W9-ASB-031

MATHEMATICAL METHODS with Applications to Problems in the Physical Sciences

MATHEMATICAL METHODS with Applications to Problems in the Physical Sciences

TED CLAY BRADBURY
Department of Physics
California State University at Los Angeles

JOHN WILEY & SONS
New York
Chichester
Brisbane
Toronto
Singapore

Library of Congress Cataloging in Publication Data:

Bradbury, T. C. (Ted Clay), 1932–
Mathematical methods with applications to problems
in the physical sciences.

Bibliography: p. 687
Includes index.
1. Mathematical physics. 2. Mathematics. I. Title.

QC20.B65 1984 530.1′5 84-3530
ISBN 0-471-88639-4

Printed in the United States of America

10 9 8 7 6 5 4 3 2 1

Preface

This book is designed primarily for physics majors who have completed the freshman-sophomore sequence in calculus, differential equations, and basic physics. It is about the application of mathematical methods to problems in physics. The basic concepts of classical mechanics, electromagnetism, and quantum mechanics are developed in the text along with the mathematics. Physical problems are used as the motivation for the development of the mathematics and then provide a continuing source of examples as the theory is further developed. The topics covered are those most likely to be needed in junior-senior level or beginning graduate courses in physics.

There is a greater emphasis on linear algebra than is customary for books at this level. This is done to give readers a better background for a later study of the more advanced algebraic methods that are now important in theoretical physics. Many topics are treated with a fair degree of mathematical rigor so that readers will have some idea of what the world looks like from the point of view of a mathematician. The treatment is intensive rather than extensive, and for this reason some topics that might have been included have been omitted. The subject of mathematical methods is voluminous, and it seemed better to give a reasonably complete treatment of the selected topics. Including more material would have resulted either in superficiality or in an unduly long book.

The chapter on numerical methods is computer-oriented and is suitable for those who are in the process of learning computer programing. It is designed for small home computing systems and can be used along with other parts of the text. In Chapters 1 through 10, problems appear at the end of each section whereas in Chapter 11 they are found at the end of the chapter.

I am indebted to Professors Donald A. Bird and Ross D. F. Thompson for their invaluable assistance in proofreading the manuscript.

Ted Clay Bradbury

Contents

MATHEMATICAL METHODS
with Applications to Problems in the Physical Sciences

$$\sum_{j=1}^{n} A_{ij}x_j = y_i$$

1 Linear Equations: An Introduction to Matrices and Determinants

The concept of linear algebraic equations is used as a vehicle to introduce the matrix concept and to develop the theory of determinants. A useful criterion is given for determining the consistency or inconsistency of a set of linear equations. Methods for solving systems of linear algebraic equations are given. The concept of a vector is introduced in an elementary way in this chapter and then developed more systematically in Chapter 2. It will be useful if the reader already has some familiarity with the vector concept.

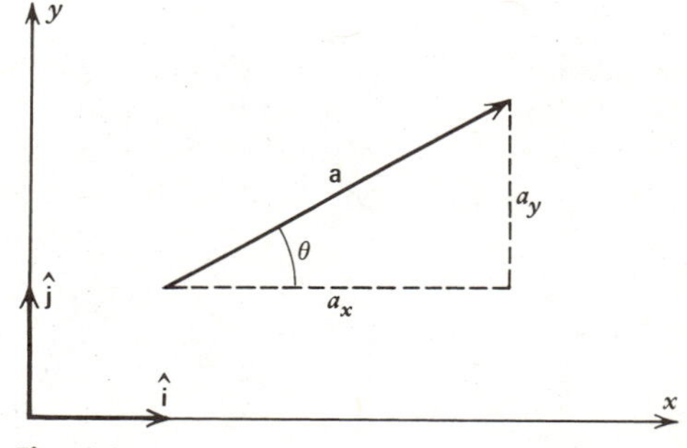

Fig. 1.1

1. VECTOR ADDITION

The notion of a vector is usually introduced in an elementary way by saying that it is a mathematical entity which has both direction and magnitude, and for the moment this will suffice. A more precise definition will be given later. In Fig. 1.1, a vector **a** lying in the x, y plane is represented by an arrow. Commonly used notations for vectors are bold-face letters or letters with arrows over them. The *components* of a vector are defined to be its projections on the axes of a rectangular Cartesian coordinate system. The components of the vector **a** of Fig. 1.1 are

$$a_x = a\cos\theta \qquad a_y = a\sin\theta \tag{1.1}$$

where θ is the angle between **a** and the x axis and a is the length or magnitude of **a** given by

$$a = |\mathbf{a}| = \sqrt{a_x^2 + a_y^2} \tag{1.2}$$

The question of defining vectors in other than rectangular coordinates will be taken up later.

A scalar is a quantity which has magnitude only and is therefore represented by an ordinary number. It is possible to multiply a vector by a scalar quantity. If α is a number and **a** is a vector, then $\alpha\mathbf{a}$ is a vector in the same direction as **a** if α is positive and in the opposite direction to **a** if α is negative. If α is zero, then $\alpha\mathbf{a}$ is the zero vector. If **a** lies in the x, y plane, the components of $\alpha\mathbf{a}$ are αa_x and αa_y.

Frequently used is the concept of a *unit vector*. Given a vector **a**, we can construct a vector of unit magnitude pointing in the same direction as **a** by means of

$$\hat{\mathbf{a}} = \frac{\mathbf{a}}{a} \tag{1.3}$$

where the hat, (̂), is used to indicate a unit vector. Even if **a** has some kind of physical units such as centimeters, the unit vector $\hat{\mathbf{a}}$, being a ratio, does not; it only indicates direction.

Figure 1.2 makes it evident that the vector **c** is equivalent to the two vectors **a** and **b**. This leads to the idea of the sum of two vectors:

$$\mathbf{c} = \mathbf{a} + \mathbf{b} \tag{1.4}$$

Figure 1.2 shows that the components of **c** are related to the components of **a** and **b** by means of

$$\begin{aligned} c_x &= a_x + b_x \\ c_y &= a_y + b_y \end{aligned} \tag{1.5}$$

Geometrically, **c** is the diagonal of the parallelogram formed by **a** and **b**.

The unit vectors $\hat{\mathbf{i}}$ and $\hat{\mathbf{j}}$ which have been included in the figures of this section are unit coordinate vectors pointing in the x and y directions. In terms of the unit coordinate vectors, the vector **a** can be represented as

$$\mathbf{a} = a_x\hat{\mathbf{i}} + a_y\hat{\mathbf{j}} \tag{1.6}$$

The vector concept can be extended to three (or more) dimensions. In Fig. 1.3, a_x, a_y, and a_z are the projections or components of the vector **a** on the x, y, and z axes. The vector **a** can therefore be expressed as

$$\mathbf{a} = a_x\hat{\mathbf{i}} + a_y\hat{\mathbf{j}} + a_z\hat{\mathbf{k}} \tag{1.7}$$

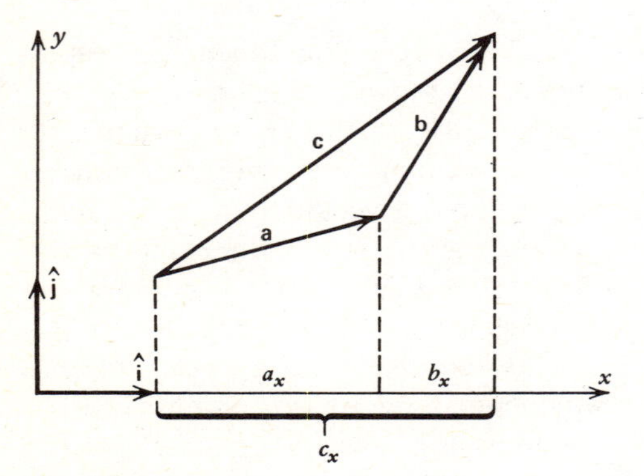

Fig. 1.2

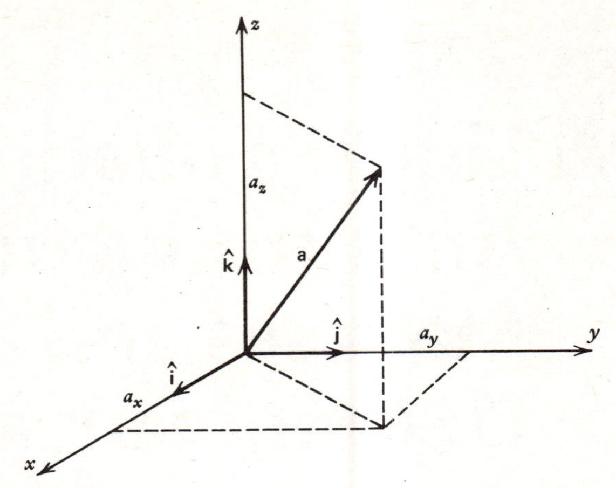

Fig. 1.3

where $\hat{\mathbf{i}}$, $\hat{\mathbf{j}}$, and $\hat{\mathbf{k}}$ are unit coordinate vectors pointing along the x, y, and z axes. Another notation which is sometimes useful is

$$\mathbf{a} = (a_x, a_y, a_z) \tag{1.8}$$

Here, the vector is thought of as an ordered triplet of numbers. This notation is especially useful when we wish to generalize the idea of vector to spaces of more than three dimensions.

Vector addition and scalar multiplication in three dimensions follow the same rules as for the two-dimensional case:

$$\begin{aligned} \mathbf{c} = \mathbf{a} + \mathbf{b} = (a_x + b_x)\hat{\mathbf{i}} + (a_y + b_y)\hat{\mathbf{j}} \\ + (a_z + b_z)\hat{\mathbf{k}} \end{aligned} \tag{1.9}$$

$$\alpha\mathbf{a} = \alpha a_x\hat{\mathbf{i}} + \alpha a_y\hat{\mathbf{j}} + \alpha a_z\hat{\mathbf{k}} \tag{1.10}$$

It is possible to solve equation (1.9) for the vector **a** in terms of **b** and **c**:

$$\mathbf{a} = \mathbf{c} - \mathbf{b} = (c_x - b_x)\hat{\mathbf{i}} + (c_y - b_y)\hat{\mathbf{j}} + (c_z - b_z)\hat{\mathbf{k}} \tag{1.11}$$

which can be viewed as the addition of the vectors **c** and $-$**b**. In this respect, vector subtraction is a special case of vector addition.

A vector is zero if, and only if, all three of its components are zero. The equation $\mathbf{a} = \mathbf{b}$ can be true if, and only if, the components of **a** are the same as those of **b**: $a_x = b_x$, $a_y = b_y$, and $a_z = b_z$.

PROBLEMS

1.1. The vectors $\mathbf{a} = -\hat{\imath} + 2\hat{\jmath} - 3\hat{\mathbf{k}}$ and $\mathbf{b} = 2\hat{\imath} + 4\hat{\jmath} - 2\hat{\mathbf{k}}$ represent displacements from the origin of coordinates (as in Fig. 1.3). What is the distance between the tips of these vectors?

1.2. Two vectors $\mathbf{a} = 5\hat{\imath} + \hat{\jmath}$ and $\mathbf{b} = \hat{\imath} + 4\hat{\jmath}$ project from the origin of coordinates and form two sides of a parallelogram as shown in Fig. 1.4. Find the two vectors which represent the two diagonals of this parallelogram and find the length of each.

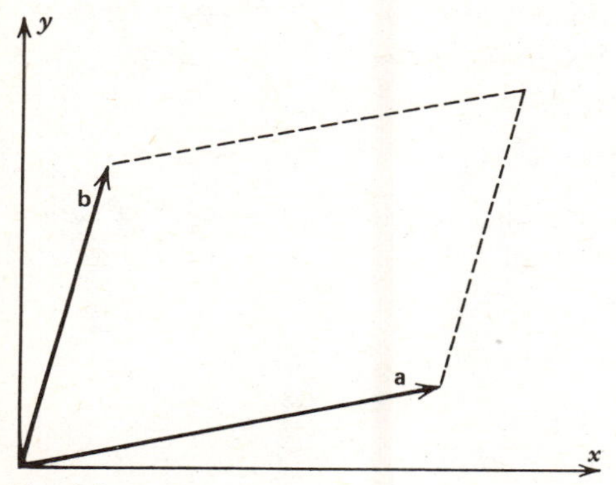

Fig. 1.4

1.3. In mechanics, a particle is in equilibrium if the vector sum of all forces acting on it is zero. A mass $m = 4$ kg is suspended by strings as shown in Fig. 1.5. The knot which is formed by the junction of the three strings is a small body in equilibrium. If $\theta_1 = 25°$ and $\theta_2 = 70°$, find the tension which exists in each of the three strings. Remember that the weight of the mass is $mg = 4\text{ kg} \times 9.80\text{ m/sec}^2 = 39.2$ newtons.

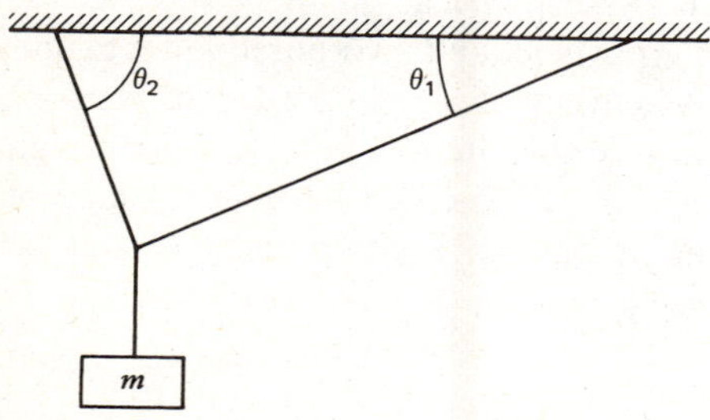

Fig. 1.5

2. EXAMPLES OF LINEAR ALGEBRAIC EQUATIONS

A direct-current circuit consists of two batteries or seats of electromotive force (emf) and three resistors as shown in Fig. 2.1. We will neglect the internal resistances of the seats of emf (or lump them with R_1 and R_2). The directions of the currents as indicated by the arrows are arbitrarily assigned; a misjudgment of the actual current direction is of no consequence since our final answers will be positive or negative numbers depending on whether a given current flows in the direction indicated or in the opposite direction. By application of Kirchoff's loop theorem to the three current loops,

$$R_1 I_1 + R_3 I_3 = \mathscr{E}_1 \tag{2.1}$$

$$R_2 I_2 - R_3 I_3 = \mathscr{E}_2 \tag{2.2}$$

$$R_1 I_1 + R_2 I_2 = \mathscr{E}_1 + \mathscr{E}_2 \tag{2.3}$$

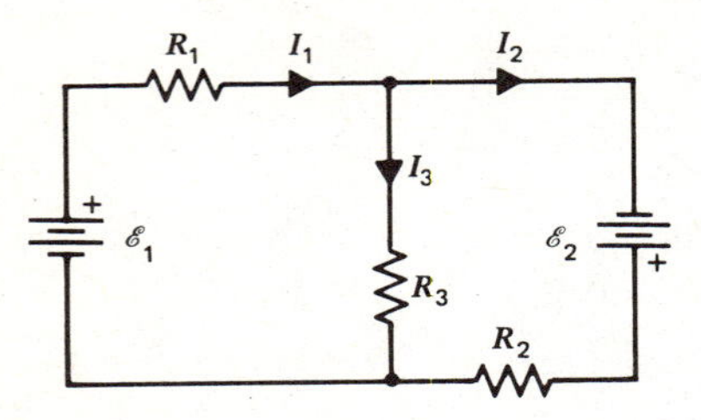

Fig. 2.1

The conservation of charge principle applied to either junction in the circuit gives a fourth equation:

$$I_1 - I_2 - I_3 = 0 \tag{2.4}$$

In the practical, or mks, system of units, current is measured in amperes, resistance is measured in ohms, and emf in volts. If we assume that the resistances R_1, R_2, and R_3 and the emfs $\mathscr{E}_1$ and $\mathscr{E}_2$ are given, then (2.1) through (2.4) are a system of four linear algebraic equations in the three unknown currents I_1, I_2, and I_3. The term *linear* means that the unknowns appear to the first power only, and not as I_1^2, I_1^3, and so forth, and not in combinations such as $I_1 I_2$, $I_1 I_3^2$, and so forth. In this example, there are more equations than unknowns. Note, however, that (2.3) is really not an independent equation since it can be obtained by simply adding together (2.1) and (2.2). It is generally true in a situation where there are more linear equations than unknowns that some of the equations are linear combinations of the others so that the number of truly independent equations equals the number of unknowns. Otherwise, there is an inconsistency.

As a numerical example, suppose that $R_1 = 10\ \Omega$, $R_2 = 12\ \Omega$, $R_3 = 14\ \Omega$, $\mathscr{E}_1 = 6$ V, and $\mathscr{E}_2 = 9$ V. Then, omitting the redundant (2.3),

$$\begin{aligned} 10I_1 + \qquad\quad 14I_3 &= 6 \\ 12I_2 - 14I_3 &= 9 \\ I_1 - \quad I_2 - \quad I_3 &= 0 \end{aligned} \tag{2.5}$$

The solution can be obtained by a method known as *Gauss elimination*. Divide each equation by the coefficient of the first unknown I_1:

$$\begin{aligned} I_1 + \qquad\quad 1.4I_3 &= 0.6 \\ 12I_2 - 14I_3 &= 9 \\ I_1 - \quad I_2 - \quad I_3 &= 0 \end{aligned}$$

Now subtract the third equation from the first to eliminate I_1 (in this example, I_1 is of course already absent from the second equation):

$$\begin{aligned} 12I_2 - 14I_3 &= 9 \\ I_2 + 2.4I_3 &= 0.6 \end{aligned}$$

Repeat the technique; this time divide each equation by the coefficient of I_2:

$$\begin{aligned} I_2 - 1.167I_3 &= 0.7500 \\ I_2 + 2.400I_3 &= 0.6000 \end{aligned}$$

Eliminate I_2 to get

$$-3.567I_3 = 0.1500 \qquad I_3 = -0.04205 \text{ A}$$

The remaining unknowns are

$$I_2 = 0.70092 \text{ A} \qquad I_1 = 0.65887 \text{ A}$$

Another example of the occurrence of linear equations is provided by the transformation of the components of a vector from one Cartesian coordinate system to another. Figure 2.2 shows a vector, which might represent some physical quantity such as the acceleration of a particle, and two possible rectangular Cartesian systems. The x', y' coordinates are obtained from the x, y coordinates by rotating through the angle θ about the z axis. If $\mathbf{a}$ lies in the plane of Fig. 2.2 so that it has no z component, it can be

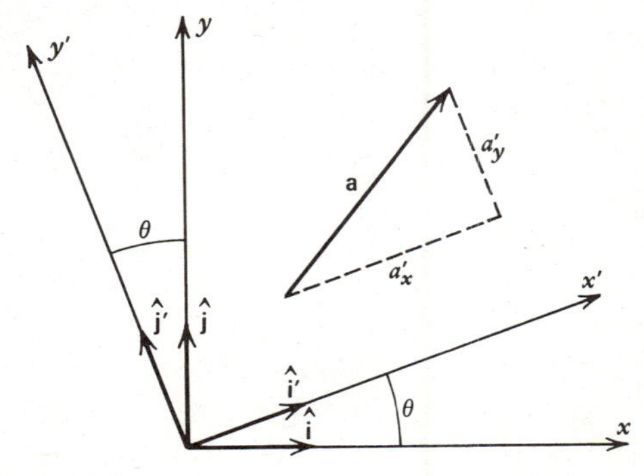

Fig. 2.2

represented in either the x, y or the x', y' coordinate systems as

$$\mathbf{a} = a'_x \hat{\imath}' + a'_y \hat{\jmath} = a_x \hat{\imath} + a_y \hat{\jmath} \tag{2.6}$$

How are the components a'_x and a'_y related to a_x and a_y? The unit vectors of the new coordinate system are given in terms of the old unit vectors by means of

$$\hat{\imath}' = \cos\theta\,\hat{\imath} + \sin\theta\,\hat{\jmath} \qquad \hat{\jmath}' = -\sin\theta\,\hat{\imath} + \cos\theta\,\hat{\jmath} \tag{2.7}$$

We substitute (2.7) into (2.6) to get

$$\begin{aligned} a'_x(\cos\theta\,\hat{\imath} + \sin\theta\,\hat{\jmath}) &+ a'_y(-\sin\theta\,\hat{\imath} + \cos\theta\,\hat{\jmath}) \\ &= a_x\hat{\imath} + a_y\hat{\jmath} \end{aligned} \tag{2.8}$$

The coefficients of $\hat{\imath}$ on the two sides of the equation must be equal and similarly for the coefficients of $\hat{\jmath}$:

$$\begin{aligned} a_x &= \cos\theta\, a'_x - \sin\theta\, a'_y \\ a_y &= \sin\theta\, a'_x + \cos\theta\, a'_y \end{aligned} \tag{2.9}$$

Thus, the components of $\mathbf{a}$ in the two coordinate systems are connected by a set of linear equations.

In electrostatics, one finds that a nonconductor polarizes in the presence of an electric field. For a crystal, which is an example of a nonisotropic medium, the polarization vector does not in general go in the same direction as the electric field, but rather its three rectangular components are related to the components of the electric field by means of the linear equations

$$\begin{aligned} \frac{1}{\varepsilon_0}P_1 &= \chi_{11}E_1 + \chi_{12}E_2 + \chi_{13}E_3 \\ \frac{1}{\varepsilon_0}P_2 &= \chi_{21}E_1 + \chi_{22}E_2 + \chi_{23}E_3 \\ \frac{1}{\varepsilon_0}P_3 &= \chi_{31}E_1 + \chi_{32}E_2 + \chi_{33}E_3 \end{aligned} \tag{2.10}$$

where the nine quantities $\chi_{11}, \chi_{12}, \ldots$ are the elements of the electric susceptibility tensor the numerical values of which are characteristic of the insulator in question; and, in mks units, ε_0 is the permittivity of free space. Frequently, it is more convenient to label the rectangular components of a vector with numerical subscripts such as P_1, P_2, and P_3 rather than P_x, P_y, and P_z. We can think of the crystal as performing a kind of operation on the given electric field $\mathbf{E}$ to produce a new vector $\mathbf{P}$ which describes the state of polarization of the medium. The mathematical representation of this operation is the electric susceptibility tensor.

A uniform plank of weight W and length s rests with one end on the ground. It is supported by a smooth rail at a point which is a distance $\frac{3}{4}s$ from the end that is on the ground, as shown in Fig. 2.3. The plank is at an angle θ with respect to the horizontal. There is no friction between the plank and the rail, but there is friction between the plank and the ground. Given the weight W of the plank and the angle θ, we want to know the forces which the rail and the ground exert on the plank. These forces are shown in Fig. 2.3. Since there is no friction between the plank and the rail, the rail can only exert a force N at right angles to the plank. Since there is friction between the ground and the plank, the force which the ground exerts has both a horizontal component H and a vertical component V. It is possible to consider the weight of the plank as a single force of magnitude W acting at the center of mass. Since the plank is uniform, this is also its geometrical center. One of the conditions for static equilibrium is that the vector sum of the forces vanishes. Setting the sum of the x and the y components separately equal to zero gives

$$H - N\sin\theta = 0 \qquad N\cos\theta + V - W = 0 \tag{2.11}$$

The second condition for equilibrium is that the sum of the torques about any point vanishes. It is convenient to calculate the torques around the point of contact O of the plank with the ground. The

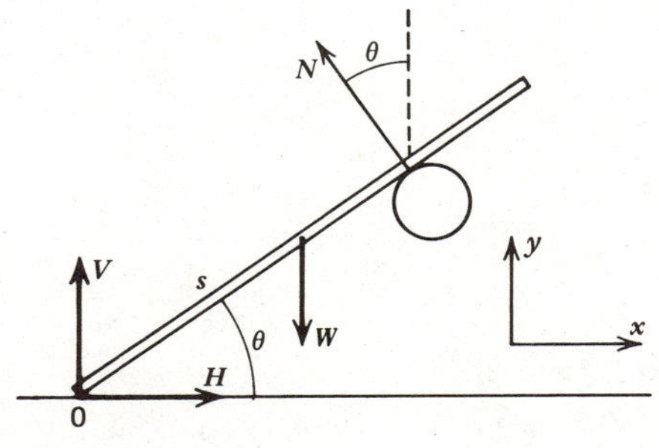

Fig. 2.3

general definition of the torque produced by a force around a point will be given later. For the present, it suffices to say that the torques produced by N and W about O are found by multiplying the force by the perpendicular distance from O to the line of action of the force. This gives $\tau_W = -W(s/2)\cos\theta$ and $\tau_N = N(3/4)s$. We have arbitrarily chosen clockwise as negative and counterclockwise as positive. Setting the sum of the torques equal to zero gives

$$\tfrac{3}{4}Ns - \tfrac{1}{2}sW\cos\theta = 0 \tag{2.12}$$

The combination of (2.11) and (2.12) provides three linear equations for the determination of the three unknowns H, V, and N.

PROBLEMS

2.1. A gate of weight W is hung by two hinges at A and B. To give extra support, a wire is run from C to D, as shown in Fig. 2.4. The forces which the hinges exert on the gate are conveniently resolved into horizontal components F_2 and F_3 and vertical components F_4 and F_5. The tension in the wire is F_1. Suppose that the dimensions of the gate are $s = 10$ ft and $h = 5$ ft and that $W = 100$ lb, $\cos\theta = 0.8$, and $\sin\theta = 0.6$. Assume that the center of mass of the gate is at its geometrical center. Using the conditions for static equilibrium of a rigid body, write down as many independent equations as you can with the idea of finding the unknown forces F_1, F_2, F_3, F_4, and F_5.

It would seem that by taking torques around different points, you can generate as many equations as you need; but in fact it does not work out that way. There is a theorem from mechanics which says that if the vector sum of all forces on a body is zero and the sum of the torques around any one point is zero, then the sum of torques around any other point will automatically be zero. Thus, taking torques around more than one point generates redundant equations. You will be able to find at most three independent equations. This means that it will not be possible to produce unique numerical values for all five of the unknown forces. The best you can do is to find some of the unknowns in terms of others.

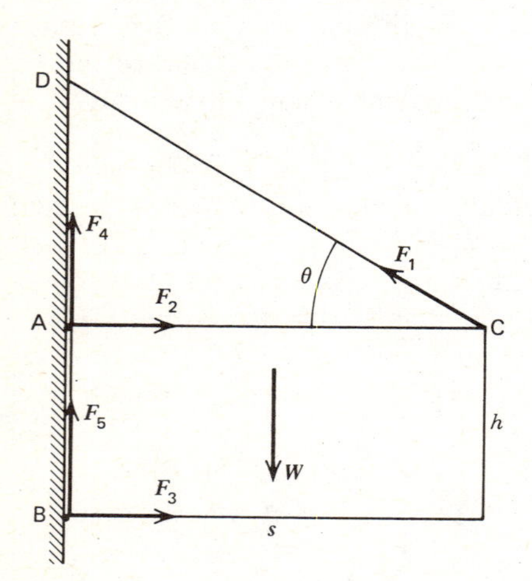

Fig. 2.4

2.2. Solve (2.9) for a'_x and a'_y in terms of a_x and a_y.

2.3. Lying in the x, y plane of a Cartesian coordinate system are two vectors **a** and **b** of the same length but differing in direction by the angle θ, as shown in Fig. 2.5. Find the linear relations connecting the components of **b** with those of **a**. Your equations will look something like (2.9), but there is a difference. In (2.9), we are talking about the same vector referred to two different coordinate systems; here, we are talking about two different vectors both referred to the same coordinate system.

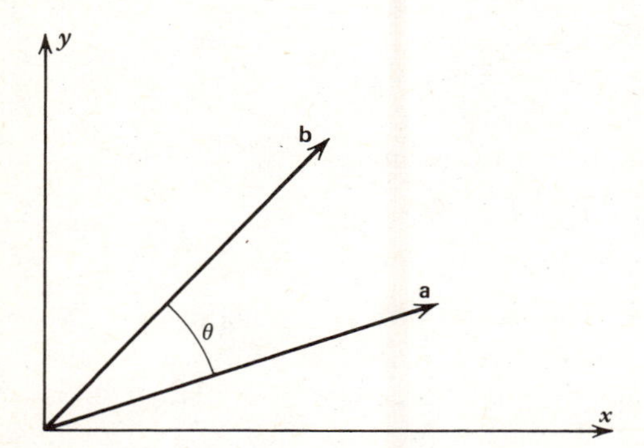

Fig. 2.5

2.4. The displacement of a particle moving on the x axis with constant acceleration is given by

$$x = x_0 + v_0 t + \tfrac{1}{2}at^2$$

where x_0 is the initial position, v_0 is the initial velocity, and a is the acceleration. The following data are taken experimentally. At $t = 1$ sec, $x = 47$ cm; at $t = 2$ sec, $x = 68$ cm; and at $t = 3$ sec, $x = 83$ cm. Find the values of x_0, v_0, and a.

2.5. In Fig. 2.6, $R = 2\ \Omega$ and the battery maintains a voltage $V = 10$ volts across the circuit. Find the current in each resistor.

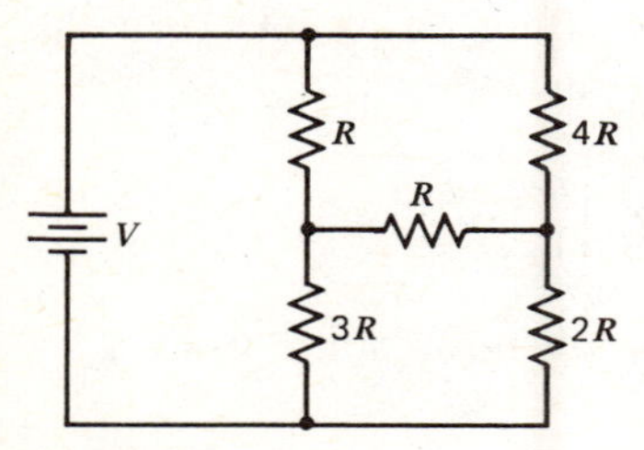

Fig. 2.6

3. THE MATRIX CONCEPT

A general system of linear equations can be represented as

$$\begin{aligned}
y_1 &= A_{11}x_1 + A_{12}x_2 + \cdots + A_{1n}x_n \\
y_2 &= A_{21}x_1 + A_{22}x_2 + \cdots + A_{2n}x_n \\
&\vdots \\
y_m &= A_{m1}x_1 + A_{m2}x_2 + \cdots + A_{mn}x_n
\end{aligned} \tag{3.1}$$

The system is said to be *homogeneous* if $y_1 = y_2 = \cdots = y_m = 0$. Otherwise, the set of equations (3.1) is *nonhomogeneous*. In (3.1), m is the number of equations and is not necessarily equal to n, the number of x values. In the circuit equations (2.1) through (2.4), there are four equations ($m = 4$) and three currents ($n = 3$). The currents correspond to the x values in (3.1). The reader who has done problem (2.1) has discovered an example where $m = 3$ and $n = 5$. Frequently, (3.1) is abbreviated as

$$y_i = \sum_{j=1}^{n} A_{ij} x_j \qquad (i = 1, 2, \ldots, m) \tag{3.2}$$

The system (3.1) is characterized, and its mathematical properties largely determined, by the values of the coefficients A_{ij}. It is convenient to display these numbers in a kind of table called a matrix:

$$A = \begin{pmatrix} A_{11} & A_{12} & \cdots & A_{1n} \\ A_{21} & A_{22} & \cdots & A_{2n} \\ \vdots & & & \\ A_{m1} & A_{m2} & \cdots & A_{mn} \end{pmatrix} \tag{3.3}$$

A matrix is basically a rectangular array of numbers. As our theory of linear equations develops, we will find it expedient to think of the matrix A as an entity in its own right which we will endow with appropriate mathematical properties.

The idea of matrix addition can be introduced as a kind of extension of vector addition. Let a vector **a** be given and suppose that a second vector **b** is constructed from it using the elements of a matrix in a set of linear equations:

$$\begin{aligned} b_1 &= A_{11} a_1 + A_{12} a_2 \\ b_2 &= A_{21} a_1 + A_{22} a_2 \end{aligned} \tag{3.4}$$

For simplicity, we suppose that our vector space, that is, the mathematical space in which the vectors exist, is two-dimensional. (A specific example of (3.4) is provided by problem 2.3.) By means of a second matrix B we can construct from the original vector a third vector **c** the components of which are

$$\begin{aligned} c_1 &= B_{11} a_1 + B_{12} a_2 \\ c_2 &= B_{21} a_1 + B_{22} a_2 \end{aligned} \tag{3.5}$$

A fourth vector **d** can now be obtained by adding together **b** and **c**:

$$\begin{aligned} d_1 &= b_1 + c_1 = (A_{11} + B_{11}) a_1 + (A_{12} + B_{12}) a_2 \\ d_2 &= b_2 + c_2 = (A_{21} + B_{21}) a_1 + (A_{22} + B_{22}) a_2 \end{aligned} \tag{3.6}$$

Observe that a direct linear connection exists between the components of **a** and **d**:

$$\begin{aligned} d_1 &= C_{11} a_1 + C_{12} a_2 \\ d_2 &= C_{21} a_1 + C_{22} a_2 \end{aligned} \tag{3.7}$$

The matrix C given by

$$C = A + B \tag{3.8}$$

is understood to have elements

$$\begin{aligned} C_{11} &= A_{11} + B_{11} \qquad & C_{12} &= A_{12} + B_{12} \\ C_{21} &= A_{21} + B_{21} \qquad & C_{22} &= A_{22} + B_{22} \end{aligned} \tag{3.9}$$

This defines matrix addition. In general, any two matrices of the same size (same number of rows and columns) can be added together in a like manner. The elements of C are then

$$C_{ij} = A_{ij} + B_{ij} \tag{3.10}$$

Matrix addition is *associative* and *commutative*:

$$(A + B) + C = A + (B + C) \tag{3.11}$$

$$A + B = B + A \tag{3.12}$$

The multiplication of a vector by a scalar is

$$\alpha \mathbf{b} = \alpha b_1 \hat{\mathbf{i}} + \alpha b_2 \hat{\mathbf{j}} \tag{3.13}$$

Each component of the vector is to be multiplied by the scalar α. Reference to (3.4) shows that the vector $\alpha\mathbf{b}$ is obtained from the vector $\alpha\mathbf{a}$ by means of the matrix A. Alternatively, one could write

$$\begin{aligned} \alpha b_1 &= (\alpha A_{11}) a_1 + (\alpha A_{12}) a_2 \\ \alpha b_2 &= (\alpha A_{21}) a_1 + (\alpha A_{22}) a_2 \end{aligned} \tag{3.14}$$

and say that the vector $\alpha\mathbf{b}$ is obtained from **a** by means of the matrix

$$\alpha A = \begin{pmatrix} \alpha A_{11} & \alpha A_{12} \\ \alpha A_{21} & \alpha A_{22} \end{pmatrix} \tag{3.15}$$

Given any matrix A, the product of A and the scalar α is a new matrix with elements αA_{ij}.

Refer again to (3.4) where the components of **b** are obtained from **a** by means of the matrix A. Suppose that a third vector **x** is obtained from **b** by means of

$$\begin{aligned} x_1 &= B_{11} b_1 + B_{12} b_2 \\ x_2 &= B_{21} b_1 + B_{22} b_2 \end{aligned} \tag{3.16}$$

How are the components of $\mathbf{x}$ related to those of $\mathbf{a}$? The elimination of b_1 and b_2 between (3.4) and (3.16) gives

$$\begin{aligned} x_1 &= (B_{11}A_{11} + B_{12}A_{21})a_1 + (B_{11}A_{12} + B_{12}A_{22})a_2 \\ x_2 &= (B_{21}A_{11} + B_{22}A_{21})a_1 + (B_{21}A_{12} + B_{22}A_{22})a_2 \end{aligned} \tag{3.17}$$

The coefficients of a_1 and a_2 which appear in (3.17) can be thought of as the elements of a matrix C which is the product of B and A:

$$\begin{aligned} C = BA &= \begin{pmatrix} B_{11} & B_{12} \\ B_{21} & B_{22} \end{pmatrix} \begin{pmatrix} A_{11} & A_{12} \\ A_{21} & A_{22} \end{pmatrix} \\ &= \begin{pmatrix} B_{11}A_{11} + B_{12}A_{21} & B_{11}A_{12} + B_{12}A_{22} \\ B_{21}A_{11} + B_{22}A_{21} & B_{21}A_{12} + B_{22}A_{22} \end{pmatrix} \end{aligned} \tag{3.18}$$

The rule is that the elements of C are found by multiplying the rows of B times the columns of A and then summing. Nonsquare matrices can be multiplied together provided the number of columns in the first matrix is the same as the number of rows in the second. In general, the elements of $C = AB$ are

$$C_{ik} = \sum_{j=1}^{n} B_{ij} A_{jk} \tag{3.19}$$

where C_{ik} is the element in the ith row and kth column of C. Matrix multiplication is in general not commutative:

$$BA \neq AB \tag{3.20}$$

This is readily verified by multiplying out the 2×2 matrices of (3.18) in the reverse order.

Special classes of matrices exist which do commute, and these are of importance in many applications. The *commutator* of two matrices is defined to be $AB - BA$ and is given the special notation

$$[A, B] = AB - BA \tag{3.21}$$

If A and B are not square, either the product will work in one order only or AB will not be the same size as BA. A commutator therefore makes sense only for square matrices.

Theorem 1. Matrix multiplication is distributive:

$$A(B + C) = AB + AC \qquad (B + C)A = BA + CA \tag{3.22}$$

Proof: The ij element of $A(B + C)$ is

$$[A(B + C)]_{ij} = \sum_k A_{ik}(B_{kj} + C_{kj})$$

Since the elements of a matrix are ordinary numbers, it is true that $A_{ik}(B_{kj} + C_{kj}) = A_{ik}B_{kj} + A_{ik}C_{kj}$. Therefore,

$$\begin{aligned} [A(B + C)]_{ij} &= \sum_k A_{ik}B_{kj} + \sum_k A_{ik}C_{kj} \\ &= (AB)_{ij} + (AC)_{ij} \end{aligned}$$

where $(AB)_{ij}$ is the ij element of AB and $(AC)_{ij}$ is the ij element of AC. Thus, $A(B + C) = AB + AC$, and the proof is complete. The proof for $(B + C)A = BA + CA$ is similar.

Theorem 2. Matrix multiplication is associative:

$$(AB)C = A(BC) \tag{3.23}$$

Proof: The method of proof is to show that the element in the ij position of $(AB)C$ is the same as the corresponding element of $A(BC)$. Expressed in words, we can form the product AB and multiply the result by C, or we can first calculate BC and then multiply by A. The same result is obtained in either case, provided that the order of the three matrices is maintained throughout:

$$\begin{aligned} [(AB)C]_{ij} &= \sum_k (AB)_{ik}C_{kj} = \sum_k \sum_n A_{in}B_{nk}C_{kj} \\ &= \sum_k \sum_n A_{in}(B_{nk}C_{kj}) = \sum_n A_{in}(BC)_{nj} \\ &= [A(BC)]_{ij} \end{aligned}$$

Thus, $(AB)C = A(BC)$. In the proof, we have used the fact that the matrix elements themselves are associative because they are ordinary numbers.

If A is a square matrix, then there exists a *zero* or *null matrix* which satisfies

$$A + O = A \qquad OA = O \qquad AO = O \tag{3.24}$$

For example, if A is a 3×3 matrix, then

$$O = \begin{pmatrix} 0 & 0 & 0 \\ 0 & 0 & 0 \\ 0 & 0 & 0 \end{pmatrix} \tag{3.25}$$

Given any matrix A, there exist *identity* or *unit matrices* which satisfy

$$AI = I'A = A \tag{3.26}$$

where $I = I'$ if A is square. In all cases, I and I' are square and have ones down the diagonal and zeros everywhere else. For example, if A is given by

$$A = \begin{pmatrix} A_{11} & A_{12} & A_{13} \\ A_{21} & A_{22} & A_{23} \end{pmatrix} \tag{3.27}$$

then

$$\begin{pmatrix} A_{11} & A_{12} & A_{13} \\ A_{21} & A_{22} & A_{23} \end{pmatrix} \begin{pmatrix} 1 & 0 & 0 \\ 0 & 1 & 0 \\ 0 & 0 & 1 \end{pmatrix} = A \tag{3.28}$$

$$\begin{pmatrix} 1 & 0 \\ 0 & 1 \end{pmatrix} \begin{pmatrix} A_{11} & A_{12} & A_{13} \\ A_{21} & A_{22} & A_{23} \end{pmatrix} = A \tag{3.29}$$

Two matrices A and B are equal if, and only if, each element of A equals the corresponding element of B:

$$A = B \Leftrightarrow A_{ij} = B_{ij}$$

Frequently used in applications are row or column matrices to represent vectors. For example, it is possible to write (2.9) in the form

$$X' = SX \tag{3.30}$$

where

$$X = \begin{pmatrix} a_x \\ a_y \end{pmatrix} \qquad S = \begin{pmatrix} \cos\theta & \sin\theta \\ -\sin\theta & \cos\theta \end{pmatrix}$$

$$X' = \begin{pmatrix} a'_x \\ a'_y \end{pmatrix} \tag{3.31}$$

Equation (3.1) can be conveniently abbreviated

$$Y = AX \tag{3.32}$$

where

$$Y = \begin{pmatrix} y_1 \\ y_2 \\ \vdots \\ y_m \end{pmatrix} \qquad X = \begin{pmatrix} x_1 \\ x_2 \\ \vdots \\ x_n \end{pmatrix} \tag{3.33}$$

It is useful to think of X and Y as vectors even if m or n is greater than 3. The fact that we happen to live in a space of three dimensions in which Euclidean geometry is valid places no restriction on a mathematical extension to spaces of higher dimension.

An example of the use of matrix multiplication is provided by the representation of rotations in three-dimensional space. In Fig. 3.1, the $\bar{x}_i$ coordinates are obtained from the x_i coordinates by rotation through the angle ϕ about the x_3 axis. We use the abbreviation x_i for (x_1, x_2, x_3). The components of the vector **a** in the two coordinate systems are related by

$$\bar{X} = S_\phi X \tag{3.34}$$

where

$$\bar{X} = \begin{pmatrix} \bar{a}_1 \\ \bar{a}_2 \\ \bar{a}_3 \end{pmatrix} \qquad S_\phi = \begin{pmatrix} \cos\phi & \sin\phi & 0 \\ -\sin\phi & \cos\phi & 0 \\ 0 & 0 & 1 \end{pmatrix}$$

$$X = \begin{pmatrix} a_1 \\ a_2 \\ a_3 \end{pmatrix} \tag{3.35}$$

In Fig. 3.2, a third coordinate system x'_i is obtained from the $\bar{x}_i$ coordinates by rotating through the angle θ about the $\bar{x}_1$ axis. The components of the vector **a** in these two coordinate systems are related by

$$X' = S_\theta \bar{X} \tag{3.36}$$

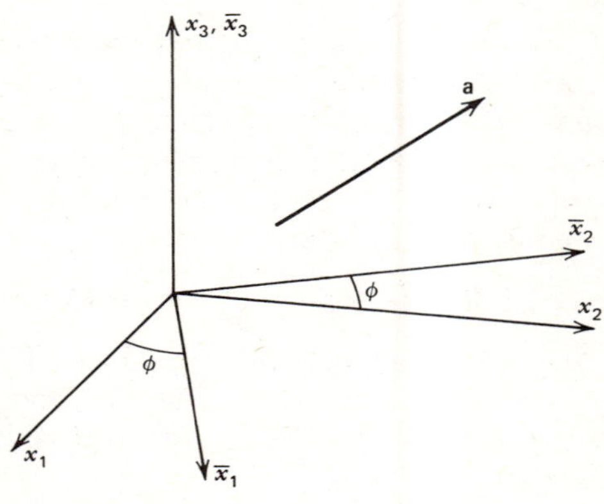

Fig. 3.1

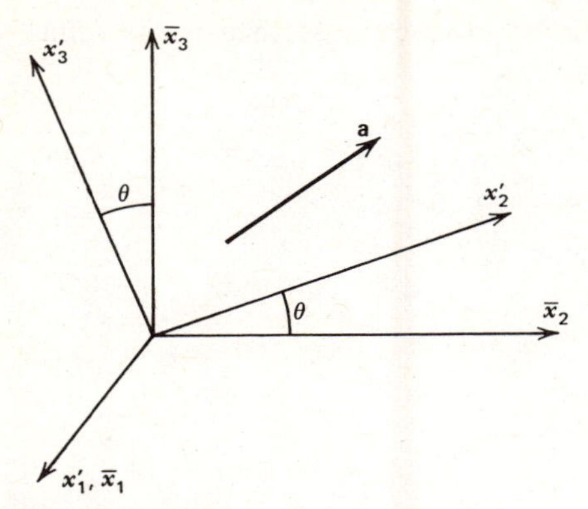

Fig. 3.2

where

$$X' = \begin{pmatrix} a_1' \\ a_2' \\ a_3' \end{pmatrix} \qquad S_\theta = \begin{pmatrix} 1 & 0 & 0 \\ 0 & \cos\theta & \sin\theta \\ 0 & -\sin\theta & \cos\theta \end{pmatrix} \tag{3.37}$$

If the transformation directly from the original x_i coordinates to the final x_i' coordinates is desired, substitute (3.34) into (3.36):

$$X' = (S_\theta S_\phi) X = SX \tag{3.38}$$

where, by carrying out the matrix multiplication, we find

$$S_\theta S_\phi = \begin{pmatrix} \cos\phi & \sin\phi & 0 \\ -\cos\theta\sin\phi & \cos\theta\cos\phi & \sin\theta \\ \sin\theta\sin\phi & -\sin\theta\cos\phi & \cos\theta \end{pmatrix} \tag{3.39}$$

The noncommutativity of matrix multiplication is illustrated in Figures 3.3 and 3.4.

Figure 3.3 shows a succession of two rotations: S_1 is a rotation by 90° about the x_3 axis and S_2 is a rotation by 90° about the x_1 axis:

$$S_1 = \begin{pmatrix} 0 & 1 & 0 \\ -1 & 0 & 0 \\ 0 & 0 & 1 \end{pmatrix}$$

$$S_2 = \begin{pmatrix} 1 & 0 & 0 \\ 0 & 0 & 1 \\ 0 & -1 & 0 \end{pmatrix} \tag{3.40}$$

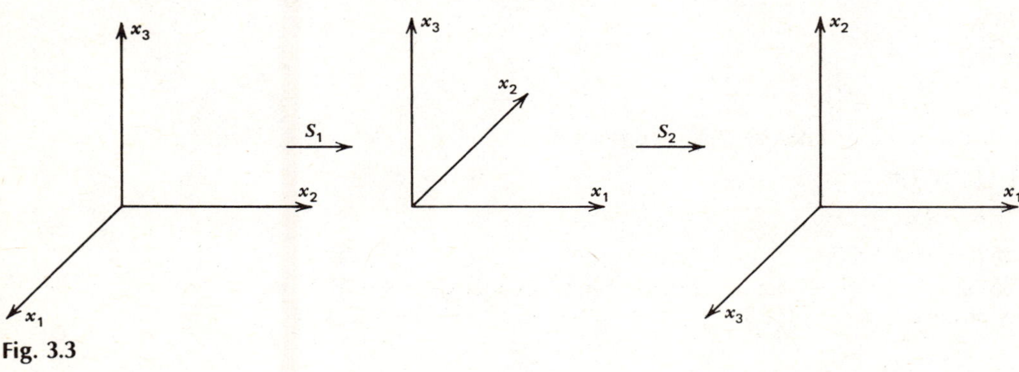

Fig. 3.3

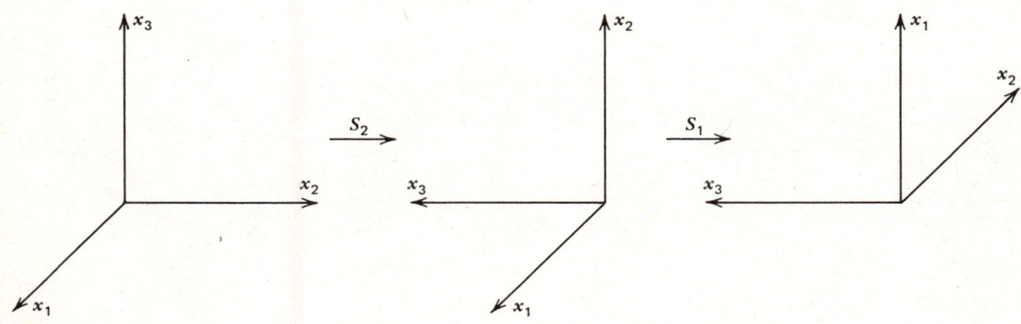

Fig. 3.4

The combined effect of S_1 and S_2 in the order illustrated by Fig. 3.3 is

$$S_2 S_1 = \begin{pmatrix} 0 & 1 & 0 \\ 0 & 0 & 1 \\ 1 & 0 & 0 \end{pmatrix} \tag{3.41}$$

Now suppose these operations are done in reverse order as illustrated in Fig. 3.4.

Note first of all the obvious difference in the final result. Also note that

$$S_1 S_2 = \begin{pmatrix} 0 & 0 & 1 \\ -1 & 0 & 0 \\ 0 & -1 & 0 \end{pmatrix} \tag{3.42}$$

which shows $S_1 S_2 \neq S_2 S_1$.

PROBLEMS

3.1. Given the two matrices

$$A = \begin{pmatrix} 1 & 2 \\ 4 & 3 \end{pmatrix} \quad B = \begin{pmatrix} 2 & 1 \\ -1 & 3 \end{pmatrix}$$

the product AB is

$$AB = \begin{pmatrix} 1 & 2 \\ 4 & 3 \end{pmatrix}\begin{pmatrix} 2 & 1 \\ -1 & 3 \end{pmatrix}$$

$$= \begin{pmatrix} 1\cdot 2 + 2(-1) & 1\cdot 1 + 2\cdot 3 \\ 4\cdot 2 + 3(-1) & 4\cdot 1 + 3\cdot 3 \end{pmatrix} = \begin{pmatrix} 0 & 7 \\ 5 & 13 \end{pmatrix}$$

Work out the product BA. If

$$C = \begin{pmatrix} -2 & 1 \\ -1 & 4 \end{pmatrix}$$

find $(AB)C$, BC, and $A(BC)$.

3.2. Verify the Jacobi identity,

$$[A,[B,C]] = [B,[A,C]] - [C,[A,B]]$$

Hint: These are commutator brackets as defined by (3.22). For instance,

$$\begin{aligned} [A,[B,C]] &= A[B,C] - [B,C]A \\ &= A(BC - CB) - (BC - CB)A \\ &= ABC - ACB - BCA + CBA \end{aligned}$$

3.3. The product of a square matrix with itself can be written $AA = A^2$. Similarly, $AAA = A^3$. Thus, A^n is defined for any positive integer n. If

$$A = \begin{pmatrix} 0 & a & b \\ 0 & 0 & c \\ 0 & 0 & 0 \end{pmatrix}$$

what are A^2 and A^3?

3.4. Suppose that

$$3\begin{pmatrix} x & y \\ z & w \end{pmatrix} = \begin{pmatrix} x & 6 \\ -1 & 2w \end{pmatrix} + \begin{pmatrix} 4 & x+y \\ z+w & 3 \end{pmatrix}$$

Find the values of x, y, z, and w.

3.5. Find AB and BA if

$$A = \begin{pmatrix} 2 & 1 & 3 \\ 1 & 0 & 4 \end{pmatrix} \quad B = \begin{pmatrix} 1 & 2 \\ 3 & 0 \\ 5 & 1 \end{pmatrix}$$

3.6. Carry through the steps to show that $(B+C)A = BA + CA$.

3.7. Find AB and BA if

$$A = (2 \quad 1 \quad 3) \qquad B = \begin{pmatrix} 1 \\ 3 \\ 5 \end{pmatrix}$$

3.8. If A and B are square matrices, what is $(A+B)(A-B)$ written out as a sum of products without parentheses?

3.9. Show that the matrices

$$S_1 = \begin{pmatrix} \cos\theta_1 & \sin\theta_1 \\ -\sin\theta_1 & \cos\theta_1 \end{pmatrix} \qquad S_2 = \begin{pmatrix} \cos\theta_2 & \sin\theta_2 \\ -\sin\theta_2 & \cos\theta_2 \end{pmatrix}$$

commute. Explain, using rotations of coordinates, why this is to be expected.

4. THE PERMUTATION SYMBOL

The permutation symbol, which we will define and study in this section, is fundamental in the definition of a determinant and also the cross product of two vectors. Basically, the permutation symbol is a function of the n integers $1,2,3,\ldots,n$. Various arrangements or permutations of n integers can be obtained by starting out with $1,2,\ldots,n$ and then transposing or exchanging pairs of integers. For example, 14325 is obtained by a transposition of 2 and 4 in the original arrangement 12345. It is in fact true that all possible arrangements of n integers (or other objects) can be obtained by a series of such transpositions. A given arrangement or permutation of the n integers is said to be *even* if an even number of transpositions is required to obtain it from the arrangement $1,2,\ldots,n$. Similarly, a given permutation is *odd* if an odd number of transpositions is needed to derive it. Formally, the permutation symbol is defined by

$$\begin{aligned} e_{ijk\ldots} &= +1 && \text{if } i, j, k, \ldots \text{ is an even permutation of } 1,2,\ldots,n \\ &= -1 && \text{if } i, j, k, \ldots \text{ is an odd permutation of } 1,2,\ldots,n \\ &= 0 && \text{if any two of the subscripts } i, j, k, \ldots \text{ are equal} \end{aligned}$$

Thus, a given permutation symbol has the three possible values $+1$, -1, and 0. For example, suppose $n=5$. Note that 42315 is obtained from 12345 by a single transposition and is therefore odd. This means that

$$e_{42315} = -1 \tag{4.1}$$

Two transpositions, such as $12345 \to 42315 \to 42513$, gives an even permutation so that

$$e_{42513} = +1 \tag{4.2}$$

Given an arbitrary arrangement of $1,2,3,\ldots,n$, how do we determine if it is even or odd? Simply restore the order to $1,2,3,\ldots,n$ and count the number of transpositions required. For example:

$$31524 \underset{(1)}{\to} 13524 \underset{(2)}{\to} 12534 \underset{(3)}{\to} 12354 \underset{(4)}{\to} 12345$$

Since four transpositions are required,

$$e_{31524} = +1 \tag{4.3}$$

Obviously, there is more than one way of arriving at a given arrangement $i, j, \ldots$ starting with $1,2,3,\ldots,n$.

Theorem 1. If any one way of changing from $1,2,\ldots,n$ involves an odd(even) number of interchanges, so does every other way of effecting the same result. This means that any arrangement of $1,2,\ldots,n$ is uniquely an even or an odd permutation.

Rather than prove this theorem in generality, we will try to convince the reader of its validity with an example. The permutation $12345 \to 42315$ involves a single transposition and is odd. Suppose that an intermediate step is introduced: 12345 → 32145. The 1 can now be brought to the final position by one more transposition: 32145 → 32415. One further exchange is necessary to get to the same final result: 32415 → 42315 giving a total of three transpositions rather than one transposition.

By a cyclic permutation is meant

$$1\,2\,3\ldots(n-1)\,n \rightarrow 2\,3\,4\ldots(n-1)\,n\,1 \qquad (4.4)$$

Theorem 2. A cyclic permutation on an odd(even) number of elements is even(odd).

Proof: We start with $1, 2, \ldots, n$ and make a series of adjacent transpositions:

	$1\,2\,3\ldots(n-1)\,n$	
Exchange 1 and 2		
	$2\,1\,3\ldots(n-1)\,n$	
Exchange 1 and 3		
	$2\,3\,1\ldots(n-1)\,n$	this involves $n-1$ transpositions
	$\vdots$	
	$2\,3\,4\ldots\quad n\,1$	
Exchange 1 and n		
	$2\,3\,4\ldots\quad 1\,n$	

Since the number of transpositions required to achieve the final result is $n-1$, the theorem follows.

Another name for the permutation symbol is *Levi–Cevita density*. Other notations commonly used are $\varepsilon_{ijk\ldots}$ or $\delta_{ijk\ldots}$. Sometimes, the number of subscripts is referred to as the *dimension* of the permutation symbol. Thus, e_{ij} is two-dimensional and e_{ijk} is three-dimensional. The reason for this is that the n-dimensional permutational symbol commonly occurs in problems which are defined in an n-dimensional space. For example, in section 5 you will see that e_{ijk} occurs in the definition of a 3×3 determinant and that 3×3 determinants occur in connection with 3×3 matrices and three-component vectors. It is easy to evaluate all the two-dimensional permutation symbols. They are

$$e_{11} = e_{22} = 0 \qquad e_{12} = +1 \qquad e_{21} = -1 \qquad (4.5)$$

Summation notation is used extensively in this book, and a short review of this subject is in order. The sum of a number of terms is abbreviated by the notation

$$a_1 + a_2 + \cdots + a_n = \sum_{i=1}^{n} a_i \qquad (4.6)$$

Note that if all the terms in the sum are equal to the same constant c, then (4.6) becomes

$$c + c + \cdots + c = \sum_{i=1}^{n} c = nc \qquad (4.7)$$

Double sums have already occurred in calculating the elements of the product of three matrices. If the matrices are 2×2, then the element in the ij position of $D = ABC$ is

$$D_{ij} = \sum_{n=1}^{2} \sum_{m=1}^{2} A_{in} B_{nm} C_{mj} \qquad (4.8)$$

The sums can be done in any order. For example, if we want to first do the sum over m followed by the sum over n, we write (4.8) as

$$\begin{aligned} D_{ij} &= \sum_{n=1}^{2} A_{in} \left(\sum_{m=1}^{2} B_{nm} C_{mj} \right) \\ &= \sum_{n=1}^{2} A_{in} (B_{n1} C_{1j} + B_{n2} C_{2j}) \\ &= \sum_{n=1}^{2} A_{in} B_{n1} C_{1j} + \sum_{n=1}^{2} A_{in} B_{n2} C_{2j} \\ &= A_{i1} B_{11} C_{1j} + A_{i2} B_{21} C_{1j} \\ &\quad + A_{i1} B_{12} C_{2j} + A_{i2} B_{22} C_{2j} \end{aligned} \qquad (4.9)$$

Alternatively, the sum over n can be done first with the result

$$\begin{aligned} D_{ij} &= \sum_{m=1}^{2} \left(\sum_{n=1}^{2} A_{in} B_{nm} \right) C_{mj} \\ &= \sum_{m=1}^{2} (A_{i1} B_{1m} + A_{i2} B_{2m}) C_{mj} \end{aligned} \qquad (4.10)$$

If you will now do the sum over m, you will again get equation (4.9).

PROBLEMS

4.1. Evaluate the following permutation symbols:

e_{132} e_{42153} e_{4113} e_{3124}

4.2. Evaluate all the nonzero three-subscript (three-dimensional) permutation symbols.

4.3. How many nonzero permutation symbols with n subscripts are there?

4.4. Complete the sum over m in (4.10) and show that the result is (4.9).

4.5. By evaluating the double sum and using the result of problem 4.2, show that

$$\sum_{i=1}^{3}\sum_{j=1}^{3} e_{1ij}a_i b_j = a_2 b_3 - b_3 a_2$$

4.6. There can be any number of sums in the same expression. Evaluate the triple sum

$$\sum_{i=1}^{3}\sum_{j=1}^{3}\sum_{k=1}^{3} e_{ijk}A_{1i}A_{2j}A_{3k}$$

The sums can be done in any order. The resulting expression has $3 \times 3 \times 3 = 27$ terms in it, but a lot of them will be zero on account of the properties of the permutation symbol. Compare your answer with equations (5.4) in the next section.

5. DETERMINANTS

Let A be an $n \times n$ square matrix. The determinant of A is defined to be

$$\det A = \sum_{i=1}^{n}\sum_{j=1}^{n}\sum_{k=1}^{n} \ldots A_{1i}A_{2j}A_{3k}\ldots e_{ijk\ldots} \tag{5.1}$$

where $e_{ijk\ldots}$ is an n-dimensional permutation symbol. Many times, a determinant is represented as a square array:

$$\det A = \begin{vmatrix} A_{11} & A_{12} & \cdots & A_{1n} \\ A_{21} & A_{22} & \cdots & A_{2n} \\ \vdots & \vdots & & \vdots \\ A_{n1} & A_{n2} & \cdots & A_{nn} \end{vmatrix} \tag{5.2}$$

A determinant however is *not* a matrix but rather a *number* derived from a matrix by means of (5.1). Straight bars rather than parentheses distinguish a determinant from a matrix. We will also use the notation $\det A = |A|$. The expansion for a 2×2 determinant is

$$|A| = \begin{vmatrix} A_{11} & A_{12} \\ A_{21} & A_{22} \end{vmatrix} = \sum_{i=1}^{2}\sum_{j=1}^{2} A_{1i}A_{2j}e_{ij}$$
$$= A_{11}A_{22}e_{12} + A_{12}A_{21}e_{21} = A_{11}A_{22} - A_{12}A_{21} \tag{5.3}$$

where use is made of (4.5). The complete expansion of a 3×3 determinant is

$$|A| = \begin{vmatrix} A_{11} & A_{12} & A_{13} \\ A_{21} & A_{22} & A_{23} \\ A_{31} & A_{32} & A_{33} \end{vmatrix}$$
$$= \sum_i \sum_j \sum_k A_{1i}A_{2j}A_{3k}e_{ijk}$$
$$= e_{123}A_{11}A_{22}A_{33} + e_{132}A_{11}A_{23}A_{32} + e_{231}A_{12}A_{23}A_{31} + e_{213}A_{12}A_{21}A_{33} + e_{312}A_{13}A_{21}A_{32} + e_{321}A_{13}A_{22}A_{31}$$
$$= A_{11}A_{22}A_{33} - A_{11}A_{23}A_{32} + A_{12}A_{23}A_{31} - A_{12}A_{21}A_{33} + A_{13}A_{21}A_{32} - A_{13}A_{22}A_{31} \tag{5.4}$$

Note that it is possible to write (5.4) as

$$|A| = A_{11}\begin{vmatrix} A_{22} & A_{23} \\ A_{32} & A_{33} \end{vmatrix} - A_{12}\begin{vmatrix} A_{21} & A_{23} \\ A_{31} & A_{33} \end{vmatrix} + A_{13}\begin{vmatrix} A_{21} & A_{22} \\ A_{31} & A_{32} \end{vmatrix} \tag{5.5}$$

This is called *Laplace's development by minors*. The 2×2 determinants which appear in (5.5) are called *cofactors* or *minors* and are defined by

$$C_{11} = \begin{vmatrix} A_{22} & A_{23} \\ A_{32} & A_{33} \end{vmatrix} \qquad C_{12} = -\begin{vmatrix} A_{21} & A_{23} \\ A_{31} & A_{33} \end{vmatrix}$$

$$C_{13} = \begin{vmatrix} A_{21} & A_{22} \\ A_{31} & A_{32} \end{vmatrix} \tag{5.6}$$

Thus, (5.5) can also be expressed as

$$|A| = \sum_{i=1}^{3} A_{1i} C_{1i} \tag{5.7}$$

The cofactors also have the representation

$$C_{1i} = \sum_{j=1}^{3} \sum_{k=1}^{3} e_{ijk} A_{2j} A_{3k} \tag{5.8}$$

where $i = 1$, 2, or 3.

Appearing in (5.6) are the cofactors of elements of the first row of det A. Cofactors corresponding to the other elements of $|A|$ can be similarly defined. They are

$$C_{21} = -\begin{vmatrix} A_{12} & A_{13} \\ A_{32} & A_{33} \end{vmatrix} \qquad C_{22} = \begin{vmatrix} A_{11} & A_{13} \\ A_{31} & A_{33} \end{vmatrix}$$

$$C_{23} = -\begin{vmatrix} A_{11} & A_{12} \\ A_{31} & A_{32} \end{vmatrix}$$

$$C_{31} = \begin{vmatrix} A_{12} & A_{13} \\ A_{22} & A_{23} \end{vmatrix} \qquad C_{32} = -\begin{vmatrix} A_{11} & A_{13} \\ A_{21} & A_{23} \end{vmatrix}$$

$$C_{33} = \begin{vmatrix} A_{11} & A_{12} \\ A_{21} & A_{22} \end{vmatrix} \tag{5.9}$$

The signs which appear in front of the determinants in (5.6) and (5.9) are found from the following scheme:

$$\begin{vmatrix} + & - & + \\ - & + & - \\ + & - & + \end{vmatrix} \tag{5.10}$$

The cofactor of a given element of det A is $\pm$ the 2×2 determinant found by erasing the row and column in which the element appears. Laplace's development by minors can be done using any row. For example,

$$|A| = \sum_i \sum_j \sum_k A_{1i} A_{2j} A_{3k} e_{ijk} = \sum_j A_{2j} C_{2j} \tag{5.11}$$

where C_{2j} are the cofactors of the second row:

$$C_{2j} = \sum_i \sum_k A_{1i} A_{3k} e_{ijk} \tag{5.12}$$

Written out in full, (5.11) is

$$|A| = -A_{21}\begin{vmatrix} A_{12} & A_{13} \\ A_{32} & A_{33} \end{vmatrix} + A_{22}\begin{vmatrix} A_{11} & A_{13} \\ A_{31} & A_{33} \end{vmatrix} - A_{23}\begin{vmatrix} A_{11} & A_{12} \\ A_{31} & A_{32} \end{vmatrix} \tag{5.13}$$

Laplace's development by minors can of course be done with determinants of any size. As an example, suppose

$$|A| = \begin{vmatrix} 2 & 0 & 0 & -2 \\ 3 & 1 & 2 & 0 \\ 4 & 0 & -1 & 3 \\ 8 & -1 & 2 & 1 \end{vmatrix} \tag{5.14}$$

Expanding by the first row gives

$$A = 2\begin{vmatrix} 1 & 2 & 0 \\ 0 & -1 & 3 \\ -1 & 2 & 1 \end{vmatrix} - (-2)\begin{vmatrix} 3 & 1 & 2 \\ 4 & 0 & -1 \\ 8 & -1 & 2 \end{vmatrix}$$

If now the 3×3 determinants are expanded, again by the first row,

$$A = 2\left\{\begin{vmatrix} -1 & 3 \\ 2 & 1 \end{vmatrix} - 2\begin{vmatrix} 0 & 3 \\ -1 & 1 \end{vmatrix}\right\}$$

$$+2\left\{3\begin{vmatrix} 0 & -1 \\ -1 & 2 \end{vmatrix} - \begin{vmatrix} 4 & -1 \\ 8 & 2 \end{vmatrix} + 2\begin{vmatrix} 4 & 0 \\ 8 & -1 \end{vmatrix}\right\}$$

$$= 2(-1 - 6 - 6) + 2(-3 - 8 - 8 - 8) = -80$$

In the following paragraphs, several theorems and properties of determinants will be discussed. The proofs are generally done using 3×3 determinants, but they are done in such a way that the generalization to determinants of arbitrary dimension is clear.

Theorem 1. Interchanging any two rows changes the sign of a determinant.

Proof: Consider

$$\begin{vmatrix} A_{21} & A_{22} & A_{23} \\ A_{11} & A_{12} & A_{13} \\ A_{31} & A_{32} & A_{33} \end{vmatrix} = A_{2i} A_{1j} A_{3k} e_{ijk} \tag{5.15}$$

which is a 3×3 determinant with rows 1 and 2 interchanged. To save time, space, writer's cramp, and expense at the printers, the *summation convention* is used. This means simply that the summation signs have been omitted in (5.15) with the understanding that subscripts which appear *in pairs* are summed. If the i and j are interchanged on the permutation symbol, the sign changes:

$$A_{2i}A_{1j}A_{3k}e_{ijk} = -A_{2i}A_{1j}A_{3k}e_{jik} \tag{5.16}$$

The i and the j are *dummy subscripts*, that is, subscripts that are summed. Thus, i and j can be interchanged everywhere in (5.16) without altering its value:

$$\begin{aligned} A_{2i}A_{1j}A_{3k}e_{ijk} &= -A_{2j}A_{1i}A_{3k}e_{ijk} \\ &= -A_{1i}A_{2j}A_{3k}e_{ijk} = -|A| \end{aligned} \tag{5.17}$$

which completes the proof.

Theorem 2. A determinant with two rows alike has the value zero.

Proof: The same kind of manipulations with subscripts that was used in the proof of theorem 1 is used here. Consider a 3×3 determinant with rows 1 and 2 alike:

$$\begin{aligned} \begin{vmatrix} A_{11} & A_{12} & A_{13} \\ A_{11} & A_{12} & A_{13} \\ A_{31} & A_{32} & A_{33} \end{vmatrix} &= A_{1i}A_{1j}A_{3k}e_{ijk} \quad \text{interchange subscripts } i \text{ and } j \text{ on } e_{ijk} \\ &= -A_{1i}A_{1j}A_{3k}e_{jik} \quad \text{exchange } i \text{ and } j \text{ everywhere} \\ &= -A_{1j}A_{1i}A_{3k}e_{ijk} \quad \text{reverse the order of } A_{1j} \text{ and } A_{1i} \\ &= -A_{1i}A_{1j}A_{3k}e_{ijk} \end{aligned} \tag{5.18}$$

From the first and fourth line of (5.18),

$$A_{1i}A_{1j}A_{3k}e_{ijk} = -A_{1i}A_{1j}A_{3k}e_{ijk}$$

which is the same as

$$2A_{1i}A_{1j}A_{3k}e_{ijk} = 0 \tag{5.19}$$

which proves the result.

The results of theorems 1 and 2 can be summarized as follows: The quantity

$$A_{mi}A_{nj}A_{pk}e_{ijk} \tag{5.20}$$

has the value $+|A|$ if m, n, p is an even permutation of 1, 2, 3 (i.e., if the rows of $|A|$ have been exchanged an even number of times); the value $-|A|$ if m, n, p is an odd permutation of 1, 2, 3; the value zero if any of the subscripts m, n, p are alike (i.e., if any two rows of $|A|$ are alike). Thus, using the definition of the permutation symbol,

$$e_{mnp}|A| = \sum_i \sum_j \sum_k A_{mi}A_{nj}A_{pk}e_{ijk} \tag{5.21}$$

Theorem 2 can also be generalized in another way:

$$\delta_{ij}|A| = \sum_k A_{ik}C_{jk} \tag{5.22}$$

where δ_{ij} is the *Kronecker delta* defined as

$$\delta_{ij} = 0 \qquad \text{if } i \neq j;\ \delta_{11} = \delta_{22} = \delta_{33} = 1 \tag{5.23}$$

To see the validity of (5.22), suppose $i = j = 1$. This is just (5.7). Now suppose $i = 1, j = 2$:

$$\begin{aligned} \sum_k A_{1k}C_{2k} &= A_{11}C_{21} + A_{12}C_{22} + A_{13}C_{23} \\ &= -A_{11}\begin{vmatrix} A_{12} & A_{13} \\ A_{32} & A_{33} \end{vmatrix} + A_{12}\begin{vmatrix} A_{11} & A_{13} \\ A_{31} & A_{33} \end{vmatrix} \\ &\quad - A_{13}\begin{vmatrix} A_{11} & A_{12} \\ A_{31} & A_{32} \end{vmatrix} \\ &= -\begin{vmatrix} A_{11} & A_{12} & A_{13} \\ A_{11} & A_{12} & A_{13} \\ A_{31} & A_{32} & A_{33} \end{vmatrix} = 0 \end{aligned} \tag{5.24}$$

The value zero is obtained because two rows of the determinant are alike.

Theorem 3. Rows and columns can be interchanged without changing the value of a determinant: $|A| = |\tilde{A}|$, where $\tilde{A}$ is the *transpose* of the matrix A defined by

$$\tilde{A} = \begin{pmatrix} A_{11} & A_{21} & A_{31} \\ A_{12} & A_{22} & A_{32} \\ A_{13} & A_{23} & A_{33} \end{pmatrix} \tag{5.25}$$

Proof: Using the summation convention,

$$e_{mnp}|A| = A_{mi}A_{nj}A_{pk}e_{ijk} \tag{5.26}$$

$$e_{ijk}|\tilde{A}| = A_{mi}A_{nj}A_{pk}e_{mnp} \tag{5.27}$$

Multiply (5.26) by e_{mnp} and (5.27) by e_{ijk}:

$$e_{mnp}e_{mnp}|A| = A_{mi}A_{nj}A_{pk}e_{mnp}e_{ijk} \tag{5.28}$$

$$e_{ijk}e_{ijk}|\tilde{A}| = A_{mi}A_{nj}A_{pk}e_{ijk}e_{mnp} \tag{5.29}$$

[In (5.28), there are sums over m, n, and p on the left side of the equation and sums over i, j, k, m, n, and p on the right side.] By working out the sums,

$$e_{mnp}e_{mnp} = e_{ijk}e_{ijk} = 6 \tag{5.30}$$

Since the right-hand sides of (5.28) and (5.29) are identical, $|A| = |\tilde{A}|$. This proves the theorem. Thus, all theorems which apply to the expansion of a determinant by rows apply to the expansion by columns:

(a) Interchange of two columns changes the sign of a determinant.
(b) If two columns are alike, the value of the determinant is zero.
(c) Laplace's expansion by minors works for columns:

$$\begin{vmatrix} A_{11} & A_{12} & A_{13} \\ A_{21} & A_{22} & A_{23} \\ A_{31} & A_{32} & A_{33} \end{vmatrix} = A_{11}\begin{vmatrix} A_{22} & A_{23} \\ A_{32} & A_{33} \end{vmatrix} - A_{21}\begin{vmatrix} A_{12} & A_{13} \\ A_{32} & A_{33} \end{vmatrix} + A_{31}\begin{vmatrix} A_{12} & A_{13} \\ A_{22} & A_{23} \end{vmatrix} \tag{5.31}$$

(d) The column version of (5.22) is

$$\delta_{ij}|A| = \sum_k A_{ki}C_{kj} \tag{5.32}$$

Theorem 4. If a linear relation exists among the rows (columns) of a determinant, the determinant is zero. By a linear relation among rows is meant

$$A_{1i} + \alpha A_{2i} + \beta A_{3i} = 0 \tag{5.33}$$

where $i = 1$, 2, or 3.

Proof: Using the summation convention,

$$|A| = A_{1i}A_{2j}A_{3k}e_{ijk} = -(\alpha A_{2i} + \beta A_{3i})A_{2j}A_{3k}e_{ijk} = 0$$

where use is made of (5.33) and theorem 2. If a relation of the form (5.33) exists, the rows are said to be *linearly dependent*. Another way of writing (5.33) is to use row matrices:

$$\begin{aligned}(A_{11} \quad A_{12} \quad A_{13}) + \alpha(A_{21} \quad A_{22} \quad A_{23}) \\ + \beta(A_{31} \quad A_{32} \quad A_{33}) = 0\end{aligned} \tag{5.34}$$

Theorem 5. The converse of theorem 4 is true. If $|A| = 0$, then a linear relation exists among the rows (columns) of $|A|$.

Proof: Assume $|A| = 0$. Then, using expansions by cofactors and (5.22),

$$\sum_i A_{1i}C_{1i} = 0 \qquad \sum_i A_{2i}C_{1i} = 0 \qquad \sum_i A_{3i}C_{1i} = 0 \tag{5.35}$$

which is equivalent to

$$\begin{pmatrix} A_{11} \\ A_{21} \\ A_{31} \end{pmatrix} C_{11} + \begin{pmatrix} A_{12} \\ A_{22} \\ A_{32} \end{pmatrix} C_{12} + \begin{pmatrix} A_{13} \\ A_{23} \\ A_{33} \end{pmatrix} C_{13} = 0 \tag{5.36}$$

which shows that a linear relation exists among the columns of $|A|$. By theorem 3, a linear relation also exists among the rows.

Theorem 6. When a determinant is multiplied by a scalar, the scalar factor multiplies the elements of any one row or column.

Proof: Using the expansion by rows and the summation convention,

$$\alpha|A| = (\alpha A_{1i})A_{2j}A_{3k}e_{ijk} = A_{1i}(\alpha A_{2j})A_{3k}e_{ijk} = \cdots$$

$$= \begin{vmatrix} \alpha A_{11} & \alpha A_{12} & \alpha A_{13} \\ A_{21} & A_{22} & A_{23} \\ A_{31} & A_{32} & A_{33} \end{vmatrix} = \begin{vmatrix} A_{11} & A_{12} & A_{13} \\ \alpha A_{21} & \alpha A_{22} & \alpha A_{23} \\ A_{31} & A_{32} & A_{33} \end{vmatrix} = \cdots \tag{5.37}$$

Similarly, by using the expansion by columns

$$|A| = A_{i1}A_{j2}A_{k3}e_{ijk} \tag{5.38}$$

we find

$$\alpha|A| = \begin{vmatrix} \alpha A_{11} & A_{12} & A_{13} \\ \alpha A_{21} & A_{22} & A_{23} \\ \alpha A_{31} & A_{32} & A_{33} \end{vmatrix} = \begin{vmatrix} A_{11} & \alpha A_{12} & A_{13} \\ A_{21} & \alpha A_{22} & A_{23} \\ A_{31} & \alpha A_{32} & A_{33} \end{vmatrix} = \cdots \tag{5.39}$$

Theorem 7. If a row (column) is multiplied by a scalar and the result is added to another row (column), the value of the determinant does not change.

Proof: We continue to use the summation convention. Since $A_{1i}A_{1j}A_{3k}e_{ijk}=0$ (because two rows are alike),

$$\begin{aligned}|A| &= A_{1i}A_{2j}A_{3k}e_{ijk}+A_{1i}(\alpha A_{1j})A_{3k}e_{ijk}\\ &= A_{1i}(A_{2j}+\alpha A_{1j})A_{3k}e_{ijk}\\ &= \begin{vmatrix} A_{11} & A_{12} & A_{13}\\ A_{21}+\alpha A_{11} & A_{22}+\alpha A_{12} & A_{23}+\alpha A_{13}\\ A_{31} & A_{32} & A_{33}\end{vmatrix}\end{aligned} \tag{5.40}$$

Theorem 7 is quite important in simplifying the process of determinant evaluation. As an example, consider

$$\underset{\substack{\llcorner -2 \uparrow \\ \llcorner \qquad -3 \uparrow}}{\begin{vmatrix} 1 & 2 & 3\\ 2 & 1 & 2\\ 3 & 3 & 2\end{vmatrix}} = \begin{vmatrix} 1 & 0 & 0\\ 2 & -3 & -4\\ 3 & -3 & -7\end{vmatrix}$$

$$= \begin{vmatrix} -3 & -4\\ -3 & -7\end{vmatrix} = 21-12=9 \tag{5.41}$$

where, as indicated, the initial simplification involves first multiplying the first column by -2 and adding the result to the second column and then multiplying the first column by -3 and adding the result to the third column. This produces two zeroes in the first row and makes the Laplace expansion by the first row easy.

Theorems 6 and 7 can be used to formally solve a system of n linear equations in n unknowns. Consider the $n=3$ case:

$$\begin{aligned} A_{11}x_1+A_{12}x_2+A_{13}x_3 &= y_1\\ A_{21}x_1+A_{22}x_2+A_{23}x_3 &= y_2\\ A_{31}x_1+A_{32}x_2+A_{33}x_3 &= y_3\end{aligned} \tag{5.42}$$

By using theorems 6 and 7,

$$x_1\begin{vmatrix} A_{11} & A_{12} & A_{13}\\ A_{21} & A_{22} & A_{23}\\ A_{31} & A_{32} & A_{33}\end{vmatrix}$$

$$= \underset{\substack{\uparrow x_2 \lrcorner \\ \llcorner x_3 \qquad \lrcorner}}{\begin{vmatrix} x_1A_{11} & A_{12} & A_{13}\\ x_1A_{21} & A_{22} & A_{23}\\ x_1A_{31} & A_{32} & A_{33}\end{vmatrix}}$$

$$= \begin{vmatrix} A_{11}x_1+A_{12}x_2+A_{13}x_3 & A_{12} & A_{13}\\ A_{21}x_1+A_{22}x_2+A_{23}x_3 & A_{22} & A_{23}\\ A_{31}x_1+A_{32}x_2+A_{33}x_3 & A_{32} & A_{33}\end{vmatrix}$$

$$= \begin{vmatrix} y_1 & A_{12} & A_{13}\\ y_2 & A_{22} & A_{23}\\ y_3 & A_{32} & A_{33}\end{vmatrix}$$

Now solve for x_1 to get

$$x_1=\frac{1}{|A|}\begin{vmatrix} y_1 & A_{12} & A_{13}\\ y_2 & A_{22} & A_{23}\\ y_3 & A_{32} & A_{33}\end{vmatrix} \tag{5.43}$$

Similarly,

$$\begin{aligned} x_2 &= \frac{1}{|A|}\begin{vmatrix} A_{11} & y_1 & A_{13}\\ A_{21} & y_2 & A_{23}\\ A_{31} & y_3 & A_{33}\end{vmatrix}\\ x_3 &= \frac{1}{|A|}\begin{vmatrix} A_{11} & A_{12} & y_1\\ A_{21} & A_{22} & y_2\\ A_{31} & A_{32} & y_3\end{vmatrix}\end{aligned} \tag{5.44}$$

A system of n equations in n unknowns always has a solution if $|A|\neq 0$. Solutions may or may not exist if $|A|=0$. The homogeneous system $y_i=0$ always has at least one solution, namely, $x_i=0$. If $|A|=0$, other solutions may exist. These questions will be taken up in more detail later. The solution as given by (5.43) and (5.44) is known as *Cramer's rule*.

Theorem 8. If A and B are $n\times n$ square matrices, then $|AB|=|A||B|$.

Proof: By use of (3.19) and the summation convention,

$$\begin{aligned} |AB| &= (A_{1m}B_{mi})(A_{2n}B_{nj})(A_{3p}B_{pk})e_{ijk}\\ &= A_{1m}A_{2n}A_{3p}(B_{mi}B_{nj}B_{pk}e_{ijk})\\ &= A_{1m}A_{2n}A_{3p}e_{mnp}|B| = |A||B|\end{aligned} \tag{5.45}$$

where use is also made of (5.21).

PROBLEMS

5.1. Evaluate the following determinants directly. Do not first simplify by using theorem 7:

$$\begin{vmatrix} 1 & 2 \\ 4 & 3 \end{vmatrix} \qquad \begin{vmatrix} 1 & 5 & 0 \\ 2 & 0 & 6 \\ 3 & 1 & 2 \end{vmatrix}$$

5.2. Evaluate (5.14) by Laplace's development by minors using the second row. Again, do not first simplify by using theorem 7.

5.3. If $e_{ijk\ldots}$ is an n-dimensional permutation symbol, what is the value of $\Sigma_i\Sigma_j\Sigma_k\cdots e_{ijk\ldots}e_{ijk\ldots}$? (The sums run from 1 to n.)

5.4. If A is an $n\times n$ matrix, prove that $|-A|=(-1)^n|A|$.

5.5. If A is a 4×4 matrix,

$$|A|=\sum_i\sum_j\sum_k\sum_m A_{1i}A_{2j}A_{3k}A_{4m}e_{ijkm}=\sum_j A_{2j}C_{2j}$$

where

$$C_{2j}=\sum_i\sum_k\sum_m A_{1i}A_{3k}A_{4m}e_{ijkm}$$

are the cofactors of the second row. Write these four cofactors as 3×3 determinants. (All of the sums run from 1 to 4 here.)

5.6. The evaluation of (5.14) given in the text is the hard way to do it. Simplify the evaluation by using theorem 7.

5.7. Suppose that $x(5 \quad 1)+y(4 \quad 3)=(1 \quad 2)$, where $(5 \quad 1)$, $(4 \quad 3)$, and $(1 \quad 2)$ are two-component row matrices. Find the values of x and y.

5.8. Determine the unknown weights x, y, and z if a weightless bar of length 11 ft with fulcrum 5 ft from one end balances in all three of the indicated positions. Do the solution using both Cramer's rule and Gauss elimination (2.5).

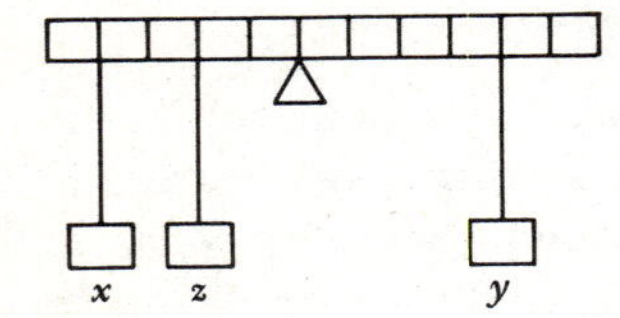

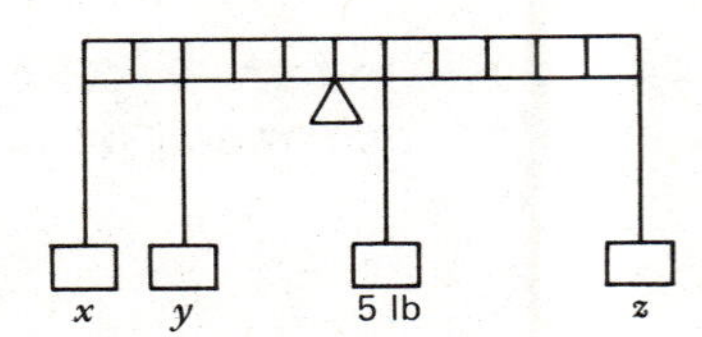

Fig. 5.1

5.9. An electrical network consisting of five resistors is connected as shown. Measurements of resistance can be made using the four terminals A, B, C, and D. The following values are obtained: $R_{AB}=8\ \Omega$, $R_{AD}=13\ \Omega$, $R_{AC}=15\ \Omega$, $R_{BD}=15\ \Omega$, $R_{BC}=17\ \Omega$, $R_{CD}=6\ \Omega$. Find the values of the five resistors. See Fig. 5.2.

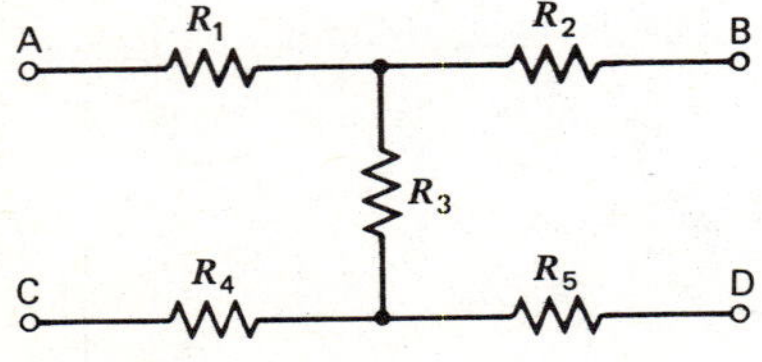

Fig. 5.2

5.10. Show that

$$\begin{vmatrix} 1 & 4 & -7 \\ 10 & 2 & 6 \\ 1 & -1 & 3 \end{vmatrix}$$

has the value zero. Find the numbers α and β in the relation

$$\begin{pmatrix} 1 \\ 10 \\ 1 \end{pmatrix} + \alpha \begin{pmatrix} 4 \\ 2 \\ -1 \end{pmatrix} + \beta \begin{pmatrix} -7 \\ 6 \\ 3 \end{pmatrix} = 0$$

5.11. Prove that

$$\begin{vmatrix} 1 & 1 & 1 \\ \alpha & \beta & \gamma \\ \beta\gamma & \gamma\alpha & \alpha\beta \end{vmatrix} = (\alpha-\beta)(\beta-\gamma)(\gamma-\alpha)$$

5.12. Suppose that the elements of a 3×3 determinant are functions of a parameter t. Express the derivative $d|A|/dt$ as the sum of three determinants in two ways: one involving derivatives of rows, the other involving derivatives of columns.

5.13. Show that

$$\begin{vmatrix} a_1 & b_1 & c_1 \\ a_2 & b_2 & c_2 \\ a_3 & b_3 & c_3 \end{vmatrix} = \begin{vmatrix} c_1 & a_1 & b_1 \\ c_2 & a_2 & b_2 \\ c_3 & a_3 & b_3 \end{vmatrix}$$

This is a cyclic permutation of columns. What happens if a cyclic permutation of the columns (or rows) of a 4×4 determinant is done?

5.14. Show that

$$\begin{vmatrix} \lambda_1 & 0 & 0 & \cdots & 0 \\ 0 & \lambda_2 & 0 & \cdots & 0 \\ 0 & 0 & \lambda_3 & \cdots & 0 \\ \vdots & \vdots & \vdots & & \vdots \\ 0 & 0 & 0 & \cdots & \lambda_n \end{vmatrix} = \lambda_1\lambda_2\lambda_3\ldots\lambda_n$$

5.15. Show that

$$\begin{vmatrix} A_{11} & A_{12}\ldots A_{1n} & 0 & 0\ldots 0 \\ A_{21} & A_{22}\ldots A_{2n} & 0 & 0\ldots 0 \\ \vdots & \vdots \quad \vdots & \vdots & \vdots \quad \vdots \\ A_{n1} & A_{n2}\ldots A_{nn} & 0 & 0\ldots 0 \\ C_{11} & C_{12}\ldots C_{1n} & B_{11} & B_{12}\ldots B_{1m} \\ C_{21} & C_{22}\ldots C_{2n} & B_{21} & B_{22}\ldots B_{2m} \\ \vdots & \vdots \quad \vdots & \vdots & \vdots \quad \vdots \\ C_{m1} & C_{m2}\ldots C_{mn} & B_{m1} & B_{m2}\ldots B_{mm} \end{vmatrix} = |A||B|$$

6. LINEAR DEPENDENCE AND INDEPENDENCE

A set of vectors $\mathbf{a}_1, \mathbf{a}_2, \ldots, \mathbf{a}_n$ is said to be *linearly dependent* if constants $\alpha_1, \alpha_2, \ldots, \alpha_n$ not all zero exist, such that

$$\alpha_1\mathbf{a}_1 + \alpha_2\mathbf{a}_2 + \cdots + \alpha_n\mathbf{a}_n = 0 \tag{6.1}$$

The concept applies to a space of any number of dimensions. In an ordinary three-dimensional Cartesian space, the linear dependence of three vectors

$$\alpha\mathbf{a} + \beta\mathbf{b} + \gamma\mathbf{c} = 0 \tag{6.2}$$

has the geometrical significance that all three of the vectors lie in the same plane.

According to theorems 4 and 5 of section 5, a determinant has the value zero *if, and only if*, a linear relation exists among its rows (columns). Thus, $|A| = 0$ implies that the rows (columns) of the square matrix A can be thought of as representing linearly dependent vectors. This provides us with a convenient test for deciding on the linear dependence or independence of vectors. Suppose that three vectors in ordinary three-dimensional space are given:

$$\begin{aligned} \mathbf{a} &= a_1\hat{\mathbf{i}} + a_2\hat{\mathbf{j}} + a_3\hat{\mathbf{k}} \\ \mathbf{b} &= b_1\hat{\mathbf{i}} + b_2\hat{\mathbf{j}} + b_3\hat{\mathbf{k}} \\ \mathbf{c} &= c_1\hat{\mathbf{i}} + c_2\hat{\mathbf{j}} + c_3\hat{\mathbf{k}} \end{aligned} \tag{6.3}$$

Construct the determinant

$$|A| = \begin{vmatrix} a_1 & b_1 & c_1 \\ a_2 & b_2 & c_2 \\ a_3 & b_3 & c_3 \end{vmatrix} \tag{6.4}$$

If $|A| = 0$, the vectors are linearly dependent; if $|A| \neq 0$, the vectors are linearly independent and no relation of the form (6.2) exists.

Theorem 1. In a three-dimensional space, any four (or more) vectors are necessarily linearly dependent.

Proof: Of the given vectors **a**, **b**, **c**, and **d**, assume that **a**, **b**, and **c** are linearly independent. (If **a**, **b**, and **c** are already linearly dependent, there is nothing to prove.) Write

$$\mathbf{d} = \alpha\mathbf{a} + \beta\mathbf{b} + \gamma\mathbf{c} \tag{6.5}$$

which is three equations in the three unknown quantities α, β, and γ. The determinant of the coefficients of (6.5) is given by (6.4), and $|A| \neq 0$ by assumption of the linear independence of **a**, **b**, and **c**. Therefore, a nonzero solution of (6.5) as found by Cramer's rule or Gauss elimination exists.

As a generalization of theorem 1, there can be at most n linearly independent vectors in an n-dimensional space. Such a set of vectors is said to *span* the space. A set of n linearly independent vectors in an n-dimensional space is also said to be *complete* in that any other vector can always be expressed as a linear combination of them.

Theorem 2. Let A be a square matrix and X be a column vector. If $|A| \neq 0$, the *only* solution of $AX = 0$ is $X = 0$.

Proof: Written out, $AX = 0$ is

$$x_1\begin{pmatrix} A_{11} \\ A_{21} \\ A_{31} \end{pmatrix} + x_2\begin{pmatrix} A_{12} \\ A_{22} \\ A_{32} \end{pmatrix} + x_3\begin{pmatrix} A_{13} \\ A_{23} \\ A_{33} \end{pmatrix} = 0 \tag{6.6}$$

Since $|A| \neq 0$, the columns of A represent three linearly independent vectors. Thus, $x_1 = x_2 = x_3 = 0$. The proof of course holds if A is an $n \times n$ matrix and X is an n-component vector.

Theorem 3. If A is a square matrix and $|A| \neq 0$, the solution of $AX = Y$ is unique.

Proof: Assume that two solutions exist: $AX = Y$ and $AX' = Y$. Then, $A(X - X') = 0$. By theorem 2, $X - X' = 0$.

Theorem 3 means that the solution of n nonhomogeneous equations in n unknowns, where $|A| \neq 0$, as given by Cramer's rule, Gauss elimination, or other means, is the only solution possible. As another consequence, the representation (6.5) of the three-dimensional vector **d** in terms of the linearly independent vectors **a**, **b**, and **c** is unique. In particular, a vector can be written in terms of its rectangular Cartesian components as

$$\mathbf{a} = a_x\hat{\mathbf{i}} + a_y\hat{\mathbf{j}} + a_z\hat{\mathbf{k}} \tag{6.7}$$

in one and only one way. An equation such as

$$a_x\hat{\mathbf{i}} + a_y\hat{\mathbf{j}} + a_z\hat{\mathbf{k}} = b_x\hat{\mathbf{i}} + b_y\hat{\mathbf{j}} + b_z\hat{\mathbf{k}} \tag{6.8}$$

implies that $a_x = b_x$, $a_y = b_y$, and $a_z = b_z$.

PROBLEMS

6.1. Decide which of the following two sets of vectors is linearly dependent:

(1) $\mathbf{a} = (4, -1, 2)$
$\mathbf{b} = (-1, 3, 1)$
$\mathbf{c} = (-3, 9, 3)$

(2) $\mathbf{a} = (1, 0, -1)$
$\mathbf{b} = (2, 2, 1)$
$\mathbf{c} = (-2, 1, 5)$

For the linearly dependent case, find the numbers α and β such that

$\mathbf{c} = \alpha\mathbf{a} + \beta\mathbf{b}$

6.2. Express the vector $\mathbf{d} = (8, -2, 5)$ as a linear combination of the set of linearly independent vectors given in problem 6.1.

6.3. Show that

$$\begin{vmatrix} 1 & -1 & 0 & 2 \\ 0 & 3 & -1 & -5 \\ -1 & 0 & 2 & 3 \\ 2 & 1 & 1 & 3 \end{vmatrix} = 0$$

Find a linear relation among the columns by finding constants α, β, γ, and δ such that

$$\alpha\begin{pmatrix} 1 \\ 0 \\ -1 \\ 2 \end{pmatrix} + \beta\begin{pmatrix} -1 \\ 3 \\ 0 \\ 1 \end{pmatrix} + \gamma\begin{pmatrix} 0 \\ -1 \\ 2 \\ 1 \end{pmatrix} + \delta\begin{pmatrix} 2 \\ -5 \\ 3 \\ 3 \end{pmatrix} = 0$$

Also find a similar linear relation among the rows.

7. THE RANK OF A MATRIX

The rank of a matrix is the dimension of the largest nonzero determinant that can be made up out of its elements. Dimension means the size (number of rows and columns) of the determinant. The term *order* is also used. (The term "rank" is used in another way in tensor analysis to be discussed later.) There are three important facts about the rank of a matrix:

1. The rank of a square matrix is the same as its dimension if, and only if, the rows (columns) are linearly independent. This follows from theorems 4 and 5 of section 5.
2. The rank of a matrix is not changed if a row (column) is multiplied by a scalar or if a row (column) is multiplied by a scalar and the result is added to another row (column). This is true because the determination of rank involves evaluating a determinant.
3. The rows (columns) of a matrix can be rearranged in any way without changing its rank.

The determination of rank is illustrated by the following example:

$$A = \begin{pmatrix} 1 & 2 & 3 \\ 5 & 2 & 1 \\ 6 & 4 & 4 \end{pmatrix} \rightarrow \quad (\text{column 1} \times -2 \text{ added to column 2}; \ \text{column 1} \times -3 \text{ added to column 3})$$

$$\begin{pmatrix} 1 & 0 & 0 \\ 5 & -8 & -14 \\ 6 & -8 & -14 \end{pmatrix} \rightarrow \quad (\text{row 1} \times -5 \text{ added to row 2}; \ \text{row 1} \times -6 \text{ added to row 3})$$

$$\begin{pmatrix} 1 & 0 & 0 \\ 0 & -8 & -14 \\ 0 & -8 & -14 \end{pmatrix} \rightarrow \quad (\text{row 2} \times -1 \text{ added to row 3})$$

$$\begin{pmatrix} 1 & 0 & 0 \\ 0 & -8 & -14 \\ 0 & 0 & 0 \end{pmatrix} \rightarrow \quad (\text{column 2} \times -\tfrac{14}{8} \text{ added to column 3})$$

$$\begin{pmatrix} 1 & 0 & 0 \\ 0 & -8 & 0 \\ 0 & 0 & 0 \end{pmatrix} \tag{7.1}$$

Arrows are used in (7.1) because the five matrices appearing there are not equal to one another, but they do all have the same rank, namely, two in this case. Here is another example:

$$A = \begin{pmatrix} 1 & -1 & 2 \\ 2 & -2 & 4 \\ -1 & 1 & -2 \end{pmatrix} \rightarrow$$

(column operations: 1 × column 1 added to column 2; −2 × column 1 added to column 3)

$$\begin{pmatrix} 1 & 0 & 0 \\ 2 & 0 & 0 \\ -1 & 0 & 0 \end{pmatrix} \rightarrow$$

(row operations: −2 × row 1 added to row 2; 1 × row 1 added to row 3)

$$\begin{pmatrix} 1 & 0 & 0 \\ 0 & 0 & 0 \\ 0 & 0 & 0 \end{pmatrix} \tag{7.2}$$

This time, the rank of A is *one*. There is really only one linearly independent vector among its rows (columns). The rows (columns) of the original matrix differ from one another by only a scalar factor. The following theorem is a summary and generalization:

Theorem 1. The rank of a matrix is equal to the number of linearly independent vectors which make up its rows (columns). The theorem holds for both square and non-square matrices.

Proof: Let the matrix have n rows and m columns. Suppose that out of the n rows there are k linearly independent vectors. Multiplying one vector by a scalar factor and adding the result to another vector does not change the number of linearly independent vectors. The application of this process to both rows and columns ultimately gives the form

$$\begin{vmatrix} \lambda_1 & 0\ldots0 & 0\ldots0 \\ 0 & \lambda_2\ldots0 & 0\ldots0 \\ \vdots & \vdots\ \ \vdots & \vdots\ \ \vdots \\ 0 & 0\ldots\lambda_k & 0\ldots0 \\ 0 & 0\ldots0 & 0\ldots0 \\ \vdots & \vdots\ \ \vdots & \vdots\ \ \vdots \\ 0 & 0\ldots0 & 0\ldots0 \end{vmatrix} \tag{7.3}$$

which makes it evident that the rank and the number of linearly independent vectors are both equal to k.

PROBLEMS

7.1. In the proof of theorem 1, it was stated that, given k linearly independent vectors, the process of multiplying one vector by a scalar and adding the result to another vector does not change the number of independent vectors. Prove this by showing that, if $\mathbf{a}_1, \mathbf{a}_2, \ldots, \mathbf{a}_k$ are k linearly independent vectors, then $\mathbf{a}_1$, $(\mathbf{a}_2 + \alpha\mathbf{a}_1), \ldots, \mathbf{a}_k$ is another set of k linearly independent vectors.

7.2. If $\mathbf{a}_1, \mathbf{a}_2, \ldots, \mathbf{a}_n$ are n linearly independent vectors and if

$$\mathbf{b}_i = \sum_{k=1}^{n} A_{ik}\mathbf{a}_k \qquad i = 1, 2, \ldots, n$$

show that the n vectors $\mathbf{b}_i$ are also linearly independent provided that $|A| \neq 0$. Here, A is the $n \times n$ matrix with elements A_{ik}. What is the matrix A for problem 7.1?

7.3. Find the rank of

$$\begin{pmatrix} 1 & 3 & -1 & 0 & 5 \\ -1 & 0 & 8 & -2 & 3 \\ 1 & -4 & 5 & -1 & 0 \\ 2 & -5 & 17 & -4 & 8 \end{pmatrix}$$

8. CONSISTENCY OF LINEAR ALGEBRAIC EQUATIONS

The solution of $AX = Y$ where A is an $n \times n$ square matrix with $|A| \neq 0$ has been discussed: The solution for X is unique; in particular, the homogeneous case $Y = 0$ has only the solution $X = 0$ (theorems 2 and 3 of section 6).

Consider now the possibility of solving $AX = Y$ where A is an $n \times n$ square matrix and $|A| = 0$. The condition $|A| = 0$ means that the rows are linearly dependent. If the dimension of A is n and its rank is k, there are $n - k$ rows which can be written as linear combinations of the k independent rows. Thus, there are actually $n-k$ relations of the form

$$\sum_{i=1}^{n} \alpha_i A_{ij} = 0 \tag{8.1}$$

Since $AX = Y$ written out is

$$\sum_{j=1}^{n} A_{ij} x_j = y_i \tag{8.2}$$

the components y_j of the vector Y must obey

$$\sum_i \sum_j \alpha_i A_{ij} x_j = \sum_i \alpha_i y_i = 0 \tag{8.3}$$

Otherwise, $AX = Y$ is inconsistent. What we are saying is simply that the same linear relations must exist among the components y_i of Y as exist among the rows of A. In other words, some of the n equations are redundant. How do we know that this is the case? Form the matrix

$$\begin{pmatrix} A_{11} & A_{12} \cdots A_{1n} & y_1 \\ A_{21} & A_{22} \cdots A_{2n} & y_2 \\ \vdots & \vdots \quad \vdots & \vdots \\ A_{n1} & A_{n2} \cdots A_{nn} & y_n \end{pmatrix} \tag{8.4}$$

This is a matrix with n rows and $n+1$ columns and is called the *augmented matrix* of the linear system $AX = Y$. If the rows of (8.4) bear the same linear relations among themselves as they did in the original matrix A, everything is OK. If not, there is trouble and $AX = Y$ is inconsistent. The test, then, is to examine the rank of (8.4) and see if it is still the same as that of the original matrix A. If the rank of (8.4) is not larger than the rank of A (adding an extra column can never cause a decrease in rank), then $AX = Y$ is consistent and a solution exists.

Theorem 1. If A is an $n \times n$ matrix and if rank $A = k$, there are $n-k$ linearly independent solutions of $AX = 0$.

Proof: Let the vectors $\mathbf{a}_1, \mathbf{a}_2, \ldots, \mathbf{a}_n$ represent the columns of A. Then, $AX = 0$ is equivalent to

$$x_1 \mathbf{a}_1 + x_2 \mathbf{a}_2 + \cdots + x_n \mathbf{a}_n = 0 \tag{8.5}$$

Since k of the n vectors are linearly independent, there are $n-k$ relations of the form (8.5), that is, $n-k$ sets of values $x_1, x_2, \ldots, x_n$. If these are written as column vectors $X_1, X_2, \ldots, X_{n-k}$, then the general solution of $AX = 0$ can be expressed as

$$X = \beta_1 X_1 + \beta_2 X_2 + \cdots + \beta_{n-k} X_{n-k} \tag{8.6}$$

where $\beta_1, \beta_2, \ldots, \beta_{n-k}$ are arbitrary constants.

Let A be an $n \times n$ matrix of rank k. Let $AX = Y$ be consistent and have as one possible solution $X = X_0$. Then, as a consequence of theorem 1, the reader can easily prove the following theorem:

Theorem 2. The general solution of $AX = Y$ is

$$X = \beta_1 X_1 + \beta_2 X_2 + \cdots + \beta_{n-k} X_{n-k} + X_0 \tag{8.7}$$

The content of theorems 1 and 2 is illustrated by finding the general solution of

$$AX = \begin{pmatrix} 1 & 0 & -2 & 1 \\ 3 & -1 & 1 & 0 \\ 4 & -1 & -1 & 1 \\ 2 & -1 & 3 & -1 \end{pmatrix} \begin{pmatrix} x_1 \\ x_2 \\ x_3 \\ x_4 \end{pmatrix} = \begin{pmatrix} 2 \\ 1 \\ 3 \\ -1 \end{pmatrix} \tag{8.8}$$

First, we examine the rank of the coefficient matrix:

$$A = \begin{pmatrix} 1 & 0 & -2 & 1 \\ 3 & -1 & 1 & 0 \\ 4 & -1 & -1 & 1 \\ 2 & -1 & 3 & -1 \end{pmatrix} \to \begin{pmatrix} 1 & 0 & 0 & 0 \\ 0 & -1 & 7 & -3 \\ 0 & -1 & 7 & -3 \\ 0 & -1 & 7 & -3 \end{pmatrix} \to \begin{pmatrix} 1 & 0 & 0 & 0 \\ 0 & -1 & 0 & 0 \\ 0 & 0 & 0 & 0 \\ 0 & 0 & 0 & 0 \end{pmatrix}$$

The rank of A is 2. Now find the rank of the augmented matrix:

$$\begin{pmatrix} 1 & 0 & -2 & 1 & 2 \\ 3 & -1 & 1 & 0 & 1 \\ 4 & -1 & -1 & 1 & 3 \\ 2 & -1 & 3 & -1 & -1 \end{pmatrix} \to \begin{pmatrix} 1 & 0 & 0 & 0 & 0 \\ 0 & -1 & 7 & -3 & -5 \\ 0 & -1 & 7 & -3 & -5 \\ 0 & -1 & 7 & -3 & -5 \end{pmatrix} \to \begin{pmatrix} 1 & 0 & 0 & 0 & 0 \\ 0 & -1 & 0 & 0 & 0 \\ 0 & 0 & 0 & 0 & 0 \\ 0 & 0 & 0 & 0 & 0 \end{pmatrix}$$

Thus, rank aug $A = 2$, and (8.8) is therefore consistent. To find the general solution of the homogeneous system $AX = 0$, pick out two columns which represent linearly independent vectors, say, the first two. Then, express the third and fourth columns in terms of the first two as

$$a_1 \begin{pmatrix} 1 \\ 3 \\ 4 \\ 2 \end{pmatrix} + a_2 \begin{pmatrix} 0 \\ -1 \\ -1 \\ -1 \end{pmatrix} + \begin{pmatrix} -2 \\ 1 \\ -1 \\ 3 \end{pmatrix} + 0 \begin{pmatrix} 1 \\ 0 \\ 1 \\ -1 \end{pmatrix} = 0 \tag{8.9}$$

$$b_1 \begin{pmatrix} 1 \\ 3 \\ 4 \\ 2 \end{pmatrix} + b_2 \begin{pmatrix} 0 \\ -1 \\ -1 \\ -1 \end{pmatrix} + 0 \begin{pmatrix} -2 \\ 1 \\ -1 \\ 3 \end{pmatrix} + \begin{pmatrix} 1 \\ 0 \\ 1 \\ -1 \end{pmatrix} = 0 \tag{8.10}$$

We of course know that there are exactly two linearly independent vectors from the fact that rank $A = 2$. We have written an apparently superfluous term in each of equations (8.9) and (8.10) because it shows that each is equivalent to $AX = 0$ if X is either $(a_1, a_2, 1, 0)$ or $(b_1, b_2, 0, 1)$. These two possibilities represent two linearly independent solutions of $AX = 0$. We find $a_1 = 2$, $a_2 = 7$, $b_1 = -1$, and $b_2 = -3$. Therefore,

$$X = \beta_1 \begin{pmatrix} 2 \\ 7 \\ 1 \\ 0 \end{pmatrix} + \beta_2 \begin{pmatrix} -1 \\ -3 \\ 0 \\ 1 \end{pmatrix} \tag{8.11}$$

represents the most general solution of $AX = 0$. In (8.11), β_1 and β_2 are arbitrary constants. Now we need to find a particular solution of $AX = Y$. (Any one will do.) Since there are really only two linearly independent equations, look at the first two:

$$\begin{aligned} x_1 \qquad - 2x_3 + x_4 &= 2 \\ 3x_1 - x_2 + x_3 \qquad &= 1 \end{aligned} \tag{8.12}$$

Choose $x_3 = x_4 = 1$. Then, $x_1 = 3$ and $x_2 = 9$. These solutions also work in the remaining two equations:

$$\begin{aligned} 4x_1 - x_2 - x_3 + x_4 &= 3 \\ 2x_1 - x_2 + 3x_3 - x_4 &= -1 \end{aligned} \tag{8.13}$$

Finally, the general solution of $AX = Y$ is

$$X = \begin{pmatrix} 3 \\ 9 \\ 1 \\ 1 \end{pmatrix} + \beta_1 \begin{pmatrix} 2 \\ 7 \\ 1 \\ 0 \end{pmatrix} + \beta_2 \begin{pmatrix} -1 \\ -3 \\ 0 \\ 1 \end{pmatrix} \tag{8.14}$$

The constants β_1 and β_2 are completely undetermined and can take on any values. (The number of undetermined constants is the difference between the dimension of A and its rank.)

Theorem 3. The general test for the consistency (existence of a solution) of a set of nonhomogeneous equations $AX = Y$ is rank aug A = rank A.

Proof: The case where A is square (same number of equations as unknowns) has already been examined in detail. Suppose that there are more equations than unknowns, for example:

$$\begin{pmatrix} A_{11} & A_{12} & A_{13} \\ A_{21} & A_{22} & A_{23} \\ A_{31} & A_{32} & A_{33} \\ A_{41} & A_{42} & A_{43} \end{pmatrix} \begin{pmatrix} x_1 \\ x_2 \\ x_3 \end{pmatrix} = \begin{pmatrix} y_1 \\ y_2 \\ y_3 \\ y_4 \end{pmatrix} \tag{8.15}$$

By theorem 1 of section 6, at least one linear relation

$$\sum_{i=1}^{4} \alpha_i A_{ij} = 0 \tag{8.16}$$

must exist among the rows. The argument from here on out is basically the same as for the case where A is square. For consistency, the same linear relation as (8.16) must exist among the components y_i of Y:

$$\sum_{i=1}^{4} \alpha_i y_i = 0 \tag{8.17}$$

This will be true if rank aug A = rank A. If there are n unknowns and m equations, where $m > n$, at least $m - n$ of the equations must be redundant. These can be discarded giving n equations in n unknowns which we already know how to solve.

An example of the case where there are fewer equations than unknowns is

$$\begin{pmatrix} A_{11} & A_{12} & A_{13} \\ A_{21} & A_{22} & A_{23} \end{pmatrix} \begin{pmatrix} x_1 \\ x_2 \\ x_3 \end{pmatrix} = \begin{pmatrix} y_1 \\ y_2 \end{pmatrix} \tag{8.18}$$

If rank $A = 2$, the two rows of A are linearly independent. We could write

$$\begin{aligned} A_{11}x_1 + A_{12}x_2 &= y_1 - A_{13}x_3 \\ A_{21}x_1 + A_{22}x_2 &= y_2 - A_{23}x_3 \end{aligned} \tag{8.19}$$

and solve for x_1 and x_2 in terms of x_3, provided of course that

$$\begin{vmatrix} A_{11} & A_{12} \\ A_{21} & A_{22} \end{vmatrix} \neq 0 \qquad (8.20)$$

[This is like (8.12).] If rank $A = 1$, it means that one row in (8.18) is a multiple of the other:

$$A_{21} = \alpha A_{11} \qquad A_{22} = \alpha A_{12} \qquad A_{23} = \alpha A_{13} \qquad (8.21)$$

For consistency, we must then have $y_2 = \alpha y_1$. In other words, rank A = rank aug A. The argument generalizes to the case of k equations in n unknowns where $k < n$.

The homogeneous system $AX = 0$ is of course always consistent, A square or not, in that necessarily rank A = rank aug A. This is because adding a column of zeros to A does not change its rank. There is always at least one solution of $AX = 0$, namely, $X = 0$.

Lastly, we add that in all cases where $AX = Y$ is consistent, A can be made square by discarding redundant equations if there are more equations than unknowns or by adding dummy equations (equations that are linear combinations of already existing equations) if the number of equations is less than the number of unknowns. We could then proceed to find the general solution in the form given by theorem 2 and illustrated by equation (8.8).

PROBLEMS

8.1. Show that

$$x_1 + 2x_2 + 3x_3 = 0$$
$$2x_1 + x_2 - x_3 = 0$$

can be solved for x_1 and x_2 in terms of x_3 for any choice of x_3.

8.2. For what value (or values) of x_3 does

$$x_1 + 2x_2 + 3x_3 = 0$$
$$2x_1 + 4x_2 - x_3 = 0$$

represent a set of consistent equations for x_1 and x_2?

8.3. Refer to problem 5.9. An experimenter reports the following data: $R_{AB} = 8\,\Omega$, $R_{AD} = 20\,\Omega$, $R_{AC} = 15\,\Omega$, $R_{BD} = 15\,\Omega$, $R_{BC} = 17\,\Omega$, and $R_{CD} = 6\,\Omega$. Show that the experimenter has made an error in his measurements.

8.4. Construct the general solution of

$$\begin{pmatrix} 5 & 2 & -3 & 0 & 1 \\ -1 & 0 & 1 & 2 & 3 \\ 4 & 2 & -2 & 2 & 4 \\ 6 & 2 & -4 & -2 & -2 \\ 3 & 2 & -5 & -4 & -5 \end{pmatrix} \begin{pmatrix} x_1 \\ x_2 \\ x_3 \\ x_4 \\ x_5 \end{pmatrix} = \begin{pmatrix} 2 \\ 1 \\ 3 \\ 1 \\ 0 \end{pmatrix}$$

as given by (8.7).

8.5. In the proof of theorem 1, it was stated that if out of n vectors k are linearly independent, then there are precisely $n - k$ relations of the form

$$x_1\mathbf{a}_1 + x_2\mathbf{a}_2 + \cdots + x_n\mathbf{a}_n = 0 \qquad (1)$$

If we assume that $\mathbf{a}_1, \mathbf{a}_2, \ldots, \mathbf{a}_k$ are the linearly independent vectors, then the $n - k$ relations can be written

$$\begin{aligned} x_{11}\mathbf{a}_1 + x_{12}\mathbf{a}_2 + \cdots + x_{1k}\mathbf{a}_k + \mathbf{a}_{k+1} &= 0 \\ x_{21}\mathbf{a}_1 + x_{22}\mathbf{a}_2 + \cdots + x_{2k}\mathbf{a}_k + \mathbf{a}_{k+2} &= 0 \\ &\vdots \\ x_{n-k,1}\mathbf{a}_1 + x_{n-k,2}\mathbf{a}_2 + \cdots + x_{n-k,k}\mathbf{a}_k + \mathbf{a}_n &= 0 \end{aligned} \qquad (2)$$

[We did this in solving (8.8).] Show that there are no more new such relations by showing that if (1) is any other possibility, then

$$\begin{pmatrix} x_1 \\ x_2 \\ \vdots \\ x_k \\ x_{k+1} \\ x_{k+2} \\ \vdots \\ x_n \end{pmatrix} = x_{k+1}\begin{pmatrix} x_{11} \\ x_{12} \\ \vdots \\ x_{1k} \\ 1 \\ 0 \\ \vdots \\ 0 \end{pmatrix} + x_{k+2}\begin{pmatrix} x_{21} \\ x_{22} \\ \vdots \\ x_{2k} \\ 0 \\ 1 \\ \vdots \\ 0 \end{pmatrix} + \cdots + x_n\begin{pmatrix} x_{n-k,1} \\ x_{n-k,2} \\ \vdots \\ x_{n-k,k} \\ 0 \\ 0 \\ \vdots \\ 1 \end{pmatrix} \tag{3}$$

meaning that (1) is really a linear combination of the equations given in (2). As a consequence, (8.6) really does represent the general solution of $AX=0$; any apparently different solution just involves a different set of constants $\beta_1, \beta_2, \ldots, \beta_{n-k}$. Note also that the $n-k$ vectors which appear on the right-hand side of (3) really are linearly independent regardless of the values of the coefficients x_{ij}, so that there are *not less* than $n-k$ linearly independent solutions of $AX=0$.

8.6. Find the general solution of

$$\begin{pmatrix} 1 & 1 & 1 \\ 0 & 0 & 0 \\ 0 & 0 & 0 \end{pmatrix}\begin{pmatrix} x_1 \\ x_2 \\ x_3 \end{pmatrix} = 0$$

as given by (8.6). This is an interesting, if somewhat trivial, example. The coefficient matrix obviously has rank 1; all the column vectors are in fact identical.

9. A LINEAR MECHANICAL SYSTEM

A mechanical system consists of two identical masses m constrained to move along a straight line and three identical springs each with force constant k. The two end springs are fastened to fixed supports as shown in Fig. 9.1. When the system is in equilibrium, the springs are extended an amount s, so that the tension in each spring is ks. When the system is displaced, the extensions of the three springs are x_1+s, x_2-x_1+s, and $s-x_2$, where x_1 and x_2 are the displacements of the two masses from equilibrium. By application of Newton's second law to each mass,

$$k(x_2-x_1+s)-k(s+x_1)=m\ddot{x}_1$$
$$k(s-x_2)-k(x_2-x_1+s)=m\ddot{x}_2 \tag{9.1}$$

where the dots refer to differentiation with respect to time: $\ddot{x}_1 = d^2x_1/dt^2$. The masses are assumed to move without friction. The factors of ks cancel out, giving

$$m\ddot{x}_1+2kx_1-kx_2=0 \tag{9.2}$$

$$m\ddot{x}_2-kx_1+2kx_2=0 \tag{9.3}$$

It is possible to express (9.2) and (9.3) as a single matrix equation:

$$M\ddot{X}+KX=0 \tag{9.4}$$

where

$$M=\begin{pmatrix} m & 0 \\ 0 & m \end{pmatrix} \qquad K=\begin{pmatrix} 2k & -k \\ -k & 2k \end{pmatrix}$$

$$X=\begin{pmatrix} x_1 \\ x_2 \end{pmatrix} \tag{9.5}$$

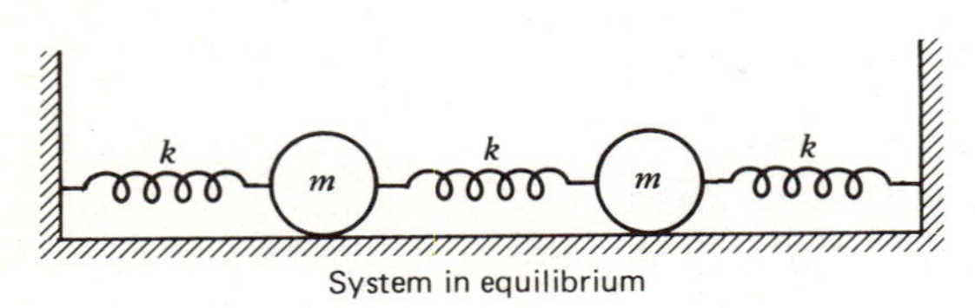

System in equilibrium

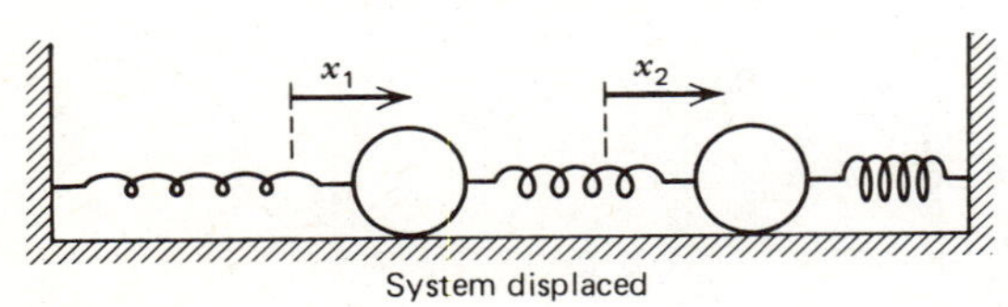

System displaced

Fig. 9.1

Our purpose is to find the possible solutions of (9.4).

We remember that the equation for a harmonic oscillator consisting of a single mass and a spring is $m\ddot{x} + kx = 0$ and that the general solution is $x = a\cos(\omega t - \phi)$, where $\omega = \sqrt{k/m}$ and a and ϕ are constants, to be determined by the initial conditions. The frequency ν of the motion is related to ω by $\omega = 2\pi\nu$. Sometimes ω is referred to as the circular frequency; it has dimensions of radians per second. We try the same approach to the solution of (9.4). The proposed solution is

$$X = A\cos(\omega t - \phi) \tag{9.6}$$

where the constant factor A is now a column matrix:

$$A = \begin{pmatrix} a_1 \\ a_2 \end{pmatrix} \tag{9.7}$$

Substitute (9.6) into (9.4) as follows:

$$\ddot{X} = -A\omega^2 \cos(\omega t - \phi)$$

$$M[-A\omega^2 \cos(\omega t - \phi)] + KA\cos(\omega t - \phi) = 0 \tag{9.8}$$

Use care to preserve the order of the factors since M, A, and K are matrices. Equation (9.8) is true for all time if

$$(K - M\omega^2)A = 0 \tag{9.9}$$

Written out in full, (9.9) is

$$\begin{pmatrix} 2k - m\omega^2 & -k \\ -k & 2k - m\omega^2 \end{pmatrix}\begin{pmatrix} a_1 \\ a_2 \end{pmatrix} = 0 \tag{9.10}$$

which is recognized as a homogeneous system of two equations in two unknowns. Such a system is always consistent and has at least one solution: $a_1 = a_2 = 0$. This corresponds to no motion of the system at all and is not very interesting. We know that (9.10) can have a nontrivial solution if, and only if,

$$\begin{vmatrix} 2k - m\omega^2 & -k \\ -k & 2k - m\omega^2 \end{vmatrix} = 0 \tag{9.11}$$

(See theorem 2, section 6, and theorem 1, section 8.) The mass m and spring constant k are fixed; however, ω is as yet undetermined. Let us then regard (9.11) as a condition on ω. By expanding the determinant, we find

$$(2k - m\omega^2)^2 - k^2 = 0 \tag{9.12}$$

The square root of (9.12) is $2k - m\omega^2 = \pm k$; therefore,

$$\omega^2 = \frac{2k \pm k}{m} \tag{9.13}$$

This means that there are actually two values of ω which will satisfy (9.11):

$$\omega_1 = \sqrt{\frac{k}{m}} \qquad \omega_2 = \sqrt{\frac{3k}{m}} \tag{9.14}$$

Equation (9.11) is called the *secular equation*, and the values (9.14) are referred to as *proper values* or *eigenvalues*.

For each eigenvalue (9.14), there is a nontrivial solution of (9.10). For $\omega = \omega_1$, (9.10) becomes

$$\begin{pmatrix} k & -k \\ -k & k \end{pmatrix}\begin{pmatrix} a_1 \\ a_2 \end{pmatrix} = 0 \tag{9.15}$$

which has the solution $a_1 = a_2$, or

$$\begin{pmatrix} a_1 \\ a_2 \end{pmatrix} = c_1\begin{pmatrix} 1 \\ 1 \end{pmatrix} \tag{9.16}$$

where c_1 is an undetermined constant. For $\omega = \omega_2$, (9.10) is

$$\begin{pmatrix} -k & -k \\ -k & -k \end{pmatrix}\begin{pmatrix} a_1 \\ a_2 \end{pmatrix} = 0 \tag{9.17}$$

which has the solution $a_1 = -a_2$, or

$$\begin{pmatrix} a_1 \\ a_2 \end{pmatrix} = c_2\begin{pmatrix} 1 \\ -1 \end{pmatrix} \tag{9.18}$$

where c_2 is another undetermined constant. The two solutions (9.16) and (9.18) are called *eigenvectors*. Thus, there are actually two solutions of the original system (9.4):

$$X_1 = c_1\begin{pmatrix} 1 \\ 1 \end{pmatrix}\cos(\omega_1 t - \phi_1) \tag{9.19}$$

$$X_2 = c_2\begin{pmatrix} 1 \\ -1 \end{pmatrix}\cos(\omega_2 t - \phi_2) \tag{9.20}$$

The solution (9.19) represents an oscillation of the two masses in phase with one another at the frequency ω_1, whereas (9.20) represents oscillations of the two masses exactly out of phase with one another at a frequency ω_2.

Since (9.4) is a linear system, the general solution is a superposition of (9.19) and (9.20):

$$\begin{pmatrix} x_1 \\ x_2 \end{pmatrix} = c_1 \begin{pmatrix} 1 \\ 1 \end{pmatrix} \cos(\omega_1 t - \phi_1) + c_2 \begin{pmatrix} 1 \\ -1 \end{pmatrix} \cos(\omega_2 t - \phi_2) \tag{9.21}$$

In (9.21), there are altogether four undetermined constants: c_1, c_2, and the two phase factors ϕ_1 and ϕ_2. These four constants must be found from specific initial conditions—usually the two positions and two velocities at $t = 0$. The two possible solutions (9.19) and (9.20) are sometimes referred to as the *normal modes* of oscillation of the system.

PROBLEMS

9.1. Show that the solution (9.21) can be expressed in the form

$$\begin{pmatrix} x_1 \\ x_2 \end{pmatrix} = \begin{pmatrix} 1 \\ 1 \end{pmatrix}(b_1 \sin \omega_1 t + b_2 \cos \omega_1 t) + \begin{pmatrix} 1 \\ -1 \end{pmatrix}(b_3 \sin \omega_2 t + b_4 \cos \omega_2 t)$$

where b_1, b_2, b_3, and b_4 are a new set of four constants. If the initial conditions (at $t = 0$) are $x_1(0) = 4$ cm, $x_2(0) = 6$ cm, $\dot{x}_1(0) = 0$, and $\dot{x}_2(0) = 0$, find the numerical values of all four constants.

9.2. Consider a one-dimensional mass–spring system consisting of two masses and three springs where the two masses are identical but the center spring has a different value from the other two. Find the general solution for the motion of this system. Evaluate all undetermined constants for the case where the initial conditions are $x_1(0) = x_0$, $x_2(0) = 0$, $\dot{x}(0) = 0$, and $\dot{x}_2(0) = 0$. Examine the case where k_2 is very weak: $k_2/k_1 = \delta \ll 1$. Make a qualitative sketch of x_1 and x_2 vs. t. (This is an example of weakly coupled oscillators.) See Fig. 9.2.

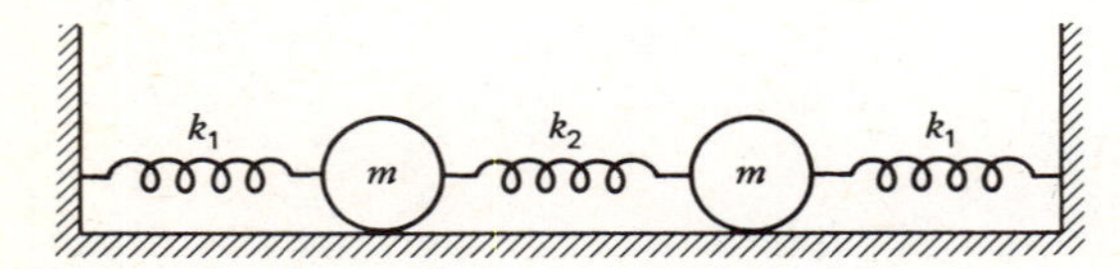

Fig. 9.2

9.3. A mass–spring system constrained to move in one dimension has three identical masses and four identical springs as shown in Fig. 9.3. Find the general solution for the motion of this system.

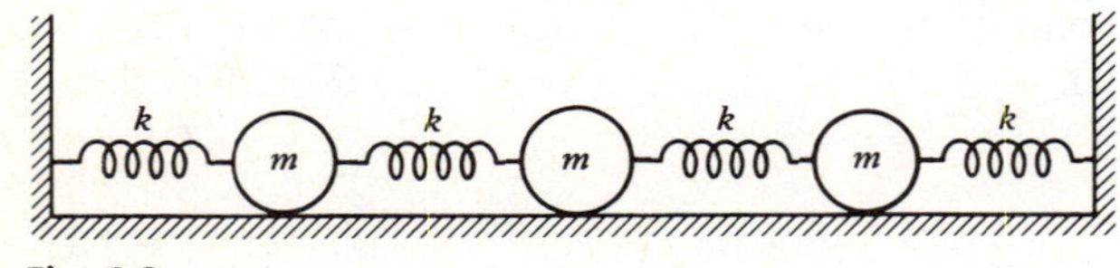

Fig. 9.3

10. THE INVERSE OF A MATRIX

Theorem 1. If A is an $n \times n$ matrix and if $X_1, X_2, \ldots, X_n$ are n linearly independent vectors, then $AX_1 = AX_2 = \cdots = AX_n = 0$ implies that $A = 0$.

Proof: Consider the n vectors

$$U_1 = \begin{pmatrix} 1 \\ 0 \\ \vdots \\ 0 \end{pmatrix} \quad U_2 = \begin{pmatrix} 0 \\ 1 \\ \vdots \\ 0 \end{pmatrix} \quad \cdots \quad U_n = \begin{pmatrix} 0 \\ 0 \\ \vdots \\ 1 \end{pmatrix} \tag{10.1}$$

which are n linearly independent unit vectors. Each of the vectors (10.1) can be expressed in a unique way in terms of the given linearly independent vectors $X_1, X_2, \ldots, X_n$, namely,

$$U_1 = \alpha_1 X_1 + \alpha_2 X_2 + \cdots + \alpha_n X_n$$

Thus, $AU_1 = AU_2 = \cdots = AU_n = 0$. If $AU_1 = 0$ is written out, the result is $A_{11} = 0, A_{21} = 0, \ldots, A_{n1} = 0$. Similarly, $AU_2 = 0$ gives $A_{12} = 0, A_{22} = 0, \ldots, A_{n2} = 0$. In this way, each element of A can be shown to be zero.

Theorem 2. If A and B are $n \times n$ square matrices and if $AX_i = BX_i$ $(i = 1, 2, \ldots, n)$ for a complete set of n linearly independent vectors $X_1, X_2, \ldots, X_n$, then $A = B$.

Proof: Theorem 2 is a direct consequence of theorem 1.

Theorem 3. Let A be an $n \times n$ matrix for which $|A| \neq 0$. If $AX_i = Y_i$ and the n vectors Y_i are known to be linearly independent, then the n solution vectors X_i are also linearly independent.

Proof: We shall show that

$$\alpha_1 X_1 + \alpha_2 X_2 + \cdots + \alpha_n X_n = 0 \tag{10.2}$$

implies that $\alpha_1 = \alpha_2 = \cdots = \alpha_n = 0$. Multiply (10.2) by the matrix A:

$$\alpha_1 AX_1 + \alpha_2 AX_2 + \cdots + \alpha_n AX_n = \alpha_1 Y_1 + \alpha_2 Y_2 + \cdots + \alpha_n Y_n = 0 \tag{10.3}$$

Since the Y_i are known to be linearly independent, the desired result follows.

Let Y_i be a given set of n linearly independent vectors. Suppose that A is an $n \times n$ matrix for which $|A| \neq 0$. Each equation

$$AX_i = Y_i \tag{10.4}$$

can be solved for X_i in terms of Y_i and the result expressed by a matrix equation as

$$X_i = BY_i \tag{10.5}$$

where B is an $n \times n$ square matrix derived from the elements of A. The solution is unique, meaning that the matrix B is unique. One way to formally construct B would be to use Cramer's rule; equation (5.44) illustrates the 3×3 case. The important properties of the matrix B are found by first multiplying (10.4) through by B:

$$BAX_i = BY_i \tag{10.6}$$

Now use (10.5) to get

$$BAX_i = X_i = IX_i \tag{10.7}$$

where I is the identity matrix. By theorem 2,

$$BA = I \tag{10.8}$$

Now multiply (10.5) by A:

$$AX_i = ABY_i = IY_i \tag{10.9}$$

Again by theorem 2,

$$AB = I \tag{10.10}$$

The matrix B is called the *inverse* of A and is generally written $B = A^{-1}$. We have proved the following theorem:

Theorem 4. A set of nonhomogeneous equations $AX = Y$ for which $|A| \neq 0$ can be solved uniquely as $X = A^{-1}Y$, where the inverse matrix A^{-1} has the property

$$A^{-1}A = AA^{-1} = I \tag{10.11}$$

If A is a square matrix, it is possible to write

$$AA = A^2 \qquad AAA = A^3 \qquad \text{and so on} \tag{10.12}$$

Thus, A^n, where n is a positive integer, is defined. If $|A| \neq 0$, it is also possible to define the negative powers of A:

$$A^{-1}A^{-1} = A^{-2} \qquad A^{-1}A^{-1}A^{-1} = A^{-3} \qquad \text{and so on} \tag{10.13}$$

Note that

$$A^{-2}A = A^{-1}A^{-1}A = A^{-1}I = A^{-1}$$

$$A^{-2}A^2 = A^{-1}A^{-1}AA = A^{-1}A = I$$

so that in general

$$A^n A^m = A^{n+m} \tag{10.14}$$

is true for n and m positive or negative integers or zero, provided that we understand

$$A^0 = I \tag{10.15}$$

The inverse of a matrix can be formally constructed by starting with equation (5.22):

$$\delta_{ij}|A| = \sum_{k=1}^{n} A_{ik}C_{jk} \tag{10.16}$$

where C_{jk} are cofactors of the elements A_{jk}. Equation (10.16) is the equivalent of (10.11), provided that

$$A^{-1} = \frac{1}{|A|}\begin{pmatrix} C_{11} & C_{21} \cdots C_{n1} \\ C_{12} & C_{22} \cdots C_{n2} \\ \vdots & \vdots \quad\quad \vdots \\ C_{1n} & C_{2n} \cdots C_{nn} \end{pmatrix} = \frac{1}{|A|}\tilde{C} \tag{10.17}$$

In (10.17), C is a matrix made up out of the cofactors of A. The matrix $\tilde{C}$ is the transpose of C and, as indicated in (10.17), results from interchanging the rows and columns of C.

Equation (10.17) provides a formal, but impractical, way of constructing A^{-1}. Another method is to use a kind of synthetic Gauss elimination procedure. Suppose that we want the inverse of

$$A = \begin{pmatrix} 1 & 2 & 1 \\ 2 & 5 & 2 \\ 3 & 8 & 4 \end{pmatrix} \tag{10.18}$$

First note that $|A| = 1$ so that A^{-1} exists. We can imagine that we are trying to solve

$$AX = IY \tag{10.19}$$

Instead of writing out the equations, we put down just the two matrices A and I:

$$A = \begin{pmatrix} 1 & 2 & 1 \\ 2 & 5 & 2 \\ 3 & 8 & 4 \end{pmatrix} \qquad I = \begin{pmatrix} 1 & 0 & 0 \\ 0 & 1 & 0 \\ 0 & 0 & 1 \end{pmatrix} \tag{10.20}$$

It is permissible in (10.19) to multiply any one of the three equations by a constant or to multiply any equation by a constant and add the result to any other equation. This is equivalent to multiplying any row of both matrices in (10.20) by a constant or multiplying any row by a constant and adding the result to another row. Column operations are not permitted here. These row operations are to be carried out until A is replaced by I:

$$A = \begin{pmatrix} 1 & 2 & 1 \\ 2 & 5 & 2 \\ 3 & 8 & 4 \end{pmatrix} \qquad \begin{pmatrix} 1 & 0 & 0 \\ 0 & 1 & 0 \\ 0 & 0 & 1 \end{pmatrix} \begin{matrix} \boxed{-2} \\ \boxed{-3} \\ \end{matrix}$$

$$\begin{pmatrix} 1 & 2 & 1 \\ 0 & 1 & 0 \\ 0 & 2 & 1 \end{pmatrix} \qquad \begin{pmatrix} 1 & 0 & 0 \\ -2 & 1 & 0 \\ -3 & 0 & 1 \end{pmatrix} \begin{matrix} \boxed{-2} \\ \\ \boxed{-2} \end{matrix}$$

$$\begin{pmatrix} 1 & 0 & 1 \\ 0 & 1 & 0 \\ 0 & 0 & 1 \end{pmatrix} \qquad \begin{pmatrix} 5 & -2 & 0 \\ -2 & 1 & 0 \\ 1 & -2 & 1 \end{pmatrix} \boxed{-1}$$

$$\begin{pmatrix} 1 & 0 & 0 \\ 0 & 1 & 0 \\ 0 & 0 & 1 \end{pmatrix} \qquad \begin{pmatrix} 4 & 0 & -1 \\ -2 & 1 & 0 \\ 1 & -2 & 1 \end{pmatrix} = A^{-1} \tag{10.21}$$

That the final matrix in (10.21) is in fact A^{-1} is verified by using (10.11).

In the remainder of this section, we state and prove a few miscellaneous theorems involving the inverse and the transpose of a matrix.

Theorem 5. If A^{-1} and B^{-1} exist, then $(AB)^{-1} = B^{-1}A^{-1}$.

Proof: $(B^{-1}A^{-1})AB = B^{-1}B = I$, $AB(B^{-1}A^{-1}) = AA^{-1} = I$, which implies that $B^{-1}A^{-1} = (AB)^{-1}$.

Theorem 6. $(\widetilde{AB}) = \tilde{B}\tilde{A}$.

Proof: In the proof, keep in mind that if A_{ik} is the ik element of A, then A_{ki} is the ik element of $\tilde{A}$: $(\tilde{A})_{ik} = A_{ki}$. We now calculate the ik element of $\widetilde{AB}$:

$$(\widetilde{AB})_{ik} = \sum_j A_{kj}B_{ji} = \sum_j (\tilde{A})_{jk}(\tilde{B})_{ij} = (\tilde{B}\tilde{A})_{ik}$$

which proves the result.

Theorem 7. If A^{-1} exists, then $(\widetilde{A^{-1}}) = (\tilde{A})^{-1}$.

Proof: Take the transpose of (10.11) and use theorem 6:

$$(\widetilde{AA^{-1}}) = (\widetilde{A^{-1}A}) = I \qquad (\widetilde{A^{-1}})\tilde{A} = \tilde{A}(\widetilde{A^{-1}}) = I$$

from which the desired result follows.

Finally, here is a bit more terminology: Matrices which do not have inverses are called *singular*. If A^{-1} exists, then A is said to be *nonsingular*.

PROBLEMS

10.1. Find the inverse of the matrix A given by (10.20) by the cofactor method as given by (10.17).

10.2. Suppose that $\mathbf{a}_1, \mathbf{a}_2, \ldots, \mathbf{a}_n$ and $\mathbf{b}_1, \mathbf{b}_2, \ldots, \mathbf{b}_n$ are both known to be linearly independent sets of vectors in an n-dimensional space. If $\mathbf{b}_i$ are written as linear combinations of $\mathbf{a}_i$ as

$$\mathbf{b}_i = \sum_{j=1}^{n} A_{ij}\mathbf{a}_j$$

show that the coefficients A_{ij} make up a nonsingular matrix. (This is the converse of the result given in problem 7.2.)

10.3. Show that theorem 7 follows directly from the representation of A^{-1} given by (10.17).

10.4. Use the synthetic Gauss elimination procedure to find the inverse of

$$\begin{pmatrix} -1 & 3 & 2 \\ 2 & 5 & -4 \\ -5 & 4 & 3 \end{pmatrix}$$

10.5. If A is nonsingular, prove that $|A^{-1}| = |A|^{-1}$.

10.6. Convince yourself that

$$D_{ij} = \sum_{k=1}^{3} A_{ik}B_{jk}$$

is the equivalent of the matrix equation $D = A\tilde{B}$. Also, find the matrix equation which is the equivalent of each of the following:

$$E_{ij} = \sum_{k=1}^{3} A_{ki}B_{kj} \qquad F_{ij} = \sum_{k=1}^{3} A_{ki}B_{jk}$$

$S\tilde{S} = I$

2 Vectors and Tensors in a Cartesian Space

The vector idea was introduced in Chapter 1. In Chapter 2, we continue to develop this concept, this time starting with a more precise definition of a vector space. The basic concepts of vector analysis including the dot product, the cross product, and vector differentiation are developed. The concept of second- and higher-rank tensors is introduced. The emphasis is on developing these ideas in a three-dimensional Cartesian space, with extensions to spaces of higher dimension where appropriate. The theory of orthogonal transformations is given, and the eigenvalue problem associated with second-rank tensors is explored.

1. VECTOR SPACES

A vector space consists of a collection of mathematical objects for which a law of addition is defined:

$$\mathbf{c} = \mathbf{a} + \mathbf{b} \tag{1.1}$$

A process of scalar multiplication is also defined: If $\mathbf{a}$ is a vector and α is a scalar (ordinary number), then $\alpha\mathbf{a}$ is also a vector. The following are the basic laws of vector algebra:

1. Closure: If $\mathbf{a}$ and $\mathbf{b}$ are vectors, $\mathbf{a} + \mathbf{b}$ is also a vector. This means that our collection of vectors is closed in the sense that if the law of addition (1.1) is performed using any two vectors, we always get another object which is recognizable as a vector and belongs to our original collection.
2. Commutative law of addition:
$$\mathbf{a} + \mathbf{b} = \mathbf{b} + \mathbf{a} \tag{1.2}$$
3. Associative law of addition:
$$(\mathbf{a} + \mathbf{b}) + \mathbf{c} = \mathbf{a} + (\mathbf{b} + \mathbf{c}) \tag{1.3}$$
4. Existence of a zero or null vector: There is a vector $\mathbf{O}$ which has the property that
$$\mathbf{a} + \mathbf{O} = \mathbf{a} \tag{1.4}$$
for all $\mathbf{a}$.
5. For a given vector $\mathbf{a}$, there is a unique vector $(-\mathbf{a})$ such that
$$\mathbf{a} + (-\mathbf{a}) = \mathbf{O} \tag{1.5}$$
6. If α_1 and α_2 are scalars,
$$(\alpha_1 + \alpha_2)\mathbf{a} = \alpha_1\mathbf{a} + \alpha_2\mathbf{a} \tag{1.6}$$
7. $\alpha(\mathbf{a} + \mathbf{b}) = \alpha\mathbf{a} + \alpha\mathbf{b}$ (1.7)
8. $\alpha_1(\alpha_2\mathbf{a}) = (\alpha_1\alpha_2)\mathbf{a}$ (1.8)

Note that in terms of the above rules, matrices could be thought of as constituting a vector space. We will, however, make our definition of vectors somewhat more restricted by requiring that their components have certain transformation properties. This question is taken up in section 2.

Section 1 of Chapter 1 discusses the ideas of vector addition and scalar multiplication as they apply to vectors defined in a three-dimensional Cartesian space. Also discussed there is the idea of the length of a vector and the concept of unit vector. We will frequently find it convenient to use the notation $\hat{\mathbf{e}}_1$, $\hat{\mathbf{e}}_2$, and $\hat{\mathbf{e}}_3$ for the three unit coordinate vectors rather than $\hat{\mathbf{i}}$, $\hat{\mathbf{j}}$, and $\hat{\mathbf{k}}$. Thus, in a three-dimensional Cartesian space, a vector can be written as

$$\mathbf{a} = a_x\hat{\mathbf{i}} + a_y\hat{\mathbf{j}} + a_z\hat{\mathbf{k}} = a_1\hat{\mathbf{e}}_1 + a_2\hat{\mathbf{e}}_2 + a_3\hat{\mathbf{e}}_3 = \sum_{i=1}^{3} a_i\hat{\mathbf{e}}_i \tag{1.9}$$

We can extend the idea of vector to an n-dimensional space simply by running the sum from 1 to n in (1.9) rather than from 1 to 3. The vectors $\hat{\mathbf{e}}_i$ are sometimes referred to as *unit basis vectors*.

The concept of linear dependence and independence of vectors has been treated in section 6 of Chapter 1. The matrix representations of the three unit coordinate vectors are

$$\begin{pmatrix}1\\0\\0\end{pmatrix} \quad \begin{pmatrix}0\\1\\0\end{pmatrix} \quad \text{and} \quad \begin{pmatrix}0\\0\\1\end{pmatrix} \tag{1.10}$$

It is clear that they are linearly independent, because

$$\alpha\begin{pmatrix}1\\0\\0\end{pmatrix} + \beta\begin{pmatrix}0\\1\\0\end{pmatrix} + \gamma\begin{pmatrix}0\\0\\1\end{pmatrix} = 0 \tag{1.11}$$

means that $\alpha = \beta = \gamma = 0$

Theorem 1. The scalar 0 multiplied by a vector $\mathbf{a}$ gives the zero vector: $0\mathbf{a} = \mathbf{O}$.

Proof: $0\mathbf{a} + \mathbf{a} = (0+1)\mathbf{a} = 1\mathbf{a} = \mathbf{a}$. Now add $(-\mathbf{a})$ to both sides to get $0\mathbf{a} = \mathbf{O}$.

Theorem 2. The scalar (-1) multiplied by a vector is the same thing as $(-\mathbf{a})$: $(-1)\mathbf{a} = (-\mathbf{a})$.

Proof: $(-1)\mathbf{a} + \mathbf{a} = [(-1)+1]\mathbf{a} = \mathbf{O}$. Now add $(-\mathbf{a})$ to both sides to get $(-1)\mathbf{a} = (-\mathbf{a})$.

Theorem 2 means that in place of writing $\mathbf{a} + (-\mathbf{b})$, we may simply write $\mathbf{a} - \mathbf{b}$.

PROBLEMS

1.1. Show that all polynomials of degree n in x,

$$P(x) = a_0 + a_1x + a_2x^2 + \cdots + a_nx^n$$

constitute a vector space. What is the dimension of this space? Find a set of basis vectors (not necessarily unit vectors) which span the space.

1.2. Consider the set of all possible pairs of real numbers (a, b). Suppose that addition of pairs and scalar multiplication are defined by

$$(a,b) + (c,d) = (a+c, b+d)$$

$$\alpha(a,b) = (\alpha a, b)$$

Show that under these conditions, the number pairs do not form a vector space.

1.3. Show that the set of all possible 2×2 matrices forms a vector space. What is the dimension of this space? Write down a set of basis vectors that spans this space.

2. ORTHOGONAL TRANSFORMATIONS

The concept of a vector as representing a physical quantity such as a force or an electric field has an existence that is independent of any particular coordinate system in which it may be represented. An important ingredient of a physical theory is its formulation in a manner which makes this independence of any particular reference frame manifest. Equivalently, basic principles can be stated in such a way that a coordinate system is implied but no particular one is singled out as being special. There is, for example, no absolute rest frame in the universe. The sets of all possible reference frames or coordinate systems are equivalent and are to be treated as comoving with respect to one another.

An important aspect of the mathematical representation of a physical quantity is its transformation property as the representation is changed from one coordinate system to another. We begin by considering *orthogonal transformations*, which are transformations among all possible rectangular Cartesian coordinate systems which involve no change of scale. The theory of orthogonal transformations can be developed in spaces of any number of dimensions, but for the most part we will think in terms of ordinary three-dimensional space. In a later chapter, we will generalize the theory to include arbitrary curvilinear coordinates. Orthogonal transformations can be broken down into three general categories as illustrated in Fig. 2.1. Figure 2.1*a* shows a simple translation of coordinate axes; Fig. 2.1*b* shows a rotation; and Fig. 2.1*c* shows a mirror reflection. Examples of rotations of coordinate axes have already been given in section 3 of Chapter 1. See, for example, equation (3.35). Each of the three categories is distinct. We cannot, for example, obtain the mirror reflection by any combination of translations or rotations.

A general orthogonal transformation connecting the components of a vector in one Cartesian coordinate system with the components of that same vector in another coordinate system can be represented as

$$X' = SX \tag{2.1}$$

where X' and X are column matrices which represent the components of the vector in the two coordinate systems and

$$S = \begin{pmatrix} S_{11} & S_{12} & S_{13} \\ S_{21} & S_{22} & S_{23} \\ S_{31} & S_{32} & S_{33} \end{pmatrix} \tag{2.2}$$

is the matrix which represents the transformation. In place of (2.1), we may also use the notation

$$a_i' = \sum_{j=1}^{3} S_{ij} a_j \tag{2.3}$$

or simply

$$a_i' = S_{ij} a_j \tag{2.4}$$

if we choose to use the summation convention.

The thing which is characteristic of an orthogonal transformation and in fact formally defines it is that it preserves the length of a vector. The square of the length of a vector can be expressed as

$$a^2 = a_1^2 + a_2^2 + a_3^2 = \tilde{X}X = (a_1 \quad a_2 \quad a_3)\begin{pmatrix} a_1 \\ a_2 \\ a_3 \end{pmatrix} \tag{2.5}$$

where the row vector $\tilde{X}$ is the transpose of the column vector X. An orthogonal transformation is therefore characterized by

$$\tilde{X}'X' = \tilde{X}X \tag{2.6}$$

The consequences of (2.6) are revealed as follows: Take the transpose of (2.1) and use theorem 6, section 10, Chapter 1:

$$\tilde{X}' = \tilde{X}\tilde{S} \tag{2.7}$$

Equation (2.6) then becomes

$$\tilde{X}\tilde{S}SX = \tilde{X}X \tag{2.8}$$

Since X can be chosen in a completely arbitrary manner, theorem 2 in section 10 of Chapter 1 implies

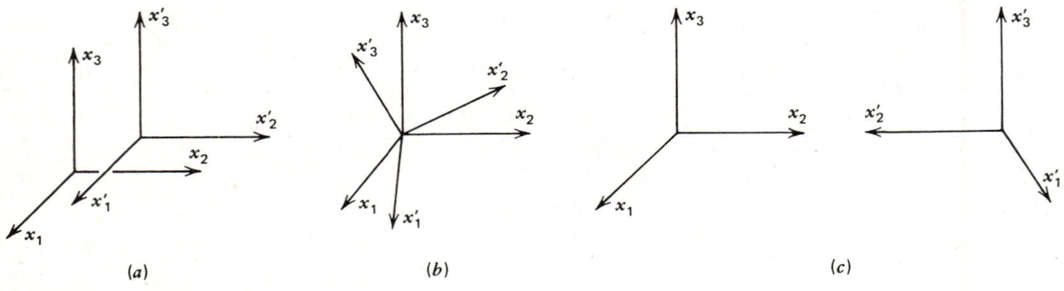

Fig. 2.1

that

$$\tilde{S}S = I \tag{2.9}$$

Therefore, the inverse of an orthogonal transformation can be found by taking its transpose:

$$S^{-1} = \tilde{S} \tag{2.10}$$

Sometimes, it is convenient to work with component equations rather than matrix equations. Note the following equivalences:

$$\tilde{S}S = I \leftrightarrow \sum_{i=1}^{3} S_{ij}S_{ik} = \delta_{jk} \tag{2.11}$$

$$S\tilde{S} = I \leftrightarrow \sum_{i=1}^{3} S_{ji}S_{ki} = \delta_{jk} \tag{2.12}$$

Equations (2.11) and (2.12) hold for all orthogonal transformations and are called *orthogonality relations.*

As an example, suppose that, as in Fig. 2.2, a coordinate system (x_i') is obtained from the original (x_i) coordinates by a rotation through an angle θ about the x_3 axis. The components of an arbitrary vector **a** in the two coordinate systems are related by

$$\begin{pmatrix} a_1' \\ a_2' \\ a_3' \end{pmatrix} = \begin{pmatrix} \cos\theta & \sin\theta & 0 \\ -\sin\theta & \cos\theta & 0 \\ 0 & 0 & 1 \end{pmatrix} \begin{pmatrix} a_1 \\ a_2 \\ a_3 \end{pmatrix} \tag{2.13}$$

Refer to equation (2.9) in Chapter 1 for a derivation of (2.13). By (2.10), the inverse of the transformation

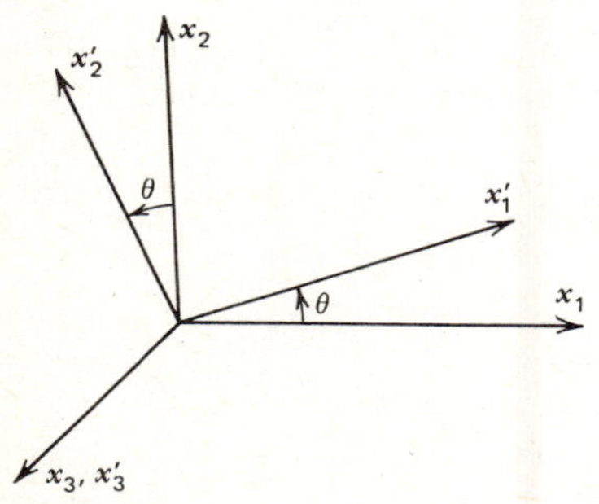

Fig. 2.2

in (2.13) is

$$S^{-1} = \tilde{S} = \begin{pmatrix} \cos\theta & -\sin\theta & 0 \\ \sin\theta & \cos\theta & 0 \\ 0 & 0 & 1 \end{pmatrix} \tag{2.14}$$

The validity of (2.9) is verified by direct multiplication.

The formal definition of a vector can now be completed. A vector (in three dimensions) is a set of three functions of position in space that transform according to $X' = SX$ under orthogonal transformations and in addition obeys the laws of vector algebra given in section 1. The three functions in question are the three components of X.

In theorem 3, section 5, Chapter 1, we proved that $|A| = |\tilde{A}|$. Theorem 8 of that same section shows that $|AB| = |A||B|$. Using these facts and (2.9) gives

$$|S|^2 = 1 \qquad |S| = \pm 1 \tag{2.15}$$

Orthogonal transformations for which $|S| = +1$ are called *proper* and include uniform translations, Fig. 2.1*a*, and rotations, Fig. 2.1*b*. Orthogonal transformations for which $|S| = -1$ are called *improper* and result from a mirror reflection as in Fig. 2.1*c*. The matrix which represents the transformation in Fig. 2.1*c* is

$$S = \begin{pmatrix} 1 & 0 & 0 \\ 0 & -1 & 0 \\ 0 & 0 & 1 \end{pmatrix} \tag{2.16}$$

and it is easy to see that $|S| = -1$. On the other hand, the matrix in (2.13) gives $|S| = +1$.

We close this section with two more definitions. A *scalar quantity* has the transformation property

$$\alpha = \alpha' \tag{2.17}$$

An *invariant* is a scalar which retains its mathematical form under transformation. An example of an invariant is the length of a vector:

$$a = \sqrt{a_1^2 + a_2^2 + a_3^2} = \sqrt{a_1'^2 + a_2'^2 + a_3'^2} = a' \tag{2.18}$$

the idea being that not only is $a = a'$, but a' bears the same functional relationship to the components a_i' as a bears to the components a_i. The following is almost obvious: If **a** is a vector and α is a scalar, then $\mathbf{b} = \alpha\mathbf{a}$ is also a vector in the sense that the transformation property of **b** is the same as that of **a**.

PROBLEMS

2.1. Show that

$$S = \begin{pmatrix} \cos\theta & \sin\theta & 0 \\ \sin\theta & -\cos\theta & 0 \\ 0 & 0 & 1 \end{pmatrix}$$

is improper. Break the transformation down into two parts as $S = S_1 S_2$, where S_1 is a mirror reflection and S_2 is a rotation.

2.2. Could

$$S = \begin{pmatrix} \frac{1}{2} & -\sqrt{\frac{1}{2}} & \frac{1}{2} \\ -\frac{1}{2} & -\sqrt{\frac{1}{2}} & -\frac{1}{2} \\ -\sqrt{\frac{1}{2}} & 0 & \sqrt{\frac{1}{2}} \end{pmatrix}$$

represent an orthogonal transformation? Could it be a rotation of coordinates?

2.3. A vector **a** has components $(1,1,1)$ in the x_i coordinates. What are its components in the x_i' coordinates obtained by a rotation through on angle ϕ about the x_2 axis? Let ϕ be given by $\cos\phi = 0.8$, $\sin\phi = 0.6$. See Fig. 2.3.

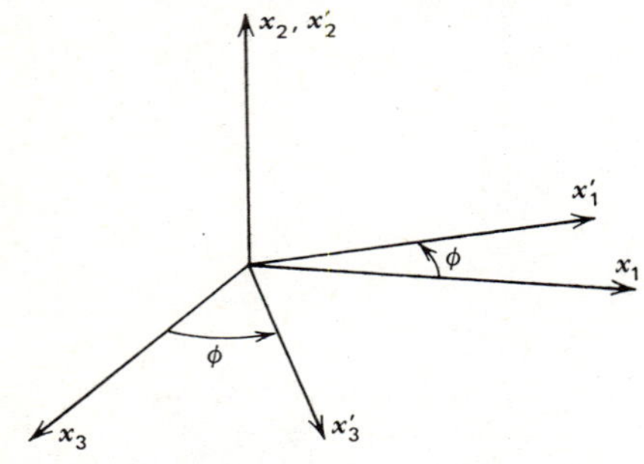

Fig. 2.3

2.4. Show that the elements of an orthogonal transformation are equal to $\pm$ their own cofactors: $S_{ij} = \pm C_{ij}$.

Hint: See equation (10.17) of Chapter 1.

2.5. If S_1 and S_2 are known to be orthogonal transformations, show that the product $S = S_1 S_2$ is also orthogonal.

2.6. A uniform circular disc is mounted on a shaft through its center, with the plane of the disc tilted at an angle θ with respect to the shaft. A rectangular Cartesian coordinate system (x_i') is fixed to the disc. The x_1' axis points along a diameter of the disc and is also perpendicular to the shaft. The x_2' axis is perpendicular to the surface of the disc and is at an angle θ with respect to the shaft. If the shaft rotates at a constant angular velocity ω, find the orthogonal transformation which gives the rotation of the x_i' coordinates as a function of time.

Hint: First rotate the coordinates $\bar{x}_i$ shown in Fig. 2.4 by an angle ωt. Then rotate the resulting coordinate system by an angle θ about the $\bar{x}_1$ axis. Combine the matrices representing these two rotations by matrix multiplication.

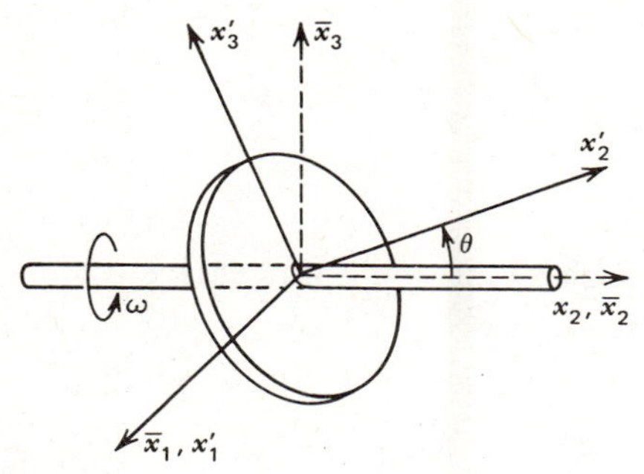

Fig. 2.4

3. DIFFERENTIATION OF VECTORS

In particle kinematics, the position of a particle can be specified by a displacement vector **r** which can be pictured as an arrow pointing from the origin of coordinates to the particle. The particle is moving along some kind of trajectory so that the three components of **r** are functions of the time:

$$\mathbf{r}(t) = x(t)\hat{\mathbf{i}} + y(t)\hat{\mathbf{j}} + z(t)\hat{\mathbf{k}} \tag{3.1}$$

In Fig. 3.1, the positions of two neighboring points on the trajectory are given by **r** and $\mathbf{r} + \Delta\mathbf{s}$ where $\Delta\mathbf{s}$ connects the two points. The instantaneous velocity of the particle is found from

$$\mathbf{u} = \lim_{\Delta t \to 0} \frac{1}{\Delta t}\Delta\mathbf{s} = \frac{d\mathbf{s}}{dt} \tag{3.2}$$

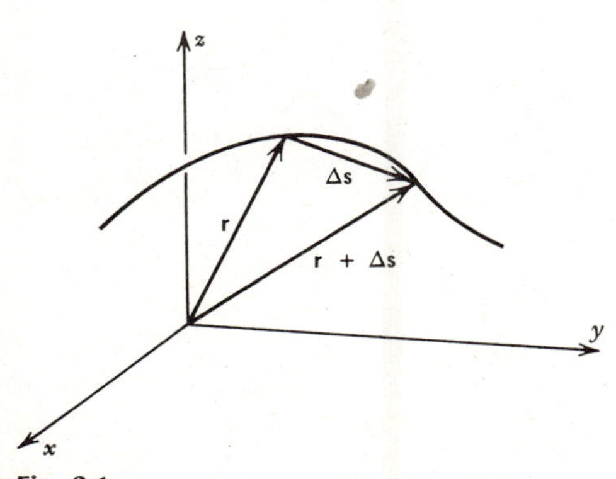

Fig. 3.1

The quantity $1/\Delta t$ is a scalar multiplying the displacement vector $\Delta\mathbf{s}$ so that it multiplies each component of $\Delta\mathbf{s}$:

$$\mathbf{u} = \lim_{\Delta t \to 0}\left(\frac{\Delta x}{\Delta t}\hat{\mathbf{i}} + \frac{\Delta y}{\Delta t}\hat{\mathbf{j}} + \frac{\Delta z}{\Delta t}\hat{\mathbf{k}}\right) = \dot{x}\hat{\mathbf{i}} + \dot{y}\hat{\mathbf{j}} + \dot{z}\hat{\mathbf{k}} \tag{3.3}$$

where Δx, Δy, and Δz are the components of $\Delta\mathbf{s}$. Since **u** is obtained from $\Delta\mathbf{s}$ by a process of scalar multiplication, it has the same transformation properties as $\Delta\mathbf{s}$ and is therefore a vector. Geometrically, in the limit as $\Delta\mathbf{s}$ and Δt approach zero, $\Delta\mathbf{s}$ and hence **u** become tangent to the trajectory at the point where **u** is computed.

The instantaneous acceleration of a particle is to be found by differentiating the velocity:

$$\mathbf{a} = \lim_{\Delta t \to 0} \frac{\Delta\mathbf{u}}{\Delta t} = \frac{d\mathbf{u}}{dt} = \ddot{x}\hat{\mathbf{i}} + \ddot{y}\hat{\mathbf{j}} + \ddot{z}\hat{\mathbf{k}} \tag{3.4}$$

The acceleration is a vector in that it has the same transformation properties as **u** and $\Delta\mathbf{s}$ have.

As an example, suppose that a particle is traveling in a circle of radius r centered at the origin as shown in Fig. 3.2. Its position vector is

$$\mathbf{r} = r\cos\theta\,\hat{\mathbf{i}} + r\sin\theta\,\hat{\mathbf{j}} \tag{3.5}$$

where θ is a function of time and r is a constant. The derivative of (3.5) is

$$\mathbf{u} = -r\omega\sin\theta\,\hat{\mathbf{i}} + r\omega\cos\theta\,\hat{\mathbf{j}} \tag{3.6}$$

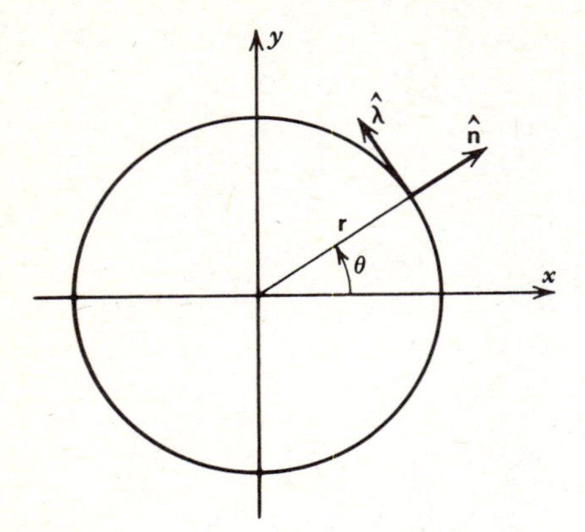

Fig. 3.2

where $\omega = \dot{\theta}$ is the angular velocity in radians/sec. From (3.6), the magnitude of $\mathbf{u}$ is easily shown to be

$$u = r\omega \tag{3.7}$$

The acceleration vector is found by differentiating (3.6):

$$\mathbf{a} = -r\alpha \sin\theta\, \hat{\mathbf{\imath}} + r\alpha\cos\theta\, \hat{\mathbf{\jmath}} - r\omega^2\cos\theta\, \hat{\mathbf{\imath}} - r\omega^2 \sin\theta\, \hat{\mathbf{\jmath}} \tag{3.8}$$

where $\alpha = \dot{\omega}$ is the angular acceleration in radians/sec^2. A unit vector tangent to the circle at the location of the particle is

$$\hat{\lambda} = \frac{\mathbf{u}}{u} = -\sin\theta\, \hat{\mathbf{\imath}} + \cos\theta\, \hat{\mathbf{\jmath}} \tag{3.9}$$

A unit vector in the same direction as $\mathbf{r}$ (perpendicular to $\hat{\lambda}$) is

$$\hat{\mathbf{n}} = \frac{\mathbf{r}}{r} = \cos\theta\, \hat{\mathbf{\imath}} + \sin\theta\, \hat{\mathbf{\jmath}} \tag{3.10}$$

Using $\hat{\lambda}$ and $\hat{\mathbf{n}}$, (3.8) can be expressed as

$$\mathbf{a} = r\alpha\hat{\lambda} - r\omega^2\hat{\mathbf{n}} \tag{3.11}$$

The term $r\omega^2$ is the familiar centripetal acceleration. From (3.7), it can be seen that

$$r\alpha = \frac{du}{dt} \tag{3.12}$$

is the rate of change of the speed of the particle.

The differentiation of matrices is similar to the differentiation of vectors in that each element of the matrix is to be differentiated. If A is a matrix the elements of which are functions of the time or some other parameter, then the derivative of A is

$$\frac{dA}{dt} = \lim_{\Delta t \to 0} \frac{1}{\Delta t}[A(t+\Delta t) - A(t)] \tag{3.13}$$

The elements of $A(t+\Delta t) - A(t)$ are $A_{ij}(t+\Delta t) - A_{ij}(t)$. Since the scalar $1/\Delta t$ multiplies each element of the matrix, the result of the limiting process in (3.13) is

$$\frac{dA}{dt} = \begin{pmatrix} \dot{A}_{11} & \dot{A}_{12} & \dot{A}_{13} \\ \dot{A}_{21} & \dot{A}_{22} & \dot{A}_{23} \\ \dot{A}_{31} & \dot{A}_{32} & \dot{A}_{33} \end{pmatrix} \tag{3.14}$$

This is to be compared and contrasted with the expression for the derivative of a determinant, which can be written in two ways:

$$\begin{aligned} \frac{d|A|}{dt} &= \begin{vmatrix} \dot{A}_{11} & \dot{A}_{12} & \dot{A}_{13} \\ A_{21} & A_{22} & A_{23} \\ A_{31} & A_{32} & A_{33} \end{vmatrix} + \begin{vmatrix} A_{11} & A_{12} & A_{13} \\ \dot{A}_{21} & \dot{A}_{22} & \dot{A}_{23} \\ A_{31} & A_{32} & A_{33} \end{vmatrix} + \begin{vmatrix} A_{11} & A_{12} & A_{13} \\ A_{21} & A_{22} & A_{23} \\ \dot{A}_{31} & \dot{A}_{32} & \dot{A}_{33} \end{vmatrix} \\ &= \begin{vmatrix} \dot{A}_{11} & A_{12} & A_{13} \\ \dot{A}_{21} & A_{22} & A_{23} \\ \dot{A}_{31} & A_{32} & A_{33} \end{vmatrix} + \begin{vmatrix} A_{11} & \dot{A}_{12} & A_{13} \\ A_{21} & \dot{A}_{22} & A_{23} \\ A_{31} & \dot{A}_{32} & A_{33} \end{vmatrix} + \begin{vmatrix} A_{11} & A_{12} & \dot{A}_{13} \\ A_{21} & A_{22} & \dot{A}_{23} \\ A_{31} & A_{32} & \dot{A}_{33} \end{vmatrix} \end{aligned} \tag{3.15}$$

You obtained this result when you did problem 5.12 of Chapter 1.

PROBLEMS

3.1. If A and B are matrices the elements of which are functions of a parameter t, show that

$$\frac{d}{dt}(AB) = A\frac{dB}{dt} + \frac{dA}{dt}B$$

Hint:

$$A(t+\Delta t)B(t+\Delta t)-A(t)B(t)=[A(t+\Delta t)-A(t)]B(t+\Delta t)$$
$$+A(t)[B(t+\Delta t)-B(t)]$$

3.2. A uniform disc of radius R and mass M is mounted on a horizontal shaft. A mass $m=\frac{1}{5}M$ hangs from a cord which is wrapped around the rim of the disc. There is no friction in the bearing. If the system is released from rest, find the acceleration vector of a particle at point P on the top of the disc at the instant that the mass m has descended a distance $y=4R$. The moment of inertia of the disc is $I=\frac{1}{2}MR^2$. Numerical answers for the two components of the acceleration vector are possible. See Fig. 3.3.

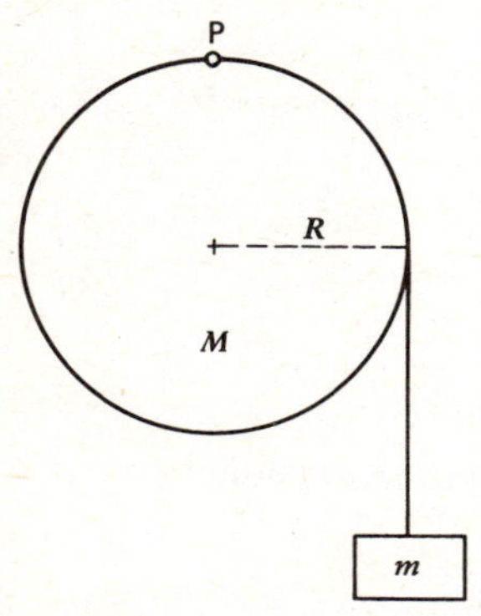

Fig. 3.3

3.3. A particle is shot from the origin of coordinates at a speed u_o and angle of elevation θ. There is a uniform gravitational field of strength g directed in the negative y direction. Find the velocity vector and the acceleration vector of the particle at point P, which is the exact top of the trajectory. Express the answers in terms of u_o, θ, and g. See Fig. 3.4.

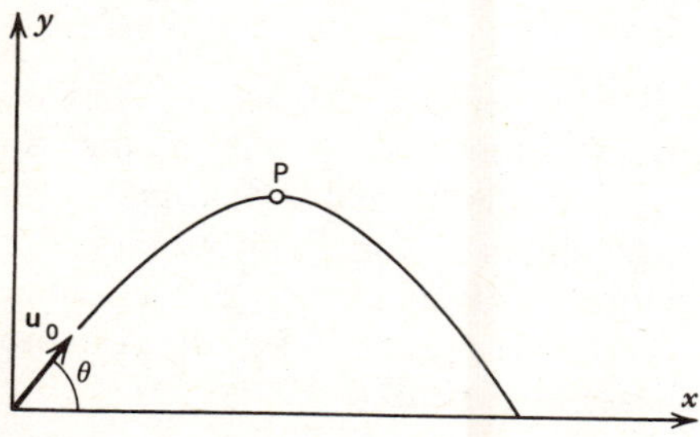

Fig. 3.4

4. THE DOT PRODUCT

Given two vectors **a** and **b**, the *dot product* is defined to be

$$\mathbf{a}\cdot\mathbf{b}=a_xb_x+a_yb_y+a_zb_z=\sum_{i=1}^{3}a_ib_i \tag{4.1}$$

This is also called the *scalar product* or *inner product*.

In matrix notation,

$$\mathbf{a}\cdot\mathbf{b}=\tilde{X}Y=\tilde{Y}X \tag{4.2}$$

where X and Y are column matrices given by

$$X=\begin{pmatrix}a_1\\a_2\\a_3\end{pmatrix}\qquad Y=\begin{pmatrix}b_1\\b_2\\b_3\end{pmatrix} \tag{4.3}$$

The square of the length of a vector is the dot product of the vector with itself:

$$a^2 = \mathbf{a} \cdot \mathbf{a} \tag{4.4}$$

Theorem 1. The dot product is an invariant scalar:

$$\mathbf{a} \cdot \mathbf{b} = \sum_{i=1}^{3} a_i b_i = \sum_{i=1}^{3} a_i' b_i' \tag{4.5}$$

where a_i, b_i and a_i', b_i' are the components of **a** and **b** expressed in two different rectangular Cartesian coordinate systems.

Proof: The proof is easy in matrix notation. The transformation of the two vectors can be expressed as

$$X' = SX \qquad Y' = SY \tag{4.6}$$

where S is an orthogonal transformation. Then,

$$\tilde{X}'Y' = (\widetilde{SX})SY = \tilde{X}\tilde{S}SY = \tilde{X}Y \tag{4.7}$$

which completes the proof.

It is also possible to do the proof using component notation. To simplify the calculation a little, the summation convention is used. In place of (4.6),

$$a_i' = S_{ij} a_j \qquad b_i' = S_{ik} b_k \tag{4.8}$$

The equivalent of (4.7) is

$$a_i' b_i' = S_{ij} S_{ik} a_j b_k = \delta_{jk} a_j b_k = a_k b_k \tag{4.9}$$

where (2.11) and the properties of the Kronecker delta as given by equation (5.23) in Chapter 1 are used. Given two vectors **a** and **b**, there are essentially three invariant scalars that can be constructed: $\mathbf{a} \cdot \mathbf{a} = a^2$, $\mathbf{b} \cdot \mathbf{b} = b^2$, and $\mathbf{a} \cdot \mathbf{b}$.

Theorem 2.

$$\mathbf{a} \cdot \mathbf{b} = ab \cos \theta \tag{4.10}$$

where θ is the angle between **a** and **b**.

Proof: Since $\mathbf{a} \cdot \mathbf{b}$ is an invariant scalar, the validity of the theorem can be established in any convenient rectangular Cartesian coordinate system. It will then be valid in any other such system. Choose a coordinate system which has its x axis in the same direction as **a**, Fig. 4.1. Then,

$$\mathbf{a} \cdot \mathbf{b} = a_x b_x = ab \cos \theta \tag{4.11}$$

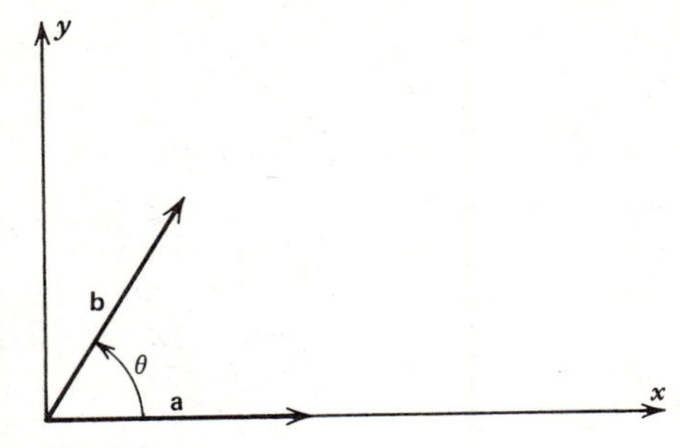

Fig. 4.1

The dot product has the following important properties:

1. $\mathbf{a} \cdot \mathbf{b} = \mathbf{b} \cdot \mathbf{a}$ (4.12)
2. $\mathbf{a} \cdot (\mathbf{b} + \mathbf{c}) = \mathbf{a} \cdot \mathbf{b} + \mathbf{a} \cdot \mathbf{c}$ (4.13)
3. If α is a scalar,
$$\alpha(\mathbf{a} \cdot \mathbf{b}) = (\alpha\mathbf{a}) \cdot \mathbf{b} = \mathbf{a} \cdot (\alpha\mathbf{b}) \tag{4.14}$$
4. If the components of **a** and **b** are functions of a parameter t,
$$\frac{d}{dt}(\mathbf{a} \cdot \mathbf{b}) = \mathbf{a} \cdot \frac{d\mathbf{b}}{dt} + \frac{d\mathbf{a}}{dt} \cdot \mathbf{b} \tag{4.15}$$

It is easy to prove each of the four properties using the definition (4.1).

The unit coordinate vectors obey

$$\hat{\imath} \cdot \hat{\imath} = \hat{\jmath} \cdot \hat{\jmath} = \hat{\mathbf{k}} \cdot \hat{\mathbf{k}} = 1 \tag{4.16}$$

$$\hat{\imath} \cdot \hat{\jmath} = \hat{\imath} \cdot \hat{\mathbf{k}} = \hat{\jmath} \cdot \hat{\mathbf{k}} = 0 \tag{4.17}$$

Equation (4.17) is true because the unit coordinate vectors are mutually orthogonal to one another so that by (4.11) $\cos \theta = 0$. The term *orthonormal* is also used for sets of mutually orthogonal unit vectors. Not only are the vectors perpendicular to one another, they have been *normalized* to unit length. A more compact way to state (4.16) and (4.17) is

$$\hat{\mathbf{e}}_i \cdot \hat{\mathbf{e}}_j = \delta_{ij} \tag{4.18}$$

The dot product can be used to prove the *cosine law* from trigonometry. Let the vectors **a**, **b**, and **c** form the three sides of a triangle. Figure 4.2 shows that

$$\mathbf{b} - \mathbf{a} = \mathbf{c} \tag{4.19}$$

Now form the dot product of (4.19) with itself:

$$(\mathbf{b} - \mathbf{a}) \cdot (\mathbf{b} - \mathbf{a}) = \mathbf{c} \cdot \mathbf{c}$$

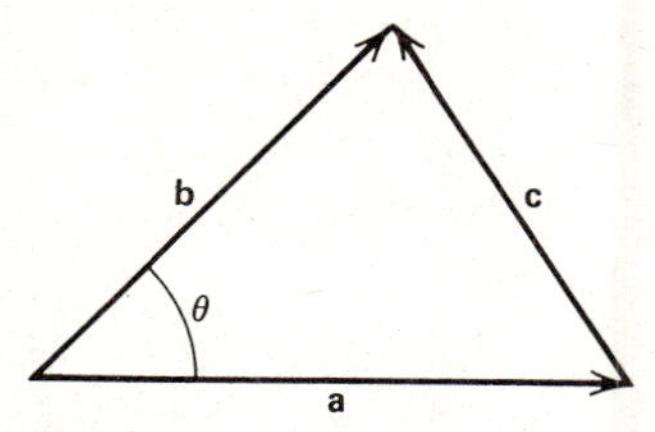

Fig. 4.2

By means of (4.4), (4.10), (4.12), and (4.13),

$$\mathbf{b}\cdot\mathbf{b}-\mathbf{b}\cdot\mathbf{a}-\mathbf{a}\cdot\mathbf{b}+\mathbf{a}\cdot\mathbf{a}=\mathbf{c}\cdot\mathbf{c}$$

$$b^2-2ab\cos\theta+a^2=c^2 \tag{4.20}$$

By means of the dot product, it is possible to construct an interesting representation of an orthogonal transformation connecting two Cartesian coordinate systems. An arbitrary vector **a** can be represented in either of the two coordinate systems in Fig. 4.3 as

$$\mathbf{a}=\sum_i a_i'\hat{\mathbf{e}}_i'=\sum_i a_i\hat{\mathbf{e}}_i \tag{4.21}$$

Form the dot product of (4.21) with $\hat{\mathbf{e}}_j'$ where $j=1$, 2, or 3:

$$\sum_i a_i'\left(\hat{\mathbf{e}}_j'\cdot\hat{\mathbf{e}}_i'\right)=\sum_i a_i\left(\hat{\mathbf{e}}_j'\cdot\hat{\mathbf{e}}_i\right) \tag{4.22}$$

By means of (4.18) and (4.10),

$$\sum_i a_i'\delta_{ij}=\sum_i(\cos\alpha_{ji})a_i \tag{4.23}$$

where, as shown in Fig. 4.3, the factors of $\cos\alpha_{ji}$ are the direction cosines of the new coordinate axes with

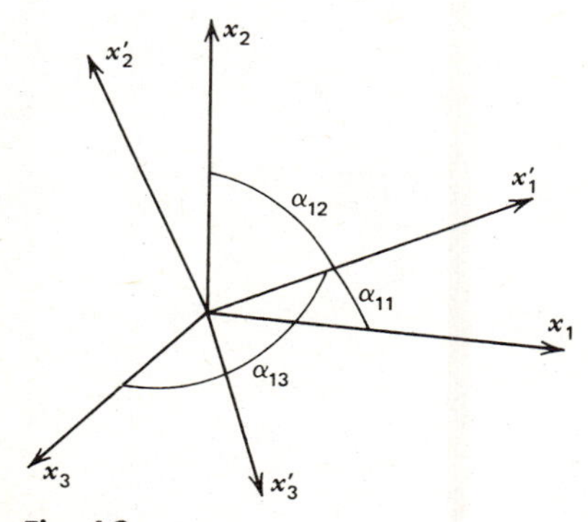

Fig. 4.3

respect to the old. Thus, (4.23) can be written as

$$a_j'=\sum_i S_{ji}a_i \tag{4.24}$$

where $S_{ji}=\cos\alpha_{ji}$. Note that the elements of the first row of S are actually the components of $\hat{\mathbf{e}}_1'$ with respect to the original x_i coordinates. Similarly, rows 2 and 3 of S are the components of $\hat{\mathbf{e}}_2'$ and $\hat{\mathbf{e}}_3'$ with respect to the x_i coordinates. The rows (or columns) of an orthogonal matrix are really a set of orthonormal vectors. This is where the term *orthogonal transformation* comes from.

The dot product occurs in the definition of work in classical mechanics. The work done on a particle by a force **F** during a displacement $d\mathbf{s}$ is

$$dW=\mathbf{F}\cdot d\mathbf{s}=F\,ds\cos\theta=F_x\,dx+F_y\,dy+F_z\,dz \tag{4.25}$$

In particular, if **F** is the net force acting on the particle, Newton's second law states that

$$\mathbf{F}=m\mathbf{a} \tag{4.26}$$

where m is the mass of the particle and **a** is the acceleration. The rate at which the force does work on the particle is

$$\frac{dW}{dt}=\mathbf{F}\cdot\frac{d\mathbf{s}}{dt}=\mathbf{F}\cdot\mathbf{u}=m\frac{d\mathbf{u}}{dt}\cdot\mathbf{u} \tag{4.27}$$

Note that by (4.15),

$$\frac{d}{dt}u^2=\frac{d}{dt}(\mathbf{u}\cdot\mathbf{u})=2\mathbf{u}\cdot\frac{d\mathbf{u}}{dt} \tag{4.28}$$

so that (4.27) becomes

$$\mathbf{F}\cdot\mathbf{u}=\frac{d}{dt}\left(\frac{1}{2}mu^2\right) \tag{4.29}$$

which says that the rate at which the net force does work on the particle is equal to the rate of change of the kinetic energy of the particle. This is called the *work–energy theorem*.

The dot product can be used to express the *flux* of a vector, which is most conveniently described by using the example of fluid flow. Suppose that a liquid or a gas is flowing through some region of space and that we desire to calculate the mass of fluid per second which flows across a surface as in Fig. 4.4. At a given instant, each point of the fluid has a velocity **u**. Figure 4.4 shows a differential parallelepiped with

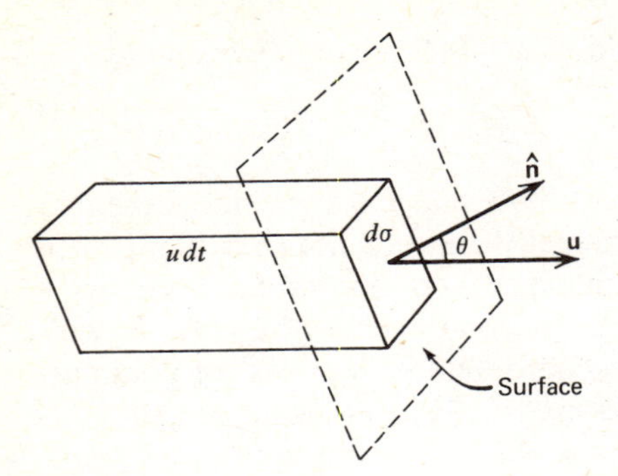

Fig. 4.4

edges parallel to **u** and with one face of area $d\sigma$ coincident with the surface across which we want to calculate the fluid flow. The volume of the parallelepiped is

$$d\Sigma = u\,dt\,d\sigma\cos\theta \tag{4.30}$$

where θ is the angle between $\hat{\mathbf{n}}$ and **u** and dt is the time for a particle of fluid to flow the length of one edge. The unit vector $\hat{\mathbf{n}}$ is perpendicular to the surface. If ρ is the density of the fluid in units of mass per unit volume, then the parallelepiped contains a mass $dm = \rho\, d\Sigma$ so that the rate at which mass crosses the differential area $d\sigma$ is

$$\frac{dm}{dt} = \rho u \cos\theta\, d\sigma = \rho\mathbf{u}\cdot\hat{\mathbf{n}}\, d\sigma = \rho\mathbf{u}\cdot d\boldsymbol{\sigma} \tag{4.31}$$

It is the usual practice to write $d\boldsymbol{\sigma} = \hat{\mathbf{n}}\, d\sigma$ and think of the element of area itself as a vector the direction of which is perpendicular to the surface. The quantity $\mathbf{u}\cdot d\boldsymbol{\sigma}$ is a differential element of the flux of the velocity vector **u**. It is possible to talk about the flux of any vector, even though it may have nothing to do with anything flowing anywhere. Thus, in electrostatics, an element of the flux of the electric field across a surface element $d\boldsymbol{\sigma}$ is defined to be

$$d\Phi_E = \mathbf{E}\cdot d\boldsymbol{\sigma} \tag{4.32}$$

We elaborate on this idea in the next section.

PROBLEMS

4.1. Find the angle between the vectors $\mathbf{a} = 2\hat{\imath} - 5\hat{\jmath} + 8\hat{\mathbf{k}}$ and $\mathbf{b} = -4\hat{\imath} + 3\hat{\jmath} - 2\hat{\mathbf{k}}$.

4.2. Prove the properties of the dot product as given by (4.12) through (4.15).

4.3. Carry out the steps leading up to (4.24) in two dimensions as follows: Write $\mathbf{a} = a'_x\hat{\imath}' + a'_y\hat{\jmath}' = a_x\hat{\imath} + a_y\hat{\jmath}$ and form the dot product first with $\hat{\imath}'$ and then with $\hat{\jmath}'$. In this manner, obtain the relations connecting a'_x and a'_y with a_x and a_y. See Fig. 4.5.

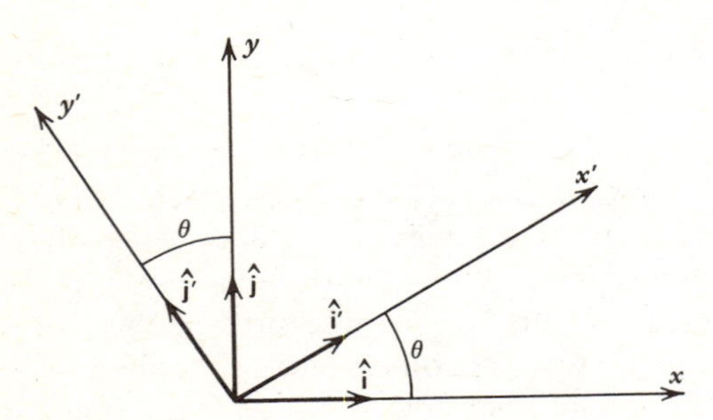

Fig. 4.5

4.4. The components of the vector **a** are constants and c is a scalar constant. If **r** is a variable displacement vector from the origin of coordinates, what kind of surfaces in three-dimensional space do $\mathbf{r}\cdot\mathbf{a} = c$ and $(\mathbf{r} - \mathbf{a})\cdot\mathbf{r} = 0$ represent?

4.5. A particle is acted on by a force $\mathbf{F} = -k\mathbf{r}$, where k is a constant. Show that $W = \frac{1}{2}mu^2 + \frac{1}{2}kr^2$ is a constant. Here, W is the total energy and $\frac{1}{2}kr^2$ is the potential energy. Such a force could be produced by a spring connecting the particle to the origin.

4.6. Figure 4.6 shows a closed surface in the shape of a hemisphere of radius R. A fluid of density ρ and constant velocity $\mathbf{u}$ flows perpendicular to the flat circular face. Calculate the total flux of the fluid (mass/sec) across the circular face and also across the spherical portion. They should be the same.

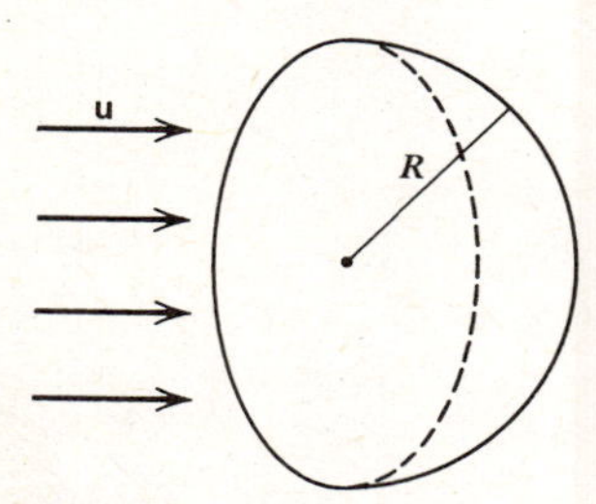

Fig. 4.6

4.7. Prove the inequalities

$$|\mathbf{a} + \mathbf{b}| \leq a + b \qquad |\mathbf{a} - \mathbf{b}| \geq a - b$$

4.8. The two vectors $\mathbf{a}$ and $\mathbf{b}$ of Fig. 4.7 lie in the x, y plane and are inclined at angles α and β with respect to the x axis. By using the dot product, establish the trigonometric identity

$$\cos(\beta - \alpha) = \cos\beta\cos\alpha + \sin\beta\sin\alpha$$

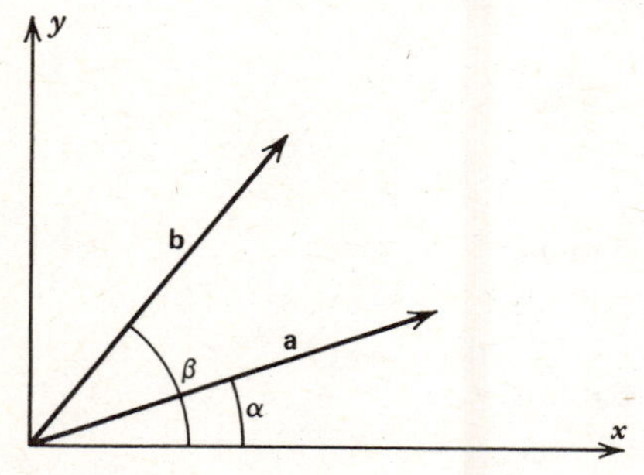

Fig. 4.7

4.9. The center of mass of a system of N particles is given by

$$\mathbf{R} = \frac{1}{M}\sum_{\alpha=1}^{N} m_\alpha \mathbf{r}_\alpha \tag{4.33}$$

where

$$M = \sum_{\alpha=1}^{N} m_\alpha \tag{4.34}$$

is the total mass of the system. The position vector $\mathbf{s}_\alpha$ of a given particle with respect to the center of mass is shown in Fig. 4.8 and is given by

$$\mathbf{s}_\alpha = \mathbf{r}_\alpha - \mathbf{R} \tag{4.35}$$

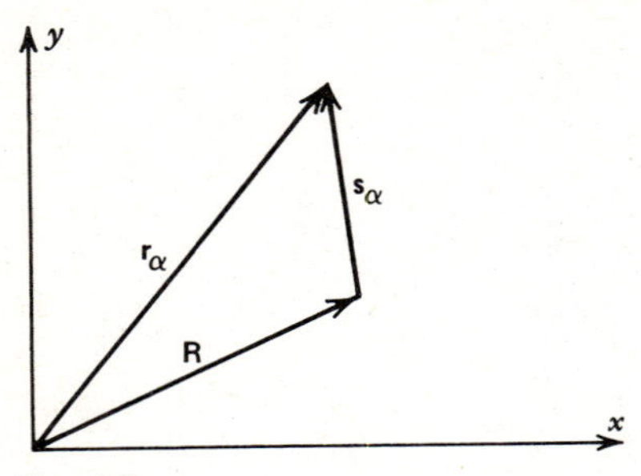

Fig. 4.8

which can be differentiated with respect to time to give

$$\mathbf{w}_\alpha = \mathbf{u}_\alpha - \mathbf{v} \tag{4.36}$$

where $\mathbf{w}_\alpha$ is the velocity of the αth particle with respect to the center of mass and $\mathbf{v} = d\mathbf{R}/dt$ is the velocity of the center of mass. Show that

$$\sum_{\alpha=1}^{N} m_\alpha \mathbf{s}_\alpha = 0 \qquad \sum_{\alpha=1}^{N} m_\alpha \mathbf{w}_\alpha = 0 \tag{4.37}$$

The total kinetic energy of the system of particles is

$$K = \sum_{\alpha=1}^{N} \tfrac{1}{2} m_\alpha (\mathbf{u}_\alpha \cdot \mathbf{u}_\alpha) \tag{4.38}$$

Prove that

$$K = \sum_{\alpha=1}^{N} \tfrac{1}{2} m_\alpha (\mathbf{w}_\alpha \cdot \mathbf{w}_\alpha) + \tfrac{1}{2} M v^2 \tag{4.39}$$

This is an important theorem in mechanics; it says that the kinetic energy of a system of particles can be divided into two parts. The first term in (4.39) is the kinetic energy due to the motion of the particles with respect to the center of mass, and the second term is due to the motion of the center of mass itself.

4.10. A wheel of radius R and mass M is rolling in a straight line along a flat surface at a velocity $\mathbf{v}$. Show, using the result of problem 4.9, that the kinetic energy of the wheel is

$$K = \tfrac{1}{2} I \omega^2 + \tfrac{1}{2} M v^2 \tag{4.40}$$

where ω is the angular velocity of rotation and the factor

$$I = \sum_{\alpha=1}^{N} m_\alpha s_\alpha^2 \tag{4.41}$$

is the moment of inertia of the wheel about its center of mass.

4.11. The kinetic energy of a two-particle system with respect to its center of mass is

$$K_{cm} = \tfrac{1}{2} m_1 w_1^2 + \tfrac{1}{2} m_2 w_2^2 \tag{4.42}$$

Show that

$$K_{cm} = \tfrac{1}{2} m u^2 \tag{4.43}$$

where m is the *reduced mass* given by

$$m = \frac{m_1 m_2}{m_1 + m_2} \tag{4.44}$$

and $\mathbf{u} = \mathbf{w}_2 - \mathbf{w}_1$ is the velocity of m_2 with respect to m_1. You will need equation (4.37).

4.12. The linear momentum of an individual particle in a system of particles is

$$\mathbf{p}_\alpha = m_\alpha \mathbf{u}_\alpha \tag{4.45}$$

Show, using equation (4.33), that the total linear momentum of a system of particles can be expressed as

$$\mathbf{p} = \sum_{\alpha=1}^{N} \mathbf{p}_\alpha = M\mathbf{v} \tag{4.46}$$

Also show that if $\mathbf{F}$ is the net external force acting on the particles, then

$$\mathbf{F} = \sum_{\alpha=1}^{N} \mathbf{F}_\alpha = M\frac{d\mathbf{v}}{dt} \tag{4.47}$$

This shows that, under the influence of external forces, the center of mass of a system of particles behaves like a single particle of mass M. Why do not the internal forces, that is, the forces which the particles exert on one another, come into (4.47)?

4.13. Show that the unit coordinate vectors themselves transform according to

$$\hat{\mathbf{e}}'_i = \sum_{j=1}^{3} S_{ij}\hat{\mathbf{e}}_j \qquad \hat{\mathbf{e}}_j = \sum_{i=1}^{3} S_{ij}\hat{\mathbf{e}}'_i \tag{4.48}$$

Hint: Equation (4.4) must hold for any vector **a**. Use the known transformation of the components of **a**.

5. SOLID ANGLE AND GAUSS' LAW

The topics of this section further illustrate the use of vectors and the dot product. Consider the point P and the surface element as shown in Fig. 5.1. The *solid angle* subtended at point P by the surface element is defined to be

$$d\Omega_P = \frac{\hat{\mathbf{r}} \cdot d\boldsymbol{\sigma}}{r^2} \tag{5.1}$$

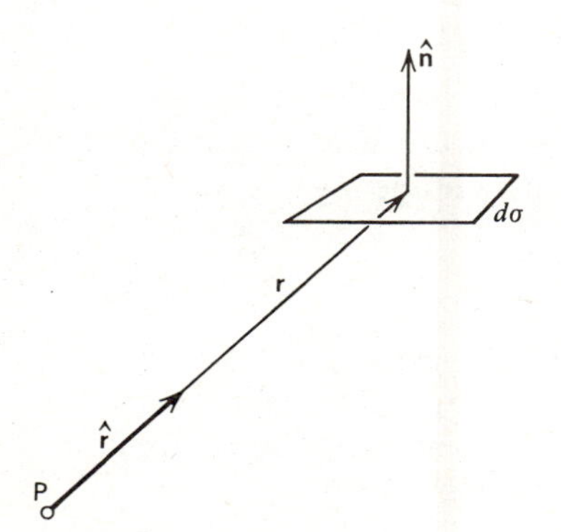

Fig. 5.1

where $\mathbf{r}$ is the displacement vector from P to the surface element and $\hat{\mathbf{r}} = \mathbf{r}/r$ is a unit vector in the same direction as $\mathbf{r}$. The unit steradian is used for solid angle, although it is really a dimensionless quantity. If P is at the exact center of a sphere of radius R and $d\sigma$ is an element of area on the sphere, then (5.1) becomes

$$d\Omega = \frac{d\sigma}{R^2} \tag{5.2}$$

The integration of (5.2) over the entire spherical surface gives

$$\Omega = \frac{1}{R^2}\int d\sigma = \frac{1}{R^2}4\pi R^2 = 4\pi \tag{5.3}$$

Note that the total solid angle (5.3) depends in no way on the radius of the sphere.

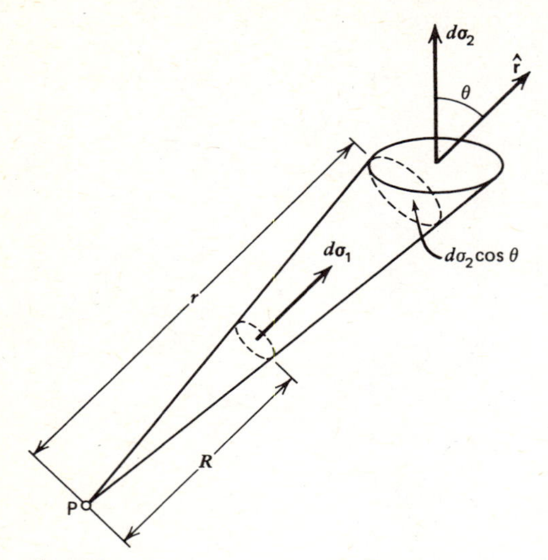

Fig. 5.2

Theorem 1. If P is a point inside a closed surface σ of any shape, then the total solid angle subtended at P by the surface is 4π steradians.

Proof: In Fig. 5.2, $d\boldsymbol{\sigma}_2$ is an element of the surface in question and $d\boldsymbol{\sigma}_1$ is an element of a spherical surface of radius R centered at P. Both of these elements of area are cross sections of a cone with its apex at P. The area $d\boldsymbol{\sigma}_1$ is parallel to $\hat{\mathbf{r}}$, but $d\boldsymbol{\sigma}_2$ is at an angle θ with respect to $\hat{\mathbf{r}}$. The solid angles subtended at P by $d\boldsymbol{\sigma}_1$ and $d\boldsymbol{\sigma}_2$ are

$$d\Omega_1 = \frac{d\sigma_1}{R^2} \qquad d\Omega_2 = \frac{d\sigma_2 \cos\theta}{r^2} \tag{5.4}$$

The area of the base of a cone is proportional to the square of its distance from the apex:

$$d\sigma_1 = \alpha R^2 \qquad d\sigma_2 \cos\theta = \alpha r^2 \tag{5.5}$$

Note that $d\sigma_2 \cos\theta$ is the projection of $d\boldsymbol{\sigma}_2$ onto a right-angle cross section of the cone. By combining (5.5) and (5.4), we find

$$d\Omega_1 = d\Omega_2 \tag{5.6}$$

By means of (5.3),

$$\int d\Omega_1 = \int d\Omega_2 = 4\pi \tag{5.7}$$

Theorem 2. If a point P is located outside a closed surface σ, Fig. 5.3, then the total solid angle subtended at P by the surface is zero.

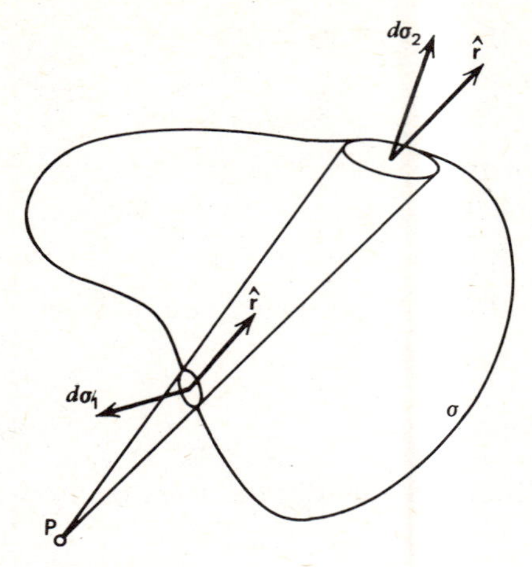

Fig. 5.3

Proof: A narrow cone with apex at P intersects σ at two locations. The solid angles subtended at P by the two surface elements $d\boldsymbol{\sigma}_1$ and $d\boldsymbol{\sigma}_2$ in Fig. 5.3 are

$$d\Omega_1 = \frac{\hat{\mathbf{r}} \cdot d\boldsymbol{\sigma}_1}{r_1^2} \qquad d\Omega_2 = \frac{\hat{\mathbf{r}} \cdot d\boldsymbol{\sigma}_2}{r_2^2} \tag{5.8}$$

Note that the angle between $d\boldsymbol{\sigma}_1$ and $\hat{\mathbf{r}}$ is greater than 90° (so that $d\Omega_1 < 0$), whereas the angle between $d\boldsymbol{\sigma}_2$ and $\hat{\mathbf{r}}$ is less than 90°. Since the magnitudes of $d\Omega_1$ and $d\Omega_2$ are both the same,

$$d\Omega_1 + d\Omega_2 = 0 \tag{5.9}$$

The argument can be extended to the entire surface with the result that

$$\int_\sigma d\Omega = 0 \tag{5.10}$$

Note that the direction of $d\boldsymbol{\sigma}$ is always outward from a closed surface. The symbol $\int_\sigma$ means integral over the entire surface.

In electrostatics, the electric field produced by a single point charge of magnitude q is

$$\mathbf{E} = \frac{1}{4\pi\varepsilon_0} \frac{q}{r^2} \hat{\mathbf{r}} \tag{5.11}$$

where $\mathbf{r}$ is the displacement vector (in units of meters) from the location of the charge to the point where the field is to be calculated, q is measured in coulombs in the mks system of units, and

$$\varepsilon_0 = 8.854 \times 10^{-12}\ \mathrm{C^2/N\text{-}m^2} \tag{5.12}$$

Here, C stands for coulombs, N for newtons, and m for meters. Suppose that the point charge q is surrounded by a closed surface σ. Equation (4.32) defines the flux of the electric field through an element $d\sigma$ of the surface. The net flux of the electric field outward through σ is

$$\Phi_E = \int_\sigma \mathbf{E} \cdot d\boldsymbol{\sigma} = \frac{q}{4\pi\varepsilon_0} \int_\sigma \frac{\hat{\mathbf{r}} \cdot d\boldsymbol{\sigma}}{r^2} \tag{5.13}$$

By means of theorem 1,

$$\int_\sigma \mathbf{E} \cdot d\boldsymbol{\sigma} = \frac{q}{\varepsilon_0} \qquad (q \text{ inside } \sigma) \tag{5.14}$$

If the charge q is outside the closed surface, theorem 2 gives

$$\int_\sigma \mathbf{E} \cdot d\boldsymbol{\sigma} = 0 \qquad (q \text{ outside } \sigma) \tag{5.15}$$

The integrals in (5.14) and (5.15) are called *surface integrals*. The content of (5.14) and (5.15) is pictured in Fig. 5.4. The electric field is represented by vectors or *lines of force* radiating outward from the charge q (assumed positive). In Fig. 5.4*a*, it is evident that each line of the electric field passes outward through the surface so that there is a net outward flux of **E**. In Fig. 5.4*b*, each line enters and leaves the closed surface giving a net zero flux. The number of lines that emanate from q can be made proportional to the magnitude of q, which can be thought of as a source for **E**. The spacing of the lines (number of lines/m^2) gives an indication of the field strength at various distances from q.

If there are two point charges inside a closed surface σ as shown in Fig. 5.5, the net outward flux of the electric field is

$$\begin{aligned}\Phi_E &= \int_\sigma \mathbf{E} \cdot d\boldsymbol{\sigma} = \int_\sigma (\mathbf{E}_1 + \mathbf{E}_2) \cdot d\boldsymbol{\sigma} \\ &= \int_\sigma \mathbf{E}_1 \cdot d\boldsymbol{\sigma} + \int_\sigma \mathbf{E}_2 \cdot d\boldsymbol{\sigma} = \frac{1}{\varepsilon_0}(q_1 + q_2)\end{aligned} \tag{5.16}$$

The argument can be extended to prove the following:

Theorem 3. Gauss' law. The net flux of the electric field outward through a closed surface is $1/\varepsilon_0$ times the total charge enclosed:

$$\int_\sigma \mathbf{E} \cdot d\boldsymbol{\sigma} = \frac{1}{\varepsilon_0} q_{\text{enclosed}} \tag{5.17}$$

The enclosed charge can be in the form of point charges of any number or can exist in the form of continuous charge.

As an example of the use of Gauss' law, consider a sphere of radius a throughout which a charge q is uniformly spread. The charge density inside the sphere

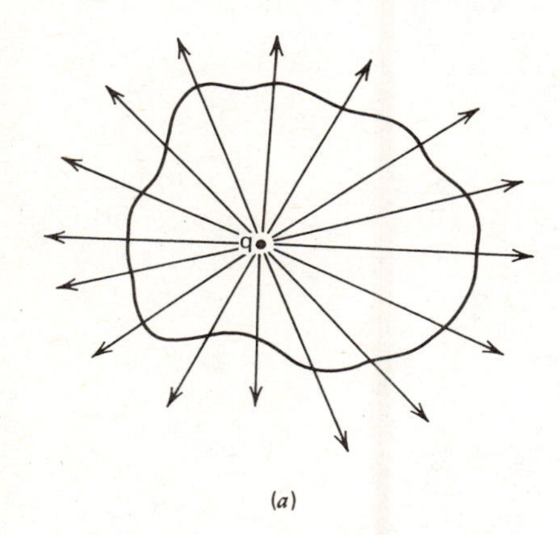

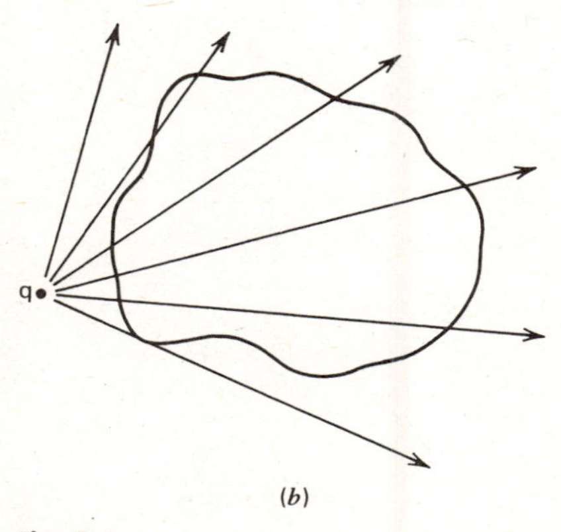

Fig. 5.4

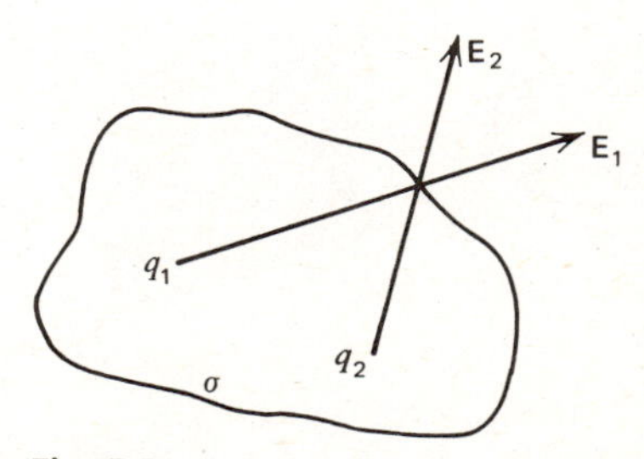

Fig. 5.5

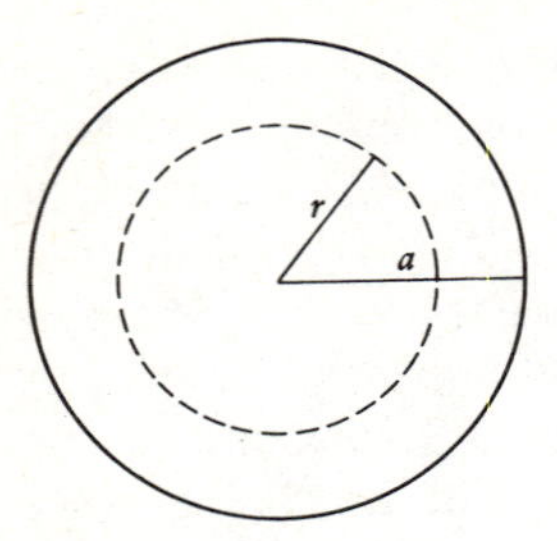

Fig. 5.6

is

$$\rho = \frac{q}{\frac{4}{3}\pi a^3} \text{ C/m}^3 \qquad (5.18)$$

Because of the simple geometry, it is possible to calculate the electric field by means of Gauss' law at points both inside and outside the sphere. Figure 5.6 shows a spherical surface of radius $r < a$ drawn inside the sphere of radius a. The total charge enclosed by this so-called *Gaussian surface* is

$$q_{\text{enclosed}} = \rho \tfrac{4}{3}\pi r^3 = q\frac{r^3}{a^3} \qquad (5.19)$$

The electric field is radially outward through the Gaussian surface and has the same strength at each point of the surface, from which we infer that

$$\int \mathbf{E} \cdot d\boldsymbol{\sigma} = E_r 4\pi r^2 \qquad (5.20)$$

This combined with (5.19) and Gauss' law (5.17) gives

$$E_r = \frac{q}{4\pi\varepsilon_0}\frac{r}{a^3} \qquad r < a \qquad (5.21)$$

where E_r is the component of **E** in the r direction (the only component in this case). By a similar argument, the field outside the sphere is

$$E_r = \frac{q}{4\pi\varepsilon_0}\frac{1}{r^2} \qquad r > a \qquad (5.22)$$

PROBLEMS

5.1. A point P is on the axis of a circular disc of radius a and is a distance z from the disc. Find the solid angle subtended at P by the disc. What is the solid angle if $a \gg z$? See Fig. 5.7.

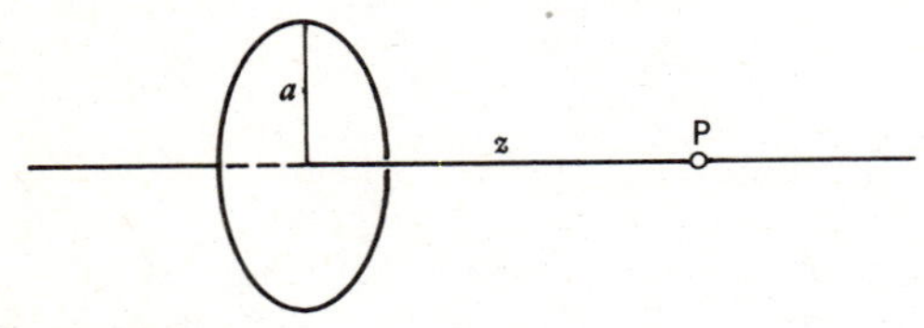

Fig. 5.7

5.2. What solid angle does the closed surface σ subtend at the points P_1 and P_2 in Fig. 5.8?

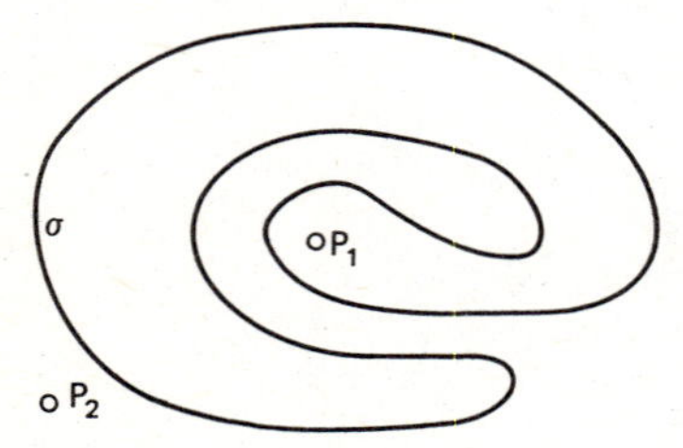

Fig. 5.8

5.3. Explain why the representation of the electric field created by a point charge by means of lines of force as pictured in Fig. 5.4 would not work if the electric field of a point charge were something other than an inverse r^2 law, for example,

$$\mathbf{E} = \frac{1}{4\pi\varepsilon_0} \frac{q}{r^3} \hat{\mathbf{r}}$$

5.4. The gravitational field produced by a point mass M is

$$\mathbf{g} = -G\frac{M}{r^2}\hat{\mathbf{r}}$$

where G is the universal gravitational constant. Develop Gauss' law for the gravitational field. Given that the field strength at the Earth's surface is $g = 9.80\ \text{m/sec}^2$, what is the field strength at a point halfway between the center of the Earth and its surface? Assume that the Earth is a sphere throughout which the total mass of the Earth is uniformly distributed.

5.5. A point charge q is at the exact center of a cylindrical surface of length L and radius a. Find the flux of the electric field through the cylindrical surface excluding the two ends.

Hint: You can use the result of problem 5.1 and Gauss' law to avoid doing an integral. See Fig. 5.9.

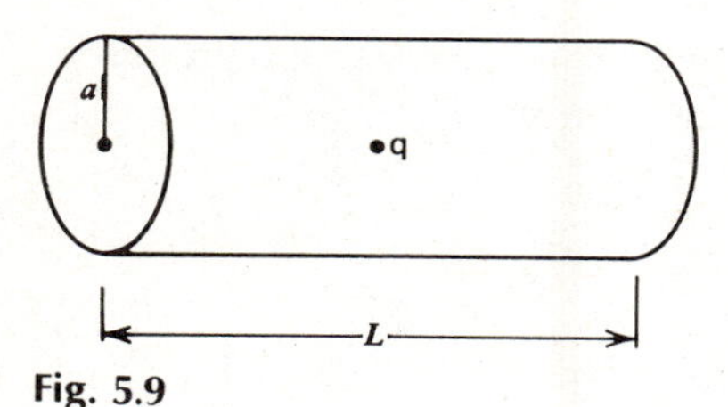

Fig. 5.9

6. TENSORS

Suppose that in a given coordinate system a relation of the form

$$Y = AX \tag{6.1}$$

exists between the vectors X and Y. The square matrix A can be applied to any vector X to produce a new vector Y. As an example, equation (9.4) of Chapter 1 can be written as

$$\ddot{X} = -AX \tag{6.2}$$

where

$$A = \frac{k}{m}\begin{pmatrix} 2 & -1 \\ -1 & 2 \end{pmatrix} \tag{6.3}$$

Figure 9.1 of Chapter 1 shows the actual physical system under consideration. It is also possible to think of (6.2) as a relation set up in a conceptual *configuration space* which is a two-dimensional rectangular Cartesian coordinate system as shown in Fig. 6.1. The motion of the two-particle system is then represented by a single point which moves along a trajectory with time. For one reason or another, it may be desirable to transform the problem to another Cartesian coordinate system, for example, by means of a rotation of coordinates.

We already know how to transform the vectors in (6.1). The transformations and their inverses are

$$\begin{aligned} Y' &= SY \qquad & Y &= \tilde{S}Y' \\ X' &= SX \qquad & X &= \tilde{S}X' \end{aligned} \tag{6.4}$$

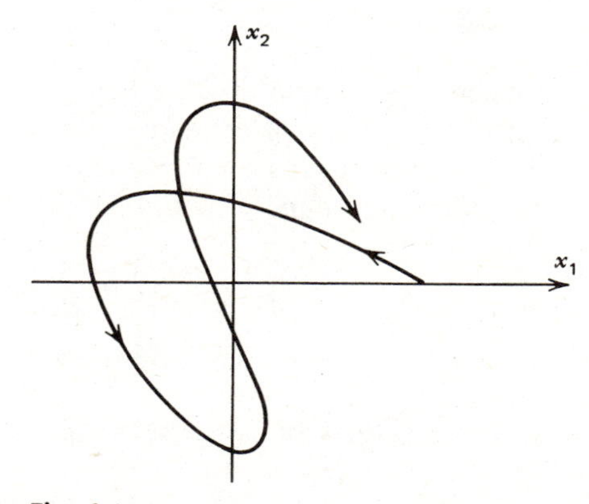

Fig. 6.1

If (6.1) is transformed, a relation

$$Y' = A'X' \tag{6.5}$$

will exist between the transformed vectors X' and Y'. To find the relation between the square matrices A and A', combine (6.4), (6.1), and (6.5):

$$\tilde{S}Y' = A\tilde{S}X'$$
$$Y' = SA\tilde{S}X' = A'X' \tag{6.6}$$

Since X' can be chosen arbitrarily,

$$A' = SA\tilde{S} \tag{6.7}$$

This is an *orthogonal similarity transformation* and is a special case of the general similarity transformation $A' = TAT^{-1}$, where T is any nonsingular matrix. It is easy to invert (6.7):

$$\tilde{S}A'S = \tilde{S}SA\tilde{S}S = IAI = A \tag{6.8}$$

The component version of (6.7) is

$$A'_{ij} = \sum_{n=1}^{3} \sum_{m=1}^{3} S_{in} S_{jm} A_{nm} \tag{6.9}$$

In three-dimensional Cartesian space, a set of nine functions which has the transformation property (6.9) is said to be a *second-rank tensor* under orthogonal transformation. The term *rank* as used here is not to be confused with the rank of a matrix. A scalar can actually be considered to be a tensor of rank zero and a vector, a tensor of rank 1. Tensors of any rank are vector fields in that they all have the algebraic properties 1 through 8 listed for vectors in section 1 of this chapter. For our purposes, we will use the term vector for a tensor of rank 1. Tensors of higher rank than 2 exist.

Sometimes, a problem can be simplified by the use of a tensor transformation. To explore this possibility, we look again at the mechanical system of section 9 in Chapter 1. The differential equations of motion are

$$m\ddot{x}_1 + 2kx_1 - kx_2 = 0 \tag{6.10}$$
$$m\ddot{x}_2 - kx_1 + 2kx_2 = 0 \tag{6.11}$$

which are the equivalent of (6.2) but are more convenient for the moment. Multiply (6.10) by $\dot{x}_1$ and (6.11) by $\dot{x}_2$ and then add the two equations:

$$m\dot{x}_1\ddot{x}_1 + m\dot{x}_2\ddot{x}_2 + 2k\dot{x}_1x_1 + 2k\dot{x}_2x_2 - k\dot{x}_1x_2 - k\dot{x}_2x_1 = 0 \tag{6.12}$$

This is equivalent to

$$\frac{d}{dt}\left[\tfrac{1}{2}m\left(\dot{x}_1^2 + \dot{x}_2^2\right) + k\left(x_1^2 + x_2^2 - x_1x_2\right)\right] = 0 \tag{6.13}$$

which states that the mechanical energy of the system is a constant. The potential energy is

$$V = k\left(x_1^2 + x_2^2 - x_1x_2\right) \tag{6.14}$$

The equipotential lines $V =$ constant when plotted in the x_1, x_2 configuration space are ellipses with their *principal axes* inclined at an angle of 45° to the x_1, x_2 coordinate axes, Fig. 6.2. This suggests that a rotation by 45° in the configuration space might be a profitable transformation to try. Such a rotation is given by

$$S = \frac{1}{\sqrt{2}}\begin{pmatrix} 1 & 1 \\ -1 & 1 \end{pmatrix} \tag{6.15}$$

and is in fact called a *principal axis transformation* because the rotated coordinates coincide with the principal axes of the ellipses $V =$ constant. By means of (6.3) and (6.7),

$$A' = \frac{k}{2m}\begin{pmatrix} 1 & 1 \\ -1 & 1 \end{pmatrix}\begin{pmatrix} 2 & -1 \\ -1 & 2 \end{pmatrix} \times \begin{pmatrix} 1 & -1 \\ 1 & 1 \end{pmatrix} = \frac{k}{m}\begin{pmatrix} 1 & 0 \\ 0 & 3 \end{pmatrix} \tag{6.16}$$

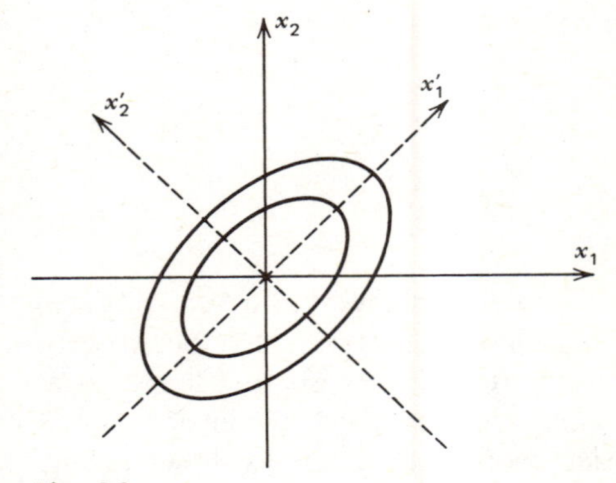

Fig. 6.2

The equations of motion in the new coordinates are $\ddot{X}' + A'X' = 0$. Written out in component form, they are

$$\ddot{x}_1' + \frac{k}{m} x_1' = 0 \tag{6.17}$$

$$\ddot{x}_2' + \frac{3k}{m} x_2' = 0 \tag{6.18}$$

which are indeed easier to solve than are (6.10) and (6.11). The solutions are

$$x_1' = a_1 \cos(\omega_1 t - \phi_1) \tag{6.19}$$

$$x_2' = a_2 \cos(\omega_2 t - \phi_2) \tag{6.20}$$

where $\omega_1 = \sqrt{k/m}$ and $\omega_2 = \sqrt{3k/m}$. The coordinates x_1', x_2' are called *normal coordinates* for the system. The original coordinates are found through $X = \tilde{S}X'$, which in this case is

$$\begin{pmatrix} x_1 \\ x_2 \end{pmatrix} = \frac{1}{\sqrt{2}} \begin{pmatrix} 1 & -1 \\ 1 & 1 \end{pmatrix} \begin{pmatrix} x_1' \\ x_2' \end{pmatrix} \tag{6.21}$$

By combining (6.19), (6.20), and (6.21),

$$x_1 = \frac{1}{\sqrt{2}} a_1 \cos(\omega_1 t - \phi_1) - \frac{1}{\sqrt{2}} a_2 \cos(\omega_2 t - \phi_2) \tag{6.22}$$

$$x_2 = \frac{1}{\sqrt{2}} a_1 \cos(\omega_1 t - \phi_1) + \frac{1}{\sqrt{2}} a_2 \cos(\omega_2 t - \phi_2) \tag{6.23}$$

which are the equivalent of (9.21) of Chapter 1. The secret of the success of the method is that the matrix A' in (6.16) comes out in *diagonal form*. In a later section, we will develop a systematic procedure for finding the principal axis transformation of a given matrix, which is essentially the transformation which renders it into a diagonal form.

Certain special types of tensors are important. A *symmetric tensor* has the property that

$$A = \tilde{A} \tag{6.24}$$

or, in component form,

$$A_{ij} = A_{ji} \tag{6.25}$$

Note that the matrix (6.3) has this property. In fact, if it were not for the symmetry property of (6.3), it could not have been put into the diagonal form as given by (6.16). We will elaborate in detail on this point later. An *antisymmetric tensor* obeys

$$A = -\tilde{A} \qquad \text{or} \qquad A_{ij} = -A_{ji} \tag{6.26}$$

Theorem 1. Symmetry and antisymmetry of second-rank tensors are invariant properties under orthogonal transformations.

Proof: Assume that $\mathbf{A} = \tilde{\mathbf{A}}$. Take the transpose of $A' = SA\tilde{S}$ to get $\tilde{A}' = \tilde{\tilde{S}}\tilde{A}\tilde{S} = SA\tilde{S} = A'$. Similarly, $A = -\tilde{A}$ implies that $A' = -\tilde{A}'$.

As was mentioned earlier, tensors of any rank can be defined. A third-rank tensor (in three-dimensional space) is a set of 27 functions of position which has the transformation property

$$A_{ijk}' = \sum_{p=1}^{3} \sum_{q=1}^{3} \sum_{r=1}^{3} S_{ip} S_{jq} S_{kr} A_{pqr} \tag{6.27}$$

and in addition obeys the algebraic properties 1 through 8 stated in section 1 of this chapter for vector fields.

There are entities which are quite important in applications which fall a bit short of the requirements for tensors. They are so-called *pseudotensors* and have the peculiarity that a minus sign appears when they are subjected to an improper orthogonal transformation. Otherwise, they behave in a normal way. Since $|S| = \pm 1$, where the plus sign applies to a proper transformation and the minus sign to an improper one, the transformation property of a pseudoscalar (pseudotensor of rank 0) is

$$\alpha' = |S|\alpha \tag{6.28}$$

A pseudovector (pseudotensor of rank 1) is characterized by

$$X' = |S|SX \qquad \text{or} \qquad a_i' = |S| \sum_{j=1}^{3} S_{ij} a_j \tag{6.29}$$

A second-rank pseudotensor has the property

$$A' = |S|SA\tilde{S} \qquad \text{or} \qquad A_{ij}' = |S| \sum_{m=1}^{3} \sum_{n=1}^{3} S_{im} S_{jn} A_{mn} \tag{6.30}$$

Finally, for a third-rank pseudotensor, the transformation is

$$A'_{ijk} = |S| \sum_{p=1}^{3} \sum_{q=1}^{3} \sum_{r=1}^{3} S_{ip} S_{jq} S_{kr} A_{pqr} \tag{6.31}$$

There are some very common examples of pseudo quantities. It turns out that the representation of area is really a pseudovector. The magnetic field and angular momentum are two other examples of pseudovectors. The permutation symbol is a third-rank pseudotensor about which we will have much more to say later.

PROBLEMS

6.1. Consider the equation of motion $M\ddot{X} + KX = 0$, where both M and K are symmetric matrices. Multiply the equation through by $\dot{X}$ and obtain the energy equation in matrix form:

$$W = \tfrac{1}{2}\tilde{\dot{X}} M \dot{X} + \tfrac{1}{2}\tilde{X} K X = \text{const.}$$

6.2. Show that equation (6.13) implies that the origin of coordinates in the configuration space is a point of stable equilibrium for the system. Prove that if the system is given an energy W, the system point in configuration space never goes outside the ellipse $V = W$.

Hint: Look ahead at problem 6.3.

6.3. Show that the potential function (6.14) can be expressed as

$$V = \tfrac{1}{2}\begin{pmatrix} x_1 & x_2 \end{pmatrix} \begin{pmatrix} 2k & -k \\ -k & 2k \end{pmatrix} \begin{pmatrix} x_1 \\ x_2 \end{pmatrix}$$

Find the expression for the potential function in normal coordinates. What are the lengths of the major and minor axes of the ellipse $V = W$, where W is the total energy of the system?

6.4. Refer to problem 9.1 of Chapter 1. If the initial conditions are $x_1(0) = 2$ cm, $x_2(0) = 0$, $\dot{x}_1(0) = 0$, and $\dot{x}_2(0) = 0$, show that the solution can be expressed

$$x_1 = \cos u + \cos\sqrt{3}\,u \qquad x_2 = \cos u - \cos\sqrt{3}\,u$$

where $u = \sqrt{k/m}\,t$. Make an accurate numerical plot of the trajectory in configuration space for $0 \le u \le 6.5$.

6.5. Show that any second-rank tensor can be resolved into two parts: $A = A_1 + A_2$, where A_1 is symmetric, $A_1 = \tilde{A}_1$, and A_2 is antisymmetric, $A_2 = -\tilde{A}_2$. Show that the resolution is unique and can be done in one and only one way.

6.6. Given that **a** and **b** are vectors, show that the *direct product*

$$\mathbf{a} \otimes \mathbf{b} = \begin{pmatrix} a_1 \\ a_2 \\ a_3 \end{pmatrix} \begin{pmatrix} b_1 & b_2 & b_3 \end{pmatrix} = \begin{pmatrix} a_1 b_1 & a_1 b_2 & a_1 b_3 \\ a_2 b_1 & a_2 b_2 & a_2 b_3 \\ a_3 b_1 & a_3 b_2 & a_3 b_3 \end{pmatrix}$$

is a second-rank tensor.

6.7. Prove that the determinant of a second-rank tensor is an invariant: $|A| = |A'|$.

6.8. The *trace* of a matrix or a second-rank tensor is defined to be the sum of its diagonal elements:

$$\operatorname{tr} A = \sum_{i=1}^{3} A_{ii}$$

Prove that the trace is an invariant: $\operatorname{tr} A = \operatorname{tr} A'$.

Hint: Use the transformation in component form, equation (6.9).

6.9. If A is known to be a second-rank tensor and X is a vector, show that $Y = AX$ is a vector. (Show that $Y' = SY$ follows from the known transformation properties of A and X.)

6.10. If A and B are second-rank tensors, show that $C = AB$ and $D = \tilde{A}B$ are also second-rank tensors. What other combinations of A and B will produce second-rank tensors? In books on tensor analysis, the component form of a product such as $C = AB$ is written out as

$$C_{ij} = \sum_k A_{ik} B_{kj}$$

and called a *contraction of indices*. Another possible contraction is

$$D_{ij} = \sum_k A_{ki} B_{kj}$$

The trace of a tensor is also considered to be a type of contraction. All second-rank tensors can be represented by matrices. The converse is not true. A matrix is not necessarily a tensor.

7. ISOTROPIC TENSORS

A second-rank isotropic tensor obeys the usual tensor transformation $A' = SA\tilde{S}$ and in addition has the property $A' = A$. This is quite a severe restriction; and as a consequence, there are not very many isotropic tensors, but the ones that do exist are very important. We will investigate the existence of isotropic tensors both in a three-dimensional space and a two-dimensional space. Since the identity tensor obeys $I = SI\tilde{S}$, it is an obvious example of an isotropic tensor. The component version of I is the Kronecker delta δ_{ij}.

Theorem 1. In three dimensions, the *only* second-rank isotropic tensor is the identity tensor I.

Proof: The condition

$$A = SA\tilde{S} \tag{7.1}$$

must hold for all possible orthogonal transformations. First, we consider only proper transformations. Let

$$S = \begin{pmatrix} -1 & 0 & 0 \\ 0 & -1 & 0 \\ 0 & 0 & 1 \end{pmatrix} \tag{7.2}$$

Equation (7.1) gives

$$\begin{pmatrix} A_{11} & A_{12} & A_{13} \\ A_{21} & A_{22} & A_{23} \\ A_{31} & A_{32} & A_{33} \end{pmatrix} = \begin{pmatrix} A_{11} & A_{12} & -A_{13} \\ A_{21} & A_{22} & -A_{23} \\ -A_{31} & -A_{32} & A_{33} \end{pmatrix} \tag{7.3}$$

By equating elements, we find, for example, that $A_{13} = -A_{13}$, which can hold only if $A_{13} = 0$. In this manner, (7.3) shows that

$$A = \begin{pmatrix} A_{11} & A_{12} & 0 \\ A_{21} & A_{22} & 0 \\ 0 & 0 & A_{33} \end{pmatrix} \tag{7.4}$$

By using

$$S = \begin{pmatrix} 1 & 0 & 0 \\ 0 & -1 & 0 \\ 0 & 0 & -1 \end{pmatrix} \tag{7.5}$$

in an exactly similar manner, it is possible to reduce A to

$$A = \begin{pmatrix} A_{11} & 0 & 0 \\ 0 & A_{22} & 0 \\ 0 & 0 & A_{33} \end{pmatrix} \tag{7.6}$$

Now try

$$S = \begin{pmatrix} 0 & 1 & 0 \\ -1 & 0 & 0 \\ 0 & 0 & 1 \end{pmatrix} \tag{7.7}$$

This time, (7.1) and (7.6) combine to give

$$\begin{pmatrix} A_{11} & 0 & 0 \\ 0 & A_{22} & 0 \\ 0 & 0 & A_{33} \end{pmatrix} = \begin{pmatrix} A_{22} & 0 & 0 \\ 0 & A_{11} & 0 \\ 0 & 0 & A_{33} \end{pmatrix} \tag{7.8}$$

which shows $A_{11} = A_{22}$. Similarly,

$$S = \begin{pmatrix} 1 & 0 & 0 \\ 0 & 0 & 1 \\ 0 & -1 & 0 \end{pmatrix} \tag{7.9}$$

shows that $A_{22} = A_{33}$. By using special proper orthogonal transformations, we have shown that $A = SA\tilde{S}$ implies that

$A = \alpha I$, where α is a scalar. Since αI is obviously isotropic under *all* orthogonal transformations, both proper and improper, it is therefore the only such tensor.

Theorem 2. There are no second-rank isotropic pseudotensors in three-dimensional space.

Proof: An isotropic pseudotensor must of course obey $A = SA\tilde{S}$ if $|S| = +1$. In the proof of theorem 1, it was possible to reduce A down to I by using only proper rotations. Since obviously I does not satisfy $I = -SI\tilde{S}$ if $|S| = -1$, there are no second-rank isotropic pseudotensors in three dimensions.

The situation in two dimensions is somewhat more complicated.

Theorem 3. Any tensor of the form

$$A = \begin{pmatrix} a & b \\ -b & a \end{pmatrix} \tag{7.10}$$

is isotropic under proper rotations in two dimensions, and in fact (7.10) is the most general such tensor.

Proof: Written out, $A = SA\tilde{S}$ is

$$\begin{pmatrix} A_{11} & A_{12} \\ A_{21} & A_{22} \end{pmatrix} = \begin{pmatrix} \cos\theta & \sin\theta \\ -\sin\theta & \cos\theta \end{pmatrix} \begin{pmatrix} A_{11} & A_{12} \\ A_{21} & A_{22} \end{pmatrix} \begin{pmatrix} \cos\theta & -\sin\theta \\ \sin\theta & \cos\theta \end{pmatrix}$$

$$= \begin{pmatrix} A_{11}\cos^2\theta + A_{22}\sin^2\theta + (A_{12}+A_{21})\sin\theta\cos\theta & (A_{22}-A_{11})\sin\theta\cos\theta - A_{21}\sin^2\theta + A_{12}\cos^2\theta \\ (A_{22}-A_{11})\sin\theta\cos\theta + A_{21}\cos^2\theta - A_{12}\sin^2\theta & A_{11}\sin^2\theta + A_{22}\cos^2\theta - (A_{12}+A_{21})\sin\theta\cos\theta \end{pmatrix} \tag{7.11}$$

The special case $\theta = \pi/2$ gives

$$\begin{pmatrix} A_{11} & A_{12} \\ A_{21} & A_{22} \end{pmatrix} = \begin{pmatrix} A_{22} & -A_{21} \\ -A_{12} & A_{11} \end{pmatrix} \tag{7.12}$$

Equation (7.10) results if $A_{11} = A_{22} = a$, $A_{12} = -A_{21} = b$. Moreover, (7.10) reduces (7.11) to an identity, which completes the proof.

Theorem 4. The only two-dimensional tensor which is isotropic under all orthogonal transformations is the identity tensor.

Proof: We start with (7.10) and require that $A = SA\tilde{S}$ for the improper two-dimensional transformation

$$S = \begin{pmatrix} -1 & 0 \\ 0 & 1 \end{pmatrix} \tag{7.13}$$

The result is

$$\begin{pmatrix} a & b \\ -b & a \end{pmatrix} = \begin{pmatrix} a & -b \\ b & a \end{pmatrix} \tag{7.14}$$

which implies $b = 0$. Thus,

$$A = aI \tag{7.15}$$

Theorem 5. The only two-dimensional pseudotensor which is isotropic under all transformations is

$$A = \begin{pmatrix} 0 & 1 \\ -1 & 0 \end{pmatrix} \tag{7.16}$$

Proof: Start with (7.10) and assume that $A = |S|SA\tilde{S}$. Use (7.13) to get

$$\begin{pmatrix} a & b \\ -b & a \end{pmatrix} = \begin{pmatrix} -a & b \\ -b & -a \end{pmatrix} \tag{7.17}$$

which implies $a = 0$. Thus, A differs from (7.16) by at most a scalar factor:

$$A = b\begin{pmatrix} 0 & 1 \\ -1 & 0 \end{pmatrix} \tag{7.18}$$

Reference to equation (4.5) of Chapter 1 shows that (7.16) is the two-dimensional permutation symbol in matrix form.

We return now to a consideration of tensors in three-dimensional space. If in equation (5.21) of Chapter 1 the elements A_{ij} of the arbitrary matrix A are replaced by the elements of an orthogonal transformation, the result is

$$e_{ijk} = |S| \sum_{p=1}^{3} \sum_{q=1}^{3} \sum_{r=1}^{3} S_{ip} S_{jq} S_{kr} e_{pqr} \tag{7.19}$$

This shows that the three-dimensional permutation symbol is a third-rank isotropic pseudotensor.

Theorem 6. In a three-dimensional space, there are no ordinary isotropic tensors of rank 3 and the only isotropic pseudotensor of rank 3 is the permutation symbol.

Proof: In the course of the proof, the summation convention will be used. Consider first of all the possibility that an ordinary isotropic tensor of rank 3 may exist. It would satisfy

$$A_{ijk} = S_{ip}S_{jq}S_{kr}A_{pqr} \tag{7.20}$$

In this case, all elements of A_{ijk} can be shown to be zero by using for S one of the following:

$$\begin{pmatrix} -1 & 0 & 0 \\ 0 & 1 & 0 \\ 0 & 0 & 1 \end{pmatrix} \quad \begin{pmatrix} 1 & 0 & 0 \\ 0 & -1 & 0 \\ 0 & 0 & 1 \end{pmatrix}$$
$$\begin{pmatrix} 1 & 0 & 0 \\ 0 & 1 & 0 \\ 0 & 0 & -1 \end{pmatrix} \quad \begin{pmatrix} -1 & 0 & 0 \\ 0 & -1 & 0 \\ 0 & 0 & 1 \end{pmatrix}$$
$$\begin{pmatrix} -1 & 0 & 0 \\ 0 & 1 & 0 \\ 0 & 0 & -1 \end{pmatrix} \quad \begin{pmatrix} 1 & 0 & 0 \\ 0 & -1 & 0 \\ 0 & 0 & -1 \end{pmatrix} \tag{7.21}$$

For any one of the matrices in (7.21), equation (7.20) reduces to

$$A_{111} = S_{11}S_{11}S_{11}A_{111}, \qquad A_{123} = S_{11}S_{12}S_{33}A_{123}, \qquad \text{and so on} \tag{7.22}$$

It is only necessary to pick out of (7.21) a transformation which will reduce (7.22) to $A_{111} = -A_{111}$, $A_{123} = -A_{123}$, and so on; and it is easy to see that this can always be done. In this way, all elements of A_{ijk} can be shown to be zero.

To investigate the possibilities for the existence of isotropic pseudotensors of rank 3, (7.20) must be replaced by

$$A_{ijk} = |S|S_{ip}S_{jq}S_{kr}A_{pqr} \tag{7.23}$$

Consider an orthogonal transformation of the form

$$S = \begin{pmatrix} 1 & 0 & 0 \\ 0 & S_{22} & S_{23} \\ 0 & S_{32} & S_{33} \end{pmatrix} \tag{7.24}$$

and let $k = 1$ in (7.23):

$$A_{ij1} = |S|S_{ip}S_{jq}S_{1r}A_{pqr} = |S|S_{ip}S_{jq}A_{pq1} \tag{7.25}$$

For i and j restricted to the values 2 or 3, (7.24) and (7.25) are the equivalent of a two-dimensional transformation. By theorem 5,

$$A_{221} = A_{331} = 0 \qquad A_{231} = -A_{321} \tag{7.26}$$

By using this same approach, it is possible to show that all elements with two subscripts alike are zero and that

$$\begin{aligned} &A_{231} = -A_{321} && A_{213} = -A_{312} && A_{123} = -A_{132} \\ &A_{132} = -A_{312} && A_{123} = -A_{321} && A_{213} = -A_{231} \\ &A_{123} = -A_{213} && A_{132} = -A_{231} && A_{312} = -A_{321} \end{aligned} \tag{7.27}$$

The elements A_{111}, A_{222}, and A_{333} can be shown to be zero by using one or another of the transformations (7.21). Then, if $A_{123} = a$, (7.27) comes down to

$$\begin{aligned} &A_{123} = a && A_{132} = -a && A_{231} = a \\ &A_{213} = -a && A_{312} = a && A_{321} = -a \end{aligned} \tag{7.28}$$

We have therefore shown that $A_{ijk} = ae_{ijk}$, which completes the proof.

The permutation symbol occurs in the definition of the *cross product* of two vectors, to be discussed in section 8 of this chapter. The cross product occurs in a wide range of physical applications, and it is significant that the permutation symbol, which has such special and unique mathematical properties, is fundamental in its definition.

PROBLEMS

7.1. Theorem 3 shows that tensors of the form

$$Z = \begin{pmatrix} x & y \\ -y & x \end{pmatrix} = x\begin{pmatrix} 1 & 0 \\ 0 & 1 \end{pmatrix} + y\begin{pmatrix} 0 & 1 \\ -1 & 0 \end{pmatrix}$$

are isotropic under proper rotations in two dimensions. Show that the algebraic properties of these tensors are identical to the algebra of complex numbers $z = x + iy$, where $i = \sqrt{-1}$.

7.2. Start with (7.20) and pick out of the transformations given in (7.21) the one which shows $A_{213} = 0$.

7.3. Start with (7.23) and pick out a transformation similar to (7.24) which will prove that $A_{121} = A_{323} = 0$ and $A_{123} = -A_{321}$.

7.4. Start with (7.23) and pick out of the transformations in (7.21) one which will show that $A_{222} = 0$.

7.5. Suppose that A_{ij} and B_{ij} are the elements of two tensors A and B in three-dimensional space. Show that the set of 81 quantities $A_{ij}B_{kl}$ form the elements of a fourth-rank tensor.

7.6. Show that the three-dimensional fourth-rank tensor with elements $\delta_{jk}\,\delta_{mn}$ is isotropic.

7.7. Prove that the tensor given in problem 7.6 is the only type of fourth-rank tensor which is possible, so that in three-dimensional space the most general fourth-rank isotropic tensor is of the form

$$A_{jkmn} = \alpha\delta_{jk}\,\delta_{mn} + \beta\delta_{jm}\,\delta_{kn} + \gamma\delta_{jn}\,\delta_{km}$$

Hint: The transformation property of a general fourth-rank tensor is

$$A'_{ijkm} = S_{ip}S_{jq}S_{kr}S_{ms}A_{pqrs}$$

(the summation convention is used). Let $k = m = 1$, use the special transformation (7.24) and invoke theorem 1.

7.8. Prove that

$$\sum_{i=1}^{3} e_{ijk}e_{imn} = \delta_{jm}\,\delta_{kn} - \delta_{jn}\,\delta_{km}$$

This is an extremely important identity involving the permutation symbol, which we will prove in another way later. You can prove it as follows: First, convince yourself that the quantity $\Sigma_i e_{ijk}e_{imn}$ is a fourth-rank isotropic tensor. Since the only possible fourth-rank isotropic tensors are of the form given in problem 7.7, it is only necessary to show that $\alpha = 0$, $\beta = 1$, $\gamma = -1$.

8. THE CROSS PRODUCT

Given two vectors **a** and **b**, it is possible to form through the use of the permutation symbol a set of three quantities:

$$c_i = \sum_{j=1}^{3}\sum_{k=1}^{3} e_{ijk}a_jb_k \tag{8.1}$$

This is the *cross product* of the two vectors **a** and **b**.

Theorem 1. The cross product of two ordinary vectors is a pseudovector. The cross product of an ordinary vector and a pseudovector is an ordinary vector. The cross product of two pseudovectors is a pseudovector.

Proof: Assume that **a** and **b** are ordinary vectors. We will use the summation convention in the proof. By using the known transformation properties of the permutation symbol and the vectors **a** and **b**,

$$\begin{aligned} c'_i = e_{ijk}a'_jb'_k &= |S|S_{ip}S_{jq}S_{kr}e_{pqr}S_{jm}a_mS_{kn}b_n \\ &= |S|S_{ip}(S_{jq}S_{jm})(S_{kr}S_{kn})e_{pqr}a_mb_n \\ &= |S|S_{ip}\,\delta_{qm}\,\delta_{rn}e_{pqr}a_mb_n = |S|S_{ip}e_{pmn}a_mb_n = |S|S_{ip}c_p \end{aligned} \tag{8.2}$$

which completes the proof. Use has been made of the orthogonality conditions (2.12) and the properties of the Kronecker delta. The proofs of the other parts of the theorem are similar. Another term for a pseudovector is *axial vector*. Ordinary vectors are also called *polar vectors*.

The commonly used vector notation for the cross product is

$$\mathbf{c} = \mathbf{a} \times \mathbf{b} = \sum_{i=1}^{3}\sum_{j=1}^{3}\sum_{k=1}^{3} e_{ijk}a_jb_k\hat{\mathbf{e}}_i \tag{8.3}$$

which can also be written out or expressed as a

determinant:

$$\mathbf{c} = (a_2b_3 - a_3b_2)\hat{\mathbf{e}}_1 + (a_3b_1 - a_1b_3)\hat{\mathbf{e}}_2 + (a_1b_2 - a_2b_1)\hat{\mathbf{e}}_3 = \begin{vmatrix} \hat{\mathbf{e}}_1 & \hat{\mathbf{e}}_2 & \hat{\mathbf{e}}_3 \\ a_1 & a_2 & a_3 \\ b_1 & b_2 & b_3 \end{vmatrix} \tag{8.4}$$

The cross product is noncommutative and obeys

$$\mathbf{a} \times \mathbf{b} = -\mathbf{b} \times \mathbf{a} \tag{8.5}$$

This is shown to be true by interchanging **a** and **b** in the determinant (8.4) and using the fact that interchanging two rows in a determinant changes its sign.

Theorem 2. The magnitude of the cross product is

$$|\mathbf{a} \times \mathbf{b}| = ab \sin\theta \tag{8.6}$$

where θ is the angle between **a** and **b**, which is less than 180°. The direction of the cross product is perpendicular to the plane formed by **a** and **b**.

Proof: Choose a coordinate system in which **a** and **b** lie in the x_1, x_2 plane and **a** points along the x_1 axis as shown in Fig. 8.1. Then, by (8.4),

$$\mathbf{a} \times \mathbf{b} = a_1 b_2 \hat{\mathbf{e}}_3 = ab \sin\theta\, \hat{\mathbf{e}}_3 \tag{8.7}$$

Saying that $\mathbf{a} \times \mathbf{b}$ is perpendicular to the plane determined by **a** and **b** is not quite enough. The coordinate system in Fig. 8.1 could as well have been constructed with its x_3 axis pointing in the opposite direction. We remove this ambiguity in the cross product by arbitrarily adopting a right-hand convention in which $\mathbf{a} \times \mathbf{b}$ points in the direction that a right-hand screw would advance if **a** were rotated into **b** through the smaller of the two angles between **a** and **b**, which is the angle θ in Fig. 8.1. The coordinate system has been drawn to conform to this rule.

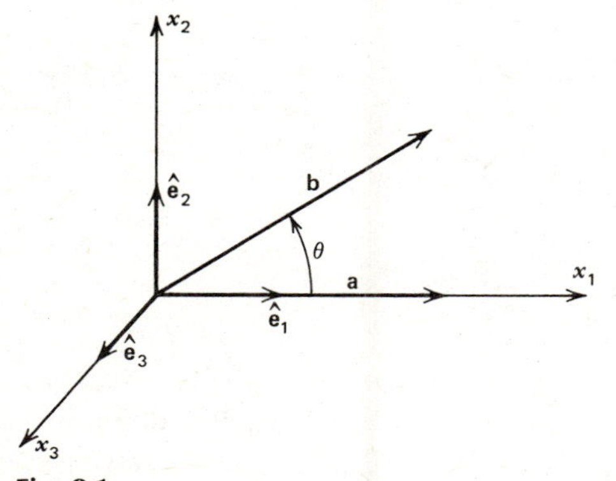

Fig. 8.1

The unit coordinate vectors themselves obey

$$\begin{aligned} &\hat{\mathbf{e}}_1 \times \hat{\mathbf{e}}_2 = \hat{\mathbf{e}}_3 \qquad \hat{\mathbf{e}}_2 \times \hat{\mathbf{e}}_3 = \hat{\mathbf{e}}_1 \qquad \hat{\mathbf{e}}_3 \times \hat{\mathbf{e}}_1 = \hat{\mathbf{e}}_2 \\ &\hat{\mathbf{e}}_1 \times \hat{\mathbf{e}}_1 = 0 \qquad \hat{\mathbf{e}}_2 \times \hat{\mathbf{e}}_2 = 0 \qquad \hat{\mathbf{e}}_3 \times \hat{\mathbf{e}}_3 = 0 \end{aligned} \tag{8.8}$$

which can be stated more compactly as

$$\hat{\mathbf{e}}_i \times \hat{\mathbf{e}}_j = \sum_{k=1}^{3} e_{ijk} \hat{\mathbf{e}}_k \tag{8.9}$$

Note that there is a definite ordering of the labeling of the coordinate axes in Fig. 8.1 to conform to the right-hand rule. This means that if $\hat{\mathbf{e}}_1$ is rotated onto $\hat{\mathbf{e}}_2$, a right-hand screw advances in the direction of $\hat{\mathbf{e}}_3$. If $\hat{\mathbf{e}}_2$ is rotated onto $\hat{\mathbf{e}}_3$, a right-hand screw advances in the direction of $\hat{\mathbf{e}}_1$. If $\hat{\mathbf{e}}_3$ is rotated into $\hat{\mathbf{e}}_1$, a right-hand screw advances in the direction of $\hat{\mathbf{e}}_2$. The equations in (8.8) and (8.9) conform to this pattern.

Let us explore what happens in a simple case when a switch is made from a right- to a left-hand coordinate system. The transformation connecting the coordinate systems in Fig. 8.2 is

$$S = \begin{pmatrix} -1 & 0 & 0 \\ 0 & 1 & 0 \\ 0 & 0 & 1 \end{pmatrix} \tag{8.10}$$

The vectors **a** and **b** are polar vectors and lie in the x_1, x_2 plane with **a** pointing along the x_1 axis. Their transformation is

$$a_1' = -a_1 \qquad b_1' = -b_1 \qquad b_2' = b_2 \tag{8.11}$$

The unit coordinate vectors are related by

$$\hat{\mathbf{e}}_1' = -\hat{\mathbf{e}}_1 \qquad \hat{\mathbf{e}}_2' = \hat{\mathbf{e}}_2 \qquad \hat{\mathbf{e}}_3' = \hat{\mathbf{e}}_3 \tag{8.12}$$

The cross product in the two systems as defined by (8.3) is

$$\mathbf{c} = a_1 b_2 \hat{\mathbf{e}}_3 \qquad \mathbf{c}' = a_1' b_1' \hat{\mathbf{e}}_3' = -a_1 b_2 \hat{\mathbf{e}}_3 = -\mathbf{c} \tag{8.13}$$

We see that in the left-hand coordinate system in Fig. 8.2*b*, the direction of $\mathbf{a} \times \mathbf{b}$ is to be found by a left-hand rule. Equation (8.13) is of course just a special case of (8.2).

An example of an axial vector is provided by the magnetic field. The magnetic field is defined by means of the Lorentz force:

$$\mathbf{F} = q\mathbf{u} \times \mathbf{B} \tag{8.14}$$

where **F** is the force on a particle of charge q moving

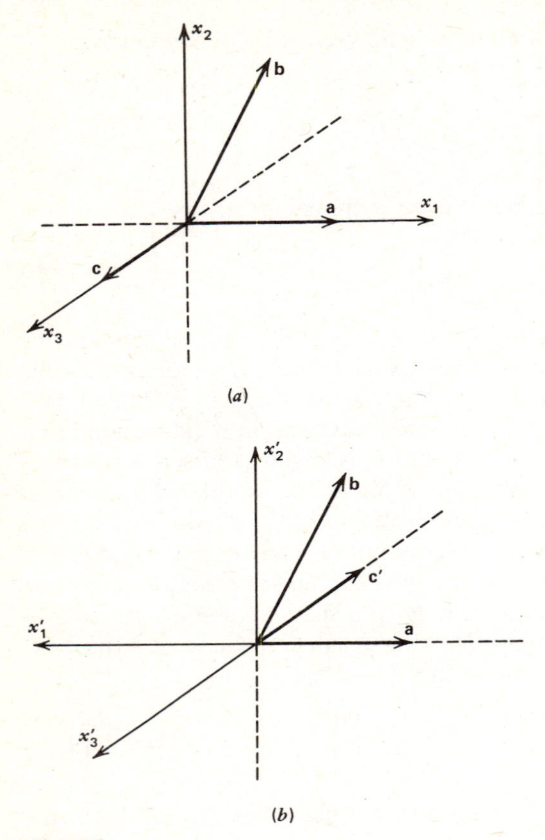

Fig. 8.2

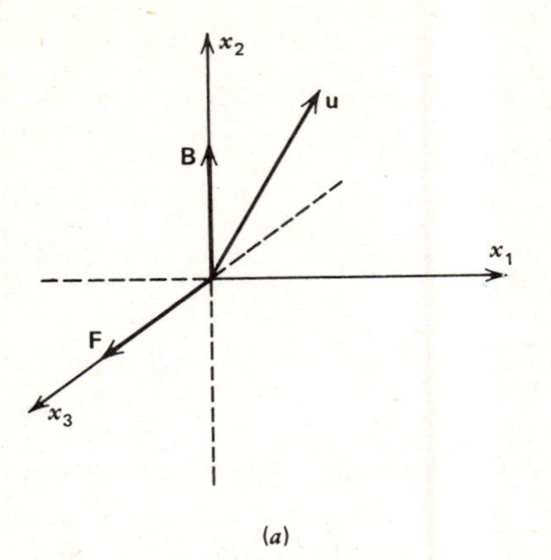

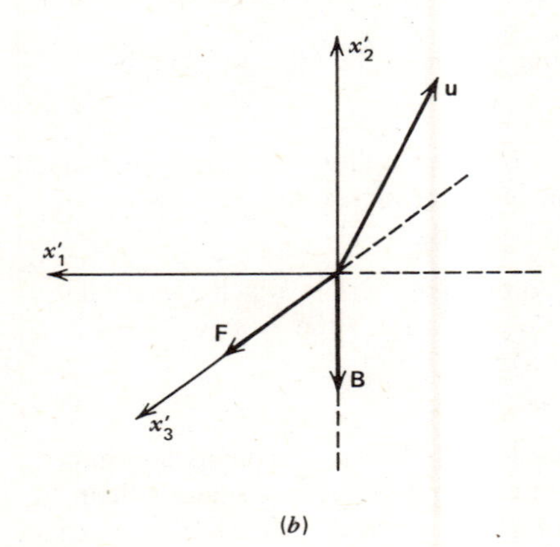

Fig. 8.3

at a velocity **u** in a region where a magnetic field **B** exists. Force and velocity are unambiguous polar vectors because they have definite directions in space. This means that **B** must be interpreted as an axial vector. Figure 8.3 shows two possible interpretations of (8.14) according as we adopt a right- or a left-hand rule. There are in fact no classical experiments based on the Lorentz force equation which allow us to remove this ambiguity in the direction of the magnetic field. We therefore arbitrarily use the right-hand convention as pictured in Fig. 8.3*a*.

As a practical matter, it is best to stick to right-handed coordinate systems when doing calculations. An unintentional use of a left-hand coordinate system is apt to introduce extraneous minus signs into a calculation. If only right-hand coordinates are employed, all pseudovectors behave as normal vectors. There are, however, areas of physics where we must study the invariance (or noninvariance) of physical processes under the full group of orthogonal transformations. In such cases, pseudo quantities cannot be avoided.

The following two rules about the cross product can easily be established using the definition (8.3). If **a** and **b** are functions of a parameter t and $\mathbf{c} = \mathbf{a} \times \mathbf{b}$, then

$$\frac{d\mathbf{c}}{dt} = \mathbf{a} \times \frac{d\mathbf{b}}{dt} + \frac{d\mathbf{a}}{dt} \times \mathbf{b} \tag{8.15}$$

Care must be used to preserve the order of the factors in (8.15) since the cross product is noncommutative. The cross product obeys the following distributive rules:

$$\mathbf{a} \times (\mathbf{b} + \mathbf{c}) = \mathbf{a} \times \mathbf{b} + \mathbf{a} \times \mathbf{c} \tag{8.16}$$

$$(\mathbf{b} + \mathbf{c}) \times \mathbf{a} = \mathbf{b} \times \mathbf{a} + \mathbf{c} \times \mathbf{a} \tag{8.17}$$

Theorem 3. The cross product is nonassociative:

$$\mathbf{a} \times (\mathbf{b} \times \mathbf{c}) \neq (\mathbf{a} \times \mathbf{b}) \times \mathbf{c} \tag{8.18}$$

The double cross product obeys

$$\mathbf{a} \times (\mathbf{b} \times \mathbf{c}) = \mathbf{b}(\mathbf{a} \cdot \mathbf{c}) - \mathbf{c}(\mathbf{a} \cdot \mathbf{b}) \tag{8.19}$$

which is sometimes called the *bac-cab rule*.

Proof: If $\mathbf{b}$ and $\mathbf{c}$ are proportional, $\mathbf{b} = \alpha\mathbf{c}$, both sides of (8.19) are zero, so the result is trivial. Assume now that $\mathbf{b}$ and $\mathbf{c}$ are not proportional. Note that $\mathbf{a} \times (\mathbf{b} \times \mathbf{c})$ is perpendicular to the plane determined by $\mathbf{a}$ and $\mathbf{b} \times \mathbf{c}$. It therefore lies in the plane determined by $\mathbf{b}$ and $\mathbf{c}$. Since $\mathbf{b}$ and $\mathbf{c}$ are linearly independent,

$$\mathbf{a} \times (\mathbf{b} \times \mathbf{c}) = \beta\mathbf{b} + \gamma\mathbf{c} \tag{8.20}$$

where β and γ are scalar factors. Choose a coordinate system such that

$$\mathbf{b} = \hat{\imath} b_x \qquad \mathbf{c} = \hat{\imath} c_x + \hat{\jmath} c_y \tag{8.21}$$

Expand both sides of (8.20) to get

$$a_y c_y b_x \hat{\imath} - a_x c_y b_x \hat{\jmath} = \beta b_x \hat{\imath} + \gamma(c_x \hat{\imath} + c_y \hat{\jmath}) \tag{8.22}$$

Equate the coefficients of $\hat{\imath}$ and $\hat{\jmath}$ on the two sides of the equation to get

$$a_y c_y b_x = \beta b_x + \gamma c_x \qquad -a_x c_y b_x = \gamma c_y \tag{8.23}$$

Solve for β and γ:

$$\gamma = -a_x b_x = -(\mathbf{a} \cdot \mathbf{b}) \qquad \beta = a_x c_x + a_y c_y = (\mathbf{a} \cdot \mathbf{c}) \tag{8.24}$$

from which (8.19) follows.

The nonassociative nature of $\mathbf{a} \times (\mathbf{b} \times \mathbf{c})$ is established by first noting that

$$(\mathbf{a} \times \mathbf{b}) \times \mathbf{c} = -\mathbf{c} \times (\mathbf{a} \times \mathbf{b}) = -\mathbf{a}(\mathbf{c} \cdot \mathbf{b}) + \mathbf{b}(\mathbf{c} \cdot \mathbf{a}) \tag{8.25}$$

Then,

$$\begin{aligned} \mathbf{a} \times (\mathbf{b} \times \mathbf{c}) - (\mathbf{a} \times \mathbf{b}) \times \mathbf{c} &= \mathbf{a}(\mathbf{b} \cdot \mathbf{c}) - \mathbf{c}(\mathbf{a} \cdot \mathbf{b}) \\ &= \mathbf{b} \times (\mathbf{a} \times \mathbf{c}) \end{aligned} \tag{8.26}$$

which, in general, is not zero. Equation (8.26) can be expressed as

$$\mathbf{a} \times (\mathbf{b} \times \mathbf{c}) = \mathbf{b} \times (\mathbf{a} \times \mathbf{c}) - \mathbf{c} \times (\mathbf{a} \times \mathbf{b}) \tag{8.27}$$

which is interesting because it has the same structure as the bac-cab rule (8.19). Another identity which has this same basic structure is the Jacobi identity involving the commutators of three matrices:

$$[A,[B,C]] = [B,[A,C]] - [C,[A,B]] \tag{8.28}$$

See problem 3.2 in Chapter 1.

Three linearly independent vectors $\mathbf{a}$, $\mathbf{b}$, and $\mathbf{c}$ can be visualized as geometrically forming three edges of a parallelepiped, as in Fig. 8.4. The cross product $\mathbf{b} \times \mathbf{c}$ has the geometrical significance that its magnitude is the area of the parallelogram determined by $\mathbf{b}$ and $\mathbf{c}$. Its direction is perpendicular to the plane of this parallelogram. In this sense, area is a pseudovector. The *triple scalar product* is defined as

$$\begin{aligned} \Sigma = \mathbf{a} \cdot (\mathbf{b} \times \mathbf{c}) &= \begin{vmatrix} a_1 & a_2 & a_3 \\ b_1 & b_2 & b_3 \\ c_1 & c_2 & c_3 \end{vmatrix} \\ &= \sum_{i=1}^{3} \sum_{j=1}^{3} \sum_{k=1}^{3} e_{ijk} a_i b_j c_k \end{aligned} \tag{8.29}$$

and is a pseudoscalar which has the geometrical significance that it is the volume of the parallelepiped in Fig. 8.4. The vectors in (8.29) can be cyclically permuted:

$$\mathbf{a} \cdot (\mathbf{b} \times \mathbf{c}) = \mathbf{c} \cdot (\mathbf{a} \times \mathbf{b}) = \mathbf{b} \cdot (\mathbf{c} \times \mathbf{a}) \tag{8.30}$$

This is because the cyclic permutation of the rows of a determinant of odd dimension does not change its sign. In other words, a cyclic permutation of an odd number of elements is even (theorem 2, section 4, Chapter 1).

Theorem 4. The permutation symbol obeys

$$\sum_{i=1}^{3} e_{ijk} e_{imn} = \delta_{jm}\delta_{kn} - \delta_{jn}\delta_{km} \tag{8.31}$$

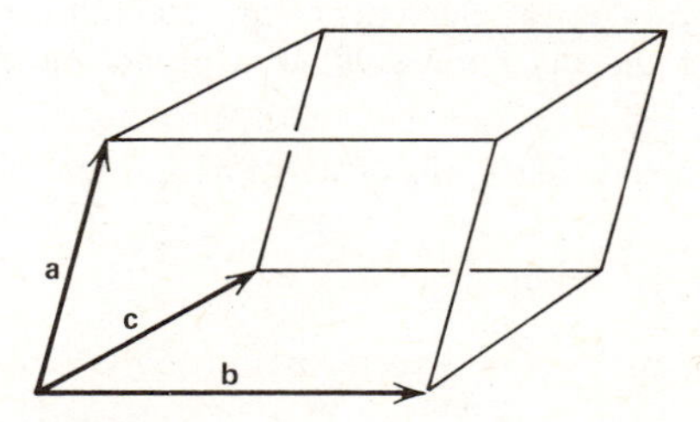

Fig. 8.4

Proof: This important identity follows from the bac-cab rule. In component form, (8.19) is

$$e_{ijk}a_j e_{kmn} b_m c_n = b_i a_n c_n - c_i a_j b_j \tag{8.32}$$

where use is made of the summation convention. By using the properties of the Kronecker delta, we can write

$$\begin{aligned} b_i &= \delta_{im} b_m \qquad & a_n &= \delta_{nj} a_j \\ c_i &= \delta_{in} c_n \qquad & b_j &= \delta_{jm} b_m \end{aligned} \tag{8.33}$$

which are used in (8.32) to get

$$e_{kij} e_{kmn} a_j b_m c_n = (\delta_{im}\delta_{nj} - \delta_{in}\delta_{jm}) a_j b_m c_n \tag{8.34}$$

Use has also been made of the fact that a cyclic permutation of the subscripts on e_{ijk} does not alter its value. Since the vectors **a**, **b**, and **c** are arbitrary, (8.34) implies

$$e_{kij} e_{kmn} = \delta_{im}\delta_{jn} - \delta_{in}\delta_{jm} \tag{8.35}$$

which, except for some changes in notation, is the same as (8.31).

Another way of proving (8.31) is based on the theory of isotropic tensors and is outlined in problems 7.6, 7.7, and 7.8 of this chapter. It would then be possible to work in the opposite direction and use (8.31) to prove the bac-cab rule. The value of (8.31) will become apparent in Chapter 3, where it is used to establish vector identities of various kinds.

PROBLEMS

8.1. Prove (8.15) and (8.16).

8.2. Prove the following:

(a)

$$\begin{aligned} (\mathbf{a}\times\mathbf{b})\times(\mathbf{c}\times\mathbf{d}) &= (\mathbf{a}\times\mathbf{b}\cdot\mathbf{d})\mathbf{c} - (\mathbf{a}\times\mathbf{b}\cdot\mathbf{c})\mathbf{d} \\ &= (\mathbf{a}\times\mathbf{c}\cdot\mathbf{d})\mathbf{b} - (\mathbf{b}\times\mathbf{c}\cdot\mathbf{d})\mathbf{a} \end{aligned}$$

(b)

$$(\mathbf{a}\times\mathbf{b})\cdot(\mathbf{c}\times\mathbf{d}) = (\mathbf{a}\cdot\mathbf{c})(\mathbf{b}\cdot\mathbf{d}) - (\mathbf{a}\cdot\mathbf{d})(\mathbf{b}\cdot\mathbf{c})$$

8.3. Show that $\hat{\mathbf{e}}_i \cdot \hat{\mathbf{e}}_j \times \hat{\mathbf{e}}_k = e_{ijk}$.

8.4. In general, how many distinct nonzero elements does a three-dimensional antisymmetric tensor have? If **a** and **b** are two polar vectors, show that $C_{ij} = a_i b_j - a_j b_i$ form the elements of an antisymmetric tensor. Show that the components of the cross product of **a** and **b** are given by

$$c_i = \frac{1}{2}\sum_j \sum_k e_{ijk} C_{jk}$$

8.5. Show that the Lorentz force equation (8.14) can be represented by

$$F_i = q \sum_{j=1}^{3} B_{ij} u_j$$

where B_{ij} are the elements of an antisymmetric tensor. Each element of the antisymmetric tensor is a component of the magnetic field. Display this tensor in the form of a matrix. Show that the transformation of B_{ij} as elements of an ordinary (not pseudo) tensor gives the same result as transforming the magnetic field as a pseudovector. You will need the result of problem 2.4.

8.6. Prove that if A_{jk} are the elements of a symmetric tensor, then

$$\sum_{j=1}^{3}\sum_{k=1}^{3} e_{ijk} A_{jk} = 0$$

In particular, the cross product of a vector with itself is zero:

$$\sum_{j=1}^{3}\sum_{k=1}^{3} e_{ijk} a_j a_k = 0$$

8.7. The three vectors

$$\mathbf{a} = \hat{\imath} + \hat{\jmath} - 2\hat{\mathbf{k}} \qquad \mathbf{b} = 2\hat{\imath} - \hat{\jmath} + \hat{\mathbf{k}} \qquad \mathbf{c} = \hat{\imath} + 3\hat{\jmath} - \hat{\mathbf{k}}$$

extend from the origin of coordinates. Their tips determine a plane. Find the perpendicular distance from the origin to this plane.

Hint: Find any vector perpendicular to the plane and then find the projection of **a** on it.

8.8. Let **a** and **b** be two vectors lying in the x, y plane at angles α and β with respect to the x axis. By calculating $\mathbf{a} \times \mathbf{b}$, prove the identity

$$\sin(\beta - \alpha) = \sin\beta\cos\alpha - \cos\beta\sin\alpha$$

8.9. The tips of the three vectors in problem 8.7 determine a triangle. Find its area.

9. ANGULAR MOMENTUM AND RIGID BODY MOTION

A discussion of angular momentum in classical mechanics provides good examples of the use of the cross product. The angular momentum of a particle of mass m and velocity **u** taken about the origin of coordinates (Fig. 9.1) is defined to be

$$\ell = \mathbf{r} \times m\mathbf{u} \tag{9.1}$$

By differentiation of (9.1) with respect to time and use of (8.15),

$$\frac{d\ell}{dt} = \mathbf{r} \times m\frac{d\mathbf{u}}{dt} + \mathbf{u} \times m\mathbf{u} \tag{9.2}$$

By use of Newton's second law and the fact that $\mathbf{u} \times \mathbf{u} = 0$,

$$\frac{d\ell}{dt} = \mathbf{r} \times \mathbf{F} = \boldsymbol{\tau} \tag{9.3}$$

where **F** is the net force on the particle. In (9.3), $\boldsymbol{\tau}$ is recognized as the net torque about the origin.

For a system of N particles, the total angular momentum is

$$\ell = \sum_{\alpha=1}^{N} \mathbf{r}_\alpha \times m_\alpha \mathbf{u}_\alpha \tag{9.4}$$

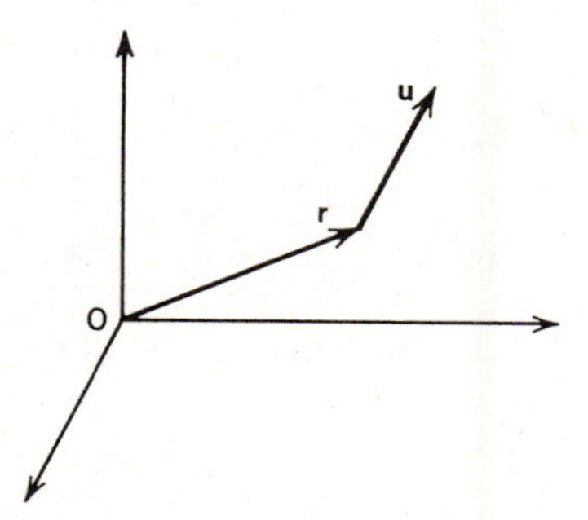

Fig. 9.1

It is possible to divide (9.4) into two parts. One part is the angular momentum due to the motion of the center of mass, and the other is the angular momentum due to the motion of the particles with respect to the center of mass. This is exactly what happens to the total kinetic energy as reference to (4.39) shows. By means of (4.35) and (4.36),

$$\ell = \sum_{\alpha=1}^{N} (\mathbf{s}_\alpha + \mathbf{R}) \times m_\alpha(\mathbf{w}_\alpha + \mathbf{v}) \tag{9.5}$$

where $\mathbf{s}_\alpha$ is the position of the αth particle with respect to the center of mass, **R** is the position vector of the center of mass, $\mathbf{w}_\alpha = d\mathbf{s}_\alpha/dt$ is the velocity of the αth particle with respect to the center of mass, and $\mathbf{v} = d\mathbf{R}/dt$ is the velocity of the center of mass. With the help of (8.16),

$$\ell = \sum_{\alpha=1}^{N} \mathbf{s}_\alpha \times m_\alpha \mathbf{w}_\alpha + \left(\sum_{\alpha=1}^{N} m_\alpha \mathbf{s}_\alpha\right) \times \mathbf{v}$$
$$+ \mathbf{R} \times \sum_{\alpha=1}^{N} m_\alpha \mathbf{w}_\alpha + \mathbf{R} \times \left(\sum_{\alpha=1}^{N} m_\alpha\right)\mathbf{v} \tag{9.6}$$

We now use (4.37) and (4.34) to get

$$\ell = \sum_{\alpha=1}^{N} \mathbf{s}_\alpha \times m_\alpha \mathbf{w}_\alpha + \mathbf{R} \times M\mathbf{v} \tag{9.7}$$

which is the desired separation of the angular momentum. The term

$$\ell_c = \sum_{\alpha=1}^{N} \mathbf{s}_\alpha \times m_\alpha \mathbf{w}_\alpha \tag{9.8}$$

is the angular momentum due to the motion of the particles with respect to the center of mass and $\mathbf{R} \times M\mathbf{v}$ is the contribution due to the motion of the center of mass itself.

The derivative of ℓ with respect to time can be found from both (9.4) and (9.7), leading to

$$\boldsymbol{\tau} = \frac{d\boldsymbol{\ell}}{dt} = \sum_{\alpha=1}^{N} \mathbf{r}_\alpha \times \mathbf{F}_\alpha = \frac{d\boldsymbol{\ell}_c}{dt} + \mathbf{R} \times \mathbf{F} \tag{9.9}$$

where $\mathbf{F}$ is the total external force as given by (4.47). Note that the net external torque $\boldsymbol{\tau}$ is the vector sum of the torques $\boldsymbol{\tau}_\alpha = \mathbf{r}_\alpha \times \mathbf{F}_\alpha$ on the individual particles. We can assume that internal torques and forces cancel in the summation process. Otherwise, an isolated system of particles would be able to exert a torque or force on itself leading to a violation of momentum conservation. For simple kinds of forces, the cancellation of internal forces and torques can be demonstrated by using Newton's third law of motion.

By substituting $\mathbf{r}_\alpha = \mathbf{s}_\alpha + \mathbf{R}$ in (9.9), we find

$$\sum_{\alpha=1}^{N} \mathbf{s}_\alpha \times \mathbf{F}_\alpha + \mathbf{R} \times \mathbf{F} = \frac{d\boldsymbol{\ell}_c}{dt} + \mathbf{R} \times \mathbf{F} \tag{9.10}$$

which shows that

$$\frac{d\boldsymbol{\ell}_c}{dt} = \sum_{\alpha=1}^{N} \mathbf{s}_\alpha \times \mathbf{F}_\alpha = \boldsymbol{\tau}_c \tag{9.11}$$

where $\boldsymbol{\tau}_c$ is the net external torque about the center of mass. This is a very useful result and allows us to calculate the rotational motion of a system with respect to the center of mass as though it were a fixed point. The motion of the center of mass itself is then found by (4.47).

It is of interest to apply these results to the kinematics and dynamics of a rigid body, which is basically an assembly of particles held rigidly with respect to one another. If the rigid body has one point fixed such as would be the case for a wheel mounted on a fixed shaft or a toy top spinning with its tip resting always at one point, we could proceed directly with (9.4) and consider the origin as the fixed point. If the motion is more complicated, we can treat the center of mass as though it were a fixed point and use (9.8). The calculation that we are about to do is basically the same either way, but we will use (9.8). The instantaneous angular velocity of the body is a pseudovector that can be defined by means of

$$\mathbf{w}_\alpha = \boldsymbol{\omega} \times \mathbf{s}_\alpha \tag{9.12}$$

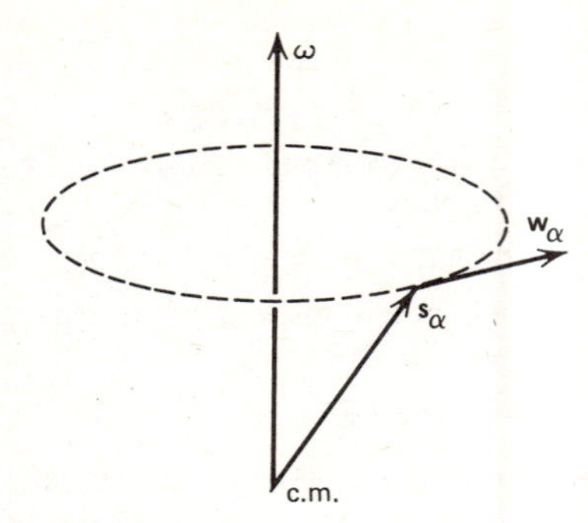

Fig. 9.2

The relationship among the vectors in (9.12) is shown in Fig. 9.2. The angular velocity vector $\boldsymbol{\omega}$ is to be visualized as defining an instantaneous axis of rotation. Both the magnitude and the direction of $\boldsymbol{\omega}$ vary as a function of time.

By combining (9.8) and (9.12) and using the bac-cab rule, we find

$$\begin{aligned} \boldsymbol{\ell}_c &= \sum_{\alpha=1}^{N} m_\alpha \mathbf{s}_\alpha \times (\boldsymbol{\omega} \times \mathbf{s}_\alpha) \\ &= \sum_{\alpha=1}^{N} m_\alpha \left[\boldsymbol{\omega} s_\alpha^2 - \mathbf{s}_\alpha (\mathbf{s}_\alpha \cdot \boldsymbol{\omega})\right] \end{aligned} \tag{9.13}$$

It is possible to write the first factor of $\boldsymbol{\omega}$ in (9.13) as

$$\boldsymbol{\omega} = \sum_{i=1}^{3} \omega_i \hat{\mathbf{e}}_i = \sum_{i=1}^{3} \sum_{j=1}^{3} \delta_{ij} \omega_j \hat{\mathbf{e}}_i \tag{9.14}$$

in which case (9.13) can be expressed in the form

$$\boldsymbol{\ell}_c = \sum_{i=1}^{3} \sum_{j=1}^{3} I_{ij} \omega_j \hat{\mathbf{e}}_i \tag{9.15}$$

where the set of nine quantities

$$I_{ij} = \sum_{\alpha=1}^{N} m_\alpha \left(\delta_{ij} s_\alpha^2 - x_{\alpha i} x_{\alpha j} \right) \tag{9.16}$$

are the elements of a second-rank symmetric tensor called the *moment of inertia tensor*. In (9.16), $x_{\alpha i}$ are the components of $\mathbf{s}_\alpha$. The tensor character of (9.16) follows from the fact that δ_{ij} and $x_{\alpha i} x_{\alpha j}$ both form the elements of a second-rank tensor.

Equation (9.15) can be set up in any number of rectangular Cartesian coordinate systems, all with

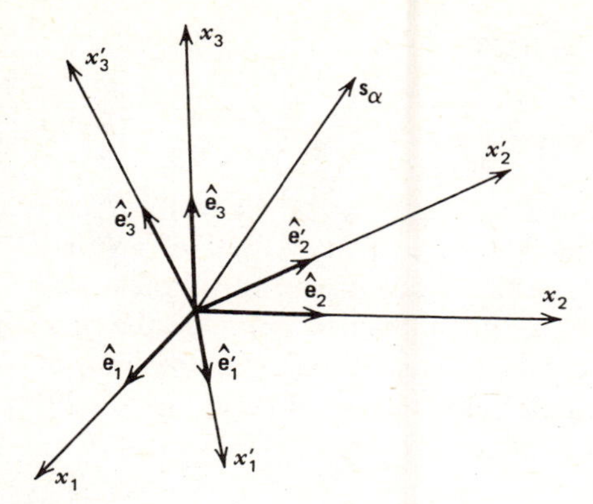

Fig. 9.3

their origins at the center of mass of the body. A coordinate system which is fixed in space, usually called a *laboratory coordinate* system, has the disadvantage that the elements of the inertia tensor (9.16) are then functions of the time due to the motion of the particles which make up the rigid body. To avoid this difficulty, it is the usual practice to use so-called *body coordinates*, which are actually fixed in the body and move with it. Assume that the x_i' coordinates in Fig. 9.3 are the body coordinates and that the x_i coordinates are stationary laboratory coordinates. The transformation of the components of $\mathbf{s}_\alpha$ in Fig. 9.3 can be expressed in matrix form as

$$X' = SX \tag{9.17}$$

The components of $\mathbf{s}_\alpha$ in the x_i' coordinates are constant in time. The elements of the transforming matrix S are however time-dependent. The reader who has done problem 2.6 has already seen an example of such a time-dependent transformation. The derivative of (9.17) with respect to time is

$$O = \dot{S}X + S\dot{X} \tag{9.18}$$

If (9.18) is multiplied through by $\tilde{S}$ and then solved for $\dot{X}$, the result is

$$\dot{X} = -\tilde{S}\dot{S}X \tag{9.19}$$

which must in fact be a matrix version of (9.12). The situation here is quite similar to the state of affairs in problem 8.5, with the angular velocity playing the same role as the magnetic field.

The quantity

$$\Omega = \tilde{S}\dot{S} \tag{9.20}$$

is an antisymmetric tensor the elements of which are the components of the angular velocity. The antisymmetry of (9.20) is established by differentiating $\tilde{S}S = I$. This gives $\dot{\tilde{S}}S + \tilde{S}\dot{S} = 0$, which is the same as $\tilde{\Omega} + \Omega = 0$. If we write

$$\Omega = \begin{pmatrix} 0 & \omega_3 & -\omega_2 \\ -\omega_3 & 0 & \omega_1 \\ \omega_2 & -\omega_1 & 0 \end{pmatrix} \tag{9.21}$$

then $\dot{X} = -\Omega X$ reproduces (9.12) exactly. It is easy to see that the elements of Ω are formally related to the angular velocity components by

$$\Omega_{ij} = \sum_{k=1}^{3} e_{ijk}\omega_k \tag{9.22}$$

which shows that Ω is in fact a tensor because e_{ijk} is a pseudotensor and $\boldsymbol{\omega}$ is a pseudovector.

As an example, suppose that the axis of rotation is the common x_3 and x_3' axes as shown in Fig. 9.4. The transformation is

$$S = \begin{pmatrix} \cos\theta & \sin\theta & 0 \\ -\sin\theta & \cos\theta & 0 \\ 0 & 0 & 1 \end{pmatrix} \tag{9.23}$$

where θ is a function of time. The derivative of (9.23) is

$$\dot{S} = \begin{pmatrix} -\dot{\theta}\sin\theta & \dot{\theta}\cos\theta & 0 \\ -\dot{\theta}\cos\theta & -\dot{\theta}\sin\theta & 0 \\ 0 & 0 & 0 \end{pmatrix} \tag{9.24}$$

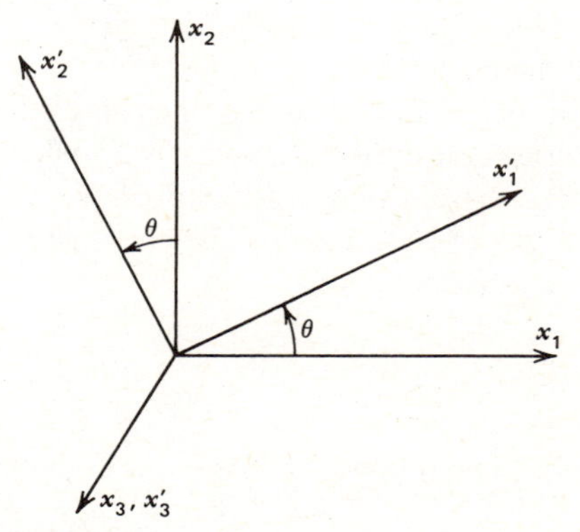

Fig. 9.4

A calculation of Ω gives

$$\Omega = \tilde{S}\dot{S} = \begin{pmatrix} 0 & \dot{\theta} & 0 \\ -\dot{\theta} & 0 & 0 \\ 0 & 0 & 0 \end{pmatrix} \tag{9.25}$$

as expected.

In Fig. 9.3, the x_i' coordinates are in motion, which means that the unit coordinate vectors $\hat{\mathbf{e}}_i'$ are functions of the time. The time rate of change of these vectors can be inferred with the help of (9.12). Let us drop the subscript α for the moment. The position vector $\mathbf{s}$, which is the position vector of any one of the particles making up the rigid body, can be expressed in either set of coordinates as

$$\mathbf{s} = \sum_{i=1}^{3} x_i \hat{\mathbf{e}}_i = \sum_{i=1}^{3} \mathbf{x}_i' \hat{\mathbf{e}}_i' \tag{9.26}$$

The derivative of (9.26) with respect to time is

$$\frac{d\mathbf{s}}{dt} = \mathbf{w} = \sum_{i=1}^{3} \dot{x}_i \hat{\mathbf{e}}_i = \sum_{i=1}^{3} x_i' \frac{d\hat{\mathbf{e}}_i'}{dt} \tag{9.27}$$

where, since $\mathbf{s}$ is fixed in the moving coordinates, x_i' are constants. By comparing (9.27) and (9.12),

$$\boldsymbol{\omega} \times \mathbf{s} = \sum_{j=1}^{3} x_j' (\boldsymbol{\omega} \times \hat{\mathbf{e}}_j') = \sum_{j=1}^{3} x_j' \frac{d\hat{\mathbf{e}}_j'}{dt} \tag{9.28}$$

we conclude that

$$\frac{d\hat{\mathbf{e}}_j'}{dt} = \boldsymbol{\omega} \times \hat{\mathbf{e}}_j' = \sum_{i=1}^{3} \omega_i' \hat{\mathbf{e}}_i' \times \hat{\mathbf{e}}_j' = \sum_{i=1}^{3} \sum_{k=1}^{3} \omega_i' e_{ijk} \hat{\mathbf{e}}_k' \tag{9.29}$$

where use has been made of (8.9).

Equation (9.15) is the total angular momentum of the rigid body about its center of mass (or, if we wish, about a fixed point), with all quantities expressed in the stationary or lab coordinates. Alternatively, it can be expressed in terms of the moving coordinates as

$$\boldsymbol{\ell}_c = \sum_{j=1}^{3} \sum_{k=1}^{3} I_{jk}' \omega_k' \hat{\mathbf{e}}_j' \tag{9.30}$$

where the elements I_{jk}' are related to I_{jk} by a second-rank tensor transformation

$$I_{jk}' = \sum_{m=1}^{3} \sum_{n=1}^{3} S_{jm} S_{kn} I_{mn} \tag{9.31}$$

The advantage of (9.30) is that the elements I_{ij}' are constant in time. The price paid is that $\hat{\mathbf{e}}_i'$ are not constant, but this can be taken care of by means of (9.29). According to (9.11), the net torque on the rigid body about its center of mass equals the time rate of change of its angular momentum. If (9.30) and (9.29) are used to calculate the torque,

$$\boldsymbol{\tau}_c = \frac{d\boldsymbol{\ell}_c}{dt} = \sum_{j=1}^{3} \sum_{k=1}^{3} I_{jk}' \dot{\omega}_k' \hat{\mathbf{e}}_j' + \sum_{i=1}^{3} \sum_{j=1}^{3} \sum_{k=1}^{3} \sum_{n=1}^{3} I_{jk}' \omega_k' \omega_i' e_{ijn} \hat{\mathbf{e}}_n' \tag{9.32}$$

The equations contained in (9.32) are called Euler's equations and can be used to determine the rotational motion of a rigid body once the torques are known. As it stands, (9.32) looks quite unwieldy. Of the many possible body coordinates which can be used, there is one in which the moment of inertia tensor is diagonal, and this results in a considerable simplification. This coordinate system is the principal axis coordinate system of the moment of inertia tensor. This idea was introduced in section 6 as a possible way of simplifying the mechanical system discussed there. See, for example, equation (6.16). The moment of inertia tensor is symmetric, and it is true that any symmetric tensor can be diagonalized by an appropriate orthogonal similarity transformation. This question is taken up in detail in section 10. Supposing that the diagonalization has been done and that the three diagonal elements are $I_1' = I_{11}'$, $I_2' = I_{22}'$, and $I_3' = I_{33}'$, the three components of (9.32) work out to

$$\tau_1' = I_1' \dot{\omega}_1' + \omega_2' \omega_3' (I_3' - I_2') \tag{9.33}$$

$$\tau_2' = I_2' \dot{\omega}_2' + \omega_3' \omega_1' (I_1' - I_3') \tag{9.34}$$

$$\tau_3' = I_3' \dot{\omega}_3' + \omega_1' \omega_2' (I_2' - I_1') \tag{9.35}$$

It is in this form that the Euler equations are usually used.

PROBLEMS

9.1. In calculating the elements of the moment of inertia tensor for a rigid body throughout which mass is continuously distributed, it is the usual practice to replace (9.16) by

$$I_{ij} = \int_{\Sigma} \rho \left(\delta_{ij} s^2 - x_i x_j \right) d\Sigma \tag{9.36}$$

where ρ is the density in units of mass/unit volume and $d\Sigma$ is a volume element. The integral is to be taken over the entire volume of the material making up the rigid body. The diagonal elements of (9.36) are referred to as *moments of inertia* and the off-diagonal elements, as *products of inertia*. Show from (9.36) that the moments and products of inertia can be written as

$$\begin{aligned} I_x &= \int_{\Sigma} \rho (y^2 + z^2)\, d\Sigma & I_{xy} &= -\int_{\Sigma} \rho xy\, d\Sigma \\ I_y &= \int_{\Sigma} \rho (x^2 + z^2)\, d\Sigma & I_{xz} &= -\int_{\Sigma} \rho xz\, d\Sigma \\ I_z &= \int_{\Sigma} \rho (x^2 + y^2)\, d\Sigma & I_{yz} &= -\int_{\Sigma} \rho yz\, d\Sigma \end{aligned} \tag{9.37}$$

9.2. A rigid body is in the form of a thin sheet lying entirely in the x, y plane as shown in Fig. 9.5. By using (9.37), show that

$$I_x + I_y = I_z \tag{9.38}$$

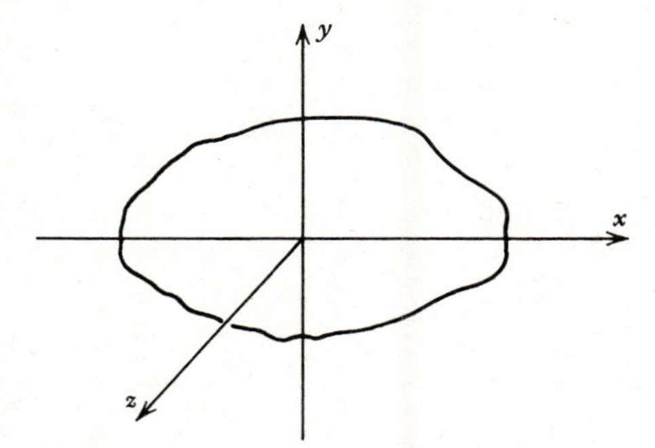

Fig. 9.5

9.3. A thin disc has radius b and total mass M uniformly distributed over it. It lies in the x, y plane with its center at the origin. Show, using (9.37) and (9.38), that

$$\begin{aligned} I_z &= \tfrac{1}{2} M b^2 \\ I_x &= I_y = \tfrac{1}{4} M b^2 \end{aligned} \tag{9.39}$$

What are the products of inertia? See Fig. 9.6.

9.4. If the moment of inertia of a body about an axis through its center of mass is I_c, show that the moment of inertia about any other parallel axis is

$$I = I_c + MR^2 \tag{9.40}$$

where M is the total mass of the body and R is the distance between the two axes. This is the

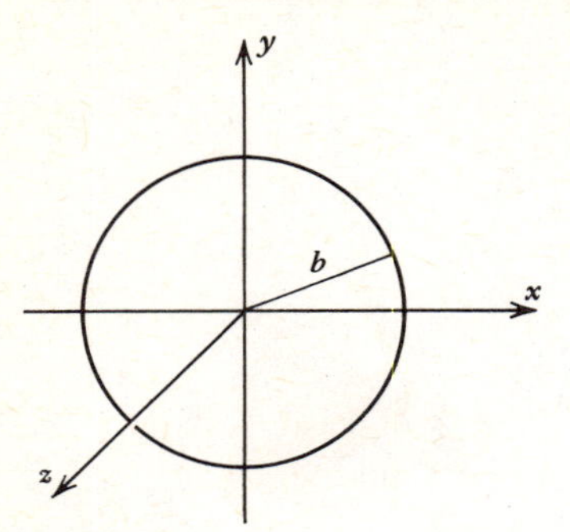

Fig. 9.6

parallel axis theorem. A uniform thin disc of radius b and mass M lies in the x, y plane, with the x axis coincident with a diameter and the y axis tangent to its rim. Use (9.39) and (9.40) to find I_z and I_y in this case. See Fig. 9.7.

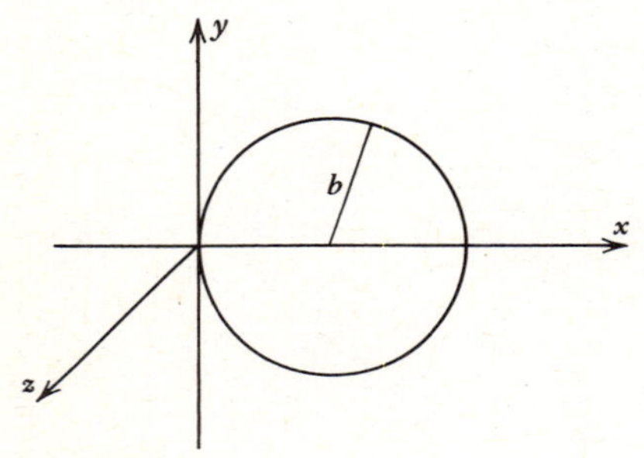

Fig. 9.7

9.5. Prove that the effect of a uniform gravitational field on a rigid body is the same as that produced by a single force equal to the weight of the body acting at the center of mass.

9.6. Prove that, if the sum of all forces acting on a body is zero and the sum of the torques on the body about any one point is zero, the sum of the torques about any other point will also be zero.

9.7. A uniform disc of radius b and mass M rolls without slipping down an inclined plane of angle θ. The three forces acting on the disc are its weight $M\mathbf{g}$, a frictional force $\mathbf{f}$ which is necessary if the disc is to roll rather than slide, and a normal force $\mathbf{N}$ exerted on the disc by the plane perpendicular to its surface. Use (4.47) and (9.11) to find the acceleration of the center of mass of the disc. See Fig. 9.8.

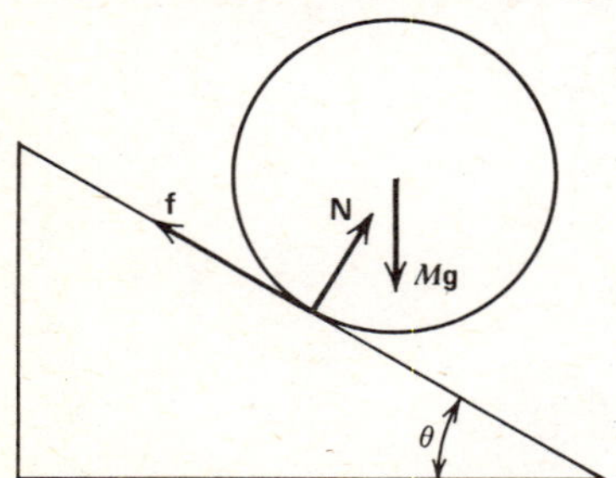

Fig. 9.8

9.8. Solve (9.22) formally for ω_k in terms of Ω_{ij} as follows: Multiply through by e_{nij}, sum over i and j and use (8.31).

9.9. According to (9.12), the velocity of any point of a rigid body with respect to the center of mass or a fixed point is $\mathbf{w} = \boldsymbol{\omega} \times \mathbf{s}$. Show that $\boldsymbol{\omega}$ defines an instantaneous axis of rotation by showing that the points for which $\mathbf{w} = 0$ fall along a straight line defined by $\mathbf{s} \propto \boldsymbol{\omega}$.

9.10. It would seem that we could differentiate (9.8) directly to get

$$\frac{d\boldsymbol{\ell}_c}{dt} = \sum_{\alpha=1}^{N} \mathbf{s}_\alpha \times m_\alpha \frac{d\mathbf{w}_\alpha}{dt} = \boldsymbol{\tau}_c$$

Explain why this is not a good idea.

9.11. If you did problem 2.6 correctly, you got

$$S = \begin{pmatrix} \cos\omega t & 0 & -\sin\omega t \\ \sin\theta\sin\omega t & \cos\theta & \sin\theta\cos\omega t \\ \cos\theta\sin\omega t & -\sin\theta & \cos\theta\cos\omega t \end{pmatrix} \tag{9.41}$$

as the transformation from laboratory coordinates to body coordinates. Use (9.20) and (9.21) to calculate the components of $\boldsymbol{\omega}$ with respect to the laboratory coordinates. Is the answer what you would expect? The transformation of the tensor Ω to body coordinates is

$$\Omega' = S\Omega\tilde{S} = S\tilde{S}\dot{S}\tilde{S} = \dot{S}\tilde{S} \tag{9.42}$$

Use (9.42) to find the components of $\boldsymbol{\omega}$ in the body coordinates. It is easy to check your result if you consider $\boldsymbol{\omega}$ to be a vector along the axis of rotation and then find its components in the x_i' coordinates. In fact, the direct use of (9.42) to find the components of $\boldsymbol{\omega}$ is the hard way to do it. Check to see that the components of $\boldsymbol{\omega}$ in the laboratory and body coordinates are related by

$$\omega_i' = \sum_{j=1}^{3} S_{ij}\omega_j$$

9.12. Use Euler's equations (9.33)–(9.35) to find the torques exerted on the rotating shaft of problem 2.6. Express the components of $\boldsymbol{\tau}$ both in terms of laboratory coordinates and body coordinates.

9.13. By differentiating the angular velocity

$$\boldsymbol{\omega} = \sum_{i=1}^{3} \omega_i \hat{\mathbf{e}}_i = \sum_{i=1}^{3} \omega_i' \hat{\mathbf{e}}_i'$$

and using (9.29), show that the angular acceleration vector can be expressed in either laboratory or body coordinates as

$$\frac{d\boldsymbol{\omega}}{dt} = \sum_{i=1}^{3} \dot{\omega}_i \hat{\mathbf{e}}_i = \sum_{i=1}^{3} \dot{\omega}_i' \hat{\mathbf{e}}_i' \tag{9.43}$$

Show from (9.43) that the components of angular acceleration transform according to

$$\dot{\omega}_i' = \sum_{j=1}^{3} S_{ij}\dot{\omega}_j \tag{9.44}$$

Hint: Remember that the unit vectors are related by

$$\hat{\mathbf{e}}_i' = \sum_{j=1}^{3} S_{ij}\hat{\mathbf{e}}_j$$

9.14. In Fig. 9.9, the x_i' coordinates are moving relative to the laboratory (x_i) coordinates in an arbitrary way involving both rotation and translation. The position vector of a particle with respect to the laboratory coordinates is **r**, and relative to the moving coordinates it is **s**. The particle is not fixed in the moving coordinates. From the figure, $\mathbf{r} = \mathbf{s} + \mathbf{R}$, which can be written as

$$\sum_{i=1}^{3} x_i \hat{\mathbf{e}}_i = \sum_{i=1}^{3} x_i' \hat{\mathbf{e}}_i' + \mathbf{R} \tag{9.45}$$

Differentiate this expression and use (9.29) to show that

$$\mathbf{u} = \mathbf{u}' + \boldsymbol{\omega} \times \mathbf{s} + \mathbf{v} \tag{9.46}$$

where

$$\mathbf{u} = \sum_{i=1}^{3} \dot{x}_i \hat{\mathbf{e}}_i \tag{9.47}$$

is the velocity of the particle with respect to the laboratory coordinates,

$$\mathbf{u}' = \sum_{i=1}^{3} \dot{x}_i' \hat{\mathbf{e}}_i' \tag{9.48}$$

is the velocity of the particle relative to an observer fixed in the moving coordinates, and $\mathbf{v} = d\mathbf{R}/dt$ is the velocity of the origin of the moving frame.

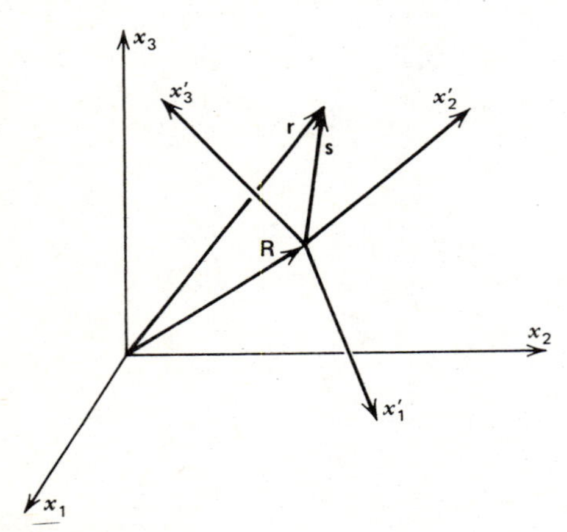

Fig. 9.9

9.15. Refer to equation (4.39). Show that the kinetic energy of a rigid body with respect to its center of mass can be expressed as

$$K_c = \sum_{\alpha=1}^{N} \tfrac{1}{2} m_\alpha (\mathbf{w}_\alpha \cdot \mathbf{w}_\alpha)$$

$$= \sum_{i=1}^{3} \sum_{j=1}^{3} \tfrac{1}{2} I_{ij} \omega_i \omega_j = \sum_{i=1}^{3} \sum_{j=1}^{3} \tfrac{1}{2} I_{ij}' \omega_i' \omega_j' \tag{9.49}$$

If the body coordinates are the principal axis coordinates of the moment of inertia tensor, then

$$K_c = \tfrac{1}{2} I_1' \omega_1'^2 + \tfrac{1}{2} I_2' \omega_2'^2 + \tfrac{1}{2} I_3' \omega_3'^2 \tag{9.50}$$

9.16. Show that the matrix version of Euler's equations is

$$\dot{L} = A\dot{W} + [A, \Omega]W = T \tag{9.51}$$

where L is a column vector representing angular momentum, A is the moment of inertia tensor, W is a column vector representing angular velocity, T represents torque, and

$$[A, \Omega] = A\Omega - \Omega A \tag{9.52}$$

is the commutator of A and Ω. Show also that

$$T' = A'\dot{W}' + [A', \Omega']W' \tag{9.53}$$

meaning that Euler's equations in exactly the same form work both in the laboratory coordinates and the body coordinates. The matrix A is of course a function of time, whereas A' is not. The proof can be done as follows: Start with equation (9.15) expressed in matrix form as

$$L = AW \tag{9.54}$$

The derivative is

$$\dot{L} = A\dot{W} + \dot{A}W \tag{9.55}$$

Now use $A = \tilde{S}A'S$ to evaluate $\dot{A}$. After this is done, express A' back in terms of A by means of $A' = SA\tilde{S}$. Equation (9.51) can be transformed into (9.53). Show that (9.33)–(9.35) can be obtained from (9.53) if A' is diagonal.

Equation (9.15) expressed in body coordinates is

$$\ell_c = \sum_{i=1}^{3} \sum_{j=1}^{3} I'_{ij}\omega'_j \hat{\mathbf{e}}'_i \tag{9.56}$$

It is possible to write a matrix version of (9.56) as

$$L' = A'W' \tag{9.57}$$

Differentiating (9.57) directly obviously does not give (9.53). The reason is that the time dependence of the unit vectors $\hat{\mathbf{e}}'_i$ is missed by this procedure. On a more fundamental level, the torque as given by $d\ell_c/dt$ is to be calculated with respect to an *inertial* (nonaccelerated) frame of reference, which is the laboratory frame in this case. After this is done, we can transform the result to the moving frame if we wish, and this is exactly what is done in going from (9.51) to (9.53). Basically, it comes down to the fact that Newton's second law is valid in its simple form only with respect to an inertial frame. This same issue is involved in problem (9.10). The trouble there is that $d\mathbf{w}_\alpha/dt$ is an acceleration measured with respect to the center of mass which itself may be undergoing acceleration. If we are to use Newton's second law, the acceleration is to be measured relative to the laboratory coordinates. This is not to say that Newtonian mechanics cannot be set up in a noninertial frame of reference by making the appropriate modifications.

10. THE EIGENVALUE PROBLEM

Given a second-rank tensor A, we now address ourselves to the general problem of constructing an orthogonal transformation such that

$$A' = SA\tilde{S} \tag{10.1}$$

is diagonal.

Theorem 1. A necessary condition for A' to be diagonal is that A be symmetric: $A = \tilde{A}$.

Proof: Solve (10.1) for A as

$$A = \tilde{S}A'S \tag{10.2}$$

Since we are assuming that A' is diagonal, it is necessarily also symmetric. By theorem 1 of section 6, A is also symmetric.

In order to consider the problem of actually constructing the diagonalizing transformation, write (10.2) as $A\tilde{S} - \tilde{S}A' = 0$. In component form, this is

$$\sum_j \left(A_{ij}S_{kj} - S_{ji}A'_{jk} \right) = 0 \tag{10.3}$$

Assume that A' is diagonal. Its elements can then be expressed as $A'_{jk} = \lambda_k \delta_{jk}$, where λ_k are the diagonal elements. This gives

$$\sum_j \left(A_{ij}S_{kj} - S_{ji}\lambda_k \delta_{jk} \right) = 0 \tag{10.4}$$

By using the properties of the Kronecker delta, we find

$$\left(\sum_j A_{ij}S_{kj} \right) - S_{ki}\lambda_k = 0 \tag{10.5}$$

Now write

$$S_{ki} = \sum_j \delta_{ij}S_{kj} \tag{10.6}$$

This gives

$$\sum_j \left(A_{ij} - \lambda_k \delta_{ij} \right) S_{kj} = 0 \tag{10.7}$$

If A is n-dimensional, there will be n different values λ_k. According to (10.7), each λ_k is associated with a row of the transforming matrix S. It is possible to write (10.7) in matrix form as

$$(\lambda_k I - A) X_k = 0 \qquad \text{or} \qquad AX_k = \lambda_k X_k \tag{10.8}$$

where the elements of the column vectors X_k make up the rows of S. Note that (10.8) represents a set of homogeneous linear equations with X_k considered as the solution vector. A nontrivial solution exists if, and only if,

$$|\lambda_k I - A| = 0 \tag{10.9}$$

which is called the *secular equation*. The matrix $\lambda I - A$ is called the *characteristic matrix* of A, and

$$f(\lambda) = |\lambda I - A| \tag{10.10}$$

is the *characteristic function* of A. Supposing that A is three-dimensional, the characteristic function is

$$f(\lambda) = \begin{vmatrix} \lambda - A_{11} & -A_{12} & -A_{13} \\ -A_{21} & \lambda - A_{22} & -A_{23} \\ -A_{31} & -A_{32} & \lambda - A_{33} \end{vmatrix} \tag{10.11}$$

If the determinant in (10.11) is multiplied out, there results a cubic equation in λ:

$$f(\lambda) = \lambda^3 + \alpha_2\lambda^2 + \alpha_1\lambda + \alpha_0 \tag{10.12}$$

where the coefficients α_0, α_1, and α_2 are functions of the elements of A. The secular equation (10.9) is $f(\lambda) = 0$, which will give three roots λ_1, λ_2, and λ_3. These are the diagonal elements of A' and are called the *eigenvalues* of the matrix A. Each eigenvalue when put back into (10.8) will yield a solution vector called an *eigenvector*. The three eigenvectors when properly normalized then make up the sought-after transforming matrix S. If A is n-dimensional, there will be n eigenvalues λ_k and n corresponding eigenvectors X_k.

The procedure just outlined is all well and good, but at this point there is no guarantee that it will always work. In what follows, the necessary theorems required to ensure the success of the method will be established. One problem has to do with the roots of (10.10). Up until now in this book, it has been tacitly assumed that all quantities (matrices, vectors, tensors, determinants, etc.) involve real numbers. The roots of a polynomial may, however, be complex even if the coefficients are real. A complex number can be written as

$$z = x + iy \tag{10.13}$$

where $i = \sqrt{-1}$. The complex conjugate of z is

$$z^* = x - iy \tag{10.14}$$

Theorem 2. If $z = z^*$, then z is real.

Proof: $x + iy = x - iy$ implies $y = -y$, which means that $y = 0$.

Theorem 3. The eigenvalues of a real symmetric matrix are real.

Proof: Assume that λ is a possible complex root of (10.10) and that X is the corresponding eigenvector. Multiply $AX = \lambda X$ by $\tilde{X}^*$ to get

$$\tilde{X}^*AX = \lambda \tilde{X}^*X \tag{10.15}$$

Now take the complex conjugate of $AX = \lambda X$:

$$AX^* = \lambda^* X^* \tag{10.16}$$

Transpose to get

$$\tilde{X}^*A = \lambda^* \tilde{X}^* \tag{10.17}$$

Now multiply by X:

$$\tilde{X}^*AX = \lambda^*\tilde{X}^*X \tag{10.18}$$

The comparison of (10.15) and (10.18) reveals that $\lambda = \lambda^*$, which means that λ is real.

Theorem 4. The eigenvectors corresponding to distinct eigenvalues are orthogonal.

Proof: Let λ_1 and λ_2 be two distinct eigenvalues and let X_1 and X_2 be the corresponding eigenvectors. Multiply $AX_1 = \lambda_1 X_1$ by $\tilde{X}_2$ and $AX_2 = \lambda_2 X_2$ by $\tilde{X}_1$:

$$\tilde{X}_2 AX_1 = \lambda_1 \tilde{X}_2 X_1 \tag{10.19}$$

$$\tilde{X}_1 AX_2 = \lambda_2 \tilde{X}_1 X_2 \tag{10.20}$$

Now transpose (10.19):

$$\tilde{X}_1 AX_2 = \lambda_1 \tilde{X}_1 X_2 \tag{10.21}$$

By equating (10.20) and (10.21),

$$(\lambda_2 - \lambda_1)\tilde{X}_1 X_2 = 0 \tag{10.22}$$

Since $\lambda_1 \neq \lambda_2$,

$$\tilde{X}_1 X_2 = 0 \tag{10.23}$$

which establishes the orthogonality.

If A is an n-dimensional real symmetric matrix with n distinct eigenvalues, it is possible to find n mutually orthogonal eigenvectors which can then be normalized and used as the rows of the orthogonal transformation S. Those cases where some of the eigenvalues are equal are said to be *degenerate*. The proof of theorem 4 breaks down if there is degeneracy.

Theorem 5. The eigenvalues of A are invariants.

Proof: Let S be any orthogonal transformation. Multiply $AX = \lambda X$ by S:

$$SAX = \lambda SX \tag{10.24}$$

Now insert $I = \tilde{S}S$ between A and X:

$$SA\tilde{S}SX = \lambda SX \tag{10.25}$$

This is equivalent to

$$A'X' = \lambda X' \tag{10.26}$$

Here, A' is not necessarily diagonal. Equation (10.26) is just $AX = \lambda X$ transformed to some other coordinates. The point is that the same value λ remains in (10.26).

Theorem 6. The symmetry of A is both a necessary and sufficient condition for the existence of an orthogonal principal axis (diagonalizing) transformation.

Proof: The theorem has already been established if A is nondegenerate. The following proof applies whether or not A is degenerate. Let λ_1 be a root of $|\lambda I - A| = 0$. Then, $(\lambda_1 I - A)X = 0$ has at least one solution X_1. Now transform the whole problem to a new frame of reference in which the x_1' axis points in the same direction as X_1. In this new coordinate system,

$$A'X_1' = \lambda_1 X_1' \qquad X_1' = \begin{pmatrix} 1 \\ 0 \\ \vdots \\ 0 \end{pmatrix} \tag{10.27}$$

Theorem 5 guarantees that the eigenvalues are not changed by this process. Written out for the case where A is 3×3, (10.27) is

$$\begin{pmatrix} A_{11}' & A_{12}' & A_{13}' \\ A_{21}' & A_{22}' & A_{23}' \\ A_{31}' & A_{32}' & A_{33}' \end{pmatrix} \begin{pmatrix} 1 \\ 0 \\ 0 \end{pmatrix} = \lambda_1 \begin{pmatrix} 1 \\ 0 \\ 0 \end{pmatrix} \tag{10.28}$$

If (10.28) is multiplied out, we find $A_{11}' = \lambda_1$, $A_{21}' = A_{31}' = 0$. Since A' is symmetric, $A_{12}' = A_{13}' = 0$. Therefore,

$$A' = \begin{pmatrix} \lambda_1 & 0 & 0 \\ 0 & A_{22}' & A_{23}' \\ 0 & A_{32}' & A_{33}' \end{pmatrix} \tag{10.29}$$

The process can be repeated. We now look for another solution of $(\lambda I - A')X' = 0$. The secular equation is

$$\begin{vmatrix} \lambda - \lambda_1 & 0 & 0 \\ 0 & \lambda - A_{22}' & -A_{23}' \\ 0 & -A_{32}' & \lambda - A_{33}' \end{vmatrix} = (\lambda - \lambda_1)\begin{vmatrix} \lambda - A_{22}' & -A_{23}' \\ -A_{32}' & \lambda - A_{33}' \end{vmatrix} = 0 \tag{10.30}$$

One solution is $\lambda = \lambda_1$, which has already been taken care of. It is evident that further solutions are involved with the two-dimensional matrix

$$\begin{pmatrix} A_{22}' & A_{23}' \\ A_{32}' & A_{33}' \end{pmatrix} \tag{10.31}$$

meaning that we can essentially work in a two-dimensional subspace which is geometrically the x_2', x_3' plane. Let λ_2 be a root of the 2×2 determinant in (10.30). This is really the secular equation associated with the 2×2 matrix (10.31). It may turn out that $\lambda_2 = \lambda_1$, but this is of no immediate consequence. Then, $(\lambda_2 I - A')X' = 0$ has at least one solution X_2'. Written out, this is

$$\begin{pmatrix} \lambda_1 & 0 & 0 \\ 0 & A'_{22} & A'_{23} \\ 0 & A'_{32} & A'_{33} \end{pmatrix} \begin{pmatrix} 0 \\ a'_2 \\ a'_3 \end{pmatrix} = \lambda_2 \begin{pmatrix} 0 \\ a'_2 \\ a'_3 \end{pmatrix} \tag{10.32}$$

Now rotate the coordinates about the x_1' axis to a new frame of reference where the x_2'' axis points in the same direction as X_2'. In this new coordinate system,

$$\begin{pmatrix} \lambda_1 & 0 & 0 \\ 0 & A''_{22} & A''_{23} \\ 0 & A''_{32} & A''_{33} \end{pmatrix} \begin{pmatrix} 0 \\ 1 \\ 0 \end{pmatrix} = \lambda_2 \begin{pmatrix} 0 \\ 1 \\ 0 \end{pmatrix} \tag{10.33}$$

which implies $A''_{22} = \lambda_2$, $A''_{23} = A''_{32} = 0$. Therefore,

$$A'' = \begin{pmatrix} \lambda_1 & 0 & 0 \\ 0 & \lambda_2 & 0 \\ 0 & 0 & A''_{33} \end{pmatrix} \tag{10.34}$$

$$X_1'' = \begin{pmatrix} 1 \\ 0 \\ 0 \end{pmatrix} \qquad X_2'' = \begin{pmatrix} 0 \\ 1 \\ 0 \end{pmatrix} \tag{10.35}$$

Obviously, the eigenvectors (10.35) are orthogonal to one another.

Suppose that the eigenvalues λ_1 and λ_2 are in fact equal. Observe that

$$\begin{pmatrix} \lambda & 0 & 0 \\ 0 & \lambda & 0 \\ 0 & 0 & A''_{33} \end{pmatrix} \begin{pmatrix} a''_1 \\ a''_2 \\ 0 \end{pmatrix} = \lambda \begin{pmatrix} a''_1 \\ a''_2 \\ 0 \end{pmatrix} \tag{10.36}$$

This means that any vector of the form $(a_1'', a_2'', 0)$ is an eigenvector corresponding to the eigenvalue λ. There are at most two vectors of this type which are linearly independent and we are at liberty to choose (10.35) as two linearly independent mutually orthogonal eigenvectors if we wish.

The proof can be extended to the case where A is $n \times n$. To summarize: If A is a real symmetric $n \times n$ matrix, it has n eigenvalues, some of which may be equal, and n linearly independent eigenvectors, which can be chosen to be mutually orthogonal.

As an example, consider the mechanical system discussed in section 6. The diagonalization of the matrix (6.3) can be done by means of the transformation (6.15), which we now show how to construct. The eigenvalues of (6.3) are found by means of the secular equation

$$\begin{vmatrix} \lambda - \dfrac{2k}{m} & \dfrac{k}{m} \\ \dfrac{k}{m} & \lambda - \dfrac{2k}{m} \end{vmatrix} = 0 \tag{10.37}$$

The roots are

$$\lambda_1 = \frac{k}{m} \qquad \lambda_2 = \frac{3k}{m} \tag{10.38}$$

The eigenvectors are found from

$$\frac{k}{m}\begin{pmatrix} 2 & -1 \\ -1 & 2 \end{pmatrix} \begin{pmatrix} a_1 \\ a_2 \end{pmatrix} = \lambda \begin{pmatrix} a_1 \\ a_2 \end{pmatrix} \tag{10.39}$$

The result is

$$a_1 = a_2 \qquad \text{if } \lambda = \lambda_1 \tag{10.40}$$

$$a_1 = -a_2 \qquad \text{if } \lambda = \lambda_2 \tag{10.41}$$

The eigenvectors must be normalized before the transformation (6.15) can be constructed. The condition $a_1^2 + a_2^2 = 1$ gives

$$a_1 = a_2 = \pm \frac{1}{\sqrt{2}} \qquad \text{if } \lambda = \lambda_1 \tag{10.42}$$

$$a_1 = -a_2 = \pm \frac{1}{\sqrt{2}} \qquad \text{if } \lambda = \lambda_2 \tag{10.43}$$

It is evident that there is some ambiguity. The reason is that there are several coordinate systems possible, each of which is inclined at an angle of 45° to the original coordinates. Some possibilities are illustrated in Fig. 10.1. The transformation (6.15) corresponds to Fig. 10.1*a*.

The eigenvalue problem has a geometrical significance which is revealed through the quadratic form

$$\phi = \tilde{X}AX = \sum_i \sum_j x_i A_{ij} x_j \tag{10.44}$$

which is called the *central quadric* of A. If the coordinates are rotated so that A becomes diagonal,

$$\phi = x_1'^2 \lambda_1 + x_2'^2 \lambda_2 + x_3'^2 \lambda_3 \tag{10.45}$$

If the eigenvalues are all positive, ϕ is a positive definite quadratic form in the sense that $\phi > 0$ for all $x_i \neq 0$. The surfaces $\phi =$ constant are ellipsoids as illustrated in Fig. 10.2*a*. The x_i' axes coincide with

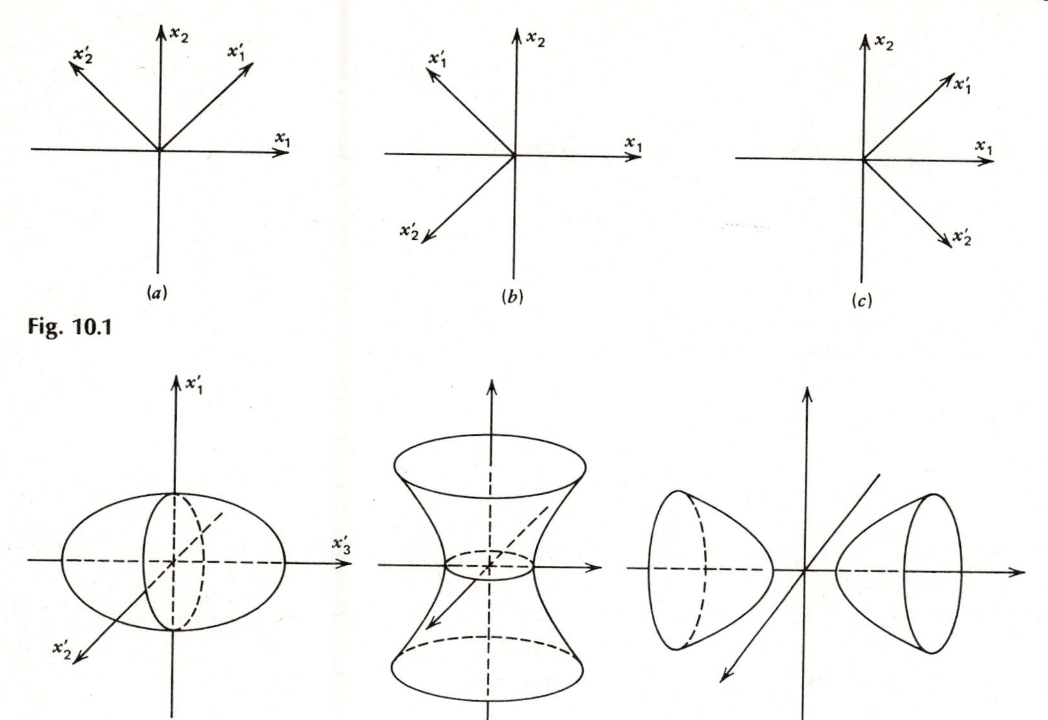

Fig. 10.1

Fig. 10.2

the principal axes of the ellipsoid, which is why the diagonalizing transformation is called the principal axis transformation. If $\lambda_1 = \lambda_2$, the cross sections of the ellipsoid parallel to the x_1', x_2' plane are circles, meaning that all directions in the x_1', x_2' plane are principal directions and all vectors lying in the x_1', x_2' plane are eigenvectors. This is the situation in (10.36). If $\lambda_1 < 0$, $\lambda_2 > 0$, and $\lambda_3 > 0$, $\phi = \text{constant}$ gives a hyperboloid of one sheet, Fig. 10.2*b*. If $\lambda_1 < 0$, $\lambda_2 < 0$, and $\lambda_3 > 0$, $\phi = \text{constant}$ gives a hyperboloid of two sheets, Fig. 10.2*c*.

We close this section with a general comment about tensor transformations. We can construct a 3×3 matrix out of any set of nine functions and then declare that it is a tensor by defining what it is in any other coordinate system by means of $A' = SA\tilde{S}$. This is useful only if A' has the same meaning in the new coordinate system as A had in the old. For example, (9.16) or (9.36) provides a prescription for the computation of the moment of inertia tensor in any rectangular Cartesian coordinate system. What is important is that the tensor has the same physical significance and is used in the Euler equations in the form (9.32) regardless of the coordinate system which is chosen. The theory is said to be *covariant* or *form invariant*, meaning that it has the same form and all quantities appearing in it have the same physical significance in all rectangular Cartesian coordinate systems. This is a general requirement of any physical theory when some kind of transformation is involved.

PROBLEMS

10.1. Any matrix, symmetric or not, has eigenvalues. If A is $n \times n$ the characteristic function will have n roots, some of which may be complex numbers. Find the eigenvalues of the two-dimensional antisymmetric matrix

$$A = \begin{pmatrix} 0 & a \\ -a & 0 \end{pmatrix}$$

where a is a real number. Find two linearly independent eigenvectors. The eigenvectors will have complex numbers as components.

10.2. Prove that the central quadric (10.44) is a scalar invariant, that is,

$$\phi = \tilde{X}AX = \tilde{X}'A'X'$$

for an arbitrary orthogonal transformation.

10.3. Consider the quadratic function $\phi = Ax^2 + 2Bxy + Cy^2$. The curves $\phi =$ constant may be ellipses, hyperbolas, or straight lines. Find the conditions on the coefficients A, B, and C which allow you to distinguish among these possibilities.

10.4. Find the eigenvalues and the normalized eigenvectors of

$$A = \begin{pmatrix} 1.36 & 0 & 0.48 \\ 0 & 3 & 0 \\ 0.48 & 0 & 1.64 \end{pmatrix}$$

Find the orthogonal transformation which diagonalizes A.

10.5. Find the eigenvalues and the normalized eigenvectors of

$$A = \begin{pmatrix} 3 & 0 & 0 & 1 \\ 0 & 2 & 0 & 0 \\ 0 & 0 & 2 & 0 \\ 1 & 0 & 0 & 3 \end{pmatrix}$$

Find an orthogonal transformation which diagonalizes A. There is a lot of degeneracy in this problem.

10.6. Given that A and D are $n \times n$ matrices and that D is diagonal with all different diagonal elements, prove that $AD = DA$ implies that A is diagonal also.

10.7. The matrices A and B are symmetric and $AB = BA$. If either A or B has all distinct eigenvalues, show that both A and B are diagonalized by the same orthogonal transformation.

10.8. If A and B are symmetric and $AB = BA$, convince yourself that an orthogonal transformation can be constructed which will diagonalize both A and B even if one or both of A and B are degenerate.

10.9. Show that the matrices

$$A = \begin{pmatrix} 2 & 0 & 0 & 0 \\ 0 & 2 & 0 & 0 \\ 0 & 0 & 2 & 1 \\ 0 & 0 & 1 & 2 \end{pmatrix} \qquad B = \begin{pmatrix} 1 & 3 & 0 & 0 \\ 3 & 1 & 0 & 0 \\ 0 & 0 & 1 & 2 \\ 0 & 0 & 2 & 1 \end{pmatrix}$$

commute. Find the eigenvalues of each. Construct a four-dimensional orthogonal transformation which diagonalizes both A and B.

10.10. In general, there are n invariants associated with an $n \times n$ matrix, which can be taken to be the n eigenvalues. This is true even if the matrix is not symmetric. All other apparently different invariants can be shown to be functions of the eigenvalues. Relate the trace and the determinant (problems 6.7 and 6.8) of a matrix to its eigenvalues. If A is 3×3, show that

$$\begin{vmatrix} A_{11} & A_{12} \\ A_{21} & A_{22} \end{vmatrix} + \begin{vmatrix} A_{11} & A_{13} \\ A_{31} & A_{33} \end{vmatrix} + \begin{vmatrix} A_{22} & A_{23} \\ A_{32} & A_{33} \end{vmatrix}$$

is an invariant.

Hint: Relate the coefficients in (10.12) both to the elements of A and to the eigenvalues.

10.11. Four point masses are located at the corners of a rectangle in the x_1, x_2 plane as follows:

Mass	$2m$	m	$2m$	m
(x_1, x_2)	$(s, 2s)$	$(-s, 2s)$	$(-s, -2s)$	$(s, -2s)$

Find the moment of inertia tensor of this distribution. Find the eigenvalues and the normalized eigenvectors of the moment of inertia tensor. Find a coordinate system in which the moment of inertia tensor is diagonal. Refer to Fig. 10.3.

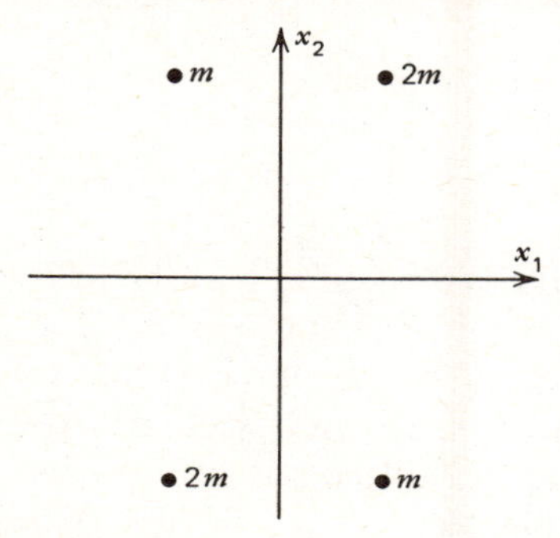

Fig. 10.3

10.12. Suppose that one eigenvalue and the corresponding eigenvector of a 3×3 symmetric matrix A have been found. Show that any orthogonal transformation the first row of which is the eigenvector will transform A into the form given by (10.29).

Hint: Consider $X = \tilde{S}X'$, where X is the eigenvector and X' is given by (10.27). This fact forms the basis of a computer program that will diagonalize a symmetric matrix. See section 10 of Chapter 11.

$\mathbf{E} = -\nabla\psi$

3 Vector Fields

We have considered problems where the components of a vector are functions of a single parameter (usually the time). By a *vector field* in three-dimensional Cartesian space is meant a vector the components of which are functions of x, y, and z. This is the basic concept of this chapter. The differential and integral calculus of vectors is developed and the operations of gradient, divergence, curl, and Laplacian are explored in some detail. Applications to electricity and magnetism, gravitation, heat flow, and diffusion are given.

1. THE GRAVITATIONAL FIELD OF A POINT MASS

The gravitational field $\mathbf{g}$ produced by a point mass M is defined by

$$\mathbf{F} = -\frac{GmM}{r^2}\hat{\mathbf{r}} = m\mathbf{g} \tag{1.1}$$

where $\mathbf{F}$ is the force of attraction between two point masses m and M (Fig. 1.1) and G is the universal gravitational constant the value of which is

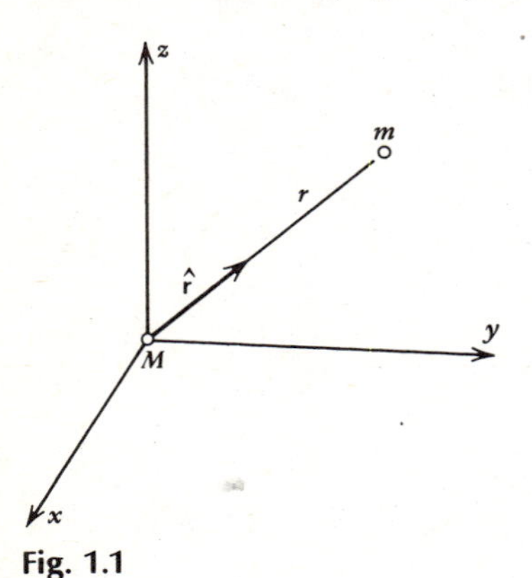

Fig. 1.1

$$G = 6.672 \times 10^{-11}\ \text{N-m}^2/\text{kg}^2$$

in the mks system of units. We visualize $\mathbf{g}$ as the field produced by the mass M and existing in the space around it. It manifests itself by the force $\mathbf{F}$ that it produces on the test mass m. (Review section 5 of Chapter 2.) The field can be written out to show its dependence on the coordinates:

$$\begin{aligned}\mathbf{g} &= -G\frac{M}{r^2}\hat{\mathbf{r}} = -G\frac{M}{r^3}\mathbf{r} \\ &= -GM\left[\frac{x\hat{\mathbf{i}}}{(x^2+y^2+z^2)^{3/2}} + \frac{y\hat{\mathbf{j}}}{(x^2+y^2+z^2)^{3/2}}\right. \\ &\qquad \left. + \frac{z\hat{\mathbf{k}}}{(x^2+y^2+z^2)^{3/2}}\right]\end{aligned} \tag{1.2}$$

The reader who has done problem 5.4 of Chapter 2 will know that the gravitational field outside of any spherically symmetric distribution of mass is the same as it is for a point mass.

In general, by a vector field we understand

$$\mathbf{a} = a_x(x, y, z)\hat{\mathbf{i}} + a_y(x, y, z)\hat{\mathbf{j}} + a_z(x, y, z)\hat{\mathbf{k}} \tag{1.3}$$

or, in more compact notation,

$$\mathbf{a} = \sum_{i=1}^{3} a_i(x_j)\hat{\mathbf{e}}_i \tag{1.4}$$

PROBLEMS

1.1. Two equal point masses are located on the x axis at $x = \pm s$. The gravitational field at the point P is the vector sum of the fields produced by the masses individually:

$$\mathbf{g} = -GM\left(\frac{\mathbf{r}_1}{r_1^3} + \frac{\mathbf{r}_2}{r_2^3}\right)$$

The x component of the field at P is

$$g_x = -GM\left\{\frac{x-s}{\left[(x-s)^2 + y^2 + z^2\right]^{3/2}} + \frac{x+s}{\left[(x+s)^2 + y^2 + z^2\right]^{3/2}}\right\}$$

Find similar expressions for the y and the z components of $\mathbf{g}$. See Fig. 1.2.

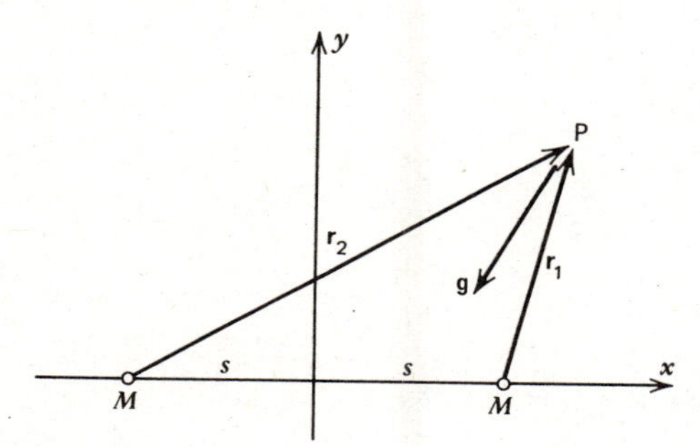

Fig. 1.2

1.2. A total mass M is distributed uniformly and continuously in a thin line along the x axis between $x = -s$ and $x = +s$. The gravitational field produced at P by a differential element dm of the mass is

$$d\mathbf{g} = -G\frac{dm}{r^3}\mathbf{r}$$

The mass dm is located in a length dx' of the distribution so that

$$dm = \frac{M}{2s}dx'$$

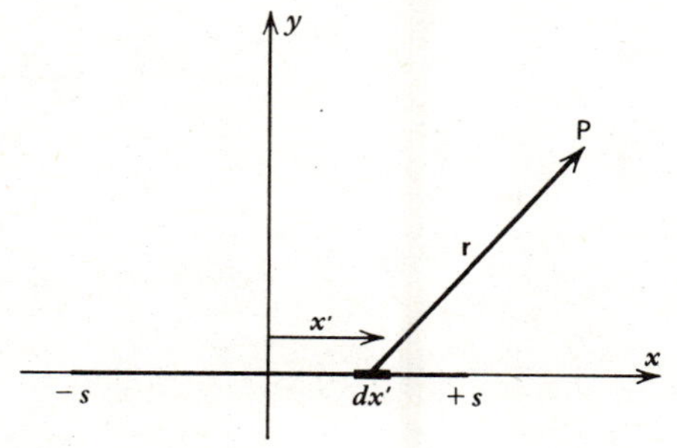

Fig. 1.3

The x component of $d\mathbf{g}$ is therefore

$$dg_x = -\frac{G(x-x')\frac{M}{2s}dx'}{\left[(x-x')^2+y^2+z^2\right]^{3/2}}$$

Integrate this expression between $x' = -s$ and $x' = +s$ to find g_x. Set up similar integrals for g_y and g_z and evaluate them. The point P (coordinates x, y, z) is called the *field point*, and the variable point of integration $(x', 0, 0)$ is the *source point*. See Fig. 1.3.

1.3. A total charge Q is distributed uniformly and continuously in a thin line along the x axis between $x = -s$ and $x = +s$. Translate the results of problem 1.2 to give the components of the resultant electric field at the point P. See equation (5.11) of Chapter 2 for the electric field produced by a single point charge.

2. THE GRADIENT OF A SCALAR FUNCTION

Given a scalar function $\psi(x_1, x_2, x_3)$, we can form three partial derivatives

$$\frac{\partial\psi}{\partial x_1} \qquad \frac{\partial\psi}{\partial x_2} \qquad \frac{\partial\psi}{\partial x_3} \tag{2.1}$$

It will be convenient at times to use the abbreviation

$$\partial_1\psi \qquad \partial_2\psi \qquad \partial_3\psi \tag{2.2}$$

for the partial derivatives. It is possible to express ψ in terms of a new set of Cartesian coordinates x_i' and again calculate the partial derivatives:

$$\partial_1'\psi \qquad \partial_2'\psi \qquad \partial_3'\psi \tag{2.3}$$

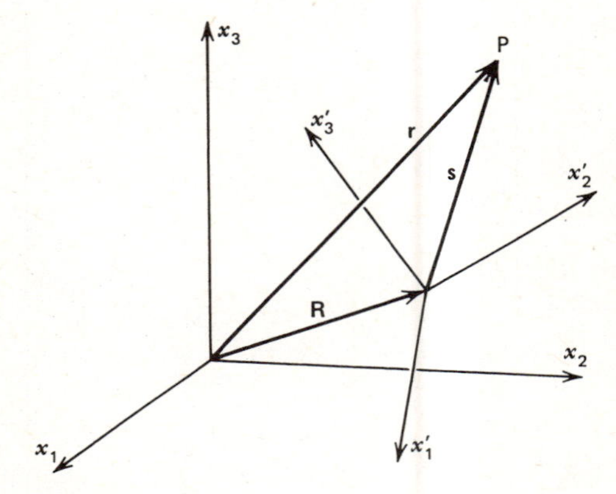

Fig. 2.1

Are the three partial derivatives the components of a vector? In order to answer this question, it is necessary to find the transformation law which connects (2.2) and (2.3).

Figure 2.1 shows two possible coordinate systems. The coordinates of the point P with respect to the two systems are connected by

$$x_j = R_j + \sum_{i=1}^{3} S_{ij}x_i' \tag{2.4}$$

where R_i are the components of the position vector $\mathbf{R}$ of the origin of the x_i' coordinate system with respect to the x_i coordinates. We can write

$$\psi = \psi\left(x_1(x_i'), x_2(x_i'), x_3(x_i')\right) \tag{2.5}$$

where it is indicated that ψ is a function of x_1, x_2, and x_3 which are in turn functions of x_1', x_2', and x_3' as given by (2.4). By means of the chain rule from differential calculus,

$$\frac{\partial\psi}{\partial x_i'} = \sum_{j=1}^{3} \frac{\partial\psi}{\partial x_j}\frac{\partial x_j}{\partial x_i'} \tag{2.6}$$

Equation (2.4) gives

$$\frac{\partial x_j}{\partial x_i'} = S_{ij} \tag{2.7}$$

so that (2.6) can be expressed

$$\partial_i'\psi = \sum_{j=1}^{3} S_{ij}\partial_j\psi \tag{2.8}$$

which is exactly the transformation law which the three partial derivatives must obey if they are to be considered as the components of a vector. The vector so obtained is given the special name *gradient* and is written

$$\text{grad}\,\psi = \nabla\psi = \frac{\partial\psi}{\partial x}\hat{\mathbf{i}} + \frac{\partial\psi}{\partial y}\hat{\mathbf{j}} + \frac{\partial\psi}{\partial z}\hat{\mathbf{k}} = \sum_{i=1}^{3} \partial_i\psi\hat{\mathbf{e}}_i \tag{2.9}$$

The symbol ∇ is called *del* or *nabla* and stands for the symbolic vector operator

$$\nabla = \hat{\mathbf{i}}\frac{\partial}{\partial x} + \hat{\mathbf{j}}\frac{\partial}{\partial y} + \hat{\mathbf{k}}\frac{\partial}{\partial z} = \sum_{i=1}^{3} \hat{\mathbf{e}}_i\,\partial_i \tag{2.10}$$

Equation (2.10) does not mean anything standing all by itself. Nevertheless, there are many formal manipulations in which ∇ can be treated like a vector. There must of course always be something in an expression which ∇ is differentiating in order for it to make any sense.

As an example, the scalar function

$$\psi = -\frac{GM}{r} \tag{2.11}$$

has the interesting feature that the gravitational field (1.2) of a point mass is obtained from it by means of

$$\mathbf{g} = -\nabla\psi \tag{2.12}$$

The function (2.11) is the *gravitational scalar potential* of a point mass.

PROBLEMS

2.1. The gravitational scalar potential (2.11) has an additive property. In problem 1.1, the potential at point P due to the two masses at $x = \pm s$ is given by

$$\psi = -\frac{GM}{r_1} - \frac{GM}{r_2}$$

Write ψ out in terms of x, y, and z. Take the gradient to find the gravitational field.

2.2. The additive property of the gravitational scalar potential can be extended to continuous distributions of mass. In problem 1.2, the contribution to the potential at P due to the element of mass dm is

$$d\psi = -\frac{G\,dm}{r}$$

Integrate this expression over the mass distribution. Take the gradient of the resulting function and verify that the correct gravitational field is obtained.

2.3. The electrostatic scalar potential of a single point charge is

$$\psi = \frac{1}{4\pi\varepsilon_0}\frac{q}{r}$$

The electric field is found by means of $\mathbf{E} = -\nabla\psi$. A total charge q is distributed uniformly and continuously in a thin line on the circumference of a circle of radius a. Find the potential at the point P which is on the axis of the circle and a distance z from its center. See Fig. 2.2. Find the electric field at P by differentiating the potential with respect to z.

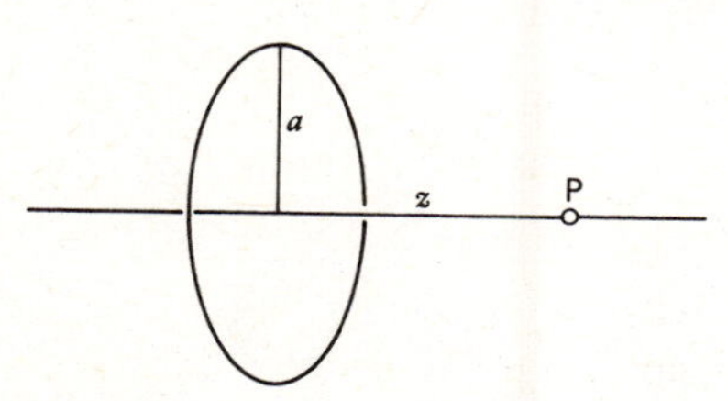

Fig. 2.2

3. REVIEW OF SOME TOPICS FROM CALCULUS

Theorem 1. The Mean Value Theorem. Let $f(x)$ be continuous in the closed interval $x_1 \le x \le x_2$ and differentiable in the open interval $x_1 < x < x_2$. Then there is at least one intermediate value ξ, $x_1 < \xi < x_2$, such that

$$f(x_2) - f(x_1) = f'(\xi)(x_2 - x_1) \tag{3.1}$$

where $f'(\xi)$ means the derivative of $f(x)$ evaluated at $x = \xi$.

Proof: The result is geometrically obvious, as Fig. 3.1 shows. It will be convenient at times to state (3.1) in the form

$$f(x + \Delta x) - f(x) = f'(\xi)\,\Delta x \tag{3.2}$$

where $x = x_1$ and $x_2 = x + \Delta x$.

Theorem 2. A function of two variables $f(x, y)$ is differentiable at x, y if it can be approximated in the neighborhood of x, y by a linear function

$$f(x + \Delta x, y + \Delta y) = f(x, y) + (A + \varepsilon_1)\,\Delta x + (B + \varepsilon_2)\,\Delta y \tag{3.3}$$

where A and B do not depend on Δx and Δy and $(\varepsilon_1, \varepsilon_2) \to (0,0)$ as $(\Delta x, \Delta y) \to (0,0)$.

Proof: Let $\Delta y = 0$. Then,

$$\frac{f(x + \Delta x, y) - f(x, y)}{\Delta x} = A + \varepsilon_1 \tag{3.4}$$

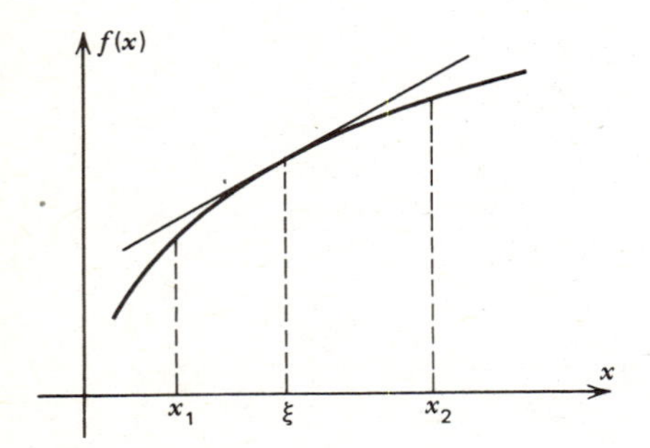

Fig. 3.1

By assumption, $\varepsilon_1 \to 0$ as $\Delta x \to 0$. Thus, in the limit $\Delta x \to 0$,

$$A = \partial_x f(x, y) \tag{3.5}$$

Similarly,

$$B = \partial_y f(x, y) \tag{3.6}$$

Theorem 3. If $\partial_x f(x, y)$ and $\partial_y f(x, y)$ exist and are continuous in a neighborhood of x, y, then $f(x, y)$ can be approximated by a linear function. This is the converse of theorem 2.

Proof: Write

$$\begin{aligned}\Delta f &= f(x + \Delta x, y + \Delta y) - f(x, y)\\ &= f(x + \Delta x, y + \Delta y) - f(x, y + \Delta y)\\ &\quad + f(x, y + \Delta y) - f(x, y)\end{aligned} \tag{3.7}$$

By use of the mean value theorem,

$$\Delta f = \partial_x f(\xi, y + \Delta y)\,\Delta x + \partial_y f(x, \eta)\,\Delta y \tag{3.8}$$

where $x < \xi < x + \Delta x$, $y < \eta < y + \Delta y$. Since we are assuming that the partial derivatives are continuous functions,

$$\partial_x f(\xi, y + \Delta y) = \partial_x f(x, y) + \varepsilon_1 \tag{3.9}$$

$$\partial_y f(x, \eta) = \partial_y f(x, y) + \varepsilon_2 \tag{3.10}$$

where $(\varepsilon_1, \varepsilon_2) \to (0,0)$ as $(\Delta x, \Delta y) \to (0,0)$. Thus,

$$\Delta f = [\partial_x f(x, y) + \varepsilon_1]\,\Delta x + [\partial_y f(x, y) + \varepsilon_2]\,\Delta y \tag{3.11}$$

It is the usual practice to write

$$df = \partial_x f\, dx + \partial_y f\, dy \tag{3.12}$$

and call df the *total differential* of the function f. Equation (3.12) has to be understood as a limiting case of (3.11) where $(\Delta x, \Delta y) \to (0,0)$. In fact, the notation dx, dy implies this kind of limit. In reality, (3.12) gives an approximation to the variation of f in the vicinity of x and y if dx and dy are thought of as small but finite increments. The concept of total differential is valid if the partial derivatives $\partial_x f$ and $\partial_y f$ exist and are continuous in a neighborhood of x, y.

PROBLEMS

3.1. The period of a simple plane pendulum is given by

$$T = 2\pi\sqrt{\frac{s}{g}}$$

where s is the length of the pendulum and g is the gravitational field strength. If s increases by 0.5% and g decreases by 0.4%, what is the approximate % change in the period?

3.2. In the x, y plane, the components of an electric field are given by

$$E_x = b - ax \qquad E_y = ay$$

where a and b are positive constants. *Field lines* or *lines of force* are curves which have the property that the field is tangent to them at each point. The field lines are therefore determined by the differential equation

$$\frac{E_x}{E_y} = \frac{dx}{dy} \qquad \text{or} \qquad E_y\, dx - E_x\, dy = 0$$

which can be thought of as the total differential of $f(x, y) = c$, where c is a constant. Integrate the differential equation to find $f(x, y)$. Each choice of the constant c yields a different field line. Make a sketch showing several different field lines.

4. GRADIENT AND DIRECTIONAL DERIVATIVE

Given a scalar function of two variables $\psi(x, y)$, it is possible to plot it vs. x and y to obtain a surface in three-dimensional space as illustrated in Fig. 4.1. The total differential of this function is

$$\begin{aligned} d\psi &= \partial_x\psi\, dx + \partial_y\psi\, dy \\ &= \nabla\psi \cdot d\mathbf{s} \end{aligned} \tag{4.1}$$

where $d\mathbf{s} = \hat{\mathbf{i}}\, dx + \hat{\mathbf{j}}\, dy$ is an infinitesimal displacement vector in the x, y plane projecting from the point x, y where the partial derivatives in (4.1) are calculated. The $d\mathbf{s}$ vector can point in a variety of directions each of which gives a possible variation of ψ in the neighborhood of x, y. It is to be emphasized that $d\psi$ depends not only on the magnitude of $d\mathbf{s}$ but on its direction as well.

If the scalar function in question depends on all three variables x, y, and z, it is not possible in a three-dimensional space to plot ψ vs. x, y, and z. What can be done is to plot the surfaces found from

$$\psi(x, y, z) = c \tag{4.2}$$

where c is a constant. Figure 4.2 shows two possible such surfaces for two different choices of the constant. The total differential of ψ is now

$$d\psi_{PQ} = \partial_x\psi\, dx + \partial_y\psi\, dy + \partial_z\psi\, dz = (\nabla\psi)_P \cdot d\mathbf{s}_{PQ} \tag{4.3}$$

which gives the variation in ψ as we move from point P to point Q. The points P and Q are separated by the infinitesimal displacement vector $d\mathbf{s}_{PQ}$, and the partial derivatives are all calculated at P. The direc-

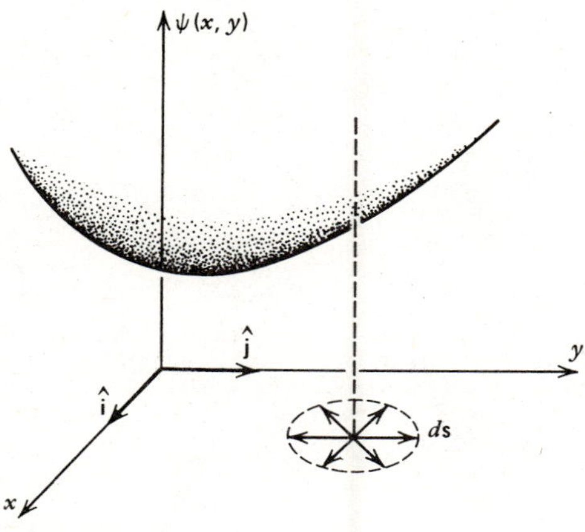

Fig. 4.1

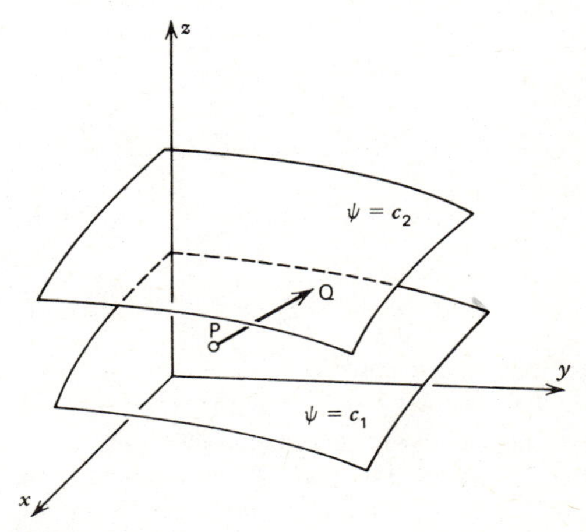

Fig. 4.2

tional derivative of ψ is defined as

$$\left(\frac{d\psi}{ds}\right)_{PQ} = \frac{(\nabla\psi)_P \cdot d\mathbf{s}_{PQ}}{ds} \tag{4.4}$$

Suppose that the point P actually lies in the surface $\psi = c_1$. If Q also lies in this surface, then $d\psi_{PQ} = 0$. It follows from this that the vector $(\nabla\psi)_P$ is perpendicular to the surface $\psi = c_1$ at the point P. Moreover, $d\psi_{PQ}$ is maximum if $d\mathbf{s}_{PQ}$ is parallel to $(\nabla\psi)_P$. Thus, $(\nabla\psi)_P$ points in the direction of maximum rate of increase of the function ψ.

As an example, suppose that we have a steady-state heat flow problem. By steady state is meant no dependence of the temperature on time. Figure 4.3 shows profiles of isotherms, namely, surfaces $T =$ constant, where T is the temperature. The heat flow vector is given by

$$\mathbf{h} = -k\,\nabla T \tag{4.5}$$

where k is the thermal conductivity of the medium. The vector $\mathbf{h}$ has units of energy per second per unit area and is always perpendicular to the isotherms. As 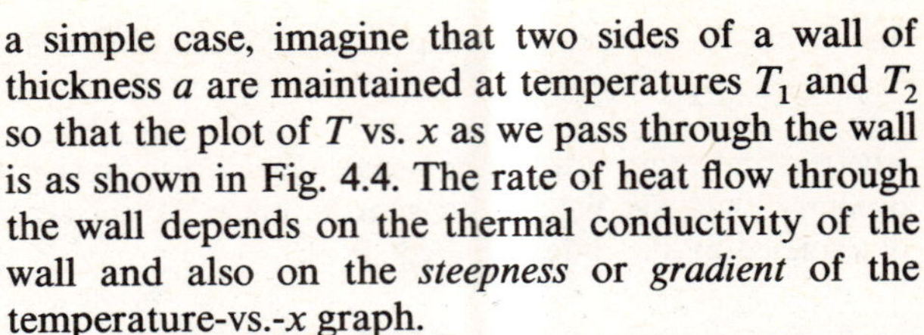 a simple case, imagine that two sides of a wall of thickness a are maintained at temperatures T_1 and T_2 so that the plot of T vs. x as we pass through the wall is as shown in Fig. 4.4. The rate of heat flow through the wall depends on the thermal conductivity of the wall and also on the *steepness* or *gradient* of the temperature-vs.-x graph.

In electrostatics, the relationship between the electric field and the electrostatic potential is

$$\mathbf{E} = -\nabla\psi \tag{4.6}$$

The surfaces $\psi =$ constant are called *equipotentials*. The field lines are perpendicular to the equipotential surfaces. As an example, the field and the potential due to a single point charge are

$$\mathbf{E} = \frac{q}{4\pi\varepsilon_0}\frac{\hat{\mathbf{r}}}{r^2} \qquad \psi = \frac{q}{4\pi\varepsilon_0 r} \tag{4.7}$$

The surfaces $\psi =$ constant are spheres with the point charge at their common centers. The field lines radiate outward from the charge and are perpendicular to the equipotential surfaces.

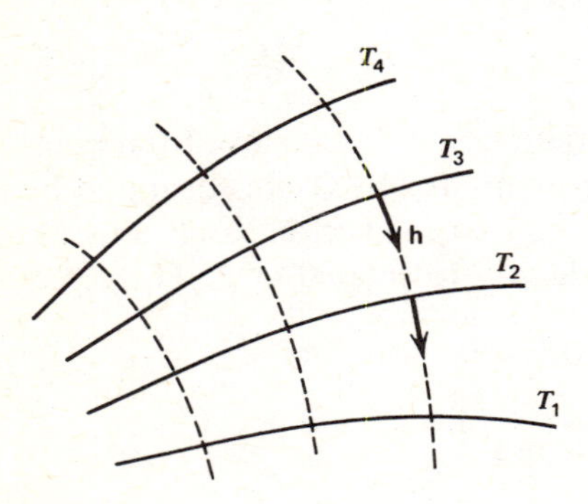

Fig. 4.3

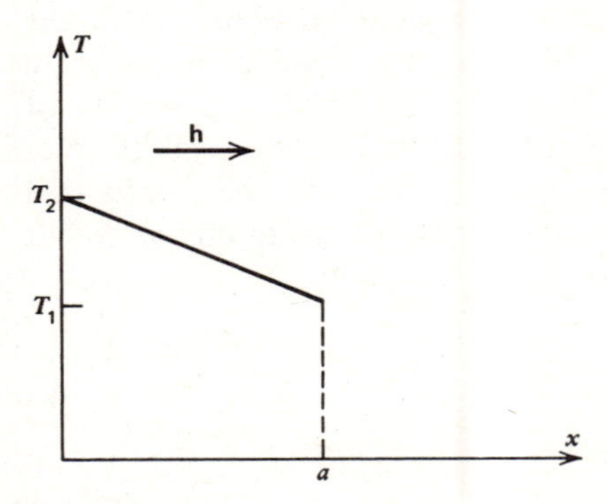

Fig. 4.4

PROBLEMS

4.1. Consider the scalar function $\psi = \mathbf{a}\cdot\mathbf{r}$, where $\mathbf{a}$ is a constant vector. The surfaces $\psi = c$ are planes. Prove that $\mathbf{a}$ is perpendicular to these planes.

4.2. Figure 4.5 shows three displacement vectors which are related by $\mathbf{r} = \mathbf{r}' + \mathbf{R}$. If a function depends on the scalar magnitude of $\mathbf{R}$, it is possible to calculate its gradient with respect to either x, y, z or x', y', z'. Prove that $\nabla f(R) = -\nabla' f(R)$.

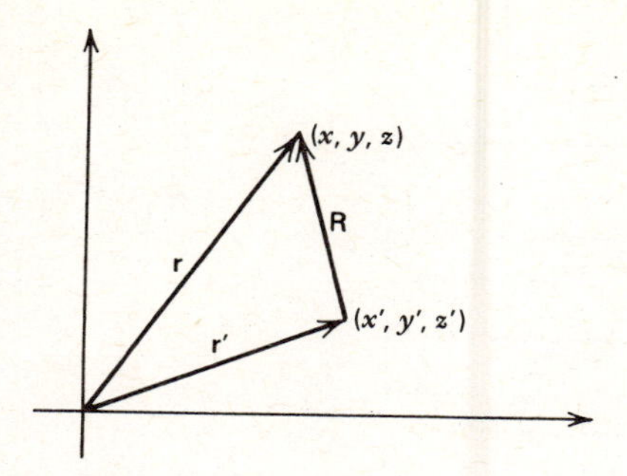

Fig. 4.5

4.3. An electrostatic field in the x, y plane is given by $E_x = -ax$, $E_y = ay$, where a is a constant. Find the electrostatic potential from $E_x = -\partial_x \psi$, $E_y = -\partial_y \psi$. Make graphs of the field lines (see problem 3.2) and the equipotential lines $\psi = c$ on the same page. Consider both positive and negative values of the constant c. Indicate the direction of the electric field on the various field lines. Note that the field lines are directed from regions of higher to lower potential.

4.4. A compound wall consists of a thickness $s_1 = 15$ cm of concrete in contact with a thickness $s_2 = 10$ cm of wood. See Fig. 4.6. The thermal conductivities are $k_1 = 0.002$ and $k_2 = 0.0003$ cal/sec-cm-°C where cal is calories and °C means centigrade or Celsius degrees. The exposed face of the concrete is at a temperature $T_1 = 0\,°\text{C}$, and the exposed face of the wood is at $T_3 = 20\,°\text{C}$. Find the heat flow through the wall in cal/sec-cm^2 and the temperature T_2 at the concrete–wood interface. Draw a graph of temperature vs. distance measured through the wall.

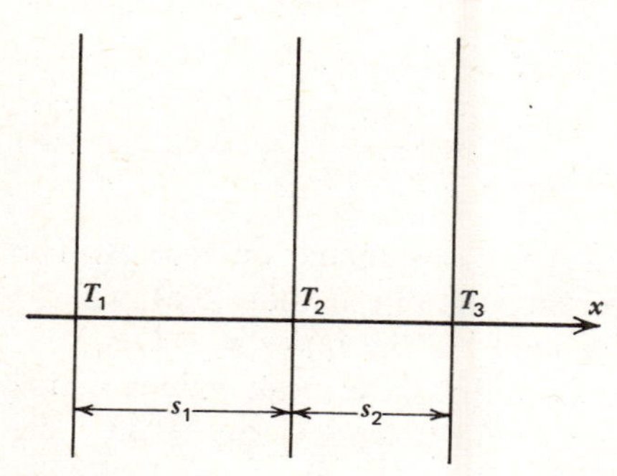

Fig. 4.6

4.5. Consider the function

$$\psi(x, y, z) = x^2 + \frac{y^2 + z^2}{4}$$

What do the surfaces $\psi = \text{constant}$ look like? At what points does the line $y = x$, $z = 0$ intersect the surface $\psi = 5$? Find unit normal vectors to the surface $\psi = 5$ at these points.

5. THE DIVERGENCE OF A VECTOR

From the components of a vector field, it is possible to form nine partial derivatives which become the elements of a second-rank tensor:

$$A = \begin{pmatrix} \partial_1 a_1 & \partial_1 a_2 & \partial_1 a_3 \\ \partial_2 a_1 & \partial_2 a_2 & \partial_2 a_3 \\ \partial_3 a_1 & \partial_3 a_2 & \partial_3 a_3 \end{pmatrix} \tag{5.1}$$

That (5.1) is in fact a second-rank tensor can be demonstrated by formally considering ∂_i as the components of a vector:

$$\partial_i' a_j' = \sum_{m=1}^{3} \sum_{n=1}^{3} S_{im} S_{jn} \partial_m a_n \tag{5.2}$$

The trace of a second-rank tensor is defined to be the sum of its diagonal elements. Reference to problems 6.8 and 10.10 of Chapter 2 shows that the trace of a second-rank tensor is an invariant. The trace of (5.1) is an invariant of some importance:

$$\operatorname{tr} A = \partial_1 a_1 + \partial_2 a_2 + \partial_3 a_3 = \nabla \cdot \mathbf{a} \tag{5.3}$$

Formally, (5.3) is the dot product of the symbolic vector ∇ and **a**. It is given the special name *divergence* and written

$$\operatorname{div} \mathbf{a} = \nabla \cdot \mathbf{a} = \sum_{i=1}^{3} \partial_i a_i \tag{5.4}$$

The meaning of the divergence of a vector and the reason for its importance become apparent in the development of the equation of continuity from fluid dynamics. If at a given point in a fluid the velocity is **u** and the density is ρ, then the vector

$$\mathbf{c} = \rho \mathbf{u} \tag{5.5}$$

represents mass per second per unit area. See equation (4.31) of Chapter 2. If $d\boldsymbol{\sigma}$ is an element of area of some surface, then

$$\mathbf{c} \cdot d\boldsymbol{\sigma} \tag{5.6}$$

gives the mass of fluid per second flowing across $d\boldsymbol{\sigma}$. Figure 5.1 shows a region of space in the form of a rectangular parallelepiped of volume

$$\Delta\Sigma = \Delta x_1 \Delta x_2 \Delta x_3 \tag{5.7}$$

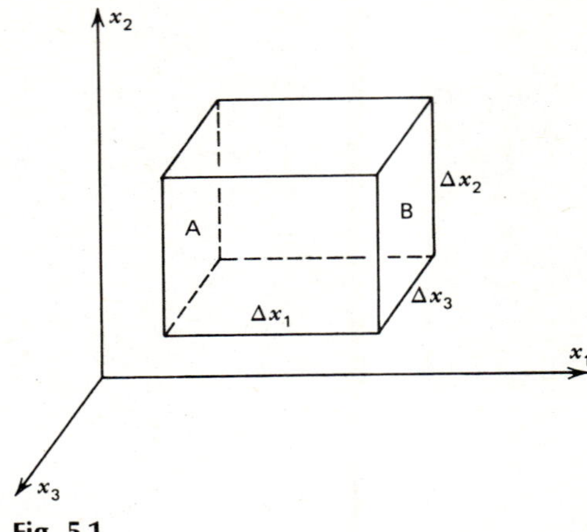

Fig. 5.1

The coordinates of the corner closest to the origin are (x_1, x_2, x_3). The fluid that we are talking about is assumed to be compressible, meaning that the total mass of fluid in $\Delta\Sigma$ can be increasing or decreasing at a given instant of time. In the following calculation, the total outflow of fluid from $\Delta\Sigma$ is evaluated. Consider the faces A and B which are parallel to the x_2, x_3 plane. By using the mean value theorem (theorem 1, section 3), it is possible to write the x_1 component of **c** on face B as

$$c_1(x_1 + \Delta x_1, \bar{x}_2, \bar{x}_3) = c_1(x_1, \bar{x}_2, \bar{x}_3) + \left(\frac{\partial c_1}{\partial x_1}\right)_{x_1 = \xi_1} \Delta x_1 \tag{5.8}$$

where $c_1(x_1, \bar{x}_2, \bar{x}_3)$ is c_1 evaluated on face A. The mean value theorem guarantees that (5.8) will be valid exactly for *some* ξ_1 in the range $x_1 < \xi_1 < x_1 + \Delta x_1$. The values $\bar{x}_2$ and $\bar{x}_3$ are mean values in the ranges $x_2 \le \bar{x}_2 \le x_2 + \Delta x_2$, $x_3 \le \bar{x}_3 \le x_3 + \Delta x_3$ defined, for example, by

$$c_1(x_1, \bar{x}_2, \bar{x}_3) = \frac{1}{\Delta x_2 \Delta x_3} \iint c_1(x_1, x_2, x_3)\, dx_2\, dx_3 \tag{5.9}$$

where the integration is over face A. By means of

(5.6),

outflow across B − inflow across A

$$= \left[c_1 + \left(\frac{\partial c_1}{\partial x_1}\right)_{\xi_1} \Delta x_1\right] \Delta x_2\, \Delta x_3 - c_1\, \Delta x_2\, \Delta x_3$$

$$= \left(\frac{\partial c_1}{\partial x_1}\right)_{\xi_1} \Delta\Sigma \tag{5.10}$$

The flow of fluid across the remaining four faces can be treated similarly with the result that

net outflow from $\Delta\Sigma$

$$= \left[\left(\frac{\partial c_1}{\partial x_1}\right)_{\xi_1} + \left(\frac{\partial c_2}{\partial x_2}\right)_{\xi_2} + \left(\frac{\partial c_3}{\partial x_3}\right)_{\xi_3}\right] \Delta\Sigma$$

$$= -\frac{d}{dt}(\bar{\rho}\,\Delta\Sigma) \tag{5.11}$$

where $x_2 \leq \xi_2 \leq x_2 + \Delta x_2$, $x_3 \leq \xi_3 \leq x_3 + \Delta x_3$, $\bar{\rho}$ is the average density in $\Delta\Sigma$, and $\bar{\rho}\,\Delta\Sigma$ represents the total mass of fluid in $\Delta\Sigma$ at a given instant. In the limit as $\Delta\Sigma \to 0$, (5.11) becomes

$$\nabla \cdot \mathbf{c} = -\frac{\partial \rho}{\partial t} \tag{5.12}$$

This is an equation of continuity, and in this case it expresses conservation (continuity) of mass. It says that the amount of mass that goes out of $d\Sigma$ is accounted for by the decrease in the total mass inside $d\Sigma$. To *diverge* means to *go out from*. Although it has not been indicated explicitly in the foregoing calculation, all quantities are in general time-dependent, meaning that $\rho = \rho(x_1, x_2, x_3, t)$, $c_1 = c_1(x_1, x_2, x_3, t)$, and so on. If the fluid is incompressible, ρ is a constant and

$$\nabla \cdot \mathbf{c} = 0 \tag{5.13}$$

A vector field which obeys (5.13) is said to be *solenoidal*.

There are other examples of equations of continuity. Let $\mathbf{j}$ represent electric current density in units of charge per second per unit area. This would be in amperes per square meter in the mks system of units. If ρ represents electric charge density (coulombs per cubic meter), then

$$\nabla \cdot \mathbf{j} + \frac{\partial \rho}{\partial t} = 0 \tag{5.14}$$

is a mathematical expression of the conservation of charge.

Theorem 1. Gauss' Divergence Theorem. Let $\mathbf{a}$ be a continuous and differentiable vector field throughout a region Σ of space as in Fig. 5.2. Then,

$$\int_\sigma \mathbf{a} \cdot d\boldsymbol{\sigma} = \int_\Sigma \nabla \cdot \mathbf{a}\, d\Sigma \tag{5.15}$$

where the surface integral is taken over the entire surface that encloses Σ. Equation (5.15) says that the net outward flux of the vector field across the closed surface is accounted for by integrating the divergence of the vector field over the volume Σ. It is possible to extend the theorem to include cases where $\mathbf{a}$ is not necessarily continuous and differentiable at every point of Σ. In fact, $\mathbf{a}$ can have certain kinds of *singularities* which are points at which $\mathbf{a}$ is undefined in some way. We will elaborate on this later.

Proof: The proof of (5.15) is best visualized in terms of the equations from fluid dynamics. Let m be the total mass of fluid in Σ:

$$m = \int_\Sigma \rho\, d\Sigma \tag{5.16}$$

If (5.16) is differentiated with respect to time,

$$\frac{dm}{dt} = \int_\Sigma \frac{\partial \rho}{\partial t} d\Sigma = -\int_\Sigma \nabla \cdot \mathbf{c}\, d\Sigma \tag{5.17}$$

where use is made of (5.12). The total rate of decrease of mass in Σ is accounted for by the total flux of the vector $\mathbf{c}$ outward through the surface σ:

$$\frac{dm}{dt} = -\int_\sigma \mathbf{c} \cdot d\boldsymbol{\sigma} = -\int_\Sigma \nabla \cdot \mathbf{c}\, d\Sigma \tag{5.18}$$

which proves the theorem.

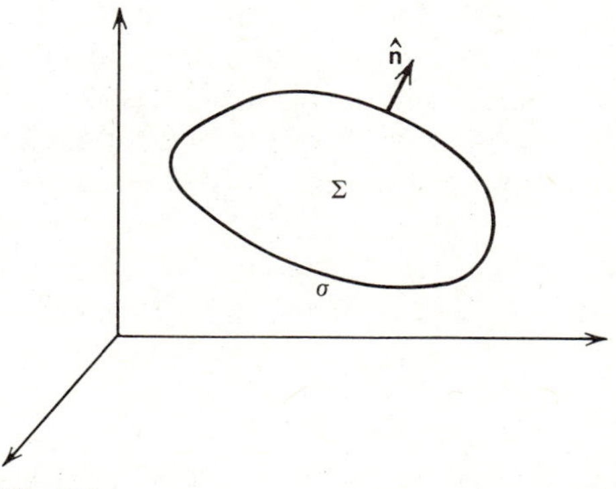

Fig. 5.2

Electrostatics provides an important application of Gauss' divergence theorem. Reference to equation (5.17) of Chapter 2 shows that Gauss' law is

$$\int_\sigma \mathbf{E} \cdot d\boldsymbol{\sigma} = \frac{1}{\varepsilon_0} q_{\text{enc}} = \frac{1}{\varepsilon_0} \int_\Sigma \rho \, d\Sigma \tag{5.19}$$

where ρ is electric charge density. Refer to Fig. 5.2 and imagine that q_{enc} is the total charge in Σ and that the surface integral in (5.19) represents the net outward flux of the electric field through the closed surface σ. By means of (5.15),

$$\int_\Sigma \nabla \cdot \mathbf{E} \, d\Sigma = \frac{1}{\varepsilon_0} \int_\Sigma \rho \, d\Sigma \tag{5.20}$$

Equation (5.20) is valid for any region Σ. We can take the limit of a very small volume, in which case (5.20) becomes

$$(\nabla \cdot \mathbf{E}) \, \Delta\Sigma = \frac{1}{\varepsilon_0} \rho \, \Delta\Sigma \tag{5.21}$$

Therefore,

$$\nabla \cdot \mathbf{E} = \frac{1}{\varepsilon_0} \rho \tag{5.22}$$

This is an extremely important and fundamental partial differential equation which all electrostatic fields must obey. In (5.22), the charge density ρ is acting as a *source* of the electric field. The field lines diverge outward from regions where there is positive charge. They converge on regions where there is negative charge, and for this reason negative charge is sometimes called a *sink* for electric field lines. This is illustrated on Fig. 5.3, which shows the field lines produced by a positive and a negative point charge which have equal magnitude. The field lines originate on the positive charge and terminate on the negative charge.

By contrast, no isolated magnetic poles or magnetic charges have so far been discovered. The consequence of this is that magnetic fields obey

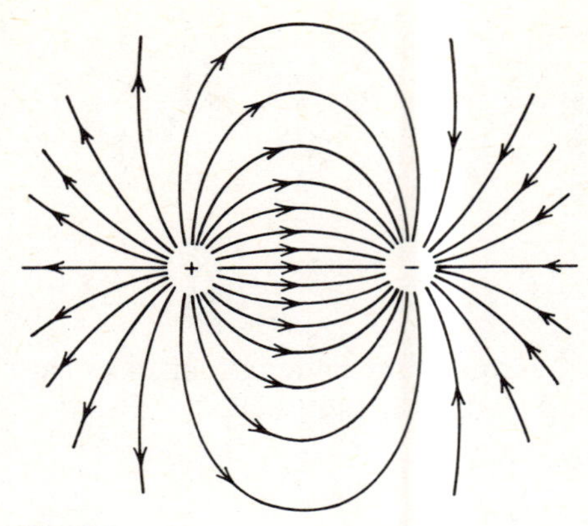

Fig. 5.3

$$\nabla \cdot \mathbf{B} = 0 \tag{5.23}$$

meaning that static magnetic fields are solenoidal. Since there are no sources or sinks on which field lines can originate or terminate, static magnetic field lines exist only in closed loops. Figure 5.4 shows the magnetic field which would be produced by a bar magnet. It is the same as the field produced by a coil of wire wound on a cylinder. Such a coil of wire is in fact called a solenoid.

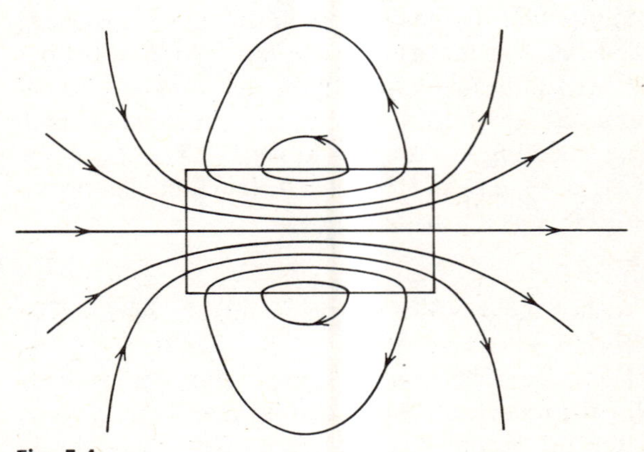

Fig. 5.4

PROBLEMS

5.1. Find the equation similar to (5.22) which the static gravitational field obeys. Refer to problem 5.4 in Chapter 2.

5.2. A long cylinder of radius a has a uniform charge density ρ throughout its volume. Find, using Gauss' law (5.19), the electric field at point P which is inside the cylinder and at point Q which is outside the cylinder. At both points, express the components of the field as functions of x and y. Calculate the divergence of the electric field so obtained at both points. See Fig. 5.5.

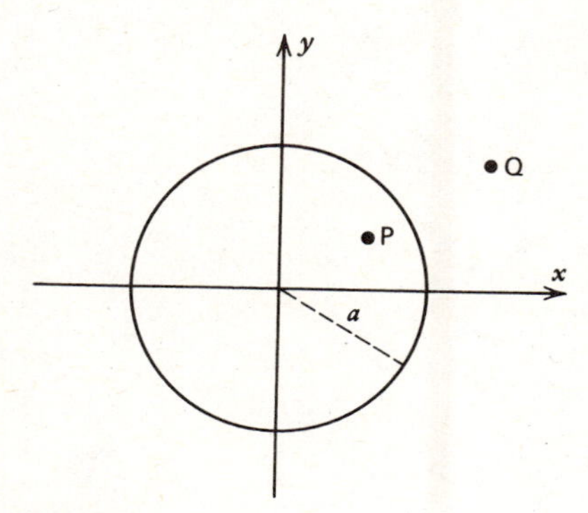

Fig. 5.5

5.3. Let $\mathbf{r}$ be the displacement vector which projects from the origin of coordinates. Find the net flux of the vector $\mathbf{r}$ through any closed surface containing a total volume Σ. See Fig. 5.2.

5.4. Inside a good conductor, the relation between the electric field and the current density can be expressed as $\mathbf{j} = \sigma \mathbf{E}$, where σ is a constant called the conductivity. By using (5.14) and (5.22), show that the charge density at any point in the conductor is

$$\rho = \rho_0 e^{-(\sigma/\varepsilon_0)t}$$

where ρ_0 is the initial charge density at the point in question. The interpretation of this result is that any excess charge placed on a conductor quickly distributes itself over the surface of the conductor, leaving no volume distribution of charge.

5.5. An electric field is given by

$$\mathbf{E} = \frac{A}{\varepsilon_0}\left(\frac{1}{3} - \frac{r^2}{5a^2}\right)\mathbf{r} \qquad r < a$$

$$\mathbf{E} = \frac{2Aa^3}{15\varepsilon_0 r^3}\mathbf{r} \qquad r > a$$

where $\mathbf{r} = x\hat{\mathbf{\imath}} + y\hat{\mathbf{\jmath}} + z\hat{\mathbf{k}}$ and A is a constant. Find the charge density at all points both inside and outside the sphere of radius a. After you have found the charge density, work backward to find the original electric field by means of Gauss' law.

5.6. A vector field is given by $\mathbf{a} = x^2\hat{\mathbf{\imath}} + xy\hat{\mathbf{\jmath}} + xz\hat{\mathbf{k}}$. Evaluate directly the net flux of $\mathbf{a}$ through a closed surface in the form of a cube of edge s with one corner at the origin. The faces of the cube are parallel to the coordinate planes. Then, evaluate $\int_\Sigma \nabla \cdot \mathbf{a}\, d\Sigma$ over the volume of the cube for comparison. See Fig. 5.6.

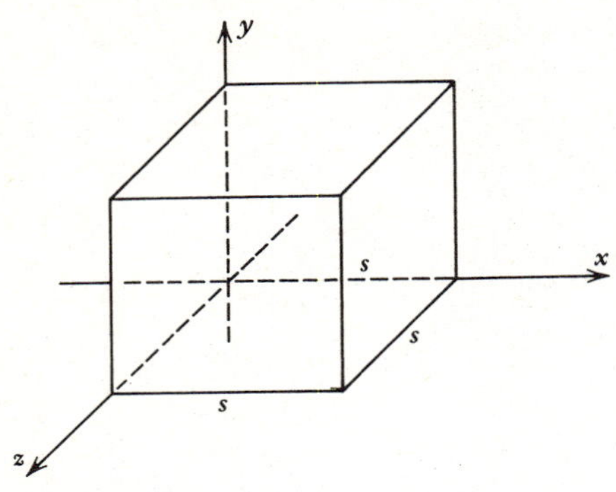

Fig. 5.6

6. HEAT CONDUCTION AND DIFFUSION

The equations of heat conduction and diffusion further illustrate the concept of divergence. In a solid or a liquid, there is little difference between the heat capacities at constant pressure and constant volume. For this reason, we can associate a definite transfer of energy in or out of a solid or liquid in the form of heat with a given temperature change regardless of how the process is carried out and write

$$dq = \rho c\, dT \tag{6.1}$$

where ρ is mass density (gm/cm^3 or kg/m^3), c is specific heat (ergs/gm-°C or joules/kg-°C), dT is temperature change (Celsius degrees), and q is heat density (ergs/cm^3 or joules/m^3). Equation (6.1) means that

$$\frac{d}{dt}(q\, d\Sigma) = \rho c \frac{\partial T}{\partial t} d\Sigma \tag{6.2}$$

is the rate of change of energy in the form of heat in the volume element $d\Sigma$. It is possible that heat energy is being generated in the solid or liquid. This could happen if an electric current is passed through the material resulting in Joule heating or if an exothermic chemical reaction were going on. Let f be the rate of generation of heat in units of ergs/sec-cm^3 or joules/sec-m^3. Then $f\, d\Sigma$ is the rate at which heat energy is created in the volume element $d\Sigma$.

Let $\mathbf{h}$ be a vector which represents the flow of heat in units of ergs/sec-cm^2 or joules/sec-m^2. Then the quantity $\nabla \cdot \mathbf{h}\, d\Sigma$ is the rate at which heat flows out of the volume element $d\Sigma$. Conservation of energy requires

$$\frac{d}{dt}(q\, d\Sigma) = f\, d\Sigma - \nabla \cdot \mathbf{h}\, d\Sigma \tag{6.3}$$

If the medium under consideration is isotropic, meaning that its physical properties are independent of direction, then the heat flow is proportional to the temperature gradient:

$$\mathbf{h} = -k \nabla T \tag{6.4}$$

where k is the thermal conductivity in units of ergs/cm-sec-°C or joules/m-sec-°C. If the medium is not isotropic, the heat may flow preferentially in one direction over another so that a more complicated relation than (6.4) is required. By combining (6.2), (6.3), and (6.4),

$$\nabla \cdot (k \nabla T) - \rho c \frac{\partial T}{\partial t} = -f \tag{6.5}$$

This is a partial differential equation the solutions of which give the temperature throughout the material as a function of position and time. Once this is done, (6.4) can be used to find the heat flow at any point. In those cases where the thermal conductivity is a constant,

$$\nabla \cdot k \nabla T = k(\nabla \cdot \nabla T) = k \nabla^2 T \tag{6.6}$$

where

$$\nabla^2 T = \frac{\partial^2 T}{\partial x^2} + \frac{\partial^2 T}{\partial y^2} + \frac{\partial^2 T}{\partial z^2} \tag{6.7}$$

The differential operator ∇^2 is formally the dot product of ∇ with itself and is called the *Laplacian*.

Thus, if k is a constant, (6.5) becomes

$$\nabla^2 T - \frac{\rho c}{k}\frac{\partial T}{\partial t} = -\frac{f}{k} \tag{6.8}$$

If a steady state has been reached, the temperature is not time dependent, and

$$\nabla^2 T = -\frac{f}{k} \tag{6.9}$$

which is called *Poisson's equation.*

Equation (6.3), which expresses conservation of energy, can be developed in another way. If (6.3) is integrated over a region of space such as the region Σ shown in Fig. 5.2, the result is

$$\frac{d}{dt}\int_\Sigma q\, d\Sigma = \int_\Sigma f\, d\Sigma - \int_\Sigma \nabla \cdot \mathbf{h}\, d\Sigma \tag{6.10}$$

The volume integral of the divergence of **h** can be converted into a surface integral by means of Gauss' theorem (5.15):

$$\frac{d}{dt}\int_\Sigma q\, d\Sigma = \int_\Sigma f\, d\Sigma - \int_\sigma \mathbf{h} \cdot d\boldsymbol{\sigma} \tag{6.11}$$

Expressed in words, (6.11) says that the rate of change of energy in the form of heat in the region Σ equals the rate at which it is being created minus the rate at which heat flows out through the surface σ that encloses Σ. For a steady state, (6.11) reads

$$\int_\sigma \mathbf{h} \cdot d\boldsymbol{\sigma} = \int_\Sigma f\, d\Sigma \tag{6.12}$$

which is Gauss' law for heat flow and should be compared to Gauss' law for electrostatic fields (5.20). In (6.12), the function f plays the role of a source for **h** and is analogous to charge density as a source for the electrostatic field. Another version of (6.3) is

$$\frac{\partial q}{\partial t} + \nabla \cdot \mathbf{h} = f \tag{6.13}$$

which is an equation of continuity with a source term. Whereas over-all energy is conserved, energy in the form of heat can be generated or converted into other forms, and this is accounted for by the function f.

Gauss' law (6.12) can be used directly to solve a few simple problems. For example, consider a long conducting wire in the form of a cylinder of radius a and length s which has an electric resistance R ohms and a thermal conductivity k_1. Surrounding the wire in the form of a cylinder of radius b is an electrically insulating material that has a thermal conductivity k_2. If an electric current I amperes flows through the wire, heat is generated in the inner cylinder so that the source function for heat flow is

$$f = \frac{I^2 R}{\pi a^2 s} \text{ watts/m}^3 \qquad 0 < r < a \tag{6.14}$$

$$f = 0 \qquad a < r < b \tag{6.15}$$

Heat flows radially outward and is dissipated at the outer surface $r = b$, which we can suppose is being maintained at some constant temperature T_0. We will assume a steady state. Application of Gauss' law to a region in the form of a cylinder of radius $r < a$, Fig. 6.1, gives

$$h_r 2\pi r s = f \pi r^2 s \qquad h_r = \tfrac{1}{2} f r \qquad 0 < r < a \tag{6.16}$$

where h_r is the radial component of **h**. Since **h** has no component other than the radial component, the heat flow vector is

$$\mathbf{h} = \tfrac{1}{2} f \mathbf{r} = \tfrac{1}{2} f (x\hat{\mathbf{i}} + y\hat{\mathbf{j}}) \tag{6.17}$$

The divergence of **h** is

$$\nabla \cdot \mathbf{h} = \frac{\partial h_x}{\partial x} + \frac{\partial h_y}{\partial y} = f \qquad 0 < r < a \tag{6.18}$$

which agrees with (6.13) for the steady state case where $\partial q/\partial t = 0$. Application of Gauss' law to a region in the form of a cylinder of radius r in the

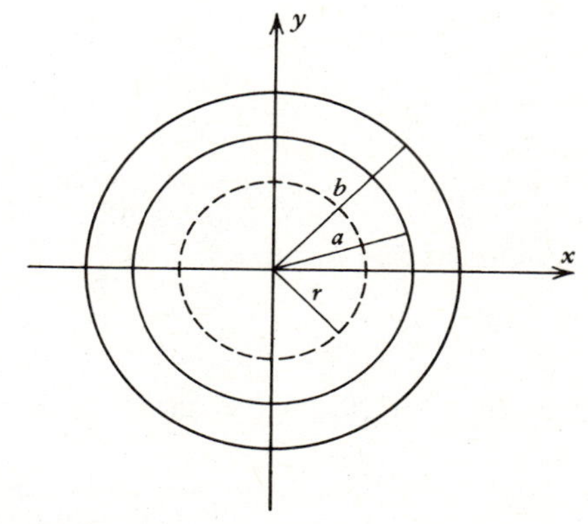

Fig. 6.1

range $a < r < b$ gives

$$h_r 2\pi rs = f\pi a^2 s \qquad h_r = \frac{1}{2} f \frac{a^2}{r}$$

$$\mathbf{h} = \frac{1}{2} f \frac{a^2}{r^2} (x\hat{\mathbf{i}} + y\hat{\mathbf{j}}) \qquad a < r < b \tag{6.19}$$

A calculation of the divergence of (6.19) gives

$$\nabla \cdot \mathbf{h} = 0 \qquad a < r < b \tag{6.20}$$

which is to be expected since no heat is generated in the region $a < r < b$. The reader who has done problem 5.2 should compare notes at this point.

The temperature can be found from (6.4):

$$h_r = \frac{1}{2} fr = -k_1 \frac{dT}{dr}$$

$$T = -\frac{1}{4}\frac{f}{k_1} r^2 + C_1 = -\frac{1}{4}\frac{f}{k_1}(x^2 + y^2) + C_1$$

$$0 < r < a \tag{6.21}$$

where C_1 is an as yet undetermined constant of integration. Note that

$$\nabla^2 T = \frac{\partial^2 T}{\partial x^2} + \frac{\partial^2 T}{\partial y^2} = -\frac{f}{k_1} \qquad 0 < r < a \tag{6.22}$$

in agreement with (6.9). For $a < r < b$,

$$h_r = \frac{1}{2} f \frac{a^2}{r} = -k_2 \frac{dT}{dr}$$

$$T = -\frac{fa^2}{2k_2} \ln r + C_2 = -\frac{fa^2}{4k_2} \ln(x^2 + y^2) + C_2$$

$$a < r < b \tag{6.23}$$

where ln is logarithm to the base e and C_2 is another constant of integration. The temperature as given by (6.23) obeys

$$\nabla^2 T = 0 \qquad a < r < b \tag{6.24}$$

The solution is completed by evaluating the constants C_1 and C_2. This is done by using the boundary condition $T = T_0$ at $r = b$. By means of (6.23),

$$T_0 = -\frac{fa^2}{2k_2} \ln b + C_2 \tag{6.25}$$

With C_2 as given by (6.25),

$$T = \frac{fa^2}{2k_2} \ln \frac{b}{r} + T_0 \qquad a < r < b \tag{6.26}$$

The temperatures as given by (6.21) and (6.26) must be equal at $r = a$:

$$-\frac{1}{4}\frac{f}{k_1} a^2 + C_1 = \frac{fa^2}{2k_2} \ln \frac{b}{a} + T_0 \tag{6.27}$$

The elimination of C_1 between (6.21) and (6.27) gives

$$T = \frac{1}{4}\frac{f}{k_1}(a^2 - r^2) + \frac{fa^2}{2k_2} \ln \frac{b}{a} + T_0 \qquad 0 < r < a \tag{6.28}$$

We shall not attempt here to give any examples of the solutions of the time-dependent equation (6.8) since the complications involved would take us too far from our main purpose, which is laying the foundations of vector analysis.

PROBLEMS

6.1. An electrical conductor in the form of a cylindrical rod of radius a and length s is thermally insulated except at its two ends, which are maintained at a temperature T_0. The cylinder has an electrical resistance R ohms and a current I amperes flows through it. In this case, heat flows parallel to the axis rather than radially outward. The steady-state temperature obeys the one-dimensional Poisson equation

$$\frac{d^2 T}{dx^2} = -\frac{f}{k}$$

Integrate Poisson's equation directly to find the steady-state temperature as a function of position along the rod.

6.2. The theory of particle diffusion is much like that of heat flow. Let n be the number of particles/unit volume and $\mathbf{c}$ be a vector which represents particles per second per unit area. If the diffusion process is isotropic, we can assume that $\mathbf{c} = -D\nabla n$, where D is a diffusion constant. Develop the theory of diffusion along the same lines as was done for heat conduction. Consider the possibility that particles can be created or destroyed in the medium through which they are diffusing.

6.3. Due to a nuclear reaction, heat is being generated at the uniform rate f watts/m^3 throughout the volume of a sphere of radius a and thermal conductivity k_1. Surrounding the sphere of radius a is a sphere of radius b and thermal conductivity k_2. No heat is being generated for $a<r<b$. If the surface of the sphere of radius b is maintained at a constant temperature T_0, find the temperature as a function of r for $0<r<b$.

7. LINE INTEGRALS

Equation (4.25) of Chapter 2 defines the work done on a particle by a force $\mathbf{F}$ acting over a displacement $d\mathbf{s}$:

$$dW = \mathbf{F}\cdot d\mathbf{s} \tag{7.1}$$

If (7.1) is integrated between any two points on the trajectory of the particle,

$$W_{12} = \int_1^2 \mathbf{F}\cdot d\mathbf{s} \tag{7.2}$$

gives the total work done by the force between points 1 and 2. This is an example of a *line integral*. By means of equation (4.29) of Chapter 2,

$$W_{\text{net}} = \int_1^2 \mathbf{F}_{\text{net}}\cdot d\mathbf{s} = \int_1^2 \mathbf{F}_{\text{net}}\cdot\mathbf{u}\,dt = \int_1^2 \frac{d}{dt}\left(\frac{1}{2}mu^2\right)dt$$
$$= \frac{1}{2}mu_2^2 - \frac{1}{2}mu_1^2 \tag{7.3}$$

where $\mathbf{F}_{\text{net}}$ is the net force acting on the particle. Equation (7.3) is a version of the work–energy theorem and says that the work done by the net force acting on a particle between any two points on its trajectory equals the change in its kinetic energy.

We can calculate the line integral of any vector. For example, suppose

$$\mathbf{a} = (xy+y^2)\hat{\mathbf{i}} + (x^2+y^2)\hat{\mathbf{j}} \tag{7.4}$$

We shall calculate the line integral of $\mathbf{a}$ along two different paths in the x, y plane connecting the points (0,0) and (2,1). Path I goes along the x axis from (0,0) to (2,0) and then parallel to the y axis to (2,1), as shown in Fig. 7.1. Path II is a straight line from (0,0) to (2,1). Along path II, $y=\frac{1}{2}x$. The line integral

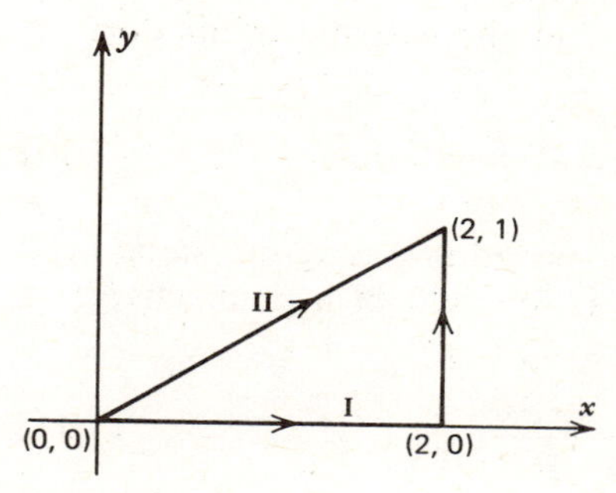

Fig. 7.1

of $\mathbf{a}$ is

$$\int_{\text{II}} \mathbf{a}\cdot d\mathbf{s} = \int_{\text{II}} (a_x\,dx + a_y\,dy)$$
$$= \int_{\text{II}} \left[(xy+y^2)\,dx + (x^2+y^2)\,dy\right]$$
$$= \int_{\text{II}} \left[\left(\tfrac{1}{2}x^2+\tfrac{1}{4}x^2\right)dx + \left(x^2+\tfrac{1}{4}x^2\right)\tfrac{1}{2}dx\right]$$
$$= \int_{x=0}^{2} \tfrac{11}{8}x^2\,dx = \tfrac{11}{3} \tag{7.5}$$

Along path I,

$$\int_{\text{I}} \mathbf{a}\cdot d\mathbf{s} = \int_{x=0}^{2} (xy+y^2)\,dx + \int_{y=0}^{1} (x^2+y^2)\,dy$$
$$= 0 + \int_{y=0}^{1} (4+y^2)\,dy = \tfrac{13}{3} \tag{7.6}$$

It is significant that two different paths connecting the initial and final points yield different values of the line integral.

A point charge Q produces an electrostatic field

$$\mathbf{E} = \frac{Q}{4\pi\varepsilon_0}\frac{\hat{\mathbf{r}}}{r^2} \tag{7.7}$$

A second point charge q placed in the field of Q experiences a force

$$\mathbf{F} = q\mathbf{E} \tag{7.8}$$

In order to hold q in static equilibrium, an external agent must exert a force

$$\mathbf{F}_{\text{ext}} = -\mathbf{F} = -q\mathbf{E} \tag{7.9}$$

Suppose that the external agent moves q quasi-statically along the path between the points a and b of Fig. 7.2. If the process is quasi-static, no kinetic energy is involved, and the external agent does an amount of work

$$W_{\text{ext}} = \int_a^b \mathbf{F}_{\text{ext}} \cdot d\mathbf{s} = -\int_a^b q\mathbf{E} \cdot d\mathbf{s} \tag{7.10}$$

which is true if q is moved in any electrostatic field. If, in particular, $\mathbf{E}$ is the field of a point charge as given by (7.7),

$$W_{\text{ext}} = -\int_a^b \frac{qQ}{4\pi\varepsilon_0} \frac{\hat{\mathbf{r}} \cdot d\mathbf{s}}{r^2} \tag{7.11}$$

As can be seen from Fig. 7.2, $\hat{\mathbf{r}} \cdot d\mathbf{s} = \cos\theta\, ds = dr$ is the change in the length of $\mathbf{r}$ as q is displaced by $d\mathbf{s}$ along the path. This gives

$$W_{\text{ext}} = -\frac{qQ}{4\pi\varepsilon_0} \int_a^b \frac{dr}{r^2} = \frac{qQ}{4\pi\varepsilon_0}\left(\frac{1}{r_b} - \frac{1}{r_a}\right) \tag{7.12}$$

This time, the value of the line integral does not depend on the path taken between the initial and the final points. Vector fields that have this property are an important subclass of general vector fields and are called *conservative*. Since the line integral (7.10) does not depend on the path, it is possible to write

$$W_{\text{ext}} = \int_a^b \mathbf{F}_{\text{ext}} \cdot d\mathbf{s} = V_b - V_a \tag{7.13}$$

where V_b and V_a depend only on the coordinates of the points a and b. By comparison of (7.12) and (7.13),

$$W_{\text{ext}} = V_b - V_a = \frac{qQ}{4\pi\varepsilon_0}\left(\frac{1}{r_b} - \frac{1}{r_a}\right) \tag{7.14}$$

The function

$$V(r) = \frac{qQ}{4\pi\varepsilon_0}\frac{1}{r} \tag{7.15}$$

is the *potential energy* of the two point charges when they are separated by a distance r and is actually the work required of an external agent to bring the charges to their final positions from an initial infinite separation. Reference to (7.10) and (7.13) shows that

$$\int_a^b q\mathbf{E} \cdot d\mathbf{s} = -(V_b - V_a) \tag{7.16}$$

The function

$$\psi(r) = \frac{1}{q}V(r) \tag{7.17}$$

is called the *electrostatic potential* of the charge Q and has units of joules/coulomb, or volts, in the mks system of units. Equation (7.16) is then

$$\int_a^b \mathbf{E} \cdot d\mathbf{s} = -(\psi_b - \psi_a) \tag{7.18}$$

which gives the line integral of $\mathbf{E}$ directly in terms of the electrostatic potential difference between the points a and b.

If $\mathbf{E}$ is the electrostatic field produced by two point charges,

$$\begin{aligned}\int_a^b \mathbf{E} \cdot d\mathbf{s} &= \int_a^b \mathbf{E}_1 \cdot d\mathbf{s} + \int_a^b \mathbf{E}_2 \cdot d\mathbf{s} \\ &= -(\psi_{b1} - \psi_{a1}) - (\psi_{b2} - \psi_{a2}) \\ &= -(\psi_{b1} + \psi_{b2}) - (\psi_{a1} + \psi_{a2}) \end{aligned} \tag{7.19}$$

which shows that the potential has a scalar addition property. The term electrostatic scalar potential is therefore appropriate. In Fig. 7.3, the potential at

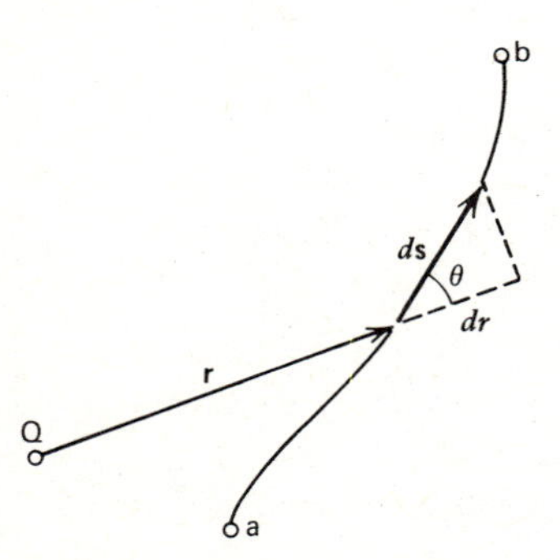

Fig. 7.2

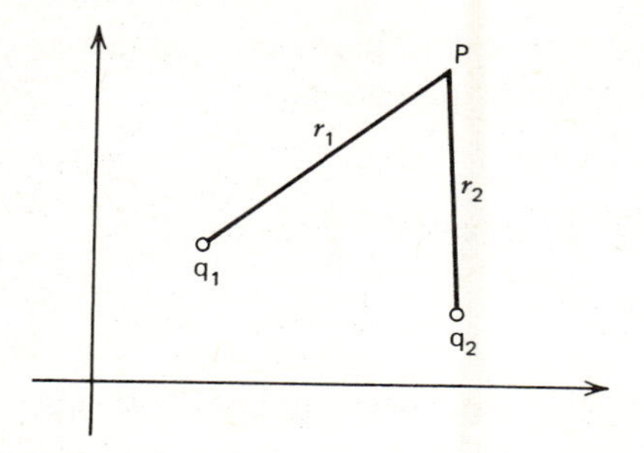

Fig. 7.3

point P due to the two charges q_1 and q_2 is

$$\psi = \frac{1}{4\pi\varepsilon_0}\left(\frac{q_1}{r_1} + \frac{q_2}{r_2}\right) \tag{7.20}$$

An external agent must do an amount of work

$$W_{\text{ext}} = q_3\psi = \frac{1}{4\pi\varepsilon_0}\left(\frac{q_1 q_3}{r_1} + \frac{q_3 q_2}{r_2}\right) \tag{7.21}$$

to move a charge q_3 from infinity to the point P. The concept can be extended to as many point charges as we like:

$$\psi = \sum_{\alpha=1}^{N} \frac{1}{4\pi\varepsilon_0}\frac{q_\alpha}{r_\alpha} \tag{7.22}$$

For a continuous distribution of charge, (7.22) is replaced by

$$\psi = \int_\Sigma \frac{1}{4\pi\varepsilon_0}\frac{1}{r}\rho\, d\Sigma \tag{7.23}$$

where the integral is to be taken over the volume Σ which contains the charge.

Since any electrostatic field is produced by a distribution of either point charges or continuous charge, all electrostatic fields are conservative, and a potential function ψ can always be constructed by means of (7.18). Any vector field defined in a region Σ of space which has the property that its line integral between any two points of Σ is independent of the path is termed conservative. It is always possible to associate a scalar potential with such a vector field by means of

$$\int_a^b \mathbf{a}\cdot d\mathbf{s} = -(\psi_b - \psi_a) \tag{7.24}$$

This is a mathematical concept; the line integral need not have the significance of work, and ψ need not be some kind of potential energy. Vector fields can be conservative in some regions of space and not in others.

Theorem 1. If a vector field is conservative in a region Σ of space, the statement that its line integral between any two points of Σ is independent of path and that

$$\oint \mathbf{a}\cdot d\mathbf{s} = 0 \tag{7.25}$$

for any closed path in Σ are equivalent. The symbol $\oint$ means integral taken around a closed path.

Proof: Observe that

$$\oint \mathbf{a}\cdot d\mathbf{s} = \underset{\text{I}}{\int_a^b} \mathbf{a}\cdot d\mathbf{s} + \underset{\text{II}}{\int_b^a} \mathbf{a}\cdot d\mathbf{s} = 0 \tag{7.26}$$

where a and b are any two points on the closed path shown in Fig. 7.4. Since reversing the limits on an integral changes its sign,

$$\underset{\text{I}}{\int_a^b} \mathbf{a}\cdot d\mathbf{s} = \underset{\text{II}}{\int_a^b} \mathbf{a}\cdot d\mathbf{s} \tag{7.27}$$

Theorem 2. The indefinite integral

$$\psi(x) = \int_a^x f(u)\, du \tag{7.28}$$

of a continuous function $f(x)$ always possesses a derivative $\psi'(x)$; moreover,

$$\psi'(x) = f(x) \tag{7.29}$$

Expressed in words, differentiation of the indefinite integral of a continuous function always gives us back that same function. This is sometimes called *the fundamental*

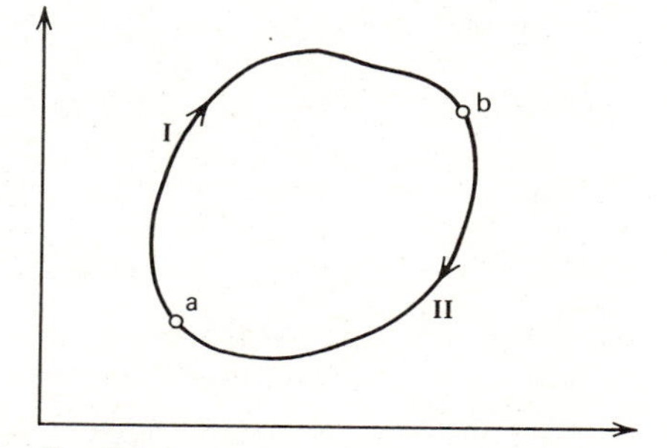

Fig. 7.4

theorem of calculus. We will not prove it here but will use it to establish the following:

Theorem 3. If a vector field is conservative in a region Σ, then at each point of Σ it can be derived from a scalar potential by means of

$$\mathbf{a} = -\nabla\psi \tag{7.30}$$

Proof: The following proof is done in two dimensions but is easily extended to three dimensions. Since $\mathbf{a}$ is conservative, its line integral between the points (x_0, y_0) and (x_1, y_1) in Fig. 7.5 is independent of the path. We can therefore write

$$\begin{aligned}\int_{(x_0,y_0)}^{(x_1,y_1)} \mathbf{a}\cdot d\mathbf{s} &= -[\psi(x_1,y_1)-\psi(x_0,y_0)]\\ &= \int_{(x_0,y_0)}^{(x_1,y_0)} a_x\,dx + \int_{(x_1,y_0)}^{(x_1,y_1)} a_y\,dy\end{aligned} \tag{7.31}$$

where

$$\int_{(x_0,y_0)}^{(x_1,y_0)} a_x\,dx = -[\psi(x_1,y_0)-\psi(x_0,y_0)] \tag{7.32}$$

$$\int_{(x_1,y_0)}^{(x_1,y_1)} a_y\,dy = -[\psi(x_1,y_1)-\psi(x_1,y_0)] \tag{7.33}$$

By application of theorem 2,

$$a_x = -\frac{\partial\psi}{\partial x} \qquad a_y = -\frac{\partial\psi}{\partial y} \tag{7.34}$$

which completes the proof.

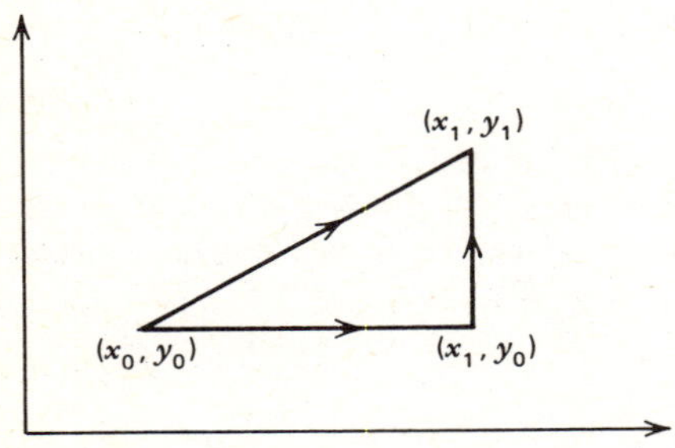

Fig. 7.5

It is possible, then, to express the electrostatic field in terms of the electrostatic potential as

$$\mathbf{E} = -\nabla\psi \tag{7.35}$$

The combination of (5.22) and (7.35) gives

$$\nabla^2\psi = -\frac{\rho}{\varepsilon_0} \tag{7.36}$$

which is recognized as Poisson's equation already encountered in the theory of heat conduction; see equation (6.9). We already know a fundamental solution of Poisson's equation; it is equation (7.23). In regions of space which are free of charge, the electrostatic potential obeys

$$\nabla^2\psi = 0 \tag{7.37}$$

which is called *Laplace's equation*. Most problems in electrostatics are solved by first finding solutions of either Poisson's or Laplace's equation and then using (7.35) to find the field.

Consider again the work–energy theorem as given by (7.3). If the net force acting on a particle is conservative, it can be derived from a potential energy function as $\mathbf{F}_{\text{net}} = -\nabla V$. Equation (7.3) then becomes

$$-\int_1^2 \nabla V\cdot d\mathbf{s} = -(V_2 - V_1) = \tfrac{1}{2}mu_2^2 - \tfrac{1}{2}mu_1^2 \tag{7.38}$$

which can be expressed as

$$W = \tfrac{1}{2}mu_2^2 + V_2 = \tfrac{1}{2}mu_1^2 + V_1 \tag{7.39}$$

The implication is that the total mechanical energy W of the particle, which is the sum of the kinetic and the potential energies, is a constant of the motion. Conservative force fields conserve mechanical energy.

PROBLEMS

7.1. An electric field is given by $\mathbf{E} = (b - ax)\hat{\mathbf{i}} + (ay)\hat{\mathbf{j}}$, where a and b are constants. Calculate the line integral of $\mathbf{E}$ between the points $(0,0)$ and (x_0, y_0).

7.2. Prove that if a vector field can be written as $\mathbf{a} = -\nabla\psi$ in some region Σ, then it is conservative in Σ. This is the converse of theorem 3.

7.3. The static gravitational field is conservative and can be derived from a scalar potential by means of $\mathbf{g} = -\nabla\psi$. Find the Poisson equation obeyed by ψ.

7.4. Find the work in joules which is required to raise a 1-kg mass from the exact center of the Earth to its surface. The radius of the Earth is $R_E = 6.37 \times 10^6$ m. Treat the Earth as a sphere throughout which the total mass is uniformly distributed. See problem 5.4 of Chapter 2. *Note*: The only numerical data needed are the Earth's radius and the fact that the gravitational field strength at the surface is 9.80 m/sec^2.

7.5. An object shot straight up from the Earth's surface rises to a height equal to 5 Earth radii above the surface. Neglecting air resistance, find its initial speed.

7.6. A proton and a helium nucleus, initially separated by a distance r, are released from rest. Show that the maximum speed attained by the proton is

$$u = \frac{2e}{\sqrt{5\pi\varepsilon_0 rm}}$$

where m and e are the mass and charge of the proton, respectively. Assume that the mass of the helium nucleus is four times the mass of the proton.

7.7. Three point charges, $q_1 = 2 \times 10^{-6}$ C, $q_2 = 5 \times 10^{-6}$ C, and $q_3 = -4 \times 10^{-6}$ C, are initially at large distances from one another and are brought to the three vertices of an equilateral triangle the sides of which are of length $s = 20$ cm. How much work in joules is required of an external agent to do this?

7.8. A thin disc of radius a has a charge q uniformly spread over it. Find the electrostatic potential at any point P on the axis of the disc. Express the potential as a function of the distance z measured from the center of the disc. Consider both positive and negative values of z. Find the electric field at any point on the z axis by means of $E_z = -\partial\psi/\partial z$. Again, consider both positive and negative values of z. Graph both the potential and the electric field as functions of z. Note that E_z is discontinuous at $z = 0$. Show that the discontinuity in E_z is equal to λ/ε_0, where λ is the charge per unit area on the disc. See Fig. 7.6.

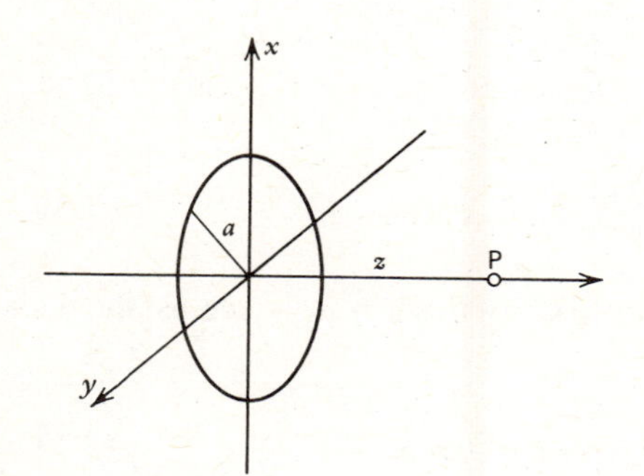

Fig. 7.6

7.9. Show that the function $\psi = 1/R$, where

$$R = \sqrt{(x-x')^2 + (y-y')^2 + (z-z')^2}$$

is a solution of Laplace's equation $\nabla^2\psi = 0$, except at the point $R = 0$. Physically, the function $1/R$ is the potential due to a point charge of magnitude $4\pi\varepsilon_0$ located at (x', y', z').

8. CIRCULATION AND CURL

The circulation of a vector field is defined to be $\oint \mathbf{a} \cdot d\mathbf{s}$. If $\mathbf{a}$ is a conservative vector field, its circulation is zero. In the following calculation, we will evaluate

$$\lim_{\Delta\sigma \to 0} \frac{1}{\Delta\sigma} \oint \mathbf{a} \cdot d\mathbf{s} \tag{8.1}$$

where, as indicated in Fig. 8.1, the circulation of $\mathbf{a}$ is calculated using a closed path bounding an area $\Delta\sigma$. Note that the direction of the unit vector $\hat{\mathbf{n}}$ which is normal to the surface follows the right-hand rule in the respect that a right-hand screw advances in the direction of $\hat{\mathbf{n}}$ if it is turned in the direction in which the line integral is calculated.

The easiest way to do the calculation is to let the area be a rectangle lying in the x_2, x_3 plane as shown in Fig. 8.2. The sides of the rectangle are parallel to the coordinate axes and are of lengths Δx_2 and Δx_3. The corner nearest to the origin has coordinates (x_2, x_3). The mean value theorem is used in very much the same way that it was used in the calculation of divergence. Thus,

$$\begin{aligned}
\oint \mathbf{a} \cdot d\mathbf{s} &= \int_{x_3}^{x_3+\Delta x_3} [-a_3(x_2, x_3)]\, dx_3 \\
&+ \int_{x_3}^{x_3+\Delta x_3} \left[a_3(x_2, x_3) + \left(\frac{\partial a_3}{\partial x_2} \right)_{x_2=\xi_2} \Delta x_2 \right] dx_3 \\
&+ \int_{x_2}^{x_2+\Delta x_2} a_2(x_2, x_3)\, dx_2 \\
&- \int_{x_2}^{x_2+\Delta x_2} \left[a_2(x_2, x_3) + \left(\frac{\partial a_2}{\partial x_3} \right)_{x_3=\xi_3} \Delta x_3 \right] dx_2 \\
&= \overline{\left(\frac{\partial a_3}{\partial x_2} \right)}_{x_2=\xi_2} \Delta x_2 \Delta x_3 \\
&- \overline{\left(\frac{\partial a_2}{\partial x_3} \right)}_{x_3=\xi_3} \Delta x_3 \Delta x_2
\end{aligned} \tag{8.2}$$

where the bar denotes a mean value. In the limit as $\Delta\sigma = \Delta x_2 \Delta x_3$ goes to zero,

$$\lim_{\Delta\sigma \to 0} \frac{1}{\Delta\sigma} \oint \mathbf{a} \cdot d\mathbf{s} = \partial_2 a_3 - \partial_3 a_2 \tag{8.3}$$

Note that (8.3) is formally the x_1 component of the cross product of ∇ and $\mathbf{a}$:

$$\mathbf{c} = \nabla \times \mathbf{a} = \sum_i \sum_j \sum_k e_{ijk} \hat{\mathbf{e}}_i \partial_j a_k \tag{8.4}$$

If $\Delta\sigma$ has an arbitrary orientation in space as shown in Fig. 8.1,

$$\lim_{\Delta\sigma \to 0} \frac{1}{\Delta\sigma} \oint \mathbf{a} \cdot d\mathbf{s} = \hat{\mathbf{n}} \cdot \nabla \times \mathbf{a} \tag{8.5}$$

The vector defined in (8.4) is called the *curl* or the *rotation* of $\mathbf{a}$. If $\mathbf{a}$ is a conservative vector field, its curl is zero. It was stated that equation (5.22) is a fundamental differential equation obeyed by the electrostatic field. We now have a second, equally important partial differential equation obeyed by $\mathbf{E}$:

$$\nabla \times \mathbf{E} = 0 \tag{8.6}$$

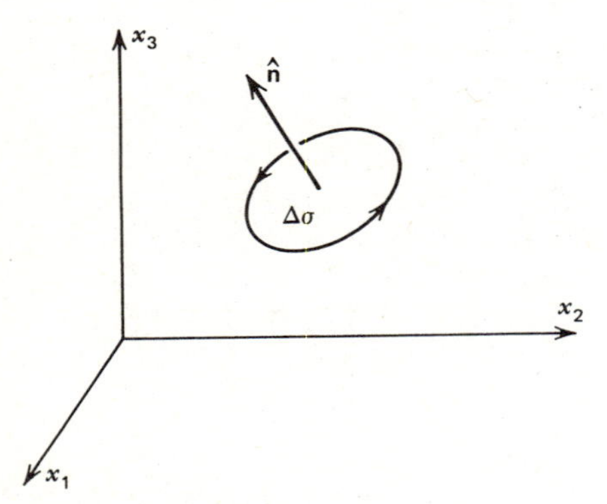

Fig. 8.1

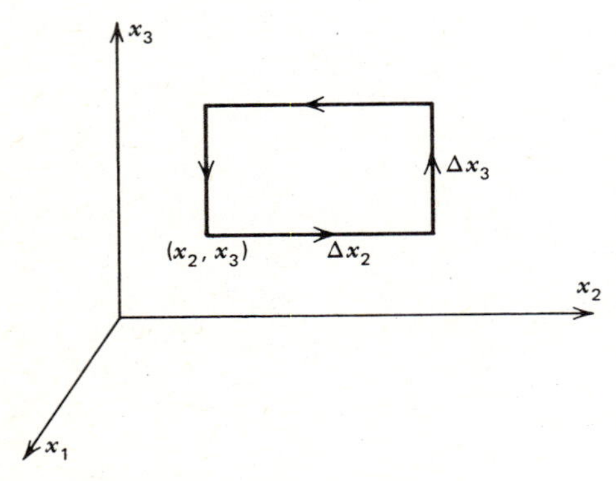

Fig. 8.2

Similarly, the static gravitational field obeys $\nabla \times \mathbf{g} = 0$. A vector field is completely determined if its divergence and curl are known and if appropriate boundary conditions are given. In this sense, the list of basic differential equations obeyed by the electrostatic field and the static gravitational field is now complete. We will take up the question of boundary conditions later.

Theorem 1. Stokes' Theorem. Let σ be a surface of any shape bounded by a closed curve C as illustrated in Fig. 8.4. If $\mathbf{a}$ is a vector field, then

$$\oint_C \mathbf{a} \cdot d\mathbf{s} = \int_\sigma \nabla \times \mathbf{a} \cdot d\boldsymbol{\sigma} \tag{8.7}$$

This important result is called Stokes' theorem.

Proof: The theorem is valid even if $\mathbf{a}$ has certain kinds of discontinuities or singularities. But for the purposes of the following proof, we can assume that $\mathbf{a}$ is continuous and differentiable at each point of σ. First, consider the three closed paths shown in Fig. 8.3. Imagine that paths 1 and 2 are expanded out until they coalesce with path 3. Since the line integrals of $\mathbf{a}$ along the portions that 1 and 2 have in common will cancel each other,

$$\oint_1 \mathbf{a} \cdot d\mathbf{s} + \oint_2 \mathbf{a} \cdot d\mathbf{s} = \oint_3 \mathbf{a} \cdot d\mathbf{s} \tag{8.8}$$

Now let the surface σ of Fig. 8.4 be divided up into a large number N of elements. The idea of equation (8.8) is extended to arrive at

$$\sum_{\alpha=1}^{N} \oint_\alpha \mathbf{a} \cdot d\mathbf{s} = \oint_C \mathbf{a} \cdot d\mathbf{s} \tag{8.9}$$

By means of (8.5),

$$\lim_{\substack{N\to\infty \\ \Delta\sigma\to 0}} \sum_{\alpha=1}^{N} \oint_\alpha \mathbf{a} \cdot d\mathbf{s} = \lim_{\substack{N\to\infty \\ \Delta\sigma\to 0}} \sum_{\alpha=1}^{N} (\nabla \times \mathbf{a} \cdot \hat{\mathbf{n}}\, \Delta\sigma)_\alpha$$

$$= \int_\sigma \nabla \times \mathbf{a} \cdot \hat{\mathbf{n}}\, d\sigma \tag{8.10}$$

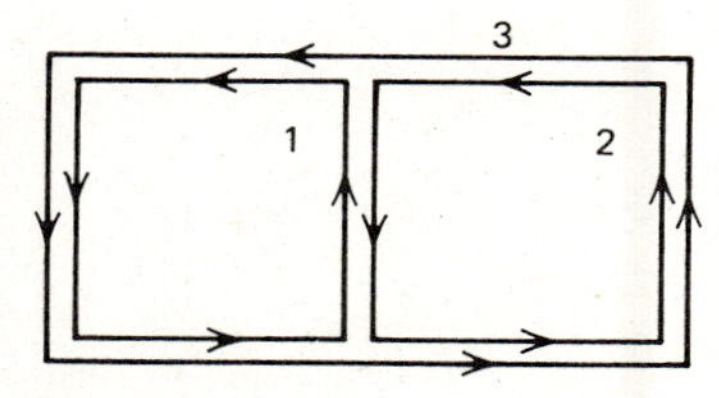

Fig. 8.3

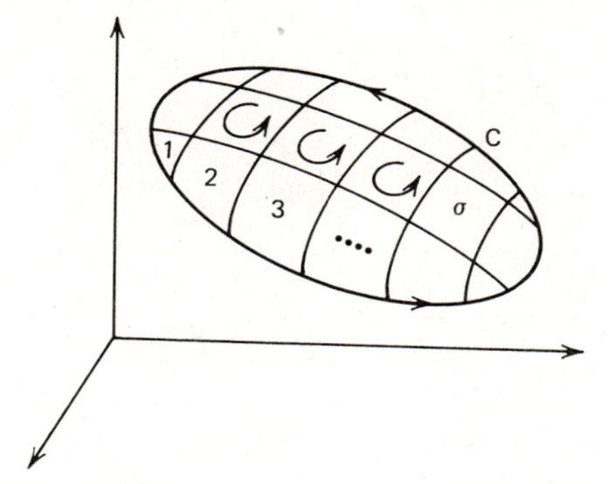

Fig. 8.4

The combination of (8.9) and (8.10) gives Stokes' theorem (8.7).

Theorem 2. If $\mathbf{a}$ obeys $\nabla \times \mathbf{a} = 0$ throughout a simply connected region Σ of space, then $\mathbf{a}$ is conservative in Σ.

Proof: A region Σ is *simply connected* if every closed path in Σ can be continuously deformed to a point without passing out of Σ. In other words, Σ should have no holes bored through it so as to form a doughnut-shaped region. The reason for this restriction is not immediately apparent but will be discussed a little later on. Pushing this question aside for the moment, the validity of the theorem is an immediate consequence of Stokes' theorem. Theorem 2 provides a very convenient test for determining whether or not a given vector field is conservative.

Note that if $\mathbf{a}$ is a polar vector, its circulation $\oint \mathbf{a} \cdot d\mathbf{s}$ is an ordinary scalar. The curl of $\mathbf{a}$ is a pseudovector, which means that the element of area $d\boldsymbol{\sigma}$ in (8.7) must be interpreted as a pseudovector in order that $\nabla \times \mathbf{a} \cdot d\boldsymbol{\sigma}$ come out to be an ordinary scalar.

One of the fundamental equations obeyed by the magnetic field is $\nabla \cdot \mathbf{B} = 0$. What about the curl of $\mathbf{B}$? One way to begin a discussion of the magnetic field is to start with the observation that the magnetic field lines produced by a current I existing in a very long, straight wire are in the form of circles surrounding the wire as pictured in Fig. 8.5. The strength of the field is given by

$$B = \frac{\mu_0 I}{2\pi r} \tag{8.11}$$

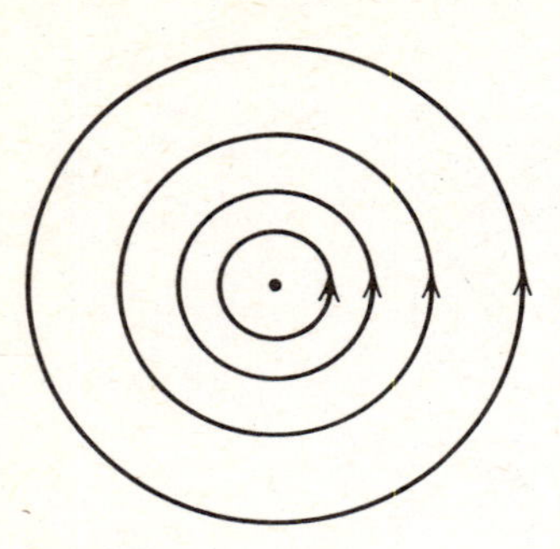

Fig. 8.5

where r is the distance from the wire. In the mks system of units, the constant μ_0 has the value

$$\mu_0 = 4\pi \times 10^{-7}\ \mathrm{N/A^2} \tag{8.12}$$

and is called the *permeability of free space*. The magnetic field itself has units of newtons/ampere-meter (N/A-m), as can be seen either from (8.11) or from the definition of magnetic field as given by the Lorentz force $\mathbf{F} = q\mathbf{u} \times \mathbf{B}$. The designation *tesla*, or *weber/m²*, is also used for the magnetic field in mks units. In Fig. 8.5, the current flows out of the plane of the page and the field lines go in a counterclockwise direction in accordance with the right-hand screw rule.

We now calculate the circulation of the magnetic field pictured in Fig. 8.5. Let the current I run along the z axis of Fig. 8.6. It is convenient to define perpendicular unit vectors $\hat{\mathbf{r}}$ and $\hat{\boldsymbol{\theta}}$, where $\hat{\boldsymbol{\theta}}$ is tangent to circles surrounding the z axis and $\hat{\mathbf{r}}$ points radially outward from the z axis. The circulation of $\mathbf{B}$ around any closed path is then

$$\oint \mathbf{B} \cdot d\mathbf{s} = \oint \frac{\mu_0 I}{2\pi r} \hat{\boldsymbol{\theta}} \cdot d\mathbf{s} \tag{8.13}$$

From the geometry of Fig. 8.6,

$$\hat{\boldsymbol{\theta}} \cdot d\mathbf{s} = ds \cos\alpha = r\, d\theta \tag{8.14}$$

Equation (8.13) becomes

$$\oint \mathbf{B} \cdot d\mathbf{s} = \frac{\mu_0 I}{2\pi} \oint d\theta \tag{8.15}$$

If the path of integration encircles the z axis, $\oint d\theta = 2\pi$; if not, $\oint d\theta = 0$. The circulation of $\mathbf{B}$ is therefore

$$\oint \mathbf{B} \cdot d\mathbf{s} = \mu_0 I_{\text{enc}} \tag{8.16}$$

where I_{enc} is the current encircled by the path of integration.

Consider now a magnetic field produced by two parallel currents I_1 and I_2 which flow perpendicular to the plane of the page in Fig. 8.7*a*. The circulation of $\mathbf{B}$ around the closed path C which encircles both I_1

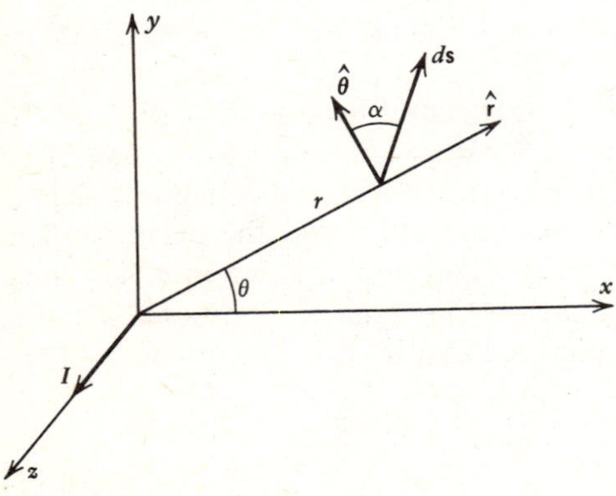

Fig. 8.6

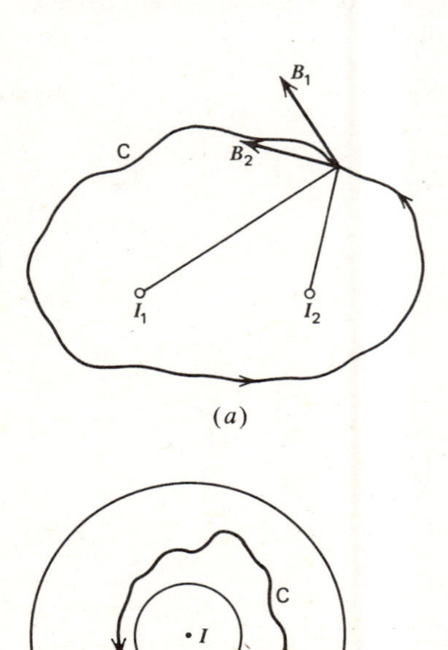

Fig. 8.7

and I_2 is

$$\oint \mathbf{B}\cdot d\mathbf{s} = \oint (\mathbf{B}_1 + \mathbf{B}_2)\cdot d\mathbf{s}$$

$$= \oint \mathbf{B}_1\cdot d\mathbf{s} + \oint \mathbf{B}_2\cdot d\mathbf{s}$$

$$= \mu_0(I_1 + I_2) \tag{8.17}$$

Equation (8.16) therefore generalizes to any number of parallel current-carrying wires encircled by the path of integration. If the current is distributed over the cross section of a conductor as a current density $\mathbf{j}$ A/m^2, then (8.16) can be expressed as

$$\oint \mathbf{B}\cdot d\mathbf{s} = \mu_0 \int_\sigma \mathbf{j}\cdot d\boldsymbol{\sigma} \tag{8.18}$$

where σ is any surface bounded by the path of integration. Equation (8.18) is known as *Ampere's law* and is valid for any static magnetic field. By means of Stokes' theorem,

$$\int_\sigma \nabla\times\mathbf{B}\cdot d\boldsymbol{\sigma} = \mu_0 \int_\sigma \mathbf{j}\cdot d\boldsymbol{\sigma} \tag{8.19}$$

The surface σ and the path of integration bounding it are completely arbitrary and can, for example, be shrunk down to differential size so that (8.19) becomes $\nabla\times\mathbf{B}\cdot\Delta\boldsymbol{\sigma} = \mu_0\mathbf{j}\cdot\Delta\boldsymbol{\sigma}$ or

$$\nabla\times\mathbf{B} = \mu_0\mathbf{j} \tag{8.20}$$

which is Ampere's law in differential form. What we have discovered is that the magnetic field has no scalar sources ($\nabla\cdot\mathbf{B} = 0$) but that it does have a *vector source* as expressed by equation (8.20).

Some insight can now be gained into the reason for restricting Σ to be a simply connected region in the proof of theorem 2. The region Σ of Fig. 8.7*b* is in the shape of a doughnut or torus and is not simply connected. Through the hole in the doughnut a current I flows. There is no current density at any point in Σ, so that the magnetic field obeys $\nabla\times\mathbf{B} = 0$ at all points of Σ. Yet the circulation of $\mathbf{B}$ calculated around the closed path C, which lies entirely in Σ, is not zero. Thus, $\mathbf{B}$ is not conservative in Σ even though $\nabla\times\mathbf{B} = 0$ there.

We close this section by presenting a summary of the conditions that a conservative vector field satisfies. In the following, let Σ be a simply connected region of space. Then, the vector field $\mathbf{a}$ is conservative in Σ if

1. $\int_1^2 \mathbf{a}\cdot d\mathbf{s}$ is independent of path for any path in Σ.
2. $\oint \mathbf{a}\cdot d\mathbf{s} = 0$ for any closed path in Σ.
3. $\nabla\times\mathbf{a} = 0$ at all points of Σ.
4. A scalar function ψ exists such that $\mathbf{a} = -\nabla\psi$.
5. $\mathbf{a}\cdot d\mathbf{s}$ is an exact differential.

The statement that $\mathbf{a}\cdot d\mathbf{s}$ is exact means that it can be written as the total differential of a scalar function: $\mathbf{a}\cdot d\mathbf{s} = -d\psi$. Each of the five conditions is both necessary and sufficient. The validity of any one of the five conditions guarantees that the other four will be valid also.

PROBLEMS

8.1. A current I flows in a conductor in the shape of a long cylinder of radius a. The current flows parallel to the axis of the cylinder and is uniformly distributed over its cross section. Use Ampere's law (8.18) to find the magnetic field at any point both inside and outside the conductor. Express the field as $\mathbf{B} = B_x\hat{\mathbf{i}} + B_y\hat{\mathbf{j}}$. Find B_x and B_y as functions of x and y. Calculate $\nabla\times\mathbf{B}$ at all points both inside and outside the conductor. See Fig. 8.8.

8.2. A rectangular area lies in the same plane as a long, straight wire carrying a current I. Two sides of the rectangle are parallel to the current. Find the flux of the magnetic field, $\int_\sigma \mathbf{B}\cdot d\boldsymbol{\sigma}$, through the rectangular area. See Fig. 8.9.

8.3. In problem 8.1, show that, for $r > a$, it is possible to write $\mathbf{B} = -\nabla\psi$, where

$$\psi = -\frac{\mu_0 I}{2\pi}\tan^{-1}\frac{y}{x} = -\frac{\mu_0 I}{2\pi}\tan^{-1}\tan\theta = -\frac{\mu_0 I}{2\pi}\theta$$

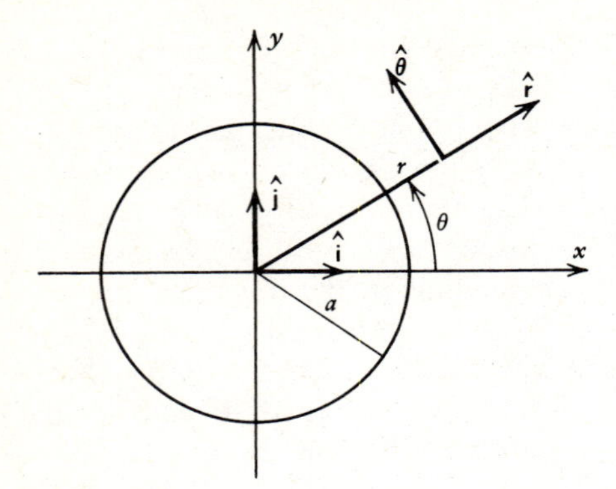

Fig. 8.8

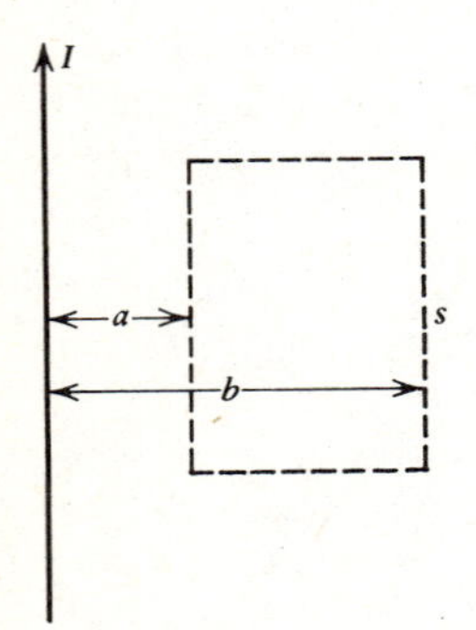

Fig. 8.9

This does not, however, mean that **B** is conservative because the function ψ is not single-valued. For example, $\psi=0$ at $\theta=0$, and $\psi=-\mu_0 I$ at $\theta=2\pi$. Multiple-valued potential functions of this type are frequently quite useful. The function ψ has nothing to do with any kind of potential energy in this example. It is a mathematical device which can be used in any region where $\nabla \times \mathbf{B}=0$. The shaded region of Fig. 8.10 is free of any current, but a current I does flow along the z axis. Is the magnetic field conservative in the shaded region?

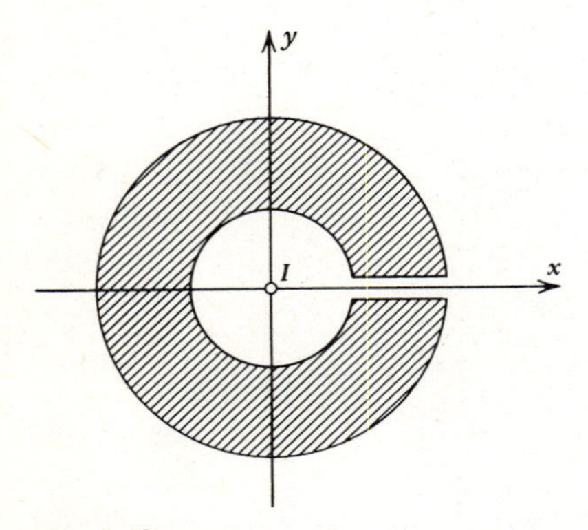

Fig. 8.10

8.4. Current flows in a sheet in the x, z plane in the z direction and has density $\mathbf{K}$ A/m. If $\mathbf{B}_1$ is the magnetic field at a point just below the x, z plane and $\mathbf{B}_2$ is the magnetic field at a point just above the x, z plane, show that

$$B_{1x} - B_{2x} = \mu_0 K \qquad B_{1y} = B_{2y} \qquad B_{1z} = B_{2z}$$

The problem can be done by evaluating the circulation of $\mathbf{B}$ around the small rectangular closed paths illustrated in Fig. 8.11. Each rectangle has one side of length Δs just above the x, z plane and the other side of length Δs just below the x, z plane. The other two sides of each rectangle are of negligible length. One rectangle is oriented parallel to the x, y plane so that the maximum possible current $K\,\Delta s$ flows through it. The other rectangle is parallel to the y, z plane so that no current flows through it. It will also be necessary to apply $\int_\sigma \mathbf{B} \cdot d\boldsymbol{\sigma} = 0$, where σ is any closed surface. This is Gauss' law for the magnetic field and is the equivalent of $\nabla \cdot \mathbf{B} = 0$. The closed surface to use is in the shape of a small box with one face of area $\Delta\sigma$ just above the x, z plane and the other face, also of area $\Delta\sigma$, just below the x, z plane. The sides of the box are of negligible area. The magnetic field lines are refracted or bent as they cross the surface where the current exists. This produces a discontinuity in the magnetic field vector.

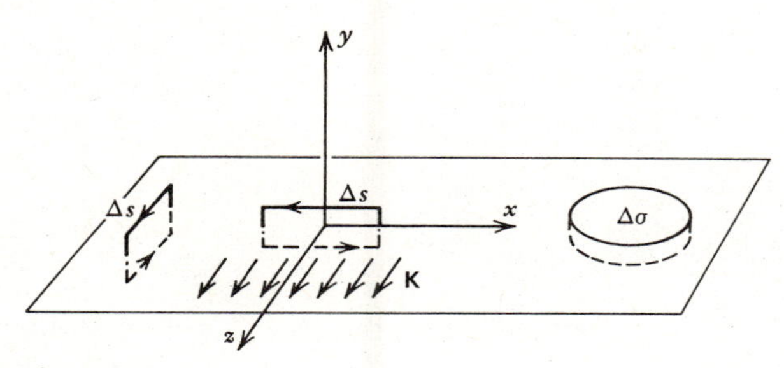

Fig. 8.11

8.5. The x, z plane of a coordinate system has a uniform surface charge density λ C/m² on it. If $\mathbf{E}_1$ is the electric field at a point just below the x, z plane and $\mathbf{E}_2$ is the electric field at a point just above the x, z plane, find the relation between $\mathbf{E}_1$ and $\mathbf{E}_2$.

8.6. Show directly that $\mathbf{a} = -\nabla\psi$ implies $\nabla \times \mathbf{a} = 0$.

8.7. Show that $\nabla \times (\mathbf{a} + \mathbf{b}) = \nabla \times \mathbf{a} + \nabla \times \mathbf{b}$.

8.8. A vector field is given by $a_x = c_1x^2 + c_2xy + c_3y^2$, $a_y = c_4x^2 + c_5xy + c_6y^2$, and $a_z = 0$. What conditions are imposed on the constants c_1 through c_6 by the requirement that $\nabla \cdot \mathbf{a} = 0$ and $\nabla \times \mathbf{a} = 0$? Write formulas for a_x and a_y which involve only c_1 and c_2 as undetermined constants. This shows that knowing the divergence and the curl is not enough to determine the vector field completely. Appropriate boundary conditions are also required.

8.9. A vector field is given by $a_x = 2bx + yz$, $a_y = xz + (2yz^2/c)$, and $a_z = xy + (2y^2z/c)$. Prove that the vector field is conservative by showing that $\nabla \times \mathbf{a} = 0$. Construct a scalar function such that $\mathbf{a} = -\nabla\psi$.

9. DIFFERENTIAL AND INTEGRAL VECTOR IDENTITIES

We present here a list of frequently used differentiation formulas. If **a** and **b** are vector fields and ψ is a scalar function, then

$$\nabla \cdot \psi \mathbf{a} = \psi \nabla \cdot \mathbf{a} + \mathbf{a} \cdot \nabla \psi \tag{9.1}$$

$$\nabla \times \psi \mathbf{a} = \psi \nabla \times \mathbf{a} + \nabla \psi \times \mathbf{a} \tag{9.2}$$

$$\nabla \cdot (\mathbf{a} \times \mathbf{b}) = \mathbf{b} \cdot (\nabla \times \mathbf{a}) - \mathbf{a} \cdot (\nabla \times \mathbf{b}) \tag{9.3}$$

$$\nabla \times (\mathbf{a} \times \mathbf{b}) = (\mathbf{b} \cdot \nabla)\mathbf{a} - (\mathbf{a} \cdot \nabla)\mathbf{b} + \mathbf{a}(\nabla \cdot \mathbf{b}) - \mathbf{b}(\nabla \cdot \mathbf{a}) \tag{9.4}$$

$$\mathbf{a} \times (\nabla \times \mathbf{b}) + \mathbf{b} \times (\nabla \times \mathbf{a}) = \nabla(\mathbf{a} \cdot \mathbf{b}) - (\mathbf{a} \cdot \nabla)\mathbf{b} - (\mathbf{b} \cdot \nabla)\mathbf{a} \tag{9.5}$$

$$\nabla \times \nabla \psi = 0 \tag{9.6}$$

$$\nabla \cdot (\nabla \times \mathbf{a}) = 0 \tag{9.7}$$

$$\nabla \times (\nabla \times \mathbf{a}) = \nabla(\nabla \cdot \mathbf{a}) - \nabla^2 \mathbf{a} \tag{9.8}$$

If $\mathbf{r} = x\hat{\mathbf{i}} + y\hat{\mathbf{j}} + z\hat{\mathbf{k}}$, then

$$\nabla \cdot \mathbf{r} = 3 \tag{9.9}$$

$$\nabla \times \mathbf{r} = 0 \tag{9.10}$$

$$(\mathbf{a} \cdot \nabla)\mathbf{r} = \mathbf{a} \tag{9.11}$$

$$\nabla r^n = n r^{n-2} \mathbf{r} \tag{9.12}$$

In the following proofs, use will be made of the summation convention in which it is understood that repeated indices in a given term are summed from 1 to 3.

Proof of (9.2):

$$\nabla \times \psi \mathbf{a} = e_{ijk} \hat{\mathbf{e}}_i \partial_j \psi a_k = e_{ijk} \hat{\mathbf{e}}_i (\partial_j \psi) a_k + e_{ijk} \hat{\mathbf{e}}_i \psi \partial_j a_k$$
$$= \nabla \psi \times \mathbf{a} + \psi \nabla \times \mathbf{a}$$

Proof of (9.4):

$$\nabla \times (\mathbf{a} \times \mathbf{b}) = e_{ijk} \hat{\mathbf{e}}_i \partial_j e_{kmn} a_m b_n = e_{kij} e_{kmn} \hat{\mathbf{e}}_i \partial_j a_m b_n$$
$$= (\delta_{im}\delta_{jn} - \delta_{in}\delta_{jm}) \hat{\mathbf{e}}_i (b_n \partial_j a_m + a_m \partial_j b_n)$$
$$= \hat{\mathbf{e}}_i b_j \partial_j a_i - \hat{\mathbf{e}}_i b_i \partial_j a_j + \hat{\mathbf{e}}_i a_i \partial_j b_j - \hat{\mathbf{e}}_i a_j \partial_j b_i$$
$$= (b_j \partial_j)(\hat{\mathbf{e}}_i a_i) - (\hat{\mathbf{e}}_i b_i)(\partial_j a_j) + (\hat{\mathbf{e}}_i a_i)(\partial_j b_j) - (a_j \partial_j)(\hat{\mathbf{e}}_i b_i)$$
$$= (\mathbf{b} \cdot \nabla)\mathbf{a} - \mathbf{b}(\nabla \cdot \mathbf{a}) + \mathbf{a}(\nabla \cdot \mathbf{b}) - (\mathbf{a} \cdot \nabla)\mathbf{b}$$

We have used $e_{ijk} = e_{kij}$ which follows from theorem 2 of section 4 in Chapter 1 and equation (8.31) of Chapter 2.

Proof of (9.7): $\nabla \cdot (\nabla \times \mathbf{a}) = e_{ijk} \partial_i \partial_j a_k = 0$ for essentially the same reason that the expansion of a determinant with two rows alike gives zero.

Proof of (9.8):

$$\nabla \times (\nabla \times \mathbf{a}) = e_{ijk} \hat{\mathbf{e}}_i \partial_j e_{kmn} \partial_m a_n = e_{kij} e_{kmn} \hat{\mathbf{e}}_i \partial_j \partial_m a_n$$
$$= (\delta_{im}\delta_{jn} - \delta_{in}\delta_{jm}) \hat{\mathbf{e}}_i \partial_j \partial_m a_n = \hat{\mathbf{e}}_i \partial_j \partial_i a_j - \hat{\mathbf{e}}_i \partial_j \partial_j a_i$$
$$= (\hat{\mathbf{e}}_i \partial_i)(\partial_j a_j) - \hat{\mathbf{e}}_i \nabla^2 a_i = \nabla(\nabla \cdot \mathbf{a}) - \nabla^2 \mathbf{a}$$

Proof of (9.11):

$$(\mathbf{a} \cdot \nabla)\mathbf{r} = (a_i \partial_i)(x_j \hat{\mathbf{e}}_j) = a_i \delta_{ij} \hat{\mathbf{e}}_j = a_i \hat{\mathbf{e}}_i = \mathbf{a}$$

In the following list of integral vector identities, Σ is a region of space bounded by a closed surface σ, **a** is a vector field, and ψ and ϕ are scalar functions.

$$\int_\Sigma (\nabla \cdot \mathbf{a}) \, d\Sigma = \int_\sigma \mathbf{a} \cdot d\boldsymbol{\sigma} \qquad \text{(Gauss' theorem)} \tag{9.13}$$

$$\int_\Sigma (\nabla \psi) \, d\Sigma = \int_\sigma \psi \, d\boldsymbol{\sigma} \tag{9.14}$$

$$\int_\Sigma (\nabla \times \mathbf{a}) \, d\Sigma = \int_\sigma (\hat{\mathbf{n}} \times \mathbf{a}) \, d\sigma \tag{9.15}$$

$$\int_\Sigma (\psi \nabla^2 \phi + \nabla \psi \cdot \nabla \phi) \, d\Sigma = \int_\sigma \psi \nabla \phi \cdot d\boldsymbol{\sigma} \qquad \text{(Green's first identity)} \tag{9.16}$$

$$\int_\Sigma (\psi \nabla^2 \phi - \phi \nabla^2 \psi) \, d\Sigma = \int_\sigma \left(\psi \frac{\partial \phi}{\partial n} - \phi \frac{\partial \psi}{\partial n} \right) d\sigma \qquad \text{(Green's second identity)} \tag{9.17}$$

Keep in mind that $\hat{\mathbf{n}}$ is always drawn outward from the closed surface, as in Fig. 5.2. The notation $\partial \psi / \partial n$ means

$$\frac{\partial \psi}{\partial n} d\sigma = \nabla \psi \cdot d\boldsymbol{\sigma} \tag{9.18}$$

In the next series of identities, σ is an open surface bounded by a closed curve C, Fig. 8.4. The direction of $\hat{\mathbf{n}}$ (or $d\boldsymbol{\sigma}$) is determined by the right-hand rule as illustrated in Fig. 8.1.

$$\oint_C \mathbf{a}\cdot d\mathbf{s} = \int_\sigma \nabla\times\mathbf{a}\cdot d\boldsymbol{\sigma} \qquad \text{(Stokes' theorem)} \tag{9.19}$$

$$\oint_C \psi\, d\mathbf{s} = \int_\sigma (\hat{\mathbf{n}}\times\nabla\psi)\, d\sigma \tag{9.20}$$

$$\oint_C d\mathbf{s}\times\mathbf{a} = \int_\sigma (\hat{\mathbf{n}}\times\nabla)\times\mathbf{a}\, d\sigma \tag{9.21}$$

Proof of (9.15): Let $\mathbf{c}$ be a constant vector. Apply Gauss' theorem to the vector $\mathbf{a}\times\mathbf{c}$ and use (9.3):

$$\int_\Sigma \nabla\cdot(\mathbf{a}\times\mathbf{c})\, d\Sigma = \int_\sigma (\mathbf{a}\times\mathbf{c})\cdot d\boldsymbol{\sigma}$$

$$\int_\Sigma [\mathbf{c}\cdot(\nabla\times\mathbf{a}) - \mathbf{a}\cdot(\nabla\times\mathbf{c})]\, d\Sigma = \int_\sigma (\hat{\mathbf{n}}\times\mathbf{a})\cdot\mathbf{c}\, d\sigma \tag{9.22}$$

Use has also been made of $\mathbf{a}\times\mathbf{c}\cdot d\boldsymbol{\sigma} = d\boldsymbol{\sigma}\times\mathbf{a}\cdot\mathbf{c} = (\hat{\mathbf{n}}\times\mathbf{a}\cdot\mathbf{c})\, d\sigma$, which follows from equation (8.30) of Chapter 2. The fact that $\mathbf{c}$ is a constant vector means that $\nabla\times\mathbf{c} = 0$ and that $\mathbf{c}$ can be taken outside the integral:

$$\mathbf{c}\cdot\int_\Sigma \nabla\times\mathbf{a}\, d\Sigma = \mathbf{c}\cdot\int_\sigma \hat{\mathbf{n}}\times\mathbf{a}\, d\sigma \tag{9.23}$$

Since $\mathbf{c}$ can be chosen arbitrarily, (9.15) follows.

Proof of (9.16): Apply Gauss' theorem to the vector $\psi\nabla\phi$:

$$\int_\sigma \psi\nabla\phi\cdot d\boldsymbol{\sigma} = \int_\Sigma \nabla\cdot(\psi\nabla\phi)\, d\Sigma \tag{9.24}$$

Now use (9.1) to get (9.16).

Proof of (9.17): Write Green's first identity (9.16) out with ϕ and ψ reversed:

$$\int_\Sigma (\phi\nabla^2\psi + \nabla\phi\cdot\nabla\psi)\, d\Sigma = \int_\sigma \phi\nabla\psi\cdot d\boldsymbol{\sigma} \tag{9.25}$$

Form the difference between (9.16) and (9.25) to get (9.17).

Proof of (9.21): Apply Stokes' theorem to $\mathbf{a}\times\mathbf{c}$, where $\mathbf{c}$ is a constant vector:

$$\oint_C \mathbf{a}\times\mathbf{c}\cdot d\mathbf{s} = \int_\sigma \hat{\mathbf{n}}\cdot\nabla\times(\mathbf{a}\times\mathbf{c})\, d\sigma \tag{9.26}$$

The integrand of the surface integral is developed as follows:

$$\begin{aligned}\hat{\mathbf{n}}\cdot\nabla\times(\mathbf{a}\times\mathbf{c}) &= n_i e_{ijk}\partial_j e_{kmn} a_m c_n = c_n e_{nkm} e_{kij} n_i \partial_j a_m \\ &= \mathbf{c}\cdot(\hat{\mathbf{n}}\times\nabla)\times\mathbf{a}\end{aligned} \tag{9.27}$$

Equation (9.26) is then

$$\oint_C (d\mathbf{s}\times\mathbf{a})\cdot\mathbf{c} = \mathbf{c}\cdot\int_\sigma (\hat{\mathbf{n}}\times\nabla)\times\mathbf{a}\, d\sigma$$

Since $\mathbf{c}$ can be chosen arbitrarily, (9.21) follows.

The proofs of the remaining identities are left as problems.

PROBLEMS

9.1. Prove (9.1), (9.3), (9.5), (9.10), and (9.12).

9.2. Prove (9.14) by application of Gauss' theorem to the vector $\mathbf{c}\psi$, where $\mathbf{c}$ is a constant vector. Use the identity (9.1).

9.3. Prove (9.20) by application of Stokes' theorem to the vector $\mathbf{c}\psi$, where $\mathbf{c}$ is a constant vector. Use the identity (9.2).

9.4. Show that Gauss' theorem and the related identities (9.14)–(9.17) can be applied to a nonsimply connected region.

Hint: Look at the figure in problem 8.3. The shaded region is simply connected so that Gauss' theorem applies. What happens when the cut along the x axis is joined so that the region becomes a torus?

9.5. A region Σ is simply connected but has a hollow or bubble in it. This is possible because any closed curve in Σ can still be continuously deformed to a point without passing out of Σ. Show that Gauss' theorem and related identities can be applied to this region, provided that the surface integral includes the surface of the hollow with $\hat{\mathbf{n}}$ drawn inward to the hollow but still directed away from Σ.

9.6. Show that Stokes' theorem and related identities can be applied to a nonsimply connected surface, provided that the line integral includes all bounding curves:

$$\oint_{C} \mathbf{a}\cdot d\mathbf{s} = \oint_{C_1} \mathbf{a}\cdot d\mathbf{s} + \oint_{C_2} \mathbf{a}\cdot d\mathbf{s} = \int_{\sigma} \nabla\times\mathbf{a}\cdot d\boldsymbol{\sigma}$$

The integrals over C_1 and C_2 must be in opposing directions as indicated in Fig. 9.1.

Hint: Make a cut in the figure which joins C_1 and C_2.

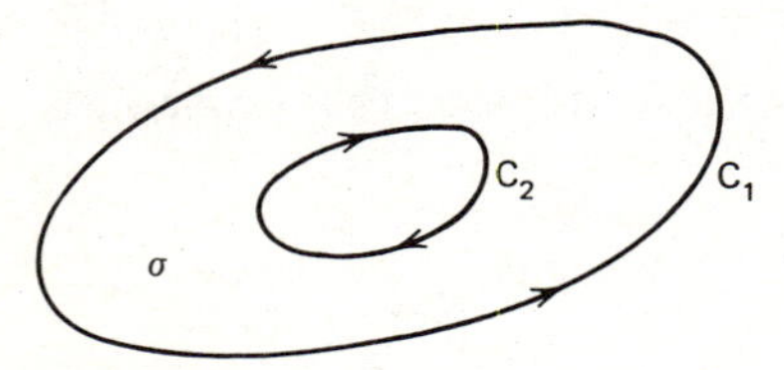

Fig. 9.1

9.7. Prove that

$$\int_{\sigma} d\boldsymbol{\sigma} = \frac{1}{2}\oint_{C} \mathbf{r}\times d\mathbf{s}$$

Hint: Use (9.21) with $\mathbf{a}=\mathbf{r}$. Prove by direct expansion using permutation symbols that $(\hat{\mathbf{n}}\times\nabla)\times\mathbf{r} = -2\hat{\mathbf{n}}$. See Fig. 9.2.

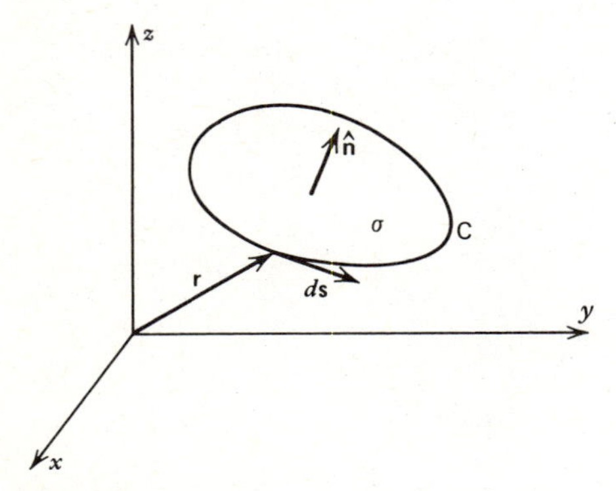

Fig. 9.2

9.8. Let $\mathbf{u}$ be the velocity of a particle in a moving medium. The instantaneous angular velocity of rotation of the particle about an axis through some point P is defined by $\mathbf{u}=\boldsymbol{\omega}\times\mathbf{r}$. Apply the identity (9.4) to show that

$$\nabla\times\mathbf{u} = 2\boldsymbol{\omega} + (\mathbf{r}\cdot\nabla)\boldsymbol{\omega} - \mathbf{r}(\nabla\cdot\boldsymbol{\omega})$$

In the limit as $\mathbf{r}\to 0$, $\boldsymbol{\omega} = \frac{1}{2}(\nabla\times\mathbf{u})$. In this sense, the curl of the linear velocity in a medium such as a flowing fluid is a kind of angular velocity density. See Fig. 9.3.

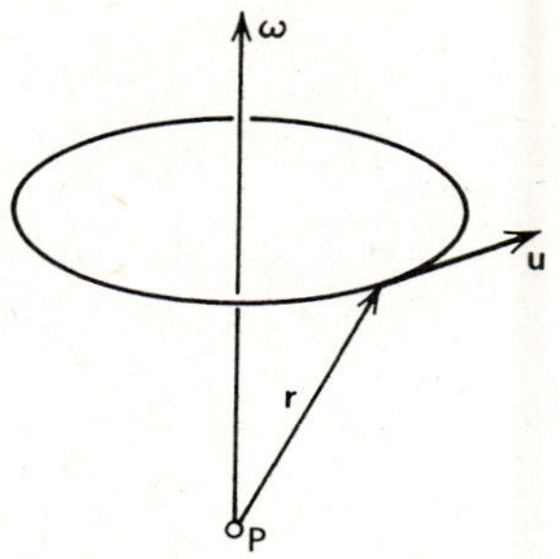

Fig. 9.3

9.9. A region of space in the shape of a toroid of rectangular cross section has an inner radius a and an outer radius b. There is a closely spaced layer of fine wire consisting of N turns and carrying a current I wound so as to form a uniform layer over the entire toroid. Use Ampere's law to find the magnetic field at all points, both inside and outside the toroid. See Fig. 9.4.

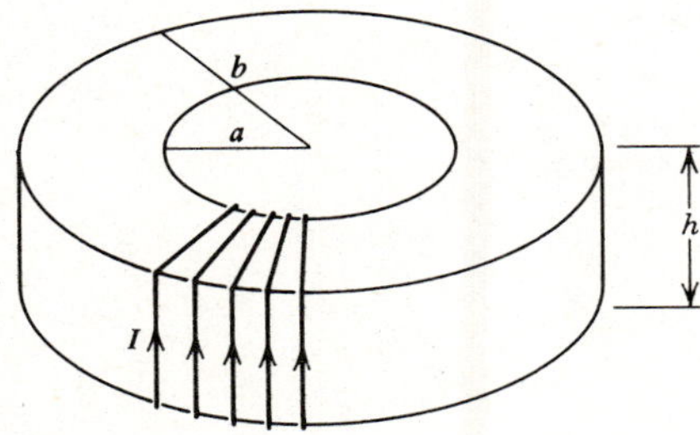

Fig. 9.4

9.10. Prove that $\nabla\psi \times \nabla\phi = \nabla \times (\psi\nabla\phi)$.

9.11. If $\mathbf{c}$ is a constant vector, prove that

$$\int_\sigma \hat{\mathbf{n}} \times (\mathbf{c} \times \mathbf{r})\, d\sigma = 2\mathbf{c}\Sigma$$

where Σ is the volume enclosed by the surface σ.

9.12. Prove that $(\mathbf{a} \times \nabla)\cdot\mathbf{r} = 0$.

9.13. Prove that

$$\int_\Sigma [(\nabla \times \mathbf{a})\cdot(\nabla \times \mathbf{b}) - \mathbf{a}\cdot\nabla \times (\nabla \times \mathbf{b})]\, d\Sigma = \int_\sigma \mathbf{a} \times (\nabla \times \mathbf{b})\cdot d\boldsymbol{\sigma}$$

Hint: Apply Gauss' theorem to the vector $\mathbf{a} \times (\nabla \times \mathbf{b})$.

9.14. Prove that

$$\int_\Sigma [\mathbf{a}\cdot\nabla \times (\nabla \times \mathbf{b}) - \mathbf{b}\cdot\nabla \times (\nabla \times \mathbf{a})]\, d\Sigma = \int_\sigma [\mathbf{b} \times (\nabla \times \mathbf{a}) - \mathbf{a} \times (\nabla \times \mathbf{b})]\cdot d\boldsymbol{\sigma}$$

9.15. Given that $\mathbf{a} = (x^2 - y^2)\hat{\imath} + xy\hat{\jmath}$, evaluate $\oint \mathbf{a}\cdot d\mathbf{s}$ in the counterclockwise sense around the square bounded by the coordinate axes and by the lines $x = c$, $y = c$ in the x, y plane. Evaluate the line integral directly and also by using Stokes' theorem.

9.16. Evaluate $\oint xy\, d\mathbf{s}$ using the path of integration in problem 9.15. Use the result to check the identity (9.20).

10. ELECTROSTATICS

Electromagnetism provides a good vehicle for the development of vector concepts. In this section, many of the results on electrostatics already presented elsewhere in this book will be rederived in a more formal fashion.

It is a direct consequence of Coulomb's law that any electrostatic field can be calculated from a known charge distribution by means of

$$\mathbf{E}(\mathbf{r}) = \frac{1}{4\pi\varepsilon_0}\int_\Sigma \frac{\rho(\mathbf{r}')\mathbf{R}}{R^3}\,d\Sigma' \tag{10.1}$$

if the charge distribution is continuous, and by means of

$$\mathbf{E}(\mathbf{r}) = \frac{1}{4\pi\varepsilon_0}\sum_{\alpha=1}^{N}\frac{q_\alpha}{R_\alpha^3}\mathbf{R}_\alpha \tag{10.2}$$

if the distribution consists of point charges. In equation (10.1), $\mathbf{r}$ is the position vector of the point where the field is to be calculated, called the *field point*, and $\mathbf{r}'$ is the position vector of the variable point of integration, called the *source point*. From Fig. 10.1,

$$\mathbf{R} = \mathbf{r} - \mathbf{r}' \tag{10.3}$$

In (10.1), the integral is extended over all space, but the charge which is producing the field is assumed to be localized in some finite region. The notation $\mathbf{E}(\mathbf{r})$ means that the electric field is a function of x, y, z. Similarly, $\rho(\mathbf{r}')$ means $\rho(x', y', z')$.

Given some function $f(R)$ of the magnitude of $\mathbf{R}$, it is possible to form its gradient either with respect

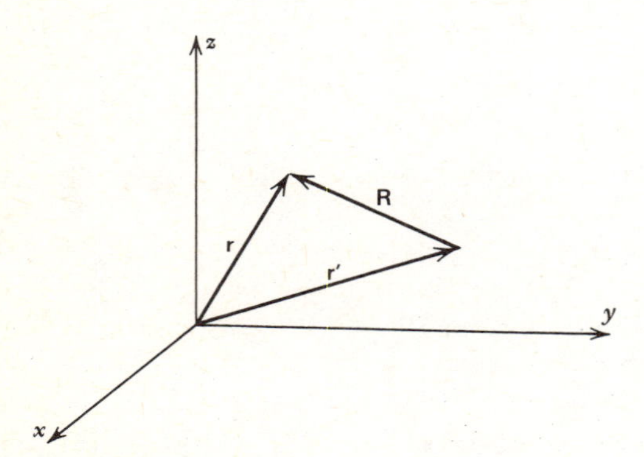

Fig. 10.1

to x, y, z or x', y', z'. By means of

$$R = \sqrt{(x-x')^2 + (y-y')^2 + (z-z')^2} \tag{10.4}$$

it is easy to show that

$$\nabla' f(R) = -\nabla f(R) \tag{10.5}$$

Also needed is the identity

$$\nabla R^n = nR^{n-2}\mathbf{R} = -\nabla' R^n \tag{10.6}$$

In particular,

$$\nabla\frac{1}{R} = -\frac{\mathbf{R}}{R^3} = -\nabla'\frac{1}{R} \tag{10.7}$$

Theorem 1.

$$\nabla\cdot\frac{\mathbf{R}}{R^3} = 0 \tag{10.8}$$

except at the singular point $R = 0$.

Proof: Equation (9.1) gives

$$\nabla\cdot\frac{\mathbf{R}}{R^3} = \frac{1}{R^3}(\nabla\cdot\mathbf{R}) + \mathbf{R}\cdot\nabla\frac{1}{R^3}$$

By means of (10.6) and $\nabla\cdot\mathbf{R} = 3$,

$$\nabla\cdot\frac{\mathbf{R}}{R^3} = \frac{3}{R^3} - \mathbf{R}\cdot\frac{3\mathbf{R}}{R^5} = \frac{3}{R^3} - \frac{3}{R^3} = 0 \qquad R \neq 0$$

The function $\mathbf{R}/R^3$ physically represents the electric field of a point charge of magnitude $4\pi\varepsilon_0$ located at $\mathbf{r}'$. This is the source point of the field, and it is desirable to include it in equation (10.8), even though the function $\mathbf{R}/R^3$ is singular (infinite) at this point. Suppose that a charge of magnitude $q = 4\pi\varepsilon_0$ is spread uniformly throughout the volume of a sphere of radius ε centered at $\mathbf{R} = 0$ as illustrated in Fig. 10.2. In place of (10.8),

$$\mathbf{E} = \frac{\mathbf{R}}{R^3} \qquad \nabla\cdot\mathbf{E} = 0 \qquad R \geq \varepsilon \tag{10.9}$$

$$\mathbf{E} = \frac{\mathbf{R}}{\varepsilon^3} \qquad \nabla\cdot\mathbf{E} = -\frac{3}{\varepsilon^3} \qquad 0 \leq R < \varepsilon \tag{10.10}$$

The electric field is now finite, continuous, and differentiable throughout any region Σ which contains the sphere of radius ε as a subregion. Gauss' theorem,

$$\int_\Sigma \nabla\cdot\mathbf{E}\,d\Sigma = \int_\sigma \mathbf{E}\cdot d\boldsymbol{\sigma} \tag{10.11}$$

applies. Both the volume integral and the surface

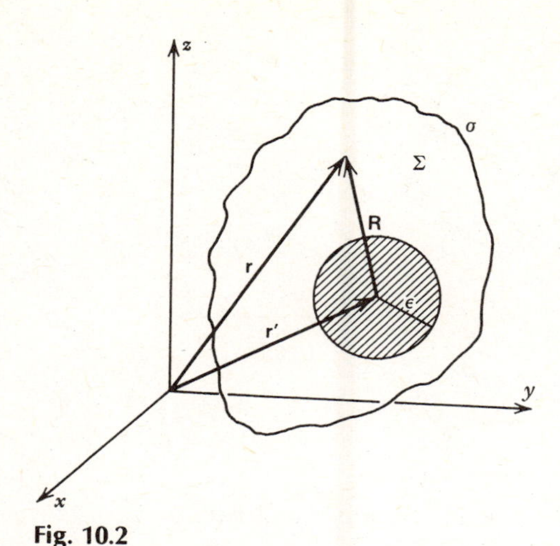

Fig. 10.2

integral in (10.11) are easily evaluated:

$$\int_\Sigma \nabla \cdot \mathbf{E}\, d\Sigma = \int_{R=0}^{\varepsilon} \frac{3}{\varepsilon^3} 4\pi R^2\, dR = 4\pi \tag{10.12}$$

$$\int_\sigma \mathbf{E} \cdot d\boldsymbol{\sigma} = \int_\sigma \frac{\mathbf{R}}{R^3} \cdot d\boldsymbol{\sigma} = 4\pi \tag{10.13}$$

Equations (10.12) and (10.13) are *independent* of the radius of the sphere which contains the charge. Thus, ε can be taken as small as we please so that the charge becomes point-like in the limit $\varepsilon \to 0$. The charge density itself becomes arbitrarily large in such a limit, but this in no way affects (10.12) and (10.13). The differentiations involved in $\nabla \cdot \mathbf{E}$ become undefined in the ordinary sense at $\mathbf{R} = 0$ as the limit is taken. It makes perfectly good sense, however, to take this limit in (10.12) and (10.13):

$$\lim_{\varepsilon \to 0} \int_\Sigma \nabla \cdot \mathbf{E}\, d\Sigma = 4\pi \tag{10.14}$$

$$\lim_{\varepsilon \to 0} \int_\sigma \mathbf{E} \cdot d\boldsymbol{\sigma} = 4\pi \tag{10.15}$$

In this way, Gauss' theorem (10.11) and the related identities (9.14)–(9.17) can be extended to include the singularities that are produced by point charges. We use a notation to indicate that such a limit has been taken and write

$$\nabla \cdot \frac{\mathbf{R}}{R^3} = 4\pi\, \delta(\mathbf{R}) \tag{10.16}$$

The quantity $\delta(\mathbf{R})$ is called a *Dirac delta function* and has the property that

$$\delta(\mathbf{R}) = 0 \qquad \mathbf{R} \neq 0 \tag{10.17}$$

$$\int_\Sigma \delta(\mathbf{R})\, d\Sigma = 1 \tag{10.18}$$

Physically, it can be thought of as representing the charge density of a point charge of unit magnitude. The charge density associated with a point charge of magnitude q is then $\rho(\mathbf{r}) = q\, \delta(\mathbf{R})$, since then $\int_\Sigma \rho(\mathbf{r})\, d\Sigma = q$.

It is possible to write the delta function as a product of one-dimensional delta functions:

$$\delta(\mathbf{R}) = \delta(x - x')\, \delta(y - y')\, \delta(z - z') \tag{10.19}$$

where, for example,

$$\int \delta(x - x')\, dx = 1 \tag{10.20}$$

provided the range of integration includes the point $x = x'$. Another extremely important property of the delta function is

$$\int_\Sigma f(\mathbf{r})\, \delta(\mathbf{R})\, d\Sigma = f(\mathbf{r}') \tag{10.21}$$

where $f(\mathbf{r})$ is any reasonably well-behaved function defined in Σ.

Theorem 2.

$$\nabla^2 \frac{1}{R} = -4\pi\, \delta(\mathbf{R}) \tag{10.22}$$

Proof: The proof is an immediate consequence of (10.7) and (10.16).

Theorem 3. The roles of the source and the field point can be exchanged with the consequence that

$$\nabla' \cdot \frac{\mathbf{R}}{R^3} = -4\pi\, \delta(\mathbf{R}) \tag{10.23}$$

$$\int_\Sigma f(\mathbf{r}')\, \delta(\mathbf{R})\, d\Sigma' = f(\mathbf{r}) \tag{10.24}$$

$$\nabla'^2 \frac{1}{R} = -4\pi\, \delta(\mathbf{R}) \tag{10.25}$$

Proof: In Fig. 10.2, imagine that the point source is situated at $\mathbf{r}$ as a fixed point rather than $\mathbf{r}'$. The variable point is now $\mathbf{r}'$ rather than $\mathbf{r}$. By Gauss' theorem,

$$\int_\Sigma \nabla' \cdot \frac{\mathbf{R}}{R^3}\, d\Sigma' = \int_\sigma \frac{\mathbf{R}}{R^3} \cdot d\boldsymbol{\sigma}' = -4\pi \tag{10.26}$$

The minus sign results because $\mathbf{R}$ now points toward the source rather than away from it. Equation (10.7) is obvious, and (10.25) follows by combining (10.7) and (10.23).

Theorem 4. If an electric field $\mathbf{E}$ is produced by a continuous distribution of charge, then

$$\nabla \cdot \mathbf{E} = \frac{1}{\varepsilon_0}\rho \tag{10.27}$$

Proof: Take the divergence of (10.1) and use (10.16) and (10.24) to get

$$\begin{aligned}\nabla \cdot \mathbf{E} &= \frac{1}{4\pi\varepsilon_0}\int_\Sigma \rho(\mathbf{r}')\, \nabla \cdot \frac{\mathbf{R}}{R^3}\, d\Sigma' \\ &= \frac{1}{4\pi\varepsilon_0}\int_\Sigma \rho(\mathbf{r}') 4\pi\, \delta(\mathbf{R})\, d\Sigma' \\ &= \frac{1}{\varepsilon_0}\rho(\mathbf{r})\end{aligned} \tag{10.28}$$

Theorem 5. A solution of Poisson's equation for a continuous distribution of charge is

$$\psi(\mathbf{r}) = \frac{1}{4\pi\varepsilon_0}\int_\Sigma \rho(\mathbf{r}')\frac{1}{R}\, d\Sigma' \tag{10.29}$$

Proof: Take the Laplacian of (10.29) and use (10.22) and (10.24):

$$\begin{aligned}\nabla^2\psi &= \frac{1}{4\pi\varepsilon_0}\int_\Sigma \rho(\mathbf{r}')\, \nabla^2 \frac{1}{R}\, d\Sigma' \\ &= -\frac{1}{4\pi\varepsilon_0}\int_\Sigma \rho(\mathbf{r}') 4\pi\, \delta(\mathbf{R})\, d\Sigma' \\ &= -\frac{1}{\varepsilon_0}\rho(\mathbf{r})\end{aligned} \tag{10.30}$$

Theorem 6. In a charge-free region of space, the potential at the center of a sphere of radius a is the average value of the potential over its surface:

$$\psi(\mathbf{r}) = \frac{1}{4\pi a^2}\int_\sigma \psi(\mathbf{r}')\, d\sigma' \tag{10.31}$$

Proof: Apply Green's second identity to the spherical region of Fig. 10.3:

$$\int_\Sigma (\phi\nabla'^2\psi - \psi\nabla'^2\phi)\, d\Sigma' = \int_\sigma (\phi\nabla'\psi - \psi\nabla'\phi) \cdot d\boldsymbol{\sigma}' \tag{10.32}$$

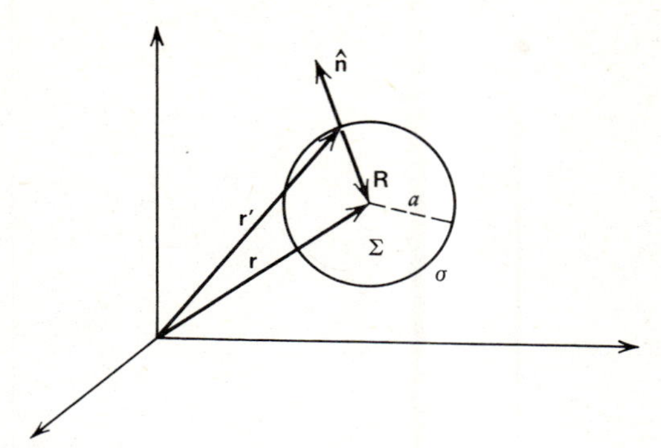

Fig. 10.3

Let $\phi = 1/R$. Since the region is charge-free, $\nabla'^2\psi = 0$:

$$\begin{aligned}\int_\Sigma \psi 4\pi\, \delta(\mathbf{R})\, d\Sigma' &= \int_\sigma \frac{1}{R}\nabla'\psi \cdot d\boldsymbol{\sigma}' - \int_\sigma \psi \frac{\mathbf{R}}{R^3} \cdot d\boldsymbol{\sigma}' \\ &= \frac{1}{a}\int_\sigma \nabla'\psi \cdot d\boldsymbol{\sigma}' + \frac{1}{a^2}\int_\sigma \psi\, d\sigma'\end{aligned} \tag{10.33}$$

Since there is no charge inside the sphere, the net flux of the electric field outward through the surface is zero:

$$\int_\sigma \nabla'\psi \cdot d\boldsymbol{\sigma}' = -\int_\sigma \mathbf{E} \cdot d\boldsymbol{\sigma}' = 0 \tag{10.34}$$

Equation (10.31) now follows by application of (10.24). A consequence of this result is that the electrostatic potential can never have a maximum or a minimum in a charge-free region of space.

Theorem 7.

$$\nabla \times \frac{\mathbf{R}}{R^3} = 0 \tag{10.35}$$

for all points, including the singular point $\mathbf{R} = 0$.

Proof: For $\mathbf{R} \neq 0$, apply (9.1) to get

$$\nabla \times \frac{\mathbf{R}}{R^3} = \left(\nabla \frac{1}{R^3}\right) \times \mathbf{R} + \frac{1}{R^3}(\nabla \times \mathbf{R})$$

By means of (10.6),

$$\nabla \times \frac{\mathbf{R}}{R^3} = -\frac{3(\mathbf{R} \times \mathbf{R})}{R^5} + \frac{\nabla \times \mathbf{R}}{R^3}$$

Since $\mathbf{R} \times \mathbf{R} = 0$ and $\nabla \times \mathbf{R} = 0$, (10.35) follows.

To extend the result to include the point $\mathbf{R} = 0$, apply (9.15) with $\mathbf{a} = \mathbf{R}/R^3$ to the spherical region of radius ε centered at $\mathbf{R} = 0$ as shown in Fig. 10.4:

$$\int_\Sigma \left(\nabla \times \frac{\mathbf{R}}{R^3}\right) d\Sigma = \int_\sigma \left(\hat{\mathbf{n}} \times \frac{\mathbf{R}}{R^3}\right) d\sigma \tag{10.36}$$

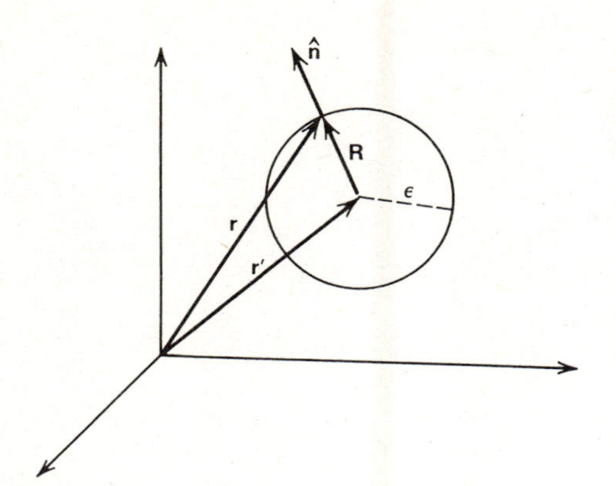

Fig. 10.4

Since $\hat{\mathbf{n}}$ and $\mathbf{R}$ are parallel,

$$\int_\Sigma \left(\nabla \times \frac{\mathbf{R}}{R^3}\right) d\Sigma = 0 \tag{10.37}$$

Equation (10.37) is independent of the value of ε:

$$\lim_{\varepsilon\to 0} \int_\Sigma \left(\nabla \times \frac{\mathbf{R}}{R^3}\right) d\Sigma = 0 \tag{10.38}$$

In this sense, (10.35) can be extended to the singular point $\mathbf{R}=0$. Just as in the case of equation (10.16), equation (10.35) is not valid in the conventional sense at $\mathbf{R}=0$. To put it another way, the divergence of $\mathbf{R}/R^3$ produces a delta function at $\mathbf{R}=0$ whereas the curl of this function does not.

Theorem 8. The curl of the electrostatic field is everywhere zero.

Proof: By taking the curl of (10.1) and using theorem 7,

$$\nabla \times \mathbf{E} = \frac{1}{4\pi\varepsilon_0}\int_\Sigma \rho(\mathbf{r}')\left(\nabla \times \frac{\mathbf{R}}{R^3}\right) d\Sigma' = 0. \tag{10.39}$$

Differentiating under the integral sign is permitted at all points of the range of the variable $\mathbf{r}'$, even at the singular point $\mathbf{R}=0$. The same result follows by taking the curl of (10.2).

Theorem 9. The electrostatic field is uniquely determined in a region Σ if the charge density is known in Σ and if the value of the potential is specified at each point of the closed surface σ which bounds Σ. This is called the *Dirichlet boundary value problem.*

Proof: In Green's first identity (9.16), let $\phi = \psi$:

$$\int_\Sigma (\phi\nabla^2\phi + |\nabla\phi|^2)\, d\Sigma = \int_\sigma \phi\, \nabla\phi \cdot d\boldsymbol{\sigma} \tag{10.40}$$

Suppose $\phi = \psi_1 - \psi_2$, where ψ_1 and ψ_2 are two possible solutions of the potential problem. At interior points,

$$\nabla^2\phi = \nabla^2\psi_1 - \nabla^2\psi_2 = -\frac{\rho}{\varepsilon_0} + \frac{\rho}{\varepsilon_0} = 0 \tag{10.41}$$

Both solutions also have to conform to the same boundary values on σ, meaning that $\phi = 0$ on σ. Equation (10.40) then becomes

$$\int_\Sigma |\nabla\phi|^2\, d\Sigma = 0 \tag{10.42}$$

Since the integrand of (10.42) is positive definite, it can only be zero, meaning that

$$\nabla\phi = \nabla\psi_1 - \nabla\psi_2 = \mathbf{E}_2 - \mathbf{E}_1 = 0 \tag{10.43}$$

which proves the uniqueness.

PROBLEMS

10.1. A function $\Delta(x)$ is defined by

$$\Delta(x) = 0 \qquad x < x_0,\ x > x_0 + \varepsilon$$

$$\Delta(x) = \frac{1}{\varepsilon} \qquad x_0 \le x \le x_0 + \varepsilon$$

and is illustrated in Fig. 10.5. Show that

$$\int_a^b \Delta(x)\, dx = 1 \qquad a < x_0,\ b > x_0 + \varepsilon$$

independently of the value of ε. In this sense,

$$\lim_{\varepsilon\to 0} \Delta(x) = \delta(x - x_0)$$

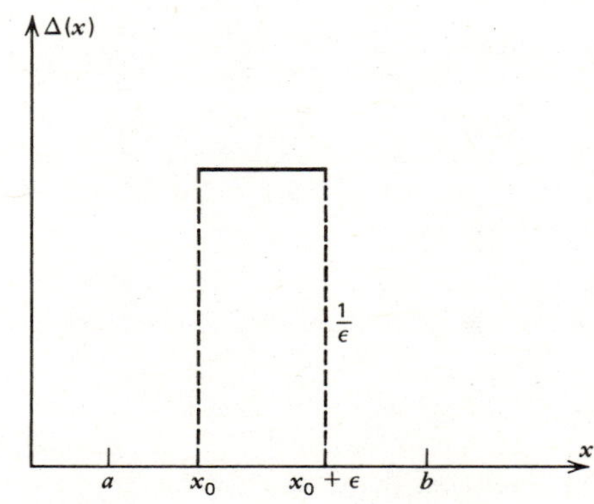

Fig. 10.5

10.2. In problem 10.1, a second function $f(x)$ is continuous in $a<x<b$ and can be expressed as $f(x)=dg(x)/dx$. Evaluate

$$\int_a^b f(x)\,\Delta(x)\,dx$$

and show that

$$\lim_{\varepsilon\to 0}\int_a^b f(x)\,\Delta(x)\,dx=\left(\frac{dg}{dx}\right)_{x_0}=f(x_0)$$

In this sense,

$$\int_a^b f(x)\,\delta(x-x_0)\,dx=f(x_0)$$

10.3. Show that the divergence of the electric field (10.2) due to a distribution of N point charges is

$$\nabla\cdot\mathbf{E}=\frac{1}{\varepsilon_0}\sum_{\alpha=1}^{N} q_\alpha\,\delta(\mathbf{R}_\alpha)$$

10.4. The one-dimensional time-dependent heat or diffusion equation studied in section 6 is of the form

$$\frac{\partial^2\psi}{\partial x^2}-\frac{1}{\alpha}\frac{\partial\psi}{\partial t}=0$$

where α is a constant and ψ can be either the temperature or the density of diffusing particles. We assume here that the source function f is zero. Show by direct substitution that

$$\psi(x,t)=\frac{1}{\sqrt{4\pi\alpha t}}\exp\left[-\frac{(x-x_0)^2}{4\alpha t}\right]$$

is a solution. Graph $\psi(x,t)$ as a function of x for $\alpha t=0.1\ \text{cm}^2$ and $\alpha t=1.0\ \text{cm}^2$. What is the behavior of the function for small values of t? Show that

$$\int_{-\infty}^{+\infty}\psi(x,t)\,dx=1$$

independently of the value of t, provided that $t>0$. The function $\psi(x,t)$ has the characteristics of a delta function centered at $x=x_0$ in the limit $t\to 0$. An approximate physical realization could be obtained by heating a thin metal foil very hot and then sandwiching it between two slabs of material at an initial uniform temperature. The temperature in the material as a function of time is then given approximately by $c\psi(x,t)$, where c is an appropriately chosen constant.

10.5. A charge of magnitude $q=4\pi\varepsilon_0$ is spread uniformly throughout the volume of a sphere of radius ε as shown in Fig. 10.6. The electric field at the center of the sphere is given by

$$\mathbf{E}=\int_\Sigma\frac{\mathbf{R}}{R^3}\,d\Sigma'$$

Use (10.7) and (9.14) to show that $\mathbf{E}=0$.

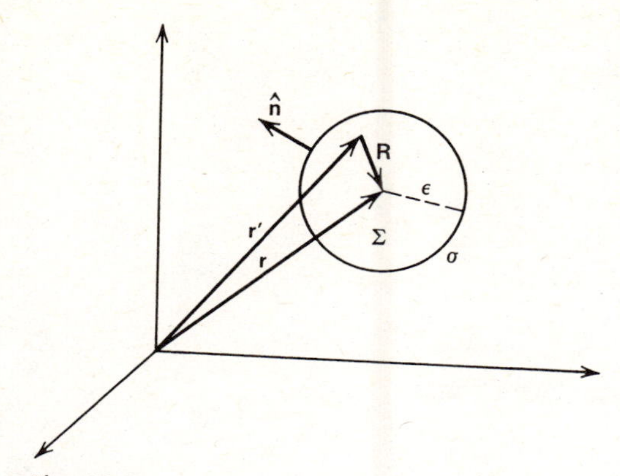

Fig. 10.6

10.6. Prove that the integral (10.1) for the electric field is convergent, that is, no problem results from the singular point $R=0$ in the integrand. Do this by evaluating the contribution to $\mathbf{E}$ from the charge inside a small sphere of radius ε centered at the field point in Fig. 10.1. Use the result of problem 5.

10.7. If σ is any closed surface, it is true that $\int_\sigma d\sigma=$ area. Show however that $\int_\sigma d\boldsymbol{\sigma}=0$.

10.8. A closed surface σ is an equipotential surface. The region Σ enclosed by σ contains no charge. If the value of the potential on σ is V_0, use theorem 9 to show that the potential at each point of Σ is also V_0. This means that the electric field at any point inside Σ is zero. As an example, the surface of a conductor in which no currents are flowing is an equipotential surface and no electric fields exist inside the conductor. Any excess charge placed on the conductor rapidly distributes itself over the surface, leaving no volume distribution of charge. Refer back to problem 5.4.

10.9. An isolated, conducting sphere of radius a is at a potential V_0 with respect to infinity. For practical purposes, this is with respect to the walls of the room considered to be grounded. What is the potential at all points of space both inside and outside the sphere? What is the electric field? Show that the electric field just as it meets the surface of the sphere is given by $E_n=\lambda/\varepsilon_0$, where λ is the surface charge density. What is the relation between λ and V_0?

10.10. Show that a static electric field is always perpendicular to the surface of a conductor of any shape in which no current is flowing and that the magnitude of the field as it meets the conducting surface is $E_n=\lambda/\varepsilon_0$.

Hint: Use a Gaussian surface in the shape of a small box as illustrated in the figure for problem 8.4.

10.11. A function ψ satisfies the boundary condition

$$\psi=\tfrac{1}{2}(x^2+y^2)$$

on the ellipse

$$\left(\frac{x}{a}\right)^2+\left(\frac{y}{b}\right)^2=1$$

and obeys Laplace's equation

$$\nabla^2\psi = \frac{\partial^2\psi}{\partial x^2} + \frac{\partial^2\psi}{\partial y^2} = 0$$

at all points interior to the ellipse. Find the function ψ in the region enclosed by the ellipse.

Hint: Look for a solution of the form

$$\psi = c_1 + c_2 x + c_3 y + c_4 x^2 + c_5 xy + c_6 y^2$$

Eliminate some of the constants by using the obvious symmetry of the problem which requires

$$\psi(x, y) = \psi(-x, y) = \psi(-x, -y) = \psi(x, -y)$$

Other constants can be eliminated by the requirement that ψ obey Laplace's equation. Finally, use the boundary conditions. Theorem 9 guarantees that the solution so obtained is the only one possible.

10.12. Prove that the electrostatic field is uniquely determined in a region Σ if the charge density is known in Σ and if the value of the normal component of the electric field is known everywhere on the surface σ which encloses Σ. This is the *Neumann boundary value problem*.

10.13. Let σ be a surface which encloses a region Σ containing a distribution of charge. Show that at any interior point,

$$\psi(\mathbf{r}) = \frac{1}{4\pi\varepsilon_0}\int_\Sigma \frac{\rho}{R}\,d\Sigma' + \frac{1}{4\pi}\int_\sigma\left(\frac{1}{R}\nabla'\psi - \psi\nabla'\frac{1}{R}\right)\cdot d\boldsymbol{\sigma}' \tag{10.44}$$

Assume that the charge is localized. Let σ be a sphere of radius a. Show that the surface integral becomes zero in the limit $a \to \infty$, so that the fundamental solution (10.29) is recovered.

Hint: At a great distance from the charge distribution, $\psi \propto 1/R$.

10.14. Show directly from (10.1) that $\mathbf{E}(\mathbf{r}) = -\nabla\psi$, where ψ is given by (10.29).

Hint: Use (10.7).

10.15. Construct an alternative proof of theorem 7 by taking the curl of the electric field given by (10.9) and (10.10) and then taking the limit $\varepsilon \to 0$.

11. MAGNETOSTATICS

The plan of this section is to reconstruct many of the results in magnetostatics previously obtained starting this time with the Biot–Savart law, which states that the magnetic field produced by a steady-state current I flowing in a thin wire so as to form a loop is

$$\mathbf{B}(\mathbf{r}) = \frac{\mu_0 I}{4\pi}\oint_C \frac{d\mathbf{s}' \times \mathbf{R}}{R^3} \tag{11.1}$$

As in the discussion of electrostatics, $\mathbf{r}'$ is the position vector of the source point or variable point of integration; $\mathbf{r}$ is the point at which the field is to be calculated, called the field point; and $\mathbf{R} = \mathbf{r} - \mathbf{r}'$. See Fig. 11.1. Note that the line element $d\mathbf{s}'$ is in the same direction as the current. All steady-state currents flow in closed loops. When we talk about the magnetic field produced by a current flowing in an infinitely long, straight wire, we really mean the magnetic field in the vicinity of a long, straight portion of a current loop, such as the point P of Fig. 11.2. The nonstraight portion, which provides the return for the current, is assumed to be more distant than the straight portion so that the contribution it makes to the field at P can be neglected.

As an example, suppose that a current I flows in a long, straight wire along the x axis of Fig. 11.3. We will calculate the contribution to the magnetic field at the point P in the x, y plane from the portion of the wire between $x = -s$ and $x = +s$. Since $\mathbf{R} = (x - x')\hat{\mathbf{i}} + y\hat{\mathbf{j}}$ and $d\mathbf{s}' = \hat{\mathbf{i}}\, dx'$, (11.1) gives

$$\mathbf{B}(x, y) = \frac{\mu_0 I}{4\pi} \int_{x'=-s}^{+s} \frac{\hat{\mathbf{k}}\, y\, dx'}{\left[(x - x')^2 + y^2\right]^{3/2}}$$

$$= -\frac{\mu_0 I \hat{\mathbf{k}}}{4\pi y} \left. \frac{x - x'}{\sqrt{(x - x')^2 + y^2}} \right|_{x'=-s}^{+s}$$

$$= \frac{\mu_0 I \hat{\mathbf{k}}}{4\pi y} \left[\frac{x + s}{\sqrt{(x + s)^2 + y^2}} - \frac{x - s}{\sqrt{(x - s)^2 + y^2}} \right] \tag{11.2}$$

In the limit $s \to \infty$,

$$\mathbf{B} = \frac{\mu_0 I \hat{\mathbf{k}}}{2\pi y} \tag{11.3}$$

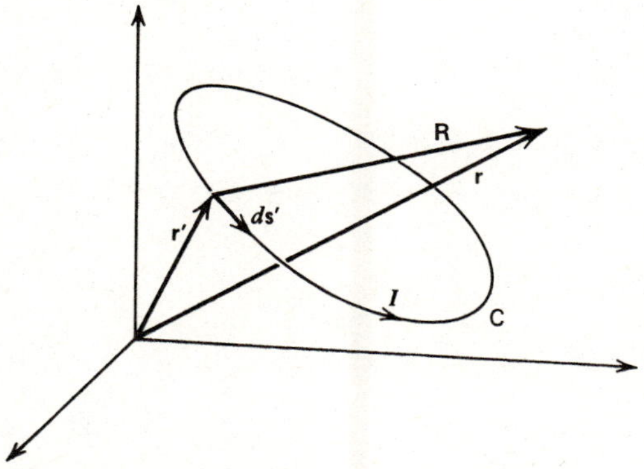

Fig. 11.1

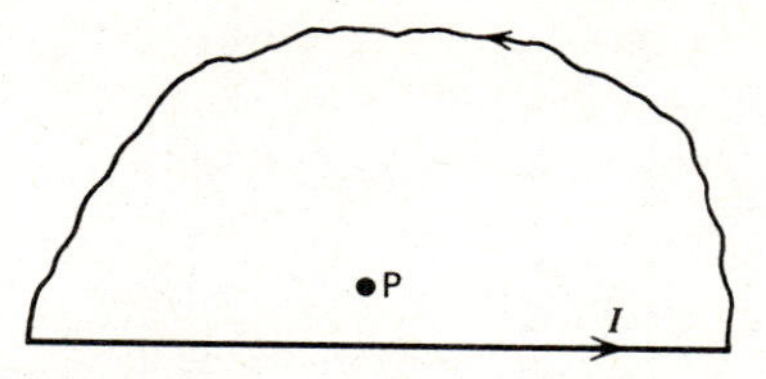

Fig. 11.2

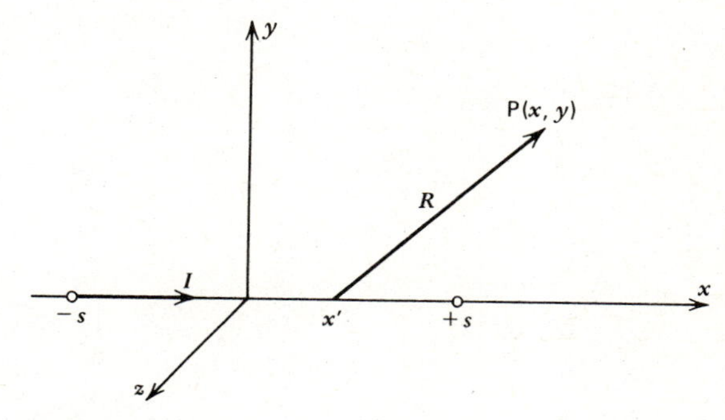

Fig. 11.3

in agreement with (8.11).

In reality, currents do not exist in wires which are mathematical lines with no cross section. Figure 11.4 shows a wire of cross-sectional area σ carrying a current I. The current density is $j = I/\sigma$. If ds' is the length of an element of the wire, $I\, ds' = j\sigma\, ds' = j\, d\Sigma'$. This shows that if current is flowing in some arbitrary way in a medium, we must replace $I\, d\mathbf{s}'$ in (11.1) by $\mathbf{j}\, d\Sigma'$. The Biot–Savart law then reads

$$\mathbf{B}(\mathbf{r}) = \frac{\mu_0}{4\pi} \int_\Sigma \frac{\mathbf{j}(\mathbf{r}') \times \mathbf{R}}{R^3}\, d\Sigma' \tag{11.4}$$

where the integral is now taken over the volume where the current exists. Since there is no time dependence, the equation of continuity (5.14) gives

$$\nabla \cdot \mathbf{j}(\mathbf{r}) = 0 \tag{11.5}$$

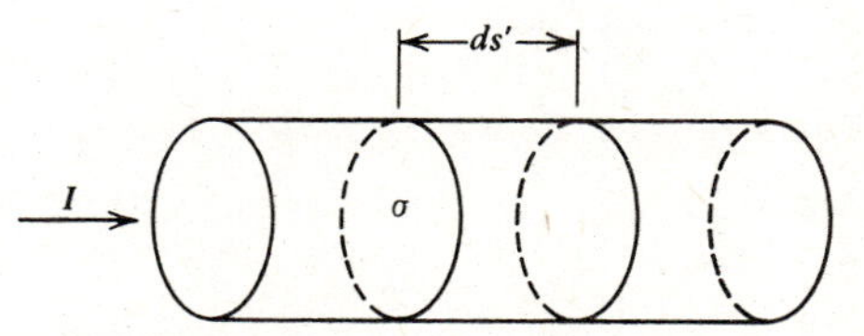

Fig. 11.4

This means that the current density vector $\mathbf{j}$ (like the static magnetic field) is solenoidal. This is another way of saying that steady-state currents always exist in closed loops.

Theorem 1. It is possible to express the magnetic field as

$$\mathbf{B} = \nabla \times \mathbf{A} \tag{11.6}$$

The vector $\mathbf{A}$ is called the *vector potential.*

Proof: Use (10.7) to express (11.4) as

$$\mathbf{B}(\mathbf{r}) = -\frac{\mu_0}{4\pi}\int_\Sigma \mathbf{j}(\mathbf{r}') \times \nabla \frac{1}{R}\, d\Sigma' \tag{11.7}$$

By means of (9.2),

$$\nabla \times \frac{1}{R}\mathbf{j}(\mathbf{r}') = \frac{1}{R}\nabla \times \mathbf{j}(\mathbf{r}') + \nabla\frac{1}{R} \times \mathbf{j}(\mathbf{r}') \tag{11.8}$$

The current density is expressed as a function of $\mathbf{r}'$. Since the differentiations in (11.8) are with respect to $\mathbf{r}$, $\nabla \times \mathbf{j}(\mathbf{r}') = 0$. If we substitute (11.8) into (11.7), we get

$$\mathbf{B} = \frac{\mu_0}{4\pi}\int_\Sigma \nabla \times \frac{1}{R}\mathbf{j}(\mathbf{r}')\, d\Sigma' = \nabla \times \frac{\mu_0}{4\pi}\int_\Sigma \frac{\mathbf{j}(\mathbf{r}')}{R}\, d\Sigma' \tag{11.9}$$

which proves the theorem and shows that a possible vector potential is

$$\mathbf{A} = \frac{\mu_0}{4\pi}\int_\Sigma \frac{\mathbf{j}(\mathbf{r}')}{R}\, d\Sigma' \tag{11.10}$$

The vector potential (11.10) is not unique. If $\mathbf{A}' = \mathbf{A} + \mathbf{a}$, then

$$\mathbf{B} = \nabla \times \mathbf{A} = \nabla \times \mathbf{A}' \tag{11.11}$$

provided that $\nabla \times \mathbf{a} = 0$. Since $\mathbf{a}$ is then essentially conservative, it can be expressed as the gradient of a scalar function. Therefore, any vector of the form

$$\mathbf{A}' = \mathbf{A} + \nabla\phi \tag{11.12}$$

is also a possible vector potential.

Theorem 2. The divergence of the magnetic field is zero.

Proof: The result is an immediate consequence of (11.6) and (9.7).

Theorem 3. A vector potential $\mathbf{A}$ can be chosen such that

$$\nabla \cdot \mathbf{A} = 0 \tag{11.13}$$

Proof: The vector potential given by (11.10) satisfies (11.13). By means of (9.1),

$$\begin{aligned}\nabla \cdot \mathbf{A} &= \frac{\mu_0}{4\pi}\int_\Sigma \left[\frac{1}{R}\nabla \cdot \mathbf{j}(\mathbf{r}') + \mathbf{j}(\mathbf{r}') \cdot \nabla \frac{1}{R}\right] d\Sigma' \\ &= \frac{\mu_0}{4\pi}\int_\Sigma \mathbf{j}(\mathbf{r}') \cdot \nabla \frac{1}{R}\, d\Sigma' \\ &= -\frac{\mu_0}{4\pi}\int_\Sigma \mathbf{j}(\mathbf{r}') \cdot \nabla' \frac{1}{R}\, d\Sigma' \end{aligned} \tag{11.14}$$

where $\nabla \cdot \mathbf{j}(\mathbf{r}') = 0$ because $\mathbf{j}$ is a function of $\mathbf{r}'$. We have also used (10.7). Now use (9.1) again, but this time differentiate with respect to $\mathbf{r}'$:

$$\nabla' \cdot \frac{1}{R}\mathbf{j}(\mathbf{r}') = \frac{1}{R}\nabla' \cdot \mathbf{j}(\mathbf{r}') + \mathbf{j}(\mathbf{r}') \cdot \nabla' \frac{1}{R} = \mathbf{j}(\mathbf{r}') \cdot \nabla' \frac{1}{R} \tag{11.15}$$

This time, $\nabla' \cdot \mathbf{j}(\mathbf{r}') = 0$ on account of (11.5). Equation (11.14) becomes

$$\nabla \cdot \mathbf{A} = -\frac{\mu_0}{4\pi}\int_\Sigma \nabla' \cdot \frac{1}{R}\mathbf{j}(\mathbf{r}')\, d\Sigma' \tag{11.16}$$

Now use Gauss' theorem (9.13) to get

$$\nabla \cdot \mathbf{A} = -\frac{\mu_0}{4\pi}\int_\sigma \frac{1}{R}\mathbf{j}(\mathbf{r}') \cdot d\boldsymbol{\sigma}' \tag{11.17}$$

The currents which are the sources of the magnetic field are assumed to be localized. The surface over which the integral in (11.17) is taken can be a sphere of sufficiently large radius so that $\mathbf{j}$ is everywhere zero on its surface. Equation (11.13) then follows.

Vector potentials can be chosen which do not obey (11.13). If ϕ in (11.12) is chosen such that $\nabla^2\phi \neq 0$, then $\nabla \cdot \mathbf{A}' \neq 0$.

Theorem 4. The curl of the magnetic field is given by

$$\nabla \times \mathbf{B} = \mu_0 \mathbf{j} \tag{11.18}$$

Proof: Take the curl of (11.6) and use (9.8):

$$\nabla \times \mathbf{B} = \nabla \times (\nabla \times \mathbf{A}) = \nabla(\nabla \cdot \mathbf{A}) - \nabla^2\mathbf{A} \tag{11.19}$$

If (11.10) is chosen as the vector potential, $\nabla \cdot \mathbf{A} = 0$ and

$$\begin{aligned}\nabla^2\mathbf{A} &= \frac{\mu_0}{4\pi}\int_\Sigma \mathbf{j}(\mathbf{r}')\nabla^2 \frac{1}{R}\, d\Sigma' \\ &= -\frac{\mu_0}{4\pi}\int_\Sigma \mathbf{j}(\mathbf{r}')4\pi\delta(\mathbf{R})\, d\Sigma' \\ &= -\mu_0\mathbf{j}(\mathbf{r})\end{aligned} \tag{11.20}$$

where use has been made of (10.21) and (10.22). Equation (11.18) then follows. Note from (11.20) that the components of **A** obey Poisson equations:

$$\nabla^2 A_x = -\mu_0 j_x \qquad \nabla^2 A_y = -\mu_0 j_y \qquad \nabla^2 A_z = -\mu_0 j_z \tag{11.21}$$

Theorem 5. If $\nabla \cdot \mathbf{B} = 0$, then there exists a vector potential **A** such that $\mathbf{B} = \nabla \times \mathbf{A}$. This is essentially the converse of theorem 2.

Proof: Assume that $\nabla \cdot \mathbf{B} = 0$ and $\nabla \times \mathbf{B} = \mu_0 \mathbf{j}$. (If both $\nabla \cdot \mathbf{B} = 0$ and $\nabla \times \mathbf{B} = 0$ everywhere, then **B** has no scalar or vector sources, $\mathbf{B} = 0$, and the problem is trivial). By means of (9.8),

$$\nabla \times (\nabla \times \mathbf{B}) = \nabla(\nabla \cdot \mathbf{B}) - \nabla^2 \mathbf{B} = -\nabla^2 \mathbf{B} = \mu_0 \nabla \times \mathbf{j} \tag{11.22}$$

This is a Poisson equation and has the solution

$$\mathbf{B}(\mathbf{r}) = \frac{\mu_0}{4\pi} \int_\Sigma \frac{\nabla' \times \mathbf{j}(\mathbf{r}')}{R} \, d\Sigma' \tag{11.23}$$

By means of (9.2) and (10.7),

$$\begin{aligned} \nabla' \times \frac{\mathbf{j}(\mathbf{r}')}{R} &= \frac{1}{R} \nabla' \times \mathbf{j}(\mathbf{r}') + \nabla' \frac{1}{R} \times \mathbf{j}(\mathbf{r}') \\ &= \frac{1}{R} \nabla' \times \mathbf{j}(\mathbf{r}') - \nabla \frac{1}{R} \times \mathbf{j}(\mathbf{r}') \end{aligned} \tag{11.24}$$

Equation (11.23) becomes

$$\mathbf{B}(\mathbf{r}) = \frac{\mu_0}{4\pi} \int_\Sigma \left[\nabla' \times \frac{\mathbf{j}(\mathbf{r}')}{R} + \nabla \frac{1}{R} \times \mathbf{j}(\mathbf{r}') \right] d\Sigma' \tag{11.25}$$

Now use (9.15) to get

$$\int_\Sigma \left[\nabla' \times \frac{\mathbf{j}(\mathbf{r}')}{R} \right] d\Sigma' = \int_\sigma \left[\hat{\mathbf{n}} \times \frac{\mathbf{j}(\mathbf{r}')}{R} \right] d\sigma' = 0 \tag{11.26}$$

The argument for setting the surface integral equal to zero is that the surface over which the integral is taken can be a sphere of sufficiently large radius that $\mathbf{j} = 0$ everywhere on its surface. This assumes that the sources, whatever they are, are localized. Now use (9.2) again, but this time differentiate with respect to **r**:

$$\nabla \times \frac{\mathbf{j}(\mathbf{r}')}{R} = \frac{1}{R} \nabla \times \mathbf{j}(\mathbf{r}') + \nabla \frac{1}{R} \times \mathbf{j}(\mathbf{r}') \tag{11.27}$$

In this case, $\nabla \times \mathbf{j}(\mathbf{r}') = 0$ because **j** is a function of $\mathbf{r}'$. Equation (11.25) is now

$$\mathbf{B}(\mathbf{r}) = \frac{\mu_0}{4\pi} \int_\Sigma \nabla \times \frac{\mathbf{j}(\mathbf{r}')}{R} \, d\Sigma' = \nabla \times \frac{\mu_0}{4\pi} \int_\Sigma \frac{\mathbf{j}(\mathbf{r}')}{R} \, d\Sigma' \tag{11.28}$$

which proves the result. We can now state that a vector potential **A** such that $\mathbf{B} = \nabla \times \mathbf{A}$ exists if, and only if, $\nabla \cdot \mathbf{B} = 0$.

Theorem 6. Ampere's Law. If σ is an open surface bounded by a closed curve C, then

$$\oint_C \mathbf{B} \cdot d\mathbf{s} = \mu_0 \int_\sigma \mathbf{j} \cdot d\boldsymbol{\sigma} \tag{11.29}$$

Proof: Integrate (11.18) over the surface and use Stokes' theorem (9.19).

Theorem 7. Uniqueness theorem for static magnetic fields. If the current density **j** is given in a region Σ and if the tangential component of the magnetic field is known over the surface σ which bounds Σ, then the magnetic field is uniquely determined at all points of Σ.

Proof: In the identity of problem (9.13), let $\mathbf{a} = \mathbf{b} = \mathbf{A}$:

$$\begin{aligned} &\int_\Sigma \left[|\nabla \times \mathbf{A}|^2 - \mathbf{A} \cdot \nabla \times (\nabla \times \mathbf{A}) \right] d\Sigma \\ &= \int_\sigma \mathbf{A} \times (\nabla \times \mathbf{A}) \cdot d\boldsymbol{\sigma} \end{aligned} \tag{11.30}$$

Let $\mathbf{B}_1$ and $\mathbf{B}_2$ be two possible solutions and let $\mathbf{A}_1$ and $\mathbf{A}_2$ be the corresponding vector potentials. Both $\mathbf{B}_1$ and $\mathbf{B}_2$ must obey $\nabla \times \mathbf{B}_1 = \nabla \times \mathbf{B}_2 = \mu_0 \mathbf{j}$ at all points of Σ. Let $\mathbf{B} = \mathbf{B}_1 - \mathbf{B}_2$ and $\mathbf{A} = \mathbf{A}_1 - \mathbf{A}_2$. Then, $\nabla \times \mathbf{A} = \mathbf{B}$ and $\nabla \times (\nabla \times \mathbf{A}) = 0$. Equation (11.30) is then

$$\int_\Sigma |\mathbf{B}|^2 \, d\Sigma = \int_\sigma (\mathbf{A} \times \mathbf{B}) \cdot \hat{\mathbf{n}} \, d\sigma = \int_\sigma \mathbf{A} \cdot (\mathbf{B} \times \hat{\mathbf{n}}) \, d\sigma \tag{11.31}$$

The magnitude of $\mathbf{B} \times \hat{\mathbf{n}}$ is

$$\begin{aligned} B \sin\theta &= B_1 \sin\theta - B_2 \sin\theta \\ &= B_{1t} - B_{2t} = 0 \end{aligned} \tag{11.32}$$

where, as is made evident by Fig. 11.5, B_{1t} and B_{2t} are the tangential components of $\mathbf{B}_1$ and $\mathbf{B}_2$. They are equal because both the solutions $\mathbf{B}_1$ and $\mathbf{B}_2$ are required to obey the same boundary conditions. Equation (11.31) becomes

$$\int_\Sigma |\mathbf{B}|^2 \, d\Sigma = 0 \tag{11.33}$$

which implies $\mathbf{B} = \mathbf{B}_1 - \mathbf{B}_2 = 0$. The uniqueness is therefore established.

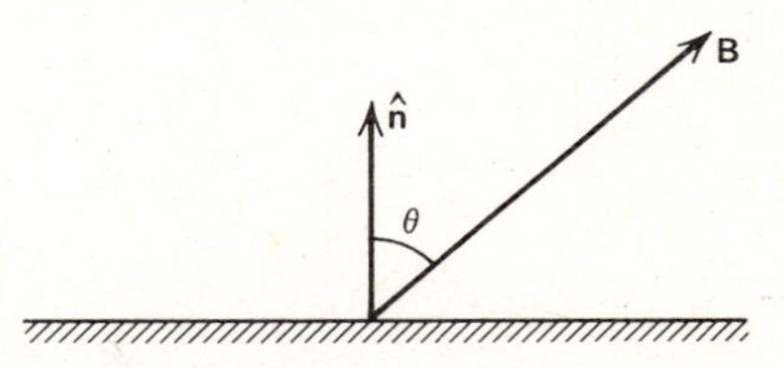

Fig. 11.5

PROBLEMS

11.1. In the proof of theorem 5, it was stated that if a vector field has zero divergence and zero curl everywhere, then the vector is zero. There is a little more to the story. Boundary conditions of some kind are implied. As a minimal requirement, we can assume that the vector field vanishes at infinity. Use either theorem 9 of section 10 or the result of problem 10.12 to show that if $\nabla \cdot \mathbf{B} = 0$ and $\nabla \times \mathbf{B} = 0$ everywhere and $\mathbf{B} = 0$ at infinity, then $\mathbf{B} = 0$ everywhere.

11.2. A current I flows in a circular loop of radius a. Find the magnetic field at any point on the axis by the following steps:

(a) Show that

$$d\mathbf{s}' \times \mathbf{R} = -\hat{\mathbf{j}} z\, dx' + z\hat{\mathbf{i}}\, dy' + \hat{\mathbf{k}}(x'\, dy' - y'\, dx')$$

(b) If $x' = a\cos\theta$ and $y' = a\sin\theta$, show that $x'\, dy' - y'\, dx' = a^2\, d\theta$. See Fig. 11.6.

(c) Use the Biot–Savart law to show that

$$\mathbf{B}(z) = \frac{\mu_0 I a^2 \hat{\mathbf{k}}}{2(a^2 + z^2)^{3/2}}$$

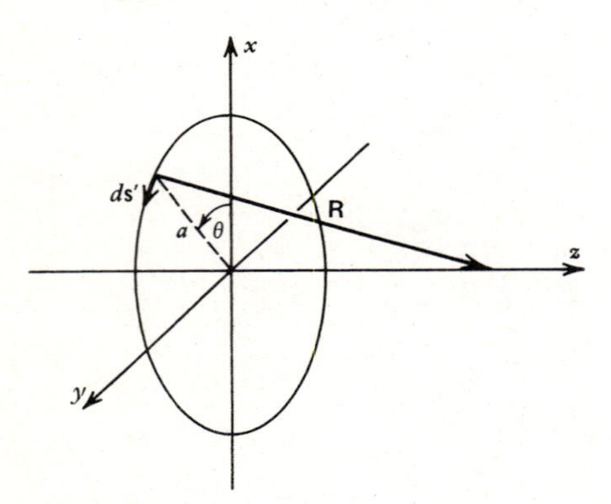

Fig. 11.6

11.3. A current I flows in a closed loop lying in the x, y plane which consists of two quarter circles of radii a and b with centers at the origin and two straight portions along the x and y axes. What is the magnetic field vector at the origin? See Fig. 11.7.

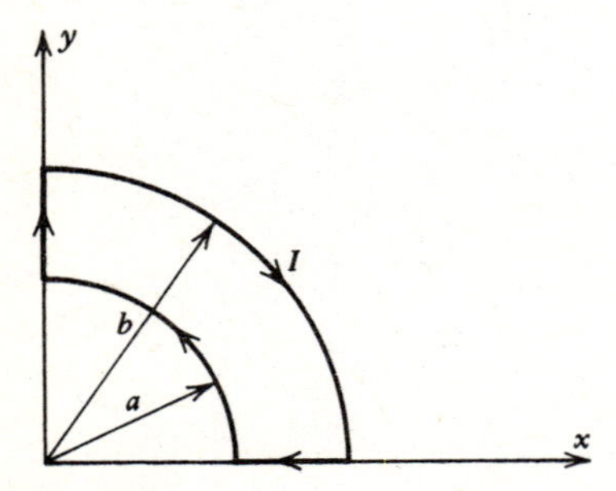

Fig. 11.7

11.4. A current loop is in a region of space where there is an externally applied uniform magnetic field. The force on an element $d\mathbf{s}$ of the loop is $d\mathbf{F} = I\, d\mathbf{s} \times \mathbf{B}$. The torque on the element $d\mathbf{s}$ about the origin is $d\boldsymbol{\tau} = \mathbf{r} \times d\mathbf{F}$. The net torque on the current loop is

$$\boldsymbol{\tau} = I \oint_C \mathbf{r} \times (d\mathbf{s} \times \mathbf{B})$$

Show that the torque can be expressed as $\boldsymbol{\tau} = \boldsymbol{\mu} \times \mathbf{B}$, where $\boldsymbol{\mu}$ is the *magnetic moment* of the loop given by

$$\boldsymbol{\mu} = I \int_\sigma d\boldsymbol{\sigma} = \tfrac{1}{2} I \oint_C \mathbf{r} \times d\mathbf{s}$$

Hint: First expand $\mathbf{r} \times (d\mathbf{s} \times \mathbf{B})$ by the bac-cab rule. Show that $\oint \mathbf{r} \cdot d\mathbf{s} = 0$. Use the identity (9.20). Refer to problem 9.7. See Fig. 11.8.

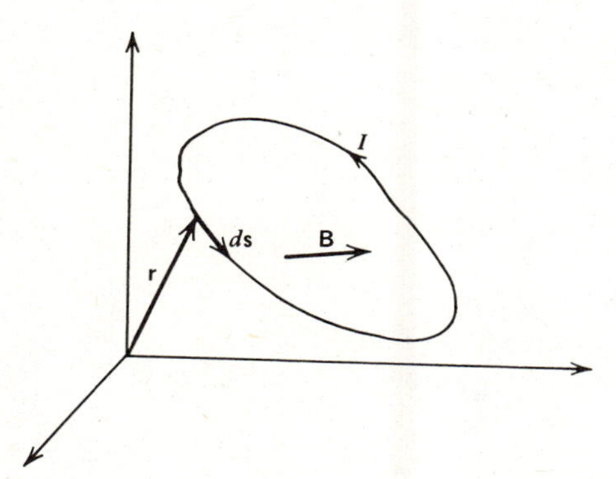

Fig. 11.8

11.5. A uniform magnetic field exists in the z direction. Find a vector potential which satisfies the condition $\nabla \cdot \mathbf{A} = 0$.

Hint: Assume $A_z = 0$.

11.6. Show that the pair of equations $\mathbf{A} = \frac{1}{2}(\mathbf{B} \times \mathbf{r})$, $\mathbf{B} = \nabla \times \mathbf{A}$ is satisfied by a constant magnetic field $\mathbf{B}$ in any orientation.

11.7. A current I flows in a thin wire along the z axis. Use (11.10) to find the vector potential at any point in the x, y plane due to the portion of the current which flows from $-s$ to $+s$. If s is very large, show that the vector potential is

$$\mathbf{A} = -\frac{\mu_0 I \hat{\mathbf{k}}}{4\pi} \ln(x^2 + y^2) + F(s)\hat{\mathbf{k}}$$

where $F(s)$ depends on s and becomes arbitrarily large as $s \to \infty$. Show however that $\nabla \times \mathbf{A}$ gives the correct magnetic field for a current flowing in a long, straight wire.

11.8. The energy stored in a magnetic field can be shown to be

$$W = \frac{1}{2\mu_0} \int_\Sigma |\mathbf{B}|^2 \, d\Sigma$$

where the integral is extended over all space. Show that

$$W = \tfrac{1}{2} \int_\Sigma \mathbf{A} \cdot \mathbf{j} \, d\Sigma$$

Hint: Write $|\mathbf{B}|^2 = \mathbf{B} \cdot \mathbf{B} = \mathbf{B} \cdot \nabla \times \mathbf{A}$. Use (9.3). Convert one of the terms in the resulting expression into a surface integral and make an argument for setting it equal to zero.

11.9. Since the vector potential is not unique, it appears from the expression given in problem 8 that different choices of the vector potential would give different values for the total energy. Show that this is not the case by demonstrating that

$$\int_{\Sigma} \nabla\phi \cdot \mathbf{j}\, d\Sigma = 0$$

for any scalar function ϕ.

11.10. Establish the equivalence of the following two ways of expressing the total energy stored in an electrostatic field:

$$W = \tfrac{1}{2}\varepsilon_0 \int_{\Sigma} |\mathbf{E}|^2\, d\Sigma = \tfrac{1}{2} \int_{\Sigma} \psi\rho\, d\Sigma$$

11.11. There are situations where a vector field has both nonzero curl and nonzero divergence: $\nabla \cdot \mathbf{a} = \rho$, and $\nabla \times \mathbf{a} = \mathbf{j}$. Prove *Helmholtz's theorem*, which states that such a vector can always be split into two parts, one of which is solenoidal (zero divergence) and the other of which is irrotational (zero curl). Construct the proof as follows: Let $\mathbf{a}$ be the given vector and let $\mathbf{b}$ be a solution of $\nabla^2\mathbf{b} = -\mathbf{a}$. Write out the identity for $\nabla \times (\nabla \times \mathbf{b})$ and show that $\mathbf{a}$ can be expressed in the form $\mathbf{a} = \nabla \times \mathbf{A} - \nabla\psi$, where $\mathbf{A}$ and ψ are suitably defined in terms of $\mathbf{b}$. Show moreover that $\nabla^2\psi = -\rho$ and $\nabla^2\mathbf{A} = -\mathbf{j}$.

11.12. Prove that a static vector field is uniquely specified in a region Σ if both its divergence and its curl are known in Σ and if either the normal or the tangential component of the field is known over the surface σ which bounds Σ.

Hint: Let $\mathbf{a}_1$ and $\mathbf{a}_2$ be two possible solutions. Let $\mathbf{a} = \mathbf{a}_1 - \mathbf{a}_2$. Then both $\nabla \cdot \mathbf{a} = 0$ and $\nabla \times \mathbf{a} = 0$. We can write either $\mathbf{a} = -\nabla\psi$ or $\mathbf{a} = \nabla \times \mathbf{A}$ and proceed as in problem 10.12 or theorem 7.

11.13. A steady current flows in a cylindrical conductor of length L and cross-sectional area A. The ends are at $x = 0$ and $x = L$ and are maintained at potentials $\psi = V_1$ and $\psi = V_2$. Show that there is a uniform electric field parallel to the axis of the cylinder of strength $E = (V_2 - V_1)/L$.

Hint: The relation between current density and electric field in the conductor can be expressed as $\mathbf{j} = \sigma\mathbf{E}$, where σ is a constant called the conductivity (not to be confused with area). Since $\mathbf{j}$ must be parallel to the sides of the conductor, $E_n = -\partial\psi/\partial n = 0$. On the ends, the potential has the known values V_1 and V_2. This is actually a mixed boundary value problem in which the potential is specified over part of the surface and its normal derivative is specified over the remainder. Find a solution of $\nabla^2\psi = 0$ which satisfies these boundary conditions.

12. MAXWELL'S EQUATIONS

It is an experimental fact that a loop of conducting wire has an electromotive force generated in it if there is a magnetic flux through it which varies with time. Quantitatively, the electromotive force is given by *Faraday's law*, which states that

$$\mathscr{E} = -\frac{d\Phi_B}{dt} \tag{12.1}$$

In (12.1), Φ_B is the flux of the magnetic field through the current loop. The electromotive force can be expressed in terms of the electric field that exists in the conductor:

$$\mathscr{E} = \oint_C \mathbf{E} \cdot d\mathbf{s} = -\frac{d}{dt}\int_{\sigma} \mathbf{B} \cdot d\boldsymbol{\sigma} = -\int_{\sigma} \frac{\partial \mathbf{B}}{\partial t} \cdot d\boldsymbol{\sigma} \tag{12.2}$$

In differentiating the flux in (12.2) with respect to the time, we have assumed that the current loop itself is not moving so that the flux changes only because the magnetic field itself varies with time. If in Fig. 12.1, **B**, and hence Φ_B, is increasing with time, the induced current flows in the direction indicated, which is opposite to the direction of $\oint \mathbf{E}\cdot d\mathbf{s}$ according to the right-hand rule if $\hat{\mathbf{n}}$ is as shown in Fig. 12.1. This is the reason for the negative signs in equations (12.1) and (12.2). Equation (12.2) is a statement about the time-dependent electromagnetic field, and as such it is true even if the path of integration does not coincide with an actual conductor. The conductor is only necessary if we want to measure the effect of the electric field which is created by the changing magnetic field. One immediate conclusion to be drawn from Faraday's law is that time-dependent electric fields are not conservative. By means of Stokes' theorem, (12.2) can be expressed as

$$\int_\sigma \nabla \times \mathbf{E}\cdot d\boldsymbol{\sigma} = -\int_\sigma \frac{\partial \mathbf{B}}{\partial t}\cdot d\boldsymbol{\sigma} \tag{12.3}$$

which must hold for an arbitrary open surface bounded by a closed curve. Therefore,

$$\nabla \times \mathbf{E} + \frac{\partial \mathbf{B}}{\partial t} = 0 \tag{12.4}$$

which can be thought of as the differential equation version of Faraday's law.

As we know, static magnetic fields obey

$$\nabla \times \mathbf{B} = \mu_0 \mathbf{j} \tag{12.5}$$

Since $\nabla\cdot\nabla\times\mathbf{B} = 0$, equation (12.5) implies that $\nabla\cdot\mathbf{j} = 0$. This presents a problem because in the time-dependent case, the charge and current density obey

$$\nabla\cdot\mathbf{j} + \frac{\partial \rho}{\partial t} = 0 \tag{12.6}$$

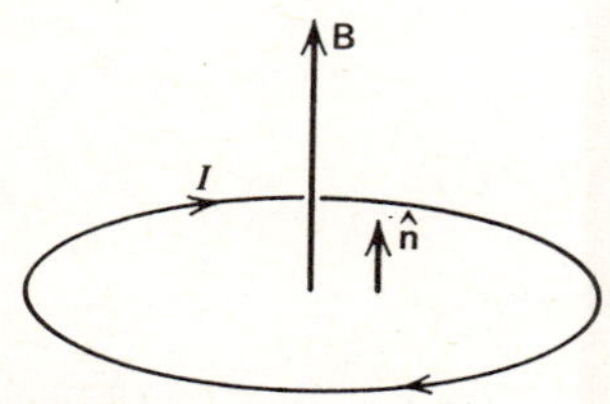

Fig. 12.1

The situation is remedied by adding another term to (12.5) so that it reads

$$\nabla \times \mathbf{B} - \mu_0\varepsilon_0 \frac{\partial \mathbf{E}}{\partial t} = \mu_0 \mathbf{j} \tag{12.7}$$

If we take the divergence of (12.7), equation (12.6) is recovered and all is well. The new term in (12.7) is called *Maxwell's displacement current*. The ultimate validity of (12.7) must of course rest on the experimental verification of the results that it predicts.

This is a propitious moment to summarize the electromagnetic field equations which we have so far obtained:

$$\nabla\cdot\mathbf{E} = \frac{1}{\varepsilon_0}\rho \tag{12.8}$$

$$\nabla \times \mathbf{E} + \frac{\partial \mathbf{B}}{\partial t} = 0 \tag{12.9}$$

$$\nabla\cdot\mathbf{B} = 0 \tag{12.10}$$

$$\nabla \times \mathbf{B} - \mu_0\varepsilon_0 \frac{\partial \mathbf{E}}{\partial t} = \mu_0 \mathbf{j} \tag{12.11}$$

Collectively, equations (12.8) through (12.11) are called *Maxwell's equations*. They are complete and general, apply to all electromagnetic fields, and require no further modification. In the following paragraphs, we will explore a few consequences of Maxwell's equations. The point of view we are taking here is that the electromagnetic field itself is created by the charges on elementary particles which are moving in some given manner. The field itself exists in the vacuum. When we talk about the field which exists inside a conduction wire or inside a volume throughout which charge exists as a density, we are really talking about a field which is the collective result of the fields produced by many individual elementary particles.

Suppose that an electromagnetic field exists in a region of space which contains no sources so that $\rho = 0$ and $\mathbf{j} = 0$. Take the curl of (12.9) and use (9.8) to get

$$\nabla(\nabla\cdot\mathbf{E}) - \nabla^2\mathbf{E} + \frac{\partial}{\partial t}(\nabla \times \mathbf{B}) = 0 \tag{12.12}$$

Now use (12.8) and (12.11) to get

$$\nabla^2\mathbf{E} - \frac{1}{c^2}\frac{\partial^2 \mathbf{E}}{\partial t^2} = 0 \tag{12.13}$$

The constant c is given by

$$c = \frac{1}{\sqrt{\mu_0 \varepsilon_0}} = 2.998 \times 10^8 \text{ m/sec} \tag{12.14}$$

and, as indicated, has the dimensions of velocity. A similar calculation gives

$$\nabla^2 \mathbf{B} - \frac{1}{c^2}\frac{\partial^2 \mathbf{B}}{\partial t^2} = 0 \tag{12.15}$$

Equations (12.13) and (12.15) are called *wave equations*.

Consider the possibility of an electromagnetic field which depends on one spatial coordinate, say z, and the time. The x component of the electric field obeys

$$\frac{\partial^2 E_x}{\partial z^2} - \frac{1}{c^2}\frac{\partial^2 E_x}{\partial t^2} = 0 \tag{12.16}$$

It is easy to show that this partial differential equation has solutions of the form

$$E_x = f(z - ct) + g(z + ct) \tag{12.17}$$

where f and g are arbitrary functions of the variables $z - ct$ and $z + ct$. A particularly important special solution is

$$E_x = E_0 \sin\left[\frac{2\pi}{\lambda}(z - ct)\right] \tag{12.18}$$

which is a *plane wave*. Figure 12.2 shows a plot of one cycle of this function at $t = 0$ and at a later time. The quantity λ is the *wavelength* and is the length of one complete cycle of the function. The constant factor E_0 is the *amplitude* of the wave, and the entire quantity $(2\pi/\lambda)(z - ct)$ is called the *phase* of the wave. During the time interval t, the wave has moved ahead a distance ct along the z axis. Thus, the constant c of equation (12.14) is the speed of propagation of the wave. More generally, the functions f and g of equation (12.16) represent disturbances of arbitrary shape propagating in both the positive and negative z directions at the speed c. The *frequency* and *period* of the wave are found from

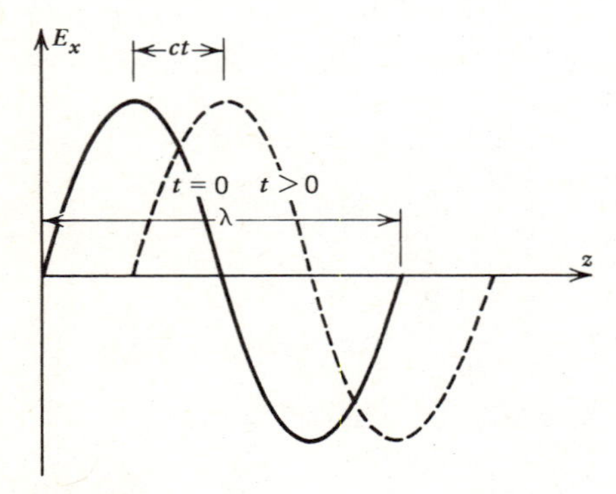

Fig. 12.2

$$\frac{2\pi c}{\lambda} = \omega = 2\pi f = \frac{2\pi}{T} \tag{12.19}$$

The frequency f represents the number of waves per second which pass a given point on the z axis, and the period T is the time for one complete wave to pass the given point. The quantity ω is called the *circular frequency* and is measured in radians/sec. It is the usual practice to define the *wave number* κ by

$$\kappa = \frac{2\pi}{\lambda} \tag{12.20}$$

The plane wave (12.18) is then

$$E_x = E_0 \sin(\kappa z - \omega t) \tag{12.21}$$

Note that

$$c = f\lambda \qquad \omega = \kappa c \tag{12.22}$$

We now represent a more general wave propagating in the z direction by

$$\mathbf{E} = \mathbf{E}_0 \sin(\kappa z - \omega t) \tag{12.23}$$

where the direction of $\mathbf{E}_0$ is as yet unspecified. Equation (12.8) gives

$$\nabla \cdot \mathbf{E} = \frac{\partial}{\partial z}\hat{\mathbf{k}} \cdot \mathbf{E}_0 \sin(\kappa z - \omega t) = 0 \tag{12.24}$$

which means that $\hat{\mathbf{k}} \cdot \mathbf{E}_0 = 0$. Thus, $\mathbf{E}_0$ is in fact perpendicular to the direction of propagation of the wave, which is the z direction in this case. For convenience, let $\mathbf{E}_0$ be in the direction of the x axis. Then,

$$\mathbf{E} = \hat{\mathbf{i}} E_0 \sin(\kappa z - \omega t) \tag{12.25}$$

Now use (12.9) to get

$$\begin{aligned}\nabla \times \mathbf{E} &= \frac{\partial}{\partial z}\hat{\mathbf{k}} \times \hat{\mathbf{i}} E_0 \sin(\kappa z - \omega t)\\ &= \hat{\mathbf{j}} E_0 \kappa \cos(\kappa z - \omega t)\\ &= -\frac{\partial \mathbf{B}}{\partial t}\end{aligned} \tag{12.26}$$

The magnetic field associated with the plane wave is therefore

$$\mathbf{B} = \hat{\mathbf{j}} E_0 \frac{1}{c} \sin(\kappa z - \omega t) \qquad (12.27)$$

In obtaining (12.27), use has been made of $\omega = \kappa c$. The reader should convince himself or herself that the solutions (12.25) and (12.27) satisfy the remaining Maxwell equations (12.10) and (12.11). Maxwell's equations thus predict the existence of plane-polarized transverse electromagnetic waves that propagate through free space at speed c. The waves are polarized in the sense that **E** and **B** vibrate in single directions; the waves are transverse in the sense that the field vectors are perpendicular to the direction of propagation. The electric and the magnetic fields are also perpendicular to one another. Figure 12.3 illustrates such a plane-polarized electromagnetic wave. Note that the wave moves in the direction of $\mathbf{E} \times \mathbf{B}$.

We will now derive another important consequence of Maxwell's equations. Form the dot product of (12.9) with **B** and the dot product of (12.11) with **E**:

$$\mathbf{B} \cdot \nabla \times \mathbf{E} + \mathbf{B} \cdot \frac{\partial \mathbf{B}}{\partial t} = 0 \qquad (12.28)$$

$$\mathbf{E} \cdot \nabla \times \mathbf{B} - \varepsilon_0 \mu_0 \mathbf{E} \cdot \frac{\partial \mathbf{E}}{\partial t} = \mu_0 \mathbf{E} \cdot \mathbf{j} \qquad (12.29)$$

Subtract (12.29) from (12.28) and use (9.3) to get

$$\nabla \cdot \mathbf{S} + \frac{\partial w}{\partial t} = -\mathbf{E} \cdot \mathbf{j} \qquad (12.30)$$

where

$$\mathbf{S} = \frac{1}{\mu_0} \mathbf{E} \times \mathbf{B} \qquad (12.31)$$

$$w = \tfrac{1}{2} \varepsilon_0 E^2 + \frac{1}{2\mu_0} B^2 \qquad (12.32)$$

The reader should have no trouble by now recognizing the significance of (12.30). Look for example at equation (6.13) which expresses conservation of energy for heat conduction.

Consider a cylindrical conductor of cross sectional area A and length s which carries a current I parallel to its axis and uniformly distributed over its cross section. The Joule heating in the conductor is given by

$$I^2 R = IV = jAEs = (jE)\Sigma \text{ watts} \qquad (12.33)$$

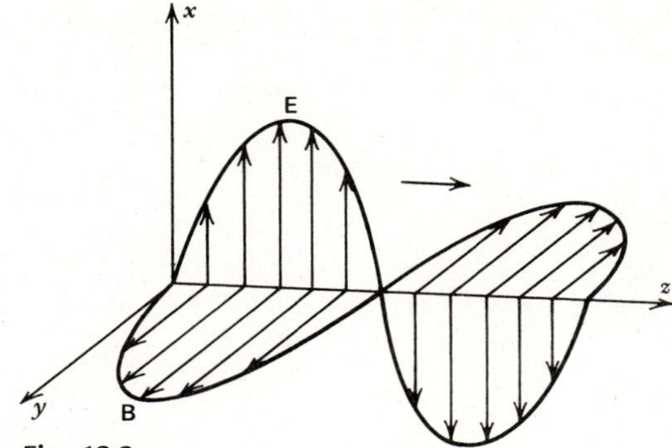

Fig. 12.3

where R is the resistance of the conductor (in ohms) and V is the potential difference between the two ends. Problem (11.13) shows that the electric field in the conductor is uniform and that the potential difference is related to the field strength by $V = Es$. Equation (12.33) reveals that jE represents power per unit volume which is drained out of the electromagnetic field to produce the Joule heating. In other circumstances, $\mathbf{E} \cdot \mathbf{j}$ may represent the reverse situation, namely, a conversion of the mechanical energy of charged particles into electromagnetic field energy.

In equation (12.30), the vector **S** is called *Poynting's vector* and has units of energy per second per unit area. The scalar w is interpreted as energy per unit volume which exists in the electromagnetic field. Equation (12.30) is therefore an equation of continuity which expresses conservation of energy. It can be integrated over a region Σ of space, with the result

$$\int_\sigma \mathbf{S} \cdot d\boldsymbol{\sigma} + \int_\Sigma \mathbf{E} \cdot \mathbf{j}\, d\Sigma = -\frac{d}{dt} \int_\Sigma w\, d\Sigma \qquad (12.34)$$

which can be taken to mean that the net outward flux of the Poynting vector through the surface σ enclosing Σ plus the total loss of energy through Joule heating gives the rate of decrease of the total field energy in Σ.

It is possible to describe a time-dependent electromagnetic field by means of potentials. The Maxwell equation (12.10) is satisfied identically if we write **B** in terms of a vector potential as

$$\mathbf{B} = \nabla \times \mathbf{A} \qquad (12.35)$$

The Maxwell equation (12.9) can be expressed as

$$\nabla \times \left(\mathbf{E} + \frac{\partial \mathbf{A}}{\partial t}\right) = 0 \tag{12.36}$$

which means that it is possible to write

$$\mathbf{E} + \frac{\partial \mathbf{A}}{\partial t} = -\nabla \psi \tag{12.37}$$

where ψ is a scalar potential. The potentials are not unique. If new potentials are defined by

$$\mathbf{A} = \mathbf{A}' + \nabla \phi \tag{12.38}$$

$$\psi = \psi' - \frac{\partial \phi}{\partial t} \tag{12.39}$$

where ϕ is any scalar function, we find that

$$\mathbf{B} = \nabla \times \mathbf{A}' \tag{12.40}$$

$$\mathbf{E} = -\nabla \psi' - \frac{\partial \mathbf{A}'}{\partial t} \tag{12.41}$$

which is to say that **E** and **B** are related to the new potentials in precisely the same way as they are to the original potentials. Equations (12.38) and (12.39) are called *gauge transformations*. Note that

$$\nabla \cdot \mathbf{A} + \frac{1}{c^2}\frac{\partial \psi}{\partial t} = \nabla \cdot \mathbf{A}' + \frac{1}{c^2}\frac{\partial \psi'}{\partial t} + \nabla^2 \phi - \frac{1}{c^2}\frac{\partial^2 \phi}{\partial t^2} \tag{12.42}$$

By an appropriate choice of ϕ it is possible to require the potentials to obey

$$\nabla \cdot \mathbf{A} + \frac{1}{c^2}\frac{\partial \psi}{\partial t} = 0 \tag{12.43}$$

which is called the *Lorentz gauge condition*. In terms of potentials, (12.8) is

$$-\nabla^2 \psi - \frac{\partial}{\partial t}(\nabla \cdot \mathbf{A}) = \frac{1}{\varepsilon_0}\rho \tag{12.44}$$

If the potentials are required to obey the Lorentz gauge condition (12.43), then

$$\nabla^2 \psi - \frac{1}{c^2}\frac{\partial^2 \psi}{\partial t^2} = -\frac{1}{\varepsilon_0}\rho \tag{12.45}$$

By a similar calculation, (12.11) leads to

$$\nabla^2 \mathbf{A} - \frac{1}{c^2}\frac{\partial^2 \mathbf{A}}{\partial t^2} = -\mu_0 \mathbf{j} \tag{12.46}$$

The vector and scalar potentials therefore obey wave equations with source terms. Equations (12.45) and (12.46) are the generalizations of Poisson's equation for time-dependent fields.

PROBLEMS

12.1. Refer to problem 8.2. Suppose that a rectangular conducting loop of resistance R coincides with the rectangular area. If the current I in the long, straight wire is reduced to zero, show that a total charge

$$q = \frac{\mu_0 I s}{2\pi R} \ln \frac{b}{a}$$

circulates in the rectangular conductor.

12.2. Derive the wave equation (12.15) from Maxwell's equations.

12.3. Show that (12.17) is a solution of the wave equation (12.16).

12.4. Show that the Maxwell equations (12.10) and (12.11), with $\mathbf{j} = 0$, are satisfied by (12.25) and (12.27).

12.5. A total charge e is uniformly distributed throughout the volume of a sphere of radius a. Calculate the total energy stored in the electrostatic field of this charge distribution. By making the assumption that $W_E = mc^2$, where m is the rest mass of the electron, it is possible to derive an expression for the "radius" of the electron. What numerical value of the radius does such a model predict? ($m = 9.11 \times 10^{-31}$ kg, $e = 1.602 \times 10^{-19}$ C.)

12.6. A uniform, constant, electric field $\mathbf{E}$ exists in the x direction and a uniform, constant, magnetic field $\mathbf{B}$ exists in the y direction. A cube of edge s has one corner at the origin and three of its faces coincident with the coordinate planes. What is the flux of the Poynting vector for each of the six faces? Is equation (12.34) valid? Is energy actually flowing through space here? See Fig. 12.4.

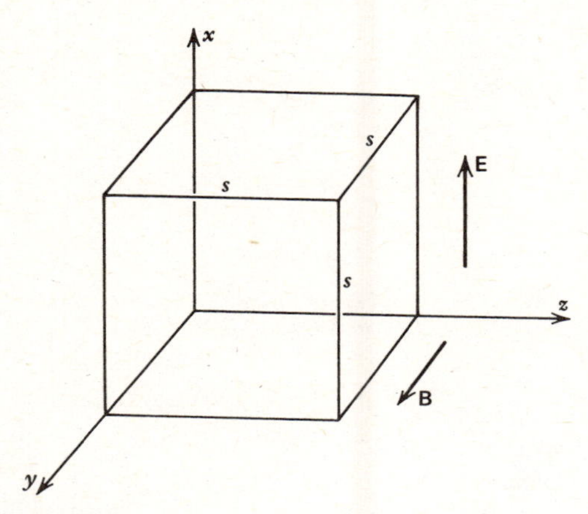

Fig. 12.4

12.7. A cylindrical conductor of cross-sectional area A, length s, and radius a carries a current I parallel to its axis and uniformly distributed over its cross section. What are the electric and the magnetic fields just outside the surface of the cylinder?

Hint: Since $\oint \mathbf{E} \cdot d\mathbf{s} = 0$ for the small rectangular path shown in Fig. 12.5, the electric field just inside the conductor is the same as it is just outside. What is the Poynting vector at the surface of the conductor? Is equation (12.34) valid? Do you believe that energy is flowing into the conductor from the surrounding space?

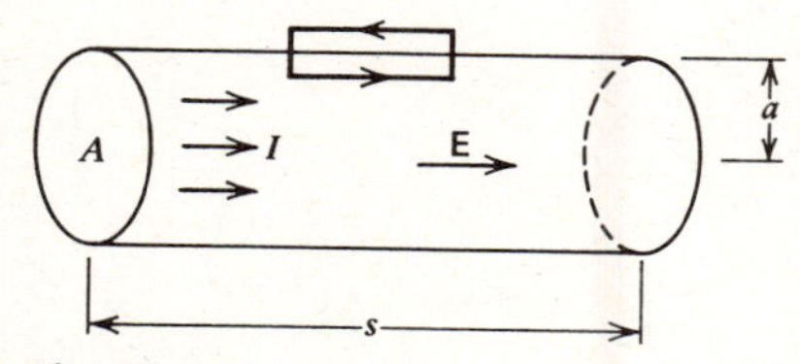

Fig. 12.5

12.8. Figure 12.3 shows one complete wave of a plane electromagnetic wave. Calculate the total electromagnetic field energy contained in the rectangular box of dimensions $\lambda \times s \times s$ shown in Fig. 12.6. In a time equal to one period, this field energy is displaced forward through the face of the box at $z = \lambda$. Show that this energy flow is correctly accounted for by the flux of the Poynting vector. The interpretation of the Poynting vector is tricky. In problems 12.6 and 12.7, the Poynting vector does not really represent a flow of energy, whereas in problem 12.8 it does. In all three cases, equations (12.30) and (12.34), being identities derived from Maxwell's equations, hold true.

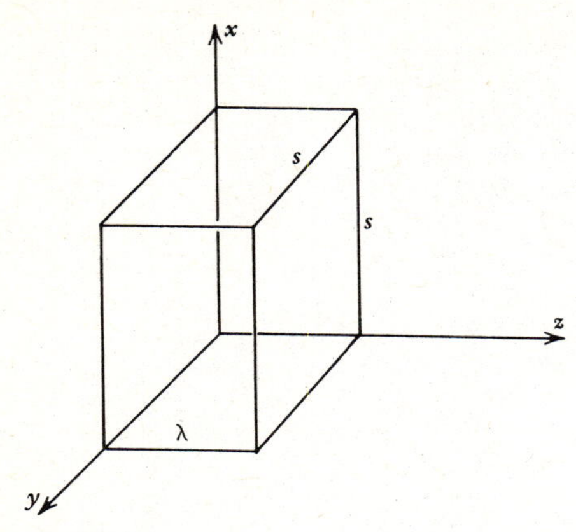

Fig. 12.6

12.9. Carry out the steps leading from (12.11) to (12.46).

12.10. The function $\mathbf{E} = \mathbf{E}_0 \sin(\boldsymbol{\kappa} \cdot \mathbf{r} - \omega t)$ describes a plane wave moving in the direction of the vector $\boldsymbol{\kappa}$. Show that this function satisfies (12.13) provided that $\omega = \kappa c$, where κ is the magnitude of $\boldsymbol{\kappa}$. The surfaces of constant phase $\boldsymbol{\kappa} \cdot \mathbf{r} - \omega t = \phi =$ constant are *planes*, hence the term *plane wave*. These surfaces are also called *wavefronts*. The vector $\boldsymbol{\kappa}$ is called the *propagation vector* and is perpendicular to the wavefronts. Consider two surfaces of constant phase $\boldsymbol{\kappa} \cdot \mathbf{r}_1 - \omega t = \phi_1$, and $\boldsymbol{\kappa} \cdot \mathbf{r}_2 - \omega t = \phi_2$. By choosing $\phi_2 - \phi_1 = 2\pi$, show that $\kappa = 2\pi/\lambda$, where λ is the wavelength or perpendicular distance between the two wavefronts.

$ds^2 = g_{ij}\,dq^i\,dq^j$

4 Curvilinear Coordinates

Up to this point, we have used rectangular Cartesian coordinates almost exclusively. Our purpose now is to extend the vector and tensor concept to arbitrary curvilinear coordinate systems, to lay the foundations of general tensor analysis, and to introduce the concept of generalized coordinates.

1. RECTANGULAR, CYLINDRICAL, AND SPHERICAL COORDINATES

In a rectangular Cartesian coordinate system, *coordinate surfaces* are defined by $x = c_1$, $y = c_2$, and $z = c_3$, where c_1, c_2, and c_3 are constants. The coordinate surfaces are planes, as shown in Fig. 1.1, over which two of the coordinates vary while the remaining one is held constant. The coordinate surfaces are mutually perpendicular and meet along straight lines which are called *coordinate lines*. Along a coordinate line, only one of the three variables changes while the other two remain constant. The intersection of three coordinate surfaces defines a point. Through this point pass three mutually orthogonal coordinate lines. At such a point, the vectors $\hat{\mathbf{i}}$, $\hat{\mathbf{j}}$, and $\hat{\mathbf{k}}$ are three mutually orthogonal unit coordinate vectors which are tangent to the coordinate lines and which point in the direction in which a given coordinate increases. The unit coordinate vectors of rectangular Cartesian coordinates are constant vectors, meaning that they have the same direction and the same magnitude throughout all space. The kinematic vectors giving the displacement, velocity, and the acceleration of a particle are easily defined:

$$\mathbf{r} = x\hat{\mathbf{i}} + y\hat{\mathbf{j}} + z\hat{\mathbf{k}} \tag{1.1}$$

$$\mathbf{u} = \frac{d\mathbf{s}}{dt} = \dot{x}\hat{\mathbf{i}} + \dot{y}\hat{\mathbf{j}} + \dot{z}\hat{\mathbf{k}} \tag{1.2}$$

$$\mathbf{a} = \frac{d\mathbf{u}}{dt} = \ddot{x}\hat{\mathbf{i}} + \ddot{y}\hat{\mathbf{j}} + \ddot{z}\hat{\mathbf{k}} \tag{1.3}$$

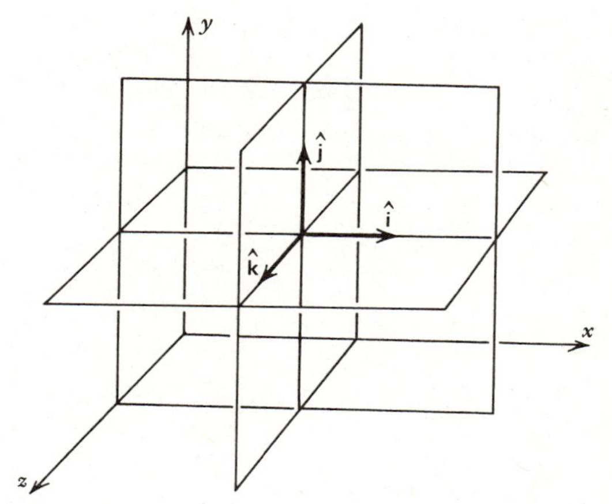

Fig. 1.1

Similarly, if a scalar function ψ and a vector field $\mathbf{a}$ are given, we know definite expressions for $\nabla\psi$, $\nabla^2\psi$, $\nabla\cdot\mathbf{a}$, and $\nabla\times\mathbf{a}$. One of the main goals of this chapter is to generalize these expressions so that they are valid in an arbitrary curvilinear coordinate system.

Cylindrical coordinates r, θ, and z can be defined in terms of rectangular coordinates by means of the transformation equations

$$x = r\cos\theta \tag{1.4}$$

$$y = r\sin\theta \tag{1.5}$$

$$z = z \tag{1.6}$$

where $0 \le \theta < 2\pi$ and $0 \le r < \infty$. An arbitrary point in space can be specified by giving either values for x, y, and z or for r, θ, and z. As Fig. 1.2 shows, the coordinate surfaces $r =$ constant are circular cylinders centered on the z axis. The coordinate surfaces $\theta =$ constant are planes which pass through the z axis and intersect the cylinders $r =$ constant along coordinate lines which run parallel to the z axis. The coordinate surfaces $z =$ constant are planes which are parallel to the x, y plane. The intersections of the planes $z =$ constant with the cylinders $r =$ constant produce coordinate lines which are circles along which only θ varies. The r, θ, and z coordinate lines which pass through any point in space are mutually orthogonal. It is possible to define three mutually orthogonal unit coordinate vectors $\hat{\mathbf{r}}$, $\hat{\boldsymbol{\theta}}$, and $\hat{\mathbf{k}}$ which are tangent to the coordinate lines and which point in the direction in which a given coordinate is increasing. A coordinate system in which the coordinate lines are mutually orthogonal is called an *orthogonal curvilinear coordinate system*. The unit coordinate vectors are not constant vectors but are vector functions which vary from point to point. From Fig. 1.2, we see that these vector functions are

$$\hat{\mathbf{r}} = \hat{\mathbf{i}} \cos\theta + \hat{\mathbf{j}} \sin\theta = \hat{\mathbf{i}} \frac{x}{\sqrt{x^2 + y^2}} + \hat{\mathbf{j}} \frac{y}{\sqrt{x^2 + y^2}} \tag{1.7}$$

$$\hat{\boldsymbol{\theta}} = -\hat{\mathbf{i}} \sin\theta + \hat{\mathbf{j}} \cos\theta = -\hat{\mathbf{i}} \frac{y}{\sqrt{x^2 + y^2}} + \hat{\mathbf{j}} \frac{x}{\sqrt{x^2 + y^2}} \tag{1.8}$$

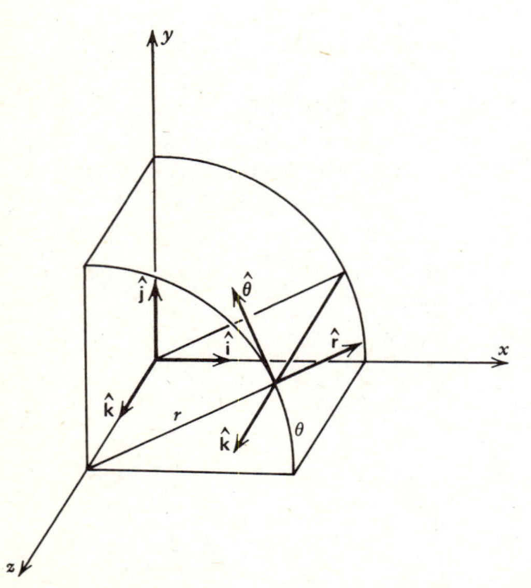

Fig. 1.2

To see what significance the dependence of $\hat{\mathbf{r}}$ and $\hat{\boldsymbol{\theta}}$ on position has, suppose that a particle is constrained to move in the x, y plane and that we want kinematic expressions for its velocity and acceleration expressed in *plane polar coordinates*, that is, cylindrical coordinates with $z = 0$. The displacement vector of the particle is

$$\mathbf{r} = r\hat{\mathbf{r}} \tag{1.9}$$

By differentiating (1.9) with respect to time and writing $d\mathbf{s}$ in place of $d\mathbf{r}$, we find

$$\mathbf{u} = \frac{d\mathbf{s}}{dt} = \dot{r}\hat{\mathbf{r}} + r\frac{d\hat{\mathbf{r}}}{dt} \tag{1.10}$$

As the particle moves along from point to point, the direction of $\hat{\mathbf{r}}$ changes so that $d\hat{\mathbf{r}}/dt$ is not zero. By means of (1.7) and (1.8), we find that

$$\frac{d\hat{\mathbf{r}}}{dt} = -\hat{\mathbf{i}}\dot{\theta}\sin\theta + \hat{\mathbf{j}}\dot{\theta}\cos\theta = \dot{\theta}\hat{\boldsymbol{\theta}} \tag{1.11}$$

Equation (1.10) becomes

$$\mathbf{u} = \dot{r}\hat{\mathbf{r}} + r\dot{\theta}\hat{\boldsymbol{\theta}} \tag{1.12}$$

which is the required expression for velocity in cylindrical coordinates. If the particle moves in the z direction as well, we have

$$\mathbf{u} = \dot{r}\hat{\mathbf{r}} + r\dot{\theta}\hat{\boldsymbol{\theta}} + \dot{z}\hat{\mathbf{k}} \tag{1.13}$$

Another way to derive (1.12) is to use a kind of intuitive differential geometry approach based on Fig. 1.3 and note that the differential line element $d\mathbf{s}$ can be expressed as

$$d\mathbf{s} = dr\hat{\mathbf{r}} + r\,d\theta\hat{\boldsymbol{\theta}} \tag{1.14}$$

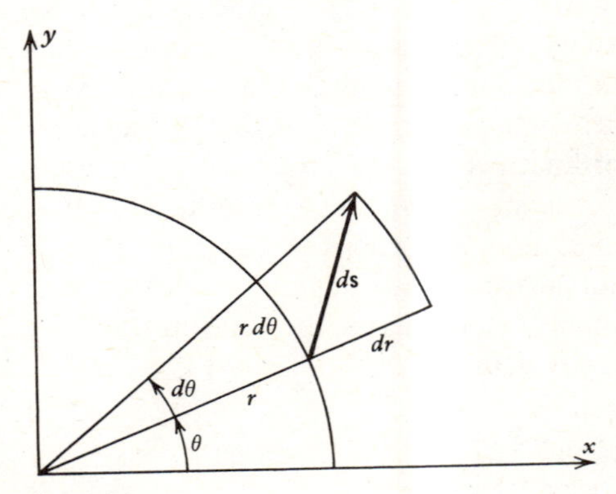

Fig. 1.3

Equation (1.12) follows if we divide through by dt. The intuitive differential geometry approach also allows us to see that a volume element expressed in cylindrical coordinates is

$$d\Sigma = r\,dr\,d\theta\,dz \tag{1.15}$$

The reader is invited to show that

$$\frac{d\hat{\boldsymbol{\theta}}}{dt} = -\dot{\theta}\hat{\mathbf{r}} \tag{1.16}$$

and that the acceleration of a particle is given by

$$\mathbf{a} = \frac{d\mathbf{u}}{dt} = (\ddot{r} - r\dot{\theta}^2)\hat{\mathbf{r}} + (r\ddot{\theta} + 2\dot{r}\dot{\theta})\hat{\boldsymbol{\theta}} \tag{1.17}$$

Spherical polar coordinates constitute a third commonly used curvilinear coordinate system. In terms of rectangular coordinates, spherical coordinates r, θ, and ϕ are defined by the transformation equations

$$x = r\sin\theta\cos\phi \tag{1.18}$$

$$y = r\sin\theta\sin\phi \tag{1.19}$$

$$z = r\cos\theta \tag{1.20}$$

where $0 \le r < \infty$, $0 \le \theta \le \pi$, and $0 \le \phi < 2\pi$. As shown in Fig. 1.4, the coordinate surfaces $r =$ constant are spheres centered at the origin, the coordinate surfaces $\theta =$ constant are cones with apex at the origin, and the coordinate surfaces $\phi =$ constant are planes which pass through the z axis. The angle θ is called the *polar angle* and ϕ is called the *azimuth angle*. Since the coordinate surfaces and coordinate lines are mutually orthogonal, spherical coordinates provide another example of an orthogonal curvilinear coordinate system. A little intuitive differential geometry allows us to see that the infinitesimal displacement vector $d\mathbf{s}$ of Fig. 1.4 can be expressed as

$$d\mathbf{s} = dr\,\hat{\mathbf{r}} + r\,d\theta\,\hat{\boldsymbol{\theta}} + r\sin\theta\,d\phi\,\hat{\boldsymbol{\phi}} \tag{1.21}$$

If (1.21) is divided by dt, the correct kinematic expression for the velocity of a particle results:

$$\mathbf{u} = \frac{d\mathbf{s}}{dt} = \dot{r}\hat{\mathbf{r}} + r\dot{\theta}\hat{\boldsymbol{\theta}} + r\sin\theta\,\dot{\phi}\hat{\boldsymbol{\phi}} \tag{1.22}$$

Volume element in spherical coordinates is given by

$$d\Sigma = (dr)(r\,d\theta)(r\sin\theta\,d\phi) = r^2\sin\theta\,dr\,d\theta\,d\phi \tag{1.23}$$

It is harder to derive the expression for acceleration, so we will put it off until later.

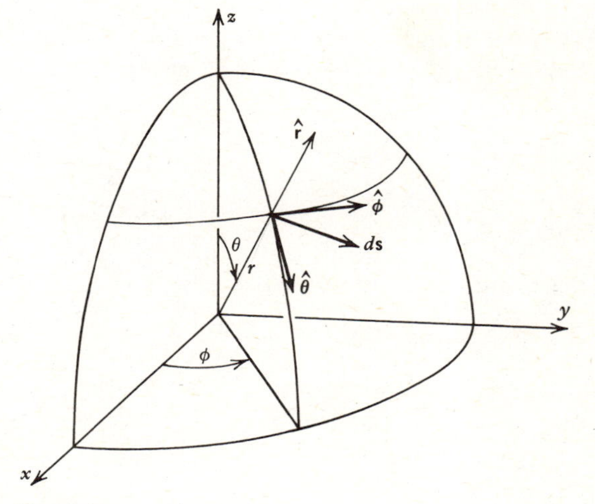

Fig. 1.4

PROBLEMS

1.1. Derive equations (1.16) and (1.17).

1.2. Show that if a particle is acted on by a force of the form $\mathbf{F} = f(r)\hat{\mathbf{r}}$, the angular momentum vector ℓ is a constant. (Refer to section 9 of Chapter 2.) Since ℓ is a constant both in direction and magnitude, the particle must move in a plane which we may take to be the x, y plane. Work out the cross product $\ell = \mathbf{r} \times m\mathbf{u}$ with $\mathbf{r}$ and $\mathbf{u}$ expressed in plane-polar coordinates. Show that the fact that the magnitude of the angular momentum vector is a constant also follows from Newton's second law written as $f(r) = ma_r$ and $F_\theta = ma_\theta = 0$, where a_r and a_θ are the r and θ components of acceleration as given by (1.17).

1.3. Evaluate the integral

$$I = \int_{-\infty}^{+\infty} e^{-x^2}\,dx$$

as follows: Write

$$I^2 = \int_{-\infty}^{+\infty} e^{-x^2}\,dx \int_{-\infty}^{+\infty} e^{-y^2}\,dy = \int_{-\infty}^{+\infty}\int_{-\infty}^{+\infty} e^{-(x^2+y^2)}\,dx\,dy$$

Reference to Fig. 1.3 shows that an element of area can be expressed as $dx\,dy = r\,dr\,d\theta$. Thus,

$$I^2 = \int_{\theta=0}^{2\pi}\int_{r=0}^{\infty} e^{-r^2} r\,dr\,d\theta$$

Note that the ranges given to the variables, $0 \le \theta < 2\pi$ and $0 \le r < \infty$, cover the entire x, y plane just as $-\infty < x < +\infty$ and $-\infty < y < +\infty$ do. Complete the evaluation of the integral.

1.4. If you are good at geometry, you can show that the unit coordinate vectors of spherical coordinates are given by

$$\hat{\mathbf{r}} = \sin\theta\cos\phi\,\hat{\mathbf{i}} + \sin\theta\sin\phi\,\hat{\mathbf{j}} + \cos\theta\,\hat{\mathbf{k}}$$
$$\hat{\boldsymbol{\theta}} = \cos\phi\cos\theta\,\hat{\mathbf{i}} + \sin\phi\cos\theta\,\hat{\mathbf{j}} - \sin\theta\,\hat{\mathbf{k}}$$
$$\hat{\boldsymbol{\phi}} = -\sin\phi\,\hat{\mathbf{i}} + \cos\phi\,\hat{\mathbf{j}}$$

1.5. Two point charges q_1 and q_2 are on the z axis at $z=0$ and $z=s$. The electric field produced by these charges is $\mathbf{E} = \mathbf{E}_1 + \mathbf{E}_2$ where

$$\mathbf{E}_1 = \frac{1}{4\pi\varepsilon_0}\frac{q_1}{R_1^3}\mathbf{R}_1$$

$$\mathbf{E}_2 = \frac{1}{4\pi\varepsilon_0}\frac{q_2}{R_2^3}\mathbf{R}_2$$

The electrostatic energy density stored in the field is

$$w = \tfrac{1}{2}\varepsilon_0 E^2 = \tfrac{1}{2}\varepsilon_0\left[E_1^2 + 2\mathbf{E}_1\cdot\mathbf{E}_2 + E_2^2\right]$$

If we integrate over all space to find the total electrostatic energy, the result is

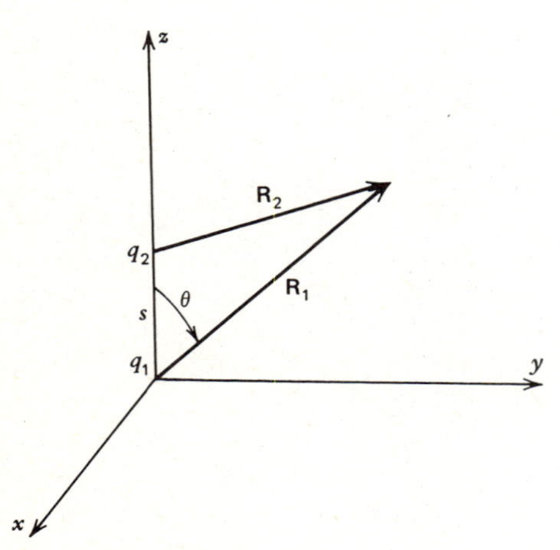

Fig. 1.5

$W = W_1 + W_2 + W_{12}$ where

$$W_1 = \int_\Sigma \frac{1}{2}\varepsilon_0 E_1^2 \, d\Sigma = \infty$$

$$W_2 = \int_\Sigma \frac{1}{2}\varepsilon_0 E_2^2 \, d\Sigma = \infty$$

$$W_{12} = \int_\Sigma \varepsilon_0 \mathbf{E}_1 \cdot \mathbf{E}_2 \, d\Sigma = \frac{q_1 q_2}{4\pi\varepsilon_0 s}$$

The quantities W_1 and W_2 are the infinite "*self-energies*" of the two charges and W_{12} is the *interaction energy* and is the work required to bring q_2 from ∞ to $z = s$. Verify all the above statements.

Hint: Let $\mathbf{R}_1 = \mathbf{r}$ and $\mathbf{R}_2 = \mathbf{r} - \hat{\mathbf{k}}s$. Do the integrals using spherical polar coordinates. See Fig. 1.5.

1.6. The displacement vectors $\mathbf{r}_1$ and $\mathbf{r}_2$ which project from the origin have polar angles θ_1 and θ_2 and azimuth angles ϕ_1 and ϕ_2. If γ is the angle between $\mathbf{r}_1$ and $\mathbf{r}_2$, show that

$$\cos\gamma = \cos\theta_1 \cos\theta_2 + \sin\theta_1 \sin\theta_2 \cos(\phi_1 - \phi_2)$$

See Fig. 1.6.

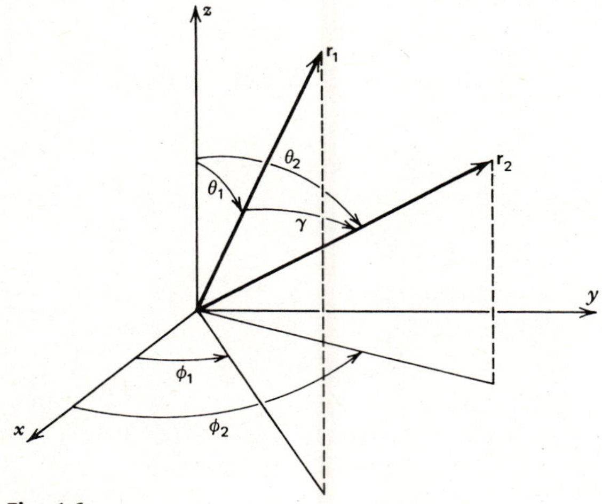

Fig. 1.6

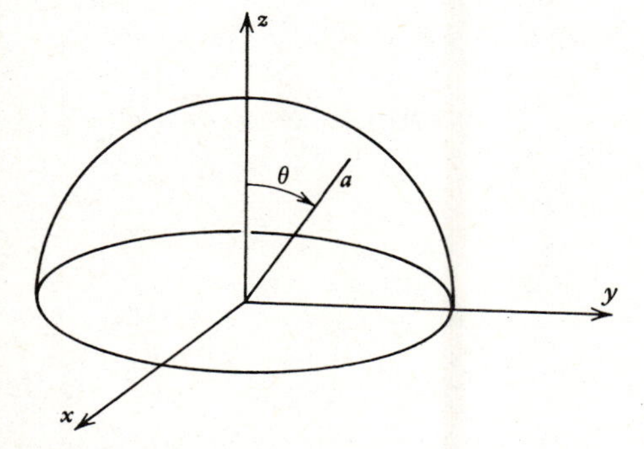

Fig. 1.7

1.7. A total charge q is spread uniformly over a hemisphere of radius a. What is the electric field at the center of the hemisphere? Use spherical coordinates; an element of area on the sphere in the figure can be expressed as $d\sigma = a^2 \sin\theta \, d\theta \, d\phi$. See Fig. 1.7.

2. ELECTRIC DIPOLES

It might seem that rectangular Cartesian coordinates are by far the simplest coordinates to use and that all other coordinate systems should be avoided as unnecessary complications. This is not the case, as the following discussion of electric dipoles shows.

The simplest way to construct an electric dipole is to place two charges of equal magnitude but opposite sign a small distance apart. For convenience, let us place a charge $-q$ at $z = -s$ and a charge $+q$ at $z = +s$ as shown in Fig. 2.1. We wish to study the resulting potential and field at distances away from the dipole which are large compared to $2s$. It is convenient to use spherical polar coordinates. Because of the rotational symmetry about the z axis, the potential will have no dependence on the azimuth angle ϕ. Thus, one advantage of spherical coordinates over rectangular coordinates is that only two variables are required rather than three. The exact potential at the point P of Fig. 2.1 is

$$\psi = \frac{q}{4\pi\varepsilon_0}\left(\frac{1}{R_1} - \frac{1}{R_2}\right) \tag{2.1}$$

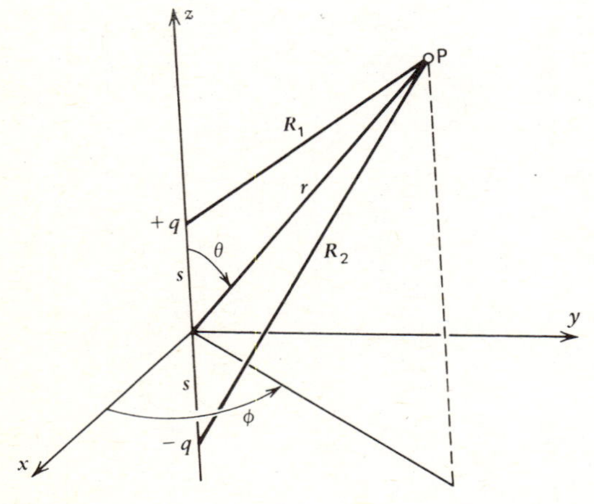

Fig. 2.1

where

$$R_1 = \sqrt{r^2 + s^2 - 2rs\cos\theta} \tag{2.2}$$

$$R_2 = \sqrt{r^2 + s^2 + 2rs\cos\theta} \tag{2.3}$$

Since $r \gg 2s$, it is possible to use the binomial expansion to get an approximation for the potential:

$$\begin{aligned}\psi &= \frac{q}{4\pi\varepsilon_0 r}\left[\left(1 + \frac{s^2}{r^2} - \frac{2s}{r}\cos\theta\right)^{-1/2} - \left(1 + \frac{s^2}{r^2} + \frac{2s}{r}\cos\theta\right)^{-1/2}\right] \\ &\simeq \frac{q}{4\pi\varepsilon_0 r}\left[1 - \frac{s^2}{2r^2} + \frac{s}{r}\cos\theta - \left(1 - \frac{s^2}{2r^2} - \frac{s}{r}\cos\theta\right)\right] \\ &= \frac{1}{4\pi\varepsilon_0}\frac{p\cos\theta}{r^2}\end{aligned} \tag{2.4}$$

where p is given by

$$p = 2qs \tag{2.5}$$

and is called the *dipole moment* of the charge distribution. We find that the potential of a dipole, when expressed in spherical coordinates, is simple and tractable. We can idealize the situation and imagine that there is such a thing as a point dipole in much the same way that we introduce the concept of a point charge. For such a point dipole, the potential given by (2.4) is exact.

Equation (2.4) for the potential can be written in the form

$$r^2 = k\cos\theta \tag{2.6}$$

where

$$k = \frac{p}{4\pi\varepsilon_0\psi} \tag{2.7}$$

If ψ is put equal to a positive constant, then $k > 0$ and necessarily $\cos\theta \geq 0$ in order that $r^2 \geq 0$. Therefore, an equipotential surface for $0 \leq \theta \leq 90°$ results. Similarly, if ψ is put equal to a negative constant, an equipotential for $90° \leq \theta \leq 180°$ results. After a plot of (2.6) is made for θ in these ranges, the figure must be rotated by 360° about the z axis to generate equipotential surfaces in three-dimensional space. Figure 2.2 shows profiles of some of these surfaces that are the intersections of the equipotentials with any plane ϕ = constant.

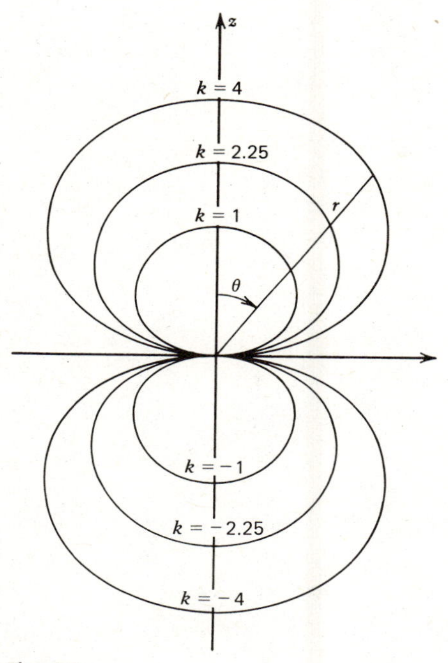

Fig. 2.2

A look at the line element (1.21) shows that displacement in the r direction is $ds_r = dr$; in the θ direction, it is $ds_\theta = r\,d\theta$; and in the ϕ direction, it is $ds_\phi = r\sin\theta\,d\phi$. It is reasonable to conclude that the gradient of a scalar function expressed in spherical coordinates is

$$\nabla\psi = \frac{\partial\psi}{\partial s_r}\hat{\mathbf{r}} + \frac{\partial\psi}{\partial s_\theta}\hat{\boldsymbol{\theta}} + \frac{\partial\psi}{\partial s_\phi}\hat{\boldsymbol{\phi}}$$

$$= \frac{\partial\psi}{\partial r}\hat{\mathbf{r}} + \frac{1}{r}\frac{\partial\psi}{\partial\theta}\hat{\boldsymbol{\theta}} + \frac{1}{r\sin\theta}\frac{\partial\psi}{\partial\phi}\hat{\boldsymbol{\phi}} \tag{2.8}$$

A more rigorous verification of the truth of (2.8) will be given later. For the moment, let us use it to find the electric field components of a dipole:

$$E_r = -\frac{\partial\psi}{\partial r} = \frac{p}{4\pi\varepsilon_0}\frac{2\cos\theta}{r^3} \tag{2.9}$$

$$E_\theta = -\frac{1}{r}\frac{\partial\psi}{\partial\theta} = \frac{p}{4\pi\varepsilon_0}\frac{\sin\theta}{r^3} \tag{2.10}$$

$$E_\phi = 0 \tag{2.11}$$

PROBLEMS

2.1. The electric field vector of a dipole and its two components E_r and E_θ are illustrated in Fig. 2.3. At a given point, the direction of a field line is determined by

$$\frac{E_r}{E_\theta} = \frac{dr}{r\,d\theta}$$

which is a first-order differential equation the solutions of which give field lines in any plane ϕ = constant. With E_r and E_θ given by (2.9) and (2.10), integrate the differential equation and express its solutions in the form $r = r_0 f(\theta)$, where r_0 is a constant. On polar coordinate graph paper, plot the field lines for $r_0 = 1$, 2, 3, 4, and 5. On the same page, make graphs of the equipotentials as given by (2.6) for $k = \pm 1$, ± 4, ± 9, ± 16, and ± 25. Put arrows on the field lines to indicate the direction of the field. Note that the field lines go from regions of higher to regions of lower potential.

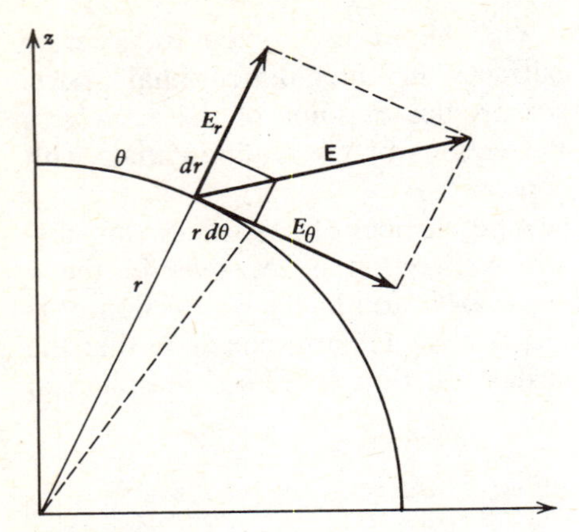

Fig. 2.3

2.2. Let **p** be a vector which has magnitude $2sq$ and which points from $-q$ to $+q$. The dipole is located at some position $\mathbf{r}'$ as illustrated in Fig. 2.4. Show that the potential at **r** is given by

$$\psi(\mathbf{r}) = \frac{1}{4\pi\varepsilon_0}\frac{\mathbf{p}\cdot\mathbf{R}}{R^3}$$

where $\mathbf{R} = \mathbf{r} - \mathbf{r}'$. Show that the electric field at **r** is given by

$$\mathbf{E} = \frac{-1}{4\pi\varepsilon_0}\left[\frac{\mathbf{p}}{R^3} - \frac{3(\mathbf{p}\cdot\mathbf{R})\mathbf{R}}{R^5}\right]$$

Hint: It is better to work out $\nabla\psi$ in rectangular coordinates here.

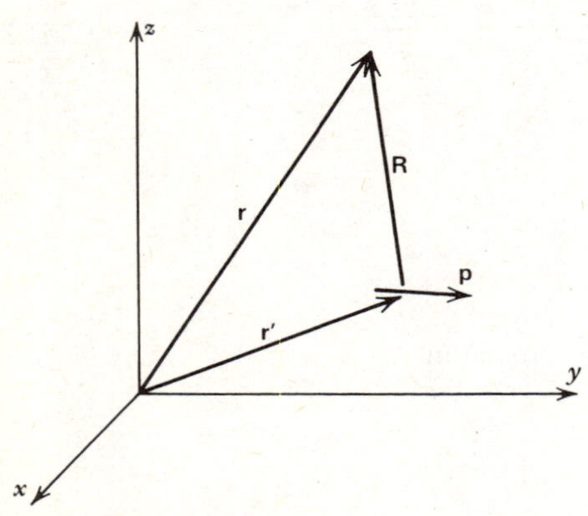

Fig. 2.4

3. ADMISSIBLE COORDINATE TRANSFORMATIONS AND JACOBIANS

In this section, we will develop a theory of curvilinear coordinates in a three-dimensional space. We will frequently use the fact that *Euclidean* geometry is valid in the space, which means that the space is *flat* and lacks any intrinsic curvature. For practical purposes, this means that a transformation from any curvilinear coordinate system back to rectangular Cartesian coordinates is always possible. A general theory of curvilinear coordinates does not require that the space be Euclidean. Much of the formalism which we will obtain remains valid for non-Euclidean spaces. The formalism may also be extended to spaces of any number of dimensions.

An example of a nonflat or non-Euclidean space is the two-dimensional space on the surface of a sphere. In this case, the two-dimensional space is embedded in a three-dimensional Euclidean space. On the surface of the sphere, the polar and azimuth angles θ and ϕ can be used as coordinates. Reference to (1.21) shows that an infinitesimal displacement vector lying in the surface of the sphere is

$$d\mathbf{s} = R\,d\theta\,\hat{\boldsymbol{\theta}} + R\sin\theta\,d\phi\,\hat{\boldsymbol{\phi}} \tag{3.1}$$

where R is now a constant. The *line element* is the magnitude of $d\mathbf{s}$ and is given by

$$ds^2 = R^2\,d\theta^2 + R^2\sin^2\theta\,d\phi^2 \tag{3.2}$$

The space is intrinsically curved. In this case, the curvature of the space is the radius R of the sphere. In the development of the theory of curvilinear coordinates, the line element is all-important since it determines not only the properties of the coordinate system being used but also the intrinsic geometry of the space. The nearest thing to a "straight line" on the surface of a sphere is a line which gives the shortest distance between two points. Such a line is called a *geodesic*. It turns out that the geodesics of a sphere are arcs of great circles. A *spherical triangle* can be constructed out of portions of arcs of great circles as shown in Fig. 3.1. One of the manifestations of the non-Euclidean nature of the space is that the three angles of a spherical triangle do not add up to 180°. This is obviously true of the triangle shown in Fig. 3.1, which has one vertex of angle α at the pole ($\theta = 0$) and one side coincident with the equator ($\theta = 90°$). The sum of the angles is $\alpha + 180°$. The intrinsic curvature of the space also manifests itself in the impossibility of introducing a two-dimensional rectangular Cartesian coordinate system which covers the entire sphere. In other words, there is no coordinate transformation $\xi = \xi(\theta, \phi)$, $\eta = \eta(\theta, \phi)$ such that the line element is

$$ds^2 = R^2\,d\theta^2 + R^2\sin^2\theta\,d\phi^2 = d\xi^2 + d\eta^2 \tag{3.3}$$

at all points of the spherical surface.

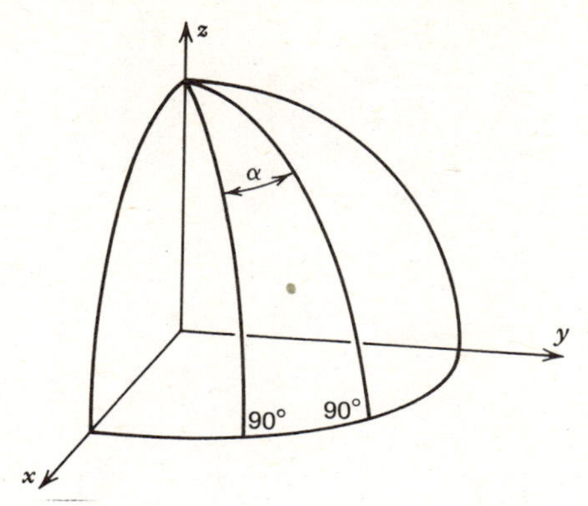

Fig. 3.1

We return now to three-dimensional Euclidean space and introduce into it a set of curvilinear coordinates defined by transformation equations of the form

$$q_1 = q_2(x_1, x_2, x_3) \qquad q_2 = q_2(x_1, x_2, x_3)$$
$$q_3 = q_3(x_1, x_2, x_3) \tag{3.4}$$

where q_1, q_2, and q_3 are the curvilinear coordinates and x_1, x_2, and x_3 are the three rectangular coordinates x, y, and z. To save writing, it is convenient to abbreviate (3.4) by the notation

$$q_i = q_i(x_j) \tag{3.5}$$

Expressed in words, (3.5) means that the three curvilinear coordinates q_i can be expressed as functions of the three rectangular coordinates x_j. For example, the cylindrical coordinates r, θ, and z as given by (1.4), (1.5), and (1.6) can be related to rectangular coordinates by means of

$$r = \sqrt{x^2 + y^2} \tag{3.6}$$

$$\theta = \cos^{-1}\left(\frac{x}{\sqrt{x^2 + y^2}}\right) \tag{3.7}$$

$$z = z \tag{3.8}$$

Coordinate surfaces are the three surfaces $q_i =$ constant, and coordinate lines are the intersections of these surfaces. Along a coordinate line, only one coordinate varies. A given point in space is specified by giving three values to q_1, q_2, and q_3. At such a point, we can construct three unit coordinate vectors

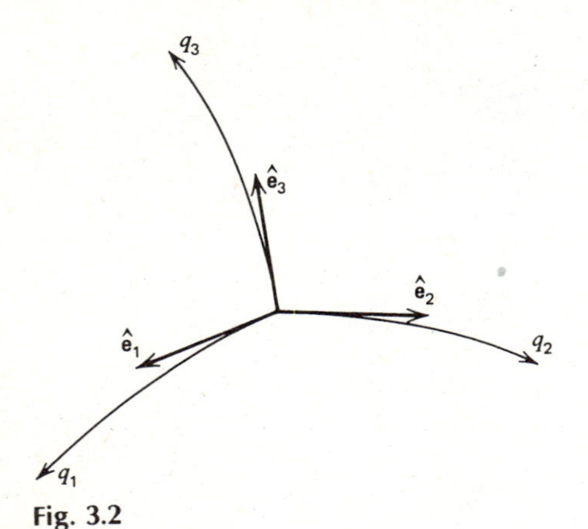

Fig. 3.2

$\hat{\mathbf{e}}_1$, $\hat{\mathbf{e}}_2$, and $\hat{\mathbf{e}}_3$ which are tangent to the three coordinate lines which pass through the point. This is illustrated in Fig. 3.2. At the given point, an infinitesimal displacement vector can be represented as

$$d\mathbf{s} = ds(1)\hat{\mathbf{e}}_1 + ds(2)\hat{\mathbf{e}}_2 + ds(3)\hat{\mathbf{e}}_3 \tag{3.9}$$

A two-dimensional version of (3.9) is illustrated in Fig. 3.3. Note that $ds(1)$ and $ds(2)$ are the lengths of the sides of a parallelogram with $d\mathbf{s}$ as the diagonal. It is not necessary that the three unit coordinate vectors be mutually orthogonal, and we will not assume that they are. It is required however that the three unit coordinate vectors be linearly independent. If it were not so, then an arbitrary infinitesimal displacement vector could not be represented as a linear combination of them, that is, the relation (3.9) would not exist. This is very important since coordinate systems are required to provide a basis in which vector quantities can be represented.

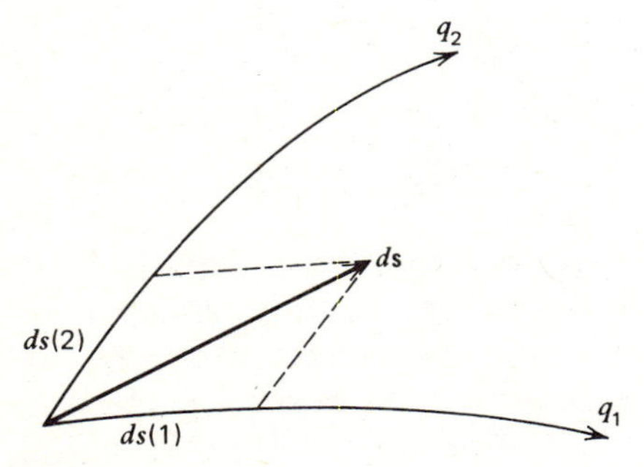

Fig. 3.3

The gradient vectors

$$\nabla q_1 = \mathbf{b}^1 \qquad \nabla q_2 = \mathbf{b}^2 \qquad \nabla q_3 = \mathbf{b}^3 \tag{3.10}$$

are perpendicular to the coordinate surfaces $q_i =$ constant. The three vectors $\mathbf{b}^i$ defined by (3.10) are called *reciprocal basis vectors*. The reason for this terminology and the use of the peculiar superscript notation will become clear as the theory is developed. The superscripts in (3.10) do not denote powers. The reciprocal base vectors are not in the same direction as the corresponding unit coordinate vectors unless the coordinate system is orthogonal. The coordinate transformations (3.5) are said to be *admissible* at those points of space for which the three unit coordinate vectors form a linearly independent set and for which the three gradient vectors (3.10) exist, are finite, and nonzero.

The *Jacobian* of the transformation is defined as follows:

$$J = \mathbf{b}^1 \cdot (\mathbf{b}^2 \times \mathbf{b}^3) = \begin{vmatrix} \partial_1 q_1 & \partial_2 q_1 & \partial_3 q_1 \\ \partial_1 q_2 & \partial_2 q_2 & \partial_3 q_2 \\ \partial_1 q_3 & \partial_2 q_3 & \partial_3 q_3 \end{vmatrix} \tag{3.11}$$

As an example, we shall calculate the Jacobian of the transformation from rectangular to cylindrical coordinates. The coordinates are identified with q_1, q_2, and q_3 as follows: $q_1 = r$, $q_2 = \theta$, and $q_3 = z$. The ordering has been chosen so that the coordinate system and the reciprocal basis vectors are right-handed. The Jacobian will then turn out to be positive. The Jacobian is

$$J = \begin{vmatrix} \dfrac{\partial r}{\partial x} & \dfrac{\partial r}{\partial y} & \dfrac{\partial r}{\partial z} \\ \dfrac{\partial \theta}{\partial x} & \dfrac{\partial \theta}{\partial y} & \dfrac{\partial \theta}{\partial z} \\ \dfrac{\partial z}{\partial x} & \dfrac{\partial z}{\partial y} & \dfrac{\partial z}{\partial z} \end{vmatrix} \tag{3.12}$$

As an alternative to differentiating (3.6) and (3.7) directly, we can proceed by taking total differentials of $x = r\cos\theta$ and $y = r\sin\theta$:

$$dx = \cos\theta\, dr - r\sin\theta\, d\theta \tag{3.13}$$

$$dy = \sin\theta\, dr + r\cos\theta\, d\theta \tag{3.14}$$

Now solve for dr and $d\theta$ to get

$$dr = \cos\theta\, dx + \sin\theta\, dy \tag{3.15}$$

$$d\theta = -\frac{1}{r}\sin\theta\, dx + \frac{1}{r}\cos\theta\, dy \tag{3.16}$$

If r and θ are considered as functions of x and y, then

$$dr = \frac{\partial r}{\partial x} dx + \frac{\partial r}{\partial y} dy \tag{3.17}$$

$$d\theta = \frac{\partial \theta}{\partial x} dx + \frac{\partial \theta}{\partial y} dy \tag{3.18}$$

The comparison of (3.15) and (3.16) with (3.17) and (3.18) allows the partial derivatives to be evaluated. The resulting Jacobian is

$$J = \begin{vmatrix} \cos\theta & \sin\theta & 0 \\ -\frac{1}{r}\sin\theta & \frac{1}{r}\cos\theta & 0 \\ 0 & 0 & 1 \end{vmatrix} = \frac{1}{r} \tag{3.19}$$

There is an important connection between the Jacobian and the unit coordinate vectors. Since the gradient vectors (3.10) are perpendicular to the coordinate surfaces,

$$\mathbf{b}^1 = k_1(\hat{\mathbf{e}}_2 \times \hat{\mathbf{e}}_3) \qquad \mathbf{b}^2 = k_2(\hat{\mathbf{e}}_3 \times \hat{\mathbf{e}}_1)$$

$$\mathbf{b}^3 = k_3(\hat{\mathbf{e}}_1 \times \hat{\mathbf{e}}_2) \tag{3.20}$$

where k_1, k_2, and k_3 are proportionality factors to be evaluated later. The condition of admissibility guarantees the existence of finite and nonzero proportionality factors. They will be positive if the coordinate system is right-handed. The Jacobian can be expressed as

$$\begin{aligned} J &= k_1k_2k_3(\hat{\mathbf{e}}_2 \times \hat{\mathbf{e}}_3)\cdot[(\hat{\mathbf{e}}_3 \times \hat{\mathbf{e}}_1)\times(\hat{\mathbf{e}}_1 \times \hat{\mathbf{e}}_2)] \\ &= k_1k_2k_3(\hat{\mathbf{e}}_2 \times \hat{\mathbf{e}}_3) \\ &\quad \cdot[\hat{\mathbf{e}}_1(\hat{\mathbf{e}}_3 \times \hat{\mathbf{e}}_1 \cdot \hat{\mathbf{e}}_2) - \hat{\mathbf{e}}_2(\hat{\mathbf{e}}_3 \times \hat{\mathbf{e}}_1 \cdot \hat{\mathbf{e}}_1)] \\ &= k_1k_2k_3(\hat{\mathbf{e}}_1 \cdot \hat{\mathbf{e}}_2 \times \hat{\mathbf{e}}_3)^2 \end{aligned} \tag{3.21}$$

where use has been made of the bac-cab rule, equation (8.19) of Chapter 3. Also used is (8.30) of Chapter 2 and $(\hat{\mathbf{e}}_3 \times \hat{\mathbf{e}}_1)\cdot\hat{\mathbf{e}}_1 = 0$. If the unit coordinate vectors are not a linearly independent set, then $\hat{\mathbf{e}}_1 \cdot \hat{\mathbf{e}}_2 \times \hat{\mathbf{e}}_3 = 0$ and $J = 0$. A coordinate transformation is admissible at a given point in space if, and only if, the Jacobian is finite and nonzero.

Commonly used coordinate systems frequently have special points or regions in which the transformation is not admissible. In the example of cylindrical coordinates, the Jacobian becomes undefined when $r = 0$. Thus, all points of the z axis are not admissible for cylindrical coordinates. The difficulty is connected with the fact that the unit vector $\hat{\boldsymbol{\theta}}$ becomes ambiguous at $r = 0$. Another problem is that specifying $x = 0$ and $y = 0$ gives $r = 0$ but leaves θ undetermined.

The transformation equations (3.5) are generally not linear. However, a type of linear transformation can be recovered if we form the total differentials of the transformation equations:

$$dq^1 = (\partial_1 q_1)\, dx^1 + (\partial_2 q_1)\, dx^2 + (\partial_3 q_1)\, dx^3 \tag{3.22}$$

$$dq^2 = (\partial_1 q_2)\, dx^1 + (\partial_2 q_2)\, dx^2 + (\partial_3 q_2)\, dx^3 \tag{3.23}$$

$$dq^3 = (\partial_1 q_3)\, dx^1 + (\partial_2 q_3)\, dx^2 + (\partial_3 q_3)\, dx^3 \tag{3.24}$$

This is obviously too much to write out, so we will use the abbreviated notation

$$dq^i = \frac{\partial q_i}{\partial x_j} dx^j \tag{3.25}$$

The summation convention is in force and will be used extensively in this chapter. In (3.25), a sum over j is implied. The index i is called a *free index* and can have the value 1, 2, or 3. The reader should have patience with the apparently unnecessary use of superscripts to label the total differentials. The Jacobian occurs here as the determinant of the coefficients of (3.25). We know from our studies of linear algebra that a unique nontrivial solution of (3.25) for the dx^j in terms of the dq^i exists if, and only if, the Jacobian is not zero. See Chapter 1, section 6, theorems 2 and 3. If the solution is carried out, the result is a set of linear equations which we can write as

$$dx^1 = f_{11}\, dq^1 + f_{12}\, dq^2 + f_{13}\, dq^3 \tag{3.26}$$

$$dx^2 = f_{21}\, dq^1 + f_{22}\, dq^2 + f_{23}\, dq^3 \tag{3.27}$$

$$dx^3 = f_{31}\, dq^1 + f_{32}\, dq^2 + f_{33}\, dq^3 \tag{3.28}$$

In rectangular Cartesian coordinates, one has

$$\int_a^b dx^1 = x_1(b) - x_2(a) \tag{3.29}$$

where a and b are any two points. The integral in (3.29) is independent of the path connecting a and b. Thus, dx^1 is an exact differential, as are dx^2 and dx^3. This implies that functions $x_i(q_j)$ exist which are the inverses of (3.5) and that the coefficients f_{ij} in (3.26), (3.27), and (3.28) are in fact partial derivatives. Thus, we may write

$$dx^i = \frac{\partial x_i}{\partial q_j} dq^j \tag{3.30}$$

Equations (3.25) and (3.30) are linear equations which are inverses of one another. The matrices which are made up out of the coefficients are likewise inverses of one another, a fact that can be stated as

$$\frac{\partial q_i}{\partial x_j}\frac{\partial x_j}{\partial q_k} = \delta^i_k \tag{3.31}$$

$$\frac{\partial x_i}{\partial q_j}\frac{\partial q_j}{\partial x_k} = \delta^i_k \tag{3.32}$$

The symbol δ^i_k is a Kronecker delta with one index written as a superscript and the other as a subscript. The determinant of the coefficients of the transformation, which is also a Jacobian, will be designated by

$$K = \left|\frac{\partial x_j}{\partial q_i}\right| \tag{3.33}$$

From what we know about matrices and determinants, we see that (3.31) and (3.32) imply that

$$JK = 1 \tag{3.34}$$

For the example of cylindrical coordinates, $J = \infty$ at $r = 0$, meaning that $K = 0$ at $r = 0$. This means that (1.4) and (1.5) cannot be solved uniquely for r and θ if $x = 0$ and $y = 0$. As previously noted, these values of x and y give $r = 0$ but leave θ undetermined. The transformation goes all right in the other direction. For example, if we say $r = 0$ and $\theta = 90°$, x and y are both unambiguously zero. This is connected with the fact that $J \neq 0$ at $r = 0$ even though it is not defined at this point.

Let q_i and q_i' be two coordinate systems that are admissible in some region. It is possible to express q_i' directly in terms of q_i, and vice versa, without going through the intermediary of rectangular coordinates. The coordinate differentials transform according to

$$dq^{i'} = \frac{\partial q_i'}{\partial q_j} dq^j \tag{3.35}$$

$$dq^i = \frac{\partial q_i}{\partial q_j'} dq^{j'} \tag{3.36}$$

It is convenient to streamline the notation further by writing the coefficients as

$$P_j^{i'} = \frac{\partial q_i'}{\partial q_j} \qquad P_{j'}^{i} = \frac{\partial q_i}{\partial q_j'} \tag{3.37}$$

The transformations are then

$$dq^{i'} = P_j^{i'} dq^j \qquad dq^i = P_{j'}^i dq^{j'} \tag{3.38}$$

The Jacobian determinants are

$$P' = |P_j^{i'}| \qquad P = |P_{j'}^i| \tag{3.39}$$

The transformation is admissible in those regions of space where both P and P' are finite and nonzero. The transformation coefficients obey

$$P_j^{i'}P_{k'}^j = \delta^i_k \qquad P_k^{j'}P_{j'}^i = \delta^i_k \tag{3.40}$$

from which

$$PP' = 1 \tag{3.41}$$

PROBLEMS

3.1. It would seem that a transformation from the coordinates θ and ϕ on the surface of a sphere to coordinates defined by $d\xi = R\,d\theta$, $d\eta = R \sin\theta\, d\phi$ would satisfy (3.3). Show, however, that no function $\eta(\theta, \phi)$ exists because $d\eta$ is not an exact differential. What happens if the coordinate transformation $\xi = R\theta$, $\eta = R\phi \sin\theta$ is made? Show that (3.3) is satisfied in the vicinity of the equator, $\theta = 90°$. An important general theorem concerning non-Euclidean spaces is that it is possible to introduce coordinates which are *locally Cartesian* in the vicinity of any point.

3.2. Calculate the Jacobian (3.11) for the transformation from rectangular to spherical coordinates as given by (1.18), (1.19), and (1.20).

3.3. A coordinate transformation is defined by means of

$$q_1 = \sqrt{x^2+y^2+z^2} \qquad q_2 = y \qquad q_3 = z$$

Draw a sketch showing coordinate surfaces and coordinate lines. Show the unit coordinate vectors at least at one point. Calculate the Jacobian of the transformation. For what region are the coordinates not admissible?

3.4. Two sets of admissible curvilinear coordinates are given in terms of rectangular coordinates by $q_i = q_i(x_j)$ and $q_i' = q_i'(x_j)$. Show that a transformation going directly from q_i to q_i' exists. Let the Jacobians be $J = |\partial_k q_j|$ and $J' = |\partial_k q_j'|$. Show that the Jacobians P and P' as given by (3.39) for the transformations connecting q_i and q_i' directly are related to J and J' by $J' = P'J$ and $J = PJ'$. Thus, if J and J' are finite and nonzero, so are P and P'.

4. VECTORS AND TENSORS

The definition of a vector space is given in section 1 of Chapter 2. In section 2 of that same chapter, a vector is further characterized by the transformation properties of its components with respect to orthogonal transformations. The vector concept will now be extended to arbitrary curvilinear coordinates in a three-dimensional space. The representation of a vector in an arbitrary curvilinear coordinate system can be done using any one of three different types of components. One type is the geometrical or physical components, which are the actual projections of the vector onto the directions of the coordinate lines at the particular point where the vector is being evaluated. An example is the representation of the infinitesimal displacement vector by means of (3.9). It turns out that the physical components $ds(1)$, $ds(2)$, and $ds(3)$ do not have simple transformation properties as a change is made from one coordinate system to another. It is much easier to think in terms of the coordinate differentials which have the simple transformation property given by (3.38). Accordingly, using (3.38) as the prototype, we define the *contravariant components* of a vector to be three functions of position which transform by means of

$$A^{i'} = P_j^{i'} A^j \qquad A^i = P_{j'}^i A^{j'} \tag{4.1}$$

We will soon discover that any vector quantity has both physical and contravariant components, as well as a third set of components, to be introduced in the next paragraph. If $\mathbf{u}$ is the velocity of a particle, then $\dot{q}^1$, $\dot{q}^2$, and $\dot{q}^3$ are seen to be its contravariant components. This follows by dividing (3.38) by dt. The components $\dot{q}^i$ are also called *generalized velocity components*. Thus, in spherical coordinates, $\dot{r}$, $\dot{\theta}$, and $\dot{\phi}$ are the generalized velocity components. The physical components of velocity are

$$u(1) = \frac{ds(1)}{dt} \qquad u(2) = \frac{ds(2)}{dt} \qquad u(3) = \frac{ds(3)}{dt} \tag{4.2}$$

In spherical coordinates, these are

$$u_r = \dot{r} \qquad u_\theta = r\dot{\theta} \qquad u_\phi = r\sin\theta\,\dot{\phi} \tag{4.3}$$

The velocity vector itself is given by (1.22).

We will now examine the behavior of the gradient of a scalar function when a transformation is made from one curvilinear coordinate system to another. The transformation properties of the partial derivatives are found by first expressing the function as

$$\psi = \psi(q_1(q_i'), q_2(q_i'), q_3(q_i')) \tag{4.4}$$

By differentiating (4.4) and using the chain rule, we find

$$\frac{\partial\psi}{\partial q_i'} = \frac{\partial\psi}{\partial q_j}\frac{\partial q_j}{\partial q_i'} = \frac{\partial\psi}{\partial q_j}P_{i'}^j \tag{4.5}$$

Equation (4.5) serves as the prototype from which we define the *covariant components* of a vector to be three functions of position which transform according to

$$A_i' = P_{i'}^j A_j \qquad A_i = P_i^{j'} A_j' \tag{4.6}$$

Compare (4.1) and (4.6) very carefully. The roles of the transformation coefficients are reversed in the two cases. At this point, the reason for using both superscripts and subscripts is made evident. Superscripts are used for contravariant quantities and subscripts are used for covariant quantities. Given some arbitrary vector, it is not clear that it should have both covariant and contravariant components, but this is in fact the case, as will be demonstrated later. At this point, we know the contravariant components of velocity and the covariant components of gradient, but not the converse. The covariant and contravariant components of vectors can be added, subtracted, and multiplied by scalars. All the conditions for vector fields as given in section 1 of Chapter 2 apply. If A^i and B^i are the contravariant components of two vectors **a** and **b** and if α and β are scalars, then the contravariant components of the vector $\mathbf{c} = \alpha\mathbf{a} + \beta\mathbf{b}$ are $C^i = \alpha A^i + \beta B^i$. Similarly, $C_i = \alpha A_i + \beta B_i$ are the covariant components of **c**. A combination such as $\alpha A^i + \beta B_i$ does not have a simple transformation property and is generally not useful. The coordinates themselves, which we have labeled as q_1, q_2, and q_3, do not have a simple covariant or contravariant transformation property. It is only the coordinate differentials which have the simple contravariant transformation property as given by (3.35) and (3.36).

Vectors are really tensors of rank one. The contravariant elements of a tensor of rank two are a set of nine functions of position which transform according to*

$$A^{ij'} = P^{i'}_{m} P^{j'}_{n} A^{mn} \tag{4.7}$$

$$A^{mn} = P^{m}_{i'} P^{n}_{j'} A^{ij'} \tag{4.8}$$

The covariant elements of a second-rank tensor have the following transformation properties:

$$A'_{ij} = P^{n}_{i'} P^{m}_{j'} A_{nm} \tag{4.9}$$

$$A_{nm} = P^{i'}_{n} P^{j'}_{m} A'_{ij} \tag{4.10}$$

Finally, a second-rank tensor has mixed elements which transform according to

$$A^{i'}{}_{j} = P^{i'}_{n} P^{m}_{j'} A^{n}{}_{m} \tag{4.11}$$

$$A^{n}{}_{m} = P^{n}_{i'} P^{j'}_{m} A^{i'}{}_{j} \tag{4.12}$$

(There is a distinction between $A^{n}{}_{m}$ and $A_{m}{}^{n}$ which will be explained later. Both have the same transformation property.) We already have an example of a second-rank mixed tensor. It is the Kronecker delta. By using (3.40), it is possible to show that

$$\delta^{i}_{j} = P^{i}_{n'} P^{m'}_{j} \delta^{n}_{m} \tag{4.13}$$

(In the case of the Kronecker delta, $\delta^{i}{}_{j}$ and δ^{i}_{j} are one and the same. In fact, $\delta^{i}_{j} = \delta^{j}_{i}$). Moreover, the Kronecker delta retains its unique property of being isotropic. See section 7 of Chapter 2.

Tensors of higher rank can be defined. Tensors of rank three have contravariant, covariant, and mixed elements which are designated by A^{ijk}, A_{ijk}, $A^{i}{}_{jk}$, $A^{ij}{}_{k}$, and so on. Examples of third-rank tensor transformations are

$$A^{ijk'} = P^{i'}_{r} P^{j'}_{s} P^{k'}_{t} A^{rst} \qquad A^{i'}{}_{jk} = P^{i'}_{r} P^{s}_{j'} P^{t}_{k'} A^{r}{}_{st} \tag{4.14}$$

We close this section by introducing a second-rank tensor which is of fundamental importance. In rectangular Cartesian coordinates, the line element, which is the length of the infinitesimal displacement vector, is given by

$$ds^2 = dx^i\, dx^i \tag{4.15}$$

By means of (3.30),

$$ds^2 = \frac{\partial x_i}{\partial q_j} \frac{\partial x_i}{\partial q_k} dq^j\, dq^k = g_{jk}\, dq^j\, dq^k \tag{4.16}$$

which expresses the line element as a quadratic form in the coordinate differentials of any curvilinear coordinate system that we may wish to use. As we will show, the set of nine functions

$$g_{jk} = \frac{\partial x_i}{\partial q_j} \frac{\partial x_i}{\partial q_k} \tag{4.17}$$

make up the covariant elements of a second-rank tensor called the metric tensor. The metric tensor is symmetric, $g_{jk} = g_{kj}$, meaning that there are really only six different elements rather than nine. As an

*Read (4.7) as $A^{ij'} = (A^{ij})'$. The entire quantity is in the primed coordinates, not just the index j.

example, the line element in spherical coordinates is found from the displacement vector (1.21):

$$ds^2 = dr^2 + r^2\,d\theta^2 + r^2 \sin^2\theta\, d\phi^2 \tag{4.18}$$

Consequently, the metric tensor is

$$(g_{ij}) = \begin{pmatrix} 1 & 0 & 0 \\ 0 & r^2 & 0 \\ 0 & 0 & r^2\sin^2\theta \end{pmatrix} \tag{4.19}$$

We can of course also calculate the metric tensor directly, using (4.17), (1.18), (1.19), and (1.20). It is significant that (4.19) is diagonal. As we shall presently show, this is connected with the fact that spherical coordinates are orthogonal.

We will now establish the tensor character of g_{ij}. Of fundamental importance in the development of the theory of orthogonal transformations is the invariance of the length of a vector when a transformation is made from one rectangular Cartesian system to another. Equally important in the development of the theory of generalized coordinates is the invariance of the line element

$$ds^2 = g_{ij}\,dq^i\,dq^j = g'_{nm}\,dq^{n'}\,dq^{m'} \tag{4.20}$$

Equation (4.20) states that the length of the infinitesimal displacement vector is the same no matter which coordinate system is used to calculate it. The term *invariant* means that a quantity is not only a scalar but that it also has the same mathematical form in all coordinate systems. The line element qualifies as an invariant as (4.20) shows. If we introduce into (4.20) the transformations (3.38) for the coordinate differentials, the result is

$$g_{ij}P^i_{n'}P^j_{m'}\,dq^{n'}\,dq^{m'} = g'_{nm}\,dq^{n'}\,dq^{m'} \tag{4.21}$$

Since (4.21) must hold for all choices of the displacement vector $d\mathbf{s}$, the implication is that

$$g'_{nm} = P^i_{n'}P^j_{m'}g_{ij} \tag{4.22}$$

which establishes that g_{ij} are elements of a second-rank tensor.

PROBLEMS

4.1. Equations (4.7) and (4.8) are inverses of one another. Obtain (4.8) from (4.7) by using (3.40).

4.2. Equation (4.17) actually represents six equations, two of which are

$$g_{11} = \left(\frac{\partial x_1}{\partial q_1}\right)^2 + \left(\frac{\partial x_2}{\partial q_1}\right)^2 + \left(\frac{\partial x_3}{\partial q_1}\right)^2$$

$$g_{12} = \frac{\partial x_1}{\partial q_1}\frac{\partial x_1}{\partial q_2} + \frac{\partial x_2}{\partial q_1}\frac{\partial x_2}{\partial q_2} + \frac{\partial x_3}{\partial q_1}\frac{\partial x_3}{\partial q_2}$$

If the transformation is from rectangular to spherical coordinates,

$$g_{11} = \left(\frac{\partial x}{\partial r}\right)^2 + \left(\frac{\partial y}{\partial r}\right)^2 + \left(\frac{\partial z}{\partial r}\right)^2$$

$$g_{12} = \frac{\partial x}{\partial r}\frac{\partial x}{\partial \theta} + \frac{\partial y}{\partial r}\frac{\partial y}{\partial \theta} + \frac{\partial z}{\partial r}\frac{\partial z}{\partial \theta}$$

Write out the remaining four elements of the metric tensor and use the results to calculate (4.19).

4.3. Find the metric tensor for the coordinate system of problem 3.3.

4.4. Prove equation (4.13).

4.5. Write out the transformation equations for the mixed elements of a third-rank tensor of the form $A^{ij}{}_k$.

4.6. Given the mixed elements of a second-rank tensor of the type $A^i{}_j$, we can calculate the quantity $\phi = A^i{}_i$. This process is called *contraction of indices*. Show that ϕ is an invariant. Given the covariant elements A_{ij} of a second-rank tensor, we could calculate the quantity $\psi = A_{ii}$. This is however not a proper contraction because ψ does not have a scalar or tensor transformation property. Show this. Such quantities are generally not useful. If $A^{ij}{}_k$ are the mixed elements of a third-rank tensor, show that $A^i = A^{ij}{}_j$ are the contravariant components of a vector. Again, $A^{ii}{}_k$ is not a proper contraction.

Given that A^{ij} are the contravariant elements of a second-rank tensor and that B_k are the covariant components of a vector, show that $C^j = B_k A^{kj}$ are the contravariant components of a vector. Note that $B^k A^{kj}$ is not a proper contraction. Both equations (4.15) and (4.16) represent examples of contractions. Equation (4.15) seems to violate the rule that proper contractions always involve a covariant and a contravariant index. Such is not the case, because in rectangular coordinates there is no distinction between covariant and contravariant components of vectors.

We believe that the laws of physics have a validity which does not depend on any particular coordinate system or frame of reference. One of the values of a tensor equation is that it has exactly the same form in all coordinate systems, provided we do not incorporate into it quantities with strange transformation properties. For example, we might express Newton's second law as $F_i = mA_i$, where F_i and A_i are the covariant components of the force vector and the acceleration vector. In this manner, a type of generalized form invariance is achieved in which the equation and the quantities in it retain the same form and the same interpretation no matter what coordinate system is used. Independence of any particular coordinate system is therefore assured.

4.7. Two particles of equal mass m are connected together by a rigid, massless rod of length s. A second rigid, massless rod also of length s is connected by means of frictionless hinges to one of the masses and to the origin as shown in Fig. 4.1. The system moves in the x, y plane so as to form a double plane pendulum. If the rectangular coordinates of the masses are designated by (x_1, y_1) and (x_2, y_2), the kinetic energy of the system is

$$K = \tfrac{1}{2}m\left(\dot{x}_1^2 + \dot{y}_1^2 + \dot{x}_2^2 + \dot{y}_2^2\right)$$

Show that the kinetic energy can also be expressed as

$$K = \tfrac{1}{2}ms^2\left[2\dot{\theta}_1^2 + \dot{\theta}_2^2 + 2\cos(\theta_1 - \theta_2)\dot{\theta}_1\dot{\theta}_2\right]$$

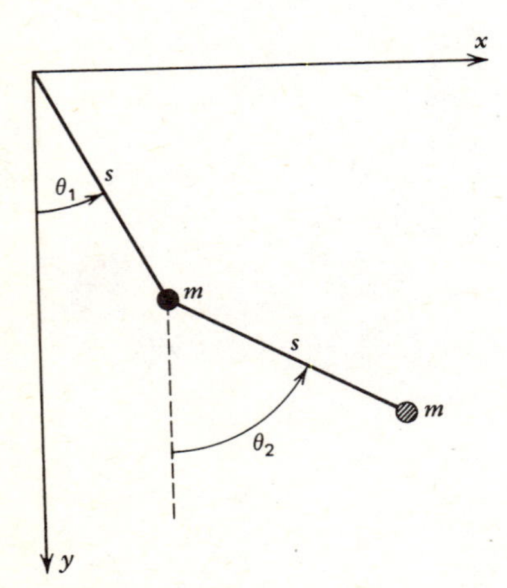

Fig. 4.1

where θ_1 and θ_2 are the angles between the two rods and the y axis.

Equation (4.16) shows that in a general curvilinear coordinate system, the kinetic energy of a particle can be expressed as

$$K = \frac{1}{2} m \left(\frac{ds}{dt} \right)^2 = \frac{1}{2} m g_{ij} \dot{q}^i \dot{q}^j$$

The motion of the double plane pendulum can therefore be represented by a single point moving in a two-dimensional configuration space characterized by a curvilinear coordinate system in which the curvilinear coordinates are θ_1 and θ_2 and the metric tensor is

$$(g_{ij}) = s^2 \begin{pmatrix} 2 & \cos(\theta_1 - \theta_2) \\ \cos(\theta_1 - \theta_2) & 1 \end{pmatrix}$$

The four variables x_1, y_1, x_2, and y_2 can be regarded as the four coordinates of a four-dimensional rectangular Cartesian coordinate system. The curvilinear coordinates θ_1 and θ_2 are then coordinates defined on a two-dimensional surface embedded in the four-space in much the same way that the angles θ and ϕ of spherical coordinates are variables defined on a sphere embedded in ordinary three-dimensional space.

5. DIFFERENTIAL GEOMETRY

If in the general expression (4.16) for the line element we allow only the coordinate differential dq^1 to be nonzero, the result is

$$ds(1)^2 = g_{11}(dq^1)^2 \tag{5.1}$$

Refer to Fig. 3.3 and equation (3.9). The quantity $ds(1)$ is an actual differential length measured along the q_1 coordinate line. With similar expressions for $ds(2)$ and $ds(3)$, equation (3.9) can be expressed as

$$d\mathbf{s} = \sqrt{g_{11}}\, dq^1 \hat{\mathbf{e}}_1 + \sqrt{g_{22}}\, dq^2 \hat{\mathbf{e}}_2 + \sqrt{g_{33}}\, dq^3 \hat{\mathbf{e}}_3 \tag{5.2}$$

The relationship between the geometrical or physical components of $d\mathbf{s}$ and the contravariant components is

$$ds(1) = \sqrt{g_{11}}\, dq^1 \qquad ds(2) = \sqrt{g_{22}}\, dq^2$$
$$ds(3) = \sqrt{g_{33}}\, dq^3 \tag{5.3}$$

From (5.3), we can infer that, given any vector $\mathbf{a}$, the relationship between its contravariant components and its physical components is

$$a(1) = \sqrt{g_{11}}\, A^1 \qquad a(2) = \sqrt{g_{22}}\, A^2$$
$$a(3) = \sqrt{g_{33}}\, A^3 \tag{5.4}$$

Equation (5.2) can be written in a much more convenient way if new basis vectors are defined by means of

$$\mathbf{b}_1 = \sqrt{g_{11}}\, \hat{\mathbf{e}}_1 \qquad \mathbf{b}_2 = \sqrt{g_{22}}\, \hat{\mathbf{e}}_2 \qquad \mathbf{b}_3 = \sqrt{g_{33}}\, \hat{\mathbf{e}}_3 \tag{5.5}$$

The infinitesimal displacement vector is then

$$d\mathbf{s} = dq^i \mathbf{b}_i \tag{5.6}$$

The new basis vectors are not unit vectors, but this inconvenience is minor compared to the advantages to be gained by their use as we will soon see. The basis vectors have a simple covariant transformation property as we now show. Equation (5.6) can be expressed in terms of any two curvilinear coordinate systems as

$$d\mathbf{s} = dq^j \mathbf{b}_j = dq^{i'} \mathbf{b}'_i \tag{5.7}$$

By means of (3.38),

$$P_{i'}^{j} dq^{i'} \mathbf{b}_j = dq^{i'} \mathbf{b}'_i \tag{5.8}$$

Since (5.8) must hold for all choices of the displacement vector $d\mathbf{s}$, we conclude that

$$\mathbf{b}'_i = P_{i'}^{j} \mathbf{b}_j \tag{5.9}$$

which proves the result.

Any vector whatsoever can be expressed either in terms of its physical components or its contravariant components as

$$\mathbf{a} = a(1)\hat{\mathbf{e}}_1 + a(2)\hat{\mathbf{e}}_2 + a(3)\hat{\mathbf{e}}_3 = A^i \mathbf{b}_i \tag{5.10}$$

The contravariant components A^i necessarily exist because the basis vectors $\mathbf{b}_i$ are linearly independent and any vector can therefore be represented as a linear combination of them.

If we start with (5.6) and calculate the line element (4.16), we find that

$$ds^2 = d\mathbf{s} \cdot d\mathbf{s} = \left(dq^i \mathbf{b}_i\right) \cdot \left(dq^j \mathbf{b}_j\right) = \left(\mathbf{b}_i \cdot \mathbf{b}_j\right) dq^i\, dq^j$$
$$= g_{ij}\, dq^i\, dq^j \qquad (5.11)$$

A very simple connection between the basis vectors and the elements of the metric tensor therefore exists:

$$\left(\mathbf{b}_i \cdot \mathbf{b}_j\right) = g_{ij} \qquad (5.12)$$

An immediate conclusion to be drawn from (5.12) is that a curvilinear coordinate system is orthogonal if, and only if, the metric tensor is diagonal. By using (5.5) and (5.12), we can calculate the angle between any two coordinate axes. For instance,

$$\sqrt{g_{11}}\,\sqrt{g_{22}}\,(\hat{\mathbf{e}}_1 \cdot \hat{\mathbf{e}}_2) = g_{12} \qquad \cos\theta_{12} = \frac{g_{12}}{\sqrt{g_{11}}\,\sqrt{g_{22}}} \qquad (5.13)$$

As you have probably guessed by now, the reciprocal basis vectors, defined in equation (3.10), have a contravariant transformation property as we now show:

$$\mathbf{b}^{i'} = \nabla q_i' = \frac{\partial q_i'}{\partial x_j}\hat{\mathbf{e}}_j = \frac{\partial q_i'}{\partial q_k}\frac{\partial q_k}{\partial x_j}\hat{\mathbf{e}}_j = P_k^{i'} \nabla q_k = P_k^{i'} \mathbf{b}^k \qquad (5.14)$$

In (5.14), $\hat{\mathbf{e}}_j$ refers to the unit coordinate vectors of rectangular coordinates. The basis vectors and the reciprocal basis vectors have been introduced separately and independently of one another. There is, however, a direct connection between them, as will be revealed in the next section. For the moment, let us note that an arbitrary vector can be represented as a linear combination of the reciprocal basis vectors since they too are a linearly independent set of vectors. Thus, in addition to (5.10), we have

$$\mathbf{a} = A_i \mathbf{b}^i \qquad (5.15)$$

Knowing that $\mathbf{b}^i$ have a contravariant transformation property, the reader can show that A_i do indeed behave covariantly. At this point, we have established that any vector has covariant, contravariant, and physical components. In the next section, we establish a direct connection between the covariant and the contravariant components of any vector.

PROBLEMS

5.1. Convince yourself of the validity of the argument leading from (5.8) to (5.9).

Hint: Since $d\mathbf{s}$ is arbitrary, it can, for instance, point along the q_1 coordinate line so that dq^2 and dq^3 are zero.

5.2. Show that the diagonal elements of the metric tensor can never be zero.

5.3. Show that (5.10) implies that A^i have a contravariant transformation property.

Hint: The vector $\mathbf{a}$ can be expressed in any two coordinate systems as $\mathbf{a} = A^j \mathbf{b}_j = A^{i'} \mathbf{b}_i'$. Use the known transformation properties of the basis vectors. Similarly, show that (5.15) implies that A_i are covariant.

5.4. In the vicinity of the origin $\theta_1 = \theta_2 = 0$, the metric tensor of problem 4.7 is

$$(g_{ij}) = s^2 \begin{pmatrix} 2 & 1 \\ 1 & 1 \end{pmatrix}$$

Explore in detail the geometry of the coordinate system defined by this metric. What is the angle between the coordinate lines? Suppose that $s = 2$ cm. Start from the origin and mark off several points along the θ_1 and θ_2 coordinate lines which are separated by 2-cm intervals. Draw coordinate lines through these points. Assign numerical values to θ_1 and θ_2 for each of the coordinate lines which you have drawn.

5.5. Let G be a matrix the elements of which are g_{ij} and let P be the matrix made up of the partial derivatives $\partial x_i/\partial q_j$. What matrix equation is equivalent to (4.17)?

5.6. A coordinate system is defined in terms of rectangular coordinates by

$$x_1 = q_1 q_2 \qquad x_2 = q_1\sqrt{1-q_2^2} \qquad x_3 = q_3$$

Find the elements of the metric tensor. What do the coordinate lines look like in the x_1, x_2 plane? What ranges should q_1 and q_2 have in order to cover the entire x_1, x_2 plane?

6. BASIS VECTORS AND RECIPROCAL BASIS VECTORS

It is possible now to evaluate the proportionality factors which appear in the expressions (3.20) for the reciprocal basis vectors. The total differential of $q_1(x_i)$ is

$$\begin{aligned} dq^1 &= \nabla q_1 \cdot d\mathbf{s} = \mathbf{b}^1 \cdot d\mathbf{s} \\ &= k_1(\hat{\mathbf{e}}_2 \times \hat{\mathbf{e}}_3) \\ &\quad \cdot\left(\sqrt{g_{11}}\, dq^1\hat{\mathbf{e}}_1 + \sqrt{g_{22}}\, dq^2\hat{\mathbf{e}}_2 + \sqrt{g_{33}}\, dq^3\hat{\mathbf{e}}_3\right) \\ &= k_1\sqrt{g_{11}}\, dq^1(\hat{\mathbf{e}}_1 \cdot \hat{\mathbf{e}}_2 \times \hat{\mathbf{e}}_3) \end{aligned} \tag{6.1}$$

where we have used $\hat{\mathbf{e}}_2 \cdot \hat{\mathbf{e}}_2 \times \hat{\mathbf{e}}_3 = \hat{\mathbf{e}}_3 \cdot \hat{\mathbf{e}}_2 \times \hat{\mathbf{e}}_3 = 0$. The expression for k_1 obtained from (6.1), along with similar expressions for k_2 and k_3, is

$$\begin{aligned} k_1 &= \frac{1}{\sqrt{g_{11}}\,(\hat{\mathbf{e}}_1 \cdot \hat{\mathbf{e}}_2 \times \hat{\mathbf{e}}_3)} \\ k_2 &= \frac{1}{\sqrt{g_{22}}\,(\hat{\mathbf{e}}_1 \cdot \hat{\mathbf{e}}_2 \times \hat{\mathbf{e}}_3)} \\ k_3 &= \frac{1}{\sqrt{g_{33}}\,(\hat{\mathbf{e}}_1 \cdot \hat{\mathbf{e}}_2 \times \hat{\mathbf{e}}_3)} \end{aligned} \tag{6.2}$$

By using (6.2) and (5.5), we can express the Jacobian (3.21) in the following more convenient form:

$$\begin{aligned} J &= \frac{1}{\sqrt{g_{11}}\sqrt{g_{22}}\sqrt{g_{33}}\,(\hat{\mathbf{e}}_1 \cdot \hat{\mathbf{e}}_2 \times \hat{\mathbf{e}}_3)} \\ &= \frac{1}{\mathbf{b}_1 \cdot \mathbf{b}_2 \times \mathbf{b}_3} \end{aligned} \tag{6.3}$$

Consider the elements of the metric tensor as given by (4.17). The determinant formed from this expression is

$$g = |g_{ij}| = \left|\frac{\partial x_i}{\partial q_j}\right|^2 = K^2 = \frac{1}{J^2} \tag{6.4}$$

where the Jacobian K is defined by (3.33). By combining (6.3) and (6.4), we find

$$K = \sqrt{g} = \mathbf{b}_1 \cdot \mathbf{b}_2 \times \mathbf{b}_3 = \frac{1}{J} \tag{6.5}$$

provided that the basis vectors form a right-handed set so that $J > 0$.

The reciprocal basis vectors (3.20) can be rewritten as follows:

$$\mathbf{b}^1 = \frac{\hat{\mathbf{e}}_2 \times \hat{\mathbf{e}}_3}{\sqrt{g_{11}}\,(\hat{\mathbf{e}}_1 \cdot \hat{\mathbf{e}}_2 \times \hat{\mathbf{e}}_3)} = \frac{\mathbf{b}_2 \times \mathbf{b}_3}{\mathbf{b}_1 \cdot \mathbf{b}_2 \times \mathbf{b}_3} = \frac{\mathbf{b}_2 \times \mathbf{b}_3}{\sqrt{g}} \tag{6.6}$$

Equation (6.6) combined with similar expressions for $\mathbf{b}^2$ and $\mathbf{b}^3$ is

$$\mathbf{b}^1 = \frac{\mathbf{b}_2 \times \mathbf{b}_3}{\sqrt{g}} \qquad \mathbf{b}^2 = \frac{\mathbf{b}_3 \times \mathbf{b}_1}{\sqrt{g}} \qquad \mathbf{b}^3 = \frac{\mathbf{b}_1 \times \mathbf{b}_2}{\sqrt{g}} \tag{6.7}$$

It is possible to solve (6.7) for the basis vectors in terms of the reciprocal basis vectors. The result is

$$\mathbf{b}_1 = \sqrt{g}\,\mathbf{b}^2 \times \mathbf{b}^3 \qquad \mathbf{b}_2 = \sqrt{g}\,\mathbf{b}^3 \times \mathbf{b}^1$$
$$\mathbf{b}_3 = \sqrt{g}\,\mathbf{b}^1 \times \mathbf{b}^2 \tag{6.8}$$

It is easy to show by using either (6.7) or (6.8) that

$$\mathbf{b}_i \cdot \mathbf{b}^j = \delta_i^j \tag{6.9}$$

Since the basis vectors and the reciprocal basis vectors are each a complete set of linearly independent vectors, relations of the form

$$\mathbf{b}_i = a_{ij}\mathbf{b}^j \tag{6.10}$$

exist. Equations (5.12) and (6.9) can be used to find the coefficients a_{ij}:

$$g_{ik} = \mathbf{b}_i \cdot \mathbf{b}_k = a_{ij}\mathbf{b}^j \cdot \mathbf{b}_k = a_{ij}\delta^j_k = a_{ik} \tag{6.11}$$

Thus, (6.10) is

$$\mathbf{b}_i = g_{ij}\mathbf{b}^j \tag{6.12}$$

The basis vectors and the reciprocal basis vectors are related through a simple linear relation in which the coefficients are the covariant elements of the metric tensor. Equation (6.5) reveals that the determinant of the metric tensor is not zero. Therefore, (6.12) can be solved for the reciprocal basis vectors in terms of the basis vectors. If the solution is expressed as

$$\mathbf{b}^i = g^{ij}\mathbf{b}_j \tag{6.13}$$

then g^{ij} are the elements of the matrix which is the inverse of the matrix made up of the elements g_{ij}. Therefore,

$$g^{ij}g_{jk} = \delta^i_k \tag{6.14}$$

By means of (6.9) and (6.13),

$$\mathbf{b}^i \cdot \mathbf{b}^k = g^{ij}\mathbf{b}_j \cdot \mathbf{b}^k = g^{ij}\delta^k_j = g^{ik} \tag{6.15}$$

Note that g^{ik} is symmetric: $g^{ik} = g^{ki}$. From the known transformation properties of the reciprocal basis vectors, it is possible to establish that

$$g^{ij\prime} = P^{i\prime}_n P^{j\prime}_m g^{nm} \tag{6.16}$$

Thus, the g^{ij} are the contravariant elements of a tensor and are in fact the contravariant version of the metric tensor.

As we have stated previously, any vector can be represented as a linear combination of either the basis vectors or the reciprocal basis vectors:

$$\mathbf{a} = A_j\mathbf{b}^j = A^i\mathbf{b}_i \tag{6.17}$$

where the components A_j are covariant and the components A^i are contravariant. By means of (6.12),

$$A_j\mathbf{b}^j = A^i g_{ij}\mathbf{b}^j \tag{6.18}$$

Since the $\mathbf{b}^j$ are linearly independent, (6.18) implies

$$A_j = g_{ij}A^i \tag{6.19}$$

If (6.19) is solved for the contravariant components in terms of the covariant components, the result is

$$A^i = g^{ij}A_j \tag{6.20}$$

Equations (6.19) and (6.20) show that the covariant and contravariant components of any vector are connected by simple linear equations. The coefficients which appear in these equations are the elements of the metric tensor.

Equations (6.19) and (6.20) are examples of *lowering and raising of indices*. This idea can be extended to tensors of higher rank. If A_{kj} are the covariant elements of any second-rank tensor, then the mixed elements of that same tensor are defined by the relations

$$A^i{}_j = g^{ik}A_{kj} \qquad A_j{}^i = g^{ik}A_{jk} \tag{6.21}$$

We find here the reason for the different possible kinds of mixed elements of tensors. In general, $A^i{}_j \neq A_j{}^i$ unless A_{kj} is symmetric. The relations between the covariant and the contravariant elements are

$$A_{ij} = g_{in}g_{jm}A^{nm} \qquad A^{nm} = g^{ni}g^{mj}A_{ij} \tag{6.22}$$

In setting up an integration in a curvilinear coordinate system, it is frequently necessary to know the expression for the volume element. This can be found by regarding the three components of the infinitesimal displacement vector $d\mathbf{s}$ to be the three edges of a differential parallelepiped. The volume of this parallelepiped is

$$\begin{aligned} d\Sigma &= (dq^1\mathbf{b}_1)\cdot(dq^2\mathbf{b}_2)\times(dq^3\mathbf{b}_3) \\ &= \sqrt{g}\,dq^1\,dq^2\,dq^3 = K\,dq^1\,dq^2\,dq^3 \end{aligned} \tag{6.23}$$

For spherical coordinates, the metric (4.19) gives

$$d\Sigma = r^2\sin\theta\,dr\,d\theta\,d\phi \tag{6.24}$$

Equation (6.23) works in any number of dimensions. For example, we can use (3.2) to find an element of surface area on a sphere:

$$d\sigma = \sqrt{g_{22}g_{33}}\,dq^2\,dq^3 = R^2\sin\theta\,d\theta\,d\phi \tag{6.25}$$

PROBLEMS

6.1. Start with (6.7) and prove (6.9). Then, solve (6.7) for the basis vectors to obtain (6.8).

6.2. In (6.21), show that if A_{ij} is symmetric, then $A^i{}_j = A_j{}^i$.

6.3. Suppose that A^j are the contravariant components of an arbitrary vector **a** and that T_{ij} are known to be the covariant elements of a second-rank tensor. Then, $B_i = T_{ij}A^j$ are necessarily the covariant components of a vector **b**. The covariant components of **a** are found from $A_i = g_{ij}A^j$, and the contravariant components of **b** are given by $B^i = g^{ij}B_j$. Show that a relation of the form $B^i = T^{ij}A_j$ exists and that necessarily $T^{ij} = g^{in}g^{jm}T_{nm}$. Solve these relations formally for T_{nm} in terms of T^{ij}.

Hint: Use (6.14).

6.4. Prove (6.16) by using (6.15) and the known transformation properties of $\mathbf{b}^i$.

6.5. Given that A_{kj} are the covariant elements of a second-rank tensor, prove that $A^i{}_j$ as given by (6.21) are in fact the mixed elements of a second-rank tensor, that is, show that $A^{i'}{}_{j'} = P^{i'}_n P^m_{j'} A^n{}_m$.

6.6. If A_{ij} and B^{ik} are the covariant and contravariant elements of second-rank tensors, show that $A_{ij}B^{ik} = A^i{}_j B_i{}^k$.

6.7. Show that the line element (4.16) can be expressed as $ds^2 = g^{ij}\,dq_i\,dq_j$, where $dq_i = g_{in}\,dq^n$.

6.8. If A^{kj} are the contravariant elements of a second-rank tensor, write down the relations between A^{kj} and the mixed elements $A^i{}_j$ and $A_j{}^i$ analogous to (6.21).

6.9. Show that $\sqrt{g'} = |P^i_{j'}|\sqrt{g}$.

6.10. In an arbitrary curvilinear coordinate system, the two-dimensional geometry of a coordinate surface q_3 = constant is determined by the two-dimensional metric tensor

$$(g_{ij}) = \begin{pmatrix} g_{11} & g_{12} \\ g_{21} & g_{22} \end{pmatrix}$$

An element of area on this coordinate surface can be expressed as $d\boldsymbol{\sigma} = \mathbf{b}_1\,dq^1 \times \mathbf{b}_2\,dq^2$. Prove that

$$|d\boldsymbol{\sigma}| = \sqrt{g_{11}g_{22} - g_{12}^2}\,dq^1\,dq^2$$

6.11. Show that the coordinate system of problem 5.6 is left-handed. Redefine it so as to eliminate this problem.

6.12. Show from (6.23) and problem 6.9 that

$$dq^{1'}\,dq^{2'}\,dq^{3'} = \left|\frac{\partial q^{i'}}{\partial q^j}\right| dq^1\,dq^2\,dq^3$$

where $q^{i'}$ and q^i are any two sets of curvilinear coordinates. This result works in any number of dimensions and is useful when variables are changed in a multiple integration.

7. DOT AND CROSS PRODUCTS

Two vectors **a** and **b** can each be represented in two ways as

$$\mathbf{a} = A^i\mathbf{b}_i = A_i\mathbf{b}^i \tag{7.1}$$

$$\mathbf{b} = B^i\mathbf{b}_i = B_i\mathbf{b}^i \tag{7.2}$$

The dot product of **a** and **b** can be calculated by using (7.1) and (7.2) and has the following forms:

$$\begin{aligned}\mathbf{a}\cdot\mathbf{b} &= (A^i\mathbf{b}_i)\cdot(B^j\mathbf{b}_j) = g_{ij}A^iB^j \\ &= (A_i\mathbf{b}^i)\cdot(B_j\mathbf{b}^j) = g^{ij}A_iB_j\end{aligned} \tag{7.3}$$

Other possible expressions for the dot product are

$$\mathbf{a}\cdot\mathbf{b} = A_i B^i = A^i B_i \tag{7.4}$$

As a preparation for the discussion of the cross product of **a** and **b**, it is convenient to first observe that the relations (6.7) and (6.8) can be written in a more compact form as

$$\mathbf{b}_i \times \mathbf{b}_j = \sqrt{g}\, e_{ijk}\mathbf{b}^k \tag{7.5}$$

$$\mathbf{b}^i \times \mathbf{b}^j = \frac{1}{\sqrt{g}} e_{ijk}\mathbf{b}_k \tag{7.6}$$

where e_{ijk} is the three-dimensional permutation symbol. We define so-called ε tensors by means of

$$\varepsilon_{ijk} = \sqrt{g}\, e_{ijk} \tag{7.7}$$

$$\varepsilon^{ijk} = \frac{1}{\sqrt{g}} e_{ijk} \tag{7.8}$$

The ε tensors are in fact third-rank pseudotensors, as we shall soon see. In terms of the ε tensors, (7.5) and (7.6) are

$$\mathbf{b}_i \times \mathbf{b}_j = \varepsilon_{ijk}\mathbf{b}^k \tag{7.9}$$

$$\mathbf{b}^i \times \mathbf{b}^j = \varepsilon^{ijk}\mathbf{b}_k \tag{7.10}$$

By calculating the dot product of (7.9) with $\mathbf{b}_n$, we find

$$\mathbf{b}_n \cdot \mathbf{b}_i \times \mathbf{b}_j = \varepsilon_{ijk}\mathbf{b}_n \cdot \mathbf{b}^k = \varepsilon_{ijk}\delta_n^k = \varepsilon_{ijn} \tag{7.11}$$

In order to establish the tensor character of the ε tensors, we can use the known transformation properties of the basis vectors as follows:

$$\begin{aligned}\varepsilon'_{ijk} &= \mathbf{b}'_i \cdot \mathbf{b}'_j \times \mathbf{b}'_k = P^r_{i'}P^s_{j'}P^t_{k'}(\mathbf{b}_r \cdot \mathbf{b}_s \times \mathbf{b}_t) \\ &= P^r_{i'}P^s_{j'}P^t_{k'}\varepsilon_{rst}\end{aligned} \tag{7.12}$$

We are assuming here that all coordinate systems are right-handed. The contravariant transformation property of (7.8) is similarly established. The ε tensors are related to one another by the normal process of raising and lowering indices as the following calculation shows:

$$\begin{aligned} g_{ir}g_{js}g_{kt}\varepsilon^{rst} &= \frac{1}{\sqrt{g}} g_{ir}g_{js}g_{kt}e_{rst} = \frac{1}{\sqrt{g}} g e_{ijk} \\ &= \sqrt{g}\, e_{ijk} = \varepsilon_{ijk}\end{aligned} \tag{7.13}$$

In (7.13), we have used the definition of a determinant as given by equation (5.21) of Chapter 1. Equation (7.13) shows that (7.7) and (7.8) are really the covariant and the contravariant elements of the same basic third-rank tensor.

It is now possible to calculate the cross product of any two vectors. By starting with (7.1) and (7.2), we have

$$\mathbf{c} = \mathbf{a} \times \mathbf{b} = \left(A^i\mathbf{b}_i\right) \times \left(B^j\mathbf{b}_j\right) = A^iB^j\varepsilon_{ijk}\mathbf{b}^k = C_k\mathbf{b}^k \tag{7.14}$$

We therefore find that the covariant components of the cross product of **a** and **b** are

$$C_i = \varepsilon_{ijk}A^jB^k \tag{7.15}$$

There are some changes in the labeling of indices between (7.14) and (7.15), but hopefully the reader will have no trouble with this by now. By an exactly similar calculation, it is easy to show that the contravariant components of $\mathbf{a} \times \mathbf{b}$ are

$$C^i = \varepsilon^{ijk}A_jB_k \tag{7.16}$$

PROBLEMS

7.1. Prove that in an orthogonal curvilinear coordinate system, the dot product of two vectors can be expressed in terms of the physical components of the vectors as

$$\mathbf{a}\cdot\mathbf{b} = a(1)b(1) + a(2)b(2) + a(3)b(3)$$

7.2. How is the result of problem 6.9 to be modified if the transformation is improper? Instead of using the method of equation (7.12) to calculate the transformation property of the ε tensors, do it directly using (7.7) in combination with the general expression for the expansion of a 3×3 determinant. Show that this leads to

$$P^i_{r'} P^j_{s'} P^k_{t'} \varepsilon_{ijk} = \pm \varepsilon'_{rst}$$

where the minus sign is to be used if the transformation is improper.

7.3. Start out with (7.1) and (7.2) and carry out the calculations leading to (7.16). Show moreover that

$$C_i = g_{ij} C^j$$

as expected, where C_i are given by (7.15).

7.4. Evaluate

$$\varepsilon^{ijk} \varepsilon_{imn}$$

in terms of Kronecker deltas.

7.5. Write out the components of (7.15) one by one and show that

$$C_1 = \sqrt{g}\,(A^2 B^3 - A^3 B^2) \qquad \text{and so on.}$$

Similarly, write out (7.16) to show that

$$C^1 = \frac{1}{\sqrt{g}}(A_2 B_3 - A_3 B_2) \qquad \text{and so on.}$$

If the coordinate system is orthogonal, show that $\mathbf{c} = \mathbf{a} \times \mathbf{b}$ in terms of the physical components of the vectors is

$$c(1) = a(2)\,b(3) - a(3)\,b(2) \qquad \text{and so on.}$$

Hint: In an orthogonal curvilinear coordinate system, (6.14) simplifies to

$$g^{11} g_{11} = 1 \qquad g^{22} g_{22} = 1 \qquad g^{33} g_{33} = 1$$

The relations between the covariant and the contravariant components of a vector also simplify considerably. For example,

$$A_1 = g_{11} A^1 \qquad A^1 = g^{11} A_1 = \frac{1}{g_{11}} A_1 \qquad \text{and so on.}$$

7.6. Prove that

$$\mathbf{b}^1 \cdot \mathbf{b}^2 \times \mathbf{b}^3 = \frac{1}{\sqrt{g}}$$

8. THE GRADIENT OPERATOR

It is convenient here to use the notation $\hat{\mathbf{u}}_1$, $\hat{\mathbf{u}}_2$, and $\hat{\mathbf{u}}_3$ for the unit coordinate vectors of Cartesian coordinates and reserve $\hat{\mathbf{e}}_1$, $\hat{\mathbf{e}}_2$, and $\hat{\mathbf{e}}_3$ for the unit coordinate vectors of an arbitrary curvilinear coordinate system. We know how to express the gradient of a scalar function in terms of rectangular Cartesian coordinates:

$$\nabla \psi = \hat{\mathbf{u}}_i \frac{\partial \psi}{\partial x_i} \tag{8.1}$$

If the rectangular coordinates are now written as functions of curvilinear coordinates, (8.1) can be expressed as

$$\nabla \psi = \hat{\mathbf{u}}_i \frac{\partial \psi}{\partial q_k} \frac{\partial q_k}{\partial x_i} = (\nabla q_k) \frac{\partial \psi}{\partial q_k} = \mathbf{b}^k \frac{\partial \psi}{\partial q_k} \tag{8.2}$$

where $\mathbf{b}^k$ are the reciprocal basis vectors as defined by (3.10). Thus, we can think of the gradient operator in an arbitrary curvilinear coordinate system as

$$\nabla = \mathbf{b}^k \frac{\partial}{\partial q_k} \tag{8.3}$$

For the special case of orthogonal curvilinear coordinates, the relations between basis vectors and reciprocal basis vectors are especially simple. From (6.10),

$$\mathbf{b}_1 = g_{11}\mathbf{b}^1 \qquad \mathbf{b}_2 = g_{22}\mathbf{b}^2 \qquad \mathbf{b}_3 = g_{33}\mathbf{b}^3 \tag{8.4}$$

Recall that

$$\mathbf{b}_1 = \sqrt{g_{11}}\,\hat{\mathbf{e}}_1 \qquad \mathbf{b}_2 = \sqrt{g_{22}}\,\hat{\mathbf{e}}_2 \qquad \mathbf{b}_3 = \sqrt{g_{33}}\,\hat{\mathbf{e}}_3 \tag{8.5}$$

Therefore, for the special case of orthogonal curvilinear coordinates, the gradient of a scalar function is

$$\nabla\psi = \frac{\hat{\mathbf{e}}_1}{\sqrt{g_{11}}}\frac{\partial\psi}{\partial q_1} + \frac{\hat{\mathbf{e}}_2}{\sqrt{g_{22}}}\frac{\partial\psi}{\partial q_2} + \frac{\hat{\mathbf{e}}_3}{\sqrt{g_{33}}}\frac{\partial\psi}{\partial q_3} \tag{8.6}$$

We already anticipated this result in Section 2 to obtain (2.8). Equation (8.6) will give us directly the physical components of $\nabla\psi$.

The gravitational scalar potential due to a point mass M situated at the origin of coordinates is

$$\psi = -\frac{GM}{r} \tag{8.7}$$

Obviously, it is much easier to calculate the gravitational field by using spherical coordinates rather than rectangular coordinates:

$$\mathbf{g} = -\nabla\psi = \hat{\mathbf{r}}\frac{\partial}{\partial r}\left(\frac{GM}{r}\right) = -\frac{GM\hat{\mathbf{r}}}{r^2}$$

PROBLEMS

8.1. Consider the two symmetric matrices

$$A = \begin{pmatrix} 1 & 1 & 0 \\ 1 & 3 & 0 \\ 0 & 0 & 1 \end{pmatrix} \qquad B = \begin{pmatrix} 1 & 2 & 0 \\ 2 & 1 & 0 \\ 0 & 0 & 1 \end{pmatrix}$$

The matrix A could represent a metric tensor whereas B could not. Why?

Hint: The quadratic form (4.16) must be positive definite, that is, $ds^2 > 0$ for all $dq^i \neq 0$. Consider the eigenvalues.

8.2. Let the matrix A of problem 8.1 represent the covariant elements of a metric tensor. Find the contravariant elements. Express the reciprocal basis vectors of the coordinate system described by this metric in terms of the basis vectors and also in terms of the unit coordinate vectors. Make a sketch showing q_1 and q_2 coordinate lines. Show the reciprocal basis vectors $\mathbf{b}^1$ and $\mathbf{b}^2$ and also the unit coordinate vectors $\hat{\mathbf{e}}_1$ and $\hat{\mathbf{e}}_2$. Find the expression for the gradient of a scalar function in terms of the unit vectors.

8.3. Find the relations between rectangular coordinates and the coordinates of problem 8.2. Start out by putting the x axis in the same direction as the q_1 axis. Use (4.17) to check out your transformation equations.

8.4. Two positive point charges are on the x axis at $x = \pm a$, and two negative point charges are on the y axis at $y = \pm a$, as shown in Fig. 8.1. All the charges have the same magnitude q. Find an approximate formula for the potential for the case where $r \gg a$. Express the potential in terms of spherical coordinates. This is an example of a *quadrupole potential*. Show that the three physical components of the electric field are

$$E_r = \frac{9qa^2}{4\pi\varepsilon_0 r^4}\sin^2\theta\cos 2\phi$$

$$E_\theta = -\frac{3qa^2}{4\pi\varepsilon_0 r^4}\sin 2\theta\cos 2\phi$$

$$E_\phi = \frac{3qa^2}{2\pi\varepsilon_0 r^4}\sin\theta\sin 2\phi$$

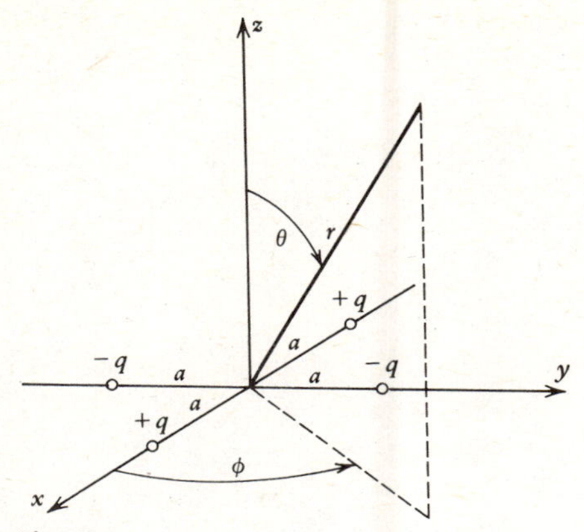

Fig. 8.1

Hint: First, write down the exact potential. It will involve terms which must be approximated by the binomial expansion. For example, one such term is

$$\frac{1}{\sqrt{(x+a)^2+y^2+z^2}}=\frac{1}{\sqrt{r^2+2xa+a^2}}=\frac{1}{r}\left(1+\frac{2xa+a^2}{r^2}\right)^{-1/2}$$

$$\simeq\frac{1}{r}\left(1-\frac{xa}{r^2}-\frac{a^2}{2r^2}+\frac{3x^2a^2}{2r^4}\right)$$

where all terms of order a^2 are retained. Higher terms involving a^3, a^4, and so on are discarded. The general form for the binomial expansion is

$$(1+x)^n=1+nx+\frac{n(n-1)}{2!}x^2+\cdots$$

8.5. Show that the relations connecting the basis vectors and the reciprocal basis vectors of a curvilinear coordinate system to the unit vectors of rectangular coordinates are

$$\mathbf{b}_i=\frac{\partial x_j}{\partial q_i}\hat{\mathbf{u}}_j \qquad \mathbf{b}^i=\frac{\partial q_i}{\partial x_j}\hat{\mathbf{u}}_j$$

$$\hat{\mathbf{u}}_i=\frac{\partial q_j}{\partial x_i}\mathbf{b}_j \qquad \hat{\mathbf{u}}_i=\frac{\partial x_i}{\partial q_j}\mathbf{b}^j$$

Note: In rectangular Cartesian coordinates, there is no distinction between covariance and contravariance. The unit coordinate vectors, basis vectors, and reciprocal basis vectors are all one and the same. If the curvilinear coordinates are orthogonal, show that

$$\hat{\mathbf{e}}_1=\frac{1}{\sqrt{g_{11}}}\frac{\partial x_j}{\partial q_1}\hat{\mathbf{u}}_j=\sqrt{g_{11}}\frac{\partial q_1}{\partial x_j}\hat{\mathbf{u}}_j$$

with similar expressions for $\hat{\mathbf{e}}_2$ and $\hat{\mathbf{e}}_3$. Also show that

$$\hat{\mathbf{u}}_i=\frac{\partial q_1}{\partial x_i}\sqrt{g_{11}}\,\hat{\mathbf{e}}_1+\frac{\partial q_2}{\partial x_i}\sqrt{g_{22}}\,\hat{\mathbf{e}}_2+\frac{\partial q_3}{\partial x_i}\sqrt{g_{33}}\,\hat{\mathbf{e}}_3$$

$$=\frac{\partial x_i}{\partial q_1}\frac{\hat{\mathbf{e}}_1}{\sqrt{g_{11}}}+\frac{\partial x_i}{\partial q_2}\frac{\hat{\mathbf{e}}_2}{\sqrt{g_{22}}}+\frac{\partial x_i}{\partial q_3}\frac{\hat{\mathbf{e}}_3}{\sqrt{g_{33}}}$$

Use these relations to find the unit coordinate vectors $\hat{\mathbf{r}}$, $\hat{\boldsymbol{\theta}}$, and $\hat{\boldsymbol{\phi}}$ of spherical coordinates in terms of $\hat{\mathbf{i}}$, $\hat{\mathbf{j}}$, and $\hat{\mathbf{k}}$. Also find the inverse of these relations.

8.6. Use the results of problem 8.5 to establish the following:

$$\frac{\partial\psi}{\partial x} = \sin\theta\cos\phi\frac{\partial\psi}{\partial r} + \cos\theta\cos\phi\frac{1}{r}\frac{\partial\psi}{\partial\theta} - \frac{\sin\phi}{r\sin\theta}\frac{\partial\psi}{\partial\phi}$$

$$\frac{\partial\psi}{\partial y} = \sin\theta\sin\phi\frac{\partial\psi}{\partial r} + \cos\theta\sin\phi\frac{1}{r}\frac{\partial\psi}{\partial\theta} + \frac{\cos\phi}{r\sin\theta}\frac{\partial\psi}{\partial\phi}$$

$$\frac{\partial\psi}{\partial z} = \cos\theta\frac{\partial\psi}{\partial r} - \sin\theta\frac{1}{r}\frac{\partial\psi}{\partial\theta}$$

Hint: Start out with

$$\nabla\psi = \frac{\partial\psi}{\partial x}\hat{\mathbf{i}} + \frac{\partial\psi}{\partial y}\hat{\mathbf{j}} + \frac{\partial\psi}{\partial z}\hat{\mathbf{k}}$$

$$= \frac{\partial\psi}{\partial r}\hat{\mathbf{r}} + \frac{1}{r}\frac{\partial\psi}{\partial\theta}\hat{\boldsymbol{\theta}} + \frac{1}{r\sin\theta}\frac{\partial\psi}{\partial\phi}\hat{\boldsymbol{\phi}}$$

8.7. The results of problem 8.5 imply that for orthogonal curvilinear coordinates

$$\frac{\partial q_1}{\partial x_i} = \frac{1}{g_{11}}\frac{\partial x_i}{\partial q_1} \qquad \frac{\partial q_2}{\partial x_i} = \frac{1}{g_{22}}\frac{\partial x_i}{\partial q_2} \qquad \frac{\partial q_3}{\partial x_i} = \frac{1}{g_{33}}\frac{\partial x_i}{\partial q_3}$$

Write out these relations for spherical coordinates. This provides another way to do problem 8.6:

$$\frac{\partial\psi}{\partial x_i} = \frac{\partial\psi}{\partial q_j}\frac{\partial q_j}{\partial x_i} = \frac{\partial\psi}{\partial q_1}\frac{1}{g_{11}}\frac{\partial x_i}{\partial q_1} + \frac{\partial\psi}{\partial q_2}\frac{1}{g_{22}}\frac{\partial x_i}{\partial q_2} + \frac{\partial\psi}{\partial q_3}\frac{1}{g_{33}}\frac{\partial x_i}{\partial q_3}$$

the advantage being that $\partial x_i/\partial q_j$ are easier to compute than are $\partial q_j/\partial x_i$.

8.8. In quantum mechanics, the operators which represent the three components of orbital angular momentum are

$$L_x\psi = -i\left(y\frac{\partial\psi}{\partial z} - z\frac{\partial\psi}{\partial y}\right)$$

$$L_y\psi = -i\left(z\frac{\partial\psi}{\partial x} - x\frac{\partial\psi}{\partial z}\right)$$

$$L_z\psi = -i\left(x\frac{\partial\psi}{\partial y} - y\frac{\partial\psi}{\partial x}\right)$$

where ψ is a wave function and $i = \sqrt{-1}$. Use the results of problem 8.6 to express these operators in spherical coordinates. See equation (4.4) in Chapter 9.

9. VELOCITY AND ACCELERATION

The velocity vector of a particle is found by dividing the infinitesimal displacement vector (5.6) by the time interval dt:

$$\mathbf{u} = \frac{d\mathbf{s}}{dt} = \dot{q}^i\mathbf{b}_i \tag{9.1}$$

The derivatives $\dot{q}^i$ are contravariant and are called *generalized velocity components.*

Obtaining the expression for acceleration is not so easy. As the particle moves along its trajectory, the basis vectors change from point to point. The acceleration vector as found by differentiating (9.1) is therefore

$$\mathbf{a} = \frac{d\mathbf{u}}{dt} = \ddot{q}^i\mathbf{b}_i + \dot{q}^i\frac{d\mathbf{b}_i}{dt} \tag{9.2}$$

We can write

$$\frac{d\mathbf{b}_i}{dt} = \frac{\partial\mathbf{b}_i}{\partial q_j}\dot{q}^j \tag{9.3}$$

The main problem, then, is to learn how to calculate the partial derivatives which appear in (9.3). We begin with the premise that the derivatives of the basis vectors are themselves vectors. Since any vector can be expressed as a linear combination of the basis vectors, relations of the form

$$\frac{\partial \mathbf{b}_i}{\partial q_j} = \begin{Bmatrix} k \\ i\,j \end{Bmatrix} \mathbf{b}_k \tag{9.4}$$

must exist. The coefficients which appear in (9.4) are designated by the peculiar notation $\begin{Bmatrix} k \\ i\,j \end{Bmatrix}$ and are given the formidable sounding name *Christoffel symbols of the second kind*. Another commonly used notation is Γ_{ij}^{k}. As you have probably guessed, there are also Christoffel symbols of the first kind; they will be defined shortly. Our goal is to evaluate the Christoffel symbols in terms of the intrinsic properties of the coordinate system in which we are working. This will be done if we can express them in terms of the metric tensor.

There must of course be similar expressions for the reciprocal basis vectors:

$$\frac{\partial \mathbf{b}^i}{\partial q_j} = C_{ijk}\mathbf{b}^k \tag{9.5}$$

The coefficients C_{ijk} which appear in (9.5) are also Christoffel symbols, as we now show. Recall that

$$\mathbf{b}^i \cdot \mathbf{b}_j = \delta_j^i \tag{9.6}$$

Differentiate (9.6) to get

$$\mathbf{b}^i \cdot \frac{\partial \mathbf{b}_j}{\partial q_k} + \frac{\partial \mathbf{b}^i}{\partial q_k} \cdot \mathbf{b}_j = 0 \tag{9.7}$$

By means of (9.4) and (9.5), we find

$$\mathbf{b}^i \cdot \begin{Bmatrix} n \\ j\,k \end{Bmatrix} \mathbf{b}_n + C_{ikn}\mathbf{b}^n \cdot \mathbf{b}_j = 0 \tag{9.8}$$

Now use (9.6) to obtain

$$\begin{Bmatrix} i \\ j\,k \end{Bmatrix} + C_{ikj} = 0 \tag{9.9}$$

Thus, (9.5) becomes

$$\frac{\partial \mathbf{b}^i}{\partial q_j} = -\begin{Bmatrix} i \\ k\,j \end{Bmatrix} \mathbf{b}^k \tag{9.10}$$

The Christoffel symbols have the following symmetry property:

$$\begin{Bmatrix} i \\ j\,k \end{Bmatrix} = \begin{Bmatrix} i \\ k\,j \end{Bmatrix} \tag{9.11}$$

This is established by first writing the basis vectors in terms of the unit coordinate vectors of rectangular coordinates:

$$\mathbf{b}_i = \frac{\partial x_k}{\partial q_i}\hat{\mathbf{u}}_k \tag{9.12}$$

The partial derivatives of (9.12) are

$$\frac{\partial \mathbf{b}_i}{\partial q_j} = \frac{\partial^2 x_k}{\partial q_j \partial q_i}\hat{\mathbf{u}}_k \tag{9.13}$$

which shows that

$$\frac{\partial \mathbf{b}_i}{\partial q_j} = \frac{\partial \mathbf{b}_j}{\partial q_i} \tag{9.14}$$

Equation (9.11) then follows.

The actual evaluation of the Christoffel symbols begins with

$$\mathbf{b}_i \cdot \mathbf{b}_k = g_{ik} \tag{9.15}$$

The partial derivatives of (9.15) are

$$\mathbf{b}_i \cdot \frac{\partial \mathbf{b}_k}{\partial q_j} + \frac{\partial \mathbf{b}_i}{\partial q_j} \cdot \mathbf{b}_k = \frac{\partial g_{ik}}{\partial q_j} \tag{9.16}$$

Two similar expressions can be found by relabeling the indices:

$$\mathbf{b}_j \cdot \frac{\partial \mathbf{b}_k}{\partial q_i} + \frac{\partial \mathbf{b}_j}{\partial q_i} \cdot \mathbf{b}_k = \frac{\partial g_{jk}}{\partial q_i} \tag{9.17}$$

$$\mathbf{b}_i \cdot \frac{\partial \mathbf{b}_j}{\partial q_k} + \frac{\partial \mathbf{b}_i}{\partial q_k} \cdot \mathbf{b}_j = \frac{\partial g_{ij}}{\partial q_k} \tag{9.18}$$

Now compute (9.16) + (9.17) − (9.18) to get

$$\frac{\partial \mathbf{b}_i}{\partial q_j} \cdot \mathbf{b}_k = \frac{1}{2}\left[\frac{\partial g_{ik}}{\partial q_j} + \frac{\partial g_{jk}}{\partial q_i} - \frac{\partial g_{ij}}{\partial q_k}\right] \equiv [ij, k] \tag{9.19}$$

In the calculation, it is necessary to use (9.14). A new symbol, $[ij, k]$, has been defined. It is the *Christoffel symbol of the first kind*. Now use (9.14) and (9.15) to get

$$\begin{Bmatrix} n \\ i\,j \end{Bmatrix} (\mathbf{b}_n \cdot \mathbf{b}_k) = \begin{Bmatrix} n \\ i\,j \end{Bmatrix} g_{nk} = [ij, k] \tag{9.20}$$

By means of

$$g_{nk}g^{mk} = \delta_n^m \tag{9.21}$$

we finally get

$$\begin{Bmatrix} m \\ i\,j \end{Bmatrix} = g^{mk}[ij, k] \tag{9.22}$$

which completes the evaluation. Equations (9.19) and (9.22) permit both types of Christoffel symbols to be computed in terms of the elements of the metric tensor.

By combining (9.2), (9.3), and (9.4), we can express the acceleration vector as

$$\mathbf{a} = A^k \mathbf{b}_k \tag{9.23}$$

where A^k are the contravariant components given by

$$A^k = \ddot{q}^k + \dot{q}^i \dot{q}^j \begin{Bmatrix} k \\ i\,j \end{Bmatrix} \tag{9.24}$$

It is interesting that $\ddot{q}^k$ by themselves do not have a contravariant transformation property. Moreover, the Christoffel symbols are not third-rank tensors. The combination (9.24) does, however, produce three quantities which transform contravariantly and which are properly identified as the contravariant components of acceleration. The covariant components of acceleration are found by lowering the index and then using (9.20):

$$A_n = g_{nk}\ddot{q}^k + \dot{q}^i \dot{q}^j [ij, n] \tag{9.25}$$

Christoffel symbols are quite messy to evaluate, so it is not surprising that everyone tries to avoid it where possible. Fortunately, the covariant form of the acceleration components as given by (9.25) can be simplified quite a bit. First, note that

$$g_{nk}\ddot{q}^k = \frac{d}{dt}(g_{nk}\dot{q}^k) - \frac{\partial g_{nk}}{\partial q_j}\dot{q}^j \dot{q}^k \tag{9.26}$$

Then, use (9.19) to get

$$\begin{aligned} A_n &= \frac{d}{dt}(g_{nk}\dot{q}^k) - \frac{\partial g_{nk}}{\partial q_j}\dot{q}^j \dot{q}^k \\ &\quad + \dot{q}^i \dot{q}^j \frac{1}{2}\left[\frac{\partial g_{in}}{\partial q_j} + \frac{\partial g_{jn}}{\partial q_i} - \frac{\partial g_{ij}}{\partial q_n}\right] \\ &= \frac{d}{dt}(g_{nk}\dot{q}^k) - \frac{1}{2}\dot{q}^i \dot{q}^j \frac{\partial g_{ij}}{\partial q_n} \end{aligned} \tag{9.27}$$

In obtaining (9.27), we have used

$$\frac{\partial g_{nk}}{\partial q_j}\dot{q}^j \dot{q}^k = \dot{q}^i \dot{q}^j \frac{\partial g_{in}}{\partial q_j} = \dot{q}^i \dot{q}^j \frac{\partial g_{jn}}{\partial q_i} \tag{9.28}$$

Equation (9.27) is fairly tractable as it stands, but an even greater simplification is possible. If we define a scalar function by means of

$$\Phi = \frac{1}{2}\left(\frac{ds}{dt}\right)^2 = \frac{1}{2} g_{ij}\dot{q}^i \dot{q}^j \tag{9.29}$$

then (9.27) is equivalent to

$$\frac{d}{dt}\left(\frac{\partial \Phi}{\partial \dot{q}^n}\right) - \frac{\partial \Phi}{\partial q_n} = A_n \tag{9.30}$$

which is really quite easy to use.

Equation (1.17) gives the components of acceleration in cylindrical coordinates. Let's see if we can get the same result from (9.30). The function Φ is

$$\Phi = \tfrac{1}{2}(\dot{r}^2 + r^2\dot{\theta}^2 + \dot{z}^2) \tag{9.31}$$

The three covariant components of acceleration are found from

$$\frac{d}{dt}\left(\frac{\partial \Phi}{\partial \dot{r}}\right) - \frac{\partial \Phi}{\partial r} = A_1 \tag{9.32}$$

$$\frac{d}{dt}\left(\frac{\partial \Phi}{\partial \dot{\theta}}\right) - \frac{\partial \Phi}{\partial \theta} = A_2 \tag{9.33}$$

$$\frac{d}{dt}\left(\frac{\partial \Phi}{\partial \dot{z}}\right) - \frac{\partial \Phi}{\partial z} = A_3 \tag{9.34}$$

We find, for example, that

$$\frac{\partial \Phi}{\partial \dot{r}} = \dot{r} \qquad \frac{d}{dt}\left(\frac{\partial \Phi}{\partial \dot{r}}\right) = \ddot{r} \qquad \frac{\partial \Phi}{\partial r} = r\dot{\theta}^2 \tag{9.35}$$

This gives

$$\ddot{r} - r\dot{\theta}^2 = A_1 \tag{9.36}$$

Similarly,

$$r^2\ddot{\theta} + 2r\dot{r}\dot{\theta} = A_2 \tag{9.37}$$

$$\ddot{z} = A_3 \tag{9.38}$$

Since the coordinate system is orthogonal, the physical components of acceleration follow from

$$a_r = \sqrt{g_{11}}\,A^1 = \frac{1}{\sqrt{g_{11}}}A_1 = \ddot{r} - r\dot{\theta}^2 \tag{9.39}$$

$$a_\theta = \frac{1}{\sqrt{g_{22}}} A_2 = r\ddot{\theta} + 2\dot{r}\dot{\theta} \tag{9.40}$$

$$a_z = \frac{1}{\sqrt{g_{33}}} A_3 = \ddot{z} \tag{9.41}$$

As another example, the function Φ in spherical coordinates is

$$\Phi = \tfrac{1}{2}(\dot{r}^2 + r^2\dot{\theta}^2 + r^2 \sin^2\theta\, \dot{\phi}^2) \tag{9.42}$$

We find

$$\frac{d}{dt}\left(\frac{\partial \Phi}{\partial \dot{r}}\right) - \frac{\partial \Phi}{\partial r} = \ddot{r} - r\dot{\theta}^2 - r\sin^2\theta\,\dot{\phi}^2 = A_1 \tag{9.43}$$

$$\frac{d}{dt}\left(\frac{\partial \Phi}{\partial \dot{\theta}}\right) - \frac{\partial \Phi}{\partial \theta} = r^2\ddot{\theta} + 2r\dot{r}\dot{\theta} - r^2 \sin\theta\cos\theta\,\dot{\phi}^2 = A_2 \tag{9.44}$$

$$\frac{d}{dt}\left(\frac{\partial \Phi}{\partial \dot{\phi}}\right) - \frac{\partial \Phi}{\partial \phi} = \frac{d}{dt}(r^2 \sin^2\theta\,\dot{\phi}) = A_3 \tag{9.45}$$

The physical components of acceleration are

$$a_r = \frac{1}{\sqrt{g_{11}}} A_1 = \ddot{r} - r\dot{\theta}^2 - r\sin^2\theta\,\dot{\phi}^2 \tag{9.46}$$

$$a_\theta = \frac{1}{\sqrt{g_{22}}} A_2 = r\ddot{\theta} + 2\dot{r}\dot{\theta} - r\sin\theta\cos\theta\,\dot{\phi}^2 \tag{9.47}$$

$$a_\phi = \frac{1}{\sqrt{g_{33}}} A_3 = \frac{1}{r\sin\theta}\frac{d}{dt}(r^2\sin^2\theta\,\dot{\phi}) \tag{9.48}$$

The significance of (9.30) goes beyond its use as a convenient way to calculate acceleration. Note that the function Φ as given by (9.29) becomes kinetic energy when it is multiplied by the mass of a particle:

$$K = m\Phi = \tfrac{1}{2}mu^2 = \tfrac{1}{2}mg_{ij}\dot{q}^i\dot{q}^j \tag{9.49}$$

Thus, (9.30) can be expressed as

$$\frac{d}{dt}\left(\frac{\partial K}{\partial \dot{q}^n}\right) - \frac{\partial K}{\partial q_n} = Q_n \tag{9.50}$$

where we define

$$Q_n = mA_n \tag{9.51}$$

as the generalized force components. They are of course the covariant components of the actual force acting on the particle and must be specified before we can proceed from (9.51) to find the equations of motion of the particle. We can think of (9.51) as a generally covariant form of Newton's second law for a single particle. The generalized force components are related to the rectangular Cartesian components of the force by the covariant transformation law:

$$Q_n = \frac{\partial x_i}{\partial q_n} F_i \tag{9.52}$$

Suppose that the forces acting on the particle are conservative. A potential function then exists, and we can write

$$Q_n = -\frac{\partial x_i}{\partial q_n}\frac{\partial V}{\partial x_i} = -\frac{\partial V}{\partial q_n} \tag{9.53}$$

Equation (9.50) can be expressed in the form

$$\frac{d}{dt}\left(\frac{\partial K}{\partial \dot{q}^n}\right) - \frac{\partial}{\partial q_n}(K - V) = 0 \tag{9.54}$$

It is convenient to define a function called the *Lagrangian* by means of

$$L = K - V \tag{9.55}$$

Since V depends on coordinates only and not the generalized velocity components,

$$\frac{\partial V}{\partial \dot{q}^n} = 0 \tag{9.56}$$

Equation (9.54) can then be written compactly as

$$\frac{d}{dt}\left(\frac{\partial L}{\partial \dot{q}^n}\right) - \frac{\partial L}{\partial q_n} = 0 \tag{9.57}$$

which are called *Lagrange's equations.*

Given a single particle moving in a conservative force field, Lagrange's equations can be used directly to find the equations of motion, thus freeing us of the necessity of dealing directly with forces. The Lagrange method is not confined to use with simple conservative forces as the derivation given here seems to indicate, but it is much more generally applicable as we will show in a later section.

PROBLEMS

9.1. How many distinct Christoffel symbols are there in a general curvilinear coordinate system? How many are there if the coordinate system is orthogonal?

9.2. Prove that

$$\frac{\partial \mathbf{b}_i}{\partial q_j} = [ij, n]\mathbf{b}^n$$

9.3. Evaluate all the Christoffel symbols of both kinds for spherical coordinates. Do the calculation directly starting with (9.19). Another way to do it is to differentiate

$$\mathbf{b}_i = \frac{\partial x_j}{\partial q_i}\hat{\mathbf{u}}_j$$

After the differentiation is done, eliminate $\hat{\mathbf{u}}_j$ by means of

$$\hat{\mathbf{u}}_j = \frac{\partial x_j}{\partial q_i}\mathbf{b}^i$$

Then, use problem 9.2. Use this second method as a check.

9.4. Let **a** be any vector field. The partial derivatives of **a** can be expressed as

$$\frac{\partial \mathbf{a}}{\partial q_j} = A^k{}_{,j}\mathbf{b}_k = A_{k,j}\mathbf{b}^k$$

Show that

$$A^k{}_{,j} = \frac{\partial A^k}{\partial q_j} + A^n\begin{Bmatrix} k \\ nj \end{Bmatrix} \qquad A_{k,j} = \frac{\partial A_k}{\partial q_j} - A_n\begin{Bmatrix} n \\ kj \end{Bmatrix}$$

These expressions are called the *covariant derivatives* of the components of **a**. What are the transformation properties of $A^k{}_{,j}$ and $A_{k,j}$? Show that $A_{k,j} = g_{kn}A^n{}_{,j}$.

9.5. A spherical pendulum consists of a particle which is constrained to move on the surface of a sphere (r = constant) and which is acted on by a uniform gravitational field. If the z axis points down as in Fig. 9.1, then the gravitational field is given by $\mathbf{g} = g\hat{\mathbf{k}}$. Use (9.47) and (9.48) to find the equations of motion of the particle. Show that the quantity

$$\ell_z = mr^2 \sin^2\theta\,\dot{\phi}$$

is a constant of the motion and prove that it is the z component of the angular momentum.

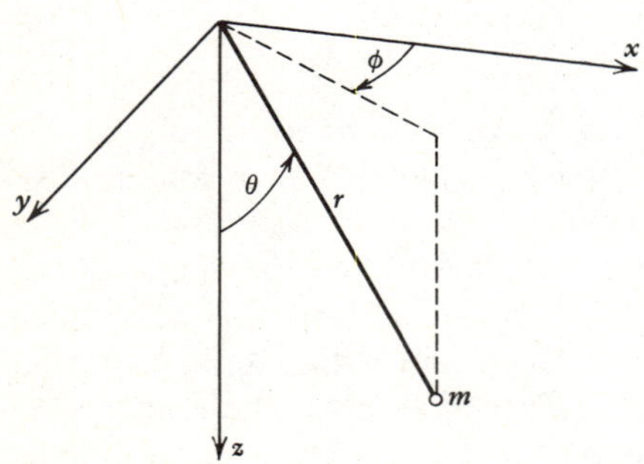

Fig. 9.1

Obtain a second-order differential equation in which $\theta(t)$ is the only unknown function. (Don't try to solve it.)

9.6. A particle moves in a uniform gravitational field and is acted on by no other forces. Express the Lagrangian for the particle in rectangular coordinates. Assume that the gravitational field points in the negative z direction. Use Lagrange's equation in the form (9.57) to obtain the equations of motion of the particle.

10. DIVERGENCE, CURL, AND LAPLACIAN

It will be convenient in this section to employ the abbreviation $\partial_k = \partial/\partial q_k$. The calculation of the divergence of a vector begins with the gradient operator as given by (8.3):

$$\nabla \cdot \mathbf{a} = \mathbf{b}^k \partial_k \cdot A^i \mathbf{b}_i = \mathbf{b}^k \cdot \mathbf{b}_i \partial_k A^i + \mathbf{b}^k A^i \cdot \partial_k \mathbf{b}_i \tag{10.1}$$

By means of (6.9) and (9.4), we get

$$\nabla \cdot \mathbf{a} = \partial_k A^k + A^i \left\{ {k \atop i\,k} \right\} \tag{10.2}$$

The contracted Christoffel symbol in (10.2) is given by

$$\left\{ {k \atop i\,k} \right\} = g^{kn} \frac{1}{2} [\partial_k g_{in} + \partial_i g_{kn} - \partial_n g_{ik}] = \frac{1}{2} g^{kn} \partial_i g_{kn} \tag{10.3}$$

where use has been made of

$$g^{kn} \partial_k g_{in} = g^{kn} \partial_n g_{ik} \tag{10.4}$$

A further simplification is possible. If g is the determinant of the metric tensor, then

$$e_{ijk} g = g_{ir} g_{js} g_{kt} e_{rst} \tag{10.5}$$

where e_{ijk} is the permutation symbol. Partial differentiation of (10.5) yields

$$e_{ijk} \partial_n g = (\partial_n g_{ir}) g_{js} g_{kt} e_{rst} + g_{ir} (\partial_n g_{js}) g_{kt} e_{rst} + g_{ir} g_{js} (\partial_n g_{kt}) e_{rst} \tag{10.6}$$

Now use (6.14) and the fact that

$$g^{1i} g^{2j} g^{3k} e_{ijk} = \det g^{ij} = \frac{1}{g} \tag{10.7}$$

to obtain

$$\begin{aligned} \frac{1}{g} \partial_n g &= (\partial_n g_{ir}) g^{1i} e_{r23} + (\partial_n g_{js}) g^{2j} e_{1s3} + (\partial_n g_{kt}) g^{3k} e_{12t} \\ &= (\partial_n g_{i1}) g^{1i} + (\partial_n g_{j2}) g^{2j} + (\partial_n g_{k3}) g^{3k} \\ &= g^{ij} \partial_n g_{ij} \end{aligned} \tag{10.8}$$

Thus, (10.3) can be written as

$$\left\{ {k \atop i\,k} \right\} = \frac{1}{2g} \partial_i g = \frac{1}{\sqrt{g}} \partial_i \sqrt{g} \tag{10.9}$$

The expression (10.2) for the divergence of a vector is then

$$\nabla \cdot \mathbf{a} = \partial_k A^k + A^k \frac{1}{\sqrt{g}} \partial_k \sqrt{g} = \frac{1}{\sqrt{g}} \partial_k (\sqrt{g} A^k) \tag{10.10}$$

which, in spite of the somewhat tedious calculation used to obtain it, is a simple and easy-to-use expression for divergence, valid in any curvilinear coordinate system. When working in orthogonal curvilinear coordinates, it is convenient to have the divergence expressed in terms of the physical components of the vector:

$$\nabla \cdot \mathbf{a} = \frac{1}{\sqrt{g_{11} g_{22} g_{33}}} \left[\partial_1 \sqrt{g_{22} g_{33}}\, a(1) + \partial_2 \sqrt{g_{11} g_{33}}\, a(2) + \partial_3 \sqrt{g_{11} g_{22}}\, a(3) \right] \tag{10.11}$$

It is easy to work (10.11) out for cylindrical and spherical coordinates. At the end of this section will be found a list containing divergence and other expressions worked out in these coordinate systems.

The expression for Laplacian follows from (10.10) by replacing $\mathbf{a}$ by $\nabla\psi$. We must remember, however, that the partial derivatives of a scalar function are covariant. Therefore, A^i must be replaced by

$$A^i = g^{ij} \partial_j \psi \tag{10.12}$$

The Laplacian is then

$$\nabla^2 \psi = \frac{1}{\sqrt{g}} \partial_i (\sqrt{g}\, g^{ij} \partial_j \psi) \tag{10.13}$$

In orthogonal coordinates, the Laplacian works out to

$$\nabla^2\psi = \frac{1}{\sqrt{g}}\left[\partial_1\left(\sqrt{\frac{g_{22}g_{33}}{g_{11}}}\,\partial_1\psi\right) + \partial_2\left(\sqrt{\frac{g_{11}g_{33}}{g_{22}}}\,\partial_2\psi\right) + \partial_3\left(\sqrt{\frac{g_{11}g_{22}}{g_{33}}}\,\partial_3\psi\right)\right] \tag{10.14}$$

The calculation of the curl of a vector proceeds as follows:

$$\begin{aligned}\mathbf{c} = \nabla\times\mathbf{a} &= \mathbf{b}^j\,\partial_j \times A_k\mathbf{b}^k \\ &= (\mathbf{b}^j\times\mathbf{b}^k)\,\partial_j A_k - A_k\mathbf{b}^j\times\left\{ {k \atop j\,n} \right\}\mathbf{b}^n \\ &= \varepsilon^{jki}\mathbf{b}_i\,\partial_j A_k - A_k\varepsilon^{jni}\mathbf{b}_i\left\{ {k \atop j\,n} \right\}\end{aligned} \tag{10.15}$$

where use has been made of (7.10). It is easy to prove that

$$\varepsilon^{jni}\left\{ {k \atop j\,n} \right\} = 0 \tag{10.16}$$

Therefore,

$$\mathbf{c} = \varepsilon^{ijk}(\partial_j A_k)\mathbf{b}_i \tag{10.17}$$

Written out, the contravariant components of $\mathbf{c}$ are

$$\begin{aligned}C^1 &= \frac{1}{\sqrt{g}}(\partial_2 A_3 - \partial_3 A_2) \\ C^2 &= \frac{1}{\sqrt{g}}(\partial_3 A_1 - \partial_1 A_3) \\ C^3 &= \frac{1}{\sqrt{g}}(\partial_1 A_2 - \partial_2 A_1)\end{aligned} \tag{10.18}$$

which are hardly more complicated than the corresponding expressions in rectangular coordinates. If the coordinates are orthogonal and all vectors are expressed in terms of physical components, we find

$$c(1) = \frac{1}{\sqrt{g_{22}g_{33}}}\left[\partial_2\sqrt{g_{33}}\,a(3) - \partial_3\sqrt{g_{22}}\,a(2)\right]$$

$$c(2) = \frac{1}{\sqrt{g_{11}g_{33}}}\left[\partial_3\sqrt{g_{11}}\,a(1) - \partial_1\sqrt{g_{33}}\,a(3)\right] \tag{10.19}$$

$$c(3) = \frac{1}{\sqrt{g_{11}g_{22}}}\left[\partial_1\sqrt{g_{22}}\,a(2) - \partial_2\sqrt{g_{11}}\,a(1)\right]$$

The following is a list of expressions for divergence, curl, and Laplacian in cylindrical and spherical coordinates. All vectors are in terms of their physical components.

Cylindrical Coordinates

$$\nabla\cdot\mathbf{a} = \frac{1}{r}\frac{\partial}{\partial r}(ra_r) + \frac{1}{r}\frac{\partial a_\theta}{\partial\theta} + \frac{\partial a_z}{\partial z} \tag{10.20}$$

$$\nabla^2\psi = \frac{1}{r}\frac{\partial}{\partial r}\left(r\frac{\partial\psi}{\partial r}\right) + \frac{1}{r^2}\frac{\partial^2\psi}{\partial\theta^2} + \frac{\partial^2\psi}{\partial z^2} \tag{10.21}$$

$$c_r = \frac{1}{r}\left[\frac{\partial a_z}{\partial\theta} - \frac{\partial}{\partial z}(ra_\theta)\right] \tag{10.22}$$

$$c_\theta = \frac{\partial a_r}{\partial z} - \frac{\partial a_z}{\partial r} \tag{10.23}$$

$$c_z = \frac{1}{r}\left[\frac{\partial}{\partial r}(ra_\theta) - \frac{\partial a_r}{\partial\theta}\right] \tag{10.24}$$

Spherical Coordinates

$$\nabla\cdot\mathbf{a} = \frac{1}{r^2}\frac{\partial}{\partial r}(r^2 a_r) + \frac{1}{r\sin\theta}\frac{\partial}{\partial\theta}(\sin\theta\, a_\theta) + \frac{1}{r\sin\theta}\frac{\partial a_\phi}{\partial\phi} \tag{10.25}$$

$$\nabla^2\psi = \frac{1}{r^2}\frac{\partial}{\partial r}\left(r^2\frac{\partial\psi}{\partial r}\right) + \frac{1}{r^2\sin\theta}\frac{\partial}{\partial\theta}\left(\sin\theta\frac{\partial\psi}{\partial\theta}\right) + \frac{1}{r^2\sin^2\theta}\frac{\partial^2\psi}{\partial\phi^2} \tag{10.26}$$

$$c_r = \frac{1}{r\sin\theta}\left[\frac{\partial}{\partial\theta}(\sin\theta\, a_\phi) - \frac{\partial a_\theta}{\partial\phi}\right] \tag{10.27}$$

$$c_\theta = \frac{1}{r\sin\theta}\left[\frac{\partial a_r}{\partial\phi} - \frac{\partial}{\partial r}(r\sin\theta\, a_\phi)\right] \tag{10.28}$$

$$c_\phi = \frac{1}{r}\left[\frac{\partial}{\partial r}(ra_\theta) - \frac{\partial a_r}{\partial\theta}\right] \tag{10.29}$$

PROBLEMS

10.1. Why is (10.4) true?

10.2. Prove that $\det g^{ij} = 1/g$.

10.3. Multiply (10.6) through by the factor $g^{1i}g^{2j}g^{3k}$ and go through the steps in detail which lead to (10.8).

10.4. Prove (10.16).

Hint: Recall that $\varepsilon^{jni} = -\varepsilon^{nji}$.

10.5. Let the metric tensor be given by the matrix A of problem 8.1. Find the expressions for the divergence and curl of a vector with the vectors expressed in terms of their physical components. Find the expression for the Laplacian of a scalar function.

10.6. Consider the wave equation given by equation (12.45) of Chapter 3. Suppose that a time-dependent point source exists at the origin but that $\rho = 0$ for $r \neq 0$. Set up the wave equation in spherical coordinates for the case where ψ depends on r and t and not on θ or ϕ. Make the substitution $\psi = (1/r)F(r, t)$. Find the partial differential equation obeyed by $F(r, t)$. What kind of solutions does it have and what do they mean?

10.7. A curvilinear coordinate system is defined by means of the transformation equations

$$x = \frac{1}{2}\left(q_1^2 - q_2^2\right) \qquad y = q_1 q_2 \qquad z = q_3$$

Verify that the coordinate system is right-handed. Find the elements of the metric tensor. Qualitatively, what do the coordinate lines $q_1 = \text{constant}$ and $q_2 = \text{constant}$ look like? Find the expressions for $\nabla \cdot \mathbf{a}$, $\nabla \times \mathbf{a}$, and $\nabla^2 \psi$. Express all vectors in terms of their physical components.

11. MAGNETIC DIPOLES

Suppose that a steady current I exists in a thin wire in the shape of a closed loop as illustrated in Fig. 11.1. Equation (11.10) of Chapter 3 gives the vector potential for a volume distribution of current. If da is the cross-sectional area of the wire, then

$$\mathbf{j}\, d\Sigma' = \mathbf{j}\, da\, ds' = I\, d\mathbf{s}' \tag{11.1}$$

where $d\mathbf{s}'$ is an infinitesimal displacement vector tangent to the wire pointing in the direction of the current. The vector potential for the current loop is then

$$\mathbf{A}(\mathbf{r}) = \frac{\mu_0 I}{4\pi} \oint \frac{d\mathbf{s}'}{R} \tag{11.2}$$

where the integral is taken around the current loop. Recall that $\mathbf{r}$ is the position vector of the field point or point at which the potential is calculated and $\mathbf{r}'$ is the position vector of the variable point of integration. Let us suppose that the current loop is in the vicinity of the origin and that the field point is at a sufficient distance from the origin so that $r \gg r'$ for all points of the loop. We then have approximately

$$\frac{1}{R} = \left(r^2 + r'^2 - 2\mathbf{r} \cdot \mathbf{r}'\right)^{-1/2}$$

$$= \frac{1}{r}\left(1 + \frac{r'^2}{r^2} - \frac{2\mathbf{r} \cdot \mathbf{r}'}{\mathbf{r}^2}\right)^{-1/2}$$

$$\simeq \frac{1}{r}\left(1 + \frac{\mathbf{r} \cdot \mathbf{r}'}{r^2}\right) \tag{11.3}$$

Since obviously $\oint d\mathbf{s}' = 0$, the vector potential is approximately

$$\mathbf{A}(\mathbf{r}) = \frac{\mu_0 I}{4\pi r^3} \oint (\mathbf{r} \cdot \mathbf{r}')\, d\mathbf{s}' \tag{11.4}$$

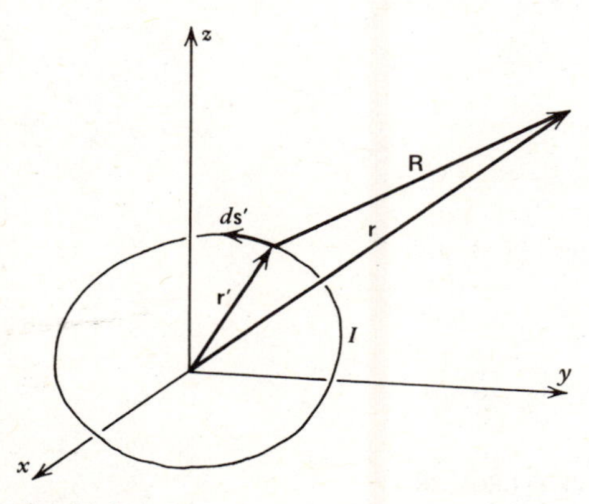

Fig. 11.1

The identity (9.20) of Chapter 3 can now be used to get

$$\oint (\mathbf{r}\cdot\mathbf{r}')\, d\mathbf{s}' = \int_\sigma \hat{\mathbf{n}} \times \nabla'(\mathbf{r}\cdot\mathbf{r}')\, d\sigma' = \int_\sigma (\hat{\mathbf{n}} \times \mathbf{r})\, d\sigma' \tag{11.5}$$

where σ is any open surface bounded by the current loop. Thus, at distances from the loop which are large compared to its dimensions, the vector potential is

$$\mathbf{A}(\mathbf{r}) = \frac{\mu_0}{4\pi}\frac{\boldsymbol{\mu}\times\mathbf{r}}{r^3} \tag{11.6}$$

where we define the *magnetic moment* of the current loop by

$$\boldsymbol{\mu} = I\int_\sigma \hat{\mathbf{n}}\, d\sigma' \tag{11.7}$$

In problem 11.4 of Chapter 3, it is shown that another form for the magnetic moment is

$$\boldsymbol{\mu} = \frac{1}{2} I \oint \mathbf{r}' \times d\mathbf{s}' \tag{11.8}$$

From (11.1) and (11.8), we can infer that the correct formula for the magnetic moment of a volume distribution of current is

$$\boldsymbol{\mu} = \frac{1}{2}\oint \mathbf{r}' \times \mathbf{j}(\mathbf{r}')\, d\Sigma' \tag{11.9}$$

The calculation of the magnetic field from (11.6) is facilitated if the magnetic moment is oriented along the z axis as shown in Fig. 11.2. Spherical coordinates are appropriate. We find

$$\mathbf{A} = \frac{\mu_0}{4\pi}\frac{\mu \sin\theta}{r^2}\hat{\boldsymbol{\phi}} = A_\phi \hat{\boldsymbol{\phi}} \tag{11.10}$$

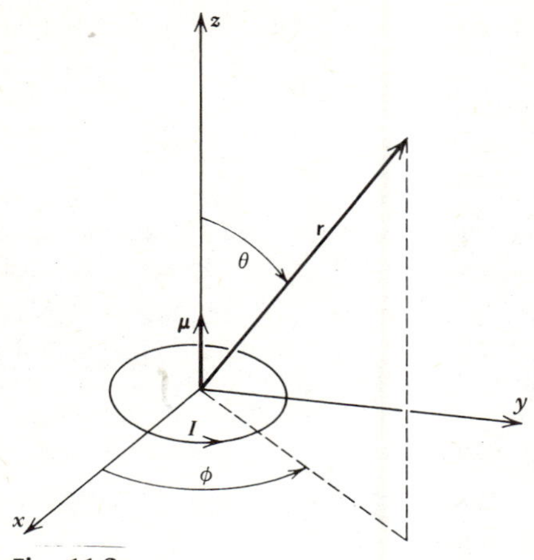

Fig. 11.2

Here, A_ϕ is a physical component. The magnetic field is found from (10.27), (10.28), and (10.29):

$$B_r = \frac{\mu_0}{2\pi}\frac{\mu\cos\theta}{r^3} \tag{11.11}$$

$$B_\theta = \frac{\mu_0}{4\pi}\frac{\mu\sin\theta}{r^3} \tag{11.12}$$

$$B_\phi = 0 \tag{11.13}$$

Reference to section 2 shows that the magnetic field at a large distance from a current loop has exactly the same structure as does the field of an electric dipole. For this reason, the magnetic field which we have obtained in this section is called a *magnetic dipole field* even though no actual *magnetic poles* are involved.

PROBLEMS

11.1. A sphere of radius a has a charge Q spread uniformly throughout its volume. Find its magnetic moment if it is set spinning about a diameter with an angular velocity ω. Do the calculation as follows: Let the z axis be the axis of spin. Show that the current density at any point inside the sphere is given by $\mathbf{j} = \rho r \sin\theta\, \omega \hat{\boldsymbol{\phi}}$, where ρ is the charge density. Show that $\mathbf{r}\times\mathbf{j} = -\rho r^2 \sin\theta\, \omega\hat{\boldsymbol{\theta}}$. Use the results of problem 8.5 to express $\hat{\boldsymbol{\theta}}$ in terms of $\hat{\mathbf{i}}$, $\hat{\mathbf{j}}$, and $\hat{\mathbf{k}}$. Use (11.9) to calculate the magnetic moment.

11.2. Calculate the divergence and the curl of the magnetic field as given by (11.11), (11.12), and (11.13).

11.3. Express the vector potential due to a current in a long, straight wire in cylindrical coordinates. See problem 11.7 of Chapter 3. Find **B** by computing the curl of **A** in cylindrical coordinates.

11.4. Show that the dipole potential (2.4) is a solution of Laplace's equation. Use (10.26).

11.5. Show that the quadrupole potential which you got for problem 8.4 is a solution of Laplace's equation. Use (10.26).

11.6. A static charge distribution is located in the vicinity of the origin of coordinates. If r is large compared to the dimensions of the distribution, show that the potential is approximately

$$\psi = \frac{1}{4\pi\varepsilon_0}\left(\frac{Q}{r} + \frac{\mathbf{p}\cdot\mathbf{r}}{r^3}\right)$$

where Q is the total charge and

$$\mathbf{p} = \int \rho \mathbf{r}' \, d\Sigma'$$

is the *dipole moment* of the distribution.

11.7. A total charge Q is distributed uniformly throughout a sphere of radius a. Set up $\nabla \cdot \mathbf{E} = (1/\varepsilon_0)\rho$ in spherical coordinates and integrate it to find the electric field for both $r < a$ and $r > a$.

11.8. A spinning object has total mass m and total charge q. The charge and mass are distributed in the same manner so that the volume densities of charge and mass are given by $\rho_q = qf(\mathbf{r})$ and $\rho_m = mf(\mathbf{r})$. Show that the magnetic moment and the angular momentum of the object are related by

$$\boldsymbol{\mu} = \frac{q}{2m}\boldsymbol{\ell}$$

A spinning charged sphere has its mass uniformly distributed throughout its volume and its charge uniformly distributed on its surface. Show that in this case

$$\boldsymbol{\mu} = \frac{5q}{6m}\boldsymbol{\ell}$$

Refer to the discussion of spin in section 6 of Chapter 9. Angular momentum is discussed in section 9 of Chapter 2.

Hint: You must replace (11.9) by

$$\boldsymbol{\mu} = \frac{1}{2}\int_\sigma \mathbf{r} \times \lambda \mathbf{u} \, d\sigma$$

where λ is the charge per unit area.

12. SPHEROIDAL COORDINATES

Rectangular, spherical, and cylindrical coordinates are probably the most widely used coordinate systems in practice. There are however other somewhat more esoteric coordinate systems which find application in special circumstances. One such coordinate system is spheroidal coordinates. There are two cases: oblate and prolate. To visualize the oblate case, imagine that a spherical coordinate system is compressed in the z direction so that the coordinate surfaces $r =$ constant are flattened into spheroids. The profiles of the coordinate surfaces in the planes $\phi =$ constant are now ellipses rather than circles. For the prolate case, spherical coordinates are imagined to be elongated in the z direction so that the new coordinate surfaces are egg-shaped.

One way to define oblate spheroidal coordinates is by means of the transformation equations

$$x = cp \sin v \cos \phi \tag{12.1}$$

$$y = cp \sin v \sin \phi \tag{12.2}$$

$$z = c\sqrt{p^2 - 1} \cos v \tag{12.3}$$

where c is a constant and the ranges of the coordinates are given by

$$1 \le p < \infty \qquad 0 \le v \le \pi \qquad 0 \le \phi \le 2\pi \tag{12.4}$$

One advantage of this definition is that it becomes identical to spherical coordinates for large values of p. We identify $cp = r$ and $v = \theta$ in the limit $p \to \infty$. A second version of the transformation from rectangular to spheroidal coordinates results if $q = \cos v$:

$$x = cp\sqrt{1-q^2}\cos\phi \tag{12.5}$$

$$y = cp\sqrt{1-q^2}\sin\phi \tag{12.6}$$

$$z = c\sqrt{p^2-1}\,q \tag{12.7}$$

where $-1 \le q \le +1$. A third way to define oblate spheroidal coordinates results if we let $p = \cosh u$:

$$x = c\cosh u \sin v \cos\phi \tag{12.8}$$

$$y = c\cosh u \sin v \sin\phi \tag{12.9}$$

$$z = c\sinh u \cos v \tag{12.10}$$

where $0 \le u < \infty$. Since not a great deal seems to be gained by the use of hyperbolic functions, the definition that will be used here is (12.1)–(12.3).

The angle ϕ is essentially the same as the azimuth angle of spherical coordinates. The coordinate system has rotational symmetry about the z axis. We will therefore find the profile of the coordinate surfaces in any plane $\phi =$ constant (i.e., any plane which passes through the z axis). The coordinate surfaces can then be generated by rotating these profiles around the z axis. It is convenient to define $s^2 = x^2 + y^2$. By means of (12.1) and (12.2),

$$s^2 = c^2p^2\sin^2 v \tag{12.11}$$

The use of (12.11) to eliminate v from (12.3) gives

$$\frac{s^2}{p^2c^2} + \frac{z^2}{(p^2-1)c^2} = 1 \tag{12.12}$$

The standard form for the equation of an ellipse is

$$\frac{s^2}{a^2} + \frac{z^2}{b^2} = 1 \tag{12.13}$$

where a and b are the lengths of the semimajor and the semiminor axes, respectively. The distance from the origin to either focus of the ellipse is given by

$$\sqrt{a^2-b^2} = \sqrt{p^2c^2-(p^2-1)c^2} = c \tag{12.14}$$

which clarifies the meaning of the constant c. Each value of p in (12.12) gives a coordinate line which is an ellipse. The family of ellipses so generated is confocal and is illustrated in Fig. 12.1 for several values of p. As $p \to \infty$, the ellipses become circles of radius $r = pc$. At the other extreme, (12.3) shows that $z = 0$ if $p = 1$. This means that the ellipse degenerates into a straight line between $s = -c$ and $s = +c$ for $p = 1$. When the family of ellipses is rotated about the z axis, coordinate surfaces $p =$ constant are generated which are oblate spheroids. The straight line between $-c$ and $+c$ corresponding to $p = 1$ generates a circular disc of radius c in the plane $z = 0$.

If p is eliminated between (12.11) and (12.3), we find

$$\frac{s^2}{c^2\sin^2 v} - \frac{z^2}{c^2\cos^2 v} = 1 \tag{12.15}$$

The standard form for the equation of a hyperbola is

$$\frac{s^2}{a^2} - \frac{z^2}{b^2} = 1 \tag{12.16}$$

A clarification of the meaning of a and b will be found in problem 12.2 at the end of this section. The distance from the origin to the focus of the hyperbola

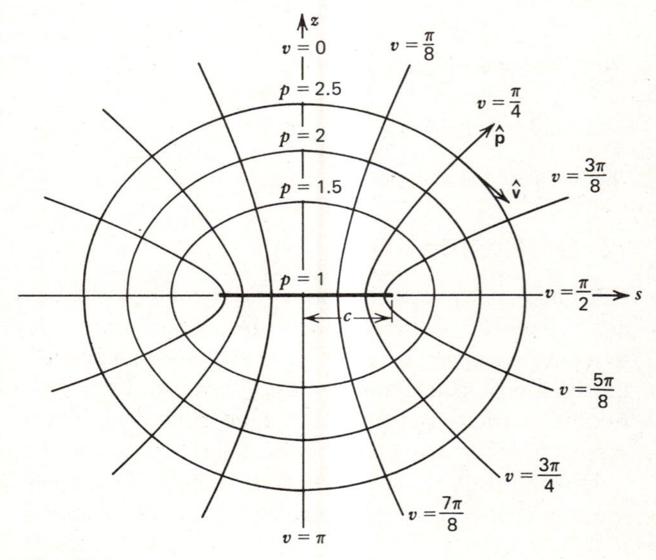

Fig. 12.1

is

$$\sqrt{a^2+b^2} = \sqrt{c^2\sin^2 v + c^2\cos^2 v} = c \tag{12.17}$$

Thus, each value of v in (12.15) gives a coordinate line which is a hyperbola. The resulting family of hyperbolas is confocal and has the same focal point as does the family of ellipses given by (12.12). If $v = \pi/2$, (12.3) shows that $z = 0$. The hyperbola degenerates into a straight line from c to ∞. The coordinate surface $v = \pi/2$ is the entire x, y plane with the exception of the circular disc $s \leq c$. Study Fig. 12.1 carefully. Note that the ranges (12.4) assigned to the variables cover the entire three-dimensional space exactly once. Note also that (12.15) does not distinguish between values of v in the ranges $0 \leq v \leq \pi/2$ and $\pi/2 \leq v \leq \pi$. Equation (12.3) makes it clear however that $z \leq 0$ requires $\pi/2 \leq v \leq \pi$.

If the coordinates are labeled $q_1 = p$, $q_2 = v$, and $q_3 = \phi$, the elements of the metric tensor are

$$g_{11} = \left(\frac{\partial x}{\partial p}\right)^2 + \left(\frac{\partial y}{\partial p}\right)^2 + \left(\frac{\partial z}{\partial p}\right)^2 = \frac{c^2(p^2-\sin^2 v)}{p^2-1} \tag{12.18}$$

$$g_{22} = \left(\frac{\partial x}{\partial v}\right)^2 + \left(\frac{\partial y}{\partial v}\right)^2 + \left(\frac{\partial z}{\partial v}\right)^2 = c^2(p^2-\sin^2 v) \tag{12.19}$$

$$g_{33} = \left(\frac{\partial x}{\partial \phi}\right)^2 + \left(\frac{\partial y}{\partial \phi}\right)^2 + \left(\frac{\partial z}{\partial \phi}\right)^2 = c^2p^2\sin^2 v \tag{12.20}$$

$$g_{12} = g_{13} = g_{23} = 0 \tag{12.21}$$

The coordinate system is therefore orthogonal. The Jacobian of the transformation is conveniently calculated from (6.5):

$$J = \frac{1}{\sqrt{g}} = \frac{\sqrt{p^2-1}}{c^3 p \sin v(p^2-\sin^2 v)} \tag{12.22}$$

Note that $J = \infty$ for $v = 0$ or $v = \pi$, which corresponds to the entire z axis. Both cylindrical and spherical coordinates have the same difficulty. Another problem region occurs when $p = 1$, since then $J = 0$. This is the circular disc $s \leq c$ in the x, y plane. The problem is connected with the fact that both the unit coordinate vectors $\hat{\mathbf{p}}$ and $\hat{\mathbf{v}}$ behave discontinuously and actually reverse direction as we cross the disc from $z > 0$ to $z < 0$. The Laplacian of a scalar function is

$$\nabla^2\psi = \frac{\sqrt{p^2-1}}{c^2p(p^2-\sin^2 v)}\frac{\partial}{\partial p}\left(p\sqrt{p^2-1}\,\frac{\partial\psi}{\partial p}\right) + \frac{1}{c^2\sin v(p^2-\sin^2 v)}\frac{\partial}{\partial v}\left(\sin v\frac{\partial\psi}{\partial v}\right) + \frac{1}{c^2p^2\sin^2 v}\frac{\partial^2\psi}{\partial\phi^2} \tag{12.23}$$

Spherical coordinates are the obvious choice for expressing the potential due to a charged conducting sphere. The potential exterior to the sphere is simply

$$\psi = \frac{Q}{4\pi\varepsilon_0 r} \tag{12.24}$$

Given the problem of finding the potential due to a charged conductor in the shape of a spheroid, we anticipate that a similar simplification results if spheroidal coordinates are used. To this end, let us look for a solution of $\nabla^2\psi = 0$ which depends only on p. Such a solution must obey

$$\frac{\partial}{\partial p}\left(p\sqrt{p^2-1}\,\frac{\partial\psi}{\partial p}\right) = 0 \tag{12.25}$$

which implies that

$$p\sqrt{p^2-1}\,\frac{\partial\psi}{\partial p} = \alpha \tag{12.26}$$

where α is a constant. Integration of (12.26) gives

$$\psi = -\alpha\sin^{-1}\left(\frac{1}{p}\right) + \beta \tag{12.27}$$

where β is a second constant. The following range of values is used for the inverse sine:

$$1 \leq p < \infty \qquad \frac{\pi}{2} \geq \sin^{-1}\left(\frac{1}{p}\right) \geq 0 \tag{12.28}$$

At a great distance from the charged spheroid, the potential is zero. This requires

$$0 = -\alpha\sin^{-1}(0) + \beta = \beta \tag{12.29}$$

If the surface of the conducting spheroid has the equation $p = p_0$, where p_0 is a constant and if the potential of the spheroid is ψ_0, then

$$\psi_0 = -\alpha \sin^{-1}\left(\frac{1}{p_0}\right) \tag{12.30}$$

The potential can then be expressed as

$$\psi = \psi_0 \frac{\sin^{-1}(1/p)}{\sin^{-1}(1/p_0)} \tag{12.31}$$

Reference to theorem 9, section 10, Chapter 3, shows that (12.31) is unique. In this case, the region Σ is the space exterior to the spheroid. The region Σ is bounded by the spheroid on which $\psi = \psi_0$ and by a surface at ∞ on which $\psi = 0$. The charge density in Σ is zero. There is, of course, charge on the spheroidal surface.

The electric field has only a component in the p direction given by

$$E_p = -\frac{1}{\sqrt{g_{11}}}\frac{\partial \psi}{\partial p} \tag{12.32}$$

By means of (12.18), (12.26), and (12.30), we find

$$E_p = \frac{\psi_0}{\sin^{-1}(1/p_0)\, pc\sqrt{p^2 - \sin^2 v}} \tag{12.33}$$

Suppose that, in Fig. 12.1, the coordinate surface $p = 1.5$ coincides with the conductor. The coordinate surfaces $p = 2$, $p = 2.5, \ldots$ then become equipotential surfaces and the coordinate lines $v =$ constant exterior to the surface $p = 1.5$ coincide with the field lines. There is, of course, no field inside the conductor. Thus, Fig. 12.1 doubles as an equipotential and field–line plot.

In problem 10.10 of Chapter 3, it is shown that the surface charge density λ is related to the field at the surface of the conductor by

$$E_n = \frac{\lambda}{\varepsilon_0} \tag{12.34}$$

The surface charge density on the surface of the charged conducting spheroid is therefore

$$\lambda = \frac{\varepsilon_0 \psi_0}{\sin^{-1}(1/p_0)\, p_0 c\sqrt{p_0^2 - \sin^2 v}} \tag{12.35}$$

To find the total charge on the spheroid, recall (see problem 6.10) that an element of area on the surface is

$$d\sigma = \sqrt{g_{22}g_{33}}\, dv\, d\phi = c^2 p_0 \sin v \sqrt{p_0^2 - \sin^2 v}\, dv\, d\phi \tag{12.36}$$

The charge on $d\sigma$ is

$$dQ = \lambda\, d\sigma = \frac{\varepsilon_0 \psi_0 c}{\sin^{-1}(1/p_0)} \sin v\, dv\, d\phi \tag{12.37}$$

The total charge is found by integrating over the entire surface of the spheroid:

$$Q = \frac{\varepsilon_0 \psi_0 c}{\sin^{-1}(1/p_0)} \int_{\phi=0}^{2\pi}\int_{v=0}^{\pi} \sin v\, dv\, d\phi$$

$$= \frac{4\pi\varepsilon_0 \psi_0 c}{\sin^{-1}(1/p_0)} \tag{12.38}$$

The expressions for the potential, field, and charge density can now be written as

$$\psi = \frac{Q}{4\pi\varepsilon_0 c} \sin^{-1}\left(\frac{1}{p}\right) \tag{12.39}$$

$$E_p = \frac{Q}{4\pi\varepsilon_0 c^2 p\sqrt{p^2 - \sin^2 v}} \tag{12.40}$$

$$\lambda = \frac{Q}{4\pi c^2 p_0 \sqrt{p_0^2 - \sin^2 v}} \tag{12.41}$$

In the limit of large values of p, $\sin^{-1}(1/p) \simeq 1/p$. If

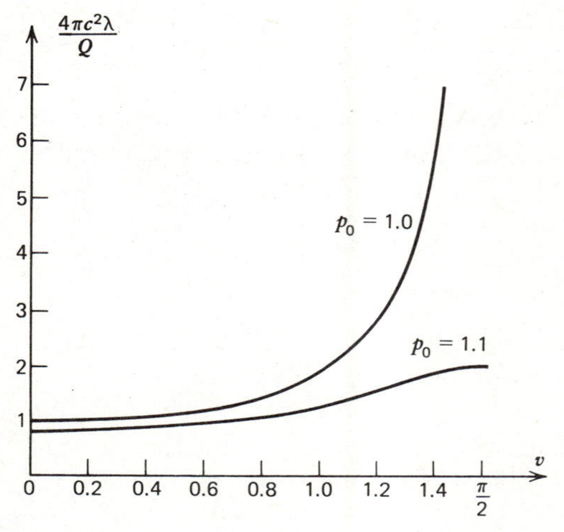

Fig. 12.2

we set $r = pc$, then

$$\psi \simeq \frac{Q}{4\pi\varepsilon_0 r} \qquad E_p \simeq \frac{Q}{4\pi\varepsilon_0 r^2} \tag{12.42}$$

as expected.

If $p_0 = 1$, the conducting spheroid degenerates into a flat disc of radius c lying in the x, y plane. With the aid of (12.11), the expression for charge density becomes

$$\lambda = \frac{Q}{4\pi c\sqrt{c^2 - s^2}} \tag{12.43}$$

Figure 12.2 shows plots of charge density as a function of v for the two cases $p_0 = 1$ and $p_0 = 1.1$. The charge density is highest at $v = \pi/2$, which is where the radius of curvature of the surface is the smallest. This is a general property of charged conductors. The field lines (coordinate lines v = constant) in Fig. 12.1 have been drawn in such a way that they are equally spaced at $p = \infty$. Note that their density is greatest where they intersect the x, y plane near the rim of the disc corresponding to the greater charge density there.

The definition of prolate spheroidal coordinates will be found in problem 12.6.

PROBLEMS

12.1. In Fig. 12.3, f_1 and f_2 are the two foci of an ellipse. The ellipse can be defined by the condition that the point P move in such a way that $r_1 + r_2 = k$, where k is a constant. Show from this that $k = 2a$ and $c = \sqrt{a^2 - b^2}$, where a and b are the semimajor and semiminor axes and c is the distance from the center to either focus. Show also that $r_1 + r_2 = 2a$ is equivalent to

$$\frac{x^2}{a^2} + \frac{y^2}{b^2} = 1$$

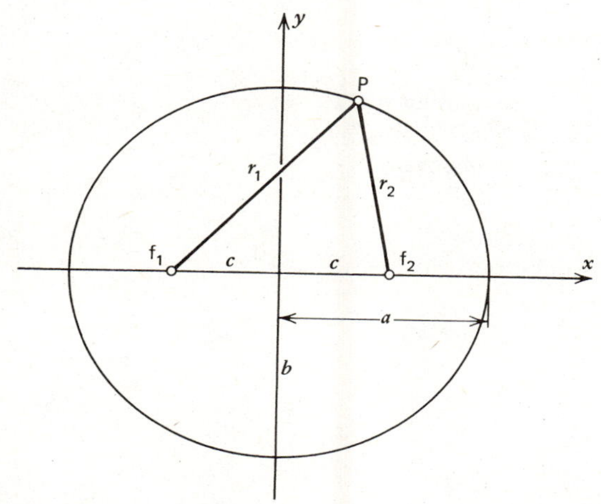

Fig. 12.3

12.2. In Fig. 12.4, f_1 and f_2 are the two foci of a hyperbola. The hyperbola can be defined by the condition that the point P move in such a way that $r_1 - r_2 = k$, where k is a constant. Show that $k = 2a$, where a is the distance from the origin to the vertex. Show from $r_1 - r_2 = 2a$ that

$$\frac{x^2}{a^2} - \frac{y^2}{b^2} = 1$$

where b is defined by $b^2 = c^2 - a^2$ and c is the distance from the origin to either focus. Show that the equations of the asymptotes are

$$y = \pm \frac{b}{a} x$$

This shows that b is the distance from the vertex to the asymptote measured in the y direction.

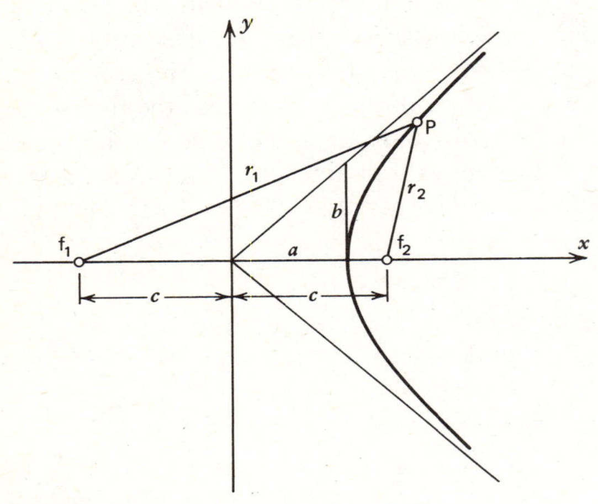

Fig. 12.4

12.3. In Fig. 12.1, $\Delta p = 0.5$ as we go from one coordinate surface $p =$ constant to the next, meaning that these coordinate surfaces are equally spaced with respect to p. Let ψ_0 be the potential of a charged disc of radius c. Plot equipotentials which differ from one another by the equal potential increments $\Delta\psi = 0.2\psi_0$. Specifically, plot the equipotentials $0.8\psi_0$, $0.6\psi_0$, $0.4\psi_0$, and $0.2\psi_0$. Use cm graph paper and let $c = 2$ cm. Where on the diagram is the gradient of the potential the greatest?

12.4. Find the unit vectors $\hat{\mathbf{p}}$, $\hat{\mathbf{v}}$, and $\hat{\boldsymbol{\phi}}$ in terms of $\hat{\mathbf{i}}$, $\hat{\mathbf{j}}$, and $\hat{\mathbf{k}}$. A charged conductor in the shape of a spheroid has the value $p_0 = 1.5$. Find the angle between the electric field line $v = \pi/8$ and the z axis at the point where the field line intersects the conductor.

12.5. Show that p is given in terms of the semimajor and semiminor axes of the ellipse to which it corresponds by

$$p = \frac{a}{\sqrt{a^2 - b^2}}$$

12.6. Prolate spheroidal coordinates can be defined by means of the transformation

$$x = c\sqrt{p^2 - 1}\, \sin v \cos \phi$$

$$y = c\sqrt{p^2 - 1}\, \sin v \sin \phi$$

$$z = cp \cos v$$

where $1 \le p < \infty, 0 \le v \le \pi, 0 \le \phi < 2\pi$. Explore the geometry of this coordinate system in detail. Construct a figure similar to Fig. 12.1 which shows the profiles of the coordinate surfaces $p =$ constant and $v =$ constant. Use cm graph paper. Let $c = 2$ cm and graph the coordinate lines for the same values of p and v which appear in Fig. 12.1.

12.7. Find the elements of the metric tensor for prolate spheroidal coordinates. Find the expression for the Laplacian of a scalar function.

12.8. Solve the potential problem for a charged conductor in the shape of a prolate spheroid. Find expressions for the potential and the electric field. Find the surface charge density on the conductor and integrate it to find the total charge. Examine the expressions for the potential and the electric field to see if they have the correct form in the limit $p \to \infty$. Show that the ratio of the maximum to the minimum surface charge density on the conductor is $\lambda_{max}/\lambda_{min} = a/b$. Show that the total charge on the conductor between z and $z + dz$ is

$$dQ_z = \frac{Q}{2cp_0} dz$$

In the limit of a thin needle-shaped conductor, $p_0 \to 1$ and

$$dQ_z = \frac{Q}{2c} dz$$

This shows that charge placed on a thin conducting wire will distribute itself uniformly. If you don't believe this, show directly that a uniformly charged thin, straight wire of length $2c$ produces the same potential as does a charged prolate spheroid.

12.9. It is desired to study the electric field and the potential in the vicinity of a circular hole of radius c which has been punched in a large charged conducting sheet. Explain why the following definition of spheroidal coordinates might be better for the study of this problem than (12.1)–(12.3):

$$x = qc \cosh u \cos \phi$$

$$y = qc \cosh u \sin \phi$$

$$z = c \sinh u \sqrt{1 - q^2}$$

$$0 \le q \le 1 \qquad -\infty < u < +\infty \qquad 0 \le \phi \le 2\pi$$

What other definitions would work? What about $p = \sinh u$?

13. LAGRANGE'S EQUATIONS

Up to this point, we have dealt almost exclusively with transformations from one curvilinear coordinate system to another which take place in the ordinary Euclidean space of everyday experience. In this section, we take up a more general, and more abstract, kind of transformation which is especially suited to the study of classical mechanics. We have already alluded to the possibility of this more general kind of transformation. For example, the double plane pendulum in problem 4.7 can be described by means of four rectangular coordinates, two being necessary for each particle. The system could be described by a single point which moves in a conceptual four-dimensional configuration space with coordinates x_1, y_1, x_2, and y_2. The condition that the two rods of length s are rigid puts mechanical constraints on the motion, the effect of which are to reduce the number of independent parameters needed to describe the motion of the system from four to two. The equations of constraint can be written down explicitly as

$$s^2 = x_1^2 + y_1^2 \tag{13.1}$$

$$s^2 = (x_2 - x_1)^2 + (y_2 - y_1)^2 \tag{13.2}$$

We could, for example, solve for y_1 and y_2 in terms of x_1 and x_2. Once x_1 and x_2 are known, the motion is then determined. A possible point of view to take is that (13.1) and (13.2) constrain the system to move in a conceptual two-dimensional subspace which is embedded in the four-dimensional hyperspace.

Using x_1 and x_2 as the independent parameters is not the cleverest way to proceed. A better pair of parameters are the two angles θ_1 and θ_2 pictured in Fig. 4.1. There exists a transformation connecting the four rectangular coordinates with these two angles:

$$x_1 = s\sin\theta_1 \qquad y_1 = s\cos\theta_1$$
$$x_2 = s\sin\theta_1 + s\sin\theta_2 \qquad y_2 = s\cos\theta_1 + s\cos\theta_2 \tag{13.3}$$

An important difference between (13.3) and transformations we have previously considered is that there are fewer (two) curvilinear coordinates than original rectangular coordinates (four). The *number of degrees of freedom* of a system is the number of independent coordinates needed to describe it, in this case, two. What we will discover is that once the kinetic and the potential energies of the double plane pendulum are correctly expressed in terms of θ_1 and θ_2, the equations of motion can be obtained from Lagrange's equation as given by (9.57):

$$\frac{d}{dt}\left(\frac{\partial L}{\partial \dot{\theta}_1}\right) - \frac{\partial L}{\partial \theta_1} = 0 \qquad \frac{d}{dt}\left(\frac{\partial L}{\partial \dot{\theta}_2}\right) - \frac{\partial L}{\partial \theta_2} = 0 \tag{13.4}$$

This is certainly as much of a simplification as we could possibly hope to obtain. The necessity of analyzing the system directly using Newton's second law with all the complications arising from the forces of constraint, which in this case are the tensions or compressions in the rods, is neatly avoided.

In order to show that Lagrange's equations are valid in as general a case as possible, let us assume that we have a system consisting of N particles. The motion of the system can be described by means of $3N$ rectangular coordinates which will be labeled consecutively as $x_1, x_2, \ldots, x_{3N}$. Newton's second law applies, and the motion of the system can, in principle, be found by means of

$$F_1 = m_1\ddot{x}_1 \qquad F_2 = m_1\ddot{x}_2 \qquad F_3 = m_1\ddot{x}_3$$
$$F_4 = m_2\ddot{x}_4 \qquad \cdots \qquad F_{3N} = m_N\ddot{x}_{3N} \tag{13.5}$$

The force components in (13.5) represent net force, including forces of constraint as well as all other forces. Equation (13.5) is clumsy as it stands, so it will be abbreviated as

$$F_i = m_{ij}\ddot{x}_j \tag{13.6}$$

where m_{ij} are the elements of the following $3N \times 3N$ matrix:

$$(m_{ij}) = \begin{pmatrix} m_1 & 0 & 0 & 0 & 0 & 0 & \cdots \\ 0 & m_1 & 0 & 0 & 0 & 0 & \cdots \\ 0 & 0 & m_1 & 0 & 0 & 0 & \cdots \\ 0 & 0 & 0 & m_2 & 0 & 0 & \cdots \\ 0 & 0 & 0 & 0 & m_2 & 0 & \cdots \\ 0 & 0 & 0 & 0 & 0 & m_2 & \cdots \\ \vdots & \vdots & \vdots & \vdots & \vdots & \vdots & \end{pmatrix} \tag{13.7}$$

The summation convention is in force. In (13.6), it is understood that j is summed from 1 to $3N$. We shall assume that a number of constraints, similar to (13.1) and (13.2), exist and can be expressed in the form

$$f_1(x_1, x_2, \ldots, x_{3N}, t) = 0$$
$$f_2(x_1, x_2, \ldots, x_{3N}, t) = 0$$
$$\vdots$$
$$f_n(x_1, x_2, \ldots, x_{3N}, t) = 0 \tag{13.8}$$

Each equation of constraint implies the existence of a force of constraint, for example, the tensions or compressions in the rods of the double plane pendulum. If there are n equations of constraint connecting the original $3N$ rectangular coordinates, there will be $f = 3N - n$ independent parameters or degrees of freedom needed to describe the motion of the system. We are getting something more in the bargain. An explicit time dependence has been included in (13.8). This would happen, for instance, if the rods of the double plane pendulum were lengthening or contracting in a known way with time. To put it another way, Lagrange's equations (13.4) still survive if $s = s(t)$ in the constraints (13.1) and (13.2) and the transformation (13.3).

We will now suppose that f parameters $q_1, q_2, \ldots, q_f$ have been chosen as suitable for the description of the motion of the system. These parameters are called *generalized coordinates*. There exist transformation equations, like (13.3), relating the original rectangular coordinates to the generalized coordinates:

$$x_i = x_i(q_\alpha, t) \tag{13.9}$$

The convention will be adopted that Latin subscripts take on the values 1 to $3N$ and Greek subscripts run from 1 to f. It will now be demonstrated in a formal way that Lagrange's equations hold. The kinetic energy of the system of particles is

$$K = \tfrac{1}{2} m_{ij} \dot{x}_i \dot{x}_j \tag{13.10}$$

The total derivative of (13.9) with respect to time is

$$\dot{x}_i = \frac{\partial x_i}{\partial q_\beta} \dot{q}_\beta + \frac{\partial x_i}{\partial t} \tag{13.11}$$

During the course of the present discussion, both covariant and contravariant quantities will be labeled with subscripts. The distinction between covariance and contravariance becomes important if a transformation from one set of generalized coordinates to another is contemplated. Should we want to do this, it will be easy enough to revive the superscript notation. The differentiation of the kinetic energy with respect to any one of the generalized coordinates gives

$$\frac{\partial K}{\partial q_\alpha} = m_{ij} \frac{\partial \dot{x}_i}{\partial q_\alpha} \dot{x}_j \tag{13.12}$$

By means of (13.11),

$$\frac{\partial K}{\partial q_\alpha} = m_{ij} \left(\frac{\partial^2 x_i}{\partial q_\alpha \, \partial q_\beta} \dot{q}_\beta + \frac{\partial^2 x_i}{\partial q_\alpha \, \partial t} \right) \dot{x}_j \tag{13.13}$$

If (13.11) is differentiated with respect to $\dot{q}_\alpha$, the result is

$$\frac{\partial \dot{x}_i}{\partial \dot{q}_\alpha} = \frac{\partial x_i}{\partial q_\alpha} \tag{13.14}$$

The derivative of the kinetic energy with respect to $\dot{q}_\alpha$ is therefore

$$\frac{\partial K}{\partial \dot{q}_\alpha} = m_{ij} \frac{\partial \dot{x}_i}{\partial \dot{q}_\alpha} \dot{x}_j = m_{ij} \frac{\partial x_i}{\partial q_\alpha} \dot{x}_j \tag{13.15}$$

The total derivative of (13.15) with respect to time is

$$\frac{d}{dt}\left(\frac{\partial K}{\partial \dot{q}_\alpha} \right) = m_{ij} \left(\frac{\partial^2 x_i}{\partial q_\beta \, \partial q_\alpha} \dot{q}_\beta + \frac{\partial^2 x_i}{\partial t \, \partial q_\alpha} \right) \dot{x}_j + m_{ij} \frac{\partial x_i}{\partial q_\alpha} \ddot{x}_j \tag{13.16}$$

The combination of (13.13) and (13.16) gives the result

$$\frac{d}{dt}\left(\frac{\partial K}{\partial \dot{q}_\alpha} \right) - \frac{\partial K}{\partial q_\alpha} = F_i \frac{\partial x_i}{\partial q_\alpha} = Q_\alpha \tag{13.17}$$

where (13.6) has been used. Reference to (9.50) and (9.52) shows that (13.17) is formally exactly the same as our previous result. It has, however, been shown to be valid under much more general conditions.

The tensions and compressions in the rods of the double plane pendulum are *workless forces of constraint*, that is, they do not in any way alter the energy balance of the system. Let us address ourselves to the general class of systems where this is true. We will also make the assumption that we have found and used all the mechanical conditions of constraint so that the f coordinates q_α truly represent the minimum number needed to describe the motion of the system and are therefore not connected by any further constraints. Under these conditions, the forces of constraint do not appear at all in the generalized force components even though they are present in the original rectangular components of force. In proving this, we will at first assume that the constraints are not time-dependent. A general displacement of the system is then given by

$$dx_i = \frac{\partial x_i}{\partial q_\alpha} dq_\alpha \tag{13.18}$$

The rectangular components of force can be divided into two parts as

$$F_i = C_i + \bar{F}_i \tag{13.19}$$

where C_i are the forces of constraint and $\bar{F}_i$ represents all other forces. Since the forces of constraint do no work,

$$0 = C_i \, dx_i = C_i \frac{\partial x_i}{\partial q_\alpha} dq_\alpha \tag{13.20}$$

The fact that the q_α are a set of independent coordinates means that the variations dq_α are linearly independent. Therefore, (13.20) implies that

$$C_i \frac{\partial x_i}{\partial q_\alpha} = 0 \tag{13.21}$$

The generalized force components are then

$$Q_\alpha = F_i \frac{\partial x_i}{\partial q_\alpha} = \bar{F}_i \frac{\partial x_i}{\partial q_\alpha} \tag{13.22}$$

and, as asserted, do not contain the forces of constraint.

If the constraints are time-dependent, a general displacement of the system is

$$dx_i = \frac{\partial x_i}{\partial q_\alpha} dq_\alpha + \frac{\partial x_i}{\partial t} dt \qquad (13.23)$$

It follows from the general nature of a total differential that the first term in (13.23) is that portion of the displacement which occurs if the coordinates are all varied at a fixed time. In some books, this is called a *virtual displacement* of the system. Equation (13.20) still holds, and (13.21) follows. This is not to say that the forces of constraint do no work. The work done by the forces of constraint is accounted for by the second term in (13.23):

$$dW_C = C_i\, dx_i = C_i \frac{\partial x_i}{\partial t} dt \qquad (13.24)$$

The example of the spherical pendulum discussed in problem 9.5 may clarify the idea of this paragraph. If r is a constant, the particle is constrained to move on a sphere. The tension in the string joining the particle to the origin is always perpendicular to the displacement of the particle and hence does no work. If someone starts pulling on the string in order to shorten it, he must do work because there will now be a component of displacement perpendicular to the original spherical surface and parallel to the tension.

It should be noted that in the event that some of the equations of constraint have been missed, the number of generalized coordinates q_α will be larger than f. Equation (13.17) is still valid, but some of the forces of constraint will now appear in the generalized force components mixed in with the other forces.

If all forces other than the forces of constraint are conservative, then

$$\bar{F}_i = -\frac{\partial V}{\partial x_i} \qquad (13.25)$$

and (13.22) becomes

$$Q_\alpha = -\frac{\partial V}{\partial x_i}\frac{\partial x_i}{\partial q_\alpha} = -\frac{\partial V}{\partial q_\alpha} \qquad (13.26)$$

From (13.17), we then obtain Lagrange's equations as

$$\frac{d}{dt}\left(\frac{\partial L}{\partial \dot{q}_\alpha}\right) - \frac{\partial L}{\partial q_\alpha} = 0 \qquad (13.27)$$

where $L = K - V$ is the Lagrangian. It is possible to construct more general types of Lagrangians which are valid for some kinds of nonconservative forces.

Some examples will show how Lagrange's equations are to be used. Suppose, as in problem 9.7 of Chapter 2 and as also illustrated in Fig. 13.1, that a disc of radius b rolls without slipping down an inclined plane. The normal force $\mathbf{N}$ and the frictional force $\mathbf{f}$ are the forces of constraint. As long as the disc rolls without slipping, the frictional force is a workless force of constraint. If x is the distance that the disc has rolled and ϕ is the angle through which it has turned, then the rolling constraint is expressed by

$$x = b\phi \qquad (13.28)$$

The system therefore has only one degree of freedom. Either x or ϕ can be used as the generalized coordinate. Reference to equation (4.40) of Chapter 2 shows that the kinetic energy is

$$K = \tfrac{1}{2} I\omega^2 + \tfrac{1}{2} Mv^2 \qquad (13.29)$$

A suitable potential function is

$$V = -Mgx\sin\theta \qquad (13.30)$$

It is necessary to express the Lagrangian entirely in terms of either x or ϕ. Choosing to use x, we have

$$\omega = \dot{\phi} = \frac{\dot{x}}{b} \qquad v = \dot{x} \qquad (13.31)$$

The Lagrangian is therefore

$$L = \frac{1}{2}\left(M + \frac{I}{b^2}\right)\dot{x}^2 + Mgx\sin\theta \qquad (13.32)$$

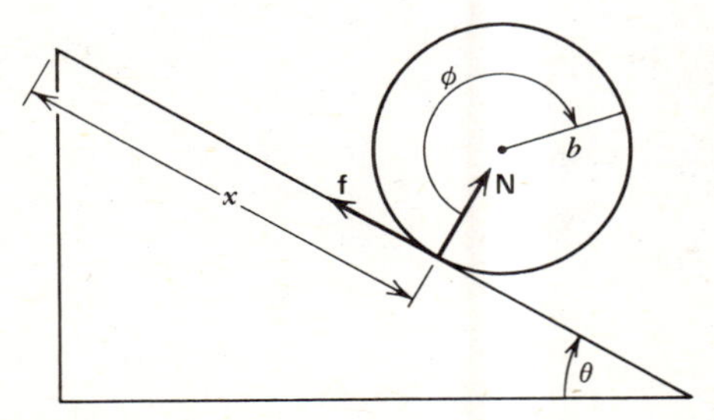

Fig. 13.1

By means of

$$\frac{d}{dt}\left(\frac{\partial L}{\partial \dot{x}}\right)-\frac{\partial L}{\partial x}=0 \tag{13.33}$$

we find

$$\ddot{x}=\frac{Mg\sin\theta}{M+(I/b^2)} \tag{13.34}$$

If the disc slips as it rolls, the frictional force is no longer a workless force of constraint. Neither is it conservative. Both coordinates ϕ and x are now needed, and the frictional force must be included in the generalized force components. Under this circumstance, the Lagrange formulation loses much of its advantage.

As a second example, suppose there is a bead on the spoke of a wheel which is rotating at a constant angular velocity ω, as shown in Fig. 13.2. The bead can slide on the spoke without friction. There is no gravity or other force on the particle. The *only* force is therefore the force of constraint between the bead and the spoke. The equation of constraint is

$$\theta=\omega t \tag{13.35}$$

This is a simple example of a time-dependent constraint. The kinetic energy of the particle is

$$K=\tfrac{1}{2}m(\dot{r}^2+r^2\dot{\theta}^2)=\tfrac{1}{2}m(\dot{r}^2+r^2\omega^2) \tag{13.36}$$

In this case, there is no potential energy function, and $K=L$. The equation of motion follows from

$$\frac{d}{dt}\left(\frac{\partial L}{\partial \dot{r}}\right)-\frac{\partial L}{\partial r}=0 \tag{13.37}$$

and is

$$\ddot{r}-r\omega^2=0 \tag{13.38}$$

As a final example, refer once more to the double plane pendulum of Fig. 4.1. The potential energy can be expressed

$$\begin{aligned}V&=-mgs\cos\theta_1-mg(s\cos\theta_1+s\cos\theta_2)\\&=-mgs(2\cos\theta_1+\cos\theta_2)\end{aligned} \tag{13.39}$$

The general motion is complicated, but a considerable simplification results if the angles are small. The potential energy is then approximately

$$V=-mgs(3-\theta_1^2-\tfrac{1}{2}\theta_2^2) \tag{13.40}$$

Discarding any constant terms, the approximate Lagrangian is

$$L=\tfrac{1}{2}ms^2(2\dot{\theta}_1^2+\dot{\theta}_2^2+2\dot{\theta}_1\dot{\theta}_2)-mgs(\theta_1^2+\tfrac{1}{2}\theta_2^2) \tag{13.41}$$

In the kinetic energy, we have used the approximation $\cos(\theta_1-\theta_2)\simeq 1$. The equations of motion follow from (13.4):

$$2\ddot{\theta}_1+\ddot{\theta}_2+\frac{2g}{s}\theta_1=0 \tag{13.42}$$

$$\ddot{\theta}_1+\ddot{\theta}_2+\frac{g}{s}\theta_2=0 \tag{13.43}$$

The nature of the approximation used here as well as the solution of (13.42) and (13.43) will be taken up in the next section.

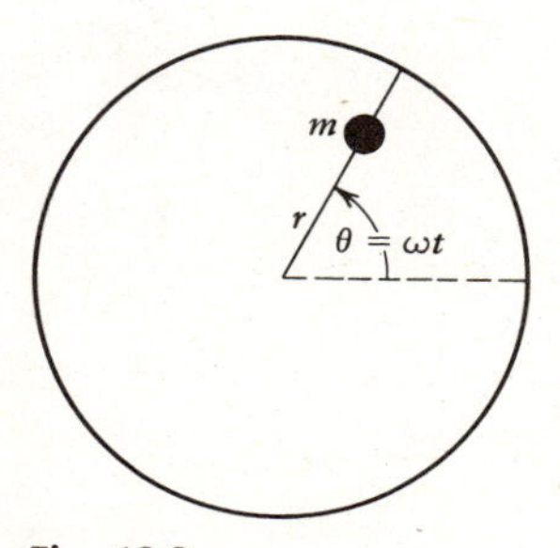

Fig. 13.2

PROBLEMS

13.1. A particle of mass m moves in the x, y plane and is joined to the origin by means of a massless rod. The rod is attached to the origin by means of a frictionless hinge. The length r of the rod is a prescribed function of the time. The force of constraint is the tension or compression **C** in the rod, as illustrated in Fig. 13.3. There is one degree of freedom and one generalized coordinate which we will take to be the angle θ. Show directly that the force of constraint cancels out of

$$Q_\theta = F_x \frac{\partial x}{\partial \theta} + F_y \frac{\partial y}{\partial \theta}$$

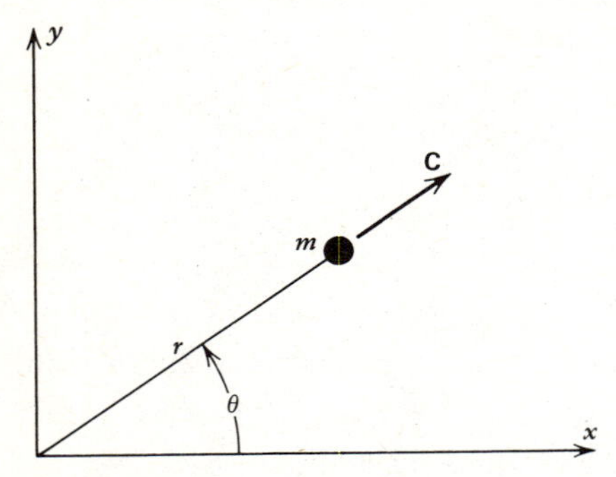

Fig. 13.3

13.2. It is the usual practice to define *generalized* or *canonical* momenta by means of

$$p_\alpha = \frac{\partial L}{\partial \dot{q}_\alpha}$$

The generalized momentum p_α is said to be *conjugate* to the coordinate q_α. Lagrange's equations then read

$$\dot{p}_\alpha - \frac{\partial L}{\partial q_\alpha} = 0$$

If a given coordinate is absent from the Lagrangian, then

$$\frac{\partial L}{\partial q_\alpha} = 0$$

and p_α is a constant of the motion. Such coordinates are called *cyclic coordinates*. Do problem 9.5 over again, this time using the Lagrange method. Show that $\ell_z = p_\phi$ is the canonical momentum conjugate to the coordinate ϕ and is a constant.

13.3. Analyze the bead on a spoke problem by means of Newton's second law in the form $F_r = ma_r$ and $F_\theta = ma_\theta$. Complete the solution of the problem, that is, integrate (13.38), for the case where the initial conditions are $r(0) = r_0$, $\dot{r}(0) = 0$. Find the force of constraint as a function of the time.

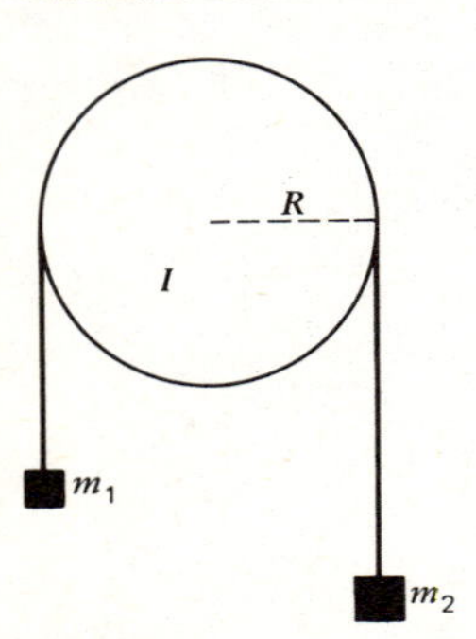

Fig. 13.4

13.4. Two masses m_1 and m_2 hang from the opposite ends of a massless string which passes over the rim of a wheel of moment of inertia I and radius R mounted on a horizontal frictionless axis. Find the acceleration of the system by the Lagrange method. Assume that the masses move either straight up or straight down and that the string does not slip on the rim of the wheel. See Fig. 13.4.

13.5. For the example of the disc rolling down an inclined plane, show that the frictional force required to produce rolling is $f = \frac{1}{3}Mg\sin\theta$, provided that the moment of inertia is $\frac{1}{2}Mb^2$. Use (4.47) and (9.11) of Chapter 2 to find the motion of the disc if the frictional force is reduced to $f = \frac{1}{6}Mg\sin\theta$. Now work backward through Lagrange's equations to show that

$$\frac{d}{dt}\left(\frac{\partial L}{\partial \dot{x}}\right) - \frac{\partial L}{\partial x} = Q_x = -f$$

$$\frac{d}{dt}\left(\frac{\partial L}{\partial \dot{\phi}}\right) - \frac{\partial L}{\partial \phi} = Q_\phi = \tau = fb$$

where

$$L = \tfrac{1}{2}M\dot{x}^2 + \tfrac{1}{4}Mb^2\dot{\phi}^2 + Mgx\sin\theta$$

In this case, the force due to the gravitational field is conservative and is taken care of by the Lagrangian. The normal force $\mathbf{N}$ is still a workless force of constraint and does not appear in the formalism. The frictional force turns up in the generalized forces. The problem with using the Lagrange formulation from the start is that it is difficult to find the form that Q_x and Q_ϕ take.

13.6. If $\phi = \phi(q_\alpha, t)$ is any scalar function of the generalized coordinates and the time, show that $d\phi/dt$ is a solution of Lagrange's equations. This means that Lagrangians are not unique. If L is a Lagrangian for a given system, then so is $L' = L + d\phi/dt$.

13.7. A time-dependent electromagnetic field is not conservative. Show however that a Lagrangian which correctly gives the motion of a charged particle in an arbitrary electromagnetic field is

$$L = K - q\psi + q\mathbf{A}\cdot\mathbf{u}$$

where ψ and $\mathbf{A}$ are the scalar and vector potentials, respectively, and q is the charge on the particle. See equations (12.35) and (12.37) of Chapter 3 for the connection between the magnetic field $\mathbf{B}$ and the electric field $\mathbf{E}$ and the potentials. Do the problem in rectangular coordinates. Specifically, show that

$$\frac{d}{dt}\left(\frac{\partial L}{\partial \dot{x}_i}\right) - \frac{\partial L}{\partial x_i} = 0$$

is the equivalent of

$$q\mathbf{E} + q\mathbf{u}\times\mathbf{B} = m\frac{d\mathbf{u}}{dt}$$

The potentials are not unique. How does a gauge transformation as given by (12.38) and (12.39) of Chapter 3 affect the Lagrangian? Does this in any way change the equations of motion?

13.8. The magnetic field of the Earth is approximately a magnetic dipole field. Use (11.10) to find a Lagrangian for a charged particle moving in the magnetic field of the Earth. Find the canonical momentum conjugate to the coordinate ϕ and show that it is a constant of the motion. A second constant of the motion is the speed of the particle. The Lagrangian of problem 13.7 does not account for any effect on the motion of the particle due to loss of energy through electromagnetic radiation.

14. LINEAR VIBRATIONS

Consider a system of N particles subject to n time-independent equations of constraint. The system can be described by $f = 3N - n$ generalized coordinates. The transformation giving the $3N$ rectangular coordinates in terms of the generalized coordinates is

$$x_i = x_i(q_\alpha) \tag{14.1}$$

The kinetic energy (13.10) can be expressed as

$$K = \tfrac{1}{2} m_{ij} \dot{x}_i \dot{x}_j = \tfrac{1}{2} g_{\alpha\beta} \dot{q}_\alpha \dot{q}_\beta \tag{14.2}$$

where

$$g_{\alpha\beta} = m_{ij} \frac{\partial x_i}{\partial q_\alpha} \frac{\partial x_j}{\partial q_\beta} \tag{14.3}$$

are the elements of a metric tensor which describes the geometry of a conceptual f-dimensional configuration space in which the behavior of the mechanical system is described by the motion of a single point. For convenience, the masses of the particles have been included as part of the metric tensor. The geometry of the configuration space may be quite complex and need not be Euclidean.

Some properties of the metric familiar from ordinary three-dimensional space are retained. For example, the metric tensor is positive definite. This is required because the quadratic form (14.2) which gives the kinetic energy must be positive for all nonzero values of the generalized velocity components.

Theorem 1. All diagonal elements of the metric tensor are positive.

Proof: If in the kinetic energy (14.2), all generalized velocity components are zero except $\dot{q}_1$, then

$$K = \tfrac{1}{2} g_{11} \dot{q}_1^2 > 0 \tag{14.4}$$

Thus, $g_{11} > 0$. Similarly, all other diagonal elements are positive.

We will assume that all forces acting on the mechanical system, other than the forces of constraint, are conservative. The origin of coordinates of the configuration space will be chosen to be a position of equilibrium for the system. This is characterized by

$$\left(\frac{\partial V}{\partial q_\alpha} \right)_{q_\beta = 0} = 0 \tag{14.5}$$

where V is the potential energy. The theory is to be developed for the case where the origin is a position of stable equilibrium and where the displacements of the system from equilibrium are not large, although there are some examples in practice where these requirements can be relaxed. The potential energy can be approximated by means of

$$V = V_0 + \tfrac{1}{2} A_{\alpha\beta} q_\alpha q_\beta \tag{14.6}$$

The linear terms in the approximation are absent because of (14.5). The constant term is of no consequence and can be discarded. The coefficients $A_{\alpha\beta}$ are the elements of a second-rank symmetric tensor. One way to see this is to view (14.6) as a Taylor's series expansion of the potential about the origin of the configuration space. Then,

$$A_{\alpha\beta} = \left(\frac{\partial^2 V}{\partial q_\alpha \, \partial q_\beta} \right)_0 = A_{\beta\alpha} \tag{14.7}$$

An appropriate Lagrangian is therefore

$$L = \tfrac{1}{2} g_{\alpha\beta} \dot{q}_\alpha \dot{q}_\beta - \tfrac{1}{2} A_{\alpha\beta} q_\alpha q_\beta \tag{14.8}$$

The elements of the metric tensor are functions of position in the configuration space with the result that the coordinate lines are curved as pictured in Fig. 14.1. A coordinate system which approximates the actual coordinate system in the vicinity of the origin is obtained by replacing the elements of the metric tensor by their values at the origin. Such a coordinate system consists of straight lines which are tangent to the actual coordinate lines at the origin as illustrated by the dashed lines in Fig. 14.1. Thus, for small displacements of the system point from the origin of the configuration space, we can consider the elements of the metric tensor in the Lagrangian (14.8) to be constants. This is basically what we did to obtain the Lagrangian (13.41) for the double plane pendulum. The equations of motion for the system then follow from Lagrange's equations (13.27) and are

$$g_{\alpha\beta} \ddot{q}_\beta + A_{\alpha\beta} q_\beta = 0 \tag{14.9}$$

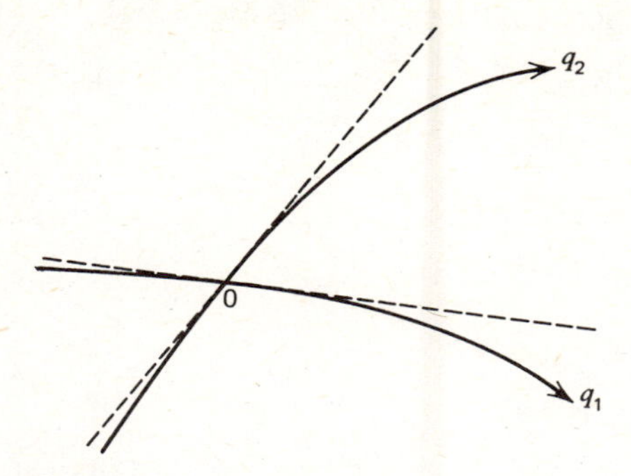

Fig. 14.1

These are coupled, linear, homogeneous equations with constant coefficients for determining the f unknown functions $q_\alpha(t)$. In matrix notation, (14.9) is

$$G\ddot{Q} + AQ = 0 \tag{14.10}$$

where G is an $f \times f$ square, symmetric, positive definite matrix with elements $g_{\alpha\beta}$; A is an $f \times f$ square symmetric matrix with elements $A_{\alpha\beta}$; and Q is a column vector the components of which are q_α. An example of a problem like (14.10) is discussed in some detail in section 6 of Chapter 2. Much of what we do in the remainder of this section parallels the discussion of the eigenvalue problem in section 10 of Chapter 2 and should be regarded as a generalization of the results obtained there.

We attempt to solve (14.10) by first looking for a transformation of coordinates which will simplify the problem. A new set of generalized coordinates can be introduced by means of

$$Q = \tilde{T}Q' \tag{14.11}$$

where T is an admissible transformation, that is, T possesses a Jacobian which is finite and nonzero. Writing the transformation as the transpose $\tilde{T}$ rather than T has no significance beyond being a notational preference. We are now working in a coordinate system where the elements of the metric tensor are constants. If the elements of T are also constants, then the transformation (14.11) involves changing the angles between coordinate lines and the scale factors for measuring distance. This is illustrated in Fig. 14.2. The transformation (14.11) is, like (3.36), contravariant. Since the coefficients in the transformation are constants, there is no distinction between the transformation of dq_α and q_α. We should, strictly speaking, be writing q^α in place of q_α; but since there are only three things we are going to be transforming, namely, Q, G, and A, it is easy enough to remember which is covariant and which is contravariant.

The combination of (14.10) and (14.11) yields

$$G\tilde{T}\ddot{Q}' + A\tilde{T}Q' = 0 \tag{14.12}$$

Multiplication from the left by T gives

$$G'\ddot{Q}' + A'Q' = 0 \tag{14.13}$$

where

$$G' = TG\tilde{T} \qquad A' = TA\tilde{T} \tag{14.14}$$

The elements of both G and A are covariant, meaning that (14.14) is the matrix version of a second-rank covariant tensor transformation. See, for example, (4.22).

Theorem 2. If G is positive definite, so is G'.

Proof: When we say that G is positive definite, we mean that

$$\phi = \tilde{Q}GQ \tag{14.15}$$

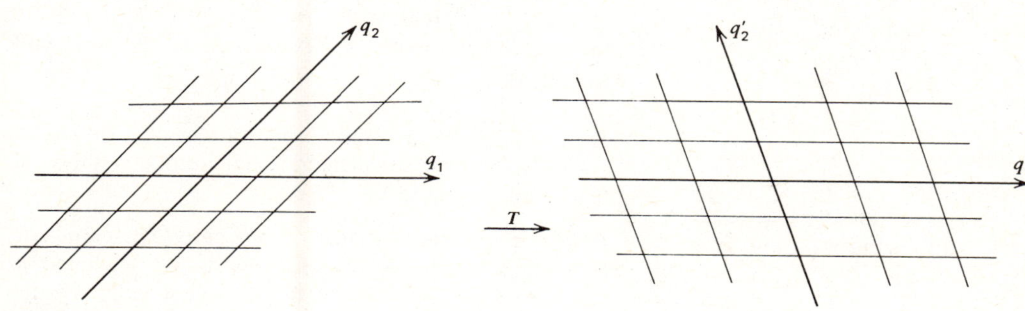

Fig. 14.2

is positive for all nonzero vectors Q. If (14.15) is transformed by means of (14.11) and (14.14), the result is

$$\phi = \left(\widetilde{TQ'}\right) G\tilde{T}Q' = \tilde{Q}'TG\tilde{T}Q' = \tilde{Q}'G'Q' \tag{14.16}$$

Since T is an admissible transformation, the coordinates Q' cover the entire space as do the coordinates Q. Thus, $\phi > 0$ for all nonzero vectors Q'. To put it another way, ϕ is a scalar invariant and retains the same range of values in all coordinate systems. Reference to (4.20) shows that ϕ is actually the square of the length of the vector Q.

Theorem 3. If A and G are symmetric, so are A' and G'.

Proof: Take the transpose of A' in (14.14):

$$\tilde{A}' = \widetilde{TA\tilde{T}} = T\tilde{A}\tilde{T} = TA\tilde{T} = A' \tag{14.17}$$

which proves the result.

A great simplification of (14.13) would result if

$$TG\tilde{T} = G' = I \qquad TA\tilde{T} = A' = \Lambda \tag{14.18}$$

where I is the identity matrix and Λ is diagonal. Equation (14.13) is then

$$\ddot{Q}' + \Lambda Q' = 0 \tag{14.19}$$

The component equations are

$$\ddot{q}'_\alpha + \lambda(\alpha) q'_\alpha = 0 \tag{14.20}$$

and are very easy to solve. The coordinates q'_α in (14.20) are called *normal coordinates*. We write the diagonal elements of Λ as $\lambda(\alpha)$ rather than λ_α because no sum over α is implied in (14.20). We will show that a transformation such that (14.18) is true does in fact exist. We first show formally how to construct the transformation and then demonstrate its existence in all cases. Something else is being accomplished here. The original configuration space is not necessarily Euclidean. This means that a transformation to f-dimensional rectangular coordinates valid simultaneously at all points of the space is not in general possible. Since $G' = I$ is possible at any one point, a transformation to coordinates which are rectangular in the neighborhood of a given point can indeed be found. Curved spaces are locally Euclidean.

In order to construct the transformation, note from (14.18) that

$$T^{-1} = G\tilde{T} \tag{14.21}$$

The transformation for A can then be expressed

$$G\tilde{T}A' - A\tilde{T} = 0 \tag{14.22}$$

The component version of (14.22) is

$$G_{\alpha\mu}T_{\nu\mu}A'_{\nu\beta} - A_{\alpha\mu}T_{\beta\mu} = 0 \tag{14.23}$$

If A' is diagonal, we can write

$$A'_{\nu\beta} = \lambda(\beta)\delta_{\nu\beta} \qquad \text{(no sum on } \beta) \tag{14.24}$$

Equation (14.23) is then

$$\left[\lambda(\beta)G_{\alpha\mu} - A_{\alpha\mu}\right]T_{\beta\mu} = 0 \tag{14.25}$$

We now go back to a matrix form and write (14.25) as

$$[\lambda(\beta)G - A]\, X_\beta = 0 \tag{14.26}$$

Each one of the diagonal elements $\lambda(\beta)$ of Λ corresponds to a vector X_β which then becomes a row of the transforming matrix T. Equation (14.26) is a homogeneous linear equation and possesses a nontrivial solution vector X_β if, and only if, the determinant of the coefficients vanishes:

$$|\lambda G - A| = 0 \tag{14.27}$$

This is an eigenvalue problem of a more general type than that discussed in section 10 of Chapter 2. Following the same terminology, we call (14.27) the *secular equation*, $\lambda G - A$ the *characteristic* matrix of A, $f(\lambda) = |\lambda G - A|$ the *characteristic function* of A, the solution vectors X *eigenvectors*, and the corresponding values λ the *eigenvalues*. The characteristic function is a polynomial of degree f in λ. Therefore, (14.27) has f roots which are the f eigenvalues. Equation (14.26) then yields an eigenvector for each one of the eigenvalues. The f rows of the transforming matrix T are formed from the eigenvectors. That this process always works is established in the following theorems.

Theorem 4. A necessary condition for the validity of (14.18) is that G be symmetric and positive definite and that A be symmetric.

Proof: The identity matrix is obviously symmetric and positive definite. The diagonal matrix Λ is symmetric. Theorems 2 and 3 then show that G is necessarily symmetric and positive definite and that A is symmetric.

Theorem 5. The eigenvalues are invariants.

Proof: Write the eigenvalue equation (14.26) as

$$AX = \lambda GX \tag{14.28}$$

Let T be any admissible transformation. If $X = \tilde{T}X'$, then (14.28) becomes

$$A\tilde{T}X' = \lambda G\tilde{T}X' \tag{14.29}$$

Multiply from the left by T to get

$$A'X' = \lambda G'X' \tag{14.30}$$

where $A' = TA\tilde{T}$ and $G' = TG\tilde{T}$. The eigenvalue λ is unaffected by this process.

Theorem 6. In order that (14.18) be valid, it is both necessary and sufficient that G be symmetric and positive definite and that A be symmetric.

Proof: The necessity of the conditions is established in theorem 4. To prove sufficiency, note that since G is symmetric, an orthogonal transformation exists which will diagonalize G:

$$G' = S_1 G\tilde{S}_1 \qquad A' = S_1 A\tilde{S}_1 \tag{14.31}$$

As indicated, it is necessary that S_1 be applied simultaneously to both G and A. A matrix T_2 exists which will transform the diagonal matrix G' into the identity matrix:

$$I = T_2 G'\tilde{T}_2 \qquad A'' = T_2 A'\tilde{T}_2 \tag{14.32}$$

For example, T_2 could be

$$T_2 = \begin{pmatrix} \frac{1}{\sqrt{g_1}} & 0 \cdots 0 & \\ 0 & \frac{1}{\sqrt{g_2}} & \cdots 0 \\ \vdots & \vdots & \vdots \\ 0 & 0 \cdots & \frac{1}{\sqrt{g_f}} \end{pmatrix} \tag{14.33}$$

where $g_1, g_2, \ldots$ are the diagonal elements of G'. Theorem 1 guarantees that (14.33) exists. Theorem 3 assures that A'' in (14.32) is symmetric. Therefore, an orthogonal matrix S_3 exists which will diagonalize A'':

$$S_3 I\tilde{S}_3 = S_3\tilde{S}_3 = I \qquad \Lambda = S_3 A\tilde{S}_3 \tag{14.34}$$

Theorem 5 assures us that we have not changed the eigenvalues. They are in fact the diagonal elements of Λ. The required transformation is therefore $T = S_3 T_2 S_1$. For example, the overall transformation of G is

$$TG\tilde{T} = S_3 T_2 S_1 G\tilde{S}_1\tilde{T}_2\tilde{S}_3 = S_3 T_2 G'\tilde{T}_2\tilde{S}_3 = S_3 I\tilde{S}_3 = S_3\tilde{S}_3 = I \tag{14.35}$$

Theorem 7. The eigenvectors can be chosen to be mutually orthogonal.

Proof: Since the rows of T are the eigenvectors, $TG\tilde{T} = I$ is the equivalent of

$$\tilde{X}_\alpha G X_\beta = \delta_{\alpha\beta} \tag{14.36}$$

which is an extension of the idea of a dot product as given by (7.3) to a general f-dimensional curvilinear coordinate system. Note also that the eigenvectors are normalized to unit length.

For a negative eigenvalue, the solution of (14.20) is

$$q'_\alpha = a_\alpha e^{-\sqrt{-\lambda(\alpha)}\,t} + b_\alpha e^{\sqrt{-\lambda(\alpha)}\,t} \tag{14.37}$$

where a_α and b_α are constants of integration. Such solutions imply that the system point moves indefinitely away from the origin in the configuration space and therefore generally violates the approximation of small displacements. If an eigenvalue is zero, the corresponding solution is

$$q'_\alpha = a_\alpha t + b_\alpha \tag{14.38}$$

which is a uniform translation of the system. Again, the small displacement approximation is violated. For positive eigenvalues, the solutions are of the form

$$q'_\alpha = c_\alpha \cos(\omega_\alpha t - \phi_\alpha) \tag{14.39}$$

where we have set $\omega_\alpha = \sqrt{\lambda(\alpha)}$. The constants of integration are c_α and ϕ_α. This is an oscillatory solution and implies that the origin is a point of stable equilibrium. If all the eigenvalues are positive, the matrix A is positive definite and the potential function (14.6) has an absolute minimum at the origin. This is the most interesting case, although occasionally a problem may occur in which we have to deal with either a zero or a negative eigenvalue. The solutions (14.39) are called the *normal modes* of oscillation, and ω_α is called a *normal frequency* or *eigenfrequency* of the system. The general solution of the problem in terms of the original coordinates is found through (14.11). Since the rows of T are the eigenvectors, we can write the solution as

$$Q = X_1 q_1' + X_2 q_2' + \cdots + X_f q_f' \tag{14.40}$$

Each normal coordinate has in it two constants of integration giving a total of $2f$ constants which must be found from the initial positions and velocities of all the particles.

Equations (13.42) and (13.43) for the double plane pendulum can be expressed in the form (14.10) with

$$G = s^2\begin{pmatrix} 2 & 1 \\ 1 & 1 \end{pmatrix} \qquad A = \begin{pmatrix} 2gs & 0 \\ 0 & gs \end{pmatrix} \tag{14.41}$$

The secular equation (14.27) yields two eigenvalues:

$$\lambda_1 = \omega_1^2 = (2 - \sqrt{2}\,)\frac{g}{s} \qquad \lambda_2 = \omega_2^2 = (2 + \sqrt{2}\,)\frac{g}{s} \tag{14.42}$$

The eigenvectors follow from (14.26) and are

$$X_1 = N_1\begin{pmatrix} 1 \\ \sqrt{2} \end{pmatrix} \qquad X_2 = N_2\begin{pmatrix} -1 \\ \sqrt{2} \end{pmatrix} \tag{14.43}$$

where N_1 and N_2 are normalization factors to be determined by (14.36). The result is

$$X_1 = \frac{1}{s\sqrt{4 + 2\sqrt{2}}}\begin{pmatrix} 1 \\ \sqrt{2} \end{pmatrix}$$

$$X_2 = \frac{1}{s\sqrt{4 - 2\sqrt{2}}}\begin{pmatrix} -1 \\ \sqrt{2} \end{pmatrix} \tag{14.44}$$

The solution can be expressed in the form

$$\begin{pmatrix} \theta_1 \\ \theta_2 \end{pmatrix} = \begin{pmatrix} 1 \\ \sqrt{2} \end{pmatrix} a_1 \cos(\omega_1 t - \phi_1) + \begin{pmatrix} -1 \\ \sqrt{2} \end{pmatrix} a_2 \cos(\omega_2 t - \phi_2) \tag{14.45}$$

The normalization factors have been absorbed into the constants a_1 and a_2.

If you did problem 5.4, you are familiar with the geometry of the θ_1, θ_2 coordinate system. By using equation (5.13), you found that the angle between coordinate lines is

$$\cos\theta_{12} = \frac{g_{12}}{\sqrt{g_{11}g_{22}}} = \frac{1}{\sqrt{2}} \qquad \theta_{12} = 45° \tag{14.46}$$

Suppose, as in Fig. 14.3, we mark off equal-distance intervals along the θ_1 and θ_2 coordinate lines. For convenience, these intervals are measured in units of s. The relation between distance intervals measured along the coordinate lines, for example, in cm and the same intervals measured in terms of θ_1 and θ_2 is determined by means of (5.1):

$$\Delta s(1) = \sqrt{g_{11}}\,\Delta\theta_1 = \sqrt{2}\,s\,\Delta\theta_1 \tag{14.47}$$

$$\Delta s(2) = \sqrt{g_{22}}\,\Delta\theta_2 = s\,\Delta\theta_2 \tag{14.48}$$

The eigenvectors (14.44) are expressed in terms of θ_1

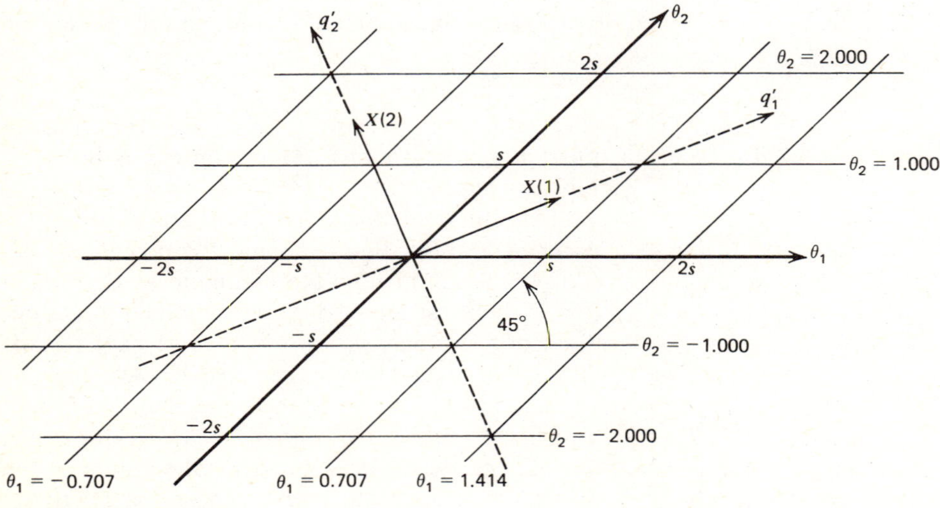

Fig. 14.3

and θ_2 units. If they are expressed in terms of their physical components, the result is

$$X(1) = N_1 \begin{pmatrix} s\sqrt{2} \\ s\sqrt{2} \end{pmatrix} \qquad X(2) = N_2 \begin{pmatrix} -s\sqrt{2} \\ s\sqrt{2} \end{pmatrix} \tag{14.49}$$

The eigenvectors are shown in Fig. 14.3. They lie along the diagonals of the parallelograms formed by the coordinate grid and are obviously perpendicular to one another. The eigenvectors point along the q_1' and q_2' axes.

The following example has some interesting features. Two particles of equal mass m are joined by two springs, each of force constant k, to a third mass M located between them. The system is constrained so that the particles all move on the x axis as shown in Fig. 14.4. The unextended length of each spring is s. A Lagrangian which will give the possible motions of this system is

$$L = \tfrac{1}{2}m(\dot{x}_1^2 + \dot{x}_3^2) + \tfrac{1}{2}M\dot{x}_2^2 - \tfrac{1}{2}k(x_2 - x_1 - s)^2 - \tfrac{1}{2}k(x_3 - x_2 - s)^2 \tag{14.50}$$

The matrix form of the equations of motion is

$$G\ddot{X} + AX = Y \tag{14.51}$$

where

$$G = \begin{pmatrix} m & 0 & 0 \\ 0 & M & 0 \\ 0 & 0 & m \end{pmatrix} \qquad X = \begin{pmatrix} x_1 \\ x_2 \\ x_3 \end{pmatrix}$$

$$A = \begin{pmatrix} k & -k & 0 \\ -k & 2k & -k \\ 0 & -k & k \end{pmatrix} \qquad Y = \begin{pmatrix} -ks \\ 0 \\ ks \end{pmatrix} \tag{14.52}$$

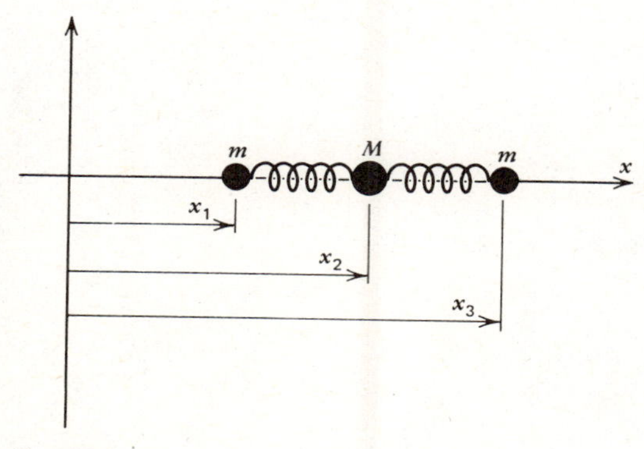

Fig. 14.4

The equation of motion (14.51) is nonhomogeneous. This difficulty is removed by first finding a particular integral. A possible particular solution is a constant vector X_0 which obeys

$$AX_0 = Y \tag{14.53}$$

We discover that det $A = 0$ and that the rank of A (section 7, Chapter 2) is 2. This means that (14.53) is consistent if, and only if, the rank of the augmented matrix

$$\text{aug } A = \begin{pmatrix} k & -k & 0 & -ks \\ -k & 2k & -k & 0 \\ 0 & -k & k & ks \end{pmatrix} \tag{14.54}$$

is also 2. (See theorem 3, section 8, Chapter 1.) We find that in fact rank aug $A = 2$. Theorem 2 of section 8 in Chapter 1 shows how to construct the general solution of (14.53). However, all we need here is a particular solution. It is easy to see that a solution of (14.53) is

$$X_0 = \begin{pmatrix} -s \\ 0 \\ s \end{pmatrix} \tag{14.55}$$

We therefore make the substitution $X = X_0 + Q$ in (14.51) to get

$$G\ddot{Q} + AQ = 0 \tag{14.56}$$

The eigenvalues follow from (14.27) and are

$$\lambda_1 = 0 \qquad \lambda_2 = \frac{k}{m} \qquad \lambda_3 = \frac{k(M+2m)}{Mm} \tag{14.57}$$

The appearance of a zero eigenvalue is not pathological in this case. No small displacement approximations were made in obtaining the Lagrangian (14.50). A uniform translation of the center of mass of the system is possible and does not interfere with the internal vibrational motion of the particles. The eigenvectors are found to be

$$X_1 = N_1 \begin{pmatrix} 1 \\ 1 \\ 1 \end{pmatrix} \qquad X_2 = N_2 \begin{pmatrix} 1 \\ 0 \\ -1 \end{pmatrix}$$

$$X_3 = N_3 \begin{pmatrix} 1 \\ -2\dfrac{m}{M} \\ 1 \end{pmatrix} \tag{14.58}$$

The general solution is therefore

$$\begin{pmatrix} x_1 \\ x_2 \\ x_3 \end{pmatrix} = \begin{pmatrix} -s \\ 0 \\ s \end{pmatrix} + \begin{pmatrix} 1 \\ 1 \\ 1 \end{pmatrix}(c_1 t + b) + \begin{pmatrix} 1 \\ 0 \\ -1 \end{pmatrix} c_2 \cos(\omega_2 t - \phi_2) + \begin{pmatrix} 1 \\ -2\dfrac{m}{M} \\ 1 \end{pmatrix} c_3 \cos(\omega_3 t - \phi_3) \tag{14.59}$$

where

$$\omega_2 = \sqrt{\frac{k}{m}} \qquad \omega_3 = \sqrt{\frac{k(M+2m)}{Mm}} \tag{14.60}$$

PROBLEMS

14.1. Let A be an arbitrary square matrix. The quadratic form associated with A is $\phi = \tilde{X}AX$. In problem 6.5 of Chapter 2, you showed that A can be resolved uniquely into a symmetric and an antisymmetric part. Show that it is only the symmetric part of A which contributes to ϕ.

14.2. Let X be a contravariant vector with the transformation property $X = \tilde{T}X'$ and let A be a square matrix. Show that if $\phi = \tilde{X}AX$ is an invariant for all choices of X, then A is a second-rank covariant tensor and transforms according to $A' = TA\tilde{T}$.

14.3. If all we had to deal with was scalars, vectors, and second-rank tensors, then all of tensor analysis could be done with matrices. We could invent a notation to keep track of covariant and contravariant quantities. Suppose we let $\bar{X}$ designate the contravariant version of a vector and $\underline{X}$ be the covariant form. Similarly, let $\bar{A}$ and $\underline{A}$ represent the contravariant and the covariant form of a second-rank tensor. For example, equation (6.14) which gives the relation between the covariant and the contravariant forms of the metric tensor would be written $\bar{G}\underline{G} = I$. The matrix version of (6.19) is $\underline{X} = \underline{G}\bar{X}$. What are the matrix versions of (6.20) and (6.22)? Equation (6.21) is a little tricky. How about $(A^i{}_j) = \bar{A}\underline{G} = \bar{G}\underline{A}$ and $(A_j{}^i) = \underline{A}\bar{G} = \underline{G}\bar{A}$? Write out the transformation properties of $\bar{X}$, $\underline{X}$, $\bar{A}$, $\underline{A}$, $(A^i{}_j)$, and $(\bar{A}_i{}^j)$ in matrix form.

If $\dot{\bar{Q}}$ represents generalized velocities, then $\underline{P} = \underline{G}\dot{\bar{Q}}$ is a covariant vector. If the Lagrangian for the system is

$$L = \tfrac{1}{2} g_{\alpha\beta}\dot{q}_\alpha \dot{q}_\beta - V(q_\alpha)$$

then show that $\underline{P}$ represents the canonical momenta. See problem 13.2.

14.4. Prove directly from $AX = \lambda GX$ that λ is real.

14.5. Start out with $AX_1 = \lambda_1 GX_1$ and $AX_2 = \lambda_2 GX_2$ and show that $\tilde{X}_2 GX_1 = 0$ if $\lambda_1 \neq \lambda_2$. *Degeneracy* occurs if $\lambda_1 = \lambda_2$, and this method of proof fails. In the text, we got around this by using the theory of orthogonal transformations. In that way, the existence of f mutually orthogonal eigenvectors is assured even if many of the eigenvalues are degenerate.

14.6. Find the most general solution of (14.53). Does the use of this more general solution rather than (14.55) alter the final solution (14.59) in any way?

14.7. Normalize (14.58) and construct the transforming matrix T.

14.8. A thin, hollow tube of mass M and length a can slide without friction on a horizontal wire as shown in Fig. 14.5. On each end of the tube hang plane pendulums of mass m and length s. The position of the end of the tube is measured by a coordinate z, and θ_1 and θ_2 are the angular displacements of the pendulums. Show that for small angular displacements of the pendulums, the approximate equations of motion of the system are

$$(2m+M)\dot{z}+ms(\dot{\theta}_1+\dot{\theta}_2)=p_z=\text{constant}$$

$$ms^2\ddot{\theta}_1+ms\ddot{z}+mgs\theta_1=0$$

$$ms^2\ddot{\theta}_2+ms\ddot{z}+mgs\theta_2=0$$

Show that the complete solution can be written in the form

$$\begin{pmatrix}\theta_1\\ \theta_2\end{pmatrix}=A\begin{pmatrix}1\\ -1\end{pmatrix}\cos(\omega_1 t-\phi_1)+B\begin{pmatrix}1\\ 1\end{pmatrix}\cos(\omega_2 t-\phi_2)$$

$$(2m+M)z+ms(\theta_1+\theta_2)=p_z t+C$$

$$\omega_1^2=\frac{g}{s}\qquad \omega_2^2=\frac{g}{s}\left(1+\frac{2m}{M}\right)$$

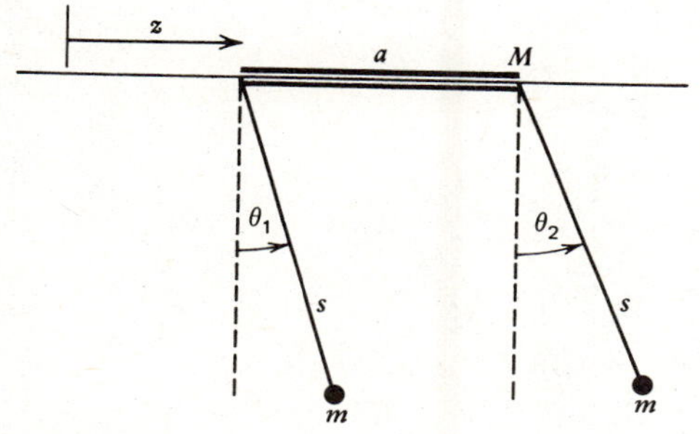

Fig. 14.5

15. EULER ANGLES AND RIGID BODY DYNAMICS

A rigid body is made up of a very large number of particles held rigidly with respect to one another by time-independent constraints. The motion of the center of mass of the body is taken care of by

$$\mathbf{F}_{\text{ext}}=M\mathbf{a}_c \tag{15.1}$$

where $\mathbf{F}_{\text{ext}}$ is the net external force acting on the body, M is its total mass, and $\mathbf{a}_c$ is the acceleration of the center of mass. See equation (4.47) of Chapter 2. The rotational motion of the body with respect to its center of mass can be specified by means of a time-dependent, proper orthogonal transformation. If the body is constrained so that one point is fixed, it is possible to discuss its rotational motion with respect to the fixed point rather than the center of mass if we so desire. The orthogonality relation

$$S\tilde{S}=\tilde{S}S=I \tag{15.2}$$

involves six conditions on the nine elements of S. This means that a general orthogonal transformation can be written as a function of three parameters. These three parameters can be used as generalized coordinates in the solution of rigid body problems either by means of Euler's equations (9.33)–(9.35) of Chapter 2 or by means of Lagrange's equations. In general, then, six coordinates are needed to specify the motion of a rigid body: three to locate the center of mass and three to specify the rotational motion with respect to the center of mass. If the body is constrained so that one point is fixed, then only the three rotational coordinates are needed.

One possible set of three parameters are the *Euler angles* which consist of an azimuth angle ϕ, a polar angle θ, and a spin angle ψ. These are illustrated in Fig. 15.1*a* as they might be used to describe the motion of a toy top with a fixed point at the origin of coordinates. The general transformation connecting the laboratory coordinates x_i with the body coordi-

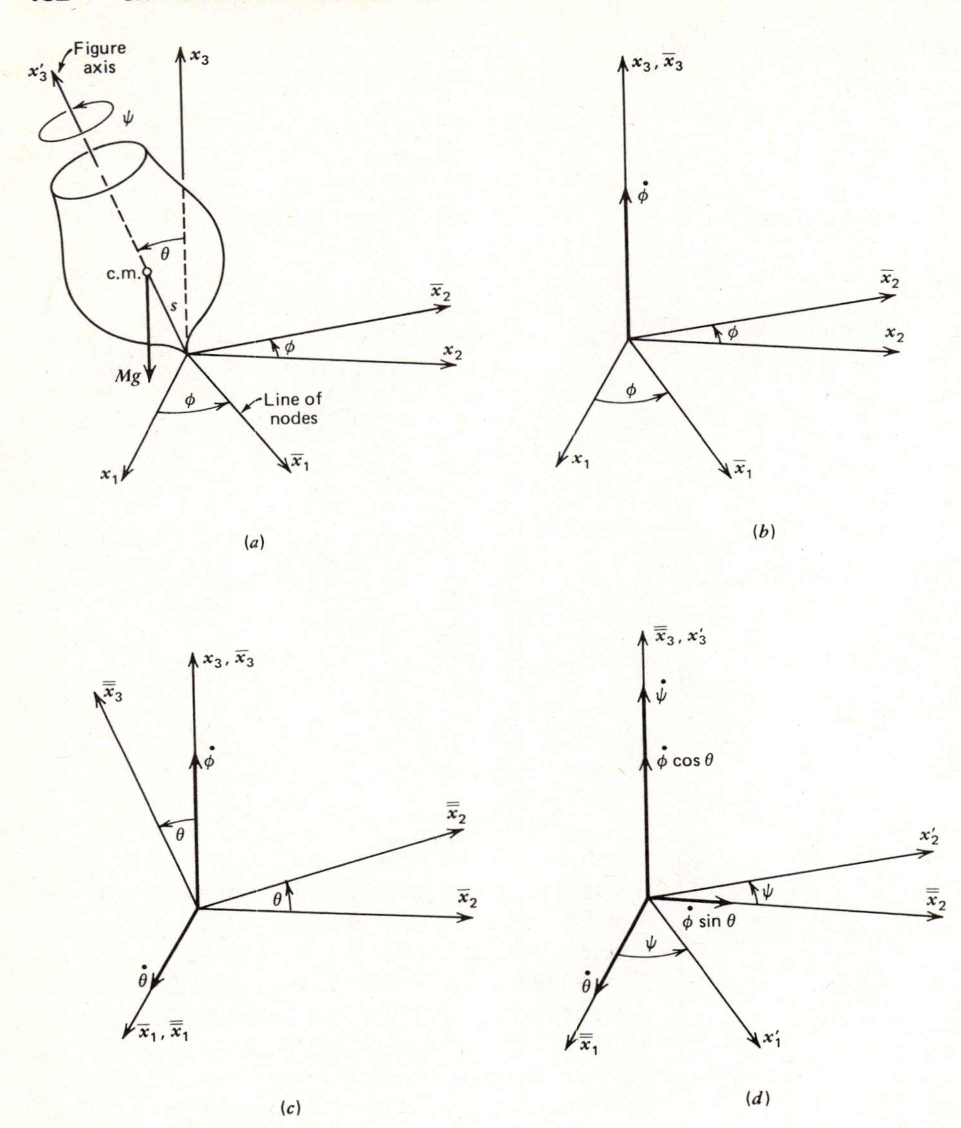

Fig. 15.1

nates x_i' can be built up as the product of three rotations. (See section 9 of Chapter 2 for a discussion of laboratory and body coordinates.) Figure 15.1*b* shows a rotation by an angle ϕ about the x_3 axis of the laboratory coordinates. The $\bar{x}_1$ axis of the new coordinate system is called the *line of nodes*. The matrix which gives this rotation is

$$S_\phi = \begin{pmatrix} \cos\phi & \sin\phi & 0 \\ -\sin\phi & \cos\phi & 0 \\ 0 & 0 & 1 \end{pmatrix} \tag{15.3}$$

A second rotation by θ about the $\bar{x}_1$ axis is shown in

Fig. 15.1*c*. The $\bar{\bar{x}}_3$ axis of this second intermediate coordinate system is usually chosen to coincide with an axis about which the toy top has rotational symmetry and is called the *figure axis* of the top. This rotation is given by

$$S_\theta = \begin{pmatrix} 1 & 0 & 0 \\ 0 & \cos\theta & \sin\theta \\ 0 & -\sin\theta & \cos\theta \end{pmatrix} \tag{15.4}$$

Now imagine that you are going to set the top into motion by spinning it about its figure axis. This involves a rotation by ψ about the $\bar{\bar{x}}_3$ axis as shown in Fig. 15.1*d* and brings us finally into coincidence with the body coordinates. Note that the x_3' axis also coincides with the figure axis of the top. The representation of this rotation is

$$S_\psi = \begin{pmatrix} \cos\psi & \sin\psi & 0 \\ -\sin\psi & \cos\psi & 0 \\ 0 & 0 & 1 \end{pmatrix} \tag{15.5}$$

The overall transformation connecting the original laboratory coordinates with the body coordinates is

$$S = S_\psi S_\theta S_\phi \tag{15.6}$$

Both Euler's equations and Lagrange's equations require that the components of the angular velocity of the body be expressed in terms of the Euler angles. A rigorous and straightforward way to find the components of the angular velocity with respect to the body coordinates is to use equation (9.42) of Chapter 2:

$$\Omega' = \dot{S}\tilde{S} \tag{15.7}$$

However, the matrix S given by (15.6) is large, and its derivative is even worse. We will therefore appeal to geometrical intuition to find the components of the angular velocity vector. In Fig. 15.1, we have kept track of the contributions made to $\boldsymbol{\omega}$ by $\dot{\phi}$, $\dot{\theta}$, and $\dot{\psi}$ in the three rotations. It is evident from Fig. 15.1*d* that

$$\omega_1' = \dot{\phi}\sin\theta\sin\psi + \dot{\theta}\cos\psi \tag{15.8}$$

$$\omega_2' = \dot{\phi}\sin\theta\cos\psi - \dot{\theta}\sin\psi \tag{15.9}$$

$$\omega_3' = \dot{\phi}\cos\theta + \dot{\psi} \tag{15.10}$$

The reader is welcome to work these out by means of (15.7) if he or she wants.

We will assume that the figure axis (the body x_3' axis) is an axis about which the toy top has rotational symmetry. Its cross sections perpendicular to the x_3' axis are circles. This means that the moment of inertia tensor with respect to body coordinates is diagonal and that two of these diagonal elements are equal: $I_1' = I_2'$. Since we are going to be working from now on in the body coordinates, primes will be dropped. The kinetic energy of the top is found from (9.50) of Chapter 2 and works out to

$$K = \tfrac{1}{2}I_1(\dot{\theta}^2 + \dot{\phi}^2\sin^2\theta) + \tfrac{1}{2}I_3(\dot{\psi} + \dot{\phi}\cos\theta)^2 \tag{15.11}$$

All that is required to complete the solution by the Lagrange method is an expression for the potential energy. If the top is acted on by a uniform gravitational field as pictured in Fig. 15.1*a*, the potential energy is

$$V = Mgs\cos\theta \tag{15.12}$$

where s is the distance from the pivot to the center of mass. The Lagrangian is then $L = K - V$.

Both ϕ and ψ are cyclic coordinates as defined in problem 13.2. The canonical momenta conjugate to these coordinates are constants of the motion which we call a and b:

$$a = \frac{\partial L}{\partial \dot{\psi}} = I_3(\dot{\psi} + \dot{\phi}\cos\theta) \tag{15.13}$$

$$b = \frac{\partial L}{\partial \dot{\phi}} = I_1\dot{\phi}\sin^2\theta + I_3(\dot{\psi} + \dot{\phi}\cos\theta)\cos\theta$$

$$= I_1\dot{\phi}\sin^2\theta + a\cos\theta \tag{15.14}$$

The solution of (15.14) for $\dot{\phi}$ yields

$$\dot{\phi} = \frac{b - a\cos\theta}{I_1\sin^2\theta} \tag{15.15}$$

The equation of motion for θ follows from

$$\frac{d}{dt}\left(\frac{\partial L}{\partial \dot{\theta}}\right) - \frac{\partial L}{\partial \theta} = 0 \tag{15.16}$$

and is

$$I_1\ddot{\theta} - I_1\dot{\phi}^2\sin\theta\cos\theta + a\dot{\phi}\sin\theta - Mgs\sin\theta = 0 \tag{15.17}$$

With the aid of (15.15) to eliminate $\dot{\phi}$, it is possible to write (15.17) as

$$\frac{I_1}{Mgs}\ddot{\theta} + \frac{dV_e}{d\theta} = 0 \tag{15.18}$$

where the function V_e is given by

$$V_e = A\left(\frac{B - \cos\theta}{\sin\theta}\right)^2 + \cos\theta \tag{15.19}$$

The constants A and B are

$$A = \frac{a^2}{2Mgs I_1} \qquad B = \frac{b}{a} \tag{15.20}$$

Equation (15.18) is a complicated differential equation the solutions of which give θ as a function of the time. A lot of insight into the nature of the motion can be found without obtaining an exact solution. Note that (15.18) implies that

$$\frac{I_1\dot{\theta}^2}{2Mgs} + V_e = \varepsilon \tag{15.21}$$

where ε is a constant. This is exactly like the statement of conservation of energy for the motion of a particle in one dimension with potential energy V_e and energy ε. Of course, V_e is not the actual potential energy of the top and ε is not its actual total energy, although (15.21) is closely related to the energy equation for the top and can be obtained from it. The function V_e is called the *effective potential.*

The effective potential can be studied numerically. Figure 15.2 shows graphs of V_e for $B = 0.75$ and three different values of A. Equations (15.20) and (15.13) show that A is determined essentially by how much spin the top is given. Large values of A mean a lot of spin. The effective potential is infinite at $\theta = 0$ and $\theta = 180°$ and has a single minimum between these values. Suppose that $A = 5.0$ and $B = 0.75$. If, as illustrated in Fig. 15.2, the initial conditions are such that $\varepsilon = 1.4$, then the figure axis of the top will move back and forth between the angles θ_1 and θ_2, called *turning points*. According to (15.21), to go outside of this range would require $\dot{\theta}^2 < 0$, which is not possible.

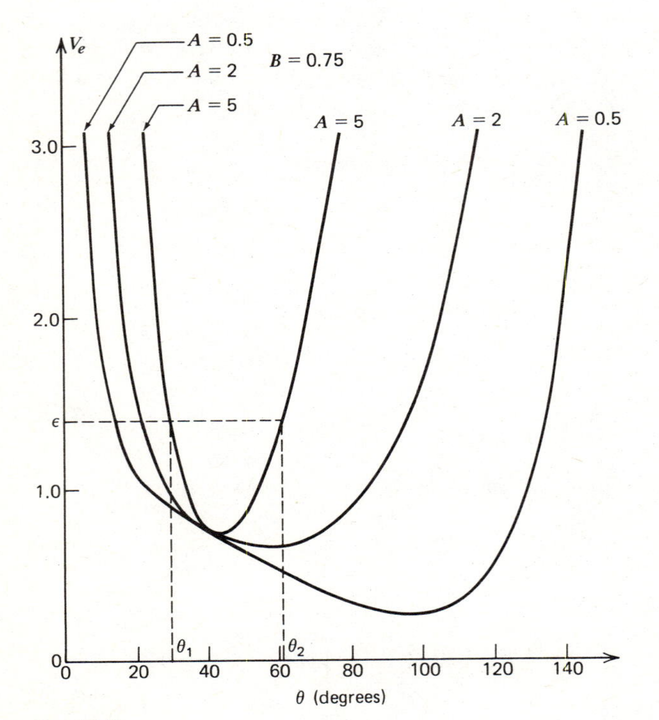

Fig. 15.2

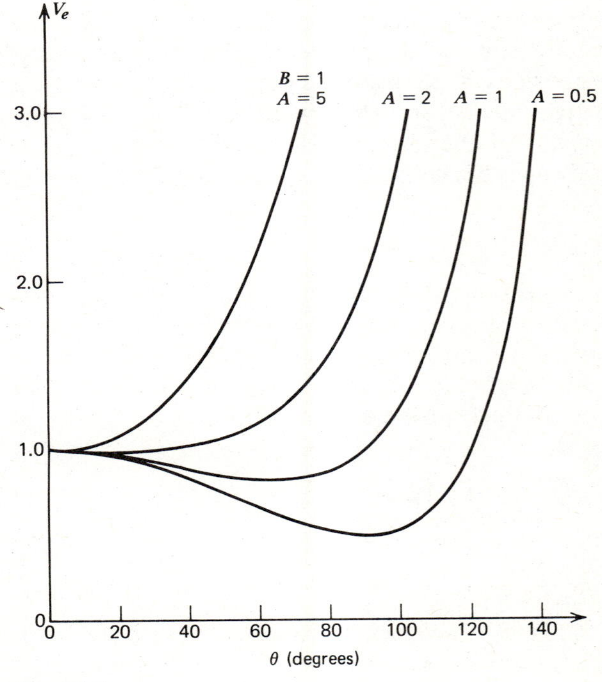

Fig. 15.3

At the turning points, $V_e = \varepsilon$ and $\dot{\theta}$ is momentarily zero. The figure axis reverses its direction of motion with respect to θ at these points. At the same time that this is going on, the figure axis advances in the ϕ direction according to equation (15.15). This part of the motion is called *precession*. The oscillation of the figure axis between θ_1 and θ_2 is called *nutation*.

To be more specific, suppose that at $t = 0$ the top is given a spin angular velocity $\dot{\psi}(0)$ about its figure axis and that $\dot{\theta}(0) = 0$, $\dot{\phi}(0) = 0$, and $\theta(0) = \theta_1$. We then have

$$a = I_3\dot{\psi}(0) \qquad b = a\cos\theta_1 \qquad B = \cos\theta_1 \tag{15.22}$$

$$\dot{\phi} = a\frac{\cos\theta_1 - \cos\theta}{I_1 \sin^2\theta} \tag{15.23}$$

Assume that the direction of spin is such that $a > 0$. The figure axis at first falls. It then picks up angular velocity in the ϕ direction as given by equation (15.23). When the figure axis reaches the turning point θ_2 in Fig. 15.2, it reverses direction and rises back to the starting angle θ_1, at which point $\dot{\theta}$ and $\dot{\phi}$ are again momentarily zero. The whole process then repeats itself. The θ motion is exactly the same kind of thing as a single particle moving in a one-dimensional potential well.

If $B = 1.0$ exactly, the qualitative nature of V_e changes. The function is now finite at the origin. Figure 15.3 shows V_e for $B = 1.0$ and several values of A. If the spin is small, V_e has a maximum at the origin and a minimum between $\theta = 0$ and $\theta = 180°$. If the spin is large, V_e has a single minimum at the origin. If the top is stood straight up and given a small spin, for example, $A = 0.5$ in Fig. 15.3, its figure axis will be in a state of unstable equilibrium and the top will fall over. If the spin is high, for example, $A = 5.0$, the equilibrium is stable and the top continues to spin quietly with its figure axis vertical.

PROBLEMS

15.1. Show that the effective potential (15.19) is finite at $\theta = 0$ if $B = 1$ exactly. Show that the effective potential has a relative maximum at $\theta = 0$ and a minimum between $\theta = 0$ and $\theta = 180°$ if $A < 2$. If $A > 2$, show that it has a single minimum at $\theta = 0$. Do this by finding the points where $dV_e/d\theta = 0$. See Fig. 15.3. It is hard to find the minimum of V_e analytically if $B \neq 1$. If you try to find the point where $dV_e/d\theta = 0$, you will get a quartic equation for $\cos\theta$.

15.2. Make an accurate plot of V_e for $B = 0.75$ and $A = 5.0$. Find the values of V_e and θ which correspond to the minimum of this function from the graph. Improve these values by numerical computation. (A computer scan of the function in the vicinity of its minimum will do it.) If the initial conditions for the top are such that $B = 0.75$, $A = 5.0$, and $\varepsilon = V_e(\text{min})$, we are exactly at the bottom of a potential well. The figure axis is in a condition of stable equilibrium with respect to its θ motion. The angle θ remains constant and the figure axis precesses on the surface of a cone at a constant angular velocity in the ϕ direction. Show that

$$\dot{\phi} = \alpha\sqrt{\frac{Mgs}{I_1}}$$

and find the numerical value of α. Assume that the direction of spin is such that $a > 0$.

15.3. Derive equation (15.21) directly from the conservation of energy equation $K + V = W$. How is the constant ε related to the actual energy W?

15.4. The angular momentum vector of the toy top expressed in body coordinates is

$$\ell = I_1'\omega_1'\hat{\mathbf{e}}_1' + I_2'\omega_2'\hat{\mathbf{e}}_2' + I_3'\omega_3'\hat{\mathbf{e}}_3'$$

Show that $I_3'\omega_3'$ is the constant a of equation (15.13). Find the component of ℓ along the laboratory x_3 axis and show that it is the constant b of equation (15.14).

15.5. A thin rod of length s and mass M is fastened by means of a frictionless pin to a vertical shaft which is rotating at a constant angular velocity $\dot{\phi} = \omega$ as pictured in Fig. 15.4. The rod can move freely in the θ direction, but the pin constrains it so that it cannot rotate about the body x_3' axis. A transformation from laboratory to body coordinates can therefore be expressed in terms of the two rotations given by (15.3) and (15.4). Find the components of the angular velocity in the body coordinates by means of (15.7). This calculation is messy, but you probably need some practice with manipulating matrices. You should get (15.8)–(15.10) with $\psi = 0$.

Hint: Write $S = S_\theta S_\phi$. Then $\dot{S} = \dot{S}_\theta S_\phi + S_\theta \dot{S}_\phi$. Finally, $\Omega' = (\dot{S}_\theta S_\phi + S_\theta \dot{S}_\phi)\tilde{S}_\phi \tilde{S}_\theta$. Wait until this last step to multiply the matrices together.

Find a Lagrangian for the motion of the rod and find a second-order differential equation obeyed by θ. Show that an effective potential exists which can be put in the form

$$V_e = (A\cos\theta - 3)\cos\theta + V_0$$

where $A = s\omega^2/g$. Plot V_e for several values of the constant A. The constant V_0 can be adjusted to any value which makes graphing convenient. There is a critical value of A where the qualitative nature of V_e changes. What is this critical value? Show that for ω above a certain critical value, it is possible for the rod to move as a conical pendulum with $\theta =$ constant and that this type of motion is not possible for smaller values of ω.

It is necessary to obtain V_e from the second-order differential equation for θ as was done in the text for the top. This is because the mechanical energy $W = K + V$ is *not* constant here. The reason for this is that the constraint, $\phi = \omega t$, is time-dependent.

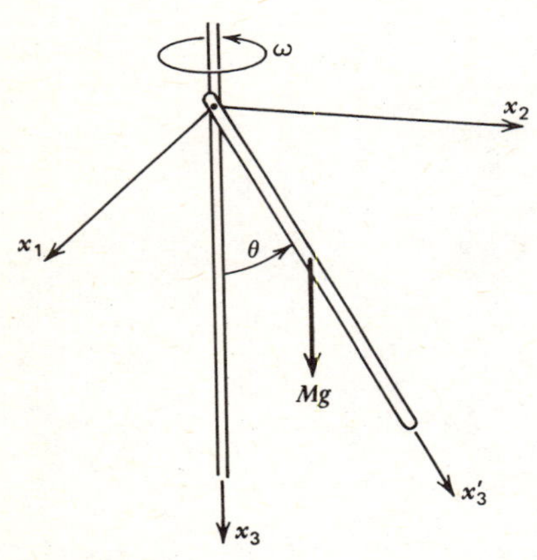

Fig. 15.4

15.6. Multiply out the matrices of equation (15.6) to show that

$$S = \begin{pmatrix} \cos\psi\cos\phi - \cos\theta\sin\phi\sin\psi & \cos\psi\sin\phi + \cos\theta\cos\phi\sin\psi & \sin\psi\sin\theta \\ -\sin\psi\cos\phi - \cos\theta\sin\phi\cos\psi & -\sin\psi\sin\phi + \cos\theta\cos\phi\cos\psi & \cos\psi\sin\theta \\ \sin\theta\sin\phi & -\sin\theta\cos\phi & \cos\theta \end{pmatrix}$$

15.7. Show that the components of $\boldsymbol{\omega}$ with respect to the laboratory coordinates are

$$\omega_1 = \dot{\theta}\cos\phi + \dot{\psi}\sin\theta\sin\phi$$

$$\omega_2 = \dot{\theta}\sin\phi - \dot{\psi}\sin\theta\cos\phi$$

$$\omega_3 = \dot{\phi} + \dot{\psi}\cos\theta$$

You can do this problem either by geometrical intuition or by using $\omega_i = S_{ji}\omega_j'$.

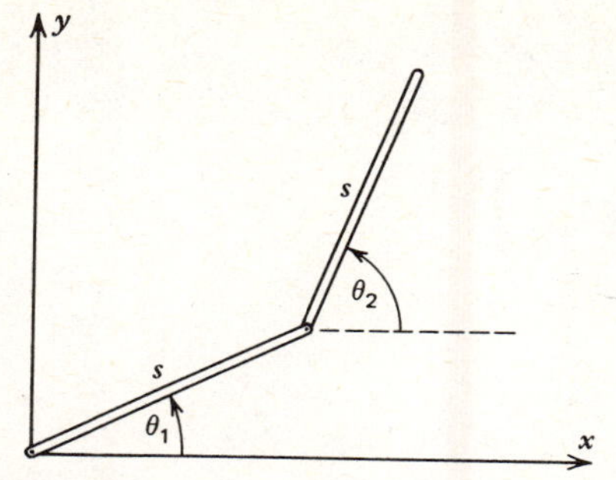

Fig. 15.5

15.8. Two identical thin, uniform rods of length s and mass M are linked together by a hinge. One end of one rod is hinged to a fixed support at the origin of coordinates as shown in Fig. 15.5. If the system is constrained to move in the x, y plane, show that the kinetic energy is

$$K = \tfrac{2}{3}Ms^2\dot{\theta}_1^2 + \tfrac{1}{6}Ms^2\dot{\theta}_2^2 + \tfrac{1}{2}Ms^2\dot{\theta}_1\dot{\theta}_2\cos(\theta_2 - \theta_1)$$

Hint: The rod which is connected to the origin is a rigid body with one point fixed. For the other rod, this is not the case, and you must express its kinetic energy as the sum of a rotational term and a translational term. See equation (4.39) of Chapter 2.

$$\frac{\partial u}{\partial x} = \frac{\partial v}{\partial y} \qquad \frac{\partial u}{\partial y} = -\frac{\partial v}{\partial x}$$

5 Complex Numbers and Variables

Up to this point, not a word has been said in this book about the concept of a number field. We started right off in Chapter 1 by making the tacit assumption that all quantities were expressed in terms of real numbers and that the reader understood the concept of a real number. It seemed best at that time not to raise the issue of the possibility of the existence of other number fields. True, the idea of complex numbers had to be dealt with briefly in connection with eigenvalue problems. This was quickly put aside when it was established that real symmetric matrices have real eigenvalues.

In this chapter, we will define the concept of a number field and then rather quickly specialize the discussion to the very important case of complex numbers. A rather complete review of the topic of infinite series is given because of its importance in the development of the theory of functions of a complex variable. Several examples will be given which show the advantages to be gained by employing complex numbers in the solution of various problems. We will find that in quantum mechanics, the employment of complex numbers is not just a convenience in facilitating calculations but is essential since the wave function is intrinsically complex.

1. THE GROUP CONCEPT

The idea of a mathematical group is basic to the definition of a number field. A group G is a set of mathematical objects $A, B, C, \ldots$ for which a single law of composition is defined. In treating groups as a mathematical abstraction, the law of composition is almost always referred to as multiplication; but in an actual realization or representation of a group, it could be many things. Thus, we assume that any two elements A and B of G, equal or unequal, possess unique products

$$AB = C \qquad BA = D \tag{1.1}$$

We do not generally require that the multiplication process be commutative. The set of elements $A, B, C, \ldots$, finite or infinite in number, forms a group if the following axioms are fulfilled:

1. Closure property: Given any two elements A and B of G, then their products C and D as given by (1.1) are also members of G.
2. Associative law: If A, B, and C are any three elements of G, then

$$A(BC) = (AB)C \tag{1.2}$$

3. Existence of an identity element: Every group contains at least one element I such that

$$IA = A \tag{1.3}$$

for every element A of G.

4. Existence of an inverse: For every A of G, there is at least one element, written A^{-1} and called the inverse of A, such that

$$A^{-1}A = I \tag{1.4}$$

Theorem 1. Axioms 3 and 4 actually define a left identity and a left inverse. These same elements are also a right identity and a right inverse, meaning that $AI = A$ and $AA^{-1} = I$.

Proof: Multiply (1.4) from the right by A^{-1} and use (1.3):

$$A^{-1}AA^{-1} = IA^{-1} = A^{-1} \tag{1.5}$$

By (1.4), A^{-1} has a left inverse which we will write as $(A^{-1})^{-1}$. Multiply (1.5) from the left by this element:

$$(A^{-1})^{-1}A^{-1}AA^{-1} = (A^{-1})^{-1}A^{-1}$$

$$IAA^{-1} = I \tag{1.6}$$

Now use (1.3) to get

$$AA^{-1} = I \tag{1.7}$$

Now consider

$$AI = A(A^{-1}A) = (AA^{-1})A = IA = A \tag{1.8}$$

which completes the proof. Similarly, a right identity and a right inverse are also necessarily a left identity and a left inverse.

Any power of a group element is defined:

$$AA = A^2 \qquad AAA = A^3 \tag{1.9}$$

$$A^{-1}A^2 = A^{-1}AA = IA = A \tag{1.10}$$

$$A^{-1}A^{-1}A^2 = A^{-1}A = I \tag{1.11}$$

Thus, we may write $A^{-1}A^{-1} = A^{-2}$. In general,

$$A^nA^m = A^{n+m} \tag{1.12}$$

where n and m are positive or negative integers or zero provided it is understood that

$$A^0 = I \tag{1.13}$$

An important class of groups is the commutative groups for which

$$AB = BA \tag{1.14}$$

for all members A and B of G. Such groups are called *Abelian*.

Theorem 2. The inverse of any element is unique.

Proof: Given the element A, suppose two elements B and C exist such that

$$BA = I \tag{1.15}$$

and

$$AC = I \tag{1.16}$$

Multiply (1.15) by C and (1.16) by B to get

$$BAC = C \qquad BAC = B \tag{1.17}$$

Thus, $B = C$.

Theorem 3. The identity element is unique. Any group contains one and only one identity.

Proof: Suppose that there are two elements I and I' such that

$$AI = A \tag{1.18}$$

$$I'B = B \tag{1.19}$$

for any two elements A and B of G. Let $A = I'$ and let $B = I$. Then,

$$I'I = I' = I \tag{1.20}$$

Given any two elements A and B of G, it is an immediate consequence of theorem 2 that

$$AX = B \qquad YA = B \tag{1.21}$$

have unique solutions for X and Y. Also, the equations

$$AX = B \qquad AY = B \tag{1.22}$$

imply that $X = Y$.

The *order* of a group is the number of distinct elements which it contains. The multiplication table of a group shows all the possible products that can be formed from the group elements. Figure 1.1 shows the multiplication table of a group of order 2. It is in fact the only possible multiplication table, as we now show. By assumption, the group contains exactly two distinct elements, one of which must be the identity. The closure property then requires either $A^2 = A$ or $A^2 = I$. In the multiplication table of any group, each element appears exactly once in each row and in each column. If the same element were duplicated, the uniqueness of solution of (1.22) would be violated. Thus, $A^2 = I$ is the only possibility, and there is really only one group of order 2. There are many representations of this group but they all have the same multiplication table. A possible representation is $I = 1$ and $A = -1$.

The same line of reasoning shows that there is only one possible group of order 3. Its multiplication table

	I	A
I	I	A
A	A	I

Fig. 1.1

	I	A	B
I	I	A	B
A	A	B	I
B	B	I	A

Fig. 1.2

is shown in Fig. 1.2. Note that the groups of orders 2 and 3 are Abelian.

An attempt to construct the possible groups of order 4 leads to four apparently different possibilities as shown in Fig. 1.3. However, if (3) is rewritten as shown in Fig. 1.4, it becomes apparent that its structure is really the same as (1) if we just exchange the elements B and C. Such groups are said to be *isomorphic* and do not really represent different groups. Similarly, it is easily shown that (1) and (4) are also isomorphic. Thus, (1) and (2) represent the only possible groups of order 4. They are both Abelian. It is not until order 6 that the first non-Abelian group appears. Group (1) can be represented by the matrices

$$I=\begin{pmatrix}1 & 0\\ 0 & 1\end{pmatrix} \qquad A=\begin{pmatrix}0 & 1\\ -1 & 0\end{pmatrix}$$

$$B=\begin{pmatrix}-1 & 0\\ 0 & -1\end{pmatrix} \qquad C=\begin{pmatrix}0 & -1\\ 1 & 0\end{pmatrix} \tag{1.23}$$

We now give some examples of groups of infinite order. All positive rational numbers form an Abelian group of infinite order with respect to ordinary multiplication as the law of combination. The number 1 is the identity, and the inverse of any element is its reciprocal.

	I	A	B	C
I	I	A	B	C
A	A	B	C	I
B	B	C	I	A
C	C	I	A	B

(1)

	I	A	B	C
I	I	A	B	C
A	A	I	C	B
B	B	C	I	A
C	C	B	A	I

(2)

	I	A	B	C
I	I	A	B	C
A	A	C	I	B
B	B	I	C	A
C	C	B	A	I

(3)

	I	A	B	C
I	I	A	B	C
A	A	I	C	B
B	B	C	A	I
C	C	B	I	A

(4)

Fig. 1.3

	I	A	C	B
I	I	A	C	B
A	A	C	B	I
C	C	B	I	A
B	B	I	A	C

Fig. 1.4

The set of all integers forms an Abelian group of infinite order if the law of combination is addition. In this case, the identity element is 0, and the inverse of an element A is $-A$ as can be seen from

$$0+A=A \tag{1.24}$$

$$(-A)+A=0 \tag{1.25}$$

PROBLEMS

1.1. Consider all possible three-dimensional orthogonal transformations. Do they form a group? Do the proper orthogonal transformations by themselves form a group? Do the improper orthogonal transformations by themselves form a group? A set of elements within a group which themselves form a group is called a *subgroup*. Can you pick out any subgroups of either of the two groups of order 4?

1.2. Construct a representation of the group of order 3 from orthogonal matrices which represent 120° rotations about the z axis.

1.3. The *order* h of a group element A is the smallest positive integer such that $A^h = I$. The identity element in any group is the only element of order 1. Show that all the elements of the group of order 3, except the identity, are of order 3. Show that all elements of the groups of order 4, other than the identity, are either of order 2 or 4. These are special cases of a general theorem: The order of any element in a finite group is a factor of the order of the group. Thus, since 5 is a prime number, all the elements in a group of order 5, except the identity, are of order 5. The entire group can be generated from a single element as $A, A^2, A^3, A^4, A^5 = I$. As a consequence, there is in fact only one group of order 5. By extending this argument, we see that there is only one group of any prime order.

1.4. Consider the set of six functions

$$f_1(z)=z \qquad f_2(z)=\frac{1}{1-z} \qquad f_3(z)=\frac{z-1}{z}$$

$$f_4(z)=\frac{1}{z} \qquad f_5(z)=1-z \qquad f_6(z)=\frac{z}{z-1}$$

Show that the set is closed if we let the law of composition be the substitution of one function into another. For example,

$$f_2f_3=f_2\{f_3(z)\}=\frac{1}{1-f_3(z)}=z=f_1(z)$$

Does the set form a group? Is it Abelian? Which is the identity? Construct the multiplication table.

2. NUMBER FIELDS

A field F is a set of mathematical objects $A, B, C, \ldots$ for which two laws of combination are defined. One of the laws of combination is required to be Abelian, but the other one need not be. All members of F are required to form a group with respect to the Abelian law of combination. It is usual to refer to the Abelian law of combination as *addition*. The following axioms therefore apply:

1. Closure: If A and B are members of F, then so is $C=A+B$.
2. Associativity:
$$A+(B+C)=(A+B)+C \tag{2.1}$$
3. Existence of an identity element: The identity element is called *zero* and has the property that
$$0+A=A \tag{2.2}$$
for all elements A of F.
4. Existence of an inverse element: For every A of F, there is an element, which is written as $(-A)$, such that
$$(-A)+A=0 \tag{2.3}$$

Conditions 1 through 4 are nothing more than a recapitulation of the group axioms 1 through 4 of section 1. All the theorems and consequences of the group axioms then follow. For example, the zero element is unique, and every element has a unique inverse. The equation

$$A+X=B \tag{2.4}$$

always has a unique solution for X. It is

$$X=B+(-A) \tag{2.5}$$

The second law of combination is called *multiplication* and is not required to be Abelian. All members of F, with the important exception of the zero element, are required to form a group with respect to the multiplicative law of combination. The continuation of the list of basic field axioms is therefore as follows:

5. Closure: If A and B are members of F, so are $C=AB$ and $D=BA$.
6. Associativity:
$$A(BC)=(AB)C \tag{2.6}$$
7. Existence of an identity element: With respect to the multiplicative law of combination, the identity element is called the unit element and has the property
$$IA=A \tag{2.7}$$
for all A of F.
8. Existence of an inverse: For every A of F, with the exception of 0, there is an element A^{-1} such that
$$A^{-1}A=I \tag{2.8}$$

Again, axioms 5 through 8 reiterate the group postulates.

It is axiom 8, which requires the existence of a multiplicative inverse, that makes trouble for the zero element necessitating our leaving it out. Zero has no multiplicative inverse. Multiplication of any element by the zero element is however defined. It is through the following, and last, axiom that the multiplicative properties of zero are revealed:

9. Distributive law: If A, B, and C are members of F, then

$$A(B+C)=AB+AC \tag{2.9}$$

$$(A+B)C=AC+BC \tag{2.10}$$

If it were not for the distributive law, there would be no connection between the two laws of combination.

Some consequences of the basic axioms are explored in the following theorems.

Theorem 1. If A is any element of F, then

$$0A=A0=0 \tag{2.11}$$

Proof: By means of axioms 7 and 9,

$$0A+A=0A+IA=(0+I)A \tag{2.12}$$

Now use axioms 3 and 7 to get

$$0A+A=IA=A \tag{2.13}$$

Add $(-A)$ to both sides of (2.13) and use axiom 4 to arrive at

$$0A=0 \tag{2.14}$$

The proof to show that $A0=0$ is similar. Theorem 1 shows that 0 can have no inverse.

Theorem 2. If A is any element of F, then

$$(-I)A=A(-I)=-A \tag{2.15}$$

Proof: The steps are similar to those of theorem 1:

$$(-I)A+A=[(-I)+I]A=0A=0$$

$$(-I)A=-A$$

Similarly, $A(-I)=-A$.

Theorem 3.

$$(-I)^2=I \tag{2.16}$$

Proof:

$$\begin{aligned}(-I)^2+(-I)&=(-I)(-I)+(-I)(I)\\&=(-I)[(-I)+I]\\&=(-I)0=0\end{aligned}$$

Now add I to both sides to get (2.16). It is the usual practice to write $A-B$ in place of $A+(-B)$. Thus, for example, $A-(-B)=A+B$.

Theorem 4. If A is not zero, then

$$AB=0 \tag{2.17}$$

implies that $B=0$.

Proof: Since $A\neq 0$, it has an inverse. Multiply (2.17) by A^{-1} to get $A^{-1}AB=A^{-1}0=0$. This is the same as $IB=B=0$.

If nonzero elements A and B existed such that (2.17) were true, it would be possible to "factor" zero into the product of two nonzero elements. To say it another way, A and B would be *divisors* of zero. Theorem 4 says that in a field, there are no divisors of zero.

The simplest field that we can think of consists of only two elements 0 and I. Since there is only one group of order 2, the addition table, which gives all possible additive combinations of the two elements, must be as given in Fig. 2.1*a*. Note that the only nonzero element must be the unit element. The unit element all by itself forms a group of order 1. The multiplication table is therefore as given in Fig. 2.1*b*. Note that for this field

$$I+I=0 \tag{2.18}$$

By virtue of the tables in Fig. 2.1, the group properties 1 through 8 are satisfied for this system. There remains the distributive property 9. Note that

$$I(I+I)=I(0)=0$$

$$I^2+I^2=I+I=0$$

+	0	I
0	0	I
I	I	0

(*a*)

×	0	I
0	0	0
I	0	I

(*b*)

Fig. 2.1

Thus, axiom 9 is satisfied and the two-element system is in fact a field.

In discussing groups in section 1, we defined powers of an element by $AA = A^2$, $AAA = A^3$, and so on. Similarly, for the additive law of combination, we can write

$$A + A = 2A \qquad A + A + A = 3A \qquad \text{and so on} \tag{2.19}$$

But be careful. Note for the field consisting of two elements that $2I = 0$ even though $2 \neq 0$ and $I \neq 0$. But then, "2" is not a member of the field.

Number fields in which the multiplicative law of combination is non-Abelian are usually called *skew fields* or *division rings*. Commutative number fields are often referred to simply as fields. Since all fields obey the same basic axioms, it is evident that vectors, matrices, and determinants with components and elements that belong to fields other than the real number field are possible. It would seem especially strange to consider the possibility of a matrix the elements of which were members of a skew field, and we shall have no reason to do so. Matrices and vectors the elements and components of which are complex numbers are useful and occur extensively in quantum theory. Probably at this point the reader is uncomfortable with the idea of a vector with complex components, so we will try to ease our way gradually into this concept.

PROBLEMS

2.1. Write out the steps to show that $A0 = 0$ and that $A(-I) = -A$.

2.2. Prove that $(-A)(-B) = AB$.

2.3. There is only one group of each of the orders 2 and 3, so that the only possibility for a three-element field has to based on the addition and multiplication tables shown in Fig. 2.2. Does this system in fact form a field? What is $3I$ in this field?

$+$	0	I	A
0	0	I	A
I	I	A	0
A	A	0	I

$\times$	0	I	A
0	0	0	0
I	0	I	A
A	0	A	I

Fig. 2.2

2.4. There are two groups of order 4. Show that it is possible to form a field with one of these groups as the additive group but that it is impossible with the other.

2.5. Let $A, B, C, \ldots$ be rational numbers. Show that all irrational numbers of the form $A + B\sqrt{2}$ form a number field with respect to ordinary addition and multiplication as the laws of combination.

3. THE COMPLEX NUMBER FIELD

The field of rational numbers consists of zero and all the possible positive and negative numbers that can be written as ratios of integers. In this number field, the equation

$$x^2 - 2 = 0 \tag{3.1}$$

has no solution. The situation is remedied by enlarging the number field to include all irrational numbers of the form $A + \sqrt{2}\,B$, where A and B are rational. A similar situation occurs when we try to solve the quadratic equation

$$ax^2 + bx + c = 0 \tag{3.2}$$

where a, b, and c are any real numbers. A formal

solution of (3.2) is

$$x = \frac{-b \pm \sqrt{b^2 - 4ac}}{2a} \tag{3.3}$$

but there is a problem if $b^2 - 4ac < 0$ because there is no way to take the square root of a negative number in the field of real numbers. We are again faced with the problem of having to enlarge the number field in order to include this possibility. Maybe there is no such number field and we are doomed to failure. It is convenient to start out by inventing a mathematical entity, usually called i, with the property that

$$i^2 = -1 \tag{3.4}$$

It is then possible, formally, to write (3.3) as

$$x = \frac{-b \pm i\sqrt{4ac - b^2}}{2a} \tag{3.5}$$

in those cases where $b^2 - 4ac < 0$. Equation (3.2) is still satisfied by (3.5), as can be checked by direct substitution. We are thus led to the possibility of an algebra of number pairs based on

$$z = x + iy \tag{3.6}$$

where x and y are real numbers and i has the property (3.4). The quantity z itself is called a complex number, or sometimes an imaginary number. The multiplicative law of combination of complex numbers is defined by

$$\begin{aligned} z_1 z_2 &= (x_1 + iy_1)(x_2 + iy_2) \\ &= x_1 x_2 + i(x_1 y_2 + y_1 x_2) + i^2 y_1 y_2 \\ &= x_1 x_2 - y_1 y_2 + i(x_1 y_2 + y_1 x_2) \end{aligned} \tag{3.7}$$

The closure property is therefore obeyed. It is easy to show that the multiplication is commutative: $z_1 z_2 = z_2 z_1$. As the additive law of combination, we define

$$z_1 + z_2 = x_1 + x_2 + i(y_1 + y_2) \tag{3.8}$$

The real numbers can be written as $x + i0$ and are a subfield of the complex number field. They can be represented by points along the x axis of a rectangular Cartesian coordinate system. It is then possible to visualize the complex numbers as being represented by points in the x, y plane. When this is done, the x axis is called the *real axis*, the y axis is called the *imaginary axis*, and the x, y plane is called the *complex plane*. The zero and the unit elements are $0 = 0 + i0$ and $1 = 1 + i0$. The reader can verify at this point that the laws of combination (3.7) and (3.8) do in fact satisfy all the postulates of a number field.

Are the laws of combination (3.7) and (3.8) the only ones that are possible? Suppose that we do not yet know the properties of i. The distributive condition permits us to write

$$\begin{aligned} z_1 z_2 &= (x_1 + iy_1)(x_2 + iy_2) \\ &= x_1 x_2 + i^2 y_1 y_2 + i(x_1 y_2 + y_1 x_2) \end{aligned} \tag{3.9}$$

Consider the four elements 1, i, -1, and $-i$. The requirements of closure, associativity, existence of an inverse, and a unit element lead to the conclusion that these four elements must form a multiplicative group of order 4. There are only two such groups. Let's try the one in Fig. 1.3(2). We make the following identification of the group elements:

$$1 \leftrightarrow I \qquad i \leftrightarrow A \qquad -1 \leftrightarrow B \qquad -i \leftrightarrow C \tag{3.10}$$

The multiplication table of the group shows that $i^2 = 1$. Equation (3.9) gives

$$z_1 z_2 = x_1 x_2 + y_1 y_2 + i(x_1 y_2 + y_1 x_2) \tag{3.11}$$

Suppose now that z_1 is given and that z_2 is required to be its inverse. Then,

$$x_1 x_2 + y_1 y_2 = 1 \qquad y_1 x_2 + x_1 y_2 = 0 \tag{3.12}$$

A solution of (3.12) for x_2 and y_2 exists if the determinant of the coefficients

$$\Delta = x_1^2 - y_1^2 \tag{3.13}$$

is not zero. There is a problem if $x_1 = \pm y_1$. Such elements do not have inverses. The field postulates are violated because all elements other than the zero element are required to have inverses. Now try the group in Fig. 1.3(1). Make the same identification as in (3.10). The multiplication table now gives $i^2 = -1$, and (3.10) becomes identical to (3.7). The existence of an inverse for z_1 requires that x_2 and y_2 satisfy

$$x_1 x_2 - y_1 y_2 = 1 \qquad y_1 x_2 + x_1 y_2 = 0 \tag{3.14}$$

The determinant of the coefficients is

$$\Delta = x_1^2 + y_1^2 \tag{3.15}$$

Thus, $\Delta \neq 0$ unless $x_1 = y_1 = 0$ and all elements except the zero element have inverses as required. The law of multiplication (3.7) is really the only one possible. This does not rule out the possibility that larger number fields exist with the complex number field as a subfield. All quadratic equations are solvable in the field of complex numbers, and there are no other similar fields which involve an algebra of number pairs.

Given the complex number $z = x + iy$, we define its *complex conjugate* by

$$z^* = x - iy \tag{3.16}$$

The product of z with its complex conjugate is

$$zz^* = x^2 + y^2 = r^2 \tag{3.17}$$

where r is the distance from the origin to the point represented by z in the complex plane. We write

$$|z| = r \tag{3.18}$$

and call r the *norm*, *modulus*, *magnitude*, or *absolute value* of z. A complex number can be represented in polar form as

$$z = r\cos\theta + ir\sin\theta \tag{3.19}$$

The angle θ is called the *argument* of z and is written $\theta = \arg z$. (Sometimes, θ is called the *amplitude* of z, but this is a bad idea since amplitude means other things in physical applications.) Complex numbers form a commutative field, so that the inverse of z can be written unambiguously as $1/z$. It is also possible to express the inverse as

$$\frac{1}{z} = \frac{z^*}{zz^*} = \frac{x - iy}{x^2 + y^2} \tag{3.20}$$

The additive law of combination (3.8) is the equivalent of two-dimensional vector addition in the x, y plane. This is illustrated in Fig. 3.1. Note that the complex number $z - z_0$ can be thought of as a vector which projects from the point z_0 to the point z. The following triangle inequality holds:

$$|z_1 + z_2| \leq |z_1| + |z_2| \tag{3.21}$$

which says simply that the length of one side of a triangle is shorter than the sum of the lengths of the other two sides.

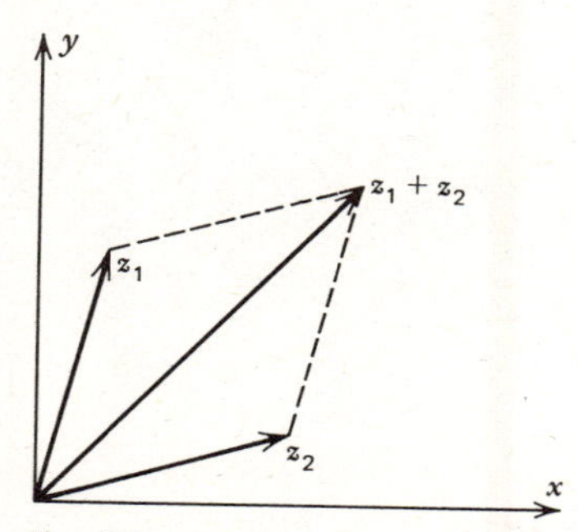

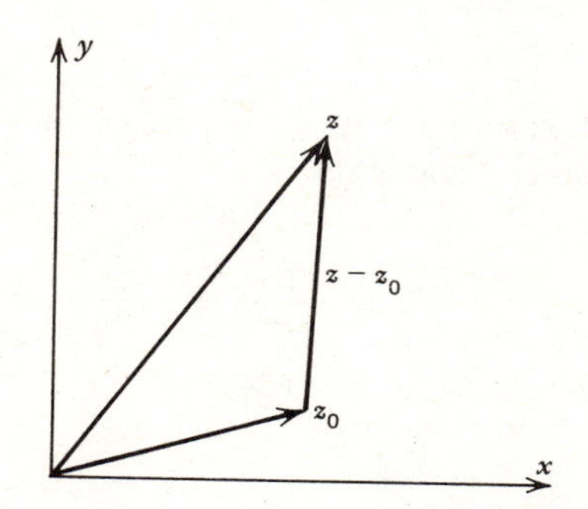

Fig. 3.1

PROBLEMS

3.1. Suppose that we try an addition law for number pairs of the form

$$(a_1, a_2) + (a_3, a_4) = \left(\sum_{j=1}^{4} \alpha_j a_j, \sum_{k=1}^{4} \beta_k a_k \right)$$

Show that the field axioms 1 through 4 reduce this to the equivalent of equation (3.8).

3.2. Solve (3.14) for x_2 and y_2 and note that the result is the same as (3.20).

3.3. Prove that if $z = z^*$, then z is real.

3.4. Any function of a complex variable z can always be separated into its real and imaginary parts. For the function

$$f(z) = z + \frac{1}{z^2} = u + iv$$

find the two functions $u(x, y)$ and $v(x, y)$. A frequently used notation is $\operatorname{Re} f(z) = u$, $\operatorname{Im} f(z) = v$.

3.5. Show that $|z_1 z_2| = |z_1||z_2| = r_1 r_2$ and that $\arg z_1 z_2 = \theta_1 + \theta_2$. In other words, if both z_1 and z_2 are written in polar form, then

$$z_1 z_2 = r_1 r_2[\cos(\theta_1 + \theta_2) + i \sin(\theta_1 + \theta_2)]$$

In particular,

$$z^2 = r^2(\cos 2\theta + i \sin 2\theta)$$

Extend this to show that

$$z^n = r^n(\cos n\theta + i \sin n\theta)$$

which is known as DeMoivre's theorem.

3.6. Let a and b be positive real numbers. The inequalities

$$|z - z_0| < a \qquad a < |z - z_0| < b$$

define certain regions of the complex plane. What is the geometry of these regions? Think of z_0 as fixed and z as a variable point.

3.7. Prove that

$$|z_1 + z_2| \geq |z_1| - |z_2|$$

Hint: Write $z_1 = z_1 + z_2 - z_2$ and use (3.21).

3.8. Suppose we try to form a number field which involves the multiplication of triplets of real numbers. Let's base it on the group of order 3 and write

$$z = a + bu + cu^2$$

where a, b, and c are real numbers and 1, u, and u^2 are identified with the group elements of Fig. 1.2 as $1 \leftrightarrow I$, $u \leftrightarrow A$, $u^2 \leftrightarrow B$. This idea doesn't work. Why? It is not until an attempt is made to multiply quadruplets that success is achieved in obtaining a larger number field. This field is called the *field of quaternions* and is noncommutative.

3.9. Show that $(z^n)^* = (z^*)^n$ and that $|z^n| = |z|^n$.

4. REVIEW OF INFINITE SERIES

Some knowledge of infinite series is necessary for the further development of the theory of functions of a complex variable. In this section, a summary of some of the important facts concerning infinite series of constant terms is presented. In section 5, we review infinite series of functions. Throughout this section, all numbers and variables are real.

Consider the finite series

$$S_n = \sum_{k=0}^{n} x^k = 1 + x + x^2 + \cdots + x^n \tag{4.1}$$

where x is any real number. This series can actually be summed. Note that

$$xS_n = x + x^2 + \cdots + x^n + x^{n+1} \tag{4.2}$$

Then,

$$S_n - xS_n = 1 - x^{n+1} \qquad S_n = \frac{1 - x^{n+1}}{1 - x} \tag{4.3}$$

If the variable x is now restricted to the range $-1 < x < +1$, then

$$\lim_{n \to \infty} x^{n+1} = 0 \tag{4.4}$$

It is therefore possible to write

$$S = \lim_{n\to\infty} S_n = \sum_{k=0}^{\infty} x^k = \frac{1}{1-x} \tag{4.5}$$

The infinite series of terms in (4.5) is called the *geometrical series*, and we have shown that it has a definite, finite value provided that $-1 < x < +1$. We say that the series *converges* to the limit $S(x)$ for every x in this interval. Obviously, the series diverges if $|x| \geq 1$. The geometrical series is important in the development of convergence tests for other series.

Let $a_1, a_2, \ldots$ be any sequence of real numbers. The nth partial sum is defined by

$$S_n = a_1 + a_2 + \cdots + a_n \tag{4.6}$$

We say that the infinite series

$$\sum_{k=1}^{\infty} a_k \tag{4.7}$$

converges and has the value S if

$$\lim_{n\to\infty} S_n = S \tag{4.8}$$

There are two basic (and equivalent) criteria for establishing the existence of the limit (4.8):

I. The series (4.7) converges to a limit S if for every positive number ε, no matter how small, there is a number $N(\varepsilon)$ such that

$$|S - S_n| < \varepsilon \tag{4.9}$$

whenever $n > N$.

II. Cauchy criterion. The series (4.7) converges to a limit if for every positive number ε, no matter how small, there is a number $N(\varepsilon)$ such that

$$|S_n - S_m| < \varepsilon \tag{4.10}$$

whenever n and m are larger than N.

In the statement of the Cauchy criterion, the existence of the limit S is implied but not specified. The criteria I and II are not theorems but are really definitions of what is meant by convergence. It is easy enough to show that II follows from I. If n and m are both larger than N, then (4.9) gives

$$\begin{aligned} |S_n - S_m| &= |S_n - S + S - S_m| \\ &\leq |S_n - S| + |S - S_m| < 2\varepsilon \end{aligned} \tag{4.11}$$

The factor of 2 is of no consequence, since ε can be made arbitrarily small. The converse, showing that I follows from II, is a little more difficult (although it seems obvious), so we do not present it here.

An obvious fact is that a necessary condition for the convergence of (4.7) is

$$\lim_{n\to\infty} a_n = 0 \tag{4.12}$$

The condition is however not sufficient. Consider, for example, the series

$$1 + \tfrac{1}{2} + \tfrac{1}{3} + \tfrac{1}{4} + \cdots \tag{4.13}$$

which satisfies (4.12). Note that

$$\begin{aligned} a_{n+1} &+ a_{n+2} + \cdots + a_{2n} \\ &= \frac{1}{n+1} + \frac{1}{n+2} + \cdots + \frac{1}{2n} \\ &> \frac{1}{2n} + \frac{1}{2n} + \cdots + \frac{1}{2n} = \frac{1}{2} \end{aligned} \tag{4.14}$$

which holds for all n, no matter how large. The Cauchy criterion for convergence is therefore violated, and (4.13) diverges.

Theorem 1. Alternating Series. The series

$$S = \sum_{k=1}^{\infty} (-1)^{k+1} a_k = a_1 - a_2 + a_3 - a_4 + \cdots \tag{4.15}$$

where (4.12) is obeyed and $0 < a_{n+1} < a_n$ for all n greater than some fixed N converges and the error in a partial sum approximation to (4.15) is less than the first neglected term.

Proof: Consider the partial sum

$$S_{2n} = a_1 - a_2 + a_3 - \cdots - a_{2n} \tag{4.16}$$

which ends in a negative term. Assume that $2n > N$. Then,

$$S - S_{2n} = (a_{2n+1} - a_{2n+2}) + (a_{2n+3} - a_{2n+4}) + \cdots \tag{4.17}$$

Since $a_{2n+1} > a_{2n+2}, a_{2n+3} > a_{2n+4}, \ldots$, (4.17) shows that

$$S - S_{2n} > 0 \tag{4.18}$$

Now write (4.17) as

$$\begin{aligned} S - S_{2n} &= a_{2n+1} - (a_{2n+2} - a_{2n+3}) \\ &\quad - (a_{2n+4} - a_{2n+5}) + \cdots \end{aligned} \tag{4.19}$$

which shows that

$$S - S_{2n} < a_{2n+1} \tag{4.20}$$

Since (4.12) is true, given a positive number ε, no matter how small, n can be made sufficiently large so that

$a_{2n+1} < \varepsilon$. Thus,

$$0 < S - S_{2n} < \varepsilon \tag{4.21}$$

The series therefore converges to the limit S and, by (4.20), the error in approximating S by S_{2n} is less than the first neglected term a_{2n+1}. To put it another way, a series converges if the terms alternate in sign and decrease in magnitude monotonically to zero. It should be pointed out that in the proof of convergence of any series, the criteria for convergence need only be met by the terms past some point in the series. In other words, a finite number of terms in any series can behave erratically as long as the series eventually settles down.

Theorem 2. Absolute Convergence. If the series

$$\sum_{k=1}^{\infty} |a_k| \tag{4.22}$$

converges, then so does

$$\sum_{k=1}^{\infty} a_k \tag{4.23}$$

The series (4.23) is then said to be *absolutely convergent*.

Proof: By the assumed convergence of (4.22),

$$|a_n + \cdots + a_m| \leq |a_n| + \cdots + |a_m| < \varepsilon \tag{4.24}$$

The series (4.23) therefore converges by the Cauchy criterion. Series which converge, but not absolutely, are said to be *conditionally convergent*. Absolute convergence is a stronger kind of convergence than conditional convergence, and it is possible to perform various kinds of operations on absolutely convergent series which are not permitted on conditionally convergent series.

Theorem 3. An absolutely convergent series can be separated into two series, one consisting of all the positive terms and the other consisting of all the negative terms:

$$\sum_{k=1}^{\infty} a_k = \sum_{k=1}^{\infty} p_k - \sum_{k=1}^{\infty} q_k \tag{4.25}$$

where $p_k > 0$ and $q_k > 0$. Such a separation is not possible for conditionally convergent series.

Proof: In the first place, both the series $\Sigma\, p_k$ and $\Sigma\, q_k$ converge by the assumption of absolute convergence of the series $\Sigma\, a_k$. To put it another way, removing part of the terms from $\Sigma\, |a_k|$ does not make it diverge. Now consider the partial sum

$$\sum_{k=1}^{n} a_k = \sum_{k=1}^{r} p_k - \sum_{k=1}^{s} q_k \tag{4.26}$$

where $r + s = n$. Equation (4.26) holds for arbitrarily large n, which proves the validity of (4.25) for absolutely convergent series. Assume now that the series $\Sigma\, a_k$ converges only conditionally and that it contains an infinite number of both positive and negative terms. Both the series $\Sigma\, p_k$ and $\Sigma\, q_k$ must then be divergent. If it were not so, then the original series would necessarily be absolutely convergent. Thus, the separation (4.25) is not valid for conditionally convergent series.

Theorem 4. In absolutely convergent series, rearrangement of the terms in any manner does not affect the convergence and the value of the sum if the series is unchanged. In conditionally convergent series, the value of the sum can be changed to any value by suitable rearrangement of the terms and the series can even be made to diverge.

Proof: By theorem 3, we need only consider a series consisting of all positive terms. Consider the partial sum

$$S_n = a_1 + \cdots + a_n \tag{4.27}$$

Let

$$T_m = b_1 + \cdots + b_m \tag{4.28}$$

be a rearrangement of a finite number of terms of $\Sigma\, a_k$. If m is large enough, T_m will contain all the terms of S_n and

$$S_n \leq T_m \tag{4.29}$$

Now consider an even larger partial sum of the original series:

$$S_{n'} = a_1 + \cdots + a_{n'} \tag{4.30}$$

Again, n' can be made sufficiently large so that all the terms of T_m are included in $S_{n'}$. This means that

$$S_n \leq T_m \leq S_{n'} \tag{4.31}$$

By the Cauchy criterion, if n is chosen sufficiently large, the difference between S_n and $S_{n'}$ can be made as small as we please. Hence, the rearranged series converges to the same limit as the original series.

Rather than construct a general proof in the case of conditionally convergent series, we will take a specific example and show that its sum can be altered by rearranging the order of the terms. By theorem 1, the *harmonic series*

$$S = 1 - \frac{1}{2} + \frac{1}{3} - \frac{1}{4} + \cdots \tag{4.32}$$

converges. It was shown in equation (4.14) that it

does not converge absolutely. Consider the following rearrangement of the terms of (4.32):

$$\begin{aligned} T &= \left(1 - \frac{1}{2}\right) - \frac{1}{4} + \left(\frac{1}{3} - \frac{1}{6}\right) \\ &\quad - \frac{1}{8} + \left(\frac{1}{5} - \frac{1}{10}\right) - \frac{1}{12} + \cdots \\ &\quad + \left(\frac{1}{2n-1} - \frac{1}{4n-2}\right) - \frac{1}{4n} + \cdots \\ &= \left(\frac{1}{2} - \frac{1}{4}\right) + \left(\frac{1}{6} - \frac{1}{8}\right) + \left(\frac{1}{10} - \frac{1}{12}\right) \\ &\quad + \cdots + \left(\frac{1}{4n-2} - \frac{1}{4n}\right) + \cdots \\ &= \frac{1}{2}\left(1 - \frac{1}{2} + \frac{1}{3} - \frac{1}{4} + \cdots \right. \\ &\quad \left. + \frac{1}{2n-1} - \frac{1}{2n} + \cdots \right) \\ &= \frac{1}{2} S \end{aligned} \tag{4.33}$$

Theorem 5. If

$$A = \sum_{k=1}^{\infty} a_k \qquad B = \sum_{k=1}^{\infty} b_k \tag{4.34}$$

are absolutely convergent series, then they can be multiplied together in any manner, and the sum of the terms will always be AB.

Proof: Consider the partial sum

$$\begin{aligned} S_n &= a_1 b_1 + a_1 b_2 + \cdots + a_1 b_n \\ &\quad + a_2 b_1 + a_2 b_2 + \cdots + a_2 b_n \\ &\quad \vdots \\ &\quad + a_n b_1 + a_n b_2 + \cdots + a_n b_n \end{aligned} \tag{4.35}$$

The finite series formed by taking the sum of the absolute values of the terms in (4.35) is

$$\sum_{k=1}^{n} |a_k| \sum_{k=1}^{n} |b_k| \tag{4.36}$$

As $n \to \infty$, (4.36) converges on account of the absolute convergence of the series (4.34). Thus, (4.35) represents a partial sum of an absolutely convergent series. By theorem 4, the terms in (4.35) can be arranged in any way without altering the sum as $n \to \infty$. In particular,

$$S_n = (a_1 + \cdots + a_n)(b_1 + \cdots + b_n) \tag{4.37}$$

Thus,

$$\lim_{n \to \infty} S_n = AB \tag{4.38}$$

which completes the proof.

The remaining theorems in this section have to do with tests for the convergence or divergence of infinite series.

Theorem 6. Comparison Test. If the series

$$\sum_{k=1}^{\infty} b_k \tag{4.39}$$

is absolutely convergent and if there exists a number N such that

$$|a_n| \le |b_n| \tag{4.40}$$

for all $n > N$, then the series

$$\sum_{k=1}^{\infty} a_k \tag{4.41}$$

is also absolutely convergent.

Proof: If both n and m are greater than N, then

$$\begin{aligned} |a_n + \cdots + a_m| &\le |a_n| + \cdots + |a_m| \\ &\le |b_n| + \cdots + |b_m| < \varepsilon \end{aligned} \tag{4.42}$$

Thus, by the Cauchy criterion, (4.41) converges absolutely. It is possible to replace the condition (4.40) by $|a_n| \le C|b_n|$, where C is any positive constant.

Theorem 7. If the series $\sum b_k$ is absolutely convergent, and if

$$\lim_{n \to \infty} \frac{|a_n|}{|b_n|} = 1 \tag{4.43}$$

then the series $\sum a_k$ is also absolutely convergent. This is a variation of the comparison test and is frequently easier to apply. If (4.43) holds, we say that the series $\sum a_k$ behaves like $\sum b_k$ for large n.

Proof: Equation (4.43) implies that there is a positive number $C > 1$ such that

$$|a_n| < C|b_n| \tag{4.44}$$

for all n greater than some fixed number N. The series

$$\sum_{k=1}^{\infty} C b_k = C \sum_{k=1}^{\infty} b_k \tag{4.45}$$

is absolutely convergent. Therefore, by theorem 6, $\sum a_k$ converges.

Theorem 8. If Σb_k converges absolutely and if there is a fixed number n such that

$$\frac{|a_{n+1}|}{|a_n|} \le \frac{|b_{n+1}|}{|b_n|} \tag{4.46}$$

for all $n > N$, then Σa_k also converges absolutely.

Proof: Start with the identity

$$|a_n| = \frac{|a_{N+1}|}{|a_N|}\frac{|a_{N+2}|}{|a_{N+1}|}\cdots\frac{|a_n|}{|a_{n-1}|}|a_N| \tag{4.47}$$

Then, use (4.46) to get

$$|a_n| \le \frac{|b_{N+1}|}{|b_N|}\frac{|b_{N+2}|}{|b_{N+1}|}\cdots\frac{|b_n|}{|b_{n-1}|}|a_N| = \frac{|a_N|}{|b_N|}|b_n| \tag{4.48}$$

The series Σa_k therefore converges absolutely by theorem 6.

Theorem 9. Ratio Test. If

$$\lim_{n\to\infty}\frac{|a_{n+1}|}{|a_n|} < 1 \tag{4.49}$$

then Σa_k converges absolutely.

Proof: The geometrical series (4.5) is known to be absolutely convergent if $|x| < 1$. Therefore, by theorem 8, Σa_k converges absolutely if

$$\frac{|a_{n+1}|}{|a_n|} \le \frac{|x^{n+1}|}{|x^n|} = |x| < 1 \tag{4.50}$$

for all n larger than some fixed N. This completes the proof. If

$$\lim_{n\to\infty}\frac{|a_{n+1}|}{|a_n|} = 1 \tag{4.51}$$

the ratio test fails. A case in point is the series

$$\sum_{k=1}^{\infty}\frac{1}{k^2} \tag{4.52}$$

which can be shown by other means to be absolutely convergent. The ratio test gives

$$\lim_{n\to\infty}\frac{n^2}{(n+1)^2} = 1 \tag{4.53}$$

The series

$$\sum_{k=1}^{\infty} k \tag{4.54}$$

is obviously divergent. The ratio test again gives

$$\lim_{n\to\infty}\frac{n+1}{n} = 1 \tag{4.55}$$

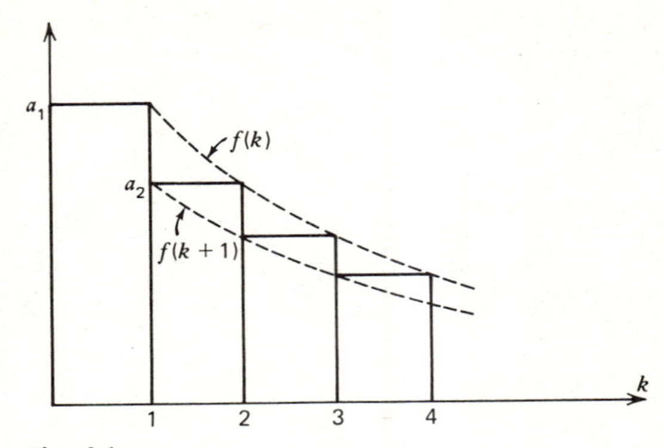

Fig. 4.1

Thus, if (4.51) holds, the series may either converge or diverge.

Theorem 10. Integral Test. Let $|a_k|$ decrease monotonically with increasing k. Define a function $f(k) = |a_k|$. The following inequalities hold:

$$|a_1| + \int_1^{\infty} f(k+1)\,dk < \sum_{k=1}^{\infty}|a_k| < |a_1| + \int_1^{\infty} f(k)\,dk \tag{4.56}$$

Proof: In Fig. 4.1, the value of the sum of the series equals the area of the rectangles. The area under the function $f(k)$ from $k = 1$ to $k = \infty$ is larger than $|a_2| + |a_3| + \cdots$. The area under the function $f(k+1)$ is smaller than $|a_2| + |a_3| + \cdots$. The inequality (4.56) then follows.

An important example of the application of theorem 10 is to the series

$$S = \sum_{k=1}^{\infty}\frac{1}{k^p} \tag{4.57*}$$

The ratio test fails on this type of series. It is apparent that (4.57) diverges if $p < 0$, so it will only be necessary to consider the case $p > 0$. The integral test (4.56) gives

$$1 + \int_1^{\infty}\frac{dk}{(k+1)^p} < S < 1 + \int_1^{\infty}\frac{dk}{k^p}$$

$$1 - \frac{1}{(p-1)(k+1)^{p-1}}\Bigg|_1^{\infty} < S < 1 - \frac{1}{(p-1)k^{p-1}}\Bigg|_1^{\infty} \tag{4.58}$$

*The function defined by (4.57) is the *Riemann zeta function* $\zeta(p)$.

The series therefore diverges if $p < 1$. If $p > 1$,

$$1 + \frac{1}{(p-1)2^{p-1}} < S < 1 + \frac{1}{(p-1)} \tag{4.59}$$

and the series converges. If $p = 1$, (4.56) gives

$$1 + \ln(k+1)\big|_1^\infty < S < 1 + \ln k\big|_1^\infty \tag{4.60}$$

and the series again diverges. In summary, (4.57) converges if $p > 1$ and diverges if $p \le 1$. It should be apparent that the crucial element in establishing the convergence or divergence of a series by the integral test is the existence or nonexistence of the integral

$$\int^\infty f(k)\,dk \tag{4.61}$$

The exact value of the lower limit is not important.

Another series to which the integral test can be successfully applied is

$$\sum_{k=2}^{\infty} \frac{1}{k \ln k} \tag{4.62}$$

The integral (4.61) is

$$\int_2^\infty \frac{dk}{k \ln k} = \ln(\ln k)\big|_2^\infty = \infty \tag{4.63}$$

and the series (4.62) therefore diverges.

The tests given in the following three theorems are quite sensitive and can be used when the ratio test fails or when the integral test is too difficult to apply.

Theorem 11. Kummer's Test. Let b_k be a sequence of positive constants. If there is a fixed number N such that

$$b_n \frac{|a_n|}{|a_{n+1}|} - b_{n+1} \ge C > 0 \tag{4.64}$$

for all $n \ge N$, then Σa_k converges absolutely. It is *not* required that Σb_k be convergent. If

$$b_n \frac{|a_n|}{|a_{n+1}|} - b_{n+1} \le 0 \tag{4.65}$$

and if $\Sigma 1/b_k$ diverges, then $\Sigma |a_k|$ diverges.

Proof: If (4.64) holds, then

$$\begin{aligned}
C|a_{N+1}| &\le b_N|a_N| - b_{N+1}|a_{N+1}| \\
C|a_{N+2}| &\le b_{N+1}|a_{N+1}| - b_{N+2}|a_{N+2}| \\
C|a_{N+3}| &\le b_{N+2}|a_{N+2}| - b_{N+3}|a_{N+3}| \\
&\vdots \\
C|a_{n+1}| &\le b_n|a_n| - b_{n+1}|a_{n+1}|
\end{aligned} \tag{4.66}$$

Addition of the equations in (4.66) gives

$$C \sum_{k=N+1}^{n+1} |a_k| \le b_N|a_N| - b_{n+1}|a_{n+1}| \le b_N|a_N| \tag{4.67}$$

Since N is fixed and since (4.67) holds for all $n \ge N$,

$$\sum_{k=N+1}^{\infty} |a_k| \le \frac{1}{C} b_N|a_N| \tag{4.68}$$

which establishes the convergence of $\Sigma |a_k|$. If (4.65) holds, then

$$b_{n+1}|a_{n+1}| \ge b_n|a_n| \ge b_{n-1}|a_{n-1}| \ge \cdots \ge b_N|a_N|$$

$$|a_{n+1}| \ge \frac{b_N}{b_{n+1}}|a_N| \tag{4.69}$$

Therefore,

$$\sum_{k=N+1}^{\infty} |a_k| \ge b_N|a_N| \sum_{k=N+1}^{\infty} \frac{1}{b_k} \tag{4.70}$$

which shows that $\Sigma |a_k|$ diverges if $\Sigma 1/b_k$ diverges.

It is usual to state Kummer's test as follows: If b_n is a sequence of positive constants and if

$$\lim_{n \to \infty} \left(b_n \frac{|a_n|}{|a_{n+1}|} - b_{n+1} \right) = C \tag{4.71}$$

then $\Sigma |a_k|$ converges if $C > 0$. It diverges if $\Sigma 1/b_k$ diverges and $C < 0$. If $C = 0$, the test fails. The problem with $C = 0$ in the limit (4.71) is that it does not allow us to distinguish between the cases (4.64) and (4.65). In other words, (4.71) by itself does not tell us whether the limit $C = 0$ was reached from positive or negative values.

Theorem 12. Raabe's Test. If

$$\lim_{n \to \infty} n \left(\frac{|a_n|}{|a_{n+1}|} - 1 \right) = P \tag{4.72}$$

then $\Sigma |a_k|$ converges if $P > 1$ and diverges if $P < 1$. The test fails if $P = 1$.

Proof: The theorem is established by setting $b_n = n$ in Kummer's test and using the fact that $\Sigma 1/k$ diverges.

Theorem 13. Gauss' Test. If

$$\lim_{n \to \infty} \frac{|a_n|}{|a_{n+1}|} = 1 + \frac{h}{n} + \frac{B(n)}{n^2} \tag{4.73}$$

where $B(n)$ is bounded, then $\Sigma |a_k|$ converges if $h > 1$ and diverges if $h \le 1$.

Proof: Convergence for $h > 1$ and divergence for $h < 1$ follows easily from Raabe's test. For the case $h = 1$, we use Kummer's test with $b_n = n \ln n$:

$$\lim_{n\to\infty}\left[n \ln n\left(1+\frac{1}{n}+\frac{B}{n^2}\right)-(n+1)\ln(n+1)\right]$$
$$=\lim_{n\to\infty}\left[(n+1)\ln n-(n+1)\ln(n+1)\right]$$
$$=\lim_{n\to\infty}\left[-(n+1)\ln\left(\frac{n+1}{n}\right)\right]$$
$$=\lim_{n\to\infty}\left[-\frac{n+1}{n}\right]=-1 \qquad (4.74)$$

We established in equation (4.64) that $\Sigma 1/(k \ln k)$ diverges. Since $-1 < 0$, the series $\Sigma |a_k|$ diverges. In (4.74), we used the fact that $\ln(1+1/n) \to 1/n$ as n becomes large. This can be seen from a Taylor's series expansion of $\ln(1+1/n)$ which will be discussed in a later section. Gauss' test can be thought of as a refinement of the ratio test and is one of the most useful tests for convergence or divergence. It is often stated in the equivalent form

$$\lim_{n\to\infty}\frac{|a_{n+1}|}{|a_n|}=1-\frac{h}{n}+\frac{B(n)}{n^2} \qquad (4.75)$$

PROBLEMS

4.1. Go through the proof of theorem 1 for the case where the partial sum (4.16) is replaced by $S_{2n+1} = a_1 - a_2 + a_3 - a_4 + \cdots + a_{2n+1}$, which ends in a positive term.

4.2. Show that any two convergent series (conditional convergence is enough) can be added term by term:

$$\sum_{k=1}^{\infty} a_k + \sum_{k=1}^{\infty} b_k = \sum_{k=1}^{\infty} (a_k + b_k)$$

4.3. If the series $\Sigma |b_k|$ diverges and if $|a_n| \geq |b_n|$ for all n greater than some fixed N, show that $\Sigma |a_k|$ diverges. Is it conceivable that Σa_k could still be conditionally convergent?

4.4. Show that if $\Sigma |b_k|$ diverges and

$$\frac{|a_{n+1}|}{|a_n|} \geq \frac{|b_{n+1}|}{|b_n|}$$

for all $n > N$, then $\Sigma |a_k|$ diverges also.

4.5. Show that if

$$\lim_{n\to\infty}\frac{|a_{n+1}|}{|a_n|} > 1$$

then $\Sigma |a_k|$ diverges.

4.6. The series Σa_k converges and has a finite number of negative terms. Show that the series must in fact converge absolutely.

4.7. Estimate the value of the sum of the series (4.52) by means of (4.59). The actual value of the series can be shown to be $\pi^2/6$.

4.8. Use a modification of (4.56) to estimate the error in approximating (4.52) by the first 100 terms.

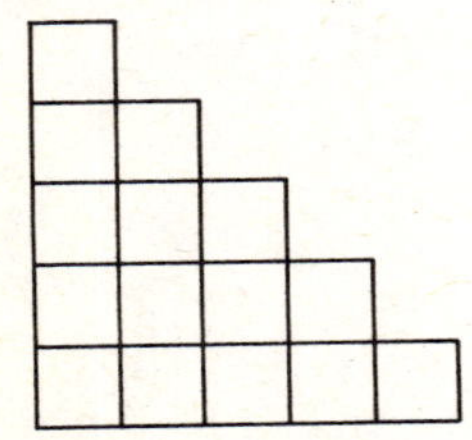

Fig. 4.2

4.9. Show that the nth partial sum of the divergent series $1+2+3+\cdots$ is

$$S_n=\sum_{k=1}^{n} k=\frac{n(n+1)}{2}$$

Hint: What is the area of the squares in Fig. 4.2?

4.10. Show that the series

$$S=\sum_{k=1}^{\infty}\frac{1}{k(k+1)}$$

converges absolutely. Show that in fact $S=1$.

Hint:

$$\frac{1}{k(k+1)}=\frac{1}{k}-\frac{1}{k+1}$$

4.11. Use Gauss' test on the series (4.54).

4.12. Prove that if $\Sigma|b_k|$ diverges and if

$$\lim_{n\to\infty}\frac{|a_n|}{|b_n|}=C$$

where C is a positive constant, then $\Sigma|a_k|$ diverges also.

4.13. Test the following series for convergence or divergence:

(a) $\displaystyle\sum_{k=0}^{\infty}\frac{1}{1+k^2}$

(b) $\displaystyle\sum_{k=1}^{\infty}\frac{k!}{k^k}$

Hint: Use the ratio test and $\lim_{n\to\infty}(1+1/n)^n=e$.

(c) $\displaystyle\sum_{k=1}^{\infty}\frac{1}{\sqrt{k(k+1)}}$

Hint: Use problem 12 with $\Sigma 1/k$ as the test series.

(d) $\displaystyle\sum_{k=2}^{\infty}\frac{1}{(\ln k)^a}$ where $a>0$.

Hint: Use the integral test. Make the substitution $\ln k=u$. It is not necessary to actually evaluate the integral in order to see that it diverges.

(e) $\displaystyle\sum_{k=1}^{\infty}\frac{k}{2^k}$

4.14. Prove that

$$\sum_{k=1}^{\infty}\sin^2\pi\left(k+\frac{1}{k}\right)$$

converges.

Hint: Expand $\sin\pi(k+1/k)$. Use the fact that $\sin(\pi/k)\to\pi/k$ as k becomes large.

4.15. Prove that

$$\sum_{k=2}^{\infty} \frac{1}{k(\ln k)^a}$$

converges when $a > 1$ and diverges when $a \leq 1$.

Hint: Follow the same procedure that was used in problem 13(d).

4.16. Consider the series

(a) $\displaystyle\sum_{k=2}^{\infty} \frac{1}{k \ln k}$ **(b)** $\displaystyle\sum_{k=2}^{\infty} \frac{1}{k(\ln k)^2}$

Show that Raabe's test gives $P = 1$ in both cases. The series (a) diverges and (b) converges.

4.17. Consider the alternating series $S = a_1 - a_2 + a_3 - a_4 + \cdots$. Assume that a_k are all positive and decrease monotonically in magnitude. Show that the sum of the series is less than its first term: $0 < S < a_1$.

5. INFINITE SERIES OF FUNCTIONS

Only real functions and variables are considered in this section. It is convenient to denote the *closed interval* $a \leq x \leq b$ by the notation $[a, b]$ and the *open interval* $a < x < b$ by (a, b). All infinite series considered in section 4, with the exception of the geometrical series (4.5), consist of terms which are constants. We take up now the case where the terms, denoted by $u_k(x)$, are functions of a variable x. If the series is convergent at every point of some interval of the variable x, then its sum is also a function of x defined over that same interval:

$$f(x) = \sum_{k=1}^{\infty} u_k(x) \tag{5.1}$$

The criteria for the convergence of (5.1) are exactly the same as for infinite series of constant terms except that they must be valid for each value of x.

Of fundamental importance is the concept of *uniform convergence*. The series $\sum u_k(x)$ is said to converge uniformly over the closed interval $[a, b]$ if for every positive number ε, no matter how small, there is a number $N(\varepsilon)$ which may depend on ε but *not* on x, such that

$$|f(x) - f_n(x)| < \varepsilon \tag{5.2}$$

for all $n \geq N$ and for all x in $[a, b]$. In (5.2), $f_n(x) = u_1(x) + \cdots + u_n(x)$ is the nth partial sum of the series. The point here is that once ε is chosen and $N(\varepsilon)$ determined, (5.2) works for *all* x in $[a, b]$. It is possible to replace (5.2) by the equivalent Cauchy criterion

$$|f_n(x) - f_m(x)| < \varepsilon \tag{5.3}$$

where n and m are any numbers larger than or equal to N.

Infinite series of functions may converge either uniformly or nonuniformly. For the geometrical series (4.5), the exact difference between the infinite series and the nth partial sum is

$$|f(x) - f_n(x)| = \left| \frac{x^{n+1}}{1 - x} \right| \tag{5.4}$$

Let $c < 1$ and let $-c \leq x \leq c$. Then,

$$|f(x) - f_n(x)| \leq \left| \frac{c^{n+1}}{1 - c} \right| \tag{5.5}$$

Define

$$\varepsilon = \frac{c^{N+1}}{1 - c} \tag{5.6}$$

As long as c is a fixed positive number less than 1, ε can be made as small as we want by making N sufficiently large. Solve for N to get

$$N(\varepsilon) = \frac{\ln \varepsilon(1 - c)}{\ln c} - 1 \tag{5.7}$$

which depends on ε but not on x. It is then true that (5.2) holds for all $n > N(\varepsilon)$. There is no dependence of N on x, so the convergence is uniform for x in the closed interval $[-c, c]$. Note that as c is taken closer and closer to 1, the situation gets "worse" in that $N(\varepsilon)$ becomes increasingly large for a given ε. It is not true that $\sum x^k$ converges uniformly for $-1 < x < +1$ because $N(\varepsilon)$ increases without limit as x is taken closer and closer to 1. In other words, N depends on x. There are no fixed values of N and ε which work for all x in $-1 < x < +1$. Thus, $\sum x^k$ converges absolutely for $-1 < x < +1$ but not uniformly. It converges both absolutely and uniformly for $-c \leq x \leq +c$ provided that c is a fixed positive number and $c < 1$.

Theorem 1. Weierstrass Test for Uniform Convergence. Let a_k be a sequence of positive constants such that $\sum a_k$ converges. If the terms of the series $\sum u_k(x)$ satisfy $|u_k(x)| \leq a_k$ for all x in $[a, b]$, then $\sum u_k(x)$ converges both uniformly and absolutely over $[a, b]$.

Proof: Given a positive number ε, which may be taken arbitrarily small, there is a fixed number $N(\varepsilon)$ such that

$$\sum_{k=n}^{m} a_k < \varepsilon \tag{5.8}$$

for all n and m larger than N. Therefore,

$$\left|\sum_{k=n}^{m} u_k(x)\right| \leq \sum_{k=n}^{m} |u_k(x)| \leq \sum_{k=n}^{m} a_k < \varepsilon \tag{5.9}$$

which is true of all x in $[a, b]$. The uniform and absolute convergence of $u_k(x)$ is therefore established.

Before proceeding to the next theorem, we remind the reader of what is meant by continuity. A function $f(x)$ is continuous at each point of $[a, b]$ if for every ε, no matter how small, there is a positive number δ such that $|f(x+h) - f(x)| < \varepsilon$ whenever $|h| < \delta$. Both x and $x + h$ must lie in the interval. Continuity at each point of a closed interval $[a, b]$ implies uniform continuity, meaning that δ does not depend on x.

Theorem 2. If a series of continuous functions converges uniformly over $[a, b]$, then its sum is also a continuous function over $[a, b]$.

Proof: Since the terms of $\sum u_k(x)$ are continuous functions, the partial sum $f_n(x)$ is also continuous. This means that

$$|f_n(x+h) - f_n(x)| < \varepsilon \tag{5.10}$$

holds throughout the interval $[a, b]$. Since $\sum u_k(x)$ is uniformly convergent, the remainder after n terms can be made to obey

$$|R_n(x)| < \varepsilon \tag{5.11}$$

throughout $[a, b]$ if n is large enough. Therefore,

$$\begin{aligned} &|f(x+h) - f(x)| \\ &\quad = |f_n(x+h) - f_n(x) + R_n(x+h) - R_n(x)| \\ &\quad \leq |f_n(x+h) - f_n(x)| + |R_n(x+h)| + |R_n(x)| < 3\varepsilon \end{aligned} \tag{5.12}$$

which holds at all points of $[a, b]$ and establishes the continuity of $f(x) = \sum u_k(x)$.

It is possible for a series of continuous functions to converge to a function which is only sectionally continuous. In the vicinity of a point of discontinuity, the convergence of the series is nonuniform.

Theorem 3. Interchange of the order of summation and integration. A series which converges uniformly over $[a, b]$ can be integrated term by term provided that the limits r and s of the integration lie in $[a, b]$:

$$\int_r^s f(x)\, dx = \sum_{k=1}^{\infty} \int_r^s u_k(x)\, dx \tag{5.13}$$

Proof: The remainder of the series after n terms is

$$f(x) - \sum_{k=1}^{n} u_k(x) = R_n(x) \tag{5.14}$$

Since the convergence is uniform, the remainder can be made to obey

$$|R_n(x)| < \varepsilon \tag{5.15}$$

throughout the interval. This means that

$$\left|\int_r^s R_n(x)\, dx\right| < \varepsilon(s - r) \tag{5.16}$$

The sum in (5.14) is finite so that it can be integrated term by term:

$$\left|\int_r^s f(x)\, dx - \sum_{k=1}^{n} \int_r^s u_k(x)\, dx\right| < \varepsilon(s - r) \tag{5.17}$$

The quantity $\varepsilon(s - r)$ can be made as small as we wish provided n is sufficiently large. This proves the result.

Theorem 3 guarantees that uniformly convergent series of continuous functions can be integrated term by term. It does not exclude the possibility that series exist which converge to discontinuous functions that can also be integrated term by term, even over a point of discontinuity. Such series require a separate investigation.

Theorem 4. Interchange of the order of differentiation and summation. If the series $f(x)=\Sigma u_k(x)$ is convergent in $[a, b]$ and if the series

$$F(x)=\sum_{k=1}^{\infty}\frac{du_k}{dx} \tag{5.18}$$

is uniformly convergent, then $F(x)=df(x)/dx$.

Proof: By assumption, (5.18) is uniformly convergent in $[a, b]$. By theorem 3, term by term integration is permitted:

$$\int_r^s F(x)\,dx=\sum_{k=1}^{\infty}u_k(s)-\sum_{k=1}^{\infty}u_k(r)=f(s)-f(r) \tag{5.19}$$

Therefore, $F(x)=df(x)/dx$, as was to be shown.

Power series are one of the most important types of series. By a power series is meant a series of the form

$$f(x)=\sum_{k=0}^{\infty}c_k(x-x_0)^k \tag{5.20}$$

where x_0 is a fixed point and c_k are constants. This is sometimes called an expansion of the function $f(x)$ about the point $x=x_0$.

Theorem 5. Power series which converge for at least one value of $x-x_0\neq 0$ and which diverge for at least one value of $x-x_0$ have a definite *interval of convergence* such that the series converges both uniformly and absolutely if $|x-x_0|\leq\xi<r$ and diverges if $|x-x_0|>r$. No general statement about the point $|x-x_0|=r$ can be made, meaning that convergence of the series at this point has to be investigated separately for each case.

Proof: Assume that the series converges for the value $|x-x_0|=\xi$. It is then true that $c_n\xi^n\to 0$ as $n\to\infty$. Let M be a fixed positive constant. The condition $|c_n\xi^n|<M$ certainly holds for n sufficiently large. If we define q by $q\xi=|x-x_0|$, then $0<q<1$ for all values of $|x-x_0|<\xi$. Therefore,

$$|c_n(x-x_0)^n|=|c_n\xi^n|q^n<Mq^n \tag{5.21}$$

The series Σq^k is the geometrical series and is convergent. By theorem 1, the series $\Sigma c_k(x-x_0)^k$ then converges both uniformly and absolutely whenever $|x-x_0|<\xi$. Now suppose that ξ is taken larger and larger until some value is reached where the series diverges. This establishes that there is some number r such that the series converges uniformly and absolutely if $|x-x_0|\leq\xi<r$ and diverges if $|x-x_0|>r$. It is possible that $r=\infty$ so that the power series converges for all finite values of $x-x_0$. As a practical matter, the value of r for a given series can be found from the ratio test (4.51). Setting $|x-x_0|=r$ gives

$$\lim_{n\to\infty}\left|\frac{c_{n+1}r^{n+1}}{c_n r^n}\right|=\lim_{n\to\infty}\left|\frac{c_{n+1}r}{c_n}\right|=1 \tag{5.22}$$

Theorem 6. Integration and differentiation of power series. Within its interval of convergence, a power series can be integrated term by term. It can also be differentiated as often as we please. The resulting series will always be convergent and have the same interval of convergence as the original series.

Proof: The integrability of a power series is an immediate consequence of theorem 3. To prove differentiability, write

$$f'(x)=\sum_{k=0}^{\infty}c_k k(x-x_0)^{k-1} \tag{5.23}$$

Both the convergence of (5.23) as required by theorem 4 and the interval of convergence can be established by using the ratio test:

$$\lim_{n\to\infty}\left|\frac{c_{n+1}(n+1)(x-x_0)^n}{c_n n(x-x_0)^{n-1}}\right|=\lim_{n\to\infty}\left|\frac{c_{n+1}(x-x_0)}{c_n}\right|=\left|\frac{x-x_0}{r}\right|<1 \tag{5.24}$$

where r is defined by (5.22). The differentiated series (5.23) thus converges uniformly and absolutely if $|x-x_0|<r$ and in fact represents the derivative of the original series (5.20).

Theorem 7. Uniqueness of power series. A function can be represented by a power series in one and only one way.

Proof: Within its interval of convergence, the series (5.20) can be differentiated as often as we please:

$$f'(x)=\sum_{k=0}^{\infty}c_k k(x-x_0)^{k-1} \tag{5.25}$$

$$f''(x)=\sum_{k=0}^{\infty}c_k k(k-1)(x-x_0)^{k-2} \tag{5.26}$$

$$\vdots$$

$$f^{(n)}(x)=\sum_{k=0}^{\infty}c_k k(k-1)(k-2)\ldots\times(k-n+1)(x-x_0)^{k-n} \tag{5.27}$$

(We use the notation $f'(x) = df/dx, \ldots, f^{(n)}(x) = d^n f/dx^n$.) By setting $x = x_0$, we find

$$f(x_0) = c_0 \qquad f'(x_0) = c_1 \qquad f''(x_0) = c_2 \cdot 2 \cdot 1$$

$$f^{(n)}(x_0) = c_n n! \tag{5.28}$$

The coefficients in the power series expansion are therefore determined uniquely by the function $f(x)$ to which the series converges. Every power series is in fact the Taylor's series expansion of the function which it represents. Different power series for the same function can be obtained by expanding the function about some point other than x_0. These series generally have different intervals of convergence. Theorem 7 says that there is only one possible power series expansion of a given function about a given point. It is possible to show that the remainder after n terms of a Taylor's expansion is

$$R_n(x) = \frac{f^{(n+1)}(\xi)}{(n+1)!}(x - x_0)^{n+1} \tag{5.29}$$

where $x_0 < \xi < x$, but we will not give a proof of this formula here. (See problem 5.16.) Generally, the exact value of ξ is not known. Nevertheless, (5.29) can be used to estimate the remainder by using the value of ξ that gives the most unfavorable value of $f^{(n+1)}(\xi)$.

The remainder of this section is concerned with some important examples of power series representations of functions. The exponential function $f(x) = e^x$ obeys the differential equation

$$f'(x) = f(x) \tag{5.30}$$

Let $f(x)$ be represented by a power series expansion taken about the point $x_0 = 0$: $f(x) = \Sigma c_k x^k$. The derivative of the power series can be written in two ways:

$$f'(x) = \sum_{k=0}^{\infty} c_k k x^{k-1} = \sum_{k=0}^{\infty} c_{k+1}(k+1)x^k \tag{5.31}$$

The differential equation (5.30) requires that

$$\sum_{k=0}^{\infty} c_{k+1}(k+1)x^k = \sum_{k=0}^{\infty} c_k x^k \tag{5.32}$$

By theorem 7, power series are unique so that the coefficients of like powers of x on the two sides of equation (5.32) must be equal. This gives

$$c_{k+1} = \frac{c_k}{k+1} \tag{5.33}$$

which is called a *recurrence formula* for the coefficients. By putting $k = 0, 1, 2, \ldots$, all coefficients in the expansion can be found in terms of c_0:

$$c_1 = c_0 \qquad c_2 = \frac{c_1}{2} = \frac{c_0}{2} \qquad c_3 = \frac{c_2}{3} = \frac{c_0}{3!}$$

$$c_4 = \frac{c_0}{4!} \qquad \ldots \qquad c_n = \frac{c_0}{n!} \tag{5.34}$$

The reason why the coefficient c_0 is not determined by this procedure is that the solution of the differential equation is only determined up to an arbitrary constant. The general solution of (5.30) is $f(x) = c_0 e^x$. The power series representation of e^x is found by setting $c_0 = 1$:

$$e^x = \sum_{k=0}^{\infty} \frac{x^k}{k!} \tag{5.35}$$

By means of the ratio test and the recurrence formula, we find that

$$\lim_{n\to\infty} \left| \frac{c_{n+1}x^{n+1}}{c_n x^n} \right| = \lim_{n\to\infty} \frac{|x|}{n+1} = 0 \tag{5.36}$$

which means that the series (5.35) converges for all finite values of x: $-\infty < x < +\infty$.

The function $f(x) = \sin x$ obeys the second-order differential equation $f''(x) = -f(x)$. Again, we look for a power series solution about the point $x_0 = 0$. The second derivative of the power series can be expressed as

$$f''(x) = \sum_{k=0}^{\infty} c_k k(k-1)x^{k-2}$$

$$= \sum_{k=0}^{\infty} c_{k+2}(k+2)(k+1)x^k \tag{5.37}$$

This time, the recurrence formula is found to be

$$c_{k+2} = -\frac{c_k}{(k+1)(k+2)} \tag{5.38}$$

As with the exponential series, an application of the ratio test shows that the series converges for all finite values of x. According to theorem 5, both series are then necessarily uniformly and absolutely convergent for all finite values of x. If we put $k = 0, 2, 4, \ldots$ in (5.38), all the even coefficients are found in terms of c_0. This determines a power series with only even powers of x in it. If $k = 1, 3, 5, \ldots$, all the odd coefficients are expressed in terms of c_1 and a power series with only odd powers of x results. In this manner, two separate solutions of $f''(x) = -f(x)$ are found, each with an arbitrary constant. This is a general feature of second-order differential equations. For the solution $f(x) = \sin x$ in particular, $f(0) = c_0 = 0$ and $f'(0) = c_1 = \cos(0) = 1$. Therefore, $\sin x$ is represented by an odd power series. This might have been anticipated on the basis that $\sin x$ is an *odd function*: $\sin(-x) = -\sin x$. A few coefficients as found from (5.38) are

$$c_3 = -\frac{c_1}{2\cdot 3} = -\frac{1}{3!} \qquad c_5 = -\frac{c_3}{4\cdot 5} = \frac{1}{5!}$$
$$c_7 = -\frac{1}{7!} \qquad \cdots \tag{5.39}$$

The function $f(x) = \sin x$ can be approximated with an increasing degree of accuracy by the polynomials

$$f_1(x) = x \qquad f_3(x) = x - \frac{x^3}{3!}$$
$$f_5(x) = x - \frac{x^3}{3!} + \frac{x^5}{5!} \qquad \cdots \tag{5.40}$$

We generally call $f_1(x)$ a *first order approximation*, $f_3(x)$ a *third order approximation*, and so on. The *even orders* are missing in this case. How good is the approximation? The answer is easy for $\sin x$ because the terms alternate in sign. Theorem 1 of section 4 says that the error is less than the first neglected term. The recent increased availability of minicomputers makes the study of the convergence properties of series easy and accessible, since such machines will add hundreds or even thousands of terms in a very short time. Table 5.1 shows a calculation of the polynomials (5.40) for $x = 1$ and $x = 2$ with a comparison to $\sin x$. It is evident that for $x = 1$ the convergence is quite rapid. It gets worse for larger values of x. It is a general feature of power series that a few terms represent the function quite well near the point about which the expansion is made. It is of course only necessary to compute $\sin x$ for $0 \le x \le \pi/2$, since all other values can then be obtained from the periodic properties of the function. The convergence of the series to $\sin x$ is further illustrated by Fig. 5.1, which shows graphs of the polynomials (5.40) for $n = 1$ through $n = 13$.

The binomial expansion of $f(x) = (1+x)^p$ about the point $x = 0$ can be found either by direct evaluation of the derivatives at $x = 0$ or by noting that $f(x)$ obeys the first-order differential equation

$$f'(x)(1+x) = pf(x) \tag{5.41}$$

The recurrence relation for the coefficients is

$$c_{k+1} = \frac{c_k(p-k)}{(k+1)} \tag{5.42}$$

Since $f(0) = c_0 = 1$, the coefficients are found to be

$$c_1 = p \qquad c_2 = \frac{p(p-1)}{2}$$
$$c_3 = \frac{p(p-1)(p-2)}{3!} \qquad \cdots$$
$$c_k = \frac{p(p-1)(p-2)\ldots(p-k+1)}{k!} \tag{5.43}$$

The notation

$$c_k = \binom{p}{k} \tag{5.44}$$

Table 5.1

n	$f_n(1)$	$f_n(1)/\sin(1)$	$f_n(2)$	$f_n(2)/\sin(2)$
3	0.833333333	0.990329255	0.666666667	0.733166780
5	0.841666667	1.00023255	0.933333333	1.02643349
7	0.841468254	0.999996755	0.907936508	0.998503329
9	0.841471010	1.00000003	0.909347443	1.00005501
11	0.841470985	1.00000000	0.909296136	0.999998580
13	0.841470985	1.00000000	0.909297451	1.00000003

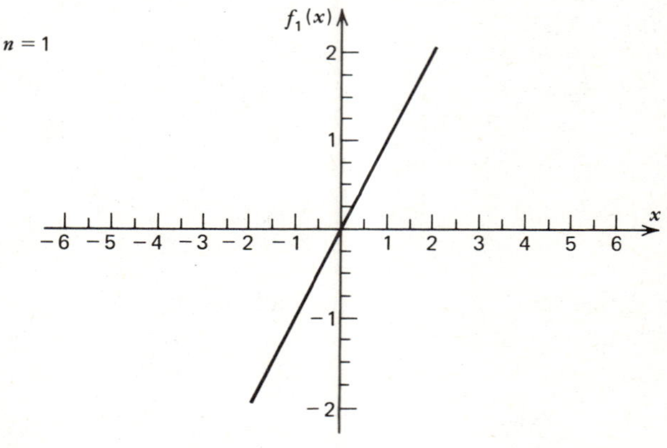

Fig. 5.1

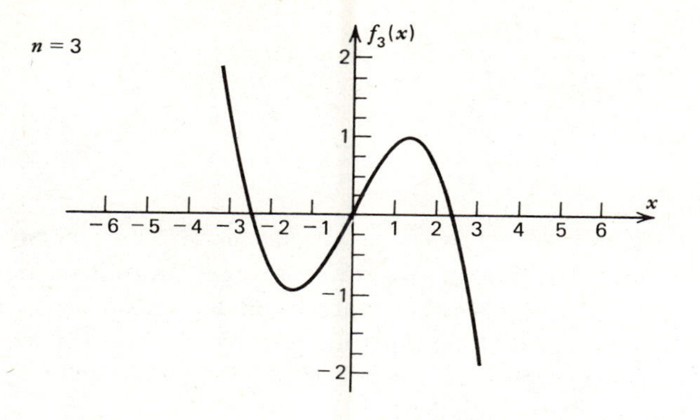

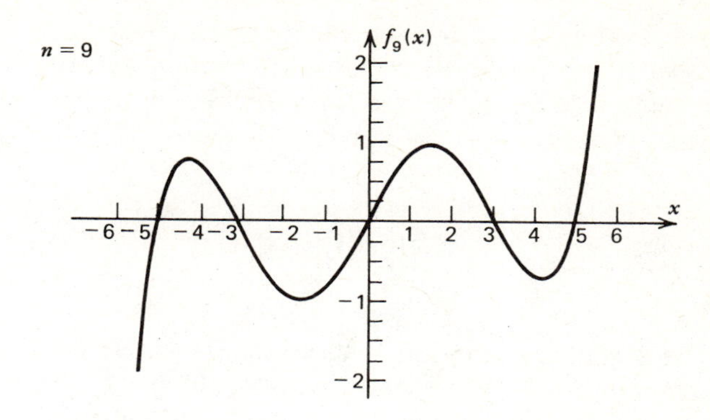

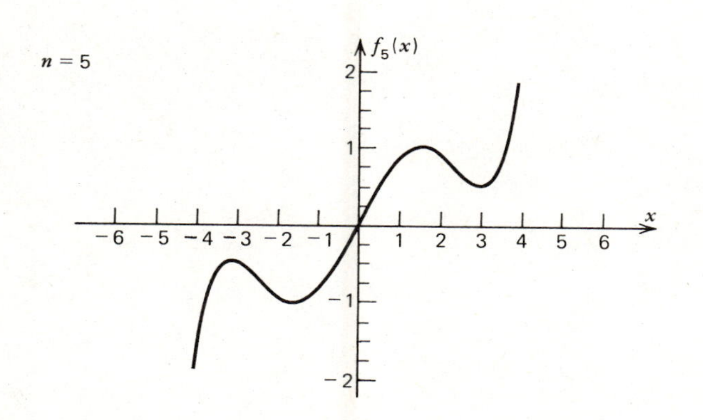

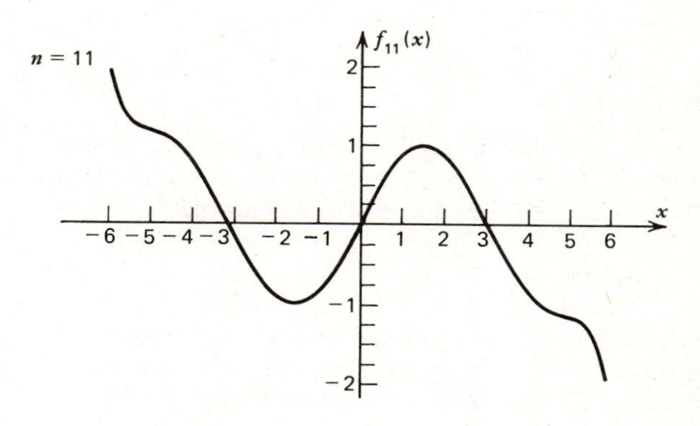

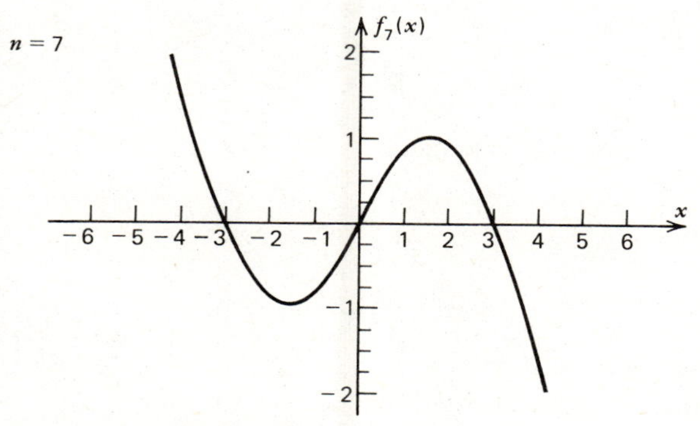

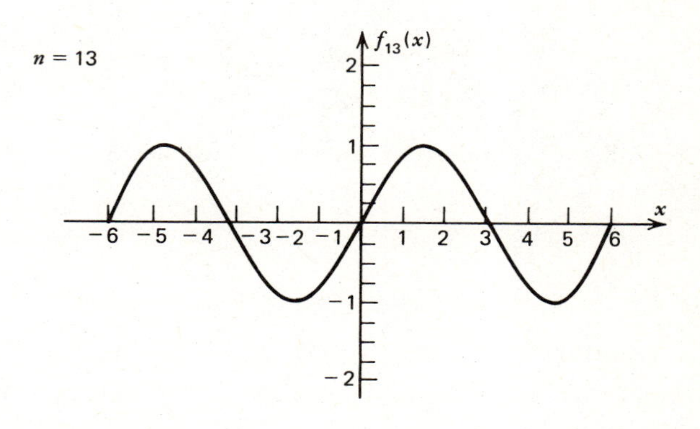

Fig. 5.1 (*Continued*)

is sometimes used for the binomial coefficients. If $p = 0, 1, 2, \ldots,$ the series terminates after a finite number of terms and there is no question of convergence. The ratio test for convergence gives

$$\lim_{n\to\infty}\left|\frac{c_{n+1}x^{n+1}}{c_n x^n}\right| = \lim_{n\to\infty}\frac{n-p}{n+1}|x| = |x| \tag{5.45}$$

The binomial expansion is therefore absolutely convergent if $-1 < x < 1$ and divergent if $|x| > 1$. The end points $x = \pm 1$ can be examined by means of Gauss' test (4.75):

$$\frac{n-p}{n+1} = \frac{1-p/n}{1+1/n} \simeq \left(1-\frac{p}{n}\right)\left(1-\frac{1}{n}\right)$$
$$= 1 - \frac{p+1}{n} + O\frac{1}{n^2} \tag{5.46}$$

The notation $O(1/n^2)$ means "terms of the order of $1/n^2$." The binomial series converges absolutely at $x = \pm 1$ if $p > 0$ and diverges at these points if $p < 0$. It turns out that the binomial series is also conditionally convergent at $x = +1$ for $-1 < p < 0$, but we will not prove this here.

PROBLEMS

5.1. Suppose that $\lim_{n\to\infty} f_n(x) = f(x)$ and that the limiting process is uniform in $[a, b]$. This means that for every positive number ε there is a fixed number $N(\varepsilon)$ such that $|f(x) - f_n(x)| < \varepsilon$ for all x in $[a, b]$ and for all $n > N$. Show that

$$\int_r^s f(x)\,dx = \lim_{n\to\infty}\int_r^s f_n(x)\,dx$$

provided that the integrals are defined and provided that r and s lie in $[a, b]$. In other words, the order of integrating and taking the limit can be interchanged. The condition of uniform convergence is *sufficient*, but not necessary.

5.2. If a power series $f(x) = \Sigma c_k x^k$ is integrated within its interval of convergence, the result is

$$\int_a^x f(t)\,dt = \sum_{k=0}^{\infty}\frac{c_k x^{k+1}}{k+1} - \sum_{k=0}^{\infty}\frac{c_k a^{k+1}}{k+1}$$

Show that the resulting power series in x has the same interval of convergence as the original series.

5.3. Prove that

$$\int_0^\infty \sin x^2\,dx$$

is convergent.

Hint: Make the change of variable $u = x^2$. Estimate the contribution to the integral from the range $n\pi \le u \le (n+1)\pi$. The integrand alternates in sign.

5.4. Show that the Taylor's series expansion of a function of two variables can be written in the form

$$f(x_1 + \Delta x_1, x_2 + \Delta x_2) = f(x_1, x_2) + (\partial_i f)\,\Delta x_i + \frac{1}{2!}(\partial_i\,\partial_j f)\,\Delta x_i\,\Delta x_j$$
$$+ \frac{1}{3!}(\partial_i\,\partial_j\,\partial_k f)\,\Delta x_i\,\Delta x_j \Delta x_k + \cdots \tag{5.47}$$

The partial derivatives are evaluated at the point x_1, x_2 and the summation convention is used. For example,

$$(\partial_i \partial_j f)\,\Delta x_i \Delta x_j = \sum_{i=1}^{2} \sum_{j=1}^{2} (\partial_i \partial_j f)\,\Delta x_i \Delta x_j$$

Hint: Define a function $F(t) = f(x_1 + t\,\Delta x_1, x_2 + t\,\Delta x_2)$ and expand it as a Taylor's series in the variable t about the point $t = 0$. What does the Taylor's series expansion for a function of several variables look like?

5.5. Find the Taylor's series expansion for $\cos x$.

Hint: $\cos x$ obeys the same differential equation as does $\sin x$. The coefficients obey the recurrence relation (5.38).

5.6. The general solution of $f''(x) = -f(x)$ can be expressed as $f(x) = c_1 \sin x + c_0 \cos x$. Show that the solution can also be written as $f(x) = A \sin(x + \phi)$ and find the constants A and ϕ in terms of c_1 and c_0.

5.7. Find the Taylor's series expansion of e^x by direct evaluation of the derivatives at $x = 0$.

5.8. Derive the recurrence relation (5.42).

5.9. Expand the function $f(x) = (1 + x)^p$ in powers of $1/x$ for the case where $|x| > 1$.

5.10. If p is an integer, show that the binomial coefficients are

$$\begin{aligned} \binom{p}{k} &= \frac{p!}{k!\,(p-k)!} \qquad 0 \le k \le p \\ &= 0 \qquad k > p \end{aligned}$$

5.11. Obtain a few terms of the Taylor's series expansion of

$$f(x) = \frac{\sin x}{1 - x}$$

about the point $x = 0$.

Hint: The expansions of $\sin x$ and $1/(1 - x)$ are known. Within their common interval of absolute convergence, the two series can be multiplied together. Be careful not to miss any terms of a given order.

5.12. Obtain a few terms of the expansion of $f(x) = \sin x$ about the point $x = \pi/2$.

Hint: Let $u = x - \pi/2$. Use the known expansion for $\cos u$.

5.13. Obtain the Taylor's series expansion of the function $f(x) = \ln(1 + x)$ about the point $x = 0$. The function $f(x)$ is undefined at $x = -1$ but is continuous and differentiable for $-1 < x < \infty$. The point $x = -1$ is called a *singularity* of $f(x)$. We anticipate that the interval of convergence of the series will be $-1 < x < 1$. Check this out. Does the series converge at $x = 1$? Show that the harmonic series (4.32) has the value $\ln 2$. The problem can be done by evaluating the derivatives of $\ln(1 + x)$ at $x = 0$. Another way to do it is to write

$$\ln(1 + x) = \int_0^x \frac{1}{1 + t}\,dt$$

and then use the known expansion for $1/(1 + t)$. Integrate the expansion term by term. Show that the remainder after n terms can be expressed as

$$R_n = (-1)^n \int_0^x \frac{t^n\,dt}{1 + t}$$

5.14. Suppose that it is necessary to calculate

$$\lim_{x \to x_0} \frac{f(x)}{g(x)} \tag{5.48}$$

and it is found that both $f(x_0)=0$ and $g(x_0)=0$. The form (5.48) is then called *indeterminate*. By expressing $f(x)$ and $g(x)$ as Taylor's series expansions about the point $x=x_0$, show that

$$\lim_{x\to x_0}\frac{f(x)}{g(x)}=\lim_{x\to x_0}\frac{f'(x)}{g'(x)} \tag{5.49}$$

provided that both $f(x)$ and $g(x)$ are continuous and differentiable at $x=x_0$. If it should turn out that both $f'(x_0)=0$ and $g'(x_0)=0$, then

$$\lim_{x\to x_0}\frac{f(x)}{g(x)}=\lim_{x\to x_0}\frac{f''(x)}{g''(x)} \tag{5.50}$$

and so on. The same formulas work if both $f(x_0)=\infty$ and $g(x_0)=\infty$.

5.15. Find the Taylor's series expansion of the function $F(t)=1/(1+t^2)$ about $t=0$. *Hint:* Set $x=-t^2$ in the geometrical series. What is the interval of convergence? Integrate the expansion term by term and use the fact that

$$\int_0^x F(t)\,dt=\arctan x$$

to find the expansion for $\arctan x$. Show that

$$\frac{\pi}{4}=1-\frac{1}{3}+\frac{1}{5}-\frac{1}{7}+\cdots \tag{5.51}$$

5.16. If a function $f(x)$ has at least n derivatives at the point $x=h$, it is possible to write

$$f(x)=f(h)+f'(h)(x-h)+\frac{1}{2!}f''(h)(x-h)^2+\cdots$$
$$+\frac{1}{n!}f^{(n)}(h)(x-h)^n+R_n(h,x) \tag{5.52}$$

which can be taken as a definition of the remainder $R_n(h,x)$. The idea is useful if R_n is small. Show that $R_n(x,x)=0$. By taking the partial derivative of (5.52) with respect to h, show that

$$\frac{\partial R_n}{\partial h}=-\frac{1}{n!}f^{(n+1)}(h)(x-h)^n \tag{5.53}$$

where now it is necessary to assume that at least $n+1$ derivatives of $f(x)$ exist at $x=h$. It is possible to integrate (5.53) from $h=x_0$ to $h=x$:

$$R_n(x,x)-R_n(x_0,x)=-\frac{1}{n!}\int_{x_0}^x f^{(n+1)}(h)(x-h)^n\,dh \tag{5.54}$$

Let's argue that it is possible to replace $f^{(n+1)}(h)$ by an appropriate average value, that is, that there is some value ξ, $x_0<\xi<x$, such that

$$-R_n(x_0,x)=-\frac{f^{(n+1)}(\xi)}{n!}\int_{x_0}^x (x-h)^n\,dh$$

Carry out the integration to get the remainder formula (5.29).

5.17. All functions which can be represented by power series are necessarily differentiable an arbitrary number of times within the interval of convergence (theorem 6), and the power series is necessarily the Taylor's series expansion of the function (theorem 7). The converse is not always true. The function

$$f(x)=e^{-1/x^2} \tag{5.55}$$

is continuous and differentiable an arbitrary number of times at $x=0$, yet it has no Taylor's series expansion about this point. Try it. It is possible to expand $f(x)$ in *inverse* powers of x by replacing x by $-1/x^2$ in the expansion for e^x. Obtain this expansion. What is its interval of convergence?

5.18. Prove that if $|x|>1$,

$$\arctan x = \frac{\pi}{2} + \sum_{k=0}^{\infty} \frac{(-1)^{k+1}}{2k+1}\left(\frac{1}{x}\right)^{2k+1} \tag{5.56}$$

Hint: Use the technique of problem 5.15. Write

$$\frac{1}{1+t^2} = \frac{1}{t^2(1+1/t^2)} \tag{5.57}$$

and expand in powers of $1/t^2$. Integrate the expansion from 1 to x. The conditions of theorem 3 are being overstepped, since the point $t=1$ does not lie in the interval of convergence of the expansion of (5.57). No general statement can be made concerning inclusion of the end points in the integration and each case has to be investigated separately. In this example, the integrated series converges conditionally at $t=1$.

5.19. When expressed in cylindrical coordinates, a static magnetic field has no θ component. The r and z components are functions of r and z and not θ. The field is symmetric with respect to the plane $z=0$, meaning that

$$\begin{aligned} B_r(r,z) &= -B_r(r,-z) \qquad & B_r(r,0) &= 0 \\ B_z(r,z) &= B_z(r,-z) \qquad & B_z(r_0,0) &= -B_0 \end{aligned} \tag{5.58}$$

A field that satisfies these requirements is shown in Fig. 5.2. Show that the Taylor's series expansions of the field components about the point $r=r_0$, $z=0$ accurate to second-order are

$$B_r = arz \qquad B_z = -B_0 - \tfrac{1}{2}ar_0^2 + a\left(\tfrac{1}{2}r^2 - z^2\right) \tag{5.59}$$

where a and B_0 are constants.

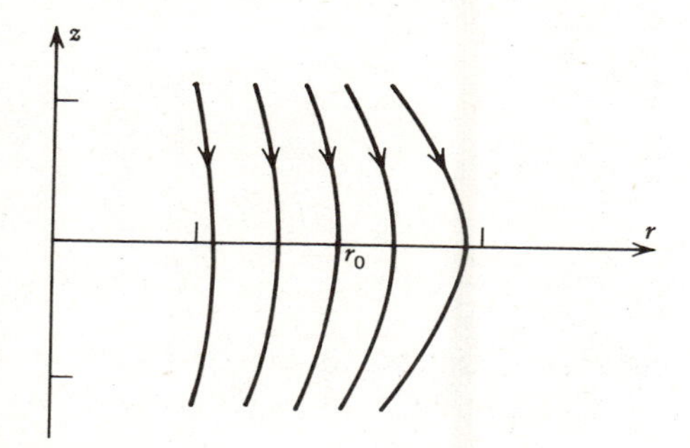

Fig. 5.2

Hint: Start out with general Taylor's series expansions of B_r and B_z considered as functions of r and z as given by (5.47). Then use (5.58), $\nabla\cdot\mathbf{B}=0$, and $\nabla\times\mathbf{B}=0$. (We assume that there are no electric currents in the vicinity of $r=r_0$, $z=0$.) Refer to equations (10.25)–(10.29) in Chapter 4. Find a vector potential of the form $\mathbf{A}=(0, A_\theta, 0)$. See equation (12.6) in Chapter 3. Show that the field lines are given by

$$\tfrac{1}{2}B_0 r^2 + \tfrac{1}{4}ar_0^2r^2 - \tfrac{1}{8}ar^4 + \tfrac{1}{2}ar^2z^2 = K \tag{5.60}$$

where K is a constant.

Hint: The differential equation of the field lines is $B_r\,dz - B_z\,dr = 0$. It can be made exact by an appropriate integrating factor.

6. PROJECTILE MOTION

If a projectile moves through air at not too great a speed, the friction produced is mainly a consequence of viscosity. The frictional force is then proportional to the velocity and oppositely directed. Assume that the motion takes place in a region near the surface of the earth so that the effect of the variation of the gravitational field with height can be neglected. If the y axis points upward and the x axis is horizontal, the equations for the motion in the x, y plane are

$$m\ddot{x} = -b\dot{x} \tag{6.1}$$

$$m\ddot{y} = -mg - b\dot{y} \tag{6.2}$$

where b is the *viscous damping constant*. In the following discussion, we assume that no motion takes place in the x direction. It is convenient to write the y equation as

$$\ddot{y} + \alpha\dot{y} = -g \tag{6.3}$$

where $\alpha = b/m$. This is a second-order, linear, nonhomogeneous differential equation with constant coefficients. A *particular integral* can be seen by inspection to be

$$y_p = -\frac{gt}{\alpha} \tag{6.4}$$

The *homogeneous equation*

$$\ddot{y} + \alpha\dot{y} = 0 \tag{6.5}$$

has the general solution

$$y_c = c_1 e^{-\alpha t} + c_2 \tag{6.6}$$

where c_1 and c_2 are constants of integration. Equation (6.6) is called the *complementary function* of (6.3). The general solution of (6.3) is

$$y = y_c + y_p = c_1 e^{-\alpha t} + c_2 - \frac{gt}{\alpha} \tag{6.7}$$

The constants c_1 and c_2 can be evaluated if the position and velocity of the projectile are known at $t = 0$:

$$y_0 = c_1 + c_2 \qquad v_0 = -\alpha c_1 - \frac{g}{\alpha} \tag{6.8}$$

The solution (6.7) is found to be

$$y = y_0 + \frac{1}{\alpha}\left(v_0 + \frac{g}{\alpha}\right)(1 - e^{-\alpha t}) - \frac{gt}{\alpha} \tag{6.9}$$

Suppose that an object is released from a height h above the ground and that t is the total time required for it to fall. Setting $y = 0$, $y_0 = h$, and $v_0 = 0$ in (6.9) gives

$$h = \frac{gt}{\alpha} - \frac{g}{\alpha^2}(1 - e^{-\alpha t}) \tag{6.10}$$

The value of the constant α for a given projectile can, in principle, be found from measurements of h and t. The problem is that (6.10) is a transcendental equation for α and is not solvable in terms of elementary functions. It is possible however to expand the exponential in (6.10) with the result

$$h = \frac{1}{2}gt^2\left(1 - \frac{\alpha t}{3} + \frac{\alpha^2 t^2}{12} - \frac{\alpha^3 t^3}{60} + \frac{\alpha^4 t^4}{360} - \cdots\right) \tag{6.11}$$

The expansion is convergent for any value of αt, but it is useful only if αt is small so that the higher-order terms become insignificant. In the limit of no friction, $\alpha = 0$ and $h = \frac{1}{2}gt^2$ as expected. It is convenient to define

$$r = \frac{h - (1/2)gt^2}{(1/2)gt^2} \qquad s = \alpha t \tag{6.12}$$

and write (6.11) as

$$r = -\frac{s}{3} + \frac{s^2}{12} - \frac{s^3}{60} + \frac{s^4}{360} - \cdots \tag{6.13}$$

Finding the constant α is reduced to solving (6.13) for s in terms of r. This is a problem in the *inversion of a power series*. In general terms, given the series

$$r = \sum_{k=1}^{\infty} a_k s^k \tag{6.14}$$

we seek to find the coefficients in the inverted series

$$s = \sum_{k=1}^{\infty} b_k r^k \tag{6.15}$$

The method is to substitute (6.15) into (6.14). Let's agree to work to fourth-order accuracy, meaning that all terms up to and including r^4 are to be retained:

$$\begin{aligned}
r &= a_1(b_1 r + b_2 r^2 + b_3 r^3 + b_4 r^4 + \cdots) \\
&\quad + a_2(b_1 r + b_2 r^2 + b_3 r^3 + \cdots)^2 \\
&\quad + a_3(b_1 r + b_2 r^2 + \cdots)^3 + a_4(b_1 r + \cdots)^4 \\
&= (a_1 b_1) r + (a_1 b_2 + a_2 b_1^2) r^2 \\
&\quad + (a_1 b_3 + 2a_2 b_1 b_2 + a_3 b_1^3) r^3 \\
&\quad + (a_1 b_4 + 2a_2 b_1 b_3 + a_2 b_2^2 + 3a_3 b_1^2 b_2 + a_4 b_1^4) r^4 \\
&\quad + \cdots
\end{aligned} \tag{6.16}$$

The uniqueness theorem for power series requires that

$$1 = a_1 b_1 \qquad 0 = a_1 b_2 + a_2 b_1^2 \qquad \text{and so on} \tag{6.17}$$

which are solved for the coefficients b_k to get

$$b_1 = \frac{1}{a_1} \qquad b_2 = -\frac{a_2}{a_1^3} \qquad b_3 = \frac{1}{a_1^5}(2a_2^2 - a_1 a_3)$$

$$b_4 = \frac{1}{a_1^7}(5a_1 a_2 a_3 - 5a_2^3 - a_1^2 a_4) \tag{6.18}$$

As an example, the inversion of the series (6.13) yields

$$s = -\tfrac{1}{3}r + 2.25r^2 - 2.025r^3 + 1.940625r^4 - \cdots \tag{6.19}$$

The coefficients are not getting small as rapidly as we might hope. It is helpful that the terms are alternating in sign, but we have no guarantee that this will continue. If the friction is not large, $h \simeq (1/2)gt^2$ and $|r| \ll 1$. The series (6.19) then gives a useful answer. Since our procedure does not yield a general expression for the recurrence formula obeyed by the coefficients, we have no way of calculating the interval of convergence of the series. There is one recourse. Any value of α that is calculated from (6.19) can be checked by means of the original equation (6.10).

PROBLEMS

6.1. When a particle is released in the atmosphere and allowed to fall, it reaches a terminal velocity. Find the terminal velocity and show its relation to the particular integral (6.4).

6.2. A projectile starts from the origin of coordinates at $t = 0$ with an angle of projection θ above the horizontal and speed v_0. If the projectile returns to $y = 0$ after a time t, show that its distance from the origin is

$$x = \frac{gtv_0 \cos\theta}{\alpha v_0 \sin\theta + g}$$

6.3. A simple plane pendulum consists of a mass m on the end of a string of length s. If the pendulum has angular amplitude θ_0, show from conservation of energy that

$$\dot{\theta}^2 = \frac{2g}{s}(\cos\theta - \cos\theta_0) \tag{6.20}$$

Show that the period of the pendulum is

$$T = 4\sqrt{\frac{s}{2g}} \int_0^{\theta_0} \frac{d\theta}{\sqrt{\cos\theta - \cos\theta_0}} \tag{6.21}$$

By making the change of variable

$$\sin\frac{\theta}{2} = \sin\frac{\theta_0}{2}\sin\phi \tag{6.22}$$

show that

$$T = 4\sqrt{\frac{s}{g}} \int_0^{\pi/2} \frac{d\phi}{\sqrt{1 - k^2 \sin^2\phi}} \tag{6.23}$$

where $k = \sin(\theta_0/2)$. The integral in (6.23) is called the *complete elliptic integral of the first kind* and cannot be evaluated in terms of elementary functions. The quantity k is called the *modulus*. Expand the integrand in (6.23) as a power series in $k \sin\phi$. Are the limits of the integration within the interval of convergence of the series? Integrate term by term to get

$$T = 2\pi\sqrt{\frac{s}{g}}\left(1 + \frac{1}{4}\sin^2\frac{\theta_0}{2} + \frac{9}{64}\sin^4\frac{\theta_0}{2} + \cdots\right) \tag{6.24}$$

6.4. A heavy object is dropped from a bridge, and the height is determined by measuring the time required for the sound of the impact to return to the bridge. If the speed of sound in air is very large and air resistance is neglected, then y is determined by $y = \frac{1}{2}gt^2$. If the finite speed of sound is taken into account, the equation for y is a quadratic which can be solved exactly. Instead of solving exactly, it is possible to obtain a useful approximate expression by writing $y = \frac{1}{2}gt^2(1+\delta)$, where δ is a small quantity. Show that $\delta \simeq -gt/u$, where u is the speed of sound. What is the approximate value of δ if $y = 30$ m and $u = 347$ m/sec?

7. ANALYTIC FUNCTIONS

In this section, we begin the discussion of functions of a complex variable and address ourselves to the concepts of limit, continuity, and differentiability. All integer powers, positive, negative, and zero, of a complex variable are defined, and this is the starting point for the construction of other functions. A polynomial function of z is

$$P(z) = a_0 + a_1 z + a_2 z^2 + \cdots + a_n z^n \tag{7.1}$$

where a_k are complex constants. Since division by any complex number is defined, the concept of function can be extended to rational functions of the form

$$f(z) = \frac{a_0 + a_1 z + \cdots + z_n z^n}{b_0 + b_1 z + \cdots + b_m z^m} \tag{7.2}$$

Other functions will be defined later. For the moment, the polynomial and rational functions will provide sufficient examples for our purposes. Functions of a complex variable can be single-valued or multiple-valued. In this section, all functions will be assumed to be defined in such a way that they are single-valued, meaning that for a given z there is only one value $f(z)$. The functions

$$f_1(z) = \frac{1}{z} \qquad f_2(z) = \frac{1}{1+z^2} \tag{7.3}$$

are single-valued. The function $f_1(z)$ is undefined at $z = 0$ and $f_2(z)$ is undefined at $z = \pm i$. These points are called *singularities* or *poles* of the functions. All functions of a complex variable can be separated into real and imaginary parts as $f(z) = u(x, y) + iv(x, y)$. For example,

$$f(z) = \frac{1}{1-z} = \frac{1-z^*}{(1-z)(1-z^*)} = \frac{1-z^*}{|1-z|^2}$$
$$= \frac{1-x}{(x-1)^2+y^2} + \frac{iy}{(x-1)^2+y^2} \tag{7.4}$$

In discussing functions of a complex variable, we speak of a *region* or *domain* of the complex plane in which the function in question is defined. For example, $|z - z_0| \le r$ is a circular region of radius r with the point z_0 at its center. It is a *closed region* because the circumference $|z - z_0| = r$ is included. The circular region $|z - z_0| < r$ is called *open* because points on the circumference are not included. A *neighborhood* of a point z_0 is the open circular region defined by $|z - z_0| < \delta$ where δ is usually thought of as being a small number.

The definitions of limit and continuity given in this paragraph parallel the corresponding definitions for real variables. The statement

$$\lim_{z \to z_0} f(z) = w \tag{7.5}$$

means that for every $\varepsilon > 0$, there is a $\delta > 0$ such that $|f(z) - w| < \varepsilon$ whenever $|z - z_0| < \delta$. If $f(z)$ is defined at every point of a region R, then $f(z)$ is continuous at a point z_0 of R if

$$\lim_{z \to z_0} f(z) = f(z_0) \tag{7.6}$$

For the idea of continuity at the point z_0 to make sense, the function $f(z)$ must at least be defined throughout some neighborhood of z_0. If a function is continuous at every point of a closed region, it is in fact uniformly continuous there. This means that once ε is chosen, the same δ works at all points z_0 of the region R. It should be emphasized that the manner in which z approaches z_0 in (7.5) and (7.6) must in no way affect the outcome. In other words, z can approach z_0 from any direction.

The derivative of a function of a complex variable is defined just as it is for real variables:

$$f'(z_0) = \lim_{z \to z_0} \frac{f(z) - f(z_0)}{z - z_0} = \lim_{\Delta z \to 0} \frac{f(z_0 + \Delta z) - f(z_0)}{\Delta z} \tag{7.7}$$

where $\Delta z = z - z_0$. A necessary condition for a function to have a derivative at a point z_0 is that it be continuous at that point, but this is not sufficient, as the following example shows. If we try to calculate the derivative of the function $f(z) = |z|^2$, we find

$$\begin{aligned} D(z_0) &= \lim_{\Delta z \to 0} \frac{|z_0 + \Delta z|^2 - |z_0|^2}{\Delta z} \\ &= \lim_{\Delta z \to 0} \frac{(z_0 + \Delta z)(z_0^* + \Delta z^*) - |z_0|^2}{\Delta z} \\ &= \lim_{\Delta z \to 0} \left(\frac{z_0 \Delta z^*}{\Delta z} + z_0^* \right) \end{aligned} \tag{7.8}$$

Let δ be the magnitude of Δz. From the geometry of Fig. 7.1, we see that $\Delta z = \delta(\cos\phi + i \sin\phi)$ and

$$\frac{\Delta z^*}{\Delta z} = \frac{(\Delta z^*)^2}{|\Delta z|^2} = \cos 2\phi - i \sin 2\phi \tag{7.9}$$

Therefore,

$$D(z_0) = z_0(\cos 2\phi - i \sin 2\phi) + z_0^* \tag{7.10}$$

The problem here is that the limit depends on the angle ϕ. In other words, $D(z_0)$ has no unique value at the point z_0 but depends on the direction ϕ with which z approaches z_0 in the limiting process. The only exception is the point $z_0 = 0$. The derivative exists at this one point and has the value $D(0) = f'(0) = 0$.

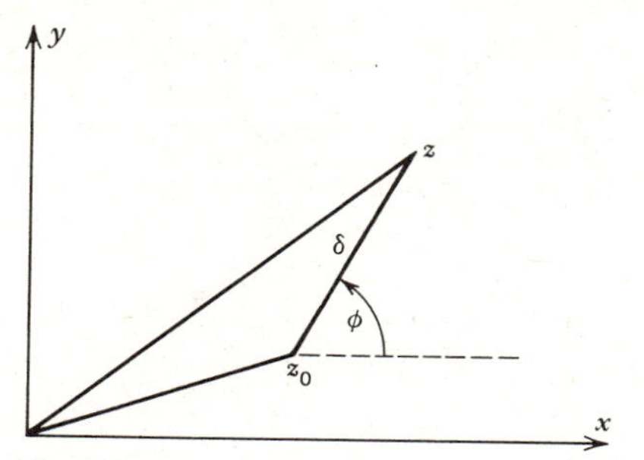

Fig. 7.1

Some fundamental formulas follow from the definition of the derivative. If c is a complex constant and if the derivatives of the functions $f(z)$ and $g(z)$ exist, then

$$\frac{dc}{dz} = 0 \tag{7.11}$$

$$\frac{d}{dz}\{cf(z)\} = cf'(z) \tag{7.12}$$

$$\frac{d}{dz} f(z)g(z) = f'(z)g(z) + f(z)g'(z) \tag{7.13}$$

$$\frac{d}{dz} f\{g(z)\} = f'\{g(z)\}g'(z) = \frac{df}{dg}\frac{dg}{dz} \tag{7.14}$$

If n is a positive or negative integer or zero, then

$$\frac{d}{dz}(z^n) = nz^{n-1} \tag{7.15}$$

The formulas (7.11) through (7.15) are familiar from real variable theory and are established in the same manner.

A single-valued function is defined to be *analytic* at a point z_0 if its derivative exists at every point in some neighborhood of z_0. The term *holomorphic* or *regular* is sometimes used in place of analytic. The function $f(z) = |z|^2$ is nowhere analytic because its derivative exists only at the isolated point $z = 0$. The function $f(z) = 1/z$ is analytic everywhere except at its singular point $z = 0$. Its derivative is $f'(z) = -1/z^2$.

If $f(z)$ is analytic, then the manner of taking the limit in (7.7) is immaterial. For example, we can write $\Delta z = \Delta x$ or $\Delta z = i\Delta y$ and approach the point at which the derivative is to be calculated either from the x direction or the y direction or even from somewhere in between if we wish. If $f(z)$ is separated into its real and imaginary parts, then

$$f'(z) = \lim_{\Delta x \to 0} \frac{u(x+\Delta x, y) - u(x, y) + i\{v(x+\Delta x, y) - v(x, y)\}}{\Delta x}$$
$$= \frac{\partial u}{\partial x} + i\frac{\partial v}{\partial x}$$
$$= \lim_{\Delta y \to 0} \frac{u(x, y+\Delta y) - u(x, y) + i\{v(x, y+\Delta y) - v(x, y)\}}{i\Delta y}$$
$$= -i\frac{\partial u}{\partial y} + \frac{\partial v}{\partial y} \tag{7.16}$$

If a, b, a', and b' are all real numbers, then the equality of the two complex numbers $a + ib = a' + ib'$ implies that $a = a'$ and $b = b'$. For the two ways of writing the derivative of $f(z)$, this implies

$$\frac{\partial u}{\partial x} = \frac{\partial v}{\partial y} \qquad \frac{\partial u}{\partial y} = -\frac{\partial v}{\partial x} \tag{7.17}$$

which are known as the *Cauchy–Riemann equations*. We have shown that their validity is a necessary condition for a function to be analytic.

Theorem 1. The Cauchy–Riemann equations are both a necessary and a sufficient condition for a function to be analytic.

Proof: To prove the sufficiency of the condition, we assume the validity of the Cauchy–Riemann equations and show that as a consequence $f'(z)$ exists and has a unique value. It is convenient to use the mean value theorem in the form (3.8) of Chapter 3 and write

$$\Delta f = \frac{\partial u}{\partial x}\Delta x + \frac{\partial u}{\partial y}\Delta y + i\left(\frac{\partial v}{\partial x}\Delta x + \frac{\partial v}{\partial y}\Delta y\right) \tag{7.18}$$

with the aid of the Cauchy–Riemann equations (7.17.),

$$\Delta f = \left(\frac{\partial u}{\partial x} + i\frac{\partial v}{\partial x}\right)\Delta z \tag{7.19}$$

where $\Delta z = \Delta x + i\,\Delta y$. Therefore,

$$f'(z) = \lim_{\Delta z \to 0} \frac{\Delta f}{\Delta z} = \frac{\partial u}{\partial x} + i\frac{\partial v}{\partial x} \tag{7.20}$$

which proves the result.

The Cauchy-Riemann equations can be differentiated to get

$$\frac{\partial^2 u}{\partial x^2} = \frac{\partial^2 v}{\partial x\,\partial y} \qquad \frac{\partial^2 v}{\partial y\,\partial x} = -\frac{\partial^2 u}{\partial y^2} \tag{7.21}$$

which can be combined to yield

$$\frac{\partial^2 u}{\partial x^2} + \frac{\partial^2 u}{\partial y^2} = \nabla^2 u = 0 \tag{7.22}$$

By a similar procedure, it is shown that

$$\nabla^2 v = 0 \tag{7.23}$$

Both the real and the imaginary part of an analytic function are solutions of Laplace's equation in two dimensions. One approach to the solution of two-dimensional potential problems is through the use of analytic functions of a complex variable. A solution of Laplace's equation is called a *harmonic function*. The functions u and v are referred to as *conjugate harmonic functions*.

Given one of the pair of conjugate harmonic functions, the other one is determined by the Cauchy–Riemann equations. For example, it is readily verified that $v(x, y) = 2xy$ is a solution of Laplace's equation. By means of (7.17), we find

$$\frac{\partial u}{\partial x} = \frac{\partial v}{\partial y} = 2x \tag{7.24}$$

$$\frac{\partial u}{\partial y} = -\frac{\partial v}{\partial x} = -2y \tag{7.25}$$

Equation (7.24) can be integrated with respect to x to find $u(x, y)$ in the form $u(x, y) = x^2 + f(y)$. The function $f(y)$ must then be determined by means of (7.25):

$$f'(y) = -2y \qquad f(y) = -y^2 + c \tag{7.26}$$

where now c is a constant. Therefore,

$$u(x, y) = x^2 - y^2 + c \tag{7.27}$$

The two conjugate harmonic functions which we have found are the real and imaginary parts of the analytic function

$$f(z) = z^2 + c \tag{7.28}$$

PROBLEMS

7.1. Go through the steps to prove (7.23).

7.2. Show that the curves $v =$ constant are perpendicular to the curves $u =$ constant. In an electrostatics problem, one set of curves gives equipotentials and the other gives the field lines.

7.3. Show that $f(z) = z^*$ and $g(z) = |z|$ are not analytic functions.

7.4. Show that a generalization of the binomial expansion is

$$(x+y)^p = x^p + px^{p-1}y + \frac{p(p-1)x^{p-2}y^2}{2!} + \cdots \tag{7.29}$$

Assume that the binomial expansion works for complex variables as well. By expanding $(z + \Delta z)^n$, prove (7.15).

7.5. Prove (7.13).

7.6. Show that a function of two variables can have neither maxima nor minima in a region of space where it is harmonic. It can have saddle points. *Hint*: Consider the Taylor's series expansion (5.47) of the function about a point where both first partial derivatives are zero. One way to look at it is to see what happens to the function first when $\Delta x = 0$, then when $\Delta y = 0$. Can you see the relation between this problem and problem 10.3 of Chapter 2? Also relevant is theorem 6 of section 10 in Chapter 3. Is your argument destroyed if the function is of the form $u = Ax^3 + Bx^2y + Cxy^2 + Dy^3$?

7.7. Show that the Cauchy–Riemann equations expressed in plane polar coordinates are

$$\frac{\partial u}{\partial r} = \frac{1}{r}\frac{\partial v}{\partial \theta} \qquad \frac{\partial v}{\partial r} = -\frac{1}{r}\frac{\partial u}{\partial \theta} \tag{7.30}$$

Hint: You can use the results of problem 8.7 in Chapter 4 to show, for example, that

$$\frac{\partial u}{\partial x} = \frac{\partial u}{\partial r}\cos\theta - \frac{\partial u}{\partial \theta}\frac{\sin\theta}{r}$$

7.8. Show that $u = 2x - x^3 + 3xy^2$ is harmonic and find its harmonic conjugate. There are two ways to do the problem. You can use the Cauchy–Riemann equations directly as was done in the text, or you can try to write down an analytic function of z that has u as its real part. The existence of such an analytic function automatically guarantees that u is harmonic and that the Cauchy–Riemann equations are satisfied.

8. COMPLEX INFINITE SERIES

Let $\phi_k(z) = u_k(x, y) + iv_k(x, y)$ be a sequence of functions of the complex variable z. The infinite series

$$f(z) = \sum_{k=0}^{\infty} \phi_k(z) = \sum_{k=0}^{\infty} u_k(x, y) + i\sum_{k=0}^{\infty} v_k(x, y) \tag{8.1}$$

exists and is convergent if the real series Σu_k and Σv_k are convergent. The series (8.1) is said to be *absolutely convergent* if $\Sigma|\phi_k|$ converges.

Theorem 1. The series $\Sigma\phi_k$ converges if $\Sigma|\phi_k|$ converges. (The converse is not necessarily true.)

Proof:

$$|\phi_k| = \sqrt{u_k^2 + v_k^2} \qquad |u_k| \le |\phi_k| \qquad |v_k| \le |\phi_k| \tag{8.2}$$

By the comparison test, Σu_k and Σv_k are both absolutely convergent and therefore convergent, which proves the theorem.

Let R be some closed region of the complex plane in which the terms $\phi_k(z)$ are defined. A region is closed if the boundary is included. Let $f_n(z)=\phi_1(z)+\cdots+\phi_n(z)$ be the nth partial sum of the series (8.1). The series converges uniformly in R if, for an arbitrarily small positive number ε, a fixed number $N(\varepsilon)$ can be found such that $|f_n(z)-f(z)|<\varepsilon$ whenever $n>N$. As with real series, the idea is that the same ε and N work for all points of R. The Weierstrass test for uniform convergence, theorem 1 of section 5, applies to series of complex terms. We need only replace $u_k(x)$ by $\phi_k(z)$ and $[a, b]$ by R in the proof. Theorem 2 of section 5 can also be restated to apply to complex infinite series. If a series of continuous functions converges uniformly at every point of R, then the sum $f(z)=\Sigma\phi_k(z)$ is also a continuous function. The proof is essentially the same as in the case of real series.

Our primary concern in this section is with power series, which are series of the form

$$f(z)=\sum_{k=0}^{\infty} c_k(z-z_0)^k \tag{8.3}$$

where c_k are complex constants. The next theorem is the most important theorem about complex power series and corresponds to theorem 5 of section 5.

Theorem 2. A power series which converges for at least one value of $z-z_0$ and diverges for at least one value of $z-z_0$ has a definite *radius of convergence* defined by $|z-z_0|=r$ such that the series is absolutely convergent if $|z-z_0|<r$ and divergent if $|z-z_0|>r$. If $r_1<r$, then the series is also uniformly convergent if $|z-z_0|\le r_1$. The reader can supply the proof by making an appropriate translation of the proof for the case of real series. The important thing to realize is that the interval of convergence on the x axis for real series is now replaced by a circular region of convergence in the complex plane with the point z_0 at its center. The radius r_1 of the region of uniform convergence must be at least slightly smaller than the radius r of the region of absolute convergence. The radius of convergence is determined by the ratio test:

$$\lim_{n\to\infty}\left|\frac{c_{n+1}r^{n+1}}{c_n r^n}\right|=1 \qquad r=\lim_{n\to\infty}\left|\frac{c_n}{c_{n+1}}\right| \tag{8.4}$$

The series may or may not converge at some or all points of the circumference of the circle. It is possible for the series to be everywhere convergent ($r=\infty$) or convergent only at $z=z_0$ in which case $r=0$.

It is possible, formally, to calculate the derivative of the series $f(z)=\Sigma c_k(z-z_0)^k$:

$$D(z)=\sum_{k=0}^{\infty} c_k k(z-z_0)^{k-1} \tag{8.5}$$

It is easy to show that the differentiated series has the same radius of convergence as the original series. Is it true that $D(z)=f'(z)$? It is certainly permissible to differentiate the nth partial sum of the original series:

$$f_n'(z)=\sum_{k=0}^{n} kc_k(z-z_0)^{k-1} \tag{8.6}$$

Consider the difference between $D(z)$ and $f_n'(z)$: .

$$|D(z)-f_n'(z)|=\left|\sum_{k=n+1}^{\infty} kc_k(z-z_0)^{k-1}\right|=|R_n| \tag{8.7}$$

The differentiated series converges uniformly and absolutely for $|z-z_0|\le r_1$. It is therefore possible to make the remainder R_n arbitrarily small if n is sufficiently large, uniformly at all points of $|z-z_0|\le r_1$. We conclude that $D(z)=\lim_{n\to\infty} f_n'(z)=f'(z)$. By the same argument, the differentiated series can also be differentiated. The following theorem is therefore established:

Theorem 3. A power series represents an analytic function. Within its radius of convergence, it can be differentiated as many times as we please. The resulting differentiated series will each have the same radius of convergence as the original series.

Although we have not yet shown it, all analytic functions can be represented by power series and can therefore be differentiated an arbitrary number of times.

The power series expansion of an analytic function about a given point is unique and is in fact the Taylor's series expansion of the function:

$$f(z)=f(z_0)+f'(z_0)(z-z_0)+\frac{1}{2!}f''(z_0)(z-z_0)^2+\cdots \tag{8.8}$$

The proof is the same as for real series. See theorem 7 of section 5.

The geometrical series can be used to illustrate an important idea. We know that $f(z)=1/(1-z)$ represents an analytic function at all points of the complex plane except at the singular point $z=1$. Let's pretend that the only thing we know about this function is its Taylor's series representation valid on the real axis for $|x|<1$:

$$f(x)=\sum_{k=0}^{\infty} x^k \tag{8.9}$$

The interval of convergence, $|x|<1$, is determined by the distance from the point about which the expansion is made, $x=0$, to the singular point, $x=1$. Suppose that we want to know an analytic function in the complex plane which matches (8.9) on the x axis. The only possible series expansion for such a function about $z=0$ is

$$f(z)=\sum_{k=0}^{\infty} z^k \tag{8.10}$$

where now the region of absolute convergence is inside the unit circle $|z|<1$. The reason why this is the only choice is because the power series expansion of an analytic function about a point is determined once all the derivatives are known at that point. All derivatives at $z=0$ are already determined by (8.9). In obtaining the expansion (8.9) by the Taylor's series method, the derivatives are calculated by taking limits along the x axis. If we want the continuation of the function into the complex plane to be analytic, then these same derivatives must be found by taking limits in any other direction.

Suppose now that we want to know what the function is outside the unit circle. We can pick a point z_0 inside $|z|=1$ and construct the Taylor's series expansion about this point. All the derivatives at $z=z_0$ can, in principle, be found from the power series (8.10) without knowing the actual closed form of the function. This extends the function in a unique manner into the circular region of radius $|z_0-1|$ with z_0 as center as shown in Fig. 8.1. This is an example

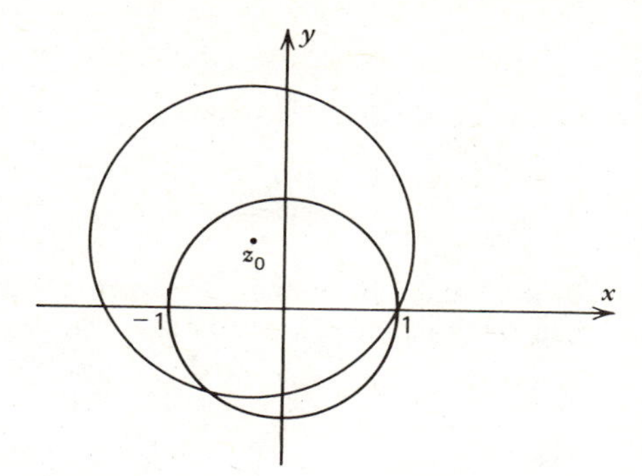

Fig. 8.1

of *analytic continuation*. There is one, and only one, possible analytic continuation of a given function. Since we really do know the closed form of the function beforehand in this case, the actual expansion about the point $z=z_0$ is easily obtained as follows:

$$\begin{aligned} f(z)=\frac{1}{1-z}&=\frac{1}{(1-z_0)-(z-z_0)}\\ &=\frac{1}{(1-z_0)\left(1-\dfrac{z-z_0}{1-z_0}\right)}\\ &=\frac{1}{1-z_0}\sum_{k=0}^{\infty}\left(\frac{z-z_0}{1-z_0}\right)^k \end{aligned} \tag{8.11}$$

which converges in the region $|z-z_0|<|1-z_0|$ as expected. Note that it is the distance $|1-z_0|$ from z_0 to the singular point $z=1$ which is the radius of convergence.

All functions cannot be continued analytically indefinitely into the complex plane. There may be certain curves which act as natural boundaries of the function across which analytic continuation does not work. Our discussion of analytic continuation is formalized in the following theorem:

Theorem 4. If a function is single-valued and analytic throughout a region R, it is uniquely determined by its values over a curve or throughout a subregion within R.

PROBLEMS

8.1. A function is defined by the binomial series

$$f(z) = \sum_{k=1}^{\infty} \binom{p}{k} z^k = 1 + pz + \frac{p(p-1)}{2!} z^2 + \cdots$$

where p is any complex number. Show that $|z| = 1$ is the radius of convergence of the series. Convince yourself that the function in question is $f(z) = (1+z)^p$.

8.2. Show that the power series expansion of $f(z) = 1/z^2$ about the point $z = -1$ is

$$f(z) = 1 + 2(z+1) + 3(z+1)^2 + \cdots + (n+1)(z+1)^n + \cdots$$

Hint: Note that

$$\frac{1}{z^2} = [1 - (z+1)]^{-2}$$

Use the binomial expansion. What is the radius of convergence of the series?

8.3. Find an expansion of $f(z) = 1/(1-z)$ in reciprocal powers of z which is valid at all points outside the unit circle $|z| = 1$.

8.4. Show that the function

$$f(z) = \frac{1}{z(1+z^2)}$$

has the expansions

$$f(z) = \frac{1}{z} + \sum_{k=1}^{\infty} (-1)^k z^{2k-1} \qquad 0 < |z| < 1$$

$$f(z) = \frac{1}{z^3} \sum_{k=0}^{\infty} \frac{(-1)^k}{z^{2k}} \qquad |z| > 1$$

Where are the singular points of the function?

8.5. Find the Taylor's series expansion of $1/(1-z)^2$ about $z = 0$ by differentiating the series for $1/(1-z)$.

8.6. An expansion of the function $f(z)$ about the point $z = 0$ is of the form

$$f(z) = \frac{b_n}{z^n} + \frac{b_{n-1}}{z^{n-1}} + \cdots + \frac{b_1}{z} + a_0 + \sum_{k=1}^{\infty} a_k z^k$$

which is valid in some region $0 < |z| < r$. If n is finite, show that the coefficients are unique. (It is in fact true that the expansion is unique even if $n \to \infty$.)

8.7. Show that the expansion

$$I(z) = \sum_{k=0}^{\infty} \frac{c_k}{k+1} (z - z_0)^{k+1}$$

has the same radius of convergence as $f(z) = \Sigma c_k (z - z_0)^k$. It is possible to think of $I(z)$ as being the indefinite integral of $f(z)$.

9. THE EXPONENTIAL FUNCTION

It was shown in section 5 that the familiar expansion (5.35) for the exponential function can be obtained as a consequence of the fact that it is a solution of the differential equation $f'(x)=f(x)$. The discussion of analytic continuation in section 8 leads to the conclusion that the appropriate extension of the exponential function into the complex plane is by means of

$$e^z = \sum_{k=0}^{\infty} \frac{z^k}{k!} \tag{9.1}$$

which is convergent everywhere in the complex plane. All the properties of the exponential function can be obtained as a consequence of its definition in terms of a power series. Of fundamental importance is the *multiplication theorem* which can be established by multiplying together two power series:

$$e^{z_1}e^{z_2} = \sum_{n=0}^{\infty} \frac{z_1^n}{n!} \sum_{m=0}^{\infty} \frac{z_2^m}{m!} \tag{9.2}$$

The multiplication of two series and the rearrangement of the terms in any manner is permitted in the common domain of absolute convergence of the two series. Therefore,

$$\begin{aligned} e^{z_1}e^{z_2} &= 1 + z_1 + \frac{1}{2!}z_1^2 + \frac{1}{3!}z_1^3 + \cdots \\ &\quad + z_2 + z_1 z_2 + \frac{1}{2!}z_1^2 z_2 + \cdots \\ &\quad + \frac{1}{2!}z_2^2 + \frac{1}{2!}z_1 z_2^2 + \cdots \\ &\quad + \frac{1}{3!}z_2^3 + \cdots \\ &\quad + \cdots \\ &= 1 + (z_1 + z_2) + \frac{1}{2!}(z_1 + z_2)^2 \\ &\quad + \frac{1}{3!}(z_1 + z_2)^3 + \cdots \\ &= e^{(z_1 + z_2)} \end{aligned} \tag{9.3}$$

which establishes the desired result.

A special case of the multiplication theorem is

$$e^z = e^{x+iy} = e^x e^{iy} \tag{9.4}$$

The power series expansion of e^{iy} is particularly interesting:

$$\begin{aligned} e^{iy} &= 1 + iy + \frac{1}{2!}(iy)^2 + \frac{1}{3!}(iy)^3 \\ &\quad + \frac{1}{4!}(iy)^4 + \frac{1}{5!}(iy)^5 + \cdots \\ &= 1 - \frac{1}{2!}y^2 + \frac{1}{4!}y^4 + \cdots \\ &\quad + i\left(y - \frac{1}{3!}y^3 + \frac{1}{5!}y^5 - \cdots\right) \\ &= \cos y + i \sin y \end{aligned} \tag{9.5}$$

An important connection between the exponential function and the trigonometric functions is revealed which is not apparent until these functions are extended into the complex plane. The complex conjugate of (9.5) is

$$(e^{iy})^* = e^{-iy} = \cos y - i \sin y \tag{9.6}$$

By solving (9.5) and (9.6) for $\sin y$ and $\cos y$, we find

$$\sin y = \frac{1}{2i}(e^{iy} - e^{-iy}) \tag{9.7}$$

$$\cos y = \frac{1}{2}(e^{iy} + e^{-iy}) \tag{9.8}$$

The exponential function provides a convenient representation of a complex number when polar coordinates are used:

$$z = x + iy = r\cos\theta + ir\sin\theta = re^{i\theta} \tag{9.9}$$

Since the magnitude of z is $|z| = r$, it follows that

$$|e^{i\theta}| = 1 \tag{9.10}$$

Each of the following special values of the exponential function represents a point on the unit circle $|z| = 1$:

$$e^{\pi i/2} = i \qquad e^{\pi i} = -1 \qquad e^{3\pi i/2} = -i \qquad e^{2\pi i} = 1 \tag{9.11}$$

The exponential function is periodic and has period 2π with respect to its argument:

$$e^{i\theta} = e^{i(\theta + 2\pi n)} \tag{9.12}$$

where n is a positive or negative integer.

The multiplication theorem (9.4) permits us to write

$$(e^{i\theta})^2 = e^{i\theta}e^{i\theta} = e^{2i\theta} = \cos 2\theta + i\sin 2\theta \qquad (9.13)$$

$$(e^{i\theta})^3 = \cos 3\theta + i\sin 3\theta \qquad (9.14)$$

$$e^{i\theta}e^{-i\theta} = e^0 = 1 \qquad (9.15)$$

$$(e^{-i\theta})^2 = e^{-i\theta}e^{-i\theta} = e^{-2i\theta} = \cos 2\theta - i\sin 2\theta \qquad (9.16)$$

The generalization is *DeMoivre's theorem* (problem 3.5):

$$z^n = r^n e^{in\theta} = r^n(\cos n\theta + i\sin n\theta) \qquad (9.17)$$

which has now been shown to hold for all positive and negative integer values of n. The possibility of extending the result to other values of n will be considered in a later section.

PROBLEMS

9.1. Show that $f(z) = e^z$ and $g(z) = e^{z+z_1}$ are both solutions of $f'(z) = f(z)$. The two solutions can at most differ by a constant: $ce^z = e^{z+z_1}$. Set $z = 0$ to get $c = e^{z_1}$. Hence, $e^{z_1}e^z = e^{z+z_1}$. This is an alternative proof of the multiplication theorem.

9.2. Show directly that the real and the imaginary parts of e^z obey the Cauchy–Riemann equations.

9.3. The rule for taking the complex conjugate of an expression is to replace i by $-i$ everywhere that it appears. Show that $(e^z)^* = e^{z^*}$. It is not always true that $f(z)^* = f(z^*)$. Show also that e^{z^*} is not analytic.

9.4. If $z = x + iy$, show that $x = \frac{1}{2}(z + z^*)$ and $y = \frac{1}{2i}(z - z^*)$.

9.5. Convince yourself of the validity of the following:

$$z_1 z_2 = r_1 r_2 e^{i(\theta_1+\theta_2)} \qquad z^{-1} = \frac{1}{z} = \frac{1}{re^{i\theta}} = \frac{1}{r}e^{-i\theta}$$

$$\frac{z_1}{z_2} = \frac{r_1}{r_2}e^{i(\theta_1-\theta_2)} \qquad |z_1 z_2| = r_1 r_2 = |z_1||z_2|$$

$$|z_1/z_2| = r_1/r_2 = |z_1|/|z_2|$$

9.6. Use DeMoivre's theorem to show that

$$\cos 3\theta = \cos^3\theta - 3\sin^2\theta\cos\theta$$

$$\sin 3\theta = -\sin^3\theta + 3\cos^2\theta\sin\theta$$

9.7. Show that $f(z) = e^{z^2}$ is analytic. What is $f'(z)$? More generally, show that an analytic function of an analytic function is analytic.

9.8. Use the exponential function to show that

$$\cos(\theta_1 + \theta_2) = \cos\theta_1\cos\theta_2 - \sin\theta_1\sin\theta_2$$

$$\sin(\theta_1 + \theta_2) = \sin\theta_1\cos\theta_2 + \cos\theta_1\sin\theta_2$$

9.9. Sometimes the evaluation of an integral can be facilitated by expressing trigonometric functions in terms of exponentials. Use (9.7) and (9.8) to show that

$$\int \sin ax \cos bx\, dx = -\frac{\cos(a-b)x}{2(a-b)} - \frac{\cos(a+b)x}{2(a+b)}$$

where a and b are real constants.

9.10. Evaluate the definite integrals

$$I_1 = \int_0^\infty e^{-ax} \cos bx \, dx \qquad I_2 = \int_0^\infty e^{-ax} \sin bx \, dx$$

where a and b are real and $a > 0$.

10. DIFFRACTION GRATING

A discussion of plane electromagnetic waves sufficient for the understanding of this section is given in section 12 of Chapter 3. A diffraction grating is made by ruling many closely spaced parallel lines on some material. As an idealization, we will consider the grating to be some opaque material in which there are many long, narrow, parallel slits. We will assume in the discussion that the slit separation d is large compared to the width of an individual slit. Suppose, as illustrated in Fig. 10.1, that a plane electromagnetic wave is incident normally on the grating. The wavefronts can be visualized as the crests of the waves of Fig. 12.2 in Chapter 3. The details of the interaction of the electromagnetic wave with the material of the grating are complicated, but the essential consequence of the interaction can be found by using Huygens' construction. According to Huygens, each slit is to be treated as a secondary source of radiation. The sources are all in phase with one another because all parts of a given wavefront arrive simultaneously at the grating. The light coming from the many inphase sources creates a characteristic interference pattern which is then displayed on the viewing screen. It is this pattern that we want to find. We will consider the case of *Fraunhofer diffraction* which occurs when the distance between the grating and the screen is large compared to the size of the grating. In this way, we can consider the interference of essentially parallel light rays as they emerge from the slits. As a practical matter, a lens can be used to focus the parallel rays onto a nearby screen. Consider the electromagnetic waves in terms of the electric field. If r_0 is the path length from the first slit to a point on the viewing screen, then the wave from this slit is

$$E_1 = A \sin(\kappa r_0 - \omega t) = A \sin\phi \tag{10.1}$$

where we have set $\phi = \kappa r_0 - \omega t$. If d is the slit separation and θ is the angle at which the radiation is being viewed, then the wave from the second slit has to travel a distance $r_1 = r_0 + d\sin\theta$ from the grating to the same point on the screen. The wave from the second slit is

$$E_2 = A \sin(\kappa r_1 - \omega t) = A \sin(\phi + \delta) \tag{10.2}$$

where $\delta = \kappa d \sin\theta$ represents the *phase difference* between the two waves. We assume that the radiation from each slit has the same amplitude A. Similarly,

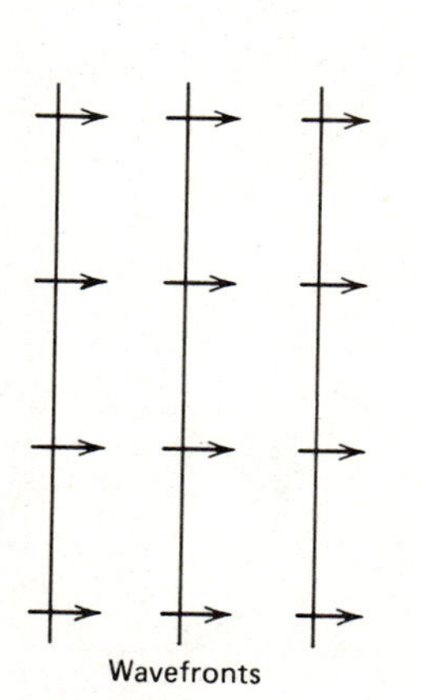

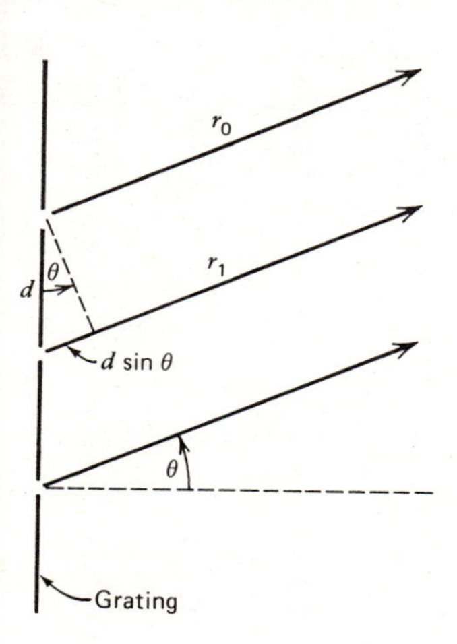

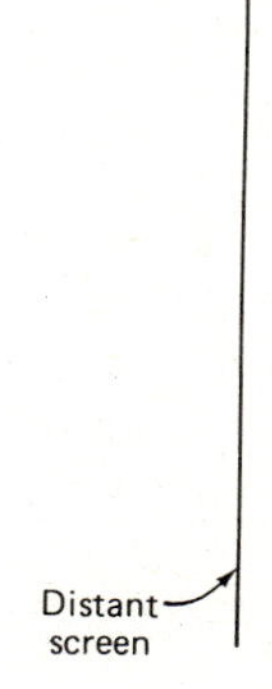

Fig. 10.1

the wave from slit 3 is

$$E_3 = A\sin(\phi + 2\delta) \tag{10.3}$$

and from slit N, it is

$$E_N = A\sin(\phi + [N-1]\delta) \tag{10.4}$$

To find the resultant electric field at the location given by the angle θ on the screen, it is necessary to add the contributions from each slit:

$$E = \sum_{n=0}^{N-1} A\sin(\phi + n\delta) \tag{10.5}$$

This is most conveniently done by evaluating the complex quantity

$$Z = \sum_{n=0}^{N-1} Ae^{i(\phi+n\delta)} = Ae^{i\phi}\sum_{n=0}^{N-1} e^{in\delta} \tag{10.6}$$

and recognizing that E is the imaginary part. There may be thousands of lines ruled on the grating so that N is large. Nevertheless, the sum is finite so there is no question of convergence. Since $|e^{i\delta}| = 1$, the series actually diverges if $N \to \infty$. The evaluation of (10.6) is easily done using the truncated form (4.3) of the geometrical series:

$$Z = Ae^{i\phi}\frac{1-e^{iN\delta}}{1-e^{i\delta}} \tag{10.7}$$

A little trick is used to extract the imaginary part. Take out a factor of $e^{iN\delta/2}$ from the numerator and a factor of $e^{i\delta/2}$ from the denominator and write (10.7) as

$$Z = Ae^{i(\phi+N\delta/2-\delta/2)}\frac{e^{-iN\delta/2}-e^{iN\delta/2}}{e^{-i\delta/2}-e^{i\delta/2}} \tag{10.8}$$

Now use (9.7) to get

$$Z = Ae^{i(\phi+N\delta/2-\delta/2)}\frac{\sin(N\delta/2)}{\sin(\delta/2)} \tag{10.9}$$

The resultant electric field is

$$E = A\sin(\phi + N\beta - \beta)\frac{\sin(N\beta)}{\sin(\beta)} \tag{10.10}$$

where

$$\beta = \frac{\delta}{2} = \frac{\kappa d\sin\theta}{2} \tag{10.11}$$

Usually what is wanted is the intensity of the radiation. For electromagnetic waves, this is given in power per unit area by the Poynting vector

$$\mathbf{S} = \frac{\mathbf{E}\times\mathbf{B}}{\mu_0} \tag{10.12}$$

In terms of the plane electromagnetic wave as given by (12.25) and (12.27) of Chapter 3, the magnitude of

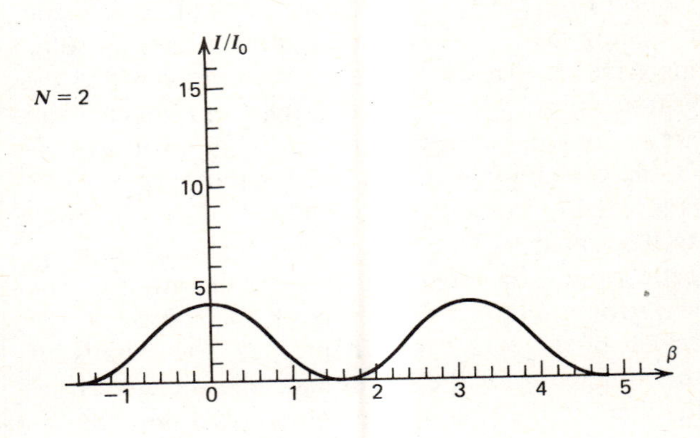

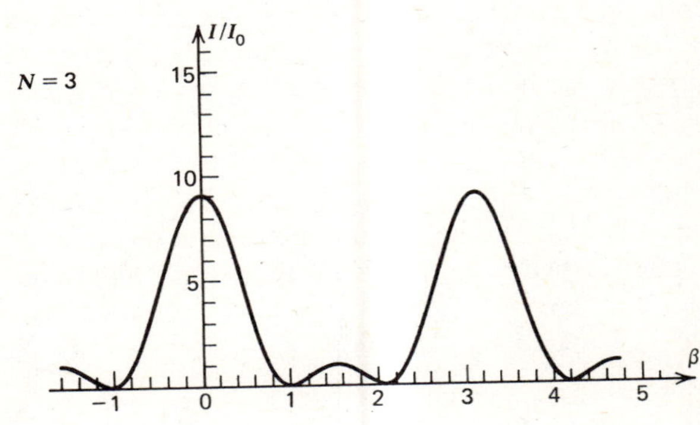

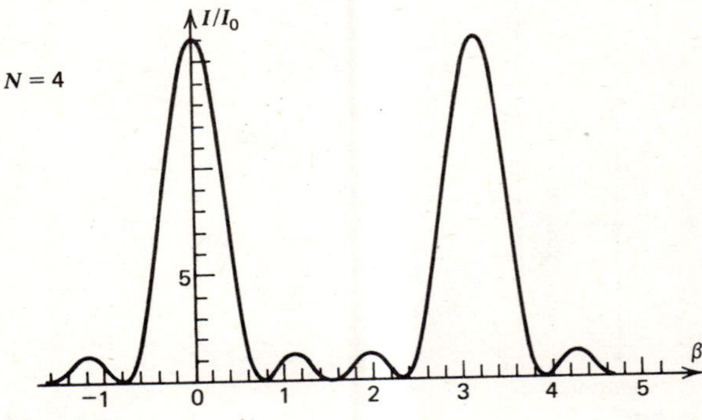

Fig. 10.2

S is

$$S = \frac{1}{\mu_0 c} E_0^2 \sin^2(\kappa z - \omega t) \tag{10.13}$$

Averaged over time, this is

$$\bar{S} = \frac{1}{2\mu_0 c} E_0^2 \tag{10.14}$$

The essential feature is that the intensity is proportional to the square of the amplitude of the wave. The intensity pattern produced by a diffraction grating can therefore be expressed as

$$I = I_0 \frac{\sin^2 N\beta}{\sin^2 \beta} \tag{10.15}$$

The function becomes indeterminate if β is a multiple of π. Near $\beta = 0$, $\sin N\beta \simeq N\beta$, $\sin\beta \simeq \beta$, and

$$\lim_{\beta \to 0} I = \lim_{\beta \to 0} I_0 \left(\frac{N\beta}{\beta} \right)^2 = I_0 N^2 \tag{10.16}$$

The same result is found for $\beta = n\pi$, where n is a positive or negative integer. These points are the locations of the *principal maxima* of the intensity pattern. Figure 10.2 shows graphs of the intensity for $N = 2$, 3, and 4. It was pointed out that the series used to calculate the intensity becomes divergent as $N \to \infty$. If N is large, the intensity pattern consists of sharp spikes resembling Dirac delta functions located at the points $\beta = n\pi$. It is this property which makes the grating useful.

PROBLEMS

10.1. Show that for $N = 2$, the intensity (10.15) is $I = 4I_0 \cos^2 \beta$.

10.2. Sketch the intensity pattern (10.15) for $N = 5$. Where are the zeros located? The height of a principal maximum is proportional to N^2. What is its width?

10.3. Show that the locations of the principal maxima on the viewing screen are given by $d \sin\theta = n\lambda$, where λ is the wavelength of the incident light and $n = 0, 1, 2, \ldots$ is an integer called the *order number*. (A given order is found for both positive and negative θ.) If there is a mixture of several wavelengths in the incident light, each wavelength produces its own principal maxima so that a *line spectrum* is produced.

10.4. Parallel light is incident normally on a single slit of width a as shown in Fig. 10.3. According to Huygens' method, each point of a wavefront as it arrives at the slit is regarded as a secondary source. For Fraunhofer diffraction, the resultant electric field is found from

$$E = \int_{x=0}^{a} A \sin(\phi + \kappa x \sin\theta)\, dx$$

which is the imaginary part of

$$Z = Ae^{i\phi} \int_{x=0}^{a} e^{i\kappa x \sin\theta}\, dx$$

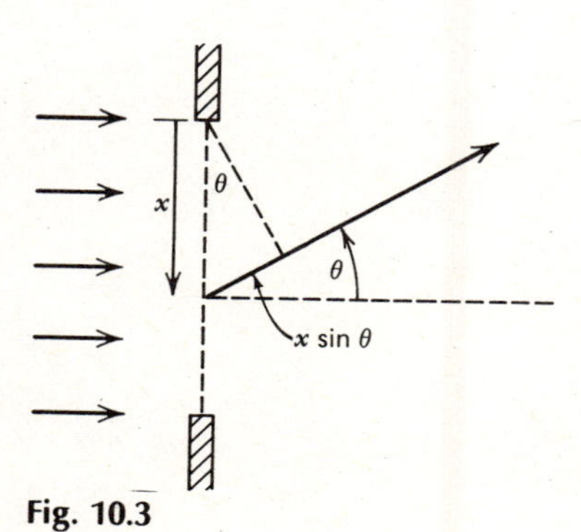

Fig. 10.3

Carry out the integration and show that the intensity is

$$I = I_0 \frac{\sin^2 \alpha}{\alpha^2} \qquad \alpha = \frac{\kappa a \sin \theta}{2}$$

Graph this function for $-3\pi < \alpha < 3\pi$.

10.5. Prove the identities

$$1 + \cos \theta + \cos 2\theta + \cdots + \cos N\theta = \frac{\cos\left(\frac{N\theta}{2}\right) \sin\left(\frac{N+1}{2}\right)\theta}{\sin \frac{\theta}{2}}$$

$$\sin \theta + \sin 2\theta + \cdots + \sin N\theta = \frac{\sin\left(\frac{N\theta}{2}\right) \sin\left(\frac{N+1}{2}\right)\theta}{\sin \frac{\theta}{2}}$$

10.6. The derivation of the grating equation is an example of a *superposition* of simple solutions of Maxwell's equations to obtain a more complicated solution. It works because Maxwell's equations are linear. Demonstrate this linearity by showing that if $\mathbf{E}_1$ and $\mathbf{E}_2$ are any two solutions of the wave equation, (12.13) of Chapter 3, then so is $\mathbf{E} = \alpha \mathbf{E}_1 + \beta \mathbf{E}_2$, where α and β are constants.

10.7. A grating consists of equally spaced slits of width a and separation d as shown in Fig. 10.4. The resultant electric field for Fraunhofer diffraction is the imaginary part of

$$Z = \sum_{n=0}^{N-1} \int_{nd}^{nd+a} A e^{i(\phi + \kappa x \sin \theta)}\, dx$$

Show that the resultant intensity of the radiation is

$$I = I_0 \left(\frac{\sin \alpha}{\alpha}\right)^2 \left(\frac{\sin N\beta}{\sin \beta}\right)^2$$

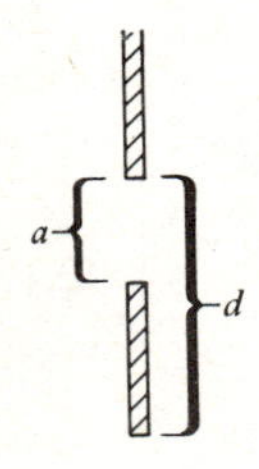

Fig. 10.4

10.8. Two waves $E_1 = A \sin \phi$ and $E_2 = B \sin(\phi + \delta)$ from two different sources are to be superimposed to obtain a resultant wave $E = E_1 + E_2 = R \sin(\phi + \gamma)$. Use complex exponential notation to show that this is the equivalent of a vector addition in two dimensions.

Specifically, show that

$$R = \sqrt{A^2 + B^2 + 2AB\cos\delta} \qquad \tan\gamma = \frac{B\sin\delta}{A + B\cos\delta}$$

Draw a diagram showing the complex numbers A, $Be^{i\delta}$, and $Re^{i\gamma}$ as two-dimensional vectors. Also show that if $A = B$, then $R^2 = 4A^2\cos^2(\delta/2)$, which is the same result as problem 10.1. Note that since the intensity of a wave is proportional to the square of the amplitude, the rule for combining the intensities of two waves of the same frequency is

$$I = I_A + I_B + 2\sqrt{I_A I_B}\cos\delta$$

11. CAPACITANCE AND INDUCTANCE

In section 12, complex numbers will be used as an aid in obtaining the solutions of linear differential equations with constant coefficients. The notion of complex impedance as it occurs in connection with alternating current circuit theory will be explained. To this end, it is a good idea to understand the behavior of capacitors and inductors, which are two of the basic circuit elements. In section 12 of Chapter 3, it is established that the density of energy in the electromagnetic field is

$$w = \frac{1}{2}\varepsilon_0 E^2 + \frac{1}{2\mu_0}B^2 \tag{11.1}$$

Consider two isolated, charged conducting bodies placed near one another. The total electrostatic energy in the field of these conductors is

$$W_E = \frac{1}{2}\varepsilon_0\int_\Sigma \mathbf{E}\cdot\mathbf{E}\,d\Sigma \tag{11.2}$$

An electrostatic field can be expressed in terms of an electrostatic potential as $\mathbf{E} = -\nabla\psi$. Therefore,

$$W_E = -\frac{1}{2}\varepsilon_0\int_\Sigma \mathbf{E}\cdot\nabla\psi\,d\Sigma \tag{11.3}$$

We now use the vector identity $\nabla\cdot\psi\mathbf{E} = \mathbf{E}\cdot\nabla\psi + \psi\nabla\cdot\mathbf{E}$ to get

$$W_E = \frac{1}{2}\varepsilon_0\int_\Sigma (\psi\nabla\cdot\mathbf{E} - \nabla\cdot\psi\mathbf{E})\,d\Sigma \tag{11.4}$$

The integral is over all space. The contribution to the integral from the interior of the conductors is of course zero because there is no field there. Imagine that the two conductors are located near the center of a very large sphere of radius R and that the interior of this sphere is the region of integration. By means of Gauss' theorem,

$$\int_\Sigma \nabla\cdot\psi\mathbf{E}\,d\Sigma = \int_\sigma \mathbf{E}\psi\cdot d\boldsymbol{\sigma} \tag{11.5}$$

where σ is the surface of the sphere. If Q is the combined charge on the two conductors, then the field and potential at a large distance away from them will be

$$\mathbf{E} = \frac{Q\hat{\mathbf{r}}}{4\pi\varepsilon_0 R^2} \qquad \psi = \frac{Q}{4\pi\varepsilon_0 R} \tag{11.6}$$

The surface integral in (11.5) is therefore

$$\int_\sigma \mathbf{E}\psi\cdot d\boldsymbol{\sigma} = \left(\frac{Q}{4\pi\varepsilon_0}\right)^2\frac{1}{R^3}\int d\sigma = \left(\frac{Q}{4\pi\varepsilon_0}\right)^2\frac{4\pi}{R} \tag{11.7}$$

which becomes zero in the limit $R \to \infty$. With the aid of $\nabla\cdot\mathbf{E} = \rho/\varepsilon_0$, the electrostatic energy can be now expressed

$$W_E = \frac{1}{2}\int_\Sigma \rho\psi\,d\Sigma \tag{11.8}$$

If you solved problem (11.10) in Chapter 3, you already did the calculation leading to (11.8).

All the charge is on the surface of the two conductors. The surface of each conductor is an equipotential surface. If λ represents surface charge density, then the electrostatic energy can be expressed in the form

$$\begin{aligned} W_E &= \frac{1}{2}\psi_1\int_{\sigma_1}\lambda_1\,d\sigma_1 + \frac{1}{2}\psi_2\int_{\sigma_2}\lambda_2\,d\sigma_2 \\ &= \frac{1}{2}\psi_1 q_1 + \frac{1}{2}\psi_2 q_2 \end{aligned} \tag{11.9}$$

where q_1 and q_2 represent the total charge on each conductor. Suppose that the two conductors are originally electrically neutral. They can be charged by connecting a battery between them which removes charge from one conductor and places it on the other. Then, $q_2 = +q$, $q_1 = -q$, and

$$W_E = \tfrac{1}{2}q(\psi_2 - \psi_1) = \tfrac{1}{2}q\psi \tag{11.10}$$

where now ψ is the potential difference between the two conductors. If an increment of charge dq is removed from one conductor and placed on the other, an external agent (the battery) must do an amount of work $dW_{\text{ext}} = \psi\, dq$. On the other hand, the total differential of W_E is

$$dW_E = \tfrac{1}{2}q\, d\psi + \tfrac{1}{2}\psi\, dq \tag{11.11}$$

Equating dW_E to dW_{ext} gives

$$q\, d\psi = \psi\, dq \tag{11.12}$$

This is a first-order differential equation with the solution

$$q = C\psi \tag{11.13}$$

where C is a constant of integration called the *capacitance* of the two-conductor system. The two conductors are collectively referred to as a capacitor. Equation (11.13) shows that the potential difference which exists between two isolated conductors is proportional to the amount of charge which has been transferred from one conductor to the other. The actual value of the capacitance depends on the geometry of the two conductors. The total energy (11.10) can be expressed as

$$W_E = \frac{1}{2C}q^2 \tag{11.14}$$

The unit of capacitance in the mks systems is the *farad*.

The concept of inductance can be arrived at by a similar calculation. The magnetostatic field energy associated with a steady current is

$$W_B = \frac{1}{2\mu_0}\int_\Sigma \mathbf{B}\cdot\mathbf{B}\, d\Sigma \tag{11.15}$$

The magnetic field can be expressed in terms of a vector potential as $\mathbf{B} = \nabla \times \mathbf{A}$. By using the vector identity $\nabla\cdot(\mathbf{B}\times\mathbf{A}) = \mathbf{A}\cdot\nabla\times\mathbf{B} - \mathbf{B}\cdot\nabla\times\mathbf{A}$, it is possible to express W_B as

$$\begin{aligned} W_B &= \frac{1}{2\mu_0}\int_\Sigma \mathbf{B}\cdot\nabla\times\mathbf{A}\, d\Sigma \\ &= \frac{1}{2\mu_0}\int_\Sigma [\mathbf{A}\cdot\nabla\times\mathbf{B} - \nabla\cdot(\mathbf{B}\times\mathbf{A})]\, d\Sigma \end{aligned} \tag{11.16}$$

The term $\int \nabla\cdot(\mathbf{B}\times\mathbf{A})\, d\Sigma$ can be converted into a surface integral and discarded. The static magnetic field obeys $\nabla\times\mathbf{B} = \mu_0\mathbf{j}$, which gives an expression for W_B which is analogous to (11.8):

$$W_B = \frac{1}{2}\int_\Sigma \mathbf{A}\cdot\mathbf{j}\, d\Sigma \tag{11.17}$$

Let us suppose that our direct current circuit is in the form of a loop of wire of small, but not zero, cross-sectional area carrying a current I. Then,

$$W_B = \frac{I}{2}\oint \mathbf{A}\cdot d\mathbf{s} \tag{11.18}$$

which is a line integral around the closed path taken by the current. Equation (11.18) is almost a swindle because it becomes infinite in the limit of a conducting wire of zero cross-sectional area. This is because the magnetic field has a singularity for a true line current, as can be seen by looking at equation (8.11) of Chapter 3. The energy as given by (11.15) is then infinite. Currents confined to mathematical lines and point charges are both convenient concepts, but we have to pay for their use by putting up with the singularities they introduce.

The flux of the magnetic field through the current loop is

$$\Phi = \int_\sigma \mathbf{B}\cdot d\boldsymbol{\sigma} = \int_\Sigma \nabla\times\mathbf{A}\cdot d\boldsymbol{\sigma} = \oint \mathbf{A}\cdot d\mathbf{s} \tag{11.19}$$

where Stokes' theorem has been used. The energy in the magnetic field can now be written

$$W_B = \tfrac{1}{2}I\Phi \tag{11.20}$$

which is the analogue of (11.10). If an external agent attempts to increase the current in the circuit, it generates a "back emf" which according to Faraday's law, equation (12.1) of Chapter 3, is

$$\mathscr{E} = -\frac{d\Phi}{dt} \tag{11.21}$$

The external agent performs an amount of work against this back emf which is given by

$$dW_{ext} = \frac{d\Phi}{dt} dq = I\,d\Phi \qquad (11.22)$$

The work done by the external agent goes to increase the magnetostatic energy. The total differential of (11.20) is

$$dW_B = \tfrac{1}{2} I\,d\Phi + \tfrac{1}{2}\Phi\,dI \qquad (11.23)$$

which is equated to dW_{ext} to get

$$I\,d\Phi = \Phi\,dI \qquad (11.24)$$

The solution of this differential equation is

$$\Phi = LI \qquad (11.25)$$

where L is a constant of integration called the *self-inductance* of the circuit. This result is quite reasonable. The magnetic field, hence the flux of the magnetic field through the circuit, is proportional to the current in the circuit. The constant of proportionality is the self-inductance. The actual value of the self-inductance depends on the geometry of the circuit in question. The magnetostatic energy can be expressed in the form

$$W_B = \tfrac{1}{2} LI^2 \qquad (11.26)$$

The back emf given by Faraday's law can now be expressed

$$\mathscr{E} = -L\frac{dI}{dt} \qquad (11.27)$$

The mks unit for inductance is the *henry*.

If there are two separate circuits near one another, as, for example, the two coaxial current loops of Fig. 11.1, then the magnetic field energy is written

$$W_B = \tfrac{1}{2} I_1\Phi_1 + \tfrac{1}{2} I_2\Phi_2 \qquad (11.28)$$

where Φ_1 is the net flux through circuit 1 and Φ_2 is the net flux through circuit 2. The net flux through either loop is due to its own magnetic field plus a contribution from the magnetic field of the other circuit:

$$\Phi_1 = L_1 I_1 + M_{12} I_2 \qquad \Phi_2 = L_2 I_2 + M_{21} I_1 \qquad (11.29)$$

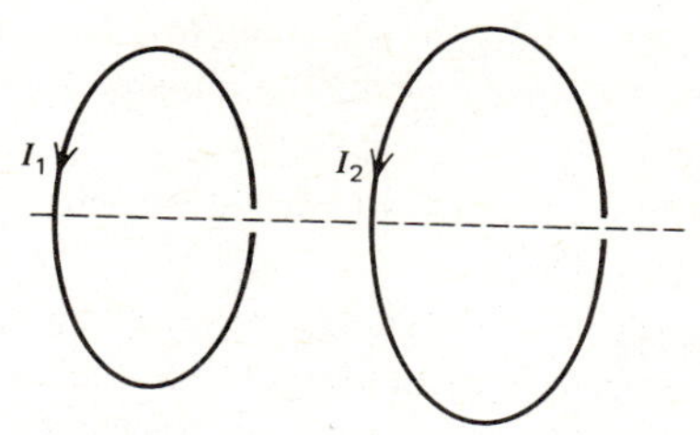

Fig. 11.1

where M_{12} and M_{21} are constants. By an argument similar to that used for a single coil, we can show that the work done by external agents in changing the currents by differential amounts is

$$dW_{ext} = I_1\,d\Phi_1 + I_2\,d\Phi_2 \qquad (11.30)$$

If the differentials $d\Phi_1$ and $d\Phi_2$ are calculated from (11.29), the result is

$$dW_{ext} = (I_1 L_1 + I_2 M_{21})\,dI_1 + (I_2 L_2 + I_1 M_{12})\,dI_2 \qquad (11.31)$$

If the amount of work done by an external agent in changing the system from one state to another is independent of how the change is brought about, in other words, independent of the path followed in a space where the coordinates are I_1 and I_2, then (11.31) is an exact differential and

$$\frac{\partial}{\partial I_2}(I_1 L_1 + I_2 M_{21}) = \frac{\partial}{\partial I_1}(I_2 L_2 + I_1 M_{12}) \qquad (11.32)$$

which shows that $M_{12} = M_{21}$. We therefore write $M = M_{12} = M_{21}$ and call M the mutual inductance of the two circuits. It is a measure of the coupling of the two circuits through the magnetic field. The energy of the magnetic field can now be written

$$W_B = \tfrac{1}{2} L_1 I_1^2 + M I_1 I_2 + \tfrac{1}{2} L_2 I_2^2 \qquad (11.33)$$

The expressions for the flux through the two loops are

$$\Phi_1 = L_1 I_1 + M I_2 \qquad \Phi_2 = L_2 I_2 + M I_1 \qquad (11.34)$$

PROBLEMS

11.1. Electric charge is spread with a uniform surface density λ over the entire y, z plane of a rectangular Cartesian coordinate system. Use Gauss' law to find the electric field everywhere.

11.2. Two large parallel sheets of area A have uniform surface charge densities $+\lambda$ and $-\lambda$. If the separation between the sheets is small compared to their dimension, find the electric field produced by the sheets.

Hint: Neglecting edge effects, the field of each sheet separately can be found from problem 11.1. Superimpose the two fields.

11.3. A parallel-plate capacitor is constructed from two parallel conducting sheets of area A separated by a distance s which is small compared to the dimensions of the sheets. Show that the expression for the capacitance is $C = \varepsilon_0 A/s$. Neglect any edge effects.

11.4. Show that the capacitance per unit length of two concentric cylindrical conductors of radii a and b is

$$C_\ell = \frac{2\pi\varepsilon_0}{\ln(b/a)} \tag{11.35}$$

11.5. Justify discarding the divergence term in (11.16) in detail. How do the magnetic field and the vector potential behave at large distances from the current loop? Look at section 11 of Chapter 4 on magnetic dipoles.

11.6. Look at problem 9.9 in Chapter 3. Show that the self-inductance of the toroid shown in Fig. 9.4 is

$$L = \frac{\mu_0 N^2 h}{2\pi} \ln\left(\frac{b}{a}\right) \tag{11.36}$$

You can do the problem either by computing the energy from (11.15) and equating it to (11.26) or by using (11.25) directly. If you use (11.25), remember that the net flux linking the circuit is $\Phi = N\Phi_0$, where Φ_0 is the flux through a cross section of the toroid.

11.7. Suppose that there are two windings on the toroid of problem 11.6, each consisting of a single layer of fine wire covering the entire toroid. One winding has a total of N_1 turns of wire and the other has N_2 turns. Show that the mutual inductance is

$$M = \frac{\mu_0 N_1 N_2 h}{2\pi} \ln\left(\frac{b}{a}\right) \tag{11.37}$$

and that $M = \sqrt{L_1 L_2}$, where L_1 and L_2 are the self-inductances of the two windings.

11.8. Consider the magnetically interacting circuits of Fig. 11.2. One is the toroid of problem 11.6, the other is a coil with N_2 turns which links the toroid but is not wound uniformly on it as in problem 11.7. Show that the mutual inductance is still given by (11.37). It is no longer true that $M = \sqrt{L_1 L_2}$, as will be evident after you do the next problem.

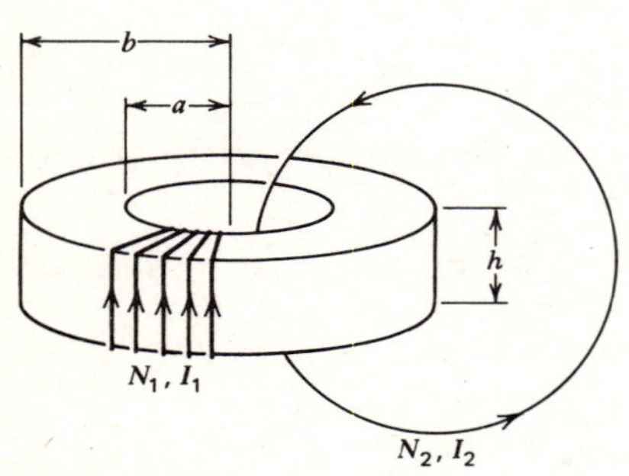

Fig. 11.2

11.9. Two coils of wire, one with N_1 turns and the other with N_2 turns, are wound on a soft iron core as illustrated in Fig. 11.3. The effect of the iron core is to concentrate the lines of the magnetic field so that they are almost entirely contained and don't wander off into the surrounding space. This means that in the most ideal case, all the lines of the magnetic field link both circuits, achieving what might be called perfect magnetic coupling of the two circuits. If Φ_0 is the flux through a cross section of the iron core, then $\Phi_1 = N_1\Phi_0$ and $\Phi_2 = N_2\Phi_0$. Use (11.34) to show that $M = \sqrt{L_1 L_2}$.

Hint: The currents I_1 and I_2 can be varied independently of one another. A relation such as $AI_1 + BI_2 = CI_1 + DI_2$ implies that $A = C$ and $B = D$. In the absence of perfect coupling, convince yourself that $M = k\sqrt{L_1 L_2}$, where $0 < k < 1$. In problem 11.7, the fields produced by both windings are entirely inside the toroid so that there is perfect coupling. In problem 11.8, all of the flux produced by the toroid goes through the second circuit, but the converse is not true.

In Fig. 11.3, the symbol ⊣||⊢ stands for a battery or source of emf. For the moment, we can think of these as being sources of direct current which we vary slightly to get the differential changes in energy talked about in various of the derivations. It becomes more interesting when these are replaced by time-varying sources of emf. Note that the convention for the positive direction of the currents both in Fig. 11.3 and Fig. 11.1 has been chosen to make M come out positive.

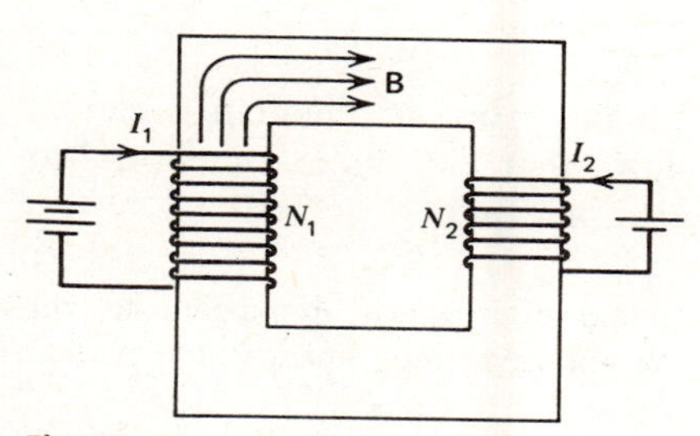

Fig. 11.3

11.10. For the case of the two interacting current loops, start with (11.18) and go through the steps in detail to get (11.28). In particular, show that

$$W_B = \frac{1}{2} I_1 \oint_1 \mathbf{A}_1 \cdot d\mathbf{s}_1 + \frac{1}{2} I_2 \oint_2 \mathbf{A}_2 \cdot d\mathbf{s}_2 \tag{11.38}$$

11.11. Equate the total differential of (11.28) to (11.30). Show that (11.29) is a solution of the resulting differential equation provided that $M_{12} = M_{21}$.

12. SECOND-ORDER LINEAR DIFFERENTIAL EQUATIONS

A few fundamental facts about the very important subject of second-order linear differential equations are reviewed in this section. Some of the ideas that will be formalized here have already been discussed to some extent. Section 6 provides a good example. By a second-order linear, nonhomogeneous differential equation is meant

$$y'' + P(x)y' + Q(x)y = F(x) \tag{12.1}$$

where $P(x)$, $Q(x)$, and $F(x)$ are any functions of x. Although we will think of x as a real variable, the theoretical development of this section applies equally well if x is replaced by the complex variable z. The function y defined by the differential equation is then viewed as an analytic function of z. The singular points of the coefficient functions P and Q as well as the function F are of special significance, and some attention will be given to them later. For the time being, assume that the range of values of x over which we are concerned has no such troublesome points. Any solution of (12.1), which may or may not represent the most general solution, is called a *particular integral*. The *homogeneous equation* is (12.1) with $F(x)=0$. The general solution of the homogeneous equation is called the *complementary function*.

Theorem 1. The most general solution of (12.1) is a linear superposition of any particular integral and the complementary function.

Proof: Let y_1 and y_2 be any two particular integrals. The function $y=y_1-y_2$ satisfies the homogeneous equation $y''+Py'+Qy=0$. Therefore, any two particular integrals differ from one another by a solution of the homogeneous equation. This means that the most general form of the solution of (12.1) is $y=y_c+y_p$, where y_c is the complementary function and y_p is any particular integral.

The idea of linear dependence is important in the theory of linear differential equations. Two functions y_1 and y_2 are defined to be linearly dependent if a relation of the form

$$c_1y_1+c_2y_2=0 \tag{12.2}$$

exists for nonzero values of the constants c_1 and c_2. A consequence of the definition is found by differentiating (12.2):

$$c_1y_1'+c_2y_2'=0 \tag{12.3}$$

A nontrivial solution of (12.2) and (12.3) for the constants c_1 and c_2 exists if the determinant

$$W(x)=y_1y_2'-y_1'y_2 \tag{12.4}$$

is zero. The function $W(x)$ is called the *Wronskian* of the functions y_1 and y_2.

Theorem 2. In order for the functions y_1 and y_2 to be linearly dependent, it is both necessary and sufficient that $W=0$.

Proof: That the condition is necessary has already been shown. To prove sufficiency, assume that $W=0$. Then, in fact (12.2) and (12.3) have nontrivial solutions for c_1 and c_2, but we have to show that these solutions are constants and not functions of x. If we divide both (12.2) and (12.3) by c_2, the result is

$$\alpha y_1+y_2=0 \qquad \alpha y_1'+y_2'=0 \qquad \alpha=\frac{c_1}{c_2} \tag{12.5}$$

If the first of these is differentiated with respect to x and combined with the second, the result is $\alpha' y_1=0$. The function y_1 may vanish at certain points, but it is not identically zero. Therefore, $\alpha'=0$ and $\alpha=$ constant, which was to be proved.

Similarly, three functions are linearly dependent if constants c_1, c_2, and c_3, not all of which are zero, exist such that

$$c_1y_1+c_2y_2+c_3y_3=0 \tag{12.6}$$

The Wronskian of the three functions is

$$W=\begin{vmatrix} y_1 & y_2 & y_3 \\ y_1' & y_2' & y_3' \\ y_1'' & y_2'' & y_3'' \end{vmatrix} \tag{12.7}$$

A proof similar to the proof of theorem 2 can be constructed to show that the three functions are linearly dependent if, and only if, $W=0$.

Theorem 3. A second-order linear, homogeneous differential equation cannot have three linearly independent solutions.

Proof: Assume that three solutions of the homogeneous equation have been found. Then,

$$\begin{aligned} y_1''+Py_1'+Qy_1&=0 \\ y_2''+Py_2'+Qy_2&=0 \\ y_3''+Py_3'+Qy_3&=0 \end{aligned} \tag{12.8}$$

which says that the three rows of the determinant W are linearly dependent. We know from the theory of determinants that whenever this happens, the determinant is zero. It has therefore been shown that a second-order linear, homogeneous equation can have at most two linearly independent solutions.

Suppose that one solution of $y''+Py'+Qy=0$ has been found. Call this solution y_1. We will try to find a second solution in the form $y_2=uy_1$. If both y_1 and y_2 are solutions, then u must obey

$$\frac{u''}{u'}+2\frac{y_1'}{y_1}+P=0 \tag{12.9}$$

which can also be expressed in the form

$$\frac{d}{dx}\ln\left(u'y_1^2\right) = -P \tag{12.10}$$

Integration yields

$$u' = \frac{c}{y_1^2}e^{-\int P\,dx} \tag{12.11}$$

where $\int P\,dx$ is the indefinite integral of $P(x)$ and c is a constant. By integrating (12.11), we find that a second solution of the homogeneous equation is

$$y_2 = uy_1 = y_1\int\frac{c}{y_1^2}e^{-\int P\,dx}\,dx \tag{12.12}$$

The Wronskian of y_1 and y_2 is

$$\begin{aligned} W &= y_1y_2' - y_1'y_2 = y_1\left(u'y_1 + uy_1'\right) - y_1'uy_1 \\ &= y_1^2u' = ce^{-\int P\,dx} \end{aligned} \tag{12.13}$$

which is not zero. We have discovered that there are two linearly independent solutions of the homogeneous equation and that, once one of them is found, the other can be constructed by means of (12.12). Equation (12.13) provides an explicit expression for the Wronskian of the two solutions. Actually, the discussion hinges on the initial assumption that at least one solution of the differential equation is known to exist, a fact which we have not rigorously demonstrated. Except for this one important point, to which we give some attention later, the following theorem has been established:

Theorem 4. Every homogeneous equation $y'' + Py' + Qy = 0$ has two linearly independent solutions y_1 and y_2. Since a third linearly independent solution does not exist, any other solution can be expressed as a linear combination of them as $y = c_1y_1 + c_2y_2$, where c_1 and c_2 are constants.

Theorem 5. The solution of the homogeneous equation is uniquely determined once the value of the solution and its first derivative are known at one point.

Proof: If y_1 and y_2 are two linearly independent solutions, then the general solution is $y = c_1y_1 + c_2y_2$, from which

$$\begin{aligned} y(x_0) &= c_1y_1(x_0) + c_2y_2(x_0) \\ y'(x_0) &= c_1y_1'(x_0) + c_2y_2'(x_0) \end{aligned} \tag{12.14}$$

The determinant of the coefficients in (12.14) is the Wronskian, which, on the account of the assumed linear independence of y_1 and y_2, is not zero. Therefore, a unique solution for c_1 and c_2 exists.

An example is provided by the one-dimensional motion of a particle of mass m which has been connected to a spring of force constant k. If the mass experiences a viscous frictional force similar to that discussed in section 6, then its equation of motion is

$$m\ddot{x} + b\dot{x} + kx = 0 \tag{12.15}$$

The coefficients are constants and the equation is homogeneous because no driving force is present. It is convenient to divide through by m and write

$$\ddot{x} + 2\beta\dot{x} + \omega_0^2x = 0 \tag{12.16}$$

where

$$\beta = \frac{b}{2m} \qquad \omega_0 = \sqrt{\frac{k}{m}} \tag{12.17}$$

The constant ω_0 will be recognized as the natural frequency of the undamped oscillator. Once two linearly independent solutions are found, we know the general solution. It is therefore legitimate to try anything to find a solution of a differential equation, including guess work. If $x = e^{\alpha t}$ is used as a trial solution, the result is

$$\alpha^2 + 2\alpha\beta + \omega_0^2 = 0 \tag{12.18}$$

which is a quadratic equation for the constant α. The two roots are

$$\alpha_1 = -\beta + \sqrt{\beta^2 - \omega_0^2} \qquad \alpha_2 = -\beta - \sqrt{\beta^2 - \omega_0^2} \tag{12.19}$$

If $\beta^2 > \omega_0^2$, we have found two linearly independent solutions. The general solution is therefore

$$x = c_1e^{\alpha_1 t} + c_2e^{\alpha_2 t} \qquad \beta^2 > \omega_0^2 \tag{12.20}$$

The solution is uniquely determined if the position and velocity are known at a particular time, which we take to be $t = 0$. The constants c_1 and c_2 are then determined by

$$\begin{aligned} x_0 &= c_1 + c_2 \\ u_0 &= \alpha_1c_1 + \alpha_2c_2 \end{aligned} \tag{12.21}$$

A special circumstance occurs when the physical constants are such that $\beta^2 = \omega_0^2$, since then $\alpha_1 = \alpha_2 = -\beta$ and there is only one exponential solution, $x_1 = e^{-\beta t}$. A second linearly independent solution can be found from (12.12):

$$x_2 = x_1 \int \frac{1}{x_1^2} e^{-\int 2\beta\, dt}\, dt = e^{-\beta t} \int e^{2\beta t} e^{-2\beta t}\, dt = te^{-\beta t} \tag{12.22}$$

The general solution is now

$$x = (c_1 + c_2 t)e^{-\beta t} \qquad \beta^2 = \omega_0^2 \tag{12.23}$$

If the solution (12.23) is valid, the oscillator is said to be *critically damped*.

If the damping is even smaller so that $\beta^2 < \omega_0^2$, the roots (12.19) become complex numbers which can be expressed in the form

$$\alpha_1 = -\beta + i\omega_1$$

$$\alpha_2 = -\beta - i\omega_1 \qquad \omega_1 = \sqrt{\omega_0^2 - \beta^2} \tag{12.24}$$

The solution for this case is

$$x = e^{-\beta t}\left(c_1 e^{i\omega_1 t} + c_2 e^{-i\omega_1 t}\right) \tag{12.25}$$

By means of (9.5), the solution can be put into the form

$$x = e^{-\beta t}(a_1 \sin \omega_1 t + a_2 \cos \omega_1 t) \tag{12.26}$$

where $a_1 = i(c_1 - c_2)$ and $a_2 = c_1 + c_2$. Since x is real, c_1 and c_2 are complex numbers related by $c_2^* = c_1$. The constants a_1 and a_2 are then real. Another form of the solution is

$$x = Ae^{-\beta t} \cos(\omega_1 t - \delta) \tag{12.27}$$

where the constants A and δ are given by

$$A = \sqrt{a_1^2 + a_2^2} \qquad \tan \delta = \frac{a_1}{a_2} \tag{12.28}$$

Any manner of expressing the solution must involve two constants which can be determined by knowing the initial position and velocity of the oscillator. The solution (12.27) consists of an oscillatory term multiplied by a damping factor $e^{-\beta t}$, which has the effect of decreasing the amplitude of the oscillations as a function of the time. The frequency ω_1 is not the same as the frequency ω_0 of the undamped oscillator. If the friction is large so that $\beta^2 > \omega_0^2$, the oscillator is said to be *overdamped*. The solution is then of the form (12.20) and no oscillations at all occur. For critical damping as given by (12.23), the friction has just reached a high enough value to prevent oscillatory motion.

The constants a_1 and a_2 in terms of the initial position and velocity are

$$a_1 = \frac{u_0 + \beta x_0}{\omega_1} \qquad a_2 = x_0 \tag{12.29}$$

As a numerical example, suppose that

$$m = 10 \text{ gm} \qquad k = 250 \text{ dynes/cm}$$

$$\frac{b}{m} = 0.5 \text{ sec}^{-1}$$

$$x_0 = 10 \text{ cm} \qquad u_0 = 0 \tag{12.30}$$

The following calculated values are found:

$$\omega_0 = 5 \text{ sec}^{-1} \qquad \beta = 0.25 \text{ sec}^{-1}$$

$$\omega_1 = \sqrt{\omega_0^2 - \beta^2} \simeq \omega_0\left(1 - \frac{\beta^2}{2\omega_0^2}\right)$$

$$= 5(1 - 0.00125) = 4.994 \text{ sec}^{-1}$$

$$a_1 = 0.5006 \qquad a_2 = 10.00$$

$$A = 10.0125 \text{ cm} \qquad \delta = 0.0500 \text{ radians} \tag{12.31}$$

The solution is then

$$x = (10.01)e^{-0.25t} \cos(4.994t - 0.0500) \tag{12.32}$$

and is shown graphed in Fig. 12.1.

Another physical system which has exactly the same behavior as the damped mechanical oscillator is the series RLC circuit shown in Fig. 12.2. RLC means resistance, inductance, and capacitance. Suppose that the capacitor initially has a charge on it.

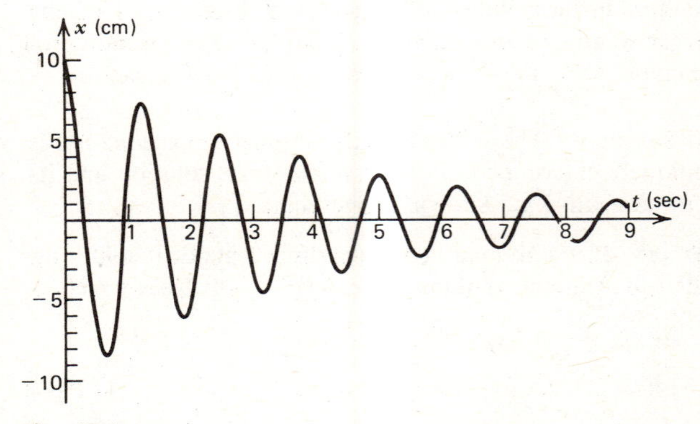

Fig. 12.1

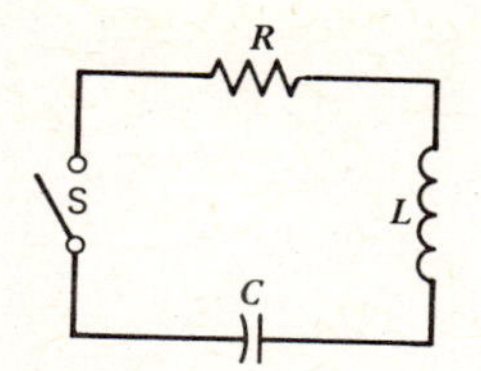

Fig. 12.2

After switch S is closed, the voltages across the three elements at any instant of time will be IR, $L(dI/dt)$, and q/C. This must add up to zero giving

$$L\ddot{q} + R\dot{q} + \frac{1}{C}q = 0 \tag{12.33}$$

Note the analogies with the mechanical oscillator: $m \leftrightarrow L$, $b \leftrightarrow R$, and $k \leftrightarrow 1/C$. The inductor is the inertial element of the circuit, and the capacitor acts like a spring.

Finally, consider the inductively coupled circuits of Fig. 12.3. By means of the flux expressions (11.34)

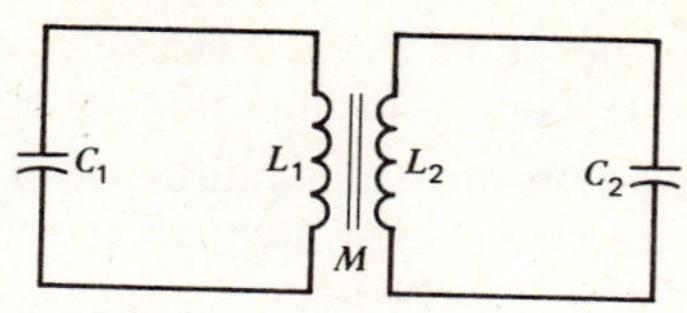

Fig. 12.3

and Faraday's law (11.21), the circuit equations are found to be

$$L_1\ddot{q}_1 + M\ddot{q}_2 + \frac{1}{C_1}q_1 = 0 \tag{12.34}$$

$$M\ddot{q}_1 + L_2\ddot{q}_2 + \frac{1}{C_2}q_2 = 0 \tag{12.35}$$

These are coupled linear differential equations similar in form to the equations of the double plane pendulum, (13.42) and (13.43) of Chapter 4. Section 14 of Chapter 4 gives a detailed treatment of the method for obtaining solutions of this type of problem.

PROBLEMS

12.1. Extend theorem 2 to the case of three linearly dependent functions, that is, prove the statement immediately following equation (12.7).

12.2. Extend theorem 5 to the nonhomogeneous case.

12.3. Show that $e^{\alpha_1 t}$ and $e^{\alpha_2 t}$ are linearly independent if $\alpha_1 \neq \alpha_2$. Also show that $e^{-\beta t}$ and $te^{-\beta t}$ are linearly independent. Do this by direct evaluation of the Wronskian determinants.

12.4. For the overdamped solution, show that α_1 and α_2 are both negative. Show that if $x_0 = 0$ and $u_0 \neq 0$, then x has a single maximum for $0 \leq t < \infty$. Make a qualitative sketch of x as a function of time.

12.5. Make the change of dependent variable $x(t) = u(t)\,v(t)$ in the oscillator equation (12.16). Determine an appropriate $v(t)$ that will make $u(t)$ obey the differential equation $\ddot{u} + Au = 0$, where A is a constant. By considering various possibilities for the constant A, obtain the three possible forms of the solution (12.20), (12.23), and (12.26).

12.6. Prove that the rate of decrease of mechanical energy of the oscillator is

$$\frac{d}{dt}\left(\frac{1}{2}m\dot{x}^2 + \frac{1}{2}kx^2\right) = -b\dot{x}^2 \tag{12.36}$$

This can be done by first multiplying the differential equation (12.15) by $\dot{x}$. What is the analogous expression for the circuit of Fig. 12.2?

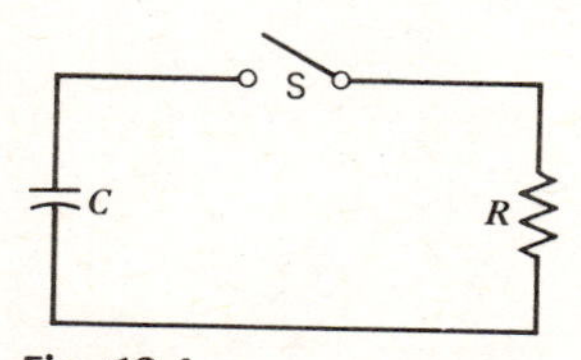

Fig. 12.4

12.7. In Fig. 12.4, the capacitor initially has a charge q_0 on it. Find the expression for the charge on the capacitor as a function of the time after switch S is closed.

12.8. In Fig. 12.5, the battery maintains a constant emf $\mathscr{E}$ across its terminals. Find the current in the circuit as a function of time after switch S is closed. Compare the differential equation that you obtain with (6.3).

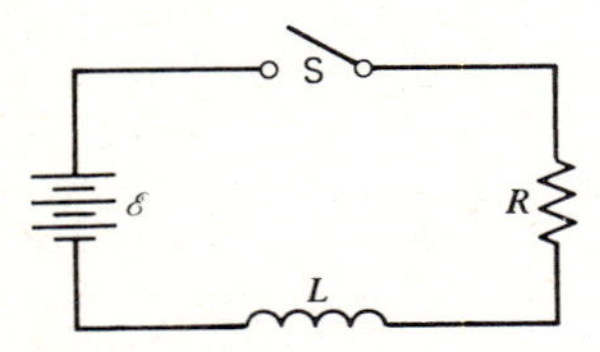

Fig. 12.5

12.9. Pretend that you don't know any simple solutions of (12.16). Try to solve it by substituting in the power series

$$x = \sum_{k=0}^{\infty} c_k t^k \tag{12.37}$$

Show that this leads to a three-term recurrence formula for the coefficients of the form

$$\alpha c_{k+2} + \beta c_{k+1} + \gamma c_k = 0 \tag{12.38}$$

and that the coefficients can all be found if x and $\dot{x}$ are known at $t = 0$. This is actually one approach to showing the existence of a solution of the general homogeneous, linear differential equation.

12.10. In the coupled circuit of Fig. 12.3, the expression

$$W_B = \tfrac{1}{2}L_1 I_1^2 + M I_1 I_2 + \tfrac{1}{2}L_2 I_2^2 \tag{12.39}$$

is the analogue of the kinetic energy in a mechanical system. Prove that this is a positive definite quadratic form only if $0 < M^2 < L_1 L_2$. We are safe because this is just the range of values for the mutual inductance we found in problem (11.9). The limit $M^2 = L_1 L_2$ gives the maximum possible coupling for the two circuits. Find the eigenfrequencies of the coupled circuit for $L_1 = L_2 = L$ and $C_1 = C_2 = C$. Note any peculiarities that occur as $M \to L$. Find the eigenvectors and the complete solution.

12.11. In setting up equations (12.34) and (12.35), we neglected the resistance of the circuit. Put two resistors in and generalize the equations appropriately. Show that the resulting pair of equations can be put into matrix form as

$$A\ddot{Q} + B\dot{Q} + KQ = 0 \tag{12.40}$$

where A, B, and K are appropriate 2×2 matrices and Q is a column matrix with elements q_1 and q_2. Solve (12.40) for the case where $L_1 = L_2 = L$, $C_1 = C_2 = C$, and $R_1 = R_2 = R$. Use a trial solution of the form $Q = Xe^{\alpha t}$, where X is a column matrix. You will get two quadratic equations for α, each one corresponding to an eigenvector. The roots of each separate quadratic equation may be real or complex, depending on the values of R, L, M, and C. Note especially what happens to the solutions as $M \to L$. The limit of perfect coupling, $M = L$, is probably not possible in practice. Can you find the solution for $M = L$? Are there critically damped solutions?

13. LINEAR DRIVEN SYSTEMS

In this section, we will consider the nonhomogeneous differential equation which results if a linear system, such as a series RLC circuit, is driven by a power source which has a periodic output voltage. We will use $V_a \cos \omega t$ as the output voltage of the power source or generator. The quantity V_a is the amplitude of the output voltage and ω is its frequency. The differential equation of the circuit shown in Fig. 13.1 is

$$L\frac{dI}{dt} + RI + \frac{q}{C} = V_a \cos \omega t \tag{13.1}$$

The symbol -(~)- in the circuit stands for the generator. Essentially what we are studying here are alternating-current (AC) circuits. In particular, we will develop the theory of *complex impedances*. The theory of AC circuits is a quasi-static theory. The theory of capacitance and inductance we have developed is based on the assumption of static fields which are allowed to vary only infinitesimally slowly. Faraday's law is acknowledged through the term $L\,dI/dt$, but no account is taken of the Maxwell displacement current term which appears in equation (12.7) of Chapter 3 and its consequent prediction of electromagnetic radiation. For practical purposes, quasi-static means that the time for one complete cycle of the term $\cos \omega t$ is long compared to the time required for an electromagnetic wave to propagate from one point in the circuit to another. We assume, for example, that, at a given time, the current I in equation (13.1) has the same value at all points of the circuit. The inductor knows instantaneously what the generator is doing. The analysis is greatly facilitated if (13.1) is replaced by

$$L\frac{dI}{dt} + RI + \frac{q}{C} = V_a e^{i\omega t} = V \tag{13.2}$$

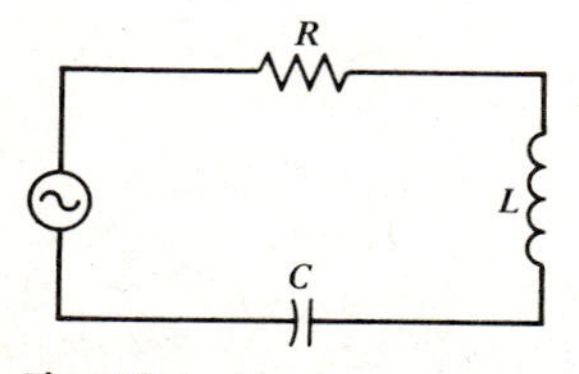

Fig. 13.1

with the understanding that the actual current in the circuit is to be found by taking the real part of the solution of (13.2). The notation $I_r = \operatorname{Re} I$ will be used. The complete solution of (13.2) consists of the complementary function and a particular integral. The complementary function involves decaying exponentials and dies out leaving only the particular integral. For this reason, it is called a *transient*. The particular integral is called the *steady-state* solution.

The steady-state solution will be of the form

$$q = q_0 e^{i\omega t} \tag{13.3}$$

where q_0 is a complex number. The complex current and its derivative can be written

$$I = \frac{dq}{dt} = i\omega q \qquad \frac{dI}{dt} = i\omega I \tag{13.4}$$

which provides the entire basis for the complex impedance method. In a steady-state linear AC circuit driven by a sinusoidal voltage consisting of a single frequency, the derivative of any time-dependent quantity is $i\omega$ times itself. Equation (13.2) can be expressed in the suggestive form

$$(Z_L + Z_R + Z_C)I = V \tag{13.5}$$

where

$$Z_L = i\omega L \qquad Z_R = R \qquad Z_C = \frac{1}{i\omega C} \tag{13.6}$$

The resemblance of (13.5) to Ohm's law for direct-current (DC) circuits is obvious and opens up the way for the analysis of complicated AC circuits by algebraic methods, using the same procedure as for DC circuits. The quantities Z_L, Z_R, and Z_C are called *complex impedances*. The procedure for finding the steady-state current in the circuit is to solve (13.5) for I and then take the real part of the resulting expression. As a first step, write I as

$$I = \frac{V}{R + i\left(\omega L - \dfrac{1}{\omega C}\right)} = \frac{\omega V_a e^{i\omega t}}{R\omega + iL\left(\omega^2 - \omega_0^2\right)} \tag{13.7}$$

where $\omega_0 = 1/\sqrt{LC}$ is the natural frequency of the free oscillations of an LC circuit with no resistance and no driving force. The most convenient way to separate the complex current into its real and imaginary parts is to write the denominator in polar form as

$$R\omega + iL(\omega^2 - \omega_0^2) = Ae^{i\phi} \tag{13.8}$$

$$A = \sqrt{R^2\omega^2 + L^2(\omega^2 - \omega_0^2)^2} \tag{13.9}$$

$$\tan\phi = \frac{L(\omega^2 - \omega_0^2)}{R\omega} \tag{13.10}$$

The various quantities are represented geometrically in Fig. 13.2. The complex current can be written conveniently as

$$I = \frac{\omega V_a}{A} e^{i(\omega t - \phi)} = I_a e^{i(\omega t - \phi)} \tag{13.11}$$

The actual current is the real part:

$$I_r = I_a \cos(\omega t - \phi) \tag{13.12}$$

The quantity $I_a = \omega V_a / A$ is the amplitude of the current. A significant feature is that the current is not in phase with the driving voltage. The phase difference is the angle ϕ. It should be noted that ϕ can be positive or negative depending on whether $\omega > \omega_0$ or $\omega < \omega_0$.

The instantaneous power being put into the circuit by the generator is

$$P = I_r V_r = I_a V_a \cos\omega t \cos(\omega t - \phi) \tag{13.13}$$

The average power is found by first writing

$$P = I_a V_a [\cos^2 \omega t \cos\phi + \cos\omega t \sin\omega t \sin\phi] \tag{13.14}$$

The average over one cycle is found in the following steps:

$$\frac{1}{2\pi}\int_0^{2\pi} \cos^2 \omega t \, d(\omega t) = \frac{1}{2}$$

$$\frac{1}{2\pi}\int_0^{2\pi} \cos\omega t \sin\omega t \, d(\omega t) = 0$$

$$\overline{P} = \frac{1}{2\pi}\int_0^{2\pi} P \, d(\omega t) = \frac{1}{2} I_a V_a \cos\phi \tag{13.15}$$

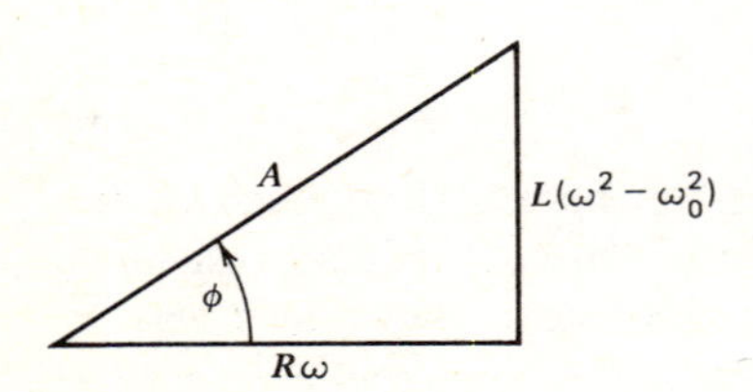

Fig. 13.2

The quantity $\cos\phi$ is called the *power factor*. Since $I_a = \omega V_a / A$ and $\cos\phi = R\omega / A$, the average power can be written in various ways as

$$\overline{P} = \frac{1}{2} I_a^2 R = \frac{\omega^2 V_a^2 R}{2A^2} = \frac{\omega^2 V_a^2 R}{2\left[R^2\omega^2 + L^2(\omega^2 - \omega_0^2)^2\right]} \tag{13.16}$$

The instantaneous power consumed by the resistor is

$$P_R = I_r^2 R = I_a^2 R \cos^2(\omega t - \phi) \tag{13.17}$$

which, if averaged over one cycle, gives (13.16). This shows that the energy put into the circuit is ultimately accounted for as heat dissipated by the resistor. The instantaneous values are different, and some energy is traded back and forth between the inductor and the capacitor, but the average energy flow is from power source to resistor.

A measure of the response of the circuit to various driving frequencies is provided by the expression for average power. The numerical values $V_a = 10$ volts, $\omega_0 = 10^3$ sec^{-1}, $L = 10^{-2}$ henry, $C = 10^{-4}$ farad give

$$\overline{P} = \frac{50R\omega^2}{R^2\omega^2 + 10^{-4}(\omega^2 - 10^6)^2} \tag{13.18}$$

This function is shown graphed in Fig. 13.3 for three different values of R. A typical resonance phenomenon is exhibited. If $R = 0$, the expression for the power becomes infinite at $\omega = \omega_0$. If R is small, the power as a function of ω has a sharp spike resembling a Dirac delta function at $\omega = \omega_0$. Recall that ω_0 is the natural frequency of the undamped oscillator. Maximum energy absorption occurs when the driving

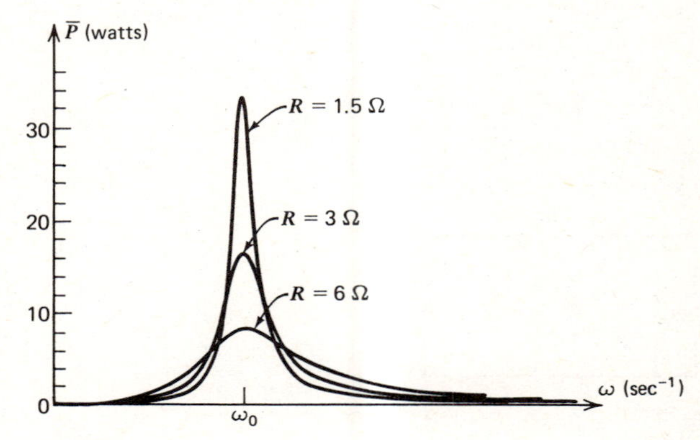

Fig. 13.3

frequency matches the natural frequency of the system. If a lot of resistance is put into the circuit, the resonance becomes broadened and less distinct. By setting $d\bar{P}/d\omega = 0$, it is possible to show formally that $\bar{P}$ has a maximum at $\omega = \omega_0$.

Complex impedances can conveniently be used to find out what is going on anywhere in the circuit. For example, suppose that we want to know what a voltmeter reads if it is connected across the capacitor in Fig. 13.1. The complex voltage is

$$V_C = IZ_C = \frac{I_a}{i\omega C} e^{i(\omega t - \phi)} = \frac{I_a}{\omega C} e^{i(\omega t - \phi - \pi/2)} \tag{13.19}$$

The actual voltmeter reading is the real part:

$$V_{Cr} = \frac{I_a}{\omega C} \cos(\omega t - \phi - \pi/2) \tag{13.20}$$

Note that the voltage across the capacitor and the current through it are 90° out of phase.

The circuit of Fig. 13.4 is sometimes called a *tank circuit*. By using complex impedances, the combined impedance of the parallel branches of the circuit can be found by the same rule used for DC circuits:

$$\frac{1}{Z_P} = \frac{1}{Z_C} + \frac{1}{R_2 + Z_L} \tag{13.21}$$

The overall impedance of the circuit is

$$Z = R_1 + Z_P = R_1 + \frac{Z_C(R_2 + Z_L)}{Z_C + Z_L + R_2}$$

$$= R_1 + \frac{\omega_0^2(R_2 + i\omega L)}{\omega_0^2 - \omega^2 + i\omega\omega_0^2 R_2 C} \tag{13.22}$$

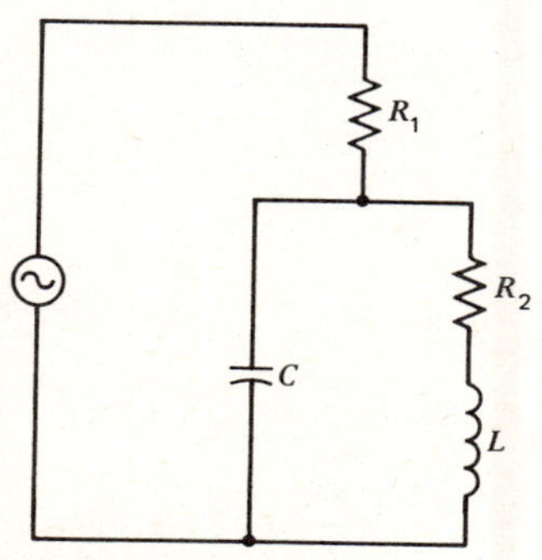

Fig. 13.4

If R_2 is small enough to be neglected, then

$$Z \simeq R_1 + \frac{\omega_0^2 i\omega L}{\omega_0^2 - \omega^2} \tag{13.23}$$

which again shows that resonance occurs if $\omega = \omega_0$. This time, it manifests itself in a large impedance.

Figure 13.5 shows a transformer circuit. The differential equations for this circuit are

$$L_1 \frac{dI_1}{dt} + M\frac{dI_2}{dt} + I_1 R_1 = V = V_a e^{i\omega t} \tag{13.24}$$

$$M\frac{dI_1}{dt} + L_2 \frac{dI_2}{dt} + I_2 Z = 0 \tag{13.25}$$

in which Z can be thought of as the load impedance in the secondary circuit, including any resistance in the wire of the transformer coils. For the steady-state, the derivatives with respect to time can be replaced by $i\omega$, giving

$$L_1 i\omega I_1 + Mi\omega I_2 + I_1 R_1 = V \tag{13.26}$$

$$Mi\omega I_1 + L_2 i\omega I_2 + I_2 Z = 0 \tag{13.27}$$

which are two linear algebraic equations for the complex currents I_1 and I_2. The formal solution is easy, but by getting bogged down in the algebra, we might overlook some simple features of this problem. Suppose that the transformer has been well designed so that $M^2 \simeq L_1 L_2$. Picture the transformer as constructed like Fig. 11.3. If perfect coupling is assumed, the flux linking the two coils can be expressed as

$$\Phi_1 = L_1 I_1 + MI_2 = N_1 \Phi_0 \tag{13.28}$$

$$\Phi_2 = MI_1 + L_2 I_2 = N_2 \Phi_0 \tag{13.29}$$

where Φ_0 is the flux through a cross section of the iron core. Let's also assume that the load impedance is purely resistive so that $Z = R_2$. Consider all quan-

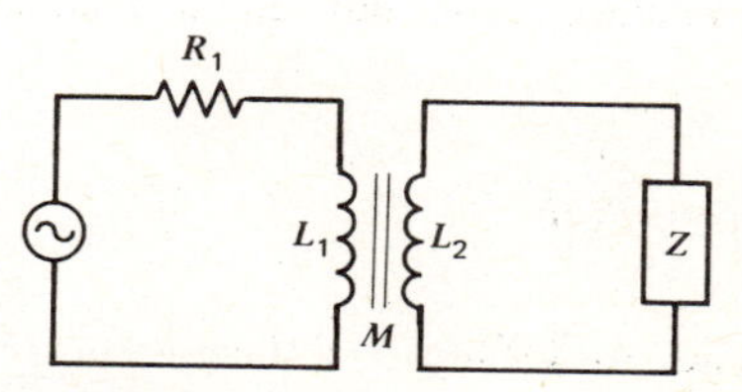

Fig. 13.5

tities to be real and express the circuit equations as

$$N_1 \frac{d\Phi_0}{dt} + I_1 R_1 = V_a \cos \omega t \tag{13.30}$$

$$N_2 \frac{d\Phi_0}{dt} + I_2 R_2 = 0 \tag{13.31}$$

Eliminate $d\Phi_0/dt$ to get

$$\frac{N_2}{N_1} = -\frac{I_2 R_2}{V_a \cos \omega t - I_1 R_1} \tag{13.32}$$

This simple result says that the ratio of the output to the input voltage is the same as the ratio of the number of turns in the two coils.

PROBLEMS

13.1. For the average power formula (13.16), show that $d\overline{P}/d\omega = 0$ gives $\omega = \omega_0$. What is $\overline{P}_{max}$? For what values of ω is $\overline{P} = \frac{1}{2}\overline{P}_{max}$? Show that these values are approximately $\omega = \omega_0 \pm R/2L$ if R is small. This problem shows that the height of the power resonance is inversely proportional to R and that its width is proportional to R.

13.2. Show that the average power formula (13.16) is given correctly by $\overline{P} = \frac{1}{2}\operatorname{Re}(IV^*)$.

13.3. What is the steady-state solution of $m\ddot{x} + kx = F_a \cos \omega t$? Is there a solution if $\omega = \omega_0 = \sqrt{k/m}$?

Hint: Try $x = At \sin \omega_0 t$.

13.4. Figure 13.6 shows a mass connected to a spring being driven by a force F. Assuming viscous friction and $F = F_a \cos \omega t$, the equation of motion is

$$m\ddot{x} + b\dot{x} + kx = F_a \cos \omega t \tag{13.33}$$

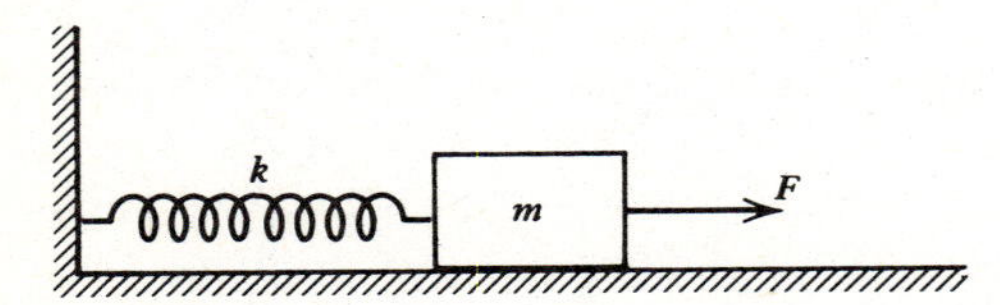

Fig. 13.6

Find the complete solution of this problem, including both the transient solution and the steady-state solution. Evaluate the constants of integration for the case where the initial position and velocity are both zero. Figure 13.7 shows graphs of the solution you will obtain for the numerical values $\omega_0 = 5\ \text{sec}^{-1}$, $\beta = b/m = 0.30\ \text{sec}^{-1}$, and $F_a/m = 30$ dynes/gm. Four different driving frequencies are shown. In Fig. 13.7*a*, the driving frequency is $\omega = 1\ \text{sec}^{-1}$ and is low compared to the resonant frequency ω_0. The higher-frequency damped oscillations of the transient solution are seen superimposed on the lower-frequency steady-state solution. The transient solution dies out after a few cycles of the steady-state solution. In Fig. 13.7*c*, the driving frequency matches the resonant frequency ω_0. The oscillations build up to a much higher steady-state amplitude than in the other three cases shown.

13.5. Show that the nonhomogeneous differential equation $\ddot{x} + 2\beta\dot{x} + \omega_0^2 x = F(t)$ can be expressed in the form

$$\left(\frac{d}{dt} - \alpha_1\right)\left(\frac{d}{dt} - \alpha_2\right)x = F(t) \tag{13.34}$$

Show that if $u(t)$ is defined by $\dot{x} - \alpha_2 x = u$, then $\dot{u} - \alpha_1 u = F(t)$. Show that the solution for

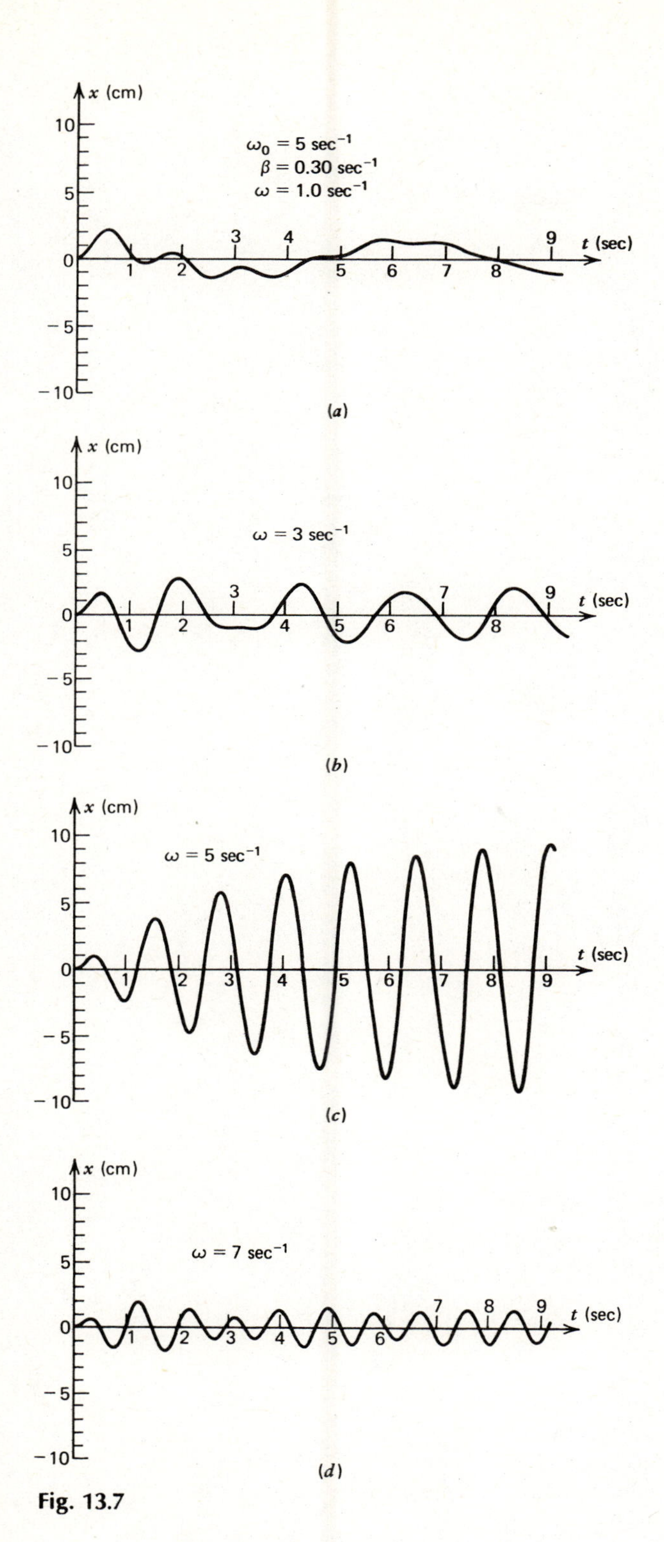
x (cm)
ω₀ = 5 sec⁻¹
β = 0.30 sec⁻¹
ω = 1.0 sec⁻¹
t (sec)
(a)
x (cm)
ω = 3 sec⁻¹
t (sec)
(b)
x (cm)
ω = 5 sec⁻¹
t (sec)
(c)
x (cm)
ω = 7 sec⁻¹
t (sec)
(d)

Fig. 13.7

$u(t)$ is

$$u(t) = e^{\alpha_1 t} \int F(t) e^{-\alpha_1 t}\, dt + c_1 e^{\alpha_1 t} \tag{13.35}$$

Hint: Multiply the differential equation obeyed by u through by the *integrating factor* $e^{-\alpha_1 t}$. Show that $x(t)$ is given by

$$x(t) = e^{\alpha_2 t} \int u(t) e^{-\alpha_2 t}\, dt + c_2 e^{\alpha_2 t} \tag{13.36}$$

13.6. Find the expression for the reading of the voltmeter in Fig. 13.8. The resistance R_2 can be thought of as the resistance of the wire making up the coil of the inductor. What is the phase relation between V and I in the limit $R_2 \to 0$? Assume that the voltmeter in no way affects the behavior of the circuit.

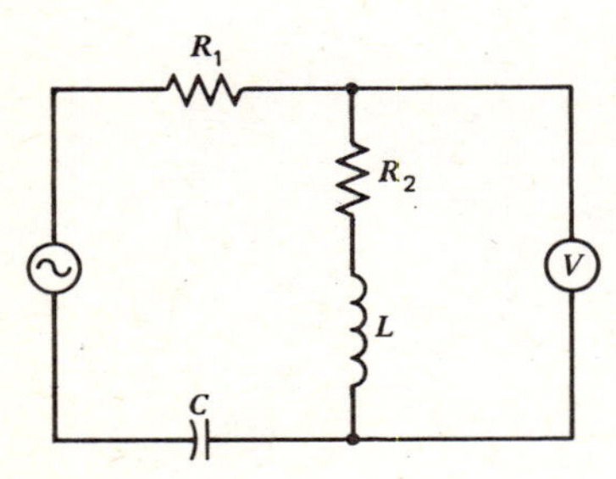

Fig. 13.8

13.7. Write down the differential equations for Fig. 13.4 and justify equation (13.21). Remember that if I_1 and I_2 are the currents in the two parallel branches, then the total current is given by $I = I_1 + I_2$. Find the expression for the total current for the case where $R_2 = 0$. If you plot the amplitude of the current as a function of ω for $\omega_0 = 1000 \text{ sec}^{-1}$, $L = 0.01$ H, $V_a = 10$ V, and $R_1 = 10\ \Omega$, it comes out like Fig. 13.9.

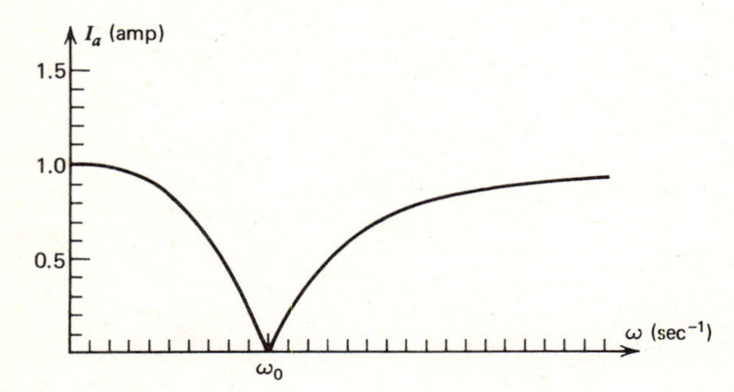

Fig. 13.9

13.8. What is the steady-state solution of $m\ddot{x} + b\dot{x} + kx = F_1 \cos \omega_1 t + F_2 \cos \omega_2 t$?

13.9. Set up equations similar to (13.26) and (13.27) for the circuit shown in Fig. 13.10. Find the expression for the average power dissipated by the resistor through which the current I_1 flows. The expression that you will get is shown graphed in Fig. 13.11 for $L = 0.01$ H,

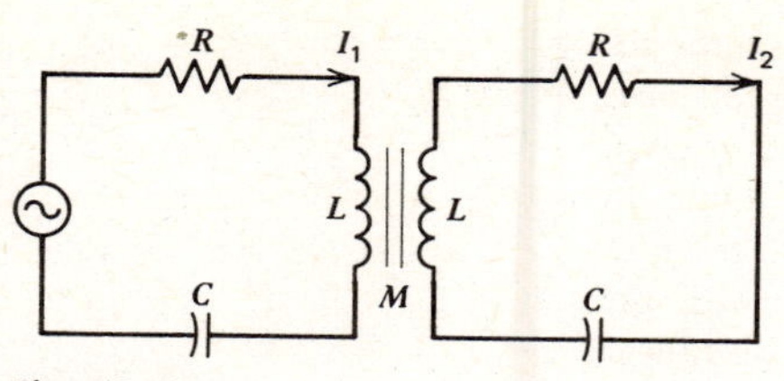

Fig. 13.10

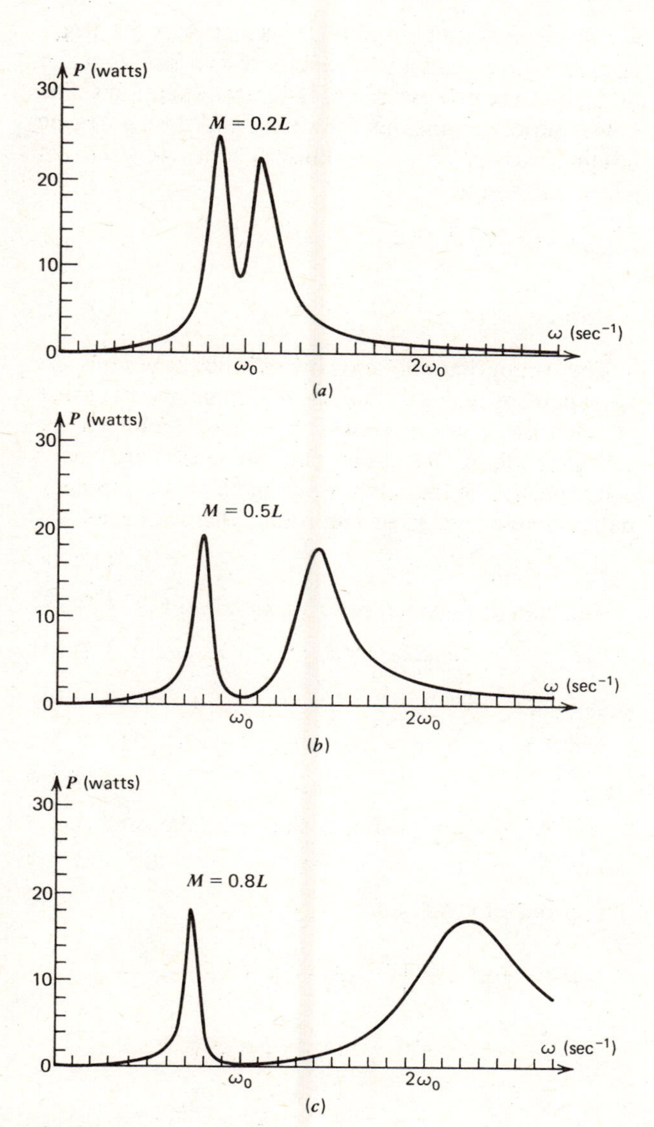

Fig. 13.11

$C=10^{-4}$ F, $\omega_0=1000\ \text{sec}^{-1}$, $V_a=10$ V, and $R=1.5\ \Omega$. If $M=0$ so that the circuits are uncoupled, there is a single resonance at $\omega=\omega_0$ as shown in Fig. 13.3. In Fig. 13.11*a*, $M=0.2L$ giving relatively weak coupling. The effect of the coupling is to split the original resonance into two peaks, one corresponding to each eigenfrequency that you found in problem 12.10. As the coupling increases, the resonances are moved farther apart as shown in Figs. 13.11*b* and 13.11*c*. Resonances can generally be expected to be associated with the natural frequencies of any system. In solving for the amplitude of the current, be careful to write the denominator in such a way that the double resonance effect is made obvious. Don't just multiply it all out in a string of terms that cannot be easily interpreted.

14. TRIGONOMETRIC AND HYPERBOLIC FUNCTIONS

In this section, the formal development of the theory of functions of a complex variable is continued. The trigonometric functions can be extended into the complex plane through the power series expansions of these functions known to be valid on the real axis just as was done for the exponential functions in section 9. Equivalently, and more conveniently, equations (9.7) and (9.8) can be used as the means of defining the functions $\sin z$ and $\cos z$:

$$\sin z=\frac{1}{2i}\left(e^{iz}-e^{-iz}\right) \tag{14.1}$$

$$\cos z=\frac{1}{2}\left(e^{iz}+e^{-iz}\right) \tag{14.2}$$

Since (14.1) and (14.2) become the familiar $\sin x$ and $\cos x$ on the real axis, theorem 4 of section 8 guarantees that they are the unique analytic continuations of $\sin z$ and $\cos z$ into the complex plane. Because of the analytic properties of the exponential function, $\sin z$ and $\cos z$ are analytic everywhere in the complex plane. The hyperbolic functions are formally defined by

$$\sinh z=\tfrac{1}{2}\left(e^{z}-e^{-z}\right) \tag{14.3}$$

$$\cosh z=\tfrac{1}{2}\left(e^{z}+e^{-z}\right) \tag{14.4}$$

These functions are also analytic everywhere in the complex plane. If z is replaced everywhere in the definitions by iz, the result is

$$\sin(iz)=\frac{1}{2i}\left(e^{-z}-e^{z}\right)=i\sinh z \tag{14.5}$$

$$\cos(iz)=\cosh z \tag{14.6}$$

$$\sinh(iz)=i\sin z \tag{14.7}$$

$$\cosh(iz)=\cos z \tag{14.8}$$

which reveals a direct connection between the trigonometric functions and the hyperbolic functions; in fact, they are not essentially different functions. The power series expansions for $\sinh z$ and $\cosh z$ can be obtained easily from the known series for the exponential function:

$$\sinh z=z+\frac{1}{3!}z^3+\frac{1}{5!}z^5+\cdots \tag{14.9}$$

$$\cosh z=1+\frac{1}{2!}z^2+\frac{1}{4!}z^4+\cdots \tag{14.10}$$

The trigonometric and hyperbolic functions are known from real variable theory to satisfy long lists of identities, most of which remain valid in the complex plane. Identities can be established as a consequence of the definitions in terms of exponentials. Some of the most important identities are

$$\sin^2 z+\cos^2 z=1 \tag{14.11}$$

$$\sin(z_1+z_2)=\sin z_1\cos z_2+\cos z_1\sin z_2 \tag{14.12}$$

$$\cos(z_1+z_2)=\cos z_1\cos z_2-\sin z_1\sin z_2 \tag{14.13}$$

$$\cosh^2 z-\sinh^2 z=1 \tag{14.14}$$

$$\sinh(z_1+z_2)=\sinh z_1\cosh z_2+\cosh z_1\sinh z_2 \tag{14.15}$$

$$\cosh(z_1+z_2)=\cosh z_1\cosh z_2+\sinh z_1\sinh z_2 \tag{14.16}$$

The proof of (14.11) is

$$\sin^2 z+\cos^2 z=\left(\frac{e^{iz}-e^{-iz}}{2i}\right)^2+\left(\frac{e^{iz}+e^{-iz}}{2}\right)^2$$

$$=-\frac{1}{4}\left(e^{2iz}-2+e^{-2iz}\right)$$

$$+\frac{1}{4}\left(e^{2iz}+2+e^{-2iz}\right)=1$$

The other identities are similarly established. Important special cases are

$$\sin z = \sin(x+iy) = \sin x \cos iy + \cos x \sin iy$$
$$= \sin x \cosh y + i \cos x \sinh y \qquad (14.17)$$
$$\cos z = \cos x \cosh y - i \sin x \sinh y \qquad (14.18)$$
$$\sinh z = \sinh x \cos y + i \cosh x \sin y \qquad (14.19)$$
$$\cosh z = \cosh x \cos y + i \sinh x \sin y \qquad (14.20)$$

Equations (14.17) through (14.20) give the separations of the various functions into their real and imaginary parts. As a consequence, the magnitudes of the functions can be calculated:

$$|\sin z|^2 = \sin^2 x \cosh^2 y + \cos^2 x \sinh^2 y$$
$$= \sin^2 x \cosh^2 y + (1 - \sin^2 x)\sinh^2 y$$
$$= \sin^2 x + \sinh^2 y \qquad (14.21)$$
$$|\cos z|^2 = \cos^2 x + \sinh^2 y \qquad (14.22)$$
$$|\sinh z|^2 = \sinh^2 x + \sin^2 y \qquad (14.23)$$
$$|\cosh z|^2 = \sinh^2 x + \cos^2 y \qquad (14.24)$$

It is well to keep in mind the behavior of the hyperbolic functions for real variables; $\sinh y$ and $\cosh y$ are graphed in Fig. 14.1. For real values of z, $|\sin z|$ and $|\cos z|$ never exceed 1. This is no longer true when these functions are extended into the complex plane. Since $|\sinh y|$ and $|\cosh y|$ become arbitrarily large as $y \to \pm\infty$, the same is true of $|\sin z|$ and $|\cos z|$.

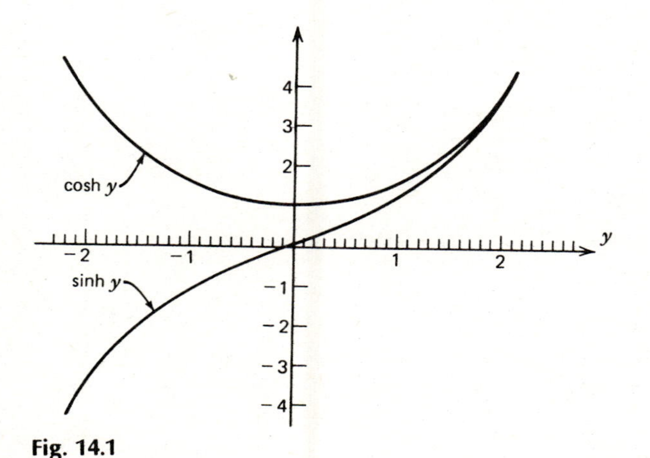

Fig. 14.1

Since the derivative of the exponential function is known, the derivatives of the trigonometric and hyperbolic functions can be calculated:

$$\frac{d}{dz}\sin z = \frac{1}{2i}(ie^{iz} + ie^{-iz}) = \cos z \qquad (14.25)$$

Similarly,

$$\frac{d}{dz}\cos z = -\sin z \qquad (14.26)$$
$$\frac{d}{dz}\sinh z = \cosh z \qquad (14.27)$$
$$\frac{d}{dz}\cosh z = \sinh z \qquad (14.28)$$

Other functions can be defined in terms of $\sin z$, $\cos z$, $\sinh z$, and $\cosh z$. The most important of these are the tangent and the hyperbolic tangent:

$$\tan z = \frac{\sin z}{\cos z} \qquad (14.29)$$
$$\tanh z = \frac{\sinh z}{\cosh z} \qquad (14.30)$$

These functions have singularities at the points where $\cos z$ and $\cosh z$ are zero; otherwise, they are analytic everywhere. The general power series expansions for $\tan z$ and $\tanh z$ are more difficult to obtain than are the other series of this section, but a few terms can be found by a process of division of one series into another. As an example, the series for $\tan z$ is found by using the known series for $\sin z$ and $\cos z$. If we agree to keep the calculation accurate to fifth order, then

$$\tan z = \frac{z - \frac{1}{3!}z^3 + \frac{1}{5!}z^5 + \cdots}{1 - \left(\frac{1}{2!}z^2 - \frac{1}{4!}z^4 + \cdots\right)}$$
$$= \left(z - \frac{1}{3!}z^3 + \frac{1}{5!}z^5 + \cdots\right)$$
$$\times\left(1 + \frac{1}{2!}z^2 - \frac{1}{4!}z^4 + \left[\frac{1}{2!}z^2\right]^2 + \cdots\right)$$
$$= z + \frac{1}{3}z^3 + \frac{2}{15}z^5 + \cdots \qquad (14.31)$$

Since the nearest zeros of $\cos z$ are at $z = \pm\pi/2$, the radius of convergence of the series for $\tan z$ can be expected to be $|z| = \pi/2$. By means of

$$\tan(iz) = \frac{\sin(iz)}{\cos(iz)} = \frac{i \sinh z}{\cosh z} = i \tanh z \tag{14.32}$$

we find that

$$\tanh z = z - \frac{1}{3}z^3 + \frac{2}{5}z^5 + \cdots \qquad |z| < \pi/2 \tag{14.33}$$

General expansions of $\tan z$ and $\tanh z$ can be found in terms of *Bernoulli numbers*, the definition and use of which is discussed in the problems.

PROBLEMS

14.1. Find the solution of $\ddot{x} + 2\beta\dot{x} + \omega_0^2 x = 0$ in terms of hyperbolic functions for the overdamped case. The solution should look like (12.26) with sin and cos replaced by sinh and cosh.

14.2. Show that $\sinh \omega t$ and $\cosh \omega t$ are linearly independent functions of t.

14.3. Show that $u = \sin x \cosh y$ and $v = \cos x \sinh y$ are harmonic functions in two different ways.

14.4. Find all the zeros of $\sin z$, $\cos z$, $\sinh z$, and $\cosh z$.

14.5. Less commonly used trigonometric and hyperbolic functions are

$$\cot z = \frac{\cos z}{\sin z} \qquad \sec z = \frac{1}{\cos z} \qquad \csc z = \frac{1}{\sin z} \tag{14.34}$$

$$\coth z = \frac{\cosh z}{\sinh z} \qquad \operatorname{sech} z = \frac{1}{\cosh z} \qquad \operatorname{csch} z = \frac{1}{\sinh z} \tag{14.35}$$

Prove the differentiation formulas

$$\frac{d}{dz}\tan z = \sec^2 z \qquad \frac{d}{dz}\cot z = -\csc^2 z$$

$$\frac{d}{dz}\sec z = \sec z \tan z \qquad \frac{d}{dz}\csc z = -\csc z \cot z \tag{14.36}$$

What are the corresponding results for the hyperbolic functions?

14.6. In special relativity, one deals with particle speeds which are less than the speed c of light. The hyperbolic tangent of a real variable has the range of values $-1 < \tanh\theta < +1$. It is therefore possible to define a function θ by $\tanh\theta = u/c$. Think, for example, of u as being the one-dimensional velocity of a particle. Show that

$$\cosh\theta = \frac{1}{\sqrt{1-(u^2/c^2)}} \qquad \sinh\theta = \frac{u/c}{\sqrt{1-(u^2/c^2)}} \tag{14.37}$$

Make a qualitative sketch of $\tanh\theta$ as a function of θ for $-\infty < \theta < +\infty$.

14.7. The Bernoulli numbers can be defined by means of the expansion

$$\frac{x}{e^x - 1} = \sum_{n=0}^{\infty} \frac{B_n}{n!} x^n \tag{14.38}$$

Calculate the values of the Bernoulli numbers at least through B_4. One way to do it is to use Taylor's series and calculate the derivatives by brute force. A better way to do it is to write

$$x = (e^x - 1)\left(B_0 + B_1 x + \frac{1}{2!}B_2 x^2 + \frac{1}{3!}B_3 x^3 + \frac{1}{4!}B_4 x^4 + \cdots\right)$$

Now use the known expansion for e^x. Multiply the series together, using care to save everything up to order x^4. If you arrange everything in a neat, sensible way, you should be able to see that the general recurrence relation is

$$1 + C_1 B_1 + C_2 B_2 + \cdots + C_{k-1} B_{k-1} = 0 \tag{14.39}$$

where $C_1, C_2, \ldots$ are binomial coefficients:

$$C_1 = k \qquad C_2 = \frac{k(k-1)}{2!} \qquad C_3 = \frac{k(k-1)(k-2)}{3!} \qquad \cdots$$

Equation (14.39) is convenient because a computer program to calculate the Bernoulli numbers can be based on it. It turns out that except for B_1, all the odd Bernoulli numbers are zero.

14.8. One reason why the Bernoulli numbers are useful is that the expansions for other functions can sometimes be expressed in terms of them. Show, for example, that

$$\coth z = 1 + \frac{1}{z} \sum_{n=0}^{\infty} \frac{B_n}{n!} (2z)^n = \frac{1}{z} \sum_{k=0}^{\infty} \frac{B_{2k}}{(2k)!} (2z)^{2k} \tag{14.40}$$

Hint:

$$\coth z = \frac{e^z + e^{-z}}{e^z - e^{-z}} = \frac{e^{2z} + 1}{e^{2z} - 1} = \frac{e^{2z} - 1 + 2}{e^{2z} - 1} = 1 + \frac{2}{e^{2z} - 1}$$

You can find an expansion for $\cot z$ by using $\coth iz = -i \cot z$. The expansion for $\tan z$ can be found by using the identity $\tan z = \cot z - 2\cot 2z$. What is the radius of convergence of (14.40)?

15. LOGARITHMS AND EXPONENTS

The exponential function $f(z) = e^z$ is uniquely defined except for an arbitrary multiplicative factor by the requirement that $f'(z) = f(z)$. All properties of e^z can be derived from this definition. We now inquire into the possibility of the existence of a function which is the inverse of the exponential, that is, a function $w(z)$ defined by the relation

$$z = e^w \tag{15.1}$$

One immediate problem is that $w(z)$ is ambiguous and is defined only up to an additive factor of the form $2\pi i n$, where n is a positive or negative integer. This is true because of the periodic nature of the exponential function as given by

$$e^w = e^{w + 2\pi i n} \tag{15.2}$$

This means that specifying z does not give $w(z)$ uniquely. To put it another way, (15.1) has many possible solutions for $w(z)$. The function $w(z)$ can be made single-valued only if conditions other than (15.1) are imposed. For the moment, let us write

$$w = \ln z \tag{15.3}$$

and pin down the solution that we are talking about by the requirement that, on the real axis, (15.3) match the conventional real function $\ln x$.

With this understanding, the properties of the logarithm familiar from real variable theory are easily extended into the complex plane. For example, if

$$z_1 = e^{w_1} \qquad z_2 = e^{w_2} \tag{15.4}$$

then by the multiplication theorem (9.3) for the exponential function,

$$z_1 z_2 = e^{w_1 + w_2}$$

Therefore,

$$w_1 + w_2 = \ln(z_1 z_2) = \ln z_1 + \ln z_2 \tag{15.5}$$

which is the *addition theorem* for the logarithm. In particular, note that $2 \ln z = \ln z^2$. These results are easily extended to

$$\ln(z_1 z_2 z_3) = \ln z_1 + \ln z_2 + \ln z_3 \tag{15.6}$$

$$n \ln z = \ln z^n \tag{15.7}$$

If we write

$$\frac{1}{z_2} = e^{-w_2} \tag{15.8}$$

then

$$\frac{z_1}{z_2} = e^{w_1} e^{-w_2} = e^{w_1 - w_2}$$

$$w_1 - w_2 = \ln(z_1 z_2^{-1}) = \ln z_1 - \ln z_2 \tag{15.9}$$

Equation (15.7) can now be seen to hold for all positive and negative integer values of n.

If z is written in polar form, then, as a special case of (15.5), we have

$$\ln z = \ln(re^{i\theta}) = \ln r + \ln e^{i\theta} = \ln r + i\theta \tag{15.10}$$

If $\theta = 0$, then $\ln z = \ln r = \ln x$, so that in fact we are talking about the solution of $z = e^w$ that matches $\ln x$ on the positive real axis. The positive real axis is also given by $\theta = 2\pi$, and this presents a problem, because then

$$\ln z = \ln r + 2\pi i \tag{15.11}$$

The function $\ln z$ is therefore not single-valued as it stands but increases by a factor of $2\pi i$ every time θ increases by 2π. An additional condition must be imposed in order that the logarithm become single-valued. One possibility is

$$f(z) = \ln z = \ln r + i\theta \qquad 0 \le \theta < 2\pi \tag{15.12}$$

When this is done, the positive real axis is said to be a *cut line* in the complex plane. The cut line must not be crossed if (15.12) is to remain single-valued. In other words, we can get off the real axis and onto the complex plane only in the direction of increasing θ. The logarithm can be made single-valued in other ways. For example, the negative real axis can be made the cut line, in which case we would write

$$f(x) = \ln z = \ln r + i\theta \qquad -\pi < \theta \le \pi \tag{15.13}$$

It is now possible to get off the positive real axis in either direction, but the only way to get on the negative real axis is through positive values of θ, and it must never be crossed. More generally, a cut line can be made at an arbitrary angle through the origin, in which case $\alpha \le \theta < \alpha + 2\pi$.

That $f(z) = \ln z$ is analytic can be shown by differentiating $z = e^{f(z)}$:

$$1 = e^{f(z)} \frac{df}{dz} \qquad \frac{df}{dz} = \frac{1}{z} \tag{15.14}$$

Thus, $f(z)$ is analytic everywhere except at $z = 0$ and along the cut line. Note that the logarithm of negative numbers is defined. For example, $\ln(-1) = \ln(1) + i\pi = i\pi$. All the solutions of the equation $z = e^w$ can be expressed as

$$w = \ln r + i\theta + 2\pi i n \qquad -\pi < \theta \le \pi \tag{15.15}$$

Different values of n are said to give different *branches* of the logarithm. The value $n = 0$ gives the *principal branch*. The singular point $z = 0$ is called the *branch point*, since encircling this point is what causes $\ln z$ to increase by the factor $2\pi i$.

It becomes possible now to define what is meant by an exponent which is an arbitrary real or complex number. If p is any such exponent and if z is expressed as $z = e^w$, then by z^p is meant

$$z^p = e^{pw} \tag{15.16}$$

Let us separate this function into its real and imaginary parts. If $p = a + ib$, then

$$\begin{aligned} z^p &= e^{(a+ib)(\ln r + i\theta)} \\ &= e^{a \ln r - b\theta} e^{i(b \ln r + a\theta)} \\ &= r^a e^{-b\theta}[\cos(b \ln r + a\theta) + i \sin(b \ln r + a\theta)] \end{aligned} \tag{15.17}$$

Just as with the logarithm, z^p is multiple-valued and in general has many branches. Since we have used the principal branch of $\ln z$ in (15.17), it represents the principal branch of z^p. The multiple-valuedness is cured just as it was for the logarithm by introducing a branch cut. The origin is the branch point and is also a singular point of z^p. There is of course no problem with multiple-valuedness if p is an integer.

If p is real, then the real and imaginary parts of z^a are

$$u = r^a \cos a\theta \qquad v = r^a \sin a\theta \tag{15.18}$$

The Cauchy–Riemann equations in polar form are

$$\frac{\partial u}{\partial r} = \frac{1}{r}\frac{\partial v}{\partial \theta} \qquad \frac{\partial v}{\partial r} = -\frac{1}{r}\frac{\partial u}{\partial \theta} \tag{15.19}$$

By calculating the derivatives, we find

$$\frac{\partial u}{\partial r} = ar^{a-1}\cos a\theta = \frac{1}{r}\frac{\partial v}{\partial \theta} \tag{15.20}$$

$$\frac{\partial v}{\partial r} = ar^{a-1}\sin a\theta = -\frac{1}{r}\frac{\partial u}{\partial \theta} \tag{15.21}$$

which seem to be valid everywhere, even including the branch point $r = 0$ in those cases where $a > 1$. But branch points are tricky. If you look back at the definition of an analytic function in section 7, you will find that in order for $f(z)$ to be analytic at a point z_0, it must be single-valued and have a derivative at every point in some neighborhood of z_0. The function z^a is not single-valued at $z = 0$. The derivatives (15.20) and (15.21) have different values if θ is increased by 2π, except for the special case where a is an integer. The basic idea of the derivative is that in calculating it at a point, we must be able to approach the point in any manner, including a spiral path that encircles the point several times if we wish. The domain in which a function is defined to be analytic is really *open*, that is, it excludes the boundaries, so that every point can be surrounded by some neighborhood at each point of which the derivative exists. The branch point, as well as the cut line, must therefore be excluded from the domain of analyticity of the function. The actual formula for the derivative of z^p follows from its definition:

$$\frac{d}{dz}z^p = \frac{d}{dz}e^{p\ln z} = \frac{p}{z}e^{p\ln z} = \frac{p}{z}z^p = pz^{p-1} \tag{15.22}$$

which is the same as for real variables.

The familiar properties of exponents follow from the definition. By means of (15.1), we can write $z' = e^{w'}$ and $z = e^w$, in which case $w' = \ln z'$ and $w = \ln z$. If $z' = z^p$, then $w' = p\ln z$, and

$$p\ln z = \ln z^p \tag{15.23}$$

which was previously shown to hold only for integer values of p. If p and q are any two complex numbers, then

$$z^p z^q = e^{pw}e^{qw} = e^{(p+q)w} = z^{p+q} \tag{15.24}$$

$$z_1^p z_2^p = e^{pw_1}e^{pw_2} = e^{p(w_1+w_2)} = (z_1 z_2)^p \tag{15.25}$$

If $w' = \ln z'$, then $(z')^q = e^{q\ln z'}$. Now let $z' = z^p$ to get

$$(z^p)^q = e^{q\ln z^p} = e^{pq\ln z} = (z)^{pq} \tag{15.26}$$

Equation (15.17) as it is written represents the principal branch of z^p. Other branches are found by increasing θ by a multiple of 2π. The existence of branches of z^p are important in finding the possible roots of algebraic equations. For example, suppose that the roots of $f(z) = z^3 - 1$ are wanted. It is possible to write the number 1 as

$$1, e^{2\pi i}, e^{4\pi i}, e^{6\pi i}, \ldots \tag{15.27}$$

The possible branches of the cube root of 1 are therefore

$$z_1 = 1 \qquad z_2 = e^{2\pi i/3} \qquad z_3 = e^{4\pi i/3} \tag{15.28}$$

There are no more because

$$e^{6\pi i/3} = e^{2\pi i} = 1 = z_1 \tag{15.29}$$

We can therefore write

$$f(z) = z^3 - 1 = (z - z_1)(z - z_2)(z - z_3) \tag{15.30}$$

Note that the three cube roots of 1 all lie on the unit circle at $\theta = 0$, 120°, and 240°. Equation (15.30) is a special case of a general theorem which states that every nth-degree polynomial

$$P(z) = z^n + a_{n-1}z^{n-1} + \cdots + a_0 \tag{15.31}$$

can be factored as

$$P(z) = (z - z_1)(z - z_2)\cdots(z - z_n) \tag{15.32}$$

In other words, every nth-degree polynomial has exactly n roots. This theorem is called *the fundamental theorem of algebra*, which is probably a more grandiose title than it deserves. In general, it is not possible to achieve such a factorization over the field of real numbers.

Some functions have more than one branch point, and care must be used in defining them in such a way that they are single-valued. An example is

$$f(z) = \sqrt{z^2 - 1} \tag{15.33}$$

From the geometry of Fig. 15.1, it becomes apparent that it is possible to write

$$z^2 - 1 = (z - 1)(z + 1) = r_1 r_2 e^{i(\theta_1 + \theta_2)} \tag{15.34}$$

If we represent the square root of each factor by its

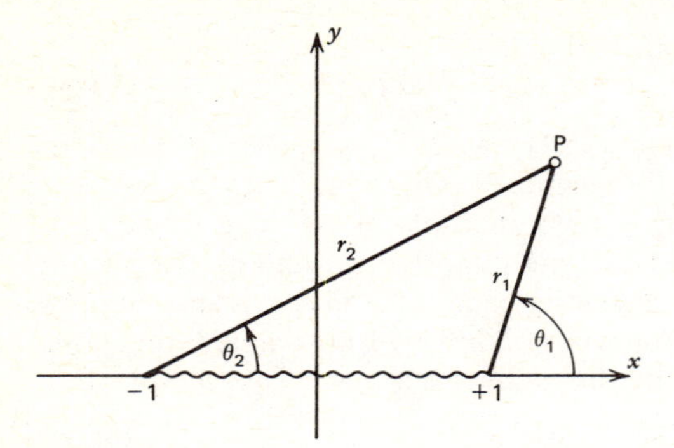

Fig. 15.1

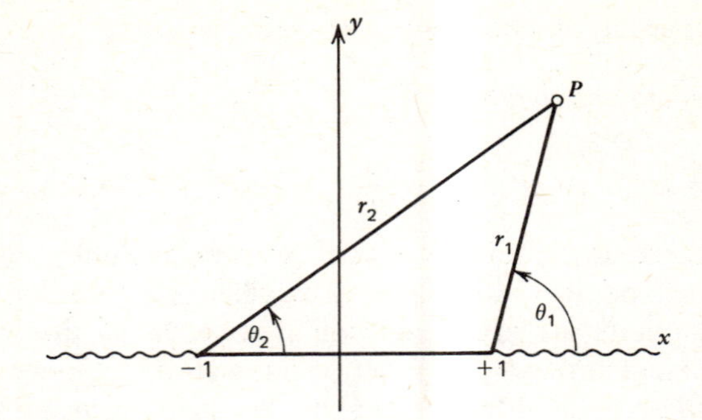

Fig. 15.2

principal value, then

$$f(z)=\sqrt{z^2-1}=\sqrt{r_1r_2}\,e^{i(\theta_1+\theta_2)/2} \tag{15.35}$$

Suppose that the point P of Fig. 15.1 moves in such a way that it crosses the real axis between -1 and $+1$ and then makes a complete circuit around $+1$ to return finally to the starting point. The angle θ_2 returns to its original value, but θ_1 increases by 2π so that the function is now

$$f(z)=\sqrt{r_1r_2}\,e^{i(\theta_1+\theta_2)/2+\pi i} \tag{15.36}$$

which is not the same as (15.35). Let's go back to the original function (15.35) and again make a complete circuit, this time enclosing both the points -1 and $+1$. Now both θ_1 and θ_2 increase by 2π so that $f(z)$ returns to its original value. The function therefore becomes single-valued if we place a branch cut between the two branch points -1 and $+1$ as indicated by the wavy line of Fig. 15.1. The point P can move in any way as long as it does not cross this line. Other solutions are possible. For example, the complex plane can be cut along the real axis from $+1$ to $+\infty$ and from $-\infty$ to -1 as shown in Fig. 15.2. The angles θ_1 and θ_2 are then restricted by $0\le\theta_1<2\pi$ and $-\pi<\theta_2\le+\pi$.

If p is an arbitrary complex number, then

$$f(z)=(1+z)^p \tag{15.37}$$

is analytic everywhere except at the branch point $z=-1$ and on any branch cuts. It can be developed into a Taylor's series about the point $z=0$, which of course must be the same as the binomial expansion because of the uniqueness of power series:

$$f(z)=1+pz+\frac{p(p-1)}{2!}z^2+\frac{p(p-1)(p-2)}{3!}z^3+\cdots \tag{15.38}$$

We talked about the convergence of this series for the case of real variables starting with equation (5.45). If you will look at the various tests for absolute convergence in section 4, you will see that they involve calculating the absolute value of the ratio of the $(n+1)$th term to the nth term. Thus, they remain valid for complex series as well. For binomial series in the complex plane, both (5.41) and (5.42) remain valid. The radius of convergence of (15.38) is determined by

$$\lim_{n\to\infty}\left|\frac{n-p}{n+1}z\right|=|z|=1 \tag{15.39}$$

which shows that it converges absolutely for $|z|<1$. Gauss' test (4.73) can be used to examine the possibility of absolute convergence on $|z|=1$:

$$\begin{aligned}\lim_{n\to\infty}\left|\frac{n-p}{n+1}\right| &= \lim_{n\to\infty}\left|\frac{1-p/n}{1+1/n}\right| \\ &= \lim_{n\to\infty}\left|\left(1-\frac{p}{n}\right)\left(1-\frac{1}{n}+\frac{1}{n^2}\right)\right| \\ &= \lim_{n\to\infty}\left|1-\frac{p+1}{n}+\frac{p+1}{n^2}\right|\end{aligned} \tag{15.40}$$

Let's drop the term of order $1/n^2$ and write $p=a+ib$.

Then,

$$\lim_{n\to\infty}\left|\frac{n-p}{n+1}\right| = \lim_{n\to\infty}\left|1-\frac{a+1}{n}-i\frac{b}{n}\right|$$
$$= \lim_{n\to\infty}\sqrt{\left(1-\frac{a+1}{n}\right)^2+\frac{b^2}{n^2}}$$
$$= \lim_{n\to\infty}\sqrt{1-\frac{2(a+1)}{n}}$$
$$= \lim_{n\to\infty}\left(1-\frac{a+1}{n}\right) \qquad (15.41)$$

From (4.7), we see that the series converges absolutely for $|z|=1$ if $a>0$. The series is not absolutely convergent if $a\leq 0$. However, it can be shown that the series converges conditionally for $|z|=1$ if $-1<a\leq 0$ provided that $z\neq -1$. Proving this requires developing tests for nonabsolute convergence of complex series, which we will not take space to carry out here.

PROBLEMS

15.1. Show that $\ln z=\ln r+i\theta$ is analytic except at $r=0$ and on any branch cuts by means of the Cauchy–Riemann equations.

15.2. In problem (5.13), you found the power series for $\ln(1+x)$. By analytic continuation, this series remains valid in the complex plane as

$$\ln(1+z)=z-\frac{1}{2}z^2+\frac{1}{3}z^3-\cdots=\sum_{k=1}^{\infty}(-1)^{k+1}\frac{1}{k}z^k \qquad |z|<1 \qquad (15.42)$$

Examine the question of absolute convergence on the circle $|z|=1$. Which branch of $\ln(1+z)$ does (15.42) represent? Obtain the following series expansions:

(a) $$\ln(z+z_0)=\ln z_0+\sum_{k=1}^{\infty}(-1)^{k+1}\frac{1}{k}\left(\frac{z}{z_0}\right)^k \qquad |z|<|z_0| \qquad (15.43)$$

(b) $$\ln\left(\frac{1+z}{1-z}\right)=2\left[z+\frac{1}{3}z^3+\frac{1}{5}z^5+\cdots\right] \qquad |z|<1 \qquad (15.44)$$

(c) $$\ln\left(\frac{z+1}{z-1}\right)=2\left[\frac{1}{z}+\frac{1}{3z^3}+\frac{1}{5z^5}+\cdots\right] \qquad |z|>1 \qquad (15.45)$$

(d) $$\ln z=2\left[\frac{z-1}{z+1}+\frac{1}{3}\left(\frac{z-1}{z+1}\right)^3+\frac{1}{5}\left(\frac{z-1}{z+1}\right)^5+\cdots\right] \qquad \operatorname{Re} z>0 \qquad (15.46)$$

15.3. By writing $1+z=re^{i\theta}$ and $p=a+ib$, give an expression for $f(z)=(1+z)^p$ which exhibits its various branches. If we want to use the expansion (15.38), which branch are we talking about? Where should a branch cut (or cuts) be made in the complex plane in order to make $f(z)$ single-valued and still have the expansion (15.38) be a valid representation for $|z|<1$?

15.4. Find all the roots of $f(z)=z^5-1$.

15.5. If a is a constant, show that

$$\frac{d}{dz}a^z=a^z\ln a \qquad (15.47)$$

15.6. Find the values of i^i, $(1+i)^i$, and $\ln(i)^{1/2}$.

15.7. Solve each of the equations $e^z=-3$ and $\ln z=i\pi/2$ for z.

15.8. If the polynomial

$$P(z)=z^n+a_{n-1}z^{n-1}+a_{n-2}z^{n-2}+\cdots+a_0 \qquad (15.48)$$

has real coefficients, then show that $\{P(z)\}^* = P(z^*)$. The roots of the polynomial are not necessarily real. Show however that if z_1 is a root of (15.48), then so is z_1^*. This means that the roots of a polynomial with real coefficients always occur in pairs that are complex conjugates of one another. A real polynomial of odd degree always has at least one real root.

15.9. Show that the function $f(z) = \sqrt{e^z + 1}$ has an infinite number of branch points as follows:

(a) Show that $e^z + 1$ is zero at the points $\pm\pi i, \pm 3\pi i, \pm 5\pi i, \ldots$.

(b) Examine the behavior of the function in the vicinity of its zeros. For example, near the point πi, write $z = \pi i + \varepsilon$, where ε is a complex number of small magnitude. Then, $e^z + 1 = e^{\pi i + \varepsilon} + 1 = -e^{\varepsilon} + 1$. In the limit $\varepsilon \to 0$, $e^{\varepsilon} \simeq 1 + \varepsilon$. Now write $\varepsilon = \rho e^{i\phi}$ and examine what happens to $f(z)$ as z moves around the point πi on a circle of infinitesimal radius ρ.

(c) Where should branch cuts be made to make $f(z)$ single-valued? Where is $f(z)$ analytic?

15.10. Define a branch of $f(z) = \sqrt{1 + \sqrt{z}}$ and show that it is analytic.

15.11. The *Riemann zeta function* is defined by

$$\zeta(z) = \sum_{k=1}^{\infty} \frac{1}{k^z} \tag{15.49}$$

In section 4, starting with equation (4.57), we examined the convergence properties of this series for the case where z is real. If z is complex, show that the series is absolutely convergent if $\text{Re } z > 1$ and not absolutely convergent if $\text{Re } z \le 1$.

15.12. If x is real, prove the inequality

$$\frac{x}{1+x} < \ln(1+x) < x \qquad x > -1 \qquad x \neq 0 \tag{15.50}$$

Hint:

$$\frac{1}{(1+t)^2} < \frac{1}{1+t} < 1 \qquad t > 0 \tag{15.51}$$

$$\frac{1}{(1+t)^2} > \frac{1}{1+t} > 1 \qquad -1 < t < 0 \tag{15.52}$$

Integrate (15.51) from 0 to x and (15.52) from x to 0.

15.13. If a and b are real numbers, $a > 0$, and $\alpha = a + ib$, establish the following limits:

$$\lim_{x \to \infty} x^{-\alpha} \ln x = 0 \qquad \lim_{x \to 0} x^{\alpha} \ln x = 0 \tag{15.53}$$

15.14. For the representation of $\ln(1+z)$ as given by equation (15.42), a branch cut can be placed along the negative real axis from $-\infty$ to -1. Show that another expansion for $\ln(1+z)$ is

$$\ln(1+z) = -\sum_{k=1}^{\infty} \frac{(2+z)^k}{k} + i\pi \qquad |2+z| < 1 \tag{15.54}$$

This time, the expansion is around the point $z = -2$ and a branch cut can be placed along the real axis from -1 to $+\infty$. Where can branch cuts be placed for equations (15.44) and (15.45)?

16. INVERSE FUNCTIONS

In section 15, the logarithm was obtained as the inverse of the exponential function. In this section, the inverses of the trigonometric and hyperbolic functions will be found. The inverse functions are multiple-valued and have many branches, so that care must be used in defining them. If we write

$$z = \frac{1}{2i}\left(e^{iw} - e^{-iw}\right) \tag{16.1}$$

then $w(z) = \arcsin z = \sin^{-1} z$ is the inverse of $\sin z$. Even in real variable theory, this function is multiple-valued. The principal branch of this function for real z is shown graphed in Fig. 16.1 and has the range of values $-\pi/2 < w < +\pi/2$. We will find its analytic continuation into the complex plane. If (16.1) is multiplied by e^{iw}, the result can be rearranged as

$$e^{2iw} - 2ize^{iw} - 1 = 0 \tag{16.2}$$

This is a quadratic equation for e^{iw} with the solutions

$$e^{iw} = iz \pm \sqrt{1 - z^2} \tag{16.3}$$

On the real axis, (16.3) is

$$e^{iw} = \cos w + i \sin w = ix \pm \sqrt{1 - x^2} \tag{16.4}$$

Since $\cos w > 0$ for $-\pi/2 < w < +\pi/2$, the plus sign is chosen. Using the principal value of the logarithm gives

$$w = \sin^{-1} z = -i \ln\left(iz + \sqrt{1 - z^2}\right) \tag{16.5}$$

There are branch points at $z = \pm 1$, and it is necessary to cut the complex plane as illustrated in Fig. 15.2. When this is done, (16.5) becomes a single-valued analytic function which matches the values of $\sin^{-1} x$ shown in Fig. 16.1 on the real axis. Differentiation of (16.5) gives

$$\frac{d}{dz} \sin^{-1} z = \frac{1}{\sqrt{1 - z^2}} \tag{16.6}$$

The binomial expansion gives

$$\frac{1}{\sqrt{1 - z^2}} = 1 + \frac{1}{2} z^2 + \frac{1 \cdot 3}{2 \cdot 4} z^4 + \frac{1 \cdot 3 \cdot 5}{2 \cdot 4 \cdot 6} z^6 + \cdots \qquad |z| < 1 \tag{16.7}$$

which can be integrated term by term, with the result that

$$\sin^{-1} z = z + \frac{1}{2 \cdot 3} z^3 + \frac{1 \cdot 3}{2 \cdot 4 \cdot 5} z^5 + \frac{1 \cdot 3 \cdot 5}{2 \cdot 4 \cdot 6 \cdot 7} z^7 + \cdots \quad |z| < 1 \tag{16.8}$$

The constant of integration is fixed by the requirement that $\sin^{-1}(0) = 0$.

The principal branch of $w = \cos^{-1} z$ is shown plotted in Fig. 16.2 for real values of z and has the range of values $0 < w < \pi$. A procedure similar to the one used for $\sin^{-1} z$ can be used to find $\cos^{-1} z$. Another way to do it is to first write $z = \cos w$ and $\sin w = (1 - z^2)^{1/2}$. This shows that $w = \sin^{-1}(1 - z^2)^{1/2}$. If in (16.5), z is replaced by $(1 - z^2)^{1/2}$, the result is

$$w = \cos^{-1} z = -i \ln\left(z + i\sqrt{1 - z^2}\right) \tag{16.9}$$

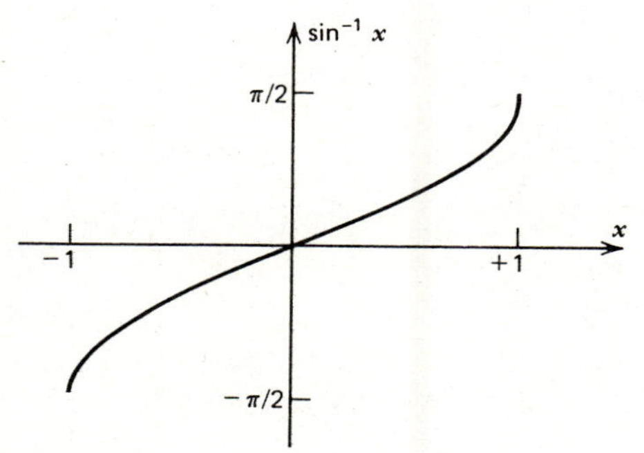

Fig. 16.1

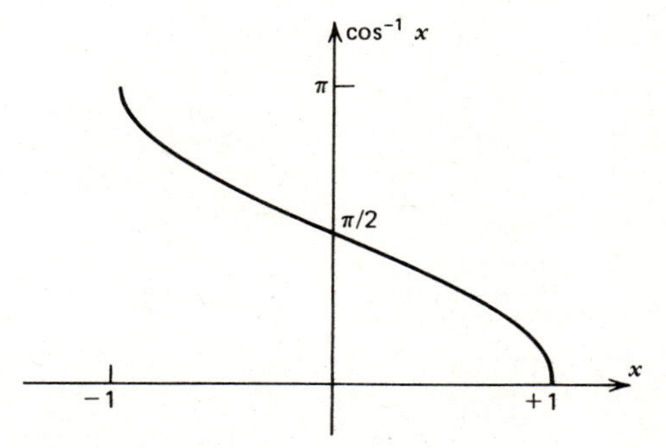

Fig. 16.2

We should make sure that this is correct for real values of z:

$$e^{iw} = \cos w + i \sin w = x + i\sqrt{1 - x^2} \tag{16.10}$$

Since $\sin w > 0$ for $0 < w < \pi$, equation (16.10) checks out, and (16.9) represents the appropriate analytic continuation into the complex plane. Again, $z = \pm 1$ are branch points. From (16.9), it follows that

$$\frac{d}{dz} \cos^{-1} z = -\frac{1}{\sqrt{1 - z^2}} \tag{16.11}$$

The series for $\cos^{-1} z$ can be obtained from the series for $\sin^{-1} z$ by means of the identity

$$\cos^{-1} z = \frac{\pi}{2} - \sin^{-1} z \tag{16.12}$$

The inverse tangent is found from

$$z = -i \frac{e^{iw} - e^{-iw}}{e^{iw} + e^{-iw}} \tag{16.13}$$

The solution for w is

$$\begin{aligned} w &= \tan^{-1} z \\ &= \frac{i}{2} \ln\left(\frac{1 - iz}{1 + iz}\right) \end{aligned} \tag{16.14}$$

The power series for $\tan^{-1} z$ can be found by replacing z by iz in equation (15.44):

$$\tan^{-1} z = z - \tfrac{1}{3}z^3 + \tfrac{1}{5}z^5 - \tfrac{1}{7}z^7 + \cdots \qquad |z| < 1 \tag{16.15}$$

Since both (16.14) and (16.15) give $\tan^{-1}(0) = 0$, they are appropriate representations of the principal branch of $\tan^{-1} z$. For real values of z, $\tan^{-1} z$ is graphed in Fig. 16.3. The derivative of (16.14) is

$$\frac{d}{dz} \tan^{-1} z = \frac{1}{1 + z^2} \tag{16.16}$$

More examples of inverse functions will be found in the problems.

A few more words about the esoteric subject of branch points and cuts need to be said. A *Riemann surface* is a generalization of the z plane into a layered surface to accommodate multiple-valued functions and make them, in effect, single-valued. The function $f(z) = \sqrt{z}$ illustrates the idea. Let a branch cut be made along the x axis as shown in Fig. 16.4. Picture the x, y plane as consisting of two layers

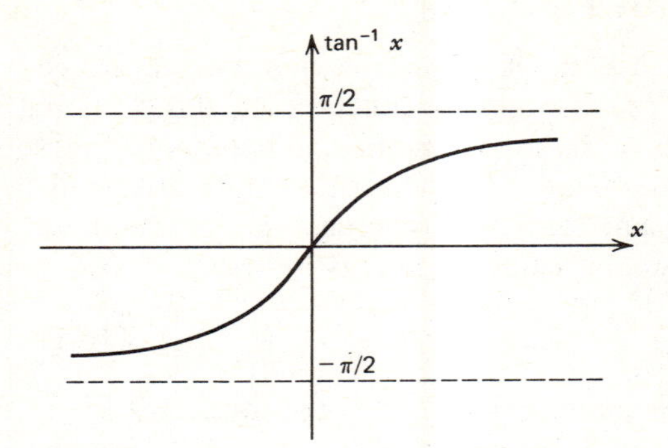

Fig. 16.3

which is called a *Riemann surface of two sheets*. The two sheets are joined along the branch cut. In polar form, the function is $f(z) = \sqrt{r}\, e^{i\theta/2}$. We will trace what happens to the function as we follow the path shown in Fig. 16.4. Start out on the x axis (the branch cut) with $\theta = 0$. Follow the upper path, which is on the upper sheet of the Riemann surface, make one complete circuit around the origin and return to the same physical location in space as the starting point. It is not really the same point because it is on another part of the double layered surface. The function is now $f(z) = \sqrt{r}\, e^{i\pi}$. Continue, this time on the lower sheet, and again encircle the origin, this time returning to the actual starting point. The angle θ has been increased from 0 to 4π and the function has its original value $f(z) = \sqrt{r}\, e^{2\pi i} = \sqrt{r}$. If we wish to further increase θ, we again pass onto the upper sheet.

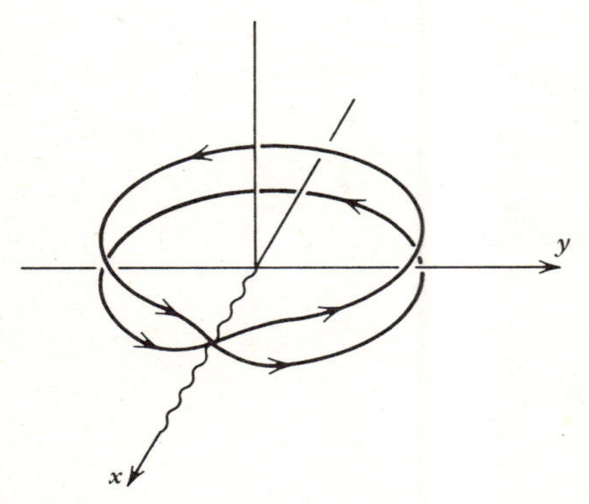

Fig. 16.4

In this manner, $f(z)$ becomes single-valued in the sense that with each point of the Riemann surface one and only one value of the function is associated. When assigning a value to the function at some point of the branch cut itself, it is necessary to remember which sheet of the surface we are on. The function $f(z)$ has now been analytically continued across the branch cut and is single-valued and analytic everywhere except at the origin. The origin itself is still a singular point because we cannot surround it by a neighborhood at each point of which the function is single-valued. Unlike the origin, a point on the branch cut itself can be chosen to be on one sheet or the other and then be completely surrounded by a neighborhood of points which is entirely on that sheet. The basic requirements of continuity and analyticity can then be met.

The imaginary part of $\ln z$ increases indefinitely as θ increases, never returning to its original value. This requires a Riemann surface like a corkscrew with an infinite number of sheets on which $\ln z$ can spiral forever.

PROBLEMS

16.1. Prove equation (16.12).

16.2. Find all the solutions of $\sin w = 2$.

16.3. Find all the possible values of $\tan^{-1}(2i)$, $\tan^{-1}(1+i)$, $\cosh^{-1}(1/2)$, and $\tanh^{-1}(0)$. (See Problem 8.)

16.4. Derive the formulas

$$\sin^{-1} z = \tan^{-1}\left(\frac{z}{\sqrt{1-z^2}}\right) \qquad \cos^{-1} z = \tan^{-1}\left(\frac{\sqrt{1-z^2}}{z}\right) \tag{16.17}$$

16.5. Show that

$$1\cdot 3\cdot 5\ldots(2k-1) = \frac{(2k)!}{2^k k!} \qquad 2\cdot 4\cdot 6\ldots 2k = 2^k k! \tag{16.18}$$

and that the expansion for $\sin^{-1} z$ can be written

$$\sin^{-1} z = \sum_{k=0}^{\infty} \frac{(2k)!z^{2k+1}}{2^{2k}(k!)^2(2k+1)} \tag{16.19}$$

16.6. Show that

$$\cot^{-1} z = \frac{i}{2}\ln\left(\frac{z-i}{z+i}\right) = \frac{1}{z} - \frac{1}{3z^3} + \frac{1}{5z^5} - \cdots \qquad |z|>1 \tag{16.20}$$

For real values of z, the principal branch of $w = \cot^{-1} z$ has the range of values $0 < w < \pi$. Make a sketch of this function. Show also that $\cot^{-1} z = \pi/2 - \tan^{-1} z$.

16.7. Show that

$$\sec^{-1} z = \cos^{-1}\frac{1}{z} \qquad \csc^{-1} z = \sin^{-1}\frac{1}{z} \tag{16.21}$$

16.8. Prove the following:

$$\sinh^{-1} z = \ln\left(z+\sqrt{z^2+1}\right) = \cosh^{-1}\sqrt{z^2+1}$$

$$= z - \frac{1}{2\cdot 3}z^3 + \frac{1\cdot 3}{2\cdot 4\cdot 5}z^5 - \frac{1\cdot 3\cdot 5}{2\cdot 4\cdot 6\cdot 7}z^7 + \cdots \qquad |z|<1 \tag{16.22}$$

$$\frac{d}{dz}\sinh^{-1} z = \frac{1}{\sqrt{z^2+1}} \tag{16.23}$$

$$\cosh^{-1} z = \ln\left(z + \sqrt{z^2 - 1}\right) = \sinh^{-1}\sqrt{z^2 - 1} \tag{16.24}$$

$$\frac{d}{dz}\cosh^{-1} z = \frac{1}{\sqrt{z^2 - 1}} \tag{16.25}$$

$$\tanh^{-1} z = \frac{1}{2}\ln\left(\frac{1+z}{1-z}\right) = z + \frac{1}{3}z^3 + \frac{1}{5}z^5 + \cdots \qquad |z| < 1 \tag{16.26}$$

$$\frac{d}{dz}\tanh^{-1} z = \frac{1}{1 - z^2} \tag{16.27}$$

Hints: In equation 16.2, $x = 2$ is on a branch cut of $(1 - z^2)^{1/2}$. A different value of w results depending on which sheet of the Riemann surface we mean by $x = 2$. Both choices give possible answers. The same thing happens in 16.3 with $\cosh^{-1}\frac{1}{2}$. In $\cosh^{-1} z$, the factor $(z^2 - 1)^{1/2}$ has branch points at $z = \pm 1$ and $z = \frac{1}{2}$ is on the cut line connecting these two points.

16.9. Show that

$$\tan^{-1} z = \frac{\pi}{2} + \frac{1}{2i}\ln\left(\frac{z - i}{z + i}\right) \tag{16.28}$$

Write the factors $z - i$ and $z + i$ in polar form. Show on a diagram how the angles are defined. Make sure that $\tan^{-1}(0) = 0$. If branch cuts are made on the imaginary axis for $|y| > 1$, show that $\tan^{-1} z$ becomes single-valued.

16.10. The function

$$w = \ln\left(z + \sqrt{z^2 - 1}\right) \tag{16.29}$$

represents $\cosh^{-1} z$. We want the function to be defined on the positive real axis for $x > 1$. The function $(z^2 - 1)^{1/2}$ has branch points at ± 1, and it might be concluded that a branch cut from -1 to $+1$ on the real axis is enough. Show however that in order to make w single-valued, the branch cut must extend from $-\infty$ to $+1$.

Hint: Use Fig. 16.5 as a guide in expressing the various factors in (16.29) in polar form.

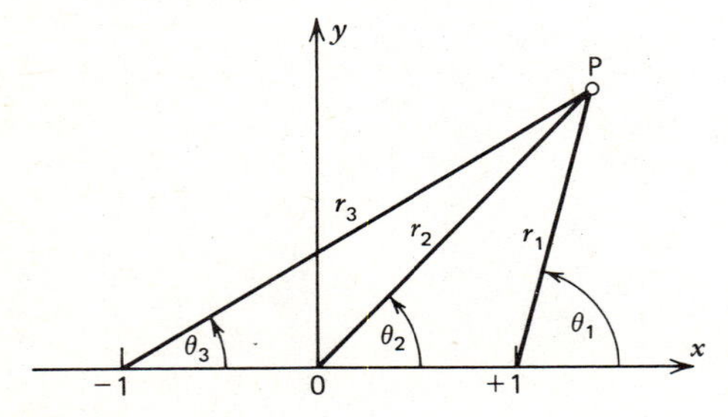

Fig. 16.5

16.11. Discuss the branch cuts which are required to make $\tanh^{-1} z$ single-valued. *Answer:* Branch cuts can be placed on the real axis for $|x| > 1$.

17. TWO-DIMENSIONAL POTENTIAL PROBLEMS

The examples of this section deal with electrostatic potentials and electric fields produced by charged conductors. The conductors are to be viewed as extending indefinitely in a direction perpendicular to the z plane without change of shape so that the potentials and fields depend on only two variables. When we speak of charge or capacitance, it is to be understood that we are talking about these quantities on a per unit length basis measured perpendicular to the z plane.

An analytic function $w(z) = u + iv$ can be viewed as a transformation from the z plane with variables x and y to a new complex plane, called the w plane, with variables u and v. This idea is central to the solution of potential problems by complex variables. An example is provided by

$$w = u + iv = z^2 = r^2 e^{2i\theta} \tag{17.1}$$

This transformation associates each point of the first quadrant of the z plane with a point of the upper half of the w plane. Such a transformation is also called a *mapping*. Corresponding points are called *images* of one another. Note the correspondence of the points (1), (2), and (3) of the z plane with the image points (1)′, (2)′, and (3)′ of the w plane. See Fig. 17.1. The positive real axis of the z plane maps onto the positive real axis of the w plane, whereas the positive imaginary axis of the z plane maps onto the negative real axis of the w plane. If the w plane is viewed as a Riemann surface of two sheets, then the upper half of the z plane maps onto the first sheet of the w plane and the lower half of the z plane maps onto the second sheet of the w plane. In this way, there is a *one-to-one* correspondence between each point of the z plane and a point of the Riemann surface of the w plane. The transformation is singular at the origin because the inverse of equation (17.1), $z = \sqrt{w}$, is not analytic at this point. Singular points have special properties when a mapping is carried out.

A mapping is said to be *conformal* at a point if the mapping does not change the angle between two lines which intersect at the point. If $w(z)$ is analytic at the point z_0 and if $w'(z_0) \neq 0$, then the mapping produced by $w(z)$ is conformal at z_0. To see this, let a line pass through z_0 and let z be another point on this line slightly removed from z_0. By Taylor's series, we have approximately

$$w(z) - w(z_0) = w'(z_0)(z - z_0) \tag{17.2}$$

The points $w(z)$ and $w(z_0)$ lie on the image of the line segment in the w plane. If all the factors in (17.2) are written in polar form, the result is

$$Re^{i\Theta} = ae^{i\alpha} re^{i\theta} \tag{17.3}$$

Thus, the angle between the image line segment and the u axis of the w plane is

$$\Theta = \theta + \alpha \tag{17.4}$$

This means that the image of each line segment which passes through the point z_0 is rotated by the same angle α. Therefore, the angle between two line segments remains unchanged by the transformation. If $w'(z_0) = 0$, then the transformed angle Θ is indeterminate. This is made evident by the transformation $w = z^2$ at the origin. The angle between the x and the y axis is 90°, and the angle between their images in the w plane is 180° so that the conformal

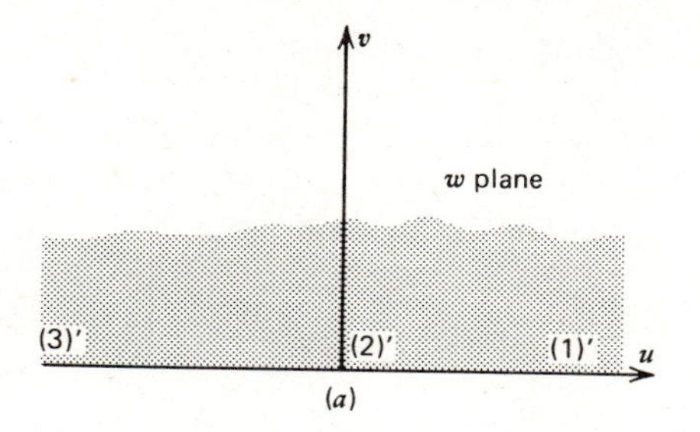

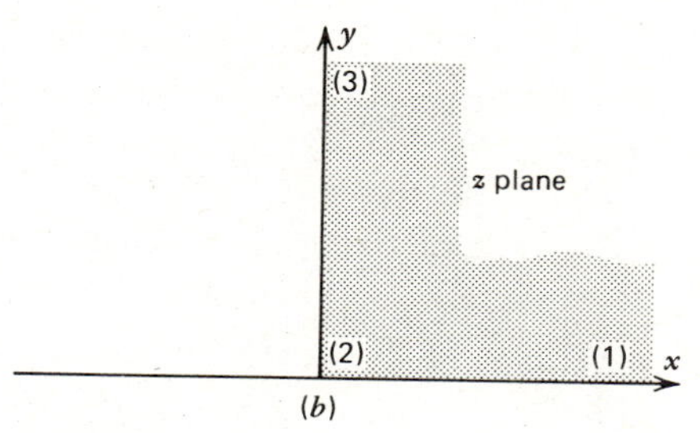

Fig. 17.1

property of the transformation does not hold at this point.

To see the relevance of mapping to the solution of potential problems, let $\psi(u, v)$ be a harmonic function expressed in terms of the variables u and v:

$$\frac{\partial^2 \psi}{\partial u^2} + \frac{\partial^2 \psi}{\partial v^2} = 0 \tag{17.5}$$

Then, there is a function of the complex variable w, $F(w) = \phi + i\psi$, such that ϕ and ψ are conjugate harmonic functions. If ψ is an electrostatic potential, then the curves $\psi =$ constant are equipotentials and the curves $\phi =$ constant represent the field lines. That these two sets of curves are mutually orthogonal is demonstrated by first writing the Cauchy–Riemann equations as

$$\frac{\partial \phi}{\partial u} = \frac{\partial \psi}{\partial v} \qquad \frac{\partial \phi}{\partial v} = -\frac{\partial \psi}{\partial u} \tag{17.6}$$

and then noting that

$$\begin{aligned} \nabla\phi \cdot \nabla\psi &= \frac{\partial \phi}{\partial u}\frac{\partial \psi}{\partial u} + \frac{\partial \phi}{\partial v}\frac{\partial \psi}{\partial v} \\ &= \frac{\partial \phi}{\partial u}\frac{\partial \psi}{\partial u} - \frac{\partial \phi}{\partial u}\frac{\partial \psi}{\partial u} = 0 \end{aligned} \tag{17.7}$$

Let a transformation to the z plane be made. Then $F(w) = F\{w(z)\}$ becomes an analytic function of z, and ϕ and ψ become harmonic functions of x and y. Herein lies the basis of using transformations to solve potential problems. The idea is to try to transform a complicated problem into a simpler one that is more easily solved. Before we see how this works, some attention must be given to boundary conditions and what happens to them when a transformation is made. Section 10 of Chapter 3 deals with the boundary conditions that are required in order to solve a potential problem uniquely. See theorem 9 and problem 10.12 of Chapter 3. We find that the solution of a potential problem in a region requires that either the potential (Dirichlet problem) or its normal derivative (Neumann problem) be known over the bounding surfaces of the region. The simplest case occurs when the bounding surfaces are conductors and are therefore equipotentials. A curve $\psi(u, v) =$ constant in the w plane transforms into the curve $\psi\{u(x, y), v(x, y)\} =$ constant in the z plane so that this type of boundary condition transforms unchanged.

The transformation of the normal (or directional) derivative is more complicated. Let $d\mathbf{s}$ be a line element which is perpendicular to a surface over which the normal derivative

$$\frac{d\psi}{ds} = \frac{\nabla\psi \cdot d\mathbf{s}}{ds} \tag{17.8}$$

has been specified. The total differential $d\psi = \nabla\psi \cdot d\mathbf{s}$ is an invariant:

$$\begin{aligned} d\psi = (\nabla\psi \cdot d\mathbf{s})_w &= \frac{\partial \psi}{\partial u} du + \frac{\partial \psi}{\partial v} dv = (\nabla\psi \cdot d\mathbf{s})_z \\ &= \frac{\partial \psi}{\partial x} dx + \frac{\partial \psi}{\partial y} dy \end{aligned} \tag{17.9}$$

The magnitude of the line element can be calculated in either the z plane or the w plane as

$$\begin{aligned} ds_w &= \sqrt{du^2 + dv^2} = |dw| \\ ds_z &= \sqrt{dx^2 + dy^2} = |dz| \end{aligned} \tag{17.10}$$

Therefore,

$$ds_w = ds_z \left| \frac{dw}{dz} \right| \tag{17.11}$$

and the transformation of the normal derivative to the surface is

$$\frac{d\psi}{ds_z} = \frac{d\psi}{ds_w} \left| \frac{dw}{dz} \right| \tag{17.12}$$

The normal derivative remains perpendicular to the surface under the transformation because of the conformal property.

A simple example is provided by $w = z^2$. Suppose that in the w plane, the entire real axis is at the constant potential V_0. This is accomplished by an infinite conducting sheet which is perpendicular to the w plane and which passes through the u axis. The solution of the potential problem in the upper half of the w plane is

$$\psi = V_0 - av \tag{17.13}$$

where a is a constant. The electric field is given by

$$E_v = -\frac{\partial \psi}{\partial v} = a \qquad E_w = -\frac{\partial \psi}{\partial w} = 0 \tag{17.14}$$

The surface charge density λ on the conductor can be found from

$$E_v = \frac{\lambda}{\varepsilon_0} = a \tag{17.15}$$

The potential is the imaginary part of the analytic function

$$F(w) = iV_0 - aw \tag{17.16}$$

The harmonic function conjugate to ψ is

$$\phi = -au \tag{17.17}$$

The equipotentials are the lines $v =$ constant, and the field lines are given by $u =$ constant. Together they form a rectangular grid in the upper half of the w plane. In order to see what happens to this very simple picture when a transformation to the z plane is made, we substitute $w = z^2$ in (17.16):

$$F(z) = iV_0 - az^2 = a(y^2 - x^2) + i(V_0 - 2axy) \tag{17.18}$$

Thus,

$$\phi = a(y^2 - x^2) \qquad \psi = V_0 - 2axy \tag{17.19}$$

which is the solution of the potential problem produced by two conducting sheets at a potential V_0 which meet at right angles. The positive x and y axes coincide with the conductors, and (17.19) gives the potential at any point in the first quadrant of the z plane. For graphing purposes, it is convenient to plot the equipotentials from $v = 2xy =$ constant and the field lines from $u = x^2 - y^2 =$ constant. Equipotentials and field lines are shown in Fig. 17.2 which result by varying u from -60 to $+60$ in 10-unit increments and by varying v from 0 to 120 also in steps of 10 units. The value $v = 0$ gives the positive x and y coordinate axes. Thus, a uniformly spaced rectangular grid of equipotentials and field lines has been folded into the first quadrant of the z plane. Because of the conformal nature of the transformation, angles between field lines and equipotentials are

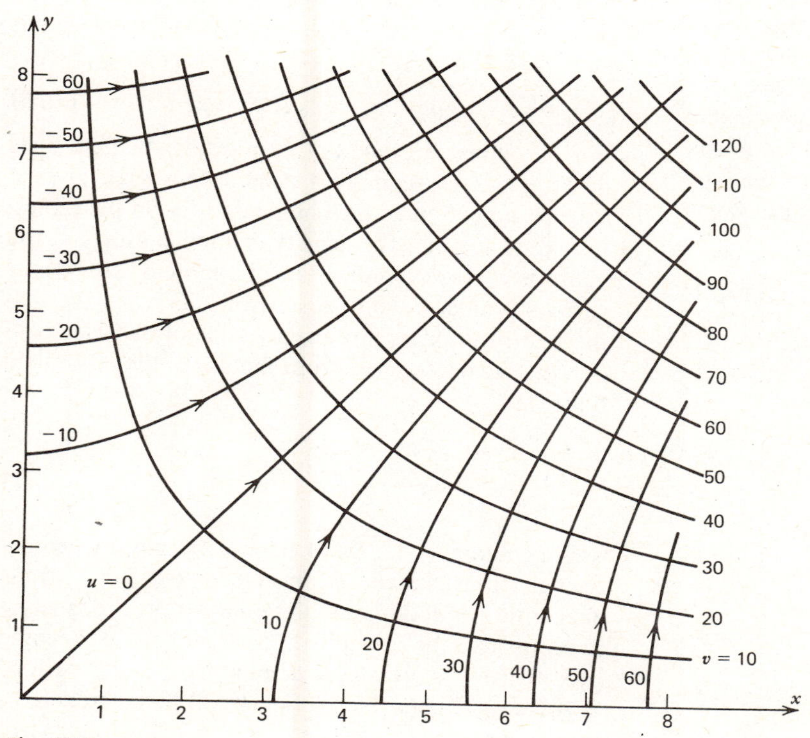

Fig. 17.2

preserved and remain at right angles. The surface charge density on the conducting plate which coincides with the x axis can be found from

$$E_y = -\frac{\partial \psi}{\partial y} = 2ax = \frac{\lambda(x)}{\varepsilon_0} \tag{17.20}$$

The surface charge density is not uniform as it was in the w plane. The nonuniformity of the charge density is also made evident by the unequal spacing of the field lines as they intersect the conducting surfaces. Recall that regions of higher field strength correspond to more densely spaced field lines. The arrows on the field lines indicate the correct direction for a postive surface density of charge. The perpendicularity of field lines and equipotentials appears to break down at the singular point $z = 0$, but the field strength is actually zero at this point. Note from equation (17.19) that the potential is V_0 on the x and y axes and, for positive charge density, decreases from this value as we move away from the conducting plates.

Before the next example is discussed, some useful relations involving charge and capacitance will be derived. Suppose that two conductors at the potentials ψ_1 and ψ_2 form a capacitor as indicated in Fig. 17.3. Two field lines $\phi = \phi_1 =$ constant and $\phi = \phi_2 =$ constant are also shown. If $d\mathbf{s}$ is a displacement vector along the surface of one of the conductors and $\hat{\mathbf{n}}$ is a unit vector normal to its surface, then by Gauss' law the total charge on the conductor between the two field lines is

$$q_{12} = \varepsilon_0 \int_1^2 \mathbf{E}\cdot\hat{\mathbf{n}}\, ds \tag{17.21}$$

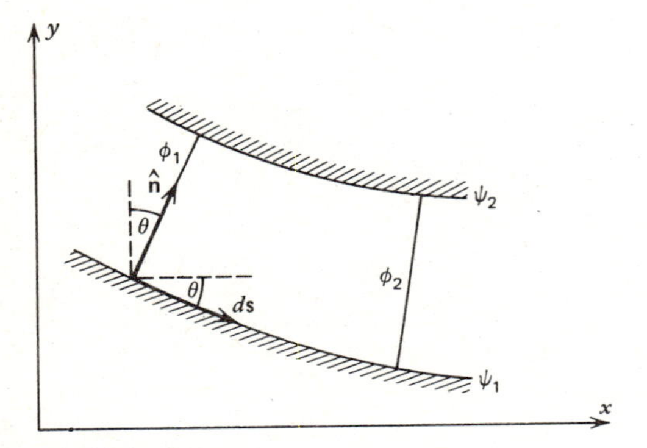

Fig. 17.3

Remember that this is actually charge on a unit length of the conductor measured perpendicular to Fig. 17.3. The electric field $\mathbf{E}$ can be written as the gradient of the potential with the result that

$$q_{12} = -\varepsilon_0 \int_1^2 \left(\frac{\partial \psi}{\partial x} n_x + \frac{\partial \psi}{\partial y} n_y \right) ds \tag{17.22}$$

By means of the Cauchy–Riemann equations (17.6), with u and v replaced by x and y, we find

$$q_{12} = \varepsilon_0 \int_1^2 \left(\frac{\partial \phi}{\partial y} n_x - \frac{\partial \phi}{\partial x} n_y \right) ds \tag{17.23}$$

The geometry of Fig. 17.3 shows that

$$n_x\, ds = \sin\theta\, ds = -dy \qquad n_y\, ds = \cos\theta\, ds = dx \tag{17.24}$$

The charge is then

$$\begin{aligned} q_{12} &= -\varepsilon_0 \int_1^2 \left(\frac{\partial \phi}{\partial y} dy + \frac{\partial \phi}{\partial x} dx \right) = -\varepsilon_0 \int_1^2 d\phi \\ &= -\varepsilon_0(\phi_2 - \phi_1) \end{aligned} \tag{17.25}$$

For example, the charge between two field lines on the conducting sheet which coincides with the x axis of Fig. 17.2 is

$$q_{12} = -\varepsilon_0 a(u_2 - u_1) = \varepsilon_0 a(x_2^2 - x_1^2) \tag{17.26}$$

For one thing, this shows that plotting the field lines of Fig. 17.2 for equally spaced values of u gives a correct indication of the charge density on the plates. It also means that the density of the field lines in the figure is a correct indication of the field strength at various points. Going back to Fig. 17.3, we see that the contribution to the capacitance of the two conductors which comes from between the two field lines is

$$C_{12} = \frac{q_{12}}{\psi_2 - \psi_1} = -\varepsilon_0 \frac{\phi_2 - \phi_1}{\psi_2 - \psi_1} \tag{17.27}$$

We could spend a lot of time exploring the geometry of transformations produced by the various elementary functions. A dictionary of such transformations can be compiled, and we could become progressively more clever at picking transformations to solve specific problems. Space will be taken here to present one more example.

Consider the transformation

$$z=\frac{a}{2\pi}(1+w+e^{w}) \tag{17.28}$$

First note that, if w is real, z is real and that $-\infty<u<+\infty$ maps onto $-\infty<x<+\infty$. The real axis of the w plane therefore maps onto the real axis of the z plane. Next, consider the horizontal line $w=u-i\pi$. The transformation is then

$$z=\frac{a}{2\pi}(1+u-i\pi-e^{u}) \tag{17.29}$$

If u is negative and of large magnitude, then

$$z=\frac{au}{2\pi}-\frac{ia}{2} \tag{17.30}$$

The point (1) of Fig. 17.4*a* has as its image (1)′ of Fig. 17.4*b*. Point (2) is given by $u=0$. Its image (2)′ is at $z=-ia/2$. We therefore find that $-\infty<u<0$, $v=-\pi$ maps onto $-\infty<x<0$, $y=-a/2$. If u is large and positive, then (17.29) gives

$$z=-\frac{a}{2\pi}e^{u}-\frac{ia}{2} \tag{17.31}$$

The point (3) therefore has as its image (3)′, and $0<u<\infty$, $v=-\pi$ also maps onto $-\infty<x<0$, $y=-a/2$. The image points (1)′ and (3)′ are at the same location on the z plane. A similar thing happens to the line $w=u+\pi i$. Note the locations of the images (4)′, (5)′, and (6)′ on the w plane. In effect, the transformation takes a large parallel-plate capacitor of plate separation 2π in the w plane, splits it in half and folds each half of each plate back on itself to form a semi-infinite parallel-plate capacitor of plate separation a in the z plane which has an edge at $x=0$. This is very useful because it will allow us to find the potential and the electric field in a region near the edge of a parallel-plate capacitor. If the plates of the capacitor in the w plane are at potentials $\pm V_0/2$, then the potential at any point between the plates is

$$\psi=\frac{1}{2\pi}V_0 v \tag{17.32}$$

assuming that the top plate is positive. This is a Dirichlet boundary value problem. Since (17.32) is a solution of Laplace's equation and satisfies the boundary conditions, it is the unique solution of the potential problem inside the capacitor. Equation (17.32) is the imaginary part of the analytic function

$$F(w)=\frac{1}{2\pi}V_0 w \tag{17.33}$$

If we were to solve (17.28) for w and substitute it into (17.33), the imaginary part of the resulting analytic function of z would be the potential due to a semiinfinite parallel-plate capacitor of plate separation a with an edge at $x=0$. Equation (17.28) is transcendental and cannot be solved for $w(z)$ in terms of elementary functions. Fortunately, there is no need to do it. The separation of (17.28) into its real and imaginary parts gives

$$x=\frac{a}{2\pi}(1+u+e^{u}\cos v) \tag{17.34}$$

$$y=\frac{a}{2\pi}(v+e^{u}\sin v) \tag{17.35}$$

The curves obtained by setting $v=$ constant and then plotting x and y parametrically as functions of u give the equipotentials. Similarly, setting $u=$ constant and

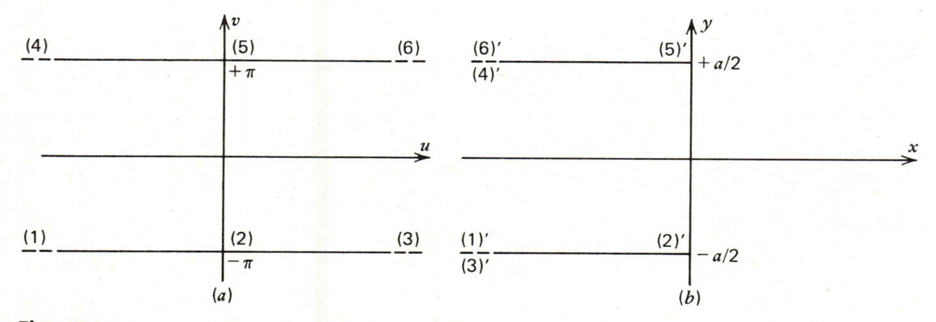

Fig. 17.4

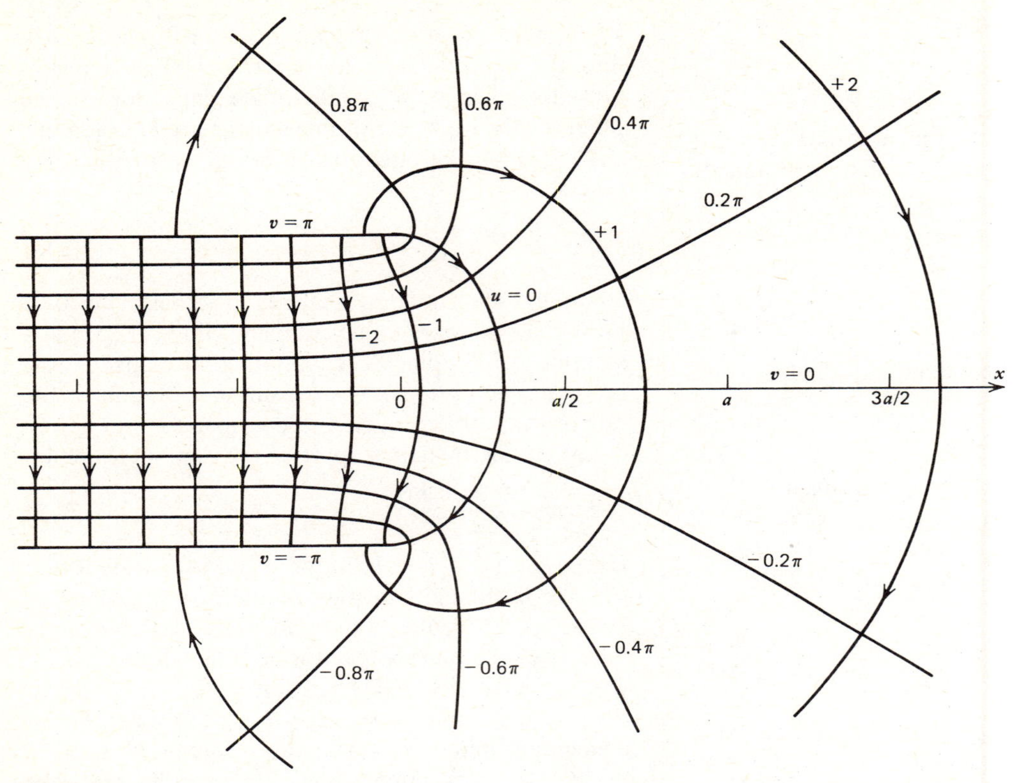

Fig. 17.5

plotting x and y as functions of v gives the field lines. Figure 17.5 was made for $-8 \leq u \leq +2$ in steps of 1 and $-\pi \leq v \leq \pi$ in steps of 0.2π. Since equally spaced values of u have been used, the density of the field lines is a correct indication of the field strength. Note that the surface charge density on the outside of the capacitor plates is quite small by comparison to the density on the inside.

Of interest is the computation of the correction to the capacitance due to the edge effect. Equation (17.33) can be written

$$\frac{1}{2\pi} V_0(u + iv) = \phi + i\psi \tag{17.36}$$

The expression (17.27) for the capacitance can therefore be replaced by

$$C_{12} = -\varepsilon_0 \frac{u_2 - u_1}{v_2 - v_1} \tag{17.37}$$

We will calculate the capacitance due to a length s of the plates as indicated in Fig. 17.6. The plates correspond to the values $v_1 = \pi$ and $v_2 = -\pi$. We start at the field line u_1 between the two plates and move along the *underside* of the positive plate to its edge

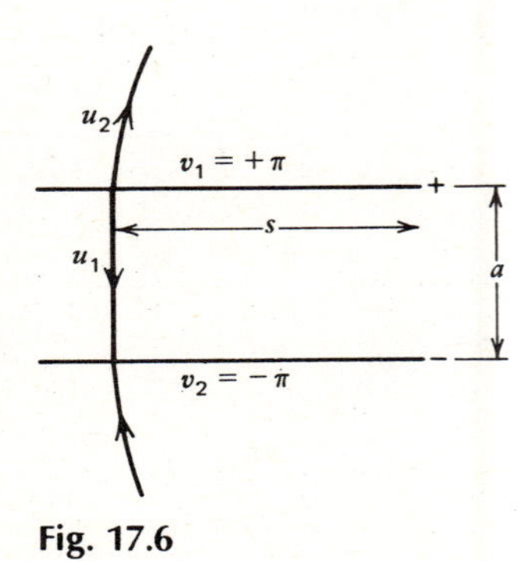

Fig. 17.6

and then back along its *top side* until we encounter the field line u_2, also at $x = -s$. The values of u_1 and u_2 are the roots of

$$-s = \frac{a}{2\pi}(1 + u - e^u) \tag{17.38}$$

which is obtained from (17.34) by setting $x = -s$ and $v = \pi$. At $x = -s$, u_1 is negative and of large magnitude so that (17.38) gives

$$u_1 \simeq -\frac{2\pi s}{a} \tag{17.39}$$

Since u_2 is large and positive, (17.38) gives in this case

$$u_2 = \ln\frac{2\pi s}{a} \tag{17.40}$$

The contribution to the capacitance from a length s of the plates is therefore

$$C \simeq \varepsilon_0 \frac{s}{a} + \frac{\varepsilon_0}{2\pi}\ln\frac{2\pi s}{a} \tag{17.41}$$

which gives the desired correction to the capacitance due to the nonuniform behavior of the field near the edge.

PROBLEMS

17.1. It is possible to represent the gradient of a scalar function in either the z plane or the w plane as

$$\text{grad}_z \psi = \frac{\partial\psi}{\partial x} + i\frac{\partial\psi}{\partial y} \qquad \text{grad}_w \psi = \frac{\partial\psi}{\partial u} + i\frac{\partial\psi}{\partial v} \tag{17.42}$$

Prove that

$$\text{grad}_z \psi = \text{grad}_w \psi\left(\frac{\partial u}{\partial x} - i\frac{\partial v}{\partial x}\right) \tag{17.43}$$

and that

$$|\text{grad}_z \psi| = |\text{grad}_w \psi|\left|\frac{dw}{dz}\right| \tag{17.44}$$

17.2. Let the function $w(z)$ be analytic at the point z_0 and assume that $w'(z_0) \neq 0$. Show that there exists a neighborhood of the point $w_0 = w(z_0)$ in the w plane in which $w(z)$ has a unique inverse $z(w)$ and that $z(w)$ is analytic at w_0.

Hint: Consider the problem of solving $u = u(x, y)$ and $v = v(x, y)$ for x and y in terms of u and v. What is the Jacobian of the transformation?

17.3. For the parallel-plate capacitor of Fig. 17.5, show that the electric field at any point is

$$E_x = \frac{V_0 \sin v}{2a(\cosh u + \cos v)} \qquad E_y = -\frac{V_0(e^{-u} + \cos v)}{2a(\cosh u + \cos v)} \tag{17.45}$$

Hint: Since (17.34) and (17.35) cannot be solved in an elementary way for u and v in terms of x and y, it is necessary to use *implicit differentiation* whereby the equations are differentiated as they stand with respect to x and y. In the differentiation, u and v are to be regarded as functions of x and y. The Cauchy–Riemann equations can be used to eliminate $\partial u/\partial x$ and $\partial u/\partial y$. The resulting equations are then solved algebraically for $\partial v/\partial x$ and $\partial v/\partial y$.

Prove that the surface charge density on the upper plate is

$$\lambda = \frac{\varepsilon_0 V_0}{a}\left|\frac{e^{-u} - 1}{e^u + e^{-u} - 2}\right| \tag{17.46}$$

What is the value of λ at $u = 0$? If you plot $\lambda a/(\varepsilon_0 V_0)$ as a function of x, it looks like Fig. 17.7. Figure 17.7*a* is for the interior side of the top plate and Fig. 17.7*b* is for the outside.

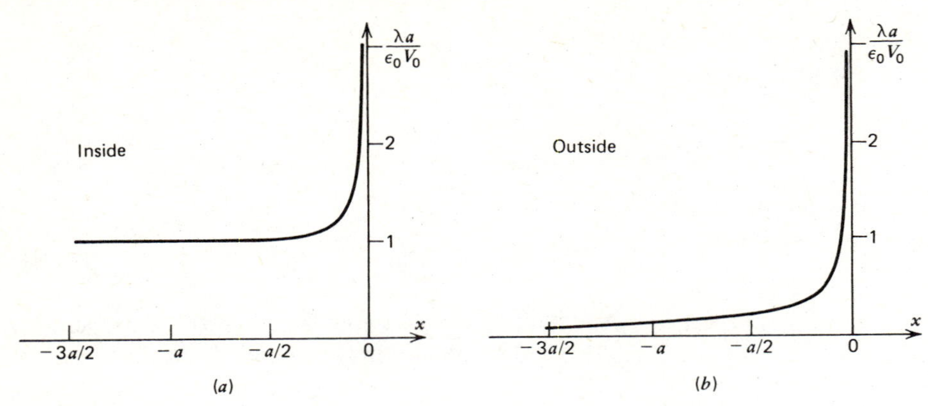

Fig. 17.7

17.4. For the parallel-plate capacitor, show that the direction of the electric field at any point with respect to the x axis is given by

$$\tan\theta = -\frac{e^{-u} + \cos v}{\sin v} \tag{17.47}$$

What is θ at $v = \pi$ if $u \neq 0$? What is it if $u = 0$? What is significant about the field line given by $u = 0$?

17.5. Either the real or the imaginary part of an analytic function can be used to represent an electrostatic potential. Derive the version of equation (17.27) that is appropriate for $F(z) = \psi + i\phi$. Remember to exchange the roles of ϕ and ψ in the Cauchy–Riemann equations (17.6).

17.6. The potential at any point between two charged coaxial conducting cylinders of radii a and b is $\psi = A \ln r + B$, where A and B are constants. Find an analytic function the imaginary part of which is ψ. Show that (17.27) gives the correct capacitance per unit length of the cylinders.

17.7. Show that the transformation

$$z = a\frac{i - w}{i + w} \tag{17.48}$$

maps the entire real axis of the w plane onto a circle of radius a centered at the origin in the z plane. The entire upper half of the w plane goes inside the circle. Suppose that the negative real axis of the w plane is at the potential V_0 and that the positive real axis is grounded. Show that in the upper half of the w plane, the imaginary part of $F(w) = A \ln w$ is an appropriate potential. Evaluate the constant A in terms of V_0. Write down an explicit formula for the transformed potential in the z plane. Where is it valid and what boundary conditions does it fit?

17.8. A parallel-plate capacitor is built out of two square conducting plates that are separated by 1 cm and are 10 cm on a side. Estimate the percent error in the capacitance if the edge effect is neglected.

18. INTEGRATION IN THE COMPLEX PLANE

Let C be a path joining the points z_0 and z_1 in the complex plane and let $f(z)$ be defined at each point of C. The definite integral of $f(z)$ along this path is

$$\int_{z_0}^{z_1} f(z)\,dz = \int_{z_0}^{z_1} (u+iv)(dx+i\,dy)$$
$$= \int_{z_0}^{z_1} (u\,dx - v\,dy) + i\int_{z_0}^{z_1} (v\,dx + u\,dy) \tag{18.1}$$

In this manner, a definite integral in the complex plane can be defined in terms of real integrals. The integral (18.1) is called a *line integral*, or a *curvilinear integral*. In general, its value depends on the path between the points z_0 and z_1.

As an example, consider the evaluation of

$$I = \int_{(0,0)}^{(1,1)} z^*\,dz = \int_{(0,0)}^{(1,1)} (x\,dx + y\,dy)$$
$$+ i\int_{(0,0)}^{(1,1)} (x\,dy - y\,dx) \tag{18.2}$$

between the points (0,0) and (1,1) over the two possible paths (a) and (b) shown in Fig 18.1. Path (a) is a straight line connecting the two points over which $y = x$ and $dy = dx$. The integral is

$$I_a = 2\int_{x=0}^{1} x\,dx = 1 \tag{18.3}$$

Path (b) consists of the x axis between (0,0) and (1,0) and a straight line parallel to the y axis between (1,0) and (1,1). Along this path, the integral is

$$I_b = \int_{x=0}^{1} x\,dx + \int_{y=0}^{1} y\,dy + i\int_{y=0}^{1} dy = 1 + i \tag{18.4}$$

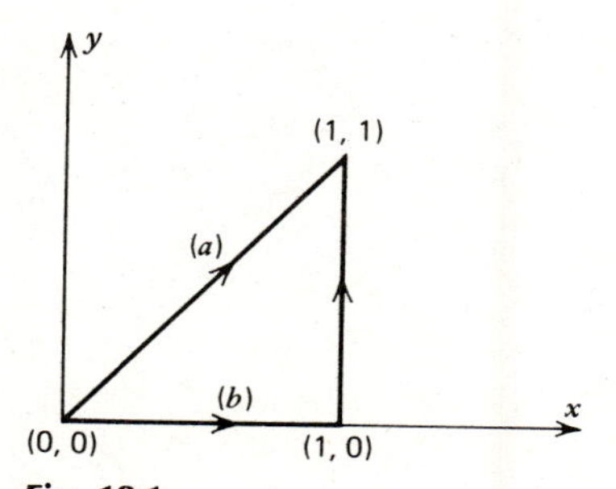

Fig. 18.1

which shows that a different value is found for each path. By contrast, one finds that

$$\int_{(0,0)}^{(1,1)} z\,dz = i \tag{18.5}$$

for either path (a) or (b). Apparently, the integral of some functions depends on the path, and for others it does not.

Theorem 1. Cauchy's Theorem. If $f(z)$ is analytic in a simple connected region R, then the integral

$$F(z_1) - F(z_0) = \int_{z_0}^{z_1} f(z)\,dz \tag{18.6}$$

is independent of the path joining z_0 and z_1, provided that it lies entirely in R. In particular, for a closed path lying entirely in R,

$$\oint f(z)\,dz = 0 \tag{18.7}$$

Moreover, $F(z)$ is an analytic function and

$$F'(z) = f(z) \tag{18.8}$$

Proof: At the end of section 8 in Chapter 3 are listed five conditions, each of which is both necessary and sufficient for a continuous and differentiable vector field to be conservative in a simply connected region of space. Review these conditions and also the discussion of simply connected regions. If in the real part of (18.1) we put $a_x = u$ and $a_y = -v$ and in the imaginary part $b_x = v$ and $b_y = u$, the result is

$$\int_{z_0}^{z_1} f(z)\,dz = \int_{z_0}^{z_1} \mathbf{a}\cdot d\mathbf{s} + i\int_{z_0}^{z_1} \mathbf{b}\cdot d\mathbf{s} \tag{18.9}$$

The integrability conditions $\nabla \times \mathbf{a} = 0$ and $\nabla \times \mathbf{b} = 0$ come out to be the Cauchy–Riemann equations which are known to be valid because of the assumption that $f(z)$ is analytic. This proves that the integral is independent of the path.

Because of the independence of path, the function $F(z)$ is uniquely defined once $F(z_0)$ is given. If we write $F(z_1) - F(z_0) = U + iV$, then

$$\int_{z_0}^{z_1} (u\,dx - v\,dy) = U \qquad \int_{z_0}^{z_1} (v\,dx + u\,dy) = V \tag{18.10}$$

It is therefore true that

$$u\,dx - v\,dy = dU = \frac{\partial U}{\partial x}dx + \frac{\partial U}{\partial y}dy$$

$$v\,dx + u\,dy = dV = \frac{\partial V}{\partial x}dx + \frac{\partial V}{\partial y}dy \qquad (18.11)$$

By comparison of the coefficients of dx and dy, we find that

$$u = \frac{\partial U}{\partial x} = \frac{\partial V}{\partial y} \qquad v = -\frac{\partial U}{\partial y} = \frac{\partial V}{\partial x} \qquad (18.12)$$

which shows that the Cauchy–Riemann equations are valid for U and V. This establishes that $F(z)$ is analytic and that

$$u + iv = f(z) = \frac{\partial U}{\partial x} + i\frac{\partial V}{\partial x} = F'(z) \qquad (18.13)$$

Consequently, as far as analytic functions are concerned, integration is the *antiderivative* or reverse of differentiation just as it is for functions of a real variable. It makes sense to think of

$$F(z) = \int f(z)\,dz + \text{constant} \qquad (18.14)$$

as the indefinite integral of $f(z)$.

As an example, since $f(z) = z$ is an analytic function, (18.5) can be evaluated as

$$\int_{(0,0)}^{(1,1)} z\,dz = \left.\frac{1}{2}z^2\right|_{(0,0)}^{(1,1)} = \frac{1}{2}(1+i)^2 = i \qquad (18.15)$$

without reference to any particular path connecting (0,0) and (1,1).

Figure 18.2b shows a multiply connected region R which has the property that every closed path in R cannot be continuously shrunk down to a point without passing out of the region. If $f(z)$ is single-valued and analytic in R as well as on the boundaries of R, then it is true that the sum of the line integrals over all boundaries of the region is zero. For the region illustrated in Fig. 18.2b, this is

$$\oint_{C_1} f(z)\,dz + \oint_{C_2} f(z)\,dz + \oint_{C_3} f(z)\,dz = 0 \qquad (18.16)$$

The truth of the statement is made evident if the multiply connected region of Fig. 18.2b is viewed as a limiting case of the simply connected region of Fig. 18.2a. Note carefully the directions in which the integrals over the various closed paths are carried out in (18.16). An integral around a closed path is generally called a *contour integral*. It is the usual practice to think of counterclockwise as the positive direction around a closed path. It is possible to rewrite (18.16) with all the integrals taken in the same direction as

$$\oint_{C_1} f(z)\,dz = \oint_{C_2} f(z)\,dz + \oint_{C_3} f(z)\,dz \qquad (18.17)$$

Integration is essentially a summation process, and as an extension of the triangle inequality $|z_1 + z_2| \le |z_1| + |z_2|$, we have

$$\left|\int f(z)\,dz\right| \le \int |f(z)|\,|dz| \qquad (18.18)$$

If $M = \max|f(z)|$ and L is the length of the path of integration, then

$$\left|\int f(z)\,dz\right| \le ML \qquad (18.19)$$

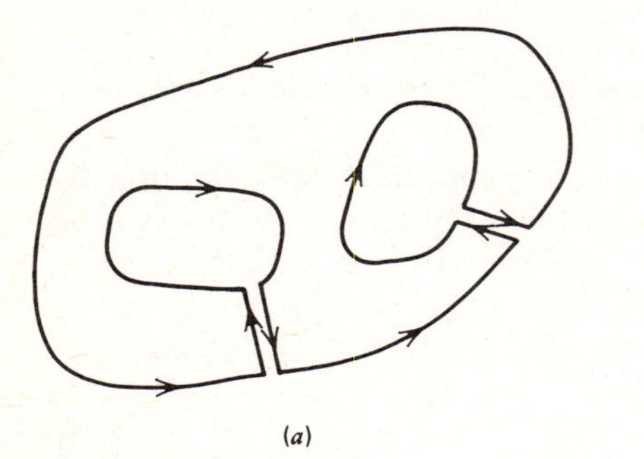
(a)

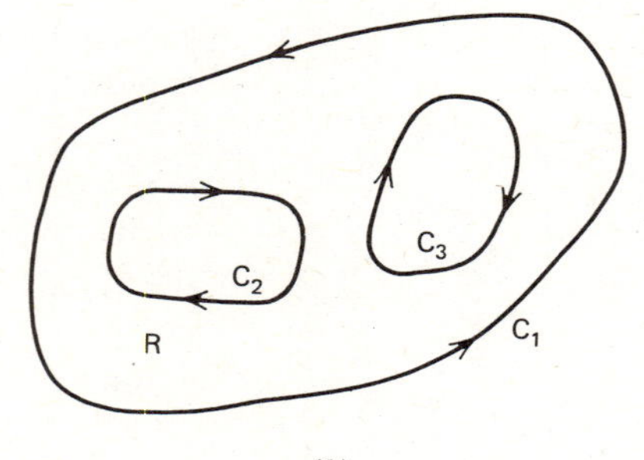

(b)

Fig. 18.2

which provides a crude but useful estimate of the upper bound of the magnitude of an integral.

Before going on to the next theorem, we will consider another simple but important example. The function $f(z)=1/z$ is single-valued and analytic throughout the simply connected region R shown in Fig. 18.3. The function is singular at the origin, but this point is not in R. Consider the two points (1) and (2), which are just slightly apart. Point (1) is just above the positive real axis at $z_1 = re^{i\varepsilon}$, and point (2) is just below it at $z_2 = re^{i(2\pi-\varepsilon)}$, where ε is a small angle. Both points are in R. Since the positive real axis has been excluded from R, it is necessary to use a path such as C in order to connect the two points by a path which lies entirely in R. We will evaluate the integral of $f(z)$ along C. Since the path lies entirely in a simply connected region in which the function is analytic, it can be circular or some other shape without affecting the value of the integral. The integral is

$$\int_1^2 \frac{dz}{z} = \ln \frac{z_2}{z_1} = \ln\left(\frac{re^{(2\pi-\varepsilon)i}}{re^{i\varepsilon}}\right) = (2\pi - 2\varepsilon)i \tag{18.20}$$

Now let the region R come together along the real axis to form a multiply connected annular region as shown in Fig. 18.4. The points (1) and (2) come together on the x axis, $\varepsilon = 0$, and

$$\oint \frac{dz}{z} = 2\pi i \tag{18.21}$$

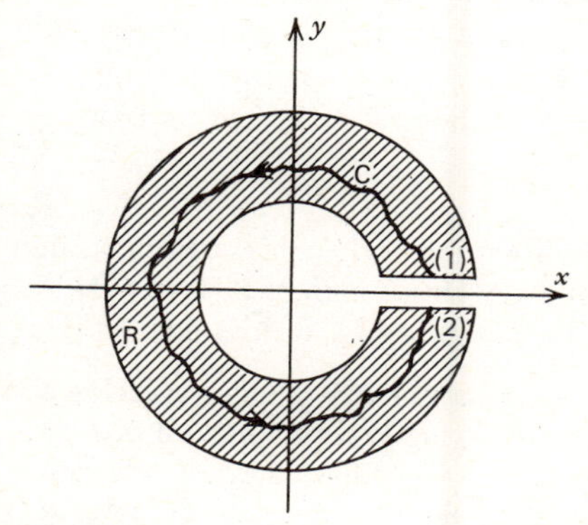

Fig. 18.3

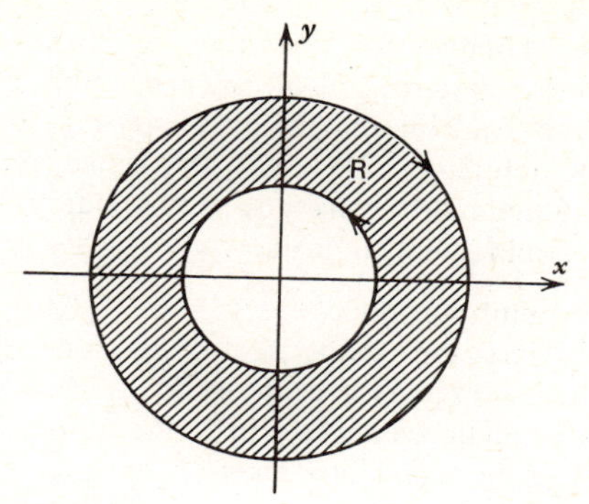

Fig. 18.4

The integral can now be on any closed path in R which encircles the origin. For any other closed path in R, $\oint(1/z)\,dz = 0$. Let's look at the problem another way. Start out by considering $\oint(1/z)\,dz$ around some contour which encloses the origin. Since $1/z$ is not analytic everywhere inside the contour, the conditions of Cauchy's theorem are violated and the integral will not necessarily be zero. Since $1/z$ is actually analytic at all points of the contour itself, it can be deformed into a circle of radius r about the origin without changing the value of the integral. It is then convenient to express z in polar form as $z = re^{i\theta}$. Since $r =$ constant, $dz = re^{i\theta} i\, d\theta$ and the integral is

$$\oint \frac{dz}{z} = \int_0^{2\pi} \frac{re^{i\theta} i\, d\theta}{re^{i\theta}} = \int_0^{2\pi} i\, d\theta = 2\pi i \tag{18.22}$$

For the multiply connected region of Fig. 18.4, $\oint(1/z)\,dz = 0$ when taken around the *entire* boundary because the inner circle gives a contribution $+2\pi i$ and the outer circle gives $-2\pi i$. This is expected because $1/z$ is single-valued and analytic everywhere on the circular boundaries and in the annular region between them.

Theorem 2. The Cauchy Integral Formula. If $f(z)$ is analytic within and on a closed contour C and if z_0 is any point inside C, then

$$f(z_0) = \frac{1}{2\pi i}\oint_C \frac{f(z)\,dz}{z - z_0} \tag{18.23}$$

where the integral is taken in the positive (counterclockwise) sense around the contour. A significant feature of the Cauchy integral formula is that it shows that, once the values of an analytic function are known on a closed contour, the function is determined at all interior points. This is no surprise, since the real and the imaginary parts of $f(z)$ are harmonic functions. We know that once a harmonic function (a potential) is known on the boundary of a region, it is determined uniquely inside.

Proof: Consider the multiply connected region R of Fig. 18.5. The outer boundary is the contour C, and the inner boundary C_0 is a circle of radius δ centered at z_0. The integrand of (18.23) is analytic everywhere in R and on the contours C and C_0. If the integrals are taken in the same sense around both C and C_0, then

$$\frac{1}{2\pi i}\oint_C \frac{f(z)\,dz}{z-z_0} = \frac{1}{2\pi i}\oint_{C_0} \frac{f(z)\,dz}{z-z_0}$$
$$= \frac{f(z_0)}{2\pi i}\oint_{C_0} \frac{dz}{z-z_0}$$
$$+\frac{1}{2\pi i}\oint_{C_0} \frac{f(z)-f(z_0)}{z-z_0}\,dz. \quad (18.24)$$

By the same procedure that was used to obtain (18.22), this time with $z-z_0=\delta e^{i\theta}$, we find

$$\oint_{C_0} \frac{dz}{z-z_0} = 2\pi i \quad (18.25)$$

The magnitude of the remaining integral can be estimated by means of (18.19):

$$\left|\oint \frac{f(z)-f(z_0)}{z-z_0}\,dz\right| = \left|\oint \{f(z)-f(z_0)\}\, i\, d\theta\right|$$
$$\leq \oint |f(z)-f(z_0)||d\theta| \leq 2\pi\varepsilon \quad (18.26)$$

where $\varepsilon = \max|f(z)-f(z_0)|$. Since $f(z)$ is continuous, $\varepsilon \to 0$ as $\delta \to 0$. The Cauchy formula (18.23) then follows.

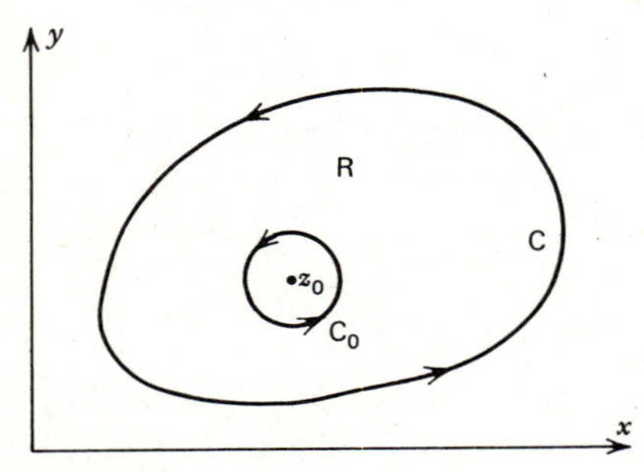

Fig. 18.5

Theorem 3. If a function is single-valued and analytic at a point, then its derivatives of all orders exist and are also analytic functions at that point.

Proof: Formally, the result follows by differentiating the Cauchy integral formula with respect to z_0:

$$f'(z_0) = \frac{1}{2\pi i}\oint_C \frac{f(z)\,dz}{(z-z_0)^2} \quad (18.27)$$

$$f''(z_0) = \frac{2!}{2\pi i}\oint_C \frac{f(z)\,dz}{(z-z_0)^3} \quad (18.28)$$

$$f^{(n)}(z_0) = \frac{n!}{2\pi i}\oint_C \frac{f(z)\,dz}{(z-z_0)^{n+1}} \quad (18.29)$$

Equation (18.27) can be justified in a somewhat more elaborate fashion. With the aid of (18.23), we can derive the identity

$$\frac{f(z_0+\Delta z_0)-f(z_0)}{\Delta z_0} - \frac{1}{2\pi i}\oint_C \frac{f(z)\,dz}{(z-z_0)^2}$$
$$= \frac{\Delta z_0}{2\pi i}\oint_C \frac{f(z)\,dz}{(z-z_0-\Delta z_0)(z-z_0)^2} \quad (18.30)$$

There is not much doubt that the last integral goes to zero in the limit $\Delta z_0 \to 0$. Therefore,

$$\lim_{\Delta z_0 \to 0} \frac{f(z_0+\Delta z_0)-f(z_0)}{\Delta z_0} = f'(z_0)$$
$$= \frac{1}{2\pi i}\oint_C \frac{f(z)\,dz}{(z-z_0)^2} \quad (18.31)$$

The higher derivatives can be similarly justified.

If $f(z)=u+iv$ is analytic, then, as a consequence of theorem 3, the existence and continuity of all the higher partial derivatives of u and v is guaranteed. In developing potential theory, we somewhat prematurely assumed the existence of the second partial derivatives.

Theorem 4. Morera's Theorem. If $f(z)$ is continuous in R and $\oint f(z)\,dz=0$ for every closed path in R, then $f(z)$ is analytic in R. This is the converse of Cauchy's theorem.

Proof: The statement that $\oint f(z)\,dz=0$ for every closed path in R is equivalent to saying that the integral of $f(z)$ between any two points in R is independent of the path which connects them. Equation (18.6) then becomes a definition of $F(z)$. Equations (18.10) through (18.13) then follow. Note that none of these equations assumes anything about $f(z)$ being analytic. Since, however, $f(z) = F'(z)$, $f(z)$ is analytic by theorem 3.

Theorem 5. Liouville's Theorem. If $f(z)$ is analytic everywhere and $|f(z)|$ is bounded for all values of z, then $f(z)$ is a constant.

Proof: Let $M = \max|f(z)|$. Let the contour in equation (18.27) be a circle of radius r centered at z_0. Write $z - z_0 = re^{i\theta}$ and use (18.19) to get

$$|f'(z_0)| \le \frac{1}{2\pi}\oint \frac{|f(z)|}{r}|d\theta| \le \frac{M}{r} \tag{18.32}$$

Now let $r \to \infty$ to get $f'(z_0) = 0$. Since this is true for all finite values of $z_0, f(z_0) =$ constant and the theorem is proved. A function which is analytic everywhere is called an *integral function*. The function $\sin z$ is an integral function, but it is not a constant because it is not bounded at ∞.

Theorem 6. The Fundamental Theorem of Algebra. A polynomial of degree n in z has exactly n roots and can be factored as

$$\begin{aligned} P_n(z) &= z^n + a_{n-1}z^{n-1} + \cdots + a_0 \\ &= (z - z_1)(z - z_2)\ldots(z - z_n) \end{aligned} \tag{18.33}$$

Proof: Consider the function

$$f(z) = \frac{1}{P_n(z)} \tag{18.34}$$

As $z \to \infty$, $|f(z)| \to 0$. Assume that $P_n(z)$ has no roots. Then, $f(z)$ is everywhere analytic and is bounded for all values of z. By theorem 5, it must be a constant. Since this is a contradiction, $P(z)$ must have at least one root z_1. The factor $z - z_1$ can be taken out of $P_n(z)$ to give

$$P_n(z) = (z - z_1)P_{n-1}(z) \tag{18.35}$$

where $P_{n-1}(z)$ is a polynomial of degree $n-1$. The process can be repeated with $P_{n-1}(z)$. The procedure is continued until (18.33) is obtained.

PROBLEMS

18.1. Prove that $\oint(1/z^2)\,dz = 0$ if the integral is taken around any closed path, including a path which encircles the singular point $z = 0$. The only restriction is that the path must not pass through $z = 0$.

18.2. Evaluate

$$I = \int_{-i}^{+i}\frac{dz}{z}$$

for paths (a) and (b) of Fig. 18.6.

Hint: Either don't cross any branch lines of $\ln z$, or if you do, remember which Riemann sheet you are on. Try putting the branch line on the positive imaginary axis. Then for path (a), $i = e^{i\pi/2}$, whereas for (b) $i = e^{-3\pi i/2}$. *Answer:* $I_a = \pi i$, $I_b = -\pi i$.

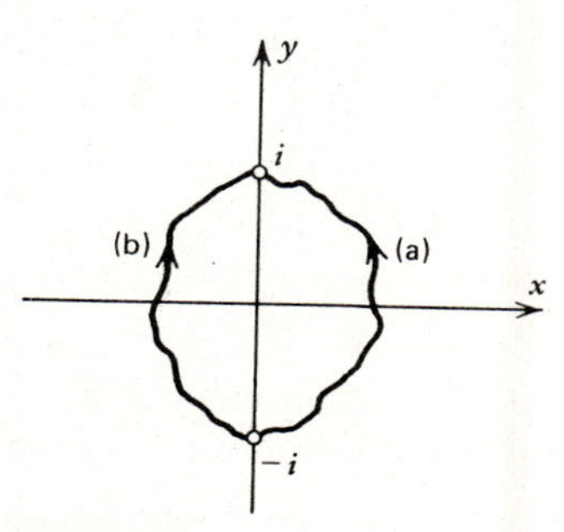

Fig. 18.6

18.3. Evaluate

$$I = \int_0^{1+i} (x - y + ix^2)\, dz$$

for paths (a) and (b) of Fig. 18.7.

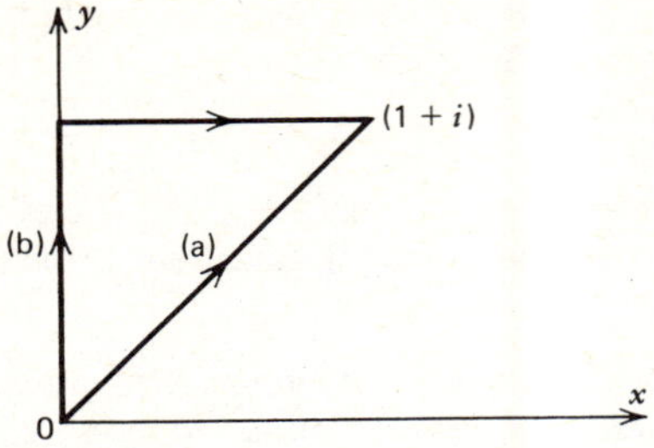

18.4. Evaluate

$$I = \int_0^{1+i} (x^2 - y^2 + 2ixy)\, dz$$

along any path joining 0 and $1 + i$.

Fig. 18.7

18.5. Find the value of

$$\oint \frac{dz}{\sin z}$$

for any closed path in the region $|z| < \pi$ which encircles the origin.

Hint: The path can be deformed into a small circle around the origin.

18.6. Find the value of

$$I = \oint \frac{2z^2 - z + 1}{z - 1}\, dz$$

for any closed path which encircles the point $z = 1$.

18.7. Use the Cauchy integral formula to show that the value of a harmonic function at the center of a circle equals its average value over the circumference of the circle, that is,

$$u_0 = \frac{1}{2\pi} \int_0^{2\pi} u\, d\theta$$

This is the two-dimensional version of theorem 6 in section 10 of Chapter 3.

18.8. Evaluate

$$I = \oint \frac{\sin z}{(z-1)^2}\, dz$$

for any closed path which encircles the point $z = 1$.

18.9. Evaluate

(a) $\displaystyle\int_0^{\pi + 2i} \cos \frac{z}{2}\, dz$

(b) $\displaystyle\int_{1-\pi i}^{1+\pi i} \sinh 2z\, dz$

18.10. Establish the indefinite integral

$$\int z^n \ln z\, dz = \frac{z^{n+1}}{n+1} \ln z - \frac{z^{n+1}}{(n+1)^2}$$

Hint: Integrate by parts.

19. TAYLOR AND LAURENT EXPANSIONS

Sections 5 and 8 of this chapter contain detailed discussions of power series. In this section, we will derive Taylor's series once again, this time in a formal manner from Cauchy's integral formula. In this manner, we are able to show that all analytic functions can be represented by a power series.

Suppose that $f(z)$ is analytic at the point z_0 and that z_s is the point nearest to z_0 where $f(z)$

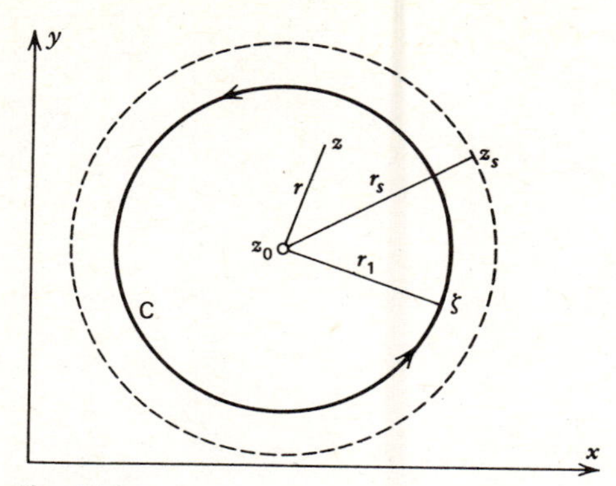

Fig. 19.1

has a singularity. The function $f(z)$ is then analytic at every point inside the circular region $|z - z_0| < |z_s - z_0|$, which is the interior of the dashed circle of Fig. 19.1. The function $f(z)$ may, of course, be analytic in other regions of the z plane as well. Let C be a circular contour of radius r_1 with center at z_0. Note that $r_1 < |z_s - z_0|$. If z is a point inside C, then by Cauchy's integral formula (18.23),

$$f(z) = \frac{1}{2\pi i}\oint_C \frac{f(\zeta)\,d\zeta}{\zeta - z} \tag{19.1}$$

where ζ is the variable point of integration on the contour C. From equations (4.1) and (4.2), the finite geometrical series plus remainder can be expressed

$$\frac{1}{1-x} = 1 + x + x^2 + \cdots + x^n + \frac{x^{n+1}}{1-x} \tag{19.2}$$

In the integrand of (19.1), it is possible to express the factor $1/(\zeta - z)$ as

$$\frac{1}{\zeta - z} = \frac{1}{\zeta - z_0 - (z - z_0)} = \frac{1}{(\zeta - z_0)\left(1 - \dfrac{z - z_0}{\zeta - z_0}\right)} \tag{19.3}$$

which can then be expanded by means of (19.2) to get

$$\frac{1}{\zeta - z} = \frac{1}{\zeta - z_0}\sum_{k=0}^{n}\left(\frac{z - z_0}{\zeta - z_0}\right)^k + \frac{1}{\zeta - z}\left(\frac{z - z_0}{\zeta - z_0}\right)^{n+1} \tag{19.4}$$

Note from the geometry of Fig. 19.1 that $|z - z_0| = r$, $|\zeta - z_0| = r_1$, and $r < r_1$. This means that an absolutely and uniformly convergent series is obtained from (19.4) in the limit $n \to \infty$. Equations (19.1) and (19.4) can be combined and term-by-term integration carried out. There is no problem about convergence because the series involved is finite. The result is

$$\begin{aligned} f(z) &= \sum_{k=0}^{n}(z - z_0)^k \frac{1}{2\pi i}\oint_C \frac{f(\zeta)\,d\zeta}{(\zeta - z_0)^{k+1}} + R_n \\ &= \sum_{k=0}^{n}\frac{f^{(k)}(z_0)}{k!}(z - z_0)^k + R_n \end{aligned} \tag{19.5}$$

where use has been made of (18.29) and

$$R_n = (z - z_0)^{n+1}\frac{1}{2\pi i}\oint_C \frac{f(\zeta)\,d\zeta}{(\zeta - z)(\zeta - z_0)^{n+1}} \tag{19.6}$$

is the remainder term. Equation (19.5) is recognized as the familiar Taylor's series expansion. One advantage that has been gained over our previous treatment of power series is an expression for the remainder which is easy to use. The magnitude of the remainder is estimated by first writing $|\zeta - z| = |\zeta - z_0 - (z - z_0)| \geq r_1 - r$ which follows from the result of problem 3.7. If $M = \max|f(\zeta)|$ on the contour C, (18.19) gives

$$|R_n| < \frac{Mr_1}{r_1 - r}\left(\frac{r}{r_1}\right)^{n+1} \tag{19.7}$$

As previously noted, $r < r_1$. This means that $R_n \to 0$ as $n \to \infty$ and

$$f(z) = \sum_{k=0}^{\infty}\frac{f^{(k)}(z_0)}{k!}(z - z_0)^k \tag{19.8}$$

We find therefore that any function which is analytic at a point z_0 can be represented by a power series which converges uniformly and absolutely in a circular region with z_0 as its center. The radius of convergence of the series is determined by the distance from z_0 to the nearest singular point of the function. Theorem 7 of section 5 shows that the power series representation of a function is unique. Review also at this time theorem 3 of section 8 and problem 8.7.

As an example, equation (14.38) is a power series about the point $x = 0$, which defines the Bernoulli numbers. Extended into the complex plane, this expansion is

$$\frac{z}{e^z - 1} = \sum_{k=0}^{\infty} \frac{B_k}{k!} z^k \qquad |z| < 2\pi \tag{19.9}$$

The origin looks like a singular point but it really isn't, because

$$\lim_{z \to 0} \frac{z}{e^z - 1} = 1 \tag{19.10}$$

and the derivatives of all orders exist at $z = 0$. The points $z = \pm 2\pi i$ are the nearest singular points and are what determine the radius of convergence of the series to be $|z| = 2\pi$.

We next suppose that $f(z)$ is analytic on the circular paths C_1 and C_2 of Fig. 19.2 and in the annular region between them. The function may have a singularity at z_0 or at other points inside C_1. This is in fact the interesting case. The Cauchy integral formula can be used to evaluate $f(z)$ at any point in the annular region:

$$f(z) = \frac{1}{2\pi i} \oint_{C_2} \frac{f(\zeta)\, d\zeta}{\zeta - z} - \frac{1}{2\pi i} \oint_{C_1} \frac{f(\zeta)\, d\zeta}{\zeta - z} \tag{19.11}$$

The integrals are taken in the positive direction about both contours. For the contour C_1, the factor $-1/(\zeta - z)$ can be expressed

$$\begin{aligned} -\frac{1}{\zeta - z} &= \frac{1}{z - z_0 - (\zeta - z_0)} \\ &= \frac{1}{(z - z_0)\left(1 - \dfrac{\zeta - z_0}{z - z_0}\right)} \\ &= \frac{1}{z - z_0} \sum_{k=0}^{n} \left(\frac{\zeta - z_0}{z - z_0}\right)^k \\ &\quad + \frac{1}{z - \zeta}\left(\frac{\zeta - z_0}{z - z_0}\right)^{n+1} \end{aligned} \tag{19.12}$$

This particular form is chosen because $|\zeta - z_0| = r_1$, $|z - z_0| = r$, and $r_1 < r$. The integral around the contour C_2 is treated in the same manner that was used

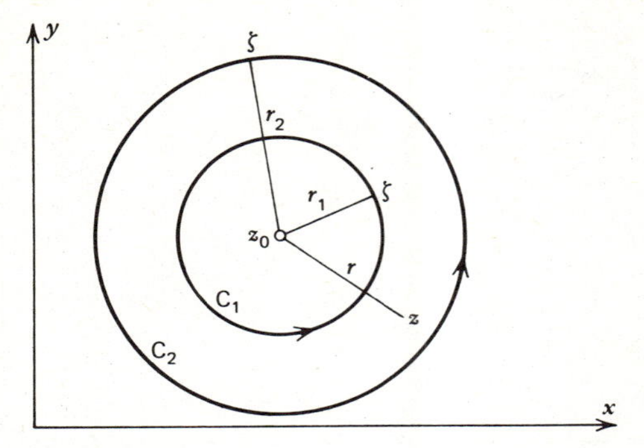

Fig. 19.2

in the derivation of Taylor's series. The result is

$$f(z) = \sum_{k=1}^{n+1} \frac{b_k}{(z - z_0)^k} + S_n + \sum_{k=0}^{n} a_k (z - z_0)^k + R_n \tag{19.13}$$

where

$$b_k = \frac{1}{2\pi i} \oint_{C_1} (\zeta - z_0)^{k-1} f(\zeta)\, d\zeta \tag{19.14}$$

$$a_k = \frac{1}{2\pi i} \oint_{C_2} \frac{f(\zeta)\, d\zeta}{(\zeta - z_0)^{k+1}} \tag{19.15}$$

The formula for R_n is given by (19.6). For S_n, we find

$$S_n = \frac{1}{2\pi i (z - z_0)^{n+1}} \oint_{C_1} \frac{f(\zeta)(\zeta - z_0)^{n+1}}{z - \zeta}\, d\zeta \tag{19.16}$$

$$|S_n| \le \frac{M r_1}{r - r_1} \left(\frac{r_1}{r}\right)^{n+1} \tag{19.17}$$

On C_1, $r_1 < r$ and $|S_n| \to 0$ as $n \to \infty$. We therefore find

$$f(z) = \sum_{k=1}^{\infty} \frac{b_k}{(z - z_0)^k} + \sum_{k=0}^{\infty} a_k (z - z_0)^k \tag{19.18}$$

which is called the *Laurent expansion* of $f(z)$ about $z = z_0$.

You will prove the uniqueness of the Laurent expansion in problem 19.4. The contours C_1 and C_2 which are used in (19.14) and (19.15) to calculate the coefficients do not have to be circles. They can be deformed in any way and can even both be the same contour provided that they are not moved across any singular points of $f(z)$ and always remain where $f(z)$ is analytic. If z_0 is the only point inside C_1 where there is a singularity and if z_1 is the next nearest singular point, then the Laurent expansion is convergent in the region $0<|z-z_0|<|z_1-z_0|$. Another Laurent expansion can be made with both z_0 and z_1 inside C_1. The region of convergence of this expansion is then $|z_1-z_0|<|z-z_0|<|z_2-z_0|$, where z_2 is the next singular point to be encountered after z_1.

If $f(z)$ is singular at z_0 and is analytic throughout some neighborhood of z_0, then $f(z)$ is said to have an *isolated singularity* at z_0. The function

$$\frac{z+2}{z^2(z^2+1)} \tag{19.19}$$

has isolated singularities at $z=0$ and $z=\pm i$. The function

$$\frac{1}{\sin(\pi/z)} \tag{19.20}$$

has an infinite number of isolated singularities at $z=\pm 1, \pm\frac{1}{2}, \pm\frac{1}{3}, \ldots$. The origin is a singular point, but it is not isolated since every neighborhood of the origin contains other singular points. The branch point $z=0$ of $\ln z$ and $\sqrt{z}$ is not classified as an isolated singularity because neighborhoods of $z=0$ cannot be constructed throughout which the functions are continuous. If z_0 is an isolated singular point of $f(z)$, then the Laurent expansion of $f(z)$ exists and is convergent in some circular region with z_0 as center. That portion of the expansion which involves negative powers of $(z-z_0)$ is called the *principal part* of the function; it is what determines the behavior of the function near $z=z_0$. If the principal part of the function terminates with the term b_n, then the singularity of $f(z)$ at z_0 is called a *pole of order n*. A pole of order one is called a *simple pole*. If $n=\infty$, then z_0 is an *essential singular point* of the function.

If z_0 is an isolated singular point of $f(z)$, then the coefficient b_1 of the term $1/(z-z_0)$ in the Laurent expansion of $f(z)$ about this point is called the *residue* of $f(z)$ at z_0. The residue is given by

$$b_1=\frac{1}{2\pi i}\oint_C f(\zeta)\,d\zeta \tag{19.21}$$

and is one if the most important results of this section because it is the basis of a method for evaluating definite integrals. If the contour C encloses several isolated singularities, it can be deformed in the manner illustrated by Fig. 19.3. In deforming the contour in Fig. 19.3*a* to that of Fig. 19.3*b*, we see that the single contour C can be replaced by three separate contours C_1, C_2, and C_3 each of which surrounds an isolated singular point. If $R_1, R_2, R_3, \ldots$ are the residues at the isolated singular points inside C, then

$$\oint_C f(z)\,dz=2\pi i(R_1+R_2+R_3+\cdots) \tag{19.22}$$

which is the *residue theorem*. The expressions (19.14) and (19.15) are seldom used as a method of obtaining the coefficients in a Laurent expansion. Rather, the expansion is found by other means, and then (19.22)

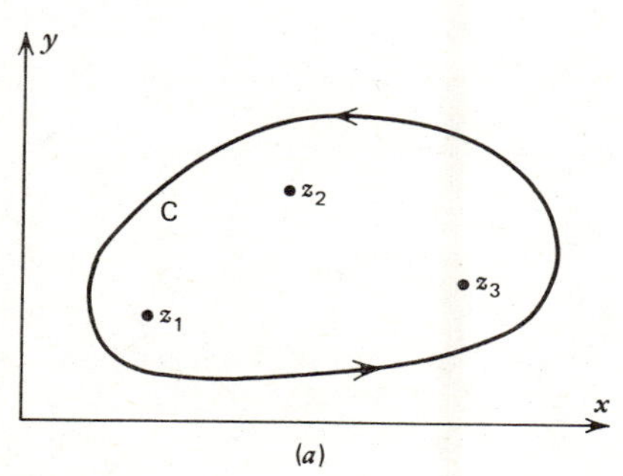

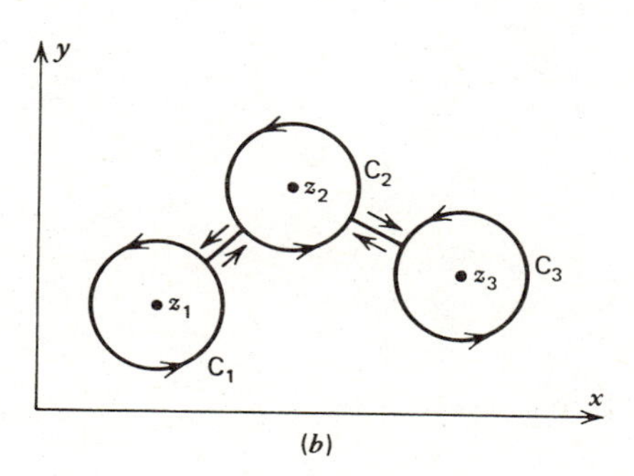

Fig. 19.3

is used as a way of evaluating integrals. For example, the Laurent expansion of

$$f(z) = \frac{\sin z}{z^4} \tag{19.23}$$

about $z = 0$ is easily found from the known Taylor's series expansion of $\sin z$:

$$\begin{aligned} f(z) &= \frac{1}{z^4}\left(z - \frac{z^3}{3!} + \frac{z^5}{5!} - \cdots\right) \\ &= \frac{1}{z^3} - \frac{1}{6z} + \frac{z}{5!} - \cdots \end{aligned} \tag{19.24}$$

The residue at $z = 0$ is therefore $-1/6$. The integral of $f(z)$ around any closed path which encircles the origin is

$$\oint \frac{\sin z}{z^4}\,dz = 2\pi i R_1 = -\frac{\pi i}{3} \tag{19.25}$$

The Laurent expansion of

$$f(z) = e^{1/z^2} \tag{19.26}$$

about the singular point $z = 0$ can be found by replacing ζ by $1/z^2$ in the Taylor's series expansion of e^{ζ}:

$$e^{1/z^2} = 1 + \frac{1}{z^2} + \frac{1}{2!z^4} + \cdots \qquad |z| > 0 \tag{19.27}$$

Therefore, $f(z)$ has an essential singularity at $z = 0$. Since the term of order $1/z$ is missing from the expansion, the line integral of $f(z)$ around a closed path which encircles the origin is zero because there is no residue there.

If in the expansion for $\sin z$ about $z = 0$ we replace z by $1/\zeta$, the result is

$$\sin\frac{1}{\zeta} = \frac{1}{\zeta} - \frac{1}{3!\zeta^3} + \frac{1}{5!\zeta^5} - \cdots \qquad |\zeta| > 0 \tag{19.28}$$

which shows that $\sin z$ has an essential singularity at $z = \infty$.

If $f(z)$ has a pole of order n at z_0, the singularity is said to be *removable* in the sense that the expansion for the function

$$\phi(z) = (z - z_0)^n f(z) \tag{19.29}$$

about z_0 is

$$\begin{aligned} \phi(z) = b_n + b_{n-1}(z - z_0) + \cdots + b_1(z - z_0)^{n-1} \\ + a_0(z - z_0)^n + a_1(z - z_0)^{n+1} + \cdots \end{aligned} \tag{19.30}$$

which is an ordinary Taylor's series expansion. The residue of the original function $f(z)$ is therefore given by

$$b_1 = \frac{1}{(n-1)!}\phi^{(n-1)}(z_0) \tag{19.31}$$

which is a useful formula for calculating residues. If $f(z)$ has a simple pole at $z = z_0$, the rule for finding the residue reduces to

$$b_1 = \lim_{z \to z_0} (z - z_0) f(z) \tag{19.32}$$

Consider the problem of evaluating

$$I = \oint \frac{dz}{\sin z} \tag{19.33}$$

around the three different contours of Fig. 19.4. The following computation shows that there are simple poles at 0, π, and 2π:

$$\begin{aligned} \lim_{z \to 0} \frac{z}{\sin z} &= 1 = R_1 \\ \lim_{z \to \pi} \frac{(z - \pi)}{\sin z} &= -1 = R_2 \\ \lim_{z \to 2\pi} \frac{(z - 2\pi)}{\sin z} &= 1 = R_3 \end{aligned} \tag{19.34}$$

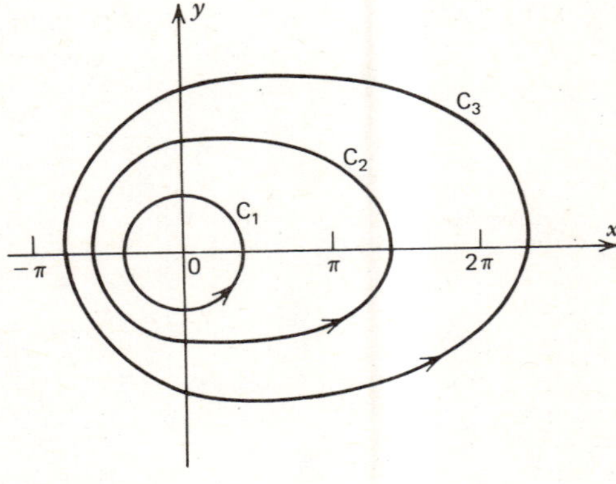

Fig. 19.4

More generally, simple poles are found at $\pm n\pi$. For the three different contours of Fig. 19.4, the integral (19.33) is

$$
\begin{aligned}
I_1 &= \oint_{C_1} \frac{dz}{\sin z} = 2\pi i R_1 = 2\pi i \\
I_2 &= \oint_{C_2} \frac{dz}{\sin z} = 2\pi i (R_1 + R_2) = 0 \\
I_3 &= \oint_{C_3} \frac{dz}{\sin z} = 2\pi i (R_1 + R_2 + R_3) = 2\pi i
\end{aligned}
\tag{19.35}
$$

The function

$$f(z) = \frac{1}{z(z+1)^2} \tag{19.36}$$

has a simple pole at $z = 0$ and a pole of order 2 at $z = -1$. The residue at $z = -1$ can be found by means of (19.29) and (19.31):

$$\phi(z) = (z+1)^2 f(z) = \frac{1}{z}$$

$$\phi'(z) = -\frac{1}{z^2} \qquad \phi'(-1) = b_1 = -1. \tag{19.37}$$

Another method is to find the Laurent expansion about $z = -1$. This can be done by using known power series expansions as the following calculation shows:

$$
\begin{aligned}
f(z) &= -\frac{1}{(z+1)^2[1-(z+1)]} \\
&= \frac{-1}{(z+1)^2}\left[1 + (z+1) + (z+1)^2 \right. \\
&\qquad \left. + (z+1)^3 + \cdots \right] \\
&= -\frac{1}{(z+1)^2} - \frac{1}{z+1} - 1 - (z+1) + \cdots
\end{aligned}
\tag{19.38}
$$

which again shows that $b_1 = -1$.

PROBLEMS

19.1. Refer to equation (19.9). Differentiate $z/(e^z - 1)$ and then take the limit as $z \to 0$ to show that $B_1 = -1/2$.

Hint: In the differentiated expression, substitute $e^z = 1 + z + \frac{1}{2}z^2 + \cdots$ before taking the limit.

19.2. For $|z| > 1$,

$$\frac{1}{1-z} = -\frac{1}{z(1-1/z)} = -\left(\frac{1}{z} + \frac{1}{z^2} + \frac{1}{z^3} + \cdots\right)$$

The point $z = 0$ is however not an essential singularity. In fact, it is not a singular point at all. Is there a contradiction here?

19.3. Make sure that you understand how (19.11) follows from the Cauchy integral formula as written in the form (18.23). Show how the integration around the circular boundaries of the annular region of Fig. 19.2 can be considered as a limiting case of a line integral around the single boundary of a simply connected region.

19.4. Let n be an integer. Show that the value of the integral

$$I = \oint (z - z_0)^n \, dz \tag{19.39}$$

when evaluated around a circular contour with z_0 at its center is zero unless $n = -1$, in which case $I = 2\pi i$. Evaluate the integral directly without reference to the residue theorem. Suppose that $f(z)$ is analytic in some annular region between concentric circles with z_0 as center and that it is represented in this region by an expansion of the form

$$f(z)=\sum_{k=-\infty}^{+\infty} A_k(z-z_0)^k \tag{19.40}$$

Multiply the expansion by the factor

$$\frac{1}{2\pi i}\frac{1}{(z-z_0)^{n+1}}$$

and integrate term by term to show that

$$A_n=\frac{1}{2\pi i}\oint\frac{f(z)\,dz}{(z-z_0)^{n+1}} \tag{19.41}$$

which gives the coefficients (19.14) and (19.15) of the Laurent expansion. Thus, (19.40) must in fact be the Laurent expansion of $f(z)$, and uniqueness is proved.

19.5. Obtain the Taylor's series expansion

$$\frac{z-1}{z^2}=\sum_{k=0}^{\infty}(-1)^k(k+1)(z-1)^{k+1} \qquad |z-1|<1 \tag{19.42}$$

19.6. If you did problem 8.4, you found two Laurent expansions for

$$f(z)=\frac{1}{z(1+z^2)} \tag{19.43}$$

one of which is valid for $0<|z|<1$ and the other for $|z|>1$. Go back and review this problem. Which version of the expansions gives the residue at $z=0$? It is important to remember that the residue at a pole of a function is found from the Laurent expansion which is valid in the *neighborhood* of the pole.

19.7. Find four terms of the Laurent expansions of the functions

$$\frac{\sinh z}{z^2} \qquad \frac{e^z}{z(z^2+1)}$$

about the point $z=0$. In each case, give the region in which the expansion converges.

19.8. Find the residue of

$$f(z)=\frac{1}{z^2(z-a)^2}$$

at the point $z=a$ by means of (19.31) and also by making a Laurent expansion of the function about $z=a$.

19.9. Find all the poles of

$$f(z)=\frac{z^2+1}{z^3+3z^2+2z}$$

Find the residue at each pole.

19.10. Find the residue of

$$f(z)=\frac{1+e^z}{\sin z+z\cos z}$$

at $z=0$.

19.11. Evaluate each of the following integrals. In each case, the direction of integration is in the positive sense:

$$\oint \frac{dz}{\sinh z} \quad \text{around the circle } |z| = 4$$

$$\oint \frac{\cosh \pi z}{z(z^2+1)}\,dz \quad \text{around the circle } |z| = 2$$

$$\oint \frac{dz}{z^2 \sin z} \quad \text{around the circle } |z| = 1$$

$$\oint z e^{1/z}\,dz \quad \text{around the circle } |z| = 1$$

20. IMPROPER INTEGRALS

In this section, we will review a few topics about integration with respect to a real variable. Special attention will be given to improper integrals. An integral is termed *improper* if the range of integration is infinite or if the integrand is singular anywhere over the range of integration. As a simple but important, example, consider

$$I = \int_a^b \frac{dx}{x^p} \tag{20.1}$$

which is not improper but becomes so if $b = \infty$. Assume that both a and b are positive. The integral is easily evaluated and is

$$I = \frac{b^{1-p}}{1-p} - \frac{a^{1-p}}{1-p} \tag{20.2}$$

which is valid if $p \neq 1$. The limit of (20.2) for $b \to \infty$ exists if $p > 1$ and does not exist if $p < 1$. We therefore say that the improper integral

$$I = \int_a^\infty \frac{dx}{x^p} \tag{20.3}$$

converges if $p > 1$ and *diverges* if $p < 1$. This fact was used in establishing the convergence properties of the Riemann zeta function defined by (4.57).

The integrand of (20.1) is singular at $x = 0$ for $p > 0$. From (20.2), the limit $a \to 0$ is possible provided that $p < 1$. Therefore,

$$I = \int_0^b \frac{dx}{x^p} \tag{20.4}$$

is convergent if $p < 1$ and divergent if $p > 1$. For the special case $p = 1$, (20.1) is

$$I = \int_a^b \frac{dx}{x} = \ln\left(\frac{a}{b}\right) \tag{20.5}$$

Neither the limit $b \to \infty$ nor $z \to 0$ is permissible. Thus, both (20.3) and (20.4) are divergent if $p = 1$. Generally, if a function $f(x)$ has a singularity at $x = x_0$, the integral of $f(x)$ over this point exists if it can be established that

$$f(x) < \frac{M}{(x - x_0)^p} \qquad p < 1 \tag{20.6}$$

for some neighborhood of the point x_0.

Improper integrals which depend on a parameter are of special interest. Such an integral is

$$F(y) = \int_a^\infty f(x, y)\,dx \tag{20.7}$$

and defines a function of y over some definite range $\alpha \leq y \leq \beta$. Just as with infinite series, the concept of *uniform convergence* is important. The integral (20.7) is said to be uniformly convergent for y in the interval $\alpha \leq y \leq \beta$, provided that for a given positive number ε, which can be arbitrarily small, there is a positive number $N(\varepsilon)$ which does not depend on y such that the remainder

$$R = \int_A^\infty f(x, y)\,dx \tag{20.8}$$

obeys $|R| < \varepsilon$ whenever $A > N$. The idea here is that the same N and ε work for all values of y in the interval $\alpha \leq y \leq \beta$. A useful test for uniform convergence is the following: The integral (20.7) converges uniformly (and absolutely) if from a fixed point $x = x_0$ onward the inequality

$$|f(x, y)| < \frac{M}{x^p} \tag{20.9}$$

holds with $M > 0$ and $p > 1$. Once chosen, x_0, M, and p must in no way depend on y. There are several important theorems about uniformly convergent integrals which parallel very closely the corresponding theorems for uniformly convergent series of functions. This is no surprise since an integral is really a limiting form of a sum. For this reason, the following theorems are quoted without proof. Reference is given to the corresponding theorems for series.

Theorem 1. A uniformly convergent integral of a continuous function is itself a continuous function (theorem 2 of section 5).

Theorem 2. A function which is defined by a uniformly convergent integral can itself be integrated and the order of the two integrals can be interchanged:

$$\begin{aligned} \int_\alpha^\beta F(y)\,dy &= \int_\alpha^\beta \int_a^\infty f(x, y)\,dx\,dy \\ &= \int_a^\infty \int_\alpha^\beta f(x, y)\,dy\,dx \end{aligned} \tag{20.10}$$

(theorem 3 of section 5). The theorem can be extended to the case where $\beta = \infty$, provided that the double improper integral

$$\int_a^\infty \int_\alpha^\infty f(x, y)\,dx\,dy$$

exists.

Theorem 3. It is possible to differentiate a uniformly convergent integral with respect to its parameter by differentiating under the integral sign, provided that the resulting integral

$$F'(f) = \int_a^\infty \frac{\partial f(x, y)}{\partial y}\,dx \tag{20.11}$$

is also uniformly convergent (theorem 4 of section 5).

The theorems just quoted, as well as those in section 5, are from classical mathematical analysis. We have already introduced a concept in this book which will allow us to go beyond what these theorems allow under certain circumstances. This concept is that of the *Dirac delta function* introduced in section 10 of Chapter 3. The Dirac delta function belongs to a class of mathematical entities called generalized functions about which we will have more to say later. We will find, for example, that there are cases in which the order of summation and integration can be interchanged which violate the conditions of theorem 3 of section 5. This is not to say that theorem 3 is wrong; it's just that it applies to ordinary mathematical functions and can be relaxed if generalized functions are involved.

Integrals which contain one or more parameters or which are used to define or represent functions are of frequent occurrence and are of considerable importance. Let's start out with the simple integral

$$\int_0^\infty e^{-at}\,dt = \frac{1}{a} \tag{20.12}$$

In order to establish uniform convergence, we calculate the remainder

$$R = \int_A^\infty e^{-at}\,dt = \frac{1}{a} e^{-aA} \tag{20.13}$$

If $\delta > 0$ and if $\delta \le a < \infty$, then

$$|R| < \frac{1}{\delta} e^{-A\delta} \tag{20.14}$$

which can be made as small as we please by taking A sufficiently large. Since the statement is true for all a in the range $\delta \le a < \infty$, the convergence is uniform for a in this range. Differentiation with respect to the parameter a is therefore permitted. Successive differentiations yield

$$\int_0^\infty te^{-at}\,dt = \frac{1}{a^2}$$

$$\int_0^\infty t^2e^{-at}\,dt = \frac{2}{a^3}$$

$$\int_0^\infty t^3e^{-at}\,dt = \frac{3!}{a^4}$$

$$\int_0^\infty t^ne^{-at}\,dt = \frac{n!}{a^{n+1}} \tag{20.15}$$

It is left as a problem to show that a uniformly convergent integral results after each differentiation.

If we set $a = 1$ in (20.15), we find

$$\int_0^\infty t^ne^{-t}\,dt = n! \tag{20.16}$$

which is an integral representation of $n!$ As a generalization, it is usual to define a function of the complex variable z by means of

$$\Gamma(z)=\int_0^{\infty} t^{z-1}e^{-t}\,dt \tag{20.17}$$

This function is called the *gamma function*. Its importance lies in the fact that it is an analytic function which has the property that $\Gamma(n)=(n-1)!$ when n is a positive integer. We give arguments later to support the claim of analyticity. The condition under which the integral converges is established by noting that

$$t^{z-1}=e^{(z-1)\ln t}=e^{(x-1)\ln t}e^{iy\ln t}=t^{x-1}e^{iy\ln t} \tag{20.18}$$

It is the behavior of (20.18) at $t=0$ that causes trouble. The integrand becomes too singular at $t=0$ unless $x>0$. Thus, (20.17) is convergent if $\operatorname{Re} z>0$.

Consider the evaluation of the definite integral

$$F(a)=\int_0^{\infty} e^{-ax}\frac{\sin x}{x}\,dx \tag{20.19}$$

In the first place, the indefinite integral of the integrand cannot be found in terms of elementary functions. The integral is absolutely convergent if $0<a<\infty$. It is also uniformly convergent if $0\le a<\infty$ and therefore defines a continuous function of a for this interval. Uniform convergence can be established by expressing the remainder as a sum:

$$R=\sum_{k=N}^{\infty}\int_{k\pi}^{(k+1)\pi} e^{-ax}\frac{\sin x}{x}\,dx \tag{20.20}$$

The terms decrease monotonically in magnitude and alternate in sign. By the result of problem 4.17, such a series has the property that the magnitude of its sum is less than the magnitude of its first term:

$$|R|<\left|\int_{N\pi}^{(N+1)\pi} e^{-ax}\frac{\sin x}{x}\,dx\right|<\int_{N\pi}^{(N+1)\pi}\frac{1}{N\pi}\,dx=\frac{1}{N} \tag{20.21}$$

Thus, R can be made as small as we wish by an appropriate choice of N independently of the value of a in the interval $0\le a<\infty$. Uniform convergence is established and differentiation of (20.19) with respect to a under the integral sign is permitted:

$$F'(a)=-\int_0^{\infty} e^{-ax}\sin x\,dx \tag{20.22}$$

which is uniformly convergent for $0<a<\infty$. Note that $a=0$ is now excluded. The evaluation of (20.22) can be done by elementary methods:

$$\begin{aligned}F'(a)&=-\operatorname{Im}\int_0^{\infty} e^{-ax+ix}\,dx\\&=-\operatorname{Im}\frac{a+i}{a^2+1}=-\frac{1}{a^2+1}\end{aligned} \tag{20.23}$$

Equation (20.23) is a differential equation for $F(a)$ with the solution

$$F(a)=C-\tan^{-1}a \tag{20.24}$$

where C is a constant. From (20.19), we see that $F(\infty)=0$. Therefore, $C=\pi/2$ and

$$F(a)=\frac{\pi}{2}-\tan^{-1}a \tag{20.25}$$

which completes the evaluation. Note that for the special case $a=0$,

$$\int_0^{\infty}\frac{\sin x}{x}\,dx=\frac{\pi}{2} \tag{20.26}$$

As a generalization of (20.26), consider the integral

$$F(a)=\int_0^{\infty}\frac{\sin ax}{x}\,dx \tag{20.27}$$

Assume that $a>0$ and make the change of variable $u=ax$:

$$F(a)=\int_0^{\infty}\frac{\sin u}{u}\,du=\frac{\pi}{2} \tag{20.28}$$

which has no apparent dependence on a. However, if $a<0$, note that

$$F(a)=-\int_0^{\infty}\frac{\sin|a|x}{x}\,dx=-\frac{\pi}{2}$$

Thus, $F(a)$ is a discontinuous function given by

$$\begin{aligned}F(a)&=-\frac{\pi}{2} &&-\infty<a<0\\&=0 &&a=0\\&=+\frac{\pi}{2} &&0<a<\infty\end{aligned} \tag{20.29}$$

The differentiation of (20.27) with respect to a yields a nonconvergent integral.

In problem 1.3 of Chapter 4, you established that

$$\int_{-\infty}^{+\infty} e^{-x^2}\,dx = \sqrt{\pi} \tag{20.30}$$

It's easy to extend this result to the integral

$$I(a) = \int_{-\infty}^{+\infty} e^{-ax^2}\,dx \tag{20.31}$$

If the change of variable $u = \sqrt{a}\,x$ is made, the result is

$$I(a) = \int_{-\infty}^{+\infty} e^{-u^2}\frac{du}{\sqrt{a}} = \sqrt{\frac{\pi}{a}} \tag{20.32}$$

Since the integral in (20.31) and all its derivatives are uniformly convergent if $0 < a < \infty$, other integrals can be found by differentiation with respect to a. For example,

$$\int_{-\infty}^{+\infty} x^2 e^{-ax^2}\,dx = \frac{\sqrt{\pi}}{2a^{3/2}} \tag{20.33}$$

The evaluation of definite integrals by the methods given in this section may seem like an exercise in frustration. You are given some integral to evaluate, and what you must do is guess what simple integral to start out with that will yield the integral you want after some clever device is used. This is one of the reasons why tables of integrals have been published. Evaluating integrals is sometimes more of an art than a science. The next section, which deals with the evaluation of definite integrals by the residue theorem, is a little more systematic in its approach. Nevertheless, a fair amount of resourcefulness will be required.

PROBLEMS

20.1. Assume that $a > 0$ and that p is any positive number. Given a positive number ε, prove that there is a number A such that

$$e^{-ax} < \frac{\varepsilon}{x^p}$$

for all $x > A$. The idea is that, even if p is a large exponent, ε can be made arbitrarily small by choosing A sufficiently large. Use this result in the next problem.

20.2. Prove that (20.15) is uniformly convergent if n is any positive integer.

20.3. Show that

$$\int_0^{\infty} x^{2n+1}e^{-x^2}\,dx = \frac{n!}{2}$$

$$\int_0^{\infty} x^{2n}e^{-x^2}\,dx = \frac{\sqrt{\pi}\,(2n)!}{2^{2n+1}n!} \tag{20.34}$$

$$\Gamma\left(n+\frac{1}{2}\right) = \frac{\sqrt{\pi}\,(2n)!}{2^{2n}n!}$$

20.4. Suppose that the limits of a definite integral as well as the integrand depend on a parameter. Prove Leibnitz's rule:

$$\frac{d}{dy}\int_{a(y)}^{b(y)} f(x,y)\,dx = \int_{a(y)}^{b(y)}\frac{\partial f(x,y)}{\partial y}\,dx + f(b,y)\frac{db}{dy} - f(a,y)\frac{da}{dy} \tag{20.35}$$

Hint: Write $f(x,y) = \partial g(x,y)/\partial x$.

20.5. Evaluate

$$\int_0^{\infty} e^{-ax}x\sin bx\,dx$$

20.6. Show that

$$\int_0^\infty e^{-ax} \frac{\sin bx}{x}\, dx = \tan^{-1}\left(\frac{b}{a}\right)$$

$$\int_0^\infty e^{-ax} \frac{1-\cos bx}{x^2}\, dx = b\tan^{-1}\left(\frac{b}{a}\right) - \frac{a}{2}\ln\left(1 + \frac{b^2}{a^2}\right)$$

Hint: Start by evaluating

$$F(a,b) = \int_0^\infty e^{-ax}\cos bx\, dx$$

Then integrate with respect to b.

20.7. Show that (20.26) is conditionally convergent.

20.8. Evaluate

$$\int_{-\infty}^\infty e^{-ax^2+bx}\, dx$$

Hint: Complete the square in the exponent.

20.9. Evaluate the following:

$$\int_0^\infty e^{-ax}\sqrt{x}\, dx = \frac{\sqrt{\pi}}{2a^{3/2}}$$

$$\int_0^\infty \frac{e^{-ax}}{\sqrt{x}}\, dx = \sqrt{\frac{\pi}{a}}$$

$$\int_0^1 (\ln x)^n\, dx = (-1)^n n!$$

$$\int_0^1 \sqrt{\ln\frac{1}{x}}\, dx = \frac{\sqrt{\pi}}{2}$$

Hint: In each case, a known integral is obtained by making an appropriate change of variable.

20.10. Prove that

$$F(b) = \int_0^\infty e^{-a^2x^2}\cos bx\, dx = \frac{\sqrt{\pi}}{2a} e^{-b^2/(4a^2)}$$

Hint: Differentiate with respect to b. Integrate the resulting integral by parts to show that $F(b)$ obeys the differential equation

$$F'(b) = -\frac{b}{2a^2}F(b)$$

20.11. Evaluate the following:

$$\int_0^\infty \frac{e^{-ax}-e^{-bx}}{x}\, dx = \ln\left(\frac{b}{a}\right) \qquad b>0,\ a>0$$

$$\int_0^1 \frac{x^b - x^a}{\ln x}\, dx = \ln\left(\frac{1+b}{1+a}\right) \qquad a>-1,\ b>-1$$

$$\int_0^\infty \frac{e^{-ax}-e^{-bx}}{x}\cos cx\, dx = \frac{1}{2}\ln\left(\frac{b^2+c^2}{a^2+c^2}\right) \qquad a>0,\ b>0$$

$$\int_0^\infty \frac{e^{-ax}-e^{-bx}}{x}\sin cx\, dx = \tan^{-1}\left(\frac{b}{c}\right) - \tan^{-1}\left(\frac{a}{c}\right)$$

Hint: In each case, a simpler integral can be integrated with respect to a parameter. Try

$$\int_0^\infty e^{-xy}\,dx \qquad \int_0^1 x^y\,dx \qquad \int_0^\infty e^{-xy}\cos cx\,dx \qquad \int_0^\infty e^{-xy}\sin cx\,dx.$$

20.12. Look back at your solution of problem 1.3 in Chapter 4 and convince yourself that (20.32) is all right if a is replaced by $a+ib$. Use this result to show that

$$\int_{-\infty}^{+\infty} e^{-ax^2}\cos bx^2\,dx = \sqrt{\frac{\pi}{\sqrt{a^2+b^2}}}\,\cos\frac{\phi}{2}$$

$$\int_{-\infty}^{+\infty} e^{-ax^2}\sin bx^2\,dx = \sqrt{\frac{\pi}{\sqrt{a^2+b^2}}}\,\sin\frac{\phi}{2}$$

where $\tan\phi = b/a$. In particular, show that

$$\int_{-\infty}^{+\infty}\cos bx^2\,dx = \int_{-\infty}^{+\infty}\sin bx^2\,dx = \sqrt{\frac{\pi}{2b}}$$

which occur in diffraction theory and are known as *Fresnel integrals*.

21. APPLICATIONS OF THE RESIDUE THEOREM

Various techniques which are based on complex variable theory that are useful in the evaluation of definite integrals are illustrated in this section by several examples. The definite integral over the real variable x,

$$I = \int_0^\infty \frac{dx}{1+x^2} = \frac{1}{2}\int_{-\infty}^{\infty}\frac{dx}{1+x^2} \tag{21.1}$$

can be evaluated through a consideration of

$$\oint \frac{dz}{1+z^2} \tag{21.2}$$

integrated around the closed contour of Fig. 21.1. The contour consists of a straight portion along the x axis from $-r$ to $+r$ and a semicircular arc of radius r centered at the origin. On the circular portion, it is convenient to use

$$z = re^{i\theta} \qquad dz = re^{i\theta}i\,d\theta \tag{21.3}$$

which converts (21.2) into

$$\oint \frac{dz}{1+z^2} = \int_{-r}^{+r}\frac{dx}{1+x^2} + \int_{\theta=0}^{2\pi}\frac{re^{i\theta}i\,d\theta}{1+r^2e^{2i\theta}} \tag{21.4}$$

The integral over the semicircular arc becomes zero in the limit $r\to\infty$. The integrand has a simple pole inside the contour at $z=i$. Therefore,

$$\int_{-\infty}^{+\infty}\frac{dx}{1+x^2} = 2\pi iR \tag{21.5}$$

The residue at $z=i$ follows from

$$R = \lim_{z\to i}(z-i)\frac{1}{1+z^2} = \lim_{z\to i}\frac{1}{z+i} = \frac{1}{2i} \tag{21.6}$$

The integral (21.1) is then

$$\int_0^\infty \frac{dx}{1+x^2} = \frac{\pi}{2} \tag{21.7}$$

If $0<k<1$, the integrand of

$$F(k) = \int_0^\infty \frac{x^{-k}\,dx}{1+x} \tag{21.8}$$

has a branch point at $x=0$. Branch points are not isolated singularities. We cannot encircle them by a contour without running into the problem of multiple-valuedness. There is no residue or Laurent expansion associated with a branch point. These difficulties

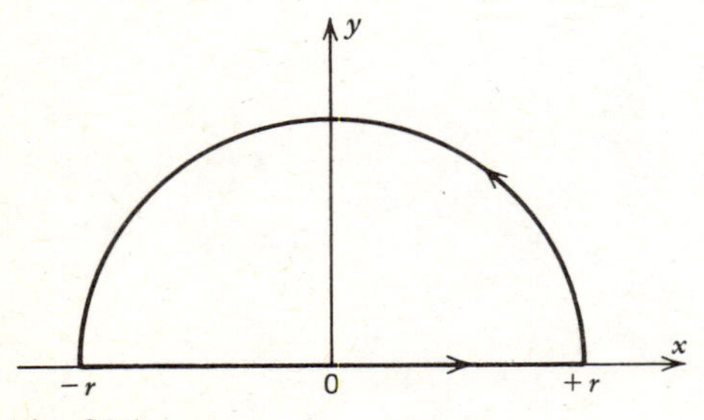

Fig. 21.1

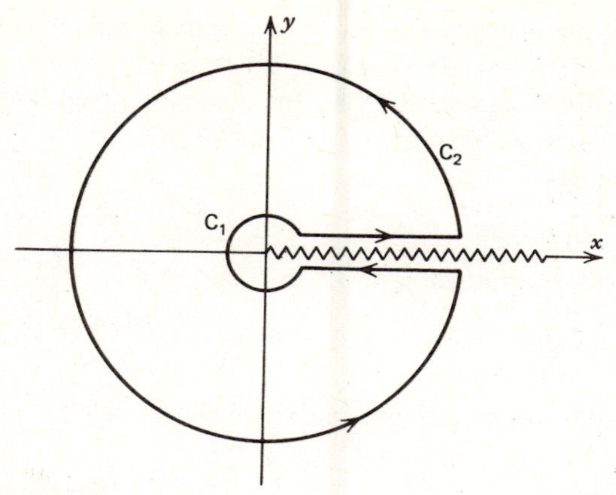

Fig. 21.2

are avoided by placing a branch cut along the positive real axis and using a contour such as that illustrated in Fig. 21.2. The integral

$$I = \oint \frac{z^{-k}\,dz}{1+z} \tag{21.9}$$

has contributions from the two circular portions C_1 and C_2 and from two straight portions, one just above the positive real axis and the other just below it. On C_1, we write $z = \delta e^{i\theta}$ and find

$$\int_{C_1} \frac{z^{-k}\,dz}{1+z} = \int_{2\pi}^{0} \frac{\delta^{1-k} e^{i\theta(1-k)} i\,d\theta}{1+\delta e^{i\theta}} \tag{21.10}$$

where δ id the radius of the contour. For $0 < k < 1$, the integral (21.10) becomes zero in the limit $\delta \to 0$. The contribution to I from the large circular arc of radius r can be expressed similarly as

$$\int_{C_2} \frac{z^{-k}\,dz}{1+z} = \int_{0}^{2\pi} \frac{r^{1-k} e^{i\theta(1-k)} i\,d\theta}{1+re^{i\theta}} \tag{21.11}$$

which becomes zero in the limit $r \to \infty$. There remain only the straight-line portions. We must be careful to remember that just above the real axis, $z = x$ whereas just below it, $z = xe^{2\pi i}$. Therefore,

$$I = \int_{\delta}^{r} \frac{x^{-k}\,dx}{1+x} + \int_{r}^{\delta} \frac{x^{-k} e^{-2\pi i k}\,dx}{1+x} \tag{21.12}$$

In the limit $\delta \to 0$ and $r \to \infty$,

$$I = F(k)(1 - e^{-2\pi i k}) \tag{21.13}$$

which, fortunately, involves the integral (21.8) that we wanted to evaluate. The integrand of (21.9) has one simple pole at $z = -1$ with residue

$$R = (-1)^{-k} = e^{-\pi i k} \tag{21.14}$$

Note that -1 must be written as $e^{\pi i}$ and not in some other way such as $e^{-\pi i}$. This is because throughout the calculation the complex variable z is written as $re^{i\theta}$ with $0 < \theta < 2\pi$. With $I = 2\pi i R$, we find that

$$F(k) = \frac{\pi}{\sin \pi k} \qquad 0 < k < 1 \tag{21.15}$$

An integral which can be related to (21.8) by a simple change of variable is

$$I(p) = \int_0^\infty \frac{dx}{1+x^p} \tag{21.16}$$

If $x^p = t$, we find

$$\begin{aligned} I(p) &= \frac{1}{p} \int_0^\infty \frac{t^{-(1-1/p)}\,dt}{1+t} \\ &= \frac{\pi}{p \sin \pi \left(\dfrac{p-1}{p} \right)} \qquad p > 1 \end{aligned} \tag{21.17}$$

An integral which involves a trigonometric function is

$$\begin{aligned} F(a,b) &= \int_0^\infty \frac{\cos ax\,dx}{b^2+x^2} = \frac{1}{2}\int_{-\infty}^{+\infty} \frac{\cos ax\,dx}{b^2+x^2} \\ &= \frac{1}{2}\int_{-\infty}^{+\infty} \frac{e^{iax}\,dx}{b^2+x^2} \end{aligned} \tag{21.18}$$

The last equality holds because

$$\int_{-\infty}^{+\infty} \frac{\sin ax\,dx}{b^2+x^2} = 0 \tag{21.19}$$

on account of the integrand being an odd function of x. It is possible to evaluate (21.18) by integrating

$$\oint \frac{e^{iaz}\,dz}{b^2+z^2} \tag{21.20}$$

around the contour of Fig. 21.1. Over the semicircular portion,

$$e^{iaz} = e^{iar\cos\theta} e^{-ar\sin\theta} \tag{21.21}$$

Over the range $0 < \theta < \pi$, $\sin\theta$ is never negative, so the exponential will not make trouble for us in the limit $r \to \infty$. The integrand of (21.20) has a simple pole inside the contour at $z = ib$ with residue

$$R = \lim_{z \to ib} (z - ib) \frac{e^{iaz}}{b^2 + z^2} = \lim_{z \to ib} \frac{e^{iaz}}{z + ib} = \frac{e^{-ab}}{2ib} \tag{21.22}$$

The integral (21.18) is therefore

$$F(a, b) = \frac{1}{2} 2\pi i R = \frac{\pi e^{-|ab|}}{2|b|} \tag{21.23}$$

The absolute value sign has been used because $F(a, b)$ as given by (21.18) is not affected by a change in sign of either a or b. Thus, (21.23) is valid for any real values of a and b.

You have already evaluated

$$F(b) = \int_0^\infty e^{-a^2x^2} \cos bx \, dx \tag{21.24}$$

in problem (20.10). Here is another way to do it:

$$F(b) = \frac{1}{2} \int_{-\infty}^{+\infty} e^{-a^2x^2 + ibx} \, dx$$

$$= \frac{1}{2} \int_{-\infty}^{+\infty} e^{-a^2\{x - ib/(2a^2)\}^2 - b^2/(4a^2)} \, dx \tag{21.25}$$

which is accomplished by completing the square of the function in the exponent. It is possible to change the variable by means of

$$z = x - \frac{ib}{2a^2} \tag{21.26}$$

and then integrate over the path of Fig. 21.3:

$$F(b) = \frac{1}{2} \int_{-\infty - ib/(2a^2)}^{+\infty - ib/(2a^2)} e^{-a^2z^2} e^{-b^2/(4a^2)} \, dz \tag{21.27}$$

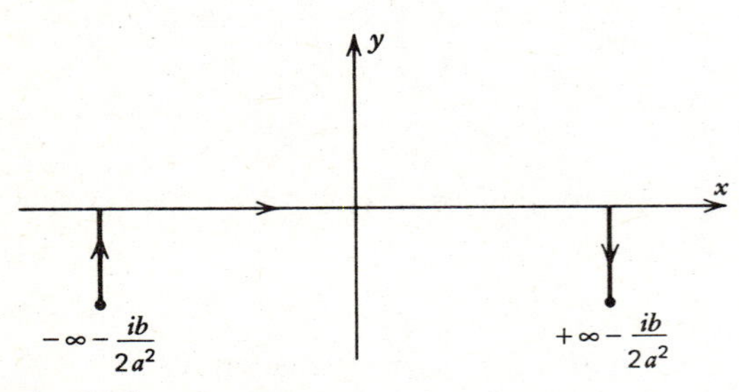

Fig. 21.3

There is no contribution to the integral from the portions which are parallel to the y axis; the contribution from the portion along the x axis is given by (20.31). The final result is

$$F(b) = \frac{\sqrt{\pi}}{2a} e^{-b^2/(4a^2)} \tag{21.28}$$

Another familiar integral is

$$\int_0^\infty \frac{\sin x}{x} dx$$

It can be found by integrating

$$I = \oint \frac{e^{iz}}{z} dz \tag{21.29}$$

around the contour of Fig. 21.4. The only singularity of the integrand in the finite z plane is at $z = 0$, and this has been avoided by placing a semicircular path C_1 around it. There are no poles inside the contour, so $I = 0$. The contribution from C_1 can be found by writing $z = \delta e^{i\theta}$ and then taking the limit $\delta \to 0$:

$$\int_{C_1} \frac{e^{iz}}{z} dz = \int_\pi^0 e^{i\delta\cos\theta - \delta\sin\theta} i \, d\theta \to \int_\pi^0 i \, d\theta = -\pi i \tag{21.30}$$

By a similar procedure, the contribution from C_2 can be shown to be zero in the limit as its radius becomes large. Equation (21.29) becomes

$$\int_{-r}^{-\delta} \frac{e^{ix}}{x} dx + \int_\delta^r \frac{e^{ix}}{x} dx - \pi i = 0 \tag{21.31}$$

If in the first integral we replace x by $-x$, the result

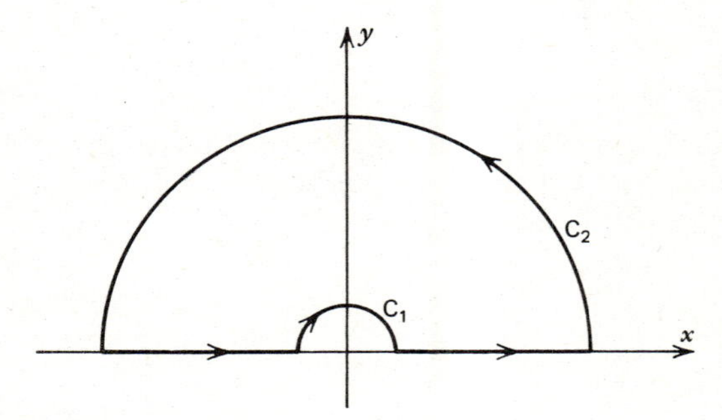

Fig. 21.4

is

$$\int_{-r}^{-\delta} \frac{e^{ix}}{x} dx = \int_{r}^{\delta} \frac{e^{-ix}}{x} dx = -\int_{\delta}^{r} \frac{e^{-ix}}{x} dx \qquad (21.32)$$

The integrals separately in (21.31) diverge in the limit $\delta \to 0$ and $r \to \infty$, but taken together they give

$$\int_{0}^{\infty} \frac{\sin x}{x} dx = \frac{\pi}{2} \qquad (21.33)$$

as expected.

Another type of definite integral involving trigonometric functions is

$$\int_0^{\pi} \frac{\cos 2\theta \, d\theta}{1 - 2a\cos\theta + a^2} = \frac{1}{2}\int_0^{2\pi} \frac{\cos 2\theta \, d\theta}{1 - 2a\cos\theta + a^2} = F(a) \qquad (21.34)$$

where a is real and $a^2 < 1$. By writing the cosine factors in exponential form, we find

$$F(a) = \frac{1}{4}\int_0^{2\pi} \frac{e^{2i\theta} + e^{-2i\theta}}{(1 - ae^{i\theta})(1 - ae^{-i\theta})} d\theta \qquad (21.35)$$

If the change of variable $z = e^{i\theta}$ is made, (21.35) becomes a contour integral around the unit circle:

$$F(a) = \frac{1}{4i}\oint \frac{z^2 + (1/z^2)}{(1 - az)(z - a)} dz \qquad (21.36)$$

Inside the contour, there is a simple pole at $z = a$ and a pole of order 2 at $z = 0$. The simplest way to find the residue at $z = 0$ is to calculate a few terms of the Laurent expansion of the integrand about this point:

$$\begin{aligned} f(z) &= -\frac{1}{4ia}\frac{z^2 + (1/z^2)}{(1 - az)(1 - z/a)} \\ &\simeq -\frac{1}{4ia}\left(z^2 + \frac{1}{z^2}\right)(1 + az + a^2z^2 + \cdots) \\ &\quad \times \left(1 + \frac{z}{a} + \frac{z^2}{a^2} + \cdots\right) \\ &\simeq -\frac{1}{4ia}\left(z^2 + \frac{1}{z^2}\right)\left(1 + \left[a + \frac{1}{a}\right]z + \cdots\right) \\ &\simeq -\frac{1}{4ia}\left(\frac{1}{z^2} + \left[a + \frac{1}{a}\right]\frac{1}{z} + \cdots\right) \end{aligned} \qquad (21.37)$$

The residue at $z = 0$ is therefore

$$R_1 = -\frac{1}{4ia}\left(a + \frac{1}{a}\right) \qquad (21.38)$$

The residue at $z = a$ is found to be

$$R_2 = \frac{1}{4i}\frac{a^2 + 1/a^2}{1 - a^2} \qquad (21.39)$$

The integral is therefore

$$F(a) = 2\pi i(R_1 + R_2) = \frac{\pi a^2}{1 - a^2} \qquad (21.40)$$

The integrand of

$$F(k) = \int_{-\infty}^{+\infty} \frac{e^{kx}\,dx}{1 + e^x} \qquad 0 < k < 1 \qquad (21.41)$$

is characterized by an infinite number of simple poles at $z = \pm\pi i, \pm 3\pi i, \ldots$ when it is extended into the complex plane. The rectangular contour of Fig. 21.5 encloses only one such pole. For this contour, we find that

$$\oint \frac{e^{kz}\,dz}{1 + e^z} = \int_{-A}^{+A} \frac{e^{kx}\,dx}{1 + e^x} + \int_{+A}^{-A} \frac{e^{k(x + 2\pi i)}}{1 + e^x} dx \qquad (21.42)$$

The contributions from the portions of the path parallel to the y axis at $x = \pm A$ become negligible as $A \to \infty$ and have been left out. Therefore, (21.42) gives, in the limit $A \to \infty$,

$$F(k)(1 - e^{2\pi ik}) = 2\pi iR \qquad (21.43)$$

where R is the residue at $z = \pi i$. The residue can be found by making a Laurent expansion about $z = \pi i$ or from

$$R = \lim_{z \to \pi i} \frac{z - \pi i}{1 + e^z} e^{kz} = \lim_{z \to \pi i} \frac{1}{e^z} e^{kz} = -e^{k\pi i} \qquad (21.44)$$

where equation (5.50) has been used. The integral is

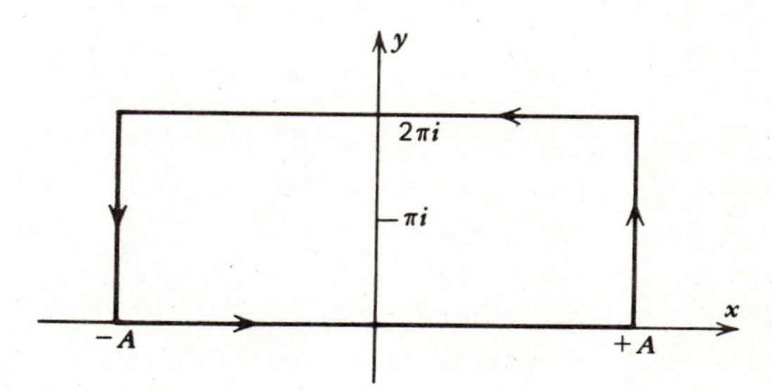

Fig. 21.5

found to be

$$F(k)=\frac{\pi}{\sin \pi k} \tag{21.45}$$

As a final example, we will use the residue theorem to evaluate the Bernoulli numbers. Recall that the Bernoulli numbers are defined as the coefficients in the expansion

$$\frac{z}{e^z-1}=\sum_{n=0}^{\infty}\frac{B_n}{n!}z^n \tag{21.46}$$

From Taylor's series as expressed by (19.5), we find

$$B_n=\frac{n!}{2\pi i}\oint_{C_0}\frac{dz}{z^n(e^z-1)} \tag{21.47}$$

where C_0 is a contour which encircles the origin and no other poles of the integrand. For $n=0$, the pole at the origin is simple, and

$$B_0=\frac{1}{2\pi i}\oint_{C_0}\frac{dz}{e^z-1}=1 \tag{21.48}$$

For $n=1$,

$$B_1=\frac{1}{2\pi i}\oint_{C_0}\frac{dz}{z(e^z-1)} \tag{21.49}$$

The integrand has a pole of order 2 at $z=0$. Its residue is found from the following Laurent expansion:

$$\begin{aligned}\frac{1}{z(e^z-1)}&\simeq\frac{1}{z(1+z+z^2/2+\cdots-1)}\\&\simeq\frac{1}{z^2\left(1+\frac{z}{2}\right)}\simeq\frac{1}{z^2}-\frac{1}{2z}+\cdots\end{aligned} \tag{21.50}$$

Thus,

$$B_1=\frac{1}{2\pi i}2\pi i\left(-\frac{1}{2}\right)=-\frac{1}{2} \tag{21.51}$$

Instead of continuing in this manner, consider the extension of the contour C_0 as shown in Fig. 21.6. Since there is no branch line on the real axis in this case, the contributions from the straight portions cancel. For $n\geq 2$, the contribution from the large circle goes to zero in the limit where its radius

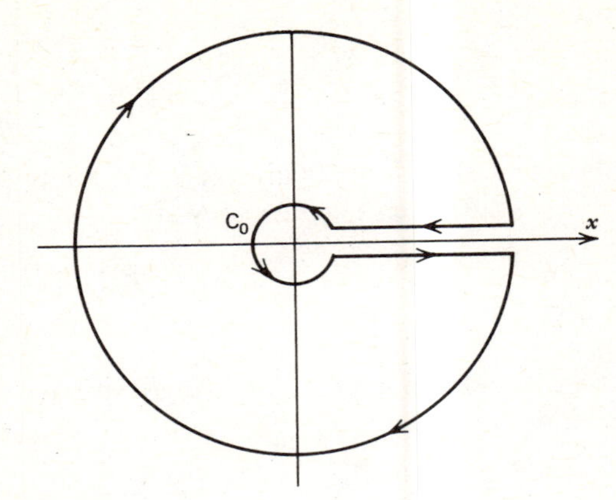

Fig. 21.6

becomes large. The added parts of the contour contribute nothing so that (21.47) is still valid if C_0 is replaced by the extended contour. The new contour encloses an infinite number of simple poles along the y axis at $\pm 2\pi i, \pm 4\pi i, \ldots$. The pole at $z=0$ is no longer inside the contour, so it contributes nothing to the integral. To put it another way, the sum of the residues at the newly enclosed poles are equivalent to the residue at $z=0$. You can show that the residue at $2\pi ik$ is

$$R_k=\frac{n!}{2\pi i(2\pi ik)^n} \tag{21.52}$$

The Bernoulli numbers are therefore

$$B_n=-n!\sum_{\substack{k=-\infty\\k\neq 0}}^{\infty}\frac{1}{(2\pi ik)^n} \tag{21.53}$$

which is valid for $n=2,3,4,\ldots$. For odd n, the contributions from positive and negative values of k cancel, giving zero. For even n, it is possible to express (21.53) as

$$B_{2p}=\frac{2(2p)!(-1)^{p+1}}{(2\pi)^{2p}}\sum_{k=1}^{\infty}\frac{1}{k^{2p}} \tag{21.54}$$

where we have put $n=2p$. The sum will be recognized as the Riemann zeta function of argument $2p$. Since, from problem 14.7, we know the values of the first few Bernoulli numbers to be

$$B_2 = \tfrac{1}{6} \qquad B_4 = -\tfrac{1}{30} \qquad B_6 = \tfrac{1}{42} \tag{21.55}$$

we can show from (21.54) that

$$\sum_{k=1}^{\infty} \frac{1}{k^2} = \frac{\pi^2}{6} \qquad \sum_{k=1}^{\infty} \frac{1}{k^4} = \frac{\pi^4}{90}$$

$$\sum_{k=1}^{\infty} \frac{1}{k^6} = \frac{\pi^6}{945} \tag{21.56}$$

What we have done is to turn (21.47) around from a way to calculate Bernoulli numbers to a way to evaluate sums.

PROBLEMS

21.1. Evaluate

$$\int_0^{\infty} \frac{x^{-k}\ln x}{1+x}\,dx = \frac{\pi^2 \cos \pi k}{\sin^2 \pi k} \qquad 0 < k < 1$$

Do it by integrating

$$\oint \frac{z^{-k}\ln z}{1+z}\,dz$$

around the contour of Fig. 21.2. Prove that the contribution to the integral from C_1 and C_2 is zero in the limits $\delta \to 0$ and $r \to \infty$. You will find the results of problem 15.13 useful.

21.2. Evaluate

$$\int_0^{\infty} \frac{dx}{x^a(1+x^b)} = \frac{\pi}{b \sin \pi\left(\dfrac{b+a-1}{b}\right)} \qquad b > 0,\ 1-b < a < 1$$

Check the convergence of the integral for the indicated ranges of a and b.

Hint: Make an appropriate change of variable and use (21.8).

21.3. Evaluate

$$\int_0^{\infty} \frac{dx}{1+x^4} = \frac{\pi\sqrt{2}}{4}$$

by residues in a manner similar to the evaluation of (21.1) and compare the result with (21.17) for $p = 4$.

21.4. Evaluate (21.41) by making the change of variable $e^x = t$. Use (21.15).

21.5. In equation (21.27), replace the upper and lower limits by $-A - ib/(2a^2)$ and $+A - ib/(2a^2)$. Go through the argument in detail to show that in the limit $A \to \infty$, there is no contribution to the integral from the portions of the path of integration in Fig. 21.3 which are parallel to the y axis. Would any other path connecting the two end points give the same answer?

21.6. Go through the details of showing that the contribution to (21.29) from the semicircle C_2 of radius r in Fig. 21.4 is zero in the limit as $r \to \infty$.

21.7. Confirm the equality of the two integrals in (21.34) by demonstrating that the same contribution to the integral comes from $\pi \le \theta \le 2\pi$ as from $0 \le \theta \le \pi$. Evaluate (21.34) for the case where $a^2 > 1$. What about $a = 1$?

21.8. An extraneous minus sign seems to have appeared in going from (21.52) to (21.53). Can you explain it? Look at the direction of integration around the contour of Fig. 21.6.

21.9. Prove that

$$\sum_{k=1}^{\infty}(-1)^{k+1}\frac{1}{k^2}=1-\frac{1}{2^2}+\frac{1}{3^2}-\cdots=\frac{1}{2}\left(1+\frac{1}{2^2}+\frac{1}{3^2}+\cdots\right)$$

$$=\frac{1}{2}\zeta(2)=\frac{\pi^2}{12}$$

By expressing $\ln(1+x)$ as a power series in x and integrating term by term, show that

$$\int_0^1\frac{\ln(1+x)}{x}dx=\frac{\pi^2}{12}$$

21.10. Evaluate

$$I=\int_0^{\infty}\frac{(\ln x)^2}{1+x^2}dx=\frac{\pi^3}{8}$$

Make the change of variable $x=e^t$ and then use the contour of Fig. 21.7. Show that the integral can be expressed as

$$I=2\int_0^{\infty}\frac{t^2e^{-t}}{1+e^{-2t}}dt$$

Expand the factor $1/(1+e^{-2t})$ in a geometrical series. Integrate term by term to show that

$$I=4\sum_{k=0}^{\infty}\frac{(-1)^k}{(1+2k)^3}\qquad \frac{\pi^3}{32}=1-\frac{1}{3^3}+\frac{1}{5^3}-\frac{1}{7^3}+\cdots$$

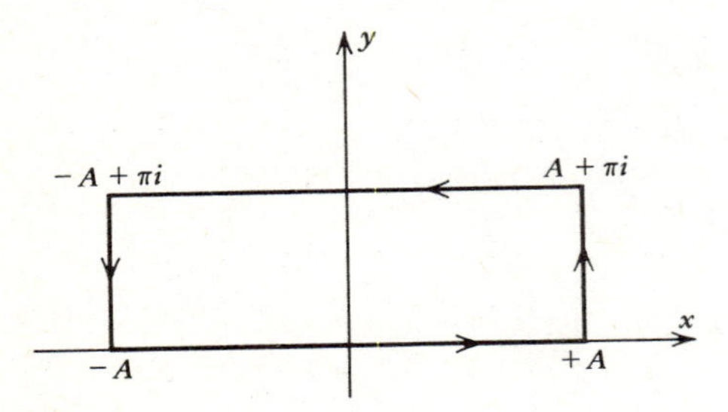

Fig. 21.7

21.11. Evaluate

$$\int_0^{\infty}\frac{x^a}{(1+x)^2}dx=\frac{\pi a}{\sin\pi a}\qquad -1<a<1$$

Hint: The Laurent expansion of the integrand around the point $z=-1$ is found by writing

$$\frac{z^a}{(1+z)^2}=\frac{(-1)^a(1-\{z+1\})^a}{(1+z)^2}$$

and using the binomial expansion on the factor $(1-\{z+1\})^a$.

21.12. Evaluate

$$\int_0^{2\pi} \frac{d\theta}{1 + a\cos\theta} = \frac{2\pi}{\sqrt{1 - a^2}} \qquad -1 < a < +1$$

$$\int_0^{\pi} \cos^{2n}\theta \, d\theta = \frac{\pi(2n)!}{2^{2n}(n!)^2} \qquad n = 0, 1, 2, \ldots$$

21.13. Evaluate

$$\int_{-1}^{+1} \frac{dx}{\sqrt{1 - x^2}\,(a + bx)} = \frac{\pi}{\sqrt{a^2 - b^2}} \qquad a > b > 0$$

Hint: As a first step, make the change of variable $x = \sin\theta$.

22. GAMMA AND BETA FUNCTIONS

Before proceeding with the discussion of the gamma and beta functions, we will say a few words about analytic functions which are defined by integrals.

Let $F(z, t)$ be an analytic function of the complex variable z in a closed region R of the complex plane for every value of the real variable t in the range $a \leq t \leq b$. Then,

$$f(z) = \int_a^b F(z, t)\, dt \tag{22.1}$$

is also an analytic function in R. The simplest way to see this is to separate (22.1) into its real and imaginary parts as

$$u(x, y) + iv(x, y) = \int_a^b U(x, y, t)\, dt + i\int_a^b V(x, y, t)\, dt \tag{22.2}$$

Then, since U and V obey the Cauchy–Riemann equations, so do u and v.

For an infinite range of integration, the integral

$$f(z) = \int_a^\infty F(z, t)\, dt \tag{22.3}$$

is said to be uniformly convergent, if for any positive number ε, which can be arbitrarily small, there is a number T such that

$$\left|\int_A^\infty F(z, t)\, dt\right| < \varepsilon \tag{22.4}$$

whenever $A > T$. This must be true for all values of z in the closed domain R, that is, the same T and ε work for all values of z. Such a uniformly convergent integral defines an analytic function. Again, this can be established by separating (22.3) into its real and imaginary parts as was done in equation (22.2). Each real integral is uniformly convergent and differentiation with respect to x or y under the integral sign is permitted. Since U and V obey the Cauchy–Riemann equations, so do u and v.

An important analytic function defined by a definite integral is the gamma function:

$$\Gamma(z) = \int_0^\infty t^{z-1}e^{-t}\, dt \tag{22.5}$$

The convergence of this integral for $\operatorname{Re} z > 0$ was established in section 20; see (20.17) and (20.18). This does not imply that the gamma function does not exist for $\operatorname{Re} z < 0$. The integral (22.5) is a valid representation only if $\operatorname{Re} z > 0$. A process of analytic continuation must be used to define $\Gamma(z)$ for other parts of the complex plane.

One of the most important relations obeyed by the gamma function is found by integrating the expression for $\Gamma(z+1)$ by parts:

$$\Gamma(z+1) = \int_0^\infty e^{-t}t^z\, dt = -\int_0^\infty \frac{d}{dt}(e^{-t})t^z\, dt = -e^{-t}t^z\Big|_0^\infty + \int_0^\infty e^{-t}zt^{z-1}\, dt = 0 + z\Gamma(z)$$

We find that

$$\Gamma(z+1) = z\,\Gamma(z) \tag{22.6}$$

which is called the *difference formula* for the gamma function.* It's easy to see from (22.5) that $\Gamma(1) = 1$.

*Equation (22.6) is also called a recurrence formula.

Thus, if n is an integer, a successive application of the difference formula gives

$$\Gamma(n+1) = n\,\Gamma(n) = n(n-1)\,\Gamma(n-1) = n(n-1)(n-2)\ldots(1) = n! \tag{22.7}$$

This result was already noted in section 20 and was a motivating factor in defining the gamma function in the first place. By expressing the difference formula as

$$\Gamma(z) = \frac{\Gamma(z+1)}{z} \tag{22.8}$$

the gamma function can be analytically continued into the region $\operatorname{Re} z > -1$ since $\Gamma(z+1)$ is defined for this range of values. Also revealed by (22.8) is the fact that there is a simple pole at $z = 0$. By a continuation of this process, we find

$$\Gamma(z) = \frac{\Gamma(z+2)}{z(z+1)} = \frac{\Gamma(z+3)}{z(z+1)(z+2)} = \frac{\Gamma(z+n+1)}{z(z+1)(z+2)\ldots(z+n)} \tag{22.9}$$

which is valid for any integer value of n and provides the analytic continuation of the gamma function into the entire z plane. It also shows that there are simple poles at $z = 0, -1, -2, -3\ldots$.

The gamma function has been extensively tabulated, and one of its uses is in the evaluation of definite integrals. For example, by making the change of variable $t = u^2$ in (22.5), we find

$$\Gamma(z) = 2\int_0^\infty e^{-u^2} u^{2z-1}\,du \tag{22.10}$$

In particular,

$$\Gamma(\tfrac{1}{2}) = 2\int_0^\infty e^{-u^2}\,du = \sqrt{\pi} \tag{22.11}$$

The change of variable $t = \ln(1/u)$ in (22.5) gives

$$\Gamma(z) = \int_0^1 \left(\ln\frac{1}{u}\right)^{z-1} du \tag{22.12}$$

The beta function is defined in terms of the gamma function as

$$B(p,q) = \frac{\Gamma(p)\,\Gamma(q)}{\Gamma(p+q)} \tag{22.13}$$

Note the symmetry property of the beta function: $B(p,q) = B(q,p)$. The motivation for this definition is that many frequently occurring definite integrals can be related to the beta function. In order to see this possibility, we calculate the factor $\Gamma(p)\,\Gamma(q)$ using the representation (22.10) for the gamma functions:

$$\Gamma(p)\,\Gamma(q) = 4\int_0^\infty e^{-u^2} u^{2p-1}\,du \int_0^\infty e^{-v^2} v^{2q-1}\,dv \tag{22.14}$$

Consider (22.14) to be a double integral over rectangular coordinates u and v. The range of integration covers only the first quadrant. Introduce plane polar coordinates by means of $u = r\cos\theta$ and $v = r\sin\theta$. The element of area is $du\,dv = r\,dr\,d\theta$. The integral becomes

$$\Gamma(p)\,\Gamma(q) = 4\int_0^\infty \int_0^{\pi/2} e^{-r^2} r^{2p+2q-1} \cos^{2p-1}\theta \sin^{2q-1}\theta\,dr\,d\theta \tag{22.15}$$

The integral over r is recognized as a gamma function of argument $p+q$. Thus,

$$\Gamma(p)\,\Gamma(q) = \Gamma(p+q)\,B(p,q)$$

$$B(p,q) = 2\int_0^{\pi/2} \cos^{2p-1}\theta \sin^{2q-1}\theta\,d\theta \tag{22.16}$$

Other integrals which can be evaluated in terms of beta functions are found by making appropriate changes of variable. For example,

$$B(p,q) = \int_0^1 t^{p-1}(1-t)^{q-1}\,dt \qquad t = \cos^2\theta \tag{22.17}$$

$$B(p,q) = \int_0^\infty \frac{u^{q-1}}{(1+u)^{p+q}}\,du \qquad t = \frac{1}{1+u} \tag{22.18}$$

A particularly important identity involving the gamma function can be obtained from (22.18) if we let $q = 1-p$:

$$B(p,1-p) = \Gamma(p)\,\Gamma(1-p) = \int_0^\infty \frac{u^{-p}}{1+u}\,du \tag{22.19}$$

This integral was evaluated in section 21 for real values of p; see (21.8) and (21.15). The same evaluation works for complex values of p as well. Thus, we evaluate

$$\oint \frac{z^{-p}}{1+z} dz \tag{22.20}$$

around the contour of Fig. 21.2. Let's check the behavior of the factor z^{-p} on the circular portions of the contour:

$$\begin{aligned} z^{-p} &= e^{-p \ln z} = e^{-(a+ib)(\ln r + i\theta)} \\ &= e^{-a \ln r + b\theta - i(b \ln r + a\theta)} \end{aligned} \tag{22.21}$$

It is only the factor $e^{-a \ln r} = r^{-a}$ which we have to worry about either as $r \to \infty$ on C_2 or as $r \to 0$ on C_1. Note that a is the real part of p. On C_1, we neglect z in the denominator of (22.20). There is an extra factor of r from dz, so that r^{1-a} must go to zero as $r \to 0$. This requires $a < 1$. On C_2, the denominator becomes essentially z, so that there is a factor of r from both the denominator and from dz. Thus, r^{-a} must go to zero as $r \to \infty$ requiring that $a > 0$. The result is that

$$\Gamma(p)\Gamma(1-p) = \int_0^\infty \frac{u^{-p}}{1+u} du = \frac{\pi}{\sin \pi p} \qquad 0 < \operatorname{Re} p < 1 \tag{22.22}$$

For future reference, let's change the notation and write

$$\Gamma(z)\,\Gamma(1-z) = \frac{\pi}{\sin \pi z} \tag{22.23}$$

which, by the process of analytic continuation, holds for all values of z except $z = 0, \pm 1, \pm 2, \ldots$.

Another useful identity involving gamma functions is obtained by starting with

$$B(z,z) = \frac{\Gamma(z)^2}{\Gamma(2z)} = \int_0^1 \{t(1-t)\}^{z-1} dt \tag{22.24}$$

Let $t = (1+s)/2$:

$$B(z,z) = \frac{1}{2^{2z-2}} \int_0^1 (1-s^2)^{z-1} ds \tag{22.25}$$

Now let $u = s^2$:

$$\begin{aligned} \frac{\Gamma(z)^2}{\Gamma(2z)} &= \frac{1}{2^{2z-1}} \int_0^1 (1-u)^{z-1} u^{-1/2} du \\ &= \frac{\Gamma(z)\Gamma(1/2)}{2^{2z-1}\Gamma(z+1/2)} \end{aligned} \tag{22.26}$$

Since $\Gamma(1/2) = \sqrt{\pi}$, we have

$$\sqrt{\pi}\,\Gamma(2z) = 2^{2z-1}\Gamma(z)\,\Gamma(z+1/2) \tag{22.27}$$

Another version is obtained by replacing z by $z + 1/2$:

$$\sqrt{\pi}\,\Gamma(2z+1) = 2^{2z}\Gamma(z+1/2)\Gamma(z+1) \tag{22.28}$$

This is known as the *Legendre duplication formula*. It can be used, for example, to compute $\Gamma(z+1/2)$. By means of $\Gamma(n+1) = n!$, (22.28) is converted to

$$\Gamma(n+1/2) = \frac{\sqrt{\pi}\,(2n)!}{2^{2n} n!} \qquad n = 0, 1, 2, 3, \ldots \tag{22.29}$$

Another form of the gamma function is found by starting with the integral

$$F(z,n) = \int_0^n \left(1 - \frac{t}{n}\right)^n t^{z-1} dt \tag{22.30}$$

If we recall that

$$\lim_{n\to\infty} \left(1 - \frac{t}{n}\right)^n = e^{-t} \tag{22.31}$$

we see that

$$\lim_{n\to\infty} F(z,n) = \Gamma(z) \tag{22.32}$$

In (22.30), let $u = t/n$. Then,

$$F(z,n) = n^z \int_0^1 u^{z-1}(1-u)^n du = \frac{n^z \Gamma(z)\,\Gamma(n+1)}{\Gamma(n+z+1)} \tag{22.33}$$

By means of the difference formula (22.6),

$$\Gamma(n+z+1) = \Gamma(z)\,z(z+1)(z+2)\ldots(z+n) \tag{22.34}$$

Therefore, (22.32) becomes

$$\Gamma(z) = \lim_{n\to\infty} \frac{n^z n!}{z(z+1)(z+2)\ldots(z+n)} \tag{22.35}$$

which is the *Euler form* of the gamma function and is

sometimes used as a definition in place of (22.5). It is useful as the basis of a method for obtaining the derivative of the gamma function, which we now do.

It is easier to differentiate $\ln \Gamma(z)$ than $\Gamma(z)$. From (22.35),

$$\ln \Gamma(z) = \lim_{n\to\infty} \left[\ln n! + z \ln n - \sum_{k=0}^{n} \ln(z+k) \right] \tag{22.36}$$

$$\psi(z) = \frac{d}{dz} \ln \Gamma(z) = \lim_{n\to\infty} \left[\ln n - \sum_{k=0}^{n} \frac{1}{z+k} \right] \tag{22.37}$$

The function $\psi(z)$ is called the *digamma function*. The series in (22.37) is divergent as $n \to \infty$, but the expression as a whole is not as we now show. It is convenient to define

$$\gamma = \lim_{n\to\infty} \left(\sum_{k=1}^{n} \frac{1}{k} - \ln n \right) \tag{22.38}$$

which is known as the *Euler constant*. Its existence is established by means of equation (4.56) with the upper limit ∞ replaced by n:

$$1 + \int_1^n \frac{dx}{x+1} < \sum_{k=1}^{n} \frac{1}{k} < 1 + \int_1^n \frac{dx}{x} \tag{22.39}$$

Carrying out the integrations yields

$$1 - \ln 2 + \ln\left(\frac{n+1}{n}\right) < \sum_{k=1}^{n} \frac{1}{k} - \ln n < 1 \tag{22.40}$$

In the limit $n \to \infty$,

$$0.30685 < \gamma < 1 \tag{22.41}$$

which not only proves the existence of γ but establishes limits on its value. The actual value is $\gamma = 0.57721566\ldots$. A minicomputer will calculate 100,000 terms of (22.38) in about 35 minutes and gives $\gamma = 0.57722$. The digamma function can now be written as

$$\begin{aligned} \psi(z) &= -\gamma - \frac{1}{z} + \sum_{k=1}^{\infty} \left(\frac{1}{k} - \frac{1}{z+k} \right) \\ &= -\gamma - \frac{1}{z} + \sum_{k=1}^{\infty} \frac{z}{k(z+k)} \end{aligned} \tag{22.42}$$

The series converges for all values of z by comparison to $\sum 1/k^2$ except at the poles $z = 0, -1, -2, \ldots$. More derivatives can be calculated and are

$$\begin{aligned} \psi'(z) &= \sum_{k=0}^{\infty} \frac{1}{(z+k)^2} \\ \psi''(z) &= -2 \sum_{k=0}^{\infty} \frac{1}{(z+k)^3} \\ \psi^{(n)}(z) &= n!(-1)^{n+1} \sum_{k=0}^{\infty} \frac{1}{(z+k)^{n+1}} \end{aligned} \tag{22.43}$$

which are known as *polygamma functions*.

PROBLEMS

22.1. Suppose that $F(z, t)$ is a continuous function of t when z is in the bounded, closed region R of the z plane and that $|F(z,t)| \le M(t)$, where $M(t) > 0$, for all z in R. Suppose also that

$$\int_a^\infty M(t)\, dt$$

converges. Show that

$$\int_a^\infty F(z,t)\, dt$$

is both uniformly and absolutely convergent in R.

22.2. Find the residue of the gamma function at $z = -n$.

22.3. Prove the Taylor's series expansion

$$\ln\Gamma(z) = -\gamma(z-1) + \frac{1}{2}\zeta(2)(z-1)^2 - \frac{1}{3}\zeta(3)(z-1)^3 + \cdots + \frac{(-1)^n}{n}\zeta(n)(z-1)^n + \cdots \qquad |z-1|<1$$

where $\zeta(n)$ is the Riemann zeta function.

22.4. Verify that

$$\lim_{n\to\infty}\left(1+\frac{t}{n}\right)^n = e^t$$

Hint: Use a binomial expansion.

22.5. Show that the binomial expansion can be expressed in the form

$$(1+x)^p = \sum_{k=0}^{\infty}\frac{\Gamma(p+1)}{\Gamma(k+1)\Gamma(p-k+1)}x^k$$

If p is an integer, the series terminates when $k=p$.

22.6. Prove each of the following:

(a) $\Gamma(z)^* = \Gamma(z^*)$.

(b) $\Gamma(iy)\,\Gamma(-iy) = |\Gamma(iy)|^2 = \dfrac{\pi}{y\sinh\pi y}$.

(c) $\Gamma(1+iy)\,\Gamma(1-iy) = |\Gamma(1+iy)|^2 = \dfrac{\pi y}{\sinh\pi y}$.

(d) $\Gamma(1/2+iy)\,\Gamma(1/2-iy) = |\Gamma(1/2+iy)|^2 = \dfrac{\pi}{\cosh\pi y}$.

(e) $\Gamma(1/2+n)\,\Gamma(1/2-n) = (-1)^n\pi$, where n is an integer.

22.7. Prove Wallis' formula

$$\int_0^{\pi/2}\cos^{2n}\theta\,d\theta = \int_0^{\pi/2}\sin^{2n}\theta\,d\theta = \frac{\Gamma(1/2)\,\Gamma(n+1/2)}{2\Gamma(n+1)} = \frac{\pi(2n)!}{2^{2n+1}(n!)^2}$$

22.8. By transforming the integrals into gamma functions, show that

$$\int_0^{\infty}e^{-x^4}\,dx = \frac{1}{4}\Gamma\left(\frac{1}{4}\right) \qquad \int_0^1 x^k\ln x\,dx = -\frac{1}{(1+k)^2} \qquad k>-1$$

22.9. Evaluate the following integrals:

$$\int_0^{\infty}e^{-t}\ln t\,dt = \Gamma'(1) = -\gamma$$

$$\int_0^{\infty}e^{-t}t\ln t\,dt = \Gamma'(2) = -\gamma+1$$

$$\int_0^{\infty}e^{-t}t^n\ln t\,dt = \Gamma'(n+1) = n!\left(1+\frac{1}{2}+\frac{1}{3}+\cdots+\frac{1}{n}-\gamma\right)$$

22.10. Prove the formulas

$$\psi(z+n)-\psi(z) = \frac{1}{z}+\frac{1}{z+1}+\cdots+\frac{1}{z+n-1} \qquad n=1,2,3,\ldots$$

$$\psi(1-z)-\psi(z) = \pi\cot\pi z$$

$$\psi(2z) = \ln 2 + \frac{1}{2}\psi(z) + \frac{1}{2}\psi(z+1/2)$$

22.11. Establish the following integrals

$$\int_{-1}^{+1}(1+x)^a(1-x)^b\,dx = 2^{a+b+1}B(a+1,b+1)$$

$$\int_t^z \frac{dx}{(z-x)^{1-\alpha}(x-t)^\alpha} = \frac{\pi}{\sin \pi\alpha} \qquad 0<\alpha<1$$

$$\int_0^{\pi/2}\sqrt{\cos\theta}\,d\theta = \frac{(2\pi)^{3/2}}{\Gamma(1/4)^2}$$

22.12. One form of exponential integral is defined by

$$E_1(x) = \int_x^\infty \frac{e^{-t}}{t}\,dt \qquad x>0$$

The integral cannot be evaluated in terms of elementary functions. Show that

$$\frac{dE_1}{dx} = -\frac{e^{-x}}{x}$$

By expanding this result in a power series in x and integrating term by term, show that

$$E_1(x) = -\ln x + C - \sum_{k=1}^{\infty}\frac{x^k(-1)^k}{kk!} \qquad x>0$$

where C is a constant of integration. Next, prove that

$$E_1(x) = -e^{-x}\ln x + \int_x^\infty e^{-t}\ln t\,dt$$

$$= -e^{-x}\ln x - \gamma - \int_0^x e^{-t}\ln t\,dt$$

where γ is Euler's constant. The constant C in the power series expansion for $E_1(x)$ is therefore $C = -\gamma$.

Hint: Write the last equation in the form $E_1(x) = -\ln x - \gamma + F(x)$. Find the first term in the Taylor's series expansion of $F(x)$.

22.13. Show that if n is an integer, then

$$\psi(n) = -\gamma + \sum_{k=1}^{n-1}\frac{1}{k} \qquad \psi(1) = -\gamma$$

23. ASYMPTOTIC SERIES

The integral

$$E_1(x) = \int_x^\infty \frac{e^{-t}}{t}\,dt \tag{23.1}$$

defines a function of x and is one form of what is known as the *exponential integral*. The integral cannot be evaluated in terms of elementary functions and so numerical procedures must be used to evaluate it. It is possible to generate a series for the exponential integral by repeated integrations by parts:

$$E_1(x) = -\int_x^\infty \frac{1}{t}\frac{d}{dt}(e^{-t})\,dt$$

$$= -\frac{1}{t}e^{-t}\Big|_x^\infty - \int_x^\infty \frac{1}{t^2}e^{-t}\,dt$$

$$= \frac{1}{x}e^{-x} - \int_x^\infty \frac{1}{t^2}e^{-t}\,dt \tag{23.2}$$

A continuation of this process yields the series

$$E_1(x)e^x = \frac{1}{x} - \frac{1}{x^2} + \frac{2!}{x^3} - \frac{3!}{x^4} + \cdots + \frac{(-1)^{n-1}(n-1)!}{x^n} + e^x \int_x^\infty \frac{(-1)^n n! e^{-t}}{t^{n+1}} dt \tag{23.3}$$

which for convenience has been expressed as a series for the function $E_1(x)e^x$. A simple application of the ratio test reveals that the series is not absolutely convergent for *any* value of x. A numerical computation of the terms in the series for a fixed value of x reveals that they get smaller for a while and then begin to increase in magnitude after a certain term.* In other words, there is a definite smallest term in the series after which the magnitudes of the terms increase without bound. The bigger x is, the greater is the number of terms before the term of minimum magnitude is reached. It may be your impression that once a series is found to be divergent, it is definitely pathological and should be discarded. Such is not the case; in fact, divergent series are frequently used in numerical work. Such series are called *semiconvergent* or *asymptotic*. Necessarily, only a finite number of terms in such a series can be used.

Suppose that

$$f(x) \simeq \sum_{k=1}^{n} u_k(x) \tag{23.4}$$

is a divergent series that we are trying to use as an approximation for the function $f(x)$ for large values of x. In order to qualify as an asymptotic series, it must satisfy the condition

$$\lim_{x\to\infty} \frac{\left[f(x) - \sum_{k=1}^{n} u_k(x)\right]}{u_n(x)} = 0 \tag{23.5}$$

which can be taken as the definition of what is meant by an asymptotic series. Note that the number of terms n in the series is fixed and that the limit in (23.5) involves taking the variable x large. In other words, the truncated series approaches the function asymptotically as x becomes large. We are thinking here of x as a real variable, but the definition (23.5) works just as well if x is replaced by the complex variable z. The definition implies that for a given positive number ε there is a fixed number A such that

$$\left| f(x) - \sum_{k=1}^{n-1} u_k(x) \right| < (\varepsilon + 1)|u_n(x)| \tag{23.6}$$

whenever $x > A$. Thus, the error in approximating $f(x)$ by its asymptotic series is actually less than the first neglected term. This is a very practical result and means that the greatest accuracy in approximating the function is achieved by cutting the series off at its smallest term. In the actual use of asymptotic series, the magnitudes of the terms might become quite small before they again increase, so that a high degree of accuracy is possible. In the case where the asymptotic series consists of reciprocal powers of x, the definition translates into

$$\lim_{x\to\infty} x^n \left[f(x) - \sum_{k=0}^{n} \frac{a_k}{x^k} \right] = 0 \tag{23.7}$$

We will now test the series (23.3) for the exponential integral to see if it is asymptotic. The series represents the function $E_1(x)e^x$ with remainder

$$|R_n| = \left| e^x \int_x^\infty \frac{n! e^{-t}}{t^{n+1}} dt \right| < \frac{e^x n!}{x^{n+1}} \int_x^\infty e^{-t} dt = \frac{n!}{x^{n+1}} \tag{23.8}$$

Since

$$\lim_{x\to\infty} x^n |R_n| = 0 \tag{23.9}$$

the series is in fact semiconvergent.

An asymptotic series for a given function is unique. If we let $n = 0$ in the form of the definition as given by (23.7), the result is

$$\lim_{x\to\infty} [f(x) - a_0] = 0$$

or

$$\lim_{x\to\infty} f(x) = a_0 \tag{23.10}$$

*See problem 20 in Chapter 11.

Similarly, $n=1$ and $n=2$ give

$$\lim_{x\to\infty} x[f(x)-a_0]=a_1$$

$$\lim_{x\to\infty} x^2\left[f(x)-a_0-\frac{a_1}{x}\right]=a_2 \tag{23.11}$$

In this way, the coefficients are found one by one. Whereas a given function has a unique asymptotic expansion, the reverse is not true, that is, the asymptotic expansion does not necessarily determine the function. More than one function may have the same asymptotic expansion. Generally, an asymptotic expression for a function is of the form

$$f(x)=g(x)\left[1+\frac{a_1}{x}+\frac{a_2}{x^2}+\cdots\right] \tag{23.12}$$

The factor $g(x)$ is called the *leading term*. For example, the leading term in the expression for $E_1(x)$ is e^{-x}. It is the factor in brackets in (23.12) that is the actual asymptotic expansion.

We now turn to the most important task of this section, which is finding an asymptotic expression for the gamma function. It is convenient to work with the expression

$$\Gamma(x+1)=\int_0^\infty e^{-t}t^x\,dt \tag{23.13}$$

We will work with real variables, although the final result is valid in the complex plane as well. We first find the value of t where the integrand of (23.13) is maximum:

$$F(t)=e^{-t}t^x \qquad \frac{dF}{dt}=-e^{-t}t^x+e^{-t}xt^{x-1}=0$$

$$t=x \tag{23.14}$$

The function $F(t)$ is shown graphed as a function of t for four different values of x in Fig. 23.1. The figure suggests that most of the contribution to the integral comes from a fairly narrow range of t centered at $t=x$. Let us write the integrand as

$$F(t)=e^{-t+x\ln t}=e^{f(t)} \tag{23.15}$$

and expand the exponent in a Taylor's series around the point $t=x$:

$$\begin{aligned}
f(t)&=-t+x\ln(x+t-x)\\
&=-t+x\ln x\left(1+\frac{t-x}{x}\right)\\
&=-t+x\ln x+x\ln\left(1+\frac{t-x}{x}\right)\\
&=-t+x\ln x+x\left[\frac{t-x}{x}-\frac{1}{2}\left(\frac{t-x}{x}\right)^2+\cdots\right]\\
&=x\ln x-x-\frac{(t-x)^2}{2x}+\cdots
\end{aligned} \tag{23.16}$$

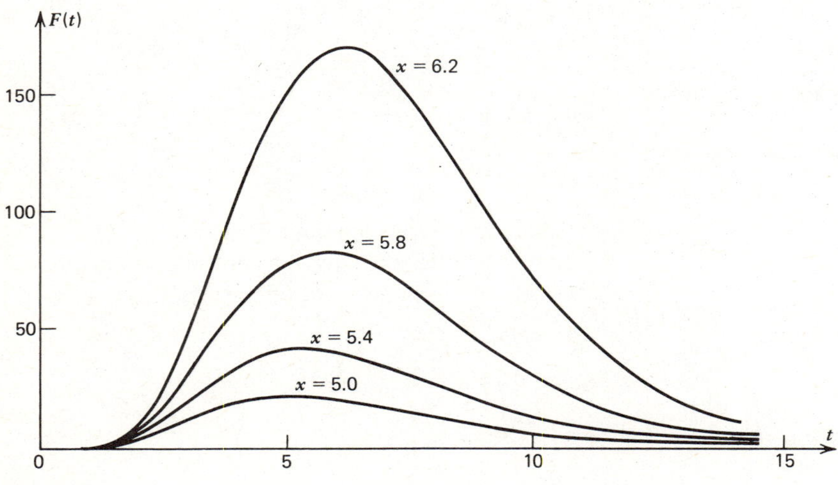

Fig. 23.1

The gamma function is then approximately

$$\Gamma(x+1) \simeq x^x e^{-x} \int_0^{\infty} e^{-(t-x)^2/(2x)}\, dt \tag{23.17}$$

If x is reasonable large, as it must be for an asymptotic expression to be valid, there is little change in the value of the integral if the range of integration is extended to $-\infty$. If this is done, we find

$$\Gamma(x+1) \simeq x^x e^{-x} \int_{-\infty}^{+\infty} e^{-u^2/(2x)}\, du = x^x e^{-x}\sqrt{2\pi x} \tag{23.18}$$

where we have put $t - x = u$ before doing the integration. What we have found here is the leading term in the asymptotic expression for $\Gamma(x+1)$. Let us assume that a series exists and write

$$\Gamma(x+1) \simeq x^x e^{-x}\sqrt{2\pi x}\left(1 + \frac{A}{x} + \frac{B}{x^2} + \frac{C}{x^3} + \cdots\right) \tag{23.19}$$

We have shown that such a series, if it exists, is unique. It is possible to find the coefficients A, B, $C, \ldots$ by means of the recurrence formula $\Gamma(x+1) = x\Gamma(x)$. The recurrence formula implies that

$$\begin{aligned} x^x e^{-x}\sqrt{2\pi x}&\left(1 + \frac{A}{x} + \frac{B}{x^2} + \frac{C}{x^3} + \cdots\right) \\ &= x(x-1)^{x-1} e^{-(x-1)}\sqrt{2\pi(x-1)} \\ &\quad\times\left(1 + \frac{A}{x-1} + \frac{B}{(x-1)^2} + \frac{C}{(x-1)^3} + \cdots\right) \end{aligned} \tag{23.20}$$

A rearrangement of factors gives

$$\begin{aligned} 1 + \frac{A}{x} + \frac{B}{x^2} + \frac{C}{x^3} + \cdots &= \left(1 - \frac{1}{x}\right)^{x-\frac{1}{2}} \\ &\quad\times e\left(1 + \frac{A}{x-1} + \frac{B}{(x-1)^2} + \frac{C}{(x-1)^3} + \cdots\right) \end{aligned} \tag{23.21}$$

Equation (23.21) contains various factors each of which must be expanded in a series of reciprocal powers of x. This is done in the following calculations in which terms to order $(1/x)^3$ are retained:

$$g(x) = \left(1 - \frac{1}{x}\right)^{x-\frac{1}{2}}$$

$$\begin{aligned} \ln g(x) &= \left(x - \frac{1}{2}\right)\ln\left(1 - \frac{1}{x}\right) = \left(x - \frac{1}{2}\right) \\ &\quad\times\left(-\frac{1}{x} - \frac{1}{2x^2} - \frac{1}{3x^3} - \frac{1}{4x^4} - \cdots\right) \\ &= -1 - \frac{1}{12x^2} - \frac{1}{12x^3} - \cdots \end{aligned}$$

$$g(x) = e^{-1}\left(1 - \frac{1}{12x^2} - \frac{1}{12x^3} - \cdots\right) \tag{23.22}$$

$$\frac{1}{x-1} = \frac{1}{x(1-1/x)} = \frac{1}{x} + \frac{1}{x^2} + \frac{1}{x^3} + \cdots \tag{23.23}$$

$$\frac{1}{(x-1)^2} = \frac{1}{x^2(1-1/x)^2} = \frac{1}{x^2} + \frac{2}{x^3} + \cdots \tag{23.24}$$

$$\frac{1}{(x-1)^3} = \frac{1}{x^3} + \cdots \tag{23.25}$$

With these results, (23.21) becomes

$$\begin{aligned} 1 + \frac{A}{x} + \frac{B}{x^2} + \frac{C}{x^3} + \cdots &= \left(1 - \frac{1}{12x^2} - \frac{1}{12x^3} - \cdots\right) \\ &\quad\times\left(1 + \frac{A}{x} + \frac{A}{x^2} + \frac{A}{x^3} + \frac{B}{x^2} + \frac{2B}{x^3} + \frac{C}{x^3} + \cdots\right) \\ &= 1 + \frac{A}{x} + \left(A + B - \frac{1}{12}\right)\frac{1}{x^2} \\ &\quad + \left(\frac{11A}{12} + 2B - \frac{1}{12} + C\right)\frac{1}{x^3} + \cdots \end{aligned} \tag{23.26}$$

Equating coefficients of like powers of $1/x$ on the two sides of (23.26) gives

$$A + B - \frac{1}{12} = B \qquad \frac{11A}{12} + 2B - \frac{1}{12} + C = C$$

Table 23.1

x	$x!$	$\Gamma(x+1)$
1	1	0.998981759
2	2	1.99896287
3	6	5.99832653
4	24	23.9958871
5	120	119.986154
6	720	719.940382
7	5040	5039.68627
8	40320	40318.0454
9	362880	362865.919
10	3628800	3628684.75

The solution for A and B is

$$A = \frac{1}{12} \qquad B = \frac{1}{288} \tag{23.27}$$

The asymptotic representation of the gamma function is therefore

$$\Gamma(x+1) = x^x e^{-x}\sqrt{2\pi x}\left(1 + \frac{1}{12x} + \frac{1}{288x^2} + \cdots\right) \tag{23.28}$$

which is actually valid even if x is replaced by the complex variable z, provided that we stay away from the negative real axis.

According to our definition (23.7), asymptotic expansions are strictly valid only in the limit $x \to \infty$. What is remarkable about (23.28) is its great accuracy, even for small values of x. Table 23.1 shows a computation of $x!$ compared with $\Gamma(x+1)$ for integer values of x as computed from (23.28). In the computations, the term $1/(288x^2)$ has been omitted.

The entire gamma function for real values of x can now be easily constructed. All we have to do is compute $\Gamma(x+1)$ for a unit interval of x values, say $5 \le x \le 6$, and then project backward along the x axis by means of the recurrence formula. Thus,

$$\Gamma(x) = \frac{\Gamma(x+1)}{x} \tag{23.29}$$

gives values of $\Gamma(x+1)$ in the range $4 \le x \le 5$;

$$\Gamma(x-1) = \frac{\Gamma(x)}{x-1} \tag{23.30}$$

gives $\Gamma(x+1)$ in the range $3 \le x \le 4$, and so on. This is the method that was used to construct the graph of $\Gamma(x+1)$ shown in Fig. 23.2. Better numerical accuracy is obtained by choosing larger initial values of x.

Expansions of functions in terms of reciprocal powers of x have been obtained at various points in this book which are actually convergent and do not

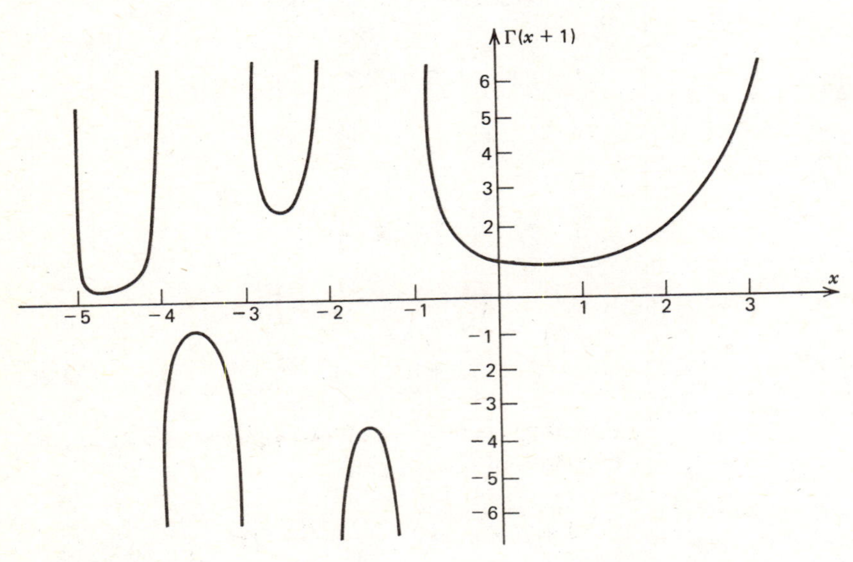

Fig. 23.2

in the strict sense qualify as asymptotic expansions. For example, the expansions (23.22) through (23.25) are all convergent for $|x| > 1$. There is really little point in making this distinction. Some expansions in powers of $1/x$ converge and others do not. Some of those that do not converge might qualify as asymptotic according to the definition (23.7) and therefore be useful as representations of functions for large values of the argument. We have not demonstrated that the expression (23.28) for the gamma function is asymptotic. There is no practical reason to worry about this since the formula can be checked against integer values of x which are easy to calculate directly. Obviously, (23.28) provides an accurate extrapolation for $\Gamma(x+1)$ between integer values of x if x is not too small.

PROBLEMS

23.1. The functions known as the sine integral and the cosine integral are defined by

$$si(x) = -\int_x^\infty \frac{\sin t}{t}\,dt \qquad Ci(x) = -\int_x^\infty \frac{\cos t}{t}\,dt \tag{23.31}$$

It is possible to establish a connection between the sine and cosine integrals and the exponential function. A general definition of the exponential integral is

$$E_1(z) = \int_z^\infty \frac{e^{-t}}{t}\,dt \tag{23.32}$$

where z is a complex variable. The series expansion obtained in problem (22.12) makes it evident that $E_1(z)$ has a logarithmic singularity at $z = 0$. It is the usual practice to put a branch cut along the negative real axis. A special case of (23.32) is

$$E_1(ix) = \int_{ix}^\infty \frac{e^{-t}}{t}\,dt \tag{23.33}$$

By making the change of variable $t = iu$, show that

$$E_1(ix) = \int_x^{-i\infty} \frac{e^{-iu}}{u}\,du \tag{23.34}$$

A possible path of integration for (23.34) is shown in Fig. 23.3a. If $R \to \infty$, then the path of Fig. 23.3b connects the same two points. Since distorting the path in this manner is done entirely in a region where the integrand of (23.34) is analytic, the two paths are equivalent for evaluating the integral. Show that the contribution to the integral from the quarter-circle arc

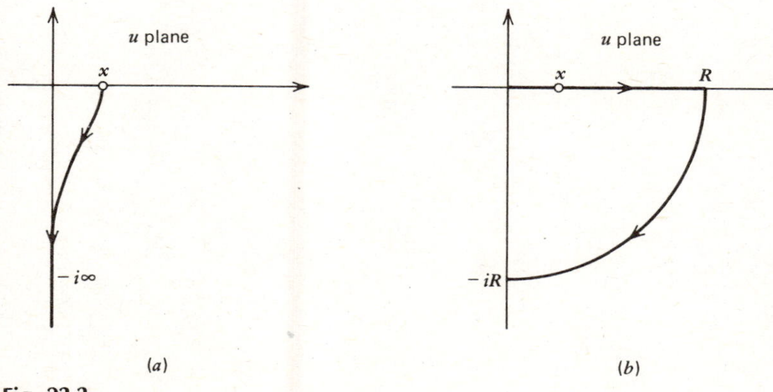

Fig. 23.3

becomes zero as $R \to \infty$. Therefore, (23.34) is equivalent to

$$E_1(ix) = \int_x^\infty \frac{e^{-iu}}{u}\,du = \int_x^\infty \frac{\cos u}{u}\,du - i\int_x^\infty \frac{\sin u}{u}\,du \tag{23.35}$$

Use this result and (23.3) to obtain the following asymptotic expansions:

$$si(x) = -\sin x \sum_{n=0}^{N} \frac{(-1)^n(2n+1)!}{x^{2n+2}} - \cos x \sum_{n=0}^{N} \frac{(-1)^n(2n)!}{x^{2n+1}} \tag{23.36}$$

$$Ci(x) = -\cos x \sum_{n=0}^{N} \frac{(-1)^n(2n+1)!}{x^{2n+2}} + \sin x \sum_{n=0}^{N} \frac{(-1)^n(2n)!}{x^{2n+1}}$$

23.2. Show that the first three terms of the asymptotic formula for the digamma function are

$$\psi(x) = \ln x - \frac{1}{2x} - \frac{1}{12x^2} + \cdots \tag{23.37}$$

23.3. The *incomplete gamma function* is defined by

$$\Gamma(n,x) = \int_x^\infty e^{-t}t^{n-1}\,dt \tag{23.38}$$

By repeated integration by parts, find an asymptotic series for $\Gamma(n,x)$, valid for $x \to \infty$.

23.4. For $x = 6$, compute the magnitudes of the terms in the asymptotic series (23.3) for $E_1(x)e^x$ up to the minimum term. Using this many terms, calculate the value of the function and estimate the error.

$$\text{Erf}(x) = \frac{2}{\sqrt{\pi}} \int_0^x e^{-t^2}\, dt$$

6 Probability, Statistics, and Experimental Errors

This is a short introductory chapter on a subject that has an extensive literature. Only those aspects of the theory which are likely to be of use to students of the physical sciences are emphasized. It is convenient to introduce the subject of probability and statistics at this point because it provides good examples of the use of some of the mathematical results of Chapter 5.

1. THE PROBABILITY CONCEPT

If someone tosses a coin, it is possible in principle to start from the initial conditions and predict by means of Newtonian mechanics the final outcome. The detailed information necessary to carry out such a program is seldom available. The best that can be done is to make a statistical statement about the possible outcome. If the coin is "fair," most everyone would agree that the probability of getting heads on a single toss is 50%. In other words, the assumption is made that equal a priori probabilities should be assigned to the possible outcomes of the experiment. As another example, a container of gas might hold 10^{23} molecules. It is *not* possible even in principle to know the exact positions and velocities of all the molecules at some given time and then predict from this information the possible future states of the gas. Statistical methods are therefore essential where the number of degrees of freedom is large. The assumption of equal a priori probabilities can be tested by tossing a coin a very large number of times and observing that, on the average, heads occurs half the time. Not exactly half the time, of course. If a coin is tossed four times, the outcome of two heads and two tails may be the most likely, but certainly three heads and one tail would occur quite frequently. One thing that we will be able to do as the theory is developed is to show that the expected *percent* deviation of the number of heads from one half of the total decreases as the number of tosses increases.

There are two formal definitions of probability. Suppose that an experiment is to be carried out consisting of many trials and that there are two possible outcomes of each trial. We will arbitrarily call one outcome "success" and the other "failure." If N is the number of trials and N_s is the number of successes, then the probability of success for any of the trials is

$$P_s = \lim_{N\to\infty} \frac{N_s}{N} \tag{1.1}$$

which is called the *a posteriori* definition of probability. As a consequence of this definition,

$$P_s + P_f = 1 \tag{1.2}$$

where P_f is the probability of failure. The a posteriori definition of probability is useful if it is suspected that a coin is bent or a pair of dice have been loaded in some unknown way. Our only recourse might then be to actually roll the dice a large number of times to measure the probability of a given outcome. If the coin is fair and the dice are not loaded and are reasonably symmetrical, another definition of probability is possible. Let n be the total number of possible outcomes of an experiment and let n_s be the number of these outcomes which result in success. Assume that each of the n outcomes is equally likely. The probability of success is defined to be

$$P_s = \frac{n_s}{n} \tag{1.3}$$

and is called *a priori* probability. In a coin tossing experiment, $n = 2$ and $n_s = 1$. In the roll of a single die, $n = 6$. If "success" is getting an ace, then $n_s = 1$. Theoretically, the two definitions of probability are equivalent and lead to the same value of P_s.

Getting a head or a tail in a single toss of a coin are mutually exclusive events in that only one of them can occur. More generally, if A, B, C,... represent all the mutually exclusive outcomes which can occur in a given experiment, then by the definition of probability,

$$P(\mathrm{A}) + P(\mathrm{B}) + P(\mathrm{C}) + \cdots = 1 \tag{1.4}$$

In the roll of a single die, there are six possible mutually exclusive outcomes. Events A, B, C,... are *statistically independent* if the occurrence of any one of them in one trial in no way affects the outcomes of subsequent trials. For example, successive tosses of a coin are independent events. If A, B, C,... are statistically independent, then the probability of the occurrence of A followed by B, then C,... is

$$P_{ABC} = P(\mathrm{A})\,P(\mathrm{B})\,P(\mathrm{C})\ldots \tag{1.5}$$

which is the multiplicative law for probabilities. We will first demonstrate the validity of this theorem by an example. Suppose that a coin is to be tossed three times and we ask for the probability of getting two heads and a tail in the order HHT. A diagram which enumerates all the $2^3 = 8$ possible outcomes of the experiment is shown in Fig. 1.1. Each of these outcomes is equally likely. Only one of the outcomes is HHT, so that the probability of getting this result is

$$P_{HHT} = \frac{1}{8} = \left(\frac{1}{2}\right)^3 \tag{1.6}$$

in agreement with (1.5). The diagram of Fig. 1.1 represents the *sample space* of the problem. There are eight points in the sample space, each of which

HHH	*HTH*	*THT*
HTT	*TTT*	*TTH*
HHT	*THH*	

Fig. 1.1

11	12	13	14	15	16
21	22	23	24	25	26
31	32	33	34	35	36
41	42	43	44	45	46
51	52	53	54	55	56
61	62	63	64	65	66

Fig. 1.2

represents a possible outcome of the experiment. To construct a more general proof of (1.5), suppose that two independent events A and B are possible. For a given trial, there are n possible outcomes of which n_A give the result A and n_B give the result B. The sample space for two such trials consists of n^2 points. The number of ways in which A can occur followed by B is $n_A \times n_B$. Therefore, the probability of getting A on the first trial and B on the second trial is

$$P_{AB} = \frac{n_A n_B}{n^2} = \frac{n_A}{n}\frac{n_B}{n} = P_A P_B \tag{1.7}$$

Suppose that two dice are rolled and we are asked to find the probability of getting at least one ace. Since the probability of getting an ace on either die is $1/6$, we are tempted to say that answer is $1/6 + 1/6 = 1/3$. But this is incorrect. The sample space for this experiment is constructed in Fig. 1.2 and consists of 36 points. The number of points which involve at least one ace is 11. The probability that one or both of the dice comes up an ace is therefore $11/36$.

The previous paragraph deals with an example of compound events which we now analyze in more general terms. Suppose that an experiment is to be performed and that there are two possible events A and B in which we are interested. The events need not be mutually exclusive or statistically independent, and they need not be the only possible events that can occur. Suppose that there are a total of n possible outcomes or points in the sample space which are all equally likely. The possibilities with respect to the events A and B are listed in the following table:

Number	Outcome
n_1	A occurs but not B
n_2	B occurs but not A
n_3	Both A and B occur
n_4	Neither A nor B occurs

Since all possibilities are exhausted,

$$n = n_1 + n_2 + n_3 + n_4 \tag{1.8}$$

It is now possible to calculate various probabilities. If $P(\mathrm{A})$ is the probability that A occurs and $P(\mathrm{B})$ is the probability that B occurs, then

$$P(\mathrm{A}) = \frac{n_1 + n_3}{n} \qquad P(\mathrm{B}) = \frac{n_2 + n_3}{n} \tag{1.9}$$

The probability of getting either A or B or both A and B together is

$$P(\mathrm{A} + \mathrm{B}) = \frac{n_1 + n_2 + n_3}{n} \tag{1.10}$$

The probability that both A and B occur together is called the *joint probability* of A and B and is

$$P(\mathrm{AB}) = \frac{n_3}{n} \tag{1.11}$$

It is now possible to combine (1.9), (1.10), and (1.11) to get

$$P(\mathrm{A} + \mathrm{B}) = P(\mathrm{A}) + P(\mathrm{B}) - P(\mathrm{AB}) \tag{1.12}$$

As an example, equation (1.12) can be used to calculate the probability of getting at least one ace in the roll of two dice. The probability that one die comes up an ace is $P(\mathrm{A}) = 1/6$. Similarly, the probability that the other die is an ace is $P(\mathrm{B}) = 1/6$. The joint probability that the two aces occur together is $P(\mathrm{AB}) = P(\mathrm{A})\,P(\mathrm{B}) = 1/36$. This gives $P(\mathrm{A} + \mathrm{B}) = 11/36$, in agreement with the result computed directly from a detailed consideration of the sample space. In general, if the two events are mutually exclusive, then $P(\mathrm{AB}) = 0$. If the two events are statistically independent, as they are in the example of the two dice, then $P(\mathrm{AB}) = P(\mathrm{A})\,P(\mathrm{B})$.

As another example, suppose (again) that two dice are rolled. Let A be the event that the sum of the faces is odd and B the event that at least one face is an ace. What are the probabilities that (a) both A and B occur, (b) either A or B or both occur, (c) A but not B occurs, and (d) B but not A occurs?

Since half of the 36 possible outcomes as displayed in the sample space of Fig. 1.2 give an odd sum for the two faces, $P(\mathrm{A}) = 0.5$. We have already shown that $P(\mathrm{B}) = 11/36$. Of the 11 points in the sample space that involve aces, 6 give an odd sum. Therefore, $P(\mathrm{AB}) = 6/36 = 1/6$. This is the joint probability of A and B and is the answer to (a). Note that $P(\mathrm{AB}) \neq P(\mathrm{A})\,P(\mathrm{B})$, so that A and B are not statistically independent. Also, since they can occur together, they are not mutually exclusive events. The answer to part (b) is $P(\mathrm{A} + \mathrm{B}) = P(\mathrm{A}) + P(\mathrm{B}) - P(\mathrm{AB}) = 23/36$. In equation (1.9), $P(\mathrm{A})$ is the probability that A occurs regardless of what else happens. If we let P_A be the probability that A occurs alone without B, then $P_A = n_1/n$ and (1.9) can be expressed as $P(\mathrm{A}) = P_A + P(\mathrm{AB})$. This gives $P_A = 1/3$. Similarly, the probability that B occurs alone is $P_B = P(\mathrm{B}) - P(\mathrm{AB}) = 5/36$.

As a somewhat different example, suppose that 30 students are in a classroom and we ask for the probability that at least two of the students have a birthday on the same day. The easiest way to do the problem is to calculate the probability that *none* of the students has a common birthday. Imagine that the 365 days of the year are bins into which the students are to be placed one by one. We first determine the total number of ways that the students can be placed in the 365 bins without regard to how many end up in the same bin. This is actually the total number of points in the sample space for the problem. There are 365 choices for the first student, 365 choices for the second student, ..., giving a total of $(365)^{30}$ ways of placing 30 students in 365 bins. We next determine the number of ways the students can be placed in 30 bins with never more than one in the same bin. The first student has 365 choices, the second 364, ..., giving a total of $(365)(365-1)(365-2)\ldots(365-29)$ choices. The probability that none of the students has common birthdays is therefore

$$P = \frac{(365)(365-1)(365-2)\ldots(365-29)}{(365)^{30}} = (1)\left(1 - \frac{1}{365}\right)\left(1 - \frac{2}{365}\right)\cdots\left(1 - \frac{29}{365}\right) = 0.2937 \tag{1.13}$$

The probability of multiple birthdays occurring is therefore $1 - P = 0.7063$.

Appearing in the birthday problem is a quantity of the form

$$ {}_nP_k = n(n-1)(n-2)\ldots(n-k+1) = \frac{n!}{(n-k)!} \tag{1.14} $$

This is generally referred to as "the number of permutations of n things taken k at a time." If k objects are selected out of n, (1.14) is the number of ways that they can be arranged side by side in a straight line. The total number of permutations of n objects is ${}_nP_n = n!$.

PROBLEMS

1.1. Two urns each contain a gold coin and two silver coins. One coin is to be drawn from each urn. Construct the sample space for the problem. What is the probability of getting a gold coin on the first draw and a silver coin on the second? What is the probability of getting a gold coin and a silver coin regardless of the order in which they are obtained?

1.2. In tossing a coin three times, what is the probability of getting two heads and one tail in *any* order?

1.3. An experiment is to be performed in which there are n possible outcomes with A and B as two possible events. Show that the probability that A or B will occur but not both is

$$ P(\text{A or B}) = P(\text{A}) + P(\text{B}) - 2P(\text{AB}) $$

1.4. A jar contains three red and five black balls. Find the probability of drawing (a) two red balls simultaneously, (b) three red balls in successive draws, with replacement after each draw, and (c) two reds and a black in a single draw of three balls. [The answer to (c) is $15/56$.]

1.5. If two dice are rolled, the sum of the two numbers obtained can be any integer from 2 to 12. List the probabilities for obtaining each sum and check that the probabilities sum to unity.

1.6. One die is rolled until an ace appears. What is the probability that this will happen on the first roll? The second? The third? The kth? Verify that the sum of these probabilities is unity in the limit $k \to \infty$.

1.7. Four cards are to be drawn from a deck of 52 playing cards. What is the probability that they are the four aces?

1.8. In a game of poker, five cards are dealt to each person. If you are in the game, what is the probability that you will be dealt four of a kind?

Solution: Since there are 52 cards in the deck, the sample space for the problem has ${}_{52}P_5 = 52 \times 51 \times 50 \times 49 \times 48$ points in it. The five different ways in which five cards can be drawn from the deck which result in four being alike and the other one being different are diagrammed in Fig. 1.3. In sequence (a), the first card drawn is to be the different one and is labeled X. Since this can be any card in the deck, there are 52 possibilities. The next card, labeled O, is to start the sequence of four like cards. Since this can be any card except one like the first one, there are 48 choices. The third card must be one of the three remaining that are like the second card. Similarly, there are two choices for the fourth card and only one choice for the fifth card. The other sequences are analyzed similarly. It turns out that the total number of ways of achieving each sequence is the same. The required probability is therefore

$$ P = \frac{5 \times (52 \times 48 \times 3 \times 2 \times 1)}{52 \times 51 \times 50 \times 49 \times 48} = \frac{1}{4165} = 2.401 \times 10^{-4} $$

(a)	X 52	O 48	O 3	O 2	O 1
(b)	O 52	X 48	O 3	O 2	O 1
(c)	O 52	O 3	X 48	O 2	O 1
(d)	O 52	O 3	O 2	X 48	O 1
(e)	O 52	O 3	O 2	O 1	X 48

Fig. 1.3

1.9. A club has 30 members. How many different committees each consisting of four persons can be formed from the members?

1.10. An experiment is to be performed in which there are n possible outcomes with A and B as two possible events. The probability that A occurs, *given that B occurs*, is called a conditional probability and is given the designation $P(\text{A}|\text{B})$. By using the notation of equations (1.8) through (1.11), show that

$$P(\text{A}|\text{B}) = \frac{n_3}{n_2 + n_3} \qquad P(\text{B}|\text{A}) = \frac{n_3}{n_1 + n_3}$$

$$P(\text{AB}) = P(\text{B})\, P(\text{A}|\text{B}) = P(\text{A})\, P(\text{B}|\text{A})$$

If C is a third possible event, show that

$$\frac{P(\text{B}|\text{A})}{P(\text{C}|\text{A})} = \frac{P(\text{B})\, P(\text{A}|\text{B})}{P(\text{C})\, P(\text{A}|\text{C})}$$

This is known as *Bayes' theorem*.

1.11. Suppose there are three jars each containing two coins. Jar A contains two gold coins, jar B contains a gold coin and a silver coin, and jar C has two silver coins. A jar is picked at random, and a coin is drawn from it and found to be gold. What is the probability that the other coin is also gold?

Hint: Let A, B, and C be the events of choosing the jars and D the event of drawing the first gold coin. Use the theorem of problem 10 to calculate $P(\text{A}|\text{D})$.

2. THE BINOMIAL DISTRIBUTION

An experiment is done in which there are n trials. Each trial has two possible outcomes one of which we will arbitrarily call "success" and the other "failure." Such trials are sometimes referred to as *Bernoulli trials*. If the probability of success for each trial is p and the probability of failure is q, then $p + q = 1$. If a coin is tossed 10 times, then $n = 10$ and $p = q = 0.5$. As another example, suppose that a particle is placed at the origin and that it can make one-unit moves in either the positive or the negative x direction. Success might be defined as a move in the positive x direction and failure as a move in the negative x direction. This is a simple statistical model of *particle diffusion* or *random walk*. It is of interest to know the probability of finding the particle at a specific location after it makes n moves. A more interesting question arises if a large number N of particles is placed at the origin and we inquire as to the *distribution* of particles around the origin after each particle has made n moves. In other words, we ask for the number of particles to be found at the end of each unit interval of the x axis.

In the particle diffusion problem, the final location of a given particle depends on the number k of moves in the positive x direction minus the number of moves in the negative x direction: $x = k - (n-k) = 2k - n$. It does not matter in which order the positive and negative moves occurred. What we therefore require is the probability of getting k successes out of n Bernoulli trials without regard to the order in which they occurred. The eight possible outcomes of tossing a coin three times are shown in Fig. 1.1. The probability of any one of the outcomes is $(1/2)^3$. There are three different ways of getting two heads and one tail. The probability of getting two heads and one tail without regard to the order in which they occur is therefore $3 \times (1/2)^3 = 3/8$. Tossing a coin n times is equivalent to tossing n coins one time. Suppose that this is done and the coins are all arranged in a straight line. The number of different possible distinguishable arrangements is

$$\frac{{}_nP_k}{k!} = {}_nC_k = \binom{n}{k} = \frac{n!}{(n-k)!k!} \tag{2.1}$$

and is called the *number of combinations of* n *things taken* k *at a time*. It is also a binomial coefficient. To understand (2.1), imagine that n objects are to be added to two jars, one labeled H and the other T. The number of ways that k objects out of n can be added to the H jar is $n(n-1)(n-2)\ldots(n-k+1) = {}_nP_k$. However, the order in which the k objects are added is immaterial. Since there are $k!$ possible arrangements of k objects, ${}_nP_k$ must be divided by $k!$. For example, the number of distinct possible arrangements of two heads and two tails is $4!/(2!2!) = 6$. Specifically, these arrangements are HHTT, HTHT, HTTH, THHT, THTH, and TTHH. We therefore find that the probability of k successes out of n Bernoulli trials is

$$P(n;k,p) = \frac{n!}{(n-k)!k!} p^k q^{n-k} \tag{2.2}$$

This is called the *binomial distribution*.

To see the meaning of the binomial distribution more clearly, suppose that a coin is tossed $n = 10$ times and that this coin tossing experiment is repeated over and over N times, where N is a very large number. In each case, the number of heads out of the 10 trials is recorded. We would expect that the number of times that k heads are obtained to be given by

$$N_k = NP(10;k,0.5) \tag{2.3}$$

In fact, according to the a posteriori definition of probability, (2.3) should become exact in the limit as $N \to \infty$. Figures 2.1*a* through 2.1*d* show plots of the binomial distribution (2.2) for $p = q = 1/2$ and $n =$ 10, 20, 30, and 40. Note the characteristic symmetric bell shape. Figures 2.1*e* and 2.1*f* are for $n = 40$ and $p = 0.1$ and 0.9. Note that the distribution is no longer symmetric around its maximum value. It is only $p = q = 1/2$ that gives a symmetric distribution.

For the symmetric distributions of Fig. 2.1, it is easy to see that the average value of k is $\bar{k} = (1/2)n = np$ and that this corresponds to the maximum of the distribution (not exactly the maximum if n is not divisible by 2). We will verify the general validity of this observation by a formal calculation. The coin-tossing experiment serves as a convenient guide. The average value or arithmetic mean of k can be defined by

$$\bar{k} = \frac{k_1 + k_2 + \cdots + k_N}{N} \tag{2.4}$$

where $k_1, k_2, \ldots k_N$ are the measured values of k for each one of the N experiments. Remember that each experiment consists of n tosses. Since we are envisioning N to be a very large number and n as small by comparison, the same value of k appears several times in (2.4). Since k can range from 0 to n, (2.4) can be rewritten as

$$\bar{k} = \frac{0 \cdot N_0 + 1 \cdot N_1 + 2 \cdot N_2 + \cdots nN_n}{N} \tag{2.5}$$

With the aid of (2.3),

$$\bar{k} = \sum_{k=0}^{n} kP(n;k,p) \tag{2.6}$$

which shows that the average value of k is to be computed using the probability of the occurrence of a given value of k as a weighting factor. Equation (2.6) is a theoretical expression; an actual experimental determination of $\bar{k}$ is based on (2.4). The two values become the same in the limit as $N \to \infty$. Since it is

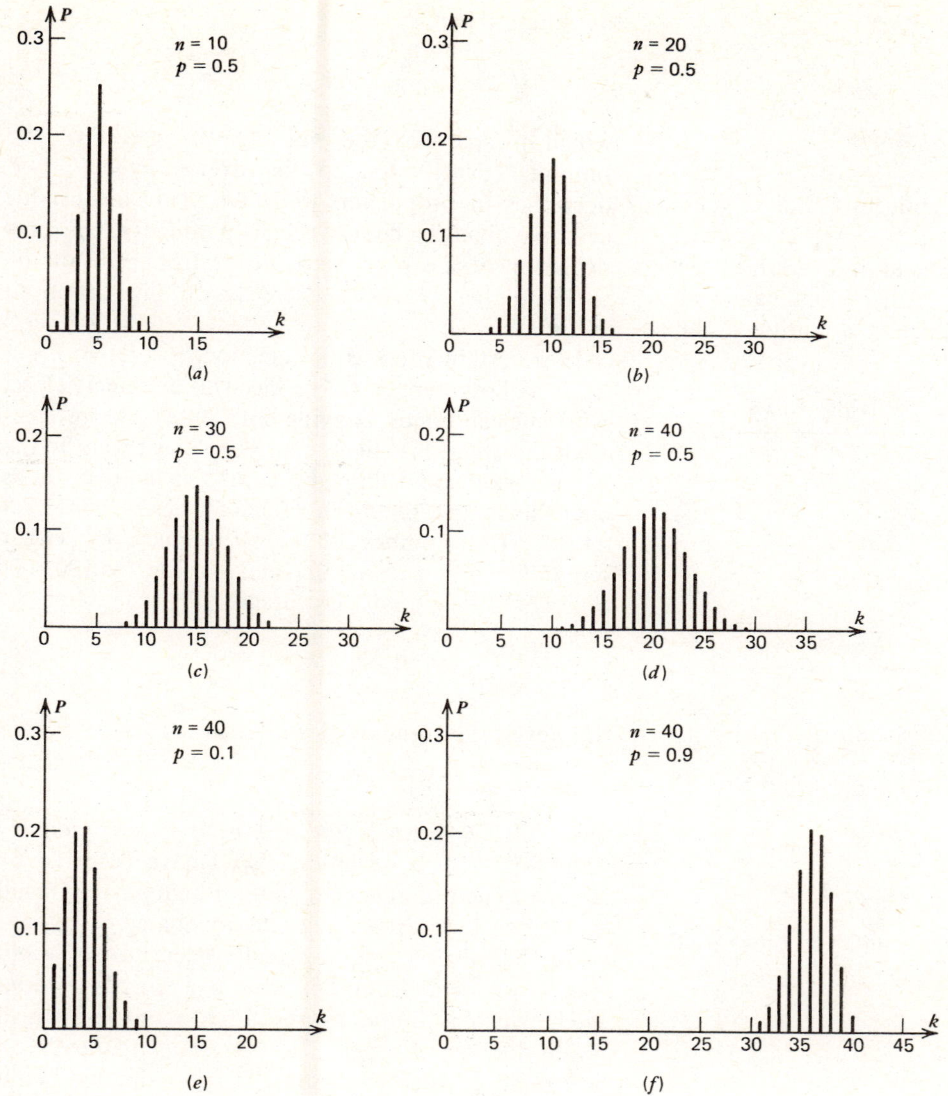

Fig. 2.1

the binomial coefficient which appears in (2.2), we have the result that

$$1 = (p+q)^n = \sum_{k=0}^{n} P(n; k, p) \tag{2.7}$$

as is to be expected since the possible k values represent the totality of the mutually exclusive events which can occur. Equation (2.6) can be expressed as

$$\begin{aligned}\bar{k} &= \sum_{k=0}^{n} \frac{n!k}{(n-k)!k!} p^k q^{n-k} \\ &= \sum_{k=1}^{n} \frac{n!}{(n-k)!(k-1)!} p^k q^{n-k}\end{aligned} \tag{2.8}$$

If k is replaced by $k+1$ and we write $n! = n(n-1)!$,

(2.8) can be rewritten as

$$\bar{k} = \sum_{k=0}^{n-1} \frac{(n-1)!}{(n-1-k)!k!} p^k q^{n-1-k} np$$

$$= (p+q)^{n-1} np = np \tag{2.9}$$

which confirms the original conjecture and is valid for $0 < p < 1$.

We need a quantity that gives an indication of the spread of either the observed or theoretical values of k around the mean value. Such a quantity is the *variance* defined by

$$v = \frac{(k_1 - \bar{k})^2 + (k_2 - \bar{k})^2 + \cdots + (k_N - \bar{k})^2}{N}$$

$$= \sum_{i=1}^{N} \frac{(k_i - \bar{k})^2}{N} \tag{2.10}$$

It is possible to express the variance as

$$v = \sum_{k=1}^{N} \frac{k_i^2 - 2k_i\bar{k} + \bar{k}^2}{N} = \overline{k^2} - 2\bar{k}^2 + \bar{k}^2 = \overline{k^2} - \bar{k}^2 \tag{2.11}$$

where $\overline{k^2}$ is the average value of k^2. By the same argument that led to (2.6), taking the limit $N \to \infty$ gives

$$\overline{k^2} = \sum_{k=0}^{n} k^2 P(n; k, p) \tag{2.12}$$

By using procedures which are similar to those used in the derivation of (2.9), we find

$$\overline{k^2} = n^2p^2 + npq \tag{2.13}$$

The variance is therefore

$$v = npq \tag{2.14}$$

The *standard deviation* is defined by

$$\sigma = \sqrt{v} = \sqrt{npq} \tag{2.15}$$

If $n = 10$ and $p = q = 1/2$, $\sigma = 1.58$. An inspection of Fig. 2.1*a* shows that almost all of the observed values of k fall in the range $\bar{k} - 2\sigma < k < \bar{k} + 2\sigma$. We will show later that if n is large and $p \simeq q$, approximately 95% of the observed values of k fall within $\pm$ two standard deviations of the mean. One of the most significant features of the standard deviation is found by noting that

$$\frac{\sigma}{\bar{k}} = \sqrt{\frac{q}{np}} \tag{2.16}$$

which means that the ratio of the width of the binomial distribution to the average value of k decreases in proportion to $1/\sqrt{n}$. This means, for example, that in a coin-tossing experiment the *percent* deviation of the number of heads from 50% of the total number of tosses decreases in proportion to $1/\sqrt{n}$.

In a certain presidential election, a total of $n = 7 \times 10^7$ ballots were cast. The race was extremely close, with one candidate receiving only 200,000 more votes than the other. Is it likely that people just went to the voting booths and flipped a coin? On the hypothesis that this is true, the mean value of k is $\bar{k} = 0.5n = 3.5 \times 10^7$. The number of votes received by either candidate differs from the mean value by 100,000, which is a very small percentage:

$$\frac{100{,}000}{3.5 \times 10^7} \times 100 = 0.29\%$$

However, the standard deviation is

$$\sigma = \sqrt{npq} = 4.18 \times 10^3$$

and 100,000 votes is about 24σ. It is very unlikely that this great a deviation would have occurred by chance. The difference, in spite of being a very small percentage, indicates a definite preference.

In this section, we have really been talking about two kinds of distributions. One is a *sample distribution*, which is found by repeating an experiment a finite number of times. The other is a mathematical abstraction which we assume is approximated better and better by the sample distribution the greater the number of times the experiment is repeated. This abstract distribution is called the *parent distribution*. In the analysis of experimental data by statistical methods, the parent distribution with its precisely defined mean value and standard deviation is presumed to exist even though it can be only approximately determined by the actual data.

An ideal gas consists of molecules which can be regarded as point particles that do not interact appreciably with one another. The particles move about at random in a container, and we expect on the average that the density of particles will be the same at any location. What kind of statistical density fluctuations might we expect to occur in a small volume element? Suppose that the container has in it a total of N particles. The container is divided conceptually into n volume elements into which the particles distribute themselves at random. The average number of particles per volume element is $\bar{k} = N/n$. Equation (2.16) can be used to estimate the expected statistical fluctuation in the number of molecules in the volume element:

$$\frac{\delta k}{\bar{k}} = \sqrt{\frac{q}{Np}} \tag{2.17}$$

Since the particles are added at random to the container, the probability that a given particle will be in the volume element in question is $p = 1/n$. The probability of it being elsewhere in the container is $q = 1 - 1/n$. If n is large, $q \simeq 1$ and

$$\frac{\delta k}{\bar{k}} = \frac{1}{\sqrt{N(1/n)}} = \frac{1}{\sqrt{\bar{k}}} \tag{2.18}$$

At one atmosphere pressure and room temperature, the number of air molecules in 1 cubic millimeter is $\bar{k} = 2.5 \times 10^{16}$. The expected statistical fluctuation of the density of the air in this volume element is

$$\frac{\delta k}{\bar{k}} = 6 \times 10^{-9} \tag{2.19}$$

which is for all practical purposes undetectable.

PROBLEMS

2.1. Show that the binomial coefficients obey the recurrence relation

$$\binom{n}{k+1} = \frac{n-k}{k+1}\binom{n}{k}$$

This is useful in computer evaluations of binomial coefficients.

2.2. Prove equation (2.13).

2.3. A coin is tossed 10 times. What is the probability that three heads and seven tails will be obtained? The experiment is repeated $N = 100$ times. What is the average or "expected" number of times that three heads and seven tails are obtained? Using the same line of reasoning that led to equation (2.18), estimate how much the number of times three heads are observed might reasonably be expected to vary from the average value. More generally, suppose that an experiment involving n trials is to be repeated N times. If the parent distribution predicts that k successes and $n - k$ failures will occur N_k times, show that the sample distribution can be expected to conform to the parent distribution to within $\delta N_k \simeq \pm\sqrt{N_k}$.

Hint: Each of the N experiments involving n subtrials is itself a Bernoulli trial where the "success" probability is $P(n; k, p)$. If n is not too small, the probability of observing a particular value of k, even one near $\bar{k}$, is small, and the failure probability is $Q = 1 - P \simeq 1$.

2.4. In what size volume element of air at a pressure of one atmosphere would a statistical density fluctuation of about 1% be expected?

2.5. A pair of dice are rolled and the sum of the numbers on the up faces are added together. Find and plot the parent probability distribution for obtaining each of the possible sums. See problem 1.5. Find (a) the most probable value of the sum, (b) the average value for a very large number of rolls, and (c) the standard deviation.

2.6. Among a large number of eggs, 1% were found to be rotten. In a dozen eggs, what is the probability that none is rotten? One? More than one?

2.7. A man stands on a street corner. He flips two coins. If he gets two tails, he walks one block south; for any other result he walks one block north. Find his possible positions after four tosses and the probability for each. What is the most probable position? If he does this experiment once a day for a large number of days, what is his average position?

2.8. Four men are sent to a jail where there are only two jail cells. As each man comes in, the officer in charge sends him to a cell at random. Find the probabilities that (a) all four men end up in the same cell, (b) three are in one cell and one is in the other, and (c) there are two men in each cell.

3. POISSON AND NORMAL DISTRIBUTIONS

We consider in this section two useful approximations to the binomial distribution. Suppose that we have a sample of radioactive material consisting of a large number of atoms, say $n \simeq 10^{20}$. In a time interval t, a few of these decay. If each atom is regarded as a Bernoulli trial and decay is a success, then the probability of k decays in the time interval can be calculated from the binomial distribution. The number of trials is very large. The probability of success is the probability that any one atom will decay and is extremely small. We therefore seek an approximation to the binomial distribution which is valid in the limit $n \to \infty$ and $p \to 0$. The mean value of k is np, and the limit is to be taken in such a way that np remains finite. The binomial distribution can be expressed as

$$P(k) = \frac{n(n-1)\ldots(n-k+1)}{k!} p^k (1-p)^{n-k} \tag{3.1}$$

Only the probabilities for observing small values of k are significant; therefore, $n(n-1)\ldots(n-k+1) \simeq n^k$ and

$$P(k) \simeq \frac{(np)^k (1-p)^{np/p}}{k!(1-p)^k} \tag{3.2}$$

In the limit $p \to 0$, $(1-p)^k \simeq 1$. It is convenient to use the notation $a = \bar{k} = np$ for the average value of k. By using

$$\lim_{p \to 0} (1-p)^{a/p} = e^{-a} \tag{3.3}$$

it is possible to express (3.2) as

$$P(k) \simeq \frac{a^k}{k!} e^{-a} \tag{3.4}$$

which is known as the *Poisson distribution*. To check that it is still correctly normalized, note that

$$\sum_{k=0}^{\infty} P(k) = \sum_{k=0}^{\infty} \frac{a^k}{k!} e^{-a} = e^a e^{-a} = 1 \tag{3.5}$$

The sum is extended to ∞ because in the derivation we assumed that the number of Bernoulli trials (atoms in the radioactive sample) is infinite.

Suppose that we have a radioactive material that contains 10^{10} atoms and that the decay rate is 3 counts per second. If counting is extended over a 100-sec interval, the depletion in the original number of atoms is negligible so that the average counting rate remains constant. It is very unlikely that exactly 3 counts would be observed in each of the 1-sec intervals. The Poisson distribution can be used to calculate the expected number of 1-sec intervals in which no counts are seen, the number in which 1 count is seen, and so on. The decay probability per atom per second is $p = 3 \times 10^{-10}$. The Poisson distribution gives the probability that k counts will be observed in a specific 1-sec interval. Table 3.1 gives values of k, $P(k)$, and $100P(k)$ for $a = 3$. The significance of $100P(x)$ is that it is the expected number of 1-sec intervals out of 100 in which k counts are seen. Note that 2 counts per second can be expected to be observed just as often as 3 counts per second even though 3 counts is the average. The Poisson distribution is not symmetric. The most probable value of k is given by the maximum of $P(k)$; because of the asymmetry in $P(k)$, this does not necessarily correspond to the average value.

Counting the decays for a single 1-sec interval would give an inaccurate value of the counting rate because of the large expected fluctuations from second to second. A measure of these fluctuations is given by the standard deviation which for the Poisson distribution is $\sigma = \sqrt{npq} = \sqrt{np} = \sqrt{a}$. This is because

Table 3.1

k	$P(k)$	$100P(k)$
0	0.0498	4.98
1	0.1494	14.94
2	0.2240	22.40
3	0.2240	22.40
4	0.1680	16.80
5	0.1008	10.08
6	0.0504	5.04
7	0.2016	2.16
8	0.00810	0.81
9	0.00270	0.27
10	0.00081	0.08

$q \simeq 1$. If $a = 3$ counts per second, then $\sigma = 1.73$. The situation is much improved if the entire 100-sec interval is used. The average number of counts over a 100-sec interval is 300, and we could determine the counting rate and its error from

$$\lambda = \frac{300 \pm \sqrt{300}}{100} = 3 \pm 0.17 \text{ counts/sec} \tag{3.6}$$

In (3.6), we have used the known value of the counting rate from the parent distribution. In actual practice, the counting rate must be determined from the experimental data. Suppose that in an actual experiment, 280 counts are observed. Let's use 2σ as an error estimate. This gives

$$\lambda = \frac{280 \pm 2\sqrt{280}}{100} = 2.8 \pm 0.33 \text{ counts/sec} \tag{3.7}$$

The actual count falls within the error range given by (3.7). Suppose that a time interval T is divided up into N subintervals of length t. How does the accuracy of the counting rate as determined from the total time T compare with that obtained from a single interval? If a is the average value of k for the subintervals, then the *relative error*, if a is determined from one subinterval, is

$$\frac{\sigma}{a} = \frac{\sqrt{a}}{a} = \frac{1}{\sqrt{a}} \tag{3.8}$$

If the total time is used, the relative error is

$$\frac{\sigma_N}{a_N} = \frac{1}{\sqrt{a_N}} = \frac{1}{\sqrt{Na}} = \frac{1}{\sqrt{N}}\frac{\sigma}{a} \tag{3.9}$$

Thus, the relative error in the counting rate found by using the total time is improved by the factor $1/\sqrt{N}$ over the error if only one subinterval is used.

The second approximation to the binomial distribution that we will consider is valid if the number of trials is large and the success probability p is not vanishingly small. Both n and k are then large, and the Stirling approximation for the factorials in the form

$$n! \simeq n^n e^{-n}\sqrt{2\pi n} \tag{3.10}$$

becomes appropriate. Since the average value of k is $\bar{k} = np$, it is convenient to introduce the notation $k = np + u$. Then, since $p + q = 1$, $n - k = nq - u$. Unless p is significantly different from 1/2, the binomial distribution has a symmetric bell shape around the value $k = \bar{k}$. For large n, the distribution is significantly different from zero only for values of u which are small compared to np or nq. The binomial distribution can now be expressed as

$$\begin{aligned} P &= \frac{n!}{(nq-u)!(np+u)!} p^{np+u} q^{nq-u} \\ &\simeq \frac{n^n e^{-n}\sqrt{2\pi n}\, p^{np+u} q^{nq-u}}{(nq-u)^{nq-u} e^{-nq+u}\sqrt{2\pi(nq-u)}\,(np+u)^{np+u} e^{-np-u}\sqrt{2\pi(np+u)}} \end{aligned} \tag{3.11}$$

Under the radicals, u can be neglected in comparison to np or nq. By using $p + q = 1$ and writing

$$n^n = n^{np+u} n^{nq-u}$$

it is possible to express (3.11) in the form

$$P = \frac{1}{\sqrt{2\pi npq}}\left(1 - \frac{u}{nq}\right)^{-nq+u}\left(1 + \frac{u}{np}\right)^{-np-u} \tag{3.12}$$

Further approximations are best done by first taking

the logarithm. In the following computation, we use the approximation $\ln(1+x) \simeq x - (1/2)x^2$. All terms of order u^3 and higher are neglected:

$$\begin{aligned}
\ln &\left(1-\frac{u}{nq}\right)^{-nq+u}\left(1+\frac{u}{np}\right)^{-np-u} \\
&= (-nq+u)\ln\left(1-\frac{u}{nq}\right) \\
&\quad -(np+u)\ln\left(1+\frac{u}{np}\right) \\
&= (-nq+u)\left(-\frac{u}{nq}-\frac{u^2}{2n^2q^2}\right) \\
&\quad -(np+u)\left(\frac{u}{np}-\frac{u^2}{2n^2p^2}\right) \\
&= -\frac{u^2}{2np}-\frac{u^2}{2nq} = -\frac{u^2}{2npq} \qquad (3.13)
\end{aligned}$$

The binomial distribution has the approximate form

$$\begin{aligned}
P &= \frac{1}{\sqrt{2\pi npq}}\exp\left[-\frac{u^2}{2npq}\right] \\
&= \frac{1}{\sqrt{2\pi}\,\sigma}\exp\left[-\frac{(k-np)^2}{2\sigma^2}\right] \qquad (3.14)
\end{aligned}$$

where $\sigma = \sqrt{npq}$ is the standard deviation. In the form (3.14), the binomial distribution is called the *normal* or *Gaussian distribution*. To this approximation, there is no asymmetry even if $p \neq 1/2$. In using the normal distribution as an approximation to the binomial distribution, we must assume only that both n and k are large.

Suppose that we are trying to measure the width of a table with a ruler. The first assumption in this endeavor is that "width" is a meaningful concept. The table is supposedly rectangular, but it may not be perfectly so. If the departure from rectangular shape is not too great, an average width may be useful. The concept of width has limits even if the table is very carefully made. The table cannot be milled to an accuracy greater than the size of the molecules which make it up. Thus, even without introducing any kind of statistical uncertainties or errors in measurement, it is meaningless to try to define the width to an accuracy beyond a certain limit. On the other hand, there are things in the world which are apparently exact. For example, two electrons have *exactly* twice as much charge as one electron. In addition to inherent limits on the accuracy to which a quantity can be measured, errors can occur in the measurement which are conveniently divided into the two categories of *systematic errors* and *random errors*. Systematic errors have to do with improperly calibrated instruments, observer bias, and so on. Statistical errors are caused by any of a large number of random variations such as temperature fluctuations or vibrations which cannot be eliminated or controlled. The random variations in the measurement are assumed to belong to some parent probability distribution which in most cases is taken to be the normal distribution. The assumption is made that the measurements of the width of a table should fall in a kind of random-walk fashion around some mean value resulting in a Gaussian distribution. This assumption is borne out in practice. We have shown that, as the number of Bernoulli trials increases, the percentage deviation of the observed value of k from $\bar{k}$ decreases. We have also shown that the arithmetic mean of the data approaches $\bar{k}$ as calculated from the parent distribution in the limit of large numbers. For these reasons, it is the arithmetic mean of the observed data which is taken as the best estimate of the value of the quantity being measured.

If some quantity represented by x is to be measured, it is necessary to make a transition between the normal distribution as given by (3.14) and some appropriate form which involves x directly. To this end, let us write $x - \bar{x} = \alpha(k - np)$, where $\bar{x}$ represents the average value of x and α is a quantity with the same dimensions as x. The exponential factor in (3.14) is then

$$\frac{(k-np)^2}{2\sigma^2} = \frac{(x-\bar{x})^2}{2\alpha^2\sigma^2} \qquad (3.15)$$

What we want is a distribution which is a continuous function of x. This can be accomplished by a limiting process where k and n become large and α becomes small in such a way that the standard deviation in x,

$\sigma_x = \alpha\sigma$, remains finite. The result is

$$P(x) = \frac{1}{\sqrt{2\pi}\,\sigma} \exp\left[-\frac{(x-\bar{x})^2}{2\sigma^2}\right] \tag{3.16}$$

The subscript x on σ has been dropped, it being understood that σ is now the standard deviation in x. The quantity $P(x)\,dx$ has the significance that it is the probability that a measurement of x will fall in the range x to $x + dx$. In future discussions, we will use the latter n to represent the number of measurements of the quantity x, but this is not the same n that appears in (3.14). The n of (3.14) has actually become infinite in the process of obtaining the parent distribution (3.16). The reader can verify that (3.16) has come through the derivation properly normalized:

$$\int_{-\infty}^{+\infty} P(x)\,dx = 1 \tag{3.17}$$

Also to be checked is the validity of

$$\sigma^2 = \int_{-\infty}^{+\infty} (x-\bar{x})^2 P(x)\,dx \tag{3.18}$$

Any apprehension about extending the limits to $\pm\infty$ is dispelled by remembering that $P(x)$ essentially dies to zero a few standard deviations away from $\bar{x}$. Figure 3.1 shows graphs of (3.16) for $\sigma = 1$, 2, and 3. For convenience, we have set $\bar{x} = 0$.

Several so-called *statistical indices* need to be defined. The *median* of a set of measurements is the middle value and is such that there are just as many measurements above it as below it. The *mode* is the most probable value and corresponds to the maximum value of the distribution. We have already defined mean, variance, and standard deviation. If a distribution is symmetric about its mode, as is the case for the normal distribution, then the median, mode, and average value are all the same.

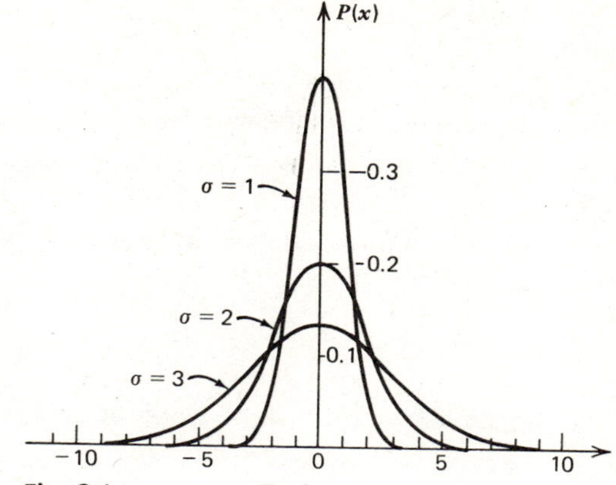

Fig. 3.1

The *error function* is defined by

$$\text{Erf}(x) = \frac{2}{\sqrt{\pi}} \int_0^x e^{-t^2}\,dt \tag{3.19}$$

and has been tabulated. Suppose that the parent distribution of a set of measurements is the normal distribution and we want to know the probability that a single measurement will fall between $\bar{x} - \sigma$ and $\bar{x} + \sigma$. This is given by

$$\begin{aligned} P(\pm\sigma) &= \int_{\bar{x}-\sigma}^{\bar{x}+\sigma} \frac{1}{\sqrt{2\pi}\,\sigma} \exp\left[-\frac{(x-\bar{x})^2}{2\sigma^2}\right] dx \\ &= \frac{2}{\sqrt{\pi}} \int_0^{1/\sqrt{2}} e^{-t^2}\,dt = \text{Erf}\left(\frac{1}{\sqrt{2}}\right) = 0.683 \end{aligned} \tag{3.20}$$

Similarly, the probability that a measurement will fall between $\bar{x} - 2\sigma$ and $\bar{x} + 2\sigma$ is

$$P(\pm 2\sigma) = \text{Erf}(\sqrt{2}) = 0.954 \tag{3.21}$$

PROBLEMS

3.1. Show directly from the Poisson distribution (3.4) that $\bar{k} = a$. Also find the average value of k^2 and from it, σ.

3.2. A counting experiment is set up to observe a very rare kind of decay. After the experiment has run for 1 hour, a single decay has been observed. The apparatus then fails and cannot be repaired for two years. Can the experimenters say anything about the expected decay rate?

Hint: How high would the counting rate have to be in order that observing only 1 count per hour is highly unlikely?

3.3. Into a batch of cookie dough are put 300 chocolate chips. If the dough makes 100 cookies, what is the expected distribution of chocolate chips among the cookies?

3.4. A fair coin is tossed 10,000 times. What is the probability that the number of heads exceeds 5100?

3.5. Suppose that a set of measurements is made and that the variance with respect to some point $x = m$ is calculated as follows:

$$v = \frac{1}{n} \sum_{k=1}^{n} (x_k - m)^2$$

Show that v is minimized if m is chosen to be the arithmetic mean of the data.

3.6. Verify (3.17) and (3.18).

3.7. The deviation of a single measurement from the mean value is $\delta x_k = x_k - \bar{x}$. The average deviation of a set of measurements is defined by

$$|\overline{\delta x}| = \frac{1}{n} \sum_{k=1}^{n} |\delta x_k|$$

The mean deviation calculated from the parent distribution is

$$\alpha = \int_{-\infty}^{+\infty} |x - \bar{x}| P(x)\, dx$$

If $P(x)$ is the normal distribution, show that

$$\alpha = \sqrt{\frac{2}{\pi}}\, \sigma = 0.79788\sigma$$

3.8. Random walk is a statistical model of heat conduction and diffusion. The diffusion equation is

$$\frac{\partial^2 \psi}{\partial x^2} - \frac{1}{\alpha} \frac{\partial \psi}{\partial t} = 0$$

where α is a constant. Show that the Gaussian (3.16) is a solution, provided that the standard deviation is an appropriate function of α and t. See problem 10.4 of Chapter 3.

3.9. Show that the power series expansion for the error function is

$$\operatorname{Erf}(x) = \frac{2}{\sqrt{\pi}} \sum_{k=0}^{\infty} \frac{(-1)^k x^{2k+1}}{k!(2k+1)}$$

The series is rapidly convergent and quite adequate for numerical evaluation of the error function. For example, 21 terms gives $\operatorname{Erf}(2) = 0.995322265$. The 21st term has the value 4×10^{-9}. Since the terms alternate in sign, the error is less than the first neglected term, provided a point has been reached in the series where the terms are monotonically decreasing in magnitude.

3.10. Show that an asymptotic expansion for the error function, valid for large x, is

$$\operatorname{Erf}(x) = 1 - \frac{e^{-x^2}}{x\sqrt{\pi}} \left[1 - \frac{1}{2x^2} + \frac{1 \cdot 3}{(2x^2)^2} - \frac{1 \cdot 3 \cdot 5}{(2x^2)^3} + \cdots \right]$$

Hint: First write

$$\text{Erf}(x)=1-\frac{2}{\sqrt{\pi}}\int_x^{\infty}e^{-t^2}\,dt$$

3.11. If you did problem 2.5, you discovered that the parent probability distribution for the possible sums of the two numbers obtained by rolling a pair of dice is *not* a binomial distribution. Suppose that a pair of dice is rolled 10,000 times. If for each roll we consider success as getting a specific sum, say 3, and failure as getting anything else, then each roll becomes a Bernoulli trial and the normal distribution applies. This is a special case of a general theorem known as the *central limit theorem*. We can calculate the expected number of times out of the 10,000 rolls that a 3 will be obtained. Not only that, we can estimate by how much the actual number of times a 3 is obtained might deviate from the average value in a given set of 10,000 rolls. Calculate the expected number of times each of the possible sums will occur and a standard deviation for each. This is a measure of how much a sample distribution obtained from 10,000 trials would reasonably be expected to deviate from the parent distribution.

4. MAXIMUM LIKELIHOOD

Suppose that the parent distribution of a set of measurements of some quantity x is the normal distribution. The probability that a given measurement will fall in the range x_k to x_k+dx_k is

$$P(x_k)\,dx_k=\frac{1}{\sqrt{2\pi}\,\sigma}\exp\left[-\frac{(x_k-\bar{x})^2}{2\sigma^2}\right]dx_k \tag{4.1}$$

Since measurements of x are mutually exclusive events, the probability that a given set of n measurements will be obtained is

$$\begin{aligned}L(x_1,x_2,\ldots,x_n)\,dx_1\,dx_2\ldots dx_n\\ &=\prod_{k=1}^{n}\frac{1}{\sqrt{2\pi}\,\sigma}\exp\left[-\frac{(x_k-\bar{x})^2}{2\sigma^2}\right]dx_k\\ &=\frac{1}{(2\pi)^{n/2}\sigma^n}\exp\left[-\sum_{k=1}^{n}\frac{(x_k-\bar{x})^2}{2\sigma^2}\right]dx_1\,dx_2\ldots dx_n\end{aligned} \tag{4.2}$$

The symbol $\prod_{k=1}^{n}$ is used to indicate a product of n factors.

Equation (4.2) has the significance that it is the probability that the first measurement will fall in the range x_1 to x_1+dx_1, the second in the range x_2 to x_2+dx_2, and so on. The function $L(x_1,\ldots,x_n)$ is called a *likelihood function*. There is good reason to believe that if a large number of measurements is taken, the actual distribution of values approximates the most likely distribution. Even if the amount of data is not large, we assume that the best way to calculate $\bar{x}$ is such as to maximize L. If we take the logarithm of L, differentiate with respect to $\bar{x}$, and then set the result equal to zero, we find

$$\frac{\partial}{\partial\bar{x}}\ln L=\frac{1}{\sigma^2}\sum_{k=1}^{n}(x_k-\bar{x})=0\qquad \bar{x}=\frac{1}{n}\sum_{k=1}^{n}x_k \tag{4.3}$$

which comes as no surprise. This illustrates the use of the maximum likelihood principle to obtain the best value of some parameter from experimental data.

Another interesting result can be obtained by changing the coordinates in (4.2) from rectangular to spherical. It is convenient to introduce the variable

$$r^2=\sum_{k=1}^{n}(x_k-\bar{x})^2 \tag{4.4}$$

and think of

$$d\Sigma=dx_1\,dx_2\ldots dx_n \tag{4.5}$$

as a volume element in an n-dimensional Cartesian space. If $\bar{x}$ is regarded as a fixed point in this n-dimensional space, then r is the distance from $\bar{x}$ to the point $(x_1,x_2,\ldots,x_n)$. Since the likelihood function depends only on r^2, it is convenient to express the volume element as the volume of an n-dimensional spherical shell of thickness dr:

$$d\Sigma = Cr^{n-1}\,dr \tag{4.6}$$

where C is a constant. The likelihood function is then

$$L(r)\,dr = \frac{1}{(2\pi)^{n/2}\sigma^n}\exp\left[-\frac{r^2}{2\sigma^2}\right]Cr^{n-1}\,dr \tag{4.7}$$

The constant C is found from the normalization condition

$$\int_0^\infty L(r)\,dr = 1 \tag{4.8}$$

The required integral is easily converted into a gamma function by the change of variable $t = r^2/(2\sigma^2)$. The result is

$$L(r)\,dr = \frac{2}{2^{n/2}\Gamma(n/2)\sigma^n}\exp\left[-\frac{r^2}{2\sigma^2}\right]r^{n-1}\,dr \tag{4.9}$$

which has the significance that $L(r)\,dr$ is the probability that a set of measurements will yield a value of r between r and $r + dr$. In the limit of a large amount of data, we assume that the actual outcome of the experiment is also the most probable outcome. Thus, the actual value of r calculated from the data should maximize $L(r)$:

$$\ln L(r) = -n\ln\sigma - \frac{r^2}{2\sigma^2} + (n-1)\ln r + \text{constant}$$

$$\frac{\partial}{\partial r}\ln L(r) = -\frac{r}{\sigma^2} + \frac{(n-1)}{r} = 0$$

$$r^2 = (n-1)\sigma^2 \tag{4.10}$$

According to (4.10), the standard deviation should be calculated from the data by

$$\sigma^2 = \frac{1}{n-1}\sum_{k=1}^{n}(x_k - \bar{x})^2 \tag{4.11}$$

which is not exactly the expected result and does not quite agree with the definition of variance as given by (2.10). One of the conditions on (2.10) is that N is large; the expression for the variance (2.14) is for the parent distribution and assumes $N \to \infty$. It is generally agreed among statisticians that (4.11) is the "best" way to estimate σ from an actual sample of data. For practical purposes, it makes little difference whether n or $n-1$ is used. If the amount of data is small, the estimate of σ is going to be bad anyway; if n is large, there is no noticeable difference between n and $n-1$.

Suppose that a quantity x is going to be measured n times and we want an estimate of the statistical error in the mean value. The uncertainty in $\bar{x}$ gets progressively smaller as the amount of data increases. As $n \to \infty$, the sample distribution approaches the parent distribution, and all statistical error in $\bar{x}$ disappears. This is true even if σ as calculated from (4.11) is large. Of necessity, x is only measured n times. But conceptualize a situation where obtaining the n measurements is repeated N times, where N is a large number. For each set of n measurements, a mean $\bar{x}_i$ is calculated. Let $\bar{\bar{x}}$ be the grand mean of all the data, or, what is equivalent, the mean of the mean values $\bar{x}_i$. The quantity

$$\sigma_m^2 = \frac{1}{N}\sum_{i=1}^{N}(\bar{x}_i - \bar{\bar{x}})^2 \tag{4.12}$$

is called the *standard deviation of the means* and is a measure of the expected variations in the values of $\bar{x}_i$. It is really (4.12) which is a measure of the statistical uncertainty in $\bar{x}$ as calculated from a single set of data. Since we are not going to repeat our experiment N times, equation (4.12) cannot be used. It is however possible to calculate σ_m another way, as we now show. The probability of measuring a particular value of $\bar{x}$ is the same as the probability of measuring the set of data from which it is calculated:

$$\frac{1}{\sqrt{2\pi}\,\sigma_m}\exp\left[-\frac{(\bar{x}-\bar{\bar{x}})^2}{2\sigma_m^2}\right]d\bar{x} = \frac{1}{\sqrt{2\pi}\,\sigma^n}\exp\left[-\sum_{k=1}^{n}\frac{(x_k-\bar{x})^2}{2\sigma^2}\right]dx_1\,dx_2\ldots dx_n \tag{4.13}$$

The logarithm of (4.13) is

$$-\ln\sigma_m - \frac{(\bar{x}-\bar{\bar{x}})^2}{2\sigma_m^2} = -n\ln\sigma - \sum_{k=1}^{n}\frac{(x_k-\bar{x})^2}{2\sigma^2} + \text{constant} \tag{4.14}$$

If (4.14) is differentiated twice with respect to $\bar{x}$, the equation so obtained can be expressed as

$$\sigma_m = \frac{\sigma}{\sqrt{n}} \tag{4.15}$$

which is one of the most useful results of this section. If a quantity is measured n times and we place equal statistical weight on all the measurements, then the standard deviation of the means as calculated by (4.15) provides an estimate of the random error. To put it another way, the uncertainty in determining the location of the mean of the parent distribution from the sample distribution decreases in proportion to $1/\sqrt{n}$. The final result of the measured value of x could be reported as $x = \bar{x} \pm 2\sigma_m$. Three or four standard deviations can be used instead of two, depending on how conservative (or pessimistic) we decide to be. Measurements are said to have high *precision* if they have small random error. The standard deviation is sometimes called a *precision index.* A measured quantity is said to have high *accuracy* if the systematic errors are small.

Consider a quantity $f(x, y)$ which is to be experimentally determined by measuring the quantities x and y on which it depends. If $\bar{x}$ and $\bar{y}$ are mean values of x and y and x_k and y_k are particular measurements, then $\Delta x_k = x_k - \bar{x}$ and $\Delta y_k = y_k - \bar{y}$ are the deviations of these measurements from their respective average values. If Δx_k and Δy_k are not too large, the deviation of $f(x_k, y_k)$ from its averge value $\bar{f} = f(\bar{x}, \bar{y})$ can be approximated by the total differential:

$$\Delta f_k = f_x \Delta x_k + f_y \Delta y_k \tag{4.16}$$

where $\Delta f_k = f_k - \bar{f}$ and f_x and f_y are the partial derivatives of f evaluated at $(\bar{x}, \bar{y})$. Conceptualize an experiment where both x and y are measured n times. If n is large, the sample distributions of both x and y approximate their respective parent distributions. The standard deviation in f can be computed from

$$\sigma_f^2 = \frac{1}{n} \sum_{k=1}^{n} \left(f_x^2 \Delta x_k^2 + 2 f_x f_y \Delta x_k \Delta y_k + f_y^2 \Delta y_k^2 \right) \tag{4.17}$$

Assume that a measurement of x in no way affects the outcome of the measurement of y. In other words, there is no dependence or correlation between x and y and a statistical error in x does not in any way bias or influence the error made in measuring y. It is then true that

$$\lim_{n \to \infty} \sum_{k=1}^{n} \Delta x_k \Delta y_k = 0 \tag{4.18}$$

because the product $\Delta x_k \Delta y_k$ has an equal probability of being either positive or negative. We conclude that

$$\sigma_f^2 = f_x^2 \sigma_x^2 + f_y^2 \sigma_y^2 \tag{4.19}$$

where σ_x and σ_y are the standard deviations in x and y. The result can be extended to more variables. For example, if $f = f(x, y, z)$, then

$$\sigma_f^2 = f_x^2 \sigma_x^2 + f_y^2 \sigma_y^2 + f_z^2 \sigma_z^2 \tag{4.20}$$

In an actual experiment, n may not be large, and x and y may not be measured the same number of times. Nevertheless, σ_x and σ_y are estimates of the standard deviations of the parent distributions of x and y, and (4.19) still serves as a useful estimate of the standard deviation in f. If the number of measurements of x and y are the same, then dividing (4.19) through by n gives

$$\sigma_{fm}^2 = f_x^2 \sigma_{xm}^2 + f_y^2 \sigma_{ym}^2 \tag{4.21}$$

If obtaining the standard deviations of the means σ_{xm} and σ_{ym} is based on different numbers of measurements, (4.21) still holds, but we will not prove this here.

PROBLEMS

4.1. In a sample of unstable particles, the number of particles which decay in a differential time dt is proportional to the number which are present:

$$dN = -\lambda N \, dt \tag{4.22}$$

where λ is a constant, usually called the decay constant. Show from this that the probability that a given particle will decay in a time dt is

$$P(t)\,dt = \lambda e^{-\lambda t}\,dt \tag{4.23}$$

Suppose that a sample consists of only a few unstable particles and that one particle decays after t_1 sec, another after t_2 sec, and so on. The probability of the experiment happening in exactly this way is proportional to the likelihood function

$$L(t_k, \lambda) = \lambda^n \exp\left[-\lambda(t_1 + \cdots + t_n)\right] \tag{4.24}$$

Use the principle of maximum likelihood to show that the best way to calculate the decay constant from the data is

$$\frac{1}{\lambda} = \frac{1}{n}\sum_{k=1}^{n} t_k \tag{4.25}$$

4.2. Carry out the integration indicated in (4.8).

4.3. If $f_k = f(x_k, y_k)$, prove that

$$\bar{f} = \frac{1}{n}\sum_{k=1}^{n} f_k$$

provided that the deviations Δx_k and Δy_k in the measurements from their mean values are small and the number of measurements is large.

4.4. The volume of an n-dimensional sphere of radius r can be calculated from

$$\Sigma_n(r) = \int\int \ldots \int dx_1\, dx_2 \ldots dx_n \tag{4.26}$$

where the region of integration is

$$x_1^2 + x_2^2 + \cdots + x_n^2 \le r^2 \tag{4.27}$$

It is possible to write (4.26) as

$$\Sigma_n(r) = \int_{-r}^{+r}\left[\int \ldots \int dx_1\, dx_2 \ldots dx_{n-1}\right] dx_n \tag{4.28}$$

where the region of integration for the quantity in brackets is

$$x_1^2 + x_2^2 + \cdots + x_{n-1}^2 \le r^2 - x_n^2 \tag{4.29}$$

which is to say, it is the volume of an $n-1$ dimensional sphere of radius $\sqrt{r^2 - x_n^2}$:

$$\Sigma_n(r) = \int_{-r}^{+r} \Sigma_{n-1}\left(\sqrt{r^2 - x_n^2}\right) dx_n \tag{4.30}$$

It is possible to write

$$\Sigma_n(r) = r^n \gamma_n \tag{4.31}$$

where γ_n is a numerical factor. Prove from (4.30) that γ_n obeys the recurrence formula

$$\gamma_n = \sqrt{\pi}\,\gamma_{n-1} \frac{\Gamma\left(\frac{n+1}{2}\right)}{\Gamma\left(\frac{n+2}{2}\right)} \tag{4.32}$$

Show from this that the volume of an n-dimensional sphere is

$$\Sigma_n = \frac{2\pi^{n/2} r^n}{n\Gamma(n/2)} \tag{4.33}$$

Use this to derive (4.9) from (4.2).

4.5. Another way to calculate the volume of an n-dimensional sphere is to use polar coordinates. Shown in Fig. 4.1 is an unconventional way to define spherical coordinates in three dimensions:

$$\begin{aligned} x_1 &= r\sin\theta_2 \sin\theta_1 \\ x_2 &= r\sin\theta_2 \cos\theta_1 \\ x_3 &= r\cos\theta_2 \end{aligned} \tag{4.34}$$

where $0 \le \theta_2 \le \pi$, $0 \le \theta_1 < 2\pi$. The generalization to four dimensions is

$$\begin{aligned} x_1 &= r\sin\theta_3 \sin\theta_2 \sin\theta_1 \\ x_2 &= r\sin\theta_3 \sin\theta_2 \cos\theta_1 \\ x_3 &= r\sin\theta_3 \cos\theta_2 \\ x_4 &= r\cos\theta_3 \end{aligned} \tag{4.35}$$

where $0 \le \theta_3 \le \pi$, $0 \le \theta_2 \le \pi$, $0 \le \theta_1 < 2\pi$. Show that the four-dimensional volume element is

$$d\Sigma_4 = r^3 \sin^2\theta_3 \sin\theta_2 \, dr\, d\theta_1\, d\theta_2\, d\theta_3 \tag{4.36}$$

Extend the transformations (4.35) to five dimensions. Show that

$$d\Sigma_5 = r^4 \sin^3\theta_4 \sin^2\theta_3 \sin\theta_2 \, dr\, d\theta_1\, d\theta_2\, d\theta_3\, d\theta_4 \tag{4.37}$$

The relevant equations from Chapter 4 are (4.17), (6.4), and (6.23) appropriately extended to more than three dimensions. The extension of (4.37) to n dimensions is

$$d\Sigma_n = r^{n-1} \sin^{n-2}\theta_{n-1} \sin^{n-3}\theta_{n-2} \ldots \sin\theta_2 \, dr\, d\theta_1\, d\theta_2 \ldots d\theta_{n-1} \tag{4.38}$$

Integrate (4.38) over a sphere of radius r and again obtain (4.33). The integrals over the angles are beta functions.

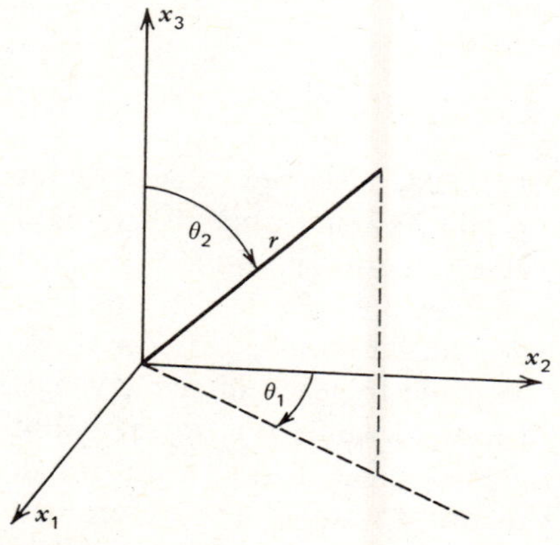

Fig. 4.1

4.6. Electrical resistors used in the laboratory are customarily given an error tolerance rating; for example, $\pm 5\%$ is common for carbon resistors. Suppose that you are given n nominally identical resistors.

(a) If the resistors are connected in series, show that the expected absolute error in the combination as measured by the standard deviation is given by

$$\sigma = \sqrt{n}\,\sigma_0$$

and that the relative error is

$$\frac{\sigma}{R} = \frac{1}{\sqrt{n}}\frac{\sigma_0}{R_0}$$

where R_0 is the nominal resistance of any one of the n resistors, σ_0 is the standard deviation in R_0, and R is the nominal resistance of the combination.

Hint: Write $R = \Sigma R_k$. Use (4.19) extended to n variables. Set $R_k = R_0$ *after* the differentiations are done.

(b) If the resistors are connected in parallel, show that

$$\sigma = \frac{1}{n^{3/2}}\sigma_0 \qquad \frac{\sigma}{R} = \frac{1}{\sqrt{n}}\frac{\sigma_0}{R_0}$$

4.7. The functional dependence of the mean value on individual measurements is $\bar{x} = (1/n)\Sigma x_k$. Use (4.19) appropriately extended to n variables to show that

$$\sigma_{\bar{x}} = \frac{1}{\sqrt{n}}\sigma_x$$

where $\sigma_{\bar{x}}$ is the standard deviation in $\bar{x}$ and σ_x is the standard deviation in any one of the measurements. This is another proof of (4.15).

4.8. A ball is loaded into a spring gun and fired horizontally across a level floor. In each of 10 trials, the horizontal range is measured yielding the following data (in cm):

$x_1 = 197.6$	$x_5 = 199.7$	$x_8 = 200.4$
$x_2 = 198.2$	$x_6 = 199.9$	$x_9 = 200.6$
$x_3 = 198.5$	$x_7 = 200.1$	$x_{10} = 202.5$
$x_4 = 199.3$		

Find $\bar{x}$, σ_x, and σ_m. The experimenters believe that their measurements are accurate to ± 2 mm. How much data should they take to reduce the statistical error to this same value? Use $\pm 2\sigma_m$ as the criterion.

5. METHOD OF LEAST SQUARES

Suppose that a quantity y is to be measured as a function of a variable x. Assume that all the statistical uncertainty occurs in y and none in x. The graph of Fig. 5.1 could represent the mean values of y from the parent distribution plotted as a function of x, or it might be a theoretical curve that we are trying to fit to experimental data. In either case, we regard the values of y which fall on the curve as mean values. If experimental data are taken, measured values of y for a given value of x will fall in some distribution around the mean value. The distribution is assumed to be normal and is indicated by the dashed curves in Fig. 5.1 for the two values x_1 and x_2. Let a single measurement of y be made for each of the x values $x_1, x_2, \ldots, x_n$. The measured value of y_1 might fall in the range dy_1 as indicated in Fig. 5.1, y_2 might fall in dy_2, and so on. The probability of the experiment occurring exactly the way it did is proportional to the

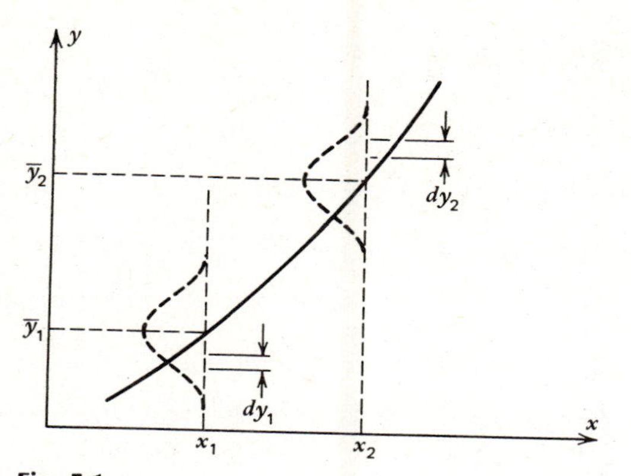

Fig. 5.1

likelihood function

$$L(y_1,\ldots,y_n)\,dy_1\,dy_2\ldots dy_n = \frac{1}{(2\pi)^{n/2}\sigma_1\sigma_2\ldots\sigma_n}$$

$$\times\,\exp\left[-\sum_{k=1}^{n}\frac{(\Delta y_k)^2}{2\sigma_k^2}\right]dy_1\,dy_2\ldots dy_n \qquad (5.1)$$

where $\Delta y_k = y_k - \bar{y}_k$ is the deviation of each measurement from its mean value. A normal distribution in y is associated with each value of x. It is not necessarily true that the standard deviation will be the same in each case, meaning that the σ_k which appear in (5.1) are not necessarily equal to one another. Specifically, $L(y_1,\ldots,y_n)\,dy_1\ldots dy_n$ is the probability that the measured value of y at x_1 will fall in the range y_1 to $y_1 + dy_1$, the value of y at x_2 will fall in the range y_2 to $y_2 + dy_2$, and so on. We introduce a variable called *chi-square* defined by

$$\chi^2 = \sum_{k=1}^{n}\frac{\Delta y_k^2}{\sigma_k^2} \qquad (5.2)$$

Suppose that we have taken some experimental data and we believe that the theoretical relation between x and y is linear: $y = a + bx$. What is the best way to determine the parameters a and b from the data? A common procedure is to plot y vs. x and draw the "best" straight line through the points by simply laying a straight edge on the page and making the best fit by eye. After this is done, the intercept a and the slope b are determined from the graph. A more high-powered way is to do a *least squares fit* whereby a and b are determined in such a way that χ^2 is minimized. This is really a version of the principle of maximum likelihood discussed in section 4. The variation of each measured value of y from its theoretical value is given by $\Delta y_k = y_k - a - bx_k$. The theoretical values $a + bx_k$ are as yet unknown and are to be estimated by minimizing χ^2 as given by

$$\chi^2 = \sum_{k=1}^{n}\frac{(y_k - a - bx_k)^2}{\sigma_k^2} \qquad (5.3)$$

In the following derivation, we will assume that the standard deviation in each value of y is the same: $\sigma_k = \sigma$. The most likely values of a and b are determined by

$$\frac{\partial\chi^2}{\partial a} = 0 \qquad \frac{\partial\chi^2}{\partial b} = 0 \qquad (5.4)$$

which gives the two algebraic equations

$$a + b\bar{x} = \bar{y} \qquad a\bar{x} + b\overline{x^2} = \overline{(xy)} \qquad (5.5)$$

The various average values are defined by

$$\bar{x} = \frac{1}{n}\sum_{k=1}^{n}x_k \qquad \bar{y} = \frac{1}{n}\sum_{k=1}^{n}y_k$$
$$\overline{x^2} = \frac{1}{n}\sum_{k=1}^{n}x_k^2 \qquad \overline{(xy)} = \frac{1}{n}\sum_{k=1}^{n}x_k y_k \qquad (5.6)$$

Note that $\overline{(xy)} \neq \bar{x}\bar{y}$. The solution of (5.5) for a and b is

$$a = \frac{\overline{x^2}\bar{y} - \bar{x}\overline{(xy)}}{\overline{x^2} - \bar{x}^2} \qquad b = \frac{\overline{(xy)} - \bar{x}\bar{y}}{\overline{x^2} - \bar{x}^2} \qquad (5.7)$$

It is possible to estimate the errors in the parameters a and b based on equation (4.19) suitably generalized to n variables:

$$\sigma_a^2 = \sum_{k=1}^{n}\left(\frac{\partial a}{\partial y_k}\right)^2\sigma_k^2 \qquad \sigma_b^2 = \sum_{k=1}^{n}\left(\frac{\partial b}{\partial y_k}\right)^2\sigma_k^2 \qquad (5.8)$$

From (5.6), we get

$$\frac{\partial\bar{y}}{\partial y_k} = \frac{1}{n} \qquad \frac{\partial\overline{(xy)}}{\partial y_k} = \frac{x_k}{n} \qquad (5.9)$$

If $\sigma_k = \sigma$, it then follows that

$$\sigma_a^2 = \sum_{k=1}^{n} \left[\frac{\overline{x^2} - \bar{x}x_k}{n\left(\overline{x^2} - \bar{x}^2\right)} \right]^2 \sigma^2$$

$$= \sum_{k=1}^{n} \left[\frac{\left(\overline{x^2}\right)^2 - 2\overline{x^2}\bar{x}x_k + \bar{x}^2 x_k^2}{n^2\left(\overline{x^2} - \bar{x}^2\right)^2} \right] \sigma^2$$

$$= \frac{n\left(\overline{x^2}\right)^2 - 2\overline{x^2}n\bar{x}^2 + \bar{x}^2 n\overline{x^2}}{n^2\left(\overline{x^2} - \bar{x}^2\right)^2} \sigma^2$$

$$= \frac{\overline{x^2}\left(\overline{x^2} - \bar{x}^2\right)\sigma^2}{n\left(\overline{x^2} - \bar{x}^2\right)^2} = \frac{\overline{x^2}\sigma^2}{n\left(\overline{x^2} - \bar{x}^2\right)}$$

The final result is

$$\sigma_a = \sqrt{\frac{\overline{x^2}}{n\left(\overline{x^2} - \bar{x}^2\right)}}\,\sigma \tag{5.10}$$

Since the standard deviation σ is the same for each y value, it can be computed from

$$\sigma^2 = \frac{1}{n-1} \sum_{k=1}^{n} \left(y_k - a - bx_k\right)^2 \tag{5.11}$$

By a similar calculation,

$$\sigma_b = \frac{}{\sqrt{n\left(\overline{x^2} - \bar{x}^2\right)}} \tag{5.12}$$

Thus, the determination of the slope and the intercept from experimental data and of the random error in these quantities is reduced to a set of algebraic equations. Calculations by hand are tedious, but the formulas are easily programmed into a computer.

By following a procedure similar to that used in section 4 in going from equation (4.4) to (4.9), it is possible to convert the likelihood function (5.1) to n-dimensional spherical coordinates. The volume of a spherical shell of radius χ can be written as

$$d\Sigma = C\chi^{n-1}\,d\chi = \tfrac{1}{2}C\left(\chi^2\right)^{(n/2)-1}\,d\chi^2 \tag{5.13}$$

It is the usual practice to write everything in terms of the variable χ^2. The constant C is to be determined by the normalization condition

$$\int_0^\infty L\left(\chi^2\right)\,d\chi^2 = 1 \tag{5.14}$$

The result is that (5.1) can be written as

$$L\left(\chi^2\right)\,d\chi^2 = \frac{1}{2^{n/2}\Gamma(n/2)}\left(\chi^2\right)^{(n/2)-1} \times \exp\left(-\frac{1}{2}\chi^2\right)\,d\chi^2 \tag{5.15}$$

which is known as the *chi-square distribution*. It is used in both the physical and the social sciences as a means of testing the *goodness of fit* of experimental data to a theoretical model. As an example, suppose that we want to test a pair of dice to find out if they are loaded. In problem (1.5), you found the a priori probabilities of obtaining the numbers 2 through 12 on a single roll, assuming fair dice. If $P(k)$ is the probability of getting the number k on a single roll of the dice and the dice are rolled N times, then $\bar{N}_k = NP(k)$ is the expected number of times that k will be obtained. Since the success probability for a given k is fairly small, the standard deviation in N_k is approximately

$$\sigma_k = \sqrt{NP(k)} = \sqrt{\bar{N}_k} \tag{5.16}$$

and is itself a measure of the expected deviation of the actual number of times k is obtained from the mean or expected value $\bar{N}_k$ (see problem 2.3). If N_k is the observed number of times k is obtained out of N rolls, then χ^2 is given by

$$\chi^2 = \sum_{k=1}^{n} \frac{\left(N_k - \bar{N}_k\right)^2}{\bar{N}_k} \tag{5.17}$$

The number n, which in this case is the total number of possible outcomes, is called the *number of degrees of freedom*. Dice have 11 degrees of freedom.* If χ^2 turns out to be large, the sample distribution does not conform well to the assumed parent distribution and the dice are suspect. More useful than the value of χ^2 itself is the integral

$$Q\left(\chi^2\right) = \int_{\chi^2}^{\infty} \frac{1}{2^{n/2}\Gamma(n/2)} t^{(n/2)-1} e^{-t/2}\,dt \tag{5.18}$$

*If (5.17) is specialized to dice, $k = 0$ and $k = 1$ never occur. The sum then runs from 2 through 12.

which is obtained from (5.14) by setting $t=\chi^2$. The integral has been tabulated and has the significance that it is the probability of observing a value of χ^2 larger than the one actually observed. If $\chi^2=0$, $Q(\chi^2)=1$, and the sample distribution matches the parent distribution perfectly. The smaller $Q(\chi^2)$ is, the greater is the discrepancy between the assumed parent distribution and the sample distribution.

Consider an experiment, such as tossing a coin, where there are two possible outcomes. The a priori assumption is that the two outcomes are equally likely, and we want to test this hypothesis by performing the experiment N times. There are two degrees of freedom. We assumed in equation (5.16) that the probability $P(k)$ is small, which will be true only if the number of degrees of freedom is fairly large. The χ^2 test gives reasonable results even if this is not the case. For two degrees of freedom with $\overline{N}_1=\overline{N}_2=\overline{N}$,

$$\chi^2=\frac{(N_1-\overline{N})^2}{\overline{N}}+\frac{(N_2-\overline{N})^2}{\overline{N}}=\frac{2(N_1-\overline{N})^2}{\overline{N}} \tag{5.19}$$

which follows because necessarily $\overline{N}=(N_1+N_2)/2$. For $n=2$, (5.18) gives

$$Q(\chi^2)=\int_{\chi^2}^{\infty}\frac{1}{2}e^{-t/2}\,dt=\exp\left(-\frac{1}{2}\chi^2\right)$$

$$\chi^2=-2\ln Q \tag{5.20}$$

By combining (5.19) and (5.20), we find

$$N_1=\overline{N}_1\pm\sqrt{-\overline{N}_1\ln Q} \tag{5.21}$$

which is not substantially different from the result found in problem 2.3. If a coin is tossed 1000 times, then $\overline{N}=500$. If we choose $Q=0.5$,

$$N_1=500\pm 19 \tag{5.22}$$

By comparison, the standard deviation is

$$\sigma=\sqrt{Npq}=\sqrt{1000\times\tfrac{1}{2}\times\tfrac{1}{2}}=16 \tag{5.23}$$

If the number of heads differs from 500 by more than, say, 20 or 30, we suspect that the initial assumption of equal a priori probabilities is not valid and that the coin is probably bent.

Consider again the measurement of y as a function of the continuous variable x. Suppose that a least squares fit of a straight line to the data has been made by means of (5.7). This procedure will give us numbers for a and b, even if the sample data are from some parent distribution that is not linear. It is possible to test the hypothesis that the relation between y and x is linear by means of a χ^2 test. There is a problem in that estimating the values of σ_k in (5.2) might be difficult. Another consideration is that (5.5) from which a and b are determined puts two conditions or constraints on a and b. To put it another way, we have forced the data to fit a straight line to some extent. Although we will not try to justify it rigorously here, the procedure that is generally followed is to reduce the number of degrees of freedom by 2 to $n-2$ when using (5.18). If an attempt is made to fit $y=a+bx+cx^2$ to the data, then three conditions must be imposed to find a, b, and c. This reduces the number of degrees of freedom to $n-3$. This is not unreasonable as the following consideration shows. If only three values of y are measured, a, b, and c can be determined so that $y=a+bx+cx^2$ fits exactly. The value of χ^2 is zero and the chi-square test is meaningless.

PROBLEMS

5.1. If the standard deviations σ_k in (5.3) are not equal, show that (5.5) is still valid provided that the mean values (5.6) are replaced by the *weighted means*

$$\bar{x}=\frac{1}{w}\sum_{k=1}^{n}\frac{1}{\sigma_k^2}x_k \qquad \bar{y}=\frac{1}{w}\sum_{k=1}^{n}\frac{1}{\sigma_k^2}y_k$$

$$\overline{x^2}=\frac{1}{w}\sum_{k=1}^{n}\frac{1}{\sigma_k^2}x_k^2 \qquad \overline{(xy)}=\frac{1}{w}\sum_{k=1}^{n}\frac{1}{\sigma_k^2}x_k y_k$$

The factor w is given by

$$w = \sum_{k=1}^{n} \frac{1}{\sigma_k^2}$$

Show that the standard deviations in a and b are now given by

$$\sigma_a = \sqrt{\frac{\overline{x^2}}{w\left(\overline{x^2} - \bar{x}^2\right)}} \qquad \sigma_b = \sqrt{\frac{1}{w\left(\overline{x^2} - \bar{x}^2\right)}}$$

5.2. Show that the maximum value of $L(\chi^2)$ as given by (5.15) occurs when $\chi^2 = n - 2$. This shows that if the number of degrees of freedom is large, χ^2 can be large and still have a value of $Q(\chi^2)$ that is acceptable in a χ^2 test.

5.3. Show that a least squares fit of the parabola $y = a + bx + cx^2$ to experimental data requires that the constants a, b, and c obey

$$a + b\bar{x} + c\overline{x^2} = \bar{y}$$

$$a\bar{x} + b\overline{x^2} + c\overline{x^3} = \left(\overline{xy}\right)$$

$$a\overline{x^2} + b\overline{x^3} + c\overline{x^4} = \left(\overline{x^2 y}\right)$$

5.4. Chickens are observed laying eggs over a period of 50 days. A total of 89 eggs are laid with the following distribution:

Eggs laid	0	1	2	3	4	5
No. of days	10	13	13	8	4	2

meaning that there were 10 days in which no eggs were laid, 13 days in which one egg was laid, and so on. Test the hypothesis that this is a Poisson distribution. *Note:* Since you have to determine the mean value from the experimental data, the number of degrees of freedom is reduced by 1.

5.5. Show that the power series expansion for $Q(x)$ is

$$Q(x) = 1 - \frac{2x^{n/2}}{2^{n/2}\Gamma(n/2)} \sum_{k=0}^{\infty} \frac{(-1)^k x^k}{k!2^k(n+2k)}$$

where $x = \chi^2$.

5.6. By making the change of variable $s = t/2$ in (5.18), relate $Q(x)$ to the incomplete gamma function defined in problem 23.3 of Chapter 5.

5.7. Show that $\overline{(x - \bar{x})^2} = \overline{x^2} - \bar{x}^2$. Since then $\overline{x^2} - \bar{x}^2 > 0$, (5.5) is always solvable for a and b.

6. LAGRANGE MULTIPLIERS

This section contains a review of a topic which will be needed in the discussion of Maxwell–Boltzmann statistics to be taken up in section 7. Given a function $\psi(x, y)$, we can locate its maxima, minima, and saddle points from the condition that its gradient vanish:

$$\frac{\partial \psi}{\partial x} = 0 \qquad \frac{\partial \psi}{\partial y} = 0 \tag{6.1}$$

If (x_0, y_0) is a point for which (6.1) is true, then the behavior of $\psi(x, y)$ in the neighborhood of this point can be determined from the first nonzero term in its Taylor's series expansion:

$$2\,\Delta\psi = a\,\Delta x^2 + 2b\,\Delta x\,\Delta y + c\,\Delta y^2 \tag{6.2}$$

where $\Delta\psi = \psi - \psi_0$, $\Delta x = x - x_0$, and $\Delta y = y - y_0$, and a, b, and c are the second-order partial deriva-

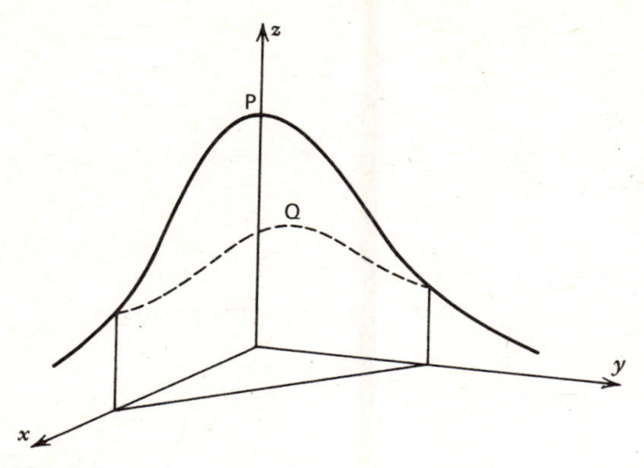

Fig. 6.1

tives

$$a = \frac{\partial^2 \psi}{\partial x^2} \qquad b = \frac{\partial^2 \psi}{\partial x\, \partial y} \qquad c = \frac{\partial^2 \psi}{\partial y^2} \tag{6.3}$$

evaluated at (x_0, y_0). If both the first and the second partial derivatives of ψ vanish at (x_0, y_0), then it is necessary to go on to higher-order terms to calculate $\Delta\psi$. Assuming nonzero second-order partial derivatives, we can determine whether (x_0, y_0) is the location of a maximum, minimum, or a saddle point by studying the quadratic form (6.2).

Our main concern is with a more complicated problem, namely, that of maximizing or minimizing a function subject to a constraint of some kind. The procedure is best explained by means of examples. The two-dimensional Gaussian function

$$\psi(x, y) = e^{-(x^2+y^2)} \tag{6.4}$$

is pictured in Fig. 6.1. Its total differential is

$$d\psi = \frac{\partial \psi}{\partial x} dx + \frac{\partial \psi}{\partial y} dy = -2x\psi\, dx - 2y\psi\, dy \tag{6.5}$$

If there are no subsidiary conditions or constraints on x and y, they can be varied independently of one another. The condition for a maximum, minimum, or a saddle point is $d\psi = 0$ for all choices of dx and dy. Since $\psi \neq 0$, the only possibility is $x = y = 0$. This gives the point P of Fig. 6.1. The thing to keep in mind is that $d\psi = \nabla\psi \cdot d\mathbf{s} = 0$ regardless of which way the infinitesimal displacement vector $d\mathbf{s} = \mathbf{i}\, dx + \mathbf{j}\, dy$ points. This can be true only if the coefficients of dx and dy in (6.5) separately vanish.

Now suppose that the problem is to find the greatest value that ψ can have if x and y are restricted by the condition

$$x + y = c \tag{6.6}$$

where c is a constant. Equation (6.6) is an *equation of constraint* and represents a plane parallel to the z axis which intersects $z = \psi(x, y)$ along the dashed curve as illustrated in Fig. 6.1. What we are trying to do is to locate the point Q on this curve where ψ is largest. At Q, it is true that $d\psi = 0$ only if $d\mathbf{s}$ is tangent to the curve where ψ and $x + y = c$ intersect. There are therefore two conditions which have to be satisfied:

$$d\psi = -2x\psi\, dx - 2y\psi\, dy = 0 \tag{6.7}$$

$$dx + dy = 0 \tag{6.8}$$

Equation (6.8) comes from forming the total differential of the equation of constraint (6.6) and is the condition that dx and dy must satisfy in order that $d\mathbf{s}$ point in the required direction. One way to solve the problem is to eliminate dy between (6.7) and (6.8) with the result that $(y - x)\, dx = 0$ or $y = x$. The equation of constraint (6.6) then gives $x = y = c/2$. The thing to emphasize is that (6.7) does not imply that the coefficients of dx and dy separately vanish as was true in the case of no constraints because dx and dy can no longer be chosen independently of one another.

Another way to solve the problem is to first multiply (6.8) by an undetermined factor λ called a *Lagrange multiplier* and then add the resulting equation to (6.7). We find

$$(\lambda - 2x\psi)\, dx + (\lambda - 2y\psi)\, dy = 0 \tag{6.9}$$

The parameter λ is then defined by the requirement that the coefficient of dy vanish. It is then necessarily true that the coefficient of dx also vanish:

$$\lambda - 2x\psi = 0 \qquad \lambda - 2y\psi = 0 \tag{6.10}$$

In other words, we can again treat the problem as though dy and dx were independent. From (6.10), we find again $x = y$.

Suppose that the three-dimensional Gaussian

$$\psi(x, y, z) = e^{-(x^2+y^2+z^2)} \tag{6.11}$$

is to be maximized subject to the two constraints

$$x + y = a \qquad y + z = b \tag{6.12}$$

Three conditions must now be met:

$$d\psi = \frac{\partial \psi}{\partial x} dx + \frac{\partial \psi}{\partial y} dy + \frac{\partial \psi}{\partial z} dz = 0 \tag{6.13}$$

$$dx + dy = 0 \tag{6.14}$$

$$dy + dz = 0 \tag{6.15}$$

The procedure is to multiply (6.14) by λ_1 and (6.15) by λ_2 and then add the resulting equations to (6.13). This gives

$$\left(\lambda_1 + \frac{\partial \psi}{\partial x}\right) dx + \left(\lambda_1 + \lambda_2 + \frac{\partial \psi}{\partial y}\right) dy + \left(\lambda_2 + \frac{\partial \psi}{\partial z}\right) dz = 0 \tag{6.16}$$

The Lagrange multipliers λ_1 and λ_2 are defined so that the coefficients of dy and dz vanish. It then follows that the coefficient of dx also vanishes, giving the three conditions

$$\lambda_1 + \frac{\partial \psi}{\partial x} = 0$$

$$\lambda_1 + \lambda_2 + \frac{\partial \psi}{\partial y} = 0 \qquad \lambda_2 + \frac{\partial \psi}{\partial z} = 0 \tag{6.17}$$

By eliminating the Lagrange multipliers, we find

$$-\frac{\partial \psi}{\partial x} - \frac{\partial \psi}{\partial z} + \frac{\partial \psi}{\partial y} = 0 \tag{6.18}$$

By using $\psi(x, y, z)$ as given by (6.11), we find $x + z - y = 0$ which when combined with the equations of constraint (6.12) gives

$$x = \frac{2a - b}{3} \qquad y = \frac{a + b}{3} \qquad z = \frac{-a + 2b}{3} \tag{6.19}$$

As another example, suppose that a trough is to be formed out of a piece of sheet metal of width w by bending up equal portions on each side by an angle θ as shown in Fig. 6.2. The dimensions x and y and the angle θ are to be chosen such that the cross-sectional area

$$A = x \sin\theta (x \cos\theta + y) = \tfrac{1}{2} x^2 \sin 2\theta + xy \sin\theta \tag{6.20}$$

is maximized. There is a single constraint given by

$$2x + y = w \tag{6.21}$$

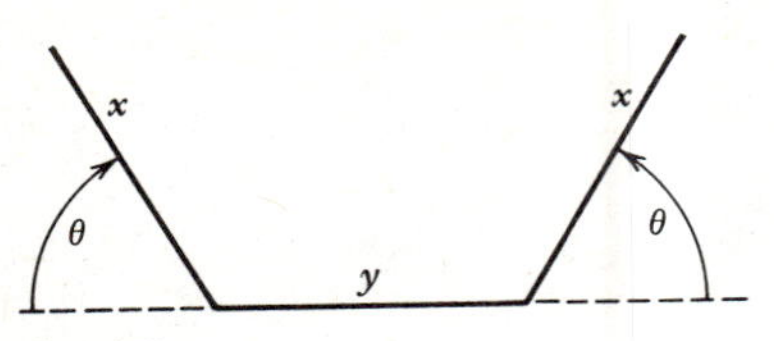

Fig. 6.2

If $2\lambda\, dx + \lambda\, dy$ is added to $dA = 0$, the result is

$$(2\lambda + x \sin 2\theta + y \sin\theta)\, dx + (\lambda + x \sin\theta)\, dy + (x^2 \cos 2\theta + xy \cos\theta)\, d\theta = 0 \tag{6.22}$$

By appropriate choice of the Lagrange multiplier λ, the coefficients of dx, dy, and $d\theta$ can all be set equal to zero. The resulting equations can be solved to yield $\theta = 60°$ and $x = y = (1/3)w$.

To summarize, a function $\psi(x_1, \ldots, x_n)$ is to be extremized (maximized or minimized) subject to a number of constraints which can be put into the form $\phi_i(x_1, \ldots, x_n) = \text{constant}$ where $i = 1, 2, \ldots, m$. Necessarily, $m \leq n$. If $n = m$, all the variables are determined without reference to the function ψ. The maxima and minima of ψ are determined by the conditions

$$d\psi = \sum_{k=1}^{n} \frac{\partial \psi}{\partial x_k} dx_k = 0 \tag{6.23}$$

$$0 = \sum_{k=1}^{n} \frac{\partial \phi_i}{\partial x_k} dx_k \tag{6.24}$$

Each of the m equations represented in (6.24) is multiplied by a Lagrange multiplier and added to (6.23) with the result

$$0 = \left(\sum_{k=1}^{n} \sum_{i=1}^{m} \lambda_i \frac{\partial \phi_i}{\partial x_k} + \sum_{k=1}^{n} \frac{\partial \psi}{\partial x_k} \right) dx_k \tag{6.25}$$

The values of the m parameters λ_i can be determined by requiring m of the coefficients of the dx_k to be zero. There then remains $n - m$ terms in (6.25). Since there are $n - m$ independent variables left, the remaining dx_k can be varied independently of one another with the consequence that the coefficients of these terms must vanish also. The result is the system of n equations

$$\sum_{i=1}^{m} \lambda_i \frac{\partial \phi_i}{\partial x_k} + \frac{\partial \psi}{\partial x_k} = 0 \tag{6.26}$$

The combination of (6.26) and the m equations of constraint give a total of $n+m$ equations for finding the n variables x_k and the m parameters λ_i.

PROBLEMS

6.1. Assume that at least one of the constants a, b, and c in the quadratic form (6.2) is not zero. Show that (x_0, y_0) gives a saddle point if $b^2 - ac > 0$, a maximum if $b^2 - ac < 0$ and $a + c < 0$, and a minimum if $b^2 - ac < 0$ and $a + c > 0$. What does the function look like in the vicinity of (x_0, y_0) if $b^2 - ac = 0$?

Hint: Find the eigenvalues of the matrix of the quadratic form. See problem 7.6 in Chapter 5.

6.2. Two positive point charges of equal magnitude are located at $x = \pm s$. If ψ is the electrostatic potential created by these charges, show that $\nabla\psi = 0$ at the origin. Find the Taylor's series expansion of ψ about the origin valid to second order in x, y, and z. You can use Taylor's series expansion for a function of many variables as given in problem 5.4 of Chapter 5, but it's easier to use a binomial expansion. Identify the origin as a maximum, minimum, or a saddle point. Again, refer to problem 7.6 in Chapter 5. Make a drawing of the curves ψ = constant in the x, y plane near the origin.

6.3. Maximize (6.11) subject to the single constraint $x + y = a$.

6.4. An aquarium with rectangular sides and bottom (and no top) is to hold 5 gallons. Find its proportions such that the least amount of material is used in its construction.

6.5. Find the shortest distance from the origin to the surface $x^2 + y^2 - 2zx = 4$.

7. MAXWELL–BOLTZMANN STATISTICS

An experiment is to be performed in which there are two trials and three possible events for each trial. The probabilities for the occurrence of each event are p_1, p_2, and p_3. An example is provided by placing two objects a and b at random into three boxes. If the two objects are identical, the possible outcomes of the experiment are illustrated in Fig. 7.1. For example, 1 1 0 means one object in box 1 and the other in box 2. This outcome is assigned a weight of 2 because it can happen in two ways: a b 0 or b a 0. Thus, Fig. 7.1 can be thought of as a weighted sample space. If each object has an equal probability of going into any one of the three boxes, $p_1 = p_2 = p_3 = 1/3$. The probability of outcome 1 1 0 is then 2/9. The probability that a definitely ends up in box 1 and b definitely ends up in box 2 is 1/9. If a and b represent molecules of a gas, they are completely indistinguishable from one another, meaning that a b 0 and b a 0 represent identical states of the system. This state is a more likely outcome of the experiment than 2 0 0 because there are more ways of achieving it. Formally, the weights of the various outcomes can be calculated from

Outcome	Weight	Outcome	Weight	Total Weight
2 0 0	1	1 1 0	2	$9 = 3^2$
0 2 0	1	1 0 1	2	
0 0 2	1	0 1 1	2	

Fig. 7.1

$$\frac{2!}{2!0!0!} = 1 \qquad \frac{2!}{1!1!0!} = 2 \tag{7.1}$$

The probabilities of the occurrence of the various states are given by

$$P(2,0,0) = \frac{2!}{2!0!0!} p_1^2 p_2^0 p_3^0$$

$$P(1,1,0) = \frac{2!}{1!1!0!} p_1^1 p_2^1 p_3^0 \quad \text{and so on} \tag{7.2}$$

If $p_1 = p_2 = p_3 = 1/3$, then $P(2,0,0) = 1/9$ and $P(1,1,0) = 2/9$. By summing over all the probabil-

(1,1)	(1,2)	(1,3)
(2,1)	(2,2)	(2,3)
(3,1)	(3,2)	(3,3)

Fig. 7.2

ities, we find that

$$\sum_{N_1=0}^{2}\sum_{N_2=0}^{2-N_1}\frac{2!}{N_1!N_2!N_3!}p_1^{N_1}p_2^{N_2}p_3^{N_3}$$

$$= p_1^2+p_2^2+p_3^2+2p_1p_2+2p_1p_3+2p_2p_3$$

$$= (p_1+p_2+p_3)^2 = 1 \tag{7.3}$$

Note that $N_1+N_2+N_3=2$. This is an example of a *multinomial distribution.*

It is possible to construct a sample space for the case where a and b are distinguishable objects. This is shown in Fig. 7.2. For example, (1,2) means a in box 1 and b in box 2. The point (2,1) means a in box 2 and b in box 1. These two outcomes are indistinguishable if a and b are identical. Figure 7.2 makes it clear that there are $3\times3=9$ possible outcomes. If $p_1=p_2=p_3=1/3$, then each point in Fig. 7.2 represents an outcome which has a probability of 1/9 of occurring.

Suppose now that three objects a, b, and c are to be placed at random in three different boxes. If a, b, and c are distinguishable objects, a sample space of the type in Fig. 7.2 can be constructed with $3^3=27$ points in it. Each point represents a possible outcome. If a, b, and c are indistinguishable, the possible outcomes and the statistical weights of each are represented by Fig. 7.3. The statistical weights can be calculated from

$$\frac{3!}{3!0!0!}=1 \qquad \frac{3!}{2!1!0!}=3 \qquad \frac{3!}{1!1!1!}=6 \tag{7.4}$$

For example, 2 1 0 means two objects in box 1, 1 in box 2, and none in box 3. In detail, the possibilities for this particular outcome are

$$\begin{matrix}(a,b)(c)(\ ) & (a,c)(b)(\ ) & (b,c)(a)(\ ) \\ (b,a)(c)(\ ) & (c,a)(b)(\ ) & (c,b)(a)(\ )\end{matrix} \tag{7.5}$$

There are $3!=6$ possibilities corresponding to the 3! ways of arranging a, b, and c in a straight line. However, (a, b)(c)( ) and (b, a)(c)( ) are one and the same outcome regardless of whether or not a, b, and c are distinguishable. Thus, there are in reality only $3!/(2!1!0!)=3$ ways of achieving 2 1 0. Note that 2! is just the number of possible arrangements of the two objects which happen to end up in box 1.

Now consider adding 5 objects to the three boxes. A possible outcome is represented by (a, b, c)(d, e)( ). There are 5! possible arrangements of 5 objects in a straight line. The 3 objects which happen to end up in box 1 can be rearranged in 3! ways without affecting the outcome. Similarly, the 2 objects in box 2 can be rearranged in 2! ways. Thus, there are $5!/(3!2!0!)=10$ ways of arriving at 3 2 0.

In the general case, N identical objects are to be placed at random in n boxes. A given outcome is sometimes referred to as a *partition* of the objects. The number of ways of obtaining a given partition where there are N_1 objects in box 1, N_2 in box 2,..., and N_n objects in box n is

$$\frac{N!}{N_1!N_2!\dots N_n!} \tag{7.6}$$

and is a *multinomial coefficient*. If p_k is the probability that an object will go into box k, then the probability of obtaining this particular partition is

$$P(N_1,N_2,\dots,N_n)=\frac{N!}{N_1!N_2!\dots N_n!}p_1^{N_1}p_2^{N_2}\dots p_n^{N_n} \tag{7.7}$$

Since (7.6) is a multinomial coefficient and since $p_1+\cdots+p_n=1$,

$$\sum_{N_1}\sum_{N_2}\dots\sum_{N_{n-1}}\frac{N!}{N_1!N_2!\dots N_n!}p_1^{N_1}p_2^{N_2}\dots p_n^{N_n}$$

$$=(p_1+p_2+\cdots+p_n)^N=1 \tag{7.8}$$

The sum is over all possible combinations such that $N_1+\cdots+N_n=N$. By giving all the probabilities the

Outcome	Weight	Outcome	Weight	Total Weight
3 0 0	1	1 2 0	3	$3^3=27$
0 3 0	1	1 0 2	3	
0 0 3	1	0 2 1	3	
2 1 0	3	0 1 2	3	
2 0 1	3	1 1 1	6	

Fig. 7.3

same value $p_k = 1/n$, we can show that

$$\sum_{N_1}\sum_{N_2}\cdots\sum_{N_{n-1}} \frac{N!}{N_1!N_2!\ldots N_n!} = n^N \tag{7.9}$$

This is the number of points there would be in the sample space if all the objects were distinguishable from one another.

The multinomial distribution forms the basis of Maxwell–Boltzmann statistics. In the following discussion, it is convenient to think in terms of an ideal monatomic gas which consists of point particles in thermal equilibrium. The particles are all identical and do not interact with one another appreciably. The only form of internal energy is kinetic energy of translation. The particles have a statistical distribution both with respect to space and velocity which will now be derived. It is convenient to introduce a six-dimensional space called *phase space* which consists of the three spatial coordinates x, y, and z and the three momentum components p_x, p_y, and p_z. Actually, we will modify this slightly and use velocity components directly. A volume element in this space is

$$\Delta\Sigma = \Delta x\,\Delta y\,\Delta z\,\Delta v_x\,\Delta v_y\,\Delta v_z \tag{7.10}$$

Imagine that the phase space is divided up into n cells of equal size. Assuming that each cell has an equal a priori probability of being occupied, the probability of a specific partition of the molecules in the phase space is

$$P(N_1, N_2, \ldots, N_n) = \frac{N!}{N_1!N_2!\ldots N_n!}\left(\frac{1}{n}\right)^N \tag{7.11}$$

The thermodynamic macroscopic state of the system is determined by how the molecules are partitioned in the phase space. We use the principle of maximum likelihood and assume that the actual state of the system is also the most likely state. The problem is however not solved by simply maximizing (7.11). There are two subsidiary conditions or constraints on the system; the total number of particles is a constant and the total energy is conserved:

$$N = \sum_{k=1}^{n} N_k = \text{constant} \tag{7.12}$$

$$E = \sum_{k=1}^{n} E_k N_k = \text{constant} \tag{7.13}$$

In (7.13), E_k is the energy of the particles in cell k of phase space. Thinking in terms of the ideal monatomic gas, different cells of the phase space correspond to different particle velocities and hence different kinetic energies. The number of particles is very large. Before maximizing (7.11), it is appropriate to use the Stirling approximation on the factorials and then take the logarithm:

$$N_k! \simeq \sqrt{2\pi N_k}\, N_k^{N_k} e^{-N_k}$$

$$\ln P = -\sum_{k=1}^{n}\left[\left(N_k + \tfrac{1}{2}\right)\ln N_k - N_k\right] + \text{constant} \tag{7.14}$$

The factor of $\frac{1}{2}$ is small in comparison to N_k and can be discarded. If the total differential of $\ln P$ is set equal to zero, the result is

$$0 = \sum_{k=1}^{n} \ln N_k\, dN_k \tag{7.15}$$

If the total differentials of the constraints (7.12) and (7.13) are multiplied by Lagrange multipliers α and β and set equal to zero, the result is

$$0 = \sum_{k=1}^{n} \alpha\, dN_k \qquad 0 = \sum_{k=1}^{n} \beta E_k\, dN_k \tag{7.16}$$

By adding (7.15) and (7.16), we find

$$0 = \sum_{k=1}^{n} (\ln N_k + \alpha + \beta E_k)\, dN_k \tag{7.17}$$

The Lagrange multipliers can be determined such that the coefficients of each dN_k vanish:

$$\ln N_k + \alpha + \beta E_k = 0$$

$$N_k = e^{-\alpha - \beta E_k} = Ce^{-\beta E_k} \tag{7.18}$$

where $C = e^{-\alpha}$ is a new constant. This is basically the Maxwell–Boltzmann distribution. The picture of the substance under consideration as ideal monatomic gas was convenient for purposes of visualization. Equation (7.18) applies however much more generally than to ideal gases. The validity of the Maxwell–Boltzmann distribution depends on the existence of some type of states that can be occupied by the particles, the correctness of (7.6) as the number of ways of obtaining a given partition, and the existence of the constraints (7.16). This in no way implies that the substance under consideration is necessarily a gas. In classical mechanics, the states occupied by the particles are cells in phase space. If the problem is being treated quantum mechanically, then the possible states are the allowed quantum states of the system. The requirements of quantum mechanics also lead to two new ways of enumerating the number of possible ways of obtaining a given partition leading to *Bose–Einstein* and *Fermi–Dirac* statistics.

The constant C and β can be found by using the equations of constraint:

$$N = C\sum_{k=1}^{n} e^{-\beta E_k} \qquad E = C\sum_{k=1}^{n} E_k e^{-\beta E_k} \tag{7.19}$$

To see how this works for an ideal monatomic gas, write the energy as $\frac{1}{2}mv_k^2$. Since the energy is a continuous function of velocity, it is appropriate to replace the sums in (7.19) by integrals over the phase space:

$$N = C\int_{\Sigma} \exp\left(-\frac{1}{2}\beta m v^2\right) d\Sigma \tag{7.20}$$

$$E = C\int_{\Sigma} \frac{1}{2}mv^2 \exp\left(-\frac{1}{2}\beta m v^2\right) d\Sigma \tag{7.21}$$

The integrands do not contain x, y, or z. This means that the particles are distributed with constant density in space. The spatial part of the integration then just gives the volume V of the container. The fact that the integrands depend only on v means that the remaining part of $d\Sigma$ can be written as the volume of a spherical shell of radius v and thickness dv. The result is

$$N = CV\int_0^{\infty} \exp\left(-\frac{1}{2}\beta m v^2\right) 4\pi v^2\, dv \tag{7.22}$$

$$E = CV\int_0^{\infty} \frac{1}{2}mv^4 \exp\left(-\frac{1}{2}\beta m v^2\right) 4\pi\, dv \tag{7.23}$$

The required integrals are

$$\int_0^{\infty} v^2 e^{-av^2}\, dv = \frac{\sqrt{\pi}}{4a^{3/2}} \qquad \int_0^{\infty} v^4 e^{-av^2}\, dv = \frac{3\sqrt{\pi}}{8a^{5/2}} \tag{7.24}$$

and can be found in equation (20.33) and problem 20.3 of Chapter 5. The result is

$$C = \frac{N}{V}\left(\frac{\beta m}{2\pi}\right)^{3/2} \qquad E = \frac{3N}{2\beta} \tag{7.25}$$

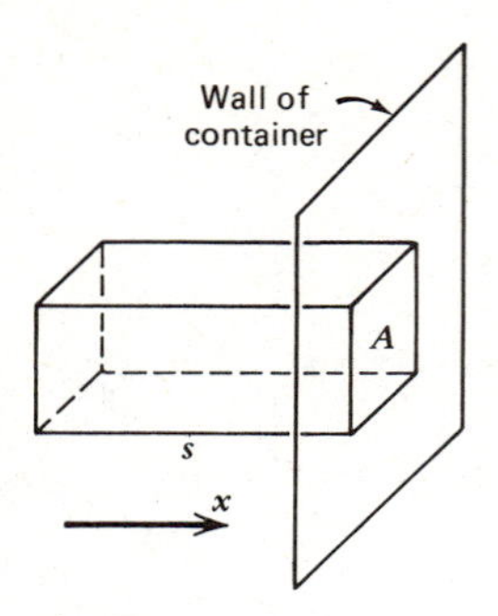

Fig. 7.4

The constant β can be related to temperature through the equation of state of an ideal gas. Suppose that an ideal gas is in thermal equilibrium inside a container of volume V. Figure 7.4 shows a volume element of the container in the shape of a rectangular parallelepiped. One face of area A coincides with the actual wall of the container. The edges which are perpendicular to the wall are of length s and are parallel to the x axis of a coordinate system. Let $dn(v_x)$ be the number of molecules per unit volume which have their x components of velocity in the range v_x to $v_x + dv_x$. Since the gas is in thermal equilibrium, the collisions of the molecules with the wall of the container will be, on the average, elastic. When the molecules inside the volume element with x components of velocity in the range v_x to $v_x + dv_x$ collide with the face of area A, they undergo a momentum change

$$\Delta p_x = 2mAs\, dn(v_x)\, v_x \tag{7.26}$$

This momentum change takes place in a time interval given by

$$\Delta t = \frac{s}{v_x} \tag{7.27}$$

The resulting force on the area A is given by

$$dF_x = \frac{\Delta p_x}{\Delta t} = 2Amv_x^2\, dn(v_x) \tag{7.28}$$

To find the net force, we must integrate (7.28) over all positive values of v_x:

$$F_x = 2Am\int_0^{\infty} v_x^2\, dn(v_x) \tag{7.29}$$

The average value of v_x^2 is given by

$$\overline{v_x^2} = \frac{\int_{-\infty}^{+\infty} v_x^2\, dn(v_x)}{\int_{-\infty}^{+\infty} dn(v_x)} = \frac{2\int_0^{\infty} v_x^2\, dn(v_x)}{N/V} \tag{7.30}$$

By combining (7.29) and (7.30), we find

$$P = \frac{F_x}{A} = \frac{mN\overline{v_x^2}}{V} \tag{7.31}$$

where P is the pressure on the container wall. Since there is no preferential direction in the container,

$$\overline{v_x^2} = \overline{v_y^2} = \overline{v_z^2} \qquad \overline{v^2} = \overline{v_x^2} + \overline{v_y^2} + \overline{v_z^2} = 3\overline{v_x^2} \tag{7.32}$$

Equation (7.31) can then be expressed in the form

$$PV = \tfrac{2}{3}N\left(\tfrac{1}{2}m\overline{v^2}\right) \tag{7.33}$$

This is to be compared with the ideal gas law

$$PV = nRT \tag{7.34}$$

where n is the number of moles, $R = 8.314$ J/mole-°K is the gas constant and T is the absolute or Kelvin temperature. The ideal gas law can be used to define temperature. Real gases approach ideal behavior at low pressures and at temperatures that are not too near the condensation point. The gas constant can be expressed as $R = N_A k$, where $N_A = 6.022 \times 10^{23}$ molecules per mole is Avogadro's number and $k = 1.381 \times 10^{-23}$ J/°K is Boltzmann's constant. Since the total number of molecules in the container is $N = nN_A$, the ideal gas law can be expressed

$$PV = nN_A kT = NkT \tag{7.35}$$

The comparison of (7.33) and (7.35) reveals that

$$\tfrac{1}{2}m\overline{v^2} = \tfrac{3}{2}kT \tag{7.36}$$

The particles of an ideal gas do not interact with one another appreciably. If the gas is monatomic, the total internal energy is essentially due to the translational kinetic energy and is

$$E = N\tfrac{1}{2}m\overline{v^2} = \tfrac{3}{2}NkT \tag{7.37}$$

Reference to equation (7.25) shows that

$$\beta = \frac{1}{kT} \tag{7.38}$$

The constant β, which was originally introduced into the formalism as a Lagrange multiplier, turns out to be related to the temperature as defined by the ideal gas law. One further point should be made. Equations (7.32) and (7.36) reveal that each of the degrees of freedom (the three rectangular coordinates x, y, and z) contribute $\frac{1}{2}kT$ to the average kinetic energy of a molecule. This is an example of the *equipartition theorem*, which states that at thermal equilibrium, there is associated with each degree of freedom an average energy of $\frac{1}{2}kT$ per molecule. By combining the expressions (7.25) and (7.38) for the constants C and β with (7.22), we find that

$$f(v)\, dv = N\left(\frac{m}{2\pi kT}\right)^{3/2} \exp\left(-\frac{mv^2}{2kT}\right) 4\pi v^2\, dv \tag{7.39}$$

represents the number of molecules in a container of ideal gas which would be expected to have velocities in the range v to $v + dv$. Figure 7.5 shows a plot of (7.39). The velocity v_0 is the *most probable* velocity and is the value of v for which $f(v)$ is a maximum. The horizontal axis in Fig. 7.5 represents v in units of v_0.

As an example of a case where the mechanical energy of the particles includes forms other than kinetic, consider a column of ideal gas of cross-sectional area A in the gravitational field of the Earth. Assume that the gas is isothermal. The energy of an individual particle is

$$E = \tfrac{1}{2}mv^2 + mgz \tag{7.40}$$

where z is the height above the Earth's surface and $g = 9.80$ m/sec² is the acceleration of a freely falling object. Assume that g is constant over the height of the column. By starting with (7.19) and replacing the

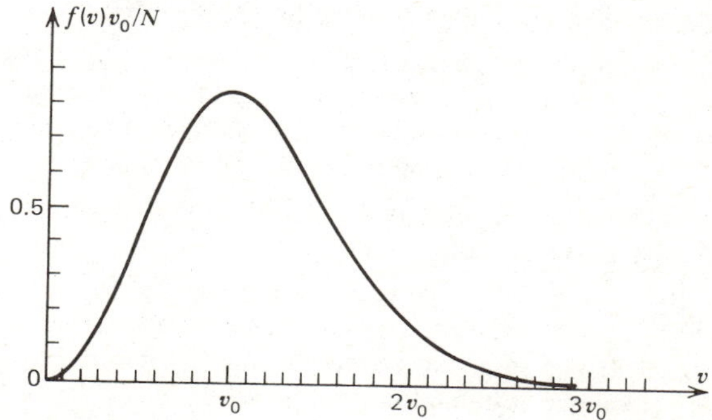

Fig. 7.5

sums by integrals, we can find the number of particles and the total energy between z and $z + dz$:

$$dN_z = AC\int_0^\infty \exp\left(-\frac{1}{2}\beta mv^2 - \beta mgz\right)4\pi v^2\,dv \tag{7.41}$$

$$dE_z = AC\int_0^\infty \left(\frac{1}{2}mv^2 + mgz\right) \times \exp\left(-\frac{1}{2}\beta mv^2 - \beta mgz\right)4\pi v^2\,dv \tag{7.42}$$

The average energy per particle is

$$\frac{dE_z}{dN_z} = \frac{\int_0^\infty (\frac{1}{2}mv^2 + mgz)\exp(-\frac{1}{2}\beta mv^2)4\pi v^2\,dv}{\int_0^\infty \exp(-\frac{1}{2}\beta mv^2)4\pi v^2\,dv}$$

$$= \frac{\int_0^\infty (\frac{1}{2}mv^4)\exp(-\frac{1}{2}\beta mv^2)\,dv}{\int_0^\infty v^2\exp(-\frac{1}{2}\beta mv^2)\,dv} + mgz$$

$$= \frac{3}{2\beta} + mgz = \frac{1}{2}m\overline{v^2} + mgz \tag{7.43}$$

The significant feature of (7.43) is that the potential energy and the average kinetic energy separate into the sum of two terms. The average kinetic energy bears the same relation to the constant β as it did when the gravitational field was absent leading to the same relation between β and the temperature as previously derived: $\beta = 1/kT$. As far as the distribution of the particles with elevation is concerned, we may write

$$f(z)\,dz = a\exp\left(-\frac{mgz}{kT}\right)dz \tag{7.44}$$

where a is a constant. If the gas is in a container of height h, the total number of particles is

$$N = \int_0^h f(z)\,dz = \frac{akT}{mg}\left[1 - \exp\left(-\frac{mgh}{kT}\right)\right] \tag{7.45}$$

If h is large,

$$f(z) \simeq \frac{mgN}{kT}\exp\left(-\frac{mgz}{kT}\right) \tag{7.46}$$

which gives an approximation to the variation of density with height in the atmosphere. The actual atmosphere of the Earth is of course not isothermal.

PROBLEMS

7.1. Three men are sent to jail where there are three jail cells. As each man comes in, the officer in charge sends him to a cell at random. Find the probabilities that (1) all three men end up in the same cell, (2) there are two men in one cell and one in one of the others, and (3) there is one man in each cell. You are man A. You like man B who is your buddy but you don't like man C. Find the probabilities that (4) you do not end up in the same cell with C, and (5) you share a cell with B, and C is in another cell.

7.2. Enumerate the possible outcomes of adding four identical objects to three boxes. Calculate the statistical weight of each. Note that the more even distribution has the highest statistical weight and hence the greatest probability of occurring. Check that the statistical weights add up to $3^4 = 81$.

7.3. For the distribution (7.39), find the most probable velocity v_0, the average velocity $\bar{v}$, and the root mean square velocity $v_{rms} = \sqrt{\overline{v^2}}$. What is the relation of $\bar{v}$ and v_{rms} to v_0? Indicate the approximate location of these three velocities on the plot of $f(v)$ vs. v given in Fig. 7.5. By writing $f(v)$ in terms of v and v_0, obtain an expression for the dimensionless quantity $f(v)v_0/N$. It is this function that is actually graphed in Fig. 7.5.

7.4. For the Maxwell–Boltzmann distribution of velocities, find the fraction of the particles which have a velocity larger than $2v_0$.

Hint: By appropriate change of variable, the required integral can be shown to be $(2/\sqrt{\pi})\Gamma(\frac{3}{2},4)$, where $\Gamma(\frac{3}{2},4)$ is an incomplete gamma function as defined in problem 23.3 of Chapter 5. It is also given by the chi-square integral (5.18) with $\chi^2 = 8$ and $n = 3$. To get a numerical value, use tables, an infinite series, or an asymptotic series.

7.5. Two systems are placed in thermal contact with one another but are otherwise thermally insulated from the environment. The probability that one system will be in a given state as characterized by the partition $N_1, N_2, \ldots, N_n$ and simultaneously the other system is in the state given by $N_1', N_2', \ldots, N_{n'}'$, is

$$P = \frac{N!}{N_1!N_2!\ldots N_n!}\left(\frac{1}{n}\right)^n \frac{N'!}{N_1'!N_2'!\ldots N_{n'}'!}\left(\frac{1}{n'}\right)^{n'}$$

Show from this that the distribution law for each system is the Maxwell–Boltzmann distribution:

$$N_i = Ce^{-\beta E_i} \qquad N_i' = C'e^{-\beta E_i'}$$

The constants C and C' are different, but the same constant β appears in both distributions. If one system is an ideal gas, then $\beta = 1/kT$. Thus, both systems are at the same temperature, and this can be taken as the temperature as defined by the ideal gas law. All systems in thermal equilibrium with one another have the same temperature. The constant β in the Maxwell–Boltzmann distribution can universally be written as $1/kT$.

7.6. Generalized coordinates and canonical momenta (problem 13.2 of Chapter 4) are the appropriate variables for a general description of phase space. If the Lagrangian for a conservative system is expressed as

$$L = \tfrac{1}{2}g_{\alpha\beta}\dot{q}^\alpha\dot{q}^\beta - V(q_\alpha)$$

we see that the generalized momenta

$$p_\alpha = \frac{\partial L}{\partial \dot{q}^\alpha} = g_{\alpha\beta}\dot{q}^\beta$$

are covariant. The coordinate differentials dq^α are by contrast contravariant. In equation (6.23) of Chapter 4, volume element in three-dimensional space is formulated as

$$d\Sigma_q = dq^1\mathbf{b}_1 \cdot dq^2\mathbf{b}_2 \times dq^3\mathbf{b}_3 = \sqrt{g}\, dq^1\, dq^2\, dq^3$$

If p_α are the canonical momenta conjugate to the coordinates q^α, then the volume element in momentum space is appropriately written as

$$d\Sigma_p = dp_1\mathbf{b}^1 \cdot dp_2\mathbf{b}^2 \times dp_3\mathbf{b}^3 = \frac{1}{\sqrt{g}} dp_1\, dp_2\, dp_3$$

In phase space, the volume therefore takes the form

$$d\Sigma = d\Sigma_q\, d\Sigma_p = dq^1\, dq^2\, dq^3\, dp_1\, dp_2\, dp_3$$

If the particles in the statistical ensemble have f degrees of freedom rather than three, the volume element in the phase space becomes $2f$-dimensional:

$$d\Sigma = dq^1 \ldots dq^f\, dp_1 \ldots dp_f$$

Suppose that a sample of gas in thermal equilibrium consists of diatomic molecules that can be considered as two point masses joined by a rigid rod. There are now three translational and two rotational degrees of freedom which we may take to be the two Euler angles θ and ϕ. The appropriate form for the energy of a single molecule is now

$$E = \tfrac{1}{2}mv^2 + \tfrac{1}{2}I(\dot{\theta}^2 + \dot{\phi}^2 \sin^2\theta)$$

as can be seen from equation (15.11) in Chapter 4 by setting $I_3 = 0$. Since the molecules are point particles, the moment of inertia I_3 about the line joining them is zero. No potential terms appear in the expression for the energy because we are assuming that the interaction between molecules is negligible. The required volume element in phase space is

$$d\Sigma = dx\, dy\, dz\, d\theta\, d\phi\, dp_x\, dp_y\, dp_z\, dp_\theta\, dp_\phi$$

Find the canonical momenta. Show that the energy can be expressed as

$$E = \frac{1}{2m}\left(p_x^2 + p_y^2 + p_z^2\right) + \frac{1}{2I}\left(p_\theta^2 + \frac{p_\phi^2}{\sin^2\theta}\right)$$

Set up and evaluate

$$\overline{E} = \frac{\int_\Sigma E e^{-E/kT}\, d\Sigma}{\int_\Sigma e^{-E/kT}\, d\Sigma}$$

for the average energy per molecule. Show that each of the five degrees of freedom contributes $\frac{1}{2}kT$ to $\overline{E}$. This is another example of the equipartition theorem for energy.

In reality, the rotational energy of a molecule is not a continuous function but is quantized. A quantum mechanical treatment therefore leads to somewhat different results and a consequent breakdown of the equipartition theorem, especially at low temperatures.

7.7. As a crude model of a crystal, assume that each atom behaves as an isotropic harmonic oscillator with a total energy given by

$$E = \tfrac{1}{2}mv^2 + \tfrac{1}{2}cr^2$$

where $v^2 = v_x^2 + v_y^2 + v_z^2$ and $r^2 = x^2 + y^2 + z^2$ gives the displacement of a given atom from its equilibrium position in the crystal lattice. We use c for the spring constant in order to avoid confusion with Boltzmann's constant. Use the procedure of problem 7.6 to show that the average energy per particle is $\overline{E} = 3kT$. The total internal energy per mole of crystal is $U = 3N_AkT = 3RT$, leading to a heat capacity of $C = dU/dT = 3R \simeq 6$ cal/mole. This is the Dulong and Petit law. It works all right at high enough temperatures, but quantum effects are important at low temperatures, leading to different results.

$\Delta p_x \Delta x \geq \frac{1}{2}\hbar$

7 Introduction to Hilbert Space and Eigenvalue Problems

A vector space of an infinite number of dimensions can be obtained by a limiting process which starts with a vector space of a finite number of dimensions. Such a program is carried out in the first two sections of this chapter by an appropriate limiting process which starts with a mechanical system that has f degrees of freedom. The infinite-dimensional space, called Hilbert space, is obtained as f is allowed to become infinite. The important concept of complete sets of orthonormal functions is discussed. Simple eigenvalue problems that occur in the mechanics of continuous media, electrodynamics, and quantum mechanics are used as examples. A substantial introduction to the theory of the Sturm–Liouville eigenvalue problem is given, although any consideration of the proof of the completeness of Sturm–Liouville systems is postponed to a later chapter.

The development of quantum mechanics leads to the necessity of complex vector spaces and to the need for Hermitian and unitary operators and matrices. We therefore take up once again the discussion of matrices which we left in Chapter 2, this time extending the theory to matrices which have complex numbers as elements. The Cayley–Hamilton theorem and the spectral theorem for matrices defined in a finite-dimensional space are discussed in detail.

1. PERIODIC SYSTEMS WITH f DEGREES OF FREEDOM

A one-dimensional mechanical system with f degrees of freedom can be constructed by connecting together f identical masses by $f+1$ identical springs. The end springs are connected to fixed supports. The masses are constrained to move along the same straight line as illustrated in Fig. 1.1. The system is periodic in the sense that it consists of many identical units connected together in a repetitive fashion. This might represent a one-dimensional model of a crystal. The solution of the system is of interest for its own sake, but a second and perhaps more important motive for studying it at this time is that it is possible by an appropriate limiting process to allow the number of degrees of freedom to become infinite. This leads to the concept of a vector space of an infinite number of dimensions. The mechanical system itself becomes a continuous medium in the limiting process.

The kinetic and potential energies of the system are

$$V = \tfrac{1}{2}ky_1^2 + \tfrac{1}{2}k(y_2 - y_1)^2 + \cdots + \tfrac{1}{2}k(y_f - y_{f-1})^2 + \tfrac{1}{2}ky_f^2 \tag{1.1}$$

$$K = \tfrac{1}{2}m\dot{y}_1^2 + \tfrac{1}{2}m\dot{y}_2^2 + \cdots + \tfrac{1}{2}m\dot{y}_f^2 \tag{1.2}$$

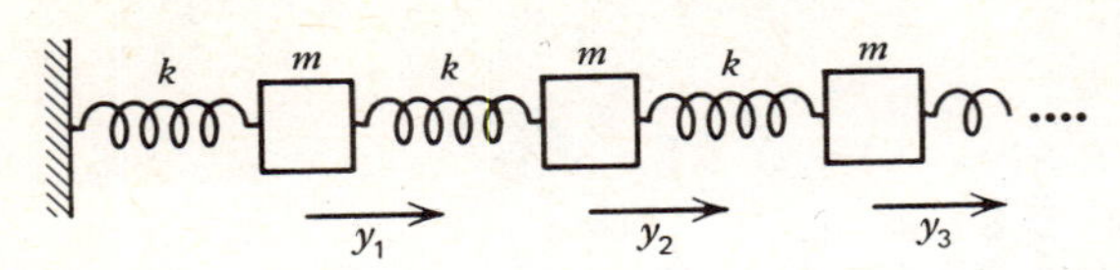

Fig. 1.1

The equations of motion are most easily found from Lagrange's equations as given by (9.57) in Chapter 4. The result is

$$m\ddot{y}_1 + 2ky_1 - ky_2 = 0$$
$$m\ddot{y}_2 - ky_1 + 2ky_2 - ky_3 = 0$$
$$\vdots$$
$$m\ddot{y}_n - ky_{n-1} + 2ky_n - ky_{n+1} = 0$$
$$\vdots$$
$$m\ddot{y}_f - ky_{f-1} + 2ky_f = 0 \tag{1.3}$$

Another system which obeys the same differential equations consists of a massless string which is stretched between fixed supports and then loaded with identical point masses at equal intervals. Figure 1.2 shows a portion of such a system. Let the x axis be the equilibrium position of the string. If the displacements in the y direction (transverse displacements) are small, the accelerations of the masses in the x direction (longitudinal displacements) will be negligible, meaning that the tension F is substantially the same everywhere in the string. In the following analysis, we assume that the system is constrained to move in the x, y plane as defined by Fig. 1.2. Small transverse displacements also mean that the angles θ_1 and θ_2 connecting a given mass to the adjacent masses will be small. Neglecting the effect of the gravitational field, the equation of motion of mass n is

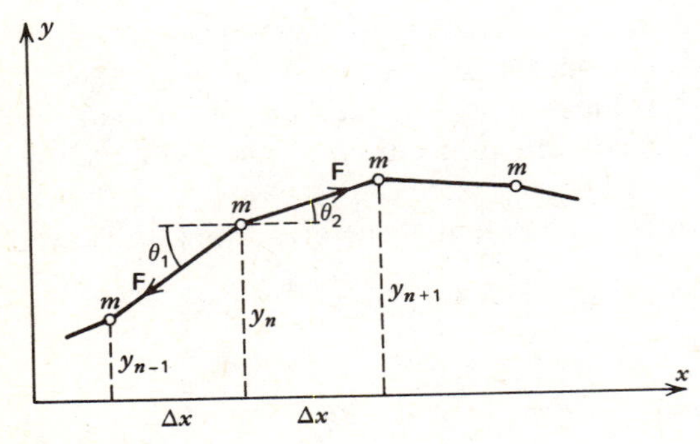

Fig. 1.2

$$F\theta_2 - F\theta_1 = m\ddot{y}_n \tag{1.4}$$

Since the angles are small, they are given to a good approximation by

$$\theta_1 = \frac{y_n - y_{n-1}}{\Delta x} \qquad \theta_2 = \frac{y_{n+1} - y_n}{\Delta x} \tag{1.5}$$

where Δx is the separation between masses. The equation of motion is then

$$m\ddot{y}_n - \frac{F}{\Delta x}y_{n-1} + \frac{2F}{\Delta x}y_n - \frac{F}{\Delta x}y_{n+1} = 0 \tag{1.6}$$

and is identical to (1.3) with $F/\Delta x$ taking the place of the spring constant. Friction has been neglected in both (1.3) and (1.6).

The general method for solving systems of the type (1.3) is discussed in section 14 of Chapter 4. By assuming solutions of the form $y_n = A_n \cos(\omega t - \phi)$, we find

$$(-m\omega^2 + 2k)A_1 - kA_2 = 0$$
$$-kA_1 + (-m\omega^2 + 2k)A_2 - kA_3 = 0$$
$$\vdots$$
$$-kA_{n-1} + (-m\omega^2 + 2k)A_n - kA_{n+1} = 0$$
$$\vdots$$
$$(-m\omega^2 + 2k)A_f - kA_{f-1} = 0 \tag{1.7}$$

The secular equation is

$$\begin{vmatrix} -m\omega^2 + 2k & -k & 0 & \cdots \\ -k & -m\omega^2 + 2k & -k & \cdots \\ 0 & -k & -m\omega^2 + 2k & \cdots \\ \vdots & \vdots & \vdots & \end{vmatrix} = 0 \tag{1.8}$$

It is convenient to divide each row of the secular determinant by k and to define

$$c = 1 - \frac{m\omega^2}{k} \tag{1.9}$$

The secular equation then reads

$$\begin{vmatrix} c+1 & -1 & 0 & \cdots \\ -1 & c+1 & -1 & \cdots \\ 0 & -1 & c+1 & \cdots \\ \vdots & \vdots & \vdots & \end{vmatrix} = \Delta_f = 0 \tag{1.10}$$

This is a polynomial of degree f in c. It has f roots which in combination with (1.9) give the f possible eigenfrequencies of the system. Note that the secular determinant obeys the recurrence relation

$$\Delta_f = (c+1)\Delta_{f-1} - \Delta_{f-2} \tag{1.11}$$

We define a parameter θ by means of

$$c + 1 = 2\cos\theta \tag{1.12}$$

Note that (1.11) is satisfied identically by

$$\Delta_f = a\sin\{(f+1)\theta\} \tag{1.13}$$

The parameter a can be found from

$$\Delta_1 = c + 1 = a\sin 2\theta = 2\cos\theta$$

$$a = \frac{2\cos\theta}{\sin 2\theta} = \frac{1}{\sin\theta} \tag{1.14}$$

The secular equation is therefore

$$\Delta_f = \frac{\sin\{(f+1)\theta\}}{\sin\theta} = 0 \tag{1.15}$$

The possible solutions are

$$(f+1)\theta = n\pi \qquad n = 1,2,3,\ldots,f \tag{1.16}$$

The cases $n = 0$ and $n = f+1$ are ruled out because (1.15) becomes indeterminant for these values. Thus, (1.16) gives all the f possible roots of the secular equation. The f possible eigenfrequencies are found from (1.9), (1.12), and (1.16) to be

$$\omega_n = 2\sqrt{\frac{k}{m}}\sin\left[\frac{n\pi}{2(f+1)}\right] \tag{1.17}$$

The eigenvectors are to be computed from (1.7) which is conveniently expressed in the form

$$\begin{aligned} A_2 &= (c+1)A_1 \\ A_3 &= (c+1)A_2 - A_1 \\ &\vdots \\ A_{j+1} &= (c+1)A_j - A_{j-1} \\ &\vdots \\ 0 &= (c+1)A_f - A_{f-1} \end{aligned} \tag{1.18}$$

The eigenvectors have to be normalized later; for the moment, it is convenient to set $A_1 = \sin\theta$. Then $A_2 = (c+1)A_1 = 2\cos\theta\sin\theta = \sin 2\theta$. The jth recurrence formula in (1.18) is satisfied by

$$A_{nj} = \sin(j\theta) = \sin\left[\frac{jn\pi}{f+1}\right] \tag{1.19}$$

where, as indicated by the double subscript, a different eigenvector, that is, a different set of solutions of (1.18), is obtained for each of the f frequencies. We already know from the general theoretical discussion in section 14 of Chapter 4 that the eigenvectors are mutually orthogonal. Since the masses are all the same, the metric tensor for this problem can be taken as δ_{ij}. The orthogonality condition therefore reads

$$\sum_{j=1}^{f}\sin\left(\frac{jn\pi}{f+1}\right)\sin\left(\frac{jn'\pi}{f+1}\right) = N\delta_{nn'} \tag{1.20}$$

The normalizing factor N is found by setting $n = n'$:

$$\sum_{j=1}^{f}\sin^2(j\theta) = N \qquad \theta = \frac{n\pi}{f+1} \tag{1.21}$$

We have encountered the problem of doing trigonometric sums of this sort before, for example, in section 10 of Chapter 5. Equation (1.21) can be written as

$$N = \sum_{j=1}^{f}\left[\frac{1}{2} - \frac{1}{2}\cos(2j\theta)\right] = \frac{f+1}{2} - \sum_{j=0}^{f}\cos(2j\theta) \tag{1.22}$$

Reference to problem 10.5 of Chapter 5 reveals that

$$N = \frac{f+1}{2} - \frac{\cos(f\theta)\sin\{(f+1)\theta\}}{\sin\theta} = \frac{f+1}{2} \tag{1.23}$$

The part of (1.23) involving θ is zero by virtue of (1.15). The elements of the normalized eigenvectors are therefore

$$S_{nj} = \sqrt{\frac{2}{f+1}}\sin\left(\frac{jn\pi}{f+1}\right) \tag{1.24}$$

and obey the orthogonality conditions

$$\sum_{j=1}^{f} S_{nj}S_{n'j} = \delta_{nn'} \qquad \sum_{n=1}^{f} S_{nj}S_{nj'} = \delta_{jj'} \tag{1.25}$$

The general solution of our mechanics problem can now be written

$$y_j = \sum_{n=1}^{f} S_{nj}(a_n \sin \omega_n t + b_n \cos \omega_n t) \tag{1.26}$$

The solution contains $2f$ arbitrary constants a_n and b_n to be determined by the initial conditions as required. Specifically, at $t = 0$,

$$y_j(0) = \sum_{n=1}^{f} S_{nj}b_n \qquad \dot{y}_j(0) = \sum_{n=1}^{f} S_{nj}\omega_n a_n \tag{1.27}$$

By application of the orthogonality conditions (1.25) we find

$$b_n = \sum_{j=1}^{f} S_{nj}y_j(0) \qquad a_n = \frac{1}{\omega_n}\sum_{j=1}^{f} S_{nj}\dot{y}_j(0) \tag{1.28}$$

Thus, all the constants are determined.

What we have done is solve a mechanics problem in its f-dimensional rectangular Cartesian configuration space. There are f mutually orthogonal eigenvectors each with f components. The S_{nj} are the elements of an f-dimensional orthogonal transformation which gives the normal coordinates of the problem through the relations

$$q_n = \sum_{j=1}^{f} S_{nj}y_j \qquad y_j = \sum_{n=1}^{f} S_{nj}q_n \tag{1.29}$$

The normal coordinates obey

$$\ddot{q}_n + \omega_n^2 q_n = 0 \tag{1.30}$$

as is made evident by the general theoretical discussion leading to equation (14.20) of Chapter 4 or by direct comparison of (1.29) to (1.26). The transformation is a rotation in the f-dimensional configuration space. The eigenvectors are *complete* in the sense that they are f linearly independent vectors which span the configuration space. This is important because it allows completely arbitrary vectors with components $y_j(0)$ and $\dot{y}_j(0)$, which represent the initial positions and velocities of all the particles, to be expressed in terms of them as given by equation (1.27).

As an example, consider the possibility that the system vibrates in a single pure mode with only one of the possible eigenfrequencies present. If the initial velocity is zero, the particle displacements as a function of time are given by

$$y_j = b_n S_{nj} \cos \omega_n t \tag{1.31}$$

Consider the case $f = 4$. There is a solution of the type (1.31) for each of the four possible eigenfrequencies ω_n as given by (1.17). The normalized eigenvectors can be displayed as the rows of the transformation matrix

$$S = \sqrt{\frac{2}{5}}\begin{pmatrix} \sin\frac{\pi}{5} & \sin\frac{2\pi}{5} & \sin\frac{2\pi}{5} & \sin\frac{\pi}{5} \\ \sin\frac{2\pi}{5} & \sin\frac{\pi}{5} & -\sin\frac{\pi}{5} & -\sin\frac{2\pi}{5} \\ \sin\frac{2\pi}{5} & -\sin\frac{\pi}{5} & -\sin\frac{\pi}{5} & \sin\frac{2\pi}{5} \\ \sin\frac{\pi}{5} & -\sin\frac{2\pi}{5} & \sin\frac{2\pi}{5} & -\sin\frac{\pi}{5} \end{pmatrix} \tag{1.32}$$

where we have written $\sin(3\pi/5) = \sin(2\pi/5)$, $\sin(8\pi/5) = -\sin(2\pi/5)$, and so on. Each row of (1.32) corresponds to one of the four eigenfrequencies. The solutions are most easily visualized for the case of transverse motion. The amplitudes and relative phases of the four particles for each of the four

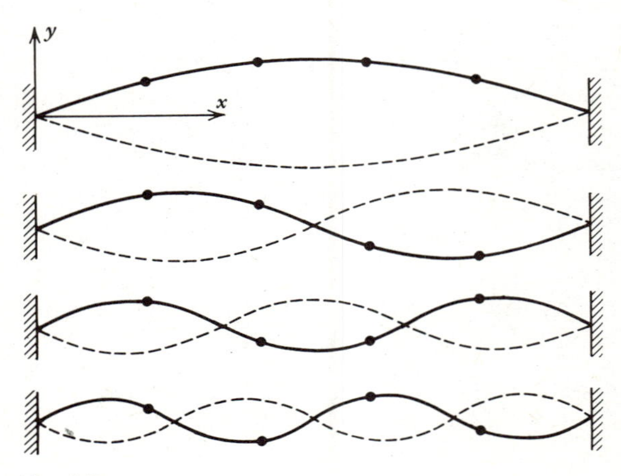

Fig. 1.3

possible modes can be read directly out of the matrix (1.32). This is facilitated by sketching the sine curves

$$\sin\left(\frac{\pi x}{5\Delta x}\right) \qquad \sin\left(\frac{2\pi x}{5\Delta x}\right) \qquad \sin\left(\frac{3\pi x}{5\Delta x}\right)$$
$$\sin\left(\frac{4\pi x}{5\Delta x}\right) \tag{1.33}$$

These are the solid curves shown in Fig. 1.3. The x coordinates of the particles are $x = \Delta x$, $2\,\Delta x$, $3\Delta x$, and $4\Delta x$. At $t = 0$, the location of each particle is on the solid curve. The dashed curves in Fig. 1.3 locate the particles after one half-cycle of the motion. The normal modes of vibration can be thought of as *standing waves*. More general motions of the system consist of mixtures or superpositions of the possible standing wave solutions.

PROBLEMS

1.1. Figure 1.4 shows a portion of a periodic mechanical system consisting of masses and springs and also a portion of a periodic electrical network consisting of inductors and capacitors. It is somet.mes convenient to consider such systems as infinitely long so that end effects can be neglected. Show that the two systems obey differential equations of the same form. The differential equation satisfied by y_n is given by (1.3). The charges q_n on the capacitors are the analogues of the displacements y_n.

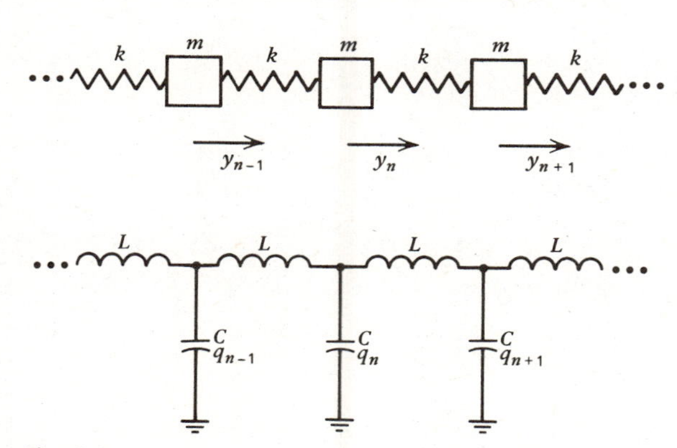

Fig. 1.4

1.2. Show that the system illustrated in Fig. 1.4 has simple harmonic or plane wave solutions of the form

$$\begin{aligned} y_{n-1} &= A\cos(\omega t - \alpha + \beta) \\ y_n &= A\cos(\omega t - \alpha) \\ y_{n+1} &= A\cos(\omega t - \alpha - \beta) \end{aligned} \tag{1.34}$$

where A and α are constants which can be chosen arbitrarily and β is dependent on the frequency. You will find it useful to use complex solutions the real parts of which are given by (1.34). Find the dependence of β on the frequency and show that plane wave solutions are not possible if the frequency exceeds the critical frequency

$$\omega_c = 2\sqrt{\frac{k}{m}} \tag{1.35}$$

This means that the system functions as a *low-pass filter*. Waves with frequencies above ω_c cannot propagate. Note that the critical frequency is essentially the same as the highest allowed eigenfrequency as given by (1.17) for standing waves, provided that the number of degrees of freedom is large.

Show that the wavelength of the plane wave is given by

$$\lambda = \frac{2\pi \Delta x}{\beta} \tag{1.36}$$

where Δx is the separation between particles as pictured for the transverse waves in Fig. 1.2. The shortest possible wavelength corresponds to the critical frequency. At this frequency, the wavelength is about two times the particle separation. Shorter waves than this cannot propagate.

The *phase velocity* is the velocity of propagation of the plane wave and is given by

$$u_p = \frac{\lambda}{T} \tag{1.37}$$

where T is the period of the motion determined by $\omega T = 2\pi$. Show that

$$u_p = \frac{\omega \Delta x}{\cos^{-1}(1 - m\omega^2/2k)} \tag{1.38}$$

Review the discussion of plane electromagnetic waves in section 12 of Chapter 3. Note especially that all electromagnetic waves propagate through the vacuum at the same speed c regardless of their frequency. This does not hold true for waves propagating through periodic structures. The phase velocity is *frequency-dependent*, meaning that plane waves of different frequency propagate at different velocities. This phenomenon is called *dispersion*. Electromagnetic waves also exhibit dispersion when they travel through material media.

If $\lambda = 8\Delta x$, make a sketch of the positions of the particles in one complete wave at $t = 0$ and also at $\omega t = 2\beta$. Set $\alpha = 0$ for convenience and consider the case of transverse motion.

1.3. Find the differential equation in which q_n appears for the electrical network of Fig. 1.5. Show that plane waves can propagate in this system if ω is *larger* than the critical frequency

$$\omega_c = \frac{1}{2\sqrt{LC}} \tag{1.39}$$

The system serves as a *high-pass* filter.

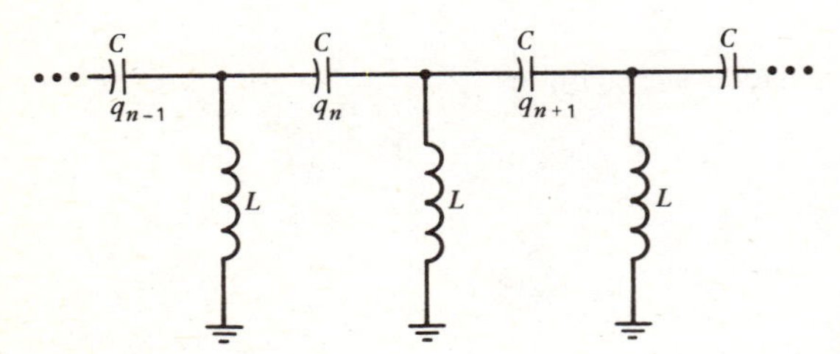

Fig. 1.5

1.4. For the periodic structures of Fig. 1.4, show that solutions exist of the form

$$\begin{aligned} y_{n-1} &= Ae^{b}e^{i(\omega t + \beta)} \\ y_n &= Ae^{i\omega t} \\ y_{n+1} &= Ae^{-b}e^{i(\omega t - \beta)} \end{aligned} \tag{1.40}$$

for $\omega > \omega_c$, provided that

$$\cosh b = \frac{m\omega^2}{2k} - 1 \qquad \beta = \pi \tag{1.41}$$

The actual solution is the real part of (1.40). This shows that, if a wave with a frequency higher than ω_c enters a periodic mechanical or electrical structure like that shown in Fig. 1.4, its amplitude will be attenuated in the direction of propagation. Note that the wavelength for such solutions is $2\,\Delta x$.

1.5. A periodic series of identical masses and springs is terminated by a spring which is connected to a fixed support as shown in Fig. 1.6. Find the differential equation obeyed by the displacement y_n of the terminal mass. Consider the possibility of solutions of the form

$$y_n = (Ae^{i\beta} + Be^{-i\beta})e^{i\omega t}$$

$$y_{n-1} = (Ae^{2i\beta} + Be^{-2i\beta})e^{i\omega t} \tag{1.42}$$

it being understood that the actual solution is the real part. In problem 1.2, you found that

$$\cos\beta = 1 - \frac{m\omega^2}{2k} \tag{1.43}$$

Use this to show that (1.42) is a solution provided that $B = -A$. The interpretation of this result is that a plane wave is totally reflected when it encounters a fixed support. The amplitude of the reflected wave is the same as that of the incident wave. The minus sign in $B = -A$ means that the reflected wave is 180° out of phase $(-1 = e^{i\pi})$ with the incident wave. Show that the real parts of (1.42) are

$$y_n = -2A\sin\beta\sin\omega t$$

$$y_{n-1} = -2A\sin 2\beta\sin\omega t \tag{1.44}$$

The incident and reflected wave combine to form a standing wave.

Any kind of interruption or change in a periodic structure can be expected to cause some degree of reflection of a plane wave.

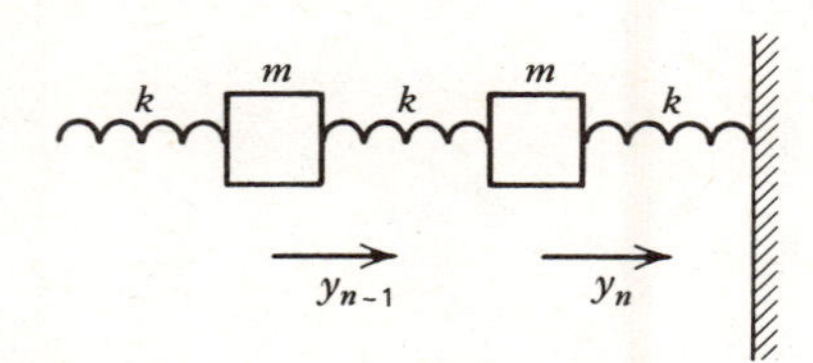

Fig. 1.6

1.6. A periodic system of identical masses and springs terminates with a mass of value $m/2$ as shown in Fig. 1.7. Find the differential equation of the terminal mass and show that solutions of the form

$$y_n = (A + B)e^{i\omega t}$$

$$y_{n-1} = (Ae^{i\beta} + Be^{-i\beta})e^{i\omega t} \tag{1.45}$$

are possible provided that $A = B$. A plane wave incident on the free end is reflected with no change of either amplitude or phase.

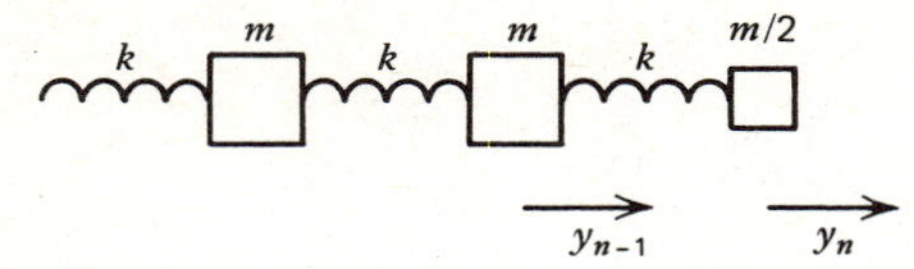

Fig. 1.7

1.7. A periodic structure consisting of identical inductors and capacitors is constructed as shown in Fig. 1.8. It is driven at one end with a harmonic driving voltage and is terminated at the other by an impedance Z. Show that the currents in the terminal elements obey

$$\frac{1}{C}(I_n - I_{n+1}) = Z\frac{dI_{n+1}}{dt} \tag{1.46}$$

Show that the traveling wave solution

$$I_n = Ae^{i\omega t} \qquad I_{n+1} = Ae^{i(\omega t - \beta)} \tag{1.47}$$

is possible, provided that

$$Z = Z_0 = \frac{e^{i\beta} - 1}{i\omega C} = i\omega\frac{L}{2} + \sqrt{\frac{L}{C}}\sqrt{1 - \omega^2 LC/4} \tag{1.48}$$

If the line is terminated by the impedance Z_0, the traveling wave is completely absorbed without any reflection. Another way to look at it is to say that Z_0 is the equivalent of the impedance of an infinitely long line of capacitors and inductors. If the terminal impedance is not Z_0, then some of the incident wave will be reflected. Show that in this case there is a solution of the form

$$I_n = (Ae^{i\beta} + Be^{-i\beta})e^{i\omega t} \qquad I_{n+1} = (A + B)e^{i\omega t} \tag{1.49}$$

The amplitude of the reflected wave is

$$B = A\frac{Z_0 - Z}{Z_0^* + Z} \tag{1.50}$$

The fixed support of problem 1.5 corresponds to $Z = \infty$. If you set $Z = i\omega L/2$, you will get the same result that you found in problem 1.6.

The impedance Z_0 is called the *characteristic impedance* of the line. It could actually be constructed from an inductor of value $L/2$ and an ordinary resistor of value

$$R_0 = \sqrt{\frac{L}{C}}\sqrt{1 - \omega^2\frac{LC}{4}} \tag{1.51}$$

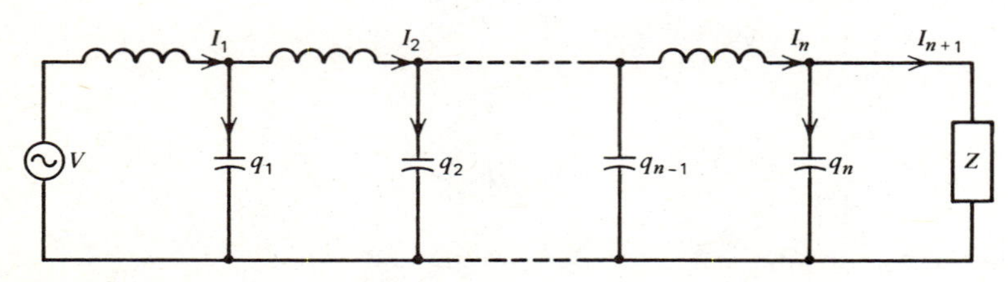

Fig. 1.8

It is possible to derive Z_0 from a consideration of the driven end of the line. Do this by showing that $V = I_1 Z_0$ provided that

$$I_1 = Ae^{i\omega t} \qquad I_2 = Ae^{i\omega t - i\beta} \qquad I_3 = Ae^{i\omega t - 2i\beta} \quad \text{and so on}$$

$$\cos\beta = 1 - \tfrac{1}{2}LC\omega^2 \tag{1.52}$$

You will have to write down the circuit equations for the first two loops of Fig. 1.8 in terms of complex impedances, for example,

$$V = I_1 i\omega L + \frac{1}{i\omega C}(I_1 - I_2) \tag{1.53}$$

for the first loop. Show that the average power which the driving voltage puts into an infinite line is

$$P = \tfrac{1}{2}|A|^2 R_0 \tag{1.54}$$

Refer to problem 13.2 in Chapter 5.

2. VIBRATING STRING

A uniform string under tension is a simple example of a one-dimensional continuous material medium. It can be considered as a limiting case of the massless string loaded at equal intervals with identical point particles. Equation (1.6) can be written in the form

$$\frac{m}{\Delta x}\ddot{y}_n - \frac{F\left(\dfrac{y_{n+1} - y_n}{\Delta x} - \dfrac{y_n - y_{n-1}}{\Delta x}\right)}{\Delta x} = 0 \tag{2.1}$$

Imagine that the point masses are placed on the string at closer and closer intervals so that $\Delta x \to 0$. At the same time, the mass of each particle approaches zero in such a way that the ratio $m/\Delta x$ approaches a finite limit:

$$\lim_{\substack{\Delta x \to 0 \\ m \to 0}} \frac{m}{\Delta x} = \mu \tag{2.2}$$

In (2.2), μ is the linear mass density of the string and is assumed here to be constant over the length of the string. By taking the limit $\Delta x \to 0$ in the equation of motion (2.1), we get the partial differential equation

$$\frac{\mu}{F}\frac{\partial^2 y}{\partial t^2} - \frac{\partial^2 y}{\partial x^2} = 0 \tag{2.3}$$

which is recognized as a one-dimensional wave equation. It might be a good idea to review the discussion in section 12 of Chapter 3 of the wave equations which occur in electromagnetic theory. The one-dimensional wave equation has solutions of the form

$$y = f(x - vt) + g(x + vt) \tag{2.4}$$

where

$$v = \sqrt{\frac{F}{\mu}} \tag{2.5}$$

is the velocity of propagation of disturbances on the string and f and g are arbitrary functions. One point to be made here is that dispersion disappears in the passage to the limit of a continuous medium from a periodic structure consisting of regularly spaced point masses. The phase velocity of a plane wave in a periodic structure is given by (1.38). The wave equation predicts that all disturbances on the string, including harmonic waves of different frequencies, will propagate at the same speed.

The wave equation for the string will now be derived again starting from the premise that it is a continuous medium of constant linear mass density. The x axis of Fig. 2.1 is the equilibrium position of the string, and we consider its transverse motion in the x, y plane. In addition to the force produced by the tension F on an element of length Δx, we assume that there is an external force per unit length in the y direction given by a function $f(x, t)$ and a viscous frictional force $b\,\partial y/\partial t$ per unit length. For small transverse displacements, the equation of mo-

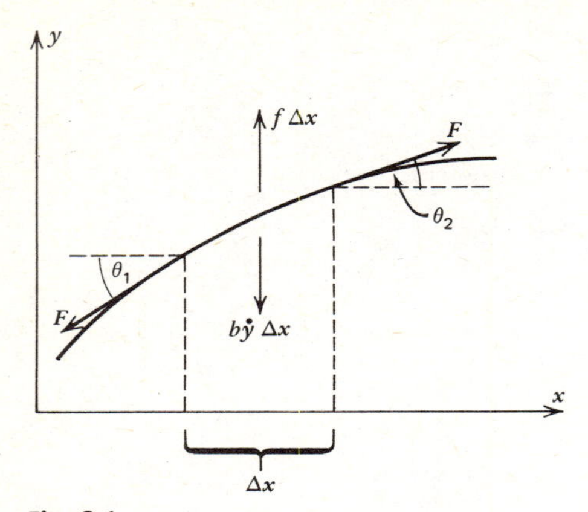

Fig. 2.1

tion of an element of length Δx is

$$F\theta_2 - F\theta_1 + f\Delta x - b\dot{y}\Delta x = \mu\,\Delta x\frac{\partial^2 y}{\partial t^2} \tag{2.6}$$

The angles θ_1 and θ_2 are approximately

$$\theta_1 = \left(\frac{\partial y}{\partial x}\right)_1 \qquad \theta_2 = \left(\frac{\partial y}{\partial x}\right)_2 \tag{2.7}$$

where the derivatives are evaluated at the two ends of the element of length Δx in Fig. 2.1. Equation (2.6) can be written

$$\frac{F\left[\left(\frac{\partial y}{\partial x}\right)_2 - \left(\frac{\partial y}{\partial x}\right)_1\right]}{\Delta x} + f(x,t) - b\frac{\partial y}{\partial t} = \mu\frac{\partial^2 y}{\partial t^2} \tag{2.8}$$

Taking the limit $\Delta x \to 0$ yields

$$\frac{\partial^2 y}{\partial x^2} - \frac{b}{F}\frac{\partial y}{\partial t} - \frac{\mu}{F}\frac{\partial^2 y}{\partial t^2} = -\frac{f(x,t)}{F} \tag{2.9}$$

In the absence of friction, $b = 0$, and (2.9) becomes a wave equation with a source term of exactly the same form as the wave equations for the vector and scalar potentials of electromagnetic theory as given by (12.45) and (12.46) of Chapter 3.

We will obtain the solution of the one-dimensional wave equation without friction or external driving forces for the example of a string under tension and fixed at its two ends. The fixed ends are taken to be at $x = 0$ and $x = s$. Imagine, for example, that a piano string has been set into vibration by being struck with a hammer. The solution will be found as the limiting form of the solution (1.26) for a massless string loaded with point particles. The components (1.24) of the eigenvectors can be written

$$\begin{aligned} S_{nj} &= \sqrt{\frac{2}{f+1}}\,\sin\left(\frac{jn\pi}{f+1}\right) \\ &= \sqrt{\Delta x}\sqrt{\frac{2}{(f+1)\,\Delta x}}\,\sin\left[\frac{\pi nj\Delta x}{(f+1)\,\Delta x}\right] \\ &= \sqrt{\Delta x}\sqrt{\frac{2}{s}}\,\sin\left(\frac{\pi n x_j}{s}\right) \end{aligned} \tag{2.10}$$

where x_j is the position of the jth particle and $s = (f+1)\,\Delta x$ is the length of the string. We will first examine the orthogonality condition

$$\begin{aligned} \sum_{j=1}^{f} S_{nj}S_{n'j} &= \sum_{j=1}^{f}\frac{2}{s}\sin\left(\frac{\pi n x_j}{s}\right)\sin\left(\frac{\pi n' x_j}{s}\right)\Delta x \\ &= \delta_{nn'} \end{aligned} \tag{2.11}$$

Recall that this is really an inner or dot product of two rows of the matrix S regarded as eigenvectors. In the limit $\Delta x \to 0$ and $f \to \infty$, (2.11) can be written

$$\int_0^s \frac{2}{s}\sin\left(\frac{\pi n x}{s}\right)\sin\left(\frac{\pi n' x}{s}\right)dx = \delta_{nn'} \tag{2.12}$$

Since x is now a continuous variable, the subscript j has been dropped. Note carefully the significance of this result. The ordinary dot product of two vectors in a space of a finite number of dimensions has been replaced by an integral in a limiting process where the dimension of the space, and hence the number of components of the vectors, becomes infinite. The discrete variable j has been replaced by a continuous variable x over which the integral is performed.

If we write $k = F/\Delta x$, the eigenfrequencies as given by (1.17) are

$$\begin{aligned} \omega_n &= 2\sqrt{\frac{F}{m\,\Delta x}}\,\sin\left[\frac{n\pi}{2(f+1)}\right] \\ &= 2\sqrt{\frac{F}{m\,\Delta x}}\,\sin\left(\frac{n\pi\Delta x}{2s}\right) \end{aligned} \tag{2.13}$$

In the limit $\Delta x \to 0$ and $f \to \infty$, an infinite number of eigenfrequencies is possible. They are

$$\omega_n = \lim_{\Delta x \to 0} 2\sqrt{\frac{F}{m\,\Delta x}}\,\frac{n\pi\Delta x}{2s} = \lim_{\Delta x \to 0} \sqrt{\frac{F\Delta x}{m}}\,\frac{n\pi}{s}$$

$$= \sqrt{\frac{F}{\mu}}\,\frac{n\pi}{s} \tag{2.14}$$

Another version of the orthogonality relations is

$$\sum_{n=1}^{f} S_{nj} S_{nj'} = \delta_{jj'} \tag{2.15}$$

which is a dot product of two columns of the matrix S. By using the expressions (2.10) for the elements S_{nJ}, (2.15) can be written

$$\sum_{n=1}^{f} \frac{2}{s} \sin\left(\frac{\pi n x_j}{s}\right) \sin\left(\frac{\pi n x_j'}{s}\right) = \frac{\delta_{jj'}}{\Delta x} \tag{2.16}$$

Even though the number of possible n values becomes infinite, the eigenfrequencies (2.14) do not become closely spaced as was the case for the positions x_j of the particles. There is no continuous variable to be associated with n in the limit of an infinite number of degrees of freedom. What then is the limiting form of (2.16)? It is zero if $j \neq j'$ and large if $j = j'$ on account of $\Delta x \to 0$. The behavior of (2.16) is like that of a Dirac delta function, and we will anticipate this identification and write for the limiting form of (2.16)

$$\sum_{n=1}^{\infty} \frac{2}{s} \sin\left(\frac{\pi n x}{s}\right) \sin\left(\frac{\pi n x'}{s}\right) = \delta(x - x') \tag{2.17}$$

The delta function is discussed in section 10 of Chapter 3.

The complete set of normalized, orthogonal eigenvectors which exist in the f-dimensional configuration space of the discrete system has been replaced in the limiting process by a complete set of normalized and orthogonal eigenfunctions

$$u_n(x) = \sqrt{\frac{2}{s}}\,\sin\left(\frac{n\pi x}{s}\right) \tag{2.18}$$

which obey the orthogonality conditions

$$\int_0^s u_n(x) u_{n'}(x)\,dx = \delta_{nn'} \tag{2.19}$$

$$\sum_{n=1}^{\infty} u_n(x) u_n(x') = \delta(x - x') \tag{2.20}$$

We will elaborate more on the concept of completeness as it applies to the eigenfunctions (2.18) a little later. It is interesting to graph a few terms of the series in (2.20). This has been done in Fig. (2.2) for the case $x' = \frac{3}{4}s$. The figures are plots of the function

$$\Delta_N\left(x - \frac{3}{4}s\right) = \frac{2}{s}\sin\left(\frac{\pi x}{s}\right)\sin\left(\frac{3\pi}{4}\right) + \frac{2}{s}\sin\left(\frac{2\pi x}{s}\right)\sin\left(\frac{3\pi}{2}\right) + \cdots + \frac{2}{s}\sin\left(\frac{N\pi x}{s}\right)\sin\left(\frac{3\pi N}{4}\right) \tag{2.21}$$

for $N = 1, 2, \ldots, 10$. The plots for $N = 3$ and $N = 4$ are both the same because the last term in the series for $N = 4$ is zero on account of $\sin(3\pi) = 0$. Similarly, $N = 7$ and $N = 8$ are the same. Constructive interference of the terms at $x = x' = \frac{3}{4}s$ occurs causing a buildup of a sharp spike at this point. Destructive interference occurs elsewhere. The actual physical system exists for $0 < x < s$. Mathematically, (2.20) can be extended to the entire x axis. The eigenfunctions are odd, meaning that $u_n(-x) = -u_n(x)$, and are of period $2s$. Any function which is expanded in terms of them will also have these properties. The plots of (2.21) in Fig. 2.2 have been extended to the range $-2s < x < 2s$ to show these features. It is evident that (2.20) really represents a series of positive delta functions at $x', x' \pm 2s, x' \pm 4s, \ldots$ and a series of negative delta functions at $-x', -x' \pm 2s, -x' \pm 4s, \ldots$. Thus,

$$\sum_{n=1}^{\infty} u_n(x) u_n(x') = \sum_{k=-\infty}^{+\infty} \left[\delta(x' + 2ks - x) - \delta(-x' + 2ks - x)\right] \tag{2.22}$$

In equations (1.27), the f coordinates $y_j(0)$ of the initial position of the system in configuration space can be thought of as the components of a vector. The eigenvectors are a complete set of orthogonal unit vectors, and the coefficients b_n are the projections or components of the initial position vector in the directions of the eigenvectors. This same concept of completeness can be preserved in the limit as $f \to \infty$. We write (1.27) as

$$y_j(0) = \sum_{n=1}^{f} b_n\sqrt{\Delta x}\sqrt{\frac{2}{s}}\sin\left(\frac{n\pi x_j}{s}\right) \tag{2.23}$$

In the limit as $\Delta x \to 0$ and $f \to \infty$,

$$y(x,0) = \sum_{n=1}^{\infty} B_n u_n(x) \tag{2.24}$$

The initial position vector of the particles has been replaced by a continuous function $y(x,0)$ which represents the initial displacement of the string. It has been necessary to redefine the coefficients as $B_n = b_n\sqrt{\Delta x}$. Examination of (1.24) shows that the numerical values of S_{nj} become small as f becomes large. The coefficients b_n in (1.27) therefore become large. It is B_n which approaches a finite limit. Note the signifi-

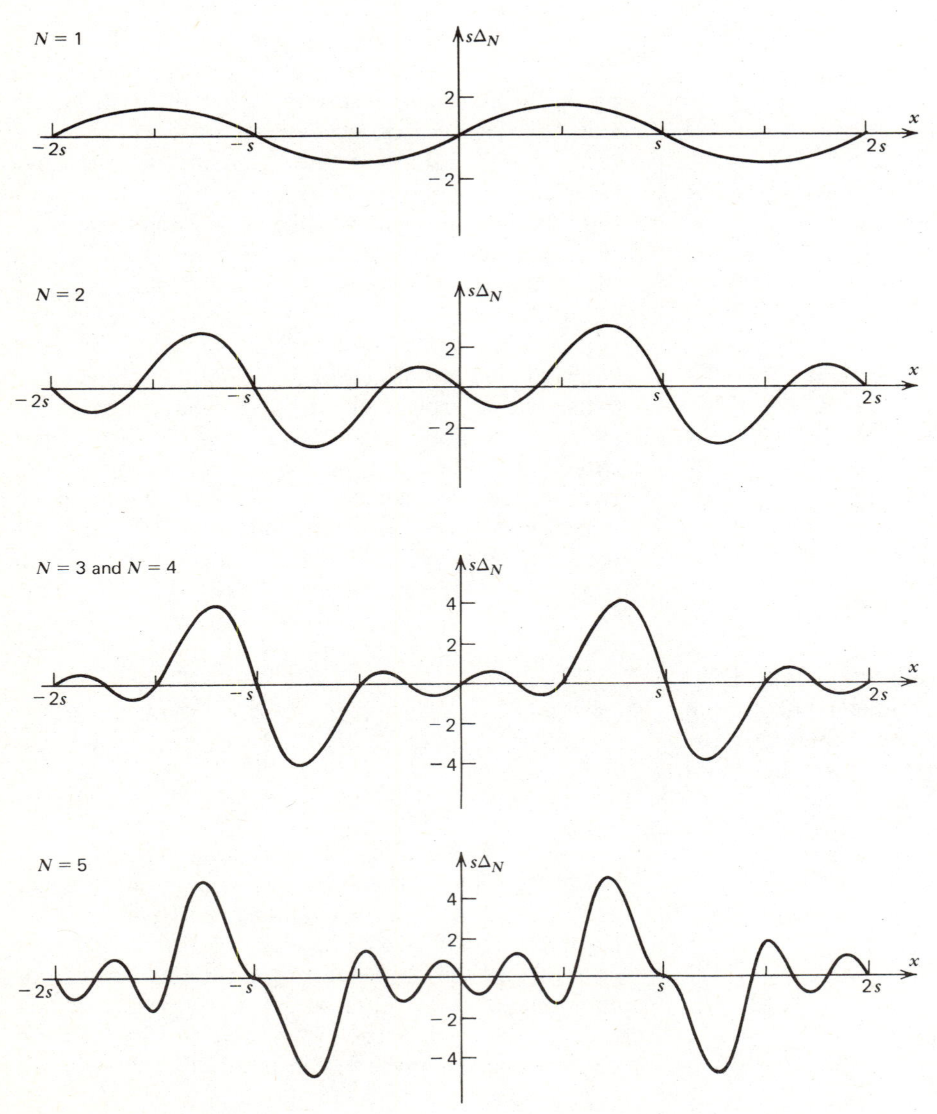

Fig. 2.2

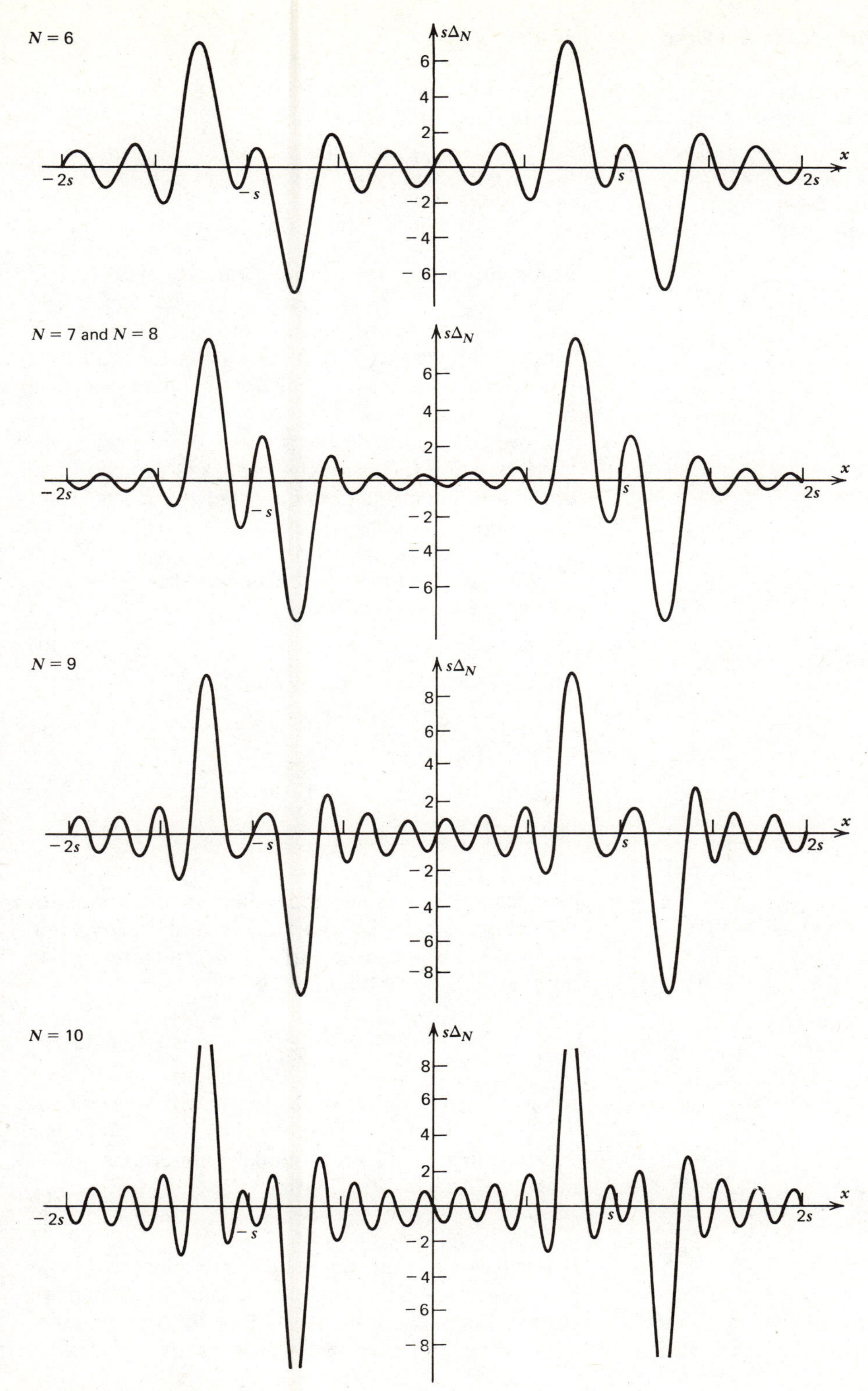

Fig. 2.2 (*continued*)

cance of (2.24). It is the expansion of an arbitrary continuous function in terms of the eigenfunctions of the string. The only restriction on $y(x,0)$ is that it obey the boundary conditions of the problem, namely, $y(0,0) = y(s,0) = 0$. It is possible to regard $y(x,0)$ as a vector in the space of infinitely many dimensions which results from the limiting process. The eigenfunctions $u_n(x)$ are a complete set of orthogonal unit vectors in this space. The components of $y(x,0)$ in the directions of $u_n(x)$ are the expansion coefficients B_n in (2.24). Equation (1.28) for the coefficients becomes

$$\begin{aligned} B_n &= \lim_{f\to\infty} \sum_{j=1}^{f} \sqrt{\Delta x}\, S_{nj} y_j(0) \\ &= \lim_{f\to\infty} \sum_{j=1}^{f} y_j(0) u_n(x_j)\,\Delta x \\ &= \int_0^s y(x,0) u_n(x)\, dx \end{aligned} \tag{2.25}$$

In a similar fashion, we find for the initial velocity of the string

$$\dot{y}(x,0) = \sum_{n=1}^{\infty} \omega_n A_n u_n(x)$$

$$A_n = \frac{1}{\omega_n} \int_0^s \dot{y}(x,0) u_n(x)\, dx \tag{2.26}$$

where $A_n = \sqrt{\Delta x}\, a_n$. The solution of the problem is

$$y(x,t) = \sum_{n=1}^{\infty} u_n(x)[A_n \sin \omega_n t + B_n \cos \omega_n t] \tag{2.27}$$

where the eigenfrequencies ω_n are given by (2.14).

We will now confirm the delta function property of (2.20). If b_n as given by (1.28) is formally substituted into (1.27), the resulting equation can be expressed in the form

$$y_j(0) = \sum_{j'=1}^{f} \left\{ \sum_{n=1}^{f} u_n(x_j) u_n(x_{j'}) \right\} y_{j'}(0)\,\Delta x \tag{2.28}$$

It has been necessary to replace the summation index j by j' in (1.28) in order to avoid confusing it with the subscript j in (1.27). We have used the fact that $S_{nj} = \sqrt{\Delta x}\, u_n(x_j)$ as is made evident by (2.10) and (2.18). The limiting form of (2.28) is

$$y(x,0) = \int_0^s \left\{ \sum_{n=1}^{\infty} u_n(x) u_n(x') \right\} y(x',0)\, dx' \tag{2.29}$$

which holds for an arbitrary function $y(x,0)$ and demonstrates the fundamental property of the delta function as given by equation (10.21) in Chapter 3. The interpretation of (2.20) as a delta function is to be understood in terms of the limiting process leading from (2.28) to (2.29).

How would we solve the problem of integrating the wave equation (2.3) directly without prior knowledge of the solution of the discrete system on which a limit can be taken? A very important technique for making a direct attack on a partial differential equation such as (2.3) is to try to separate the variables. To do this, we assume a solution of the form

$$y(x,t) = u(x)g(t) \tag{2.30}$$

The wave equation requires that

$$u''(x)g(t) - \frac{1}{v^2} u(x) g''(t) = 0 \tag{2.31}$$

Dividing by $u(x)g(t)$ and rearranging gives

$$\frac{u''(x)}{u(x)} - \frac{1}{v^2}\frac{g''(t)}{g(t)} = 0 \tag{2.32}$$

The separation has succeeded because the terms involving x are grouped separately from those involving t. Since x and t can be varied independently of one another, (2.32) can hold only if

$$\frac{u''(x)}{u(x)} = -\kappa^2 \qquad \frac{1}{v^2}\frac{g''(t)}{g(t)} = -\kappa^2 \tag{2.33}$$

where κ^2 is a constant called the *separation constant*. The choice to write it as $-\kappa^2$ rather than some other way is dictated by the boundary conditions which must be satisfied. The solution of

$$u''(x) + \kappa^2 u(x) = 0 \tag{2.34}$$

is oscillatory in character, which is required to get the solution to vanish at both $x = 0$ and $x = s$. Had we chosen the separation constant such that $u''(x) - \kappa^2 u(x) = 0$, the solutions would be exponential in character. The solution which vanishes at $x=0$ is then $u(x)=A\sinh(\kappa x)$. There is no way to make this solution vanish at $x=s$. The solution of (2.34) which vanishes at $x=0$ is

$$u(x)=A\sin(\kappa x) \tag{2.35}$$

The requirement that $u(s)=0$ means that the separation constant must satisfy

$$\kappa_n s=n\pi \qquad n=1,2,3,\ldots \tag{2.36}$$

The value $n=0$ is excluded because it gives $u(x)=0$, which is a possible but not interesting solution. Negative integer values of n give the same set of functions. Thus, there is not one separation constant but an infinite number and therefore an infinite number of solutions given by

$$u_n(x)=A\sin\left(\frac{n\pi x}{s}\right) \tag{2.37}$$

This is the set of eigenfunctions (2.18) already found. The eigenfunctions are appropriately normalized if $A=\sqrt{2/s}$. The differential equation for $g(t)$ is conveniently written

$$g_n''(t)+\omega_n^2 g_n(t)=0 \qquad \omega_n=\kappa_n v=\frac{n\pi v}{s} \tag{2.38}$$

The subscript n indicates that there is not one solution but many depending on the choice of the separation constant. The general solution of (2.38) is

$$g_n(t)=A_n\sin\omega_n t+B_n\cos\omega_n t \tag{2.39}$$

The general solution of the problem is a superposition of all the eigenfunctions, which is equation (2.27).

Formally, the coefficients B_n in (2.27) can be found by first setting $t=0$ and then multiplying through by $u_{n'}(x)$:

$$y(x,0)u_{n'}(x)=\sum_{n=0}^{\infty}u_n(x)u_{n'}(x)B_n \tag{2.40}$$

By integrating over $0<x<s$ and using the orthogonality relations (2.19), we find

$$\int_0^s y(x,0)u_{n'}(x)\,dx=\sum_{n=0}^{\infty}\delta_{nn'}B_n=B_{n'} \tag{2.41}$$

in agreement with (2.25). The coefficients A_n are found by first differentiating (2.28) with respect to t and then proceeding similarly.

In our future study of eigenvalue problems associated with differential equations, we will for the most part proceed formally from the differential equation itself and not search each time for a discrete system on which we can take appropriate limits to obtain the required solution. The use of the limiting procedure as an introduction to the subject has several advantages. It makes the idea of an eigenfunction as some kind of unit vector in an abstract space of an infinite number of dimensions seem plausible. The orthogonality relations (2.19) and (2.20) are seen as natural consequences of the more familiar orthogonality relations obeyed by complete sets of orthogonal unit vectors in a space of a finite number of dimensions. Most important of all, the completeness of the eigenfunctions is made apparent. In a space of f dimensions, any f linearly independent vectors forms a basis, meaning that any other vector can be represented as a linear combination of them. This idea carries over by an appropriate limiting process into a space of an infinite number of dimensions as is exemplified by the relations (2.25) and (2.26) where arbitrary functions representing the initial position and velocity of the string are represented by expansions in terms of the eigenfunctions. We have established that the expansion works for functions $y(x,0)$ which are continuous and obey the same boundary conditions as are required of the solutions of the physical problem under consideration. We will find that the expansion theorem can be pushed beyond these requirements and that it is possible, for example, to expand functions which are only piecewise continuous in terms of complete sets of eigenfunctions. The only price paid is that such expansions are not uniformly convergent in the vicinity of the discontinuities. The expansion (2.20) for the Dirac delta function does not of course converge in the classical mathematical sense.

The space of an infinite number of dimensions which we have utilized in the discussion of the vibrating string is a special case of what is called *Hilbert space*. More generally, the vectors and matrices of Hilbert space can have components and elements that are complex numbers. This generalization is required when problems in quantum mechanics are to be considered.

PROBLEMS

2.1. Verify the orthogonality condition (2.19) by direct evaluation of the integrals.

2.2. Consider the case where the initial velocity of a string under tension and fixed at both ends is zero. The solution (2.27) can then be written

$$y(x,t) = \sum_{n=1}^{\infty} B_n y_n(x,t) \tag{2.42}$$

where

$$y_n(x,t) = \sqrt{\frac{2}{s}} \sin\left(\frac{n\pi x}{s}\right) \cos \omega_n t \tag{2.43}$$

can be thought of as a time-dependent eigenfunction. It is also a *standing wave*. The relation between the velocity of propagation of the waves, the frequency of vibration, and the wavelength is $v = \sqrt{F/\mu} = f_n \lambda_n$. Use it to find the allowed wavelengths of the standing waves. The frequency f_n in cycles per second is related to the circular frequency ω_n by $\omega_n = 2\pi f_n$. It is possible for the string to vibrate in such a way that only one of the solutions (2.43) is present. Make a free-hand sketch of (2.43) for $n = 1$, 2, 3, and 4 at $t = 0$ and also at $\omega_n t = \pi$. Locate the points of the string that are always stationary. These points are called *nodes*. The standing wave for $n = 1$ is called the *fundamental* or *first harmonic*. The second harmonic occurs for $n = 2$, the third for $n = 3$, and so on.

2.3. In the expression (1.38) for the phase velocity of a harmonic wave in a periodic structure, replace the spring constant k by $F/\Delta x$ and show that

$$\lim_{\substack{\Delta x \to 0 \\ m \to 0}} u_p = v = \sqrt{\frac{F}{\mu}}$$

as expected. In the derivation of the wave equation (2.9), we assumed that the string is completely flexible, meaning that it has no resistance to being bent. If the string has stiffness, such as would be the case for a piano wire, some dispersion remains even in the limit of a continuous medium.

2.4. A plane harmonic wave traveling in the positive x direction on a string can be represented in various ways, for example,

$$y = A \cos\left[\frac{2\pi}{\lambda}(vt - x)\right] = A\cos(\omega t - \kappa x) \tag{2.44}$$

Sometimes it is easier to do calculations if the waves are written in complex form as

$$y = Ae^{i(\omega t - \kappa x)} \tag{2.45}$$

it being understood that the actual wave is the real part. Suppose that two flexible strings with linear mass densities μ_1 and μ_2 are joined together. The string of density μ_1 lies along the negative x axis, and the string of density μ_2 lies along the positive x axis. The junction is at $x = 0$. When a harmonic wave traveling in the positive x direction on the string of density μ_1 reaches the origin, it will generate a wave of the same frequency in the string of density μ_2. Part of the incident wave is transmitted into the second string and part of it is reflected. The waves on the two strings can then be expressed as

$$y_1 = Ae^{i(\omega t - \kappa_1 x)} + Be^{i(\omega t + \kappa_1 x)}$$

$$y_2 = Ce^{i(\omega t - \kappa_2 x)} \tag{2.46}$$

The velocity of propagation on the two strings is not the same:

$$v_1 = \frac{\omega}{\kappa_1} = \sqrt{\frac{F}{\mu_1}} \qquad v_2 = \frac{\omega}{\kappa_2} = \sqrt{\frac{F}{\mu_2}} \tag{2.47}$$

The wave numbers κ_1 and κ_2 therefore obey

$$\frac{\kappa_1}{\kappa_2} = \frac{v_2}{v_1} = \sqrt{\frac{\mu_1}{\mu_2}} \tag{2.48}$$

At the junction, the solutions (2.46) obey the conditions

$$y_1(0) = y_2(0) \qquad \left.\frac{\partial y_1}{\partial x}\right|_{x=0} = \left.\frac{\partial y_2}{\partial x}\right|_{x=0} \tag{2.49}$$

Show that the amplitudes of the reflected and transmitted waves are given by

$$B = \frac{A(\kappa_1 - \kappa_2)}{\kappa_1 + \kappa_2} \qquad C = \frac{2\kappa_1 A}{\kappa_1 + \kappa_2} \tag{2.50}$$

The behavior of plane waves at the interface between two media is important in many applications, for example, in the derivation of the laws of reflection and refraction of electromagnetic waves.

If the second string is extremely dense, $\mu_2 \to \infty$, and it becomes the equivalent of a fixed support for the first string. Show that in this limit, $B = -A$. The reflected wave has the same amplitude as the incident wave but is 180° out of phase with it. Show that there is now a standing wave on the first string given by

$$y_1 = 2A \sin \omega t \sin \kappa_1 x \qquad x < 0 \tag{2.51}$$

Note that the wave number κ_1 is the separation constant for the wave equation. It is not restricted by the condition (2.36) because we have imposed boundary conditions only on one end of the string. Compare this result to the result of problem 1.5.

2.5. We have noted in equation (1.6) that $F/\Delta x$ is analogous to a spring constant for the transverse motion of the massless string loaded with point masses at equal intervals. This means that the energy associated with the jth particle is

$$W_j = \frac{1}{2} m \dot{y}_j^2 + \frac{1}{2}\left(\frac{F}{\Delta x}\right)(y_j - y_{j-1})^2 \tag{2.52}$$

The linear energy density of the continuous flexible string can be found by means of

$$w = \lim_{\Delta x \to 0} \frac{W_j}{\Delta x} \tag{2.53}$$

Show that

$$w = \frac{1}{2}\mu\left(\frac{\partial y}{\partial t}\right)^2 + \frac{1}{2}F\left(\frac{\partial y}{\partial x}\right)^2 \tag{2.54}$$

For a plane harmonic wave, show that the average energy density is

$$\overline{w} = \tfrac{1}{2}\mu\omega^2 A^2 \tag{2.55}$$

The power carried by the wave is the average energy per second which passes a given point:

$$P = \overline{w}v = \tfrac{1}{2}\mu v \omega^2 A^2 = \tfrac{1}{2}\sqrt{F\mu}\,\omega^2 A^2 \tag{2.56}$$

The intensity of the wave, defined as the power carried by the wave, is proportional to the square of the amplitude of the wave. The same is true of an electromangetic wave. See equation (10.14) of Chapter 5.

Refer to problem 2.4. The fraction of the power carried by the incident wave which is transmitted into the second string is called the *transmission coefficient*, and the fraction of the power reflected is called the *reflection coefficient*. Show that these coefficients are

$$R = \left(\frac{\kappa_1 - \kappa_2}{\kappa_1 + \kappa_2}\right)^2 \qquad T = \frac{4\kappa_1\kappa_2}{(\kappa_1 + \kappa_2)^2} \tag{2.57}$$

Show that $R + T = 1$.

2.6. A flexible string of constant mass density has a point mass m on it at $x = 0$. Show that the presence of the point mass causes a discontinuity in the derivative given by

$$m\ddot{y} = F\left(\frac{\partial y_2}{\partial x} - \frac{\partial y_1}{\partial x}\right)_{x=0} \tag{2.58}$$

where y_1 is the solution for $x < 0$ and y_2 is the solution for $x > 0$. By the same method that was used in problem 2.4, show that a plane wave incident on the point mass will be partially reflected and partially transmitted. You can set $\kappa_1 = \kappa_2$ and use (2.46) as a starting point. The real parts of the resulting solutions will be

$$y_1 = A\cos(\omega t - \kappa x) + \frac{A\alpha}{\sqrt{1+\alpha^2}}\cos(\omega t + \kappa x - \phi - \pi/2) \qquad x < 0$$

$$y_2 = \frac{A}{\sqrt{1+\alpha^2}}\cos(\omega t - \kappa x - \phi) \qquad x > 0$$

$$\alpha = \frac{m\kappa}{2\mu} \qquad \tan\phi = \alpha \tag{2.59}$$

Do these solutions behave as you would expect if m becomes large?

2.7. If there is viscous friction, the wave equation for the flexible string can be expressed in the form

$$v^2\frac{\partial^2 y}{\partial x^2} - a\frac{\partial y}{\partial t} - \frac{\partial^2 y}{\partial t^2} = 0 \tag{2.60}$$

where we have replaced the constant b of (2.9) by a new constant $a = b/\mu$. Assume a solution of the form

$$y = Ae^{i(\kappa x - \omega t)} \tag{2.61}$$

and show that κ is a complex number with real and imaginary parts given by

$$\operatorname{Re}\kappa = \kappa_1 = \frac{\omega}{\sqrt{2}\,v}\sqrt{1 + \sqrt{1 + \frac{a^2}{\omega^2}}}$$

$$\operatorname{Im}\kappa = \kappa_2 = \frac{a\omega}{2\kappa_1 v^2} \tag{2.62}$$

The solution (real part) therefore takes the form

$$y = Ae^{-\kappa_2 x}\cos(\kappa_1 x - \omega t) \tag{2.63}$$

The relation between ω and κ_1 is called a *dispersion relation*. The effect of the friction is to produce an attenuation of the wave with distance as it moves along the string and also dispersion. The phase velocity u_p is no longer v but is a function of the frequency. Show that if a is small, the phase velocity is approximately

$$u_p = v\left(1 - \frac{a^2}{8\omega^2}\right) \tag{2.64}$$

2.8. A flexible string under tension and fixed at its two ends is set into motion by being drawn aside a distance h at its center and then released. The initial shape of the string is shown in Fig. 2.3. The initial velocity is zero, which means that all the coefficients A_n in (2.27) are zero. Show that $B_n = 0$ if n is even, without actually evaluating any integrals. Evaluate B_n if n is odd, and show that the motion of the string is given by

$$y(x,t) = \frac{8h}{\pi^2}\left[\sin\left(\frac{\pi x}{s}\right)\cos\omega_1 t - \frac{1}{9}\sin\left(\frac{3\pi x}{s}\right)\cos(3\omega_1 t) + \frac{1}{25}\sin\left(\frac{5\pi x}{s}\right)\cos(5\omega_1 t) - + \cdots\right] \tag{2.65}$$

Set $t = 0$, $x = s/2$, and $y = h$ to show that

$$\frac{\pi^2}{8} = 1 + \frac{1}{3^2} + \frac{1}{5^2} + \cdots \tag{2.66}$$

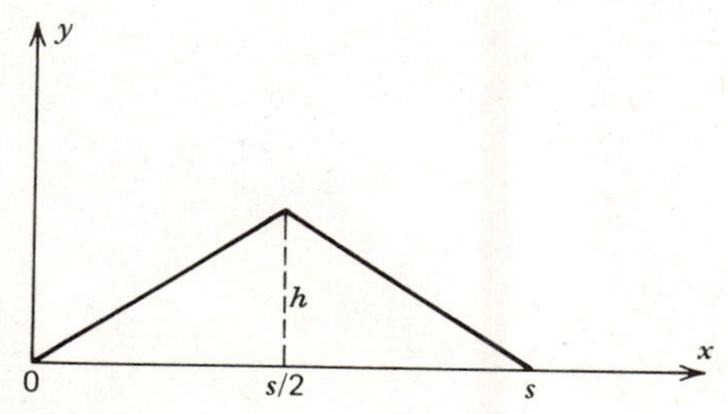

Fig. 2.3

2.9. A wave pulse on a string under tension given by $F(x - vt)$ and shown in Fig. 2.4 is traveling on the negative x axis toward the origin. A second wave pulse $G(x + vt)$ is traveling on the positive x axis also toward the origin. The displacement of the string at any time and position is given by $y = F(x - vt) + G(x + vt)$. Show that $y = 0$ at $x = 0$ for all times, provided that $G(x + vt) = -F(-x - vt)$. Add a sketch of $G(x + vt)$ to Fig. 2.4. What happens as time progresses? Since $y = 0$ at $x = 0$ for all time, $x = 0$ could be replaced by a fixed support. The solution $y = F(x - vt) - F(-x - vt)$ then describes a wave pulse which is reflected from a rigid support at $x = 0$. Note that in the reflection process the pulse gets turned over. This same effect is responsible for the 180° phase shift that occurs when a harmonic wave is reflected from a fixed support as noted in problems 1.5 and 2.4.

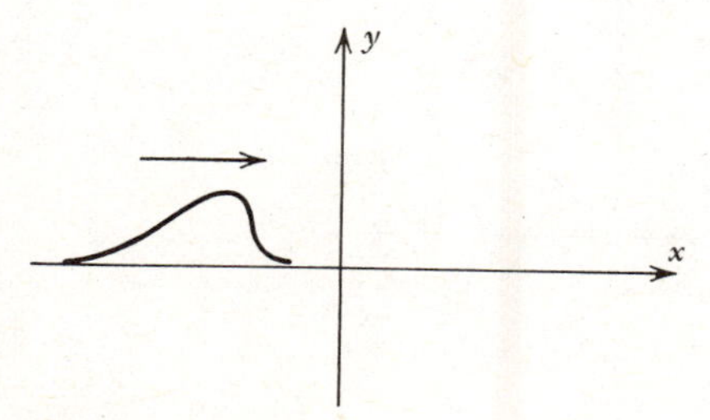

Fig. 2.4

2.10. A heat-conducting cylindrical rod is thermally insulated except at its two ends, which are maintained at a temperature of 0°C. Let the two ends of the rod be at $x=0$ and $x=s$. The problem is one-dimensional in that the temperature depends only on x and t. Assume that at $t=0$ the temperature is a known function $T(x,0)$. The theory of heat conduction is developed in section 6 of Chapter 3. The one-dimensional heat conduction equation is

$$\frac{\partial^2 T}{\partial x^2} - \frac{1}{\alpha}\frac{\partial T}{\partial t} = 0 \tag{2.67}$$

where α is a constant. Separate the variables by writing the temperature as $T(x,t)=u(x)f(t)$. Find the complete solution for the temperature in the rod as a function of x and t as an expansion in terms of a complete set of eigenfunctions. Evaluate all the expansion coefficients in terms of $T(x,0)$. If the two ends of the rod are at temperatures T_1 and T_2, show that the function

$$F(x,t) = T(x,t) - T_1 - \frac{(T_2 - T_1)x}{s} \tag{2.68}$$

obeys (2.67) with $F(0,t)=F(s,t)=0$. Comment on the similarities and differences between the solutions of the one-dimensional heat equation of this problem and the solution of the one-dimensional wave equation for the string with both ends fixed.

3. VIBRATING MEMBRANE

Suppose that a flexible membrane is stretched tight over a frame in the shape of a closed curve lying in the x, y plane as illustrated in Fig. 3.1. The stretching is done in such a way that the tension is the same at all points of the membrane. A drumhead is constructed in this manner. It will be our purpose to find the possible transverse vibrations of the membrane perpendicular to the x, y plane. Let F be the tension in the membrane in units of force per unit length. The forces on the four edges of the element of area $\Delta x\,\Delta y$ shown in Fig. 3.1 are then $F\Delta x$ and $F\Delta y$. Figure 3.2 shows a profile of the membrane seen edge on and looking in the y direction. The displacement of the membrane perpendicular to the x, y plane is represented by $\eta(x, y, t)$. The derivation of the equation of motion of the membrane parallels the analysis of the flexible string quite closely. The components of the forces shown in Fig. 3.2 in the direction of the displacement are

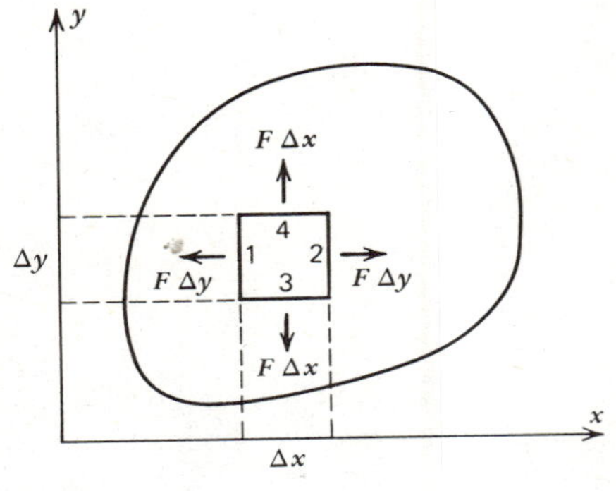

Fig. 3.1

$$F\Delta y\theta_2 - F\Delta y\theta_1 = F\Delta y\left(\left.\frac{\partial \eta}{\partial x}\right|_2 - \left.\frac{\partial \eta}{\partial x}\right|_1\right) \tag{3.1}$$

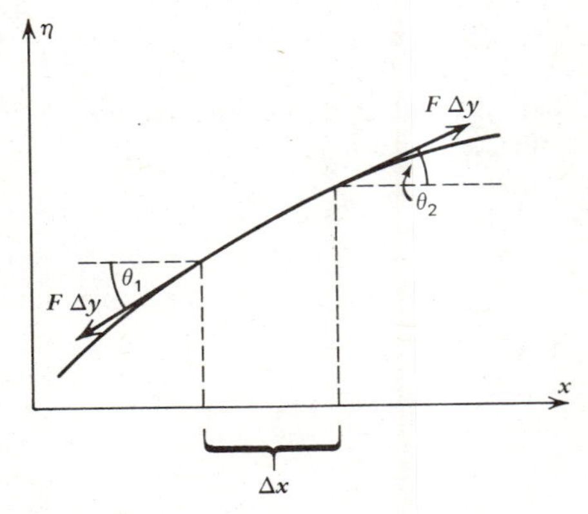

Fig. 3.2

where small displacements and angles are assumed. If the forces on the other two edges of the element are taken into account, its equation of motion can be written

$$F\Delta y\left(\left.\frac{\partial\eta}{\partial x}\right|_2-\left.\frac{\partial\eta}{\partial x}\right|_1\right)+F\Delta x\left(\left.\frac{\partial\eta}{\partial y}\right|_4-\left.\frac{\partial\eta}{\partial y}\right|_3\right)$$
$$=\nu\,\Delta x\,\Delta y\frac{\partial^2\eta}{\partial t^2} \tag{3.2}$$

where ν is the surface mass density of the membrane in units of mass per unit area. Frictional forces and external driving forces are not included in (3.2). If we divide (3.2) by $\Delta x\,\Delta y$ and then take the limit as $\Delta x\to 0$ and $\Delta y\to 0$, the result is

$$\frac{\partial^2\eta}{\partial x^2}+\frac{\partial^2\eta}{\partial y^2}-\frac{1}{v^2}\frac{\partial^2\eta}{\partial t^2}=0 \tag{3.3}$$

where the constant v has the units of velocity and is given by

$$v=\sqrt{\frac{F}{\nu}} \tag{3.4}$$

Equation (3.3) is a homogeneous, two-dimensional wave equation. Plane wave solutions of (3.3) exist of the form

$$\eta=A\sin(\boldsymbol{\kappa}\cdot\mathbf{r}-\omega t) \tag{3.5}$$

provided that the magnitude of the propagation vector $\boldsymbol{\kappa}$ obeys $\omega=\kappa v$. Review problem 12.10 of Chapter 3 for a discussion of this type of solution. Plane transverse waves can move on the membrane in the direction of $\boldsymbol{\kappa}$ with phase velocity given by v. Since all such plane waves move at the same speed v regardless of their frequency, there is no dispersion. In this section we will concern ourselves with the solutions of (3.3) which conform to the boundary conditions $\eta=0$ on $x=0$, $y=0$, $x=a$, and $y=b$. In other words, we will find the possible motions of a rectangular drumhead. This can be regarded as an extension of the problem of the vibrations of the one-dimensional string into two dimensions.

The two-dimensional wave equation can be solved by separating the variables. As a first step, the time dependence will be separated out by expressing the solution as

$$\eta(x,y,t)=u(x,y)g(t) \tag{3.6}$$

The wave equation (3.3) requires that

$$\frac{1}{u}\left(\frac{\partial^2 u}{\partial x^2}+\frac{\partial^2 u}{\partial y^2}\right)-\frac{1}{v^2 g}\frac{d^2 g}{dt^2}=0 \tag{3.7}$$

The same argument holds here that was used in the separation of the one-dimensional wave equation. See equations (2.32) and (2.33). We introduce the separation constant κ^2 with the result

$$\frac{\partial^2 u}{\partial x^2}+\frac{\partial^2 u}{\partial y^2}+\kappa^2 u=0 \tag{3.8}$$

$$\frac{d^2 g}{dt^2}+\omega^2 g=0 \qquad \omega=\kappa v \tag{3.9}$$

The separation constant therefore appears as the wave number. Equation (3.8) is called the *Helmholtz equation*. It can be further separated by writing its possible solutions as $u(x,y)=u_1(x)u_2(y)$. The result is

$$\frac{1}{u_1}\frac{d^2u_1}{dx^2}+\frac{1}{u_2}\frac{d^2u_2}{dy^2}+\kappa^2=0 \tag{3.10}$$

which can hold only if

$$\frac{1}{u_1}\frac{d^2u_1}{dx^2}=-\kappa_x^2 \qquad \frac{1}{u^2}\frac{d^2u_2}{dy^2}=-\kappa_y^2 \tag{3.11}$$

where κ_x^2 and κ_y^2 are new separation constants which must satisfy

$$\kappa_x^2+\kappa_y^2=\kappa^2 \tag{3.12}$$

The functions $u_1(x)$ and $u_2(y)$ obey the differential equations

$$\frac{d^2u_1}{dx^2}+\kappa_x^2u_1=0 \qquad \frac{d^2u_2}{dy^2}+\kappa_y^2u_2=0 \tag{3.13}$$

Just as with the solution of the vibrating string problem, the choice of the form of the separation constants is guided by the boundary conditions. Solutions are required which vanish at $x=0$, $y=0$, $x=a$, and $y=b$. The separation constants κ_x and κ_y can be regarded as the x and y components of the propagation vector. We are in the process of finding two-dimensional standing wave solutions which can be regarded as being built up out of plane wave solutions which are propagating in various directions on the membrane and which are undergoing reflections at the boundaries. The solutions of (3.13) that meet the boundary conditions are

$$u_1(x) = \sqrt{\frac{2}{a}} \sin\left(\frac{m\pi x}{a}\right)$$

$$u_2(y) = \sqrt{\frac{2}{b}} \sin\left(\frac{n\pi y}{b}\right) \tag{3.14}$$

where $m = 1, 2, 3, \ldots$ and $n = 1, 2, 3, \ldots$. The multiplicative constants have been chosen to make the solutions properly normalized. The separation constants therefore obey

$$\kappa_{xm} = \frac{m\pi}{a} \qquad \kappa_{yn} = \frac{n\pi}{b} \qquad \kappa_{mn} = \pi\sqrt{\frac{m^2}{a^2} + \frac{n^2}{b^2}} \tag{3.15}$$

Subscripts have been added to the separation constants to indicate that many are possible. The frequencies of oscillation of the normal modes are given by the eigenfrequencies

$$\omega_{mn} = \kappa_{mn} v \tag{3.16}$$

The normalized eigenfunctions for the rectangular membrane are

$$u_{mn}(x, y) = \frac{2}{\sqrt{ab}} \sin\left(\frac{m\pi x}{a}\right) \sin\left(\frac{n\pi y}{b}\right) \tag{3.17}$$

The eigenfunctions obey the orthogonality conditions

$$\int_0^a \int_0^b u_{mn}(x, y) u_{m'n'}(x, y)\, dy\, dx = \delta_{mm'}\delta_{nn'} \tag{3.18}$$

The general solution of the problem is a linear superposition of all the possible solutions which conform to the boundary conditions:

$$\eta(x, y, t) = \sum_{m=1}^{\infty} \sum_{n=1}^{\infty} u_{mn}(A_{mn} \sin \omega_{mn} t + B_{mn} \cos \omega_{mn} t) \tag{3.19}$$

The solution is uniquely determined if, in addition to the boundary conditions, the functions η and $\partial\eta/\partial t$ are known over the entire membrane at $t = 0$. If we set $t = 0$ in (3.19), multiply by $u_{m'n'}$, and then integrate over the entire membrane, we find

$$\begin{aligned} \int_0^a \int_0^b \eta(x, y, 0) u_{m'n'}(x, y)\, dy\, dx &= \sum_{m=1}^{\infty} \sum_{n=1}^{\infty} \int_0^a \int_0^b B_{mn} u_{mn} u_{m'n'}\, dy\, dx \\ &= \sum_{m=1}^{\infty} \sum_{n=1}^{\infty} B_{mn} \delta_{mm'} \delta_{nn'} = B_{m'n'} \end{aligned} \tag{3.20}$$

which determines the coefficients B_{mn}. By a similar procedure,

$$A_{mn} = \frac{1}{\omega_{mn}} \int_0^a \int_0^b \left.\frac{\partial \eta}{\partial t}\right|_{t=0} u_{mn}\, dy\, dx \tag{3.21}$$

As an example, suppose that the membrane is vibrating in a pure mode given by

$$\eta(x, y, t) = \frac{2A}{\sqrt{ab}} \sin\left(\frac{2\pi x}{a}\right) \sin\left(\frac{3\pi y}{b}\right) \cos(\omega_{23} t) \tag{3.22}$$

The straight lines $x = a/2$, $y = b/3$, and $y = 2b/3$ are always at rest and are called *nodal lines*. The portions of the membrane bounded by the nodal lines oscillate in a direction perpendicular to the plane of the membrane at the frequency ω_{23}.

It seems as though the solution of the rectangular membrane is little more than an extension of the solution of the vibrating string into two dimensions. This is true to a large extent, but there is an interesting feature of the solution of the membrane problem which is not present in the vibrating string. Suppose that the membrane is square. The possible eigenfrequencies are then given by

$$\omega_{mn} = \frac{\pi v}{a}\sqrt{m^2 + n^2} \tag{3.23}$$

Since now $\omega_{mn} = \omega_{nm}$, there are at least two combinations of m and n that give the same frequency, with the exception of the case $m = n$. This means that there are at least two eigenfunctions corresponding to a given frequency if $m \neq n$. This phenomenon is called *degeneracy*. We have already studied degeneracy in connection with the eigenvalue problem associated with a symmetric matrix. Review the discussion in section 10 of Chapter 2. For example, the two eigenfunctions u_{12} and u_{21} both correspond to the same eigenfrequency

$$\omega_{12} = \omega_{21} = \frac{\pi v}{a}\sqrt{5} \tag{3.24}$$

Any linear combination of u_{12} and u_{21} such as

$$\begin{aligned} u &= Au_{21} - Bu_{12} \\ &= \frac{2}{a}\left[A\sin\left(\frac{2\pi x}{a}\right)\sin\left(\frac{\pi y}{a}\right)\right. \\ &\quad \left. - B\sin\left(\frac{\pi x}{a}\right)\sin\left(\frac{2\pi y}{a}\right)\right] \end{aligned} \tag{3.25}$$

is also an eigenfunction corresponding to the eigenfrequency (3.24). What does the membrane look like if it is oscillating in the mode given by (3.25)? The possible nodal lines are found by setting (3.25) equal to zero:

$$0 = \sin\left(\frac{\pi x}{a}\right)\sin\left(\frac{\pi y}{a}\right)\left[\frac{A}{B}\cos\left(\frac{\pi x}{a}\right) - \cos\left(\frac{\pi y}{a}\right)\right] \tag{3.26}$$

where we have used $\sin(2\pi x/a) = 2\sin(\pi x/a)\cos(\pi x/a)$. The nodal lines are therefore found from

$$y = \frac{a}{\pi}\cos^{-1}\left[R\cos\left(\frac{\pi x}{a}\right)\right] \tag{3.27}$$

where $R = A/B$. Nodal lines for various mixtures of the two eigenfunctions u_{12} and u_{21} are shown in Fig. 3.3. The diagonal and curved nodal lines are possible because of the degeneracy and do not occur if the degeneracy is removed by making $a \neq b$.

For $m \neq n$, each eigenfrequency is at least twofold degenerate. An examination of the possible solutions of $m^2 + n^2 = c$ for various choices of c reveals that more degeneracy is possible, especially if c is large. The first six possible cases of more than twofold degeneracy are listed in Table 3.1. The eigenfrequency

$$\omega_{17} = \omega_{71} = \omega_{55} = \frac{\pi v}{a}\sqrt{50} \tag{3.28}$$

is threefold degenerate, meaning that there are three linearly independent eigenfunctions which give this

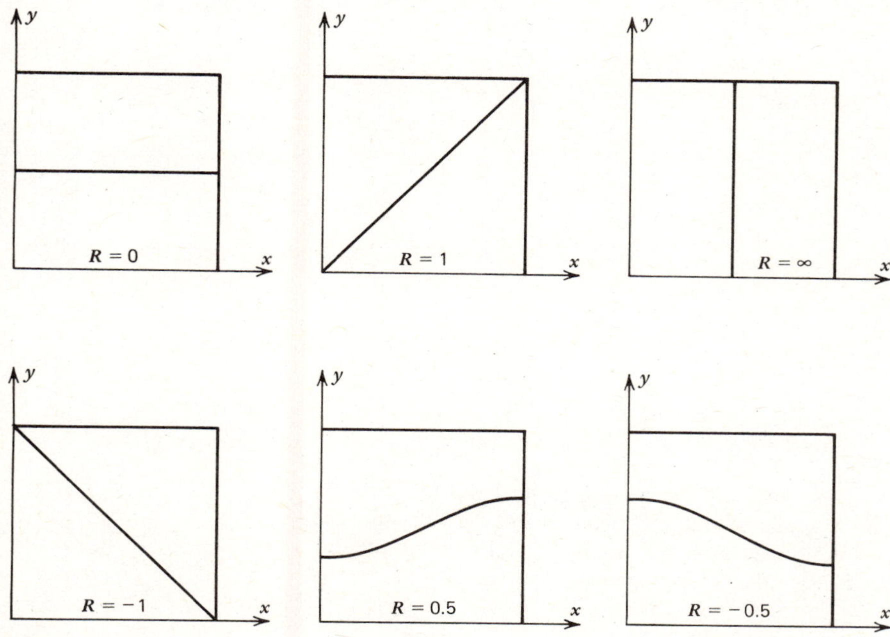

Fig. 3.3

Table 3.1

m^2+n^2	m	n	m^2+n^2	m	n	m^2+n^2	m	n
50	1	7	65	1	8	85	2	9
	5	5		4	7		6	7
	7	1		7	4		7	6
				8	1		9	2
125	2	11	130	3	11	145	1	12
	5	10		7	9		8	9
	10	5		9	7		9	8
	11	2		11	3		12	1

eigenfrequency. Similarly,

$$\omega_{18}=\omega_{47}=\omega_{74}=\omega_{81}=\frac{\pi v}{a}\sqrt{65} \tag{3.29}$$

is fourfold degenerate.

Perturbation theory deals with what happens when small changes are made in a physical system, for example, the addition of a small point mass to the drumhead. If the unperturbed state is degenerate, this must be taken into account. Typically what happens is that degeneracy is partially or totally removed when perturbations are added to a system.

PROBLEMS

3.1. A stretched flexible membrane with uniform mass density and tension lies in the x, y plane in the half space $x<0$. The membrane is fastened rigidly to the y axis as a boundary. The angle of incidence or reflection is taken to be the angle between the propagation vector and a normal drawn to the boundary. Suppose that a plane wave with propagation vector $\boldsymbol{\kappa}_1$ is incident on the boundary at the angle of incidence θ_1 and is reflected with propagation vector $\boldsymbol{\kappa}_2$ and angle of reflection θ_2 as shown in Fig. 3.4. Since both incident and reflected waves travel in the same medium, $|\boldsymbol{\kappa}_1|=|\boldsymbol{\kappa}_2|=\kappa$. The combination of the incident and reflected wave can be written in complex form as

$$\eta=A_1\exp\left[i\left(\kappa_{1x}x+\kappa_{1y}y-\omega t\right)\right]+A_2\exp\left[i\left(\kappa_{2x}x+\kappa_{2y}y-\omega t\right)\right] \tag{3.30}$$

Use the condition that $\eta=0$ at $x=0$ for all values of y to show that $A_2=-A_1$, $\kappa_{1y}=\kappa_{2y}$, and $\theta_1=\theta_2$. The requirement $|\boldsymbol{\kappa}_1|=|\boldsymbol{\kappa}_2|$ then means that $\kappa_{1x}=\pm\kappa_{2x}$. Since the second wave is reflected, the choice $\kappa_{1x}=-\kappa_{2x}$ is required.

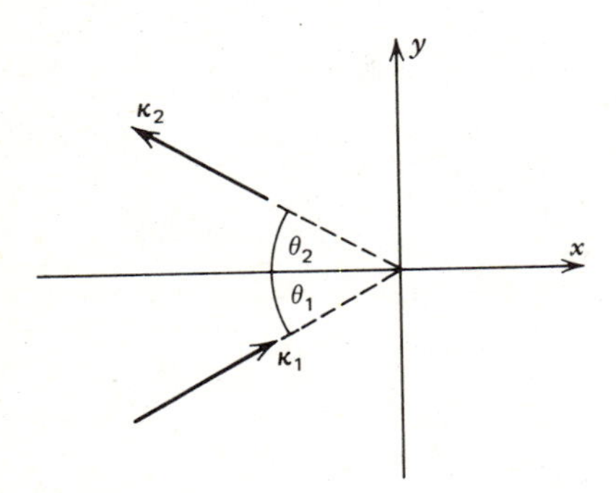

Fig. 3.4

3.2. A stretched flexible membrane is in the shape of a very long strip of width b between $y=0$ and $y=b$ as shown in Fig. 3.5. The boundaries are rigid. Separate out the y dependence of the wave equation by writing its solution as $\eta(x,y,t)=u_2(y)f(x,t)$. What are the possible solutions of the y equation? Show that the remaining equation for $f(x,t)$ has solutions of the form

$$f(x,t)=\exp\left[i\left(\kappa_x x-\omega t\right)\right] \tag{3.31}$$

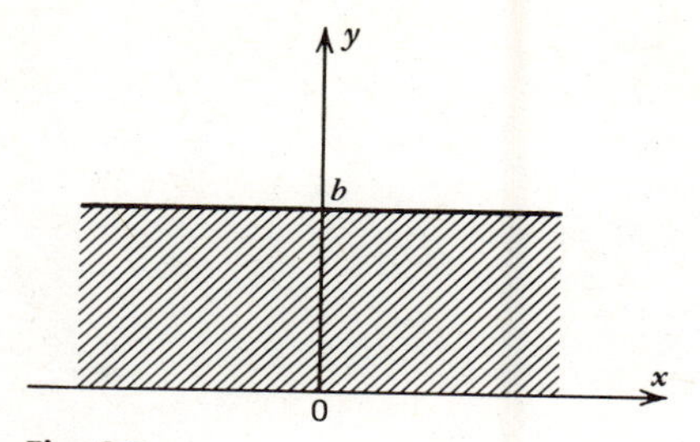

Fig. 3.5

provided that

$$\kappa_x^2 = \frac{\omega^2}{v^2} - \left(\frac{n\pi}{b}\right)^2 \qquad n = 1,2,3,\ldots \tag{3.32}$$

The phase velocity of the wave represented by (3.31) is

$$u_{px} = \frac{\omega}{\kappa_x} = \frac{\omega v}{\sqrt{\omega^2 - (n\pi v/b)^2}} \tag{3.33}$$

which is interesting because the membrane strip is now behaving like a medium with dispersion. The wavelength as defined by $\lambda_x = 2\pi/\kappa_x$ is

$$\lambda_x = \frac{2\pi v}{\sqrt{\omega^2 - (n\pi v/b)^2}} \tag{3.34}$$

Suppose that someone at a large distance down the negative x axis is applying a periodic driving force to the surface of the membrane at a frequency in the range $\pi v/b < \omega < 2\pi v/b$. Waves of the form (3.31) can then propagate in the positive x direction on the strip. The phase velocity and wavelength are given by (3.33) and (3.34) with $n = 1$. If ω is near to $\pi v/b$, both u_{px} and λ_x are large. This can be understood in terms of a plane wave propagating on the strip at an angle with respect to the y axis and undergoing multiple reflections from the boundaries at $y = 0$ and $y = b$. Figure 3.6 shows such a wave. The wavefronts shown can be thought of as the crests of two successive waves. For a given driving frequency ω and a given value of n, only certain directions of propagation are possible because κ_x is then fixed by (3.32). As ω is lowered to $\pi v/b$, θ and κ_x become small. The "component" λ_x of the

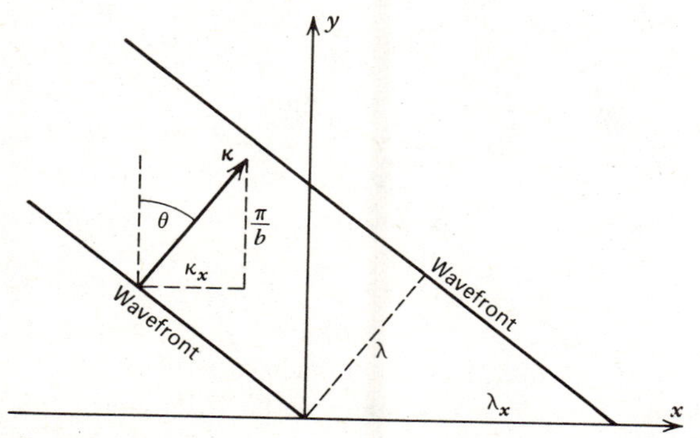

Fig. 3.6

wavelength in the x direction becomes large. If $\omega = \pi v/b$, $\lambda_x = \infty$, and it is only possible to have a standing wave in the y direction and no propagation in the x direction. If $\omega < \pi v/b$, κ_x becomes imaginary and the wave is attenuated in the x direction. If $2\pi v/b < \omega < 3\pi v/b$, propagation for both $n=1$ and $n=2$ is possible. This means that waves in this frequency range can be propagated in two different modes. The strip membrane acts as a *mechanical wave guide*.

3.3. A drum head is made in the shape of an isosceles triangle as shown in Fig. 3.7. What is the lowest eigenfrequency and the corresponding eigenfunction?

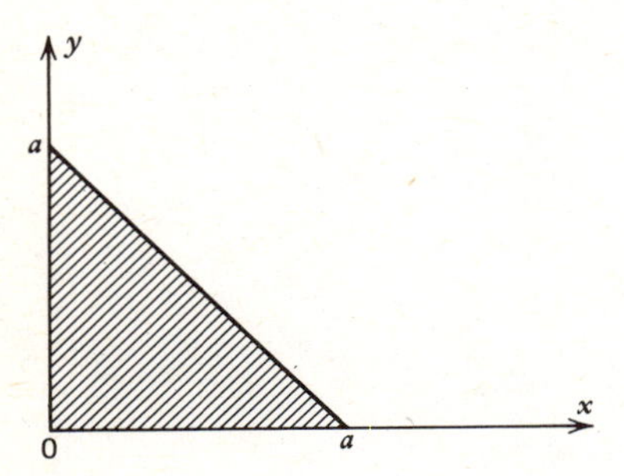

Fig. 3.7

3.4. A rectangular membrane has dimensions $x = 2b$ and $y = b$. Find two degenerate eigenvalues and the corresponding eigenfunctions.

3.5. The wave equation (3.3) for the membrane can be expressed in the form

$$F(\nabla \cdot \nabla \eta) = \nu \frac{\partial^2 \eta}{\partial t^2} \tag{3.35}$$

Multiply (3.35) through by $\partial\eta/\partial t$ and use identities (9.1) and (9.13) of Chapter 3 to show that

$$\oint F \frac{\partial \eta}{\partial t} \nabla\eta \cdot \hat{\mathbf{n}}\, ds = \frac{d}{dt} \int \left[\frac{1}{2} F \nabla\eta \cdot \nabla\eta + \frac{1}{2} \nu \left(\frac{\partial \eta}{\partial t} \right)^2 \right] d\sigma \tag{3.36}$$

What does this result mean? See equation (2.54) for the flexible string. Identity (9.13) is Gauss' theorem. You have to apply it to a two-dimensional region with the volume element replaced by $d\sigma$ and the element of area replaced by ds. In (3.36), the surface integral is over any region of the membrane, and the line integral is over the curve bounding the region.

4. ELECTROMAGNETIC WAVES IN METALLIC CONDUCTORS

It might be a good idea to review Maxwell's equations as they are developed in section 12 of Chapter 3. In this section, we will give some attention to the behavior of time-dependent electromagnetic fields inside metallic conductors. As a set of partial differential equations, Maxwell's equations are considerably more complicated than are the partial differential equations which describe the motion of a flexible string or a membrane. Nevertheless, solutions can be found by similar mathematical techniques, and it will be our purpose in this and the next section to exploit some of these similarities. If an excess charge is placed on a metallic conductor, it quickly distributes itself over the surface. See problem 5.4 of Chapter 3. We are not primarily concerned here with the fields produced by such excess charge. The conductor under consideration is essentially electrically neutral.

The important effect on an electromagnetic field in such a conductor is due to the electric current produced by a flow of conduction electrons which are essentially the valence electrons in the case of a metallic conductor. The electrons move from atom to atom, and there is no buildup anywhere of an excess volume density of charge. There will however be a time-dependent induced surface density of charge when an electromagnetic wave is incident on the surface of a conductor. To a good approximation, the current density in a metal is proportional to the electric field:

$$\mathbf{j} = \sigma \mathbf{E} \tag{4.1}$$

where σ is a constant which is characteristic of the metal and is called the *conductivity*. Actually, the conductivity of a metal is frequency-dependent, but this effect does not become important until frequencies above the infrared region of the electromagnetic spectrum are considered. We will therefore assume that (4.1) holds with σ a constant. (The letter σ has been used to represent area elsewhere in this book. Hopefully, no confusion about notation will result.) Equation (4.1) is really a version of *Ohm's law*. Maxwell's equations, (12.8) through (12.11) of Chapter 3, appropriately stated for electromagnetic fields inside metallic conductors, are therefore

$$\nabla \cdot \mathbf{E} = 0 \tag{4.2}$$

$$\nabla \cdot \mathbf{B} = 0 \tag{4.3}$$

$$\nabla \times \mathbf{E} + \frac{\partial \mathbf{B}}{\partial t} = 0 \tag{4.4}$$

$$\nabla \times \mathbf{B} - \frac{1}{c^2}\frac{\partial \mathbf{E}}{\partial t} = \mu_0 \sigma \mathbf{E} \tag{4.5}$$

where c is the speed of electromagnetic waves in vacuum as defined by equation (12.14) of Chapter 3. Any electric or magnetic polarization effects are neglected. This means essentially that the dielectric constant of the metal is taken to be unity and the magnetic susceptibility as zero.

It is possible to derive wave equations for the fields by a procedure similar to that used in deriving equations (12.13) and (12.15) of Chapter 3. Through the use of identity (9.8) of Chapter 3, the curl of (4.5) can be written as

$$\nabla(\nabla \cdot \mathbf{B}) - \nabla^2 \mathbf{B} - \frac{1}{c^2}\frac{\partial}{\partial t}(\nabla \times \mathbf{E}) = \mu_0 \sigma (\nabla \times \mathbf{E}) \tag{4.6}$$

which in combination with (4.3) and (4.4) gives

$$\nabla^2 \mathbf{B} - \mu_0 \sigma \frac{\partial \mathbf{B}}{\partial t} - \frac{1}{c^2}\frac{\partial^2 \mathbf{B}}{\partial t^2} = 0 \tag{4.7}$$

By a similar procedure, we find that the electric field obeys

$$\nabla^2 \mathbf{E} - \mu_0 \sigma \frac{\partial \mathbf{E}}{\partial t} - \frac{1}{c^2}\frac{\partial^2 \mathbf{E}}{\partial t^2} = 0 \tag{4.8}$$

A similar wave equation (2.9) is found for the vibrating string when the viscous frictional effect of the string moving in air is taken into account. We will look for plane harmonic wave solutions of (4.7) and (4.8).

Electric and magnetic fields are real quantities. It is nevertheless a convenient mathematical device to express the solutions of (4.7) and (4.8) in complex form. We look for a solution of (4.7) of the form

$$\mathbf{E} = \mathbf{E}_0 e^{i(\omega t - \kappa z)} \tag{4.9}$$

which represents a plane harmonic wave of frequency ω propagating in the z direction. We assume that (4.9) has no dependence on x or y. The amplitude vector $\mathbf{E}_0$ is therefore constant. The condition $\nabla \cdot \mathbf{E} = 0$ tells us that

$$(\hat{\mathbf{e}}_3 \cdot \mathbf{E}_0)\frac{\partial}{\partial z} e^{i(\omega t - \kappa z)} = 0 \tag{4.10}$$

which means that $\mathbf{E}_0$ is at right angles or transverse to the direction of propagation of the wave. For convenience, the direction of $\mathbf{E}_0$ is taken along the x axis. It is probably a good idea to try and dispel any uneasiness which the reader may have about representing fields which he or she knows to be real by complex numbers. Maxwell's equations are *linear* partial differential equations for the six functions E_x, E_y, E_z, B_x, B_y, and B_z which represent the components of the electric and magnetic fields. The equations are linear because combinations such as E_x^2, $E_x E_y$, $E_x(\partial E_x/\partial x)$, and so on, do not appear. If each of the six functions which represent the components of $\mathbf{E}$ and $\mathbf{B}$ are assumed to be complex, then the real parts and imaginary parts each separately constitute a set of possible solutions of Maxwell's equations. This is true because the vector components never get squared or multiplied together in any way so as to cause the real and imaginary parts to get mixed up. Thus, either the real or the imaginary part of (4.9) can be used as the actual solution. We have chosen to write the exponential factor as $\exp\{i(\omega t - \kappa z)\}$ rather than $\exp\{i(\kappa z - \omega t)\}$ because it will be more convenient later when the incident, reflected, and transmitted waves at the surface of a conductor are considered.

The substitution of (4.9) into the wave equation (4.8) yields

$$-\kappa^2 - \mu_0 \sigma i\omega + \frac{\omega^2}{c^2} = 0 \tag{4.11}$$

The constant κ is therefore more than just a wave number. It is a complex number which is conveniently expressed in the form $\kappa = \alpha - i\beta$. The minus sign is used because β then comes out as a positive number. By separating (4.11) into its real and imaginary parts, we find

$$\alpha^2 - \beta^2 = \frac{\omega^2}{c^2} \qquad \alpha\beta = \frac{\mu_0\sigma\omega}{2} \tag{4.12}$$

The elimination of β gives

$$\alpha^4 - \left(\frac{\omega\alpha}{c}\right)^2 - \left(\frac{\mu_0\sigma\omega}{2}\right)^2 = 0 \tag{4.13}$$

Since α^2 is positive, the positive root of (4.13) is required. Also, α itself is positive for propagation of waves in the positive z direction. We therefore find

$$\alpha = \frac{\omega}{c\sqrt{2}}\sqrt{1 + \sqrt{1 + (\mu_0\sigma c^2/\omega)^2}} \qquad \beta = \frac{\mu_0\sigma\omega}{2\alpha} \tag{4.14}$$

Compare these results with equation (2.62). The electric field can now be expressed as

$$\mathbf{E} = E_0\hat{\mathbf{e}}_1 e^{-\beta z} e^{i(\omega t - \alpha z)} \tag{4.15}$$

The associated magnetic field is found from (4.4) to be

$$\mathbf{B} = E_0\hat{\mathbf{e}}_2 \frac{(\alpha - i\beta)}{\omega} e^{-\beta z} e^{i(\omega t - \alpha z)} \tag{4.16}$$

The magnetic field is perpendicular to the electric field, and both are perpendicular to the direction of propagation just as in the case of plane waves moving in vacuum. The direction of propagation is the same as the direction of $\mathbf{E} \times \mathbf{B}$. A possible physical electromagnetic wave moving in the conductor is found by taking the real parts of (4.15) and (4.16):

$$\mathbf{E} = E_0\hat{\mathbf{e}}_1 e^{-\beta z}\cos(\omega t - \alpha z) \tag{4.17}$$

$$\mathbf{B} = E_0\hat{\mathbf{e}}_2 \frac{1}{\omega}\sqrt{\alpha^2 + \beta^2}\, e^{-\beta z}\cos(\omega t - \alpha z - \phi) \tag{4.18}$$

$$\tan\phi = \frac{\beta}{\alpha} \tag{4.19}$$

The wave is attenuated in the direction of propagation, as would be expected since energy is being removed from the field in the form of Joule heating. The wave number for propagation in the conductor is the parameter α. There is dispersion because the phase velocity

$$u_p = \frac{\omega}{\alpha} = \frac{c\sqrt{2}}{\sqrt{1 + \sqrt{1 + (\mu_0\sigma c^2/\omega)^2}}} \tag{4.20}$$

is frequency dependent. The magnetic field is out of phase with the electric field by the angle ϕ as given by (4.19).

It is of interest to find out what happens when a plane-polarized electromagnetic wave traveling in a vacuum is incident on the surface of a conductor. We will consider only the simplest case of normal incidence. It is first necessary to consider the boundary conditions obeyed by the field vectors at the surface of the conductor. Figure 4.1 shows a differential closed path of integration in the shape of a rectangle with one side (2) of length Δs in the conductor and the opposite side (1) in the vacuum. The two ends of length Δh are perpendicular to the surface of the conductor. By integrating equation (4.4) over the surface bounded by the path of integration (this surface is at right angles to the physical surface of the conductor) and using Stokes' theorem (9.19) of Chapter 3, we find

$$\oint \mathbf{E}\cdot d\mathbf{s} + \frac{\partial}{\partial t}\int \mathbf{B}\cdot d\boldsymbol{\sigma} = 0 \tag{4.21}$$

In the limit $\Delta h \to 0$, there will be no contribution to

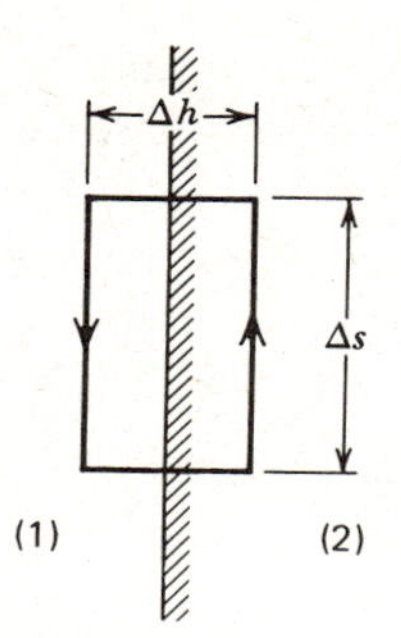

Fig. 4.1

the line integral from the two sides of length Δh. Equation (4.21) then gives

$$(E_{2t} - E_{1t})\,\Delta s + \left(\frac{\partial \mathbf{B}}{\partial t}\right)\cdot\hat{\mathbf{n}}\,\Delta s\,\Delta h = 0 \tag{4.22}$$

where E_{1t} and E_{2t} are the components of **E** in the two media which are tangent to the interface and $\hat{\mathbf{n}}$ is a unit vector normal to the surface of integration. In the limit $\Delta h \to 0$, we find $E_{1t} = E_{2t}$. Thus, the tangential component of the electric field is continuous as we cross the boundary. This result was previously obtained for a static electric field in problem 12.7 of Chapter 3. A similar calculation based on equation (4.5) gives $B_{1t} = B_{2t}$.

Let a conducting medium occupy the half-space $z > 0$. The x, y plane then coincides with its surface. A plane-polarized electromagnetic wave traveling on the negative z axis toward the conductor can be expected to be partially reflected and partially transmitted into the conductor. In the vacuum, the electromagnetic field is therefore

$$\mathbf{E}_1 = E_{00}\hat{\mathbf{e}}_1 e^{i(\omega t - \kappa z)} + E_{01}\hat{\mathbf{e}}_1 e^{i(\omega t + \kappa z)} \tag{4.23}$$

$$\mathbf{B}_1 = \frac{1}{c}E_{00}\hat{\mathbf{e}}_2 e^{i(\omega t - \kappa z)} - \frac{1}{c}E_{01}\hat{\mathbf{e}}_2 e^{i(\omega t + \kappa z)} \tag{4.24}$$

Recall that in the vacuum, $\omega = \kappa c$. Consult equations (12.25) and (12.27) of Chapter 3 to see the correct relation between **E** and **B** for a plane-polarized electromagnetic wave moving in vacuum. Keep in mind that both the incident and reflected waves move in the direction of $\mathbf{E} \times \mathbf{B}$. Inside the conductor, the fields are

$$\mathbf{E}_2 = E_{02}\hat{\mathbf{e}}_1 e^{-\beta z} e^{i(\omega t - \alpha z)} \tag{4.25}$$

$$\mathbf{B}_2 = E_{02}\hat{\mathbf{e}}_2 \frac{\alpha - i\beta}{\omega} e^{-\beta z} e^{i(\omega t - \alpha z)} \tag{4.26}$$

By imposing the continuity conditions on the fields at the surface of the conductor at $z = 0$, we find

$$E_{00} + E_{01} = E_{02} \tag{4.27}$$

$$E_{00} - E_{01} = \frac{c(\alpha - i\beta)}{\omega} E_{02} \tag{4.28}$$

In the case of normal incidence, the field vectors are entirely parallel to the surface and have no normal components. It is possible to solve (4.27) and (4.28) for the reflected and transmitted amplitudes E_{01} and E_{02} in terms of the amplitude E_{00} of the incident wave:

$$E_{01} = \frac{\kappa - \alpha + i\beta}{\kappa + \alpha - i\beta} E_{00} \qquad E_{02} = \frac{2\kappa}{\kappa + \alpha - i\beta} E_{00} \tag{4.29}$$

where $\omega = \kappa c$ has been used. The fact that (4.29) involves complex numbers simply means that the transmitted and reflected waves are not in phase with the incident wave. For a very good conductor, α and β are large. The limiting form of (4.29) for a perfect conductor is

$$E_{01} = -E_{00} \qquad E_{02} = 0 \tag{4.30}$$

This is similar to the result found for the reflection from a fixed support of a harmonic wave traveling on a string. In the limit of a perfect conductor, the tangential component of **E** vanishes. Only a normal component of **E** is possible.

It is of interest to calculate the average energy flow in the waves from Poynting's vector

$$\mathbf{S} = \frac{1}{\mu_0}\mathbf{E} \times \mathbf{B} \tag{4.31}$$

We cannot just put the complex forms of **E** and **B** into (4.31) and then take the real part. This is because (4.31) is not linear in the fields and the real and imaginary parts would get mixed up. There is however a way to calculate the average energy flow in a harmonic wave using the complex form of the fields.

To consider a fairly general case, suppose that

$$\mathbf{E} = E_x \hat{\mathbf{e}}_1 e^{i\omega t} \qquad \mathbf{B} = B_y \hat{\mathbf{e}}_2 e^{i\gamma} e^{i\omega t} \tag{4.32}$$

where E_x and B_y are real. By using the real parts of (4.32), we find

$$\begin{aligned} \operatorname{Re}\mathbf{E} \times \operatorname{Re}\mathbf{B} &= E_x B_y \cos \omega t \cos(\omega t + \gamma) \hat{\mathbf{e}}_3 \\ &= E_x B_y (\cos^2 \omega t \cos \gamma - \cos \omega t \sin \omega t \sin \gamma) \hat{\mathbf{e}}_3 \end{aligned} \tag{4.33}$$

The time average of this expression is

$$\overline{\operatorname{Re}\mathbf{E} \times \operatorname{Re}\mathbf{B}} = \tfrac{1}{2} E_x B_y \cos \gamma \hat{\mathbf{e}}_3 \tag{4.34}$$

Now use the complex forms of the fields and note that

$$\begin{aligned} \mathbf{E} \times \mathbf{B}^* &= E_x B_y e^{-i\gamma} \hat{\mathbf{e}}_3 \\ \operatorname{Re}(\mathbf{E} \times \mathbf{B}^*) &= E_x B_y \cos \gamma \hat{\mathbf{e}}_3 = 2\overline{\operatorname{Re}\mathbf{E} \times \operatorname{Re}\mathbf{B}} \end{aligned} \tag{4.35}$$

Therefore, the time-averaged Poynting vector for a harmonic wave expressed in complex form is

$$\overline{\mathbf{S}} = \frac{1}{2\mu_0} \operatorname{Re}(\mathbf{E} \times \mathbf{B}^*) \tag{4.36}$$

By means of (4.36), we can avoid finding the real parts of $\mathbf{E}$ and $\mathbf{B}$ before calculating energy flow. By a similar computation, the time-averaged energy density can be shown to be

$$w = \frac{1}{4} \varepsilon_0 \mathbf{E} \cdot \mathbf{E}^* + \frac{1}{4\mu_0} \mathbf{B} \cdot \mathbf{B}^* \tag{4.37}$$

The energy flow in the incident wave is found by application of (4.36) to the incident part of the wave as given by (4.23) and (4.24):

$$\overline{S}_I = \frac{1}{2\mu_0 c} E_{00}^2 \tag{4.38}$$

The energy flow in the reflected wave is

$$\begin{aligned} \overline{S}_R &= \frac{1}{2\mu_0 c} \operatorname{Re}(E_{01} E_{01}^*) \\ &= \frac{1}{2\mu_0 c} \operatorname{Re}\left(\frac{\kappa - \alpha + i\beta}{\kappa + \alpha - i\beta} \right) \left(\frac{\kappa - \alpha - i\beta}{\kappa + \alpha + i\beta} \right) E_{00}^2 \\ &= \frac{1}{2\mu_0 c} \left[\frac{(\kappa - \alpha)^2 + \beta^2}{(\kappa + \alpha)^2 + \beta^2} \right] E_{00}^2 \end{aligned} \tag{4.39}$$

The energy flow transmitted into the conductor is found from (4.25) and (4.26) applied at $z = 0$:

$$\begin{aligned} \overline{S}_T &= \frac{1}{2\mu_0} \operatorname{Re}\left[E_{02} E_{02}^* \frac{(\alpha + i\beta)}{\kappa c} \right] \\ &= \frac{2\kappa\alpha}{\mu_0 c} \left[\frac{1}{(\kappa + \alpha)^2 + \beta^2} \right] E_{00}^2 \end{aligned} \tag{4.40}$$

It is easy to check that $\overline{S}_I = \overline{S}_R + \overline{S}_T$. For a very good conductor, α and β are large, and the energy loss to the conductor as given by (4.40) becomes insignificant.

PROBLEMS

4.1. If the conductivity σ is large, show that

$$\alpha \simeq \beta \simeq \sqrt{\frac{\mu_0 \sigma \omega}{2}} \tag{4.41}$$

4.2. A plane harmonic electromagnetic wave moving in an arbitrary direction in vacuum can be represented by

$$\mathbf{E} = \mathbf{E}_0 e^{i(\omega t - \boldsymbol{\kappa} \cdot \mathbf{r})} \qquad \mathbf{B} = \mathbf{B}_0 e^{i(\omega t - \boldsymbol{\kappa} \cdot \mathbf{r})} \tag{4.42}$$

Show that for such waves it is possible to replace ∇ by $-i\boldsymbol{\kappa}$ and $\partial/\partial t$ by $i\omega$. Make these replacements in the four Maxwell equations and interpret the result. Show that

$$\boldsymbol{\kappa} = \frac{\omega(\mathbf{E} \times \mathbf{B})}{E^2} \tag{4.43}$$

4.3. A perfect conductor ($\sigma = \infty$) occupies the half-space $z > 0$. A plane wave traveling in the x, z plane at an angle of incidence θ_1 with respect to the z axis is incident on the surface of the conductor as shown in Fig. 4.2. If the incident wave is polarized such that its magnetic field is parallel to the surface of the conductor, then the electric field of the combined incident and reflected waves is

$$\mathbf{E} = (E_{1x}\hat{\mathbf{e}}_1 + E_{1z}\hat{\mathbf{e}}_3)\exp[i(\omega t - \kappa_{1x}x - \kappa_{1z}z)]$$
$$+ (E_{2x}\hat{\mathbf{e}}_1 + E_{2y}\hat{\mathbf{e}}_2 + E_{2z}\hat{\mathbf{e}}_3)\exp\left[i(\omega t - \kappa_{2x}x - \kappa_{2y}y + \kappa_{2z}z)\right] \tag{4.44}$$

By imposing the condition that the tangential component of $\mathbf{E}$ vanishes at $z=0$ for all values of x and y, show that $E_{2x} = -E_{1x}$, $E_{2y}=0$, $E_{2z}=E_{1z}$, $\kappa_{1x}=\kappa_{2x}$, $\kappa_{2y}=0$, $\kappa_{2z}=\kappa_{1z}$, and $\theta_1=\theta_2$. Note that the electric field in (4.44) has a component normal to the conducting surface. The electric field is zero inside the conductor. Since the electric field lines must end on charge, there is a time-dependent induced surface charge on the conductor. What happens if the incident wave is polarized with its electric field parallel to the surface of the conductor?

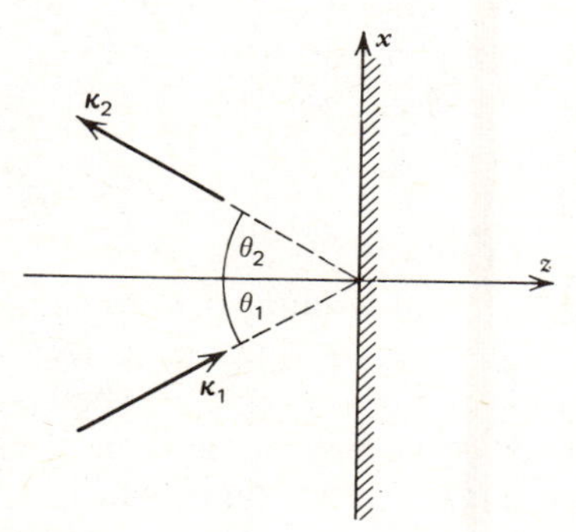

Fig. 4.2

4.4. The Joule heating per unit volume in the conductor is given by $\mathbf{j}\cdot\mathbf{E} = \sigma E^2$. The time-averaged power dissipated per unit volume is $\frac{1}{2}\mathrm{Re}(\sigma\mathbf{E}\cdot\mathbf{E}^*)$. Show that this correctly accounts for the average energy transmitted into the conductor as given by (4.40). The quantity $\delta = 1/\beta$ is called the *skin depth*. Note that the fields and currents essentially exist in a layer of thickness δ inside the conductor. As σ becomes large, δ approaches zero and the currents become surface currents.

5. ELECTROMAGNETIC WAVES IN HOLLOW CONDUCTORS

Electromagnetic waves can propagate inside hollow pipes made out of good conducting material. When used in this manner, a hollow metal pipe is called a *wave guide*. In this section, detailed results will be obtained for the propagation of waves in wave guides which are rectangular in cross section. We will assume that the wave guide is made out of a perfect electrical conductor. The discussion of section 4 makes it clear that the appropriate boundary condition is the vanishing of the tangential component of the electric field at the surface of the conductor. The space inside the pipe is assumed to be vacuum, although the presence of air in the pipe makes little difference. The appropriate form of Maxwell's equations is therefore (4.2) through (4.5) with $\sigma=0$. We look for solutions of a single frequency ω of the form

$$\mathbf{E} = (E_1\hat{\mathbf{e}}_1 + E_2\hat{\mathbf{e}}_2 + E_3\hat{\mathbf{e}}_3)e^{i(\kappa_3 z - \omega t)}$$
$$\mathbf{B} = (B_1\hat{\mathbf{e}}_1 + B_2\hat{\mathbf{e}}_2 + B_3\hat{\mathbf{e}}_3)e^{i(\kappa_3 z - \omega t)} \tag{5.1}$$

The z axis is parallel to the axis of the pipe which is assumed to be either infinitely long or to have a source of electromagnetic waves of frequency ω at one end and an absorber at the other. Equation (5.1) therefore describes some sort of wave which propagates in the z direction in the space inside the pipe. The field components E_1, E_2, E_3, B_1, B_2, and B_3 are expected to be functions of x and y. By direct substitution of (5.1) into the four Maxwell equations, we find

$$\frac{\partial E_1}{\partial x} + \frac{\partial E_2}{\partial y} + i\kappa_3 E_3 = 0 \tag{5.2}$$

$$\frac{\partial B_1}{\partial x} + \frac{\partial B_2}{\partial y} + i\kappa_3 B_3 = 0 \tag{5.3}$$

$$\frac{\partial E_3}{\partial y} - i\kappa_3 E_2 - i\omega B_1 = 0 \tag{5.4}$$

$$i\kappa_3 E_1 - \frac{\partial E_3}{\partial x} - i\omega B_2 = 0 \tag{5.5}$$

$$\frac{\partial E_2}{\partial x} - \frac{\partial E_1}{\partial y} - i\omega B_3 = 0 \tag{5.6}$$

$$\frac{\partial B_3}{\partial y} - i\kappa_3 B_2 + \frac{i\omega}{c^2} E_1 = 0 \tag{5.7}$$

$$i\kappa_3 B_1 - \frac{\partial B_3}{\partial x} + \frac{i\omega}{c^2} E_2 = 0 \tag{5.8}$$

$$\frac{\partial B_2}{\partial x} - \frac{\partial B_1}{\partial y} + \frac{i\omega}{c^2} E_3 = 0 \tag{5.9}$$

The task before us is finding the solutions of the set of partial differential equations given by (5.2) through (5.9). If (5.4) and (5.8) are solved for B_1 and E_2, the result is

$$B_1 = \frac{i}{\gamma^2}\left[\kappa_3 \frac{\partial B_3}{\partial x} - \frac{\omega}{c^2}\frac{\partial E_3}{\partial y}\right] \tag{5.10}$$

$$E_2 = \frac{i}{\gamma^2}\left[\kappa_3 \frac{\partial E_3}{\partial y} - \omega \frac{\partial B_3}{\partial x}\right] \tag{5.11}$$

where we have set

$$\gamma^2 = \frac{\omega^2}{c^2} - \kappa_3^2 \tag{5.12}$$

If (5.5) and (5.7) are similarly solved for E_1 and B_2, the result is

$$E_1 = \frac{i}{\gamma^2}\left[\omega \frac{\partial B_3}{\partial y} + \kappa_3 \frac{\partial E_3}{\partial x}\right] \tag{5.13}$$

$$B_2 = \frac{i}{\gamma^2}\left[\kappa_3 \frac{\partial B_3}{\partial y} + \frac{\omega}{c^2}\frac{\partial E_3}{\partial x}\right] \tag{5.14}$$

The field components E_3 and B_3 are in the direction of propagation and are called the *longitudinal components*. Equations (5.10), (5.11), (5.13), and (5.14) express the *transverse components* E_1, E_2, B_1, and B_2 in terms of the longitudinal components. If we can succeed in solving for the functions E_3 and B_3, the entire electromagnetic field will be determined.

If E_1 and E_2 as given by (5.11) and (5.13) are substituted into (5.2), the result is

$$\frac{\partial^2 E_3}{\partial x^2} + \frac{\partial^2 E_3}{\partial y^2} + \gamma^2 E_3 = 0 \tag{5.15}$$

Exactly the same equation is found if B_1 and B_2 as given by (5.10) and (5.14) are substituted into (5.9). Similarly, either (5.3) or (5.6) leads to

$$\frac{\partial^2 B_3}{\partial x^2} + \frac{\partial^2 B_3}{\partial y^2} + \gamma^2 B_3 = 0 \tag{5.16}$$

Equations (5.15) and (5.16) are quite encouraging because each is a partial differential equation, in fact, a Helmholtz equation involving a single unknown function. The waves which can propagate in the wave guide naturally divide into two parts, one of which is determined by E_3 and the other by B_3. It is convenient to consider these separately. The part of the field which is determined by E_3 alone and for which $B_3 = 0$ is called a *transverse magnetic* (TM) field. Similarly, the part of the field determined by B_3 alone and for which $E_3 = 0$ is called a *transverse electric* (TE) field. Both types of waves can be present simultaneously. Appropriate boundary conditions for E_3 are already known. Since E_3 is a component of the electric field which is always tangent to the walls of the guide, it must be zero on the boundaries. The boundary conditions for B_3 are found from (5.11) and (5.13) with $E_3 = 0$:

$$E_2 = -\frac{i\omega}{\gamma^2}\frac{\partial B_3}{\partial x} \qquad E_1 = \frac{i\omega}{\gamma^2}\frac{\partial B_3}{\partial y} \tag{5.17}$$

If $\hat{\mathbf{n}}$ is a unit vector drawn to the surface of the conductor, then the vanishing of the tangential component of $\mathbf{E}$ of the boundary translates into $\hat{\mathbf{n}} \times \mathbf{E} = n_1 E_2 - n_2 E_1 = 0$. Equation (5.17) gives

$$n_1 \frac{\partial B_3}{\partial x} + n_2 \frac{\partial B_3}{\partial y} = \hat{\mathbf{n}} \cdot \nabla B_3 = \frac{\partial B_3}{\partial n} = 0 \tag{5.18}$$

The Helmholtz equation for B_3 is therefore to be

solved subject to the boundary condition that the normal derivative of B_3 vanish on the walls of the wave guide.

All results so far obtained apply to wave guides with cross sections of arbitrary shape. The discussion will now be specialized to the case of a guide with a rectangular cross section bounded by $x=0$, $x=a$, $y=0$, and $y=b$. The reader who has done problem 3.2 already has considerable insight into the wave guide problem. The first step in finding the possible TM waves is to obtain the solutions of (5.15) which satisfy the boundary conditions $E_3=0$ on $x=0$, $x=a$, $y=0$, and $y=b$. The mathematics has already been done in section 3 in connection with the rectangular membrane. The solutions are the eigenfunctions

$$E_3 = E_0 \sin\left(\frac{m\pi x}{a}\right)\sin\left(\frac{n\pi y}{b}\right) \tag{5.19}$$

with $m=1,2,3,\ldots$ and $n=1,2,3,\ldots$. The parameter γ is restricted to the eigenvalues

$$\gamma^2 = \pi^2\left(\frac{m^2}{a^2}+\frac{n^2}{b^2}\right) = \frac{\omega^2}{c^2} - \kappa_3^2 \tag{5.20}$$

This means that for a given frequency, the possible values of κ_3 are also restricted. The phase velocity is given by

$$u_p = \frac{\omega}{\kappa_3} = \frac{\omega}{\sqrt{\omega^2/c^2-\gamma^2}} \tag{5.21}$$

Note that it is quite possible for the phase velocity to become infinite. The remaining field components for the TM waves are found from (5.10), (5.11), (5.13), and (5.14) with E_3 as given by (5.19) and $B_3=0$:

$$\begin{aligned}
E_1 &= \frac{im\pi\kappa_3 E_0}{\gamma^2 a}\cos\left(\frac{m\pi x}{a}\right)\sin\left(\frac{n\pi y}{b}\right)\\
E_2 &= \frac{in\pi\kappa_3 E_0}{\gamma^2 b}\sin\left(\frac{m\pi x}{a}\right)\cos\left(\frac{n\pi y}{b}\right)\\
B_1 &= -\frac{\omega}{\kappa_3 c^2}E_2\\
B_2 &= \frac{\omega}{\kappa_3 c^2}E_1
\end{aligned} \tag{5.22}$$

Each pair of values of the integers m and n gives a possible mode for the propagation of a TM wave. For a given frequency, only those modes are possible for which κ_3^2 is positive as determined by (5.20).

It is possible to calculate the average energy flow in the wave guide for a given mode by means of the Poynting vector in the form (4.36). The fields (5.22) give for the z component of the Poynting vector

$$\begin{aligned}
\overline{S}_3 &= \frac{\omega}{2\mu_0\kappa_3 c^2}(E_1E_1^* + E_2E_2^*)\\
&= \frac{\omega\pi^2\kappa_3 E_0^2}{2\mu_0 c^2\gamma^4}\left[\frac{m^2}{a^2}\cos^2\left(\frac{m\pi x}{a}\right)\sin^2\left(\frac{n\pi y}{b}\right)\right.\\
&\quad\left.+\frac{n^2}{b^2}\sin^2\left(\frac{m\pi x}{a}\right)\cos^2\left(\frac{n\pi y}{b}\right)\right]
\end{aligned} \tag{5.23}$$

The power flow is found by integrating (5.23) over the cross section of the wave guide. Needed are the integrals

$$\int_0^a \cos^2\left(\frac{m\pi x}{a}\right)dx = \int_0^a \sin^2\left(\frac{m\pi x}{a}\right)dx = \frac{a}{2} \tag{5.24}$$

The integrals over y are similar. The result is

$$\overline{P}_{TM} = \int_0^a\int_0^b \overline{S}_3\,dy\,dx = \frac{\omega\kappa_3 E_0^2 ab}{8\mu_0 c^2\gamma^2} \tag{5.25}$$

It is of interest to calculate the velocity of energy transport. For this purpose, we need the time-averaged energy density:

$$\overline{w} = \frac{\varepsilon_0\mathbf{E}\cdot\mathbf{E}^*}{4} + \frac{\mathbf{B}\cdot\mathbf{B}^*}{4\mu_0} = \frac{1}{4\mu_0 c^2}(\mathbf{E}\cdot\mathbf{E}^* + c^2\mathbf{B}\cdot\mathbf{B}^*) \tag{5.26}$$

where $\varepsilon_0 = 1/(\mu_0 c^2)$ has been used. The magnetic field can be written in terms of the electric field by means of (5.22) with the result

$$\overline{w} = \frac{1}{4\mu_0 c^2}\left[(E_1E_1^* + E_2E_2^*)\left(2+\frac{\gamma^2}{\kappa_3^2}\right) + E_3E_3^*\right] \tag{5.27}$$

The relation (5.12) has also been used in obtaining (5.27). The average energy density per unit length is found by integrating (5.27) over the cross section of the guide. The result is

$$\overline{w}_L = \int_0^a\int_0^b \overline{w}\,dy\,dx = \frac{E_0^2 ab\omega^2}{8\mu_0\gamma^2 c^4} \tag{5.28}$$

The velocity at which energy is transported is given

by

$$u_w = \frac{\bar{P}}{\bar{w}_L} = \frac{\kappa_3 c^2}{\omega} \tag{5.29}$$

The *group velocity* of the waves is defined by

$$u_g = \frac{d\omega}{d\kappa_3} \tag{5.30}$$

By writing (5.12) as

$$\omega = c\sqrt{\gamma^2 + \kappa_3^2} \tag{5.31}$$

we find

$$u_g = \frac{c\kappa_3}{\sqrt{\gamma^2 + \kappa_3^2}} = \frac{\kappa_3 c^2}{\omega} = u_w \tag{5.32}$$

Thus, for waves propagating in the evacuated space in a hollow conducting pipe, the velocity of energy transport is the same as the group velocity. Group velocity and phase velocity are related by

$$u_p u_g = c^2 \tag{5.33}$$

If the group velocity is expressed as

$$u_g = c\sqrt{1 - (\gamma c/\omega)^2} \tag{5.34}$$

it becomes evident that it can never be larger than c.

To find the possible TE waves, it is first necessary to solve (5.16) for B_3 subject to the boundary condition $\partial B_3/\partial n = 0$. For a rectangular wave guide, this translates into

$$\frac{\partial B_3}{\partial x} = 0 \quad \text{on } x = 0 \text{ and } x = a$$
$$\frac{\partial B_3}{\partial y} = 0 \quad \text{on } y = 0 \text{ and } y = b \tag{5.35}$$

The appropriate solutions are

$$B_3 = B_0 \cos\left(\frac{m\pi x}{a}\right)\cos\left(\frac{n\pi y}{b}\right) \tag{5.36}$$

where $m = 0, 1, 2, \ldots$ and $n = 0, 1, 2, \ldots$ but m and n are not both simultaneously zero. Equation (5.20) for γ^2 holds for both TE and TM waves. All previously obtained eigenfunctions and eigenvalues have occurred when the function in question vanishes on the boundaries of the region. The solutions given by (5.36) provide an example of a set of eigenfunctions which result from the requirement that the normal derivative vanish on the boundaries. Later on in this chapter, we will discover that eigenfunctions can occur for a variety of different boundary conditions.

The complete solution for TE waves is obtained in essentially the same manner as was done for TM waves. The results are summarized in the following equations:

$$E_1 = -\frac{i\omega B_0 n\pi}{\gamma^2 b}\cos\left(\frac{m\pi x}{a}\right)\sin\left(\frac{n\pi y}{b}\right) \tag{5.37}$$

$$E_2 = \frac{i\omega B_0 m\pi}{\gamma^2 a}\sin\left(\frac{m\pi x}{a}\right)\cos\left(\frac{n\pi y}{b}\right) \tag{5.38}$$

$$B_1 = -\frac{\kappa_3}{\omega}E_2 \qquad B_2 = \frac{\kappa_3}{\omega}E_1 \tag{5.39}$$

$$\bar{P}_{TE} = \frac{\kappa_3 \omega B_0^2 ab}{8\mu_0\gamma^2} \tag{5.40}$$

Since (5.20) holds for TE waves, the expressions (5.21) and (5.32) are valid for phase and group velocities. Again, the group velocity is the velocity of energy transport.

PROBLEMS

5.1. Consider the possibility that transverse electromagnetic (TEM) waves can propagate inside a hollow cylindrical conductor. These are waves which are purely transverse to the walls of the wave guide and are therefore characterized by both $E_3 = 0$ and $B_3 = 0$. Equations (5.10) through (5.14) show that in order for such waves to exist, it is necessary that $\gamma = 0$. This means that $\omega = \kappa_3 c = \kappa c$. Write out equations (5.2) through (5.9) for TEM waves. Show that it is possible to express the electric field as the gradient of a scalar function by means of

$$E_1 = -\frac{\partial\psi}{\partial x} \qquad E_2 = -\frac{\partial\psi}{\partial y} \tag{5.41}$$

and that ψ obeys Laplace's equation

$$\frac{\partial^2\psi}{\partial x^2}+\frac{\partial^2\psi}{\partial y^2}=0 \tag{5.42}$$

with the boundary condition $\psi=$ constant on the walls of the conductor. What is the unique solution of Laplace's equation for this boundary condition if the region inside the guide is simply connected? Can TEM waves exist? Suppose that the cross section of the guide is not simply connected. This would be true, for example, if the guide were constructed by placing a solid conducting cylinder inside a hollow one. Show that it is in fact possible for TEM waves to exist in the space between the two conductors.

5.2. In working this problem, it is not necessary to assume that the cross section of the guide is rectangular. For TM waves, the function E_3 obeys the Helmholtz equation (5.15) and is real. In Green's first identity, equation (9.16) of Chapter 3, let $\phi=\psi=E_3$. Since E_3 depends only on x and y, (9.16) can be applied to a cross section of the wave guide with the volume element replaced by an element of area and the element of area replaced by a line element:

$$\int_\sigma\left(E_3\nabla^2E_3+\nabla E_3\cdot\nabla E_3\right)d\sigma=\oint E_3(\nabla E_3\cdot\hat{\mathbf{n}})\,ds \tag{5.43}$$

The line integral coincides with the wall of the wave guide, and $\hat{\mathbf{n}}$ is a unit vector drawn outward from its cylindrical surface. The gradient operator in (5.43) is two-dimensional:

$$\nabla=\hat{\mathbf{e}}_1\frac{\partial}{\partial x}+\hat{\mathbf{e}}_2\frac{\partial}{\partial y} \tag{5.44}$$

By application of (5.43) to TM waves, show that

$$\int_\sigma\nabla E_3\cdot\nabla E_3\,d\sigma=\int_\sigma\gamma^2E_3^2\,d\sigma \tag{5.45}$$

Show that equation (5.26) for the energy density can be written in the form

$$\overline{w}=\frac{1}{4\mu_0c^2}\left[\left(\frac{1}{\gamma^2}+\frac{2\kappa_3^2}{\gamma^4}\right)(\nabla E_3\cdot\nabla E_3)+E_3^2\right] \tag{5.46}$$

By integrating (5.46) over a cross section of the wave guide, show that the average energy density per unit length is

$$\overline{w}_L=\frac{\omega^2}{2\mu_0\gamma^2c^4}\int_\sigma E_3^2\,d\sigma \tag{5.47}$$

Check to see that (5.47) gives (5.28) for a rectangular wave guide.

5.3. TM waves are propagating in a wave guide of arbitrary cross-sectional shape. Show that the z component of the time-averaged Poynting vector is

$$\overline{S}_3=\frac{\omega\kappa_3}{2\mu_0\gamma^4c^2}\nabla E_3\cdot\nabla E_3 \tag{5.48}$$

and that the average power flow is

$$\overline{P}_{TM}=\frac{\omega\kappa_3}{2\mu_0\gamma^2c^2}\int_\sigma E_3^2\,d\sigma \tag{5.49}$$

5.4. TE waves are propagating in a wave guide of arbitrary cross-sectional shape. Show that the average energy density per unit length can be expressed as

$$\overline{w}_L=\frac{\omega^2}{2\mu_0c^2\gamma^2}\int_\sigma B_3^2\,d\sigma \tag{5.50}$$

and that the average power flow is

$$\overline{P}_{TE} = \frac{\omega \kappa_3}{2\mu_0 \gamma^2} \int_\sigma B_3^2 \, d\sigma \tag{5.51}$$

5.5. Two plane waves of slightly different frequency and the same amplitude are given by

$$\begin{aligned} \eta_1 &= A \cos(\kappa x - \omega t) \\ \eta_2 &= A \cos(\{\kappa + \Delta\kappa\} x - \{\omega + \Delta\omega\} t) \end{aligned} \tag{5.52}$$

Show that

$$\eta = \eta_1 + \eta_2 = 2A \cos(\kappa x - \omega t) \cos\left(\frac{\Delta\kappa x - \Delta\omega t}{2}\right) \tag{5.53}$$

provided that $\Delta\kappa$ and $\Delta\omega$ are small compared to κ and ω. The waves can be electromagnetic waves or any other type of waves which obey a linear superposition principle. At $t = 0$, (5.53) can be written as

$$\eta = 2A \cos\left(\frac{2\pi x}{\lambda}\right) \cos\left(\frac{\pi x}{\lambda_B}\right) \tag{5.54}$$

where we have put $\kappa = 2\pi/\lambda$ and $\Delta\kappa = 2\pi/\lambda_B$. Figure 5.1 shows a plot of (5.54) with $\lambda_B = 8\lambda$. The longer wavelength part of (5.54) is called a *beat*. The beats move with the group velocity $u_g = d\omega/d\kappa$. What is u_g for plane electromagnetic waves moving in vacuum?

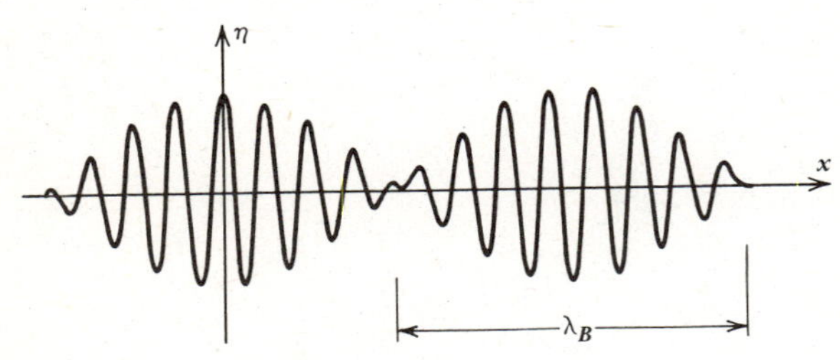

Fig. 5.1

5.6. Write out the complete electromagnetic field for TE waves in a rectangular wave guide, including the z dependence and the time dependence, for the case $m = 1$ and $n = 0$. Show that the real part is

$$\begin{aligned} B_x &= \frac{\kappa_3 B_0 a}{\pi} \sin\left(\frac{\pi x}{a}\right) \sin(\kappa_3 z - \omega t) \\ B_y &= 0 \\ B_z &= B_0 \cos\left(\frac{\pi x}{a}\right) \cos(\kappa_3 z - \omega t) \\ E_x &= 0 \qquad E_y = -\frac{\omega}{\kappa_3} B_x \qquad E_z = 0 \end{aligned} \tag{5.55}$$

The magnetic field lines are determined by

$$\frac{B_x}{B_z} = \frac{dx}{dz} \tag{5.56}$$

By integrating this differential equation, show that the magnetic field lines at $t = 0$ are given

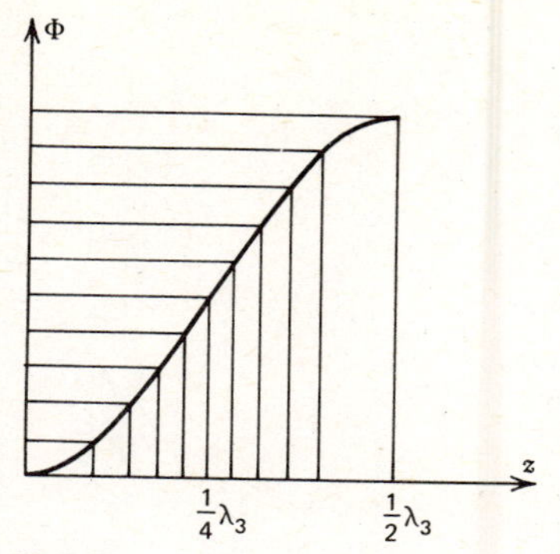

Fig. 5.2

by

$$\sin\left(\frac{\pi x}{a}\right)\cos(\kappa_3 z)=\alpha \tag{5.57}$$

where α is a constant. Figure 5.3 shows the magnetic field in the wave guide at $t=0$. The field lines for $-\lambda_3/4<z<\lambda_3/4$ are given by $\alpha=0.2$, 0.4, 0.6, and 0.8. The field lines for $\lambda_3/4<z<3\lambda_3/4$ are for negative values of α. The wave length is determined by $\lambda_3=2\pi/\kappa_3$.

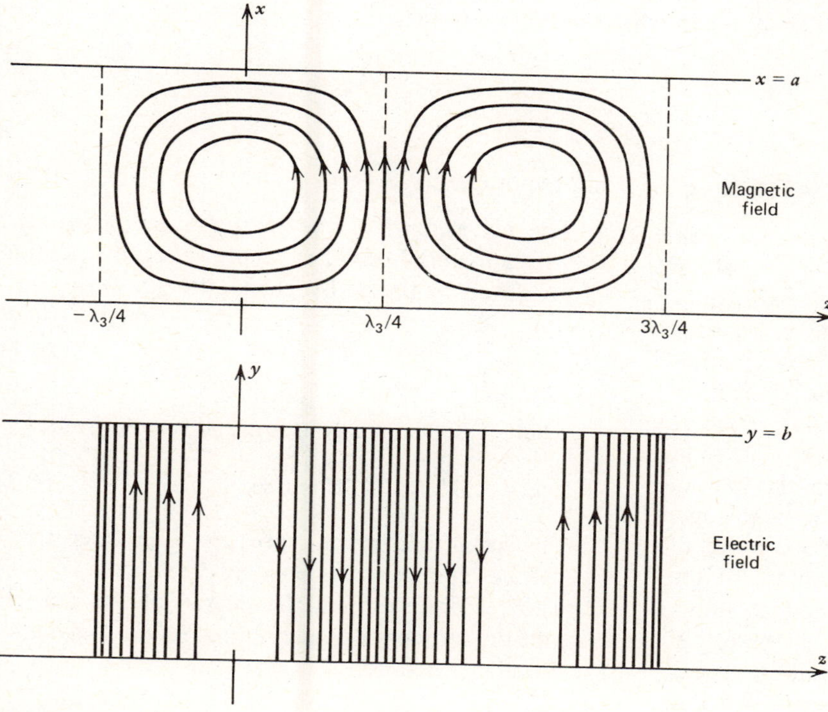

Fig. 5.3

Show that the flux of the magnetic field through the plane $x = a/2$ is

$$\Phi(z) = \frac{B_0 ab}{\pi}\left[1 - \cos\left(\frac{2\pi z}{\lambda_3}\right)\right] \tag{5.58}$$

By plotting $\Phi(z)$ vs. z as shown in Fig. 5.2 and then drawing equally spaced lines representing equal increments of flux and then projecting the intersections of these lines with $\Phi(z)$ onto the z axis, we can see how to space the field lines so that their density is a correct representation of the field strength. Equation (5.58) in combination with (5.57) shows that equal increments of flux in the plane $x = a/2$ correspond to equal increments for the parameter α. The magnetic field lines at $z = -\lambda_3/4$, $+\lambda_3/4$, and so on, are peculiar. They terminate on the walls of the wave guide at a point where the field strength is zero. Use equations (5.55) to check this. Also check to see that the arrows on the field lines point in the correct direction. The electric field lines are perpendicular to the magnetic field lines and are also shown in Fig. 5.3 as they would exist in the plane $x = a/2$. The electric field is of course zero at $x = 0$ and $x = a$.

5.7. Electromagnetic waves exist inside a closed rectangular box made of perfectly conducting material. The box is bounded by the coordinate planes and by $x = a$, $y = b$, and $z = s$. What are the possible standing waves which can exist in the box? Show that the possible frequencies are given by

$$\omega^2 = c^2\pi^2\left(\frac{m^2}{a^2} + \frac{n^2}{b^2} + \frac{p^2}{s^2}\right) \tag{5.59}$$

Hint: The problem can be solved by superposition of the solutions already found in the text with similar waves moving in the negative z direction. Both TM and TE waves should be considered. The superposition is done in such a way as to produce nodes at $z = 0$ and $z = s$ in the transverse components of the electric field.

6. THE SCHRÖDINGER WAVE EQUATION

The mathematical techniques which have so far been developed in this book are quite adequate to allow some simple solutions of the Schrödinger wave equation to be obtained. The wave function is a scalar field and as such is in many respects simpler mathematically than the Maxwell field. The difficulties which students have with quantum mechanics are at least initially more apt to be conceptual than mathematical. We assume that the reader has had an introduction to modern physics and at least to some extent has been led to believe in the existence of "matter waves." The basic physical content of the Schrödinger wave equation as well as the plausibility arguments leading to it will be reviewed here, but space will not be taken to go over in detail the crucial experiments needed to support the theory. Our discussion of quantum mechanics is based on the following hypotheses:

1. The wavelength of the matter waves associated with a material particle moving with a momentum p is given by the de Broglie relation

$$\lambda = \frac{h}{p} \tag{6.1}$$

where h is Planck's constant and has the value

$$h = 6.6251 \times 10^{-27} \text{ erg-sec} \tag{6.2}$$

The scalar field which represents the matter waves will be represented by $\psi(x, y, z, t)$.

2. The relation between the energy of the particle and the frequency of the matter waves is given by the Planck relation

$$W = hf \tag{6.3a}$$

3. Matter waves obey a linear superposition principle. The differential equation which they obey is therefore expected to be linear.
4. The intensity of the matter waves is given by $|\psi|^2$ and has the significance that $|\psi|^2\, d\Sigma$ is the probability of finding the particle in the volume element $d\Sigma$. The wave function itself is not observable.
5. All the observable quantities of a system can be calculated from a knowledge of its wave function.

We will explore a simple consequence of the hypothesis of linear superposition. In problem 10.4 of Chapter 5, you showed that the intensity of the Fraunhofer diffraction pattern of a single slit is given by

$$I = I_0 \frac{\sin^2 \alpha}{\alpha^2} \qquad \alpha = \frac{\kappa a \sin\theta}{2} \tag{6.3b}$$

The intensity of the Maxwell field as given by the Poynting vector provides a probability distribution function for photons in much the same way that $|\psi|^2$ gives the probability density for material particles. If the intensity of the light illuminating the slit is made very low, it is possible to record individual interactions of the electromagnetic radiation on the viewing screen in energy units of hf. This could be done, for example, by observing the ejection of photoelectrons from specific points on the screen. As time goes on and the number of observed interactions becomes large, the intensity pattern as given by (6.3) emerges. The derivation of (6.3) depends on the linear superposition property of electromagnetic waves. It is therefore reasonable to assume that it applies equally well to matter waves. Suppose that electrons or some other material particles traveling in the x direction with a definite energy and momentum are incident on a slit of width a as shown in Fig. 6.1. The particles are assumed not to interact with one another so that the wave function under discussion is a single-particle wave function. Let y be the coordinate on the viewing screen which measures position from the center of the intensity pattern. The value of $|\psi(x, y)|^2$ as calculated on the viewing screen is a probability density function for a single particle, meaning that $|\psi(x, y)|^2\, dy$ gives the probability that a particle will land on the screen between y and $y + dy$. The intensity pattern can be experimentally observed by allowing many particles to come through the apparatus over a period of time. The slit itself can be thought of as an instrument for determining to within an accuracy $\Delta y = a$ the y coordinate of an individual particle. Because of the interference of the matter waves, the exact direction of motion of the particle after it leaves the slit is no longer known. A measurement of the y coordinate of the particle to within an accuracy Δy has introduced an uncertainty in the y component of the momentum. Most of the particles fall in the region of the central maximum of the intensity pattern, which is between $\alpha = -\pi$ and $\alpha = +\pi$. If the angle θ corresponding to $\alpha = \pi$ is assumed to be small, then the uncertainty in the y component of the momentum is approximately

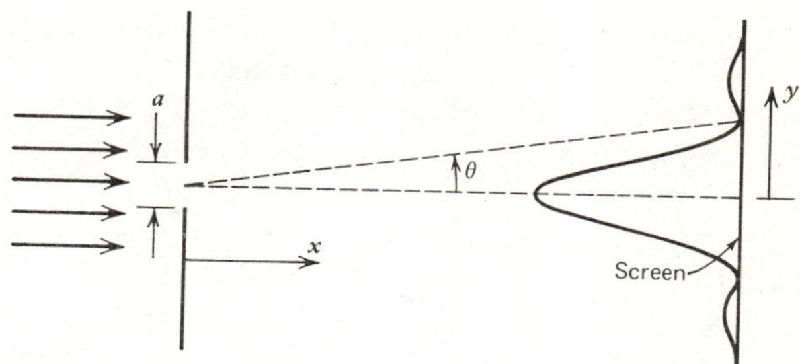

Fig. 6.1

$$\pi = \frac{\kappa a \theta}{2} \simeq \frac{\kappa \Delta y \Delta p_y}{4 p_x} \tag{6.4}$$

The de Broglie relation (6.1) gives $p_x = h/\lambda = h\kappa/(2\pi)$, which allows (6.4) to be put into the form

$$\Delta y \Delta p_y \simeq 2h \tag{6.5}$$

This is an imprecise statement of the *Heisenberg uncertainty relation*. A more precise form of it can be given only after exact definitions are given for Δy and Δp_y. After this is done, the accurate form of the uncertainty relation is found to be

$$\Delta y \Delta p_y \geq \tfrac{1}{2}\hbar \tag{6.6}$$

where we have introduced the commonly used notation $\hbar = h/(2\pi)$. More generally, the uncertainty relation holds between generalized coordinates and their canonically conjugate momenta as defined in problem 13.2 of Chapter 4:

$$\Delta q_\alpha \Delta p_\alpha \geq \tfrac{1}{2}\hbar \tag{6.7}$$

The uncertainty relation states that the y coordinate and its canonically conjugate momentum p_y cannot be measured simultaneously with unlimited accuracy. The limitation on the accuracy which is possible is given by (6.6).

Suppose that charged particles are accelerated through a known potential difference and then collimated into a beam. The energy and the momentum are known to within a very small error. The beam may be several centimeters or even meters long. The location of any one of the particles in the beam is not known so that the uncertainty in the position is large as is required since the momentum is accurately known. The wave function of a particle in the beam is fairly accurately represented by a plane wave of the form

$$\psi = Ae^{i(\boldsymbol{\kappa}\cdot\mathbf{r}-\omega t)} \tag{6.8}$$

The probability density for the particle is then

$$|\psi|^2 = \psi^*\psi = |A|^2 \tag{6.9}$$

and is a constant as would be expected since the particle is equally likely to be anywhere in the beam. A plane wave such as $\psi = A\sin(\boldsymbol{\kappa}\cdot\mathbf{r} - \omega t)$ would not work because it gives a nonconstant probability density. We therefore choose the complex form of a plane wave and write the probability density as $\psi^*\psi$. In discussing solutions of Maxwell's equations and the wave equation for elastic waves moving on a stretched string or membrane, we find that expressing the solutions in complex form is a great convenience. The final solutions of such problems are however always real functions. We are now asserting that the wave function for matter waves is intrinsically complex. The observable quantity is the probability density $\psi^*\psi$ and is real.

To discover the partial differential equation that the wave function obeys, first write the de Broglie and Planck relations in the form

$$\mathbf{p} = \hbar\boldsymbol{\kappa} \qquad W = \hbar\omega \tag{6.10}$$

where $|\boldsymbol{\kappa}| = 2\pi/\lambda$ and $\omega = 2\pi f$. For a free particle, the relation between energy and momentum is

$$W = \frac{p^2}{2m} \tag{6.11}$$

By combining (6.10) and (6.11), we find

$$\hbar\omega = \frac{\hbar^2\kappa^2}{2m} \tag{6.12}$$

There is dispersion in the matter waves because the phase velocity

$$u_p = \frac{\omega}{\kappa} = \frac{\hbar\kappa}{2m} \tag{6.13}$$

depends on κ. The group velocity is

$$u_g = \frac{d\omega}{d\kappa} = \frac{\hbar\kappa}{m} = \frac{p}{m} \tag{6.14}$$

and is identified as the classical particle velocity. The phase velocity has no direct physical significance. A wave equation obeyed by the plane wave (6.8) can be found by noting that

$$\nabla\psi = i\boldsymbol{\kappa}\psi \qquad \nabla^2\psi = -\kappa^2\psi \qquad \frac{\partial\psi}{\partial t} = -i\omega\psi \tag{6.15}$$

The dispersion relation (6.12) then implies that

$$i\hbar\frac{\partial\psi}{\partial t} = -\frac{\hbar^2}{2m}\nabla^2\psi \tag{6.16}$$

This is the Schrödinger wave equation for a free particle of mass m. If forces are present which can be derived from a potential energy function $V(\mathbf{r}, t)$, then the expression (6.11) for the total energy is

$$W = \frac{p^2}{2m} + V(\mathbf{r}, t) \tag{6.17}$$

The wave equation which is appropriate for the description of a particle moving in a conservative force field is

$$i\hbar\frac{\partial\psi}{\partial t} = -\frac{\hbar^2}{2m}\nabla^2\psi + V\psi \tag{6.18}$$

The steps leading to the Schrödinger wave equation (6.18) should in no way be construed to be a derivation. It is not possible to derive the wave equation except insofar as it may be a special case of some more general theory. The form of the wave equation has been shown to be plausible, but its validity must rest on the experimental verification of the results it predicts. An important feature of the wave equation is the appearance of $i = \sqrt{-1}$. There is no classical theory in which of necessity the imaginary unit i appears in a fundamental equation.

Formally, the total energy and linear momentum of the particle are represented by the operators

$$W \leftrightarrow i\hbar\frac{\partial}{\partial t} \qquad \mathbf{p} \leftrightarrow -i\hbar\nabla \tag{6.19}$$

The operator concept is of great importance in the development of quantum mechanics. The *Hamiltonian operator* for a particle moving in a conservative force field is defined to be

$$H = -\frac{\hbar^2}{2m}\nabla^2 + V \tag{6.20}$$

The Schrödinger equation expressed in terms of the Hamiltonian is

$$i\hbar\frac{\partial\psi}{\partial t} = H\psi \tag{6.21}$$

In the form (6.21), the Schrödinger equation is quite general and applies to systems other than a single particle moving in a conservative force field, provided that a Hamiltonian appropriate to the physical system under consideration is used.

With every physical theory we have so far developed, there has been associated a conservation theorem. In Chapter 3, equation (5.12) expresses conservation of mass in fluid flow; (6.13) expresses conservation of energy in heat flow; (12.6) expresses conservation of charge; (12.30) expresses conservation of energy in the Maxwell field. In this chapter, equation (3.36) expresses conservation of mechanical energy in a vibrating membrane. It is not surprising that there is a conservation theorem associated with the Schrödinger wave equation. We begin its discussion by establishing an identity. The wave equation multiplied by ψ^* is

$$i\hbar\psi^*\frac{\partial\psi}{\partial t} = \psi^* H\psi \tag{6.22}$$

By taking the complex conjugate of (6.22), we find

$$-i\hbar\psi\frac{\partial\psi^*}{\partial t} = (H\psi)^*\psi \tag{6.23}$$

If (6.23) is subtracted from (6.22), the resulting equation can be written as

$$i\hbar\frac{\partial}{\partial t}(\psi^*\psi) = \psi^* H\psi - (H\psi)^*\psi \tag{6.24}$$

In terms of the Hamiltonian (6.20), this is

$$\begin{aligned} i\hbar\frac{\partial}{\partial t}(\psi^*\psi) &= -\frac{\hbar^2}{2m}(\psi^*\nabla^2\psi - \psi\nabla^2\psi^*) \\ &= -\frac{\hbar^2}{2m}\nabla\cdot(\psi^*\nabla\psi - \psi\nabla\psi^*) \end{aligned} \tag{6.25}$$

provided that $V(\mathbf{r}, t)$ is real. Equation (6.25) can be put in the form

$$\nabla\cdot\mathbf{S} + \frac{\partial P}{\partial t} = 0 \tag{6.26}$$

where

$$P = \psi^*\psi \qquad \mathbf{S} = \frac{\hbar}{2im}(\psi^*\nabla\psi - \psi\nabla\psi^*) \tag{6.27}$$

The function P is the probability density for the particle and $\mathbf{S}$ represents probability current. The physical content of (6.26) is conservation of probability. As the wave function varies in time, P increases in some regions and decreases in others while the total probability of finding the particle somewhere never changes. We are assuming here that the discussion is confined to stable particles which do not decay into other quantities. It is usual to normalize the wave function so that

$$\int_\Sigma \psi^*\psi\, d\Sigma = 1 \tag{6.28}$$

The integral is extended over a region of space which is sufficiently large so that the probability of finding the particle outside the region is nil. In other words, the particle is localized in space. Mathematically speaking, the wave function is such that (6.28) converges when the integral is extended over all space. We have already encountered a wave function for which (6.28) does not hold, namely, the plane wave (6.8). No particle is actually described by a plane wave stretching from $-\infty$ to $+\infty$. If the particle exists in the beam of a linear accelerator which is 1 mile long, the plane wave description is useful and accurate but not exact. The wave train describing the particle eventually falls to zero *somewhere* and is really a *wave packet*. More will be said about wave packets later.

The quantity in parentheses in (6.27) is the difference between a number and its complex conjugate. The probability current vector therefore comes out to be a real quantity. It is of interest to calculate it for the plane wave (6.8). The result is

$$\mathbf{S} = \frac{\hbar \boldsymbol{\kappa}}{m}|A|^2 = \frac{\mathbf{p}}{m}|A|^2 = \mathbf{u}|A|^2 \tag{6.29}$$

where $\mathbf{u}$ is the classical velocity of the particle. Some enlightenment on the meaning of this result and how it can be used will be found in the next section.

Another result of interest can be obtained from (6.24) by integrating it over a region Σ of space. We find

$$i\hbar \frac{d}{dt} \int_\Sigma \psi^* \psi \, d\Sigma = \int_\Sigma \left[\psi^* H \psi - (H\psi)^* \psi \right] d\Sigma \tag{6.30}$$

If the region of integration is taken sufficiently large, then (6.28) holds, and

$$\int_\Sigma (H\psi)^* \psi \, d\Sigma = \int_\Sigma \psi^* H \psi \, d\Sigma \tag{6.31}$$

Operators that obey this condition are said to be *Hermitian*. The hermiticity property of the Hamiltonian depends on two factors: the Schrödinger equation (6.21) and the convergence of the integral (6.28). In other words, the hermiticity property of an operator depends not only on the form of the operator but also on the class of functions on which it operates. Hermitian operators play a fundamental role in quantum mechanics.

PROBLEMS

6.1. Electrons are accelerated through a potential difference of 25 volts. What is their de Broglie wave length? In one experiment, the electrons are projected into a region of space where there is a magnetic field of strength 0.001 tesla and in another, they are allowed to scatter from a crystal in which the separation of the atoms is about 3 Å (Å $= 10^{-8}$ cm). Explain qualitatively what happens in each of the two experiments. In which of the two are quantum effects important? (For the electron, $m = 9.11 \times 10^{-31}$ kg and $e = 1.602 \times 10^{-19}$ C.)

6.2. The wave function $\psi = A \sin(\boldsymbol{\kappa} \cdot \mathbf{r} - \omega t)$ obeys the wave equation

$$\nabla^2 \psi - \frac{1}{v^2} \frac{\partial^2 \psi}{\partial t^2} = 0 \tag{6.32}$$

with $v = \omega / \kappa$. Why is this not a satisfactory wave equation for matter waves that describe a free particle?

6.3. Prove that the linear momentum operator $\mathbf{p} = -i\hbar \nabla$ is Hermitian:

$$\int_\Sigma \psi^* \mathbf{p} \psi \, d\Sigma = \int_\Sigma (\mathbf{p}\psi)^* \psi \, d\Sigma \tag{6.33}$$

In doing the proof, assume that Σ is sufficiently large so that $\psi = 0$ over its bounding surface. *Hint:* The identity (9.14) of Chapter 3 is useful.

6.4. If the intensity function

$$I = I_0 \frac{\sin^2 y}{y^2} \tag{6.34}$$

is to be used as a probability distribution function, it should be normalized so that

$$\int_{-\infty}^{+\infty} I(y) \, dy = 1 \tag{6.35}$$

Determine I_0 so that (6.35) is satisfied. (Evaluate the integral; don't just look it up in a table.)

Show that

$$\frac{\sin^2 y}{y^2} = 2 \sum_{k=0}^{\infty} \frac{(-1)^k (2y)^{2k}}{(2k+2)!} \tag{6.36}$$

$$\int_0^x \frac{\sin^2 y}{y^2} dy = \sum_{k=0}^{\infty} \frac{(-1)^k (2x)^{2k+1}}{(2k+1)(2k+2)!} \tag{6.37}$$

A numerical evaluation using 12 terms of (6.37) gives for $x = \pi$

$$\int_0^{\pi} \frac{\sin^2 y}{y^2} dy = 1.41815157 \pm 8.9 \times 10^{-9} \tag{6.38}$$

The error estimate is the value of the first neglected term. In a single slit diffraction experiment involving material particles, what fraction of the particles falls in the region of the central maximum?

6.5. The Hamiltonian appears also in classical mechanics. It is defined by

$$H = p_\alpha \dot{q}_\alpha - L \tag{6.39}$$

where the sum over α runs from 1 to f, p_α is the canonical momentum conjugate to the coordinate q_α defined by

$$p_\alpha = \frac{\partial L}{\partial \dot{q}_\alpha} \tag{6.40}$$

and L is the Lagrangian. The Lagrangian for a conservative system with f degrees of freedom is

$$L = \tfrac{1}{2} g_{\alpha\beta} \dot{q}_\alpha \dot{q}_\beta - V(q_\alpha) \tag{6.41}$$

Show that for this Lagrangian,

$$H = \tfrac{1}{2} g_{\alpha\beta} \dot{q}_\alpha \dot{q}_\beta + V = K + V \tag{6.42}$$

For a conservative system, the Hamiltonian is the total energy.

6.6. Suppose that the Hamiltonian (or any other operator) is Hermitian with respect to the two functions ψ and ϕ:

$$\int_\Sigma \psi^* H\psi \, d\Sigma = \int_\Sigma (H\psi)^* \psi \, d\Sigma \qquad \int_\Sigma \phi^* H\phi \, d\Sigma = \int_\Sigma (H\phi)^* \phi \, d\Sigma \tag{6.43}$$

Show that as a consequence

$$\int_\Sigma \psi^* H\phi \, d\Sigma = \int_\Sigma (H\psi)^* \phi \, d\Sigma \tag{6.44}$$

Note that conversely (6.44) implies (6.43).

Hint: Let λ be any complex number. Equation (6.43) implies that

$$\int_\Sigma (\phi + \lambda\psi)^* H(\phi + \lambda\psi) \, d\Sigma = \int_\Sigma [H(\phi + \lambda\psi)]^* (\phi + \lambda\psi) \, d\Sigma \tag{6.45}$$

Expand (6.45). In the resulting equation, first let $\lambda = a$, then let $\lambda = ib$. Add the two equations.

7. POTENTIAL BARRIER

If the Hamiltonian in the Schrödinger equation (6.21) is independent of the time, it is possible to separate the space and time variables by writing the wave function as $\psi(\mathbf{r}, t) = u(\mathbf{r})f(t)$. The wave equation gives

$$i\hbar\frac{f'(t)}{f(t)} = \frac{Hu(\mathbf{r})}{u(\mathbf{r})} \tag{7.1}$$

Each side can be set equal to a separation constant W giving

$$i\hbar f'(t) = Wf(t) \tag{7.2}$$

$$Hu(\mathbf{r}) = Wu(\mathbf{r}) \tag{7.3}$$

The solution of (7.2) is

$$f(t) = e^{-iWt/\hbar} = e^{-i\omega t} \tag{7.4}$$

The identification of the separation constant as the total energy is consistent with the Planck relation $W = \hbar\omega$. Equation (7.3) is an eigenvalue problem in a form first encountered as the matrix equation (10.8) of Chapter 2. We have already seen several examples in this chapter of eigenvalue problems that come from differential equations when certain boundary conditions are imposed. In (7.3), the Hamiltonian is the total energy operator and W is a possible energy eigenvalue. A definite energy eigenvalue is possible only if the Hamiltonian is not time-dependent.

We will consider a one-dimensional problem in which the potential function is the step function shown in Fig. 7.1 and given by

$$\begin{aligned} V(x) &= 0 \qquad x < 0 \\ V(x) &= V_0 \qquad x \geq 0 \end{aligned} \tag{7.5}$$

From (7.3) and (6.20), we find

$$\frac{d^2u}{dx^2} + \frac{2mW}{\hbar^2}u = 0 \qquad x < 0 \tag{7.6}$$

$$\frac{d^2u}{dx^2} + \frac{2m(W - V_0)}{\hbar^2}u = 0 \qquad x \geq 0 \tag{7.7}$$

Assume that initially particles with definite energy and momentum are traveling toward the origin on the negative x axis. The appropriate solution of (7.6) to represent such particles is

$$u(x) = Ae^{i\kappa_1 x} \qquad \kappa_1 = \sqrt{\frac{2mW}{\hbar^2}} \tag{7.8}$$

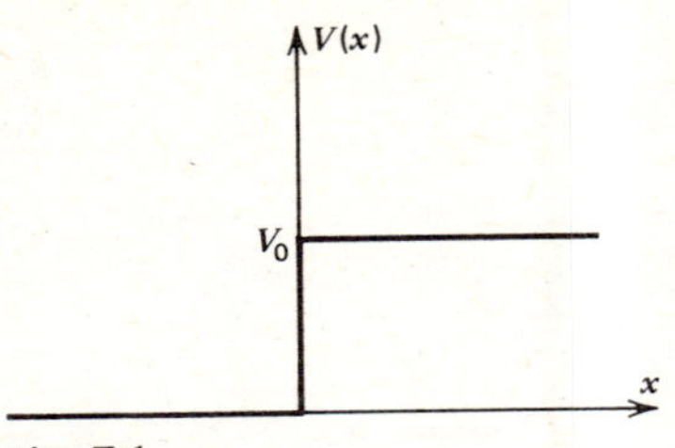

Fig. 7.1

If the time-dependent part is included, the wave function is

$$\psi(x, t) = Ae^{i(\kappa_1 x - \omega t)} \tag{7.9}$$

which is the familiar plane wave. We have been adequately prepared by examples from classical physics to correctly guess that the plane wave is partially transmitted and partially reflected by the potential step at the origin. If $W > V_0$, then the complete time-independent wave function which describes this reflection and transmission is

$$u_1 = Ae^{i\kappa_1 x} + Be^{-i\kappa_1 x} \qquad x < 0 \tag{7.10}$$

$$u_2 = Ce^{i\kappa_2 x} \qquad \kappa_2 = \sqrt{\frac{2m(W - V_0)}{\hbar^2}} \qquad x \geq 0 \tag{7.11}$$

It should be emphasized that these are single-particle wave functions. A single particle of course cannot be both reflected and transmitted. The correct interpretation comes from the probability current vector as given by (6.29) for plane waves. If there are a lot of particles in the beam, each described by the wave functions (7.10) and (7.11), it is possible to set $|A|^2$ equal to the particle density calculated as number of particles per unit volume in the incident beam. The probability current vector $S_A = \mathbf{u}|A|^2$ then gives the incident particle flux calculated as particles per second per unit area. The ratio

$$P_R = \frac{S_B}{S_A} = \frac{|B|^2}{|A|^2} \tag{7.12}$$

gives the fraction of the incident particles which are reflected and

$$P_T = \frac{S_C}{S_A} = \frac{\kappa_2|C|^2}{\kappa_1|A|^2} \tag{7.13}$$

gives the fraction transmitted into the region $x > 0$. In terms of a single-particle interpretation, P_R is the probability that a given particle will be reflected and P_T is the probability that it will be transmitted. Another way to look at it is to say that a particle which is originally in the state described by $Ae^{i\kappa_1 x}$ can decay into two possible states, one given by $Be^{-i\kappa_1 x}$ and the other by $Ce^{i\kappa_2 x}$. It is only possible to give the relative probability of the two final outcomes. We cannot make a definite prediction as to the final state of the particle. This is in sharp contrast to classical physics where, in principle, the exact outcome of any experiment can be predicted. In the present example, classical mechanics predicts that all the particles are transmitted into the region $x > 0$.

What kind of boundary conditions are to be imposed on the wave functions at $x = 0$? Part of the answer comes from the time-independent wave equation

$$\frac{d^2u}{dx^2} + \frac{2m\{W - V(x)\}}{\hbar^2}u = 0 \tag{7.14}$$

By integrating over the range $x = 0 - \varepsilon$ to $x = 0 + \varepsilon$, we get

$$\left.\frac{du_2}{dx}\right|_{+\varepsilon} - \left.\frac{du_1}{dx}\right|_{-\varepsilon} + \frac{2m}{\hbar^2}\int_{-\varepsilon}^{+\varepsilon}\{W - V(x)\}u\,dx \tag{7.15}$$

If the functions in the integrand are piecewise continuous, then in the limit $\varepsilon \to 0$

$$\left.\frac{du_2}{dx}\right|_0 = \left.\frac{du_1}{dx}\right|_0 \tag{7.16}$$

More insight into the boundary conditions can be obtained from the conservation of probability theorem. If (6.26) is integrated over a finite region Σ of space and Gauss' divergence theorem is used, the result is

$$\int_\sigma \mathbf{S}\cdot d\boldsymbol{\sigma} = -\frac{d}{dt}\int_\Sigma P\,d\Sigma \tag{7.17}$$

Equation (7.17) can be applied to a small pillbox with one face at $x = \varepsilon/2$ and the other at $x = -\varepsilon/2$. We find

$$(S_{2x} - S_{1x})\,\Delta\sigma = -\frac{d}{dt}P\,\Delta\sigma\varepsilon \tag{7.18}$$

where $\Delta\sigma$ is the area of the faces of the box and ε is its thickness in the x direction. We have used the pillbox idea before; see problems 8.4 and 10.10 in Chapter 3. In the limit as $\varepsilon \to 0$, $S_{2x} = S_{1x}$. Thus,

$$S_x = \frac{\hbar}{2im}\left(\psi^*\frac{\partial\psi}{\partial x} - \psi\frac{\partial\psi^*}{\partial x}\right) \tag{7.19}$$

is continuous. A discontinuity in S_x at $x = 0$ would imply some kind of a surface source or sink density for probability, meaning that particles could spontaneously appear or disappear. The continuity of (7.19) is assured if both the wave function and its gradient are continuous at $x = 0$:

$$u_1(0) = u_2(0) \qquad \left.\frac{du_1}{dx}\right|_0 = \left.\frac{du_2}{dx}\right|_0 \tag{7.20}$$

Application of these boundary conditions to the wave functions (7.10) and (7.11) yields

$$A + B = C \qquad \kappa_1(A - B) = \kappa_2 C \tag{7.21}$$

By solving for B and C in terms of A, we find

$$B = \frac{\kappa_1 - \kappa_2}{\kappa_1 + \kappa_2}A \qquad C = \frac{2\kappa_1 A}{\kappa_1 + \kappa_2} \tag{7.22}$$

The reflection and transmission probabilities as defined by (7.12) and (7.13) come out to be

$$P_R = \left(\frac{\kappa_1 - \kappa_2}{\kappa_1 + \kappa_2}\right)^2 \qquad P_T = \frac{4\kappa_1\kappa_2}{(\kappa_1 + \kappa_2)^2} \tag{7.23}$$

Note that $P_R + P_T = 1$, as required. The potential step problem should be compared to the reflection and transmission of a plane wave at the junction between two stretched strings of unequal density as given in problem 2.4. Even though the two physical situations and the interpretation of the solutions are entirely different, the mathematics is practically the same in the two cases.

If the energy of the incident particles is less than V_0, equation (7.7) can be expressed in the form

$$\frac{d^2u}{dx^2} - \alpha^2 u = 0 \qquad \alpha = \sqrt{\frac{2m(V_0 - W)}{\hbar^2}} \tag{7.24}$$

The wave function for $x < 0$ is still given by (7.10); but for $x \geq 0$, the appropriate solution is

$$u_2 = Ce^{-\alpha x} \qquad x \geq 0 \tag{7.25}$$

Another possible solution is $e^{+\alpha x}$, but this is discarded because it becomes arbitrarily large as x increases. By application of the boundary conditions (7.20), we find

$$A + B = C \qquad i\kappa_1(A - B) = -\alpha C \tag{7.26}$$

The solution for B and C in terms of A is

$$B = \frac{\kappa_1 - i\alpha}{\kappa_1 + i\alpha} A \qquad C = \frac{2A\kappa_1}{\kappa_1 + i\alpha} \tag{7.27}$$

The probability current for the wave function in the region $x \geq 0$ is found from (7.19) to be zero. This is confirmed by computing the reflection probability from (7.27):

$$P_R = \left| \frac{A}{B} \right|^2 = 1 \tag{7.28}$$

If $W \leq V_0$, all particles are reflected from the potential step, as is also true classically. The region $x \geq 0$ is a classically forbidden region, since $W = K + V_0$ implies that the kinetic energy of the particles is negative. In the quantum mechanical treatment, something very nonclassical happens. A wave function exists for $x \geq 0$. There is a finite probability of finding the particle in the classically forbidden region.

Something analogous happens when an electromagnetic wave is incident on a conductor. The wave penetrates the conductor with an amplitude which decreases exponentially with distance into the conductor. The reflected intensity as measured by the Poynting vector is less than the incident intensity because the conductor soaks up some of the energy in the form of Joule heating. Nothing analogous to this happened in our treatment of matter waves. The probability of finding the particle is conserved. If we think of the interaction of an electromagnetic wave with a conductor in terms of photons, then we do have a situation where the photons, considered as particles, are annihilated. This does not happen with material particles until the energy exchanges become of the same order of magnitude as the rest energies of the particles. A treatment by relativistic quantum mechanics is then required. Photons are particles with zero rest mass traveling at speed c and are intrinsically relativistic.

Another mathematically analogous situation can be found by joining together two periodic structures consisting of masses and springs or inductors and capacitors as shown in Fig. 7.2. The first line is effectively terminated by the characteristic impedance of the second line so that the amplitude of the reflected wave at the junction as given by equation (1.50) is

$$B = A\frac{Z_0 - Z_0'}{Z_0^* + Z_0'} \tag{7.29}$$

The characteristic impedance of the second section of the periodic structure is

$$Z_0' = i\omega\frac{L'}{2} + \sqrt{\frac{L'}{C'}}\sqrt{1 - \omega^2\frac{L'C'}{4}} \tag{7.30}$$

The critical frequency for the second section is

$$\omega_c = \frac{2}{\sqrt{L'C'}} \tag{7.31}$$

If $\omega < \omega_c$, then a plane wave incident on the junction

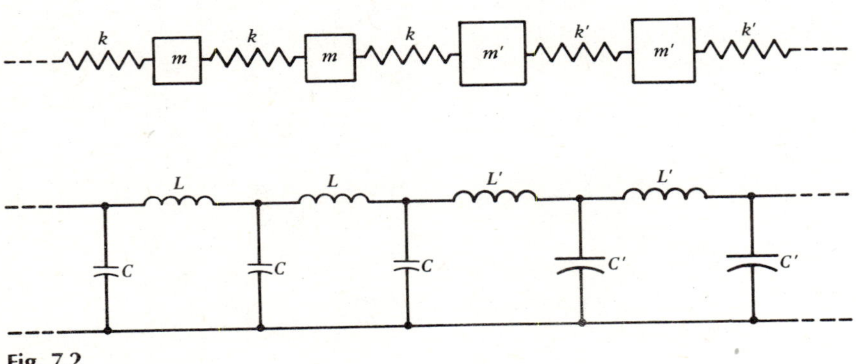

Fig. 7.2

from the left is partially transmitted and partially reflected. If $\omega > \omega_c$, then Z_0' is purely imaginary and $|B/A| = 1$. All of the energy in the incident wave is reflected. There is a solution in the second section which decays exponentially with distance as discussed in problem 1.4. It is quite interesting that the behavior of a plane wave at the junction of two periodic lattices is so similar to the behavior of a matter wave when it encounters a potential step. This in no way implies that the classical system is some kind of a model for the quantum mechanical system. The two situations are quite different.

PROBLEMS

7.1. Show that the probability density function for $W < V_0$ is

$$\begin{aligned} \psi\psi^* &= 4|A|^2 \cos^2 \phi \, e^{-2\alpha x} \qquad & x \geq 0 \\ &= 4|A|^2 \cos^2 (\kappa_1 x + \phi) \qquad & x < 0 \end{aligned} \tag{7.32}$$

where $\tan \phi = \alpha/\kappa_1$. If you graph this function for $W = 0.5V_0$, it comes out like Fig. 7.3. The wavelength is defined by $\lambda = 2\pi/\kappa_1$.

Express the ratio α/κ_1 in terms of V_0 and W. Qualitatively, what happens to the probability distribution function in the two limiting cases $W \to 0$ and $W \to V_0$? Make sketches for the two cases.

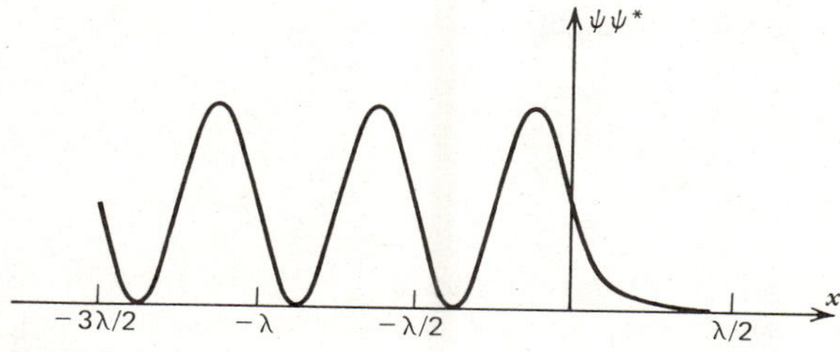

Fig. 7.3

7.2. It is of interest to consider what happens if a plane wave encounters a potential barrier as illustrated in Fig. 7.4 and defined by

$$\begin{aligned} V &= 0 \qquad & x < 0 \text{ and } x > a \\ V &= V_0 \qquad & 0 \leq x \leq a \end{aligned} \tag{7.33}$$

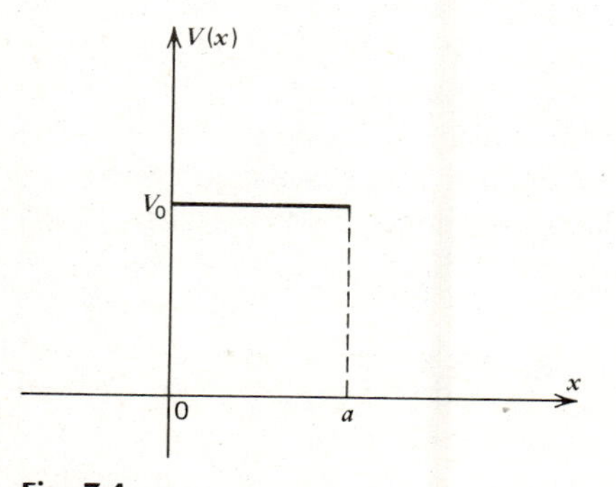

Fig. 7.4

To make the problem a little simpler, let's make the potential into the delta function

$$V(x) = U\delta(x) \tag{7.34}$$

where U is a constant. Show from (7.15) that

$$\left.\frac{du_2}{dx}\right|_0 - \left.\frac{du_1}{dx}\right|_0 = \frac{2mU}{\hbar^2}u(0) \tag{7.35}$$

Show that the probability current vector (7.19) is continuous provided that $u_2(0) = u_1(0)$. Find the transmission and reflection probabilities for a particle moving on the x axis with a definite energy. In the calculation, it is convenient to define $\kappa_0 = mU/\hbar^2$.

8. PARTICLE IN A BOX

If the potential V_0 of the potential step in Fig. 7.1 is large, then the parameter α defined by equation (7.24) is also large. Think of an idealization where V_0 and α become infinite. The wave function (7.25) for $x > 0$ then vanishes and the infinitely high potential wall at $x = 0$ becomes a perfect reflector for matter waves of any energy. The boundary condition at $x = 0$ becomes simply $\psi = 0$. Consider a rectangular parallelepiped-shaped region of space bounded by $x = 0$, $y = 0$, $z = 0$, $x = a$, $y = b$, and $z = c$. Suppose that there is a particle in the box and that $V(\mathbf{r}) = 0$ inside and $V(\mathbf{r}) = \infty$ outside. The walls of the box are infinite potential steps. Even though this is artificial physically, the idea is useful as a first approximation in the analysis of the free electron model of a metallic conductor. The boundaries of the conductor can be regarded as potential step functions. The problem is simple mathematically and gives some insight into the kinds of solutions to be expected from the Schrödinger equation. The time-independent part of the Schrödinger equation as given by (7.3) becomes

$$\nabla^2 u + \kappa^2 u = 0 \qquad \kappa^2 = \frac{2mW}{\hbar^2} \tag{8.1}$$

and is to be solved subject to the boundary condition $u = 0$ on the walls of the box. Equation (8.1) is a Helmholtz equation. We have already looked at its solutions in one dimension for the vibrating string with both ends fixed and in two dimensions for the vibrating rectangular membrane with fixed edges. Its solutions in three dimensions are obtained by extending the method that was used to solve (3.8) for the membrane. The result is the set of three-dimensional normalized eigenfunctions

$$u_{n_1 n_2 n_3} = \sqrt{\frac{8}{abc}}\,\sin\left(\frac{n_1\pi x}{a}\right)\sin\left(\frac{n_2\pi y}{b}\right)\sin\left(\frac{n_3\pi z}{c}\right) \tag{8.2}$$

where n_1, n_2, and n_3 can have the integer values $1, 2, 3, \ldots$. The possible wave numbers and energies are given by the eigenvalues

$$\kappa^2 = \frac{2mW}{\hbar^2} = \pi^2\left(\frac{n_1^2}{a^2} + \frac{n_2^2}{b^2} + \frac{n_3^2}{c^2}\right) \tag{8.3}$$

Even though the mathematics for the membrane and the particle in a box is the same, the implications of the final results are profoundly different. For the particle in a box, the possible energies are *quantized* and can have only the discrete values permitted by equation (8.3). For the vibrating string or membrane, the wave number is quantized just as it is in equation (8.3), but there is no direct connection between this fact and the energy of the system. If a string fixed at both ends is vibrating in its fundamental mode, its energy can be varied continuously by changing the amplitude of the oscillations. You should also compare (8.2) and (8.3) to your solution of problem 5.7 for electromagnetic waves inside a box made out of a perfect metallic conductor. There is a lot of similarity although the electromagnetic field is more complicated than the scalar field of the Schrödinger equation.

The solution of the problem of the potential step in section 7 did not produce discrete energy eigenvalues. The energy of the incoming wave can take on any value, meaning that the energy eigenvalues form a continuum. The solution of the eigenvalue problem as given by equation (7.3) is capable of yielding either a continuous or a discrete spectrum of eigenvalues depending on the boundary conditions which are imposed. It is possible for both to occur in the same problem.

The general solution of the Schrödinger equation for a particle in a box is a linear superposition of the possible eigenfunctions:

$$\psi(\mathbf{r},t)=\sum_{n_1=1}^{\infty}\sum_{n_2=1}^{\infty}\sum_{n_3=1}^{\infty}A_{n_1n_2n_3}u_{n_1n_2n_3}\times\exp\left(-iW_{n_1n_2n_3}t/\hbar\right) \tag{8.4}$$

The eigenfunctions obey the orthogonality relations

$$\int_0^a\int_0^b\int_0^c u_{n_1n_2n_3}u_{n_1'n_2'n_3'}dx\,dy\,dz=\delta_{n_1n_1'}\delta_{n_2n_2'}\delta_{n_3n_3'} \tag{8.5}$$

By setting $t=0$ in (8.4), we find

$$\psi(\mathbf{r},0)=\sum_{n_1=1}^{\infty}\sum_{n_2=1}^{\infty}\sum_{n_3=1}^{\infty}A_{n_1n_2n_3}u_{n_1n_2n_3} \tag{8.6}$$

By the same procedure that was used to obtain (3.20), the coefficients are found to be

$$A_{n_1n_2n_3}=\int_0^a\int_0^b\int_0^c\psi(\mathbf{r},0)\,u_{n_1n_2n_3}dx\,dy\,dz \tag{8.7}$$

In the case of either the string or the membrane, it was necessary to know both the function and its first derivative with respect to time in order to complete the solution. For the Schrödinger equation, only the function itself need be known at $t=0$. The difference is due to the fact that the wave equations for the string and the membrane contain a second derivative with respect to time, whereas the Schrödinger equation contains a first derivative.

To simplify the discussion in the remainder of this section, we will look at a one-dimensional version of the particle in a box. Consider a particle moving on the x axis in the potential defined by $V(x)=0$ if $0<x<a$ and $V(x)=\infty$ outside of this range. This is a *square well potential* with infinitely high walls. The wave function is

$$\psi(x,t)=\sum_{n=1}^{\infty}A_nu_n(x)\exp(-i\omega_nt)$$
$$u_n(x)=\sqrt{\frac{2}{a}}\sin\left(\frac{n\pi x}{a}\right) \tag{8.8}$$

$$W_n=\hbar\omega_n=\frac{n^2\pi^2\hbar^2}{2ma^2} \tag{8.9}$$

The normalization requirement puts a condition on the expansion coefficients:

$$\int_0^a\psi^*\psi\,dx=\int_0^a\sum_{n=1}^{\infty}\sum_{n'=1}^{\infty}A_n^*u_n^*A_{n'}u_{n'}\,dx$$
$$=\sum_{n=1}^{\infty}\sum_{n'=1}^{\infty}A_n^*A_{n'}\delta_{nn'}=\sum_{n=1}^{\infty}|A_n|^2=1 \tag{8.10}$$

(In this example, the eigenfunctions are real, but this is not usually true.) The wave function is a superposition of eigenfunctions of all the possible allowed energies. If the energy of the system is measured, the value obtained must be one of the eigenvalues as given by (8.9). The expansion coefficients have the significance that $|A_n|^2$ is the probability that the energy eigenvalue W_n will be measured. To take a simple example, suppose that there are only two terms in the eigenfunction expansion (8.8) so that the wave function is

$$\psi=A_1u_1e^{-i\omega_1t}+A_2u_2e^{-i\omega_2t} \tag{8.11}$$

Equation (8.10) requires

$$|A_1|^2+|A_2|^2=1 \tag{8.12}$$

If we choose $A_1=A_2=1/\sqrt{2}$, then

$$\psi=\frac{1}{\sqrt{2}}u_1e^{-i\omega_1t}+\frac{1}{\sqrt{2}}u_2e^{-i\omega_2t} \tag{8.13}$$

The system has been prepared in such a way that its wave function is a mixture of equal parts of the first two eigenfunctions. The exact energy is not known, but there is an equal probability of measuring either W_1 or W_2 and a zero probability of measuring any other energy. If a determination of the energy is made and found to be W_1, then the wave function is no longer (8.13) but is

$$\psi = u_1 e^{-i\omega_1 t} \tag{8.14}$$

Because the Hamiltonian for the particle in a box is independent of the time, the wave function, once established, remains the same superposition of eigenfunctions until the system is disturbed in some way. The disturbance could be someone making a measurement on the system. The disturbance introduces time-dependent terms in the Hamiltonian, and transitions can occur from one energy level to another. The stationary states of the time-independent Hamiltonian are now only approximately stationary.

An atom, such as the hydrogen atom which we will discuss later, has a definite set of possible energy levels. It is usual to refer to the lowest possible energy level as the *ground state* and the other states as excited states. According to the simple picture of quantum mechanics that we have developed, an atom in one of its excited states would just sit there indefinitely. What actually happens is that the atom sooner or later makes a transition to a lower energy state with the emission of a photon. One mechanism for the decay is *stimulated emission* caused by a time-dependent perturbation term in the Hamiltonian as explained in the last paragraph. The perturbation could be caused, for example, by an external electromagnetic field. The atom can also absorb energy from the electromagnetic wave and go into a higher excited state. There is another important process called *spontaneous emission* which happens even in the absence of external effects. Spontaneous emission cannot be properly understood without a development of the theory of creation and annihilation of photons.

PROBLEMS

8.1. The one-dimensional square well shown in Fig. 8.1 is defined by

$$V(x) = 0 \qquad |x| \geq a/2$$
$$V(x) = -V_0 \qquad -a/2 < x < a/2$$

A quantum mechanical particle moving in this potential can have a continuum of energy eigenvalues as long as $W > 0$. Plane waves incident on the well with $W > 0$ are partially reflected and partially transmitted. If $-V_0 < W < 0$, then *bound states* occur with only certain energy eigenvalues allowed. This is an example in which the Hamiltonian has both a continuous and a discrete spectrum of energy eigenvalues. The wave function is oscillatory in the range $-a/2 < x < a/2$ and decays exponentially if $|x| \geq a/2$, so that the probability of finding the particle is substantial only in the vicinity of the well. What is the behavior of a particle according to *classical* mechanics in this potential? Consider both positive and negative values of the energy of the particle. Show that if $-V_0 < W < 0$, a classical particle oscillates back and forth between $-a/2$ and $a/2$ with the frequency

$$\omega = \frac{\pi}{a}\sqrt{\frac{2(W+V_0)}{m}} \tag{8.15}$$

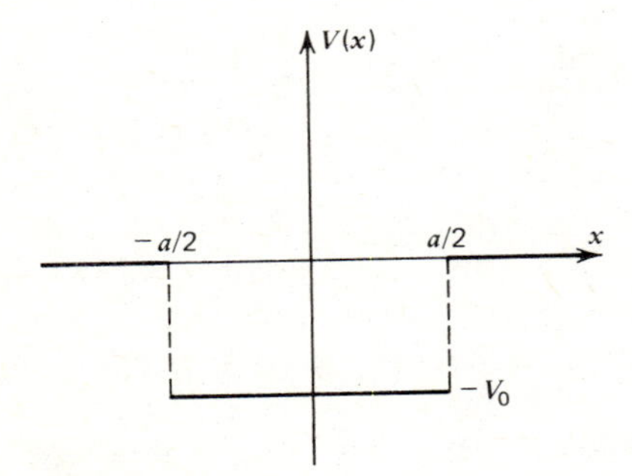

Fig. 8.1

8.2. Show that the probability density computed from the wave function (8.13) is

$$\psi^*\psi = \frac{1}{a}\sin^2\left(\frac{\pi x}{a}\right) + \frac{1}{a}\sin^2\left(\frac{2\pi x}{a}\right) + \frac{2}{a}\sin\left(\frac{\pi x}{a}\right)\sin\left(\frac{2\pi x}{a}\right)\cos(3\omega_1 t) \tag{8.16}$$

The time-dependent term $\cos(3\omega_1 t)$ goes through a complete cycle for $0 < \omega_1 t < 2\pi/3$. Figure 8.2 shows graphs of (8.16) for five equally spaced times covering one half of a cycle of the motion. The wave function has been constructed so that initially there is a high probability of finding the particle in the first half of the well. As time goes on, the probability density oscillates back and forth. Show that if $W = (W_1 + W_2)/2$ and $\omega = 3\omega_1$, then

$$\omega = \frac{3\pi}{\sqrt{10}\,a}\sqrt{\frac{2W}{m}} \tag{8.17}$$

which corresponds roughly to the classical frequency (8.15).

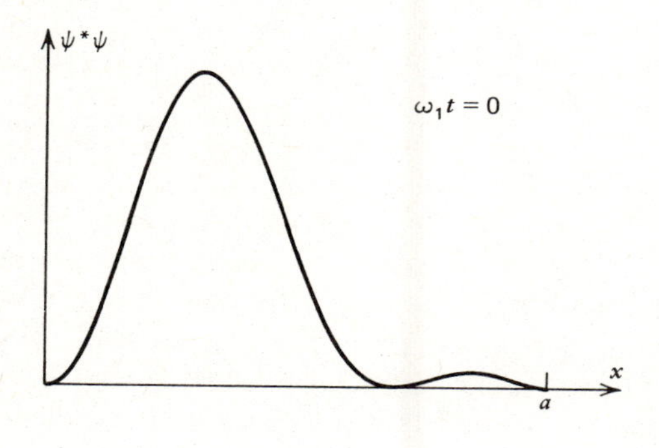

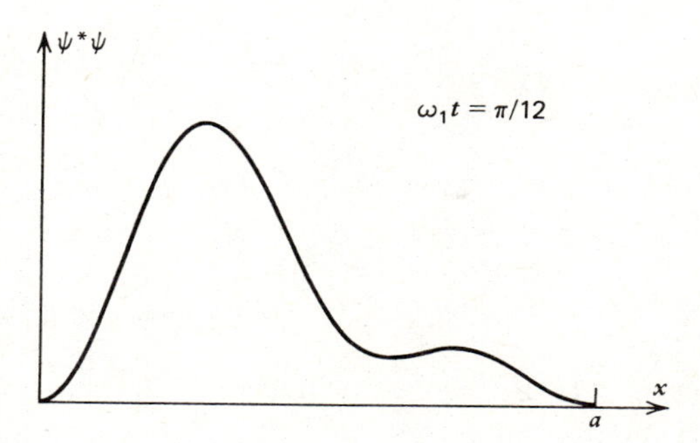

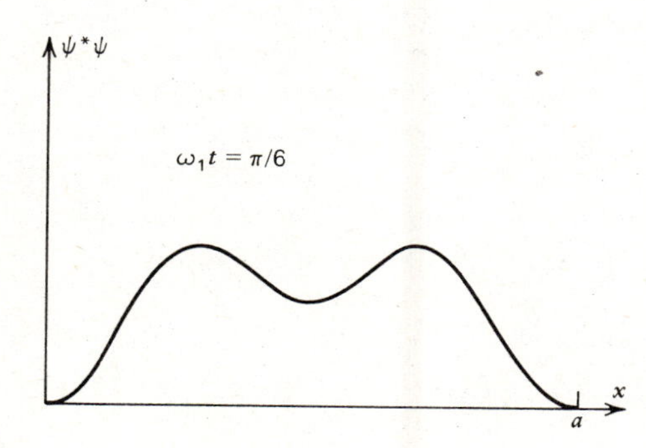

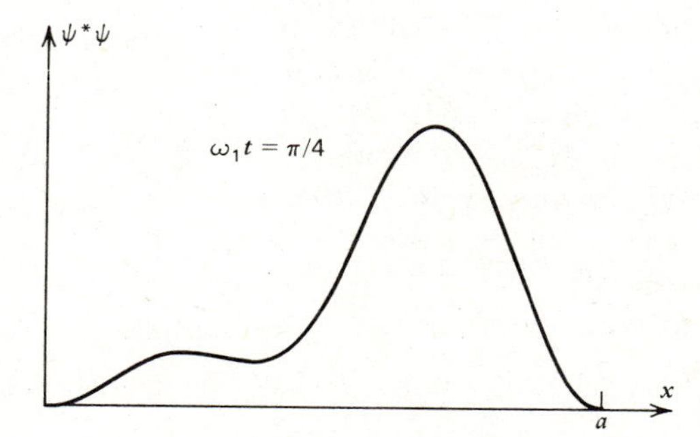

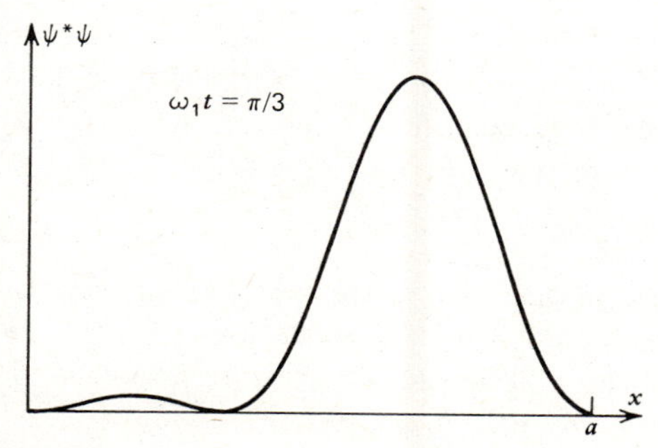

Fig. 8.2

8.3. For a particle in a three-dimensional box, there is a lot of degeneracy if $a=b=c$. Enumerate all the states which give $n_1^2+n_2^2+n_3^2=6$, 9, and 14.

8.4. The potential function for a particle constrained to move in one dimension is given by $V(x)=-U\delta(x)$, where U is a positive constant. Show that only one bound state is possible and that the energy for this state is given by

$$E=-\frac{mU^2}{2\hbar^2} \tag{8.18}$$

Show that the normalized time-independent wave function is given by

$$u(x)=\sqrt{\kappa}\,e^{-\kappa|x|} \qquad \kappa=\sqrt{\frac{mU}{\hbar^2}} \tag{8.19}$$

See equation (7.35).

9. ONE-DIMENSIONAL WAVES IN ELASTIC MEDIA

If a material medium is put under simple tension or compression as illustrated in Fig. 9.1, it will undergo a distortion which to a first approximation is directly proportional to the applied force F. The material can be visualized as a bar of length s and cross-sectional area σ. The *stress* is defined to be the ratio of the tension (or compression) to the cross-sectional area: F/σ. The stress is a pure tension or compression and involves no twisting or bending. The strain is a measure of the distortion of the medium and is defined to be the ratio of the change in the length of the rod to its original length: $\Delta s/s$. We assume that the material is *elastic* and will return to its original shape after the stress is removed. This is true of many substances provided that the stress is not so large that there is permanent distortion. Under these circumstances, stress and strain are proportional:

$$\frac{F}{\sigma}=Y\frac{\Delta s}{s} \tag{9.1}$$

The constant of proportionality Y has units of force per unit area and is called *Young's modulus*. Its value is characteristic of the material.

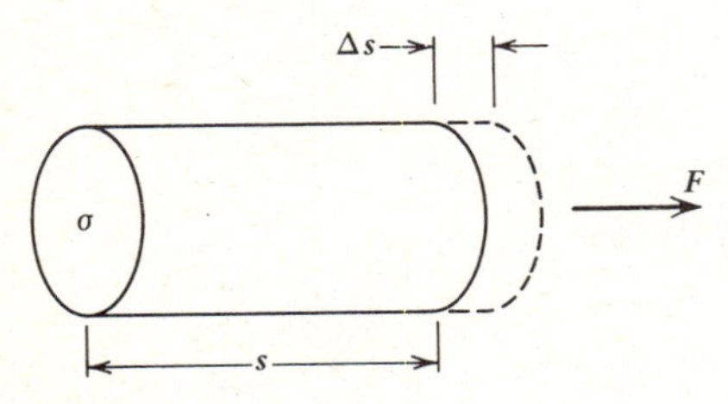

Fig. 9.1

It is possible to construct a model of an elastic material as a limiting case of the system of masses connected by springs as shown in Fig. 9.2. The coordinates η_n are measured from the equilibrium positions of the masses. The masses all move along the same straight line so that the system is one-dimensional. The distances Δx_n are the equilibrium separations between the masses. In the limiting process, the separations Δx_n are allowed to approach zero at the same time that the number of masses approaches infinity. The masses and springs are not identical. We will assume that the difference between force constants of adjacent springs approaches zero in the limiting process so that the force constant approaches a continuous function of x. Similarly, the masses are assumed to approach a continuous density function. The equation of motion of mass m_n is

$$k_{n+1}(\eta_{n+1}-\eta_n)-k_n(\eta_n-\eta_{n-1})=m_n\ddot{\eta}_n \tag{9.2}$$

The force which exists in a given spring is

$$F_n=k_n(\eta_n-\eta_{n-1})=k_n\,\Delta\eta_n \tag{9.3}$$

The strain is $\Delta\eta_n/\Delta x_n$. According to the definition of Young's modulus as given by (9.1),

$$\frac{k_n\,\Delta\eta_n}{\sigma}=Y_n\frac{\Delta\eta_n}{\Delta x_n} \qquad Y_n=\frac{k_n\,\Delta x_n}{\sigma} \tag{9.4}$$

In a limiting process where the number of masses increases and $\Delta x_n\to 0$, it is Y_n as given by (9.4) which approaches a finite limit and has the interpretation of Young's modulus for a continuous medium. We

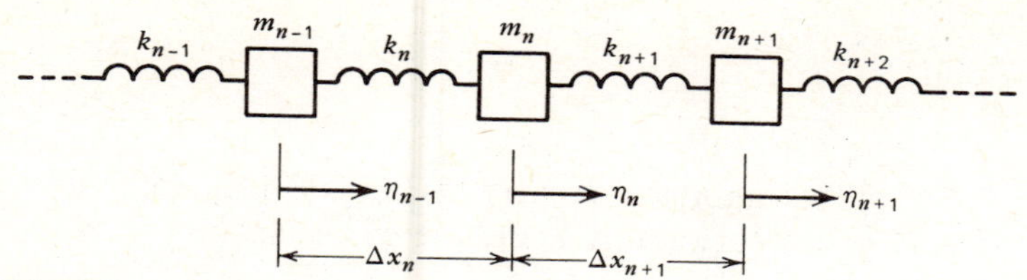

Fig. 9.2

therefore use (9.4) to eliminate the spring constants from (9.2) and express the resulting equation as

$$\frac{Y_{n+1}\left(\frac{\eta_{n+1}-\eta_n}{\Delta x_{n+1}}\right)-Y_n\left(\frac{\eta_n-\eta_{n-1}}{\Delta x_n}\right)}{\Delta x_n}=\frac{m_n}{\sigma\Delta x_n}\ddot{\eta}_n \tag{9.5}$$

As the limit is taken, the density of the material in units of mass per unit volume is

$$\rho(x)=\lim_{x_n\to 0}\frac{m_n}{\sigma\Delta x_n} \tag{9.6}$$

The limiting form of (9.5) is

$$\frac{\partial}{\partial x}\left(Y\frac{\partial\eta}{\partial x}\right)=\rho\frac{\partial^2\eta}{\partial t^2} \tag{9.7}$$

which is an appropriate differential equation for describing longitudinal or compressional waves in a one-dimensional elastic medium. The medium is not necessarily homogeneous. Both Young's modulus and the density can be functions of x. When general three-dimensional distortions of an elastic medium are analyzed, the situation is more complicated. Both longitudinal and transverse waves can exist.

The time and space variables in (9.7) can be separated by writing $\eta(x,t)=u(x)f(t)$. The result is

$$\frac{1}{\rho u}\frac{d}{dx}\left(Y\frac{du}{dx}\right)=\frac{1}{f}\frac{d^2f}{dt^2}=-\omega^2 \tag{9.8}$$

where ω^2 is the separation constant. The function $u(x)$ obeys

$$\frac{d}{dx}\left(Y\frac{du}{dx}\right)+\omega^2\rho u=0 \tag{9.9}$$

It would be a formidable task to proceed as we did in section 1 and find the exact equations of motion of the discrete system shown in Fig. 9.2. Think of trying to solve equation (14.10) in Chapter 4 if G and A are f-dimensional matrices containing as elements masses and spring constants which are all different. Fortunately, we know qualitatively what would happen without actually finding the exact solution. There would be f eigenvalues giving f normal modes of oscillation and f mutually orthogonal eigenvectors. We therefore anticipate that a differential equation of the form (9.9), which is a limiting case of such a system, will yield an infinite number of eigenvalues and mutually orthogonal eigenfunctions if the boundary conditions are appropriately chosen. Because of their physical interpretation, the functions $Y(x)$ and $\rho(x)$ are positive throughout the domain of the variable x.

It is fairly easy to derive (9.7) by a direct consideration of a continuous elastic medium. The medium is assumed to be in the shape of a rod of cross-sectional area σ. We consider only compressions and tensions in the rod and omit the possibility of bending or twisting. Figure 9.3a shows an element of the rod of

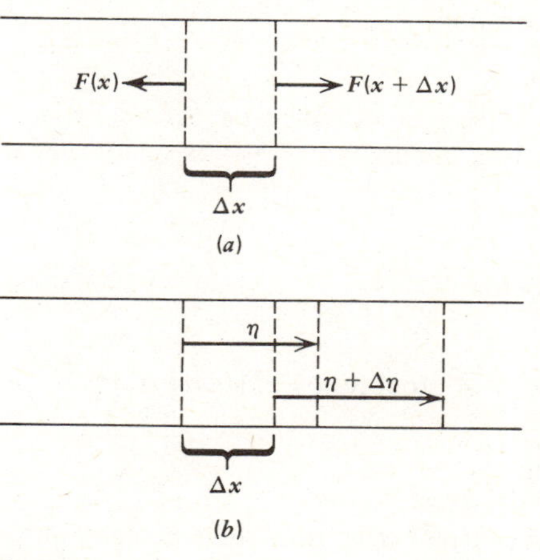

Fig. 9.3

length Δx. Its equation of motion is

$$F(x+\Delta x)-F(x)=\rho\,\Delta x\,\sigma\frac{\partial^2\eta}{\partial t^2} \tag{9.10}$$

In the limit as $\Delta x \to 0$, this is

$$\frac{\partial F}{\partial x}=\rho\sigma\frac{\partial^2\eta}{\partial t^2} \tag{9.11}$$

Figure 9.3*b* shows a much exaggerated picture of the distortion of the element as it is displaced from its equilibrium position. The strain is $\Delta\eta/\Delta x$. By means of (9.1),

$$\frac{F}{\sigma}=Y\frac{\partial\eta}{\partial x} \tag{9.12}$$

The combination of (9.11) and (9.12) again gives (9.7). Note that if Y and ρ are constants,

$$\frac{\partial^2\eta}{\partial x^2}=\frac{\rho}{Y}\frac{\partial^2\eta}{\partial t^2} \tag{9.13}$$

which predicts that longitudinal elastic waves will propagate through the medium with the velocity

$$v=\sqrt{\frac{Y}{\rho}} \tag{9.14}$$

PROBLEMS

9.1. Longitudinal waves can propagate through a liquid or a gas. The required elastic properties of such media can be described by the *bulk modulus* defined by

$$P-P_0=-B\frac{\Delta V}{V} \tag{9.15}$$

where V is the original volume of the material and ΔV is the volume decrease caused by the pressure increase $P-P_0$. The minus sign is used in (9.15) because ΔV is negative when $P-P_0$ is positive. We can think of P_0 as being the ambient pressure, for example, atmospheric pressure, if the propagation of sound waves is to be considered. By an argument similar to that used in deriving (9.11), (9.12), and (9.13) for one-dimensional plane waves, show that

$$P-P_0=-B\frac{\partial\eta}{\partial x} \qquad -\frac{\partial P}{\partial x}=\rho\ddot{\eta} \qquad B\frac{\partial^2\eta}{\partial x^2}=\rho\frac{\partial^2\eta}{\partial t^2} \tag{9.16}$$

9.2. The propagation of sound through an ideal gas is essentially an adiabatic process governed by the equation

$$PV^{\gamma}=\text{constant} \tag{9.17}$$

where $\gamma=C_P/C_V$ is the ratio of the heat capacity at constant pressure to the heat capacity at constant volume. By assuming that $P-P_0$ is small compared to P_0, we can express (9.15) as

$$\frac{dP}{dV}=-\frac{B}{V} \tag{9.18}$$

Show that for an ideal gas

$$B=\gamma P_0 \tag{9.19}$$

and that the velocity of propagation of sound waves is

$$v=\sqrt{\frac{\gamma P_0}{\rho}}=\sqrt{\frac{\gamma RT}{M}} \tag{9.20}$$

where R is the gas constant, T is the absolute temperature, and M is the mass of 1 mole of the gas.

9.3. If we wish to consider the standing waves or eigenfunctions which can exist in the air in a pipe of length s which is closed at one end and open at the other as shown in Fig. 9.4, then the appropriate boundary conditions to use are $\eta=0$ at $x=0$ and $P=P_0$ at $x=s$. Find the eigenfunctions of the wave equation (9.16) for these boundary conditions.

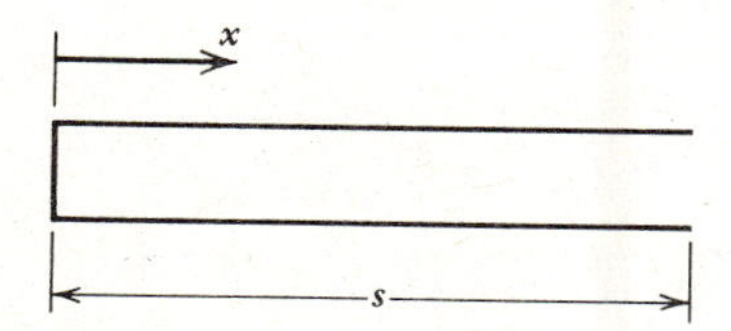

Fig. 9.4

9.4. Starting with equation (9.7), show that

$$\frac{\partial}{\partial x}\left[Y\frac{\partial \eta}{\partial t}\frac{\partial \eta}{\partial x}\right]=\frac{\partial}{\partial t}\left[\frac{1}{2}Y\left(\frac{\partial \eta}{\partial x}\right)^2+\frac{1}{2}\rho\left(\frac{\partial \eta}{\partial t}\right)^2\right] \tag{9.21}$$

Do the calculation as follows: First multiply (9.7) by $\partial\eta/\partial t$. Then use the identities

$$\frac{\partial}{\partial x}\left[\frac{\partial \eta}{\partial t}\left(Y\frac{\partial \eta}{\partial x}\right)\right]=\frac{\partial^2 \eta}{\partial x\,\partial t}\left(Y\frac{\partial \eta}{\partial x}\right)+\frac{\partial \eta}{\partial t}\frac{\partial}{\partial x}\left(Y\frac{\partial \eta}{\partial x}\right) \tag{9.22}$$

$$\frac{\partial}{\partial t}\left[\frac{1}{2}Y\left(\frac{\partial \eta}{\partial x}\right)^2\right]=Y\frac{\partial \eta}{\partial x}\frac{\partial^2 \eta}{\partial x\,\partial t} \tag{9.23}$$

Show from (9.21) that

$$I_2-I_1=-\frac{d}{dt}\int_1^2\left[\frac{1}{2}Y\left(\frac{\partial \eta}{\partial x}\right)^2+\frac{1}{2}\rho\left(\frac{\partial \eta}{\partial t}\right)^2\right]dx \tag{9.24}$$

where

$$I=-Y\frac{\partial \eta}{\partial t}\frac{\partial \eta}{\partial x} \tag{9.25}$$

What does (9.24) mean? Compare these results with problem 3.5. For a plane harmonic wave given by

$$\eta=A\sin(\kappa x-\omega t) \tag{9.26}$$

show that

$$\bar{I}=\tfrac{1}{2}\rho v A^2\omega^2 \text{ watts/m}^2 \tag{9.27}$$

This is the correct formula for the intensity of a sound wave.

10. THE STURM–LIOUVILLE EIGENVALUE PROBLEM

In this section, we study the eigenvalues and eigenfunctions which come from the second-order differential equation

$$\frac{d}{dx}\left[p(x)\frac{du(x)}{dx}\right]-q(x)u(x)+\lambda\rho(x)u(x)=0 \tag{10.1}$$

when certain boundary conditions are imposed. For one-dimensional longitudinal waves in an elastic medium, $p(x)$ is Young's modulus and $\rho(x)$ is the mass density of the medium. The parameter λ is the square of the frequency. The functions $p(x)$ and $\rho(x)$ are assumed to be positive throughout the domain of the variable x, although there are problems of interest where $p(x)$ is zero at one or both of the end points. In Sturm–Liouville theory, $\rho(x)$ is referred to as the *weight function*. The interval over which the problem is defined is taken to be $a \leq x \leq b$. Most of the proofs of this section assume that a and b are finite. Infinite intervals do however occur in many of the eigenvalue problems which we will encounter. If the one-dimensional Schrödinger wave equation for a conservative system is written as

$$\frac{d^2u}{dx^2} - \frac{2mV(x)}{\hbar^2}u(x) + \frac{2mW}{\hbar^2}u(x) = 0 \quad (10.2)$$

we see that the function $q(x)$ corresponds to the potential function. Thus, (10.1) is sufficiently general to encompass all of the eigenvalue problems of interest that have so far been discussed in this book.

In studying the Sturm–Liouville equation, it is convenient to write it in operator form as

$$Lu + \lambda\rho u = 0 \quad (10.3)$$

where L is the operator

$$L = \frac{d}{dx}p(x)\frac{d}{dx} - q(x) \quad (10.4)$$

It is of interest to compare (10.3) to the algebraic eigenvalue equation (14.28) in Chapter 4. Note that the density function $\rho(x)$ is a continuous version of the metric tensor. As a second-order differential equation, (10.3) can be solved for any value of the parameter λ. Our primary concern in this section is with those types of boundary conditions that require λ to be restricted to certain values which we refer to as eigenvalues. In most cases, the eigenfunctions are real. Since the exceptions are important, we will develop most of the theory assuming complex functions. This also has the advantage that the formalism is appropriate for quantum mechanics. It should be noted that a complex function is not necessarily an analytic function. For example, the wave function (8.8) is complex, but it is not an analytic function of a complex variable z as defined in section 7 of Chapter 5.

With respect to any two functions u_1 and u_2, the Sturm–Liouville operator obeys the important identity

$$u_1^*Lu_2 - (Lu_1)^*u_2 = \frac{d}{dx}\left[p\left(u_1^*\frac{du_2}{dx} - u_2\frac{du_1^*}{dx}\right)\right] \quad (10.5)$$

The identity is an immediate consequence of (10.4).

Theorem 1. The Sturm–Liouville operator is Hermitian provided that the functions on which it operates are restricted by the boundary condition

$$\left[p\left(u_1^*\frac{du_2}{dx} - u_2\frac{du_1^*}{dx}\right)\right]_{x=b} - \left[p\left(u_1^*\frac{du_2}{dx} - u_2\frac{du_1^*}{dx}\right)\right]_{x=a} = 0 \quad (10.6)$$

Proof: By integrating (10.5) over the range $a \leq x \leq b$, we find

$$\int_a^b \left[u_1^*Lu_2 - (Lu_1)^*u_2\right] dx = 0 \quad (10.7)$$

which shows that L is Hermitian in the generalized sense as explained in problem 6.6. The Sturm–Liouville operator is real if $p(x)$ and $q(x)$ are real. The operators of quantum mechanics, for example, the linear momentum operator $\mathbf{p} = -i\hbar\nabla$, are in general complex. Throughout this section, L is assumed to be real.

Theorem 2. If an operator is Hermitian, its eigenvalues are real.

Proof: In (10.7), let $u_1 = u_2 = u$. By means of $Lu = -\lambda\rho u$,

$$(\lambda - \lambda^*)\int_a^b uu^*\rho\, dx = 0 \quad (10.8)$$

Since the integral cannot be zero, $\lambda = \lambda^*$, and the proof is complete.

The fact that Hermitian operators have real eigenvalues is of crucial importance in quantum mechanics. To the list of fundamental hypotheses of quantum mechanics given at the beginning of section 6 may be added the following: Dynamical variables in quantum mechanics are represented by Hermitian operators. One or another of the eigenvalues of the operator is the only possible result of the measurement of the dynamical variable.

Theorem 3. Eigenfunctions corresponding to distinct eigenvalues are orthogonal.

Proof: If u_1 and u_2 are two possible eigenfunctions, then

$$Lu_1 + \lambda_1\rho u_1 = 0 \qquad Lu_2 + \lambda_2\rho u_2 = 0 \quad (10.9)$$

By means of the Hermitian property of L as given by (10.7),

$$(\lambda_2 - \lambda_1)\int_a^b u_1^*u_2\rho\, dx = 0 \quad (10.10)$$

Since $\lambda_1 \neq \lambda_2$,

$$\int_a^b u_1^* u_2 \rho \, dx = 0 \tag{10.11}$$

which establishes the orthogonality.

It was demonstrated in section 1 that the orthogonality condition could be arrived at by a limiting process which starts with an ordinary dot product in a space of a finite number of dimensions. A new feature has crept into our formalism. Equation (10.11), considered as a dot product of two unit vectors, involves complex quantities. In the next section, we will formally generalize the vector concept to take this into account.

Theorem 4. If an eigenfunction of the Sturm–Liouville equation is complex and its real and imaginary parts are linearly independent, then it is twofold degenerate. This is the greatest degree of degeneracy which is possible.

Proof: If $u(x)$ is a complex eigenfunction, it can be written as $u(x) = f(x) + ig(x)$. The Sturm–Liouville equation then gives

$$L(f + ig) + \lambda\rho(f + ig) = 0 \tag{10.12}$$

The real and imaginary parts can be written separately as

$$Lf + \lambda\rho f = 0 \qquad Lg + \lambda\rho g = 0 \tag{10.13}$$

No greater than twofold degeneracy can occur because not more than two linearly independent solutions exist for a given value of λ. The proof is therefore complete.

You may have noticed two things about degeneracy. One is that the possibilities for degeneracy increase as the number of dimensions involved in the problem increases. The Sturm–Liouville equation, being only one-dimensional, can at most produce twofold degeneracy. Another feature of degeneracy is that it is usually connected with some kind of symmetry. A square membrane produces degenerate eigenfunctions, whereas no degeneracy occurs if the lengths of the sides of the membrane are not related in any particular way.

Theorem 5. Let u_1 and u_2 be two real eigenfunctions which satisfy the boundary condition

$$p\left(u_1 u_2' - u_2 u_1'\right) = 0 \tag{10.14}$$

at both end points $x = a$ and $x = b$ *separately*. Then u_1 and u_2 cannot be degenerate.

Proof: If u_1 and u_2 are degenerate, then

$$Lu_1 + \lambda\rho u_1 = 0 \qquad Lu_2 + \lambda\rho u_2 = 0 \tag{10.15}$$

This means that

$$u_1 L u_2 - u_2 L u_1 = 0 \tag{10.16}$$

for all values of x in $a \le x \le b$. The identity (10.5) then requires

$$\begin{gathered} \frac{d}{dx} p\left(u_1 u_2' - u_2 u_1'\right) = 0 \\ W(x) = u_1 u_2' - u_2 u_1' = \frac{C}{p(x)} \end{gathered} \tag{10.17}$$

where C is a constant. The function $W(x)$ is recognized as the *Wronskian* of the two functions u_1 and u_2 (see section 12 in Chapter 5). The boundary condition (10.14) requires that

$$W(a) = W(b) = \frac{C}{p(a)} = \frac{C}{p(b)} = 0 \tag{10.18}$$

which means $C = 0$ and $W(x) = 0$. The functions u_1 and u_2 are therefore linearly dependent and can differ by at most a constant. For degeneracy to occur at all, it is necessary for (10.6) to hold *without* the quantities in brackets separately vanishing. This type of boundary condition is called *periodic*.

Theorem 6. The eigenvalues of the Sturm–Liouville equation form a monotonically increasing sequence in which $\lambda_n \to \infty$ as $n \to \infty$.

Proof: The proof will be done for the boundary condition $u(a) = u(b) = 0$. By theorem 5, the eigenfunctions are necessarily real and nondegenerate. The proof can be extended to other boundary conditions as well. It will also be assumed that the end points a and b are finite.

The eigenfunctions can be arranged in a sequence determined by how many times the function crosses the x axis between a and b. Viewed as standing waves, the eigenfunctions are to be classified according to the number of nodes which they have. By an appropriate choice of a multiplicative factor, all the eigenfunctions can be made to have a positive tangent at $x = a$. Figure 10.1 shows the first two eigenfunctions in the classification scheme. The function $u_2(x)$ crosses the x axis at $x = \zeta$. By utilizing the identity (10.5) and the boundary condition at $x = a$, we can show that

$$\int_a^\zeta \left(u_1 L u_2 - u_2 L u_1\right) dx = p\left(u_1 u_2' - u_2 u_1'\right)_{x=\zeta} \tag{10.19}$$

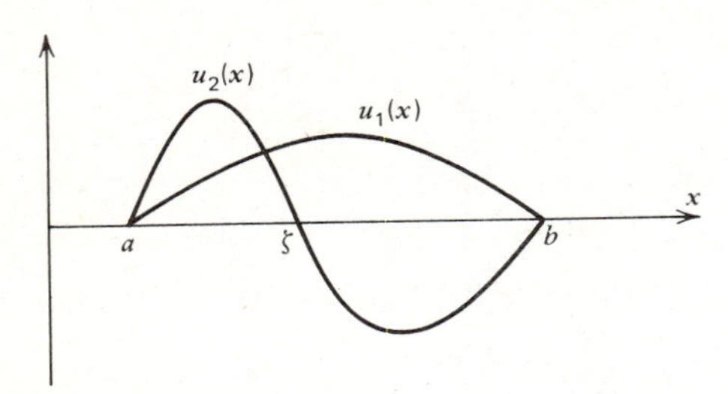

Fig. 10.1

By (10.9) and the fact that $u_2(\zeta)=0$, this reduces to

$$(\lambda_2-\lambda_1)\int_a^\zeta u_1 u_2 \rho \, dx = -pu_1 u_2'|_{x=\zeta} \tag{10.20}$$

Figure 10.1 shows that $u_1(\zeta)>0$ and $u_2'(\zeta)<0$. Therefore,

$$(\lambda_2-\lambda_1)\int_a^\zeta u_1 u_2 \rho \, dx > 0 \tag{10.21}$$

The functions u_1 and u_2 are both positive over $a<x<\zeta$. The integral is therefore positive and

$$\lambda_2-\lambda_1>0 \tag{10.22}$$

Now consider the possibility that any two of the eigenfunctions could have the behavior illustrated in Fig. 10.2. By a procedure similar to that used in deriving (10.20), we can show that

$$(\lambda_{n+1}-\lambda_n)\int_1^2 u_n u_{n+1}\rho \, dx = -pu_n u_{n+1}'|_2 + pu_n u_{n+1}'|_1 \tag{10.23}$$

where the integral is taken between the points (1) and (2) of Fig. 10.2. The signs of the various terms in (10.23) are indicated by

$$(\lambda_{n+1}-\lambda_n)(+) = -(-)+(+) = (+)+(+) > 0 \tag{10.24}$$

from which we conclude

$$\lambda_{n+1}-\lambda_n>0 \tag{10.25}$$

Similarly, integration between (3) and (4) gives

$$(\lambda_{n+1}-\lambda_n)\int_3^4 u_n u_{n+1}\rho \, dx = pu_{n+1}u_n'|_4 - pu_{n+1}u_n'|_3 \tag{10.26}$$

The signs of the terms are indicated by

$$(\lambda_{n+1}-\lambda_n)(+) = (-)-(+) = (-) < 0 \tag{10.27}$$

which implies that

$$\lambda_{n+1}-\lambda_n<0 \tag{10.28}$$

Since (10.25) and (10.28) are contradictory, the behavior indicated in Fig. 10.2 is not possible. If u_{n+1} has one more zero or node than u_n, the zeros must interleave with one another as shown in Fig. 10.3. By integrating between a and the first node of u_{n+1}, we get a result similar to (10.20) and conclude that

$$\lambda_{n+1}-\lambda_n>0 \tag{10.29}$$

The eigenvalues can therefore be arranged in an increasing sequence. The more nodes an eigenfunction has, the greater is its eigenvalue.

Now integrate (10.5) between $x=a$ and the last node of u_{n+1} to get

$$(\lambda_{n+1}-\lambda_n)\int_0^\zeta u_n u_{n+1}\rho \, dx = -pu_n u_{n+1}'|_\zeta \tag{10.30}$$

The values of p, u_n, and u_{n+1}' all remain finite as $n\to\infty$. However,

$$\lim_{n\to\infty}\int_0^\zeta u_n u_{n+1}\rho \, dx = 0 \tag{10.31}$$

This is because $\zeta\to b$ as $n\to\infty$ and the integral in (10.31) vanishes on account of the orthogonality condition. There-

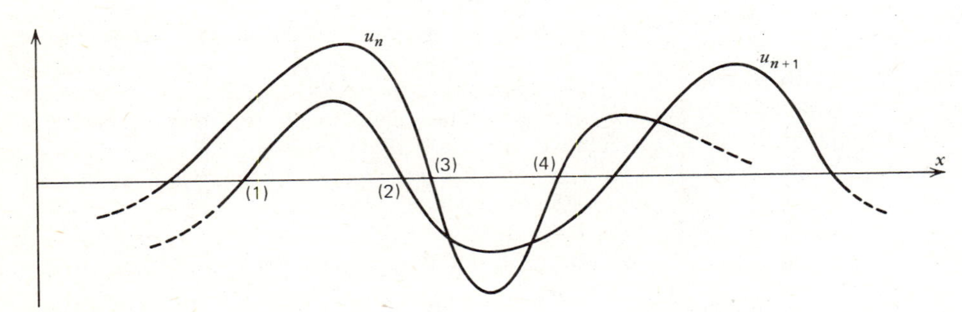

Fig. 10.2

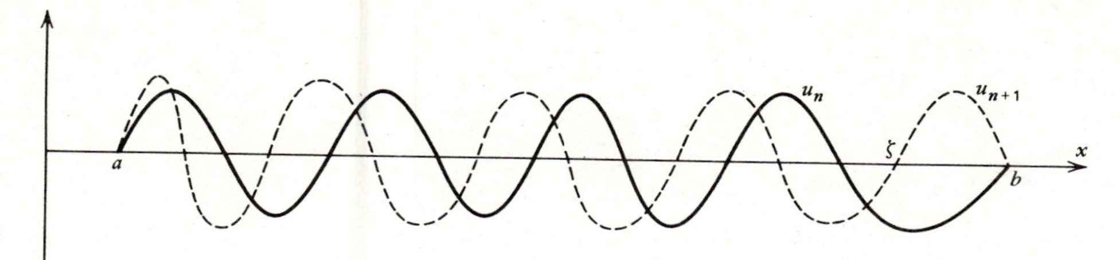

Fig. 10.3

fore,

$$\lim_{n\to\infty} (\lambda_{n+1} - \lambda_n) = \infty \tag{10.32}$$

which completes the proof of theorem 6. As an example, observe that the eigenvalue κ^2 in equation (2.34) is, by equation (2.38), proportional to n^2.

Of great importance is the question of completeness of any set of eigenfunctions. In section 1, a particular example of a set of Sturm–Liouville eigenfunctions was shown to be complete by a limiting process which started with a finite-dimensional vector space. We will now approach the idea of completeness from a different point of view. Let $f(x)$ be a function which is at least piecewise continuous over the interval $a \le x \le b$. Consider the quantity

$$\delta_n^2 = \int_a^b \left| f(x) - \sum_{k=1}^n b_k u_k(x) \right|^2 \rho(x)\, dx \tag{10.33}$$

For the sake of generality, we will allow for the possibility that $f(x)$ and $u_k(x)$ are complex. We are interested in approximating $f(x)$ by the partial sum

$$\sum_{k=1}^n b_k u_k(x) \tag{10.34}$$

in such a way that δ_n is minimized. What is the best way to choose the expansion coefficients b_k in order to do this? If the integrand in (10.33) is multiplied out, the result is

$$\begin{aligned}\delta_n^2 = \int_a^b |f|^2 \rho\, dx &- \sum_{k=1}^n b_k \int_a^b f^* u_k \rho\, dx \\ &- \sum_{k=1}^n b_k^* \int_a^b f u_k^* \rho\, dx \\ &+ \sum_{k=1}^n \sum_{k'=1}^n b_k^* b_{k'} \int_a^b u_k^* u_{k'} \rho\, dx\end{aligned} \tag{10.35}$$

If we define

$$c_k = \int_a^b u_k^* f \rho\, dx \tag{10.36}$$

and use the orthogonality relation, (10.35) can be expressed as

$$\begin{aligned}\delta_n^2 &= \int_a^b |f|^2 \rho\, dx - \sum_{k=1}^n (b_k c_k^* + b_k^* c_k) + \sum_{k=1}^n |b_k|^2 \\ &= \int_a^b |f|^2 \rho\, dx + \sum_{k=1}^n |b_k - c_k|^2 - \sum_{k=1}^n |c_k|^2\end{aligned} \tag{10.37}$$

Since δ_n^2 is positive, δ_n is minimized if the choice $b_k = c_k$ is made. The expansion coefficients are therefore to be calculated by (10.36). The eigenfunctions are said to be complete if

$$\lim_{n\to\infty} \delta_n^2 = 0 \tag{10.38}$$

for every function $f(x)$ which is piecewise continuous. It then follows from (10.37) that

$$\int_a^b |f|^2 \rho\, dx = \sum_{k=1}^\infty |c_k|^2 \tag{10.39}$$

which is called the *completeness relation*. Look at equation (8.10). It is really the completeness relation for the eigenfunctions of a particle in a one-dimensional square well with perfectly reflecting walls. We then write

$$f(x) = \sum_{k=1}^\infty c_k u_k(x) \tag{10.40}$$

We have here quite a different definition of convergence than was given for infinite series in section 4 of Chapter 5. It is possible for (10.38) to hold even if the convergence of the expansion (10.40) breaks down at isolated points of the interval. If $f(x)$ is continuous, has piecewise continuous first and second derivatives, and satisfies the same boundary conditions as the eigenfunctions, then the expansion (10.40) is in fact uniformly and absolutely convergent. If $f(x)$ is piecewise continuous, then the expansion converges uniformly in all subintervals which are free of points of discontinuity. At the points of discontinuity, the series converges but exhibits what is called *Gibbs' phenomenon*. The value to which the series converges is not precisely defined at the points of discontinuity. Gibbs' phenomenon will be explained more fully later in connection with Fourier series. It is actually possible to carry out the expansion even if $f(x)$ does not obey the same boundary conditions as the eigenfunctions. It can be shown that the Sturm–Liouville eigenfunctions are complete. The proof will be given in a later chapter.

Another relation of some importance can be derived by starting with the finite sum

$$f_n(x) = \sum_{k=1}^{n} c_k u_k(x) \tag{10.41}$$

By substituting the expansion coefficients given by (10.36) into (10.41), we find

$$f_n(x) = \int_a^b f(x') \left[\sum_{k=1}^{n} u_k^*(x') u_k(x) \rho(x') \right] dx' \tag{10.42}$$

In the limit $n \to \infty$, $f_n(x)$ becomes $f(x)$ and the expression in brackets therefore becomes a Dirac delta function:

$$\sum_{k=1}^{\infty} u_k^*(x') u_k(x) = \frac{\delta(x - x')}{\rho(x)} \tag{10.43}$$

This is the expansion of a Dirac delta function in terms of Sturm–Liouville eigenfunctions and is called the *closure relation*. We have therefore shown that the completeness relation implies the closure relation. Conversely, the closure relation implies the completeness relation. The completeness of a set of eigenfunctions is therefore established if it can be shown that either (10.39) or (10.43) is true. Equation (2.20) is a special case of the closure relation.

PROBLEMS

10.1. Let u_1 and u_2 be degenerate eigenfunctions. It is not necessarily true that they are orthogonal. Show however that linear combinations of u_1 and u_2 exist which are mutually orthogonal.

10.2. Show formally that the expansion coefficients in (10.40) are given by (10.36) by using the orthogonality property of the eigenfunctions.

10.3. Suppose that the potential function in the one-dimensional Schrödinger wave equation (10.2) is a potential well of some sort so that the particle is bound as explained in problem 8.1. The boundary condition on the wave function is $u(x) \to 0$ as $x \to \pm\infty$. Will this boundary condition produce eigenfunctions? Is degeneracy to be expected? It will be necessary to consider the limiting case where the end points a and b of equation (10.6) become infinite.

A potential function which obeys $V(x) = V(-x)$ is said to be *symmetric*. Show that for symmetric potentials, $u(x)$ and $u(-x)$ are eigenfunctions corresponding to the same energy. If there is no degeneracy, then $u(x) = Cu(-x)$, where C is a constant. Show that $C = \pm 1$. The even wave functions for which $C = +1$ are said to have *even parity* and the odd wave functions for which $C = -1$ are said to have *odd parity*.

Consider the square well potential shown in Fig. 8.1. If we set $W = -E$, then bound states occur when $0 < E < V_0$. The wave equation is

$$u'' - \alpha^2 u = 0 \qquad \alpha^2 = \frac{2mE}{\hbar^2} \qquad |x| > \frac{a}{2} \tag{10.44}$$

$$u'' + \kappa^2 u = 0 \qquad \kappa^2 = \frac{2m(V_0 - E)}{\hbar^2} \qquad -\frac{a}{2} \le x \le \frac{a}{2} \tag{10.45}$$

The even solutions are

$$
\begin{aligned}
u &= Ae^{\alpha x} && x < -\frac{a}{2} \\
u &= Ae^{-\alpha x} && x > \frac{a}{2} \\
u &= B\cos\kappa x && -\frac{a}{2} \le x \le \frac{a}{2}
\end{aligned}
\tag{10.46}
$$

The odd solutions are

$$
\begin{aligned}
u &= Ae^{\alpha x} && x < -\frac{a}{2} \\
u &= -Ae^{-\alpha x} && x > \frac{a}{2} \\
u &= B\sin\kappa x && -\frac{a}{2} \le x \le \frac{a}{2}
\end{aligned}
\tag{10.47}
$$

By using appropriate boundary conditions at $x = \pm a/2$, show that

$$\alpha = \kappa\tan\left(\frac{\kappa a}{2}\right) \qquad \text{even solutions} \tag{10.48}$$

$$-\alpha = \kappa\cot\left(\frac{\kappa a}{2}\right) \qquad \text{odd solutions} \tag{10.49}$$

These are transcendental equations the roots of which produce the possible energy eigenvalues. They cannot be solved for E in terms of elementary functions but must be solved by numerical methods. There are only a finite number of roots possible and therefore only a finite number of bound states. The actual number of bound states can be anything from zero on up depending on the well depth V_0.

If $W > 0$, the wave equation is

$$u'' + \kappa_1^2 u = 0 \qquad \kappa_1^2 = \frac{2mW}{\hbar^2} \qquad |x| > \frac{a}{2} \tag{10.50}$$

$$u'' + \kappa_2^2 u = 0 \qquad \kappa_2^2 = \frac{2m(V_0 + W)}{\hbar^2} \qquad -\frac{a}{2} < x < \frac{a}{2} \tag{10.51}$$

Harmonic wave solutions with a continuum of energies are possible. Such solutions do not vanish at $\pm\infty$. The conditions for a Sturm–Liouville eigenvalue problem as we have developed them break down. It is however possible to regard harmonic waves as eigenfunctions with a continuum of energy eigenvalues through an appropriate limiting process. The method is to define a Sturm–Liouville eigenvalue problem over a finite interval and then allow the interval to become infinite. This procedure will be taken up later.

10.4. An ordinary second-order linear homogeneous differential equation

$$P(x)u'' + Q(x)u' + R(x)u = 0 \tag{10.52}$$

is said to be *self-adjoint* if $Q(x) = P'(x)$. Show that a self-adjoint differential equation can be expressed in Sturm–Liouville form as given by (10.1). Show that if (10.52) is not self-adjoint, it can be put into that form by multiplying through by the factor

$$\exp\left[\int \frac{Q - P'}{P}\,dx\right] \tag{10.53}$$

This shows that any differential equation of the type (10.52) can be put into Sturm–Liouville form.

10.5. Let $u(x)$ be any solution of the Sturm–Liouville equation (10.1). Show that $u(x)=0$ does not have any repeated roots. By this is meant that there is no point $x=\alpha$ such that both $u(\alpha)=0$ and $u'(\alpha)=0$.

Hint: Show from the differential equation that $u(\alpha)=0$ and $u'(\alpha)=0$ implies that all derivatives of $u(x)$ at $x=\alpha$ vanish. By Taylor's theorem, $u(x)$ is then identically zero. [An exception occurs if $p(\alpha)=0$.]

10.6. Show that the formula for the coefficients (10.36) and the closure relation (10.43) imply the validity of the completeness relation (10.39).

11. HILBERT SPACES AND UNITARY SPACES

A wave function in quantum mechanics can be expanded in terms of the eigenfunctions of the Hamiltonian (or some other Hermitian operator) as

$$\psi(\mathbf{r},t)=\sum_{k=1}^{\infty} A_k(t)u_k(\mathbf{r}) \tag{11.1}$$

The wave function (8.4) for a particle in a box has three sums. In such cases, imagine that the sum in (11.1) is a kind of supersum where each value of k stands for just one of the possible eigenfunctions. In general, the coefficients in (11.1) can be time-dependent. Another example of the same kind of thing is the expansion of an arbitrary function $f(x)$ in terms of the eigenfunctions of a Sturm–Liouville operator as given by (10.40). The point was made in section 1 that complete sets of normalized orthogonal eigenfunctions can be obtained as a limiting form of orthogonal unit vectors defined in a space of a finite number of dimensions when the dimension of the space is allowed to approach infinity. The expansion coefficients A_k are to be regarded as the components of the "vector" ψ in the abstract vector space of infinitely many dimensions, called Hilbert space, which results from the limiting process. These components are in general complex numbers. A finite-dimensional complex vector space with the same algebraic properties as Hilbert space is referred to as a *unitary space*.

The *norm* of a vector is the square of its length and is defined by

$$(\psi,\psi)=\sum_{k=1}^{\infty} A_k^* A_k \tag{11.2}$$

This is a natural definition and is nothing more than the completeness relation as given, for example, by (8.10) or (10.39). Note that since (ψ,ψ) consists of a series of nonnegative terms, $(\psi,\psi)\geq 0$. If $(\psi,\psi)=0$, then each A_k must individually vanish, meaning that $\psi=0$. The idea of a vector represented by an ordered sequence of complex numbers as

$$\psi=(A_1, A_2,\ldots,A_n,\ldots) \tag{11.3}$$

makes sense if the infinite series given by (11.2) converges. This could be established, for example, in Sturm–Liouville theory by the convergence of the integral which appears in (10.39). If ψ is a wave function defined over a region Σ of space, then the convergence requirement can be expressed as

$$(\psi,\psi)=\int_{\Sigma}\psi^*\psi\, d\Sigma=1 \tag{11.4}$$

which is a statement of conservation of probability. If functions are used as a representation of the vectors in Hilbert space, they must be square integrable over the domain in question. The component representation (11.3) makes it evident that the postulates of a vector space as given by 1 through 8 in section 1 of Chapter 2 are satisfied. All vector components and scalars must now be regarded as complex numbers. The notation (ψ,ψ) is used to represent the scalar or dot product of ψ with itself. Equations (11.2) and (11.4) are two possible realizations or representations of the scalar product depending on whether we want a vector component picture or a function picture of ψ. The idea of representing a dot product as an integral of the product of two functions is not new. Look at (2.12) and review how it was derived. Another commonly used notation for the dot product is $\langle\psi|\psi\rangle$ and is due to Dirac.

A set of unit base vectors or coordinate vectors can be written as

$$u_1 = (1,0,0,\ldots)$$
$$u_2 = (0,1,0,\ldots)$$
$$u_3 = (0,0,1,\ldots)$$
$$\vdots \qquad\qquad (11.5)$$

Any vector in the space can be written as a linear combination of the base vectors:

$$\psi = \sum_{k=1}^{\infty} A_k u_k \qquad (11.6)$$

A set of vectors ψ_k, finite or infinite in number, is said to be linearly dependent if there exists scalars $\alpha_1, \alpha_2, \ldots$, not all zero, such that

$$\alpha_1\psi_1 + \alpha_2\psi_2 + \cdots + \alpha_n\psi_n + \cdots = 0 \qquad (11.7)$$

If no such relation exists, then ψ_k are linearly independent. There is scarcely any difference between the theory of unitary spaces and the theory of finite-dimensional vector spaces as developed in Chapters 1 and 2. This is to be expected since complex numbers constitute a number field with all the same algebraic properties as the real numbers. Thus, matrices, determinants, and linear algebraic equations all make sense if the number field is enlarged to include complex numbers. There are important differences between a space of a finite number of dimensions and one of an infinite number of dimensions. In a space of a finite number of dimensions, there is no problem with the convergence of the expression for the scalar product since there is only a finite number of terms in the sum. In a space of dimension n, any set of n linearly independent vectors is complete and spans the space. In a space of infinitely many dimensions, an infinite set of linearly independent vectors is not necessarily complete. For example, if u_1 is omitted in the set of mutually orthogonal unit vectors (11.5), there is still an infinite number of mutually orthogonal unit vectors left, but the set is no longer complete.

The scalar product of the two vectors $\psi = (A_1, A_2, \ldots)$ and $\phi = (B_1, B_2, \ldots)$ is defined by

$$(\psi, \phi) = \sum_{k=1}^{\infty} A_k^* B_k \qquad (11.8)$$

If a functional representation of ψ and ϕ exists, as would be the case if they were wave functions, then expansions in terms of a complete set of eigenfunctions is possible:

$$\psi = \sum_{k=1}^{\infty} A_k u_k(\mathbf{r}) \qquad \phi = \sum_{j=1}^{\infty} B_j u_j(\mathbf{r}) \qquad (11.9)$$

We note that

$$\int_\Sigma \psi^*\phi \, d\Sigma = \sum_{k=1}^{\infty}\sum_{j=1}^{\infty} A_k^* B_j \int_\Sigma u_k^*(\mathbf{r}) u_j(\mathbf{r}) \, d\Sigma$$
$$= \sum_{k=1}^{\infty}\sum_{j=1}^{\infty} A_k^* B_j \delta_{kj} = \sum_{k=1}^{\infty} A_k^* B_k \qquad (11.10)$$

The scalar product is therefore also represented by

$$(\psi, \phi) = \int_\Sigma \psi^*\phi \, d\Sigma \qquad (11.11)$$

The domain Σ over which the functions are defined can be finite or infinite. It may involve more than three spatial coordinates. For example, a two-particle system in quantum mechanics requires at least six coordinates for a complete description. A little later on in this section, we will show that the convergence of (ψ, ϕ) is guaranteed if (ψ, ψ) and (ϕ, ϕ) converge. If α is a complex number, then the scalar product has the following properties:

$$(\psi, \alpha\phi) = \alpha(\psi, \phi) \qquad (11.12)$$
$$(\alpha\psi, \phi) = \alpha^*(\psi, \phi) \qquad (11.13)$$
$$(\psi, \phi)^* = (\phi, \psi) \qquad (11.14)$$
$$(\psi + \phi, \chi) = (\psi, \chi) + (\phi, \chi) \qquad (11.15)$$

In Dirac's notation, the scalar product is written $\langle\psi|\phi\rangle$.

Theorem 1. Triangle inequality. We use the notation $|\psi| = \sqrt{(\psi, \psi)}$ to indicate the "length" of a vector. If ψ and ϕ are any two vectors, then

$$|\psi + \phi| \le |\psi| + |\phi| \qquad (11.16)$$

Proof: Let x be any real number. Then,

$$0 \le (\psi + x\phi, \psi + x\phi)$$
$$= |\psi|^2 + x[(\psi, \phi) + (\phi, \psi)] + x^2|\phi|^2 \quad (11.17)$$

If $x = -|\psi|/|\phi|$, then (11.17) becomes

$$(\psi, \phi) + (\phi, \psi) \le 2|\phi||\psi| \quad (11.18)$$

We can now write

$$|\psi + \phi|^2 = |\psi|^2 + (\psi, \phi) + (\phi, \psi) + |\phi|^2$$
$$\le |\psi|^2 + 2|\psi||\phi| + |\phi|^2 = (|\psi| + |\phi|)^2 \quad (11.19)$$

from which (11.16) follows.

Theorem 2. Schwartz Inequality.

$$|\psi||\phi| \ge |(\psi, \phi)| \quad (11.20)$$

The equality holds if, and only if, ψ and ϕ are proportional: $\psi = \alpha\phi$.

Proof:

$$0 \le \sum_{j=1}^{\infty} \sum_{k=1}^{\infty} (A_j^* B_k^* - A_k^* B_j^*)(A_j B_k - A_k B_j)$$
$$= 2|\psi|^2|\phi|^2 - 2(\psi, \phi)(\phi, \psi) \quad (11.21)$$

Since $(\psi, \phi)(\phi, \psi) = (\psi, \phi)(\psi, \phi)^* = |(\psi, \phi)|^2$, (11.20) follows. The equality holds if, and only if,

$$A_j B_k - A_k B_j = 0 \quad (11.22)$$

for all values of j and k. This means that

$$A_1 B_2 - A_2 B_1 = 0 \qquad A_2 B_3 - A_3 B_2 = 0, \ldots$$
$$\frac{A_1}{B_1} = \frac{A_2}{B_2} = \frac{A_3}{B_3} = \cdots = \alpha \qquad \psi = \alpha\phi \quad (11.23)$$

As a consequence of the Schwartz inequality, the convergence of the scalar product (ψ, ϕ) is established if it is known that (ψ, ψ) and (ϕ, ϕ) converge. If ψ and ϕ are functions defined over a region Σ of space, then the Schwartz inequality can be stated in the form

$$\int_\Sigma \psi^* \psi \, d\Sigma \int_\Sigma \phi^* \phi \, d\Sigma \ge \left| \int_\Sigma \psi^* \phi \, d\Sigma \right|^2 \quad (11.24)$$

Two functions can differ from one another at isolated points and still give the same value for their scalar product defined as an integral. Their expansions in terms of complete sets of vectors are the same since the expansion coefficients are evaluated by means of integrals. In Hilbert space theory, such functions are regarded as being essentially the same.

Two vectors ψ and ϕ are defined to be orthogonal if $(\psi, \phi) = 0$. The unit vectors (11.5) are normalized and orthogonal:

$$(u_j, u_k) = \delta_{jk} \quad (11.25)$$

It is not necessary that the unit vectors "point" along the "coordinate axes." They could be any complete set of mutually orthogonal unit vectors. The orthogonality of the Sturm–Liouville eigenfunctions as given by (10.11) provides a concrete example.

The calculation leading to the completeness relation (10.39) can be repeated using the abstract inner product notation. The vector ψ is to be approximated by the finite sum

$$\psi_n = \sum_{k=1}^{n} A_k u_k \quad (11.26)$$

The mean square error in the approximation is

$$|\psi - \psi_n|^2 = |\psi|^2 - (\psi, \psi_n) - (\psi_n, \psi) + |\psi_n|^2$$
$$= |\psi|^2 - \left(\psi, \sum A_k u_k\right)$$
$$- \left(\sum A_k u_k, \psi\right) + \sum |A_k|^2$$
$$= |\psi|^2 - \sum A_k (u_k, \psi)^*$$
$$- \sum A_k^* (u_k, \psi) + \sum |A_k|^2$$
$$= |\psi|^2 + \sum |A_k - (u_k, \psi)|^2 - \sum |(u_k, \psi)|^2$$
$$(11.27)$$

Since $|\psi - \psi_n|^2$ is positive, the mean square error is minimized if

$$A_k = (u_k, \psi) \quad (11.28)$$

which is the familiar expression for the expansion coefficients. Equation (11.27) can be expressed as

$$|\psi|^2 \ge \sum_{k=1}^{n} |A_k|^2 \quad (11.29)$$

which is called *Bessel's inequality*. The set of orthogonal vectors u_k is, by definition, complete if

$$|\psi|^2 = \sum_{k=1}^{\infty} |A_k|^2 \quad (11.30)$$

If the space is a finite-dimensional unitary space, equality is obtained if n equals the dimension of the space. The completeness relation (11.30) also goes by the name of *Parseval's relation.*

Sometimes we are given a complete set of linearly independent vectors in a Hilbert space or a unitary space which are not orthogonal. Linear combinations of the vectors that are mutually orthogonal can be chosen in a variety of ways. We will give one particular method of choosing the linear combinations which is called *Schmidt orthogonalization.* Let $\phi_1, \phi_2, \ldots, \phi_n, \ldots$ be a set of linearly independent vectors. Let a new set of vectors be chosen by means of

$$\begin{aligned} \psi_1 &= \phi_1 \\ \psi_2 &= \phi_2 + \alpha_{21}\psi_1 \\ \psi_3 &= \phi_3 + \alpha_{31}\psi_1 + \alpha_{32}\psi_2 \\ &\vdots \\ \psi_n &= \phi_n + \alpha_{n1}\psi_1 + \cdots + \alpha_{n,n-1}\psi_{n-1} \end{aligned} \tag{11.31}$$

We require the new set to be mutually orthogonal:

$$0 = (\psi_1, \psi_2) = (\psi_1, \phi_2) + \alpha_{21}|\psi_1|^2$$

$$\alpha_{21} = -\frac{(\psi_1, \phi_2)}{|\psi_1|^2} \tag{11.32}$$

$$0 = (\psi_1, \psi_3) = (\psi_1, \phi_3) + \alpha_{31}|\psi_1|^2$$

$$\alpha_{31} = -\frac{(\psi_1, \phi_3)}{|\psi_1|^2} \tag{11.33}$$

$$0 = (\psi_2, \psi_3) = (\psi_2, \phi_3) + \alpha_{32}|\psi_2|^2$$

$$\alpha_{32} = -\frac{(\psi_2, \phi_3)}{|\psi_2|^2} \tag{11.34}$$

The general coefficient is

$$\alpha_{jk} = -\frac{(\psi_k, \phi_j)}{|\psi_k|^2} \tag{11.35}$$

Neither the original set nor the new orthogonal set is normalized.

PROBLEMS

11.1. Prove that $|\psi| - |\phi| \leq |\psi + \phi|$.

11.2. If ψ and ϕ are vectors in Hilbert space, then (ψ, ψ) and (ϕ, ϕ) are convergent. If α and β are complex numbers and $\chi = \alpha\psi + \beta\phi$, prove that (χ, χ) is also convergent. This demonstrates the closure property of vectors in Hilbert space. Closure is explained in section 1 of Chapter 2.

11.3. Let the domain of x be $-1 \leq x \leq +1$. We know from the theory of Taylor's series that the functions $\phi_0 = 1$, $\phi_1 = x$, $\phi_2 = x^2, \ldots, \phi_n = x^n, \ldots$ are complete. It is possible to construct a set of mutually orthogonal polynomials defined over $-1 \leq x \leq +1$ by Schmidt orthogonalization. Find the first four of these. The inner product is written as

$$(\psi_n, \phi_m) = \int_{-1}^{+1} \psi_n x^m \, dx$$

Except for a multiplicative factor, the polynomials generated in this fashion are the *Legendre polynomials*. They occur in potential theory when Laplace's equation is separated in spherical coordinates and also in the two-body central force problem in quantum mechanics.

11.4. Prove that the expansion of a function in terms of a given complete set of normalized orthogonal unit vectors is unique, that is,

$$\psi = \sum_{k=1}^{\infty} A_k u_k = \sum_{k=1}^{\infty} B_k u_k$$

implies that $A_k = B_k$.

11.5. If ψ and ϕ are both functions of the time, show that

$$\frac{d}{dt}(\psi, \phi) = \left(\frac{\partial \psi}{\partial t}, \phi\right) + \left(\psi, \frac{\partial \phi}{\partial t}\right)$$

11.6. What is the statement of the Schwartz inequality for two ordinary vectors in three-dimensional space and what does it mean geometrically?

12. HERMITIAN AND UNITARY MATRICES

In the theory of vectors and tensors in a Cartesian space, transformations from one rectangular Cartesian coordinate system to another are important. The same is true in unitary and Hilbert spaces. Assume that we are given a set u_j of orthonormal basis vectors which is known to be complete. It is possible to define a new set of orthonormal vectors by means of the linear transformation

$$u'_n = \sum_j S^*_{nj} u_j \tag{12.1}$$

which should be compared to equation (4.48) in Chapter 2. Writing the transformation coefficients as the complex conjugate is a notational preference. The expansions given are possible because of the known completeness of the set u_j. The sum over j is finite for a unitary space and infinite for a Hilbert space. The properties of the expansion coefficients are found by utilizing the orthonormality of the unit vectors:

$$\begin{aligned}(u'_n, u'_m) &= \left(\sum_j S^*_{nj} u_j, \sum_k S^*_{mk} u_k\right) \\ &= \sum_j \sum_k S_{nj} S^*_{mk} (u_j, u_k) \\ &= \sum_j \sum_k S_{nj} S^*_{mk}\, \delta_{jk} = \sum_k S_{nk} S^*_{mk} = \delta_{nm}\end{aligned} \tag{12.2}$$

For $n = m$, the sums are completeness relations and are therefore convergent. For $n \neq m$, the sums converge as a consequence of the Schwartz inequality as explained at the end of the proof of theorem 2 in section 11.

Let S be a matrix with elements S_{jk}. It is infinite-dimensional for a Hilbert space. The complex conjugate of the transpose of S is written as

$$S^\dagger = \tilde{S}^* \tag{12.3}$$

and called the *adjoint* of S. In matrix notation, (12.2) is

$$SS^\dagger = I \tag{12.4}$$

Up to this point, everything that we have said is valid for both unitary and Hilbert spaces. In a unitary space, (12.4) implies that

$$SS^\dagger = S^\dagger S = I \tag{12.5}$$

since in a space of a finite number of dimensions a right inverse is also a left inverse. This is not necessarily true in Hilbert space. The source of the difficulty is best illustrated by an example. Suppose that the new set of orthonormal unit vectors is related to the old by

$$u'_n = u_{n+1} = S^*_{n+1,\,n} u_n \tag{12.6}$$

The matrix which represents this transformation is

$$S = \begin{pmatrix} 0 & 1 & 0 & 0 & \cdots \\ 0 & 0 & 1 & 0 & \cdots \\ 0 & 0 & 0 & 1 & \cdots \\ 0 & 0 & 0 & 0 & \cdots \\ \vdots & \vdots & \vdots & \vdots & \end{pmatrix} \tag{12.7}$$

We find that $SS^\dagger = I$, but

$$S^\dagger S = \begin{pmatrix} 0 & 0 & 0 & 0 & \cdots \\ 0 & 1 & 0 & 0 & \cdots \\ 0 & 0 & 1 & 0 & \cdots \\ 0 & 0 & 0 & 1 & \cdots \\ \vdots & \vdots & \vdots & \vdots & \end{pmatrix} \neq I \tag{12.8}$$

Consequently, there is no guarantee that the new set of orthonormal vectors u'_n given by (12.1) is complete. If the completeness of u'_n can be established, then everything that we have said works in reverse. A linear relation for the u_j in terms of the u'_n exists, or, what is the same thing, (12.1) is solvable for the u_j in terms of the u'_n. The validity of (12.5) can then be assumed. In problems of practical importance, this will be the case. Note that the component version of $S^\dagger S = I$ is

$$\sum_k S^*_{kn} S_{km} = \delta_{nm} \tag{12.9}$$

A transformation with the property (12.5) is called *unitary*. Unitary transformations are a generalization of orthogonal transformations and include them as a special case. The transformation (12.1) can be inverted by multiplying by S_{nk} and summing over n:

$$\sum_n S_{nk} u'_n = \sum_j \sum_n S_{nk} S^*_{nj} u_j = \sum_j \delta_{kj} u_j = u_k \tag{12.10}$$

In the remaining discussion in this section, both u_k and u'_n are assumed to be complete.

An arbitrary vector in the space can be expanded either in terms of the primed or unprimed sets of unit vectors:

$$\psi = \sum_k A_k u_k = \sum_j A'_j u'_j \tag{12.11}$$

By means of (12.10) we find

$$\sum_j \sum_k A_k S_{jk} u'_j = \sum_j A'_j u'_j \tag{12.12}$$

As shown in problem (11.4), the expansion coefficients are unique. This requires that

$$A'_j = \sum_k S_{jk} A_k \tag{12.13}$$

If the two representations of the vector are expressed as column matrices, then (12.13) is equivalent to the matrix equation

$$\psi' = S\psi \qquad \psi' = \begin{pmatrix} A'_1 \\ A'_2 \\ \vdots \end{pmatrix} \qquad \psi = \begin{pmatrix} A_1 \\ A_2 \\ \vdots \end{pmatrix} \tag{12.14}$$

Strictly speaking, we should use different letters such as X and X' to stand for specific matrix representations of the abstract vector ψ. To avoid proliferation of notation, we have chosen not to do this, hoping that the meaning will be clear from the context. The complex conjugate transpose of ψ is the row matrix

$$\psi^\dagger = (A_1^* \quad A_2^* \quad A_3^* \ldots) \tag{12.15}$$

In theorem 6 in section 10 of Chapter 1, it is shown that the rule for forming the transpose of the product of two matrices is $\widetilde{AB} = \tilde{B}\tilde{A}$. The complex conjugate transpose goes by the same rule: $(AB)^\dagger = B^\dagger A^\dagger$. The norm of ψ' is

$$\begin{aligned}(\psi', \psi') &= \psi'^\dagger \psi' = (S\psi)^\dagger S\psi = \psi^\dagger S^\dagger S\psi \\ &= \psi^\dagger \psi = (\psi, \psi)\end{aligned} \tag{12.16}$$

and is therefore preserved by the unitary transformation. The idea is familiar from the theory of orthogonal transformations as developed in section 2 of Chapter 2. In quantum mechanics, conservation of probability requires that the functions which describe a physical system be normalized so that $(\psi, \psi) = 1$. Since this normalization must be maintained if a change of representation is made, the allowed transformations of quantum mechanics are unitary.

An operator in the vector space changes one vector into another:

$$H\psi = \phi \tag{12.17}$$

Operators may appear in fundamental equations such as the Schrödinger equation (6.21). The operators of the theory are *linear*, which is defined to mean

$$H(\alpha\psi + \beta\phi) = \alpha H\psi + \beta H\phi \tag{12.18}$$

for any vectors ψ and ϕ and any complex numbers α and β. Suppose that a specific representation is chosen and that the vectors are expanded in terms of a complete orthonormal system. The operator equation (12.17) can then be expressed as

$$H\Big(\sum_k A_k u_k\Big) = \sum_k B_k u_k \tag{12.19}$$

By forming the inner product of (12.19) with any one of the unit vectors, we find

$$\Big(u_j, H\sum_k A_k u_k\Big) = \Big(u_j, \sum_k B_k u_k\Big) \tag{12.20}$$

The *matrix elements* of the operator are defined to be

$$H_{jk} = (u_j, Hu_k) \tag{12.21}$$

Equation (12.20) can then be written

$$\sum_k H_{jk} A_k = B_j \tag{12.22}$$

which is a matrix representation of equation (12.17). If we start with a representation in terms of functions, then the calculation of matrix elements involves evaluating integrals. In a quantum mechanical application, the wave function and the complete sets of eigenfunctions in terms of which it is expanded will conform to the boundary conditions of the problem. These boundary conditions are then automatically built into the matrix representation. In another "coordinate system," the matrix elements are

$$H'_{jk} = (u'_j, Hu'_k) \tag{12.23}$$

By means of the transformation (12.1), we find

$$\begin{aligned} H'_{jk} &= \left(\sum_n S^*_{jn} u_n,\ H \sum_m S^*_{km} u_m\right) \\ &= \sum_n \sum_m S_{jn} S^*_{km} (u_n, Hu_m) \\ &= \sum_n \sum_m S_{jn} H_{nm} S^*_{km} \end{aligned} \tag{12.24}$$

Translated into a matrix equation, this is

$$H' = SHS^\dagger \tag{12.25}$$

Compare this result to the development in section 6 of Chapter 2. Specifically, look at equation (6.7).

Given an operator H, its *adjoint* (also called *Hermitian adjoint* or *Hermitian conjugate*) is defined by

$$(A\psi, \phi) = (\psi, H\phi) \tag{12.26}$$

We have already used this term in connection with matrices. We will find that the general definition (12.26) leads to an equivalent result when the operators are represented by matrices. If the operator is Hermitian, then $A = H$. Specific examples of Hermitian operators are provided by the linear momentum (6.33) and Hamiltonian (6.44) operators in quantum mechanics and by the Sturm–Liouville operator (10.7). The matrix elements of the adjoint operator are found from

$$\begin{aligned} (Au_n, u_m) &= (u_n, Hu_m) \\ (u_m, Au_n) &= (u_n, Hu_m)^* \\ A_{mn} &= H^*_{nm} \end{aligned} \tag{12.27}$$

In matrix notation, this is

$$A = H^\dagger \tag{12.28}$$

which is our previous definition of the adjoint of a matrix. If the operator is Hermitian, then it is represented by a *Hermitian matrix*:

$$H = H^\dagger \tag{12.29}$$

Hermitian matrices contain real symmetric matrices as a subclass. The theory of the eigenvalue problem for Hermitian and real symmetric matrices is quite similar. The following development parallels quite closely the discussion in section 10 of Chapter 2.

The eigenfunctions and eigenvalues of an operator are defined by

$$Hu = \lambda u \tag{12.30}$$

For the concept of an eigenvalue and corresponding stationary states to make sense in a quantum mechanical application, the operator must be independent of the time. In the following theorems, a lot of results which you already know will be derived again, this time from a more general point of view.

Theorem 1. The eigenvalues of a Hermitian operator are real.

Proof: Assume that the eigenfunctions are normalized. By calculating the inner product of (12.30) with u, we find $\lambda = (u, Hu)$. The complex conjugate of this equation is $\lambda^* = (Hu, u)$. Then, because H is Hermitian, $\lambda^* = (u, Hu) = \lambda$, and the proof is complete.

Theorem 2. The Hermitian property of an operator is preserved when a unitary transformation is made.

Proof: Assume that H is Hermitian. The adjoint of the transformation $H' = SHS^\dagger$ is $H'^\dagger = SH^\dagger S^\dagger = SHS^\dagger = H'$, which proves the theorem.

Theorem 3. If H has all real eigenvalues, then a necessary condition for the existence of a unitary matrix which will transform it into a diagonal matrix with the eigenvalues appearing as the diagonal elements is that it be Hermitian.

Proof: If the diagonalizing unitary transformation exists, then $SHS^\dagger = \Lambda$, where Λ is real and diagonal. Since Λ is then Hermitian, theorem 2 guarantees that H is also. Q.E.D.

The formal construction of the diagonalizing unitary matrix is done by first writing the transformation in the form $HS^\dagger = S^\dagger\Lambda$. The component version is

$$\begin{aligned} \sum_k H_{jk} S^*_{nk} &= \sum_k S^*_{kj} \Lambda_{kn} = \sum_k S^*_{kj} \delta_{kn} \lambda_n = S^*_{nj} \lambda_n \\ &= \sum_k S^*_{nk} \delta_{jk} \lambda_n \end{aligned} \tag{12.31}$$

Equation (12.31) can be rewritten in the following

equivalent forms:

$$\sum_k (H_{jk} - \lambda_n \delta_{jk}) S^*_{nk} = 0 \qquad (H - \lambda_n I) u_n = 0$$

$$H u_n = \lambda_n u_n \tag{12.32}$$

For the finite-dimensional case, a nontrivial solution exists if, and only if,

$$|H - \lambda_n I| = 0 \tag{12.33}$$

which is the secular equation. The characteristic function is

$$f(\lambda) = |\lambda I - H| \tag{12.34}$$

The roots of $f(\lambda)$ are the eigenvalues and are guaranteed to be real if H is Hermitian. For each eigenvalue, an eigenvector can be constructed by means of (12.32). After normalization, the components of the eigenvectors give the required transformation by means of

$$u_n = (S^*_{n1}, S^*_{n2}, S^*_{n3}, \ldots) \tag{12.35}$$

The concept of determinant and characteristic function have no satisfactory generalization which is valid in the infinite-dimensional case. Instead, the eigenfunctions might be found as solutions of a differential equation as was done for the vibrating string in section 2 and the particle in a box in section 8. Infinite-dimensional problems can also be solved by alternate algebraic methods. This approach will be used in later sections to solve the harmonic oscillator problem and also to find the eigenvalues of the angular momentum operator.

Theorem 4. The eigenvectors corresponding to distinct eigenvalues are orthogonal.

Proof: The Hermiticity property of H is expressed by

$$(u_m, Hu_n) - (Hu_m, u_n) = 0 \tag{12.36}$$

where u_n and u_m are any two eigenfunctions of H. By means of (12.32), we find

$$(\lambda_n - \lambda_m)(u_m, u_n) = 0 \tag{12.37}$$

Thus, $(u_m, u_n) = 0$ if $\lambda_n \neq \lambda_m$, and the proof is complete. The orthogonality of the eigenvectors assures us that the unitary transformation can be constructed.

For unitary spaces, the degenerate case can be handled by theorem 6 in section 10 of Chapter 2. No essential modification is required if the matrix is Hermitian rather than symmetric. Thus, if H has real eigenvalues, its hermiticity is both necessary and sufficient for the existence of a diagonalizing unitary transformation. The eigenvalues are specified to be real because there is a class of non-Hermitian operators which have complex eigenvalues and which can be diagonalized by a unitary transformation. These are the *normal operators* defined by the condition $AA^\dagger = A^\dagger A$. Hermitian matrices are a subclass of the normal matrices.

Theorem 5. If H and J are two operators which are known to commute in one representation, then they commute in all representations.

Proof: The relation $HJ = JH$ is preserved in a unitary transformation:

$$\begin{aligned} SHJS^\dagger &= SJHS^\dagger \\ SHS^\dagger SJS^\dagger &= SJS^\dagger SHS^\dagger \\ H'J' &= J'H' \end{aligned} \tag{12.38}$$

Theorem 6. If the matrices H and J are diagonalized by the same transformation, they commute.

Proof: Assume that $SHS^\dagger = \Lambda_1$ and $SJS^\dagger = \Lambda_2$ where Λ_1 and Λ_2 are diagonal. Then $\Lambda_1 \Lambda_2 = \Lambda_2 \Lambda_1$ because any two diagonal matrices commute. By theorem 5, $HJ = JH$.

Theorem 7. If H and J are commuting Hermitian operators, then a unitary transformation can always be found which diagonalizes both.

Proof: Let S be a unitary transformation which diagonalizes H: $SHS^\dagger = \Lambda_1$. Since $HJ = JH$, theorem 5 shows that $\Lambda_1 J' = J' \Lambda_1$, where $J' = SJS^\dagger$. In component form, this is

$$\begin{aligned} \sum_k \Lambda_{nk} J'_{km} &= \sum_k J'_{nk} \Lambda_{km} \\ \sum_k \lambda_n \delta_{nk} J'_{km} &= \sum_k J'_{nk} \lambda_m \delta_{km} \\ \lambda_n J'_{nm} &= \lambda_m J'_{nm} \\ (\lambda_n - \lambda_m) J'_{nm} &= 0 \end{aligned} \tag{12.39}$$

Thus, if $\lambda_n \neq \lambda_m$, $J'_{nm} = 0$. The theorem is therefore established if there is no degeneracy. As usual, degeneracy requires special consideration. Suppose that the matrices are 5×5 and that H has two distinct eigenvalues. Equation

(12.39) then shows that Λ_1 and J' will be of the form

$$\Lambda_1 = \begin{pmatrix} \lambda_1 & 0 & 0 & 0 & 0 \\ 0 & \lambda_1 & 0 & 0 & 0 \\ 0 & 0 & \lambda_2 & 0 & 0 \\ 0 & 0 & 0 & \lambda_2 & 0 \\ 0 & 0 & 0 & 0 & \lambda_2 \end{pmatrix}$$

$$J' = \begin{pmatrix} J'_{11} & J'_{12} & 0 & 0 & 0 \\ J'_{21} & J'_{22} & 0 & 0 & 0 \\ 0 & 0 & J'_{33} & J'_{34} & J'_{35} \\ 0 & 0 & J'_{43} & J'_{44} & J'_{45} \\ 0 & 0 & J'_{53} & J'_{54} & J'_{55} \end{pmatrix} \tag{12.40}$$

A second unitary transformation can now be found of the form

$$S = \begin{pmatrix} S_{11} & S_{12} & 0 & 0 & 0 \\ S_{21} & S_{22} & 0 & 0 & 0 \\ 0 & 0 & S_{33} & S_{34} & S_{35} \\ 0 & 0 & S_{43} & S_{44} & S_{45} \\ 0 & 0 & S_{53} & S_{54} & S_{55} \end{pmatrix} \tag{12.41}$$

which will complete the diagonalization of J' and at the same time will not disrupt the diagonal form of Λ_1. To see this, note that the 2×2 submatrix

$$S_2 = \begin{pmatrix} S_{11} & S_{12} \\ S_{21} & S_{22} \end{pmatrix} \tag{12.42}$$

is itself unitary:

$$S_2 S_2^\dagger = \begin{pmatrix} 1 & 0 \\ 0 & 1 \end{pmatrix} = I_2 \tag{12.43}$$

When the transformation $S\Lambda_1 S^\dagger$ is carried out, S_2 works only on the 2×2 submatrix

$$\begin{pmatrix} \lambda_1 & 0 \\ 0 & \lambda_1 \end{pmatrix} = \lambda_1 I_2 \tag{12.44}$$

Then, since $S_2 \lambda_1 I_2 S_2^\dagger = \lambda_1 I_2$, nothing happens to Λ_1. The matrix S_2 can therefore be chosen so that it will diagonalize

$$\begin{pmatrix} J'_{11} & J'_{12} \\ J'_{21} & J'_{22} \end{pmatrix} \tag{12.45}$$

The remaining 3×3 submatrices can be treated in the same way.

We have shown that Hermitian operators can be diagonalized by the same unitary transformation if, and only if, they commute. This theorem is of fundamental importance in quantum mechanics. In a given quantum mechanical system, there are a number of dynamical variables each of which is represented by a Hermitian operator. For example, in the two-body central force problem, these variables are the total energy, the magnitude of the angular momentum, and the three components of the angular momentum. Some (not all) of these operators commute with one another. We have shown that a representation exists in which all the operators that commute with one another are diagonal. For a given group of such operators, there exists a wave function which is simultaneously an eigenfunction of each. This eigenfunction represents a possible stationary state of the system in which each of the dynamical variables represented by the commuting operators can have precise values. The values of each of these variables can be simultaneously accurately measured and known.

Let H be a Hermitian operator which represents a dynamical variable in quantum mechanics and suppose that a system is described by the wave function ψ. The quantity defined by

$$\langle H \rangle = (\psi, H\psi) \tag{12.46}$$

is of some importance. If ψ is a normalized eigenfunction of H with the eigenvalue λ, then the system is in a pure state and

$$\langle H \rangle = (u, Hu) = (u, \lambda u) = \lambda \tag{12.47}$$

A more general wave function can be expanded in terms of the eigenfunctions of H with the result that

$$\begin{aligned} \langle H \rangle &= \left(\sum_k A_k u_k, \sum_j A_j H u_j \right) \\ &= \sum_k \sum_j A_k^* A_j (u_k, Hu_j) \\ &= \sum_k \sum_j A_k^* A_j \lambda_j \delta_{kj} = \sum_k \lambda_k |A_k|^2 \end{aligned} \tag{12.48}$$

Given that the system is in the state described by the wave function ψ, $|A_k|^2$ is the probability that a measurement of the dynamical variable λ represented by H will yield the value λ_k. Now imagine a very large number of identical physical systems each described by the wave function ψ. If λ is measured for each of these systems and an average calculated, the result will be $\langle H \rangle$. For this reason, $\langle H \rangle$ is called the *expectation value* of the operator H. Conversely, we could begin the theory of operators in quantum mechanics by using (12.46) as the definition of expectation value. The requirement that $\langle H \rangle$ be real leads at once to the conclusion that H must be Hermitian.

Theorem 8. The eigenfunctions of a Hermitian operator are complete.

At the time that this book was written, a general proof of this theorem was not known, although everyone believes that it is true. Completeness can be established in special cases. For example, it is known that the eigenfunctions of the Sturm–Liouville differential equation are complete. The theorems of this section show that any Hermitian matrix of dimension N has N mutually orthogonal eigenvectors which span the space and are therefore complete. We assume that this remains valid as $N \to \infty$.

We will give an example of a Hermitian operator and demonstrate its completeness. In problem 10.3, the concept of the parity of a wave function is introduced. In one dimension, the *parity operator* is defined by

$$P\psi(x) = \psi(-x) \tag{12.49}$$

where the domain of ψ is $-\infty < x < \infty$. It is possible to write any function as a linear combination of an even function and an odd function:

$$\begin{aligned} \psi(x) &= \tfrac{1}{2}[\psi(x) + \psi(-x)] + \tfrac{1}{2}[\psi(x) - \psi(-x)] \\ &= \psi_e(x) + \psi_0(x) \end{aligned} \tag{12.50}$$

Note that $\psi_e(-x) = \psi_e(x)$ and $\psi_0(-x) = -\psi_0(x)$. That P is Hermitian is established by calculating its expectation value for an arbitrary function:

$$\begin{aligned} \langle P \rangle &= \int_{-\infty}^{+\infty} \psi^* P \psi \, dx = \int_{-\infty}^{+\infty} [\psi_e^*(x) + \psi_0^*(x)] \\ &\quad \times P[\psi_e(x) + \psi_0(x)] \, dx \\ &= \int_{-\infty}^{+\infty} [\psi_e^*(x) + \psi_0^*(x)][\psi_e(x) - \psi_0(x)] \, dx \\ &= \int_{-\infty}^{+\infty} |\psi_e|^2 \, dx - \int_{-\infty}^{+\infty} |\psi_0|^2 \, dx \\ &\quad + \int_{-\infty}^{+\infty} (\psi_0^* \psi_e - \psi_e^* \psi_0) \, dx \\ &= \int_{-\infty}^{+\infty} |\psi_e|^2 \, dx - \int_{-\infty}^{+\infty} |\psi_0|^2 \, dx \end{aligned} \tag{12.51}$$

The last step is true because $\psi_0^*\psi_e - \psi_e^*\psi_0$ is an odd function of x, and its integral is therefore zero. Thus, $\langle P \rangle$ is real and P is Hermitian. The eigenfunctions and eigenvalues of P are to be found from $Pu(x) = \lambda u(x)$. By the defining property of P as given by (12.49), this becomes $u(-x) = \lambda u(x)$. Multiplication by P and use of (12.49) again results in $u(x) = \lambda^2 u(x)$. Thus, $\lambda = \pm 1$ are the two possible eigenvalues of P. For $\lambda = +1$, $Pu_1(x) = u_1(x)$ or $u_1(-x) = u_1(x)$. For $\lambda = -1$, $Pu_2(x) = -u_2(x)$ or $u_2(-x) = -u_2(x)$. Thus, any even function is an eigenfunction if $\lambda = +1$ and any odd function is an eigenfunction if $\lambda = -1$. There is an infinite amount of degeneracy. Equation (12.50) gives the expansion of an arbitrary function in terms of even and odd functions.

The linear momentum operator in quantum mechanics is given by $\mathbf{p} = -i\hbar\nabla$. Its eigenfunctions and possible eigenvalues are to be found from

$$-i\hbar\nabla u(\mathbf{r}) = \hbar\boldsymbol{\kappa} u(\mathbf{r}) \tag{12.52}$$

The eigenfunctions are defined over all space and are

$$u_\kappa(\mathbf{r}) = Ce^{i\boldsymbol{\kappa}\cdot\mathbf{r}} \tag{12.53}$$

There are no restrictions on the possible values of $\boldsymbol{\kappa}$. The requirements of Hilbert space theory as developed in this section are overstepped because (12.53) is not square-integrable. The inner product of two "eigenfunctions" is

$$(u_{\kappa'}, u_\kappa) = \int_\Sigma |C|^2 e^{i(\boldsymbol{\kappa}-\boldsymbol{\kappa}')\cdot\mathbf{r}} \, d\Sigma \tag{12.54}$$

and does not seem to make any sense. In fact, this does not seem like a legitimate eigenvalue problem at all. The respect in which (12.53) may be regarded as a complete set of eigenfunctions with a continuum of possible eigenvalues will be explored in a later section. We will also show that (12.54) can be interpreted as a delta function.

The Hamiltonian for a free particle is

$$H = -\frac{\hbar^2}{2m}\nabla^2 \tag{12.55}$$

and commutes with the linear momentum operator because the derivatives of a function can be taken in any order. The eigenfunctions and eigenvalues of H

are found from $Hu_\kappa = W_\kappa u_\kappa$, which can be expressed as

$$\nabla^2 u_\kappa + \kappa^2 u_\kappa = 0 \qquad \kappa^2 = \frac{2mW_\kappa}{\hbar^2} \tag{12.56}$$

The momentum eigenfunctions (12.53) are also solutions of (12.56) on account of the relation

$$W_\kappa = \frac{\hbar^2\kappa^2}{2m} = \frac{p_\kappa^2}{2m} \tag{12.57}$$

Thus, the functions (12.53) are simultaneously eigenfunctions of the linear momentum operator and the free-particle Hamiltonian.

PROBLEMS

12.1. Show that the inner product (ψ, ϕ) of two different vectors is invariant under unitary transformations.

12.2. Theorem 5 is a special case of a more general theorem which states that any algebraic relation involving vectors and matrices is preserved when a unitary transformation is carried out. If $\alpha, \beta, \gamma, \ldots$ are constants and $H, J, K, \ldots$ are operators, show that the relation $\alpha H + \beta HJ + \gamma HJK + \cdots = 0$ implies that the transformed operators will obey $\alpha H' + \beta H'J' + \gamma H'J'K' + \cdots = 0$.

12.3. Prove that the eigenvalues of any operator are invariants when a unitary transformation is carried out.

12.4. Show that the non-Hermitian matrix

$$A = \begin{pmatrix} 4 & 1 \\ 3 & 2 \end{pmatrix}$$

has two distinct real eigenvalues and two linearly independent eigenvectors. The eigenvectors are not orthogonal. Normalize them and calculate their inner product. If ψ is a possible two-component wave function in a quantum mechanics problem, then it can be represented as a linear combination of the two eigenvectors U_1 and U_2 as $\psi = c_1U_1 + c_2U_2$, where c_1 and c_2 will in general be complex numbers. Show that the expectation value $(\psi, A\psi)$ is not in general real. Another problem with trying to use non-Hermitian operators in quantum mechanics is that they cannot be diagonalized by unitary transformations. Transformations are restricted to be unitary because of the requirement that wave functions preserve their normalization when a transformation is made.

12.5. Show that the non-Hermitian matrix

$$B = \begin{pmatrix} 4 & 1 \\ -5 & 2 \end{pmatrix}$$

has two distinct complex eigenvalues and two linearly independent but nonorthogonal eigenvectors.

12.6. Show that the non-Hermitian matrix

$$C = \begin{pmatrix} 4 & 1 \\ -1 & 2 \end{pmatrix}$$

is degenerate and has only one eigenvalue. Show that only one eigenvector exists.

12.7. Show that if H is an $n \times n$ Hermitian matrix with only one eigenvalue, then it must be a constant times the identity matrix: $H = \lambda I$. Does H have n linearly independent eigenvectors?

12.8. In a quantum mechanics problem, let u_k be the eigenfunctions of the Hermitian operator K. The Hamiltonian H does not commute with K so that the eigenfunctions u_k cannot be eigenfunctions of H. The wave function which describes the system can be expressed as

$$\psi = \sum_{k=1}^{\infty} A_k(t) u_k$$

Find the differential equations obeyed by the expansion coefficients $A_k(t)$.

12.9. If A and B are any operators, then their adjoints are defined by the relations $(\psi, A\phi) = (A^\dagger\psi, \phi)$ and $(\psi, B\phi) = (B^\dagger\psi, \phi)$. Show that the relation $(AB)^\dagger = B^\dagger A^\dagger$ follows as a consequence and can be proved without reference to a matrix representation.

Hint: Consider $(\psi, AB\phi)$.

12.10. Show that the most general form of a 2×2 unitary matrix is

$$S = \begin{pmatrix} e^{i\alpha}\cos\theta & e^{i\beta}\sin\theta \\ -e^{i\gamma}\sin\theta & e^{i\delta}\cos\theta \end{pmatrix}$$

where $\alpha + \delta - \beta - \gamma = 2n\pi$.

12.11. If S is a finite-dimensional unitary matrix, show that $|S||S|^* = 1$.

12.12. Show that the determinant of a finite-dimensional matrix is invariant under a unitary transformation. In problem 10.10 of Chapter 2, you showed that the determinant of a matrix equals the product of its eigenvalues. Show that the trace of a matrix is an invariant and that it equals the sum of the eigenvalues.

Hint: The result is most easily obtained by first expressing $H' = SHS^\dagger$ in component form as

$$H'_{nm} = \sum_j \sum_k S_{nj} H_{jk} S^*_{mk}$$

12.13. Let A be a 3×3 real orthogonal matrix: $A\tilde{A} = \tilde{A}A = I$. If A is applied to a vector in three-dimensional space, it rotates the vector without changing its length. Consider the eigenvalue problem defined by $AX = \lambda X$. Show that $|\lambda| = 1$. Since the characteristic function is a cubic with real coefficients, one of the eigenvalues is real and must be given by $\lambda_1 = \pm 1$. The other two roots are complex conjugates of one another and can be written as $\lambda_2 = e^{i\phi}$ and $\lambda_3 = e^{-i\phi}$. If A is proper, use the result of problem 12 to show that $\lambda_1 = +1$. If X_1 is the eigenvector associated with λ_1, then $AX_1 = X_1$. Thus, X_1 is unaffected by the action of A and must therefore be an axis of rotation. The trace of A is the sum of its three eigenvalues:

$$\operatorname{tr} A = 1 + e^{i\phi} + e^{-i\phi} = 1 + 2\cos\phi$$

If a coordinate system is chosen in which the x_3 axis points in the same direction as X_1, then S must be

$$S = \begin{pmatrix} \cos\phi & \sin\phi & 0 \\ -\sin\phi & \cos\phi & 0 \\ 0 & 0 & 1 \end{pmatrix}$$

Thus, ϕ is the angle of rotation about the eigenvector X_1. Since the general displacement of a rigid body can be described by an orthogonal transformation, we have proven *Euler's theorem*, which states that the most general displacement of a rigid body with one point fixed is a rotation about some axes.

12.14. Show that the expectation value of the linear momentum for a system described by the wave function (8.13) is

$$\langle p_x \rangle = (\psi, p_x\psi) = \int_0^a \psi^* \left(-i\hbar \frac{\partial\psi}{\partial x} \right) dx = \frac{8\hbar}{3a} \sin(3\omega_1 t)$$

12.15. The Hamiltonian for a particle moving in one dimension is

$$H = -\frac{\hbar^2}{2m}\frac{d^2}{dx^2} + V(x)$$

where $V(x)$ is even: $V(-x) = V(x)$. If P is the parity operator, show that $PH = HP$. If the potential is given by $V(x) = 0$ for $-a/2 < x < a/2$ and $V = \infty$ for $|x| > a/2$, show that the normalized eigenfunctions can be expressed as

$$\text{Even states: } u_n(x) = \sqrt{\frac{2}{a}}\cos\left(\frac{\pi n x}{a}\right) \qquad n = 1, 3, 5, \ldots$$

$$\text{Odd states: } u_n(x) = \sqrt{\frac{2}{a}}\sin\left(\frac{\pi n x}{a}\right) \qquad n = 2, 4, 6, \ldots$$

Show that the matrix representation of the parity operator with respect to this set of functions is

$$P = \begin{pmatrix} 1 & 0 & 0 & 0 & \cdots \\ 0 & -1 & 0 & 0 & \cdots \\ 0 & 0 & 1 & 0 & \cdots \\ 0 & 0 & 0 & -1 & \cdots \\ \vdots & \vdots & \vdots & \vdots & \end{pmatrix}$$

12.16. A *normal matrix* is defined by the condition $AA^\dagger = A^\dagger A$. Prove that a necessary condition for a matrix A to be diagonalized by a unitary transformation is that it be normal. The elements of the diagonalized matrix are not necessarily real.

12.17. Prove that any matrix can be expressed in the form $A = B + iC$, where both B and C are Hermitian. Prove that if A is normal, then $BC = CB$. By theorems 6 and 7, a unitary transformation exists which will diagonalize both B and C and therefore A. In problems 12.16 and 12.17, we have shown that a matrix can be diagonalized by a unitary transformation if, and only if, it is normal. Hermitian matrices are the subset of all normal matrices which have real eigenvalues. The normal matrix is the most general type of matrix that can be brought into diagonal form by a unitary transformation.

12.18. A quantum mechanical particle moves in one dimension under the influence of the potential $V(x)$. If H is the Hamiltonian and p_x is the linear momentum operator, show that

$$p_x H - H p_x = -i\hbar\frac{\partial V}{\partial x}$$

Thus, we do not expect to find a set of functions which are simultaneously eigenfunctions of both p_x and H.

Hint: Consider the action of $p_x H - H p_x$ on a function $\psi(x)$. Classically speaking, the momentum is not constant because of the force $F_x = -\partial V/\partial x$. In quantum mechanics, the fact that the momentum is not constant means that there are no stationary momentum states. Thus, dynamical variables in quantum mechanics with stationary states correspond to conserved quantities in classical mechanics.

13. THE HEISENBERG UNCERTAINTY RELATION

Consider a quantum mechanical system for which there are a number of Hermitian operators each of which corresponds to a measurable quantity. There are subsets of these operators which commute with one another, and the dynamical variables which they represent can therefore be simultaneously determined. If the wave function for the system is not an eigenfunction of a given operator, the associated dynamical variable can still be measured and the value obtained will be one of the eigenvalues of the operator. Repeated measurements on identically prepared systems yield different values the average of which is the expectation value of the operator. If P is any Hermitian operator and the system is in the state ψ, we define the uncertainty in P by means of

$$\Delta P^2 = (\psi, [P - \langle P \rangle]^2 \psi) \tag{13.1}$$

Equation (3.18) in Chapter 6 shows that ΔP corresponds to the conventional standard deviation. It is therefore an indication of the expected spread in the measured values of the dynamical variable corresponding to P with respect to the mean value $\langle P \rangle$. If ψ is an eigenfunction of P corresponding to the eigenvalue λ, then $\langle P \rangle = \lambda$ and $\Delta P = 0$.

Let P and Q be any two Hermitian operators. It is convenient to define new operators

$$A = P - \langle P \rangle \qquad B = Q - \langle Q \rangle \tag{13.2}$$

which are also Hermitian. It should be understood that $P - \langle P \rangle$ really means $P - \langle P \rangle I$, where I is the identity operator. In a matrix formulation, I is the identity matrix. If P is a differential operator, then the identity operator is just the number 1. By using the Hermitian property of A and B, it is possible to write

$$\Delta P^2 = (\psi, A^2\psi) = (A\psi, A\psi) = |A\psi|^2$$
$$\Delta Q^2 = |B\psi|^2 \tag{13.3}$$

Let μ be a real variable. We start with the obvious inequality

$$F(\mu) = ([A + i\mu B]\psi, [A + i\mu B]\psi) \geq 0 \tag{13.4}$$

The equality can hold if, and only if, $A\psi + i\mu B\psi = 0$. By again using the Hermitian property of A and B, (13.4) can be expressed as

$$\begin{aligned} F(\mu) &= (\psi, [A + i\mu B]^\dagger [A + i\mu B]\psi) \\ &= (\psi, [A - i\mu B][A + i\mu B]\psi) \\ &= (\psi, [A^2 + \mu i(AB - BA) + \mu^2 B^2]\psi) \\ &= |A\psi|^2 + \mu^2 |B\psi|^2 + \mu \langle i(AB - BA)\rangle \end{aligned} \tag{13.5}$$

We note that the operator $i(AB - BA)$ is Hermitian:

$$\begin{aligned} &[i(AB - BA)]^\dagger \\ &\quad = -i(B^\dagger A^\dagger - A^\dagger B^\dagger) = i(AB - BA) \end{aligned} \tag{13.6}$$

Thus, $\langle i(AB - BA)\rangle$ is a real number. This is also obvious from the fact that $F(\mu)$ as defined by (13.4) is real. By using the definitions (13.2) for A and B, it is possible to show that

$$AB - BA = PQ - QP \tag{13.7}$$

Thus,

$$F(\mu) = \Delta P^2 + \mu^2 \Delta Q^2 + \mu \langle i[P, Q]\rangle \geq 0 \tag{13.8}$$

We now choose the parameter μ such that $F(\mu)$ is a minimum:

$$\frac{dF}{d\mu} = 2\mu \, \Delta Q^2 + \langle i[P, Q]\rangle = 0,$$
$$\mu = -\frac{\langle i[P, Q]\rangle}{2\,\Delta Q^2} \tag{13.9}$$

If (13.8) and (13.9) are combined, the result is

$$\Delta P \Delta Q \geq \tfrac{1}{2}\langle i[P, Q]\rangle \tag{13.10}$$

which is the precise form of the Heisenberg uncertainty relation for any two operators P and Q.

As an example, suppose that P and Q are the x components of linear momentum and position:

$$P = p_x = -i\hbar \frac{\partial}{\partial x} \qquad Q = x \tag{13.11}$$

These operators do not commute. In fact,

$$(p_x x - x p_x)\psi = -i\hbar \frac{\partial}{\partial x}(x\psi) + i\hbar x \frac{\partial \psi}{\partial x} = -i\hbar\psi \tag{13.12}$$

Symbolically, this is

$$[p_x, x] = -i\hbar \tag{13.13}$$

This means that there is no set of functions which are simultaneously eigenfunctions of both p_x and x. The uncertainties Δp_x and Δx are never both simultaneously zero. The uncertainty relation (13.10) gives

$$\Delta p_x \Delta x \geq \tfrac{1}{2}\hbar \tag{13.14}$$

which was derived in an approximate form in equation (6.5). The equality in (13.14) holds if, and only if,

$$(p_x - \langle p_x \rangle)\psi = -i\mu(x - \langle x \rangle)\psi \qquad \mu = -\frac{\hbar}{2\Delta x^2} \tag{13.15}$$

which is a differential equation for $\psi(x)$. Assume a solution of the form

$$\psi = N \exp\left[\alpha(x - \langle x \rangle)^2 + \beta x\right] \tag{13.16}$$

By substituting (13.16) into (13.15), we find

$$-i\hbar[2\alpha(x - \langle x \rangle) + \beta] - \langle p_x \rangle = -i\mu(x - \langle x \rangle) \tag{13.17}$$

which can be true for all values of x if the constants α and β are

$$\alpha = \frac{\mu}{2\hbar} \qquad \beta = \frac{i\langle p_x \rangle}{\hbar} \tag{13.18}$$

The constant N is found from the normalization condition

$$\int_{-\infty}^{+\infty} \psi^*\psi \, dx = 1 \tag{13.19}$$

The required integral can be found in equation (20.31) in Chapter 5. The final result is

$$\psi = \frac{1}{(2\pi\Delta x^2)^{1/4}} \exp\left[-\frac{(x - \langle x \rangle)^2}{4\Delta x^2} + \frac{i\langle p_x \rangle x}{\hbar}\right] \tag{13.20}$$

The equality holds in the uncertainty relation (13.14) only if the system is described by the wave function (13.20).

If the position and momentum of the particle are measured to as great an accuracy as the uncertainty relation allows, the wave function for the system at the instant of measurement is given by (13.20). If $t = 0$ is taken as the time of measurement, then (13.20) is interpreted as the wave function at $t = 0$. The wave function $\psi(x, t)$ at some later time must be found by integrating Schrödinger's equation (6.21). Since Δx, Δp_x, $\langle x \rangle$, and $\langle p_x \rangle$ are in general functions of the time, the values of these quantities which appear in (13.20) are really initial values. The wave function (13.20) is called the *minimum uncertainty wave packet*.

PROBLEMS

13.1. If ψ is an eigenfunction of P, show that $\Delta P = 0$.

13.2. Show that $\Delta P^2 = \langle P^2 \rangle - \langle P \rangle^2$. Compare this result to problem 5.7 in Chapter 6.

13.3. In general, the expectation value of an operator is given by $\langle P \rangle = (\psi, P\psi)$. If the operator has no explicit time dependence, show, using the Schrödinger equation (6.21), that the time rate of change of the expectation value is

$$\frac{d}{dt}\langle P \rangle = \frac{i}{\hbar}\langle HP - PH \rangle \tag{13.21}$$

If P commutes with H, then $\langle P \rangle$ is a constant. For example, if $P = p_x$ and H is the free-particle Hamiltonian, then $[H, p_x] = 0$ and $\langle p_x \rangle$ is a constant. Since now there is no potential function, there is no force on the particle, and the expectation value of its momentum is constant in time. Show that $\langle P \rangle$ is also constant when H is independent of the time and ψ is an eigenfunction of H. Under these conditions, the system is said to be in a *stationary state*.

14. THE CAYLEY–HAMILTON THEOREM

In this and the next section, some further developments in the theory of matrices are presented.

Theorem 1. If the eigenvalues of an n-dimensional matrix A are all different, then the eigenvectors form a linearly independent set.

Proof: The validity of this theorem has already been established for real symmetric, Hermitian, and normal matrices. For these classes of matrices, the theorem holds even if there is degeneracy. There are however matrices which do not fall into any of these categories which also have complete sets of eigenvectors. The proof will be written for the case where A is 3×3, but it is easily extended to n dimensions. Let X_1, X_2, and X_3 be the three eigenvectors corresponding to the eigenvalues λ_1, λ_2, and λ_3. The eigenvalues may be real or complex. If the eigenvectors are linearly dependent, then there exists a relation of the form

$$\alpha X_1 + \beta X_2 + \gamma X_3 = 0 \tag{14.1}$$

where α, β, and γ are not all zero. If (14.1) is multiplied by A, the result is

$$\alpha\lambda_1 X_1 + \beta\lambda_2 X_2 + \gamma\lambda_3 X_3 = 0 \tag{14.2}$$

A second multiplication by A yields

$$\alpha\lambda_1^2 X_1 + \beta\lambda_2^2 X_2 + \gamma\lambda_3^2 X_3 = 0 \tag{14.3}$$

Let the three eigenvectors be written as

$$X_1 = \begin{pmatrix} x_1 \\ y_1 \\ z_1 \end{pmatrix} \qquad X_2 = \begin{pmatrix} x_2 \\ y_2 \\ z_2 \end{pmatrix} \qquad X_3 = \begin{pmatrix} x_3 \\ y_3 \\ z_3 \end{pmatrix} \tag{14.4}$$

Each of equations (14.1), (14.2), and (14.3) is really three equations. If we pick out of each set the equation which involves only the first components of the eigenvectors, we get three equations which can be expressed in the form

$$\begin{pmatrix} 1 & 1 & 1 \\ \lambda_1 & \lambda_2 & \lambda_3 \\ \lambda_1^2 & \lambda_2^2 & \lambda_3^2 \end{pmatrix} \begin{pmatrix} \alpha x_1 \\ \beta x_2 \\ \gamma x_3 \end{pmatrix} = 0 \tag{14.5}$$

There are similar equations for the second and third components of the eigenvectors. The determinant of the square matrix which appears in (14.5) is

$$D = (\lambda_1 - \lambda_2)(\lambda_1 - \lambda_3)(\lambda_2 - \lambda_3) \tag{14.6}$$

since the eigenvalues are all different, $D \neq 0$. This means that the only solution of (14.5) is the zero solution. Since all the components of all the eigenvectors are not zero, we conclude that $\alpha = \beta = \lambda = 0$. Thus, no relation of the form (14.1) exists and the eigenvectors are linearly independent. Theorem 1 does not rule out the possible existence of matrices which are not real symmetric, Hermitian, or normal and yet have complete sets of eigenvectors in spite of degeneracy.

Theorem 2. The eigenvalues of a matrix are unaffected by a general similarity transformation.

Proof: By a general similarity transformation is meant $A' = T^{-1}AT$. (By a redefinition of T, a similarity transformation can also be expressed as $A' = TAT^{-1}$, but this is notationally less convenient for our present purposes.) If $T = S^\dagger$, where S is unitary, the transformation is a unitary similarity transformation. The eigenvalue equation $AX = \lambda X$ can be expressed as

$$T^{-1}ATT^{-1}X = \lambda T^{-1}X \qquad \text{or} \qquad A'X' = \lambda X' \tag{14.7}$$

where $X' = T^{-1}X$. Thus, the eigenvalues remain invariant when a similarity transformation is carried out.

Theorem 3. Any matrix can be brought into triangular form with the eigenvalues along the main diagonal and zeros everywhere below the main diagonal by means of a similarity transformation. The transformation can be chosen to be unitary, but it need not be.

Proof: Let λ_1 be an eigenvalue of the n-dimensional matrix A and let X_1 be the corresponding eigenvector. Choose a transformation T_1 such that

$$T_1 X_1' = \begin{pmatrix} 11 & 12 & \cdots \\ 21 & 22 & \cdots \\ \vdots & \vdots & \end{pmatrix} \begin{pmatrix} 1 \\ 0 \\ \vdots \end{pmatrix} = X_1 = \begin{pmatrix} x_1 \\ x_2 \\ \vdots \end{pmatrix} \tag{14.8}$$

To save some space, the elements of matrices will sometimes be indicated by number pairs in those cases where no confusion will result. By multiplying (14.8) out, we find

$$(11) = x_1 \qquad (21) = x_2 \qquad (31) = x_3 \qquad \cdots \tag{14.9}$$

Thus, only the first column of T_1 is determined. If X_1 is normalized, the undetermined part of T_1 can be chosen such as to make it unitary if we wish. In any event, T_1 must be nonsingular. Let T_1 be applied to A. By theorem 2, this does not affect the eigenvalues. In the transformed reference frame, the eigenvalue equation is

$$AX_1 = \begin{pmatrix} 11 & 12 & 13 & \cdots \\ 21 & 22 & 23 & \cdots \\ 31 & 32 & 33 & \cdots \\ \vdots & \vdots & \vdots & \end{pmatrix} \begin{pmatrix} 1 \\ 0 \\ 0 \\ \vdots \end{pmatrix} = \lambda_1 \begin{pmatrix} 1 \\ 0 \\ 0 \\ \vdots \end{pmatrix} \tag{14.10}$$

We really should label A, X_1, and all the elements of A with primes, but these are left out to simplify the notation. Let's just remember that we have done a transformation. If (14.10) is multiplied out, the result is

$$(11) = \lambda_1 \qquad (21) = 0 \qquad (31) = 0 \qquad \cdots \tag{14.11}$$

Thus, the transformed matrix must be

$$A = \begin{pmatrix} \lambda_1 & 12 & 13 & \cdots \\ 0 & 22 & 23 & \cdots \\ 0 & 32 & 33 & \cdots \\ \cdot & \cdot & \cdot & \end{pmatrix} \tag{14.12}$$

Let us now construct a second transformation of the form

$$T_2 = \begin{pmatrix} 1 & 0 & 0 & \cdots \\ 0 & 22 & 23 & \cdots \\ 0 & 32 & 33 & \cdots \\ \vdots & \vdots & \vdots & \end{pmatrix} \tag{14.13}$$

By analyzing what happens when $T_2^{-1}AT_2$ is carried out, you can convince yourself that the first column of A is not disturbed. Another important fact is that the elements

$$\begin{pmatrix} 22 & 23 & \cdots \\ 32 & 33 & \cdots \\ \vdots & \vdots & \end{pmatrix} \tag{14.14}$$

in A are affected only by the corresponding elements in T_2. Thus, by working in the $n-1$ dimensional subspace defined by leaving out the first columns and first rows of all matrices, we can transform A into the form

$$A = \begin{pmatrix} \lambda_1 & 12 & 13 & \cdots \\ 0 & \lambda_2 & 23 & \cdots \\ 0 & 0 & 33 & \cdots \\ \vdots & \vdots & \vdots & \end{pmatrix} \tag{14.15}$$

In constructing the $n-1 \times n-1$ transformation, an $n-1$ dimensional eigenvector of the $n-1$ dimensional subspace of A is used. This is not necessarily an eigenvector of the full matrix A. This is of no consequence since we are mainly interested in constructing the transformation (14.13). By continuing in this manner, A can be brought into triangular form with the eigenvalues appearing on the main diagonal and zeros everywhere below the main diagonal. This process works whether or not there is degeneracy. We have just proved a more general form of theorem 6 in section 10 of Chapter 2.

Theorem 4. The Cayley-Hamilton Theorem. Since any positive power of a matrix is defined, it is possible to construct polynomial functions of matrices:

$$f(A) = A^n + a_1 A^{n-1} + a_2 A^{n-2} + \cdots + a_{n-1}A + a_n I \tag{14.16}$$

In particular, if $f(\lambda)$ is the characteristic function of A given by

$$f(\lambda) = |\lambda I - A| = \lambda^n + a_1\lambda^{n-1} + \cdots + a_{n-1}\lambda + a_n \tag{14.17}$$

then

$$f(A) = 0 \tag{14.18}$$

Proof: Consider first the case where A has n linearly independent eigenvectors. If X_k is any one of the eigenvectors, then

$$AX_k = \lambda_k X_k \qquad A^2X = \lambda_k^2 X_k \qquad \cdots \qquad A^n X_k = \lambda_k^n X_k \tag{14.19}$$

$$\begin{aligned} f(A)X_k &= \left(A^n + a_1 A^{n-1} + \cdots + a_{n-1}A + a_n I\right)X_k \\ &= \left(\lambda_k^n + a_1\lambda_k^{n-1} + \cdots + a_{n-1}\lambda_k + a_n\right)X_k \\ &= f(\lambda_k)X_k = 0 \end{aligned} \tag{14.20}$$

The last step is true because the eigenvalues are roots of $f(\lambda)$. Since (14.20) holds for each of the n linearly independent eigenvectors, $f(A)=0$ by theorem 1 in section 10 of Chapter 1. If A does not have n linearly independent eigenvalues, it can still be brought into triangular form by an appropriate similarity transformation. Consider the 3×3 case. With A given by

$$A = \begin{pmatrix} \lambda_1 & 12 & 13 \\ 0 & \lambda_2 & 23 \\ 0 & 0 & \lambda_3 \end{pmatrix} \tag{14.21}$$

it is possible to write the matrix polynomial derived from the characteristic function in factored form as

$$\begin{aligned} f(A) &= (A-\lambda_1 I)(A-\lambda_2 I)(A-\lambda_3 I) \\ &= \begin{pmatrix} 0 & 12 & 13 \\ 0 & \lambda_2-\lambda_1 & 23 \\ 0 & 0 & \lambda_3-\lambda_1 \end{pmatrix} \\ &\times \begin{pmatrix} \lambda_1-\lambda_2 & 12 & 13 \\ 0 & 0 & 23 \\ 0 & 0 & \lambda_3-\lambda_2 \end{pmatrix} \\ &\times \begin{pmatrix} \lambda_1-\lambda_3 & 12 & 13 \\ 0 & \lambda_2-\lambda_3 & 23 \\ 0 & 0 & 0 \end{pmatrix} \end{aligned} \tag{14.22}$$

By multiplying the matrices out, we find $f(A)=0$.

Theorem 5. If A is an $n \times n$ matrix and if $t_1 = \operatorname{tr} A$, $t_2 = \operatorname{tr} A^2, \ldots t_n = \operatorname{tr} A^n$ are the traces of the various powers of A, then the coefficients in the characteristic function (14.17) are given by

$$\begin{aligned} a_1 &= -t_1 \\ a_2 &= -\tfrac{1}{2}(t_2 + a_1 t_1) \\ a_3 &= -\tfrac{1}{3}(t_3 + a_1 t_2 + a_2 t_1) \\ &\vdots \\ a_k &= -\frac{1}{k}(t_k + a_1 t_{k-1} + a_2 t_{k-2} + \cdots + a_{k-1}t_1) \end{aligned} \tag{14.23}$$

Proof: The characteristic function can be expressed in factored or expanded form as

$$f(\lambda)=(\lambda-\lambda_1)(\lambda-\lambda_2)\ldots(\lambda-\lambda_n)$$
$$=\lambda^n-(\lambda_1+\lambda_2+\cdots+\lambda_n)\lambda^{n-1}$$
$$+\cdots+(-1)^n\lambda_1\lambda_2\ldots\lambda_n \tag{14.24}$$

Thus, in general,

$$a_1=-(\lambda_1+\lambda_2+\cdots+\lambda_n)=-t_1 \tag{14.25}$$

Suppose that A is a 2×2 matrix. By the Cayley–Hamilton theorem,

$$A^2+a_1A+a_2I=0 \tag{14.26}$$

By taking the trace and solving for a_2, we find

$$a_2=-\tfrac{1}{2}(t_2+a_1t_1) \tag{14.27}$$

We have already shown that $a_1=-t_1$ if A is a matrix of any finite dimension. Since the coefficients in the characteristic function as well as the traces are invariants, we can expect that the same relation between the coefficients and the traces will prevail independently of the dimension of the matrix. Thus, (14.27) holds generally. If A is 3×3, the Cayley–Hamilton theorem gives

$$A^3+a_1A^2+a_2A+a_3I=0 \tag{14.28}$$

Taking the trace and solving for a_3 gives

$$a_3=-\tfrac{1}{3}(t_3+a_1t_2+a_2t_1) \tag{14.29}$$

By continuing in this manner, we can show that (14.23) gives the general formula for the coefficients.

The value of theorem 5 lies in the fact that it is quite easy to write a computer program that uses (14.23) to compute the coefficients in the characteristic function. The eigenvalues can then be found by numerical computation of the roots of $f(\lambda)$. If A is nonsingular, then the Cayley–Hamilton theorem can be used to find its inverse by means of

$$A^{-1}=\frac{-1}{a_n}(A^{n-1}+a_1A^{n-2}+\cdots$$
$$+a_{n-2}A+a_{n-1}I) \tag{14.30}$$

Another result of interest is that the determinant of A is given by

$$|A|=(-1)^n a_n \tag{14.31}$$

PROBLEMS

14.1. Given the matrices

$$A=\begin{pmatrix}\lambda & a & 0\\ 0 & \lambda & b\\ 0 & 0 & \mu\end{pmatrix}\qquad B=\begin{pmatrix}\lambda & 0 & c\\ 0 & \lambda & 0\\ 0 & 0 & \mu\end{pmatrix}$$

show that A has two linearly independent eigenvectors and that B has three. Assume that a, b, and c are not zero and that $\lambda\neq\mu$.

14.2. Go through the steps used in proving theorem 3 to reduce

$$A=\begin{pmatrix}2 & 0 & -1\\ -1 & 1 & 1\\ -1 & 0 & 3\end{pmatrix}$$

to triangular form.

Hint: One of the roots of the characteristic function is $\lambda_1=1$.

14.3. If A has n linearly independent eigenvectors, show that a similarity transformation exists which will bring it into diagonal form by means of $T^{-1}AT=\Lambda$.

Hint: First write the transformation as $AT=T\Lambda$. Follow the same steps that led from equation (12.31) to (12.32).

14.4. If we were to laboriously expand the characteristic function as given by (14.17), we would find that

$$a_1=-\operatorname{tr}A=-(A_{11}+A_{22}+\cdots+A_{nn})\qquad a_n=(-1)^n|A|$$

Comparison to (14.24) shows that

$$\operatorname{tr} A = \lambda_1 + \lambda_2 + \cdots + \lambda_n \qquad |A| = \lambda_1 \lambda_2 \ldots \lambda_n$$

The remaining coefficients in the characteristic function are more complicated functions of the elements of A. Show however that these coefficients are invariants under a general similarity transformation.

14.5. If A is 3×3, use (14.24) to show that

$$a_2 = \lambda_1 \lambda_2 + \lambda_1 \lambda_3 + \lambda_2 \lambda_3$$

By using the fact that

$$t_1 = \lambda_1 + \lambda_2 + \lambda_3 \qquad t_2 = \lambda_1^2 + \lambda_2^2 + \lambda_3^2$$

show that

$$a_2 = -\tfrac{1}{2}(a_1 t_1 + t_2)$$

14.6. Given the matrix

$$A = \begin{pmatrix} 1 & 2 & 0 \\ 2 & 3 & 0 \\ 0 & 0 & 4 \end{pmatrix}$$

compute the traces t_1, t_2, and t_3. Compute the coefficients in the characteristic polynomial by means of (14.23). Use (14.30) and (14.31) to compute A^{-1} and $|A|$.

14.7. In this and the next problem, assume that the matrices are 3×3. The results are easily generalized to the $n \times n$ case. A possible transformation of the type used in the proof of theorem 3 is

$$T_1 = \begin{pmatrix} x_1 & 0 & 0 \\ x_2 & 1 & 0 \\ x_3 & 0 & 1 \end{pmatrix}$$

What is the inverse of this transformation? Assume that $x_1 \neq 0$.

14.8. If $A' = T^{-1}AT$, where T is any transformation of the type given in problem 14.7, show that A and A' have the same rank. This means that the triangularization procedure carried out in the proof of theorem 3 yields a transformed matrix with the same rank as the original matrix. Thus, the number of eigenvalues of A that are zero equals the difference between its dimension and its rank.

14.9. Every element of an $n \times n$ matrix equals 1. Show that one eigenvalue equals n and that all other eigenvalues are zero.

15. SPECTRAL THEOREM

Many of the results of this section will be stated for 3×3 matrices, but in all cases the generalization to the $n \times n$ case is possible and the reader should have no trouble making this extension. Let A be a 3×3 matrix with three distinct eigenvalues λ_1, λ_2, and λ_3. We define three auxiliary matrices associated with A by means of

$$G_1 = \frac{(\lambda_2 I - A)(\lambda_3 I - A)}{(\lambda_2 - \lambda_1)(\lambda_3 - \lambda_1)}$$
$$G_2 = \frac{(\lambda_1 I - A)(\lambda_3 I - A)}{(\lambda_1 - \lambda_2)(\lambda_3 - \lambda_2)} \tag{15.1}$$
$$G_3 = \frac{(\lambda_1 I - A)(\lambda_2 I - A)}{(\lambda_1 - \lambda_3)(\lambda_2 - \lambda_3)}$$

By theorem 1 in section 14, A has three linearly independent eigenvectors X_1, X_2, and X_3. We will determine the effect of the matrices (15.1) on each of the three eigenvectors. Note that

$$\begin{aligned}(\lambda_2 I - A)(\lambda_3 I - A)X_1 &= (\lambda_2 I - A)(\lambda_3 - \lambda_1)X_1 \\ &= (\lambda_3 - \lambda_1)(\lambda_2 I - A)X_1 \\ &= (\lambda_3 - \lambda_1)(\lambda_2 - \lambda_1)X_1\end{aligned} \tag{15.2}$$

$$\begin{aligned}(\lambda_2 I - A)(\lambda_3 I - A)X_2 &= (\lambda_3 I - A)(\lambda_2 I - A)X_2 \\ &= 0\end{aligned} \tag{15.3}$$

By use of (15.2) and (15.3) and other similarly derived relations, it is established that

$$\begin{array}{lll} G_1X_1 = X_1 & G_1X_2 = 0 & G_1X_3 = 0 \\ G_2X_1 = 0 & G_2X_2 = X_2 & G_2X_3 = 0 \\ G_3X_1 = 0 & G_3X_2 = 0 & G_3X_3 = X_3 \end{array} \tag{15.4}$$

Let X be any vector. Since the three eigenvectors are linearly independent, X can be expressed as a linear combination of them:

$$X = \alpha_1 X_1 + \alpha_2 X_2 + \alpha_3 X_3 \tag{15.5}$$

By means of (15.4), the effect of the three matrices G_1, G_2, and G_3 on the arbitrary vector (15.5) is determined to be

$$G_1 X = \alpha_1 X_1 \qquad G_2 X = \alpha_2 X_2 \qquad G_3 X = \alpha_3 X_3 \tag{15.6}$$

The effect of each of the three matrices G_1, G_2, and G_3 on X is to compute its projection or component in the direction of the corresponding eigenvector. For this reason, these matrices are called *projection operators*. Other important properties of the projection operators are found by considering the effect of various combinations of them on an arbitrary vector. For instance,

$$G_1 G_2 X = G_1 \alpha_2 X_2 = 0 \tag{15.7}$$

Theorem 2 in section 10 of Chapter 1 shows that $G_1G_2 = 0$. Similarly,

$$G_1^2 X = G_1 \alpha_1 X_1 = \alpha_1 X_1 = G_1 X \tag{15.8}$$

proves that $G_1^2 = G_1$, and

$$(G_1 + G_2 + G_3)X = \alpha_1 X_1 + \alpha_2 X_2 + \alpha_3 X_3 = X \tag{15.9}$$

shows that $G_1 + G_2 + G_3 = I$. These properties are summarized as follows:

$$\begin{array}{lll} G_1^2 = G_1 & G_1G_2 = 0 & G_1G_3 = 0 \\ G_2G_1 = 0 & G_2^2 = G_2 & G_2G_3 = 0 \\ G_3G_1 = 0 & G_3G_2 = 0 & G_3^2 = G_3 \\ G_1 + G_2 + G_3 = I & & \end{array} \tag{15.10}$$

If A is a 3×3 matrix and one eigenvalue is twofold degenerate, it is still possible to find three linearly independent eigenvectors if A is normal and even in some cases where A is not normal, as problem 14.1 reveals. Normal matrices are defined in problem 12.16. Recall that Hermitian and real symmetric matrices are special types of normal matrices. Assume that λ_1 is the eigenvalue corresponding to the eigenvectors X_1 and X_2 and that λ_3 corresponds to X_3. Two projection operators can be defined as

$$G_1 = \frac{\lambda_3 I - A}{\lambda_3 - \lambda_1} \qquad G_3 = \frac{\lambda_1 I - A}{\lambda_1 - \lambda_3} \tag{15.11}$$

Application of G_1 and G_3 to the three eigenvectors gives

$$\begin{array}{lll} G_1X_1 = X_1 & G_1X_2 = X_2 & G_1X_3 = 0 \\ G_3X_1 = 0 & G_3X_2 = 0 & G_3X_3 = X_3 \end{array} \tag{15.12}$$

If X is an arbitrary vector expressed as a linear combination of the eigenvectors as given in equation (15.5), then

$$G_1 X = \alpha_1 X_1 + \alpha_2 X_2 \tag{15.13}$$

Thus, G_1 yields the projection of X onto the subspace determined by X_1 and X_2. By calculating the effect of various combinations of G_1 and G_3 on an arbitrary vector which has been expressed as a linear combination of the eigenvectors, we can show that

$$\begin{array}{lll} G_1^2 = G_1 & G_3^2 = G_3 & G_1G_3 = 0 \\ G_1 + G_3 = I & & \end{array} \tag{15.14}$$

For example,

$$G_1G_3X = G_1G_3(\alpha_1X_1 + \alpha_2X_2 + \alpha_3X_3)$$
$$= G_1(\alpha_3X_3) = 0 \quad (15.15)$$

Since X is arbitrary, $G_1G_3 = 0$. This result is interesting because it means that

$$(A - \lambda_1 I)(A - \lambda_3 I) = A^2 - (\lambda_1 + \lambda_3)A + \lambda_1\lambda_3 I$$
$$= 0 \quad (15.16)$$

This shows that there is a matrix polynomial of lower degree than the degree of the matrix polynomial derived from the characteristic function, which is zero. The result generalizes. If A is $n \times n$ and has n linearly independent eigenvectors and f distinct eigenvalues, then

$$P_f(A) = (A - \lambda_1 I)(A - \lambda_2 I)\ldots(A - \lambda_f I) = 0 \quad (15.17)$$

where $\lambda_1, \lambda_2, \ldots,$ and λ_f are the distinct eigenvalues. The polynomial $P_f(\lambda)$ is called the *minimum polynomial* of A and is the polynomial of lowest degree such that $P_f(A) = 0$. Necessarily, $f \le n$.

If A is a 3×3 matrix and has a single threefold degenerate eigenvalue and three linearly independent eigenvectors, then the effect of A on an arbitrary vector is

$$AX = A(\alpha_1X_1 + \alpha_2X_2 + \alpha_3X_3)$$
$$= \alpha_1\lambda X_1 + \alpha_2\lambda X_2 + \alpha_3\lambda X_3$$
$$= \lambda X = \lambda IX \quad (15.18)$$

Thus, every vector is an eigenvector and A is a multiple of the identity matrix: $A = \lambda I$.

If X_1 is an eigenvector of A, then repeated application of A to X_1 shows that

$$A^kX_1 = \lambda^kX_1 \quad (15.19)$$

If $P(A)$ is any polynomial function of A, then

$$P(A)X_1 = P(\lambda_1)X_1 \quad (15.20)$$

Let A be a 3×3 matrix with three linearly independent eigenvectors. If X is an arbitrary vector, then

$$P(A)X = P(A)[\alpha_1X_1 + \alpha_2X_2 + \alpha_3X_3]$$
$$= \alpha_1P(\lambda_1)X_1 + \alpha_2P(\lambda_2)X_2 + \alpha_3P(\lambda_3)X_3 \quad (15.21)$$

By using the properties of the projection operators as given by (15.4), we can put (15.21) into the form

$$P(A)X = [P(\lambda_1)G_1 + P(\lambda_2)G_2 + P(\lambda_3)G_3]X \quad (15.22)$$

Because X is arbitrary, (15.22) implies that

$$P(A) = P(\lambda_1)G_1 + P(\lambda_2)G_2 + P(\lambda_3)G_3 \quad (15.23)$$

In particular, if $P(A) = A$,

$$A = \lambda_1G_1 + \lambda_2G_2 + \lambda_3G_3 \quad (15.24)$$

If A has only two distinct eigenvalues but still has three linearly independent eigenvectors, then the corresponding result is

$$P(A) = P(\lambda_1)G_1 + P(\lambda_3)G_3 \quad (15.25)$$
$$A = \lambda_1G_1 + \lambda_3G_3 \quad (15.26)$$

where now G_1 and G_3 are given by (15.10). Equations (15.24) and (15.25) are referred to as the *spectral resolution* of a matrix and constitute the main result of what is known as the *spectral theorem*. Fortunately, this terminology is consistent with the use of the word *spectrum* in physics, since many times what a physicist calls a spectrum is actually a set of eigenvalues of a Hermitian operator.

We will now summarize and extend the results so far obtained in this section to the case where A is $n \times n$. Assume that A has n linearly independent eigenvectors and f distinct eigenvalues. There are f projection operators which can be defined by

$$G_k = \prod_{\substack{j=1 \\ j \ne k}}^{f} \frac{\lambda_j I - A}{\lambda_j - \lambda_k} \quad (15.27)$$

For example,

$$G_1 = \frac{(\lambda_2 I - A)(\lambda_3 I - A)\ldots(\lambda_f I - A)}{(\lambda_2 - \lambda_1)(\lambda_3 - \lambda_1)\ldots(\lambda_f - \lambda_1)} \quad (15.28)$$

The projection operators obey

$$G_jG_k = 0 \text{ if } j \ne k \qquad G_k^2 = G_k \qquad \sum_{k=1}^{f} G_k = I \quad (15.29)$$

If X_k is any one of the eigenvectors corresponding to the eigenvalue λ_k, then $G_kX_k = X_k$. If X_j is any other eigenvector, then $G_kX_j = 0$.

If A is a 3×3 degenerate matrix with fewer than three linearly independent eigenvectors, the spectral theorem breaks down. We will now show why this happens. Theorem 3 of section 14 shows that any matrix is similar to a triangular matrix, that is, it can be brought into the form

$$A=\begin{pmatrix}\lambda & a & c\\ 0 & \lambda & b\\ 0 & 0 & \mu\end{pmatrix} \tag{15.30}$$

by a similarity transformation. We are assuming here that the eigenvalue λ is twofold degenerate. Further reduction of the matrix by similarity transformations is possible. Consider the transformation given by

$$T=\begin{pmatrix}1 & 0 & 0\\ 0 & 1 & x\\ 0 & 0 & 1\end{pmatrix} \qquad T^{-1}=\begin{pmatrix}1 & 0 & 0\\ 0 & 1 & -x\\ 0 & 0 & 1\end{pmatrix}$$

$$T^{-1}AT=\begin{pmatrix}\lambda & a & c+ax\\ 0 & \lambda & \lambda x+b-\mu x\\ 0 & 0 & \mu\end{pmatrix} \tag{15.31}$$

Thus, if x is chosen by the condition $\lambda x+b-\mu x=0$, the transformed matrix has a zero in the 23 position. This always works because $\lambda\neq\mu$. Moreover, we have not ruined the previously accomplished triangularization. Now consider the transformation

$$T=\begin{pmatrix}1 & 0 & x\\ 0 & 1 & 0\\ 0 & 0 & 1\end{pmatrix} \qquad T^{-1}=\begin{pmatrix}1 & 0 & -x\\ 0 & 1 & 0\\ 0 & 0 & 1\end{pmatrix} \tag{15.32}$$

With A given by (15.30), we find

$$T^{-1}AT=\begin{pmatrix}\lambda & a & c-\mu x+\lambda x\\ 0 & \lambda & b\\ 0 & 0 & \mu\end{pmatrix} \tag{15.33}$$

If x is chosen by the condition $c-\mu x+\lambda x=0$, then the element in the 13 position of the transformed matrix is zero. None of the other elements in the matrix is disturbed. For example, b could have been made zero by (15.31) before (15.32) is applied. The result is that A can always be reduced to

$$A=\begin{pmatrix}\lambda & a & 0\\ 0 & \lambda & 0\\ 0 & 0 & \mu\end{pmatrix} \tag{15.34}$$

If $a=0$, A is in diagonal form and there is no point in continuing. If $a\neq 0$, a further simplification is possible. Let T be

$$T=\begin{pmatrix}1 & 0 & 0\\ 0 & x & 0\\ 0 & 0 & 1\end{pmatrix} \qquad T^{-1}=\begin{pmatrix}1 & 0 & 0\\ 0 & \dfrac{1}{x} & 0\\ 0 & 0 & 1\end{pmatrix} \tag{15.35}$$

The effect of T on A as given by (15.34) is

$$T^{-1}AT=\begin{pmatrix}\lambda & ax & 0\\ 0 & \lambda & 0\\ 0 & 0 & \mu\end{pmatrix} \tag{15.36}$$

If we choose x by the condition $ax=1$, then

$$A=\begin{pmatrix}\lambda & 1 & 0\\ 0 & \lambda & 0\\ 0 & 0 & \mu\end{pmatrix} \tag{15.37}$$

This is known as the *Jordan canonical form* and is essentially the greatest degree of simplification that is possible by means of similarity transformations.

If A is given by (15.37), there are only two linearly independent eigenvectors. They are

$$X_1=\begin{pmatrix}1\\0\\0\end{pmatrix} \qquad X_2=\begin{pmatrix}0\\0\\1\end{pmatrix} \tag{15.38}$$

where X_1 corresponds to λ and X_2 to μ. If we define

$$G_1=\frac{\mu I-A}{\mu-\lambda} \qquad G_2=\frac{\lambda I-A}{\lambda-\mu} \tag{15.39}$$

we find that G_1 and G_2 have some, but not all, of the properties of projection operators. For example, it is true that

$$\begin{aligned}&G_1X_1=X_1 \qquad G_1X_2=0 \qquad G_2X_1=0\\ &G_2X_2=X_2 \qquad G_1+G_2=I\end{aligned} \tag{15.40}$$

However,

$$G_1G_2\neq 0 \qquad G_1^2\neq G_1 \qquad G_2^2\neq G_2 \tag{15.41}$$

Also, we find that

$$G_1\begin{pmatrix}0\\1\\0\end{pmatrix}=\begin{pmatrix}\dfrac{1}{\lambda-\mu}\\ 1\\ 0\end{pmatrix} \tag{15.42}$$

which shows that G_1 does not give the projection of

an arbitrary vector onto X_1. By direct computation, it can be shown that

$$A = \lambda G_1 + \mu G_2 \tag{15.43}$$

We find however that

$$A^2 \neq \lambda^2 G_1 + \mu^2 G_2 \tag{15.44}$$

which means that there is no simple representation for an arbitrary polynomial function of A such as is given by (15.25). We will say that A has a spectral decomposition only if the matrices G_1 and G_2 are true projection operators. Otherwise, the spectral theorem fails.

We have shown that it is possible to reduce an arbitrary matrix to Jordan form by a similarity transformation. Could it not be that an even greater simplification would result if a more general transformation were used? Possibly so. The reason for using a similarity transformation is that it preserves the essential features of the matrix in which we are interested. For example, the eigenvalues are not changed by a similarity transformation. Also, algebraic equations involving matrices retain their form when a similarity transformation is made. You investigated this property for the special case of unitary similarity transformations in problem 12.2.

Let us now return to the case of a 3×3 matrix with three distinct eigenvalues. Consider the matrix polynomial

$$\begin{aligned} P_N(A) &= \sum_{k=0} \frac{1}{k!} A^k \\ &= \sum_{k=0}^{N} \frac{1}{k!}\left(\lambda_1^k G_1 + \lambda_2^k G_2 + \lambda_3^k G_3\right) \end{aligned} \tag{15.45}$$

The limit $N \to \infty$ exists and gives the result

$$e^A = \sum_{k=0}^{\infty} \frac{1}{k!} A^k = e^{\lambda_1} G_1 + e^{\lambda_2} G_2 + e^{\lambda_3} G_3 \tag{15.46}$$

In this way, functions of matrices can be defined. If one of the eigenvalues is twofold degenerate but three linearly independent eigenvectors still exist, then (15.25) is valid and

$$e^A = e^{\lambda_1} G_1 + e^{\lambda_3} G_3 \tag{15.47}$$

It is in fact true that e^A can be defined for any square matrix. If no spectral theorem exists for A, then some other means must be found to establish convergence of the series which defines e^A. Consider again the polynomial of degree N:

$$P_N(A) = \sum_{k=0}^{N} \frac{1}{k!} A^k \tag{15.48}$$

If a similarity transformation is applied, the result is

$$T^{-1} P_N(A) T = \sum_{k=0}^{N} \frac{1}{k!} T^{-1} A^k T = \sum_{k=0}^{N} \frac{1}{k!} (T^{-1} A T)^k \tag{15.49}$$

But we know that A can be brought into the Jordan canonical form (15.37) if T is appropriately chosen. Assuming that this has been done, the powers of A are

$$A^k = \begin{pmatrix} \lambda^k & k\lambda^{k-1} & 0 \\ 0 & \lambda^k & 0 \\ 0 & 0 & \mu^k \end{pmatrix} \tag{15.50}$$

Thus, in the limit $N \to \infty$, (15.48) becomes

$$e^A = \begin{pmatrix} e^\lambda & e^\lambda & 0 \\ 0 & e^\lambda & 0 \\ 0 & 0 & e^\mu \end{pmatrix} \tag{15.51}$$

and the convergence is established.

Let A be an $n \times n$ matrix which has a spectral resolution. We can define

$$\cos A = \sum_k G_k \cos \lambda_k \tag{15.52}$$

$$\sin A = \sum_k G_k \sin \lambda_k \tag{15.53}$$

There may be degeneracy, meaning that there are only as many terms in the sums as there are distinct eigenvalues. There must be n linearly independent eigenvectors. By using the properties of the projection operators, we can show that

$$\cos^2 A = \sum_k G_k \cos^2 \lambda_k \qquad \sin^2 A = \sum_k G_k \sin^2 \lambda_k$$

$$\sin^2 A + \cos^2 A = I \tag{15.54}$$

Generally, if A has a spectral resolution, any function of it can be defined by

$$f(A) = \sum_k f(\lambda_k) G_k \tag{15.55}$$

If no spectral resolution exists, $f(A)$ can be defined by an appropriate series expansion, provided that convergence can be established by other means.

Suppose that we are required to solve the system of n first-order linear differential equations in n unknown functions given by

$$\frac{dx_j}{dt} = \sum_{k=1}^{n} A_{jk} x_k \tag{15.56}$$

This is equivalent to the matrix equation

$$\dot{X} = AX \tag{15.57}$$

If A has no dependence on the variable t, the formal solution is

$$X = e^{At} X_0 \tag{15.58}$$

where each component of the column vector X_0 is a constant. If the spectral theorem is valid for A, then the solution can be expressed as

$$X = \sum_{k=1}^{f} \exp(\lambda_k t) G_k X_0 \tag{15.59}$$

where λ_k are the eigenvalues of A.

PROBLEMS

15.1. Let A be a 4×4 matrix with four linearly independent eigenvectors. Write out the projection operators in the form (15.1) for the cases where A has four, three, and two distinct eigenvalues.

15.2. Let A be an $n \times n$ matrix with n linearly independent eigenvectors and f distinct eigenvalues. Consider the possibility that there are two sets of projection operators G_k and H_k each of which obeys the conditions (15.29). The matrix A can then be expressed as

$$A = \sum_k \lambda_k G_k = \sum_k \lambda_k H_k$$

where the sums run from 1 to f. Show that

$$H_j A = A H_j = \lambda_j H_j \qquad G_k A = A G_k = \lambda_k G_k$$

$$G_k H_j A = A G_k H_j = \lambda_j G_k H_j \qquad G_k A H_j = A G_k H_j = \lambda_k G_k H_j$$

$$G_k H_j = 0 \qquad H_k = H_k \sum_j G_j = H_k G_k = \sum_j H_j G_k = G_k$$

This shows that the projection operators for a given matrix are unique.

15.3. Let A be a 3×3 normal matrix with three distinct eigenvalues and three mutually orthogonal eigenvectors X_1, X_2, and X_3. Show that the three projection operators can be written

$$G_1 = X_1 X_1^\dagger \qquad G_2 = X_2 X_2^\dagger \qquad G_3 = X_3 X_3^\dagger \tag{15.60}$$

provided that the eigenvectors are normalized.

Hint: Since we have shown that the projection operators are unique, it is only necessary to show that the conditions (15.4) and (15.10) are satisfied. The matrix A can therefore be expressed in the form

$$A = \lambda_1 X_1 X_1^\dagger + \lambda_2 X_2 X_2^\dagger + \lambda_3 X_3 X_3^\dagger \tag{15.61}$$

If $\lambda_1 = \lambda_2$, there still exist three mutually orthogonal eigenvectors. Show that in this case

$$G_1 = X_1 X_1^\dagger + X_2 X_2^\dagger \qquad G_3 = X_3 X_3^\dagger \tag{15.62}$$

15.4. If A is $n \times n$ and has n mutually orthogonal eigenvectors and f distinct eigenvalues, show that the rank of G_k equals the multiplicity of the corresponding eigenvalue. (The multiplicity of an eigenvalue is the number of times it is repeated as a root of the secular equation; it is also equal to the number of linearly independent eigenvectors which correspond to it in those cases where A is normal.)

Hint: Use (15.62) appropriately extended to the case where A is $n \times n$. Calculate the projection operators in a representation in which A is diagonal.

15.5. Let A and B be two commuting Hermitian matrices. Discuss the relation between the projection operators of the two matrices. Consider first the case where neither A nor B has degenerate eigenvalues and then the case where there is degeneracy.

15.6. The matrix (15.37) has a twofold degenerate eigenvalue and only two linearly independent eigenvectors. Show by direct computation that (15.16) does not work for this matrix.

15.7. Show that the symmetry of a matrix is *not* preserved by a general similarity transformation. The fact that some matrices can be brought into diagonal form by a similarity transformation does not prove that they were originally symmetric. Recall that symmetry *is* an invariant property of matrices under *unitary* transformations.

15.8. Write out the two operators G_1 and G_2 defined by (15.39) with A given by (15.37) and show by direct computation that (15.43) is valid.

15.9. By use of the series expansions for e^A and e^B, show that

$$e^A e^B = e^{A+B}$$

provided that $AB = BA$.

15.10. If A is given by (15.37), convince yourself that (15.50) is correct by computing A^2 and A^3.

15.11. Show that similarity transformations exist which will reduce the following matrices to Jordan canonical form:

$$(1)\begin{pmatrix} \lambda & a & c \\ 0 & \lambda & b \\ 0 & 0 & \lambda \end{pmatrix} \qquad (2)\begin{pmatrix} \lambda & 0 & c \\ 0 & \lambda & b \\ 0 & 0 & \lambda \end{pmatrix} \qquad (3)\begin{pmatrix} \lambda & 0 & c \\ 0 & \lambda & 0 \\ 0 & 0 & \lambda \end{pmatrix}$$

A matrix is in Jordan canonical form when the eigenvalues appear on the main diagonal and the elements immediately above the main diagonal are either 1 or 0. All other elements are 0. How many linearly independent eigenvectors does each of the three matrices have?

15.12. For the matrix A defined by

$$A = \begin{pmatrix} \lambda & 1 & 0 \\ 0 & \lambda & 1 \\ 0 & 0 & \lambda \end{pmatrix}$$

show that

$$A^k = \begin{pmatrix} \lambda^k & \dfrac{d}{d\lambda}\lambda^k & \dfrac{1}{2}\dfrac{d^2}{d\lambda^2}\lambda^k \\ 0 & \lambda^k & \dfrac{d}{d\lambda}\lambda^k \\ 0 & 0 & \lambda^k \end{pmatrix}$$

Show that if the series expansion for $f(\lambda)$ is defined, then

$$f(A) = \begin{pmatrix} f(\lambda) & f'(\lambda) & \frac{1}{2}f''(\lambda) \\ 0 & f(\lambda) & f'(\lambda) \\ 0 & 0 & f(\lambda) \end{pmatrix}$$

15.13. Use the series expansion for e^A to show that

$$T^{-1}e^A T = e^{T^{-1}AT}$$

15.14. Find the Jordan canonical form of

$$A = \begin{pmatrix} 4 & 1 & 1 \\ -9 & -2 & -4 \\ 0 & 0 & 2 \end{pmatrix}$$

15.15. For the matrix A given by

$$A = \begin{pmatrix} \frac{2}{3} & 1 & 0 \\ 1 & \frac{2}{3} & 0 \\ 0 & 0 & 1 \end{pmatrix}$$

find the eigenvalues and the normalized eigenvectors. Construct the three projection operators. Check the validity of (15.24). Compute A^{-1} by means of

$$A^{-1} = \frac{1}{\lambda_1} G_1 + \frac{1}{\lambda_2} G_2 + \frac{1}{\lambda_3} G_3$$

Find a matrix B which has the property that $B^2 = A$. In this way, it is possible to define $\sqrt{A}$. How many such matrices are there?

15.16. Prove (15.17) by showing that $P_f(A)X = 0$ for any vector X. Give an argument supporting the claim that there is no matrix polynomial of lower degree than $P_f(A)$ that is zero.

15.17. If H is Hermitian, show that e^{iH} is unitary.

15.18. If the elements of the matrix A are constants, show that

$$\frac{d}{dt} e^{At} = A e^{At}$$

Hint: Differentiate the series expansion for e^{At}.

15.19. If A is normal and has eigenvalues λ_k and projection operators G_k, show that

$$AA^\dagger = \sum_k |\lambda_k|^2 G_k$$

The eigenvalues of A are in general complex.

15.20. Find e^A if

$$A = \begin{pmatrix} 0 & x & y \\ 0 & 0 & z \\ 0 & 0 & 0 \end{pmatrix}$$

15.21. For the matrix (15.37), show that

$$\sin A = \begin{pmatrix} \sin\lambda & \cos\lambda & 0 \\ 0 & \sin\lambda & 0 \\ 0 & 0 & \sin\mu \end{pmatrix} \qquad \cos A = \begin{pmatrix} \cos\lambda & -\sin\lambda & 0 \\ 0 & \cos\lambda & 0 \\ 0 & 0 & \cos\mu \end{pmatrix}$$

Verify that $\sin^2 A + \cos^2 A = I$.

15.22. If A is an $n \times n$ matrix, use the series expansion for e^{iA} to show that

$$e^{iA} = \cos A + i \sin A$$

15.23. If $AB = BA$, show that

$$\sin(A + B) = \sin A \cos B + \cos A \sin B$$

15.24. Find the eigenvalues of the matrix

$$A = \begin{pmatrix} -a & 0 & 0 \\ a & -b & 0 \\ 0 & b & -c \end{pmatrix}$$

Compute the projection operators G_1, G_2, and G_3. The nuclei of three radioactive substances decay successively so that the numbers $N_1(t)$, $N_2(t)$, and $N_3(t)$ of the three types of nuclei obey the equations

$$\frac{dN_1}{dt} = -aN_1 \qquad \frac{dN_2}{dt} = aN_1 - bN_2 \qquad \frac{dN_3}{dt} = bN_2 - cN_3$$

where the rate constants a, b, and c are all different. If initially $N_1 = N$, $N_2 = 0$, and $N_3 = n$, use (15.58) to find N_1, N_2, and N_3 as functions of the time.

15.25. If A is a 4×4 matrix with two distinct eigenvalues λ and μ, theorem 3 in section 14 guarantees that a similarity transformation exists that will bring it into the triangular form

$$A = \begin{pmatrix} \lambda & a & b & c \\ 0 & \lambda & d & e \\ 0 & 0 & \mu & f \\ 0 & 0 & 0 & \mu \end{pmatrix}$$

Show that a similarity transformation exists which will bring it into the Jordan canonical form

$$A = \begin{pmatrix} \lambda & 1 & 0 & 0 \\ 0 & \lambda & 0 & 0 \\ 0 & 0 & \mu & 1 \\ 0 & 0 & 0 & \mu \end{pmatrix}$$

Consider the case where none of the elements a, b, c, d, e, or f is zero.

$$A(\kappa) = \frac{1}{\sqrt{2\pi}} \int_{-\infty}^{+\infty} f(x)\, e^{-i\kappa x}\, dx$$

8 Fourier Series and Integrals

1. DEFINITION AND CONVERGENCE OF FOURIER SERIES

Consider the Sturm–Liouville problem defined by the differential equation

$$u'' + \kappa^2 u = 0 \qquad a \le x \le b \tag{1.1}$$

The most general boundary conditions which will produce eigenfunctions are

$$\left(u_1^* \frac{du_2}{dx} - u_2 \frac{du_1^*}{dx} \right)_{x=a} = \left(u_1^* \frac{du_2}{dx} - u_2 \frac{du_1^*}{dx} \right)_{x=b} \tag{1.2}$$

where u_1 and u_2 are two possible eigenfunctions. If the solutions of (1.1) are represented in complex form as

$$u_1 = N_1 e^{i\kappa_1 x} \qquad u_2 = N_2 e^{i\kappa_2 x} \tag{1.3}$$

then the boundary conditions (1.2) require

$$e^{i(\kappa_1 - \kappa_2)(b-a)} = 1 \qquad \kappa_2 - \kappa_1 = \frac{2n\pi}{b-a}$$

$$n = 0, \pm 1, \pm 2, \dots \tag{1.4}$$

The eigenvalues are therefore given by

$$\kappa_k = \frac{2k\pi}{b-a} \qquad k = 0, \pm 1, \pm 2, \dots \tag{1.5}$$

Note that for a given eigenfunction

$$\begin{aligned} u_k(b) &= N \exp\left[\frac{2k\pi i b}{b-a}\right] \\ &= N \exp\left[\frac{2k\pi i(b-a) + 2k\pi i a}{b-a}\right] \\ &= N \exp(2k\pi i) \exp\left[\frac{2k\pi i a}{b-a}\right] = u_k(a) \end{aligned} \tag{1.6}$$

Boundary conditions of this type are said to be *periodic*. If the eigenfunctions are required to obey the normalization condition

$$\int_a^b |u_k|^2\, dx = 1 \tag{1.7}$$

then

$$u_k(x) = \frac{1}{\sqrt{b-a}} \exp\left[\frac{2k\pi i x}{b-a}\right] \tag{1.8}$$

The eigenfunctions are known as *Fourier functions*, and series representations of functions in terms of them are called *Fourier series*. The eigenfunctions are twofold degenerate in that both $u_k(x)$ and $u_{-k}(x)$ correspond to the same eigenvalue. (It is κ^2 which is the eigenvalue λ in Sturm–Liouville theory.) Theorem 4 in section 10 of Chapter 7 shows that twofold degeneracy is the greatest degree of degeneracy which is possible in a one-dimensional Sturm–Liouville system.

It is known that Sturm–Liouville eigenfunctions are complete and that sectionally continuous functions can be expanded in terms of them. We will prove the convergence theorems in detail for the special case of Fourier series. It is convenient to use as the fundamental interval $-\pi \le x \le \pi$. The Fourier functions are then

$$u_k(x) = \frac{1}{\sqrt{2\pi}} e^{ikx} \tag{1.9}$$

The Fourier functions are periodic of period 2π:

$$u_k(x) = u_k(x + 2\pi) \tag{1.10}$$

For this reason, if a function $f(x)$ is defined over

$-\pi \le x \le \pi$ and represented as the Fourier series

$$f(x) = \sum_{k=-\infty}^{+\infty} c_k u_k(x) \tag{1.11}$$

it is automatically continued out of the fundamental interval in a periodic fashion: $f(x) = f(x+2\pi)$. The coefficients in the expansion are calculated by means of

$$c_k = \int_{-\pi}^{+\pi} u_k^*(t)\, f(t)\, dt \tag{1.12}$$

which follows from the general expression for the coefficients in any eigenfunction expansion as given by (10.36) in Chapter 7.

If $f(x)$ has a jump discontinuity at $x = x_0$, it is convenient to indicate its value just to the right of the point of discontinuity by $f(x_0+0)$ and its value just to the left by $f(x_0-0)$. In the discussion of this section, the value of $f(x)$ at x_0 will be defined by

$$f(x_0) = \tfrac{1}{2}[f(x_0+0) + f(x_0-0)] \tag{1.13}$$

If $f(x)$ is continuous at the point $x = x_0$, then $f(x_0+0) = f(x_0-0) = f(x_0)$. The left and right derivatives of $f(x)$ at x_0 are defined by

$$f_L'(x_0) = \lim_{\varepsilon \to 0} \frac{f(x_0-0) - f(x_0-\varepsilon)}{\varepsilon} \tag{1.14}$$

$$f_R'(x_0) = \lim_{\varepsilon \to 0} \frac{f(x_0+\varepsilon) - f(x_0+0)}{\varepsilon} \tag{1.15}$$

At any point where $f'(x)$ is continuous, there is no distinction between the left and right derivatives.

If the expansion coefficients (1.12) are substituted into (1.11) and the series truncated at $k = n$, the result is

$$f_n(x) = \int_{-\pi}^{+\pi} f(t)\, \Delta_n(x-t)\, dt \tag{1.16}$$

where

$$\Delta_n(x-t) = \sum_{k=-n}^{+n} u_k^*(t)\, u_k(x) = \frac{1}{2\pi} \sum_{k=-n}^{+n} e^{ik(x-t)} \tag{1.17}$$

The convergence of the Fourier series to $f(x)$ is established if it can be shown that

$$\lim_{n \to \infty} f_n(x) = f(x) \tag{1.18}$$

It is convenient to define $\theta = t - x$. Note that $\Delta_n(-\theta) = \Delta_n(\theta)$ and that $\Delta_n(\theta)$ is periodic of period 2π. This means that the integral of $\Delta_n(\theta)$ over any interval of length 2π gives the same value. It is possible to write $\Delta_n(\theta)$ as

$$\begin{aligned} \Delta_n(\theta) &= \frac{1}{2\pi}\left(1 + \sum_{k=1}^{n} e^{ik\theta} + \sum_{k=1}^{n} e^{-ik\theta}\right) \\ &= \frac{1}{2\pi}\left(1 + 2\sum_{k=1}^{n} \cos k\theta\right) \end{aligned} \tag{1.19}$$

The sum is evaluated in problem 10.5 of Chapter 5. Through the use of the identity

$$\sin\left(n\theta + \frac{\theta}{2}\right) + \sin\frac{\theta}{2} = 2\sin\left(\frac{n+1}{2}\theta\right)\cos\frac{n\theta}{2} \tag{1.20}$$

we find that the sum can be expressed as

$$\Delta_n(\theta) = \frac{\sin\left(n\theta + \dfrac{\theta}{2}\right)}{2\pi \sin(\theta/2)} \tag{1.21}$$

From the series representation (1.19), it is possible to show that

$$\int_{-\pi}^{0} \Delta_n(\theta)\, d\theta = \frac{1}{2} \qquad \int_{0}^{\pi} \Delta_n(\theta)\, d\theta = \frac{1}{2}$$
$$\int_{-\pi}^{\pi} \Delta_n(\theta)\, d\theta = 1 \tag{1.22}$$

We anticipate that

$$\lim_{n \to \infty} \Delta_n(\theta) = \delta(t - x) \tag{1.23}$$

which is the closure relation as defined by equation (10.42) in Chapter 7. The proof of the convergence of Fourier series will also prove (1.23).

In the following discussion, a sectionally continuous (or piecewise continuous) function is defined to be a function which is continuous except at a finite number of isolated points.

Theorem 1. Let $s(t)$ be sectionally continuous over $a \le x \le b$ and let the points where it is discontinuous be simple jump discontinuities. If

$$I = \int_a^b s(t) \sin \lambda t\, dt \tag{1.24}$$

then

$$\lim_{\lambda \to \infty} I = 0 \tag{1.25}$$

Proof: The contribution to I from an interval containing a jump discontinuity in $s(t)$ can be made arbitrarily small by making the interval small. We may therefore regard $a \le x \le b$ as a subinterval over which $s(t)$ is continuous. The theorem is true because a small change in t produces many oscillations of $\sin \lambda t$ when λ is large. Since $s(t)$ does not change much over one period of $\sin \lambda t$, the contributions to the integral from adjacent oscillations cancel. To be more rigorous, let $t = \tau + \pi/\lambda$. Then,

$$I = -\int_{a-\pi/\lambda}^{b-\pi/\lambda} s(\tau + \pi/\lambda) \sin \lambda\tau \, d\tau \tag{1.26}$$

In the limit $\lambda \to \infty$, the limits in (1.26) can be replaced by a and b. If we set $t = \tau$ in (1.26) and add it to (1.24), the result is

$$2I = \int_a^b [s(t) - s(t + \pi/\lambda)] \sin \lambda t \, dt \tag{1.27}$$

As $\lambda \to \infty$, $I \to 0$ on account of the continuity of $s(t)$.

To continue with the proof of the convergence of Fourier series, set $\theta = t - x$ in (1.16) to get

$$\begin{aligned} f_n(x) &= \int_{-\pi-x}^{\pi-x} f(x+\theta)\, \Delta_n(\theta)\, d\theta \\ &= \int_{-\pi}^{\pi} f(x+\theta)\, \Delta_n(\theta)\, d\theta \end{aligned} \tag{1.28}$$

The last step is true because the integrand is periodic with a period 2π. Now write (1.28) as

$$\begin{aligned} f_n(x) &= \int_{-\pi}^{0} f(x+\theta)\, \Delta_n(\theta)\, d\theta \\ &\quad + \int_0^{\pi} f(x+\theta)\, \Delta_n(\theta)\, d\theta \end{aligned} \tag{1.29}$$

In the first integral, replace θ by $-\theta$:

$$\begin{aligned} f_n(x) &= \int_0^{\pi} f(x-\theta)\, \Delta_n(\theta)\, d\theta \\ &\quad + \int_0^{\pi} f(x+\theta)\, \Delta_n(\theta)\, d\theta \end{aligned} \tag{1.30}$$

By means of the expression for $\Delta_n(\theta)$ given by (1.21), we can express $f_n(x)$ as

$$\begin{aligned} f_n(x) &= \int_0^{\pi} \frac{f(x-\theta) - f(x-0)}{2\pi \sin(\theta/2)} \sin \lambda\theta \, d\theta \\ &\quad + f(x-0) \int_0^{\pi} \frac{\sin \lambda\theta}{2\pi \sin(\theta/2)} \, d\theta \\ &\quad + \int_0^{\pi} \frac{f(x+\theta) - f(x+0)}{2\pi \sin(\theta/2)} \sin \lambda\theta \, d\theta \\ &\quad + f(x+0) \int_0^{\pi} \frac{\sin \lambda\theta}{2\pi \sin(\theta/2)} \, d\theta \end{aligned} \tag{1.31}$$

where $\lambda = n + \frac{1}{2}$. We now make the assumption that $f(x)$ has both a right and a left derivative at x. It is not necessary that they be equal. We also assume that $f(x)$ is sectionally continuous with isolated points where there are jump discontinuities. The integrands exist at $\theta = 0$ because of the assumed existence of the right and left derivatives of $f(x)$ at x. By theorem 1, the first and the third integrals in (1.31) vanish in the limit as $n \to \infty$. The remaining integrals are evaluated in (1.22). The result is

$$\lim_{n \to \infty} f_n(x) = \frac{1}{2}[f(x+0) + f(x-0)] \tag{1.32}$$

The following theorem has therefore been established:

Theorem 2. If $f(x)$ is sectionally continuous except at a finite number of points where it has jump discontinuities, then its Fourier series converges to the value (1.32) at every point where it has both a right and a left derivative. This is true even at points where the function has jump discontinuities. There are however subtleties about the convergence at the points of discontinuity (Gibbs phenomenon) which we will explain later. Our proof of convergence also verifies the closure property of the Fourier eigenfunctions:

$$\sum_{k=-\infty}^{+\infty} u_k^*(t) u_k(x) = \delta(x-t) \tag{1.33}$$

Theorem 3. Let $f(x)$ be continuous over $-\pi \le x \le \pi$ and let $f'(x)$ be sectionally continuous over this interval. If the discontinuities in $f'(x)$ are jump discontinuities and if $f(\pi) = f(-\pi)$, then the Fourier series representation of $f(x)$ converges uniformly and absolutely.

Proof: Since $f'(x)$ satisfies the conditions of theorem 2, it has the Fourier series representation

$$f'(x) = \sum_{k=-\infty}^{+\infty} b_k u_k(x) \qquad b_k = \int_{-\pi}^{\pi} u_k^*(t) f'(t)\, dt \tag{1.34}$$

The expansion coefficients can be integrated by parts to yield

$$b_k = u_k^*(t) f(t)\Big|_{-\pi}^{+\pi} + ik\int_{-\pi}^{\pi} u_k^*(t) f(t)\, dt = ikc_k \tag{1.35}$$

where c_k are the Fourier coefficients in the expansion for $f(x)$. The last step in (1.35) is true because $u_k^*(\pi) f(\pi) = u_k^*(-\pi) f(-\pi)$. The condition $f(\pi) = f(-\pi)$ ensures that the periodic extension of $f(x)$ out of the fundamental interval will be continuous at $x = \pm\pi$.

Now consider the partial sum

$$\sqrt{2\pi} \sum_{k=-n}^{+n} |c_k u_k| = \sum_{k=-n}^{+n} |c_k| = \sum_{\substack{k=-n \\ k\neq 0}}^{+n} \left|\frac{1}{k}\right| |kc_k| + |c_0| \tag{1.36}$$

If A_k and B_k are a series of positive constants, then

$$\sum A_k B_k \le \sqrt{\sum A_k^2}\sqrt{\sum B_k^2} \tag{1.37}$$

which is a special case of the Schwartz inequality (11.20) of Chapter 7. Applied to (1.36), the Schwartz inequality gives

$$\sqrt{2\pi} \sum_{k=-n}^{+n} |c_k u_k| \le \sqrt{\sum \frac{1}{k^2}}\sqrt{\sum k^2 |c_k|^2} + |c_0| \tag{1.38}$$

Bessel's inequality, (11.29) of Chapter 7, applied to the expansion (1.34) gives

$$|f'|^2 = \int_{-\pi}^{\pi} |f'(x)|^2\, dx \ge \sum_{k=-n}^{+n} |kc_k|^2 \tag{1.39}$$

We therefore have the result

$$\sqrt{2\pi} \sum_{k=-n}^{+n} |c_k u_k| \le \sqrt{\sum \frac{1}{k^2}}\, |f'| + |c_0| \tag{1.40}$$

Since $\Sigma 1/k^2$ is convergent, the limit $n \to \infty$ exists. Since (1.40) is true for all x in the interval $-\pi \le x \le +\pi$, the uniform and absolute convergence of the Fourier series for $f(x)$ is established.

If a series of continuous functions converges uniformly over an interval $a \le x \le b$, then its sum is also a continuous function over this interval. This is shown in theorem 2 in section 5 of Chapter 5. A Fourier series therefore does not converge uniformly over any interval which contains a discontinuity of the function it represents. It can be shown that the Fourier series representation of a discontinuous function converges uniformly (not necessarily absolutely) over any subinterval which does not contain a discontinuity. We will not prove this result here.

Theorem 4. Let the Fourier series for $f(x)$ be uniformly and absolutely convergent over $-\pi \le x \le \pi$. If the Fourier series for $f'(x)$ exists, then it can be obtained by differentiating the series for $f(x)$ term by term.

Proof: The Fourier series for $f'(x)$ is given by (1.34), and the coefficients obey (1.35). Therefore,

$$f'(x) = \sum_{k=-\infty}^{+\infty} ikc_k u_k = \sum_{k=-\infty}^{+\infty} c_k u_k' \tag{1.41}$$

which completes the proof.

Theorem 5. The Fourier series representation of a sectionally continuous function with isolated jump discontinuities can be integrated term by term.

Proof: Let $F(x)$ be defined by

$$F(x) = \int_{-\pi}^{x} f(t)\, dt - \frac{x+\pi}{\sqrt{2\pi}} c_0 \qquad c_0 = \frac{1}{\sqrt{2\pi}} \int_{-\pi}^{\pi} f(t)\, dt \tag{1.42}$$

Since $f(x)$ is sectionally continuous with no more than jump discontinuities, $F(x)$ is continuous. It is also true that

$$F(\pi) = F(-\pi) = 0 \qquad F'(x) = f(x) - \frac{c_0}{\sqrt{2\pi}} \tag{1.43}$$

By theorem 3, the Fourier series expansion

$$F(x) = \sum_{k=-\infty}^{+\infty} a_k u_k(x) \qquad a_k = \int_{-\pi}^{\pi} u_k^*(t) F(t)\, dt \tag{1.44}$$

is both uniformly and absolutely convergent. From (1.42), the integral of $f(x)$ has the form

$$\int_{-\pi}^{x} f(x)\, dx = \frac{x+\pi}{\sqrt{2\pi}} c_0 + \sum_{k=-\infty}^{+\infty} a_k u_k(x) \tag{1.45a}$$

To complete the proof, it is necessary to establish that

(1.45) is identical to the series which would be obtained through term-by-term integration of the original series for $f(x)$. This is left as a problem.

It will not be possible to show in this book that every conceivable set of orthonormal functions defined by a Sturm–Liouville eigenvalue problem is complete. We have shown it in detail for the Fourier functions. The convergence properties of other Sturm–Liouville eigenfunctions are similar. Functions with even fewer continuity properties than those assumed in the proofs of the theorems in this section can be expanded in Fourier series. For example, the closure relation (1.33) is the Fourier series expansion of a Dirac delta function.

Frequently, it is desirable to have Fourier series expressed in terms of trigonometric functions. The series (1.11) can be expressed as

$$\begin{aligned} f(x) &= \frac{c_0}{\sqrt{2\pi}} + \sum_{n=1}^{\infty} c_n u_n(x) + \sum_{n=1}^{\infty} c_{-n} u_{-n}(x) \\ &= \frac{c_0}{\sqrt{2\pi}} + \sum_{n=1}^{\infty} \frac{c_n}{\sqrt{2\pi}} (\cos nx + i \sin nx) \\ &\quad + \sum_{n=1}^{\infty} \frac{c_{-n}}{\sqrt{2\pi}} (\cos nx - i \sin nx) \\ &= b_0 + \sum_{n=1}^{\infty} (a_n \sin nx + b_n \cos nx) \end{aligned} \tag{1.45b}$$

where

$$\begin{aligned} b_0 &= \frac{c_0}{\sqrt{2\pi}} = \frac{1}{2\pi} \int_{-\pi}^{+\pi} f(x)\, dx \\ a_n &= \frac{i}{\sqrt{2\pi}} (c_n - c_{-n}) = \frac{1}{\pi} \int_{-\pi}^{+\pi} f(x) \sin nx\, dx \\ b_n &= \frac{1}{\sqrt{2\pi}} (c_n + c_{-n}) = \frac{1}{\pi} \int_{-\pi}^{+\pi} f(x) \cos nx\, dx \end{aligned} \tag{1.46}$$

The Fourier series expansion of a function defined over the interval $a \le x \le b$ is

$$\begin{aligned} f(x) &= b_0 + \sum_{n=1}^{\infty} \left[a_n \sin\left(\frac{2\pi n x}{s}\right) + b_n \cos\left(\frac{2\pi n x}{s}\right) \right] \\ b_0 &= \frac{1}{s} \int_a^b f(x)\, dx \\ a_n &= \frac{2}{s} \int_a^b f(x) \sin\left(\frac{2\pi n x}{s}\right) dx \\ b_n &= \frac{2}{s} \int_a^b f(x) \cos\left(\frac{2\pi n x}{s}\right) dx \end{aligned} \tag{1.47}$$

where $s = b - a$.

PROBLEMS

1.1. If the Fourier series (1.11) can be differentiated term by term n times, show that

$$\sum_{k=-\infty}^{+\infty} k^{2n} |c_k|^2$$

remains below a fixed bound. This gives some indication as to how rapidly the Fourier series representation of a given function can be expected to converge. The more derivatives that a function has, the more rapidly does its Fourier series converge.

1.2. Show that the boundary conditions on the derivative of the Fourier functions are also periodic: $u_k'(a) = u_k'(b)$.

1.3. Complete the proof of theorem 5 as follows: Integrate the expression (1.44) for the coefficients a_k by parts to get

$$F(x) = a_0 u_0 + \sum_{k \ne 0} \frac{c_k}{ik} u_k(x)$$

Then, since

$$F(-\pi)=0=a_0u_0+\sum_{k\neq 0}\frac{c_k}{ik}u_k(-\pi)$$

we can write $F(x)$ in the form

$$F(x)-F(-\pi)=F(x)=\sum_{k\neq 0}\frac{c_k}{ik}[u_k(x)-u_k(-\pi)]$$

Now show that the same expression for $F(x)$ is obtained from (1.42) by direct integration of the Fourier series for $f(x)$.

1.4. Show that the appropriate orthogonality relations for use with the Fourier series in the form (1.47) are

$$\int_0^s \sin\left(\frac{2\pi nx}{s}\right)\sin\left(\frac{2\pi n'x}{s}\right)dx=\frac{s}{2}\delta_{nn'}$$

$$\int_0^s \sin\left(\frac{2\pi nx}{s}\right)\cos\left(\frac{2\pi n'x}{s}\right)dx=0$$

$$\int_0^s \cos\left(\frac{2\pi nx}{s}\right)\cos\left(\frac{2\pi n'x}{s}\right)dx=\frac{s}{2}\delta_{nn'}$$

For convenience, we have set $a=0$ and $b=s$.

2. EXAMPLES OF FOURIER SERIES

A sectionally continuous function is defined by

$$\begin{aligned} f(x)&=-1 \qquad -\pi<x<0 \\ f(x)&=+1 \qquad 0<x<\pi \end{aligned} \tag{2.1}$$

This function and its periodic extension is pictured in Fig. 2.1. It does not matter what value we assign to the function at the points of discontinuity $x=0, \pm\pi, \pm 2\pi, \ldots$ since theorem 2 in section 1 predicts that the Fourier series will converge to the arithmetic mean of the two values of the function on either side of the discontinuity, zero in this case. The Fourier series is given by (1.45). The coefficient b_0 is given by

$$b_0=\frac{1}{2\pi}\int_{-\pi}^{+\pi}f(x)\,dx=0 \tag{2.2}$$

which is true because the average value of the function is zero. For b_n, the result is

$$b_n=\frac{1}{\pi}\int_{-\pi}^{+\pi}f(x)\cos nx\,dx=0 \tag{2.3}$$

which follows because $f(x)$ is odd and $\cos nx$ is even. The integrand of (2.3) is therefore odd. In evaluating Fourier coefficients, it is important to make the job easier by exploiting whatever symmetry properties the function may have. The remaining coefficients are

$$\begin{aligned} a_n&=-\frac{1}{\pi}\int_{-\pi}^{0}\sin nx\,dx+\frac{1}{\pi}\int_0^{\pi}\sin nx\,dx \\ &=\frac{2}{\pi n}[1-(-1)^n] \\ &=\frac{4}{\pi n} \qquad \text{if } n \text{ is odd} \\ &=0 \qquad \text{if } n \text{ is even} \end{aligned} \tag{2.4}$$

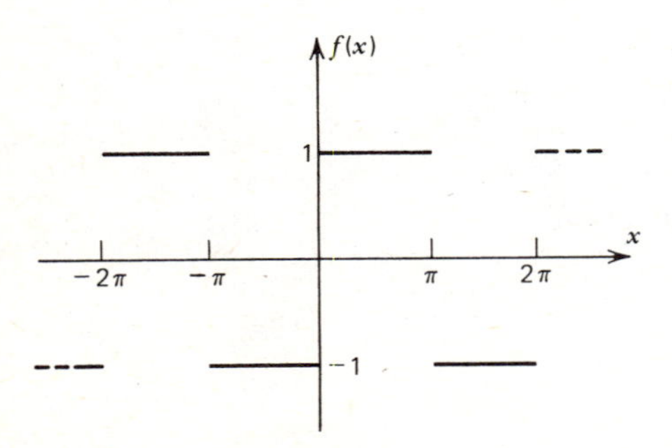

Fig. 2.1

The Fourier series is therefore

$$f(x) = \frac{4}{\pi}\left(\sin x + \frac{\sin 3x}{3} + \frac{\sin 5x}{5} + \cdots\right)$$
$$= \frac{4}{\pi}\sum_{k=0}^{\infty}\frac{\sin(2k+1)x}{2k+1} \tag{2.5}$$

By setting $x = \pi/2$, we find

$$\frac{\pi}{4} = 1 - \frac{1}{3} + \frac{1}{5} - \frac{1}{7} + - \cdots \tag{2.6}$$

which is called Gregory's series. This series also appears in problem 5.15 of Chapter 5. Equation (2.5) is a *Fourier sine series*. This is not the first place in this book that such a series has been encountered. If we set $t = 0$ in (2.65) of Chapter 7, the result is the Fourier sine series

$$y(x) = \frac{8h}{\pi^2}\left[\sin\left(\frac{\pi x}{s}\right) - \frac{1}{9}\sin\left(\frac{3\pi x}{s}\right) + \frac{1}{25}\sin\left(\frac{5\pi x}{s}\right) - \cdots\right] \tag{2.7}$$

Figure 2.2 shows $y(x)$ periodically extended out of its fundamental interval $0 \le x \le s$. The original use of (2.7) was as the series representation of the initial displacement of a string which physically exists between $x = 0$ and $x = s$. The physical system does not exist outside of this interval but the mathematical extension of the Fourier series expansion (2.7) does. Equation (2.5) represents a discontinuous function and converges more slowly than does (2.7), which represents a continuous function. Equation (2.7) converges absolutely, and (2.5) does not. These convergence properties are predicted by theorem 3 in section 1.

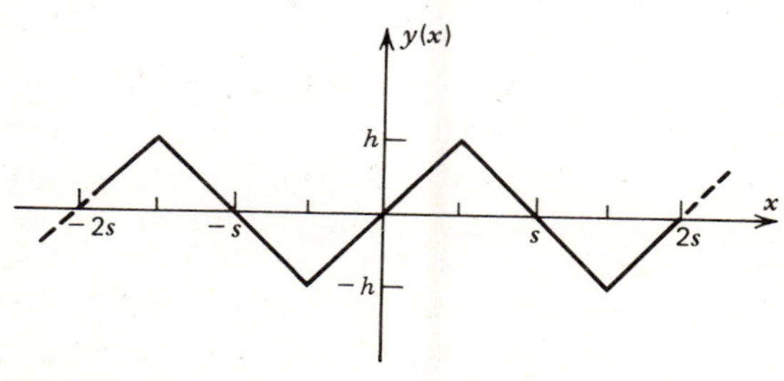

Fig. 2.2

To explore more fully the properties of the series (2.5), it is convenient to use the finite series

$$f_n(x) = \frac{4}{\pi}\sum_{k=0}^{n-1}\frac{\sin(2k+1)x}{2k+1} \tag{2.8}$$

The convergence properties of the series are illustrated in Fig. 2.3, where $f_n(x)$ is graphed for $n = 1$, 2, 3, 4, 5, and 10. A peculiarity becomes evident as n increases. The series representation overshoots the original function on either side of the points of discontinuity, resulting in the small spikes especially evident for $n = 10$. This is the *Gibbs phenomenon*. In spite of this, the series actually converges to 0 at the points of discontinuity, as (2.5) makes evident. The amount of overshoot can be computed. The finite series found by differentiating $f_n(x)$ is

$$f_n'(x) = \frac{4}{\pi}\sum_{k=0}^{n-1}\cos(2k+1)x = \operatorname{Re}\frac{4}{\pi}\sum_{k=0}^{n-1}e^{i(2k+1)x}. \tag{2.9}$$

This series can be summed as follows:

$$f_n'(x) = \frac{4}{\pi}\operatorname{Re} e^{ix}\sum_{k=0}^{n-1}e^{2ikx} = \frac{4}{\pi}\operatorname{Re} e^{ix}\left(\frac{1-e^{2inx}}{1-e^{2ix}}\right)$$
$$= \frac{4}{\pi}\operatorname{Re}\frac{e^{inx}(e^{-inx}-e^{inx})}{e^{-ix}-e^{ix}}$$
$$= \frac{4}{\pi}\operatorname{Re}\frac{e^{inx}(-2i\sin nx)}{-2i\sin x}$$
$$= \frac{4\cos nx\sin nx}{\pi\sin x} = \frac{2\sin 2nx}{\pi\sin x} \tag{2.10}$$

In the calculation, use has been made of (4.3) in Chapter 5. The finite sum (2.8) therefore has the representation

$$f_n(x) = \frac{2}{\pi}\int_0^x \frac{\sin 2nt}{\sin t}\,dt \tag{2.11}$$

We are interested in the behavior of $f_n(x)$ near the point of discontinuity at $x = 0$. If x is small (n is assumed large), then

$$f_n(x) \simeq \frac{2}{\pi}\int_0^x \frac{\sin 2nt}{t}\,dt = \frac{2}{\pi}\int_0^{2nx}\frac{\sin\theta}{\theta}\,d\theta \tag{2.12}$$

where $\theta = 2nt$. The function

$$\operatorname{Si}(u) = \int_0^u \frac{\sin\theta}{\theta}\,d\theta \tag{2.13}$$

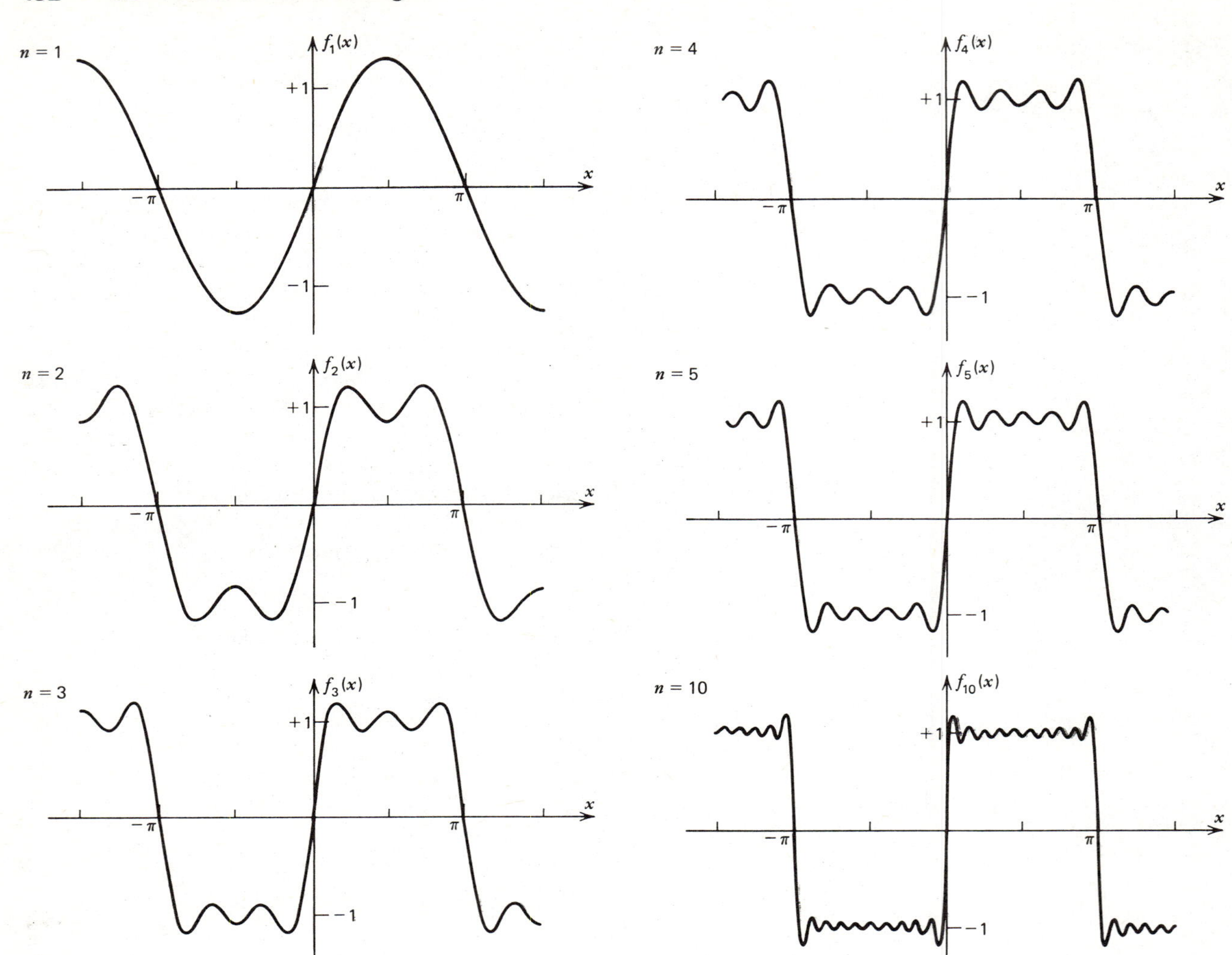

Fig. 2.3

is one of the *sine integral functions* and is tabulated. These functions are introduced in problem 23.1 of Chapter 5. Note that the function $\mathrm{si}(u)$ defined in equation (23.31) of Chapter 5 is related to $\mathrm{Si}(u)$ by

$$\mathrm{Si}(u) = \mathrm{si}(u) + \frac{\pi}{2} \tag{2.14}$$

From tables, we find that the maximum value of the sine integral function is

$$\max \mathrm{Si}(u) = 1.852 \qquad u = 2nx_{\max} = 3.1 \tag{2.15}$$

Thus, as $n \to \infty$, $x_{\max} \to 0$ from positive values and

$$\lim_{n\to\infty} f_n\left(\frac{3.1}{2n}\right) = \frac{2}{\pi} \times 1.852 = 1.18 \tag{2.16}$$

If we set $x = 0$ in the series (2.5), we get $f(0) = 0$. However, (2.16) shows that as $n \to \infty$, there are thin spikes on either side of $x = 0$ which are about 18% larger in magnitude than the original function. In an interval where there is a jump discontinuity, the convergence is nonuniform. The order of taking the limits $n \to \infty$ and $x \to 0$ matters. If we set $x = 0$ and then take the limit $n \to \infty$, we get $f(0) = 0$. If, instead, x is constrained by the relation $2nx = 3.1$ as given by (2.15), then $f_n(x)$ has the limit 1.18 as $n \to \infty$. In this sense, the value to which the series converges at $x = 0$ is undefined to within the range $-1.18 \leq f(0) \leq 1.18$. At any rate, functions can be assigned different values at isolated points and still have the same Fourier series expansion since this does not affect the values of the coefficients. In section 1, we *defined* the function to have the value (1.32) at points where a jump discontinuity exists.

Another result of some interest can be obtained by noting that

$$\begin{aligned} \lim_{n\to\infty} \int_{-\pi/2}^{0} f_n'(x)\,dx &= f(0) - f\left(-\frac{\pi}{2}\right) = 1 \\ \lim_{n\to\infty} \int_{0}^{\pi/2} f_n'(x)\,dx &= f\left(\frac{\pi}{2}\right) - f(0) = 1 \end{aligned} \tag{2.17}$$

Let $F(x)$ be a function which has both a right and a left derivative at $x = 0$. By using an adaptation of the proof of theorem 2 in section 1, we find

$$\begin{aligned} \int_{-\pi/2}^{+\pi/2} F(x)\, f_n'(x)\,dx &= \frac{2}{\pi}\int_{-\pi/2}^{0} \frac{F(x) - F(-0)}{\sin x}\sin 2nx\,dx \\ &\quad + F(-0)\int_{-\pi/2}^{0} f_n'(x)\,dx \\ &\quad + \frac{2}{\pi}\int_{0}^{\pi/2} \frac{F(x) - F(+0)}{\sin x}\sin 2nx\,dx \\ &\quad + F(+0)\int_{-\pi/2}^{0} f_n'(x)\,dx \end{aligned} \tag{2.18}$$

In the limit $n \to \infty$, theorem 1 in section 1 gives

$$\lim_{n\to\infty} \int_{-\pi/2}^{+\pi/2} F(x)\, f_n'(x)\,dx = F(-0) + F(+0) \tag{2.19}$$

The choice of the limits $\pm\pi/2$ is not critical as long as the lower limit is somewhere in $-\pi < x < 0$ and the upper limit is in $0 < x < \pi$. Taking into account the other jump discontinuities, the interpretation is that

$$\begin{aligned} \lim_{n\to\infty} f_n'(x) &= 2\sum_{k=-\infty}^{+\infty} (-1)^k\,\delta(x - k\pi) \\ &= \sum_{k=0}^{\infty} \frac{4}{\pi}\cos(2k+1)x \end{aligned} \tag{2.20}$$

We have therefore pushed beyond what theorem 4 in section 1 allows and have shown that the Fourier series representation of the sectionally continuous function (2.1) can be differentiated term by term. The resulting series is not convergent in the ordinary sense but represents a series of Dirac delta functions. This nonconvergent series can be integrated term by term to obtain the original series.

As another example, consider the function

$$f(x) = \frac{x^2}{2\pi} \qquad -\pi \leq x \leq \pi \tag{2.21}$$

This function and its periodic extension is shown in Fig. 2.4. Since $f(x)$ is an even function and since $\sin nx$ is odd, all the coefficients a_n in the Fourier series are zero. To evaluate the remaining coefficients, consider the integral

$$\int_{-\pi}^{+\pi} \cos ax\,dx = \frac{2\sin \pi a}{a} \tag{2.22}$$

By differentiating twice with respect to the parameter a, we find

$$\int_{-\pi}^{+\pi} x \sin ax\,dx = \frac{2\sin\pi a}{a^2} - \frac{2\pi\cos\pi a}{a} \tag{2.23}$$

$$\int_{-\pi}^{+\pi} x^2 \cos ax\,dx = \left(\frac{2\pi^2}{a} - \frac{4}{a^3}\right)\sin \pi a + \frac{4\pi\cos\pi a}{a^2} \tag{2.24}$$

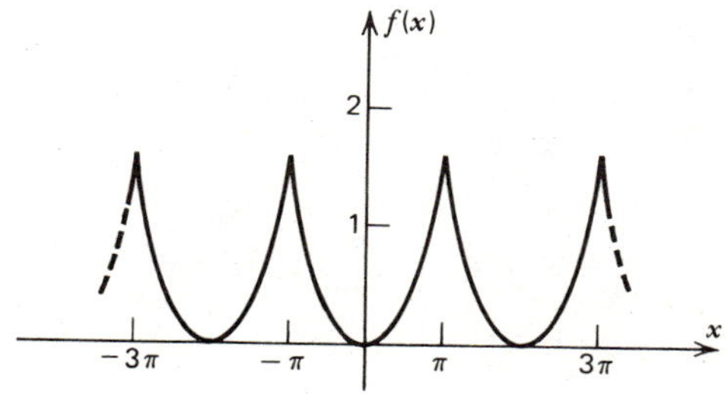

Fig. 2.4

If n is an integer and $a = n$, then

$$\int_{-\pi}^{+\pi} x^2 \cos nx \, dx = \frac{4\pi(-1)^n}{n^2} \tag{2.25}$$

The nonzero coefficients in the Fourier series are

$$b_0 = \frac{1}{2\pi} \int_{-\pi}^{+\pi} \frac{1}{2\pi} x^2 \, dx = \frac{\pi}{6} \tag{2.26}$$

$$b_n = \frac{1}{\pi} \int_{-\pi}^{+\pi} \frac{1}{2\pi} x^2 \cos nx \, dx = \frac{2(-1)^n}{\pi n^2} \tag{2.27}$$

The Fourier series is therefore

$$\frac{x^2}{2\pi} = \frac{\pi}{6} + \frac{2}{\pi} \sum_{n=1}^{\infty} \frac{(-1)^n}{n^2} \cos nx \tag{2.28}$$

This is the Fourier series representation of a continuous periodic function and is absolutely and uniformly convergent. By setting $x = 0$ we find

$$\frac{\pi^2}{12} = 1 - \frac{1}{2^2} + \frac{1}{3^2} - \frac{1}{4^2} + \cdots \tag{2.29}$$

This series also appears in problem 21.9 of Chapter 5.

Let $f(x)$ be given by

$$f(x) = \frac{x}{\pi} \qquad -\pi \le x \le \pi \tag{2.30}$$

Figure 2.5 shows $f(x)$ and its periodic extension. Its Fourier series can be found by differentiating (2.28) term by term:

$$\frac{x}{\pi} = -\frac{2}{\pi} \sum_{n=1}^{\infty} \frac{(-1)^n}{n} \sin nx \tag{2.31}$$

This series represents a discontinuous function and is not absolutely convergent. If we set $x = \pi/2$, we again find Gregory's series (2.6).

Our study of the sectionally continuous function (2.1) has revealed some properties of the Dirac delta function. Pictured in Fig. 2.6 is a step function with steps of heights a and b at the points x_1 and x_2. Note that there is a step up at x_1 and a step down at x_2. The derivative of this function can be expressed in terms of Dirac delta functions as

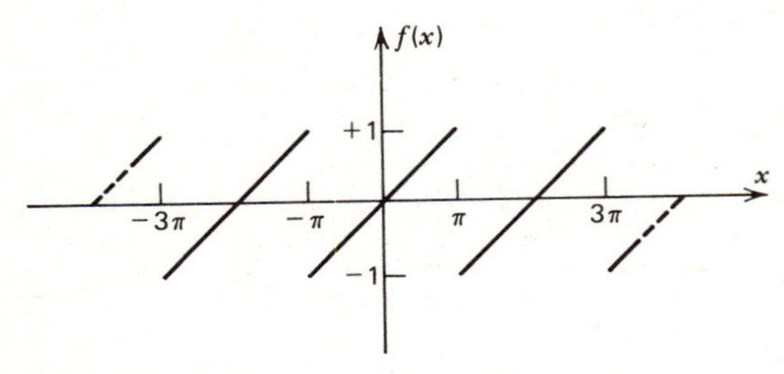

Fig. 2.5

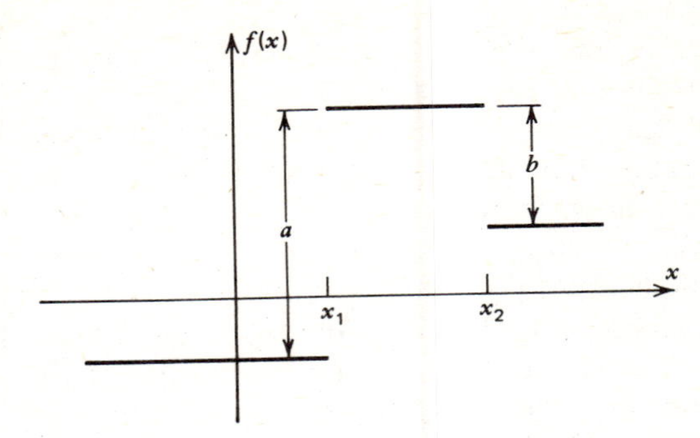

Fig. 2.6

$$f'(x) = a\delta(x - x_1) - b\delta(x - x_2) \tag{2.32}$$

If $f'(x)$ is integrated over an interval containing the point x_1, the result is

$$\int_{x_1-\varepsilon}^{x_1+\varepsilon} f'(x) \, dx = a \int_{x_1-\varepsilon}^{x_1+\varepsilon} \delta(x - x_1) \, dx$$
$$= a = f(x_1 + \varepsilon) - f(x_1 - \varepsilon) \tag{2.33}$$

as expected. It is convenient to define a function

$$\begin{aligned} H(x - x_1) &= 0 \qquad x < x_1 \\ &= 1 \qquad x \ge x_1 \end{aligned} \tag{2.34}$$

which is called the Heaviside unit step function. Its derivative is

$$H'(x - x_1) = \delta(x - x_1) \tag{2.35}$$

Let $F(x)$ be a sectionally continuous function with a jump discontinuity of height h at $x = x_1$ as pictured in Fig. 2.7*a*. As shown in Fig. 2.7*b*, $F(x)$ can be expressed as

$$F(x) = f(x) + hH(x - x_1) \tag{2.36}$$

where $f(x)$ is continuous. Provided that $F(x)$ has both a right and a left derivative at $x = x_1$, its derivative everywhere can be expressed as

$$F'(x) = f'(x) + h\delta(x - x_1) \tag{2.37}$$

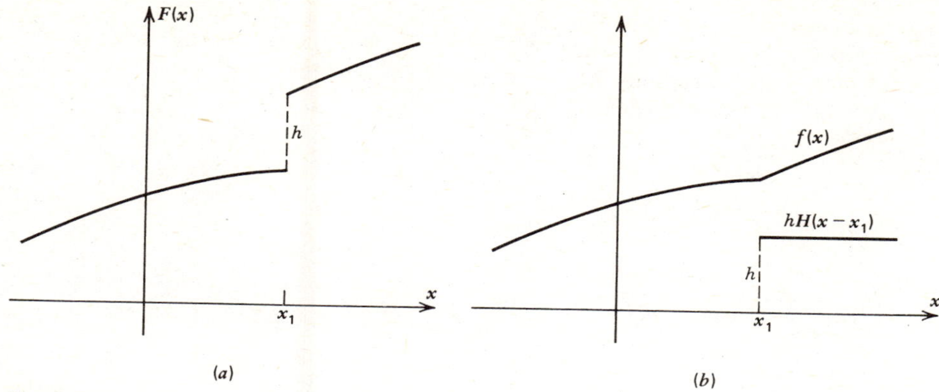

Fig. 2.7

The delta function has the property that

$$\int_{x_1-\varepsilon}^{x_1+\varepsilon} F(x)\,\delta(x-x_1)\,dx = \frac{1}{2}[F(x_1+0)+F(x_1-0)] \tag{2.38}$$

Thus if $F(x)$ is discontinuous at x_1, (2.38) represents the arithmetic mean of the values of $F(x)$ on either side of x_1. The delta function is an even function, meaning that

$$\int_{x_1-\varepsilon}^{x_1} F(x)\,\delta(x-x_1)\,dx = \frac{1}{2}F(x_1-0)$$
$$\int_{x_1}^{x_1+\varepsilon} F(x)\,\delta(x-x_1)\,dx = \frac{1}{2}F(x_1+0) \tag{2.39}$$

We have assumed throughout that $F(x)$ has both a right and a left derivative at all points.

PROBLEMS

2.1. If $f(x)$ is defined by

$$f(x) = \frac{x}{\pi} \qquad 0 \le x \le \pi \tag{2.40}$$

show that its Fourier series is

$$f(x) = \frac{1}{2} - \frac{1}{\pi}\sum_{n=1}^{\infty}\frac{\sin 2nx}{n} \tag{2.41}$$

This function and its periodic extension are shown in Fig. 2.8. By differentiating (2.41), show that

$$\sum_{k=-\infty}^{+\infty}\delta(x-k\pi) = \frac{1}{\pi} + \frac{2}{\pi}\sum_{n=1}^{\infty}\cos(2nx) \tag{2.42}$$

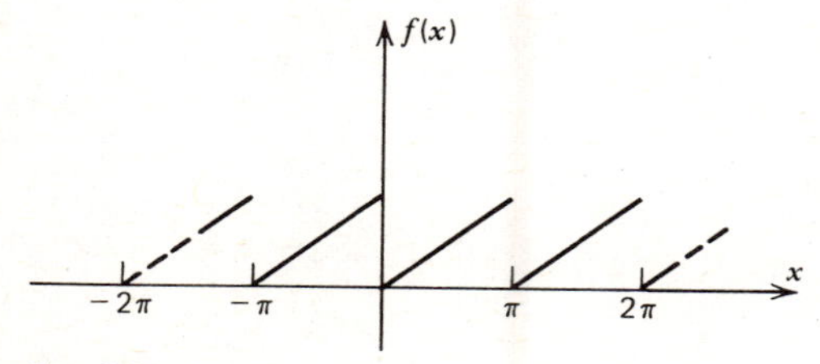

Fig. 2.8

2.2. The function defined by equation (2.40) can be extended over the interval $-\pi \le x \le \pi$ as given by equation (2.30) and shown in Fig. 2.5. It is then represented by the Fourier sine series (2.31). Another possibility is to extend it so that it becomes the even function shown in Fig. 2.9. Show that in this case it is represented by the *Fourier cosine series*

$$f(x) = \frac{1}{2} - \frac{4}{\pi^2} \sum_{k=0}^{\infty} \frac{\cos(2k+1)x}{(2k+1)^2} \tag{2.43}$$

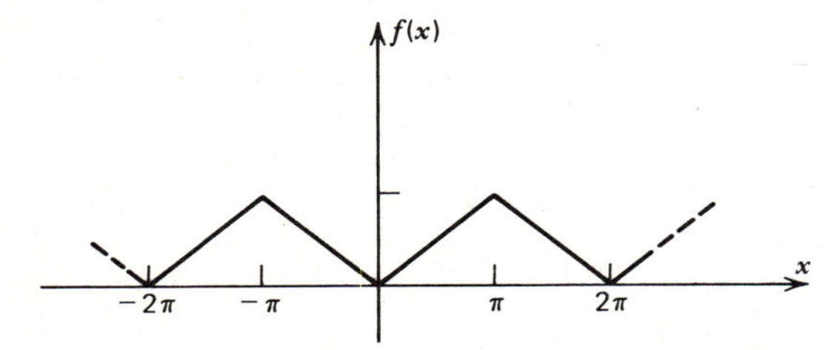

Fig. 2.9

2.3. If the closure relation (1.33) is extended out of its fundamental interval, it gives a series of delta functions. In particular, if we set $t = 2n\pi$, we find

$$\frac{1}{2\pi} \sum_{k=-\infty}^{+\infty} e^{ikx} = \sum_{n=-\infty}^{+\infty} \delta(x - 2n\pi) \tag{2.44}$$

Multiply this expression through by e^{-x} and integrate from 0 to ∞ to show that

$$\sum_{k=0}^{\infty} \frac{1}{1+k^2} = \frac{1}{2} - \frac{\pi}{2} + \frac{\pi}{1 - e^{-2\pi}} \tag{2.45}$$

Watch out for the terms $k = 0$ and $n = 0$. They need to be treated as special cases.

2.4. By integrating (2.44) over the range $-t \le x \le t$, show that

$$\phi(t) = \frac{t}{\pi} + \frac{2}{\pi} \sum_{k=1}^{\infty} \frac{\sin kt}{k} \tag{2.46}$$

represents the stair step function shown in Fig. 2.10. Be careful to treat the $k = 0$ term as a special case when integrating.

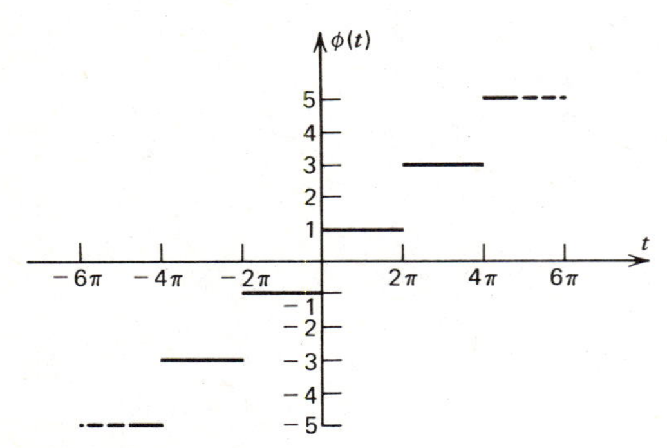

Fig. 2.10

2.5. A few values of $\mathrm{Si}(u)$ are given in the following table:

u	0	1	2	3	3.1	4	5
$\mathrm{Si}(u)$	0	0.946	1.605	1.849	1.852	1.758	1.550

If $f_n(x)$ is given by (2.12), convince yourself that

$$\lim_{\substack{n\to\infty \\ x\to 0}} f_n(x) \qquad x>0$$

can be made to take on any value between 0 and 1.18 depending on how the limits are taken.

2.6. Obtain the Fourier series

$$\cos xt = \frac{\sin \pi t}{\pi t} + \frac{2t\sin \pi t}{\pi}\sum_{n=1}^{\infty}\frac{(-1)^n}{t^2-n^2}\cos nx \qquad -\pi\le x\le \pi \tag{2.47}$$

By setting $x=\pi$, show that

$$\cot \pi t - \frac{1}{\pi t} = -\frac{2t}{\pi}\sum_{n=1}^{\infty}\frac{1}{n^2-t^2} \tag{2.48}$$

By integrating this expression with respect to t from 0 to x, obtain the series

$$\ln\left(\frac{\sin \pi x}{\pi x}\right) = \sum_{n=1}^{\infty}\ln\left(1-\frac{x^2}{n^2}\right) \tag{2.49}$$

Show that this series is absolutely convergent for all x. Equation (2.49) can be expressed as

$$\frac{\sin \pi x}{\pi x} = \prod_{n=1}^{\infty}\left(1-\frac{x^2}{n^2}\right) \tag{2.50}$$

which is the *infinite product* representation for $\sin \pi x$. Equation (2.48) is sometimes called the *resolution of the cotangent into partial fractions.*

2.7. Show that

$$\begin{aligned} &u''+\kappa^2 u = 0 \qquad 0\le x\le s \\ &u'(0) = u'(s) = 0 \end{aligned} \tag{2.51}$$

defines a Sturm–Liouville eigenvalue problem. Show that the expansion of an arbitrary function in terms of the resulting eigenfunctions is a Fourier cosine series. Ordinarily, in an application we would be interested in expanding functions which also conform to the boundary conditions (2.51). As problem 2.2 shows, functions which do not satisfy these boundary conditions can also be expanded in terms of Fourier cosine series.

3. PERIODIC SOLUTIONS OF NONLINEAR DIFFERENTIAL EQUATIONS

A simple plane pendulum consists of a particle of mass m on one end of a massless rigid rod of length s. The rod is pivoted at the other end and the system moves in a plane. Neglecting friction, the equation of motion of the pendulum is

$$\ddot{\theta} + \frac{g}{s}\sin\theta = 0 \tag{3.1}$$

where θ is the angular displacement measured from the vertical and g is the acceleration of a freely falling object. The differential equation for θ as a function of time is *nonlinear*. If the pendulum is given an initial angular displacement θ_0 and then released, it oscillates periodically with time between the limits $\pm\theta_0$. It is possible to solve (3.1) in terms of Jacobian elliptic functions. Since θ is a periodic function of the time, another possibility is to find an approximate solution in terms of a Fourier series. If $\sin\theta$ is expanded in a

Taylor's series around the equilibrium position $\theta = 0$, the result is the approximate differential equation

$$\ddot{\theta} + \frac{g}{s}\left(\theta - \frac{1}{3!}\theta^3 + \frac{1}{5!}\theta^5 - \cdots\right) = 0 \tag{3.2}$$

The familiar small-angle approximation to the motion is found if only the linear term in θ is retained.

More generally, consider the motion of a mechanical system which depends on one coordinate x and obeys the differential equation

$$\ddot{x} + f(x) = 0 \tag{3.3}$$

If we write $f(x) = dV/dx$ and multiply the equation of motion by $\dot{x}$, we find

$$\dot{x}\ddot{x} + \dot{x}\frac{dV}{dx} = 0$$

$$\frac{d}{dt}\left(\frac{1}{2}\dot{x}^2 + V\right) = 0 \tag{3.4}$$

$$\frac{1}{2}\dot{x}^2 + V = W = \text{constant}$$

Thus, V represents potential energy per unit mass and W is the total energy per unit mass. Since there is no friction and no external driving force, W is a constant. Suppose that $V(x)$ has the form indicated in Fig. 3.1 so that the origin is a point of stable equilibrium. If the particle is given an energy W, then it moves periodically back and forth between the turning points x_1 and x_2 as indicated in Fig. 3.1. It is not possible classically for the particle to go outside these limits since to do so would mean $V(x) > W$ and $\dot{x}^2 < 0$, which is impossible. Quantum mechanically it is possible for the particle to penetrate the classically forbidden region as the discussion in section 7 of Chapter 7 shows. Another example of a physical system which obeys a differential equation of the type (3.3) is provided by the symmetric top with one point fixed, as reference to equation (15.18) of Chapter 4 shows.

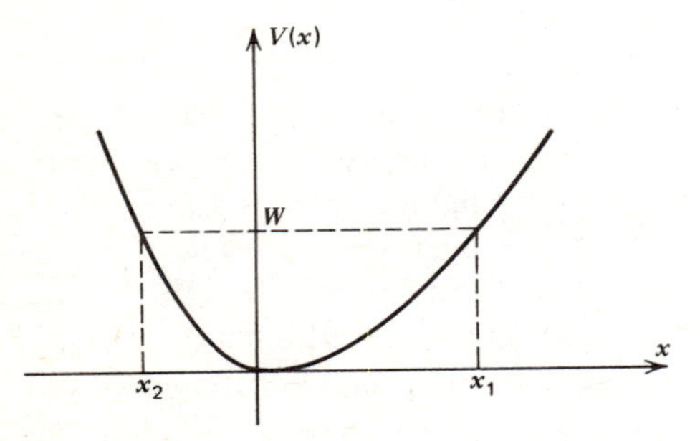

Fig. 3.1

It is convenient to express the Taylor's series expansion of the potential function about $x = 0$ in the form

$$V(x) = \tfrac{1}{2}\alpha x^2 + \tfrac{1}{3}\beta x^3 + \tfrac{1}{4}\gamma x^4 + \tfrac{1}{5}\delta x^5 + \tfrac{1}{6}\varepsilon x^6 + \cdots \tag{3.5}$$

The approximate equation of motion is then

$$\ddot{x} + \alpha x + \beta x^2 + \gamma x^3 + \delta x^4 + \varepsilon x^5 = 0 \tag{3.6}$$

The validity of the Fourier series method of solution depends only on the fact that x is a periodic function of the time. We will however obtain the solution with the additional restriction that the higher-order terms in (3.6) become progressively smaller. The Fourier series solution can be expressed in the form

$$x = b_0 + \sum_{n=1}^{\infty} (a_n \sin n\omega t + b_n \cos n\omega t) \tag{3.7}$$

The fundamental interval for the expansion is the period or time for one complete cycle. The period and the circular frequency are related by

$$\omega P = 2\pi \tag{3.8}$$

The first term in the series, $a_1 \sin \omega t + b_1 \cos \omega t$, is the fundamental or first harmonic, and the other terms are the higher harmonics. The coefficients in the Fourier series are

$$b_0 = \frac{1}{P}\int_0^P x(t)\,dt$$

$$a_n = \frac{2}{P}\int_0^P x(t)\sin n\omega t\,dt \tag{3.9}$$

$$b_n = \frac{2}{P}\int_0^P x(t)\cos n\omega t\,dt$$

Assume that the particle is drawn aside a distance x_1 and then released with zero initial velocity. Qualitatively, one cycle of the motion looks like Fig. 3.2. One thing to note is that the graph for $P/2 < t < P$ is a mirror image of the graph for $0 < t < P/2$. This is because the differential equation (3.3) is invariant with respect to time reversal: $t \to -t$. In other words,

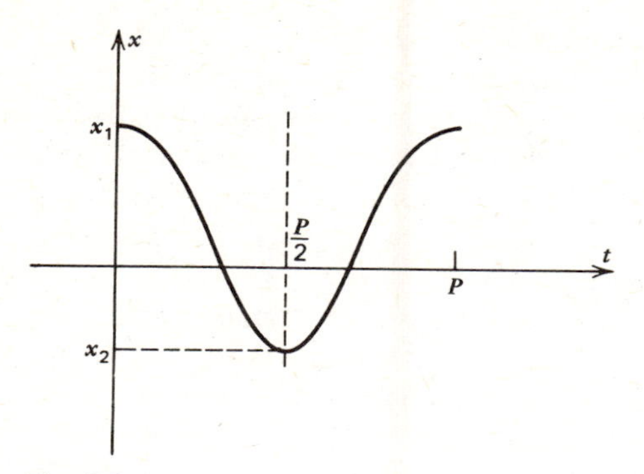

Fig. 3.2

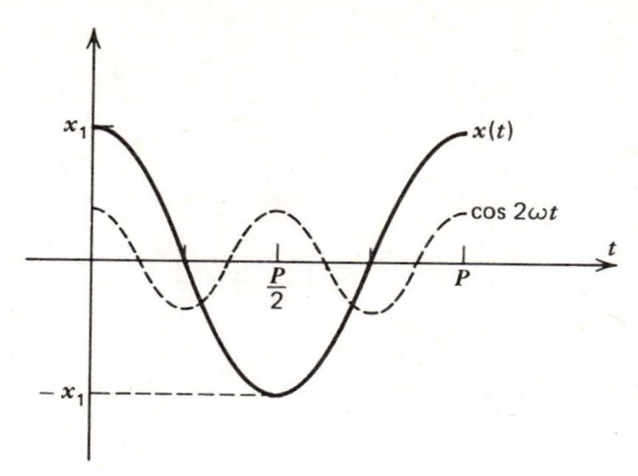

Fig. 3.3

both $x(t)$ and $x(-t)$ obey the same differential equation. A moving picture of the motion of the particle could be run in reverse and no one would know the difference. Since the periodic extension of $x(t)$ is an even function of t, all the sine terms in the Fourier series will be missing. The motion can therefore be represented by the series

$$x = b_0 + \sum_{n=1}^{\infty} b_n \cos n\omega t \tag{3.10}$$

Should the velocity be other than zero at $t = 0$, the solution can be conveniently reexpressed as

$$x = b_0 + \sum_{n=1}^{\infty} b_n \cos n(\omega t - \phi) \tag{3.11}$$

where ϕ is a constant.

We will begin with a simple case by assuming that $V(x)$ is symmetric: $V(-x) = V(x)$. Then all the odd terms in the expansion (3.5) are missing. If the fifth-order term is neglected, the equation of motion is

$$\ddot{x} + \alpha x + \gamma x^3 = 0 \tag{3.12}$$

The differential equation is now invariant both with respect to time reversal and with respect to space inversion: $x \to -x$. This means that the portion of the graph of $x(t)$ in Fig. 3.3 for $x < 0$ has exactly the same shape as the portion for $x > 0$. Also shown in Fig. 3.3 is a plot of the term $\cos 2\omega t$. It is evident that in the calculation of the coefficient b_2, there will be equal positive and negative contributions to the integral giving $b_2 = 0$. Similarly, all the other even coefficients can be shown to be zero. We are left with

$$x = b_1 \cos \omega t + b_3 \cos 3\omega t + \cdots \tag{3.13}$$

as the Fourier series representation of the solution. The fact that we have represented the potential function by a Taylor's series means that it is continuous and has derivatives of all orders. The solution $x(t)$ therefore also has these properties. Reference to problem 1.1 indicates that rapid convergence of the Fourier series can be expected.

It is not possible to evaluate the coefficients from (3.9) because $x(t)$ is not known beforehand. It is therefore necessary to proceed by substituting the series (3.13) directly into the differential equation. In the following computation, we will assume that the nonlinear term γx^3 is small by comparison to the linear term αx. Because of the convergence properties of the Fourier series, $b_3 < b_1$. In computing x^3, a sufficient approximation is

$$x^3 \simeq b_1^3 \cos^3 \omega t = b_1^3 \left(\tfrac{3}{4} \cos \omega t + \tfrac{1}{4} \cos 3\omega t\right) \tag{3.14}$$

In doing computations of this sort, it is necessary to express each term as a linear combination of harmonic terms of the form $\cos n\omega t$. A list of trigonometric identities useful for this purpose is found at the end of this section. The differential equation gives

$$\left[(\alpha - \omega^2) b_1 + \tfrac{3}{4}\gamma b_1^3\right] \cos \omega t + \left[(\alpha - 9\omega^2) b_3 + \tfrac{1}{4}\gamma b_1^3\right] \cos 3\omega t = 0 \tag{3.15}$$

The coefficients of $\cos \omega t$ and $\cos 3\omega t$ must separately vanish. This gives the nonlinear algebraic equations

$$\alpha - \omega^2 + \tfrac{3}{4}\gamma b_1^2 = 0 \tag{3.16}$$

$$(\alpha - 9\omega^2)b_3 + \tfrac{1}{4}\gamma b_1^3 = 0 \tag{3.17}$$

From (3.16), we find that

$$\omega^2 = \alpha + \tfrac{3}{4}\gamma b_1^2 \tag{3.18}$$

In (3.17), it is sufficient to use the approximation $\omega^2 = \alpha$. The result is

$$b_3 = \frac{\gamma b_1^3}{32\alpha} \tag{3.19}$$

The approximate Fourier series solution is therefore

$$x = b_1 \cos \omega t + \frac{\gamma b_1^3}{32\alpha} \cos 3\omega t + \cdots \tag{3.20}$$

Thus, the frequency and the coefficients of all the higher harmonics are found as functions of the single parameter b_1 which must be found from the initial conditions. If x_1 is the initial displacement, then b_1 is found from

$$x_1 = b_1 + \frac{\gamma b_1^3}{32\alpha} \tag{3.21}$$

An important feature of a periodic solution of a nonlinear differential equation is that the frequency is *amplitude-dependent*.

Some general features of the solution and the method used to find it should be noted. The coefficients $b_3, b_5, \ldots$ of the higher harmonics and the frequency ω are found as power series in the coefficient b_1 of the fundamental. The leading term in the expression for b_3 is b_1^3. We anticipate that the leading term in the series for b_5 is b_1^5. The higher harmonics are all odd functions of b_1, and ω is an even function of b_1. These features persist when the solution is carried to a higher order of accuracy. The steps leading to the solution (3.20) were all consistently carried out to third-order accuracy. Thus, in computing x^3, only the third-order term proportional to b_1^3 was retained. A term such as $b_1^2 b_3$ is fifth order, $b_1 b_3^2$ is seventh order, and b_3^3 is ninth order. These terms are all discarded. There is not much point in retaining them since fifth- and higher-order terms have already been neglected in the differential equation (3.12). In substituting ω^2 as given by (3.18) into (3.17), the term proportional to $b_1^2 b_3$ is discarded because it is fifth order. The assumption throughout is that the higher-order terms are smaller in magnitude than the lower-order terms. The method of solution of the nonlinear algebraic equations (3.16) and (3.17) is that of *iteration* or *successive approximation*.

If the potential function is not symmetric and if we again work to third-order accuracy, the approximate differential equation of motion is

$$\ddot{x} + \alpha x + \beta x^2 + \gamma x^3 = 0 \tag{3.22}$$

Both even and odd harmonics will now be present in the solution:

$$x = b_0 + b_1 \cos \omega t + b_2 \cos 2\omega t + b_3 \cos 3\omega t + \cdots \tag{3.23}$$

where b_0, b_2, b_3, and ω are to be regarded as functions of b_1. We anticipate that b_0 and b_2 will be second order in b_1 and that b_3 will be third order. For x^2, we find

$$\begin{aligned} x^2 &= b_1^2 \cos^2 \omega t + 2b_1 \cos \omega t (b_0 + b_2 \cos 2\omega t) + \cdots \\ &= \tfrac{1}{2} b_1^2 (1 + \cos 2\omega t) + 2b_1 b_0 \cos \omega t \\ &\quad + 2b_1 b_2 (\tfrac{1}{2} \cos 3\omega t + \tfrac{1}{2} \cos \omega t) + \cdots \end{aligned} \tag{3.24}$$

Terms proportional to b_0^2, b_2^2, and $b_0 b_2$ are fourth order; these as well as higher-order terms have been neglected. Equation (3.14) for x^3 is still valid. If the Fourier series (3.23) is substituted into the differential equation (3.22) and the coefficients of each harmonic set equal to zero, the result is the system of nonlinear algebraic equations

$$b_0 = -\frac{\beta b_1^2}{2\alpha} \tag{3.25}$$

$$\omega^2 - \alpha = \tfrac{3}{4}\gamma b_1^2 + 2\beta b_0 + \beta b_2 \tag{3.26}$$

$$(4\omega^2 - \alpha) b_2 = \tfrac{1}{2}\beta b_1^2 \tag{3.27}$$

$$(9\omega^2 - \alpha) b_3 = \tfrac{1}{4}\gamma b_1^3 + \beta b_1 b_2 \tag{3.28}$$

Equation (3.26) shows that $\omega^2 = \alpha +$ terms of order two. Thus, in maintaining third-order accuracy in (3.27) and (3.28), the approximation $\omega^2 = \alpha$ is sufficient. We find that the solution is

$$x = -\frac{\beta b_1^2}{2\alpha} + b_1 \cos \omega t + \frac{\beta b_1^2}{6\alpha} \cos 2\omega t + \left(\frac{\gamma b_1^3}{32\alpha} + \frac{\beta^2 b_1^3}{48\alpha^2}\right) \cos 3\omega t \tag{3.29}$$

$$\omega^2 = \alpha\left[1 + \frac{3\gamma b_1^2}{4\alpha} - \frac{5\beta^2 b_1^2}{6\alpha^2} + \cdots\right] \tag{3.30}$$

In equation (3.30) for the frequency, there are two second-order terms, one depending on γ and the other on β. In applications, these two terms are frequently of the same order of magnitude. It is therefore necessary to express the differential equation accurate to third order if all second-order contributions to ω^2 are to be obtained. If higher approximations are considered, no further second-order contributions to ω^2 are found. If we make the assumption that all terms beyond $\omega^2 = \alpha$ go to zero in the limit as $\gamma, \beta, \ldots$ go to zero, then the coefficient of b_1^2 must involve factors such as γ/α, with α appearing in the denominator. But γ/α and β^2/α^2 are the only such factors which can be constructed out of the coefficients $\alpha, \beta, \gamma, \ldots$ that have dimensions of $1/(\text{length})^2$ and hence are the only factors which will appear in the second-order correction to ω^2.

Trigonometric Identities

1. $\cos^2\theta = \frac{1}{2}(1 + \cos 2\theta)$
2. $\cos^3\theta = \frac{3}{4}\cos\theta + \frac{1}{4}\cos 3\theta$
3. $\cos^4\theta = \frac{3}{8} + \frac{1}{2}\cos 2\theta + \frac{1}{8}\cos 4\theta$
4. $\cos^5\theta = \frac{5}{8}\cos\theta + \frac{5}{16}\cos 3\theta + \frac{1}{16}\cos 5\theta$
5. $\cos\theta\cos 2\theta = \frac{1}{2}\cos 3\theta + \frac{1}{2}\cos\theta$
6. $\cos\theta\cos 3\theta = \frac{1}{2}\cos 4\theta + \frac{1}{2}\cos 2\theta$
7. $\cos\theta\cos 4\theta = \frac{1}{2}\cos 3\theta + \frac{1}{2}\cos 5\theta$
8. $\cos\theta\cos 5\theta = \frac{1}{2}\cos 4\theta + \frac{1}{2}\cos 6\theta$
9. $\cos^2\theta\cos 2\theta = \frac{1}{4} + \frac{1}{2}\cos 2\theta + \frac{1}{4}\cos 4\theta$
10. $\cos^2\theta\cos 3\theta = \frac{1}{4}\cos\theta + \frac{1}{2}\cos 3\theta + \frac{1}{4}\cos 5\theta$
11. $\cos^2\theta\cos 4\theta = \frac{1}{4}\cos 2\theta + \frac{1}{2}\cos 4\theta + \frac{1}{4}\cos 6\theta$
12. $\cos^2\theta\cos 5\theta = \frac{1}{4}\cos 3\theta + \frac{1}{2}\cos 5\theta + \frac{1}{4}\cos 7\theta$
13. $\cos\theta\cos^2 2\theta = \frac{1}{2}\cos\theta + \frac{1}{4}\cos 3\theta + \frac{1}{4}\cos 5\theta$
14. $\cos\theta\cos^2 3\theta = \frac{1}{2}\cos\theta + \frac{1}{4}\cos 5\theta + \frac{1}{4}\cos 7\theta$
15. $\cos 2\theta\cos^3\theta = \frac{1}{2}\cos\theta + \frac{3}{8}\cos 3\theta + \frac{1}{8}\cos 5\theta$
16. $\cos 2\theta\cos 3\theta = \frac{1}{2}\cos\theta + \frac{1}{2}\cos 5\theta$
17. $\cos^2 2\theta\cos 3\theta = \frac{1}{4}\cos\theta + \frac{1}{2}\cos 3\theta + \frac{1}{4}\cos 7\theta$

PROBLEMS

3.1. In the process of substituting the Fourier series (3.10) into the differential equation (3.6), the result is always reduced to a harmonic series of the form

$$B_0 + \sum_{n=1}^{\infty} B_n \cos n\omega t = 0$$

Why does this imply that $B_0 = 0$ and $B_n = 0$?

Hint: What are the coefficients in the expansion of a function which is identically zero?

3.2. Find the solution of (3.21) for b_1 in terms of x_1 correct to third-order accuracy.

3.3. Is $m\ddot{x} + b\dot{x} + kx = 0$ invariant with respect to time reversal? Could you tell if a motion picture of a particle obeying this differential equation were being run backward?

3.4. Show that the solution of

$$\ddot{x} + \alpha x + \gamma x^3 + \varepsilon x^5 = 0 \tag{3.31}$$

correct to fifth order is

$$x = b_1 \cos \omega t + \left(\frac{\gamma b_1^3}{32\alpha} - \frac{21\gamma^2 b_1^5}{1024\alpha^2} + \frac{5\varepsilon b_1^5}{128\alpha}\right) \cos 3\omega t + \left(\frac{\gamma^2 b_1^5}{1024\alpha^2} + \frac{\varepsilon b_1^5}{384\alpha}\right) \cos 5\omega t + \cdots \tag{3.32}$$

$$\omega^2 = \alpha + \frac{3}{4}\gamma b_1^2 + \left(\frac{3\gamma^2}{128\alpha} + \frac{5\varepsilon}{8}\right) b_1^4 + \cdots \tag{3.33}$$

Hint: The three equations you will get when you substitute the Fourier series into (3.31) are

$$(\alpha - \omega^2) b_1 + \tfrac{3}{4}\gamma(b_1^3 + b_1^2 b_3) + \tfrac{5}{8}\varepsilon b_1^5 = 0 \tag{3.34}$$

$$(\alpha - 9\omega^2) b_3 + \tfrac{1}{4}\gamma b_1^3 + \tfrac{3}{2}\gamma b_1^2 b_3 + \tfrac{5}{16}\varepsilon b_1^5 = 0 \tag{3.35}$$

$$(\alpha - 25\omega^2) b_5 + \tfrac{3}{4}\gamma b_1^2 b_3 + \tfrac{1}{16}\varepsilon b_1^5 = 0 \tag{3.36}$$

In equation (3.36), fifth-order accuracy is maintained if we set $\omega^2 = \alpha$ and use the known third-order approximation for b_3:

$$b_3 = \frac{\gamma}{32\alpha} b_1^3 \tag{3.37}$$

In (3.35), it is necessary to use ω^2 as given by (3.18) to obtain b_3 to fifth order. Finally, ω^2 as given by (3.33) is found from (3.34).

3.5. In the solutions we have so far obtained, the linear term has been regarded as more important than the higher-order terms in the differential equation. Suppose that the linear term is missing altogether and that the exact equation of motion is

$$\ddot{x} + \gamma x^3 = 0 \qquad \gamma > 0 \tag{3.38}$$

The potential function is symmetric, the motion is periodic, and a Fourier series solution therefore exists. Show that it is approximately

$$x = b_1(\cos \omega t + 0.04496 \cos 3\omega t + 0.00204 \cos 5\omega t + \cdots) \tag{3.39}$$

$$\omega^2 = 0.7867 \gamma b_1^2 \tag{3.40}$$

Caution: You will have to retain seventh-order terms in the computation to get this result.

3.6. The equation of motion of a particle is given by

$$\ddot{x} - \alpha x + \gamma x^3 = 0 \qquad \alpha > 0, \gamma > 0 \tag{3.41}$$

Find the potential function and sketch it. Describe qualitatively the kinds of motion that the particle can have. If the energy of the particle is exactly zero, show that the exact solution is

$$x = \frac{\sqrt{2\alpha}}{\sqrt{\gamma} \cosh \sqrt{\alpha}\, t} \tag{3.42}$$

This is known as the *sticking solution*. Does the name seem appropriate?

3.7. The equation of motion of a particle is given by

$$\ddot{x} + \alpha x - \beta x^2 = 0 \qquad \alpha > 0, \beta > 0 \tag{3.43}$$

Find the potential function and sketch it. Describe qualitatively the kinds of motion that the particle can have. Show that if the energy is in the range

$$0 < W < \alpha^3/6\beta^2 \tag{3.44}$$

the particle can have both bound periodic motion and also unbounded motion where it can escape to infinity.

3.8. One distinguishing characteristic of a nonlinear differential equation is the failure of the superposition principle. Show that if x_1 and x_2 are each separately solutions of (3.43), then $x = Ax_1 + Bx_2$ is not in general a solution.

3.9. Show that the equation which expresses conservation of energy for the plane pendulum can be expressed in the form

$$W = \tfrac{1}{2} m s^2 \dot{\theta}^2 + mgs(1 - \cos \theta) \tag{3.45}$$

where W is the total energy. The potential energy is

$$V(\theta) = mgs(1 - \cos\theta) = 2mgs\sin^2\frac{\theta}{2} \tag{3.46}$$

Sketch $V(\theta)$ and show that oscillatory motion results if $0 \leq W \leq 2mgs$. The turning points are at $\pm\theta_0$, where θ_0 is determined by

$$W = 2mgs\sin^2\frac{\theta_0}{2} \tag{3.47}$$

The *modulus* k is a dimensionless parameter defined by

$$k^2 = \frac{W}{2mgs} = \sin^2\frac{\theta_0}{2} \tag{3.48}$$

The modulus has the range of values $0 \leq k \leq 1$. By making the change of dependent variable

$$y = \frac{1}{k}\sin\frac{\theta}{2} \tag{3.49}$$

show that

$$\dot{y}^2 = \frac{g}{s}(1 - y^2)(1 - k^2y^2) \tag{3.50}$$

Note that y has the range of values $-1 \leq y \leq +1$. Now make the change of independent variable $u = \sqrt{g/s}\,t$ to get

$$\left(\frac{dy}{du}\right)^2 = (1 - y^2)(1 - k^2y^2) \tag{3.51}$$

By differentiating (3.51), obtain the second-order differential equation

$$\frac{d^2y}{du^2} + (1 + k^2)y - 2k^2y^3 = 0 \tag{3.52}$$

Equations (3.51) and (3.52) are nonlinear and do not have solutions in terms of elementary functions. The solution of either (3.51) or (3.52) is one of the *Jacobian elliptic functions* and is generally written as

$$y = \operatorname{sn}(u - u_0) \tag{3.53}$$

where u_0 is a constant of integration which we will set equal to zero in the remaining discussion. We will take the point of view that (3.51) along with the condition $\operatorname{sn}(0) = 0$ *defines* the function $y = \operatorname{sn}(u)$. Qualitatively, $\operatorname{sn}(u)$ resembles $\sin u$ and in fact becomes $\sin u$ if $k = 0$. Show from (3.51) that

$$u = \int_0^y \frac{dy}{\sqrt{(1 - y^2)(1 - k^2y^2)}} \tag{3.54}$$

This result is valid if $0 \leq y \leq 1$, for example, over the first quarter cycle of $\operatorname{sn}(u)$. Equation (3.54) is called the *elliptic integral of the first kind*. If $y = 1$, we get the *complete elliptic integral of the first kind*

$$K = \int_0^1 \frac{dy}{\sqrt{(1 - y^2)(1 - k^2y^2)}} \tag{3.55}$$

By making the change of variable $y = \sin\phi = \operatorname{sn}(u)$, show that

$$u = \int_0^\phi \frac{d\phi}{\sqrt{1 - k^2\sin^2\phi}} \qquad K = \int_0^{\pi/2} \frac{d\phi}{\sqrt{1 - k^2\sin^2\phi}} \tag{3.56}$$

See problem 6.3 in Chapter 5. The elliptic integrals have been tabulated. Use the result of problem 3.4 to show that

$$\mathrm{sn}(u) \simeq \left(1 + \frac{a}{32} + \frac{23a^2}{1024}\right)\sin(\omega u) + \left(\frac{a}{32} + \frac{3a^2}{128}\right)\sin(3\omega u) + \frac{a^2}{1024}\sin(5\omega u) \quad (3.57)$$

where

$$a = \frac{2k^2}{1+k^2} \qquad \omega^2 = (1+k^2)\left(1 - \frac{3a}{4} - \frac{3a^2}{128}\right) \quad (3.58)$$

Since $\mathrm{sn}(u)$ has been defined in such a way that $\mathrm{sn}(0) = 0$, you will have to shift the time axis of the Fourier series (3.32) by a factor of $\pi/2$: $\omega u \to \omega u - \pi/2$. It is necessary to determine the constant b_1 by the condition $-1 \leq \mathrm{sn}(u) \leq 1$. If you set $y_{max} = A$ in (3.32), you will get the result

$$A = b_1 - \frac{ab_1^2}{32} - \frac{20a^2b_1^5}{1024} \quad (3.59)$$

You can solve for b_1 in terms of A by an iterative procedure or you can use the inversion of power series method of section 6 in Chapter 5. After this is done, set $A = 1$. The solution should work well if k is small and should be poor if k is near unity. If $k = 0.5$ and $u = 0.5251$, (3.57) gives $\mathrm{sn}(u) \simeq 0.4996$. The value from tables is $\mathrm{sn}(u) = 0.5000$. By expressing everything in terms of k, show from (3.58) that

$$\frac{1}{\omega} \simeq 1 + \frac{k^2}{4} + \frac{9k^4}{64} \quad (3.60)$$

Compare this result to equation (6.24) in Chapter 5.

3.10. In problem 3.9, $k = 1$ gives the sticking solution. The pendulum has exactly the energy $W = 2mgs$ and sticks in unstable equilibrium at $\theta = 180°$. Find the exact solution of (3.51) for this case in terms of elementary functions.

4. CLASSICAL CENTRAL FORCE PROBLEM

If a force acting on a particle has the form

$$\mathbf{F} = \hat{\mathbf{r}}F(r) \qquad F(r) = -\frac{dV}{dr} \quad (4.1)$$

it is said to be a *central force*. The gravitational attraction between two particles is an example. The force is directed along the line joining the two particles, and its magnitude depends only on the scalar distance between them. In a two-body system, such as the Earth and the moon, the two bodies rotate around their common center of mass. It is shown in problem 4.11 of Chapter 2 that the kinetic energy of a two-particle system with respect to its center of mass can be expressed as

$$K_c = \tfrac{1}{2}mu^2 \quad (4.2)$$

where $\mathbf{u}$ is the velocity of one particle referred to the other as an origin and

$$m = \frac{m_1 m_2}{m_1 + m_2} \quad (4.3)$$

is the reduced mass of the system. The center-of-mass angular momentum of a two-particle system can be similarly expressed. Equation (9.8) in Chapter 2 shows that

$$\ell_c = \mathbf{s}_1 \times m_1\mathbf{w}_1 + \mathbf{s}_2 \times m_2\mathbf{w}_2 \quad (4.4)$$

where, as shown in Fig. 4.1, $\mathbf{s}_1$ and $\mathbf{s}_2$ are the position vectors of the particles with respect to the center of mass and

$$\mathbf{w}_1 = \frac{d\mathbf{s}_1}{dt} \qquad \mathbf{w}_2 = \frac{d\mathbf{s}_2}{dt} \quad (4.5)$$

are the particle velocities with respect to the center of mass. The position vector of m_2 referred to m_1 as origin is given by

$$\mathbf{r} = \mathbf{s}_2 - \mathbf{s}_1 \quad (4.6)$$

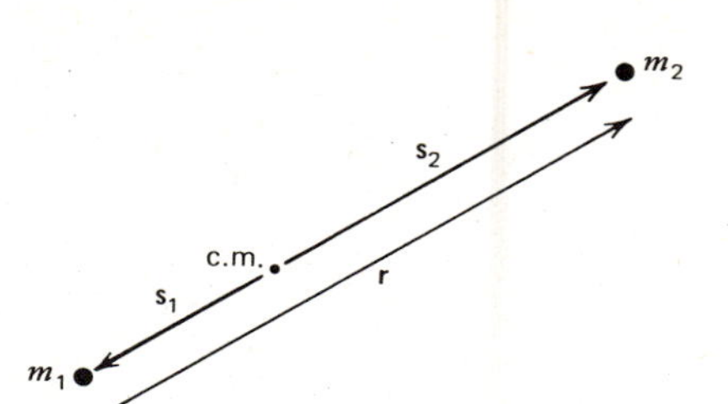

Fig. 4.1

The velocity of m_2 with respect to m_1 is therefore

$$\mathbf{u} = \frac{d\mathbf{r}}{dt} = \mathbf{w}_2 - \mathbf{w}_1 \tag{4.7}$$

By equation (4.37) in Chapter 2,

$$m_1\mathbf{w}_1 + m_2\mathbf{w}_2 = 0 \tag{4.8}$$

If (4.7) and (4.8) are solved for $\mathbf{w}_1$ and $\mathbf{w}_2$, the result is

$$\mathbf{w}_2 = \frac{m_1}{m_1 + m_2}\mathbf{u} \qquad \mathbf{w}_1 = -\frac{m_2}{m_1 + m_2}\mathbf{u} \tag{4.9}$$

The angular momentum (4.3) can therefore be expressed as

$$\ell_c = \mathbf{r} \times m\mathbf{u} \tag{4.10}$$

If you will review your solution of problem 4.11 in Chapter 2, you will find that it is quite similar to the derivation of (4.10). The essential feature of the two-body central force problem that is revealed by the expressions for center of mass kinetic energy and angular momentum is that it is possible to treat the motion of the two particles with respect to their common center of mass as though one particle moved about the other as a fixed origin. It is necessary to use the reduced mass rather than the actual mass of one of the particles.

If there are no external forces acting on the system, its angular momentum is a constant. This means that the angular momentum always points in the same direction in space and that the motion of the two-particle system takes place in a plane. It is convenient to introduce polar coordinates in this plane and to write the Lagrangian for the system as

$$L = \tfrac{1}{2}m(\dot{r}^2 + r^2\dot{\theta}^2) - V(r) \tag{4.11}$$

A discussion of Lagrange's equations is to be found in section 13 of Chapter 4. The equations of motion follow from

$$\frac{d}{dt}\frac{\partial L}{\partial \dot{r}} - \frac{\partial L}{\partial r} = 0 \qquad \frac{d}{dt}\frac{\partial L}{\partial \dot{\theta}} - \frac{\partial L}{\partial \theta} = 0 \tag{4.12}$$

The resulting equations are

$$m(\ddot{r} - r\dot{\theta}^2) = -\frac{\partial V}{\partial r} = F(r) \tag{4.13}$$

$$\frac{\partial L}{\partial \dot{\theta}} = mr^2\dot{\theta} = \ell = \text{constant} \tag{4.14}$$

We find that θ is a cyclic coordinate as explained in problem 13.2 of Chapter 2. The constant ℓ is the magnitude of the angular momentum. It is possible to eliminate $\dot{\theta}$ between (4.13) and (4.14) to get a nonlinear differential equation for r as a function of the time:

$$m\ddot{r} - \frac{\ell^2}{mr^3} = F(r) \tag{4.15}$$

In addition to the angular momentum, the total energy is also a constant of the motion. It can be expressed in the form

$$W = \tfrac{1}{2}m(\dot{r}^2 + r^2\dot{\theta}^2) + V(r) = \tfrac{1}{2}m\dot{r}^2 + V_e(r)$$

$$V_e(r) = \frac{\ell^2}{2mr^2} + V(r) \tag{4.16}$$

where $V_e(r)$ is an *effective potential*. The equation of motion (4.15) can be expressed in the form

$$m\ddot{r} + \frac{dV_e(r)}{dr} = 0 \tag{4.17}$$

which is the equivalent of a one-dimensional equation of motion in which $V_e(r)$ is the actual potential. In those cases where the radial motion is periodic, the solution can be found by the method developed in section 3. An effective potential was also used in the discussion of the motion of a toy top in section 15 of Chapter 4.

Rather than solve equation (4.15) for r as a function of time, it is convenient to work with a differential equation which gives directly the shape of the orbit. To this end, a new variable x is introduced by means of $r = 1/x$. The derivative of r with respect to time can be expressed as

$$\dot{r} = -\frac{1}{x^2}\frac{dx}{dt} = -\frac{1}{x^2}\frac{dx}{d\theta}\dot{\theta} = -\frac{\ell}{m}\frac{dx}{d\theta} \tag{4.18}$$

where $\dot{\theta}$ has been eliminated by means of $\ell = mr^2\dot{\theta}$. A second differentiation gives

$$\ddot{r} = -\frac{\ell}{m}\frac{d^2x}{d\theta^2}\dot{\theta} = -\frac{\ell^2}{m^2}x''x^2 \tag{4.19}$$

where $x'' = d^2x/d\theta^2$. The differential equation (4.15) can therefore be expressed as

$$x'' + x = -\frac{m}{\ell^2 x^2}F\left(\frac{1}{x}\right) \tag{4.20}$$

If a charged particle moves in the space between charged concentric cylindrical electrodes, it experiences a force

$$F(r) = -\frac{C}{r} \tag{4.21}$$

where C is a constant and r is measured from the common axes of the two electrodes. This is to be regarded as the z axis of a cylindrical coordinate system. If $C > 0$, then $F(r)$ is a force of attraction toward the inner electrode. If $C < 0$, the particles are repelled from the inner electrode. This is not a true central force problem since the force is not directed toward the origin. However, all the results of this section apply to the motion in the r, θ plane. There can at most be an additional motion at constant velocity in the z direction. The effective potential (4.16) is

$$V_e(r) = \frac{\ell^2}{2mr^2} + C\ln\frac{r}{k} \tag{4.22}$$

where k is a constant. We find that $V_e(r)$ has a single minimum at

$$r_0 = \frac{\ell}{\sqrt{mC}} \tag{4.23}$$

provided that $C > 0$. It is convenient to let $k = r_0$ and write $V_e(r)$ as

$$V_e(r) = C\left(\frac{r_0^2}{2r^2} + \ln\frac{r}{r_0}\right) \tag{4.24}$$

Figure 4.2 shows a plot of $V_e(r)/C$. It is evident that periodic solutions for r as a function of time exist if $C > 0$, and this is the case which we will consider. It is possible for the particle to have a stable circular

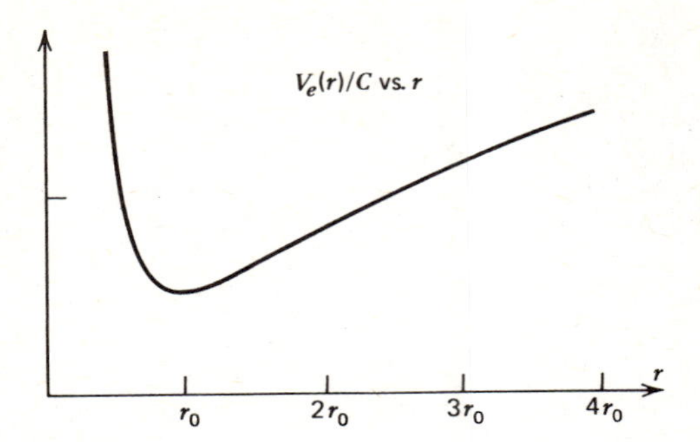

Fig. 4.2

orbit of a given radius r_0 provided that the initial conditions are properly determined. If the particle is disturbed slightly, r will oscillate with a small amplitude about r_0 as the equilibrium point.

If we set $x_0 = 1/r_0$, then equation (4.20) for the orbit can be expressed as

$$x'' + x - \frac{x_0^2}{x} = 0 \tag{4.25}$$

In spite of its simple appearance, this nonlinear differential equation is not solvable in terms of elementary functions. We will therefore find its approximate solution in terms of a Fourier series. If the function $f(x) = x - x_0^2/x$ appearing in the differential equation is expanded in a Taylor's series about x_0, the result is the approximate differential equation

$$x'' + 2(x - x_0) - \frac{1}{x_0}(x - x_0)^2 + \frac{1}{x_0^2}(x - x_0)^3 - \cdots = 0 \tag{4.26}$$

The interval of convergence of the Taylor's series is $|x - x_0| < x_0$. In terms of r and r_0, this is $|r - r_0| < r$, which is equivalent to

$$\tfrac{1}{2}r_0 < r < \infty \tag{4.27}$$

If the coefficients in (4.26) are identified as

$$\alpha = 2 \qquad \beta = \frac{-1}{x_0} \qquad \gamma = \frac{1}{x_0^2} \tag{4.28}$$

the Fourier series solution can be obtained from

(3.29) and (3.30) as

$$\frac{r_0}{r} = 1 + \frac{(b_1r_0)^2}{4} + b_1r_0\cos\lambda\theta - \frac{(b_1r_0)^2}{12}\cos 2\lambda\theta + \frac{(b_1r_0)^3}{48}\cos 3\lambda\theta + \cdots \tag{4.29}$$

$$\lambda^2 = 2\left[1 + \tfrac{1}{6}(b_1r_0)^2\right] \tag{4.30}$$

We have used λ in place of the frequency ω. Note that if $b_1 = 0$, then (4.29) reduces to a circular orbit of radius r_0. Both r_0 and b_1 are determined by the initial conditions.

As a numerical example, suppose that the initial conditions are such that $b_1r_0 = 0.400$. The approximate solution is then

$$\frac{r_0}{r} = 1.040 + 0.400\cos\lambda\theta - 0.0133\cos 2\lambda\theta + 0.0013\cos 3\lambda\theta$$

$$\lambda = 1.433 \tag{4.31}$$

Figure 4.3 is obtained from this equation. The particle starts out at an inner turning point at A where $r = 0.700r_0$ and moves to an outer turning point at B where $r = 1.600r_0$. In orbit theory, a turning point is a maximum or a minimum value of r and is called an *apse* (sometimes *apsis*). The plural form is *apsides*. The angle between the inner and the outer apsides is found from

$$\lambda\theta = 180° \qquad \theta = 125.6° \tag{4.32}$$

The shape of the orbit from the outer apse at B to the next inner apse at C can be found by a mirror

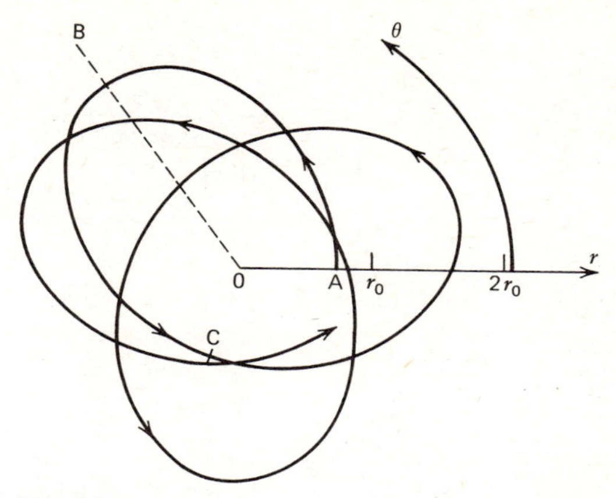

Fig. 4.3

reflection of the orbit from A to B in the apsidal line OB. By continuing in this manner, the entire figure can be produced.

An estimate of error can be made by first writing the differential equation as

$$x'' + 2(x - x_0)\left[1 - \frac{1}{2}\left(\frac{x - x_0}{x_0}\right) + \frac{1}{2}\left(\frac{x - x_0}{x_0}\right)^2 - \frac{1}{2}\left(\frac{x - x_0}{x_0}\right)^3 + \cdots\right] = 0 \tag{4.33}$$

For $0.700r_0 < r < 1.600r_0$, the first neglected term has the range of values

$$-0.026 < \frac{1}{2}\left(\frac{x - x_0}{x_0}\right)^3 < 0.039 \tag{4.34}$$

PROBLEMS

4.1. A particle of charge q moves in the space between concentric cylindrical electrodes. The inner electrode has radius a and is at a potential ϕ. The outer electrode has radius b and is grounded. Relate the constant C of equation (4.21) to a, b, q, and ϕ.

4.2. The electrostatic force between charged particles and the gravitational force between point masses are both inverse r^2 forces:

$$F(r) = -\frac{C}{r^2} \tag{4.35}$$

where $C > 0$ for attraction and $C < 0$ for repulsion. If $C > 0$, show that the effective potential

(4.16) is

$$V_e(r) = \frac{C}{r_0}\left[\frac{r_0^2}{2r^2} - \frac{r_0}{r}\right] \tag{4.36}$$

where

$$r_0 = \frac{\ell^2}{mC} \tag{4.37}$$

is the location of the minimum of $V_e(r)$ and is the radius of a stable circular orbit of angular momentum ℓ. Graph the effective potential. Show that the radial motion is periodic if the energy is in the range

$$-\frac{C}{2r_0} < W < 0 \tag{4.38}$$

4.3. Continuing with the inverse r^2 force, show that for bound orbits

$$x'' + x = x_0 \qquad x_0 = \frac{1}{r_0}$$
$$x = x_0 + b_1 \cos\theta \tag{4.39}$$

where b_1 is a constant of integration. Show that in terms of the apsidal distances r_1 and r_2 the equation for the orbit is

$$\frac{1}{r} = \frac{1}{2}\left(\frac{1}{r_1} + \frac{1}{r_2}\right) + \frac{1}{2}\left(\frac{1}{r_1} - \frac{1}{r_2}\right)\cos\theta \tag{4.40}$$

When the orbits of planets around the sun are under discussion, the distance of closest approach or inner apse r_1 is called *perihelion* and the outer apse is called *aphelion*. The inverse r^2 force has the unusual property of producing an orbit which closes on itself. In fact, the shape of the orbit as given by (4.40) is that of an ellipse with the center of force at one focus. The isotropic harmonic oscillator force, $F(r) = -Cr$, also gives a closed orbit.

4.4. Show that the turning points for bound orbits in an inverse r^2 force field are given by

$$r_1 = \frac{C}{2|W|} - \sqrt{\left(\frac{C}{2W}\right)^2 - \frac{\ell^2}{2m|W|}} \tag{4.41}$$

$$r_2 = \frac{C}{2|W|} + \sqrt{\left(\frac{C}{2W}\right)^2 - \frac{\ell^2}{2m|W|}} \tag{4.42}$$

4.5. A discussion of the geometry of an ellipse is to be found in problem 12.1 of Chapter 4. As shown in Fig. 4.4, an ellipse can be defined by the condition that the point P move in such a way that

$$r + s = 2a \tag{4.43}$$

where a is the length of the semimajor axis. Show that the equation of the ellipse can be put into the form

$$\frac{1}{r} = \frac{a}{b^2} + \sqrt{\left(\frac{a}{b^2}\right)^2 - \frac{1}{b^2}}\cos\theta \tag{4.44}$$

where b is the length of the semiminor axis. The orbit equation (4.40) is written in such a way that the focus f_2 is the center of force. Show that a and b can be evaluated in terms of the

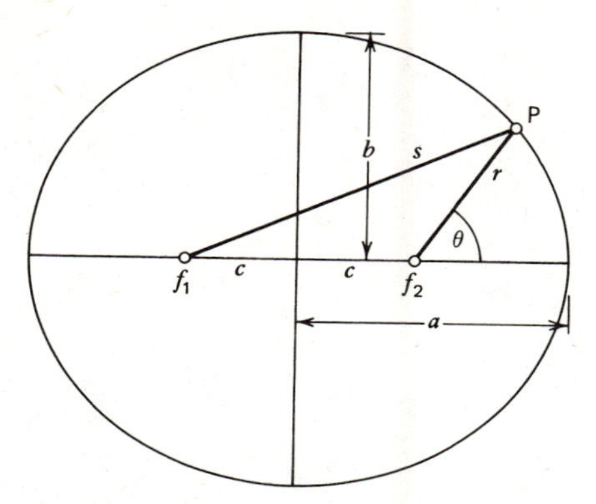

Fig. 4.4

energy and angular momentum as

$$b = \frac{\ell}{\sqrt{2m|W|}} \qquad a = \frac{C}{2|W|} \tag{4.45}$$

4.6. Show that as a consequence of conservation of angular momentum, the time required for the radius vector to sweep out an area $d\sigma$ is

$$dt = \frac{2m}{\ell} d\sigma \tag{4.46}$$

4.7. The equation of an ellipse in parametric form is

$$x = a\cos\phi \qquad y = b\sin\phi \tag{4.47}$$

By expressing an element of the area of the ellipse as $d\sigma = x\,dy$ and integrating, show that its total area is

$$\sigma = \pi ab \tag{4.48}$$

4.8. Show that for the inverse r^2 force law, the period of a bound orbit is

$$P = 2\pi a^{3/2}\sqrt{\frac{m}{C}} \tag{4.49}$$

4.9. By a consideration of the effective potential, show that the force law

$$F(r) = -\frac{C}{r^n} \qquad 1 < n < 3 \tag{4.50}$$

is qualitatively similar to the inverse r^2 force law. Show that the differential equation of the orbit is

$$x'' + x - r_0^{n-3}x^{n-2} = 0 \tag{4.51}$$

where r_0 is the location of the minimum of $V_e(r)$ and is given by

$$r_0^{n-3} = \frac{mC}{\ell^2} \tag{4.52}$$

provided that $C > 0$. Show that for the case of bound orbits, an approximate solution in terms of a Fourier series is

$$x - x_0 = \frac{(2-n)r_0 b_1^2}{4} + b_1 \cos \lambda\theta - \frac{(2-n)r_0 b_1^2}{12} \cos 2\lambda\theta$$

$$+ \frac{(2-n)(3-n)r_0^2 b_1^3}{96} \cos 3\lambda\theta + \cdots \tag{4.53}$$

$$\lambda^2 = (3-n)\left[1 - \frac{(n-2)(n+1)}{12} b_1^2 r_0^2\right] \tag{4.54}$$

If the law of force is almost an inverse r^2 force, then $n = 2 + \varepsilon$, where $\varepsilon \ll 1$. Show that

$$\lambda \simeq 1 - \tfrac{1}{2}\varepsilon \tag{4.55}$$

The orbit does not quite close on itself and can be pictured as an ellipse the axes of which slowly turn. This phenomenon is called *precession*. Show that for each revolution, the perihelion advances by an approximate amount

$$\Delta\theta = \varepsilon\pi \tag{4.56}$$

and that the precessional angular velocity or rate of advance of the perihelion is

$$\omega_P = \frac{\varepsilon\pi}{P} = \frac{\varepsilon}{2a^{3/2}}\sqrt{\frac{C}{m}} \tag{4.57}$$

The planet Mercury exhibits an anomalous precessional effect. This is however a relativistic effect and is not due to a failure of the inverse r^2 law of force. See also problem 12.12 of Chapter 9.

4.10. Find the effective potential for

$$F(r) = -\frac{C}{r^4} \qquad C > 0 \tag{4.58}$$

and show that circular orbits exist but that they are unstable. Describe qualitatively the type of bound motion which can occur. Note that the particle can pass through the center of force even if it has angular momentum. Comment on the significant differences between the bound motion for (4.58) and the inverse r^2 force.

4.11. If you did problem 5.19 of Chapter 5, you obtained the vector potential

$$A_\theta = -\tfrac{1}{2}arz^2 + \tfrac{1}{8}ar^3 - \tfrac{1}{2}rB_0 - \tfrac{1}{4}arr_0^2 \tag{4.59}$$

Find a Lagrangian which gives the equations of motion of a charged particle moving in the magnetic field given by equation (5.59) of Chapter 5. Refer to problem 13.7 in Chapter 4. The kinetic energy of the particle is a constant of the motion because the force exerted on it by the magnetic field is at right angles to the direction of motion. What is another constant of the motion? Refer to problem 13.2 in Chapter 4. Find an effective potential for the particle if its motion is confined to the plane $z = 0$. Show that stable circular orbits of radius r_0 about the z axis are possible if $a < B_0/r_0^2$. This problem does not answer the question of stability with respect to motion in the z direction.

5. THE FOURIER INTEGRAL

In section 2 of Chapter 7, the eigenfunctions for a flexible string were arrived at by a limiting process which began with a discrete system consisting of a massless string loaded at equal intervals by identical point masses. We are going to continue the limiting process in another way by allowing the interval over which a set of eigenfunctions is defined to become infinite. Consider the Fourier eigenfunctions

$$u_k(x) = \frac{1}{\sqrt{2s}} e^{ik\pi x/s} \qquad -s \le x \le s \tag{5.1}$$

and the expansion of an arbitrary function in terms

of them:

$$f(x)=\sum_{k=-\infty}^{\infty} c_k\frac{1}{\sqrt{2s}}e^{ik\pi x/s} \tag{5.2}$$

$$c_k=\int_{-s}^{s}\frac{1}{\sqrt{2s}}e^{-ik\pi x/s}f(x)\,dx \tag{5.3}$$

If we define $\kappa=k\pi/s$, then it is possible to write (5.2) and (5.3) in the form

$$f(x)=\frac{1}{\sqrt{2\pi}}\sum_{k=-\infty}^{\infty}\sqrt{\frac{s}{\pi}}\,c_k e^{i\kappa x}\frac{\pi}{s} \tag{5.4}$$

$$\sqrt{\frac{s}{\pi}}\,c_k=\frac{1}{\sqrt{2\pi}}\int_{-s}^{s}e^{-i\kappa x}f(x)\,dx \tag{5.5}$$

The reason for this apparently peculiar arrangement of the factors will soon become clear. As s becomes large, the possible values of κ become more closely spaced; in the limit $s\to\infty$, κ becomes a continuous variable. When k changes by one integer, the change in κ is $d\kappa=\pi/s$. The limiting forms of (5.4) and (5.5) are therefore

$$f(x)=\frac{1}{\sqrt{2\pi}}\int_{-\infty}^{\infty}A(\kappa)e^{i\kappa x}\,d\kappa \tag{5.6}$$

$$A(\kappa)=\frac{1}{\sqrt{2\pi}}\int_{-\infty}^{\infty}f(x)e^{-i\kappa x}\,dx \tag{5.7}$$

where we have defined

$$A(\kappa)=\lim_{s\to\infty}\sqrt{\frac{s}{\pi}}\,c_k \tag{5.8}$$

Equations (5.6) and (5.7) constitute what is known as *Fourier's integral theorem*. Equation (5.7) is called the *Fourier transform* of $f(x)$. Equation (5.6) is the inverse transformation. Another way to look at it is to regard (5.6) as the expansion of an arbitrary function $f(x)$ in terms of the eigenfunctions

$$u_\kappa(x)=\frac{1}{\sqrt{2\pi}}e^{i\kappa x}\qquad -\infty<x<\infty \tag{5.9}$$

where κ now represents a continuum of eigenvalues. The coefficients in the expansion are given by the function $A(\kappa)$. Equations (5.6) and (5.7) are symmetric; we could equally regard (5.7) as an expansion of the function $A(\kappa)$.

The closure relation for the eigenfunctions (5.1) can be written

$$\sum_{k=-\infty}^{\infty}u_k^*(x')u_k(x)=\sum_{k=-\infty}^{\infty}\frac{1}{2\pi}e^{ik\pi(x-x')/s}\frac{\pi}{s}$$
$$=\delta(x-x') \tag{5.10}$$

In the limit $s\to\infty$,

$$\frac{1}{2\pi}\int_{-\infty}^{\infty}e^{i\kappa(x-x')}\,d\kappa=\delta(x-x') \tag{5.11}$$

A corresponding result can be found by carrying the orthogonality relation for the eigenfunctions through a similar set of steps:

$$\int_{-s}^{s}u_k(x)u_{k'}^*(x)\,dx=\frac{1}{2s}\int_{-s}^{s}e^{i(k-k')\pi x/s}\,dx=\delta_{kk'}$$
$$\frac{1}{2\pi}\int_{-s}^{s}e^{i(\kappa-\kappa')x}\,dx=\frac{s}{\pi}\delta_{kk'}=\frac{\delta_{kk'}}{\Delta\kappa} \tag{5.12}$$
$$\frac{1}{2\pi}\int_{-\infty}^{\infty}e^{i(\kappa-\kappa')x}\,dx=\delta(\kappa-\kappa')$$

We will now undertake the task of establishing the Fourier integral theorem and the closure relations in a somewhat more rigorous fashion. To prove the closure relation (5.11), consider the finite integral

$$\Delta_a(u)=\frac{1}{2\pi}\int_{-a}^{a}e^{i\kappa u}\,d\kappa \tag{5.13}$$

where $u=x-x'$. The integral is evaluated to give

$$\Delta_a(u)=\frac{\sin au}{\pi u} \tag{5.14}$$

The function $\Delta_a(u)$ gives a thin high spike at $u=0$ and is small elsewhere when a is large. These are the required properties of a function if its limit is to be a delta function. Note that

$$\int_{-\infty}^{\infty}\Delta_a(u)\,du=\int_{-\infty}^{\infty}\frac{\sin au}{\pi u}du$$
$$=\frac{1}{\pi}\int_{-\infty}^{\infty}\frac{\sin y}{y}dy=1 \tag{5.15}$$

The integral is evaluated in equation (20.26) of Chapter 5. Note that (5.15) is independent of the value of a. Note also that because $\Delta_a(u)$ is an even function of u,

$$\int_{-\infty}^{0}\Delta_a(u)\,du=\frac{1}{2}\qquad\int_{0}^{\infty}\Delta_a(u)\,du=\frac{1}{2} \tag{5.16}$$

The proof that (5.14) becomes a delta function in the limit $a \to \infty$ is almost identical to the proof of (1.23). Let $F(u)$ be a sectionally continuous function which has both a right and a left derivative everywhere. Consider

$$\begin{aligned}
&\int_{-\infty}^{\infty} F(u)\,\Delta_a(u)\,du \\
&= \int_{-\infty}^{0} F(u)\,\Delta_a(u)\,du + \int_{0}^{\infty} F(u)\,\Delta_a(u)\,du \\
&= \int_{0}^{\infty} F(-u)\,\Delta_a(u)\,du + \int_{0}^{\infty} F(u)\,\Delta_a(u)\,du \\
&= \int_{0}^{\infty} \frac{F(-u)-F(-0)}{\pi u}\sin au\,du \\
&\quad + F(-0)\int_{0}^{\infty} \Delta_a(u)\,du \\
&\quad + \int_{0}^{\infty} \frac{F(u)-F(+0)}{\pi u}\sin au\,du \\
&\quad + F(+0)\int_{0}^{\infty} \Delta_a(u)\,du
\end{aligned} \tag{5.17}$$

By means of theorem 1 in section 1 and equation (5.16),

$$\lim_{a\to\infty} \int_{-\infty}^{\infty} F(u)\,\Delta_a(u)\,du = \frac{1}{2}[F(+0)+F(-0)] \tag{5.18}$$

Thus, (5.11) and the companion result (5.12) are established.

We will now confirm the validity of the Fourier integral transform pair (5.6) and(5.7). Given the function $f(x)$, define

$$A_s(\kappa) = \frac{1}{\sqrt{2\pi}} \int_{-s}^{s} f(x')e^{-i\kappa x'}\,dx' \tag{5.19}$$

Multiply by

$$u_\kappa(x)\,d\kappa = \frac{1}{\sqrt{2\pi}} e^{i\kappa x}\,d\kappa \tag{5.20}$$

and integrate κ over the range $-a$ to $+a$:

$$\begin{aligned}
&\frac{1}{\sqrt{2\pi}} \int_{-a}^{a} A_s(\kappa)e^{i\kappa x}\,d\kappa \\
&= \int_{-s}^{s} f(x')\left[\frac{1}{2\pi}\int_{-a}^{a} e^{i\kappa(x-x')}\,d\kappa\right] dx'
\end{aligned} \tag{5.21}$$

In the limit $a \to \infty$,

$$\begin{aligned}
\frac{1}{\sqrt{2\pi}} \int_{-\infty}^{\infty} A_s(\kappa)e^{i\kappa x}\,d\kappa &= \int_{-s}^{s} f(x')\,\delta(x-x')\,dx' \\
&= f(x)
\end{aligned} \tag{5.22}$$

The result is independent of s as long as $-s < x < s$. In the limit as $s \to \infty$, (5.19) and (5.22) give (5.6) and (5.7). The integrals do not have to be convergent in the ordinary sense. For example, if $f(x) = c =$ constant for $-\infty < x < \infty$, then (5.7) gives

$$A(\kappa) = c\sqrt{2\pi}\,\delta(\kappa) \tag{5.23}$$

Putting (5.23) back into (5.6) gives

$$f(x) = \frac{1}{\sqrt{2\pi}} \int_{-\infty}^{\infty} c\sqrt{2\pi}\,\delta(\kappa)e^{i\kappa x}\,d\kappa = c \tag{5.24}$$

Let $f(x)$ and $g(x)$ be two functions which are square-integrable. This means that

$$(f, f) = \int_{-\infty}^{\infty} f^* f\,dx \qquad (g, g) = \int_{-\infty}^{\infty} g^* g\,dx \tag{5.25}$$

are both convergent. As a consequence of the Schwartz inequality (11.20) of Chapter 7,

$$(f, g) = \int_{-\infty}^{\infty} f^* g\,dx \tag{5.26}$$

is also convergent. If the Fourier transforms for $f(x)$ and $g(x)$ are substituted into (5.26), the result is

$$\begin{aligned}
&\int_{-\infty}^{\infty} f^* g\,dx \\
&= \int_{-\infty}^{\infty} \frac{1}{2\pi} \int_{-\infty}^{\infty} A^*(\kappa')e^{-i\kappa' x}\,d\kappa' \int_{-\infty}^{\infty} B(\kappa)e^{i\kappa x}\,d\kappa\,dx \\
&= \int_{-\infty}^{\infty}\int_{-\infty}^{\infty} \left[\frac{1}{2\pi}\int_{-\infty}^{\infty} e^{i(\kappa-\kappa')x}\,dx\right] A^*(\kappa')\,B(\kappa)\,d\kappa'\,d\kappa \\
&= \int_{-\infty}^{\infty}\int_{-\infty}^{\infty} \delta(\kappa-\kappa')\,A^*(\kappa')\,B(\kappa)\,d\kappa'\,d\kappa \\
&= \int_{-\infty}^{\infty} A^*(\kappa)\,B(\kappa)\,d\kappa
\end{aligned} \tag{5.27}$$

This result is analogous to (11.10) of Chapter 7 except that the "components" of f and g now form a continuum. If $f = g$, equation (5.27) becomes Parseval's relation

$$|f|^2 = \int_{-\infty}^{\infty} f^* f\,dx = \int_{-\infty}^{\infty} A^* A\,d\kappa \tag{5.28}$$

As an example of a Fourier transform, suppose

$$f(x)=c \qquad -s\le x\le s$$
$$f(x)=0 \quad |x|>s \tag{5.29}$$

The Fourier transform of this function is

$$A(\kappa)=\frac{1}{\sqrt{2\pi}}\int_{-s}^{s} ce^{-i\kappa x}\,dx=\sqrt{\frac{2}{\pi}}\,\frac{c\sin\kappa s}{\kappa} \tag{5.30}$$

The inverse is

$$f(x)=\frac{1}{\pi}\int_{-\infty}^{\infty}\frac{c\sin\kappa s}{\kappa}e^{i\kappa x}\,d\kappa \tag{5.31}$$

Since $(\sin\kappa s)/\kappa$ is an even function of κ and $\sin\kappa x$ is odd, it is only the cosine part of $e^{i\kappa x}$ which contributes to the integral. Therefore,

$$f(x)=\frac{1}{\pi}\int_{-\infty}^{\infty}\frac{c\sin\kappa s\cos\kappa x}{\kappa}\,d\kappa \tag{5.32}$$

The value of this integral is the original function as given by (5.29). Thus, one possible use of Fourier transforms is in the evaluation of definite integrals. The reader can show that at the points of discontinuity $x=\pm s$, the integral actually converges to the value $c/2$.

All the results of this section can be extended to three dimensions:

$$f(\mathbf{r})=\frac{1}{(2\pi)^{3/2}}\int_{\Sigma_\kappa} A(\boldsymbol{\kappa})e^{i\boldsymbol{\kappa}\cdot\mathbf{r}}\,d\Sigma_\kappa \tag{5.33}$$

$$A(\boldsymbol{\kappa})=\frac{1}{(2\pi)^{3/2}}\int_{\Sigma_r} f(\mathbf{r})e^{-i\boldsymbol{\kappa}\cdot\mathbf{r}}\,d\Sigma_r \tag{5.34}$$

We have used the notation

$$d\Sigma_\kappa=d\kappa_x\,d\kappa_y\,d\kappa_z \qquad d\Sigma_r=dx\,dy\,dz \tag{5.35}$$

The integrals are to be extended over all space. The closure relations are

$$\delta(\mathbf{r}-\mathbf{r}')=\frac{1}{(2\pi)^3}\int_{\Sigma_\kappa} e^{i\boldsymbol{\kappa}\cdot(\mathbf{r}-\mathbf{r}')}\,d\Sigma_\kappa \tag{5.36}$$

$$\delta(\boldsymbol{\kappa}-\boldsymbol{\kappa}')=\frac{1}{(2\pi)^3}\int_{\Sigma_r} e^{i(\boldsymbol{\kappa}-\boldsymbol{\kappa}')\cdot\mathbf{r}}\,d\Sigma_r \tag{5.37}$$

PROBLEMS

5.1. The Fourier sine integral and its inverse are given by

$$f(x)=\sqrt{\frac{2}{\pi}}\int_0^{\infty}A(\kappa)\sin\kappa x\,d\kappa \tag{5.38}$$

$$A(\kappa)=\sqrt{\frac{2}{\pi}}\int_0^{\infty}f(x)\sin\kappa x\,dx \tag{5.39}$$

Derive this result from the eigenfunctions for a flexible string with both ends fixed by taking the limit $s\to\infty$. Show that the closure relation is

$$\frac{2}{\pi}\int_0^{\infty}\sin\kappa x\sin\kappa x'\,d\kappa=\delta(x-x') \tag{5.40}$$

5.2. The Fourier cosine integral and its inverse are given by

$$f(x)=\sqrt{\frac{2}{\pi}}\int_0^{\infty}A(\kappa)\cos\kappa x\,d\kappa \tag{5.41}$$

$$A(\kappa)=\sqrt{\frac{2}{\pi}}\int_0^{\infty}f(x)\cos\kappa x\,dx \tag{5.42}$$

Obtain them from (5.6) and (5.7) by assuming that $f(x)$ is an even function. Show that the appropriate closure relation is

$$\delta(x-x')=\frac{2}{\pi}\int_0^{\infty}\cos\kappa x\cos\kappa x'\,d\kappa \tag{5.43}$$

5.3. It is convenient to denote the Fourier transform of a function by

$$F[f(x)] = A(\kappa) = \frac{1}{\sqrt{2\pi}} \int_{-\infty}^{\infty} f(x) e^{-i\kappa x} \, dx \tag{5.44}$$

Establish the following properties of the Fourier transform:

$$F[xf(x)] = i \frac{d}{d\kappa} F[f(x)] \tag{5.45}$$

$$F[f(x+a)] = e^{i\kappa a} F[f(x)] \tag{5.46}$$

If $f(x)$ vanishes at $\pm\infty$, then

$$F[f'(x)] = i\kappa F[f(x)] \tag{5.47}$$

5.4. The *convolution* of two functions $g(x)$ and $h(x)$ is defined as

$$f(x) = \int_{-\infty}^{\infty} g(y) h(x-y) \, dy \tag{5.48}$$

Show that

$$F[f(x)] = \sqrt{2\pi} \, F[g(x)] F[h(x)] \tag{5.49}$$

5.5. If $F[f_1] = g_1$ and $F[f_2] = g_2$, show that

$$F[f_1 f_2] = \frac{1}{\sqrt{2\pi}} \int_{-\infty}^{\infty} g_1(y) g_2(\kappa - y) \, dy \tag{5.50}$$

5.6. Show that the integral given by (5.32) has the value $c/2$ if $x = \pm s$.

5.7. By writing the volume integral as the product of three integrals, show that (5.36) is equivalent to

$$\delta(\mathbf{r} - \mathbf{r}') = \delta(x - x') \, \delta(y - y') \, \delta(z - z') \tag{5.51}$$

5.8. The Dirac delta function belongs to a group of mathematical entities called *generalized functions*. Another generalized function can be found by differentiating the delta function. If $f(x)$ is an arbitrary function defined over $a \le x \le b$, prove that

$$\int_a^b \delta'(x - x_0) f(x) \, dx = -f'(x_0) \tag{5.52}$$

Hint: Let

$$\delta(x - x_0) = \lim_{n \to \infty} \Delta_n(x - x_0) \tag{5.53}$$

and consider

$$\lim_{n \to \infty} \int_a^b \Delta'_n(x - x_0) f(x) \, dx \tag{5.54}$$

The members of the sequence $\Delta_n(x - x_0)$ can be chosen to be continuous and differentiable. One example of such a sequence is given by (5.14). It is however not necessary to use a specific formula for $\Delta_n(x - x_0)$ in this problem. Note that since $\delta(x)$ is even, $\delta'(x)$ is odd.

5.9. Prove that

$$F[1] = \sqrt{2\pi} \, \delta(\kappa) \qquad F[x] = \sqrt{2\pi} \, i\delta'(\kappa) \tag{5.55}$$

5.10. Prove that

$$\int_a^b \delta''(x - x_0) f(x) \, dx = f''(x_0) \tag{5.56}$$

$$F[x^2] = -\sqrt{2\pi} \, \delta''(\kappa) \tag{5.57}$$

5.11. Prove that

$$\int_a^b \delta[c(x-x_0)]f(x)\,dx = \frac{1}{|c|}f(x_0) \tag{5.58}$$

where c is a nonzero constant.

5.12. If $g(x)$ vanishes at x_0 and at no other point, show that

$$\int_a^b \delta[g(x)]f(x)\,dx = \frac{f(x_0)}{|g'(x_0)|} \tag{5.59}$$

Use a modification of the same idea to show that

$$\delta(x^2-c^2) = \frac{1}{2|c|}[\delta(x-c)+\delta(x+c)] \tag{5.60}$$

5.13. Evaluate the following integrals:

$$\int_0^\infty \frac{\cos\kappa x}{a^2+\kappa^2}\,d\kappa = \frac{\pi}{2a}e^{-ax} \tag{5.61}$$

$$\int_0^\infty \frac{\kappa\sin\kappa x}{a^2+\kappa^2}\,d\kappa = \frac{\pi}{2}e^{-ax} \tag{5.62}$$

where $a>0$ and $x>0$.

Hint: Start out by finding the Fourier sine and cosine transforms of e^{-ax}. The integral in (5.61) is evaluated by another method in equation (21.23) of Chapter 5.

6. THE FREE PARTICLE IN QUANTUM MECHANICS

Some discussion of the free-particle Schrödinger equation is given in section 12 of Chapter 7; the relevant equations are (12.52) through (12.57). The development of this section will be done in one dimension. The Hamiltonian and linear momentum operators are

$$H = -\frac{\hbar^2}{2m}\frac{\partial^2}{\partial x^2} \qquad p_x = -i\hbar\frac{\partial}{\partial x} \tag{6.1}$$

These operators commute with one another and have the common eigenfunctions

$$u_\kappa(x) = \frac{1}{\sqrt{2\pi}}e^{i\kappa x} \tag{6.2}$$

The momentum eigenvalues are $\hbar\kappa$, and the energy eigenvalues are

$$W_\kappa = \frac{\hbar^2\kappa^2}{2m} \tag{6.3}$$

The eigenvalues form a continuum and the eigenfunctions are defined over all space. Since they do not vanish at infinity, they cannot be normalized in the conventional sense. As we have discovered in section 5, delta function normalization is possible so that

$$\int_{-\infty}^{\infty} u_\kappa^*(x)u_{\kappa'}(x)\,dx = \delta(\kappa-\kappa') \tag{6.4}$$

An arbitrary wave function can be expanded in terms of the eigenfunctions (6.2) as

$$\psi(x,t) = \frac{1}{\sqrt{2\pi}}\int_{-\infty}^{\infty}\phi(\kappa,t)e^{i\kappa x}\,d\kappa \tag{6.5}$$

$$\phi(\kappa,t) = \frac{1}{\sqrt{2\pi}}\int_{-\infty}^{\infty}\psi(x,t)e^{-i\kappa x}\,dx \tag{6.6}$$

which is nothing more than Fourier's integral theorem. It is valid for both the free particle and for the case where there is a force. There is a symmetry here leading to the interpretation that $\phi(\kappa,t)$ is the wave function in momentum space just as $\psi(x,t)$ is the wave function in coordinate space. Equation (6.6) is then a transformation or change in representation of the theory. The preservation of normalization requires that such transformations be unitary, and in fact the condition of unitarity in this case is

$$\frac{1}{2\pi}\int_{-\infty}^{\infty} e^{-i\kappa x'}e^{i\kappa x}\,d\kappa = \delta(x-x') \tag{6.7}$$

We know that $|\psi|^2\,dx$ has the interpretation that it is the probability that the particle will be found in the range x to $x + dx$; similarly, $|\phi|^2\, d\kappa$ is the probability that a measurement of momentum will yield a value of κ in the range κ to $\kappa + d\kappa$. If ψ is normalized, then Parseval's relation (5.28) guarantees that ϕ will also be normalized:

$$\int_{-\infty}^{\infty} |\psi|^2\, dx = \int_{-\infty}^{\infty} |\phi|^2\, d\kappa = 1 \tag{6.8}$$

Thus, the entire formal structure of quantum mechanics is maintained when plane wave eigenfunctions are used even though these functions do not vanish at ∞.

The expectation value of the momentum, if the state of the system is described by the wave function $\phi(\kappa, t)$ in momentum space, is

$$\langle p_x \rangle = \int_{-\infty}^{\infty} \phi^*(\kappa, t)\hbar\kappa\phi(\kappa, t)\, d\kappa \tag{6.9}$$

If one of the factors of ϕ in (6.9) is eliminated by means of (6.6) we find

$$\begin{aligned}\langle p_x \rangle &= \int_{-\infty}^{\infty} \phi^* \hbar\kappa \left[\frac{1}{\sqrt{2\pi}} \int_{-\infty}^{\infty} \psi e^{-i\kappa x}\, dx \right] d\kappa \\ &= \int_{-\infty}^{\infty} \phi^* \left[\frac{1}{\sqrt{2\pi}} \int_{-\infty}^{\infty} \psi i\hbar \frac{d}{dx} e^{-i\kappa x}\, dx \right] d\kappa\end{aligned} \tag{6.10}$$

If the integral over x is integrated by parts, the result is

$$\langle p_x \rangle = \int_{-\infty}^{\infty} \phi^* \left[\frac{1}{\sqrt{2\pi}} \int_{-\infty}^{\infty} \left(-i\hbar \frac{\partial \psi}{\partial x} \right) e^{-i\kappa x}\, dx \right] d\kappa \tag{6.11}$$

where we have assumed that ψ vanishes at $\pm\infty$. The integral over κ in (6.11) is just ψ^*:

$$\langle p_x \rangle = \int_{-\infty}^{\infty} \psi^* \left(-i\hbar \frac{\partial \psi}{\partial x} \right) dx \tag{6.12}$$

Thus, the definition of expectation value of momentum as a simple average in momentum space leads to the operator form in coordinate space as given by (6.12).

Now assume that no forces act on the particle so that its wave function obeys

$$i\hbar \frac{\partial \psi}{\partial t} = -\frac{\hbar^2}{2m} \frac{\partial^2 \psi}{\partial x^2} \tag{6.13}$$

We will find the most general solution of this partial differential equation. The entire problem can be transformed into a momentum space representation by means of (6.5). We find

$$\frac{1}{\sqrt{2\pi}} \int_{-\infty}^{\infty} \left(i\hbar \frac{\partial \phi}{\partial t} - \frac{\hbar^2 \kappa^2}{2m} \phi \right) e^{i\kappa x}\, d\kappa = 0 \tag{6.14}$$

If the Fourier transform of a function is identically zero, then the function itself is zero. Thus,

$$i\hbar \frac{\partial \phi}{\partial t} - \frac{\hbar^2 \kappa^2}{2m} \phi = 0 \tag{6.15}$$

which is the free-particle Schrödinger wave equation expressed in momentum space. The advantage gained is that the momentum space version is easier to solve. In fact, the solution of (6.15) is readily seen to be

$$\phi(\kappa, t) = A(\kappa) \exp\left[-\frac{i\hbar\kappa^2 t}{2m} \right] \tag{6.16}$$

The wave function in coordinate space is therefore

$$\psi(x, t) = \frac{1}{\sqrt{2\pi}} \int_{-\infty}^{\infty} A(\kappa) \exp\left[i\left(\kappa x - \frac{\hbar\kappa^2}{2m} t \right) \right] d\kappa \tag{6.17}$$

We recognize

$$\omega(\kappa) = \frac{\hbar\kappa^2}{2m} \tag{6.18}$$

as the circular frequency of a plane wave of propagation number κ. Thus, (6.17) mathematically represents a wave packet, that is, a wave function defined over a limited region of space that has been built up out of a large number of plane waves. Each plane wave represents an eigenstate of precisely defined energy and momentum and hence completely undefined position. If the particle is described by the wave function (6.17), its position is determined to within an uncertainty Δx. The spread in possible momentum values in (6.17) represents the corresponding uncertainty in the momentum as is required by the uncertainty relation. At $t = 0$,

$$\psi(x, 0) = \frac{1}{\sqrt{2\pi}} \int_{-\infty}^{\infty} A(\kappa) e^{i\kappa x}\, d\kappa \tag{6.19}$$

In the extreme case where $\psi(x, 0)$ represents a particle with a precisely defined position,

$$\psi(x,0)=\delta(x-x_0)$$
$$=\frac{1}{2\pi}\int_{-\infty}^{\infty}\exp\left[i\kappa(x-x_0)\right]d\kappa \tag{6.20}$$
$$A(\kappa)=\frac{1}{\sqrt{2\pi}}\exp(-i\kappa x_0) \qquad -\infty<\kappa<\infty \tag{6.21}$$

To obtain such a state, it is necessary to superimpose all momentum states for $-\infty<\kappa<+\infty$ with equal weight. In other words, the momentum of the state becomes completely indeterminate. In reality, it is not possible to measure the position of a particle to delta function accuracy. The inversion of (6.19) yields

$$A(\kappa)=\frac{1}{\sqrt{2\pi}}\int_{-\infty}^{\infty}\psi(x',0)e^{-i\kappa x'}dx' \tag{6.22}$$

The wave function (6.17) becomes

$$\psi(x,t)$$
$$=\frac{1}{2\pi}\int_{-\infty}^{\infty}\psi(x',0)\int_{-\infty}^{\infty}\exp\left[i\kappa(x-x')-\frac{i\hbar\kappa^2t}{2m}\right]d\kappa\,dx' \tag{6.23}$$

The definite integral over κ can be evaluated. It is quite similar to the integrals in problem 20.12 of Chapter 5. Briefly,

$$\int_{-\infty}^{\infty}\exp(-ia\kappa^2+ib\kappa)\,d\kappa$$
$$=\int_{-\infty}^{\infty}\exp\left[ia\left(\kappa-\frac{b}{2a}\right)^2+i\frac{b^2}{4a}\right]d\kappa$$
$$=\exp\left(\frac{ib^2}{4a}\right)\int_{-\infty}^{\infty}\exp(-iau^2)\,du$$
$$=\sqrt{\frac{\pi}{ia}}\exp\left(\frac{ib^2}{4a}\right) \tag{6.24}$$

The resulting wave function can be expressed as

$$\psi(x,t)=\int_{-\infty}^{\infty}G(x,t;x',0)\psi(x',0)\,dx' \tag{6.25}$$

where

$$G(x,t;x',0)=\sqrt{\frac{m}{2\pi i\hbar t}}\exp\left[\frac{im(x-x')^2}{2\hbar t}\right] \tag{6.26}$$

is called the *propagator*. The wave function for all time is determined once it is known at $t=0$.

As an example, we can take for the initial wave function the minimum uncertainty wave packet as given by equation (13.20) of Chapter 7. It will be convenient to designate the expectation values of x, p_x, and Δx at $t=0$ by

$$\langle x\rangle_{t=0}=x_0 \qquad \langle p_x\rangle_{t=0}=p_0 \qquad \Delta x_{t=0}=\Delta \tag{6.27}$$

The result of problem 13.3 of Chapter 7 shows that $\langle p_x\rangle$ is a constant for a free particle. Needed in the calculation is the definite integral

$$\int_{-\infty}^{\infty}\exp(-ax^2+bx)\,dx=\sqrt{\frac{\pi}{a}}\exp\left(\frac{b^2}{4a}\right) \tag{6.28}$$

The computation is tedious but straightforward and yields

$$\psi=\frac{1}{(2\pi)^{1/4}}\frac{1}{\sqrt{\Delta+\dfrac{i\hbar t}{2m\Delta}}}$$
$$\times\exp\left[-\frac{(x-x_0)^2+\dfrac{2tx_0p_0}{m}-\dfrac{4i\Delta^2p_0}{\hbar}\left(x-\dfrac{p_0t}{2m}\right)}{4\Delta^2+\dfrac{2i\hbar t}{m}}\right] \tag{6.29}$$

The probability density is given by

$$|\psi|^2$$
$$=\frac{1}{\sqrt{2\pi}}\frac{1}{\sqrt{\Delta^2+\dfrac{\hbar^2t^2}{4m^2\Delta^2}}}\exp\left[-\frac{\left(x-x_0-\dfrac{p_0t}{m}\right)^2}{2\Delta^2+\dfrac{\hbar^2t^2}{2m^2\Delta^2}}\right] \tag{6.30}$$

The center of the initial wave packet is at x_0; and its initial width, consistent with the accuracy of the determination of the position of the particle at $t=0$, is Δ. As time progresses, the center of the wave packet moves at the velocity p_0/m, and its width increases according to

$$\Delta(t)=\sqrt{\Delta^2+\frac{\hbar^2t^2}{4m^2\Delta^2}} \tag{6.31}$$

Once located by measurement to within an uncertainty Δ, the position probability density of the particle does not remain stationary in time but spreads or diffuses. The longer the time lapse before another measurement is made, the less certain will be the outcome. This is in contrast to a particle described by the wave function

$$\psi(x,t)=\frac{1}{\sqrt{2\pi}}e^{i(\kappa x-\omega t)} \tag{6.32}$$

which is an eigenfunction of a free particle corresponding to precise values of energy and momentum. Once in this state, the particle stays there until measured again. The reason there is no stationary position state is that x does not commute with the Hamiltonian. At the instant of measurement, the position is known to within Δ, which may be very small, so that the initial wave is nearly a delta function. The expectation value of x is not a constant, and the particle does not remain in a pure position state. As time progresses, the wave function becomes a superposition or mixture of more and more possible position states so that the actual position of the particle becomes less and less certain.

PROBLEMS

6.1. Show that for a free particle moving in the x direction

$$\frac{d}{dt}\langle x\rangle=\frac{1}{m}\langle p_x\rangle \tag{6.33}$$

Hint: Use (13.21) in Chapter 7.

6.2. Show that the momentum space representation of $\langle x\rangle$ is

$$\langle x\rangle=\int_{-\infty}^{\infty}\phi^*(\kappa,t)\left(i\frac{d}{d\kappa}\right)\phi(\kappa,t)\,d\kappa \tag{6.34}$$

6.3. Show that the general solution of

$$i\hbar\frac{\partial\psi}{\partial t}=H\psi \tag{6.35}$$

can be put into the propagator form

$$\psi(\mathbf{r},t)=\int_\Sigma G(\mathbf{r},t;\mathbf{r}',0)\psi(\mathbf{r}',0)\,d\Sigma' \tag{6.36}$$

where

$$G(\mathbf{r},t;\mathbf{r}',0)=\sum_{k=1}^{\infty}u_k^*(\mathbf{r}')u_k(\mathbf{r})\exp\left(-\frac{iW_k t}{\hbar}\right) \tag{6.37}$$

The functions $u_k(\mathbf{r})$ are eigenfunctions of the Hamiltonian:

$$Hu_k(\mathbf{r})=W_k u_k(\mathbf{r}) \tag{6.38}$$

The Hamiltonian is assumed to be independent of the time, and the energy eigenvalues W_k form a discrete set.

6.4. The one-dimensional equation of heat conduction is

$$\frac{\partial^2\psi}{\partial x^2}-\frac{1}{\alpha}\frac{\partial\psi}{\partial t}=0 \tag{6.39}$$

where ψ is the temperature and α is a constant. If the initial temperature is given as $\psi(x,0)$, show that the temperature at later times is given by

$$\psi(x,t)=\frac{1}{\sqrt{4\pi\alpha t}}\int_{-\infty}^{\infty}\psi(x',0)\exp\left[-\frac{(x-x')^2}{4\alpha t}\right]dx' \tag{6.40}$$

Use the same method that was used to solve the one-dimensional free-particle Schrödinger equation. See problem 3.8 in Chapter 6.

6.5. Find the Fourier transform of

$$f(x)=\exp(i\kappa_0 x) \qquad -\Delta x<x<\Delta x$$
$$f(x)=0 \qquad |x|\geq \Delta x \tag{6.41}$$

Show that $A(\kappa)$ is large for

$$-\pi<\Delta x(\kappa_0-\kappa)<\pi \tag{6.42}$$

and is small outside this range. If we define $\Delta\kappa=\kappa_0-\kappa$, then (6.42) is equivalent to $\Delta x\,\Delta\kappa \simeq 2\pi$. Note that if $p_x=\hbar\kappa$, then $\Delta x\,\Delta p_x \simeq 2\pi\hbar$.

7. SOLUTION OF THE WAVE EQUATION

The solution of the classical wave equation provides further examples of the use of the Fourier transform and the Dirac delta function. Consider first the one-dimensional homogeneous case

$$\frac{\partial^2\psi}{\partial x^2}-\frac{1}{c^2}\frac{\partial^2\psi}{\partial t^2}=0 \tag{7.1}$$

The function ψ could represent an electromagnetic potential, the transverse displacement of a string, or the longitudinal displacement of an acoustic wave. If we assume that $\psi(x,t)$ is defined over the entire x axis, then it can be represented as a Fourier integral:

$$\psi(x,t)=\frac{1}{\sqrt{2\pi}}\int_{-\infty}^{\infty}\phi(\kappa,t)e^{i\kappa x}\,d\kappa \tag{7.2}$$

The combination of (7.1) and (7.2) yields the result

$$\frac{1}{\sqrt{2\pi}}\int_{-\infty}^{\infty}\left[\kappa^2\phi(\kappa,t)+\frac{1}{c^2}\frac{\partial^2\phi(\kappa,t)}{\partial t^2}\right]e^{i\kappa x}\,d\kappa=0 \tag{7.3}$$

The integrand of (7.3) must vanish identically:

$$\frac{\partial^2\phi}{\partial t^2}+\kappa^2c^2\phi=0 \tag{7.4}$$

The general solution of (7.4) is

$$\phi(\kappa,t)=A(\kappa)e^{i\kappa ct}+B(\kappa)e^{-i\kappa ct} \tag{7.5}$$

The solution (7.2) can therefore be written

$$\psi=\frac{1}{\sqrt{2\pi}}\int_{-\infty}^{\infty}A(\kappa)e^{i\kappa(x+ct)}\,d\kappa+\frac{1}{\sqrt{2\pi}}\int_{-\infty}^{\infty}B(\kappa)e^{i\kappa(x-ct)}\,d\kappa \tag{7.6}$$

The significance of (7.6) is that it represents two wave pulses or wave packets, one traveling in the positive x direction and the other traveling in the negative x direction. Both wave packets travel at the speed c. We assume that $\psi(x,0)=f(x)$ and $(\partial\psi/\partial t)_{t=0}=g(x)$ are known functions. Equation (7.6) then gives

$$f(x)=\frac{1}{\sqrt{2\pi}}\int_{-\infty}^{\infty}(A+B)e^{i\kappa x}\,d\kappa \tag{7.7}$$

$$g(x)=\frac{1}{\sqrt{2\pi}}\int_{-\infty}^{\infty}i\kappa c(A-B)e^{i\kappa x}\,d\kappa \tag{7.8}$$

The inversion of (7.7) and (7.8) can be accomplished by means of Fourier's integral theorem (5.7):

$$A+B=\frac{1}{\sqrt{2\pi}}\int_{-\infty}^{\infty}f(x)e^{-i\kappa x}\,dx \tag{7.9}$$

$$i\kappa c(A-B)=\frac{1}{\sqrt{2\pi}}\int_{-\infty}^{\infty}g(x)e^{-i\kappa x}\,dx \tag{7.10}$$

Simultaneous solution for A and B gives

$$A=\frac{1}{2\sqrt{2\pi}}\int_{-\infty}^{\infty}\left[f(x)+\frac{g(x)}{i\kappa c}\right]e^{-i\kappa x}\,dx \tag{7.11}$$

$$B=\frac{1}{2\sqrt{2\pi}}\int_{-\infty}^{\infty}\left[f(x)-\frac{g(x)}{i\kappa c}\right]e^{-i\kappa x}\,dx \tag{7.12}$$

The solution (7.6) can therefore be expressed in the form

$$\psi=\frac{1}{4\pi}\int_{-\infty}^{\infty}\int_{-\infty}^{\infty}\left[f(x')+\frac{g(x')}{i\kappa c}\right]e^{i\kappa(x-x'+ct)}\,d\kappa\,dx'+\frac{1}{4\pi}\int_{-\infty}^{\infty}\int_{-\infty}^{\infty}\left[f(x')-\frac{g(x')}{i\kappa c}\right]e^{i\kappa(x-x'-ct)}\,d\kappa\,dx' \tag{7.13}$$

By means of the delta function in the form (5.11), it is possible to write (7.13) as

$$\begin{aligned}\psi &= \frac{1}{2}[f(x+ct)+f(x-ct)] \\ &\quad + \frac{1}{4\pi}\int_{-\infty}^{\infty}\int_{-\infty}^{\infty} g(x') \\ &\quad \times \left[\frac{e^{i\kappa(x-x'+ct)}-e^{i\kappa(x-x'-ct)}}{i\kappa c}\right] d\kappa\, dx' \\ &= \frac{1}{2}[f(x+ct)+f(x-ct)] \\ &\quad + \frac{1}{4\pi}\int_{-\infty}^{\infty}\int_{-\infty}^{\infty} g(x')\int_{u=x-x'-ct}^{x-x'+ct}\frac{1}{c}e^{i\kappa u}\, du\, d\kappa\, dx' \\ &= \frac{1}{2}[f(x+ct)+f(x-ct)] \\ &\quad + \frac{1}{2c}\int_{-\infty}^{\infty} g(x')\int_{u=x-x'-ct}^{x-x'+ct}\delta(u)\, du\, dx' \qquad (7.14)\end{aligned}$$

It is convenient to use the fact that the delta function is the derivative of the Heaviside unit step function: $\delta(u)=H'(u)$. The Heaviside function is defined in equation (2.34). We then find that

$$\begin{aligned}\psi &= \frac{1}{2}[f(x+ct)+f(x-ct)] \\ &\quad + \frac{1}{2c}\int_{-\infty}^{\infty} g(x')F(x,x',t)\, dx' \qquad (7.15)\end{aligned}$$

where $F(x,x',t)=H(x-x'+ct)-H(x-x'-ct)$. This function has the property that

$$\begin{aligned} F(x,x',t)&=1 \qquad x-ct<x'<x+ct \\ F(x,x',t)&=0 \qquad x'<x-ct \text{ or } x'>x+ct \end{aligned} \qquad (7.16)$$

We therefore conclude that the complete formal solution of the one-dimensional homogeneous wave equation is

$$\begin{aligned}\psi &= \frac{1}{2}[f(x+ct)+f(x-ct)] \\ &\quad + \frac{1}{2c}\int_{x'=x-ct}^{x+ct} g(x')\, dx' \qquad (7.17)\end{aligned}$$

An important qualitative difference between the solution of the free-particle Schrödinger equation (6.13) and the wave equation (7.1) should be pointed out. In the case of matter waves, the phase velocity of an individual harmonic wave is $u_p=\omega/\kappa=\hbar\kappa/(2m)$. This phase velocity is a function of κ. For light waves traveling in vacuum or for waves traveling without dissipation on a stretched flexible string, the phase velocity is $u_p=\omega/\kappa=\kappa c/\kappa=c=$ constant. For matter waves, there is dispersion because each individual harmonic wave that makes up the wave packet moves at a different velocity. This is the reason why a matter wave packet does not stay together but rather spreads as it moves. Initially, the waves may be all in phase at one point causing the wave function to be large there and small elsewhere. As times goes on, the waves get out of step because of the dispersion. This is not true of the solutions of (7.1) since all harmonic wave solutions have the same phase velocity c. An initial wave packet therefore propagates without change of shape.

Partial differential equations of the form (7.1) are called *hyperbolic differential equations*. By contrast, Laplace's equation is called an *elliptic differential equation*. The lines $x \pm ct =$ constant when graphed in an x, t coordinate system are called the *characteristics* of the hyperbolic differential equations. These are shown in Fig. 7.1. The boundary conditions $f(x)=\psi(x,0)$, $g(x)=(\partial\psi/\partial t)_{t=0}$ are called *Cauchy conditions*. If Cauchy boundary conditions are given only over the portion $a \le x \le b$ of the x axis, then the solution is determined only in the shaded portion of Fig. 7.1. This region is bounded by the x axis and two of the characteristics. For example, suppose that $\psi(x,t)$ is desired at the time and position indicated

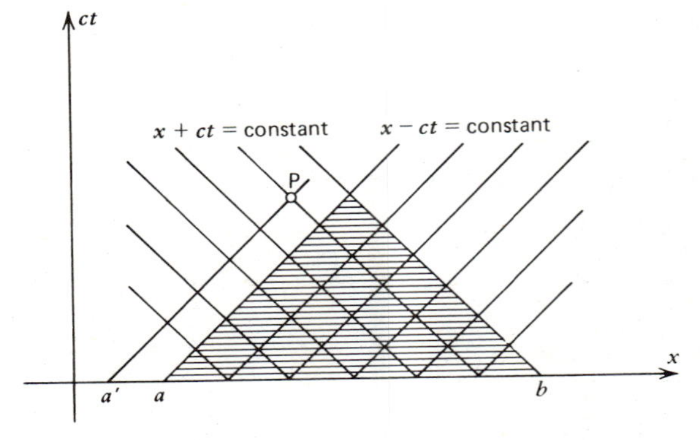

Fig. 7.1

by the point P in Fig. 7.1. If the characteristics are drawn through P, it is found that one of them passes through the point a' on the x axis. This means that a wave pulse originally at a' has had time to reach the x coordinate of P and is therefore influencing the motion there. Since initial conditions are known only for $a \le x \le b$, we cannot predict $\psi(x, t)$ at P. The solution can be made unique for all time in $a \le x \le b$ if end conditions are given where $\psi(a, t)$ and $\psi(b, t)$ are specified for all t. For example, in problem 2.8 of Chapter 7, you found the solution of the wave equation in terms of a Fourier series for the case where the initial shape of a string fixed at its two ends is as shown in Fig. 2.3. The solution can also be found from (7.17). In order to do so, the function $f(x)$ must first be periodically extended outside the fundamental interval $0 \le x \le s$ as shown in Fig. 7.2a. This is because the validity of (7.17) depends on $f(x)$ and $g(x)$ being known for the entire x axis. If the initial velocity of the string is zero, then

$$\psi(x, t) = \tfrac{1}{2}[f(x+ct) + f(x-ct)] \tag{7.18}$$

In order to construct the solution for all times, imagine that $\frac{1}{2}f(x+ct)$ propagates in the negative x direction and that $\frac{1}{2}f(x-ct)$ propagates in the positive x direction. This is indicated by the succession of diagrams in Fig. 7.2. The two dashed curves, one representing $\frac{1}{2}f(x+ct)$ and the other $\frac{1}{2}f(x-ct)$, always add to zero at $x = 0$ and $x = s$ as is required by the boundary conditions. The solution is actually a standing wave produced by two identical sawtooth waves moving in opposite directions on the x axis.

The three-dimensional nonhomogeneous wave equation satisfied by the electromagnetic scalar potential is

$$\Box\psi = \frac{-\rho}{\varepsilon_0} \tag{7.19}$$

The operator $\Box$ is called the *D'Alembertian* and is defined by

$$\Box = \nabla^2 - \frac{1}{c^2}\frac{\partial^2}{\partial t^2} \tag{7.20}$$

The wave equation (7.19) as well as a similar one for the vector potential is developed in section 12 of Chapter 3. We know that the solution of Poisson's

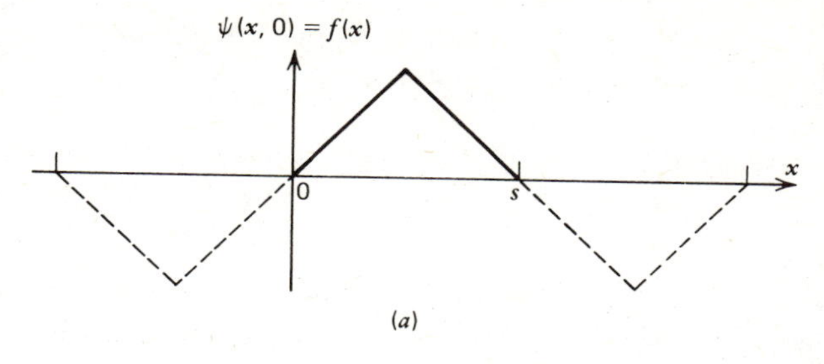

(a)

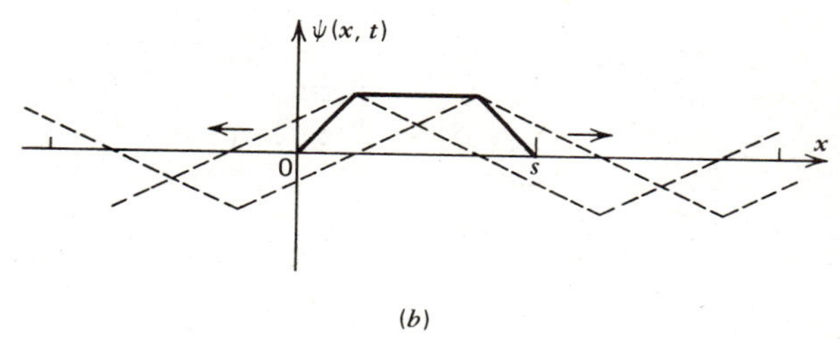

(b)

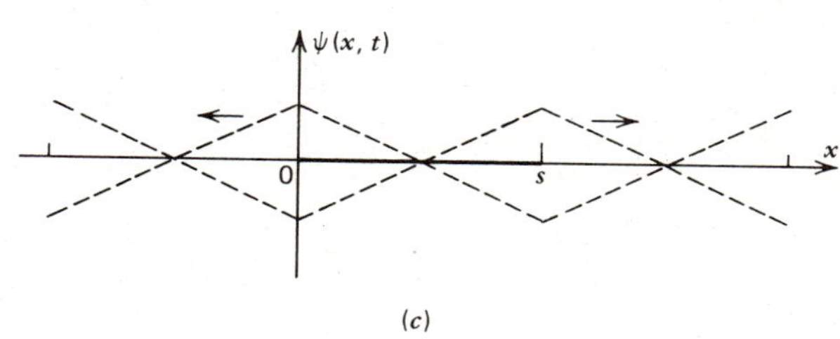

(c)

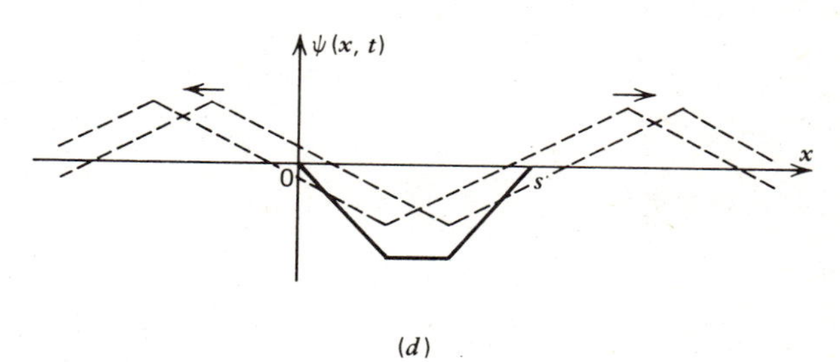

(d)

Fig. 7.2

equation, $\nabla^2\psi = -\rho/\varepsilon_0$, can be put into the form

$$\psi(\mathbf{r}) = \frac{1}{4\pi\varepsilon_0}\int_\Sigma G(\mathbf{r}, \mathbf{r}')\rho(\mathbf{r}')\,d\Sigma' \tag{7.21}$$

where the integral is over all space and

$$G(\mathbf{r}, \mathbf{r}') = \frac{1}{|\mathbf{r} - \mathbf{r}'|} \tag{7.22}$$

Equation (10.22) of Chapter 3 shows that

$$\nabla^2 G = -4\pi\delta(\mathbf{R}) \tag{7.23}$$

where $\mathbf{R} = \mathbf{r} - \mathbf{r}'$. It is possible to put the solution of the wave equation in a similar form:

$$\psi(\mathbf{r}, t) = \frac{1}{4\pi\varepsilon_0} \int_\Sigma \int_{-\infty}^{\infty} G(\mathbf{r}, t; \mathbf{r}', t') \rho(\mathbf{r}', t')\, dt'\, d\Sigma' \tag{7.24}$$

There is a similarity between the approach being taken here and the propagator approach to the solution of the Schrödinger equation that was used in section 6. See equation (6.25) and also problem (6.3). The functions $G(x, t; x', 0)$ of equation (6.25), $G(\mathbf{r}, \mathbf{r}')$ of equation (7.22), and $G(\mathbf{r}, t; \mathbf{r}', t')$ of equation (7.24) are called *Green's functions*. In quantum mechanics, the "source" of the wave function is its value at $t = 0$; the sources of the electromagnetic potentials are charge and current.

If (7.24) is substituted into the wave equation, the result is

$$\Box\psi = \frac{1}{4\pi\varepsilon_0} \int_\Sigma \int_{-\infty}^{\infty} (\Box G) \rho\, dt'\, d\Sigma' = -\frac{\rho}{\varepsilon_0} \tag{7.25}$$

The implication is that the Green's function satisfies

$$\Box G = -4\pi\delta(\mathbf{R})\, \delta(t - t') \tag{7.26}$$

This is analogous to (7.23). We will look for a solution of (7.26). We begin by "expanding" the time dependence of the Green's function as a Fourier integral:

$$G(\mathbf{r}, t; \mathbf{r}', t') = \frac{1}{2\pi} \int_{-\infty}^{\infty} g(\mathbf{r}, \mathbf{r}', \omega) e^{i\omega(t-t')}\, d\omega \tag{7.27}$$

(It is convenient here to use a factor $1/(2\pi)$ rather than $1/\sqrt{2\pi}$.) The delta function $\delta(t - t')$ can be similarly represented:

$$\delta(t - t') = \frac{1}{2\pi} \int_{-\infty}^{\infty} e^{i\omega(t-t')}\, d\omega \tag{7.28}$$

Equation (7.26) then gives

$$\int_{-\infty}^{\infty} \left[\nabla^2 g + \kappa^2 g + 4\pi\delta(\mathbf{R})\right] e^{i\omega(t-t')}\, d\omega = 0 \tag{7.29}$$

where $\kappa = \omega/c$. The function g therefore obeys the nonhomogeneous Helmholtz equation

$$\nabla^2 g + \kappa^2 g = -4\pi\delta(\mathbf{R}) \tag{7.30}$$

We will find a solution of (7.30) which depends only on the magnitude of $\mathbf{R}$. To this end, it is convenient to use spherical coordinates and write

$$\frac{1}{R^2} \frac{\partial}{\partial R}\left(R^2 \frac{\partial g}{\partial R}\right) + \kappa^2 g = -4\pi\delta(\mathbf{R}) \tag{7.31}$$

The possible solutions of (7.31) valid at all points except $R = 0$ are

$$g_1 = \frac{e^{i\kappa R}}{R} \qquad g_2 = \frac{e^{-i\kappa R}}{R} \tag{7.32}$$

These solutions actually remain valid even at $R = 0$. To see that this is true, multiply (7.30) by $d\Sigma$ and integrate over a small sphere of radius ε centered at $R = 0$:

$$\int_\Sigma \nabla^2 g_1\, d\Sigma + \int_0^\varepsilon \kappa^2 g_1 4\pi R^2\, dR = -4\pi \tag{7.33}$$

In the limit $\varepsilon \to 0$, $g_1 \to 1/R$. The first term in (7.33) becomes

$$\int_\Sigma \nabla^2 \frac{1}{R}\, d\Sigma = -4\pi \int_\Sigma \delta(R)\, d\Sigma = -4\pi \tag{7.34}$$

The second term gives zero. Thus, (7.30) is satisfied at all points by the function g_1. The same holds true for the function g_2.

If we use the function g_2, the Green's function (7.27) becomes

$$\begin{aligned} G(\mathbf{r}, t; \mathbf{r}', t') &= \frac{1}{2\pi} \int_{-\infty}^{\infty} \frac{1}{R} e^{i\omega(t-t'-R/c)}\, d\omega \\ &= \frac{1}{R}\delta(t - t' - R/c) \end{aligned} \tag{7.35}$$

The solution (7.24) can then be expressed as

$$\psi = \frac{1}{4\pi\varepsilon_0} \int_\Sigma \int_{-\infty}^{\infty} \frac{1}{R}\delta(t - t' - R/c) \rho(\mathbf{r}', t')\, dt'\, d\Sigma' \tag{7.36}$$

By utilizing the properties of the delta function, the integral over t' can be carried out with the result

$$\psi(\mathbf{r}, t) = \frac{1}{4\pi\varepsilon_0} \int_\Sigma \frac{\rho(\mathbf{r}', t - R/c)}{R}\, d\Sigma' \tag{7.37}$$

This solution of the nonhomogeneous wave equation looks very much like the solution of Poisson's equation. The difference is that the value of charge which is used at the source point is not its value at the time t of observation but rather is its value at the earlier time $t-R/c$. The time lag R/c is just the time required for an electromagnetic signal to traverse the distance R. For this reason, (7.37) is called a *retarded potential.* Had we used the function g_1 rather than g_2, we would have obtained the solution

$$\psi(\mathbf{r}, t)=\frac{1}{4\pi\varepsilon_0}\int_\Sigma \frac{\rho(\mathbf{r}', t+R/c)}{R}\,d\Sigma' \tag{7.38}$$

This is called an *advanced potential* because the value of charge used in the integrand is a future value. This is unreasonable from the point of view of causality. Charge and current values which are going to occur in the future presumably cannot affect present values of an electromagnetic field. If we take the point of view that the electromagnetic signals which we are observing were generated at earlier times, then the appropriate solution to use is the retarded potential (7.37). There is a similar solution for the vector potential:

$$\mathbf{A}=\frac{\mu_0}{4\pi}\int_\Sigma \frac{\mathbf{j}(\mathbf{r}', t-R/c)}{R}\,d\Sigma' \tag{7.39}$$

PROBLEMS

7.1. Show that in terms of the variables $u=x+ct$ and $v=x-ct$, the wave equation (7.1) is

$$\frac{\partial^2\psi}{\partial u\,\partial v}=0 \tag{7.40}$$

Show that any function of the form $\psi=F(u)+G(v)$ is a solution.

7.2. Solve the differential equation (7.31) for the case $R\neq 0$.

Hint: Let $g(R)=u(R)v(R)$. In the resulting differential equation, determine $v(R)$ by the condition that the coefficient of $u'(R)$ vanish.

7.3. If a point particle of charge q moves along a trajectory given by $\mathbf{s}(t)$, then the resulting charge density and current are given by

$$\rho(\mathbf{r}, t)=q\delta(\mathbf{r}-\mathbf{s}(t)) \qquad \mathbf{j}(\mathbf{r}, t)=\rho(\mathbf{r}, t)\mathbf{u}(t) \tag{7.41}$$

where $\mathbf{u}=d\mathbf{s}/dt$ is the particle velocity. By starting out with the scalar and vector potential both written in the form (7.36) and then doing the integral over all space, show that

$$\psi(\mathbf{r}, t)=\frac{1}{4\pi\varepsilon_0}\int_{-\infty}^{\infty}\frac{q}{R}\delta(t-t'-R/c)\,dt' \tag{7.42}$$

$$\mathbf{A}(\mathbf{r}, t)=\frac{\mu_0}{4\pi}\int_{-\infty}^{\infty}\frac{q}{R}\delta(t-t'-R/c)\mathbf{u}(t')\,dt' \tag{7.43}$$

As shown in Fig. 7.3, $\mathbf{R}=\mathbf{r}-\mathbf{s}(t')$. Note that $\mathbf{s}(t')$ is the position vector of the particle and also the source point of the radiation. The vector $\mathbf{r}$ is the field point or point at which the

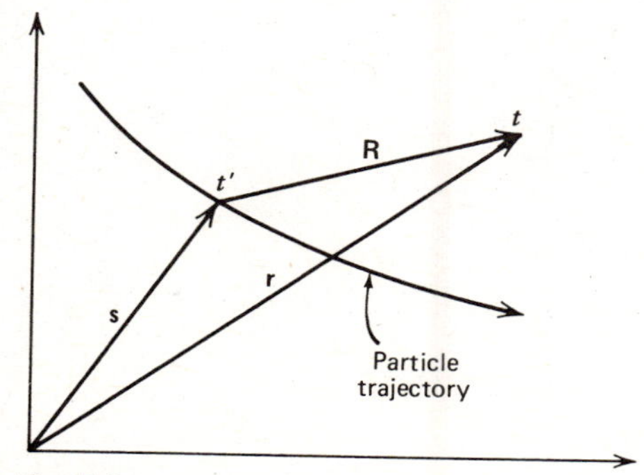

Fig. 7.3

radiation from the charge is observed. The integrals of (7.42) and (7.43) are over all values of t' and are therefore extended over the entire trajectory of the particle. However, if t is the time of observation of the radiation, there is only one point on the trajectory which contributes to the integrals. This point is determined by $t - t' - R/c = 0$ and is called the *retardation condition*. In other words, the radiation which is observed at $\mathbf{r}$ at the time t was emitted at the time $t' = t - R/c$. Equations (7.42) and (7.43) are one form of what are commonly called the *Lienard–Wiechert potentials*.

7.4. It is shown in section 12 of Chapter 3 that the electromagnetic field is to be obtained from the potentials by means of

$$\mathbf{E} = -\nabla\psi - \frac{\partial \mathbf{A}}{\partial t} \qquad \mathbf{B} = \nabla \times \mathbf{A} \tag{7.44}$$

Show that the electromagnetic field produced by a single point charge moving in any prescribed manner is

$$\mathbf{E} = \frac{q}{4\pi\varepsilon_0}\int_{-\infty}^{\infty}\left[\left(\frac{\mathbf{u}}{Rc^2} - \frac{\mathbf{R}}{R^2c}\right)\delta'(T) + \frac{\mathbf{R}\delta(T)}{R^3}\right] dt' \tag{7.45}$$

$$\mathbf{B} = \frac{\mu_0 q}{4\pi}\int_{-\infty}^{\infty}\frac{\mathbf{R}\times\mathbf{u}}{R}\left[\frac{\delta'(T)}{Rc} - \frac{\delta(T)}{R^2}\right] dt' \tag{7.46}$$

where $T = t' - t + R/c$ and $\delta'(T)$ is the derivative of the delta function with respect to T.
Hint: Since the integrands of (7.42) and (7.43) depend on $\mathbf{r}$ through the scalar R, it is possible to express the gradient as

$$\nabla = \frac{\mathbf{R}}{R}\frac{\partial}{\partial R} \tag{7.47}$$

Be careful to observe that the differentiations when carried out under the integral sign are with respect to the field point variables $\mathbf{r}$ and t. The variable t occurs only through T. Remember also that $c = 1/\sqrt{\mu_0\varepsilon_0}$. The derivative of the delta function is discussed in problem (5.8).

7.5. Show that

$$\frac{dT}{dt'} = \frac{K}{R} \qquad K = R - \frac{\mathbf{R}\cdot\mathbf{u}}{c} \qquad \mathbf{u} = \frac{d\mathbf{s}(t')}{dt'} \tag{7.48}$$

The electromagnetic field of a point charge can therefore be expressed as

$$\mathbf{E} = \frac{q}{4\pi\varepsilon_0}\int_{-\infty}^{\infty}\left[\left(\frac{\mathbf{u}}{Kc^2} - \frac{\mathbf{R}}{RKc}\right)\delta'(T) + \frac{\mathbf{R}}{KR^2}\delta(T)\right] dT \tag{7.49}$$

$$\mathbf{B} = \frac{\mu_0 q}{4\pi}\int_{-\infty}^{\infty}\frac{\mathbf{R}\times\mathbf{u}}{K}\left[\frac{\delta'(T)}{Rc} - \frac{\delta(T)}{R^2}\right] dT \tag{7.50}$$

7.6. Evaluate the integrals in (7.49) and (7.50) to show that

$$\mathbf{E} = \frac{q}{4\pi\varepsilon_0}\left[\frac{R}{K}\frac{d}{dt'}\left(\frac{\mathbf{R}c - R\mathbf{u}}{RKc^2}\right) + \frac{\mathbf{R}}{KR^2}\right]_{T=0} \tag{7.51}$$

$$\mathbf{B} = \frac{\mu_0 q}{4\pi}\left[\frac{R}{K}\frac{d}{dt'}\left(\frac{\mathbf{u}\times\mathbf{R}}{RKc}\right) + \frac{\mathbf{u}\times\mathbf{R}}{KR^2}\right]_{T=0} \tag{7.52}$$

7.7. Evaluate the derivatives with respect to t' in (7.51) and (7.52) to show that

$$\mathbf{E} = \frac{q}{4\pi\varepsilon_0}\left[-\frac{R\mathbf{a}}{K^2c^2} + \frac{\mathbf{R}\cdot\mathbf{a}}{K^3c^2}\left(\mathbf{R} - R\frac{\mathbf{u}}{c}\right) + \frac{1}{K^3}\left(\mathbf{R} - R\frac{\mathbf{u}}{c}\right)\left(1 - \frac{u^2}{c^2}\right)\right]_{T=0} \tag{7.53}$$

$$\mathbf{B} = \frac{\mu_0 q}{4\pi}\left[\frac{\mathbf{a}\times\mathbf{R}}{K^2c} + \frac{(\mathbf{u}\times\mathbf{R})(\mathbf{R}\cdot\mathbf{a})}{K^3c^2} + \frac{\mathbf{u}\times\mathbf{R}}{K^3}\left(1 - \frac{u^2}{c^2}\right)\right]_{T=0} \tag{7.54}$$

where $\mathbf{a} = d\mathbf{u}/dt'$ is the acceleration of the particle. One of the basic assumptions of special relativity is that Maxwell's equations are valid in all Lorentz frames of reference. The fields given by (7.53) and (7.54) are therefore relativistically correct.

8. THE LAPLACE TRANSFORM

Many integral transforms other than the Fourier transform are in common use. The study of integral transform methods is called the *operational calculus*. A general integral transform of a function might be represented as

$$\phi(s) = T[F(x)] = \int_a^b K(x,s)F(x)\,dx \tag{8.1}$$

The function $\phi(s)$ is the *transform* or *image* of the function $F(x)$. The function $K(x,s)$ is called the *kernel* of the transformation. We are going to give a brief account in this section of the Laplace transform and some of its applications.

Let $f(x)$ be an arbitrary function and let $H(x)$ be the Heaviside unit step function as defined by equation (2.34). Consider the Fourier transform

$$A(y) = \int_{-\infty}^{\infty} f(x)H(x)e^{-cx}e^{-ixy}\,dx \tag{8.2}$$

where c is a positive constant. The inverse of (8.2) is

$$f(x)e^{-cx}H(x) = \frac{1}{2\pi}\int_{-\infty}^{\infty} A(y)e^{ixy}\,dy \tag{8.3}$$

For convenience, a placement of the factor of 2π is being used here which is different from that originally used in equations (5.6) and (5.7). Next, we introduce the new variable $s = c + iy$ and write $A(y) = \phi(s)$. The result is

$$\phi(s) = \int_0^{\infty} f(x)e^{-sx}\,dx \qquad \operatorname{Re} s > 0 \tag{8.4}$$

$$f(x) = \frac{1}{2\pi i}\int_{c-i\infty}^{c+i\infty} \phi(s)e^{sx}\,ds \tag{8.5}$$

We have dropped the factor of $H(x)$ with the understanding that $f(x) = 0$ if $x < 0$. Equation (8.4) formally defines the Laplace transform. Sometimes we will use the notation $L[f(x)] = \phi(s)$. Extensive tables of Laplace transforms exist in mathematical handbooks. In the usual application of the theory, $\phi(s)$ is known and $f(x)$ must be found. Either tables or the *inversion theorem* (8.5) can be used. Without the existence of an inverse transformation, an integral transform is of little use. As an example, suppose it is known that

$$\phi(s) = \frac{s}{s^2 + a^2} \tag{8.6}$$

Since $x \geq 0$, it is possible to complete the contour of the integral in (8.5) by a large semicircle of radius R to the left of the line $\operatorname{Re} s = c$ as shown in Fig. 8.1. Over this semicircle,

$$e^{sx} = e^{R(\cos\theta + i\sin\theta)x + cx} \tag{8.7}$$

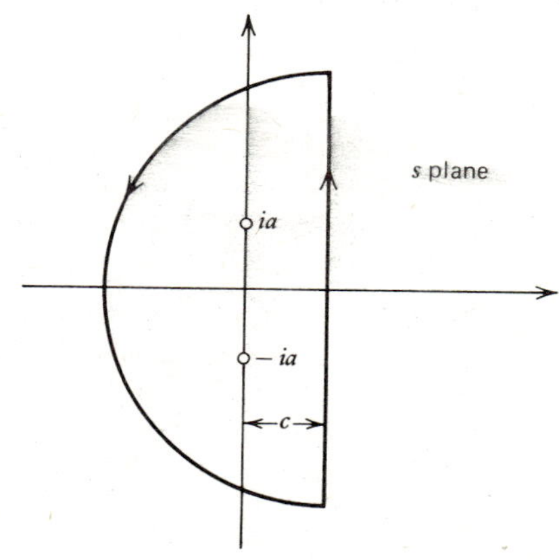

Fig. 8.1

Since $\pi/2 \le \theta \le 3\pi/2$, $\cos\theta \le 0$ and the contribution to the integral from the semicircle vanishes as $R \to \infty$. The integrand of (8.5) can be written as

$$\phi(s)e^{sx} = \frac{se^{sx}}{(s+ia)(s-ia)} \tag{8.8}$$

which reveals that there are two simple poles at $s = \pm ia$ inside the contour. The reader may wish to review the residue theorem in section 19 of Chapter 5. See equations (19.22) and (19.31). Section 21 of Chapter 5 discusses the evaluation of definite integrals by the residue theorem. The residues of (8.8) at the two poles are

$$\begin{aligned} R_1 &= \lim_{s\to ia}(s-ia)\phi(s)e^{sx} = \frac{iae^{iax}}{2ia} = \frac{1}{2}e^{iax} \\ R_2 &= \tfrac{1}{2}e^{-iax} \end{aligned} \tag{8.9}$$

It is therefore found that

$$f(x) = \frac{1}{2\pi i}2\pi i(R_1 + R_2) = \cos ax \tag{8.10}$$

Note that if $x < 0$ the contour would have to be completed by a semicircle to the right of $\operatorname{Re} s = c$ in order to ensure convergence of the integral. Since this contour encloses no poles, $f(x) = 0$ if $x < 0$.

The following is a list of some of the basic properties of the Laplace transform.

$$L[f'(x)] = s\phi(s) - f(0) \tag{8.11}$$

$$L[f''(x)] = s^2\phi(s) - sf(0) - f'(0) \tag{8.12}$$

$$L[f(x-a)] = e^{-as}L[f(x)] \qquad a > 0 \tag{8.13}$$

$$L[f(x+a)] = e^{as}\int_a^\infty f(x)e^{-sx}\,dx \qquad a > 0 \tag{8.14}$$

$$L[f(x)e^{ax}] = \phi(s-a) \qquad \operatorname{Re}(s-a) > 0 \tag{8.15}$$

$$L\left[\int_0^x f(t)\,dt\right] = \frac{1}{s}\phi(s) \tag{8.16}$$

$$L[xf(x)] = -\frac{d}{ds}\phi(s) \tag{8.17}$$

Equations (8.11) and (8.12) are established by integrating by parts. For example,

$$\begin{aligned} \int_0^\infty f'(x)e^{-sx}\,dx &= f(x)e^{-sx}\Big|_0^\infty - \int_0^\infty f(x)(-se^{-sx})\,dx \\ &= -f(0) + s\phi(s) \end{aligned} \tag{8.18}$$

To prove (8.13), make the change of variable $u = x - a$ in the integral

$$\int_0^\infty f(x-a)e^{-sx}\,dx = \int_{u=-a}^\infty f(u)e^{-s(u+a)}\,du \tag{8.19}$$

Now remember that $f(u) = 0$ if $u < 0$. Therefore,

$$L[f(x-a)] = e^{-as}\int_0^\infty f(u)e^{-su}\,du = e^{-as}\phi(s) \tag{8.20}$$

Now consider (8.16):

$$L\left[\int_0^x f(t)\,dt\right] = \int_{x=0}^\infty e^{-sx}\int_{t=0}^x f(t)\,dt\,dx \tag{8.21}$$

Note that the region of integration is the shaded portion of Fig. 8.2. If we reverse the order of integration, then this same region is covered if we write

$$\begin{aligned} L\left[\int_0^x f(t)\,dt\right] &= \int_{t=0}^\infty \int_{x=t}^\infty e^{-sx}\,dx\,f(t)\,dt \\ &= \int_{t=0}^\infty \frac{1}{s}e^{-st}f(t)\,dt = \frac{1}{s}\phi(s) \end{aligned} \tag{8.22}$$

The proofs of the remaining properties are left as problems.

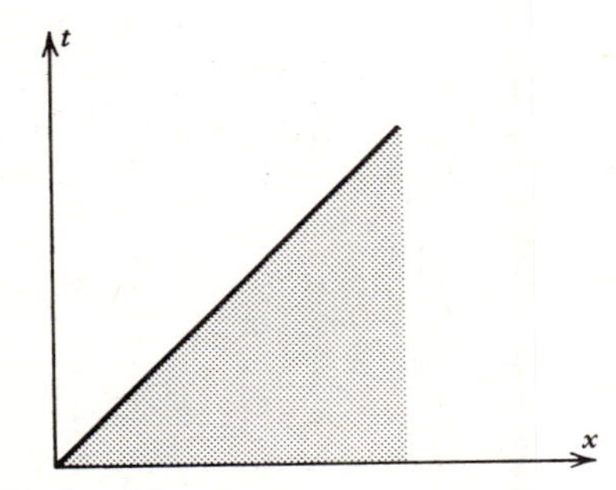

Fig. 8.2

Given two functions $f_1(x)$ and $f_2(x)$, we define their *convolution* by means of

$$g(x) = \int_0^x f_1(t)\, f_2(x-t)\, dt \tag{8.23}$$

The convolution of two functions in connection with Fourier transforms is defined by equation (5.48). The Laplace transform of (8.23) is

$$L[g(x)] = \int_{x=0}^{\infty} e^{-sx} \int_{t=0}^{x} f_1(t) f_2(x-t)\, dt\, dx \tag{8.24}$$

Now use the same inversion of the order of integration that was used on equation (8.21):

$$L[g(x)] = \int_{t=0}^{\infty} f_1(t) \int_{x=t}^{\infty} e^{-sx} f_2(x-t)\, dx\, dt \tag{8.25}$$

In the integral over x, make the change of variable $u = x - t$:

$$\begin{aligned} L[g(x)] &= \int_{t=0}^{\infty} f_1(t) \int_{u=0}^{\infty} e^{-s(t+u)} f_2(u)\, du\, dt \\ &= \int_0^{\infty} f_1(t) e^{-st}\, dt \int_0^{\infty} f_2(u) e^{-su}\, du \\ &= L[f_1] L[f_2] \end{aligned} \tag{8.26}$$

This result is known as the *convolution theorem*.

The Laplace transform is widely used to solve both ordinary and partial linear differential equations with constant coefficients. As a simple example, consider the series RLC circuit shown in Fig. 8.3. The switch S is closed at $t = 0$ so that the differential equation for the charge on the capacitor is

$$\begin{aligned} L\ddot{q} + R\dot{q} + \frac{q}{C} &= 0 \qquad t < 0 \\ &= V_0 \qquad t \geq 0 \end{aligned} \tag{8.27}$$

Assume that the initial conditions are $q(0) = 0$ and $\dot{q}(0) = 0$. By using (8.11) and (8.12) with t rather than x as the independent variable, we find that the Laplace transform of (8.27) is

$$Ls^2\phi + Rs\phi + \frac{\phi}{C} = \frac{V_0}{s} \qquad \phi(s) = \int_0^{\infty} q(t) e^{-st}\, dt \tag{8.28}$$

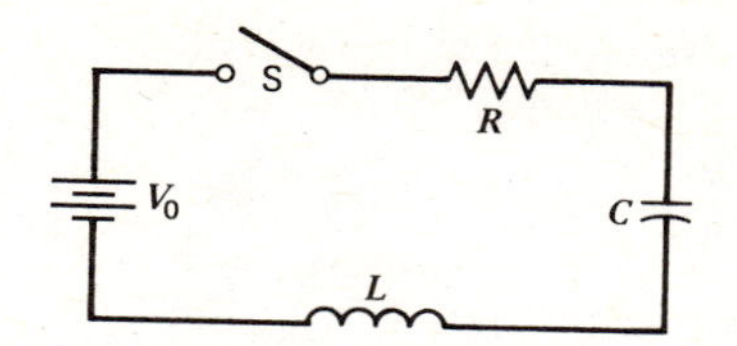

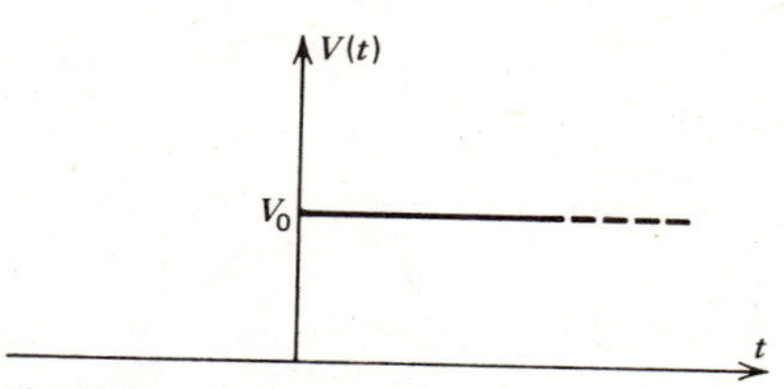

Fig. 8.3

By solving for ϕ we find

$$\phi(s) = \frac{V_0}{sL(s^2 + 2\beta s + \omega_0^2)} = \frac{V_0}{sL(s - s_1)(s - s_2)} \tag{8.29}$$

where we have defined

$$\beta = \frac{R}{2L} \qquad \omega_0^2 = \frac{1}{LC} \tag{8.30}$$

If the damping is less than critical, then the roots s_1 and s_2 are

$$s_1 = -\beta + i\omega \qquad s_2 = -\beta - i\omega \qquad \omega = \sqrt{\omega_0^2 - \beta^2} \tag{8.31}$$

The function $\phi(s)$ has three simple poles at $s = 0$, $s = s_1$, and $s = s_2$. The inversion can be accomplished with the same contour that appears in Fig. 8.1. The constant c must be large enough so that all three of the poles fall on the left side of the line $\text{Re}\, s = c$. By evaluating the three residues, we find

$$q(t) = \frac{V_0}{L}\left[\frac{e^{s_1 t}}{s_1(s_1 - s_2)} + \frac{e^{s_2 t}}{s_2(s_2 - s_1)} + \frac{1}{s_1 s_2}\right] \tag{8.32}$$

The current in the circuit is given by

$$\dot{q} = \frac{V_0}{L}\left[\frac{e^{s_1 t} - e^{s_2 t}}{s_1 - s_2}\right] = \frac{V_0}{L} e^{-\beta t} \frac{\sin \omega t}{\omega} \tag{8.33}$$

Note that the steady-state portion of (8.32) is

$$\frac{V_0}{Ls_1s_2} = \frac{V_0}{L\omega_0^2} = V_0C \tag{8.34}$$

as expected.

The Laplace transform can be used to evaluate definite integrals. For example, if

$$f(t) = \int_0^\infty \frac{\sin tx}{x}\,dx \tag{8.35}$$

then

$$\begin{aligned}\phi(s) &= \int_0^\infty \frac{1}{x}\int_0^\infty \sin tx\, e^{-st}\,dt\,dx\\ &= \int_0^\infty \frac{1}{s^2+x^2}\,dx = \frac{1}{s}\tan^{-1}\left(\frac{x}{s}\right)\Big|_0^\infty = \frac{\pi}{2s}\end{aligned} \tag{8.36}$$

The inversion of (8.36) gives

$$f(t) = \frac{\pi}{2} \qquad t>0 \tag{8.37}$$

This integral also appears in equation (20.27) of Chapter 5.

As an example of the use of the Laplace transform in solving partial differential equations, we will find the solution of the equation of heat conduction

$$\frac{\partial^2\psi}{\partial x^2} - \frac{1}{\alpha}\frac{\partial\psi}{\partial t} = 0 \tag{8.38}$$

for a semi-infinite slab of material which occupies the space $x>0$. The initial temperature in the slab is everywhere zero. The temperature is maintained at T_0 at all points of the y, z plane at all times:

$$\psi(0,t) = T_0 \tag{8.39}$$

The Laplace transform of (8.38) with respect to the variable t is

$$\frac{\partial^2\phi}{\partial x^2} - \frac{1}{\alpha}s\phi = 0 \qquad \phi(s,x) = \int_0^\infty \psi(x,t)e^{-st}\,dt \tag{8.40}$$

The differential equation (8.40) is easily integrated:

$$\phi(s,x) = Ae^{-x\sqrt{s/\alpha}} + Be^{x\sqrt{s/\alpha}} \tag{8.41}$$

The positive exponent part of the solution is eliminated as being unsuitable since it becomes infinite as $x\to\infty$. By use of the boundary condition (8.39) in the Laplace transform (8.40), we find

$$\phi(s,0) = \int_0^\infty T_0e^{-st}\,dt = \frac{T_0}{s} \tag{8.42}$$

This allows the constant A in (8.41) to be evaluated with the result

$$\phi(x,s) = T_0\frac{1}{s}e^{-x\sqrt{s/\alpha}} \tag{8.43}$$

The inversion theorem (8.5) gives for the temperature, at any time and position in the half space $x\geq 0$,

$$\psi(x,t) = \frac{1}{2\pi i}\int_{c-i\infty}^{c+i\infty}\frac{T_0}{s}e^{-x\sqrt{s/\alpha}+st}\,ds \tag{8.44}$$

It turns out that it is a little easier to evaluate the integral which is found by differentiating (8.44) once with respect to x:

$$\frac{\partial\psi}{\partial x} = -\frac{T_0}{2\pi i\sqrt{\alpha}}\int_{c-i\infty}^{c+i\infty}\frac{1}{\sqrt{s}}e^{-x\sqrt{s/\alpha}+st}\,ds \tag{8.45}$$

Because of the factor $\sqrt{s}$, the integrand has a branch point at $s=0$. If we place a branch cut along the negative real axis, then the contour can be completed as shown in Fig. 8.4. Since there are no poles inside

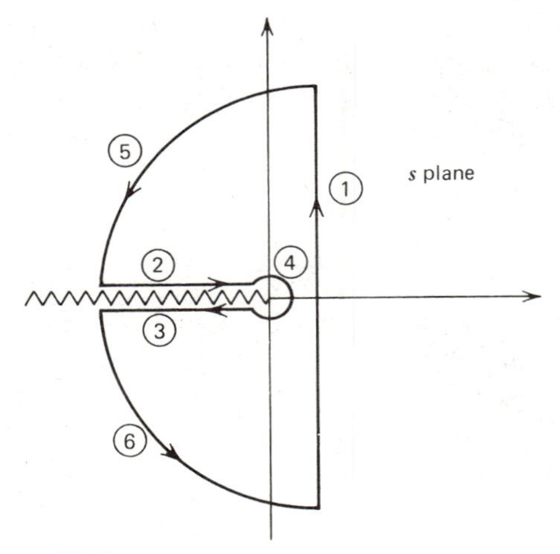

Fig. 8.4

this contour,

$$\oint \frac{1}{\sqrt{s}} e^{-x\sqrt{s/\alpha}+st}\, ds = 0 \tag{8.46}$$

By writing $s = re^{i\theta}$, the reader can show that there is no contribution to the integral from the small circle ④ about the origin in the limit $r \to 0$. Similarly, you can show that in the limit $r \to \infty$, there is no contribution from the large quarter arcs ⑤ and ⑥. The remaining contributions must therefore obey

$$① = -② - ③ \tag{8.47}$$

It is the integral ① which appears in (8.45). The integrals along the straight portions ② and ③ can be evaluated. Along ②, $s_2 = re^{i\pi}$, $\sqrt{s_2} = i\sqrt{r}$. Along ③, $s_3 = re^{-i\pi}$, $\sqrt{s_3} = -i\sqrt{r}$. We therefore find that

$$\begin{aligned} ① &= -\int_\infty^0 \frac{1}{i\sqrt{r}} e^{-ix\sqrt{r/\alpha}-rt}(-dr) \\ &\quad -\int_0^\infty \frac{1}{-i\sqrt{r}} e^{ix\sqrt{r/\alpha}-rt}(-dr) \\ &= \int_0^\infty \frac{i}{\sqrt{r}} e^{-rt}\left(e^{ix\sqrt{r/\alpha}} + e^{-ix\sqrt{r/\alpha}}\right) dr \end{aligned} \tag{8.48}$$

Next, make the change of variable $z^2 = r$ to get

$$① = 2i\int_0^\infty \left(e^{-z^2t+izx/\sqrt{\alpha}} + e^{-z^2t-izx/\sqrt{\alpha}}\right) dz \tag{8.49}$$

The integrals in (8.49) can be evaluated by completing the square:

$$\begin{aligned} &\int_0^\infty e^{-z^2t+iaz}\, dz \\ &= \exp\left(-\frac{a^2}{4t}\right)\int_0^\infty \exp\left[-t\left(z-\frac{ia}{2t}\right)\right]^2 dz \\ &= \exp\left(-\frac{a^2}{4t}\right)\int_0^\infty e^{-tu^2}\, du = \frac{\sqrt{\pi}}{2\sqrt{t}}\exp\left(-\frac{a^2}{4t}\right) \end{aligned} \tag{8.50}$$

We therefore find that

$$① = 2i\sqrt{\frac{\pi}{t}}\exp\left(-\frac{x^2}{4\alpha t}\right) \tag{8.51}$$

Equation (8.45) becomes

$$\frac{\partial\psi}{\partial x} = -\frac{T_0}{\sqrt{\pi\alpha t}}\exp\left(-\frac{x^2}{4\alpha t}\right) \tag{8.52}$$

Integration with respect to x yields

$$\psi = T_0\left[1 - \frac{1}{\sqrt{\pi\alpha t}}\int_0^x \exp\left(-\frac{x^2}{4\alpha t}\right) dx\right] \tag{8.53}$$

The constant of integration has been adjusted so that $\phi(0, t) = T_0$ as required by the boundary conditions. By making the change of variable

$$u = \frac{x}{\sqrt{4\alpha t}} \tag{8.54}$$

we get as a final result

$$\begin{aligned} \psi &= T_0\left[1 - \frac{2}{\sqrt{\pi}}\int_0^{x/(2\sqrt{\alpha t})} e^{-u^2}\, du\right] \\ &= T_0\left[1 - \operatorname{erf}\left(\frac{x}{2\sqrt{\alpha t}}\right)\right] \end{aligned} \tag{8.55}$$

The error function is defined by equation (3.19) of Chapter 6.

PROBLEMS

8.1. Prove (8.12), (8.14), (8.15), and (8.17).

8.2. Show that the expression (8.32) for the charge can be written as

$$q(t) = V_0C - \frac{V_0e^{-\beta t}}{L\omega\omega_0^2}(\beta\sin\omega t + \omega\cos\omega t) \tag{8.56}$$

8.3. Show that both $F[f(x)]$ and $L[f(x)]$ are linear operators. The definition of a linear operator is given in equation (12.18) of Chapter 7.

8.4. If $L[f_1]=\phi_1(s)$ and $L[f_2]=\phi_2(s)$, show that

$$L[f_1 f_2]=\frac{1}{2\pi i}\int_{c-i\infty}^{c+i\infty}\phi_1(u)\phi_2(u-s)\,du \tag{8.57}$$

8.5. The Laplace transform can be used to solve systems of linear differential equations. Use this method to solve the system of differential equations given in problem 15.24 of Chapter 7. By taking the Laplace transform of the three differential equations, you will get three linear algebraic equations for the three functions $\phi_1(s)$, $\phi_2(s)$, and $\phi_3(s)$, which are the Laplace transforms of $N_1(t)$, $N_2(t)$ and $N_3(t)$, respectively. Solve these three equations and then use the inversion theorem.

8.6. Solve the linear mechanical system given by equations (9.2) and (9.3) in Chapter 1 by the Laplace transform method. Use as initial conditions $x_1(0)=a$, $\dot{x}_1(0)=0$, $x_2(0)=0$, and $\dot{x}_2(0)=0$. See problem 6.4 in Chapter 1. *Caution*: Arrange the calculation carefully. The inversion integral has four simple poles inside of the contour.

8.7. A linear mechanical system initially at rest is struck a sharp blow. If the force is approximated by a delta function, the equation of motion of the system is

$$\ddot{x}+2\beta\dot{x}+\omega_0^2 x=J\delta(t) \tag{8.58}$$

where J is the impulse per unit mass delivered by the force. Show that the solution is

$$x(t)=\frac{J}{\omega}e^{-\beta t}\sin\omega t \tag{8.59}$$

Assume that the damping is less than critical.

8.8. A series RLC circuit is driven by a time-dependent voltage so that the differential equation for the current is

$$\begin{aligned} L\dot{I}+IR+\frac{1}{C}\int_0^t I\,dt &= 0 \qquad t<0 \\ &= V(t) \qquad t\geq 0 \end{aligned} \tag{8.60}$$

If $I(0)=0$, show that $L[I(t)]=L[f(t)]L[V(t)]$, where

$$L[f(t)]=\frac{s}{L(s^2+2\beta s+\omega_0^2)} \tag{8.61}$$

Find $f(t)$ and show by means of the convolution theorem that

$$I(t)=\int_0^t e^{-\beta u}\left(\frac{\cos\omega u}{L}-\frac{\beta\sin\omega u}{L\omega}\right)V(t-u)\,du \tag{8.62}$$

This gives a formal solution for an arbitrary driving voltage.

8.9. Show that

$$L[x^n]=\frac{\Gamma(n+1)}{s^{n+1}} \tag{8.63}$$

The gamma function is defined by equation (22.5) in Chapter 5.

8.10. Prove the following:

$$L[x^n f(x)] = (-1)^n \frac{d^n\phi}{ds^n} \tag{8.64}$$

$$L[x^2 f'(x)] = s\phi''(s) + 2\phi'(s) \tag{8.65}$$

$$L[xf''(x)] = -s^2\phi'(s) - 2s\phi(s) + f(0) \tag{8.66}$$

8.11. If $f(x)$ is periodic of period a, then $f(x+a) = f(x)$. Show that

$$L[f(x)] = \frac{1}{1-e^{-as}} \int_0^a f(x) e^{-sx}\, dx \tag{8.67}$$

$$A^{\dagger}u_n = N_+ u_{n+1}$$

9 Partial Differential Equations and Special Functions

1. SERIES SOLUTIONS OF ORDINARY DIFFERENTIAL EQUATIONS

Partial differential equations can sometimes be solved by separating the variables. This is illustrated for the one-dimensional wave equation in section 2 of Chapter 7, starting with equation (2.30). Separating the variables in a partial differential equation leads to ordinary differential equations in each of the variables. Many times, it is necessary to solve these ordinary differential equations by expressing the solution as a power series. We will investigate in this section the important features of this approach, which is sometimes called the *Frobenius method*. It might be a good idea at this point to review the introduction to the theory of second-order linear differential equations which is given in section 12 of Chapter 5.

A general second-order homogeneous differential equation defined in the complex plane can be expressed as

$$w''(z) + P(z)w'(z) + Q(z)w(z) = 0 \tag{1.1}$$

A point z_0 in the complex z plane is said to be an *ordinary point* of the differential equation (1.1) if both $P(z)$ and $Q(z)$ are analytic in a neighborhood of z_0. All other points are *singular points* of the differential equation.

Theorem 1. There exists a unique solution of (1.1) in the neighborhood of an ordinary point which satisfies the conditions $w(z_0) = a_1$ and $w'(z_0) = a_2$. This shows that the only possible poles of $w(z)$ are the poles of the coefficients $P(z)$ and $Q(z)$. Moreover, the radius of convergence of the power series representation of the solution is the distance between z_0 and the nearest pole of either $P(z)$ or $Q(z)$. For a discussion of the concept of radius of convergence of a power series, see theorem 2 in section 8 of Chapter 5.

Proof: Rather than prove the theorem in the general case, we will use a particular differential equation which is of some importance. It is called *Legendre's differential equation* and is given by

$$(1 - z^2)w''(z) - 2zw'(z) + \lambda w(z) = 0 \tag{1.2}$$

where λ is a constant. The coefficient functions are

$$P(z) = -\frac{2z}{1 - z^2} \qquad Q(z) = \frac{\lambda}{1 - z^2} \tag{1.3}$$

and have simple poles at $z = \pm 1$. Legendre's differential equation occurs in the separation of Laplace's equation and the Schrödinger wave equation in spherical coordinates. The origin is an ordinary point, and we will attempt to find a solution in terms of a power series expansion about the origin of the form

$$w(z) = \sum_{k=0}^{\infty} c_k z^{s+k} \tag{1.4}$$

where s is a constant to be determined. If theorem 1 is correct, the power series will have a radius of convergence $|z| = 1$. By direct substitution of the power series into the differential equation, we find

$$\sum_{k=0}^{\infty} c_k[\lambda-(s+k)(s+k+1)]z^{s+k}$$
$$+\sum_{k=0}^{\infty} c_k(s+k)(s+k-1)z^{s+k-2}=0 \qquad (1.5)$$

In order to compare the coefficients of like powers of z, it is convenient to reexpress the last term by replacing k everywhere by $k+2$. Equation (1.5) then reads

$$\sum_{k=0}^{\infty} c_k[\lambda-(s+k)(s+k+1)]z^{s+k}$$
$$+\sum_{k=-2}^{\infty} c_{k+2}(s+k+2)(s+k+1)z^{s+k}=0 \qquad (1.6)$$

Note that the sum in the last term now runs from -2 to ∞ thus making it equivalent to its original form in equation (1.5). For a power series expansion to be identically zero, the coefficients of the like powers of z must all separately vanish. For $k=-2$ and $k=-1$, we find

$$c_0 s(s-1)=0 \qquad c_1(s+1)s=0 \qquad (1.7)$$

which are called *indicial equations*. The indicial equations present us with three choices:

$$s=0 \qquad c_0\neq 0 \qquad c_1\neq 0 \qquad (1.8)$$
$$s=1 \text{ or } 0 \qquad c_0\neq 0 \qquad c_1=0 \qquad (1.9)$$
$$s=-1 \text{ or } 0 \qquad c_0=0 \qquad c_1\neq 0 \qquad (1.10)$$

It turns out that each of the three choices leads ultimately to the same two linearly independent solutions of Legendre's equation. Since it is a little simpler, we will set $s=0$ and use (1.8). By setting the coefficients of the remaining powers of z equal to zero, we find

$$c_{k+2}=-c_k\frac{\lambda-k(k+1)}{(k+1)(k+2)} \qquad (1.11)$$

which is a recurrence formula that allows all the even coefficients to be found in terms of c_0 and all the odd coefficients to be found in terms of c_1. Two power series are therefore found, one of which is even and the other is odd:

$$w_1(z)=c_0+c_2z^2+\cdots+c_{2n}z^{2n}+\cdots \qquad (1.12)$$
$$w_2(z)=c_1z+c_3z^3+\cdots+c_{2n+1}z^{2n+1}+\cdots \qquad (1.13)$$

These are two linearly independent solutions of the differential equation.

The general term of the even series is $u_n=c_{2n}z^{2n}$. The convergence properties of this series are determined by computing

$$\lim_{n\to\infty}\left|\frac{u_{n+1}}{u_n}\right|=\lim_{n\to\infty}\left|\frac{\lambda-2n(2n+1)}{(2n+1)(2n+2)}\right||z|^2 \qquad (1.14)$$

where we have set $k=2n$. In evaluating the limit, we will keep all terms of order $1/n$ and discard terms of order $1/n^2$ and higher:

$$\lim_{n\to\infty}\left|\frac{u_{n+1}}{u_n}\right|$$
$$=\lim_{n\to\infty}\left|\frac{4n^2\left(1+\frac{1}{2n}\right)}{4n^2\left(1+\frac{1}{2n}\right)\left(1+\frac{2}{2n}\right)}\right||z|^2$$
$$=\lim_{n\to\infty}\left|\left(1+\frac{1}{2n}\right)\left(1-\frac{1}{2n}\right)\left(1-\frac{1}{n}\right)\right||z|^2$$
$$=\left|1-\frac{1}{n}\right||z|^2 \qquad (1.15)$$

By theorem 2 of section 8 in Chapter 5, the radius of convergence of the series is $|z|=1$. By Gauss' test, equation (4.75) in Chapter 5, the series diverges on the unit circle $|z|=1$. The convergence properties of the odd series are the same. Since two linearly independent solutions have been found, theorem 5 in section 12 of Chapter 5 shows that a unique solution is determined once the value of the solution and its first derivative are known at a point. Theorem 1 is therefore established for Legendre's equation. The proof for the general case proceeds along similar lines.

By using the recurrence formula (1.11) to work out the coefficients, we find for the two series

$$w_1(z)=c_0\left[1-\frac{\lambda}{2!}z^2+\frac{\lambda(\lambda-2\cdot 3)}{4!}z^4-\frac{\lambda(\lambda-2\cdot 3)(\lambda-4\cdot 5)}{6!}z^6+\cdots\right] \qquad (1.16)$$

$$w_2(z)=c_1\left[z-\frac{(\lambda-1\cdot 2)}{3!}z^3+\frac{(\lambda-1\cdot 2)(\lambda-3\cdot 4)}{5!}z^5-\cdots\right] \qquad (1.17)$$

If the constant λ is written as $l(l+1)$ and l has integer values, one or the other of the series (1.16) or (1.17) terminates after a finite number of terms giving a polynomial solution called a *Legendre polynomial.* The first few Legendre polynomials are

$$\begin{aligned} &P_0(z)=1 && P_3(z)=\tfrac{1}{2}(5z^3-3z) \\ &P_1(z)=z && P_4(z)=\tfrac{1}{8}(35z^4-30z^2+3) \\ &P_2(z)=\tfrac{1}{2}(3z^2-1) && P_5(z)=\tfrac{1}{8}(63z^5-70z^3+15z) \end{aligned} \tag{1.18}$$

The constants c_0 and c_1 have been chosen in each case to conform to the conventional definition of the Legendre polynomials. These solutions of the Legendre equation are of considerable importance, since they are the only solutions which are finite at the singular points $z=\pm 1$. When you solved problem 11.3 in Chapter 7, you found the first four Legendre polynomials by Schmidt orthogonalization.

Reference to section 10 of Chapter 7 shows that the Legendre equation can be expressed in Sturm–Liouville form as

$$\frac{d}{dz}\left[(1-z^2)\frac{dw}{dz}\right]+\lambda w(z)=0 \tag{1.19}$$

We identify

$$\begin{aligned} &p(z)=1-z^2 && q(z)=0 \\ &\lambda=l(l+1) && \rho(z)=1 \end{aligned} \tag{1.20}$$

The Legendre differential equation defines an eigenvalue problem over $-1\le x\le +1$ provided that the boundary conditions as given by (10.6) in Chapter 7 are met. Since $p(x)$ vanishes at $x=\pm 1$, the only requirement is that the solutions be finite at $x=\pm 1$. We have already found that these solutions are the Legendre polynomials. The Legendre polynomials therefore define a complete set of orthogonal functions over $-1\le x\le +1$. Sturm–Liouville theory shows that

$$\int_{-1}^{+1} P_l(x)P_{l'}(x)\,dx=N(l)\delta_{ll'} \tag{1.21}$$

where $N(l)$ is a normalizing factor.

For integer values of l, there is of course a second linearly independent solution of Legendre's equation which is usually denoted by $Q_l(x)$ in the literature. Many times, the two linearly independent solutions of Legendre's equation are denoted $P_l(z)$ and $Q_l(z)$ for general values of l and are called Legendre functions of the first and second kind. Reference to equation (10.17) in Chapter 7 shows that the Wronskian of the Legendre equation is

$$W(x)=P_l(x)Q_l'(x)-Q_l(x)P_l'(x)=\frac{C}{1-x^2} \tag{1.22}$$

where C is a nonzero constant. Since the Legendre polynomials are finite at $x=\pm 1$, (1.22) shows that $Q_l(x)$ must be singular at these points. For this reason, they cannot be eigenfunctions for the interval $-1\le x\le +1$, and the Legendre polynomials are therefore guaranteed to be complete. It is possible to find $Q_l(x)$ in closed form by using equation (12.12) of Chapter 5. For example, if $l=0$,

$$Q_0(x)=C\int e^{-\int P(x)\,dx}\,dx$$

$$\begin{aligned} -\int P(x)\,dx &= \int \frac{2x}{1-x^2}\,dx=\int\left(\frac{1}{1-x}-\frac{1}{1+x}\right)dx \\ &= -\ln(1-x)-\ln(1+x) \\ &= -\ln(1-x^2) \end{aligned}$$

$$\begin{aligned} Q_0(x) &= C\int\frac{dx}{1-x^2}=\frac{1}{2}C\int\left(\frac{1}{1+x}+\frac{1}{1-x}\right)dx \\ &= \tfrac{1}{2}C[\ln(1+x)-\ln(1-x)] \end{aligned} \tag{1.23}$$

It is conventional to define $Q_0(x)$ by choosing $C=1$:

$$Q_0(x)=\tfrac{1}{2}\ln\left(\frac{1+x}{1-x}\right) \tag{1.24}$$

The Taylor's series expansion of $Q_0(x)$ is

$$Q_0(x)=x+\tfrac{1}{3}x^3+\tfrac{1}{5}x^5+\cdots \tag{1.25}$$

Note that this same expansion is found from (1.17) if $\lambda=0$. Thus, for a given integer value of l, one of the solutions (1.16) or (1.17) gives a Legendre polynomial and the other supplies the second linearly independent solution as a series expansion. When (1.24) is extended into the complex plane, it is frequently written as

$$Q_0(z)=\frac{1}{2}\ln\left(\frac{z+1}{z-1}\right) \tag{1.26}$$

The singular points $z = \pm 1$ are branch points, and a cut is placed along the real axis between -1 and $+1$. Look at Fig. 15.1 in Chapter 5. If we write

$$z - 1 = r_1 e^{i\theta_1} \qquad z + 1 = r_2 e^{i\theta_2} \tag{1.27}$$

and then approach the real axis between -1 and $+1$ from above,

$$z - 1 = (1 - x)e^{i\pi} \qquad z + 1 = 1 + x \tag{1.28}$$

Equation (1.26) becomes

$$Q_0(z) = \frac{1}{2} \ln\left(\frac{1+x}{1-x} e^{-i\pi}\right) = -\frac{i\pi}{2} + \frac{1}{2} \ln\left(\frac{1+x}{1-x}\right) \tag{1.29}$$

which agrees with (1.24) except for an additive constant. The constant is a solution of Legendre's equation for $l = 0$ since $i\pi/2 = (i\pi/2)P_0(x)$. There is of course nothing to stop us from expressing the solution in the complex plane as

$$Q_0(z) = \frac{1}{2} \ln\left(\frac{1+z}{1-z}\right) \tag{1.30}$$

if that should prove more convenient in some application. Branch cuts can then be placed as illustrated in Fig. 15.2 of Chapter 5.

Because of their great importance, we will study the Legendre polynomials in more detail later. For the present, we continue with the main theme of this section. Consider again the general differential equation (1.1). Let z_0 be a singular point. If $(z - z_0)P(z)$ and $(z - z_0)^2 Q(z)$ are analytic functions, then z_0 is called a *regular singular point* of the differential equation. Otherwise, z_0 is called an *irregular singular point*. To put it another way, z_0 is a regular singular point if the behavior of $P(z)$ is no worse than $1/(z - z_0)$ and the behavior of $Q(z)$ is no worse than $1/(z - z_0)^2$ in the vicinity of z_0.

Theorem 2. In the neighborhood of a regular singular point, a second-order differential equation has two linearly independent solutions. One or both of them may be singular at z_0. The content of theorems 1 and 2 are generally referred to as *Fuch's theorem*.

Proof: The proof will again be done using a specific example. We choose another differential equation of considerable importance called Bessel's equation. It is given by

$$z^2 w''(z) + z w'(z) + (z^2 - m^2) w(z) = 0 \tag{1.31}$$

where m is a constant. Bessel's equation occurs, for example, in the separation of Laplace's equation in cylindrical coordinates. Note that the origin is a regular singular point of (1.31) and that there are no other singular points in the finite complex plane. Again, we try a power series solution of the form (1.4). This time, we find

$$\sum_{k=0}^{\infty} c_k \left[(s+k)^2 - m^2\right] z^{s+k} + \sum_{k=2}^{\infty} c_{k-2} z^{s+k} = 0 \tag{1.32}$$

Indicial equations are obtained for $k = 0$ and $k = 1$:

$$c_0(s^2 - m^2) = 0 \qquad c_1\left[(s+1)^2 - m^2\right] = 0 \tag{1.33}$$

We will assume $c_0 \neq 0$ and use $s = \pm m$. This requires $c_1 = 0$. The other choice is $c_0 = 0$, $c_1 \neq 0$, and $s + 1 = \pm m$. The second choice leads ultimately to the same solution. The recurrence formula for the other coefficients is

$$c_k = -\frac{c_{k-2}}{(s+k)^2 - m^2} = -\frac{c_{k-2}}{(s+k+m)(s+k-m)} \tag{1.34}$$

Choosing $s = +m$ and computing the coefficients from the recurrence relation yields the solution

$$w(z) = c_0 z^m \left[1 - \frac{1}{m+1}\left(\frac{z}{2}\right)^2 + \frac{1}{2!(m+1)(m+2)}\left(\frac{z}{2}\right)^4 - \frac{1}{3!(m+1)(m+2)(m+3)}\left(\frac{z}{2}\right)^6 + \cdots\right] \tag{1.35}$$

The solution for $s = -m$ can be found by changing m to $-m$ in (1.35). If m is not an integer, two linearly independent solutions are found for the two cases $s = \pm m$. Both have branch points at $z = 0$. If m is an integer, then $s = -m$ gives no solution since the coefficients in the series become undefined. A second solution must then be found by other means, for example, by means of equation (12.12) in Chapter 5. It is usual to define the *Bessel functions* by choosing the constant c_0 in (1.35) to be

$$c_0 = \frac{1}{\Gamma(m+1)2^m} \tag{1.36}$$

The gamma function $\Gamma(m+1)$ is defined in section 22 of Chapter 5. The series representation of the Bessel functions is then

$$J_m(z) = \left(\frac{z}{2}\right)^m \sum_{n=0}^{\infty} \frac{(-1)^n}{n!\Gamma(m+n+1)}\left(\frac{z}{2}\right)^{2n} \tag{1.37}$$

The reader can show that the series converges if $|z| < \infty$ as is to be expected since there are no singularities other than $z = 0$ in the finite complex plane. If m is a positive integer, then

$$J_m(z) = \left(\frac{z}{2}\right)^m \sum_{n=0}^{\infty} \frac{(-1)^n}{n!(m+n)!}\left(\frac{z}{2}\right)^{2n} \tag{1.38}$$

Note that this solution is analytic at $z = 0$. Because of their importance, more space will be devoted later to the study of the solutions of Bessel's equation. The constant m is called the *order* of the Bessel function.

In the vicinity of an irregular singularity of the differential equation (1.1), the behavior of the solution is more complicated, and there is no guarantee that the method of solution by series will work.

PROBLEMS

1.1. It is possible for the point at infinity to be an ordinary point, a regular singularity, or an irregular singularity of a second-order differential equation. To study the point at infinity, make the change of variable $\zeta = 1/z$ in the differential equation (1.1) and show that

$$\frac{d^2w}{d\zeta^2} + \left[\frac{2}{\zeta} - \frac{1}{\zeta^2}P\left(\frac{1}{\zeta}\right)\right]\frac{dw}{d\zeta} + \frac{1}{\zeta^4}Q\left(\frac{1}{\zeta}\right)w = 0 \tag{1.39}$$

Find the transformed versions of Legendre's and Bessel's equations and show that Legendre's equation has a regular singularity at infinity and that Bessel's equation has an irregular singularity at infinity.

1.2. By making the change of dependent variable $w(z) = u(z)v(z)$, show that the differential equation (1.1) becomes

$$u'' + \left(\frac{2v'}{v} + P\right)u' + \left(\frac{v''}{v} + \frac{Pv'}{v} + Q\right)u = 0 \tag{1.40}$$

If $v(z)$ is determined by the condition that the coefficient of $u'(z)$ is zero, show that

$$u'' - \left(\tfrac{1}{4}P^2 + \tfrac{1}{2}P' - Q\right)u = 0 \tag{1.41}$$

This is called the *normal form* of the original differential equation. As an example, reduce the differential equation

$$w'' - \frac{2}{z}w' + \left(a^2 + \frac{2}{z^2}\right)w = 0 \tag{1.42}$$

and show that its general solution is

$$w(z) = c_1 z\cos az + c_2 z\sin az \tag{1.43}$$

Note that the solution is analytic at $z = 0$ even though this is a regular singular point of the differential equation. Thus, differential equations exist where both linearly independent solutions are analytic in the neighborhood of a regular singular point.

1.3. If $z = 0$ is a regular singular point of (1.1), then $P(z)$ and $Q(z)$ can be expanded in Laurent series about $z = 0$ as

$$P(z) = \frac{p_1}{z} + p_2 + p_3 z + \cdots$$
$$Q(z) = \frac{q_1}{z^2} + \frac{q_2}{z} + q_3 + q_4 z + \cdots \tag{1.44}$$

The normal form of the differential equation can therefore be expressed as

$$z^2u'' - (a_0 + a_1z + a_2z^2 + \cdots)u = 0 \tag{1.45}$$

By assuming a power series solution for u of the form (1.4), show that the indicial equation gives

$$s = \tfrac{1}{2} \pm \sqrt{\tfrac{1}{4} + a_0} \tag{1.46}$$

and that the coefficients in the series can all be found in terms of c_0.

1.4. The differential equation

$$z^3w'' - w = 0 \tag{1.47}$$

has an irregular singular point at $z = 0$. What happens if you try to solve it by a power series expansion about $z = 0$?

1.5. Show that (1.9) gives the solutions (1.16) and (1.17).

1.6. Establish the convergence of the series (1.37) for Bessel functions.

1.7. For $l = 1$, show that the second linearly independent solution of Legendre's equation is

$$Q_1(x) = \frac{x}{2}\ln\left(\frac{1+x}{1-x}\right) - 1 \tag{1.48}$$

Assume that x is real.

1.8. Find the Sturm–Liouville form of Bessel's differential equation. If the two linearly independent solutions are denoted $J_m(z)$ and $N_m(z)$, show that the Wronskian takes the form

$$W(z) = N_mJ_m' - J_mN_m' = \frac{C}{z} \tag{1.49}$$

where C is a constant. If m is a positive integer, J_m is analytic at $z = 0$. The function $N_m(z)$, called a Neumann function, is therefore singular at $z = 0$.

1.9. If $m = 0$, show that the second solution of Bessel's equation has a logarithmic singularity at $z = 0$.

Hint: Use equation (12.12) of Chapter 5. It is only necessary to use the leading term in the expansion for $J_0(z)$ to find the behavior of $N_0(z)$ at $z = 0$.

2. THE HARMONIC OSCILLATOR

It is shown in section 7 of Chapter 7 that the time-dependent part of the wave function in the Schrödinger wave equation

$$i\hbar\frac{\partial\psi}{\partial t} = H\psi \tag{2.1}$$

can be separated out by writing $\psi(\mathbf{r}, t) = u(\mathbf{r})f(t)$ in those cases where the Hamiltonian operator does not depend on the time. The result is

$$Hu(\mathbf{r}) = Wu(\mathbf{r}) \qquad \psi(\mathbf{r}, t) = u(\mathbf{r})e^{-iWt/\hbar} \tag{2.2}$$

where W is the energy of the particle. If a particle of mass m moves in a harmonic oscillator potential, the time-independent part of the wave function obeys

$$\left(\frac{p^2}{2m} + \frac{1}{2}K_1x_1^2 + \frac{1}{2}K_2x_2^2 + \frac{1}{2}K_3x_3^2\right)u(\mathbf{r}) = Wu(\mathbf{r}) \tag{2.3}$$

where p^2 is the square of the magnitude of the total momentum operator and K_1, K_2, and K_3 are force constants which, in the most general case of an anisotropic oscillator, are not equal. If we view (2.3) as a partial differential equation in the three rectangular coordinates x_1, x_2, and x_3, it can be further separated by writing $u(\mathbf{r}) = u_1(x_1)u_2(x_2)u_3(x_3)$. We find

$$\left(-\frac{\hbar^2}{2m}\frac{\partial^2}{\partial x_1^2}+\frac{1}{2}K_1x_1^2-\frac{\hbar^2}{2m}\frac{\partial^2}{\partial x_2^2}+\frac{1}{2}K_2x_2^2 -\frac{\hbar^2}{2m}\frac{\partial^2}{\partial x_3^2}+\frac{1}{2}K_3x_3^2\right)u_1u_2u_3 = Wu_1u_2u_3 \tag{2.4}$$

If (2.4) is divided through by $u_1u_2u_3$, the result is

$$\frac{1}{u_1}\left(-\frac{\hbar^2}{2m}\frac{d^2u_1}{dx_1^2}+\frac{1}{2}K_1x_1^2u_1\right) + \frac{1}{u_2}\left(-\frac{\hbar^2}{2m}\frac{d^2u_2}{dx_2^2}+\frac{1}{2}K_2x_2^2u_2\right) + \frac{1}{u_3}\left(-\frac{\hbar^2}{2m}\frac{d^2u_3}{dx_3^2}+\frac{1}{2}K_3x_3^2u_3\right) = W \tag{2.5}$$

Since each of the three terms depends on only one of the independent variables, each must be a constant. This gives

$$-\frac{\hbar^2}{2m}\frac{d^2u_1}{dx_1^2}+\frac{1}{2}K_1x_1^2u_1 = W_1u_1 \tag{2.6}$$

$$-\frac{\hbar^2}{2m}\frac{d^2u_2}{dx_2^2}+\frac{1}{2}K_2x_2^2u_2 = W_2u_2 \tag{2.7}$$

$$-\frac{\hbar^2}{2m}\frac{d^2u_3}{dx_3^2}+\frac{1}{2}K_3x_3^2u_3 = W_3u_3 \tag{2.8}$$

where W_1, W_2, and W_3 are separation constants which must add up to the total energy:

$$W_1 + W_2 + W_3 = W \tag{2.9}$$

We find that the original partial differential equation separates into three ordinary equations, each of which is equivalent to a one-dimensional harmonic oscillator. The separation succeeds because the potential energy is a sum of three terms each of which depends on only one of the independent variables.

In the remainder of this section, the solution of the one-dimensional harmonic oscillator given by

$$-\frac{\hbar^2}{2m}u'' + \frac{1}{2}Kx^2u = Wu \tag{2.10}$$

will be analyzed in detail. In classical mechanics, the circular frequency of the harmonic oscillator is given by

$$\omega = \sqrt{\frac{K}{m}} \tag{2.11}$$

and this same parameter will be used in the quantum mechanical solution. It is convenient to express the differential equation in terms of dimensionless variables. If (2.10) is divided through by $\hbar\omega$, the result is

$$-\frac{\hbar}{2m\omega}u'' + \frac{K}{2\hbar\omega}x^2u = \frac{W}{\hbar\omega}u \tag{2.12}$$

We then define

$$\lambda = \frac{W}{\hbar\omega} \qquad y = \beta x \qquad \beta = \sqrt{\frac{K}{\hbar\omega}} = \sqrt{\frac{m\omega}{\hbar}} \tag{2.13}$$

The differential equation is converted to

$$\frac{d^2u}{dy^2} - y^2u + 2\lambda u = 0 \tag{2.14}$$

Since $\hbar\omega$ has dimensions of energy, the parameter λ is dimensionless. Since y^2 also appears as a coefficient of u in (2.14), it too is dimensionless. Equation (2.14) is in Sturm–Liouville form. Look at the boundary conditions given by equation (10.6) in Chapter 7. The interval over which the problem is defined is infinite: $a = \infty$, $b = -\infty$. The boundary conditions are satisfied by solutions which vanish at infinity. This is reasonable physically because the motion is bound both classically and quantum mechanically. Review the discussion of Hermitian operators in section 6 of Chapter 7. If the wave function vanishes sufficiently rapidly at infinity so that $\int |u|^2\,dx$ exists, then the Hamiltonian is Hermitian and the eigenvalues, which are the possible energies, are real.

Another point to be made is that the harmonic oscillator potential is symmetric: $V(-x) = V(x)$. The discussion in problem 10.3 of Chapter 7 shows that the eigenfunctions can be expected to have either even parity or odd parity. There is a discussion of the parity operator just after theorem 8 in section 12 of Chapter 7.

The solution of (2.14) is best done by first investigating its behavior at infinity. If y^2 is large, then

$$u'' - y^2 u \simeq 0 \tag{2.15}$$

which, in the limit $y^2 \to \infty$, has the approximate solutions

$$u(y) \simeq e^{\pm y^2/2} \tag{2.16}$$

The negative exponent gives acceptable behavior at infinity. We therefore "factor out" the behavior of the wave function at infinity by making the change of dependent variable

$$u(y) = w(y) e^{-y^2/2} \tag{2.17}$$

The resulting differential equation for $w(y)$ is

$$w'' - 2yw' + (2\lambda - 1)w = 0 \tag{2.18}$$

This is known as *Hermite's differential equation*. It has no singularities for finite values of y. We use the power series method as discussed in section 1 and express the solution as

$$w(y) = \sum_{k=0}^{\infty} c_k y^{s+k} \tag{2.19}$$

The differential equation gives

$$\sum_{k=-2}^{\infty} c_{k+2}(s+k+2)(s+k+1) y^{s+k} - \sum_{k=0}^{\infty} c_k [2(s+k) - (2\lambda - 1)] y^{s+k} = 0 \tag{2.20}$$

Indicial equations result for $k = -2$ and $k = -1$:

$$c_0 s(s-1) = 0 \qquad c_1(s+1)s = 0 \tag{2.21}$$

These are both satisfied if $s = 0$. The remaining coefficients obey the recurrence formula

$$c_{k+2} = c_k \frac{2k - (2\lambda - 1)}{(k+1)(k+2)} \tag{2.22}$$

Just as in the example of Legendre's equation, an even solution is obtained by evaluating all the even coefficients in terms of c_0 and an odd solution is found by computing the odd coefficients in terms of c_1. To investigate the convergence of the solutions, we write $k = 2j$ and evaluate the ratio of successive terms in the series for $j \to \infty$:

$$\lim_{j\to\infty} \left| \frac{c_{2j+2}}{c_{2j}} \right| |y|^2 = \lim_{j\to\infty} \left| \frac{4j}{(2j+1)(2j+2)} \right| |y|^2 = \lim_{j\to\infty} \frac{1}{j} |y|^2 \tag{2.23}$$

The series therefore converges for all finite values of y. The behavior of the solution at infinity is found by comparing it to the series expansion

$$e^{y^2} = \sum_{j=1}^{\infty} \frac{1}{j!} y^{2j} \tag{2.24}$$

The ratio of successive terms of this series is

$$\left| \frac{y^{2j}}{j!} \frac{(j-1)!}{y^{2j-2}} \right| = \frac{1}{j} |y|^2 \tag{2.25}$$

and is the same as the limiting form (2.23). The reason for setting $k = 2j$ in (2.23) is that j increases by one for each successive term just as it does in (2.24). Thus, the power series solutions behave asymptotically like e^{y^2} at infinity. This is to be expected since we removed a factor of $e^{-y^2/2}$ from the solution of (2.14). Since the possible forms of the solution at infinity are $e^{\pm y^2/2}$, the series solutions must behave like e^{y^2}. Since this behavior is not acceptable, the series must be terminated after a finite number of terms by an appropriate choice of the possible eigenvalues to give a polynomial solution. An examination of the recurrence formula (2.22) shows that this is accomplished if

$$2\lambda - 1 = 2n \tag{2.26}$$

where n is an integer which we identify as a quantum number. The eigenvalues of the Sturm–Liouville differential equation (2.14) are $2\lambda = 2n + 1$. From (2.13) we see that the possible energy levels of the harmonic oscillator are

$$W_n = \hbar\omega\left(n + \tfrac{1}{2}\right) \qquad n = 0, 1, 2, 3, \ldots \tag{2.27}$$

The even and odd solutions work out to

$$w_1(y) = c_0\left[1 - ny^2 + \frac{2^2}{4!}n(n-2)y^4 - \frac{2^3}{6!}n(n-2)(n-4)y^6 + \cdots\right] \quad (2.28)$$

$$w_2(y) = c_1\left[y - \frac{2}{3!}(n-1)y^3 + \frac{2^2}{5!}(n-1)(n-3)y^5 - \cdots\right] \quad (2.29a)$$

The polynomial solutions which result from integer values of n are

$$H_0(y) = 1 \qquad H_2(y) = 4y^2 - 2$$
$$H_1(y) = 2y \qquad H_3(y) = 8y^3 - 12y$$
$$H_4(y) = 12 - 48y^2 + 16y^4 \quad (2.29b)$$
$$H_5(y) = 120y - 160y^3 + 32y^5$$
$$H_6(y) = -120 + 720y^2 - 480y^4 + 64y^6$$

These are known as *Hermite polynomials*. In each case, the constants c_0 and c_1 have been chosen to conform to the conventional definition of these functions. The harmonic oscillator wave functions are

$$u_n(x) = N(n)H_n(\beta x)e^{-\beta^2x^2/2} \quad (2.30)$$

where $N(n)$ is a normalizing factor.

We are now going to solve the harmonic oscillator problem all over again, this time using an approach which uses only the algebraic properties of the position, momentum, and Hamiltonian operators. In the following derivation, x, p, and H are to be regarded as Hermitian operators which could be represented by matrices or by differential operators, as in the analytic approach to the solution just completed. The Hamiltonian is

$$H = \frac{p^2}{2m} + \frac{1}{2}Kx^2 \quad (2.31)$$

The position and momentum obey the commutator relation

$$[x, p] = i\hbar \quad (2.32)$$

The problem can be completely solved using nothing more than (2.31) and (2.32). Again, it is convenient to use dimensionless quantities and write the Hamiltonian as

$$H = \tfrac{1}{2}\hbar\omega(y^2 + q^2) \quad (2.33)$$

where $y = \beta x$ as already defined by (2.13) and

$$q = \frac{p}{\sqrt{m\omega\hbar}} \quad (2.34)$$

The commutator for y and q is

$$[y, q] = i \quad (2.35)$$

The algebraic solution is facilitated by introducing a new non-Hermitian operator and its Hermitian adjoint by

$$A = \frac{1}{\sqrt{2}}(y + iq) \qquad A^\dagger = \frac{1}{\sqrt{2}}(y - iq) \quad (2.36)$$

Remember that y and q are themselves Hermitian so that $y^\dagger = y$ and $q^\dagger = q$. By working out $AA^\dagger$, we find

$$AA^\dagger = \frac{1}{2}(y^2 - i[yq - qy] + q^2) = \frac{H}{\hbar\omega} + \frac{1}{2} \quad (2.37)$$

where use is made of (2.35). Similarly,

$$A^\dagger A = \frac{H}{\hbar\omega} - \frac{1}{2} \quad (2.38)$$

Thus, the commutator of A and $A^\dagger$ is

$$AA^\dagger - A^\dagger A = 1 \quad (2.39)$$

and the Hamiltonian can be written in two ways as

$$H = \hbar\omega AA^\dagger - \tfrac{1}{2}\hbar\omega \qquad H = \hbar\omega A^\dagger A + \tfrac{1}{2}\hbar\omega \quad (2.40)$$

The eigenvectors and eigenvalues are to be found from $Hu_n = W_n u_n$. This can be expressed in two ways as

$$(AA^\dagger - \tfrac{1}{2})u_n = \lambda_n u_n \quad (2.41)$$

$$(A^\dagger A + \tfrac{1}{2})u_n = \lambda_n u_n \quad (2.42)$$

The relation of the eigenvalue λ_n to the energy is given by (2.13). If (2.42) is multiplied from the left by A, the resulting equation can be expressed in the form

$$(AA^\dagger + \tfrac{1}{2})Au_n = \lambda_n Au_n \quad (2.43)$$

By using the commutator relation (2.39), we find

$$(A^\dagger A + \tfrac{1}{2})Au_n = (\lambda_n - 1)Au_n \tag{2.44}$$

By comparing (2.42) and (2.44), we see that if u_n is an eigenvector with eigenvalue λ_n, then Au_n is an eigenvector with eigenvalue $\lambda_n - 1$. Thus, the significance of the operator A is that it is a *lowering operator*. By doing a similar calculation starting with (2.41), the reader can show that

$$(AA^\dagger - \tfrac{1}{2})A^\dagger u_n = (\lambda_n + 1)A^\dagger u_n \tag{2.45}$$

This shows that $A^\dagger$ is a *raising operator*. The vector $A^\dagger u_n$ is an eigenvector with the eigenvalue $\lambda_n + 1$. The results can be summarized by

$$Au_n = N_- u_{n-1} \tag{2.46}$$

$$A^\dagger u_n = N_+ u_{n+1} \tag{2.47}$$

where N_- and N_+ are factors which are required to preserve the normalization of the eigenvectors.

The expectation value of the Hamiltonian is

$$\langle H \rangle = \frac{1}{2m}(\psi, p^2\psi) + \frac{K}{2}(\psi, x^2\psi) \tag{2.48}$$

We can use a functional representation and write

$$(\psi, p^2\psi) = \int_{-\infty}^{+\infty} \psi^* p^2 \psi \, dx = \int_{-\infty}^{+\infty} (p\psi)^* p\psi \, dx$$

$$= \int_{-\infty}^{+\infty} |p\psi|^2 \, dx \tag{2.49}$$

This shows that $(\psi, p^2\psi) = (p\psi, p\psi) \geq 0$. Similarly, $(\psi, x^2\psi) \geq 0$. Refer to equation (6.31) in Chapter 7 for the general definition of a Hermitian operator. The inner product notation is introduced by equation (11.11) of Chapter 7. See also equation (12.26) of Chapter 7 and the ensuing discussion. The matrix notation version of (2.49) is

$$(\psi, p^2\psi) = \psi^\dagger p^2 \psi = \psi^\dagger p^\dagger p \psi = (p\psi)^\dagger p\psi \tag{2.50}$$

In either case, it can be concluded that $\langle H \rangle \geq 0$. The eigenvalues of H are therefore positive. The significance of this is that there must be a lowest eigenvector and a lowest eigenvalue. Physically, this is the ground state of the system. If the lowering operator is applied to the lowest eigenvector, the result must be zero:

$$Au_0 = 0 \tag{2.51}$$

Equation (2.42) then gives the result

$$\lambda_0 = \frac{1}{2} \tag{2.52}$$

Equations (2.45) and (2.47) show that the eigenvalue of u_1 is $\lambda_1 = 1 + \frac{1}{2}$. By applying the raising operator over and over again, the eigenvalue of u_n is found to be $\lambda_n = n + \frac{1}{2}$, where $n = 0, 1, 2, \ldots$. This is the same result that we found by the analytical method.

Let us now find the normalizing factors which appear in (2.46) and (2.47). By forming the inner product of (2.47) with itself and requiring that both u_n and u_{n+1} be normalized, we get

$$|N_+|^2(u_{n+1}, u_{n+1}) = (A^\dagger u_n, A^\dagger n_n) = (u_n, AA^\dagger u_n)$$

$$|N_+|^2 = \left(u_n, \left[\frac{H}{\hbar\omega} + \frac{1}{2}\right]u_n\right)$$

$$= (n+1)(u_n, u_n) = n + 1$$

$$N_+ = \sqrt{n+1} \tag{2.53}$$

We might have assumed that N_+ is complex and written $N_+ = \sqrt{n+1}\, e^{i\delta}$. However, the presence of the phase factor δ in no way affects the probability density or other measurable quantities that are calculated from the eigenvector, and so we have set it equal to zero. Similarly, by starting with (2.46), the reader can show that

$$N_- = \sqrt{n} \tag{2.54}$$

Starting with the lowest eigenvector u_0, all normalized eigenvectors can be generated by using the raising operator:

$$u_1 = \frac{1}{\sqrt{1}} A^\dagger u_0 \qquad u_2 = \frac{1}{\sqrt{2}} A^\dagger u_1 = \frac{1}{\sqrt{2!}} (A^\dagger)^2 u_0$$

$$u_3 = \frac{1}{\sqrt{3}} A^\dagger u_2 = \frac{1}{\sqrt{3!}} (A^\dagger)^3 u_0$$

$$u_n = \frac{1}{\sqrt{n!}} (A^\dagger)^n u_0 \tag{2.55}$$

In addition to providing a second method for solving problems in quantum mechanics, the algebraic solution provides a means of deriving some important properties of the solution obtained by the analytical method. In terms of a functional representation, the operator q of equation (2.34) is

$$q = \frac{p}{\sqrt{m\omega\hbar}} = -\frac{i\hbar}{\sqrt{m\omega\hbar}}\frac{\partial}{\partial x} = -i\frac{\partial}{\partial y} \tag{2.56}$$

The raising and lowering operators are then

$$A^\dagger = \frac{1}{\sqrt{2}}\left(y - \frac{\partial}{\partial y}\right) \qquad A = \frac{1}{\sqrt{2}}\left(y + \frac{\partial}{\partial y}\right) \tag{2.57}$$

Equation (2.51) becomes a differential equation for the determination of the lowest eigenfunction:

$$yu_0 + \frac{du_0}{dy} = 0 \tag{2.58}$$

The solution is

$$u_0 = Ne^{-y^2/2} = Ne^{-\beta^2x^2/2} \tag{2.59}$$

The normalizing factor is found from

$$\int_{-\infty}^{+\infty} |u_0|^2\, dx = |N|^2 \int_{-\infty}^{+\infty} e^{-\beta^2x^2}\, dx = |N|^2\frac{\sqrt{\pi}}{\beta} = 1 \tag{2.60}$$

Note that the normalization is carried out with respect to the variable x, not y. The normalized ground-state wave function is then

$$u_0 = \sqrt{\frac{\beta}{\sqrt{\pi}}}\, e^{-\beta^2x^2/2} = \sqrt{\frac{\beta}{\sqrt{\pi}}}\, e^{-y^2/2} \tag{2.61}$$

All of the normalized wave functions can now be found by using the raising operator as given by equation (2.55):

$$u_n(y) = \sqrt{\frac{\beta}{\sqrt{\pi}\, n!2^n}}\left(y - \frac{\partial}{\partial y}\right)^n e^{-y^2/2} \tag{2.62}$$

It is possible to express (2.62) in another form by noting that, if $f(y)$ is any function of y, then

$$\left(y - \frac{\partial}{\partial y}\right) f(y)e^{y^2/2} = -f'(y)e^{y^2/2}$$
$$\left(y - \frac{\partial}{\partial y}\right)^n f(y)e^{y^2/2} = (-1)^n f^{(n)}(y)e^{y^2/2} \tag{2.63}$$

If $f(y) = e^{-y^2}$, then

$$\left(y - \frac{\partial}{\partial y}\right)^n e^{-y^2/2} = (-1)^n\left(\frac{d^n}{dy^n}e^{-y^2}\right)e^{y^2/2} \tag{2.64}$$

The normalized wave functions (2.62) can therefore be expressed as

$$u_n(y) = \sqrt{\frac{\beta}{\sqrt{\pi}\, n!2^n}}\,(-1)^n e^{y^2}\left(\frac{d^n}{dy^n}e^{-y^2}\right)e^{-y^2/2} \tag{2.65}$$

which should be compared to (2.30).

When we made up the list of Hermite polynomials (2.29), the multiplicative constants c_0 and c_1 were chosen so as to conform to some as yet undisclosed definition of these functions. At some point, a formal definition must be given, and this is as good a place to do it as any. The definition is

$$H_n(y) = (-1)^n e^{y^2}\frac{d^n}{dy^n}e^{-y^2} \tag{2.66}$$

Relations of this type are known as *Rodrigues formulas*. Many of the special functions of mathematical physics can be expressed in this manner. The normalized harmonic oscillator wave functions are now expressed in their final form as

$$u_n(y) = \sqrt{\frac{\beta}{\sqrt{\pi}\, n!2^n}}\, H_n(y)e^{-y^2/2} \tag{2.67}$$

Figure 2.1 shows graphs of the probability densities $|u_n|^2$ for several of the wave functions. The fact that the wave functions are normalized shows that

$$\int_{-\infty}^{+\infty} H_n(y)H_m(y)e^{-y^2}\, dy = n!2^n\sqrt{\pi}\,\delta_{nm} \tag{2.68}$$

The Hermite polynomials are therefore orthogonal with respect to the weight function e^{-y^2}.

Another important result can be derived from the Rodrigues formula. Consider the Taylor's series expansion

$$e^{-(y-t)^2} = \sum_{n=0}^{\infty}\frac{t^n}{n!}\left[\frac{d^n}{dt^n}e^{-(y-t)^2}\right]_{t=0} \tag{2.69}$$

in which the expansion parameter is t. Let $f(y-t)$ be

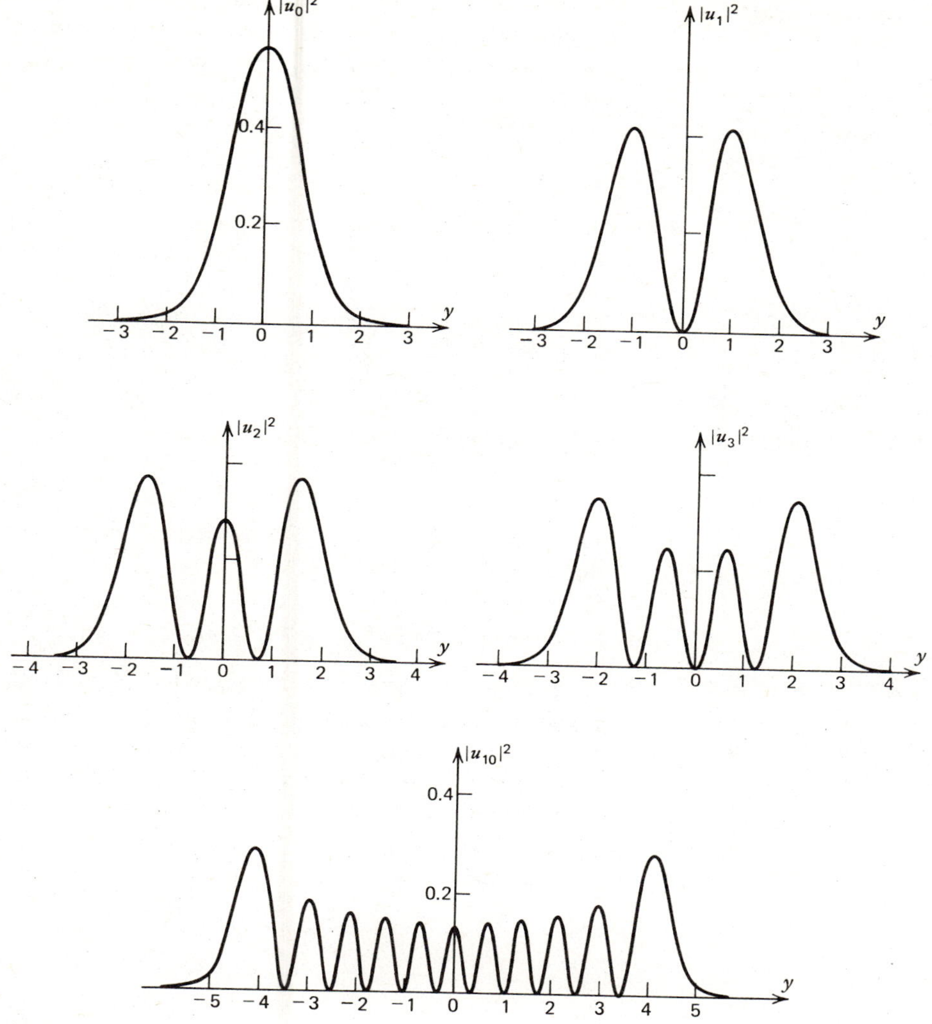

Fig. 2.1

any function of the variable $y-t$. Note that

$$\frac{d}{dt}f(y-t)\bigg|_{t=0} = f'(y-t)(-1)|_{t=0} = (-1)f'(y)$$

$$\frac{d^n}{dt^n}f(y-t)\bigg|_{t=0} = (-1)^n f^{(n)}(y) \tag{2.70}$$

The Taylor's series expansion (2.69) can then be written

$$e^{-(t-y)^2} = \sum_{n=0}^{\infty} \frac{(-1)^n t^n}{n!} \frac{d^n}{dy^n} e^{-y^2} \tag{2.71}$$

If (2.71) is multiplied through by a factor e^{y^2} and (2.66) is used, the result is

$$e^{2yt-t^2} = \sum_{n=0}^{\infty} H_n(y)\frac{t^n}{n!} \tag{2.72}$$

which is a *generating function* for the Hermite polynomials.

One of the uses of the generating function is in the evaluation of definite integrals involving the Hermite polynomials. As an example, we will again evaluate the integral (2.68). Write down the generating function twice and multiply the two equations together to get

$$e^{2yt-t^2}e^{2ys-s^2} = \sum_{n=0}^{\infty}\sum_{m=0}^{\infty}\frac{t^n s^m}{n!m!}H_n(y)H_m(y) \tag{2.73}$$

Now multiply through by e^{-y^2} and integrate both sides of the equation:

$$e^{-s^2-t^2}\int_{-\infty}^{+\infty}e^{-y^2+y(2t+2s)}\,dy = \sum_{n=0}^{\infty}\sum_{m=0}^{\infty}\frac{t^n s^m}{n!m!}I_{nm} \tag{2.74}$$

where I_{nm} is the same integral which appears in (2.68). The evaluation of the integral on the left side of (2.74) gives

$$e^{-s^2-t^2}\sqrt{\pi}\,e^{(t+s)^2} = \sqrt{\pi}\,e^{2st} \tag{2.75}$$

Now expand (2.75) as a power series and equate it to (2.74):

$$\sqrt{\pi}\sum_{n=0}^{\infty}\frac{2^n}{n!}s^n t^n = \sum_{n=0}^{\infty}\sum_{m=0}^{\infty}\frac{t^n s^m}{n!m!}I_{nm} \tag{2.76}$$

Compare the coefficients of like powers of t and s on the two sides of the equation to conclude that

$$I_{nm} = \sqrt{\pi}\,2^n n!\delta_{nm} \tag{2.77}$$

in agreement with (2.68).

The generating function can be used to derive recurrence formulas for the Hermite polynomials. If the generating function (2.72) is differentiated with respect to t, the result is

$$(2y-2t)e^{2yt-t^2} = \sum_{n=1}^{\infty}\frac{nt^{n-1}}{n!}H_n(y) \tag{2.78}$$

Now use the generating function to eliminate the exponential factor. After a little manipulation, we find

$$\sum_{n=0}^{\infty}\frac{2yH_n}{n!}t^n - \sum_{n=1}^{\infty}\frac{2H_{n-1}}{(n-1)!}t^n = \sum_{n=0}^{\infty}\frac{H_{n+1}}{n!}t^n \tag{2.79}$$

Equating the like powers of t gives

$$2yH_n(y) - 2nH_{n-1}(y) = H_{n+1}(y) \tag{2.80}$$

This can be converted into a recurrence formula for the wave functions to give

$$u_{n+1} = \sqrt{\frac{2}{n+1}}\,yu_n - \sqrt{\frac{n}{n+1}}\,u_{n-1} \tag{2.81}$$

Recurrence formulas are extremely valuable in numerical computations. Starting only with the wave functions u_0 and u_1, a computer program can easily be written which computes numerical values of the higher wave functions by repeated use of the recurrence formula. This was the method that was used to produce the graphs of Fig. 2.1.

PROBLEMS

2.1. Write Hermite's differential equation (2.18) in Sturm–Liouville form and show that the functions $p(y)$, $q(y)$, and $\rho(y)$ are all given by e^{-y^2}. The Sturm–Liouville differential equation is defined by equation (10.1) of Chapter 7. The functions $p(y)$ and $q(y)$ are of course not to be confused with momentum and the operator q of equation (2.34). Show that the Wronskian of Hermite's differential equation is

$$w_1w_2' - w_2w_1' = Ce^{y^2} \tag{2.82}$$

where C is a constant. If w_1 is a Hermite polynomial, then the other linearly independent solution necessarily becomes arbitrarily large as $y \to \infty$ and is not suitable as an eigenfunction.

2.2. Prove equation (2.45).

2.3. Prove equation (2.54).

2.4. Find the first three Hermite polynomials from the Rodrigues formula (2.66).

2.5. Prove the differential recurrence formula

$$H_n'(y) = 2nH_{n-1}(y) \tag{2.83}$$

in two ways, once by differentiating the generating function with respect to y and then again by using the lowering operator and equation (2.46).

2.6. If $n = 3$, show that the amplitude of a classical harmonic oscillator in terms of y units is $y_0 = \sqrt{7} = 2.65$. Look at the $n = 3$ graph of Fig. 2.1. Note that the wave function is oscillatory in the classically allowed region and that it falls off exponentially in the classically forbidden region.

2.7. Show that the probability density function for a classical harmonic oscillator is

$$P(y) = \frac{1}{\pi\sqrt{y_0^2 - y^2}} \tag{2.84}$$

where y_0 is the classical amplitude.

Hint: The probability that the particle is found between y and $y + dy$ is proportional to how much time it spends there: $P(y)\,dy = C\,dt$, where C is a constant. Find y_0 for the case where $n = 10$ and graph $P(y)$. Compare your graph to the $n = 10$ graph in Fig. 2.1.

2.8. It was not necessary to use a specific representation in the algebraic solution of the harmonic oscillator. A specific matrix representation can however be used if we wish. Section 12 in Chapter 7 explains how the matrix elements of an operator are to be found. The eigenvectors of the Hamiltonian can be represented by

$$u_0 = \begin{pmatrix} 1 \\ 0 \\ 0 \\ 0 \\ \vdots \end{pmatrix} \quad u_1 = \begin{pmatrix} 0 \\ 1 \\ 0 \\ 0 \\ \vdots \end{pmatrix} \quad u_2 = \begin{pmatrix} 0 \\ 0 \\ 1 \\ 0 \\ \vdots \end{pmatrix} \cdots \tag{2.85}$$

The matrix elements of the lowering operator are given by

$$A_{mn} = (u_m, Au_n) = u_m^{\dagger} A u_n \tag{2.86}$$

With the aid of (2.46), show that

$$A = \begin{pmatrix} 0 & \sqrt{1} & 0 & 0 & \cdots \\ 0 & 0 & \sqrt{2} & 0 & \cdots \\ 0 & 0 & 0 & \sqrt{3} & \cdots \\ \vdots & \vdots & \vdots & \vdots & \end{pmatrix} \tag{2.87}$$

You can now find the matrix representation of the Hamiltonian by means of (2.40) and, as you will see, it comes out diagonal with the energy eigenvalues as the diagonal elements. Solve equations (2.36) for the operators y and q and find their matrix representations.

2.9. Use the generating function to evaluate the integral

$$I_{mn} = \int_{-\infty}^{+\infty} H_m(y) H_n(y)\, y e^{-y^2}\, dy \tag{2.88}$$

Use the result to find the matrix elements

$$y_{mn} = (u_m, yu_n) = \int_{-\infty}^{+\infty} y u_m(y) u_n(y)\, dy \tag{2.89}$$

Check the answer you got for problem 2.8.

2.10. According to equations (2.6) through (2.9), the energy levels of a three-dimensional isotropic harmonic oscillator are given by

$$W_{n_1 n_2 n_3} = \hbar\omega\left(n_1 + n_2 + n_3 + \tfrac{3}{2}\right) \qquad \omega = \sqrt{\frac{K}{m}} \tag{2.90}$$

Review your solutions to problems 7.6 and 7.7 in Chapter 6. Again assume that each atom of a crystal is an isotropic harmonic oscillator. Since the possible energies are no longer being treated as a continuum, it is now necessary to calculate the average energy per atom by means of

$$\overline{W} = \frac{\sum We^{-W/kT}}{\sum e^{-W/kT}} \tag{2.91}$$

where the sums are over all possible energy levels. By combining (2.90) and (2.91),

$$\frac{\overline{W}}{\hbar\omega} = \frac{\sum_{n_1}\sum_{n_2}\sum_{n_3}(n_1 + n_2 + n_3 + 3/2)\,e^{-\alpha(n_1+n_2+n_3+3/2)}}{\sum_{n_1}\sum_{n_2}\sum_{n_3} e^{-\alpha(n_1+n_2+n_3+3/2)}} \tag{2.92}$$

where $\alpha = \hbar\omega/kT$. Evaluate the sums to show that

$$\overline{W} = \frac{3}{2}\hbar\omega + \frac{3\hbar\omega}{e^{\hbar\omega/kT} - 1} \tag{2.93}$$

Hint: You know how to evaluate the sum $\Sigma e^{-\alpha n}$. Differentiate with respect to α to find $\Sigma ne^{-\alpha n}$. Show that the heat capacity per mole of crystal is

$$C = N_A \frac{d\overline{W}}{dT} = 3N_A k\left(\frac{\hbar\omega}{kT}\right)^2 \frac{e^{\hbar\omega/kT}}{\left(e^{\hbar\omega/kT} - 1\right)^2} \tag{2.94}$$

Show that the Dulong and Petit law results if the temperature is high.

Equation (2.94) explains the general features of the specific heat of a crystal quite well. It deviates the most from experiment at low temperatures. The reason for the discrepancy is that we have assumed only one natural frequency ω for the system. We know from the discussion in section 1 of Chapter 7 that a periodic structure, such as a crystal lattice, has many normal modes of oscillation and an improvement in the theory results if we take this into account by averaging over the possible normal modes.

According to equation (2.93), the crystal has a vibrational energy $\frac{3}{2}N_A\hbar\omega$ per mole even at absolute zero temperature. This is called the *zero point energy* and results because the lowest possible state of a harmonic oscillator does not correspond to zero energy. The theory leading to (2.93) and (2.94) was first given by Einstein. It is usual to introduce a characteristic temperature called the *Einstein temperature* by $\theta = \hbar\omega/k$. Thermodynamically speaking, (2.94) actually gives the molar specific heat at constant volume. It can be written

$$C_V = 3R\left(\frac{\theta}{T}\right)^2 \frac{e^{\theta/T}}{\left(e^{\theta/T} - 1\right)^2} = 3RF_E\left(\frac{\theta}{T}\right) \tag{2.95}$$

where $F_E(\theta/T)$ is called the *Einstein function* and R is the gas constant. At moderate temperatures, there is little difference between C_V and the specific heat at constant pressure. The Einstein temperature has a characteristic value for a given material. For diamond, $\theta = 1320$ degrees Kelvin. Graph the Einstein function as a function of the variable $x = T/\theta$. Some mathematical handbooks have tables of this function.

An approximate theory can be constructed to take into account the fact that there are many normal modes of vibration. If the crystal is treated as a continuum, elastic waves propagating in it obey the wave equation $\Box\psi = 0$. If the crystal is a cube of edge s and we use as boundary conditions the vanishing of the wave, then the acoustical eigenfunctions are

$$u(\mathbf{r}) = \sqrt{\frac{8}{s^3}} \sin\left(\frac{n_1 \pi x_1}{s}\right) \sin\left(\frac{n_2 \pi x_2}{s}\right) \sin\left(\frac{n_3 \pi x_3}{s}\right) \tag{2.96}$$

The eigenfrequencies are given by

$$\left(\frac{\omega}{v}\right)^2 = \frac{\pi^2}{s^2}\left(n_1^2 + n_2^2 + n_3^2\right) \tag{2.97}$$

where v is the velocity of the wave and n_1, n_2, and n_3 are positive integers. They are of course not quantum numbers. Review the discussion of the vibrating membrane in section 3 of Chapter 7. Thus, instead of a single characteristic frequency ω as we assumed in deriving (2.94), there are many as given by (2.97). The number of possible frequencies is however not infinite. The crystal is really a periodic structure and, as is revealed in section 1 of Chapter 7, there is a definite maximum frequency ω_c above which waves will not propagate. See especially problem 1.2 of Chapter 7. To carry out an averaging process, we need to know the number of modes of vibration between ω and $\omega + d\omega$. It is convenient to define

$$R^2 = n_1^2 + n_2^2 + n_3^2 \tag{2.98}$$

and write (2.97) as

$$\omega = \frac{v\pi R}{s} \tag{2.99}$$

Since n_1, n_2, n_3 are all positive, the number of modes between R and $R + dR$ is just one eighth of the volume of a spherical shell:

$$N(\omega)\, d\omega = \tfrac{1}{8} 4\pi R^2\, dR \tag{2.100}$$

In making such an approximation, we assume that the values of n_1, n_2, and n_3 which are important are large so that they can be treated as almost a continuum. By using (2.99) to express R in terms of ω, we find

$$N(\omega)\, d\omega = \frac{1}{2\pi^2}\left(\frac{s}{v}\right)^3 \omega^2\, d\omega \tag{2.101}$$

The total number of modes between $\omega = 0$, and the cutoff frequency ω_c is

$$N = \int_0^{\omega_c} N(\omega)\, d\omega = \frac{1}{6\pi^2}\left(\frac{s}{v}\right)^3 \omega_c^3 \tag{2.102}$$

The average energy per atom in the crystal as given by (2.93) is now replaced by

$$\begin{aligned} \overline{W} &= \frac{1}{N}\int_0^{\omega_c} \frac{3\hbar\omega}{e^{\hbar\omega/kT} - 1} N(\omega)\, d\omega \\ &= \frac{9}{\omega_c^3}\int_0^{\omega_c} \frac{\hbar\omega^3}{e^{\hbar\omega/kT} - 1}\, d\omega \end{aligned} \tag{2.103}$$

We have omitted the zero point energy since it does not contribute to the specific heat. Show from (2.103) that the molar specific heat is

$$C_V = 9R\left(\frac{T}{\theta_c}\right)^3 \int_0^{\theta_c/T} \frac{x^4 e^x}{(e^x - 1)^2}\, dx = 3RF_D\left(\frac{\theta_c}{T}\right) \tag{2.104}$$

where θ_c is defined by

$$\theta_c = \frac{\hbar\omega_c}{k} \tag{2.105}$$

and is called the *Debye temperature*. The function $F_D(\theta_c/T)$ which appears in (2.105) is one form of the *Debye function*. Show also that (2.103) can be put into the form

$$\overline{W} = \frac{9kT^4}{\theta_c^3}\int_0^{\theta_c/T}\frac{x^3\,dx}{e^x - 1} \tag{2.106}$$

The Debye temperature is characteristic of a given material.

The Debye theory we have just presented is a significant improvement over the Einstein theory, especially at low temperatures. A further refinement results if we take into account the fact that there are both transverse and longitudinal acoustical waves in an elastic solid, each with its own characteristic velocity of propagation. There is another improvement if the crystal is treated properly as a periodic structure rather than as a continuum when the normal modes of vibration are found.

2.11. Show that the Debye function as defined by (2.104) can be expressed as

$$F_D(t) = \frac{3}{t^3}\int_0^t \frac{x^4 e^x}{(e^x - 1)^2}\,dx = \frac{-3t}{e^t - 1} + \frac{12}{t^3}\int_0^t \frac{x^3}{e^x - 1}\,dx \tag{2.107}$$

The last integral which appears in (2.107) is found in tables.

2.12. Show that

$$\int_0^\infty \frac{x^n}{e^x - 1}\,dx = n!\,\zeta(n+1) \tag{2.108}$$

where $\zeta(n+1)$ is the Riemann zeta function defined by equation (4.57) of Chapter 5. See also equations (21.54) and (21.56) of Chapter 5.

Hint: Use the expansion

$$\frac{1}{1 - e^{-x}} = \sum_{j=0}^{\infty} e^{-jx} \tag{2.109}$$

Show that, at very low temperatures,

$$\overline{W} = \frac{3}{5}\pi^4 k\frac{T^4}{\theta_c^3} \qquad C_V = \frac{12}{5}\pi^4 R\left(\frac{T}{\theta_c}\right)^3 \tag{2.110}$$

Thus, the specific heat of a crystal is proportional to T^3 at low temperatures.

2.13. Use the definite integral

$$\int_{-\infty}^{+\infty} e^{-\frac{1}{4}t^2 + iyt}\,dt = 2\sqrt{\pi}\,e^{-y^2} \tag{2.111}$$

to show that the harmonic oscillator wave functions can be expressed as

$$u_n(y) = \frac{(-i)^n}{2\sqrt{\pi}\sqrt{\sqrt{\pi}\,n!2^n}}e^{y^2/2}\int_{-\infty}^{+\infty} t^n e^{-t^2/4 + iyt}\,dt \tag{2.112}$$

For convenience, we have omitted the factor of β. The wave functions as given by (2.112) are then normalized with respect to the variable y. Write down (2.112) twice, multiply the factors together, and then sum over n to show that

$$\sum_{n=0}^{\infty} u_n(y)u_n(y') = \frac{e^{(y^2+y'^2)/2}}{4\pi\sqrt{\pi}} \int_{-\infty}^{+\infty}\int_{-\infty}^{+\infty} e^{-(s+t)^2/4+i(yt+y's)}\, ds\, dt \tag{2.113}$$

Make the change of variables in the integrand given by

$$s = v + w \qquad t = v - w \tag{2.114}$$

Caution: Recall that

$$ds\, dt = \begin{vmatrix} \dfrac{\partial s}{\partial w} & \dfrac{\partial s}{\partial v} \\ \dfrac{\partial t}{\partial w} & \dfrac{\partial t}{\partial v} \end{vmatrix} dw\, dv \tag{2.115}$$

See problem 6.12 in Chapter 4. Evaluate the integrals over w and v to show that

$$\sum_{n=0}^{\infty} u_n(y)u_n(y') = e^{(y-y')^2/4}\delta(y-y') = \delta(y-y') \tag{2.116}$$

We have verified that the closure relation for the harmonic oscillator wave functions holds. Now let $f(y)$ be any function and consider the expansion coefficients

$$c_n = \int_{-\infty}^{+\infty} u_n(y) f(y)\, dy \tag{2.117}$$

Use the closure relation to show that

$$\sum_{n=0}^{\infty} |c_n|^2 = \int_{-\infty}^{+\infty} |f(y)|^2\, dy \tag{2.118}$$

We have therefore shown that the harmonic oscillator wave functions obey the completeness relation defined by equation (10.39) of Chapter 7. We have said elsewhere in this book that general proofs of completeness covering wide classes of problems are difficult. So far, we have demonstrated in a rigorous way the completeness of the Fourier eigenfunctions and now the harmonic oscillator eigenfunctions.

2.14. Show that

$$H_{2n}(0) = (-1)^n \frac{(2n)!}{n!} \qquad H_{2n+1}(0) = 0 \tag{2.119}$$

Hint: Set $y = 0$ in the generating function (2.72). Expand e^{-t^2} and compare the coefficients of the like powers of t on the two sides of the equation.

2.15. The series expansion of the generating function (2.72) converges for all t in the finite complex plane. Multiply (2.72) by the factor t^{-m-1} and then integrate the resulting equation around any contour which encircles the origin. Use the residue theorem (19.22) of Chapter 5 to show that

$$H_n(y) = \frac{n!}{2\pi i} \oint t^{-n-1} e^{2yt - t^2}\, dt \tag{2.120}$$

This is an *integral representation* for the Hermite polynomials.

2.16. Show that the matrix elements of y^2 are

$$\begin{aligned}(y^2)_{nm} &= \int_{-\infty}^{+\infty} y^2 u_n(y) u_m(y)\, dy \\ &= \tfrac{1}{2}\delta_{n+2,\,m}\sqrt{(n+1)m} + \delta_{nm}\left(n+\tfrac{1}{2}\right) + \tfrac{1}{2}\delta_{n,\,m+2}\sqrt{n(m+1)}\end{aligned} \tag{2.121}$$

You can do this problem by using the generating function or you can simply take the matrix representation for y you found in problem 2.8 and square it. The matrix elements of y can be conveniently expressed as

$$y_{nm} = \frac{1}{\sqrt{2}}\left(\delta_{n+1,m}\sqrt{m} + \delta_{n,m+1}\sqrt{n}\right) \tag{2.122}$$

3. ALGEBRA OF ANGULAR MOMENTUM

In section 9 of Chapter 2, we defined the angular momentum of a single particle by $\boldsymbol{\ell} = \mathbf{r} \times \mathbf{p}$, where $\mathbf{p} = m\mathbf{u}$ is its linear momentum. In our quantum mechanical discussion, we will be concerned with two kinds of angular momentum. One is the orbital angular momentum of particles such as that of electrons revolving around the nucleus of atoms, and the other is the intrinsic spin angular momentum of the particles themselves. Our initial discussion of angular momentum will be based on extending the concept of orbital angular momentum from classical to quantum mechanics. Since it is convenient to reserve the letter l for other purposes, the expression for the orbital angular momentum of a particle will be written

$$\mathbf{P} = \mathbf{r} \times \mathbf{p} \tag{3.1}$$

This expression is valid quantum mechanically if the linear momentum is represented by its operator as $\mathbf{p} = -i\hbar\nabla$.

It is a good idea to review the commutator relations between linear momentum and position. It is shown in section 13 of Chapter 7 that $[x, p_x] = i\hbar$. By proceeding in a similar manner, the other possible commutators between the components of $\mathbf{p}$ and $\mathbf{r}$ are found to be

$$\begin{array}{lll} [x, p_x] = i\hbar & [x, p_y] = 0 & [x, p_z] = 0 \\ [y, p_x] = 0 & [y, p_y] = i\hbar & [y, p_z] = 0 \\ [z, p_x] = 0 & [z, p_y] = 0 & [z, p_z] = i\hbar \end{array} \tag{3.2}$$

The commutator relations can be written compactly as

$$[x_j, p_k] = i\hbar\delta_{jk} \tag{3.3}$$

It is convenient to work with a dimensionless operator $\mathbf{L}$ defined by

$$\mathbf{L} = \frac{1}{\hbar}\mathbf{P} = \mathbf{r} \times \frac{1}{i}\nabla \tag{3.4}$$

The components of $\mathbf{L}$ work out to

$$\begin{aligned} L_x &= -i\left(y\frac{\partial}{\partial z} - z\frac{\partial}{\partial y}\right) \\ L_y &= -i\left(z\frac{\partial}{\partial x} - x\frac{\partial}{\partial z}\right) \\ L_z &= -i\left(x\frac{\partial}{\partial y} - y\frac{\partial}{\partial x}\right) \end{aligned} \tag{3.5}$$

We will now prove that the angular momentum operator as given by (3.4) is Hermitian. According to equation (6.31) of Chapter 7, we are required to show that

$$\langle \mathbf{L} \rangle = \int_\Sigma \psi^*\left(\mathbf{r} \times \frac{1}{i}\nabla\psi\right) d\Sigma = \int_\Sigma \left(\mathbf{r} \times \frac{1}{i}\nabla\psi\right)^* \psi \, d\Sigma \tag{3.6}$$

where the integral is taken over all space. Look at the identity given by equation (9.2) in Chapter 3. If $\mathbf{a}$ is replaced by $\psi^*\mathbf{r}$, the result is

$$\nabla \times \psi\psi^*\mathbf{r} = \psi\nabla \times \psi^*\mathbf{r} + \nabla\psi \times \psi^*\mathbf{r} \tag{3.7}$$

The same identity can be used again to get

$$\nabla \times \psi^*\mathbf{r} = \psi^*\nabla \times \mathbf{r} + \nabla\psi^* \times \mathbf{r} = \nabla\psi^* \times \mathbf{r} \tag{3.8}$$

The last step is true because $\nabla \times \mathbf{r} = 0$. Equation (3.7) can therefore be written

$$\nabla \times \psi\psi^*\mathbf{r} = -(\mathbf{r} \times \nabla\psi^*)\psi - \psi^*\mathbf{r} \times \nabla\psi \tag{3.9}$$

By using the integral identity given by equation (9.15) of Chapter 3, we find

$$\int_\Sigma (\nabla \times \psi\psi^*\mathbf{r}) \, d\Sigma = \int_\sigma (\hat{\mathbf{n}} \times \psi\psi^*\mathbf{r}) \, d\sigma \tag{3.10}$$

As is usually the case when a functional representation of quantum mechanics is used, the Hermiticity property of the operators depends not only on the form of the operator but also on the boundary conditions that are imposed on the wave function as well. In the present example, it is necessary to impose the condition that $\psi\psi^*$ vanish sufficiently rapidly at infinity to make the surface integral in (3.10) zero. The surface can be taken to be a sphere of arbitrarily large radius. Thus, if (3.9) is integrated over all space, the result is

$$\int_\Sigma \psi^*(\mathbf{r}\times\nabla\psi)\,d\Sigma = -\int_\Sigma(\mathbf{r}\times\nabla\psi^*)\psi\,d\Sigma \qquad (3.11)$$

If the factors of i are now put in the right places, (3.11) becomes identical to (3.6) and the hermiticity property of $\mathbf{L}$ is established. We therefore have a quantum mechanical operator that is Hermitian and which corresponds to classical angular momentum. Since it is the more general theory, it is not possible to derive quantum mechanics from classical mechanics. We must adopt procedures to infer the form of quantum mechanical operators which seem reasonable and then test the predictions of the theory by experiment.

The eigenvectors and eigenvalues of angular momentum can be found by purely algebraic methods once the basic commutation relations are known. The commutation relations obeyed by the components of $\mathbf{L}$ can be found by using

$$\left[\frac{\partial}{\partial z}, x\right]=0 \qquad \left[\frac{\partial}{\partial z}, y\right]=0 \qquad \left[\frac{\partial}{\partial z}, z\right]=1 \qquad (3.12)$$

with similar relations for $\partial/\partial x$ and $\partial/\partial y$. These commutators are really equivalent to those given by (3.2). We find, for example, that

$$\begin{aligned}[L_x, L_y] &= -\left[\left(y\frac{\partial}{\partial z}-z\frac{\partial}{\partial y}\right),\left(z\frac{\partial}{\partial x}-x\frac{\partial}{\partial z}\right)\right]\\ &= -\left[y\frac{\partial}{\partial z}, z\frac{\partial}{\partial x}\right]+\left[y\frac{\partial}{\partial z}, x\frac{\partial}{\partial z}\right]\\ &\quad+\left[z\frac{\partial}{\partial y}, z\frac{\partial}{\partial x}\right]-\left[z\frac{\partial}{\partial y}, x\frac{\partial}{\partial z}\right]\\ &= -y\frac{\partial}{\partial x}\left[\frac{\partial}{\partial z}, z\right]+0+0-x\frac{\partial}{\partial y}\left[z, \frac{\partial}{\partial z}\right]\\ &= -y\frac{\partial}{\partial x}+x\frac{\partial}{\partial y}=iL_z \end{aligned} \qquad (3.13)$$

The other components of $\mathbf{L}$ obey similar relations. They are summarized by

$$[L_y, L_z]=iL_x \qquad [L_z, L_x]=iL_y$$
$$[L_x, L_y]=iL_z \qquad (3.14)$$

Formally, the commutation relations can be written compactly as

$$\mathbf{L}\times\mathbf{L}=i\mathbf{L} \qquad (3.15)$$

Of course, no ordinary vector would obey a relation such as this. The operator

$$L^2=\mathbf{L}\cdot\mathbf{L}=L_x^2+L_y^2+L_z^2 \qquad (3.16)$$

is the quantum mechanical version of the square of the magnitude of $\mathbf{L}$. To find the commutation relations between L^2 and the components of $\mathbf{L}$, it is useful to first prove an identity. If A and B are any two operators, then

$$\begin{aligned}[A, B^2] &= AB^2-B^2A\\ &= [AB-BA]B+BAB-B^2A\\ &= [A, B]B+B[A, B]\end{aligned} \qquad (3.17)$$

By application of (3.17), we find

$$\begin{aligned}[L_x, L^2] &= [L_x, L_x^2]+[L_x, L_y^2]+[L_x, L_z^2]\\ &= 0+[L_x, L_y]L_y+L_y[L_x, L_y]\\ &\quad+[L_x, L_z]L_z+L_z[L_x, L_z]\\ &= iL_zL_y+iL_yL_z-iL_yL_z-iL_zL_y=0\end{aligned} \qquad (3.18)$$

where use of (3.14) has been made. Similarly,

$$[L_y, L^2]=0 \qquad [L_z, L^2]=0 \qquad (3.19)$$

What we have found are four operators, L^2, L_x, L_y, and L_z, which are the quantum mechanical generalizations of classical orbital angular momentum. These operators are Hermitian, and L^2 commutes with any one of the components L_x, L_y, and L_z but the components do not commute among themselves. This means that it is possible to find a set of eigenvectors which are simultaneously eigenvectors of L^2 and any one of the three components. We will arbitrarily give L_z this favored status. The eigenvectors so found will then not be eigenvectors of L_x and L_y. Any set of Hermitian operators that obey the commutation relations (3.14) are defined in quantum mechanics to be angular momentum operators. We will find that such a set of operators can represent not only the orbital angular momentum but also the intrinsic spin angular momentum of particles. The properties of angular momentum can be derived by starting with $\mathbf{r} \times \mathbf{p}$ as orbital angular momentum. The operators of spin angular momentum have no such representation but still obey the basic commutation relations (3.14) and (3.16).

The eigenvectors and eigenvalues of L^2 and L_z are to be determined by

$$L^2 u_{\lambda m} = \lambda u_{\lambda m} \tag{3.20}$$

$$L_z u_{\lambda m} = m u_{\lambda m} \tag{3.21}$$

The procedure is very much like that used in the algebraic solution of the harmonic oscillator. We define a non-Hermitian operator and its Hermitian conjugate by

$$A = L_x - iL_y \qquad A^\dagger = L_x + iL_y \tag{3.22}$$

These operators will turn out to be raising and lowering operators like those defined by (2.36). The following commutation relations and identities are easily established:

$$L_z A - AL_z = -A \tag{3.23}$$

$$L_z A^\dagger - A^\dagger L_z = A^\dagger \tag{3.24}$$

$$A^\dagger A = L^2 - L_z^2 + L_z \tag{3.25}$$

$$AA^\dagger = L^2 - L_z^2 - L_z \tag{3.26}$$

$$A^\dagger A - AA^\dagger = 2L_z \tag{3.27}$$

If (3.21) is multiplied from the left by A and (3.23) is used, the result is

$$\begin{aligned} AL_z u_{\lambda m} &= mAu_{\lambda m} \\ (L_z A + A) u_{\lambda m} &= mAu_{\lambda m} \\ L_z(Au_{\lambda m}) &= (m-1)(Au_{\lambda m}) \end{aligned} \tag{3.28}$$

In a similar manner, it is shown that

$$L_z(A^\dagger u_{\lambda m}) = (m+1)(A^\dagger u_{\lambda m}) \tag{3.29}$$

This verifies that $A^\dagger$ and A have the desired raising and lowering properties with respect to the z component of angular momentum. By writing

$$L^2 - L_z^2 = L_x^2 + L_y^2 \tag{3.30}$$

and then calculating the expectation value with respect to one of the eigenstates, we find

$$\begin{aligned} &(u_{\lambda m}, L^2 u_{\lambda m}) - (u_{\lambda m}, L_z^2 u_{\lambda m}) \\ &\qquad = (u_{\lambda m}, L_x^2 u_{\lambda m}) + (u_{\lambda m}, L_y^2 u_{\lambda m}) \geq 0 \\ &\lambda - m^2 \geq 0 \end{aligned} \tag{3.31}$$

The expectation value of the square of any Hermitian operator is never negative. See equation (2.49). Equation (3.31) shows that m has a maximum and a minimum value. Since the choice of the direction of the positive z axis is arbitrary and $-z$ could as well be substituted for z, these maximum and minimum values are the same. We therefore can write

$$-l \leq m \leq +l \qquad l \geq 0 \tag{3.32}$$

If $m = l$, then $u_{\lambda l}$ is the highest possible eigenstate, and $A^\dagger u_{\lambda l} = 0$. By using (3.26) we find

$$\begin{aligned} AA^\dagger u_{\lambda l} &= L^2 u_{\lambda l} - L_z^2 u_{\lambda l} - L_z u_{\lambda l} \\ &= (\lambda - l^2 - l) u_{\lambda l} = 0 \end{aligned} \tag{3.33}$$

Therefore,

$$\lambda = l(l+1) \tag{3.34}$$

The same result is found from $Au_{\lambda, -l} = 0$. Equations (3.28) and (3.29) show that m must vary by integer steps from $-l$ to $+l$. The only way this can happen is for l to have either integer or half-integer values. Thus, if $l = \frac{1}{2}$, $m = -\frac{1}{2}$ or $+\frac{1}{2}$. If $l = 1$, then $m = -1$, 0, or $+1$. It is found that half-integer values of l occur only as the intrinsic spin angular momentum of particles. For example, $l = \frac{1}{2}$ for the spin of the electron. Integer values of l occur in orbital angular momentum, although there are particles that have integer values of intrinsic spin. For example, the photon has unit spin.

The raising operator has the property that

$$A^\dagger u_{lm} = N_+ u_{l,m+1} \tag{3.35}$$

where N_+ is a factor which preserves the normalization of the eigenvectors. It is convenient now to write the eigenvectors as u_{lm} rather than $u_{\lambda m}$. To find N_+, form the inner product of (3.35) with itself:

$$(A^\dagger u_{lm}, A^\dagger u_{lm}) = (u_{lm}, AA^\dagger u_{lm}) = |N_+|^2 \tag{3.36}$$

Now use (3.26) to get

$$\left(u_{lm}, \left[L^2 - L_z^2 - L_z\right] u_{lm}\right) = l(l+1) - m^2 - m$$
$$= |N_+|^2$$
$$N_+ = \sqrt{l(l+1) - m(m+1)}$$
$$= \sqrt{(l-m)(l+m+1)} \tag{3.37}$$

Similarly,

$$Au_{lm} = N_- u_{l,m-1}$$
$$N_- = \sqrt{l(l+1) - m(m-1)}$$
$$= \sqrt{(l+m)(l-m+1)} \tag{3.38}$$

PROBLEMS

3.1. Classically, it is true that $\mathbf{P} = \mathbf{r} \times \mathbf{p} = -\mathbf{p} \times \mathbf{r}$. Show that this is also true for orbital angular momentum in quantum mechanics.

Hint: It is necessary to show that

$$\left(x\frac{\partial}{\partial y} - y\frac{\partial}{\partial x}\right)\psi = -\left(\frac{\partial}{\partial x}y - \frac{\partial}{\partial y}x\right)\psi \tag{3.39}$$

3.2. Suppose that there are two particles in a system with angular momenta $\mathbf{L}_1$ and $\mathbf{L}_2$. Show that if we define $\mathbf{L} = \mathbf{L}_1 + \mathbf{L}_2$, then the components of $\mathbf{L}$ also obey the angular momentum commutation relations; for example,

$$\left[L_x, L_y\right] = \left[L_{1x} + L_{2x}, L_{1y} + L_{2y}\right] = i(L_{1z} + L_{2z}) = iL_z \tag{3.40}$$

Hint: There are two sets of coordinates, (x_1, y_1, z_1) and (x_2, y_2, z_2), one set for each particle. Write out the components of $\mathbf{L}_1$ and $\mathbf{L}_2$, each in terms of its own coordinates, and note that the components of $\mathbf{L}_1$ commute with those of $\mathbf{L}_2$. This result is valid for systems of many particles and also holds for spin angular momentum. This is an important result because it shows that the angular momenta of the individual parts of a complex system can be added to get a resultant angular momentum.

3.3. Prove (3.23) through (3.27).

3.4. Verify (3.29).

3.5. Show that $Au_{\lambda,-l} = 0$ also leads to (3.34).

3.6. Verify that N_- is given correctly by (3.38).

3.7. Show that the Hermitian property of $P_z = xp_y - yp_x$ follows directly from the known Hermitian properties of x, y, p_y, and p_x.

3.8. For a given value of l, the matrix elements of A are given by

$$A_{mn} = (u_{lm}, Au_{ln}) = u_{lm}^\dagger A u_{ln} \tag{3.41}$$

Consider the case $l = 1$. If the elements of the matrix are labeled by values of m and n according to the scheme

$$\begin{matrix} 1,1 & 1,0 & 1,-1 \\ 0,1 & 0,0 & 0,-1 \\ -1,1 & -1,0 & -1,-1 \end{matrix} \tag{3.42}$$

show that the matrix representations of A and $A^\dagger$ are

$$A = \begin{pmatrix} 0 & 0 & 0 \\ \sqrt{2} & 0 & 0 \\ 0 & \sqrt{2} & 0 \end{pmatrix} \qquad A^\dagger = \begin{pmatrix} 0 & \sqrt{2} & 0 \\ 0 & 0 & \sqrt{2} \\ 0 & 0 & 0 \end{pmatrix} \tag{3.43}$$

Also show that

$$L_x = \begin{pmatrix} 0 & \frac{1}{\sqrt{2}} & 0 \\ \frac{1}{\sqrt{2}} & 0 & \frac{1}{\sqrt{2}} \\ 0 & \frac{1}{\sqrt{2}} & 0 \end{pmatrix} \qquad L_y = i \begin{pmatrix} 0 & -\frac{1}{\sqrt{2}} & 0 \\ \frac{1}{\sqrt{2}} & 0 & -\frac{1}{\sqrt{2}} \\ 0 & \frac{1}{\sqrt{2}} & 0 \end{pmatrix}$$

$$L_z = \begin{pmatrix} 1 & 0 & 0 \\ 0 & 0 & 0 \\ 0 & 0 & -1 \end{pmatrix} \qquad L^2 = \begin{pmatrix} 2 & 0 & 0 \\ 0 & 2 & 0 \\ 0 & 0 & 2 \end{pmatrix} \tag{3.44}$$

3.9. Find the eigenvalues and the normalized eigenvectors of the matrix representation of L_x given by (3.44). Let us suppose that the system is known to have its angular momentum in the positive z direction so that $l = 1$, $m = 1$. The system is then described by the eigenvector of L_z given by

$$u_{1,1} = \begin{pmatrix} 1 \\ 0 \\ 0 \end{pmatrix} \tag{3.45}$$

Since the eigenvectors of L_x are also a complete set of eigenvectors for the three-dimensional space defined by $l = 1$, it is possible to expand (3.45) in terms of them as

$$u_{1,1} = c_1 u'_{1,1} + c_0 u'_{1,0} + c_{-1} u'_{1,-1} \tag{3.46}$$

Find the expansion coefficients c_1, c_0, and c_{-1}.

The system is known to be in the state given by the wave function (3.45). This means that if the z component of the angular momentum is measured, we are certain to get the value $m = 1$. (In terms of angular momentum units, this would be $\hbar$.) There is nothing to stop us from measuring the component of the angular momentum in the x direction. The outcome of the measurement will be one of the eigenvalues $\hbar$, 0, or $-\hbar$, but we cannot say for sure which one it will be. It is however possible to say what the probabilities of measuring each of the three eigenvalues will be. These probabilities are $|c_1|^2$, $|c_0|^2$, and $|c_{-1}|^2$. See equation (12.48) of Chapter 7 and the ensuing discussion for an explanation of this concept. The discussion in connection with equation (8.13) of Chapter 7 gives another relevant example. Note that the three probabilities add to unity.

After the measurement is made and the x component of $\mathbf{L}$ is definitely known, the system is no longer described by the representation (3.44) and the wave function (3.45). It is necessary to go to a representation in which L_x is diagonal. You can do this by constructing a unitary transformation out of the eigenvectors of L_x. See equation (12.35) of Chapter 7. Transform each of the matrices in (3.44) by means of this unitary transformation.

3.10. If a particle is in an angular momentum state described by the eigenvector u_{lm}, show that

$$\langle L_x \rangle = 0 \qquad \langle L_x^2 \rangle = \Delta L_x^2 = \frac{l(l+1) - m^2}{2} \tag{3.47}$$

See section 13 of Chapter 7 for the definition and meaning of ΔL_x. What is ΔL_z? The easy way to do the problem is to first recognize that

$$\langle L_x^2 \rangle = \langle L_y^2 \rangle \tag{3.48a}$$

Another way to do it is to first prove the identity

$$L_x^2 = \tfrac{1}{4}\left[(A^\dagger)^2 + A^\dagger A + AA^\dagger + A^2\right] \tag{3.48b}$$

Since the effects of A and $A^\dagger$ on u_{lm} are known, $\langle L_x^2 \rangle$ can be calculated.

4. ORBITAL ANGULAR MOMENTUM AND SPHERICAL HARMONICS

As we know from problem (3.8), the angular momentum operators have matrix representations. For integer values of l, the angular momentum operators can describe orbital angular momentum, and a representation exists in terms of functions. In this section, we will find this representation in spherical coordinates. If you will look at equation (8.3) in Chapter 4, you will be reminded of how the gradient operator can be represented in an arbitrary curvilinear coordinate system. In terms of spherical coordinates, the operator $\mathbf{L}$ is

$$\begin{aligned}\mathbf{L} &= \frac{1}{i}\mathbf{r}\times\nabla = \frac{1}{i}r\mathbf{b}^1\times\mathbf{b}^k\partial_k \\ &= \frac{r}{i}(\mathbf{b}^1\times\mathbf{b}^2)\frac{\partial}{\partial\theta} + \frac{r}{i}(\mathbf{b}^1\times\mathbf{b}^3)\frac{\partial}{\partial\phi}\end{aligned} \tag{4.1}$$

where $\mathbf{b}^k$ are the reciprocal base vectors. According to equation (6.8) of Chapter 4, $\mathbf{L}$ can be written as

$$\mathbf{L} = \frac{r}{i\sqrt{g}}\left(\mathbf{b}_3\frac{\partial}{\partial\theta} - \mathbf{b}_2\frac{\partial}{\partial\phi}\right) \tag{4.2}$$

Note that in spherical coordinates, $\mathbf{L}$ has only two components. Equations relating the base vectors of an arbitrary curvilinear coordinate system to the unit vectors of Cartesian coordinates are displayed in problem (8.5) of Chapter 4. The relevant transformations are

$$\begin{aligned}\mathbf{b}_2 &= r\cos\theta\cos\phi\hat{\mathbf{e}}_1 + r\cos\theta\sin\phi\hat{\mathbf{e}}_2 - r\sin\theta\hat{\mathbf{e}}_3 \\ \mathbf{b}_3 &= -r\sin\theta\sin\phi\hat{\mathbf{e}}_1 + r\sin\theta\cos\phi\hat{\mathbf{e}}_2\end{aligned} \tag{4.3}$$

The rectangular components of $\mathbf{L}$ work out to

$$\begin{aligned}L_x &= i\left(\sin\phi\frac{\partial}{\partial\theta} + \cot\theta\cos\phi\frac{\partial}{\partial\phi}\right) \\ L_y &= -i\left(\cos\phi\frac{\partial}{\partial\theta} - \cot\theta\sin\phi\frac{\partial}{\partial\phi}\right) \\ L_z &= -i\frac{\partial}{\partial\phi}\end{aligned} \tag{4.4}$$

The raising and lowering operators are

$$\begin{aligned}A^\dagger &= e^{i\phi}\left(\frac{\partial}{\partial\theta} + i\cot\theta\frac{\partial}{\partial\phi}\right) \\ A &= -e^{-i\phi}\left(\frac{\partial}{\partial\theta} - i\cot\theta\frac{\partial}{\partial\phi}\right)\end{aligned} \tag{4.5}$$

The eigenfunctions of orbital angular momentum are called *spherical harmonics* and are designated by the notation $Y_{lm}(\theta,\phi)$. The eigenfunctions of L_z are found from

$$\begin{aligned}L_z u_m(\phi) &= -i\frac{\partial}{\partial\phi}u_m(\phi) = m u_m(\phi) \\ u_m(\phi) &= Ne^{im\phi}\end{aligned} \tag{4.6}$$

The spherical harmonics can therefore be written in the form

$$Y_{lm}(\theta,\phi) = F_{lm}(\theta)e^{im\phi} \tag{4.7}$$

where $F_{lm}(\theta)$ is a function of θ only. In many references, the argument is made that m must be restricted to integer values to make the wave function single-valued. This is not enough since it is only the measurable quantities which are calculated from a wave function that must be unambiguous. A probability density calculated from the multiple-valued function $f(\theta)e^{i\phi/2}$ is single-valued and quite respectable. One problem with wave functions which are a superposition of possible half-integral angular momentum states is that they produce a probability current which has a θ component. This means that the positive z axis acts as a source for particles while the negative z axis is a sink. This unphysical situation is avoided by using only integral values of l for orbital angular momentum.*

A second-order differential equation obeyed by $F_{lm}(\theta)$ can be found by expressing L^2 in terms of differential operators and then using the eigenvalue

*In problem 4.4, it is shown conclusively that half-integer values of l must be excluded from orbital angular momentum.

equation (3.20). Probably the easiest way to work out L^2 is to use the identity (3.26). In calculating $AA^\dagger$, consider its action on some function ψ:

$$AA^\dagger\psi = -e^{-i\phi}\left(\frac{\partial}{\partial\theta} - i\cot\theta\frac{\partial}{\partial\phi}\right)e^{i\phi}\left(\frac{\partial\psi}{\partial\theta} + i\cot\theta\frac{\partial\psi}{\partial\phi}\right) \tag{4.8}$$

Remember that the partial derivatives differentiate everything to their right. The final result is that the eigenvalue equation $L^2Y_{lm} = l(l+1)Y_{lm}$ can be expressed as

$$\frac{1}{\sin\theta}\left(\frac{\partial}{\partial\theta}\sin\theta\frac{\partial Y_{lm}}{\partial\theta}\right) + \frac{1}{\sin^2\theta}\frac{\partial^2 Y_{lm}}{\partial\phi^2} + l(l+1)Y_{lm} = 0 \tag{4.9}$$

The ϕ dependence of the spherical harmonics is already known through equation (4.7). The θ dependence can be found by expressing the spherical harmonics as

$$Y_{lm} = N_{lm}P_{lm}(\cos\theta)e^{im\phi} \tag{4.10}$$

where N_{lm} is a normalizing factor. The functions $P_{lm}(\cos\theta)$ are called *associated Legendre functions*. It is conventional to express them as a function of the argument $\cos\theta$. By combining (4.9) and (4.10), we find that the associated Legendre functions obey the ordinary differential equation

$$\frac{1}{\sin\theta}\frac{d}{d\theta}\left(\sin\theta\frac{dP_{lm}}{d\theta}\right) - \frac{m^2}{\sin^2\theta}P_{lm} + l(l+1)P_{lm} = 0 \tag{4.11}$$

For most purposes, we will want to express the associated Legendre equation in terms of the variable $x = \cos\theta$. There is a risk of notational problems here since x is conventionally used to represent one of the rectangular coordinates. There are only so many letters in the alphabet, and we trust that the reader can make the distinction when necessary. In terms of x, (4.11) is

$$\frac{d}{dx}\left[(1-x^2)\frac{dP_{lm}}{dx}\right] - \frac{m^2}{1-x^2}P_{lm} + l(l+1)P_{lm} = 0 \tag{4.12}$$

If $m = 0$, we find the familiar Legendre equation already discussed in section 1.

Since Y_{ll} is the spherical harmonic with the highest m value for a given l, the application of the raising operator to it must give zero. If we write $Y_{ll} = F(\theta)e^{il\phi}$, then $A^\dagger Y_{ll} = 0$ gives the differential equation

$$\frac{dF}{d\theta} - l\cot\theta F = 0 \tag{4.13}$$

By solving the differential equation, we find that

$$Y_{ll} = N\sin^l\theta e^{il\phi} \tag{4.14}$$

The normalizing factor N is determined by the condition

$$\int_{\theta=0}^{\pi}\int_{\phi=0}^{2\pi} Y_{ll}^* Y_{ll}\sin\theta\, d\theta\, d\phi = 1 \tag{4.15}$$

Recall that the expression for volume element in spherical coordinates is $d\Sigma = r^2\sin\theta\, d\theta\, d\phi$. The factor $\sin\theta\, d\theta\, d\phi$ is an element of solid angle. Evaluation of the integral over ϕ gives

$$2\pi|N|^2\int_{\theta=0}^{\pi}\sin^{2l+1}\theta\, d\theta = 1 \tag{4.16}$$

The integral over θ can be evaluated by means of the beta function given by equation (22.13) of Chapter 5. Also needed is the Legendre duplication formula given by equation (22.28) in Chapter 5. The final result is

$$Y_{ll}(\theta,\phi) = \frac{(-1)^l\sqrt{(2l+1)!}}{2^l l!\sqrt{4\pi}}\sin^l\theta e^{il\phi} \tag{4.17}$$

The factor of $(-1)^l$ is inserted to make (4.17) agree with the conventional definition of the spherical harmonics. It in no way affects the normalization.

It is now possible to generate all the spherical harmonics by a successive application of the lowering operator to (4.17). As a first step, the correct normalization can be found by repeated application of (3.38). The result is

$$\begin{aligned}
AY_{ll} &= \sqrt{(2l)(1)}\;Y_{l,l-1}\\
A^2Y_{ll} &= \sqrt{(2l)(1)(2l-1)(2)}\;Y_{l,l-2}\\
A^3Y_{ll} &= \sqrt{(2l)(1)(2l-1)(2)(2l-2)(3)}\;Y_{l,l-3}\\
A^nY_{ll} &= \sqrt{(2l)(2l-1)\dots(2l-n+1)n!}\;Y_{l,l-n}\\
&= \sqrt{\frac{(2l)!n!}{(2l-n)!}}\;Y_{l,l-n}
\end{aligned} \tag{4.18}$$

If we substitute $n = l - m$, we find

$$Y_{lm} = \sqrt{\frac{(l+m)!}{(2l)!(l-m)!}}\, A^{l-m} Y_{ll} \tag{4.19}$$

The task of finding the effect of A on Y_{ll} is best done in terms of the variable $x = \cos\theta$. The lowering operator is

$$A = e^{-i\phi}\left(\sqrt{1-x^2}\,\frac{\partial}{\partial x} + \frac{ix}{\sqrt{1-x^2}}\,\frac{\partial}{\partial \phi}\right) \tag{4.20}$$

If the spherical harmonic is written in abbreviated form as $Y_{lm} = F(x)e^{im\phi}$, then

$$\begin{aligned} A Y_{lm} &= e^{-i\phi}\left(\sqrt{1-x^2}\,\frac{dF}{dx} - \frac{mx}{\sqrt{1-x^2}}\,F\right)e^{im\phi} \\ &= (1-x^2)^{(1-m)/2}\frac{d}{dx}\left[(1-x^2)^{m/2}F\right]e^{i(m-1)\phi} \end{aligned} \tag{4.21}$$

It is now necessary to apply (4.21) over and over to Y_{ll}. First set

$$F(x) = N\sin^l\theta = N(1-x^2)^{l/2} \tag{4.22}$$

and then set $m = l$ in (4.21) to get

$$A Y_{ll} = N(1-x^2)^{(1-l)/2}\left[\frac{d}{dx}(1-x^2)^l\right]e^{i(l-1)\phi} \tag{4.23}$$

In the next step, $F(x)$ is the coefficient of $e^{i(l-1)\phi}$ in (4.23) and $m = l-1$ in (4.21):

$$A^2 Y_{ll} = N(1-x^2)^{(2-l)/2}\left[\frac{d^2}{dx^2}(1-x^2)^l\right]e^{i(l-2)\phi} \tag{4.24}$$

A continuation of this process leads to

$$A^{l-m} Y_{ll} = N(1-x^2)^{-m/2}\left[\frac{d^{l-m}}{dx^{l-m}}(1-x^2)^l\right]e^{im\phi} \tag{4.25}$$

Now use (4.19) and the normalizing factor which appears in (4.17) to get

$$\begin{aligned} Y_{lm} = \frac{(-1)^l}{2^l l!}\sqrt{\frac{2l+1}{4\pi}\,\frac{(l+m)!}{(l-m)!}} \\ \times(1-x^2)^{-m/2}\left[\frac{d^{l-m}}{dx^{l-m}}(1-x^2)^l\right]e^{im\phi} \end{aligned} \tag{4.26}$$

In particular, if $m = 0$, we find

$$Y_{l,0} = \frac{(-1)^l}{2^l l!}\sqrt{\frac{2l+1}{4\pi}}\,\frac{d^l}{dx^l}(1-x^2)^l \tag{4.27}$$

Apart from a normalizing factor, $Y_{l,0}$ is a Legendre polynomial. We will define the Legendre polynomials by the Rodrigues formula

$$P_l(x) = \frac{(-1)^l}{2^l l!}\,\frac{d^l}{dx^l}(1-x^2)^l \tag{4.28}$$

If you will calculate a few polynomials from this formula, you will see that they agree with the list given in equation (1.18). It must be so because, as already pointed out, the differential equation (4.12) reduces to the Legendre equation if $m = 0$. The spherical harmonic $Y_{l,0}$ is normalized. This means that

$$\begin{aligned} &\int_{\theta=0}^{\pi}\int_{\phi=0}^{2\pi} Y_{l,0}^* Y_{l,0}\sin\theta\, d\theta\, d\phi \\ &\quad = 2\pi\int_{-1}^{+1}\frac{2l+1}{4\pi}[P_l(x)]^2\,dx = 1 \end{aligned} \tag{4.29}$$

$$\int_{-1}^{+1}[P_l(x)]^2\,dx = \frac{2}{2l+1}$$

This is the normalization integral for the Legendre polynomials.

By using the same procedure, we can start with $Y_{l,0}$ and work our way up by using the raising operator. The following formulas summarize the results:

$$Y_{lm} = \sqrt{\frac{(l-m)!}{(l+m)!}}\,(A^\dagger)^m Y_{l,0} \tag{4.30}$$

$$A^\dagger = e^{i\phi}\left(-\sqrt{1-x^2}\,\frac{\partial}{\partial x} + \frac{ix}{\sqrt{1-x^2}}\,\frac{\partial}{\partial\phi}\right) \tag{4.31}$$

$$Y_{lm} = F(x)e^{im\phi} \qquad x = \cos\theta \tag{4.32}$$

$$\begin{aligned} A^\dagger Y_{lm} = &-(1-x^2)^{(m+1)/2} \\ &\times\frac{d}{dx}\left[(1-x^2)^{-m/2}F(x)\right]e^{i(m+1)\phi} \end{aligned} \tag{4.33}$$

$$(A^\dagger)^m Y_{l,0} = (-1)^m(1-x^2)^{m/2}\frac{d^m Y_{l,0}}{dx^m}e^{im\phi} \tag{4.34}$$

$$Y_{lm} = \sqrt{\frac{2l+1}{4\pi}\,\frac{(l-m)!}{(l+m)!}}\,P_{lm}(x)e^{im\phi} \tag{4.35}$$

$$P_{lm}(x) = (-1)^m(1-x^2)^{m/2}\frac{d^m P_l(x)}{dx^m} \tag{4.36}$$

Equation (4.36) gives a method for calculating the associated Legendre functions. There is no universal agreement as to where to place the factor of $(-1)^m$. In many references, it is omitted from (4.36) and included in the formula (4.35) for Y_{lm}.

Because L^2 and L_z are Hermitian, the spherical harmonics are a complete set of orthonormal functions with respect to the variables θ and ϕ. They obey the orthogonality condition

$$\int_{\theta=0}^{\pi}\int_{\phi=0}^{2\pi} Y_{lm}^* Y_{l'm'} \sin\theta \, d\theta \, d\phi = \delta_{ll'}\delta_{mm'} \tag{4.37}$$

An arbitrary function can be expanded in terms of them:

$$f(\theta,\phi) = \sum_{l=0}^{\infty}\sum_{m=-l}^{+l} c_{lm} Y_{lm}(\theta,\phi)$$

$$c_{lm} = \int_{\theta=0}^{\pi}\int_{\phi=0}^{2\pi} Y_{lm}^*(\theta,\phi) f(\theta,\phi) \sin\theta \, d\theta \, d\phi \tag{4.38}$$

We can use equation (4.26) to continue the lowering process into negative values of m with the result that

$$\begin{aligned} Y_{l,-m} &= \frac{(-1)^l}{2^l l!}\sqrt{\frac{2l+1}{4\pi}\frac{(l-m)!}{(l+m)!}}(1-x^2)^{m/2} \\ &\quad \times\left[\frac{d^{l+m}}{dx^{l+m}}(1-x^2)^l\right]e^{-im\phi} \\ &= \sqrt{\frac{2l+1}{4\pi}\frac{(l-m)!}{(l+m)!}}(1-x^2)^{m/2} \\ &\quad \times \frac{d^m}{dx^m}P_l(x)e^{-im\phi} \\ &= (-1)^m Y_{lm}^* \end{aligned} \tag{4.39}$$

If (4.26) and (4.35) are equated, we find another formula for calculating the associated Legendre functions:

$$\begin{aligned} P_{lm}(x) &= \frac{(-1)^l}{2^l l!}\frac{(l+m)!}{(l-m)!}(1-x^2)^{-m/2} \\ &\quad \times \frac{d^{l-m}}{dx^{l-m}}(1-x^2)^l \end{aligned} \tag{4.40}$$

If we replace m by $-m$ we get

$$\begin{aligned} P_{l,-m}(x) &= \frac{(l-m)!}{(l+m)!}(1-x^2)^{m/2}\frac{d^m}{dx^m}P_l(x) \\ &= (-1)^m\frac{(l-m)!}{(l+m)!}P_{lm}(x) \end{aligned} \tag{4.41}$$

Apart from a constant, $P_{lm}(x)$ and $P_{l,-m}(x)$ are the same function. This result could have been anticipated from the fact that the differential equation (4.12) depends on m^2. Thus, the solutions for $+m$ and $-m$ are essentially the same.

Associated Legendre functions and Spherical Harmonics

$P_{0,0} = 1$	$Y_{0,0} = \sqrt{\frac{1}{4\pi}}$
$P_{1,1} = -\sin\theta$	$Y_{1,1} = -\frac{1}{2}\sqrt{\frac{3}{2\pi}}\sin\theta e^{i\phi}$
$P_{1,0} = \cos\theta$	$Y_{1,0} = \frac{1}{2}\sqrt{\frac{3}{\pi}}\cos\theta$
$P_{1,-1} = \frac{1}{2}\sin\theta$	$Y_{1,-1} = \frac{1}{2}\sqrt{\frac{3}{2\pi}}\sin\theta e^{-i\phi}$
$P_{2,2} = 3\sin^2\theta$	$Y_{2,2} = \frac{1}{4}\sqrt{\frac{15}{2\pi}}\sin^2\theta e^{2i\phi}$
$P_{2,1} = -3\sin\theta\cos\theta$	$Y_{2,1} = -\frac{1}{2}\sqrt{\frac{15}{2\pi}}\sin\theta\cos\theta e^{i\phi}$
$P_{2,0} = \frac{1}{2}(3\cos^2\theta - 1)$	$Y_{2,0} = \frac{1}{4}\sqrt{\frac{5}{\pi}}(3\cos^2\theta - 1)$
$P_{2,-1} = \frac{1}{2}\sin\theta\cos\theta$	$Y_{2,-1} = \frac{1}{2}\sqrt{\frac{15}{2\pi}}\sin\theta\cos\theta e^{-i\phi}$
$P_{2,-2} = \frac{1}{8}\sin^2\theta$	$Y_{2,-2} = \frac{1}{4}\sqrt{\frac{15}{2\pi}}\sin^2\theta e^{-2i\phi}$

PROBLEMS

4.1. Carry out the normalization integral for Y_{ll} given by equation (4.16).

4.2. Equation (4.12) for the associated Legendre functions is in Sturm–Liouville form. What is the Wronskian for this differential equation? See equation (10.17) in Chapter 7. For each value of m, the associated Legendre functions are a complete set of orthogonal functions over the interval $-1 \le x \le +1$. Show that

$$\int_{-1}^{+1} P_{lm} P_{l'm} \, dx = \frac{2}{2l+1} \frac{(l+m)!}{(l-m)!} \delta_{ll'} \tag{4.42}$$

4.3. Show from the Rodrigues formula (4.28) that $P_l(-x) = (-1)^l P_l(x)$. Thus, the Legendre polynomials are even or odd according as l is even or odd. This fact is also evident from the derivation of the Legendre polynomials in section 1.

4.4. Assume that a functional representation of the angular momentum eigenfunctions exists for $l = \frac{1}{2}$. Write $Y_{1/2,1/2} = f_1(\theta) e^{(1/2)i\phi}$ and use the condition $A^\dagger Y_{1/2,1/2} = 0$ to find $f_1(\theta)$. Now write $Y_{1/2,-1/2} = f_2(\theta) e^{-(1/2)i\phi}$ and apply the lowering operator to $Y_{1/2,1/2}$ to find $f_2(\theta)$. The functions $f_1(\theta)$ and $f_2(\theta)$ are not the same, thus breaking the symmetry with respect to the transformation $z \to -z$. Show that f_1 and f_2 are the two linearly independent solutions of the associated Legendre equation for $m^2 = \frac{1}{4}$. Note that $f_2(\theta)$ is singular at $\theta = 0$ and π. The singularity can be tolerated because the wave function is still square-integrable, that is, the normalization integral

$$\int_0^\pi \int_0^{2\pi} |Y_{1/2,-1/2}|^2 \sin\theta \, d\theta \, d\phi \tag{4.43}$$

exists. Verify this statement. Also verify that $Y_{1/2,1/2}$ and $Y_{1/2,-1/2}$ are orthogonal to one another. Show that the lowering operator when applied to $Y_{1/2,-1/2}$ does not give zero, thus violating the condition $-l \le m \le +l$. Half-integral values of l are therefore rejected as possibilities for orbital angular momentum. What happens if you determine $f_2(\theta)$ by the condition $A Y_{1/2,-1/2} = 0$?

4.5. Show that

$$P_{lm}(-x) = (-1)^{m+l} P_{lm}(x) \tag{4.44}$$

4.6. Prove from the Rodrigues formula (2.28) that

$$P_{2n}(0) = \frac{(-1)^n (2n)!}{2^{2n}(n!)^2} \qquad P_{2n+1}(0) = 0 \tag{4.45}$$

Hint: Expand the factor $(1 - x^2)^l$ in a binomial series. Use the differentiation formula

$$\frac{d^l}{dx^l} x^{2k} = \frac{(2k)!}{(2k-l)!} x^{2k-l} \tag{4.46}$$

4.7. Prove that

$$P_{lm}(0) = \frac{(-1)^{(l+m)/2}}{2^l} \frac{(l+m)!}{\left(\frac{l-m}{2}\right)! \left(\frac{l+m}{2}\right)!} \qquad l+m \text{ even}$$

$$P_{lm}(0) = 0 \qquad l+m \text{ odd} \tag{4.47}$$

4.8. Show that

$$\int_{-1}^{+1} x^{2n} P_{2l}(x) \, dx = \frac{(2n)! 2^{2l+1} (n+l)!}{(n-l)!(2n+2l+1)!} \qquad n \ge l$$

$$= 0 \qquad n < l \tag{4.48}$$

Hint: Use the Rodrigues formula and integrate by parts. You will also need the beta function and the Legendre duplication formula. See equations (22.17) and (22.28) in Chapter 5.

4.9. Prove directly that L_z as given by equation (4.4) is Hermitian by showing that

$$\int_0^{2\pi} \psi^* L_z \psi \, d\phi = \int_0^{2\pi} (L_z \psi)^* \psi \, d\phi \tag{4.49}$$

where ψ is any wave function such that $\psi^*\psi$ is single-valued. Note that hermiticity does not require ψ itself to be single-valued.

5. QUANTUM CENTRAL FORCE PROBLEM

If the interaction between two particles can be described by a scalar potential, then the appropriate Schrödinger wave equation is

$$-\frac{\hbar^2}{2m_1}\nabla_1^2\psi - \frac{\hbar^2}{2m_2}\nabla_2^2\psi + V(\mathbf{r})\psi = i\hbar\frac{\partial\psi}{\partial t} \tag{5.1}$$

The position coordinates of the two particles are $\mathbf{r}_1$ and $\mathbf{r}_2$ and m_1 and m_2 are their masses. The operator ∇_1^2 operates on $\mathbf{r}_1$ and ∇_2^2 operates on $\mathbf{r}_2$. The potential function depends on the relative coordinates of the two particles defined by $\mathbf{r} = \mathbf{r}_1 - \mathbf{r}_2$. We will assume that no external forces act on the system. Equation (5.1) is a partial differential equation in the six coordinates needed to specify the positions of the particles plus the time, giving a total of seven independent variables. In this section, we will find bound-state solutions of (5.1) for the case where the potential depends only on the scalar magnitude of $\mathbf{r}$.

In section 4 of Chapter 8, we treated the two-body central force problem classically. We found that a great simplification resulted from the use of the coordinates of the center of mass and the coordinates of one particle referred to the other as origin. These coordinates are defined by

$$x = x_2 - x_1 \qquad x_c = \frac{m_1 x_1 + m_2 x_2}{m_1 + m_2} \tag{5.2}$$

with similar equations for the y and z components. Fortunately, the wave equation (5.1) can be similarly simplified. The problem is to change the variables in the partial differential equation from x_1, y_1, z_1, x_2, y_2, and z_2 to x, y, z, x_c, y_c, and z_c. If ψ is expressed in terms of the new variables, then by the chain rule for taking partial derivatives,

$$\frac{\partial\psi}{\partial x_1} = \frac{\partial\psi}{\partial x}\frac{\partial x}{\partial x_1} + \frac{\partial\psi}{\partial x_c}\frac{\partial x_c}{\partial x_1} = -\frac{\partial\psi}{\partial x} + \frac{\partial\psi}{\partial x_c}\left(\frac{m_1}{m_1+m_2}\right) \tag{5.3}$$

If the process is repeated once more, the result is

$$\frac{\partial^2\psi}{\partial x_1^2} = \frac{\partial}{\partial x}\left[-\frac{\partial\psi}{\partial x} + \frac{\partial\psi}{\partial x_c}\frac{m_1}{m_1+m_2}\right]\frac{\partial x}{\partial x_1} + \frac{\partial}{\partial x_c}\left[-\frac{\partial\psi}{\partial x} + \frac{\partial\psi}{\partial x_c}\frac{m_1}{m_1+m_2}\right]\frac{\partial x_c}{\partial x_1}$$
$$= \frac{\partial^2\psi}{\partial x^2} - 2\frac{\partial^2\psi}{\partial x\,\partial x_c}\frac{m_1}{m_1+m_2} + \frac{\partial^2\psi}{\partial x_c^2}\left(\frac{m_1}{m_1+m_2}\right)^2 \tag{5.4}$$

Similarly,

$$\frac{\partial^2\psi}{\partial x_2^2} = \frac{\partial^2\psi}{\partial x^2} + 2\frac{\partial^2\psi}{\partial x\,\partial x_c}\left(\frac{m_2}{m_1+m_2}\right) + \frac{\partial^2\psi}{\partial x_c^2}\left(\frac{m_2}{m_1+m_2}\right)^2 \tag{5.5}$$

Therefore,

$$\frac{1}{m_1}\frac{\partial^2\psi}{\partial x_1^2} + \frac{1}{m_2}\frac{\partial^2\psi}{\partial x_2^2} = \frac{1}{M}\frac{\partial^2\psi}{\partial x^2} + \frac{1}{m_1+m_2}\frac{\partial^2\psi}{\partial x_c^2} \tag{5.6}$$

where

$$M = \frac{m_1 m_2}{m_1 + m_1} \tag{5.7}$$

is the reduced mass. We must not use m for the reduced mass as was done in section 4 of Chapter 8 to avoid confusion with the quantum number for the z component of angular momentum. The wave equation (5.1) can therefore be put into the form

$$-\frac{\hbar^2}{2M}\nabla^2\psi - \frac{\hbar^2}{2(m_1+m_2)}\nabla_c^2\psi + V(r)\psi = i\hbar\frac{\partial\psi}{\partial t} \tag{5.8}$$

where ∇^2 operates on the relative coordinates and ∇_c^2 operates on the coordinates of the center of mass.

The three components of the angular momentum operator commute with the Hamiltonian of equation (5.8). The two particles are being treated as structureless, each with mass and charge but no spin. It is therefore only the orbital form of angular momentum as given by (3.5) with which we are currently concerned. It is readily verified, for example, that

$$L_z\nabla^2 - \nabla^2 L_z = 0 \tag{5.9}$$

If the angular momentum operators are put in spherical coordinates as given by (4.4), it becomes obvious that they commute with $V(r)$, since it depends only on r and not θ and ϕ. We therefore expect that it will be possible to find eigenfunctions of (5.8) which are simultaneous eigenfunctions of H, L^2, and L_z.

The method of solution of (5.8) is by separation of variables. As a first step, the time dependence is separated out by writing the wave function as

$$\psi(\mathbf{r},\mathbf{r}_c,t) = \phi(\mathbf{r},\mathbf{r}_c)f(t) \tag{5.10}$$

The result is

$$\frac{1}{\phi}\left[-\frac{\hbar^2}{2M}\nabla^2\phi - \frac{\hbar^2}{2(m_1+m_2)}\nabla_c^2\phi\right] + V(r) = \frac{i\hbar}{f}\frac{df}{dt} = W + W_c \tag{5.11}$$

The separation constant is the total energy, and it has been written as $W+W_c$ in anticipation of the next step in the separation process. The time-dependent part of the wave function is

$$f(t) = e^{-i(W+W_c)t/\hbar} \tag{5.12}$$

To continue the separation, write

$$\phi(\mathbf{r},\mathbf{r}_c) = u(\mathbf{r})u_c(\mathbf{r}_c) \tag{5.13}$$

We find

$$-\frac{\hbar^2}{2Mu}\nabla^2 u + V(r) - \frac{\hbar^2}{2(m_1+m_2)u_c}\nabla_c^2 u_c = W + W_c \tag{5.14}$$

If all the terms which depend only on r are equated to W and those which depend on r_c to W_c, the resulting equations are

$$-\frac{\hbar^2}{2M}\nabla^2 u + V(r)u = Wu \tag{5.15}$$

$$-\frac{\hbar^2}{2(m_1+m_2)}\nabla_c^2 u_c = W_c u_c \tag{5.16}$$

The separation has succeeded and gives results which are analogous to the classical treatment of the problem. According to (5.15), the relative motion of the two particles can be treated as though one of them were a fixed center of force as long as the mass of the other particle is replaced by the reduced mass of the system. Equation (5.16) is a free-particle Schrödinger equation for the motion of the center of mass of the system. The separation constant W is the center-of-mass energy and W_c is the energy due to the motion of the center of mass. In the remainder of this section, we will concentrate on finding the bound-state solutions of (5.15).

Spherical coordinates will be used. Refer to equation (10.26) in Chapter 4 for the appropriate expression for the Laplacian. The r dependence is separated out by writing

$$u(\mathbf{r}) = F(r)Y(\theta,\phi) \tag{5.17}$$

The result is

$$\frac{1}{F}\frac{d}{dr}\left(r^2\frac{dF}{dr}\right) + \frac{1}{Y\sin\theta}\left(\frac{\partial}{\partial\theta}\sin\theta\frac{\partial Y}{\partial\theta}\right) + \frac{1}{Y\sin^2\theta}\frac{\partial^2 Y}{\partial\phi^2} + \frac{2Mr^2}{\hbar^2}[W - V(r)] = 0 \tag{5.18}$$

Look at equation (4.9). As expected, the angular dependence has separated out as an eigenfunction of

L^2 and L_z. We can therefore write

$$\frac{d}{dr}\left(r^2\frac{dF}{dr}\right) - l(l+1)F - \frac{2MV(r)}{\hbar^2}r^2F + \frac{2MW}{\hbar^2}r^2F = 0 \tag{5.19}$$

Note that $l(l+1)$ is the separation constant. The solutions of the angular part of (5.18) are given in section 4. Equation (5.19) is called the *radial equation*, and there remains the task of finding its solutions.

If you will look at equation (10.1) in Chapter 7, you will see that (5.19) defines a Sturm–Liouville problem for $0 \le r \le \infty$. In the notation of Chapter 7, $p(r) = r^2$. Inspection of equation (10.6) of Chapter 7 shows that Sturm–Liouville boundary conditions are met provided that the behavior of any two eigenfunctions $F_1(r)$ and $F_2(r)$ is such that

$$\left[r^2(F_1F_2' - F_2F_1')\right]_{r=0} = 0 \tag{5.20}$$

$$\left[r^2(F_1F_2' - F_2F_1')\right]_{r=\infty} = 0 \tag{5.21}$$

It is also evident from (5.19) that potentials which are more strongly singular than $V(r) \propto 1/r^2$ at the origin may cause trouble because the origin is then not a regular singular point of the differential equation. See theorem 2 in section 1. The energy appears in (5.19) as an eigenvalue. The weight function is $\rho(r) = r^2$. Sturm–Liouville theory tells us that $F_1(r)$ and $F_2(r)$ are orthogonal:

$$\int_0^\infty F_1(r)F_2(r)r^2\,dr = 0 \qquad W_1 \neq W_2 \tag{5.22}$$

The factor r^2 is to be expected because volume element in spherical coordinates is $d\Sigma = r^2\sin\theta\, d\theta\, d\phi$. Theorem 5 in section 10 of Chapter 7 shows that $F_1(r)$ and $F_2(r)$ cannot be degenerate.

In solving the radial equation (5.19), it is convenient to use a dimensionless variable ρ defined by $r = a\rho$. The parameter a has the dimensions of length and will be chosen to suit our convenience at the appropriate time. We will also make the change of dependent variable

$$v(\rho) = rF(r) = a\rho F(a\rho) \tag{5.23}$$

The resulting differential equation for $v(\rho)$ is

$$v'' - \frac{l(l+1)}{\rho^2}v + \frac{2Ma^2W}{\hbar^2}v - \frac{2Ma^2V(a\rho)}{\hbar^2}v = 0 \tag{5.24}$$

By writing the limiting form of $v(\rho)$ as ρ^p as $\rho \to 0$, we find that the normalization integral converges at the origin if $p > -\frac{1}{2}$. The normalization integral is (5.22) with $F_1(r) = F_2(r)$. As shown in problem 5.1, a further restriction is put on p by the boundary condition (5.20). A potential of considerable interest is the Coulomb potential between two charged particles. If one of the particles is an electron with charge $-e$ and the other is an atomic nucleus with charge $+Ze$, the potential is

$$V(a\rho) = -\frac{Ze^2}{4\pi\varepsilon_0 a\rho} \tag{5.25}$$

The radial equation (5.24) is then appropriate for the hydrogen atom ($Z = 1$), single ionized helium, ($Z = 2$), or any system consisting of two charges of opposite sign. We will find the possible bound states of such a system and the corresponding energy eigenvalues. The coefficients of v in (5.24) are all dimensionless. It is convenient to define the parameter a in terms of the energy by the condition

$$\frac{2Ma^2W}{\hbar^2} = -\frac{1}{4} \tag{5.26}$$

A minus sign is used because bound-state energies are negative. If a dimensionless parameter λ is defined by

$$\lambda = \frac{aZMe^2}{2\pi\varepsilon_0\hbar^2} \tag{5.27}$$

then the energy can be expressed as

$$W = -\frac{Z^2Me^4}{2(4\pi\varepsilon_0)^2\hbar^2\lambda^2} \tag{5.28}$$

The differential equation is

$$v'' + \left[-\frac{l(l+1)}{\rho^2} - \frac{1}{4} + \frac{\lambda}{\rho}\right]v = 0 \tag{5.29}$$

At large distances from the origin, the behavior of

$v(\rho)$ is determined by

$$v'' - \tfrac{1}{4}v \simeq 0 \qquad v \simeq e^{\pm\rho/2} \tag{5.30}$$

The negative exponential is appropriate, and we therefore make the change of dependent variable

$$v(\rho) = w(\rho)e^{-\rho/2} \tag{5.31}$$

The differential equation for $w(\rho)$ is

$$w'' - w' - \frac{l(l+1)}{\rho^2}w + \frac{\lambda}{\rho}w = 0 \tag{5.32}$$

The origin is a regular singular point. The behavior of $w(\rho)$ near the origin can be found by looking at the leading term in its series expansion. If

$$\lim_{\rho\to 0} w(\rho) = \rho^s \tag{5.33}$$

then the differential equation gives

$$s(s-1) - l(l+1) = 0 \tag{5.34}$$

which is really the indicial equation. Thus, $s = -l$ or $s = l+1$. Convergence of the normalization integral is required. The possible wave functions must also satisfy the boundary condition (5.20). It might be well at this point to remind the reader that requiring $F(r)$ to satisfy Sturm–Liouville boundary conditions is equivalent to the basic requirement that the Hamiltonian operator be Hermitian. The choice is $s = l+1$. Accordingly, we make a final transformation of the differential equation by writing

$$w(\rho) = \rho^{l+1}L(\rho) \tag{5.35}$$

The function $L(\rho)$ obeys

$$\rho L'' + (k+1-\rho)L' + nL = 0 \tag{5.36}$$

To agree with conventional notation, we have defined

$$k = 2l+1 \qquad n = \lambda - l - 1 \tag{5.37}$$

The differential equation (5.36) is called the *associated Laguerre equation*. Since the singularity at the origin has already been removed, its power series solution can be represented as

$$L(\rho) = \sum_{j=0}^{\infty} c_j\rho^j \tag{5.38}$$

It is found that the coefficients obey the recurrence relation

$$c_{j+1} = -c_j\frac{n-j}{(j+1)(j+1+k)} \tag{5.39}$$

For large j,

$$\frac{c_{j+1}}{c_j} \simeq \frac{1}{j} \tag{5.40}$$

which is the same behavior exhibited by the coefficients in the series

$$e^{\rho} = \sum_{j=0}^{\infty} \frac{1}{j!}\rho^j \tag{5.41}$$

This is to be expected because $v(\rho)$ is given by

$$v(\rho) = \rho^{l+1}L(\rho)e^{-\rho/2} \tag{5.42}$$

and equation (5.30) shows that the behavior of $v(\rho)$ at infinity is either $e^{+\rho/2}$ or $e^{-\rho/2}$. To prevent the unacceptable positive exponential behavior at infinity, it is necessary to terminate the series generated by (5.39) by making n a nonnegative integer: $n = 0,1,2,\ldots$. The function (5.38) then becomes a polynomial. Since $n = \lambda - l - 1$, λ is also an integer: $\lambda = N = n + l + 1$. We know that l is the orbital angular momentum quantum number and that $l = 0,1,2,\ldots$. Therefore, the possible values of N are $1,2,3,\ldots$. From (5.28), the energy eigenvalues are given by

$$W_N = -\frac{Z^2Me^4}{2(4\pi\varepsilon_0)^2\hbar^2N^2} \tag{5.43}$$

By writing $l = N - n - 1$, we see that for a given N, the possible l values are $l = 0,1,2,\ldots,N-1$. It is usual to refer to n as the *radial quantum number* and to N as the *principal quantum number*.

The solutions of the associated Laguerre equation are usually designated by $L_n^k(\rho)$. By repeated application of the recurrence relation (5.39), we can show that

$$L_n^k(\rho) = c_0\sum_{j=0}^{n}\frac{(-1)^j n!k!\rho^j}{(n-j)!(k+j)!j!} \tag{5.44}$$

This provides an explicit representation of the solutions for integer values of n and k. It is possible to solve the associated Laguerre equation by other methods. If $k=0$, the differential equation (5.36) becomes

$$\rho L_n'' + (1-\rho)L_n' + nL_n = 0 \tag{5.45}$$

and is called the *Laguerre equation*. The solutions of the associated Laguerre equation can be found from the solutions of the Laguerre equation in much the same way that the associated Legendre functions can be generated from the Legendre polynomials. The fact that $L_n(\rho)$ is defined for $0 \le \rho < \infty$ suggests that the method of Laplace transforms might be useful. The necessary Laplace transforms are worked out in problem 8.10 of Chapter 8. By taking the Laplace transform of (5.45), we find that

$$s(s-1)\phi'(s) + (s-n-1)\phi(s) = 0 \tag{5.46}$$

$$\phi(s) = \int_0^\infty L_n(\rho)e^{-\rho s}\,d\rho \tag{5.47}$$

The differential equation for $\phi(s)$ can be written

$$\frac{d\phi}{\phi} + \left[\frac{n+1}{s} - \frac{n}{s-1}\right]ds = 0 \tag{5.48}$$

Its solution is found to be

$$\phi(s) = \frac{c(s-1)^n}{s^{n+1}} \tag{5.49}$$

where c is a constant of integration. We will determine the multiplicative factor in the definition of $L_n(\rho)$ by setting $c=1$. For integer values of n, (5.49) has a pole of order $n+1$ at $s=0$. The inversion of the Laplace transform gives

$$L_n(\rho) = \frac{1}{2\pi i}\oint \frac{(s-1)^n e^{s\rho}}{s^{n+1}}\,ds \tag{5.50}$$

See equation (8.5) and Fig. 8.1 in Chapter 8. The path of integration can be deformed into any contour which encircles the origin. Equation (5.50) provides us with an integral representation of the Laguerre polynomials. The change of the variable of integration $s = 1 - z/\rho$ gives

$$L_n(\rho) = \frac{e^\rho}{2\pi i}\oint \frac{z^n e^{-z}\,dz}{(z-\rho)^{n+1}} \tag{5.51}$$

In the complex z plane, the contour of integration must encircle the point $z=\rho$. The integral representations provide compact formulas for the solution of the Laguerre equation and have the advantage that they are sometimes easier to work with than the series representation.

Taylor's series for a function of a complex variable is discussed in section 19 of Chapter 5. Relevant to the following discussion are equations (18.29) and (19.8). An analytic function $f(t)$ can be represented by its Taylor's series expansion about the point $t=\rho$:

$$f(t) = \sum_{n=0}^{\infty} a_n(\rho)(t-\rho)^n \tag{5.52}$$

$$a_n(\rho) = \frac{1}{2\pi i}\oint \frac{f(z)\,dz}{(z-\rho)^{n+1}} = \frac{1}{n!}\left[\frac{d^n}{dt^n}f(t)\right]_{t=\rho} \tag{5.53}$$

If we put

$$f(t) = e^\rho t^n e^{-t}$$

then (5.53) gives

$$a_n(\rho) = \frac{e^\rho}{2\pi i}\oint \frac{z^n e^{-z}\,dz}{(z-\rho)^{n+1}} = \frac{e^\rho}{n!}\frac{d^n}{d\rho^n}(\rho^n e^{-\rho}) \tag{5.54}$$

By using (5.51), we find

$$L_n(\rho) = \frac{e^\rho}{n!}\frac{d^n}{d\rho^n}(\rho^n e^{-\rho}) \tag{5.55}$$

which is a Rodrigues formula for the Laguerre polynomials.

Next, we will find a generating function for the Laguerre polynomials of the form

$$G(\rho,t) = \sum_{n=0}^{\infty} L_n(\rho)t^n \tag{5.56}$$

By regarding (5.56) as a Taylor's series expansion of $G(\rho,t)$ about $t=0$, we can express the coefficients as

$$L_n(\rho) = \frac{1}{2\pi i}\oint \frac{G(\rho,z)}{z^{n+1}}\,dz \tag{5.57}$$

Now go back to equation (5.50) and try to make a change of variable which will make the dependence of the integrand on n appear only through the factor z^{n+1} in the denominator. Success is achieved by

$$z = \frac{s}{s-1} \qquad s = \frac{z}{z-1} \qquad ds = -\frac{dz}{(z-1)^2} \tag{5.58}$$

The result is

$$L_n(\rho)=\frac{1}{2\pi i}\oint\frac{e^{-z\rho/(1-z)}}{(1-z)z^{n+1}}dz \tag{5.59}$$

Note that $s=0$ corresponds to $z=0$ whereas $s=\infty$ corresponds to $z=1$. The contour of equation (5.59) must therefore encircle the point $z=0$ but not $z=1$. The comparison of (5.57) and (5.59) shows that the generating function is

$$G(\rho,t)=\frac{e^{-\rho t/(1-t)}}{1-t}=\sum_{n=0}^{\infty}L_n(\rho)t^n \qquad |t|<1 \tag{5.60}$$

In some references, the constant c of equation (5.49) is chosen to be $n!$. A factor of $1/n!$ then appears in front of $L_n(\rho)$ in (5.60). The reader can show that the generating function obeys the partial differential equations

$$\frac{\partial G}{\partial\rho}=-\frac{t}{1-t}G \tag{5.61}$$

$$(1-t)^2\frac{\partial G}{\partial t}=(1-t-\rho)G \tag{5.62}$$

from which follow the recurrence relations

$$L'_{n-1}-L'_n=L_{n-1} \tag{5.63}$$

$$(n+1)L_{n+1}=(2n+1-\rho)L_n-nL_{n-1} \tag{5.64}$$

Other recurrence formulas can be found by combining (5.63) and (5.64) in various ways. For example, if (5.64) is differentiated, we find

$$(n+1)L'_{n+1}=(2n+1-\rho)L'_n-L_n-nL'_{n-1} \tag{5.65}$$

Replace n by $n+1$ in (5.63) to get

$$L'_n-L'_{n+1}=L_n \tag{5.66}$$

Now use (5.63) and (5.66) to eliminate L'_{n+1} and L'_{n-1} from (5.65). The result is

$$\rho L'_n=nL_n-nL_{n-1} \tag{5.67}$$

If we set $\rho=0$ in the generating function (5.60), we find

$$\frac{1}{1-t}=\sum_{n=0}^{\infty}L_n(0)t^n=\sum_{n=0}^{\infty}t^n \tag{5.68}$$

Therefore,

$$L_n(0)=1 \tag{5.69}$$

We will now find various representations for the associated Laguerre polynomials. If $k=0$ in the series (5.44),

$$L_n(\rho)=c_0\sum_{j=0}^{n}\frac{(-1)^j n!\rho^j}{(n-j)!j!j!} \tag{5.70}$$

By setting $\rho=0$ and comparing (5.69) and (5.70), we conclude that $c_0=1$. If (5.70) is differentiated with respect to ρ, the result is

$$\begin{aligned}\frac{d}{d\rho}L_n(\rho)&=\sum_{j=0}^{n}\frac{(-1)^j n!j\rho^{j-1}}{(n-j)!j!j!}\\&=\sum_{j=1}^{n}\frac{(-1)^j n!\rho^{j-1}}{(n-j)!(j-1)!j!}\\&=\sum_{j=0}^{n-1}\frac{(-1)^{j+1}n!\rho^j}{(n-j-1)!j!(j+1)!}\end{aligned} \tag{5.71}$$

Replacing n by $n+1$ yields

$$\frac{d}{d\rho}L_{n+1}(\rho)=\sum_{j=0}^{n}\frac{(-1)^{j+1}(n+1)!\rho^j}{(n-j)!j!(j+1)!} \tag{5.72}$$

Repeat the process k times to get

$$\begin{aligned}\frac{d^k}{d\rho^k}L_{n+k}(\rho)&=(-1)^k\sum_{j=0}^{n}\frac{(-1)^j(n+k)!\rho^j}{(n-j)!(j+k)!j!}\\&=(-1)^kL_n^k(\rho)\end{aligned} \tag{5.73}$$

By comparing (5.44) and (5.73), you will see that they can be made the same by choosing

$$c_0=\frac{(n+k)!}{n!k!}=L_n^k(0) \tag{5.74}$$

The associated Laguerre polynomials are therefore found by differentiating the Laguerre polynomials. By differentiating the generating function (5.60) k times, you can show that

$$\frac{e^{-\rho t/(1-t)}}{(1-t)^{k+1}}=\sum_{n=0}^{\infty}L_n^k(\rho)t^n \tag{5.75}$$

To find an integral representation for $L_n^k(\rho)$, write down equation (5.50) with n replaced by $n+k$ and then differentiate k times with respect to ρ. The result is

$$L_n^k(\rho)=\frac{(-1)^k}{2\pi i}\oint\frac{(s-1)^{n+k}e^{s\rho}}{s^{n+1}}ds \tag{5.76}$$

The change of variable $s=1-z/\rho$ yields

$$L_n^k(\rho)=\frac{e^\rho\rho^{-k}}{2\pi i}\oint\frac{e^{-z}z^{n+k}\,dz}{(z-\rho)^{n+1}} \tag{5.77}$$

By exactly the same procedure that led to (5.55), you can derive the Rodrigues formula

$$L_n^k(\rho)=\frac{e^\rho\rho^{-k}}{n!}\frac{d^n}{d\rho^n}(e^{-\rho}\rho^{n+k}) \tag{5.78}$$

Recurrence relations for $L_n^k(\rho)$ can be derived from the generating function or by differentiating the recurrence relations for $L_n(\rho)$. Some possibilities are

$$(n+1)L_{n+1}^k=(2n+k+1-\rho)L_n^k-(n+k)L_{n-1}^k \tag{5.79}$$

$$\rho(L_n^k)'=nL_n^k-(n+k)L_{n-1}^k \tag{5.80}$$

By tracing back through the various transformations which have been made, we find that the radial part of the hydrogen atom wave function is

$$F(r)=\frac{1}{a}\rho^l L_n^k(\rho)e^{-\rho/2} \tag{5.81}$$

In order to normalize the wave function, we must calculate

$$\int_0^\infty[F(r)]^2r^2\,dr=a\int_0^\infty\rho^{k+1}[L_n^k(\rho)]^2e^{-\rho}\,d\rho \tag{5.82}$$

where we have used $k+1=2l+2$. The evaluation of the integral (5.82) is somewhat involved. Before carrying it out, we first point out a complication. The parameter a, which has the dimensions of length, is not the same for all wave functions. As (5.27) reveals, it depends on the principal quantum number N and can be expressed as

$$a=Nb \qquad b=\frac{2\pi\varepsilon_0\hbar^2}{ZMe^2}=\frac{a_0}{2Z} \tag{5.83}$$

The characteristic length

$$a_0=\frac{4\pi\varepsilon_0\hbar^2}{Me^2}=0.529177\times10^{-8}\text{ cm} \tag{5.84}$$

is called the *Bohr radius*. This means that the orthogonality condition (5.22) comes out as

$$\int_0^\infty r^{k+1}L_n^k\left(\frac{r}{Nb}\right)L_{n'}^k\left(\frac{r}{N'b}\right)\times\exp\left[-\frac{r}{2b}\left(\frac{1}{N}+\frac{1}{N'}\right)\right]dr=0 \qquad N\neq N' \tag{5.85}$$

where $n=N-l-1$, $n'=N'-l-1$, and N and N' are two possible values of the principal quantum number. In the normalization integral (5.82), $N=N'$ and there is no such complication. We now proceed with its evaluation.

Problem 10.4 of Chapter 7 shows that any differential equation can be converted to Sturm–Liouville form. In particular, if we multiply the associated Laguerre differential equation (5.36) by the factor $\rho^k e^{-\rho}$, we find that it can be expressed as

$$\frac{d}{d\rho}(\rho^{k+1}e^{-\rho}L')+n\rho^k e^{-\rho}L=0 \tag{5.86}$$

Thus, the Laguerre polynomials are a complete set of orthogonal polynomials with respect to the weight function $\rho^k e^{-\rho}$ and obey*

$$\int_0^\infty L_n^k(\rho)L_{n'}^k(\rho)\rho^k e^{-\rho}\,d\rho=I_{nk}\delta_{nn'} \tag{5.87}$$

where I_{nk} is an appropriate normalizing factor. Note that (5.87) is *not* the same as (5.85). In the following, we will evaluate both (5.87) and (5.82). We begin by considering the integral

$$I_{nk}(\alpha)=\int_0^\infty[L_n^k(\alpha\rho)]^2\rho^k e^{-\alpha\rho}\,d\rho \tag{5.88}$$

which becomes the normalization integral of (5.87) if $\alpha=1$. To evaluate (5.88), square the generating func-

*Note that a different set is obtained for each value of k.

tion (5.75), multiply by $\rho^k e^{-\alpha\rho}$, and integrate to get

$$F(\alpha, t) = \int_0^\infty \frac{\rho^k}{(1-t)^{2k+2}} \exp\left[-\alpha\rho\left(\frac{1+t}{1-t}\right)\right] d\rho$$
$$= \sum_{n=0}^{\infty} \sum_{m=0}^{\infty} t^{n+m} \int_0^\infty L_n^k(\alpha\rho) L_m^k(\alpha\rho) \rho^k e^{-\alpha\rho} d\rho$$
$$= \sum_{n=0}^{\infty} \sum_{m=0}^{\infty} t^{n+m} \delta_{nm} I_{nk}(\alpha) = \sum_{n=0}^{\infty} t^{2n} I_{nk}(\alpha) \tag{5.89}$$

The integral is evaluated by making the change of variable

$$x = \alpha\rho\left(\frac{1+t}{1-t}\right) \qquad \rho = \frac{x}{\alpha}\left(\frac{1-t}{1+t}\right) \tag{5.90}$$

The result is

$$F(\alpha, t) = \frac{1}{\alpha^{k+1}(1-t^2)^{k+1}} \int_0^\infty e^{-x} x^k \, dx$$
$$= \frac{k!}{\alpha^{k+1}(1-t^2)^{k+1}}$$
$$= \frac{1}{\alpha^{k+1}} \sum_{n=0}^{\infty} \frac{(k+n)!}{n!} t^{2n} \tag{5.91}$$

The final form of (5.91) is obtained by a binomial expansion. The comparison of (5.89) and (5.91) shows that

$$I_{nk}(\alpha) = \int_0^\infty [L_n^k(\alpha\rho)]^2 \rho^k e^{-\alpha\rho} \, d\rho = \frac{1}{\alpha^{k+1}} \frac{(k+n)!}{n!} \tag{5.92}$$

If (5.92) is differentiated with respect to the parameter α, the result is

$$-\int_0^\infty [L_n^k(\alpha\rho)]^2 \rho^{k+1} e^{-\alpha\rho} \, d\rho$$
$$+ \int_0^\infty 2L_n^k(\alpha\rho)[L_n^k(\alpha\rho)]' \rho^{k+1} e^{-\alpha\rho} \, d\rho$$
$$= -\frac{(k+1)(k+n)!}{\alpha^{k+2} n!} \tag{5.93}$$

where

$$[L_n^k(\alpha\rho)]' = \frac{d}{d(\alpha\rho)} L_n^k(\alpha\rho) \tag{5.94}$$

In terms of the variable $\alpha\rho$, the recurrence relation (5.80) is

$$\alpha\rho[L_n^k(\alpha\rho)]' = nL_n^k(\alpha\rho) - (n+k)L_{n-1}^k(\alpha\rho) \tag{5.95}$$

When it is used in (5.93), the term involving $L_{n-1}^k(\alpha\rho)$ gives zero on account of the orthogonality condition (5.87). The remaining terms can be regrouped to give

$$\int_0^\infty [L_n^k(\alpha\rho)]^2 \rho^{k+1} e^{-\alpha\rho} \, d\rho$$
$$= \frac{(k+1)(k+n)!}{\alpha^{k+2} n!} + \frac{2n}{\alpha} \int_0^\infty [L_n^k(\alpha\rho)]^2 \rho^k e^{-\alpha\rho} d\rho$$
$$= \frac{(k+n)!(k+2n+1)}{\alpha^{k+2} n!} \tag{5.96}$$

The last step follows because of (5.92). Setting $\alpha = 1$ gives the required integral in (5.82).

By using the normalization integral (5.96), we can express the normalized hydrogen atom eigenfunctions as

$$u_{Nlm}(r, \theta, \phi) = \frac{1}{a^{3/2}} \sqrt{\frac{(N-l-1)!}{2N(N+l)!}} \left(\frac{r}{a}\right)^l$$
$$\times L_n^k\left(\frac{r}{a}\right) e^{-r/(2a)} Y_{lm}(\theta, \phi)$$
$$a = \frac{Na_0}{2Z} \qquad k = 2l+1 \qquad n = N-l-1 \tag{5.97}$$

Laguerre Polynomials

$L_0 = 1$
$L_1 = -\rho + 1$
$2!L_2 = \rho^2 - 4\rho + 2$
$3!L_3 = -\rho^3 + 9\rho^2 - 18\rho + 6$
$4!L_4 = \rho^4 - 16\rho^3 + 72\rho^2 - 96\rho + 24$

Associated Laguerre Polynomials

$L_0^1 = 1$
$L_1^1 = -\rho + 2$
$L_2^1 = \frac{1}{2}\rho^2 - 3\rho + 3$
$L_3^1 = -\frac{1}{6}\rho^3 + 2\rho^2 - 6\rho + 4$
$L_0^2 = 1$
$L_1^2 = -\rho + 3$
$L_2^2 = \frac{1}{2}\rho^2 - 4\rho + 6$
$L_0^3 = 1$
$L_1^3 = -\rho + 4$
$L_0^4 = 1$

Hydrogen Atom Wave Functions

$N=1 \qquad a=\dfrac{a_0}{2Z} \qquad \rho=\dfrac{2Z}{a_0}r$

$$u_{100}=\frac{1}{a^{3/2}\sqrt{8\pi}}e^{-\rho/2}$$

$N=2 \qquad a=\dfrac{a_0}{Z} \qquad \rho=\dfrac{Z}{a_0}r$

$$u_{200}=\frac{1}{a^{3/2}\sqrt{32\pi}}(2-\rho)e^{-\rho/2}$$

$$u_{211}=-\frac{1}{a^{3/2}8\sqrt{\pi}}\rho e^{-\rho/2}\sin\theta e^{i\phi}$$

$$u_{210}=\frac{1}{a^{3/2}\sqrt{32\pi}}\rho e^{-\rho/2}\cos\theta$$

$$u_{21,-1}=\frac{1}{a^{3/2}8\sqrt{\pi}}\rho e^{-\rho/2}\sin\theta e^{-i\phi}$$

PROBLEMS

5.1. Consider the function $v(\rho)$ defined by equation (5.23). Assume that near the origin its behavior is given by ρ^p. Show that the Sturm–Liouville boundary condition (5.20) is satisfied if $p>\frac{1}{2}$.

5.2. In our approximate treatment of the hydrogen atom, the energy levels depend only on the principal quantum number N. There is degeneracy because there are different possible angular momentum states for a given N. Show that the degree of the degeneracy is N^2, that is, for a given N there are actually N^2 linearly independent wave functions. If we take into account the two spin states of the electron, the total number of states for a given N is actually $2N^2=2,8,18,32,\ldots$.

5.3. Derive the generating function (5.75).

5.4. Start with (5.77) and carry out the derivation of (5.78).

5.5. Given two functions $f(\rho)$ and $g(\rho)$, show that the rule for differentiating their product k times is

$$d^k(fg)=(d^kf)g+k(d^{k-1}f)(dg)+\frac{k(k-1)}{2!}(d^{k-2}f)(d^2g)+\cdots$$

$$=\sum_{m=0}^{k}\frac{k!}{(k-m)!m!}(d^{k-m}f)(d^mg) \tag{5.98}$$

where $d=d/d\rho$. Note the resemblance to a binomial expansion.

5.6. Prove the recurrence relations (5.79) and (5.80) by differentiating (5.64) and (5.67) k times.

5.7. Work out the binomial expansion of the factor $(1-t^2)^{-k-1}$ that was used in obtaining (5.91).

5.8. If a hydrogen atom is in the ground state u_{100}, find the probability density function $P(r)$ which has the property that $P(r)\,dr$ is the probability that the electron is found between r and $r+dr$. Find the expectation value of r and also the most probable value of r. Express the answers in terms of the Bohr radius a_0.

5.9. Show that the formula for finding the expectation value of r in any hydrogen atom state is

$$\langle r\rangle=\frac{n!a}{2N(N+l)!}\int_0^\infty \rho^{k+2}\left[L_n^k(\rho)\right]^2e^{-\rho}\,d\rho \tag{5.99}$$

5.10. Prove the differentiation formula

$$\frac{d^m}{d\rho^m}(\rho^n)=\frac{n!}{(n-m)!}\rho^{n-m} \tag{5.100}$$

5.11. Evaluate the normalization integral (5.88) again, using the following steps. Replace one of the factors of L_n^k with the Rodrigues formula (5.78). Now integrate by parts n times to get

$$\int_0^\infty \rho^{k+1}\left[L_n^k(\rho)\right]^2 e^{-\rho}\,d\rho = \frac{(-1)^n}{n!}\int_0^\infty \left[\frac{d^n}{d\rho^n}(\rho L_n^k)\right] e^{-\rho}\rho^{n+k}\,d\rho \tag{5.101}$$

Since L_n^k is a polynomial of degree n, not too many terms of ρL_n^k survive n differentiations. Use the series representation (5.73) to show that

$$\frac{d^n}{d\rho^n}(\rho L_n^k) = (-1)^n[(n+1)\rho - n(n+k)] \tag{5.102}$$

Now complete the evaluation of the integral. If you feel up to it, you can evaluate the integral in problem 9 by the same technique.

5.12. Show that the probability current vector for any eigenstate of the hydrogen atom is

$$\mathbf{S}_{Nlm} = \frac{\hbar m}{Mr\sin\theta}|u_{Nlm}|^2\hat{\boldsymbol{\phi}} \tag{5.103}$$

See equation (6.27) of Chapter 7. Don't confuse the mass of the particle with the quantum number m. *Suggestion*: Represent the gradient operator in spherical coordinates as given by equation (2.8) in Chapter 4.

Since $|u|^2$ is a probability density function for the electron, $-e|u|^2$ can be interpreted as charge density. The equation of continuity for probability can be expressed as

$$\nabla\cdot(e\mathbf{S}) + \frac{\partial}{\partial t}(e|u|^2) = 0 \tag{5.104}$$

Thus, $\mathbf{j} = -e\mathbf{S}$ is the electrical current density. Use equation (11.9) of Chapter 4 to show that the magnetic moment associated with any eigenstate of the hydrogen atom is

$$\boldsymbol{\mu} = -\frac{\hbar m e}{2M}\hat{\mathbf{k}} \tag{5.105}$$

Caution: Before you do the integration, you will have to express the unit vectors of spherical coordinates in terms of the unit vectors of rectangular coordinates. See problem 1.4 in Chapter 4.

5.13. Show that for any eigenstate of the hydrogen atom

$$\langle V\rangle = 2W_N \qquad \langle K\rangle = -W_N \tag{5.106}$$

where $\langle V\rangle$ and $\langle K\rangle$ are the expectation values of the potential and the kinetic energies. These same relations between the average potential and kinetic energies also hold classically. In classical mechanics, this result is known as the *virial theorem*.

5.14. The total angular momentum of a two-particle system can be written as $\mathbf{L}_T = \mathbf{L}_1 + \mathbf{L}_2$. Show that, just as in classical mechanics, the total angular momentum separates into two parts as $\mathbf{L}_T = \mathbf{L}_c + \mathbf{L}$, where $\mathbf{L}_c$ represents the angular momentum due to the motion of the center of mass and $\mathbf{L}$ is the contribution from the relative motion of the two particles. Do this by showing that

$$\left(y_1\frac{\partial\psi}{\partial z_1} - z_1\frac{\partial\psi}{\partial y_1}\right) + \left(y_2\frac{\partial\psi}{\partial z_2} - z_2\frac{\partial\psi}{\partial y_2}\right) = \left(y\frac{\partial\psi}{\partial z} - z\frac{\partial\psi}{\partial y}\right) + \left(y_c\frac{\partial\psi}{\partial z_c} - z_c\frac{\partial\psi}{\partial y_c}\right) \tag{5.107}$$

6. SPIN AND QUATERNIONS

Many of the elementary particles have an intrinsic angular momentum called spin. The term no doubt originates from attempts to construct mechanical models in which the particle is pictured as spinning on an axis. Frequently, a magnetic moment is associated with the spin. Consider a mechanical model of the electron in which it is visualized as a little sphere of spinning charge. Classically, its angular momentum and magnetic moment can be expressed as

$$\mathbf{P} = \int_\Sigma \mathbf{r} \times \rho_m \mathbf{u}\, d\Sigma \tag{6.1}$$

$$\boldsymbol{\mu} = \frac{1}{2}\int_\Sigma \mathbf{r} \times \rho_e \mathbf{u}\, d\Sigma \tag{6.2}$$

where ρ_m and ρ_e are mass and charge densities, respectively. Refer to section 11 in Chapter 4 for a discussion of magnetic moment. Angular momentum is defined by equation (9.4) in Chapter 2. Suppose that the charge and mass are both distributed in the same way throughout the volume of the electron so that the mass and charge densities can be expressed as

$$\rho_m = Mf(\mathbf{r}) \qquad \rho_e = -ef(\mathbf{r}) \tag{6.3}$$

This means that the ratio of charge density to mass density has the constant value e/M throughout the volume of the electron. The magnetic moment and angular momentum are then related by

$$\boldsymbol{\mu} = -\frac{e\mathbf{P}}{2M} \tag{6.4}$$

Reference to equation (5.105) shows that this agrees exactly with the expression for the magnetic moment due to the orbital motion of an electron in a hydrogen atom. It is however too small by a factor of 2 for the magnetic moment due to the intrinsic spin of the electron. Other models can be constructed. For example, if we assume that the mass is uniformly distributed throughout the volume of a sphere and that the charge is uniformly distributed over the surface, the result is

$$\boldsymbol{\mu} = -\frac{5e\mathbf{P}}{6M} \tag{6.5}$$

Refer to problem 11.1 in Chapter 4 for the details of this type of calculation. The pursuit of such mechanical models is however unproductive, and one must ultimately rely on the empirically determined ratio between magnetic moment and angular momentum, which is

$$\boldsymbol{\mu} = \gamma\mathbf{P} \qquad \gamma = -1.00116\frac{e}{M} \tag{6.6}$$

The quantity γ is called the *gyromagnetic ratio.* Its value is predicted to a very high accuracy by relativistic quantum mechanics.

Let us pursue the classical model a bit further and imagine that a rotating negatively charged object is placed in a uniform external magnetic field. According to problem 11.4 in Chapter 3, such a rotating object experiences a torque given by $\boldsymbol{\mu} \times \mathbf{B}$. From equation (9.11) in Chapter 2, we see that its equation of motion is

$$-\frac{e}{2M}\mathbf{P} \times \mathbf{B} = \frac{d\mathbf{P}}{dt} \tag{6.7}$$

Remember that we are now using $\mathbf{P}$ for angular momentum rather than ℓ. To solve (6.7), note that

$$\frac{d}{dt}P^2 = \frac{d}{dt}(\mathbf{P}\cdot\mathbf{P}) = 2\mathbf{P}\cdot\frac{d\mathbf{P}}{dt} = -\frac{e}{M}\mathbf{P}\cdot\mathbf{P}\times\mathbf{B} = 0 \tag{6.8}$$

$$\frac{d}{dt}(\mathbf{P}\cdot\mathbf{B}) = \mathbf{B}\cdot\frac{d\mathbf{P}}{dt} = -\frac{e}{2M}\mathbf{B}\cdot\mathbf{P}\times\mathbf{B} = 0 \tag{6.9}$$

This means that the magnitude of $\mathbf{P}$ and its component in the direction of $\mathbf{B}$ are constants. Let $\mathbf{B}$ define the z axis. Then θ and ϕ of spherical coordinates give the orientation of $\mathbf{P}$. The angle θ is a constant, and $\mathbf{P}$ can move only in the ϕ direction. If we write

$$\mathbf{P} = \hat{\mathbf{r}}P = P(\sin\theta\cos\phi\,\hat{\mathbf{i}} + \sin\theta\sin\phi\,\hat{\mathbf{j}} + \cos\theta\,\hat{\mathbf{k}}) \tag{6.10}$$

then

$$\frac{d\mathbf{P}}{dt} = P(-\sin\phi\,\hat{\mathbf{i}} + \cos\phi\,\hat{\mathbf{j}})\dot{\phi}\sin\theta = P\dot{\phi}\sin\theta\,\hat{\boldsymbol{\phi}} \tag{6.11}$$

The equation of motion (6.7) then gives

$$\frac{e}{2M}PB\sin\theta\,\hat{\boldsymbol{\phi}} = P\dot{\phi}\sin\theta\,\hat{\boldsymbol{\phi}} \qquad \dot{\phi} = \frac{eB}{2M} \tag{6.12}$$

The quantity $eB/2M$ is called the *Larmor frequency.*

Classically, the angular momentum vector precesses at the Larmor frequency about the magnetic field vector. It moves on the surface of a cone at the angle θ with respect to the magnetic field. Something like this also happens quantum mechanically except that the only orientations of **P** with respect to **B** that are possible are given by $P_z = m\hbar$.

Examination of the spectral lines of hydrogen and the hydrogen-like spectra of the alkali metals reveals that states of nonzero angular momentum are split into doublets. This can be explained if the electron has two possible spin states corresponding to $l = \frac{1}{2}$, $m = \pm\frac{1}{2}$. The splitting occurs because the orbital motion of the electron creates a magnetic field. The two possible orientations of the spin with respect to the magnetic field have slightly different energies. The magnitude of the observed splitting is one way to determine the gyromagnetic ratio of the electron. In problem 4.4, we gave the coup de grace to half-integer values of l for orbital angular momentum. Empirical evidence demands that we now revive half-integer values of l to describe the spin of the elementary particles. The $l = \frac{1}{2}$ case will be considered in detail.

In all our previous considerations of quantum mechanical problems, there was always the choice of working with either a functional or a matrix representation. For spin, there is no functional representation, and we must either work with spin operators and eigenvectors considered as abstract entities in a vector space or choose a matrix representation. Following conventional notation, we will use S_x, S_y, and S_z to represent the dimensionless spin angular momentum operators, reserving L_x, L_y, and L_z for orbital angular momentum. The algebra of angular momentum is discussed in section 3. In particular, the algebraic properties of spin are contained in the relations

$$[S_y, S_z] = iS_x \qquad [S_z, S_x] = iS_y \qquad [S_x, S_y] = iS_z \tag{6.13}$$

$$[S^2, S_x] = [S^2, S_y] = [S^2, S_z] = 0 \tag{6.14}$$

Any set of operators which obey (6.13) and (6.14) and which also have a vector addition property allowing them to be combined with other similar operators are qualified to be representations of angular momentum. The spin operators exist in a two-dimensional unitary space. If the eigenvectors of S_z and S^2 are denoted by χ_+ and χ_-, then

$$S^2\chi_+ = \tfrac{1}{2}(\tfrac{1}{2}+1)\chi_+ = \tfrac{3}{4}\chi_+ \qquad S^2\chi_- = \tfrac{3}{4}\chi_- \tag{6.15}$$

$$S_z\chi_+ = \tfrac{1}{2}\chi_+ \qquad S_z\chi_- = -\tfrac{1}{2}\chi_- \tag{6.16}$$

The eigenvectors are called *spinors*, and it is usual to think of χ_+ as representing "spin up" and χ_- as representing "spin down." An arbitrary spin state of the electron, or another spin $\frac{1}{2}$ particle, can be represented by the linear combination

$$\chi = \alpha\chi_+ + \beta\chi_- \tag{6.17}$$

If a measurement is made on the system, then $|\alpha|^2$ is the probability that the spin of the particle is found pointing in the positive z direction and $|\beta|^2$ is the probability that it is found pointing in the negative z direction. The algebraic properties of the spin operators are most conveniently determined by using the raising and lowering operators

$$A^\dagger = S_x + iS_y \qquad A = S_x - iS_y \tag{6.18}$$

Reference to equations (3.35) through (3.38) shows that

$$\begin{aligned} A^\dagger\chi_+ &= 0 \qquad & A^\dagger\chi_- &= \chi_+ \\ A\chi_+ &= \chi_- \qquad & A\chi_- &= 0 \end{aligned} \tag{6.19}$$

If (6.18) is solved for S_x and S_y, the result is

$$S_x = \frac{1}{2}(A^\dagger + A) \qquad S_y = \frac{1}{2i}(A^\dagger - A) \tag{6.20}$$

It is now possible to determine the effect of the spin operators on any spinor:

$$\begin{aligned} S_x\chi_+ &= \frac{1}{2}\chi_- \qquad & S_x\chi_- &= \frac{1}{2}\chi_+ \\ S_y\chi_+ &= \frac{i}{2}\chi_- \qquad & S_y\chi_- &= -\frac{i}{2}\chi_+ \end{aligned} \tag{6.21}$$

The matrix elements of A are

$$\begin{aligned} A_{11} &= \chi_+^\dagger A\chi_+ = 0 \qquad & A_{12} &= \chi_+^\dagger A\chi_- = 0 \\ A_{21} &= \chi_-^\dagger A\chi_+ = 1 \qquad & A_{22} &= \chi_-^\dagger A\chi_- = 0 \end{aligned} \tag{6.22}$$

Thus,

$$A = \begin{pmatrix} 0 & 0 \\ 1 & 0 \end{pmatrix} \qquad A^\dagger = \begin{pmatrix} 0 & 1 \\ 0 & 0 \end{pmatrix} \tag{6.23}$$

It is usual to define *Pauli spin matrices* by $S_x = \frac{1}{2}\sigma_x$, $S_y = \frac{1}{2}\sigma_y$, and $S_z = \frac{1}{2}\sigma_z$. By means of (6.20) and the fact that S_z is diagonal, we find

$$\sigma_x = \begin{pmatrix} 0 & 1 \\ 1 & 0 \end{pmatrix} \quad \sigma_y = \begin{pmatrix} 0 & -i \\ i & 0 \end{pmatrix} \quad \sigma_z = \begin{pmatrix} 1 & 0 \\ 0 & -1 \end{pmatrix} \tag{6.24}$$

One of the conveniences of the Pauli matrices over the S matrices is that all the factors of $\frac{1}{2}$ are eliminated. As a direct consequence of the commutation relations (6.13),

$$[\sigma_y, \sigma_z] = 2i\sigma_x \quad [\sigma_z, \sigma_x] = 2i\sigma_y \quad [\sigma_x, \sigma_y] = 2i\sigma_z \tag{6.25}$$

The effects of multiplying the eigenspinors by the Pauli matrices are summarized as

$$\begin{aligned} \sigma_x\chi_+ &= \chi_- & \sigma_x\chi_- &= \chi_+ \\ \sigma_y\chi_+ &= i\chi_- & \sigma_y\chi_- &= -i\chi_+ \\ \sigma_z\chi_+ &= \chi_+ & \sigma_z\chi_- &= -\chi_- \end{aligned} \tag{6.26}$$

The following properties of the Pauli matrices can be established either by means of (6.26) or by direct multiplication of the matrices:

$$\begin{aligned} \sigma_x\sigma_y &= -\sigma_y\sigma_x = i\sigma_z \\ \sigma_y\sigma_z &= -\sigma_z\sigma_y = i\sigma_x \\ \sigma_z\sigma_x &= -\sigma_x\sigma_z = i\sigma_y \end{aligned} \tag{6.27}$$

Because of (6.27), the Pauli matrices are said to *anticommute*. A further property is

$$\sigma_x^2 = \sigma_y^2 = \sigma_z^2 = \begin{pmatrix} 1 & 0 \\ 0 & 1 \end{pmatrix} = I \tag{6.28}$$

The eigenspinors are given by

$$\chi_+ = \begin{pmatrix} 1 \\ 0 \end{pmatrix} \quad \chi_- = \begin{pmatrix} 0 \\ 1 \end{pmatrix} \tag{6.29}$$

The Pauli matrices have an interesting and significant mathematical property. If an isomorphic relation between the symbols 1, k_1, k_2, and k_3 and the Pauli matrices is established by

$$1 \leftrightarrow I \quad k_1 \leftrightarrow i\sigma_x \quad k_2 \leftrightarrow i\sigma_y \quad k_3 \leftrightarrow -i\sigma_z \tag{6.30}$$

then the algebraic properties of 1, k_1, k_2, and k_3 are

$$\begin{aligned} k_1^2 = k_2^2 = k_3^2 &= -1 & k_1k_2 &= -k_2k_1 = k_3 \\ k_2k_3 = -k_3k_2 &= k_1 & k_3k_1 &= -k_1k_3 = k_2 \end{aligned} \tag{6.31}$$

One set of mathematical elements is said to be *isomorphic* to another set if there is a one-to-one correspondence between the elements of one set and those of the other and if the algebraic properties of both sets are identical. Let q, x, y, and z be real numbers. Then,

$$Q = q + xk_1 + yk_2 + zk_3 \tag{6.32}$$

is called a *quaternion*. The significance of quaternions is that they form a number field, sometimes called a *hypercomplex number field*. It was shown in problem 3.8 of Chapter 5 that a number field based on three unit elements does not exist. We find here that an algebraic system based on the four quantities 1, k_1, k_2, and k_3 does in fact form a noncommutative number field. The unit element "1" can of course be thought of as the ordinary real number 1.

PROBLEMS

6.1. Establish the anticommutation properties (6.27) without reference to any matrix representation.

6.2. Establish the following isomorphic relations:

$$1 \leftrightarrow \begin{pmatrix} 1 & 0 \\ 0 & 1 \end{pmatrix} \leftrightarrow \begin{pmatrix} 1 & 0 & 0 & 0 \\ 0 & 1 & 0 & 0 \\ 0 & 0 & 1 & 0 \\ 0 & 0 & 0 & 1 \end{pmatrix}$$

$$k_1 \leftrightarrow \begin{pmatrix} 0 & i \\ i & 0 \end{pmatrix} \leftrightarrow \begin{pmatrix} 0 & 0 & 0 & 1 \\ 0 & 0 & -1 & 0 \\ 0 & 1 & 0 & 0 \\ -1 & 0 & 0 & 0 \end{pmatrix}$$

$$k_2 \leftrightarrow \begin{pmatrix} 0 & 1 \\ -1 & 0 \end{pmatrix} \leftrightarrow \begin{pmatrix} 0 & 0 & 1 & 0 \\ 0 & 0 & 0 & 1 \\ -1 & 0 & 0 & 0 \\ 0 & -1 & 0 & 0 \end{pmatrix} \tag{6.33}$$

$$k_3 \leftrightarrow \begin{pmatrix} -i & 0 \\ 0 & i \end{pmatrix} \leftrightarrow \begin{pmatrix} 0 & -1 & 0 & 0 \\ 1 & 0 & 0 & 0 \\ 0 & 0 & 0 & 1 \\ 0 & 0 & -1 & 0 \end{pmatrix}$$

Recall the isomorphic relation between 1 and $i = \sqrt{-1}$ and the matrices

$$\begin{pmatrix} 1 & 0 \\ 0 & 1 \end{pmatrix} \quad \begin{pmatrix} 0 & 1 \\ -1 & 0 \end{pmatrix} \tag{6.34}$$

Equation (6.33) is a generalization of this idea to the quaternion number field.

6.3. Show that the product of the two quaternions

$$Q_1 = x_0 + x_1 k_1 + x_2 k_2 + x_3 k_3$$
$$Q_2 = y_0 + y_1 k_1 + y_2 k_2 + y_3 k_3 \tag{6.35}$$

is

$$\begin{aligned} Q_1 Q_2 = x_0 y_0 - x_1 y_1 - x_2 y_2 - x_3 y_3 + x_0 (y_1 k_1 + y_2 k_2 + y_3 k_3) \\ + y_0 (x_1 k_1 + x_2 k_2 + x_3 k_3) + (x_2 y_3 - x_3 y_2) k_1 \\ + (x_3 y_1 - x_1 y_3) k_2 + (x_1 y_2 - x_2 y_1) k_3 \end{aligned} \tag{6.36}$$

Note that parts of the expression are like dot and cross products. If we define vector-like quantities by

$$\underset{\sim}{a} = x_1 k_1 + x_2 k_2 + x_3 k_3 \qquad \underset{\sim}{b} = y_1 k_1 + y_2 k_2 + y_3 k_3 \tag{6.37}$$

then it is possible to write formally

$$Q_1 Q_2 = x_0 y_0 - \underset{\sim}{a} \cdot \underset{\sim}{b} + x_0 \underset{\sim}{b} + y_0 \underset{\sim}{a} + \underset{\sim}{a} \times \underset{\sim}{b} \tag{6.38}$$

The quaternions themselves can be expressed as

$$Q_1 = x_0 + \underset{\sim}{a} \qquad Q_2 = y_0 + \underset{\sim}{b} \tag{6.39}$$

While $\underset{\sim}{a}$ and $\underset{\sim}{b}$ retain many of the characteristics of ordinary vectors, there are important differences. For example, the product **ab** makes no sense for ordinary vectors, but for quaternion vectors, it is

$$\underset{\sim}{a}\underset{\sim}{b} = -\underset{\sim}{a} \cdot \underset{\sim}{b} + \underset{\sim}{a} \times \underset{\sim}{b} \tag{6.40}$$

6.4. The appropriate generalization of the idea of complex conjugate to quaternions is

$$Q^c = q - xk_1 - yk_2 - zk_3 \tag{6.41}$$

which is called the *quaternion conjugate* of Q. Show that

$$QQ^c = Q^c Q = q^2 + x^2 + y^2 + z^2 \tag{6.42}$$

We define the *magnitude* or *modulus* of a quaternion by

$$|Q| = \sqrt{QQ^c} \tag{6.43}$$

This leads to the important result that any nonzero quaternion has the inverse

$$Q^{-1} = \frac{Q^c}{|Q|^2} \tag{6.44}$$

The addition of two quaternions is defined by

$$Q_1 + Q_2 = (x_0 + y_0) + (x_1 + y_1)k_1 + (x_2 + y_2)k_2 + (x_3 + y_3)k_3 \tag{6.45}$$

Refer to section 2 of Chapter 5 and verify that quaternions satisfy all the number field axioms. Note that the quaternion vectors (6.37) add like ordinary vectors, which is at least one justification for calling them vectors in the first place.

6.5. Show that

$$(Q_1Q_2)^c = Q_2^cQ_1^c \tag{6.46}$$

6.6. Show that the eight elements 1, k_1, k_2, k_3, -1, $-k_1$, $-k_2$, and $-k_3$ form a group of order 8 under multiplication.

6.7. Let the dimensionless angular momentum operators of a spin $\frac{3}{2}$ particle be denoted by J_x, J_y, and J_z. The matrix representation is four-dimensional. The four eigenvectors of J_z obey

$$J_z\chi_1 = \tfrac{3}{2}\chi_1 \qquad J_z\chi_2 = \tfrac{1}{2}\chi_2 \qquad J_z\chi_3 = -\tfrac{1}{2}\chi_3 \qquad J_z\chi_4 = -\tfrac{3}{2}\chi_4 \tag{6.47}$$

If χ is any one of the four eigenvectors, then

$$J^2\chi = \tfrac{3}{2}(\tfrac{3}{2}+1)\chi = \tfrac{15}{4}\chi \tag{6.48}$$

Show that

$$J_x = \frac{1}{2}\begin{pmatrix} 0 & \sqrt{3} & 0 & 0 \\ \sqrt{3} & 0 & 2 & 0 \\ 0 & 2 & 0 & \sqrt{3} \\ 0 & 0 & \sqrt{3} & 0 \end{pmatrix} \qquad J_y = \frac{1}{2i}\begin{pmatrix} 0 & \sqrt{3} & 0 & 0 \\ -\sqrt{3} & 0 & 2 & 0 \\ 0 & -2 & 0 & \sqrt{3} \\ 0 & 0 & -\sqrt{3} & 0 \end{pmatrix}$$

$$J_z = \frac{1}{2}\begin{pmatrix} 3 & 0 & 0 & 0 \\ 0 & 1 & 0 & 0 \\ 0 & 0 & -1 & 0 \\ 0 & 0 & 0 & -3 \end{pmatrix} \tag{6.49}$$

7. ANGULAR MOMENTUM AND ROTATION

If a complex number is multiplied by the factor $e^{i\theta}$, the result is a rotation of the complex number by an angle θ:

$$\begin{aligned} z' &= e^{i\theta}z = e^{i\theta}re^{i\alpha} \\ &= re^{i(\theta+\alpha)} \end{aligned} \tag{7.1}$$

This is illustrated in Fig. 7.1. Note that the same result could be achieved by rotating the coordinates in the opposite direction.

A similar transformation can be done with quaternions. A rotation of the quaternion Q by an angle ϕ about the axis defined by k_3 is accomplished by

$$Q' = e^{\phi k_3/2}Qe^{-\phi k_3/2} \tag{7.2}$$

as we now show. First note that

$$e^{\phi k_3/2} = \cos\frac{\phi}{2} + k_3 \sin\frac{\phi}{2} \tag{7.3}$$

The truth of (7.3) is established by making a Taylor's series expansion of the exponential and using $k_3^2 =$

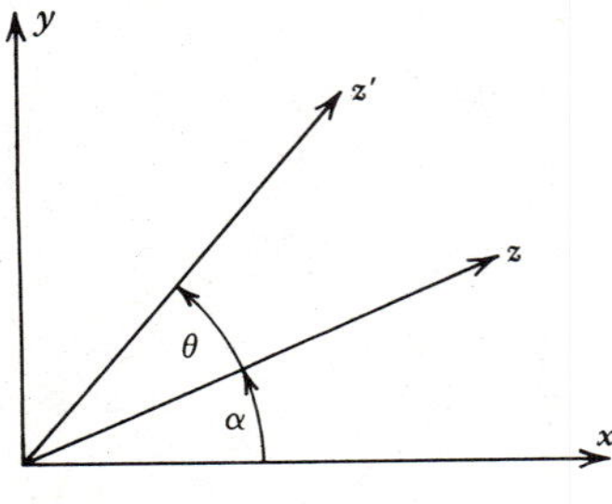

Fig. 7.1

-1. In this respect, it is the same as $e^{i\phi} = \cos\phi + i\sin\phi$ for ordinary complex numbers. The terms in Q are to be transformed one-by-one. By using (6.31), we find

$$
\begin{aligned}
e^{\phi k_3/2}k_1e^{-\phi k_3/2}
&= \left(\cos\frac{\phi}{2} + k_3\sin\frac{\phi}{2}\right)k_1\left(\cos\frac{\phi}{2} - k_3\sin\frac{\phi}{2}\right)\\
&= \left(\cos\frac{\phi}{2} + k_3\sin\frac{\phi}{2}\right)\left(k_1\cos\frac{\phi}{2} + k_2\sin\frac{\phi}{2}\right)\\
&= \left(\cos^2\frac{\phi}{2} - \sin^2\frac{\phi}{2}\right)k_1 + \left(2\cos\frac{\phi}{2}\sin\frac{\phi}{2}\right)k_2\\
&= k_1\cos\phi + k_2\sin\phi
\end{aligned}
\tag{7.4}
$$

Similarly,

$$e^{\phi k_3/2}k_2e^{-\phi k_3/2} = -k_1\sin\phi + k_2\cos\phi \tag{7.5}$$

$$e^{\phi k_3/2}k_3e^{-\phi k_3/2} = k_3 \tag{7.6}$$

If $Q = q + xk_1 + yk_2 + zk_3$, its transformation is

$$
\begin{aligned}
Q' &= q + x(k_1\cos\phi + k_2\sin\phi)\\
&\quad + y(-k_1\sin\phi + k_2\cos\phi) + zk_3\\
&= q + (x\cos\phi - y\sin\phi)k_1\\
&\quad + (x\sin\phi + y\cos\phi)k_2 + zk_3\\
&= q + x'k_1 + y'k_2 + z'k_3
\end{aligned}
\tag{7.7}
$$

Therefore,

$$
\begin{aligned}
x' &= x\cos\phi - y\sin\phi\\
y' &= x\sin\phi + y\cos\phi \qquad z' = z
\end{aligned}
\tag{7.8}
$$

The vector part of the quaternion has components which in fact transform like a vector under the rotation given by (7.2). Just as in the case of a complex number, a positive rotation of the quaternion corresponds to a negative rotation of the coordinate axes as pictured in Fig. 7.2. Note from (7.4) and (7.5) that the unit elements k_1 and k_2 also transform vectorially except that they are rotated in the opposite sense to that shown in Fig. 7.2. The transformation (7.8) gives more credibility to labeling the quantities defined by (6.37) as vectors. A succession of transformations can be built up. For example, the transformation corresponding to the three Euler angles pictured in Fig. 15.1 of Chapter 4 is

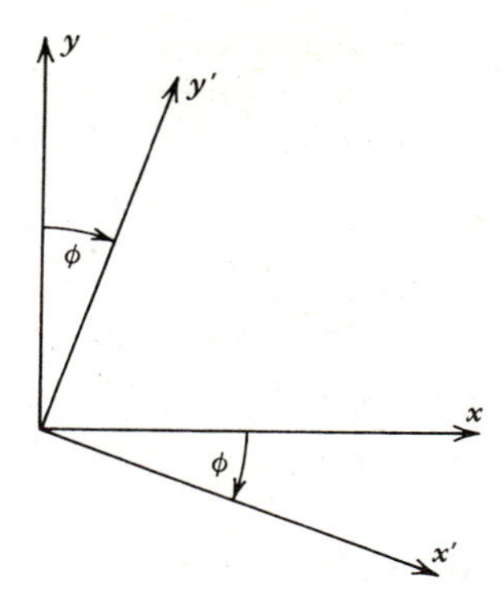

Fig. 7.2

$$Q' = e^{-\psi k_3/2}e^{-\theta k_2/2}e^{-\phi k_3/2}Qe^{\phi k_3/2}e^{\theta k_2/2}e^{\psi k_3/2} \tag{7.9}$$

The signs of the angles in (7.9) are opposite to those in (7.2) in order to conform to the fact that the Euler angles have been defined as positive rotations of coordinate axes. It should be noted that

$$e^{\theta k_2/2}e^{\psi k_3/2} \neq e^{(\theta k_2/2)+(\psi k_3/2)} \tag{7.10}$$

because k_2 and k_3 do not commute. See problem 15.9 in Chapter 7.

Consider the quaternion defined by

$$V = \cos\frac{\alpha}{2} + \underline{n}\sin\frac{\alpha}{2} \tag{7.11}$$

$$\underline{n} = n_xk_1 + n_yk_2 + n_zk_3 \tag{7.12}$$

The quaternion vector $\underline{n}$ is defined to have unit length:

$$\underline{n}\cdot\underline{n} = \underline{n}\underline{n}^c = n_x^2 + n_y^2 + n_z^2 = 1 \tag{7.13}$$

It also has the property that

$$\underline{n}\underline{n} = \underline{n}^2 = -\underline{n}\cdot\underline{n} + \underline{n}\times\underline{n} = -1 \tag{7.14}$$

This permits us to write

$$V = e^{\alpha\underline{n}/2} \tag{7.15}$$

The transformation

$$Q' = e^{\alpha\underline{n}/2}Qe^{-\alpha\underline{n}/2} \tag{7.16}$$

rotates the quaternion Q through an angle α about the direction defined by $\underline{n}$.

Equations (7.4) and (7.5) show that the unit elements themselves have a vector transformation property. Accordingly, one sometimes writes

$$\mathbf{k} = k_1\hat{\mathbf{e}}_1 + k_2\hat{\mathbf{e}}_2 + k_3\hat{\mathbf{e}}_3 \tag{7.17}$$

$$\mathbf{a} = x\hat{\mathbf{e}}_1 + y\hat{\mathbf{e}}_2 + z\hat{\mathbf{e}}_3 \tag{7.18}$$

where $\hat{\mathbf{e}}_1$, $\hat{\mathbf{e}}_2$, and $\hat{\mathbf{e}}_3$ are the ordinary unit vectors of Cartesian coordinates. The quaternion vector $\underline{a}$ is then

$$\underline{a} = \mathbf{k}\cdot\mathbf{a} \tag{7.19}$$

If $\mathbf{a}$ and $\mathbf{b}$ are any two ordinary vectors, then

$$(\mathbf{k}\cdot\mathbf{a})(\mathbf{k}\cdot\mathbf{b}) = -\mathbf{a}\cdot\mathbf{b} + \mathbf{k}\cdot(\mathbf{a}\times\mathbf{b}) \tag{7.20}$$

is the equivalent of (6.40). Note that there is no distinction between $\underline{a}\cdot\underline{b}$ and $\mathbf{a}\cdot\mathbf{b}$.

We return now to the discussion of angular momentum in quantum mechanics. Consider the inverse of the transformation given by (7.8) and pictured in Fig. 7.2. If the angle ϕ is very small, we have

$$x = x' + y'\,d\phi \qquad y = -x'\,d\phi + y' \qquad z = z' \tag{7.21}$$

This is an example of an *infinitesimal transformation*. At a fixed point in space, the value of an arbitrary function can be computed in terms of the new coordinates by means of

$$f(x,y,z) = f(x',y',z') + \frac{\partial f}{\partial x'}(x - x') + \frac{\partial f}{\partial y'}(y - y') \tag{7.22}$$

By utilizing (7.21) and (3.5), we find

$$\begin{aligned} f(x,y,z) &= f(x',y',z') + i\,d\phi\left[i\left(x'\frac{\partial}{\partial y'} - y'\frac{\partial}{\partial x'}\right)\right]f(x',y',z') \\ &= [1 - i\,d\phi L_z]\,f(x',y',z') \end{aligned} \tag{7.23}$$

For this reason, the angular momentum operators are said to be *generators* of rotations. The proper transformations of ordinary three-dimensional space are either translations or rotations. Linear momentum generates translations and angular momentum generates rotations. This is why linear and angular momentum are such fundamental quantities. A finite transformation can be built up by multiplying together many infinitesimal transformations. If ϕ is a finite angle and we write $d\phi = \phi/N$, then a finite rotation about the z axis is given by the unitary transformation

$$U_\phi = \lim_{N\to\infty}\left(1 - \frac{i\phi}{N}L_z\right)^N = e^{-i\phi L_z} \tag{7.24}$$

If you are uncomfortable with the appearance of an operator in the exponent, review section 15 in Chapter 7.

The derivation leading to (7.23) only works if angular momentum can be represented by $\mathbf{r}\times\mathbf{p}$. If we extend the idea to spin space where $L_z = S_z = \frac{1}{2}\sigma_z$, then (7.24) becomes

$$U_\phi = e^{-\phi i\sigma_z/2} \tag{7.25}$$

Our experience with quaternions assures us that (7.25) is correct. To investigate the problem further, we will construct a succession of two rotations which will transform the spinor χ_+ into a spinor which represents spin in an arbitrary direction. Useful in keeping track of the geometry as we go along is the Hermitian matrix

$$\underline{R} = x\sigma_x + y\sigma_y + z\sigma_z \tag{7.26}$$

Note that $\underline{R}$ is essentially the same thing as the quaternion vector defined by equation (6.37). Consider first the transformation given by

$$U_\theta = e^{-\theta i\sigma_y/2} = I\cos\frac{\theta}{2} - i\sigma_y\sin\frac{\theta}{2} = \begin{pmatrix} \cos\frac{\theta}{2} & -\sin\frac{\theta}{2} \\ \sin\frac{\theta}{2} & \cos\frac{\theta}{2} \end{pmatrix} \tag{7.27}$$

Writing the exponential in terms of sines and cosines is justified because $(i\sigma_y)^2 = -I$. The transformation of the Pauli matrices by (7.27) can be worked out by direct matrix multiplication or by using the algebraic properties of the Pauli matrices as given by (6.27). For example,

$$\begin{aligned} \bar{\sigma}_x &= U_\theta\sigma_x U_\theta^\dagger \\ &= \left(I\cos\frac{\theta}{2} - i\sigma_y\sin\frac{\theta}{2}\right)\sigma_x\left(I\cos\frac{\theta}{2} + i\sigma_y\sin\frac{\theta}{2}\right) \end{aligned}$$

$$= \left(I\cos\frac{\theta}{2} - i\sigma_y \sin\frac{\theta}{2}\right)\left(\sigma_x \cos\frac{\theta}{2} - \sigma_z \sin\frac{\theta}{2}\right)$$

$$= \sigma_x\left(\cos^2\frac{\theta}{2} - \sin^2\frac{\theta}{2}\right) - \sigma_z\left(2\sin\frac{\theta}{2}\cos\frac{\theta}{2}\right)$$

$$= \sigma_x \cos\theta - \sigma_z \sin\theta \tag{7.28}$$

Similarly,

$$\bar{\sigma}_y = U_\theta \sigma_y U_\theta^\dagger = \sigma_y \tag{7.29}$$

$$\bar{\sigma}_z = U_\theta \sigma_z U_\theta^\dagger = \sigma_z \cos\theta + \sigma_x \sin\theta \tag{7.30}$$

This is of course the same type of calculation that we did with quaternions in equations (7.4), (7.5), and (7.6). If the unitary matrix $\underline{R}$ given by equation (7.26) is transformed, the result is

$$U_\theta \underline{R} U_\theta^\dagger = x\bar{\sigma}_x + y\bar{\sigma}_y + z\bar{\sigma}_z$$

$$= x(\sigma_x \cos\theta - \sigma_z \sin\theta) + y\sigma_y$$

$$+ z(\sigma_x \sin\theta + \sigma_z \cos\theta)$$

$$= \sigma_x(x\cos\theta + z\sin\theta) + y\sigma_y$$

$$+ \sigma_z(-x\sin\theta + z\cos\theta)$$

$$= \bar{x}\sigma_x + \bar{y}\sigma_y + \bar{z}\sigma_z \tag{7.31}$$

We discover that

$$\bar{x} = x\cos\theta + z\sin\theta \qquad \bar{y} = y$$
$$\bar{z} = -x\sin\theta + z\cos\theta \tag{7.32}$$

which shows that the coordinate transformation which corresponds to the unitary transformation (7.27) is as pictured in Fig. 7.3. Note from (7.28), (7.29), and (7.30) that the Pauli matrices also have a vector transformation property, although they are rotated in the opposite sense.

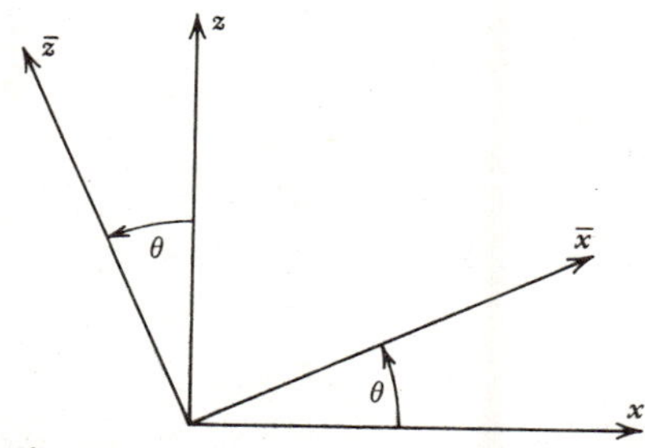

Fig. 7.3

As the second transformation, we will use

$$U_\phi = e^{-\phi i\sigma_z/2} = \begin{pmatrix} e^{-i\phi/2} & 0 \\ 0 & e^{i\phi/2} \end{pmatrix} \tag{7.33}$$

By following the same procedure that we used with U_θ, we find

$$\sigma_x' = U_\phi \sigma_x U_\phi^\dagger = \sigma_x \cos\phi + \sigma_y \sin\phi$$

$$\sigma_y' = -\sigma_x \sin\phi + \sigma_y \cos\phi \qquad \sigma_z' = \sigma_z \tag{7.34}$$

$$x' = \bar{x}\cos\phi - \bar{y}\sin\phi$$

$$y' = \bar{x}\sin\phi + \bar{y}\cos\phi \qquad z' = \bar{z} \tag{7.35}$$

Since it is σ_x, σ_y, and σ_z which appear in (7.31), we have transformed them rather than using $\bar{\sigma}_x$, $\bar{\sigma}_y$, and $\bar{\sigma}_z$ in (7.34). The application of U_ϕ to (7.31) then gives (7.35). The three-dimensional coordinate transformation which corresponds to the two-dimensional unitary transformation (7.33) is shown in Fig. 7.4. In Fig. 7.3, the y axis points into the page and in Fig. 7.4, the $\bar{z}$ axis points out of the page. Thus, both of the rotations are negative.

Frequently, the three Pauli spin matrices are represented as the components of a vector:

$$\boldsymbol{\sigma} = \sigma_x \hat{\mathbf{e}}_1 + \sigma_y \hat{\mathbf{e}}_2 + \sigma_z \hat{\mathbf{e}}_3 \tag{7.36}$$

The reader is probably uncomfortable with matrices as the components of a vector. The idea is credible because two basic requirements are satisfied. For one thing, the transformation properties of the Pauli matrices are correct for a vector. Secondly, angular

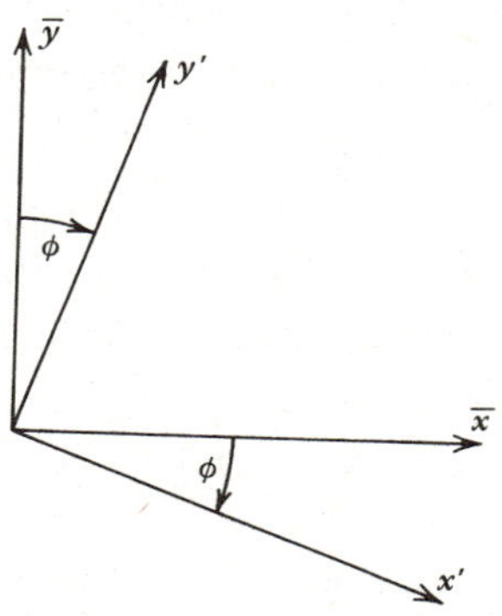

Fig. 7.4

momenta have a vector addition property although the details of actually carrying it out are more complicated than for ordinary vectors. Operators such as (7.36) are called *vector operators*. The corresponding equation for quaternions is (7.17). Note that the Hermitian matrix (7.26) can be expressed as

$$\underline{R} = \boldsymbol{\sigma}\cdot\mathbf{r} \tag{7.37}$$

The combination of the two rotations given by (7.27) and (7.33) is

$$U = U_\phi U_\theta = \begin{pmatrix} e^{-i\phi/2}\cos\frac{\theta}{2} & -e^{-i\phi/2}\sin\frac{\theta}{2} \\ e^{i\phi/2}\sin\frac{\theta}{2} & e^{i\phi/2}\cos\frac{\theta}{2} \end{pmatrix} \tag{7.38}$$

and we will now investigate its significance. The eigenspinor χ_+ represents spin pointing in the positive z direction. Its transformation by means of (7.38) is

$$\chi'_+ = U\chi_+ = \begin{pmatrix} e^{-i\phi/2}\cos\frac{\theta}{2} \\ e^{i\phi/2}\sin\frac{\theta}{2} \end{pmatrix} \tag{7.39}$$

The expectation value of the spin vector (7.36) in this state is

$$\begin{aligned} \langle\boldsymbol{\sigma}\rangle &= (\chi'_+)^\dagger \boldsymbol{\sigma}\chi'_+ \\ &= \hat{\mathbf{e}}_1 \sin\theta\cos\phi + \hat{\mathbf{e}}_2 \sin\theta\sin\phi + \hat{\mathbf{e}}_3\cos\theta \end{aligned} \tag{7.40}$$

This is a unit vector pointing in the direction (θ, ϕ). The transformation of the eigenspinor χ_- and the Pauli matrix σ_z gives

$$\chi'_- = U\chi_- = \begin{pmatrix} -e^{-i\phi/2}\sin\frac{\theta}{2} \\ e^{i\phi/2}\cos\frac{\theta}{2} \end{pmatrix} \tag{7.41}$$

$$\sigma'_z = U\sigma_z U^\dagger = \begin{pmatrix} \cos\theta & e^{-i\phi}\sin\theta \\ e^{i\phi}\sin\theta & -\cos\theta \end{pmatrix} \tag{7.42}$$

The spinors χ'_+ and χ'_- are eigenspinors of σ'_z; χ'_+ represents spin in the direction (θ, ϕ), and χ'_- represents spin in the opposite direction. The transformations were chosen so as to rotate the coordinates in the negative sense. The spinor χ_+ is rotated in the positive sense so that its transformation gives an eigenspinor which represents spin in the direction (θ, ϕ).

We have established that there is a correspondence between two-dimensional unitary transformations in the spin space and ordinary three-dimensional rotations in coordinate space. Because of the half-angle dependence of the unitary transformations, the correspondence is not one-to-one but is double-valued. A given rotation in three dimensions, say, θ about the y axis, corresponds to the two possibilities $\theta/2$ and $(\theta + 2\pi)/2 = \theta/2 + \pi$ in spin space. The transformations that we have derived all have determinant $+1$ and constitute a representation of what is known as the *unimodular group*. This group is also referred to as SU(2).

Another significant result is that the spinors are double-valued. For example, if ϕ is replaced by $\phi + 2\pi$ in (7.39), the eigenspinor becomes $-\chi'_+$. The double-valuedness is permissible because eigenvalues and probability densities that are calculated from the eigenvectors are unaffected.

PROBLEMS

7.1. Let the difference between the coordinate systems (x', y', z') and (x, y, z) be a uniform infinitesimal translation without rotation. Show that a function evaluated at a fixed point with respect to the new coordinates is

$$f(x', y', z') = \left(1 + \frac{i}{\hbar} d\mathbf{s}\cdot\mathbf{p}\right) f(x, y, z) \tag{7.43}$$

where $\mathbf{p} = -i\hbar\nabla$ is the linear momentum operator. In this sense, linear momentum generates translations.

7.2. Show that the operator given by equation (7.24) is unitary and that its determinant is +1.

Hint: Assume that L_z is diagonal. Show that

$$|e^{-i\phi L_z}| = \prod_{m=-l}^{+l} e^{-im\phi} = 1 \tag{7.44}$$

7.3. Let $(x, y, z) \rightarrow (x', y', z')$ be the three-dimensional coordinate transformation which corresponds to the two-dimensional unitary transformation U in spin space. Show that the Hermitian matrix $\underline{R}$ defined by (7.26) is invariant in form, i.e. that

$$\underline{R} = \begin{pmatrix} z & x - iy \\ x + iy & -z \end{pmatrix}$$

$$\underline{R}' = U\underline{R}U^\dagger = \begin{pmatrix} z' & x' - iy' \\ x' + iy' & -z' \end{pmatrix} \tag{7.45}$$

You can verify the validity of (7.45) by using specific examples of rotations such as U_θ and U_ϕ. You can also argue the point more generally by noting that the determinant and trace, being invariants, must be the same for both $\underline{R}$ and $\underline{R}'$. These two invariants plus the fact that $\underline{R}'$ must also be Hermitian are enough to establish the validity of (7.45). Calculate the determinant of $\underline{R}$ and comment on its significance.

7.4. Show that the transformation of the vector operator (7.36) by U_ϕ yields

$$U_\phi \boldsymbol{\sigma} U_\phi^\dagger = \sigma_x \hat{\mathbf{e}}_1' + \sigma_y \hat{\mathbf{e}}_2' + \sigma_z \hat{\mathbf{e}}_3' \tag{7.46}$$

7.5. Calculate directly the eigenvalues and the normalized eigenvectors of the matrix (7.42).

7.6. Prove that any 2×2 matrix can be expressed as

$$A = a_0 I + \mathbf{a} \cdot \boldsymbol{\sigma} \tag{7.47}$$

where a_0 and the components of $\mathbf{a}$ are, in general, complex numbers. This is the equivalent of (6.39) for a quaternion.

7.7. If $\mathbf{a}$ and $\mathbf{b}$ are any two vectors, show that

$$(\mathbf{a} \cdot \boldsymbol{\sigma})(\mathbf{b} \cdot \boldsymbol{\sigma}) = \mathbf{a} \cdot \mathbf{b} + i\boldsymbol{\sigma} \cdot (\mathbf{a} \times \mathbf{b}) \tag{7.48}$$

Compare this result to (7.20). In particular, if $\hat{\mathbf{n}}$ is a real unit vector, then

$$(\hat{\mathbf{n}} \cdot \boldsymbol{\sigma})^2 = 1 \tag{7.49}$$

7.8. An arbitrary unitary transformation can be expressed in the form

$$U = e^{i\gamma/2} e^{(i\alpha/2)(\hat{\mathbf{n}} \cdot \boldsymbol{\sigma})} \tag{7.50}$$

where $\hat{\mathbf{n}}$ is a real unit vector. What is its geometrical significance? What is $\det U$?

7.9. The two-dimensional unitary transformation which corresponds to a succession of three rotations by the Euler angles is given by

$$U = e^{i\psi\sigma_z/2} e^{i\theta\sigma_x/2} e^{i\phi\sigma_z/2} \tag{7.51}$$

Refer to Fig. 15.1 in Chapter 4 for a definition of the Euler angles. Show that

$$U = \begin{pmatrix} e^{i(\psi+\phi)/2} \cos\dfrac{\theta}{2} & ie^{i(\psi-\phi)/2} \sin\dfrac{\theta}{2} \\ ie^{-i(\psi-\phi)/2} \sin\dfrac{\theta}{2} & e^{-i(\psi+\phi)/2} \cos\dfrac{\theta}{2} \end{pmatrix} \tag{7.52}$$

Note that the Euler angle θ does not correspond exactly to the angle θ of spherical coordinates. The angle ψ is superfluous in discussing the spin of a particle because a spin angle or spin angular velocity is meaningless. It is only the orientation of the spin which is important, and this is taken care of by two angles.

7.10. The group SU(2) is a representation of the rotation group in the respect that there is a correspondence between its elements and rotations in ordinary three-dimensional Cartesian space. Representations of the rotation group of any dimension exist. For example, we can generate a representation of SU(3) by means of rotations of the type

$$U_\theta = e^{-i\theta L_y} \qquad U_\phi = e^{-i\phi L_z} \tag{7.53}$$

where L_y and L_z are angular momentum operators for $l = 1$ as given by equation (3.44). Since $(iL_y)^2 \neq -I$, the matrix representation of U_θ is not given by a formula like that in equation (7.27). One way to find matrix representations of the rotations is to use the spectral resolution of a matrix as given by equation (15.55) in Chapter 7. Thus,

$$U_\theta = \sum_{m=-l}^{+l} e^{-im\theta} G_m = e^{-i\theta} G_1 + G_2 + e^{i\theta} G_3 \tag{7.54}$$

where G_1, G_2, and G_3 are the projection operators of L_y. Find the normalized eigenvectors of L_y and use equation (15.59) of Chapter 7 to construct the projection operators. Do the same computation for U_ϕ and show that

$$U = U_\phi U_\theta$$

$$= \begin{pmatrix} \frac{1}{2}e^{-i\phi}(1+\cos\theta) & -\frac{1}{\sqrt{2}}e^{-i\phi}\sin\theta & \frac{1}{2}e^{-i\phi}(1-\cos\theta) \\ \frac{1}{\sqrt{2}}\sin\theta & \cos\theta & -\frac{1}{\sqrt{2}}\sin\theta \\ \frac{1}{2}e^{i\phi}(1-\cos\theta) & \frac{1}{\sqrt{2}}e^{i\phi}\sin\theta & \frac{1}{2}e^{i\phi}(1+\cos\theta) \end{pmatrix} \tag{7.55}$$

This matrix does the same job for a spin 1 particle as (7.38) does for a spin $\frac{1}{2}$ particle. The representation is not double-valued as it is in the $l = \frac{1}{2}$ case, meaning that there is a one-to-one correspondence between rotations in Cartesian space and transformations in the three-dimensional unitary space. The eigenvectors of a system with angular momentum $l = 1$ are single-valued. Double-valuedness occurs for all half-integer values of l.

8. SPIN $\frac{1}{2}$ PARTICLE IN A MAGNETIC FIELD

If the effects of spin are to be included in a description of the mechanical behavior of a particle when it is acted on by an external force field, then it is necessary to modify the Schrödinger equation to include the spin variables. The Hamiltonian will generally be a function of the time, the three position coordinates x, y, and z, and a spin variable s. The position coordinates are continuous, whereas s can have only the values $\pm\frac{1}{2}$. In the simple examples to be worked out in this section, the Hamiltonian can be separated into two parts, one involving the position coordinates of the particle and the other its spin variables:

$$H(\mathbf{r}, s, t) = IH_r + H_s \tag{8.1}$$

In (8.1), I is the 2×2 identity matrix and H_r is the part of the Hamiltonian involving the kinetic energy of translation and the potential energy of position. The term H_s is a 2×2 matrix and gives the contribution to the energy due to the spin. The Hamiltonian (8.1) would not be suitable for finding the fine structure splitting of the spectral lines of hydrogen due to spin because no interaction terms between spin and space variables have been included. It is suitable for problems where the interaction of the spin with an external field is the dominant feature and effects due to translational motion are of secondary importance.

If the wave function is expressed as a product of a spatial part and a spin part, then the wave equation is

$$(IH_r + H_s)\psi(\mathbf{r}, t)\chi(s, t) = i\hbar\left(\frac{\partial \psi}{\partial t}\chi + \psi\frac{\partial \chi}{\partial t}\right) \tag{8.2}$$

where $\chi(s, t)$ is a two-component spinor. If (8.2) is multiplied through by $\chi^\dagger$ and $\chi^\dagger\chi = 1$ is used, the resulting equation can be put into the form

$$\frac{1}{\psi}\left(H_r\psi - i\hbar\frac{\partial \psi}{\partial t}\right) = -\chi^\dagger\left(H_s\chi - i\hbar\frac{\partial \chi}{\partial t}\right) \tag{8.3}$$

The separation of space and spin variables is therefore achieved. Both sides of (8.3) can at most be a function of t. Setting this function equal to zero gives the ordinary Schrödinger equation

$$H_r\psi = i\hbar\frac{\partial \psi}{\partial t} \tag{8.4}$$

for the translational motion and

$$H_s\chi = i\hbar\frac{\partial \chi}{\partial t} \tag{8.5}$$

for the time dependence of the spinor. It is to be emphasized that this separation procedure works only if H_s has no dependence on position or if the particle is more or less stationary so that translational motion does not enter into the picture at all. In the remainder of this section, we will find solutions of (8.5) for the case where a spin ½ particle has been placed in a magnetic field.

Classically, the potential energy of a magnetic dipole of moment $\boldsymbol{\mu}$ when placed in an external magnetic field is $-\boldsymbol{\mu}\cdot\mathbf{B}$. We will assume that this same relation carries over into quantum mechanics and express the spin Hamiltonian as

$$H_s = -\gamma\mathbf{P}\cdot\mathbf{B} = -\tfrac{1}{2}\gamma\hbar(\boldsymbol{\sigma}\cdot\mathbf{B})$$

$$= -\tfrac{1}{2}\gamma\hbar\begin{pmatrix} B_z & B_x - iB_y \\ B_x + iB_y & -B_z \end{pmatrix} \tag{8.6}$$

Equation (6.6) relating magnetic moment and intrinsic spin angular momentum has been used. Different spin ½ particles have different values of the gyromagnetic ratio γ; the value given in equation (6.6) applies to the electron. If the magnetic field is constant in time and points in the z direction, then (8.6) simplifies to

$$H_s = -\tfrac{1}{2}\hbar\omega_0\sigma_z \qquad \omega_0 = \gamma B \tag{8.7}$$

where B is the strength of the magnetic field. Classically, ω_0 is the Larmor frequency. The Schrödinger equation is then

$$-\frac{1}{2}\omega_0\sigma_z\chi = i\frac{\partial \chi}{\partial t} \tag{8.8}$$

Since the spinors χ_+ and χ_- as given by (6.29) are a complete set, the solution of (8.8) can be expressed as a linear combination of them as

$$\chi = C_1(t)\chi_+ + C_2(t)\chi_- \tag{8.9}$$

where $C_1(t)$ and $C_2(t)$ are functions of the time. Since χ_+ and χ_- are eigenspinors of σ_z with eigenvalues $+1$ and -1, (8.8) gives

$$-\tfrac{1}{2}\omega_0 C_1 = i\dot{C}_1 \qquad \tfrac{1}{2}\omega_0 C_2 = i\dot{C}_2 \tag{8.10}$$

The solutions are

$$C_1 = \alpha e^{i\omega_0 t/2} \qquad C_2 = \beta e^{-i\omega_0 t/2} \tag{8.11}$$

The normalization condition $\chi^\dagger\chi = 1$ requires that the constants α and β obey

$$|\alpha|^2 + |\beta|^2 = 1 \tag{8.12}$$

There is always an arbitrary phase factor which is not determined by the normalization and which has no effect on the measurable quantities that can be calculated from the wave function. We are at liberty to choose

$$\alpha = \cos\frac{\theta}{2} \qquad \beta = \sin\frac{\theta}{2} \tag{8.13}$$

The solution of (8.8) can now be written as

$$\chi = e^{i\omega_0 t/2}\cos\frac{\theta}{2}\chi_+ + e^{-i\omega_0 t/2}\sin\frac{\theta}{2}\chi_-$$

$$= \begin{pmatrix} e^{i\omega_0 t/2}\cos\dfrac{\theta}{2} \\ e^{-i\omega_0 t/2}\sin\dfrac{\theta}{2} \end{pmatrix} \tag{8.14}$$

Note that χ_+ and χ_- are eigenspinors of the Hamiltonian. They correspond to the energy eigenvalues

$$W_+ = -\tfrac{1}{2}\hbar\omega_0 = -\tfrac{1}{2}\hbar\gamma B \qquad W_- = \tfrac{1}{2}\hbar\omega_0 = \tfrac{1}{2}\hbar\gamma B \tag{8.15}$$

A spinning particle has its lowest energy when its magnetic moment points in the same direction as the magnetic field. A positively charged particle has its magnetic moment and spin in the same direction. Hence, χ_+ corresponds to the lower energy. For a negatively charged particle, $\gamma < 0$ and everything is reversed.

If a measurement is made on a system described by the state (8.14), the probability of finding the spin pointing in the direction of the magnetic field is $\cos^2(\theta/2)$ and the probability of finding the spin pointing in the opposite direction is $\sin^2(\theta/2)$. An examination of equation (7.39) shows that (8.14) describes a state where the spin is at an angle θ with respect to the direction of the magnetic field and precesses at the angular rate ω_0 in the negative ϕ direction if $\gamma > 0$ and in the positive ϕ direction if $\gamma < 0$. This is not at all unlike the classical behavior of a spinning charged particle in a magnetic field as described at the beginning of section 6.

Almost all problems involving time-dependent Hamiltonians are too difficult to solve exactly. The following example provides an exception. Assume that a spin $\frac{1}{2}$ particle is placed in a magnetic field which has a constant and uniform component in the z direction and a transverse time-dependent part given by

$$B_x = b\cos\omega t \qquad B_y = -b\sin\omega t \tag{8.16}$$

where b is a constant and ω is the frequency of the applied field. The particular form of the time dependence is chosen to make the analysis easy. Note that the field still has no dependence on position so that the conditions required for the validity of the separation of the Schrödinger equation (8.2) still hold. Note that

$$B_x - iB_y = be^{i\omega t} \tag{8.17}$$

The equation of motion (8.5) for the spinor is

$$-\frac{1}{2}\gamma\hbar\begin{pmatrix} B & be^{i\omega t} \\ be^{-i\omega t} & -B \end{pmatrix}\begin{pmatrix} C_1 \\ C_2 \end{pmatrix} = i\hbar\begin{pmatrix} \dot{C}_1 \\ \dot{C}_2 \end{pmatrix} \tag{8.18}$$

If we define $\omega_0 = \gamma B$ and $\omega_b = \gamma b$, (8.18) works out to

$$\begin{aligned} \omega_0 C_1 + \omega_b C_2 e^{i\omega t} &= -2i\dot{C}_1 \\ \omega_b C_1 e^{-i\omega t} - \omega_0 C_2 &= -2i\dot{C}_2 \end{aligned} \tag{8.19}$$

A solution can be found of the form

$$C_1 = \alpha e^{i\omega_1 t} \qquad C_2 = \beta e^{i\omega_2 t} \tag{8.20}$$

where α, β, ω_1, and ω_2 are constants. We find that

$$\begin{aligned} (\omega_0 - 2\omega_1)\alpha + \omega_b \beta e^{i(\omega+\omega_2-\omega_1)t} &= 0 \\ \omega_b \alpha e^{i(\omega_1-\omega_2-\omega)t} - (\omega_0 + 2\omega_2)\beta &= 0 \end{aligned} \tag{8.21}$$

To eliminate the time dependence from (8.21), ω_1 and ω_2 are required to satisfy

$$\omega + \omega_2 - \omega_1 = 0 \tag{8.22}$$

With this condition, (8.21) becomes a set of two homogeneous linear equations for α and β. A nontrivial solution exists if the determinant of the coefficients vanishes:

$$(\omega_0 - 2\omega_1)(\omega_0 + 2\omega_2) + \omega_b^2 = 0 \tag{8.23}$$

The elimination of ω_1 by means of (8.23) gives a quadratic equation for ω_2. Its solutions are

$$\omega_2 = -\tfrac{1}{2}\omega \pm \nu \qquad \nu = \tfrac{1}{2}\sqrt{(\omega-\omega_0)^2 + \omega_b^2} \tag{8.24}$$

The general solution for C_2 is therefore

$$C_2(t) = e^{-i\omega t/2}\left(\beta_1 e^{i\nu t} + \beta_2 e^{-i\nu t}\right) \tag{8.25}$$

If we assume as an initial condition that the spin is definitely in the positive z direction at $t = 0$, then $C_2(0) = 0$ and $\beta_2 = -\beta_1$. Equation (8.25) can then be written

$$C_2(t) = 2i\beta_1 e^{-i\omega t/2}\sin\nu t \tag{8.26}$$

It is now possible to find $C_1(t)$ directly from (8.19). The result is

$$C_1(t) = \frac{2i\beta_1}{\omega_b}e^{i\omega t/2}\left[(\omega_0 - \omega)\sin\nu t - 2i\nu\cos\nu t\right] \tag{8.27}$$

By using the fact that $C_1(0) = 1$, we find that

$$\beta_1 = \frac{\omega_b}{4\nu} \tag{8.28}$$

Thus, (8.26) can be expressed as

$$C_2(t) = \frac{i\omega_b}{2\nu} e^{-i\omega t/2} \sin \nu t \tag{8.29}$$

The reader should check that

$$|C_1|^2 + |C_2|^2 = 1 \tag{8.30}$$

The significance of

$$|C_2|^2 = A(\omega)\sin^2 \nu t \qquad A(\omega) = \frac{\omega_b^2}{(\omega-\omega_0)^2 + \omega_b^2} \tag{8.31}$$

is that it is the probability that the particle will be found with its spin pointing in the negative z direction at the time t. Equation (8.31) exhibits a typical resonance phenomenon. A plot of the amplitude $A(\omega)$ as a function of the driving frequency ω appears in Fig. 8.1 for three values of ω_b. The resonance becomes very sharp if $\omega_b \ll \omega_0$, and its location can be accurately measured. The resonant frequency is ω_0, and since $\omega_0 = \gamma B$, the gyromagnetic ratio of the particle can be determined. To determine γ for a proton, a sample containing a high density of protons, such as water or paraffin, is used. Equation (8.31) then gives the fraction of the total number of particles which have flipped their spins at a given time. At resonance, essentially all the particles are flipping their spins. The resonance is detected either by measuring the power absorbed from the oscillating field or by the rapidly changing magnetization of the sample.

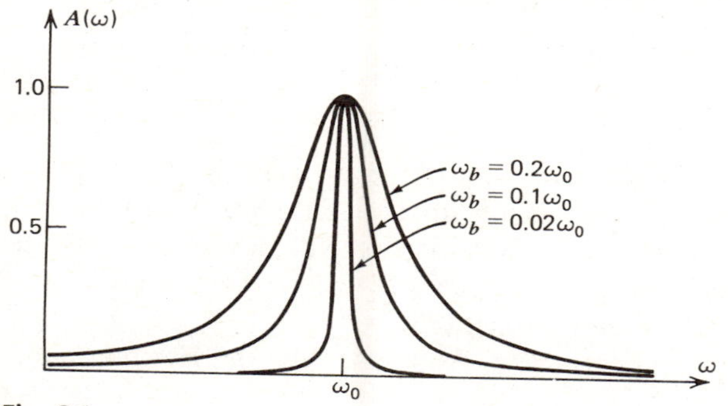

Fig. 8.1

The time-independent Hamiltonian (8.7) has the stationary states given by χ_+ and χ_-. The probability amplitudes as given by (8.11) have the property that $|C_1|^2$ and $|C_2|^2$ are constants. Once the particle is in the state described by (8.14), it stays there until disturbed by some external influence. By contrast, the time-dependent Hamiltonian has no stationary states. The solution can still be written in the form (8.9), but $|C_1|^2$ and $|C_2|^2$ are now time-dependent. If the amplitude b of the time-dependent part of the field is small compared to the magnitude B of the constant field in the z direction, then the main effect of the time dependence is to induce transitions between the stationary states of the time-independent part of the Hamiltonian. Note that the difference between the energy eigenvalues of the time-independent Hamiltonian is

$$W_- - W_+ = \hbar\omega_0 \tag{8.32}$$

Thus, at resonance, the energy of the photons absorbed from the field exactly matches the energy difference between the two eigenstates. Nuclear magnetic resonance experiments are an important application of the theory just discussed.

The example just considered presents a somewhat more realistic view of quantum mechanics than can be obtained from a consideration of time-independent Hermitian operators and their mathematically precise eigenvalues. We do not really ever observe these eigenvalues with arbitrary accuracy. The resonance of Fig. 8.1 is more typical of what is actually observed. No spectral line is precisely sharp. When you examine a spectral line through a spectroscope, what you really see is a resonance phenomenon giving a spectral line of finite width.

PROBLEMS

8.1. If there is a uniform magnetic field at an arbitrary direction, then

$$B_x = B\sin\theta\cos\phi \qquad B_y = B\sin\theta\sin\phi \qquad B_z = B\cos\theta \tag{8.33}$$

Show that the spin Hamiltonian (8.5) is

$$H_s = -\tfrac{1}{2}\gamma\hbar B\sigma(\theta,\phi) \tag{8.34}$$

where $\sigma(\theta, \phi)$ is the spin operator given by equation (7.42). Complete the solution of (8.5) for the Hamiltonian (8.34).

8.2. Find the expectation value of **P** in the state (8.14) and show that

$$\frac{d\langle \mathbf{P} \rangle}{dt} = \gamma \langle \mathbf{P} \rangle \times \mathbf{B} \tag{8.35}$$

Compare this to the classical result (6.7).

8.3. Verify (8.30).

8.4. What happens to the solutions of (8.18) when $B = 0$? Sketch the amplitude $A(\omega)$ as given by (8.31) for this case. Consider only positive values of ω. What is the significance of the solution when $\omega = 0$?

9. LAPLACE'S EQUATION IN SPHERICAL COORDINATES

The Laplacian in spherical coordinates is given by equation (10.26) in Chapter 4. Laplace's equation from electrostatics is $\nabla^2 \psi = 0$, and if its solution is separated as $\psi(r, \theta, \phi) = F(r)Y(\theta, \phi)$, we find

$$\frac{1}{F}\frac{d}{dr}\left(r^2 \frac{dF}{dr}\right) + \frac{1}{Y \sin\theta}\frac{\partial}{\partial\theta}\left(\sin\theta \frac{\partial Y}{\partial\theta}\right) + \frac{1}{Y \sin^2\theta}\frac{\partial^2 Y}{\partial\phi^2} = 0 \tag{9.1}$$

The reader will have to shift notational gears and remember that ψ is now an electrostatic potential and not a wave function. If the separation constant is written as $l(l+1)$, then the radial equation is

$$\frac{d}{dr}\left(r^2 \frac{dF}{dr}\right) - l(l+1)F = 0 \tag{9.2}$$

Reference to equation (4.9) shows that $Y = Y_{lm}(\theta, \phi)$ is a spherical harmonic.

The development of the theory of spherical harmonics in section 4 connected them intimately with angular momentum in quantum mechanics. The reader is perhaps surprised (maybe even shocked) to find them asserting themselves once again in potential theory. Spherical harmonics were known and studied long before the advent of quantum mechanics. Introducing these functions as angular momentum eigenfunctions has the advantage that the raising and lowering operators as developed through the algebra of angular momentum provide the most economical and straightforward way of deriving many of their properties. In potential theory, l and m are separation constants and have nothing to do with angular momentum. If the potential is to be defined over the whole range of values of θ, then l must be an integer in order to prevent the solution from becoming divergent at $\theta = 0$ and π. The potential must be single-valued, and this is enough to require that m be restricted to integer values. Fortunately, the solution of the radial equation (9.2) is much easier than in quantum mechanics. It is

$$F(r) = Ar^l + \frac{B}{r^{l+1}} \tag{9.3}$$

where A and B are constants. The complete solution of Laplace's equation is

$$\psi(r, \theta, \phi) = \sum_{l=0}^{\infty} \sum_{m=-l}^{+l} \left(a_{lm} r^l + \frac{b_{lm}}{r^{l+1}}\right) Y_{lm}(\theta, \phi) \tag{9.4}$$

Applications of (9.4) to the solution of specific problems will be taken up in later sections. In the remainder of this section, we will derive more of the mathematical properties of the Legendre and associated Legendre functions.

The Legendre polynomials, like the Hermite polynomials and the Laguerre polynomials, have a generating function. The easiest way to find it is to use a specific solution of Laplace's equation. Suppose, as illustrated in Fig. 9.1, that a point charge of magnitude $4\pi\varepsilon_0$ has been placed on the z axis at $z = a$. The resulting potential is

$$\psi = \frac{1}{R} = \frac{1}{\sqrt{r^2 + a^2 - 2ra\cos\theta}} \tag{9.5}$$

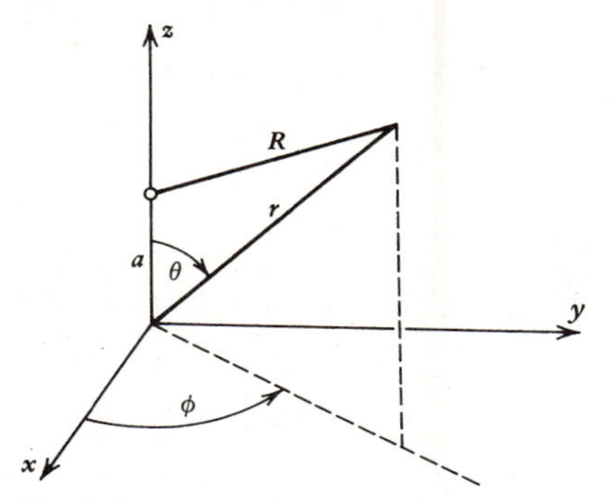

Fig. 9.1

We know that it must also be possible to express this potential in the form (9.4). Since there is no dependence on the azimuth angle ϕ, only $m=0$ appears in the expansion. Also, the coefficients a_{lm} must all be zero to prevent the potential from becoming arbitrarily large as $r \to \infty$. Thus,

$$\psi = \sum_{l=0}^{\infty} \frac{b_l}{r^{l+1}} P_l(\cos\theta) \tag{9.6}$$

It is possible to expand (9.5) directly by using a binomial expansion:

$$\psi = \frac{1}{R} = \frac{1}{r\sqrt{1+\dfrac{a^2}{r^2}-\dfrac{2a}{r}\cos\theta}}$$

$$= \frac{1}{r}\sum_{l=0}^{\infty}\left(\frac{a}{r}\right)^l P_l(\cos\theta) \tag{9.7}$$

where we have used a/r as the expansion parameter. In so doing, the coefficients of the different powers of a/r will come out to be polynomials in $\cos\theta$. By comparison to (9.6), these polynomials must be at least proportional to the Legendre polynomials. There is still the question of a possible difference by a multiplicative factor. It is convenient to write $h=a/r$ and express the generating function as

$$G(h,x) = \frac{1}{\sqrt{1+h^2-2xh}} = \sum_{l=0}^{\infty} h^l P_l(x) \tag{9.8}$$

where $x=\cos\theta$. We will show a little later that the series converges absolutely if $h<1$. By carrying out a few terms of the expansion indicated in (9.8), you can show that the first two or three polynomials generated are in fact the same as those listed in equation (1.18). Review section 1 and make sure that you understand why the Legendre polynomials are the only solutions of the Legendre equation that are finite at $x=\pm 1$.

By calculating the normalization integral directly from (9.8), we can confirm that the polynomials generated by it can differ from the Legendre polynomials by at most a factor of ± 1. By squaring (9.8) and integrating over the range $-1 \le x \le +1$, we get

$$\int_{-1}^{+1} \frac{1}{1+h^2-2xh}\,dx = \sum_{l=0}^{\infty}\sum_{n=0}^{\infty} h^{l+n}\int_{-1}^{+1} P_l P_n\,dx$$

$$\frac{1}{h}\ln\left(\frac{1+h}{1-h}\right) = \sum_{l=0}^{\infty} N_l h^{2l} \tag{9.9}$$

where we have used

$$\int_{-1}^{+1} P_l P_n\,dx = N_l \delta_{ln} \tag{9.10}$$

If the logarithm in (9.9) is expanded in a power series, the result is

$$\sum_{l=0}^{\infty} \frac{2}{2l+1} h^{2l} = \sum_{l=0}^{\infty} N_l h^{2l} \tag{9.11}$$

Thus,

$$N_l = \frac{2}{2l+1} \tag{9.12}$$

in agreement with (4.29).

The generating function can be used to derive recurrence formulas obeyed by the Legendre polynomials. The technique is the same as that used to derive (2.80) for the Hermite polynomials. The reader can show that the generating function obeys

$$h\frac{\partial G}{\partial h} = (x-h)\frac{\partial G}{\partial x} \tag{9.13}$$

$$(1+h^2-2xh)\frac{\partial G}{\partial h} = (x-h)G \tag{9.14}$$

from which follow the two recurrence formulas

$$lP_l = xP_l' - P_{l-1}' \tag{9.15}$$

$$(l+1)P_{l+1} = x(1+2l)P_l - lP_{l-1} \tag{9.16}$$

The last of these is quite important because it allows all the polynomials to be generated once it is known

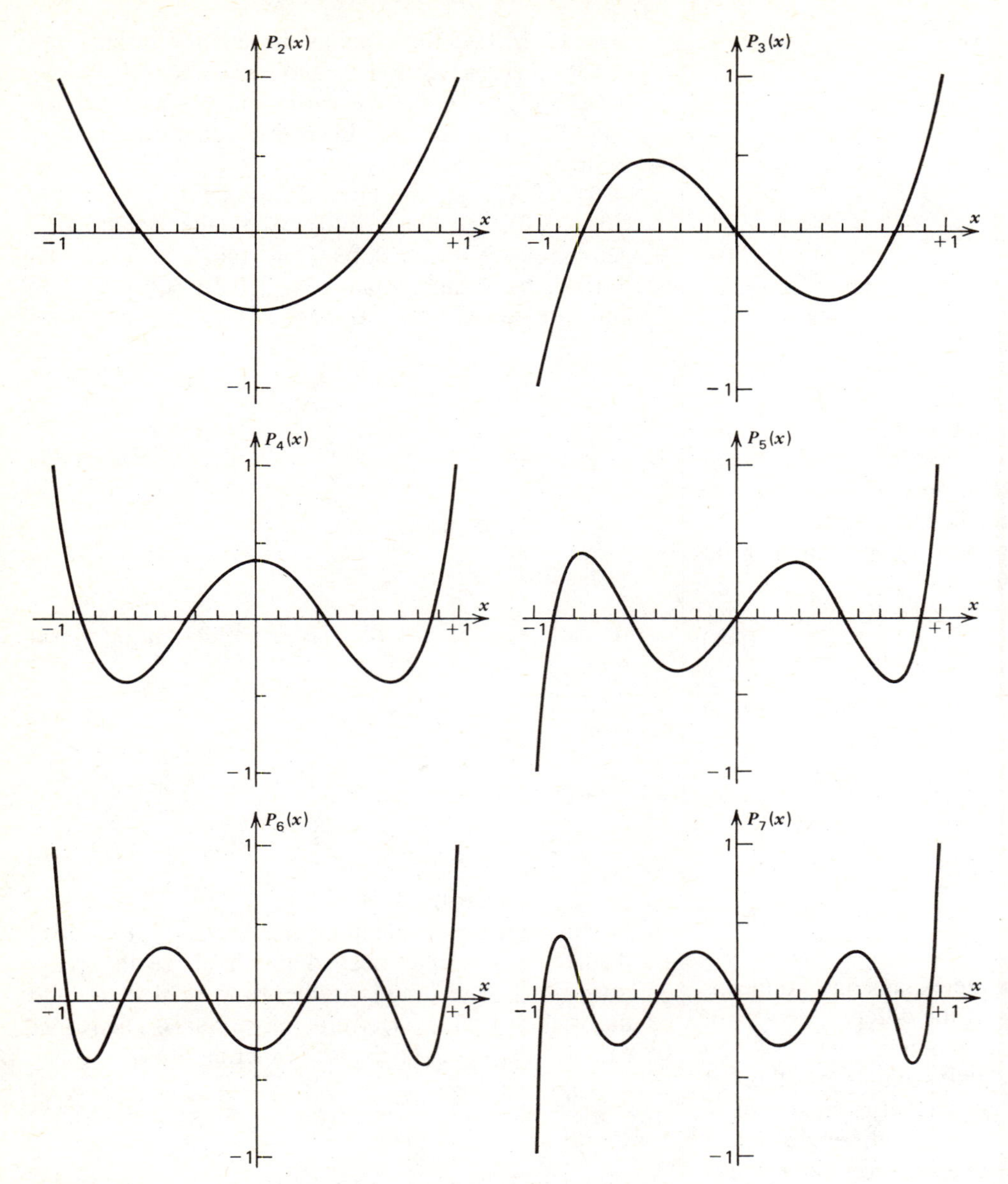

Fig. 9.2

that $P_0 = 1$ and $P_1 = x$. For any value of x, a computer program can be written that uses (9.16) to calculate the numerical value of any $P_l(x)$. This is how the graphs of the Legendre polynomials shown in Fig. 9.2 were obtained. The following recurrence relations can be found by combining (9.15) and (9.16) in various ways:

$$P'_{l+1} - P'_{l-1} = (2l+1)P_l \tag{9.17}$$

$$P'_{l+1} - xP'_l = (l+1)P_l \tag{9.18}$$

$$(x^2-1)P'_l - lxP_l + lP_{l-1} = 0 \tag{9.19}$$

For example, (9.17) is derived by first differentiating (9.16) and then using (9.15) to eliminate P_l'.

If we set $x=1$ in the generating function, we find

$$\frac{1}{1-h}=\sum_{l=0}^{\infty} h^l=\sum_{l=0}^{\infty} h^l P_l(1) \tag{9.20}$$

Therefore, $P_l(1)=1$. You can verify that the polynomials listed in equation (1.18) satisfy this condition. The polynomials found from the generating function therefore agree completely with those defined by the Rodrigues formula (4.28). As a special case of the general relation $P_l(-x)=(-1)^l P_l(x)$, we have $P_l(-1)=(-1)^l$. The convergence properties of the generating function can be established by setting $x=\cos\theta$ and writing

$$\begin{aligned}
\sum_{l=0}^{\infty} h^l P_l(\cos\theta) &= (1+h^2-2h\cos\theta)^{-1/2} \\
&= (1+h^2-he^{i\theta}-he^{-i\theta})^{-1/2} \\
&= (1-he^{i\theta})^{-1/2}(1-he^{-i\theta})^{-1/2} \\
&= \left(1+\tfrac{1}{2}he^{i\theta}+\tfrac{3}{8}h^2e^{2i\theta}+\cdots\right) \\
&\quad \times\left(1+\tfrac{1}{2}he^{-i\theta}+\tfrac{3}{8}h^2e^{-2i\theta}+\cdots\right) \\
&= 1+\tfrac{1}{2}he^{-i\theta}+\tfrac{3}{8}h^2e^{-2i\theta}+\cdots \\
&\quad +\tfrac{1}{2}he^{i\theta}+\tfrac{1}{4}h^2+\cdots+\tfrac{3}{8}h^2e^{2i\theta}+\cdots \\
&= 1+h\cos\theta+h^2\left(\tfrac{1}{4}+\tfrac{3}{4}\cos 2\theta\right)+\cdots
\end{aligned} \tag{9.21}$$

In this manner, it is shown that

$$P_l(\cos\theta)=\sum_{n=0}^{l} a_n\cos n\theta \tag{9.22}$$

where the coefficients a_n are all nonnegative. The maximum value of $P_l(\cos\theta)$ therefore occurs when $\theta=0$. Since $P_l(1)=1$, we have the inequality

$$|P_l(\cos\theta)|\le 1 \tag{9.23}$$

This shows that the series expansion (9.8) for the generating function converges absolutely when $h<1$, for example, by comparison to the series (9.20) which is known to converge for $h<1$. The generating function actually converges for $h=1$ if $x\neq\pm 1$.

There is a second set of solutions of the Legendre equation for integer values of l denoted by $Q_l(x)$ which are singular at $x=\pm 1$. The first two of these, $Q_0(x)$ and $Q_1(x)$, are given by equations (1.24) and (1.48). Starting with these two functions, it can be shown by a process of mathematical induction that the $Q_l(x)$ satisfy the same set of recurrence relations as the $P_l(x)$. Thus, all the $Q_l(x)$ can be generated from (9.16) rewritten as

$$(l+1)Q_{l+1}=x(1+2l)Q_l-lQ_{l-1} \tag{9.24}$$

Solutions of the Legendre equation for arbitrary values of l are given by the series expansions (1.16) and (1.17). We will not study these solutions further.

Formulas for calculating the associated Legendre functions are given by (4.36) and (4.40). One way to find recurrence relations for the associated functions is to use the raising and lowering operators. If equation (3.35) is written out for spherical harmonics with the raising operator given by (4.31), the result is

$$\begin{aligned}
&e^{i\phi}\left(-\sqrt{1-x^2}\,\frac{\partial}{\partial x}+\frac{ix}{\sqrt{1-x^2}}\,\frac{\partial}{\partial\phi}\right)Y_{lm} \\
&\qquad=\sqrt{(l-m)(l+m+1)}\,Y_{l,m+1}
\end{aligned} \tag{9.25}$$

Now use the formula for the spherical harmonics given by (4.35) to get

$$(1-x^2)P_{lm}'+mxP_{lm}+\sqrt{1-x^2}\,P_{l,m+1}=0 \tag{9.26}$$

Similarly, using the lowering operator gives

$$\begin{aligned}
&(1-x^2)P_{lm}'-mxP_{lm} \\
&\qquad-\sqrt{1-x^2}\,(l+m)(l-m+1)P_{l,m-1}=0
\end{aligned} \tag{9.27}$$

By combining (9.26) and (9.27) in various ways, we find

$$\begin{aligned}
&\sqrt{1-x^2}\,P_{lm}'+\tfrac{1}{2}P_{l,m+1} \\
&\qquad-\tfrac{1}{2}(l+m)(l-m+1)P_{l,m-1}=0
\end{aligned} \tag{9.28}$$

$$\begin{aligned}
&2mxP_{lm}+\sqrt{1-x^2}\,P_{l,m+1} \\
&\qquad+\sqrt{1-x^2}\,(l+m)(l-m+1)P_{l,m-1}=0
\end{aligned} \tag{9.29}$$

Other recurrence formulas can be found by differentiating the recurrence formulas for the Legendre polynomials.

The solutions of the associated Legendre equation which are singular at $x = \pm 1$ can be found by generalizing (4.36):

$$Q_{lm}(x) = (-1)^m (1-x^2)^{m/2} \frac{d^m Q_l(x)}{dx^m} \tag{9.30}$$

The justification for this is that once any solution of the Legendre equation is found, a solution of the associated equation can be obtained by (9.30). The proof can be done directly by differentiating the Legendre equation m times. This has the interesting consequence that the raising and lowering operators can be used on the functions $Q_{lm}e^{im\phi}$. These functions are not eigenfunctions and are not normalizable. Moreover, the raising and lowering process does not terminate at $m = \pm l$ as it does with the spherical harmonics.

PROBLEMS

9.1. Prove that $P_l(-x) = (-1)^l P_l(x)$ directly from the generating function (9.8).

9.2. Use the generating function (9.8) to obtain another proof of equation (4.45).

9.3. The first few Legendre polynomials are easily obtained by direct expansion of the generating function, for example as was done to obtain equation (9.21). The recurrence relation (9.16) is also derived directly from the generating function and can be used to continue the process. Start with $P_0(x) = 1$ and $P_1(x) = x$ and use (9.16) to obtain $P_2(x)$ and $P_3(x)$.

9.4. The expansion given by (9.7) works if $r > a$. Find a similar expansion which is valid if $r < a$. Note that this expansion is also a special case of (9.4).

9.5. Consider a region of space in the form of a rectangular box with edges a, b, and c as pictured in Fig. 9.3. Suppose that the potential is the known function

$$\psi(x, y, 0) = f(x, y) \tag{9.31}$$

over the face which lies in the x, y plane and that each of the other five faces is at zero potential. Find the expression for the potential at any point inside the box.

As a first step in the solution of the problem, separate Laplace's equation in rectangular coordinates by writing

$$\psi(x, y, z) = f_x(x) f_y(y) f_z(z) \tag{9.32}$$

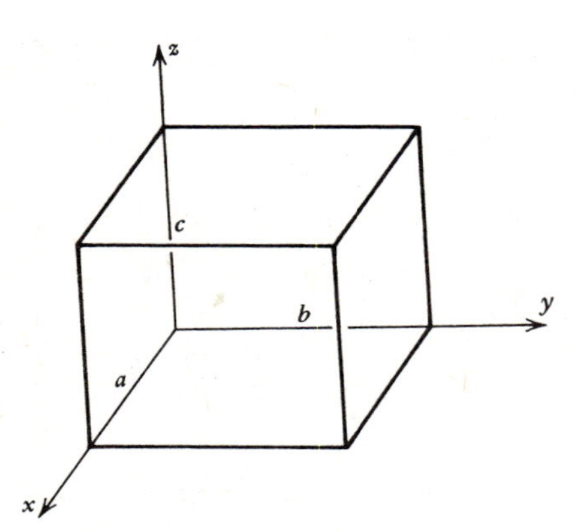

Fig. 9.3

If separation constants α, β, and γ are appropriately chosen, then

$$f_x'' + \alpha^2 f_x = 0 \qquad f_y'' + \beta^2 f_y = 0 \qquad f_z'' - \gamma^2 f_z = 0 \tag{9.33}$$

where $\gamma^2 = \alpha^2 + \beta^2$. The separation constants are chosen in this particular way because eigenfunctions are needed for the expansion of the function (9.31). The final form of the solution is

$$\psi(x, y, z) = \sum_{m=0}^{\infty} \sum_{n=0}^{\infty} A_{mn} u_{mn}(x, y) \sinh[\gamma_{mn}(c - z)] \tag{9.34}$$

$$u_{mn}(x, y) = \frac{2}{\sqrt{ab}} \sin\left(\frac{m\pi x}{a}\right) \sin\left(\frac{n\pi y}{b}\right) \tag{9.35}$$

$$\gamma_{mn}^2 = \pi^2 \left(\frac{m^2}{a^2} + \frac{n^2}{b^2}\right) \tag{9.36}$$

$$A_{mn} = \frac{1}{\sinh(\gamma_{mn} c)} \int_0^a \int_0^b u_{mn}(x, y) f(x, y)\, dx\, dy \tag{9.37}$$

The eigenfunctions (9.35) have come up in several other problems. See equations (2.18), (3.14), (5.19), and (8.2) in Chapter 7.

9.6. Let $w(x)$ be any solution of the Legendre equation

$$(1 - x^2) w'' - 2xw' + l(l+1) w = 0 \tag{9.38}$$

Show that the function

$$v(x) = (1 - x^2)^{m/2} f(x) \qquad f(x) = \frac{d^m w}{dx^m} \tag{9.39}$$

is a solution of the associated Legendre equation (4.12). This leaves no doubt that (9.30) gives the second linearly independent solution of the associated equation for integer values of l and m. The proof can be done by first differentiating (9.38) m times. The calculation is facilitated by equation (5.98). This leads to a second-order differential equation obeyed by $f(x)$.

9.7. Show that

$$P_{ll}(\cos\theta) = \frac{(2l)!(-1)^l}{2^l l!} \sin^l \theta \tag{9.40}$$

See equations (4.17) and (4.35). What is $P_{l,-l}$?

9.8. By differentiating (9.17) m times, show that

$$P_{l+1,m+1} - P_{l-1,m+1} = -\sqrt{1 - x^2}\,(2l+1) P_{l,m} \tag{9.41}$$

9.9. The first two Legendre functions of the second kind for integer values of l are given by

$$Q_0(x) = \frac{1}{2} \ln\left(\frac{1+x}{1-x}\right) \qquad Q_1(x) = \frac{1}{2} x \ln\left(\frac{1+x}{1-x}\right) - 1 \tag{9.42}$$

See equations (1.24) and (1.48). Show that these functions satisfy the recurrence relations (9.15) and (9.19) for $l = 1$. Let the function $Q_2(x)$ be defined in terms of $Q_0(x)$ and $Q_1(x)$ by setting $l = 1$ in equation (9.24). Show that

$$Q_2(x) = \left(\tfrac{3}{2}x^2 - \tfrac{1}{2}\right) Q_0(x) - \tfrac{3}{2}x \tag{9.43}$$

Show that the recurrence relations (9.15) and (9.19) are valid for Legendre functions of the second kind if $l = 2$.

9.10. Assume that all the Legendre functions of the second kind are known up to a given value of l and that

$$(x^2-1)Q_l' - lxQ_l + lQ_{l-1} = 0 \tag{9.44}$$

$$lQ_l - xQ_l' + Q_{l-1}' = 0 \tag{9.45}$$

are valid. Define Q_{l+1} by means of (9.24). Show as a consequence that (9.44) and (9.45) are valid if l is replaced by $l+1$. This shows by mathematical induction that the functions generated by (9.24) obey the recurrence relations (9.44) and (9.45) for arbitrary l. All other recurrence relations can be found by combining (9.44), (9.45), and (9.24) in various ways. Thus, $Q_l(x)$ and $P_l(x)$ obey the same recurrence relations. The functions $Q_{lm}(x)$ are given in terms of $Q_l(x)$ by (9.30), which is the same as the relation between $P_{lm}(x)$ and $P_l(x)$. Thus, $Q_{lm}(x)$ and $P_{lm}(x)$ also obey the same recurrence relations.

Assume that Q_l and Q_{l-1} are known to be solutions of the Legendre equation (9.38). In the case of Q_{l-1}, we must of course replace l by $l-1$. Show that as a consequence, Q_{l+1} obeys (9.38), with l replaced by $l+1$.

9.11. Show that an integral representation of the Legendre polynomials is

$$P_l(z) = \frac{1}{2\pi i 2^l} \oint \frac{(\zeta^2-1)^l \, d\zeta}{(\zeta - z)^{l+1}} \tag{9.46}$$

where the contour of integration encircles the point $\zeta = z$. This is called the *Schlaefi integral representation* and is valid if z is a complex variable. Refer to equations (5.52) and (5.53) for a hint as to the procedure. By making the change of variable

$$\zeta = z + \sqrt{z^2-1}\, e^{i\phi} \tag{9.47a}$$

show that

$$P_l(z) = \frac{1}{2\pi} \int_0^{2\pi} \left(z + \sqrt{z^2-1} \cos\phi \right)^l d\phi \tag{9.47b}$$

which is called *Laplace's integral representation.*

9.12. Prove that

$$P_{lm}(z) = \frac{(l+m)!(-1)^m e^{-im\pi/2}}{2\pi l!} \int_0^{2\pi} \left(z + \sqrt{z^2-1} \cos\phi \right)^l \cos m\phi \, d\phi \tag{9.48}$$

provided that

$$(z^2-1)^{-m/2} = (1-z^2)^{-m/2} e^{-im\pi/2} \tag{9.49}$$

Hint: At some point in the calculation, your expression will contain the factor

$$(1-z^2)^{m/2}(z^2-1)^{-m/2} \tag{9.50}$$

There are branch points at $z = \pm 1$. There are two choices as to where to put branch cuts, either of which can be used. These are illustrated in Figs. 15.1 and 15.2 of Chapter 5. Review the discussion of equation (15.33) in Chapter 5.

9.13. Prove the formulas

$$\frac{\partial}{\partial z} r^l P_l(\cos\theta) = l r^{l-1} P_{l-1}(\cos\theta) \tag{9.51}$$

$$\frac{\partial}{\partial z} r^{-l-1} P_l(\cos\theta) = -(l+1) r^{-l-2} P_{l+1}(\cos\theta) \tag{9.52}$$

Hint: First show that

$$\frac{\partial r}{\partial z} = \cos\theta \qquad \frac{\partial \theta}{\partial z} = -\frac{\sin\theta}{r} \tag{9.53}$$

9.14. Prove the recurrence formula

$$(l+1-m)P_{l+1,m} = (2l+1)xP_{lm} - (m+l)P_{l-1,m} \tag{9.54}$$

For a given m, the associated Legendre functions P_{lm} for $l \geq m$ form a complete set of orthogonal functions. See equation (4.42). Equation (9.54) is a generalization of (9.16) and is useful for generating these functions.

10. EXAMPLES OF SOLUTIONS OF LAPLACE'S EQUATION

Assume that the potential is the known function

$$\psi(a,\theta,\phi) = f(\theta,\phi) \tag{10.1}$$

over the surface of a sphere of radius a and that the potential is zero at infinity. If there is no charge in the region $r > a$, then, from (9.4), the potential exterior to the sphere is

$$\psi_e(r,\theta,\phi) = \sum_{l=0}^{\infty}\sum_{m=-l}^{+l} \frac{b_{lm}}{r^{l+1}} Y_{lm}(\theta,\phi) \tag{10.2}$$

Terms proportional to r^l have been discarded because they do not become zero at infinity. At the surface of the sphere,

$$f(\theta,\phi) = \sum_{l=0}^{\infty}\sum_{m=-l}^{+l} \frac{b_{lm}}{a^{l+1}} Y_{lm}(\theta,\phi) \tag{10.3}$$

The spherical harmonics are complete and obey the orthogonality relation (4.37). For practical purposes, this means that any function $f(\theta,\phi)$ which could represent a physically realizable potential on the surface of the sphere can be expanded in terms of them as given by (10.3). Reference to equation (4.38) shows that the coefficients in (10.3) are given by

$$b_{lm} = a^{l+1}c_{lm}$$

$$c_{lm} = \int_{\theta=0}^{\pi}\int_{\phi=0}^{2\pi} Y^*(\theta,\phi) f(\theta,\phi) \sin\theta \, d\theta \, d\phi \tag{10.4}$$

The exterior solution is therefore

$$\psi_e = \sum_{l=0}^{\infty}\sum_{m=-l}^{+l} \left(\frac{a}{r}\right)^{l+1} c_{lm} Y_{lm}(\theta,\phi) \tag{10.5}$$

By a similar procedure, the potential at points interior to the sphere is found to be

$$\psi_i = \sum_{l=0}^{\infty}\sum_{m=-l}^{+l} \left(\frac{r}{a}\right)^{l} c_{lm} Y_{lm}(\theta,\phi) \tag{10.6}$$

We have assumed that there is no charge in the region $r < a$. Terms proportional to r^{-l-1} have been discarded because they are singular at $r = 0$. We know from theorem 9 in section 10 of Chapter 3 that the potentials given by (10.5) and (10.6) are unique. In the case of the exterior problem, the region Σ of theorem 9 is the region exterior to the sphere. This region is also bounded by a second closed surface at infinity. The potential must be specified over both of these surfaces to make the solution unique. We see that the interior and exterior solutions are related by

$$\begin{aligned} \psi_e(r,\theta,\phi) &= \frac{a}{r}\psi_i\left(\frac{a^2}{r},\theta,\phi\right) \\ \psi_i(r,\theta,\phi) &= \frac{a}{r}\psi_e\left(\frac{a^2}{r},\theta,\phi\right) \end{aligned} \tag{10.7}$$

This result is called *inversion in the sphere* or sometimes the *inversion theorem*.

As an example, suppose that the potential over the surface of the sphere is known to be

$$f(\theta,\phi) = V\sin\theta\cos\phi \tag{10.8}$$

The coefficients can be formally calculated by means of (10.4). Because (10.8) is a simple function, it is easier to express it directly as a linear combination of spherical harmonics by noting that

$$Y_{11}(\theta,\phi) = -\sqrt{\frac{3}{8\pi}}\sin\theta e^{i\phi}$$

$$Y_{1,-1}(\theta,\phi) = \sqrt{\frac{3}{8\pi}}\sin\theta e^{-i\phi}$$

$$f(\theta,\phi) = \frac{V}{2}\sin\theta(e^{i\phi}+e^{-i\phi})$$

$$= \frac{V}{2}\sqrt{\frac{8\pi}{3}}(Y_{1,-1}-Y_{11}) \tag{10.9}$$

The exterior solution must therefore be

$$\psi_e = \left(\frac{a}{r}\right)^2 \frac{V}{2}\sqrt{\frac{8\pi}{3}}(Y_{1,-1}-Y_{11})$$

$$= \left(\frac{a}{r}\right)^2 V\sin\theta\cos\phi \tag{10.10}$$

The interior solution can be found from the inversion theorem (10.7):

$$\psi_i = \frac{r}{a}V\sin\theta\cos\phi \tag{10.11}$$

In the remainder of this section, several specific solutions will be worked out. In all the remaining cases, the potentials are axially symmetric, meaning that they have no dependence on the azimuth angle ϕ. The general form of the solution is therefore

$$\psi(r,\theta) = \sum_{l=0}^{\infty}\left(A_l r^l + \frac{B_l}{r^{l+1}}\right)P_l(\cos\theta) \tag{10.12}$$

Conducting Sphere in a Uniform Field

Imagine that an uncharged conducting sphere of radius a is placed in a region of space where prior to the introduction of the sphere a uniform electric field existed. The problem is to find how the field is modified by the presence of the sphere. Far from the sphere, the field is still uniform and we will take it to be in the positive z direction. The potential at infinity is therefore

$$\psi_\infty = -Ez = -Er\cos\theta \tag{10.13}$$

where E is the field strength. The general expression for the potential can be represented as

$$\psi = -Er\cos\theta + \frac{B_0}{r} + \frac{B_1}{r^2}\cos\theta + \sum_{l=2}^{\infty}\frac{B_l}{r^{l+1}}P_l(\cos\theta) \tag{10.14}$$

All terms proportional to r^l with the exception of $l=1$ have been discarded so that (10.14) agrees with (10.13) in the limit $r\to\infty$. The center of the sphere is at the origin of coordinates; and if it has no net charge on it, the potential of its surface is zero. Setting $r=a$ in (10.14) therefore gives

$$0 = -Ea\cos\theta + \frac{B_0}{a} + \frac{B_1}{a^2}\cos\theta + \sum_{l=2}^{\infty}\frac{B_l}{a^{l+1}}P_l(\cos\theta) \tag{10.15}$$

which is an expansion of the zero function in terms of the Legendre polynomials. Since the Legendre polynomials are complete, this means that the coefficients of each Legendre polynomial must separately vanish. Thus,

$$-Ea + \frac{B_1}{a^2} = 0 \qquad B_0 = 0 \qquad B_l = 0 \text{ if } l \geq 2 \tag{10.16}$$

The required potential is therefore

$$\psi = E\left(\frac{a^3}{r^2} - r\right)\cos\theta \qquad r > a \tag{10.17}$$

The electric field components are given by

$$E_r = -\frac{\partial\psi}{\partial r} = E\left(\frac{2a^3}{r^3}+1\right)\cos\theta$$

$$E_\theta = -\frac{1}{r}\frac{\partial\psi}{\partial\theta} = E\left(\frac{a^3}{r^3}-1\right)\sin\theta \tag{10.18}$$

The differential equation for the field lines is $E_r r\,d\theta = E_\theta\,dr$ from which we find

$$\frac{\cos\theta}{\sin\theta}d\theta = \frac{a^3-r^3}{2a^3r+r^4}dr = \frac{dr}{2r} - \frac{3r^2\,dr}{2(r^3+2a^3)} \tag{10.19}$$

Integration yields

$$\ln(\sin\theta) - \ln ak = \tfrac{1}{2}\ln r - \tfrac{1}{2}\ln(r^3+2a^3)$$

$$\sin\theta = k\sqrt{\frac{ra^2}{2a^3+r^3}} \tag{10.20}$$

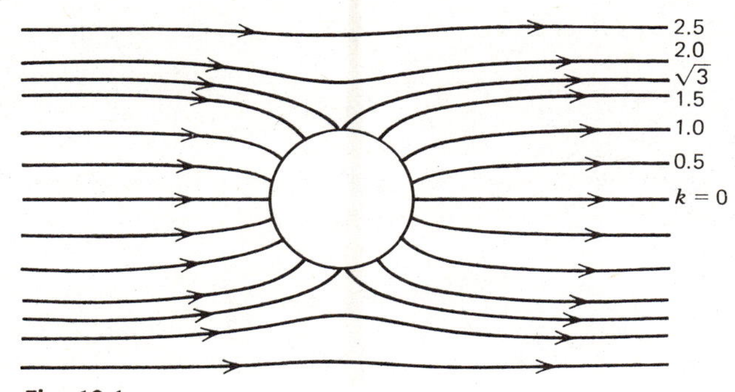

Fig. 10.1

where k is a constant of integration each choice of which gives a different field line. As $r \to \infty$, we find the limiting form

$$r \sin \theta = ka \tag{10.21}$$

Since (10.21) represents a straight line parallel to the z axis, equal increments of the constant k give equally spaced field lines at infinity. Figure 10.1 shows the field lines which are obtained from (10.20). By setting $r = a$ in (10.20), we find that the particular choice $k = \sqrt{3}$ gives $\theta = \pi/2$. This field line represents the division between the field lines which terminate on the sphere and those which pass it by. This is the one field line which does not intersect the sphere at right angles in apparent violation of the well-known fact that the electric field is always at right angles to the equipotential surfaces. Note however from (10.18) that the field strength is actually zero at the point of intersection.

Point Dipole at the Center of a Grounded Conducting Sphere

Electric dipoles are discussed in section 2 of Chapter 4. A point dipole represents a singularity at $r = 0$ and when the general solution (10.12) is specialized to fit this case, the term corresponding to a dipole singularity is retained and all other singular terms are discarded. The result is

$$\psi = \frac{p \cos \theta}{4\pi\varepsilon_0 r^2} + \sum_{l=0}^{\infty} A_l r^l P_l(\cos \theta) \tag{10.22}$$

On the surface of the sphere, $\psi = 0$ and $r = a$, giving

$$0 = \frac{p \cos \theta}{4\pi\varepsilon_0 a^2} + A_0 + A_1 a \cos \theta + \sum_{l=2}^{\infty} A_l a^l P_l(\cos \theta) \tag{10.23}$$

Therefore,

$$A_0 = 0 \qquad \frac{p}{4\pi\varepsilon_0 a^2} + A_1 a = 0 \qquad A_l = 0 \text{ if } l \geq 2 \tag{10.24}$$

The potential is then

$$\psi = \frac{p}{4\pi\varepsilon_0 a^3}\left(\frac{a^3}{r^2} - r\right) \cos \theta \qquad r < a \tag{10.25}$$

The term proportional to $r \cos \theta$ is the contribution to the potential resulting from the surface density of charge which is induced on the inner surface of the sphere. Note that (10.17) and (10.25) are the same if

$$\frac{p}{4\pi\varepsilon_0 a^3} = E \tag{10.26}$$

Both potentials represent the superposition of a dipole field and a uniform field, with the dipole vector **p** pointing in the same direction as the uniform field. In the example of a conducting sphere placed in a uniform field, there is of course no point dipole at its center. However, the induced surface charge density on the sphere produces the same field external to the sphere as would a point dipole. If a point dipole is placed in a uniform field of strength E and aligned with it, a sphere of radius a as determined by (10.26) centered on the dipole is a surface of zero potential of the combined fields. The field lines interior to the sphere are given by (10.20). Figure 10.2 shows these field lines for the same values of the parameter k that were used to obtain Fig. 10.1. Again, the field line given by $k = \sqrt{3}$ is the boundary between the field lines which intersect the spherical surface and those which do not. The point $\theta = \pi/2$ where this field line intersects the spherical surface is a saddle point of the potential function. An electrostatic potential can have saddle points but no maxima or minima in a charge-free region of space. Refer to theorem 6 in section 10 of Chapter 3.

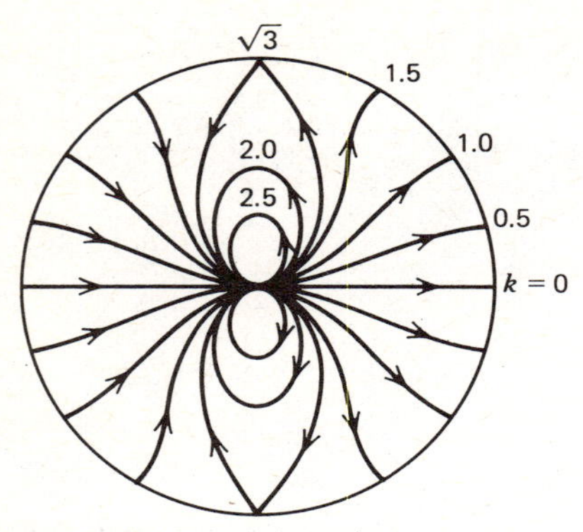

Fig. 10.2

Conducting Hemispheres at ± V

A conducting sphere of radius a with its center at the origin is cut in half at the x, y plane. The two halves are slightly separated to permit insulation. The top half ($z > 0$) is charged to a potential $+V$ and the bottom half is charged to a potential $-V$. The problem is to find the potential everywhere exterior to the sphere. If the potential is zero at infinity, then all terms in (10.12) proportional to r^l are discarded. The resulting potential is

$$\psi(r, \theta) = \sum_{l=0}^{\infty} \frac{B_l}{r^{l+1}} P_l(\cos\theta) \tag{10.27}$$

On the surface of the sphere, the potential is a known function of θ:

$$f(x) = \sum_{l=0}^{\infty} \frac{B_l}{a^{l+1}} P_l(x) \qquad x = \cos\theta \tag{10.28}$$

The coefficients are evaluated using the orthogonality of the Legendre polynomials and the normalization integral (4.29):

$$B_l = a^{l+1}(l + \tfrac{1}{2}) \int_{-1}^{+1} f(x) P_l(x)\, dx \tag{10.29}$$

The function $f(x)$ is given by

$$\begin{aligned} f(x) &= -V \qquad -1 < x < 0 \qquad \frac{\pi}{2} < \theta < \pi \\ f(x) &= +V \qquad 0 < x < 1 \qquad 0 < \theta < \frac{\pi}{2} \end{aligned} \tag{10.30}$$

Since $f(x)$ is odd and the since the Legendre polynomials are even if l is even and odd if l is odd, the even coefficients are zero. The odd coefficients are

$$B_l = Va^{l+1}(2l + 1) \int_0^1 P_l(x)\, dx \qquad l \text{ odd} \tag{10.31}$$

The integral can be evaluated by using the recurrence relation (9.17) with the result

$$\begin{aligned} B_l &= Va^{l+1} \int_0^1 \left[P'_{l+1} - P'_{l-1} \right] dx \\ &= Va^{l+1}\left[P_{l-1}(0) - P_{l+1}(0) \right] \end{aligned} \tag{10.32}$$

where use is made of $P_l(1) = 1$. By setting $l = 2n + 1$, we can express the potential as

$$\begin{aligned} \psi(r, \theta) = V \sum_{n=0}^{\infty} \left(\frac{a}{r}\right)^{2n+2} &\left[P_{2n}(0) \right. \\ &\left. - P_{2n+2}(0) \right] P_{2n+1}(\cos\theta) \end{aligned} \tag{10.33}$$

The values of $P_{2n}(0)$ are given by equation (4.45).

A finite number of terms of the expansion of the step function (10.30) is

$$f_N(x) = V \sum_{n=0}^{N} \left[P_{2n}(0) - P_{2n+2}(0) \right] P_{2n+1}(x) \tag{10.34}$$

It is of interest to examine the convergence properties of this series by graphing $f_N(x)$ for different values of N. This has been done in Fig. 10.3 for $N = 5$ and $N = 10$. Figure 2.3 of Chapter 8 shows a Fourier series expansion of a step function and its obvious resemblance to Fig. 10.3 should be noted. In particular, note the appearance of the Gibbs phenomenon in the Legendre series expansion. The Gibbs phenomenon is discussed in connection with Fourier series in section 2 of Chapter 8.

Uniformly Charged Ring

If a charge q is distributed uniformly on a thin wire in the shape of a circle of radius a, it is easy to show that the potential at any point on the axis at a distance z from the center of the circle is

$$\psi(z) = \frac{q}{4\pi\varepsilon_0} \frac{1}{\sqrt{a^2 + z^2}} \tag{10.35}$$

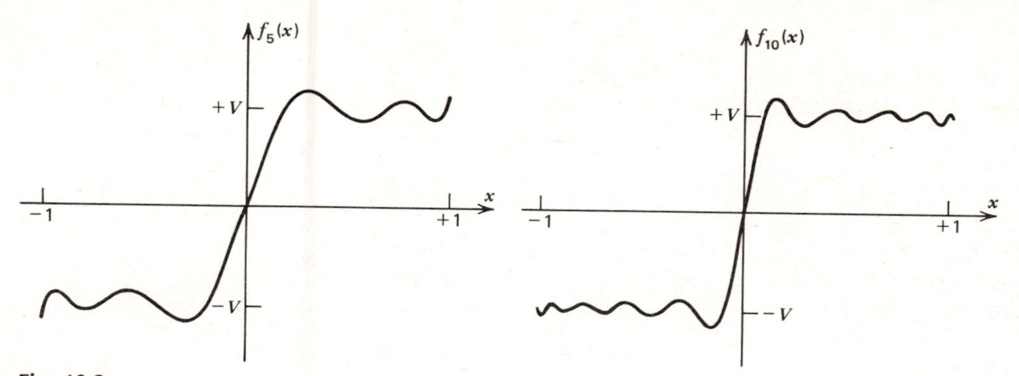

Fig. 10.3

By means of a binomial expansion,

$$\psi(z)=\frac{q}{4\pi\varepsilon_0 z}\left[1+\left(\frac{a}{z}\right)^2\right]^{-1/2}$$

$$=\frac{q}{4\pi\varepsilon_0 z}\sum_{n=0}^{\infty}\frac{(-1)^n(2n)!}{2^{2n}(n!)^2}\left(\frac{a}{z}\right)^{2n} \tag{10.36}$$

which is valid if $z>a$. The general form of the potential for $r>a$ is given by (10.27). By setting $\cos\theta=1$ and then comparing (10.27) and (10.36), the coefficients can be evaluated with the result that

$$\psi(r,\theta)=\frac{q}{4\pi\varepsilon_0 r}\sum_{n=0}^{\infty}\frac{(-1)^n(2n)!}{2^{2n}(n!)^2}\left(\frac{a}{r}\right)^{2n}P_{2n}(\cos\theta)$$

$$r>a \tag{10.37}$$

Note that the odd coefficients are zero. This same technique can be used to solve any axially symmetric potential problem once the potential is known along the axis. The reader can find an expansion similar to (10.37) which is valid if $r<a$.

Axially Symmetric Problems in Cylindrical Coordinates

From equation (10.21) in Chapter 4, we see that Laplace's equation in cylindrical coordinates for an axially symmetric potential problem is

$$\frac{1}{r}\frac{\partial}{\partial r}\left(r\frac{\partial\psi}{\partial r}\right)+\frac{\partial^2\psi}{\partial z^2}=0 \qquad r=\sqrt{x^2+y^2} \tag{10.38}$$

where r is now measured perpendicular to the z axis and ψ depends only on r and z. Suppose that the potential along the z axis is the known function $\psi_0(z)$. At points off the axis, the potential can be expanded in a Taylor's series with respect to the variable r:

$$\psi(r,z)=\psi_0(z)+f_1(z)r+f_2(z)r^2$$
$$+f_3(z)r^3+f_4(z)r^4+\cdots \tag{10.39}$$

The coefficients of the various powers of r are functions of z which we will now determine. By direct substitution of (10.39) into (10.38), we find that

$$f_1+4f_2r+9f_3r^2+16f_4r^3+\cdots$$
$$+\frac{d^2\psi_0}{dz^2}r+\frac{d^2f_1}{dz^2}r^2+\frac{d^2f_2}{dz^2}r^3+\cdots=0 \tag{10.40}$$

Since this relation must hold for all values of r, the coefficients of the like powers of r must separately vanish:

$$f_1=0 \qquad 4f_2+\frac{d^2\psi_0}{dz^2}=0$$

$$9f_3+\frac{d^2f_1}{dz^2}=0 \qquad 16f_4+\frac{d^2f_2}{dz^2}=0 \tag{10.41}$$

By solving for the unknown functions of z, we find that the potential can be expressed as

$$\psi(r,z)=\psi_0-\frac{1}{4}\frac{d^2\psi_0}{dz^2}r^2+\frac{1}{64}\frac{d^4\psi_0}{dz^4}r^4-\cdots \tag{10.42}$$

This result is especially useful if the potential is needed at points which are not too far removed from the axis of symmetry.

PROBLEMS

10.1. Separate Laplace's equation in cylindrical coordinates for the case where the potential has no dependence on z and show that the general solution for the case where $0 \le \theta \le 2\pi$ can be put in the form

$$\psi(r,\theta) = C_0 + C_1 \ln r + \sum_{n=1}^{\infty} (A_n r^n + B_n r^{-n})(a_n \sin n\theta + b_n \cos n\theta) \tag{10.43}$$

A uniform electric field of strength E exists in the x direction. An infinitely long uncharged conducting cylinder of radius a is placed in this field. If the axis of the cylinder coincides with the z axis, show that the resulting potential is

$$\psi(r,\theta) = E\left(\frac{a^2}{r} - r\right)\cos\theta \qquad r > a \tag{10.44}$$

10.2. Find the expression for the induced surface charge density on an initially uncharged conducting sphere after it is placed in a uniform field. Show that the integral over the sphere of the surface charge density is zero. How is the potential (10.17) modified if the sphere initially has a net charge Q on it before it is placed in the field?

10.3. Show that the field line given by $k = \sqrt{3}$ in Fig. 10.1 intersects the surface of the conducting sphere at an angle of 45°.

10.4. Use the inversion theorem to find the potential due to a charged ring of radius a which is valid for $r < a$.

10.5. Show that the potential at points slightly off the axis of a uniformly charged ring is

$$\psi \simeq \frac{q}{4\pi\varepsilon_0}\frac{1}{\sqrt{a^2+z^2}}\left[1 - \frac{(2z^2-a^2)r^2}{4(a^2+z^2)^2}\right] \tag{10.45}$$

10.6. Two spheres of radii a and b are concentric. Over the surface of the inner sphere of radius a, the potential is given by

$$\psi(a,\theta) = f(\cos\theta) = f(x) \tag{10.46}$$

where $f(x)$ is a known function. There is no dependence on the azimuth angle ϕ. The outer sphere of radius b is grounded. Show that the potential at any point between the spheres is given by

$$\psi(r,\theta) = \sum_{l=0}^{\infty} A_l \left[r^l - \frac{b^{2l+1}}{r^{l+1}}\right] P_l(\cos\theta) \tag{10.47}$$

where the coeffients are determined by

$$A_l a^l \left[1 - \left(\frac{b}{a}\right)^{2l+1}\right] = \left(l + \frac{1}{2}\right)\int_{-1}^{+1} f(x) P_l(x)\,dx \tag{10.48}$$

10.7. The surface charge density on a charged conducting disc of radius c is given by equation (12.43) in Chapter 4. Counting both sides of the disc, the total surface charge density is

$$\lambda(s) = \frac{Q}{2\pi c\sqrt{c^2 - s^2}} \tag{10.49}$$

Show that for $r > c$, the potential of the disc can be expressed as

$$\psi = \frac{Q}{4\pi\varepsilon_0 r}\sum_{n=0}^{\infty}\frac{(-1)^n}{2n+1}\left(\frac{c}{r}\right)^{2n} P_{2n}(\cos\theta) \tag{10.50}$$

You can do the calculation by first using (10.37) to get an expression for the contribution to the potential from a ring of charge between s and $s+ds$. You will need equations (22.17) and (22.29) of Chapter 5. The inversion theorem cannot be used in (10.50) to find the potential for $r<c$. Why not?

Another way to do the problem is to show from equation (12.39) of Chapter 4 that the potential on the axis of the disc is

$$\psi=\frac{Q}{4\pi\varepsilon_0 c}\sin^{-1}\left(\frac{c}{\sqrt{c^2+r^2}}\right)=\frac{Q}{4\pi\varepsilon_0 c}\tan^{-1}\left(\frac{c}{r}\right) \tag{10.51}$$

Now use equation (16.15) of Chapter 5 to expand the inverse tangent. If you want an expansion of the potential in terms of Legendre polynomials valid for $r<c$, you can start out with

$$\tan^{-1}\left(\frac{c}{r}\right)=\frac{\pi}{2}-\tan^{-1}\left(\frac{r}{c}\right) \tag{10.52}$$

Show that the capacitance of the disc is

$$C=8\varepsilon_0 c \tag{10.53}$$

10.8. Evaluate the following integral:

$$\begin{aligned}\int_{-1}^{+1} xP_l P_n\,dx &= \frac{2(l+1)}{(2l+1)(2l+3)} \qquad n=l+1\\ &= \frac{2l}{(2l+1)(2l-1)} \qquad n=l-1\\ &=0 \qquad n\neq l\pm 1\end{aligned} \tag{10.54}$$

11. PROBLEMS IN MAGNETOSTATICS

If an electric dipole of moment $\mathbf{p}$ is located at $\mathbf{r}'$, it creates an electrostatic potential at $\mathbf{r}$ given by

$$\psi=\frac{1}{4\pi\varepsilon_0}\frac{\mathbf{p}\cdot\mathbf{R}}{R^3} \tag{11.1}$$

Electric dipoles are discussed in section 2 of Chapter 4; see especially problem 2.2 and Fig. 2.4. The molecules of an insulating material can be regarded as point dipoles which are randomly oriented in the absence of any external field. When such a material is placed in an electric field, it polarizes and thereby modifies the original field because the individual dipoles which make up the medium are making a contribution to the potential given by (11.1). The net contribution due to the polarization of the insulator can be written as

$$\psi=\frac{1}{4\pi\varepsilon_0}\int_\Sigma \frac{\mathbf{P}\cdot\mathbf{R}}{R^3}\,d\Sigma' \tag{11.2}$$

where $\mathbf{P}$ is the polarization vector of the medium and the integral is over its volume. This is the starting point for the development of the theory of dielectric materials. The significance of the polarization vector is that it is the dipole moment per unit volume. Thus, $\mathbf{P}\,d\Sigma'$ is the dipole moment of the volume element $d\Sigma'$.

A surface layer of dipoles can be constructed by taking two surfaces with surface charge densities of equal magnitude but opposite sign and then placing them close together, as illustrated in Fig. 11.1. The potential created by such a surface is

$$\psi=\frac{1}{4\pi\varepsilon_0}\int_\sigma \frac{N\mathbf{R}\cdot d\boldsymbol{\sigma}'}{R^3} \tag{11.3}$$

where N is the strength of the dipole layer in units of dipole moment per unit area and $N\,d\boldsymbol{\sigma}'$ is the dipole moment of the area $d\boldsymbol{\sigma}'$. The individual dipoles which make up the layer are assumed to be aligned perpendicular to the surface. If N is constant, then

$$\psi=\frac{N}{4\pi\varepsilon_0}\int_\sigma \frac{\mathbf{R}\cdot d\boldsymbol{\sigma}'}{R^3} \tag{11.4}$$

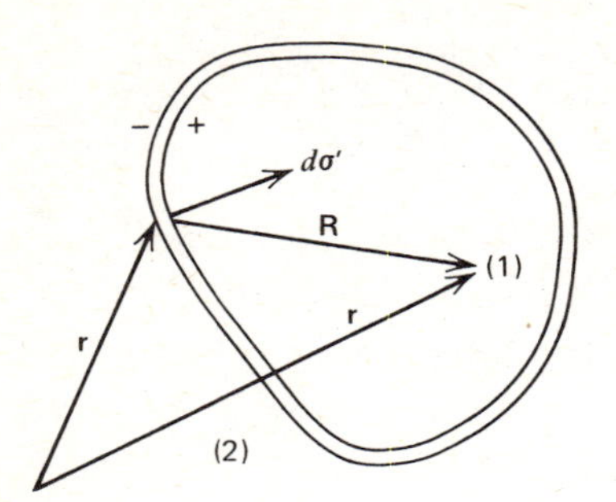

Fig. 11.1

which is interesting because the integral is the solid angle subtended by the surface at the point where the potential is being evaluated. If, as indicated in Fig. 11.1, the surface is closed and the dipoles point inward, then (11.4) gives

$$\psi_1 = \frac{N}{\varepsilon_0} \qquad \psi_2 = 0 \tag{11.5}$$

where ψ_1 is the potential at any point inside the surface and ψ_2 is the potential at any point outside. A significant feature of this result is that the potential behaves discontinuously at the surface. The amount of the discontinuity is

$$\psi_1 - \psi_2 = \frac{N}{\varepsilon_0} \tag{11.6}$$

As an example, consider a surface layer of dipoles on a circular disc of radius a as pictured in Fig. 11.2. If the density of dipoles is uniform, then the potential at points along the axis is given by

$$\begin{aligned}\psi &= \frac{N}{4\pi\varepsilon_0}\int_0^a \frac{\cos\theta\, 2\pi r'\,dr'}{R^2}\\ &= \frac{N}{2\varepsilon_0}\int_0^a \frac{zr'\,dr'}{(r'^2+z^2)^{3/2}}\\ &= \frac{Nz}{2\varepsilon_0}\left[-\frac{1}{\sqrt{r'^2+z^2}}\right]_0^a = \frac{N}{2\varepsilon_0}\left[\frac{z}{|z|} - \frac{z}{\sqrt{a^2+z^2}}\right]\end{aligned} \tag{11.7}$$

The result of graphing ψ as a function of z along the axis of the disc is shown in Fig. 11.3. Note that at the surface of the disc ($z=0$), there is a discontinuity in the potential function which is again given by (11.6). All potentials we have previously studied are due to volume or surface densities of charge and are continuous functions of position, except at the singularities which are produced either by point charges or by line charges. By means of a dipole layer, we are able to create a discontinuity in the potential function. As unlikely as it may seem, dipole layers can be used as a method of solving problems in magnetostatics as we now show. Before continuing to read this section, you may wish to review section 11 in Chapter 3.

Figure 11.4 shows a loop of wire carrying a current I. Everywhere except at the location of the wire it is true that $\nabla \times \mathbf{B} = 0$; and in any simply connected region of space through which the wire does not pass, we can write $\mathbf{B} = -\nabla\psi$. Then since $\nabla\cdot\mathbf{B} = 0$ everywhere, we have $\nabla^2\psi = 0$. We will refer to ψ as a *magnetic scalar potential*. Since vector potentials are

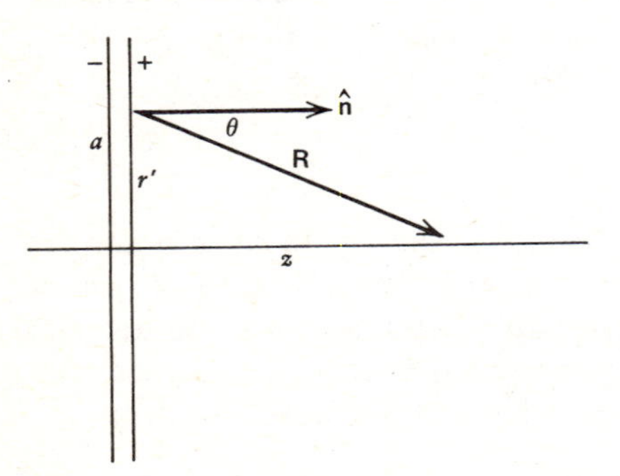

Fig. 11.2

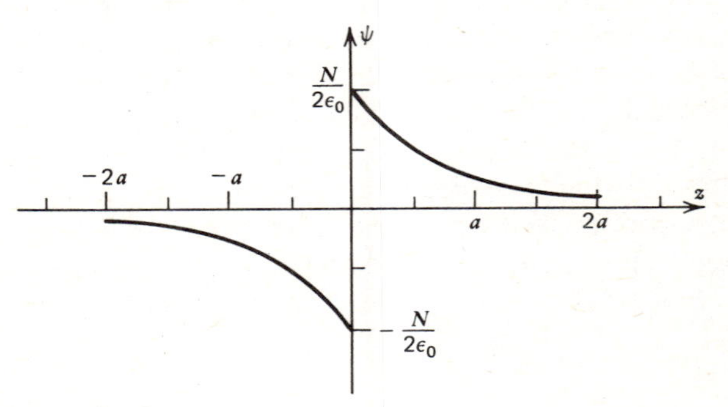

Fig. 11.3

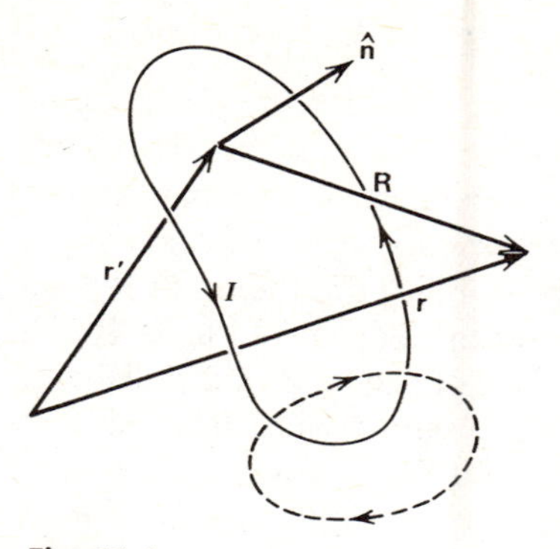

Fig. 11.4

harder to use than scalar potentials, we are motivated to exploit the possibility of a magnetic scalar potential as a way to calculate static magnetic fields. A clue as to how to relate such a potential to the source of the field, which is really a current rather than a scalar source, is provided by the discussion of dipole layers in electrostatics. If the line integral of the magnetic field is calculated around a closed path which links the current loop, such as the dashed path of Fig. 11.4, then Ampere's law gives

$$\oint \mathbf{B}\cdot d\mathbf{s} = \mu_0 I \tag{11.8}$$

If we write $\mathbf{B} = -\nabla\psi$, then

$$\int_1^2 \mathbf{B}\cdot d\mathbf{s} = -\psi_2 + \psi_1 \tag{11.9}$$

The magnetic scalar potential can be made consistent with (11.8) provided that it is multiple-valued and has a jump discontinuity given by

$$\psi_1 - \psi_2 = \mu_0 I \tag{11.10}$$

at an open surface which is bounded by the current loop. The comparison of (11.10) and (11.6) suggests that it is possible to replace the current loop by an equivalent dipole layer the strength of which is determined by

$$N = \mu_0 \varepsilon_0 I \tag{11.11}$$

The magnetic scalar potential is then to be calculated by means of

$$\psi = \frac{\mu_0 I}{4\pi}\int_\sigma \frac{\mathbf{R}\cdot d\sigma'}{R^3} \tag{11.12}$$

The surface over which the integral is carried out can be any open surface bounded by the current loop. The equivalent dipole layer is called a *magnetic double layer* or *magnetic shell.*

In order to test this idea out, let us replace N by $\mu_0\varepsilon_0 I$ in equation (11.7) and then calculate the field along the z axis. We find

$$\begin{aligned} \psi &= \frac{\mu_0 I}{2}\left[\frac{z}{|z|} - \frac{z}{\sqrt{a^2+z^2}}\right] \\ B_z &= -\frac{\partial\psi}{\partial z} = \frac{\mu_0 I}{2}\frac{a^2}{(a^2+z^2)^{3/2}} \end{aligned} \tag{11.13}$$

which agrees exactly with the magnetic field calculated from the Biot–Savart law. The potential is discontinuous at $z=0$, but the field evaluated at $z-0$ is the same as at $z+0$ because ψ has the same slope at these points. We remark that in the case of (11.7), which gives the potential of an actual dipole disc, we would get

$$E_z = \frac{N}{2\varepsilon_0}\frac{a^2}{(a^2+z^2)^{3/2}} - \frac{N}{\varepsilon_0}\delta(z) \tag{11.14}$$

in place of (11.13). Recall that $\delta(z)$ is a Dirac delta function. See equation (2.33) in Chapter 8. The reason for the appearance of the delta function is that in the limit as the distance between the two discs in Fig. 11.2 goes to zero, the field at $z=0$ becomes arbitrarily large. There is no delta function in the case of the magnetic field because the magnetic shell has no actual physical existence.

The formal confirmation of the validity of an equivalent magnetic shell as a method for calculating magnetic fields starts with the Biot–Savart law for the magnetic field of a single current loop:

$$\mathbf{B} = \frac{\mu_0 I}{4\pi}\oint \frac{d\mathbf{s}'\times\mathbf{R}}{R^3} \tag{11.15}$$

By means of the integral vector identity (9.21) in Chapter 3,

$$\mathbf{B} = \frac{\mu_0 I}{4\pi}\int_\sigma (\mathbf{n}\times\nabla')\times\frac{\mathbf{R}}{R^3}d\sigma' \tag{11.16}$$

where σ is any open surface bounded by the current loop. We temporarily set $\mathbf{a} = \mathbf{R}/R^3$ and rewrite the integrand by expressing the cross products in terms of permutation symbols as given by equation (8.4) in Chapter 3:

$$(\hat{\mathbf{n}} \times \nabla') \times \mathbf{a} = e_{ijk}\left(e_{jrs} n_r \partial'_s\right) a_k \hat{\mathbf{e}}_i = e_{jki} e_{jrs} n_r \partial'_s a_k \hat{\mathbf{e}}_i \tag{11.17}$$

The various components of **a** are functions only of the difference between **r** and **r**′:

$$a_k = a_k\left(x_1 - x'_1, x_2 - x'_2, x_3 - x'_3\right) \tag{11.18}$$

Therefore,

$$\partial'_s a_k = -\partial_s a_k \tag{11.19}$$

Now use the identity given by equation (8.31) in Chapter 2 to get

$$\begin{aligned}(\hat{\mathbf{n}} \times \nabla') \times \mathbf{a} &= -(\delta_{kr}\delta_{is} - \delta_{ks}\delta_{ir}) n_r \partial_s a_k \hat{\mathbf{e}}_i \\ &= (-n_k \partial_i a_k + n_i \partial_k a_k)\hat{\mathbf{e}}_i \end{aligned} \tag{11.20}$$

Since n_k is a function of **r**′ and not **r**, we can move it past the symbol ∂_i and write (11.20) as

$$\begin{aligned}(\hat{\mathbf{n}} \times \nabla') \times \frac{\mathbf{R}}{R^3} &= -\nabla\left(\hat{\mathbf{n}} \cdot \frac{R}{R^3}\right) + \hat{\mathbf{n}}\left(\nabla \cdot \frac{\mathbf{R}}{R^3}\right) \\ &= -\nabla\left(\hat{\mathbf{n}} \cdot \frac{\mathbf{R}}{R^3}\right) \end{aligned} \tag{11.21}$$

where we have used equation (10.8) of Chapter 3. The Biot–Savart law (11.15) is therefore equivalent to

$$\mathbf{B} = -\frac{\mu_0 I}{4\pi} \int_\sigma \nabla\left(\hat{\mathbf{n}} \cdot \frac{\mathbf{R}}{R^3}\right) d\sigma' = -\nabla\left(\frac{\mu_0 I}{4\pi} \int_\sigma \frac{d\sigma' \cdot \mathbf{R}}{R^3}\right) \tag{11.22}$$

which leaves no doubt as to the validity of the magnetic shell method for calculating a magnetic field. The equivalent dipole layer has no actual physical reality. The idea is that a vector source (the current) can be replaced by an equivalent, though physically nonexistent, scalar source which ultimately produces the same field. One way to extend the method to volume distributions of current is to first find the contribution to the potential due to a single loop and then integrate the loop over the volume. The remainder of this section contains examples of magnetic field calculations.

Magnetic Field of a Circular Current Loop

It is possible to find expressions for the components of the magnetic field due to a circular current loop which are valid at all points in space as expansions in terms of Legendre functions about an arbitrary point on the axis of the loop. In Fig. 11.5, the point P is on the axis of a circular current loop of radius a. It is a distance s from the center of the loop and a distance b from each point of its circumference. Imagine that a spherical surface of radius b has its center at P. Let r and θ be polar coordinates with respect to P as origin. The spherical cap defined by $0 \le \theta \le \alpha$ will be chosen as the location of the equivalent magnetic shell bounded by the current loop. Since the magnetic scalar potential is a solution of Laplace's equation, it can be represented as an expansion in terms of Legendre polynomials:

$$\begin{aligned} \psi_1 &= \sum_{l=0}^{\infty} r^l A_l P_l(\cos\theta) \qquad r < b \\ \psi_2 &= \sum_{l=0}^{\infty} \frac{B_l}{r^{l+1}} P_l(\cos\theta) \qquad r > b \end{aligned} \tag{11.23}$$

The potential is discontinuous at the location of the dipole layer and continuous at other points of the spherical surface $r = b$:

$$\begin{aligned} \psi_2 - \psi_1 &= \mu_0 I \qquad 0 \le \theta < \alpha \\ \psi_2 - \psi_1 &= 0 \qquad \alpha \le \theta \le \pi \end{aligned} \tag{11.24}$$

Application of this condition to the potential (11.23)

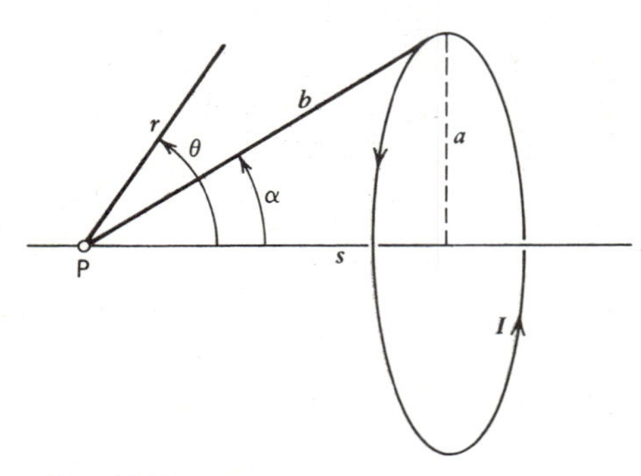

Fig. 11.5

gives

$$\sum_{l=0}^{\infty}\left(\frac{B_{l+1}}{b^{l+1}}-A_l b^l\right)P_l(x)=\mu_0 I \qquad \cos\alpha\le x\le 1$$
$$=0 \qquad -1\le x<\cos\alpha \tag{11.25}$$

Equation (11.25) is the expansion of a step function in terms of Legendre polynomials. The coefficients are

$$\frac{B_l}{b^{l+1}}-A_l b^l=\mu_0 I\left(l+\frac{1}{2}\right)\int_{\cos\alpha}^{1}P_l(x)\,dx \tag{11.26}$$

A second condition on the potential is provided by the requirement that the magnetic field itself be continuous everywhere on the surface $r=b$:

$$\left(\frac{\partial\psi_1}{\partial r}\right)_b=\left(\frac{\partial\psi_2}{\partial r}\right)_b \tag{11.27}$$

Therefore,

$$lb^{l-1}A_l=-\frac{l+1}{b^{l+2}}B_l \tag{11.28}$$

The solution of (11.36) and (11.28) for the coefficients is

$$A_l=-\frac{1}{2}\mu_0 I(l+1)b^{-l}\int_{\cos\alpha}^{1}P_l(x)\,dx \tag{11.29}$$

$$B_l=\frac{1}{2}\mu_0 I l b^{l+1}\int_{\cos\alpha}^{1}P_l(x)\,dx \tag{11.30}$$

The integral which appears in (11.29) and (11.30) can be evaluated with the aid of the recurrence relation (9.17):

$$\int_{\cos\alpha}^{1}P_l(x)\,dx=\frac{1}{2l+1}\int_{\cos\alpha}^{1}\left[P'_{l+1}-P'_{l-1}\right]dx$$
$$=\frac{1}{2l+1}\left[P_{l-1}(\cos\alpha)-P_{l+1}(\cos\alpha)\right] \tag{11.31}$$

This can be used as it stands for numerical computation. However, another form is possible. By setting $m=-1$ in (9.41) and then using (4.41), we find

$$P_{l-1}-P_{l+1}=\sqrt{1-x^2}\,(2l+1)P_{l,-1}$$
$$=\sqrt{1-x^2}\,(2l+1)(-1)\frac{(l-1)!}{(l+1)!}P_{l,1}$$
$$=-\sqrt{1-x^2}\,\frac{2l+1}{l(l+1)}P_{l,1} \tag{11.32}$$

Thus,

$$\int_{\cos\alpha}^{1}P_l(x)\,dx=-\frac{\sin\alpha}{l(l+1)}P_{l,1}(\cos\alpha) \tag{11.33}$$

Equations (11.31) and (11.33) do not work for $l=0$. By direct evaluation of the integral for this case, we find

$$\int_{\cos\alpha}^{1}P_0(x)\,dx=1-\cos\alpha \tag{11.34}$$

Evaluation of some of the coefficients gives

$$A_0=-\tfrac{1}{2}\mu_0 I(1-\cos\alpha)$$
$$A_1=-\frac{\mu_0 I}{2b}\sin^2\alpha$$
$$A_2=\frac{3\mu_0 I}{4b^2}\sin^2\alpha\cos\alpha \tag{11.35}$$
$$A_3=\frac{\mu_0 I}{4b^3}\sin^2\alpha(5\cos^2\alpha-1)$$
$$A_4=-\frac{5\mu_0 I}{16b^4}\sin^2\alpha(7\cos^3\alpha-3\cos\alpha)$$
$$A_5=-\frac{3\mu_0 I}{16b^5}\sin^2\alpha(21\cos^4\alpha-14\cos^2\alpha+1)$$

The field components are given by

$$B_r=-\frac{\partial\psi}{\partial r}=-\sum_{l=0}^{\infty}A_l l r^{l-1}P_l(\cos\theta) \tag{11.36}$$

$$B_\theta=-\frac{1}{r}\frac{\partial\psi}{\partial\theta}=-\sum_{l=0}^{\infty}A_l r^{l-1}P'_l(\cos\theta)(-\sin\theta) \tag{11.37}$$

By means of equation (4.36),

$$P_{l,1}=-\sqrt{1-x^2}\,P'_l(x)=-\sin\theta P'_l(\cos\theta)$$
$$B_\theta=-\sum_{l=0}^{\infty}A_l r^{l-1}P_{l,1}(\cos\theta) \tag{11.38}$$

Helmholtz Coils

Two circular coils of radius a are arranged coaxially and are separated by a distance $2s$ as illustrated in Fig. 11.6. The current has the same magnitude and circulates in the same direction in both coils. The magnetic field created by the coils is to be utilized in

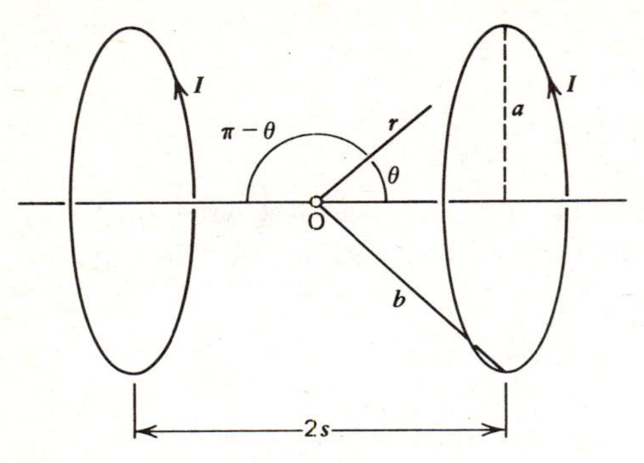

Fig. 11.6

the vicinity of the point O which is halfway between the coils and on their common axis. We therefore want an expansion of the potential about O as origin. The potentials of the two coils separately can be expressed as

$$\psi_1 = \sum_{l=0}^{\infty} A_l r^l P_l(\cos\theta)$$

$$\psi_2 = -\sum_{l=0}^{\infty} A_l r^l P_l(\cos[\pi-\theta])$$

$$= -\sum_{l=0}^{\infty} A_l r^l (-1)^l P_l(\cos\theta) \tag{11.39}$$

The signs of ψ_1 and ψ_2 are opposite because ψ_1 is on the negative side of the magnetic shell of the first coil and ψ_2 is on the positive side of the magnetic shell of the second coil. The combined potential is a superposition of ψ_1 and ψ_2:

$$\psi = \psi_1 + \psi_2 = \sum_{l=0}^{\infty} A_l r^l \left[1-(-1)^l\right] P_l(\cos\theta)$$

$$= 2A_1 r P_1(\cos\theta) + 2A_3 r^3 P_3(\cos\theta)$$

$$+ 2A_5 r^5 P_5(\cos\theta) + \cdots \tag{11.40}$$

Note that all the even terms cancel out. The first term, if present alone, would give a uniform magnetic field. In many applications, it is desirable to have a field in the vicinity of O which is as uniform as possible. The best we can do with two coils is to choose their separation such that the third-order term in (11.40) is zero. From (11.35), we see that $A_3 = 0$ requires

$$\cos^2\alpha = \frac{1}{5} = \frac{s^2}{s^2+a^2} \qquad a = 2s \tag{11.41}$$

The separation of the coils is therefore chosen to be equal to their radius. With this choice, the expression for the potential becomes

$$\psi = -\frac{4\mu_0 I}{5b} r\cos\theta + 0.288\mu_0 I\left(\frac{r}{b}\right)^5 P_5(\cos\theta) + \cdots \tag{11.42}$$

Charged Conducting Sphere Rotating About a Diameter

If an excess charge is placed on a conducting sphere, it appears as a uniform surface charge density of magnitude λ. If the sphere rotates around a diameter which coincides with the z axis as shown in Fig. 11.7, then a surface current is produced which is given by

$$\mathbf{K} = \lambda\mathbf{u} = \lambda a \sin\theta\,\omega\,\hat{\boldsymbol{\phi}} \tag{11.43}$$

where $\mathbf{u}$ is the velocity of the surface of the sphere, a is its radius, and ω is its angular velocity of rotation. If λ is in units of coulombs per square meter, then $\mathbf{K}$ has units of amperes per meter. The magnetic field produced by the surface current has r and θ components and no ϕ component. Reference to problem 8.4 in Chapter 3 shows that the tangential component of the magnetic field is discontinuous at the surface of the sphere. If $B_{1\theta}$ is the θ component of the magnetic field just inside the surface and $B_{2\theta}$ is its value just

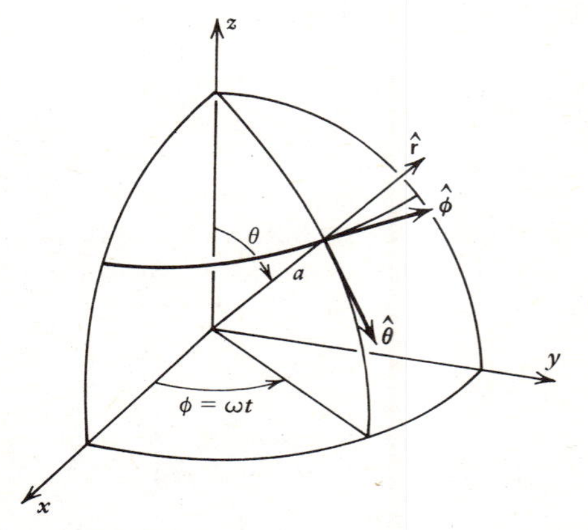

Fig. 11.7

outside, then

$$B_{2\theta} - B_{1\theta} = \mu_0 K = \mu_0 \lambda a \omega \sin\theta \tag{11.44}$$

We will assume that the sphere is not made out of any ferromagnetic material and that it has no significant magnetic properties. Since there is no current density either inside or outside the sphere, it is permissible to use scalar potentials in these regions. The appropriate potentials are given by (11.23), with b replaced by a. Since $\mathbf{B} = -\nabla\psi$, (11.44) gives

$$-\frac{1}{a}\frac{\partial\psi_2}{\partial\theta} + \frac{1}{a}\frac{\partial\psi_1}{\partial\theta} = \mu_0 \lambda a \omega \sin\theta \tag{11.45}$$

The requirement that the normal component of the field be continuous at the surface gives equation (11.28) with b replaced by a. Both boundary conditions can be met if we retain the $l = 1$ term and discard all others. We find that

$$\begin{aligned} \psi_1 &= -\tfrac{2}{3} a \mu_0 \lambda \omega r \cos\theta = -B_1 z \\ \psi_2 &= \frac{1}{3} a^4 \mu_0 \lambda \omega \frac{\cos\theta}{r^2} \end{aligned} \tag{11.46}$$

The field inside the sphere is uniform and of strength B_1. Outside, it is

$$B_r = \frac{\mu\mu_0}{2\pi}\frac{\cos\theta}{r^3} \qquad B_\theta = \frac{\mu\mu_0}{4\pi}\frac{\sin\theta}{r^3} \tag{11.47}$$

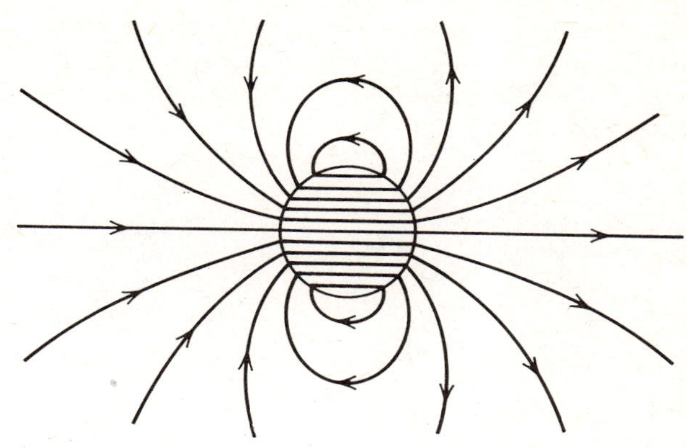

Fig. 11.8

where

$$\mu = \frac{4\pi}{3} a^4 \lambda \omega \tag{11.48}$$

is the dipole moment of the sphere. The comparison of (11.47) with equations (11.11) and (11.12) of Chapter 4 shows that the field external to the sphere is a dipole field. You can use equation (11.9) of Chapter 4 to calculate the dipole moment (11.48) directly. Look at your solution of problem 11.1 in Chapter 4. Figure 11.8 shows the field lines of the rotating sphere. It turns out that a uniformly magnetized sphere produces exactly the same field.

PROBLEMS

11.1. Show that the potential at any point on the axis of a single disc of radius a on which there is a uniform surface charge density λ is given by

$$\psi = \frac{\lambda}{2\varepsilon_0}\left[\sqrt{a^2 + z^2} - |z|\right] \tag{11.49}$$

where z is measured along the axis from the center of the disc. Figure 11.9 shows a graph of this potential.

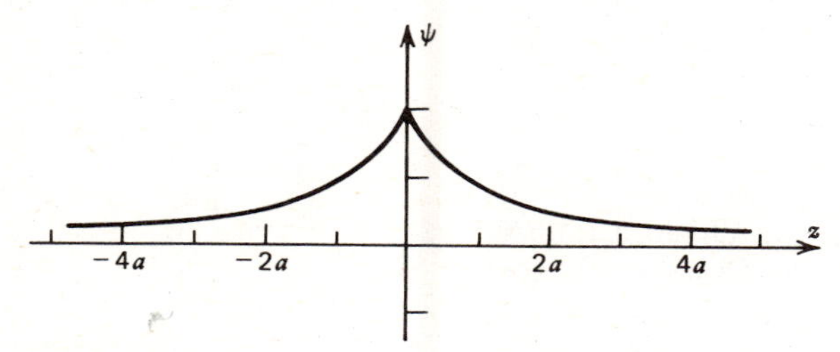

Fig. 11.9

11.2. Two discs of radius a each carry a uniform surface charge density of magnitude λ and are located at $z = \pm s$. The disc at $z = +s$ is positively charged while that at $z = -s$ is negatively charged. Show that the electrostatic potential along the common axis of the two discs is given by

$$\begin{aligned} \psi &= \frac{\lambda}{2\varepsilon_0}[2s - F(z)] \qquad z > s \\ \psi &= \frac{\lambda}{2\varepsilon_0}[2z - F(z)] \qquad -s < z < s \\ \psi &= \frac{\lambda}{2\varepsilon_0}[-2s - F(z)] \qquad z < -s \end{aligned} \tag{11.50}$$

where

$$F(z) = \sqrt{a^2 + (z+s)^2} - \sqrt{a^2 + (z-s)^2} \tag{11.51}$$

Figure 11.10 shows a graph of the potential given by (11.50) for $s = 2a$.

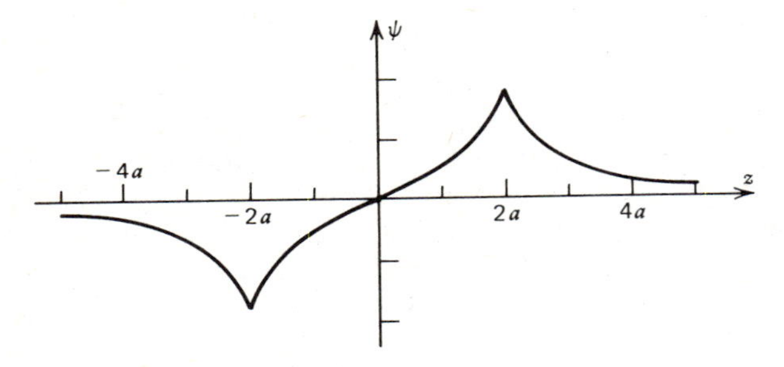

Fig. 11.10

11.3. We can produce a dipole layer by pushing the two discs of problem 11.2 close together. Show that in the limit $s \to 0$, (11.50) gives, for the case $z > 0$,

$$\psi = \frac{2s\lambda}{2\varepsilon_0}\left[1 - \frac{z}{\sqrt{a^2 + z^2}}\right] \tag{11.52}$$

Note that this agrees with (11.7) if the dipole moment per unit area is $N = 2s\lambda$. The limit must be taken in such a way that N approaches a finite value.

Hint: Write $F(z) = 2s[F(z)/2s]$. What is $F(z)/2s$ in the limit $s \to 0$?

11.4. A solenoid consists of a single layer of insulated wire wound on a cylinder of radius a and length $2s$. The axis of the solenoid coincides with the z axis and its ends are at $z = \pm s$. Start with equation (11.13) and show that the magnetic field along the axis of the solenoid is

$$B_z = \frac{\mu_0 nI}{4s}\left[\frac{z+s}{\sqrt{a^2 + (z+s)^2}} - \frac{z-s}{\sqrt{a^2 + (z-s)^2}}\right] \tag{11.53}$$

where n is the total number of turns of wire. Show that (11.53) can be obtained from the potential (11.50) for the case $z > 0$ if $2s\lambda = \varepsilon_0\mu_0 nI$. The field outside the solenoid is therefore the same as that produced by two charged discs of opposite sign at its two ends. Equation (11.53) is correct for the axial field inside the solenoid but is *not* the same as the field found from a potential of the form (11.50) for the case $-s < z < s$.

11.5. Use the recurrence relation (9.54) to generate the associated Legendre functions $P_{l,1}$ through $l = 5$. Start with $P_{1,1}$ and $P_{2,2}$ as given in the list at the end of section 4. The result is

$$\begin{aligned} P_{3,1} &= \tfrac{3}{2}\sin\theta\,(1 - 5\cos^2\theta) \\ P_{4,1} &= \tfrac{1}{2}\sin\theta\,(15\cos\theta - 35\cos^3\theta) \\ P_{5,1} &= \tfrac{15}{8}\sin\theta\,(-21\cos^4\theta + 14\cos^2\theta - 1) \end{aligned} \tag{11.54}$$

11.6. Show that the z component of the magnetic field in the vicinity of the point O of Fig. 11.6 is approximately

$$B_z = B_0\left[1 - 1.8\left(\frac{r}{b}\right)^4 P_4(\cos\theta)\right] \tag{11.55}$$

where B_0 is the field strength at O.

Hint: Use equation (9.51). Show that, in the plane defined by $\theta = \pi/2$,

$$B_z = B_0\left[1 - 0.675\left(\frac{r}{b}\right)^4\right] = B_0\left[1 - 0.432\left(\frac{r}{a}\right)^4\right] \tag{11.56}$$

In particular, if $r = 0.50a$, then $B_z = 0.973B_0$.

11.7. If in Fig. 11.5 P is at the center of the loop so that $a = b$ and $\alpha = \pi/2$, show that the components of the magnetic field for $r > a$ can be expressed as

$$\begin{aligned} B_r &= \sum_{n=0}^{\infty} \mu_0 I \frac{(-1)^n(2n+1)!}{2^{2n+1}(n!)^2}\frac{a^{2n+2}}{r^{2n+3}} P_{2n+1}(\cos\theta) \\ B_\theta &= -\sum_{n=0}^{\infty} \mu_0 I \frac{(-1)^n(2n+1)!}{2^{2n+2}(n+1)!n!}\frac{a^{2n+2}}{r^{2n+3}} P_{2n+1,1}(\cos\theta) \end{aligned} \tag{11.57}$$

11.8. Show that the potential function given by (11.46) is discontinuous at $r = a$.

11.9. Reference to problem 2.1 in Chapter 4 shows that dipole field lines are given by

$$r = c_n \sin^2\theta \tag{11.58}$$

where each value of c_n gives a different field line. Let the spacing between the uniform field lines inside the sphere of Fig. 11.8 be s. In constructing the figure, it is necessary to choose the constants c_n such that the dipole field lines join onto the uniform field lines at the surface of the sphere. Show that this can be done if the constants obey the recurrence relation

$$c_{n+1} = \frac{c_n a^3}{\left(a^{3/2} + s\sqrt{c_n}\right)^2} \qquad c_1 = \frac{a^3}{s^2} \tag{11.59}$$

11.10. It is shown in section 7 of Chapter 7 that both the real and the imaginary parts of an analytic function $F(z)$ of the complex variable z are solutions of Laplace's equation in two dimensions. Separate $F(z) = \ln z$ into its real and imaginary parts. Show by appropriate choices of the multiplicative constants that the real part is the potential of an infinitely long line charge and that the imaginary part is a multiple-valued function which can be used as the magnetic scalar potential of a current flowing in a long straight wire. See problem 8.3 in Chapter 3. Note that the magnetic shell for this potential is a plane bounded by the wire and extending to infinity in the x direction. Show that these same solutions can be found by separating Laplace's equation in cylindrical coordinates and that they correspond to the case where the separation constant is zero. The multiple-valued part of the solution was excluded from (10.43).

11.11. The vector potential can be used to solve problems in magnetostatics. Consider once more the problem of the circular current loop of radius a. Let the origin of coordinates be at the center of the loop. From section 11 of Chapter 3, we find that the vector potential obeys

$$\nabla \times \mathbf{B} = \nabla \times (\nabla \times \mathbf{A}) = \mu_0 \mathbf{j} \tag{11.60}$$

where $\mathbf{j}$ is the current density. In the problem under consideration, $\mathbf{j}$ is zero everywhere except at points on the current loop. The use of the vector identity (9.8) and the reduction of (11.60) to equation (11.3) in Chapter 3 require that $\mathbf{A}$ be expressed in terms of its rectangular components. If we want $\mathbf{A}$ expressed in curvilinear coordinates, then we must work with (11.60) directly. We already know that $\mathbf{B}$ has only an r and a θ component. Assume that the vector potential has only a ϕ component, that is, $\mathbf{A} = A_\phi(r,\theta)\hat{\boldsymbol{\phi}}$. At all points of space where $\mathbf{j} = 0$, show that A_ϕ obeys the differential equation

$$\frac{1}{r}\frac{\partial^2}{\partial r^2}(rA_\phi) + \frac{1}{r^2}\frac{\partial^2 A_\phi}{\partial \theta^2} + \frac{1}{r^2}\frac{\partial}{\partial \theta}(\cot\theta A_\phi) = 0 \tag{11.61}$$

Show that the general solution is

$$A_\phi = \sum_{l=0}^{\infty}\left(A_l r^l + \frac{B_l}{r^{l+1}}\right)P_{l,1}(\cos\theta) \tag{11.62}$$

For the case $r > a$, show that the field components are

$$B_r = -\sum_{l=0}^{\infty}\frac{B_l}{r^{l+2}}l(l+1)P_l(\cos\theta) \qquad r > a$$

$$B_\theta = \sum_{l=0}^{\infty}\frac{B_l}{r^{l+2}}lP_{l,1}(\cos\theta) \qquad r > a \tag{11.63}$$

Hint: The recurrence relation (9.27) is useful. The problem can be completed by expanding the relation for B_z on the axis as given by (11.13) and then comparing the result to (11.63) for $\theta = 0$. The required expansion can be found by differentiating (10.36) with respect to z.

12. ADDITION THEOREM FOR SPHERICAL HARMONICS

Let $\mathbf{r}$ and $\mathbf{r}'$ be two displacement vectors projecting from the origin of coordinates, and let $\mathbf{R} = \mathbf{r} - \mathbf{r}'$ as shown in Fig. 12.1. Equation (9.7) is an expansion of the scalar function $G(\mathbf{r},\mathbf{r}') = 1/R$ and is important because it leads to a generating function for the Legendre polynomials. The main goal of this section is to generalize this idea by finding an expansion for $G(\mathbf{r},\mathbf{r}')$ in terms of spherical harmonics. The function $G(\mathbf{r},\mathbf{r}')$ is actually a special case of a Green's function. We introduced the concept of a Green's function in connection with the solution of partial differential equations in section 7 of Chapter 8; see

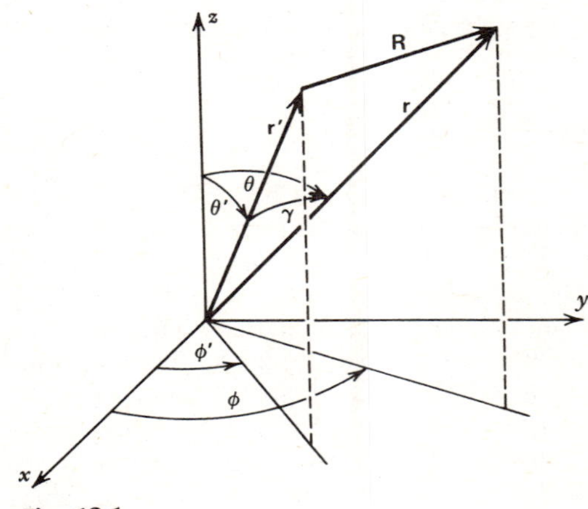

Fig. 12.1

equation (7.24). We will elaborate more on the Green's function approach to the solution of partial differential equations later on in this chapter. According to equation (10.22) of Chapter 3, $G(\mathbf{r},\mathbf{r}')$ obeys the Poisson equation

$$\nabla^2 G(\mathbf{r},\mathbf{r}') = -4\pi\delta(\mathbf{R}) \tag{12.1}$$

where $\delta(\mathbf{R})$ is a Dirac delta function. One of its basic properties is

$$\int_\Sigma \delta(\mathbf{R})\,d\Sigma = 1 \tag{12.2}$$

where it is assumed that the region of integration includes the point $\mathbf{r}=\mathbf{r}'$.

There is symmetry with respect to $\mathbf{r}$ and $\mathbf{r}'$. The Green's function obeys

$$G(\mathbf{r},\mathbf{r}') = G(\mathbf{r}',\mathbf{r}) \tag{12.3}$$

Equations (12.1) and (12.2) remain valid if $\mathbf{r}'$ is considered to be the variable point and $\mathbf{r}$ the fixed point:

$$\nabla'^2 G(\mathbf{r},\mathbf{r}') = -4\pi\delta(\mathbf{R}) \qquad \int_\Sigma \delta(\mathbf{R})\,d\Sigma' = 1 \tag{12.4}$$

Recall also that the delta function is even:

$$\delta(\mathbf{R}) = \delta(\mathbf{r}-\mathbf{r}') = \delta(\mathbf{r}'-\mathbf{r}) \tag{12.5}$$

Delta functions with respect to each of the three variables r, θ, and ϕ have the property

$$\begin{aligned} &\int_0^\infty \delta(r-r')\,dr = 1 \\ &\int_0^\pi \delta(\theta-\theta')\,d\theta = 1 \\ &\int_0^{2\pi} \delta(\phi-\phi')\,d\phi = 1 \end{aligned} \tag{12.6}$$

It is convenient to use equation (5.59) of Chapter 8 to reexpress the θ delta function as

$$\frac{\delta(\theta-\theta')}{\sin\theta} = \delta(\cos\theta - \cos\theta') \tag{12.7}$$

The three integrals in (12.6) can be combined to give

$$\begin{aligned} 1 &= \int_0^\infty \int_0^\pi \int_0^{2\pi} \delta(r-r')\delta(\theta-\theta')\delta(\phi-\phi')\,dr\,d\theta\,d\phi \\ &= \int_\Sigma \delta(\mathbf{R})\,d\Sigma \end{aligned} \tag{12.8}$$

Since the volume element in spherical coordinates is $r^2 \sin\theta\,dr\,d\theta\,d\phi$, the delta function must be expressed as

$$\delta(\mathbf{R}) = \frac{\delta(r-r')}{r^2}\delta(\cos\theta - \cos\theta')\delta(\phi-\phi') \tag{12.9}$$

Note that the region Σ of integration is all space.

If the Poisson equation (12.1) is written out in spherical coordinates, the result is

$$\begin{aligned} &\frac{1}{r^2}\frac{\partial}{\partial r}\left(r^2\frac{\partial G}{\partial r}\right) + \frac{1}{r^2\sin\theta}\frac{\partial}{\partial\theta}\left(\sin\theta\frac{\partial G}{\partial\theta}\right) \\ &\quad + \frac{1}{r^2\sin^2\theta}\frac{\partial^2 G}{\partial\phi^2} \\ &\quad = -4\pi\frac{\delta(r-r')}{r^2}\delta(\cos\theta-\cos\theta')\delta(\phi-\phi') \end{aligned} \tag{12.10}$$

The spherical harmonics are a complete set of orthonormal functions. It is therefore possible to expand the Green's function in terms of them as

$$G(\mathbf{r},\mathbf{r}') = \sum_{l=0}^{\infty}\sum_{m=-l}^{+l} F_l(r) A_{lm} Y_{lm}(\theta,\phi) \tag{12.11}$$

Moreover, the spherical harmonics obey the closure relation

$$\begin{aligned} &\sum_{l=0}^{\infty}\sum_{m=-l}^{+l} Y_{lm}^*(\theta',\phi')Y_{lm}(\theta,\phi) \\ &\quad = \delta(\cos\theta-\cos\theta')\delta(\phi-\phi') \end{aligned} \tag{12.12}$$

By combining (12.10), (12.11), and (12.12) and also using (4.9), we find that

$$\begin{aligned} &\sum_{l=0}^{\infty}\sum_{m=-l}^{+l}\left[\frac{1}{r^2}\frac{d}{dr}\left(r^2\frac{dF_l}{dr}\right) - \frac{F_l}{r^2}l(l+1)\right]A_{lm}Y_{lm}(\theta,\phi) \\ &\quad = -4\pi\frac{\delta(r-r')}{r^2}\sum_{l=0}^{\infty}\sum_{m=-l}^{+l} Y_{lm}^*(\theta',\phi')Y_{lm}(\theta,\phi) \end{aligned} \tag{12.13}$$

Since $Y_{lm}(\theta,\phi)$ are a complete set of functions, their coefficients on the two sides of the equation are equal. This yields

$$A_{lm} = Y_{lm}^*(\theta', \phi') \tag{12.14}$$

$$\frac{d}{dr}\left(r^2 \frac{dF_l}{dr}\right) - F_l l(l+1) = -4\pi\delta(r - r') \tag{12.15}$$

The symmetry with respect to **r** and **r′** could have been used to anticipate (12.14). The procedure being used here is quite similar to that followed in solving the wave equation (7.19) of Chapter 8.

For $r \neq r'$, the solution of (12.15) which is finite at $r = 0$ and $r = \infty$ is

$$\begin{aligned} F_l(r) &= Ar^l \qquad r < r' \\ F_l(r) &= Br^{-l-1} \qquad r > r' \end{aligned} \tag{12.16}$$

where A and B are constants. The effect of the delta function is found by integrating (12.15) over the range $r' - \varepsilon < r < r' + \varepsilon$. In the limit $\varepsilon \to 0$, we get

$$\left(r^2 \frac{dF_l}{dr}\right)_{r'+0} - \left(r^2 \frac{dF_l}{dr}\right)_{r'-0} = -4\pi \tag{12.17}$$

This means that dF_l/dr is discontinuous at $r = r'$. Figure 12.2 shows a plot of $F_l(r)$ for the case $l = 2$. The function $F_l(r)$ itself is continuous at $r = r'$. Otherwise, dF_l/dr would be proportional to a delta function at $r = r'$, and (12.15) would contain a term involving the derivative of the delta function.

Equation (12.17) and the continuity of $F_l(r)$ give the two conditions

$$\begin{aligned} (l+1)(r')^{-l}B + l(r')^{l+1}A &= 4\pi \\ (r')^{-l-1}B - (r')^l A &= 0 \end{aligned} \tag{12.18}$$

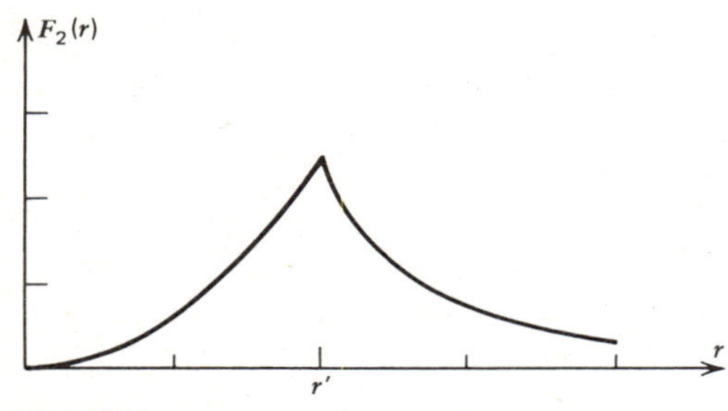

Fig. 12.2

By solving for A and B, we find that

$$\begin{aligned} F_l(r) &= \frac{4\pi}{2l+1} \frac{r^l}{(r')^{l+1}} \qquad r < r' \\ F_l(r) &= \frac{4\pi}{2l+1} \frac{(r')^l}{r^{l+1}} \qquad r > r' \end{aligned} \tag{12.19}$$

The following expansions have therefore been established:

$$\frac{1}{R} = \sum_{l=0}^{\infty} \sum_{m=-l}^{+l} \frac{4\pi}{2l+1} \frac{r^l}{(r')^{l+1}} Y_{lm}^*(\theta', \phi') Y_{lm}(\theta, \phi) \qquad r < r' \tag{12.20}$$

$$\frac{1}{R} = \sum_{l=0}^{\infty} \sum_{m=-l}^{+l} \frac{4\pi}{2l+1} \frac{(r')^l}{r^{l+1}} Y_{lm}^*(\theta', \phi') Y_{lm}(\theta, \phi) \qquad r > r' \tag{12.21}$$

If the radial function given in (12.19) is regarded as a function of the two variables r and r', then $F_l(r, r') = F_l(r', r)$. In demonstrating this symmetry property, you must exchange r and r' throughout (12.19), including where they appear in the inequalities. Symmetry with respect to the two sets of variables (r, θ, ϕ) and (r', θ', ϕ') is therefore maintained.

The angle γ between **r** and **r′** in Fig. 12.1 is given by

$$\cos\gamma = \cos\theta\cos\theta' + \sin\theta\sin\theta'\cos(\phi - \phi') \tag{12.22}$$

Reference to the generating function (9.5) shows that it is possible to expand $1/R$ in terms of Legendre polynomials of argument $\cos\gamma$ by writing

$$\frac{1}{R} = \frac{1}{\sqrt{r^2 + r'^2 - 2rr'\cos\gamma}} = \sum_{l=0}^{\infty} \frac{(r')^l}{r^{l+1}} P_l(\cos\gamma) \tag{12.23}$$

By equating the coefficients of like powers of r in (12.21) and (12.23), we find that

$$P_l(\cos\gamma) = \sum_{m=-l}^{+l} \frac{4\pi}{2l+1} Y_{lm}^*(\theta', \phi') Y_{lm}(\theta, \phi) \tag{12.24}$$

This result is known as the *addition theorem* for spherical harmonics.

One way to find the solution of Poisson's equation

$$\nabla^2\psi = -\frac{\rho}{\varepsilon_0} \tag{12.25}$$

for a known charge distribution is to carry out the integral

$$\psi = \frac{1}{4\pi\varepsilon_0}\int_\Sigma \frac{\rho(r')}{R}\,d\Sigma' \tag{12.26}$$

The integral is almost always too difficult to evaluate directly, but headway can sometimes be made by using the expansions (12.20) and (12.21). Suppose that the charge distribution in question is localized in the vicinity of the origin as shown in Fig. 12.3 and that it is required to find the potential at points sufficiently far removed from the distribution that $r > r'$ for the entire range of the variable of integration r'. By substituting (12.21) into (12.26) and integrating term by term, we find an expansion for the potential in terms of spherical harmonics given by

$$\psi = \frac{1}{\varepsilon_0}\sum_{l=0}^{\infty}\sum_{m=-l}^{+l}\frac{q_{lm}}{2l+1}\frac{1}{r^{l+1}}Y_{lm}(\theta,\phi) \tag{12.27}$$

$$q_{lm} = \int_\Sigma \rho(r',\theta',\phi')(r')^l Y_{lm}^*(\theta',\phi')\,d\Sigma \tag{12.28}$$

The quantities q_{lm} are called the *multipole moments* of the charge distribution. Because of the relation (4.39), they obey

$$q_{l,-m} = (-1)^m q_{lm}^* \tag{12.29}$$

The spherical harmonics used in the following computations are listed at the end of section 4. The first term in the expansion (12.27) is given by $l=0$ and is called the *monopole term*. The monopole moment is

$$q_{00} = \int_\Sigma \rho Y_{00}\,d\Sigma' = \frac{q}{\sqrt{4\pi}} \tag{12.30}$$

where q is the net charge in the distribution. It is of course possible to have equal numbers of positive and negative charges in the distribution so that the monopole term is absent. Assuming that $q \neq 0$, the leading

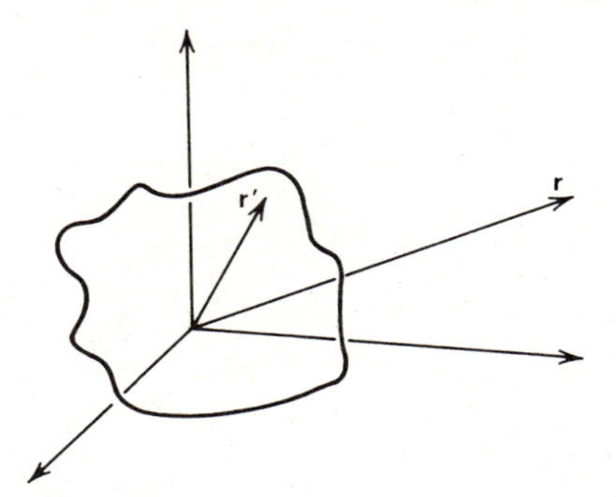

Fig. 12.3

term in the expansion for the potential is

$$\psi_0 = \frac{q}{4\pi\varepsilon_0 r} \tag{12.31}$$

Retaining only this term in the expansion amounts to approximating the charge distribution by a point charge.

It is sometimes useful to calculate the first few multipole moments in terms of rectangular rather than spherical coordinates. The spherical harmonics for $l=1$ can be expressed as

$$\begin{aligned} Y_{1,1} &= -\sqrt{\frac{3}{8\pi}}\left(\frac{x+iy}{r}\right) \\ Y_{1,0} &= \sqrt{\frac{3}{4\pi}}\frac{z}{r} \\ Y_{1,-1} &= -Y_{1,1}^* \end{aligned} \tag{12.32}$$

Thus,

$$\begin{aligned} q_{11} &= -\sqrt{\frac{3}{8\pi}}\int_\Sigma \rho(x'-iy')\,d\Sigma' \\ &= -\sqrt{\frac{3}{8\pi}}(p_x - ip_y) \\ q_{10} &= \sqrt{\frac{3}{8\pi}}\int_\Sigma \rho z'\,d\Sigma' = \sqrt{\frac{3}{4\pi}}p_z \\ q_{1,-1} &= -q_{11}^* \end{aligned} \tag{12.33}$$

where p_x, p_y, and p_z are the rectangular components of the *dipole moment* of the charge distribution defined by

$$\mathbf{p} = \int_\Sigma \rho\mathbf{r}'\,d\Sigma' \tag{12.34}$$

The contribution to the potential from the dipole terms is

$$\psi_1 = \frac{1}{3\varepsilon_0 r^2} \sum_{m=-1}^{+1} q_{1,m} Y_{1,m}$$

$$= \frac{1}{3\varepsilon_0 r^2} [q_{11} Y_{11} + q_{10} Y_{10} + q_{1,-1} Y_{1,-1}]$$

$$= \frac{1}{3\varepsilon_0 r^2} [2 \operatorname{Re} q_{11} Y_{11} + q_{10} Y_{10}]$$

$$= \frac{1}{3\varepsilon_0 r^3} \left[\frac{3}{4\pi} \operatorname{Re}(p_x - ip_y)(x + iy) + \frac{3}{4\pi} p_z z \right]$$

$$= \frac{1}{4\pi\varepsilon_0} \frac{\mathbf{p} \cdot \mathbf{r}}{r^3} \tag{12.35}$$

which is the familiar dipole potential.

The $l = 2$ terms in (12.27) constitute what is called the *quadrupole potential*. We will show that the quadrupole potential can be expressed in terms of a second-rank symmetric Cartesian tensor, called the *quadrupole tensor*, and defined by

$$Q_{ij} = \int_\Sigma \rho \left(3x_i' x_j' - \delta_{ij} r'^2\right) d\Sigma' \tag{12.36}$$

There is some similarity between the quadrupole tensor and the moment of inertia tensor defined in Chapter 2 by equation (9.36). The quadrupole tensor has the property that its trace is zero:

$$Q_{11} + Q_{22} + Q_{33} = 0 \tag{12.37}$$

The $l = 2$ spherical harmonics expressed in rectangular coordinates are

$$Y_{22} = \sqrt{\frac{15}{32\pi}} \frac{x^2 + 2ixy - y^2}{r^2}$$

$$Y_{21} = -\sqrt{\frac{15}{8\pi}} \frac{xz + iyz}{r^2} \tag{12.38}$$

$$Y_{20} = \sqrt{\frac{5}{16\pi}} \frac{3z^2 - r^2}{r^2}$$

The evaluation of the $l = 2$ multipole moments in terms of the elements of the quadrupole tensor gives

$$q_{22} = \sqrt{\frac{15}{32\pi}} \int_\Sigma \rho(x'^2 - 2ix'y' - y'^2)\, d\Sigma'$$

$$= \frac{1}{3} \sqrt{\frac{15}{32\pi}} \int_\Sigma [\rho(3x'^2 - r'^2) - \rho(3y'^2 - r'^2) - 6i\rho x'y']\, d\Sigma'$$

$$= \sqrt{\frac{5}{96\pi}} (Q_{11} - Q_{22} - 2iQ_{12}) \tag{12.39}$$

$$q_{21} = -\sqrt{\frac{15}{8\pi}} \int_\Sigma \rho(x'z' - iy'z')\, d\Sigma'$$

$$= -\sqrt{\frac{5}{24\pi}} (Q_{13} - iQ_{23})$$

$$q_{20} = \sqrt{\frac{5}{16\pi}} \int_\Sigma \rho(3z'^2 - r'^2)\, d\Sigma' = \sqrt{\frac{5}{16\pi}} Q_{33}$$

The quadrupole potential is

$$\psi_2 = \frac{1}{5\varepsilon_0 r^3} [2 \operatorname{Re}(q_{22} Y_{22} + q_{21} Y_{21}) + q_{20} Y_{20}] \tag{12.40}$$

After a little algebra, we find that

$$\psi_2 = \frac{1}{8\pi\varepsilon_0 r^5} \sum_{j=1}^{3} \sum_{k=1}^{3} Q_{jk} x_j x_k \tag{12.41}$$

In the course of the calculation, it is necessary to use the fact that the trace of the quadrupole tensor is zero. The combination of the monopole, dipole, and quadrupole terms gives the approximate potential

$$\psi = \frac{1}{4\pi\varepsilon_0} \left(\frac{q}{r} + \frac{\mathbf{p} \cdot \mathbf{r}}{r^3} + \frac{1}{2r^5} \sum_j \sum_k Q_{jk} x_j x_k + \cdots \right) \tag{12.42}$$

When you did problem (11.6) in Chapter 4, you obtained the first two terms of (12.42). An example of a quadrupole potential is provided by problem 8.4 in Chapter 4. In that example, there is no monopole or dipole term. The monopole term is zero because there are equal numbers of positive and negative charges. The dipole term is zero because the array consists of two dipoles of equal strength oppositely aligned.

The $l=3$ terms in the expansion for the potential make up the *octupole potential*. We could define a third-rank Cartesian octupole tensor and reexpress the octupole potential in terms of it, but this probably has no advantage over the original form of the expansion as given by (12.27).

We will now examine the quadrupole term in some detail. In section 10 of Chapter 2, we demonstrated that a Cartesian coordinate system exists in which any second-rank symmetric tensor is diagonal. Therefore, by an appropriate choice of coordinates, the quadrupole tensor can be made diagonal. Since its trace is zero, only two elements have to be evaluated. If the z axis is an axis of symmetry of the charge distribution so that the potential does not depend on the azimuth angle ϕ, then $Q_{11}=Q_{22}$. It is then necessary to calculate only the single element $Q=Q_{33}$. The quadrupole potential is then

$$\psi_2=\frac{Q}{8\pi\varepsilon_0 r^5}\left(-\frac{1}{2}x^2-\frac{1}{2}y^2+z^2\right)$$

$$=\frac{Q}{16\pi\varepsilon_0 r^3}(3\cos^2\theta-1) \tag{12.43}$$

It is usual to call Q the *quadrupole moment* of the distribution, but in doing so we must remember that it is only in the case of an axially symmetric charge distribution that the quadrupole potential can be characterized by a single constant. Imagine that an idealized point quadrupole exists and that its potential is given by (12.43) for all values of the coordinates. The resulting field components are

$$E_r=\frac{3Q}{16\pi\varepsilon_0 r^4}(3\cos^2\theta-1)$$

$$E_\theta=\frac{3Q}{16\pi\varepsilon_0 r^4}\sin 2\theta \tag{12.44}$$

The field lines are solutions of the differential equation

$$\frac{E_r}{E_\theta}=\frac{dr}{r\,d\theta}=\frac{3\cos\theta}{2\sin\theta}-\frac{1}{\sin 2\theta} \tag{12.45}$$

Fig. 12.4

Integration yields

$$\ln\frac{r}{k}=\frac{3}{2}\ln\sin\theta-\frac{1}{2}\ln\tan\theta$$

$$r=k\sin\theta\sqrt{\cos\theta} \qquad 0<\theta<\frac{\pi}{2} \tag{12.46}$$

$$r=k\sin\theta\sqrt{-\cos\theta} \qquad \frac{\pi}{2}<\theta<\pi$$

Field lines obtained for different values of the constant of integration k are shown in Fig. 12.4. You can see where the term "quadrupole" comes from. Figure 12.4 has four lobes. Also, the quadrupole potential of problem 8.4 in Chapter 4 is produced by four charges or "poles."

The gravitational scalar potential obeys the Poisson equation

$$\nabla^2\psi=4\pi G\rho \tag{12.47}$$

where now ρ is the mass density. See section 1 and problem 5.1 of Chapter 3. The gravitational potential can therefore be represented by a multipole expansion either in the form (12.27) or (12.42). It is merely necessary to replace $1/\varepsilon_0$ by $-4\pi G$. It is of interest to find the gravitational potential produced by a distribution of mass which deviates only slightly from being spherically symmetric. Assume, for example, that the Earth is a spheroid with semimajor and

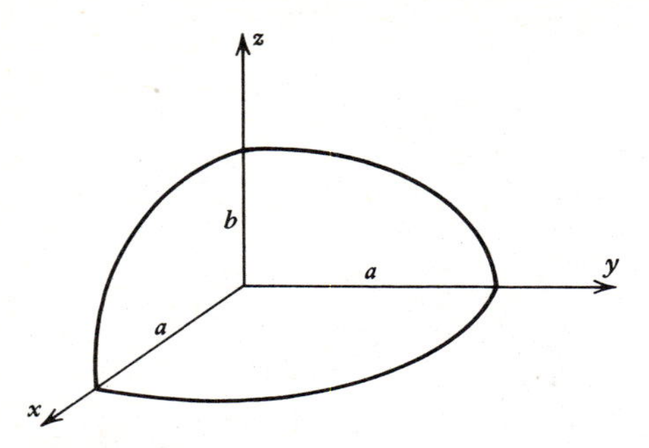

Fig. 12.5

semiminor axes a and b, respectively, throughout which a mass M is uniformly distributed. As pictured in Fig. 12.5, the origin of coordinates is at the center of the spheroid and the z axis is the axis of symmetry. The dipole moment of this mass distribution as given by (12.34) is zero. The quadrupole moment is

$$Q = \int_{\Sigma} (3z^2 - r^2)\rho \, d\Sigma = \int_{\Sigma} (2z^2 - x^2 - y^2)\rho \, d\Sigma \tag{12.48}$$

The equation of the spheroid can be expressed as

$$\frac{R^2}{a^2} + \frac{z^2}{b^2} = 1 \qquad R^2 = x^2 + y^2 \tag{12.49}$$

The integral (12.48) can be evaluated by expressing the variables of integration in cylindrical coordinates. The result is

$$\begin{aligned} Q &= 2\pi\rho \int_{z=-b}^{+b} \int_{R=0}^{a\sqrt{1-(z/b)^2}} (2z^2 - R^2) R \, dR \, dz \\ &= 2\pi\rho \int_{z=-b}^{+b} \left[z^2 a^2 (1 - z^2/b^2) - \frac{1}{4} a^4 (1 - z^2/b^2)^2 \right] dz \\ &= \frac{8}{15} \pi\rho a^2 b (b^2 - a^2) \end{aligned} \tag{12.50}$$

The volume and total mass of the spheroid are, respectively,

$$\Sigma = \frac{4}{3}\pi a^2 b \qquad M = \rho\Sigma \tag{12.51}$$

The quadrupole moment is therefore

$$Q = \frac{2}{5} M (b^2 - a^2) \tag{12.52}$$

The approximate potential is

$$\psi = -GM \left[\frac{1}{r} + \frac{b^2 - a^2}{10r^3} (3\cos^2\theta - 1) + \cdots \right] \tag{12.53}$$

This result is valid if r is larger than both a and b. Note that Q can be positive or negative depending on whether the spheroid is prolate (cigar-shaped) or oblate (pancake-shaped). If the spheroid deviates only slightly from a perfect sphere, then Q is small. The actual value of Q, as well as the higher multipoles in the expansion, can be found from measurements on the orbits of artificial satellites.

PROBLEMS

12.1. Show that the potential that you found in problem 8.4 of Chapter 4 can be expressed as the following linear combination of $l = 2$ spherical harmonics:

$$\psi = \frac{qa^2}{\varepsilon_0 r^3} \sqrt{\frac{3}{10\pi}} (Y_{22} + Y_{2,-2}) \tag{12.54}$$

12.2. For axially symmetric distributions of charge, show that the multipole expansion (12.27) can be expressed as

$$\psi = \frac{1}{4\pi\varepsilon_0} \sum_{l=0}^{\infty} \frac{q_l}{r^{l+1}} P_l(\cos\theta)$$

$$q_l = q_{l0} \sqrt{\frac{4\pi}{2l+1}} = \int_{\Sigma} \rho r'^l P_l(\cos\theta') \, d\Sigma' \tag{12.55}$$

Refer to equation (4.35). What is the relation between q_2 and the quadrupole moment Q?

12.3. A point charge $2q$ is located on the z axis at $z = +s$, and a second point charge q is on the z axis at $z = -2s$. Show that the multipole moments of this charge distribution as given by (12.55) are

$$q_l = qs^l\left[2 + 2^l(-1)^l\right] \tag{12.56}$$

Note that the dipole ($l = 1$) term is zero. If all the charges in a distribution are of the same sign, it is always possible to choose the origin of coordinates such that the dipole term is zero. *Hint:* Represent the charge density as

$$\rho = q[2\delta(z - s) + \delta(z + 2s)]\,\delta(x)\,\delta(y) \tag{12.57}$$

Do the integral in rectangular coordinates.

12.4. Verify that the trace of the quadrupole tensor is zero.

12.5. Show that the potential at a point inside of a known charge distribution can be expressed as

$$\psi = \frac{1}{\varepsilon_0}\sum_{l=0}^{\infty}\sum_{m=-l}^{+l}\frac{1}{2l+1}\left[A_{lm}(r)r^l + \frac{B_{lm}(r)}{r^{l+1}}\right]Y_{lm}(\theta,\phi)$$

$$A_{lm}(r) = \int_{r'>r}\frac{\rho(r',\theta',\phi')}{(r')^{l+1}}Y_{lm}^*(\theta',\phi')\,d\Sigma' \tag{12.58}$$

$$B_{lm}(r) = \int_{r'<r}\rho(r',\theta',\phi')(r')^l Y^*{}_{lm}(\theta',\phi')\,d\Sigma'$$

The integral for $A_{lm}(r)$ is over that portion of the charge which lies outside the sphere of radius r. For $B_{lm}(r)$, the integral is over the charge inside the sphere of radius r. Since the limits of integration depend on r, the integrals come out as functions of r. If the potential has no dependence on ϕ, then it can be written as

$$\psi = \frac{1}{4\pi\varepsilon_0}\sum_{l=0}^{\infty}\left(A_l r^l + \frac{B_l}{r^{l+1}}\right)P_l(\cos\theta)$$

$$A_l(r) = \int_{r'>r}\frac{\rho(r',\theta')}{(r')^{l+1}}P_l(\cos\theta')\,d\Sigma' \tag{12.59}$$

$$B_l(r) = \int_{r'<r}\rho(r',\theta')(r')^l P_l(\cos\theta')\,d\Sigma'$$

12.6. The charge density inside a sphere of radius a is given by $\rho = k\cos\theta$, where k is a constant. There is no charge outside the sphere. Show that the electrostatic potential is

$$\psi = \frac{k}{12\varepsilon_0}(4ar - 3r^2)\cos\theta \qquad 0 < r < a$$

$$\psi = \frac{ka^4}{12\varepsilon_0}\frac{\cos\theta}{r^2} \qquad r > a \tag{12.60}$$

12.7. A total charge q is spread uniformly throughout the volume of a cylinder of radius a and length $2s$. As shown in Fig. 12.6, the axis of the cylinder coincides with the z axis and its center is at the origin. Find a formula for the potential expressed in spherical coordinates which is valid for r large compared to either s or a and which is accurate through the quadrupole term. Show that the quadrupole term is zero if $s = (\sqrt{3}/2)a$. Note that all the odd multipoles ($l = 1, 3, 5, \ldots$) are missing from the expansion.

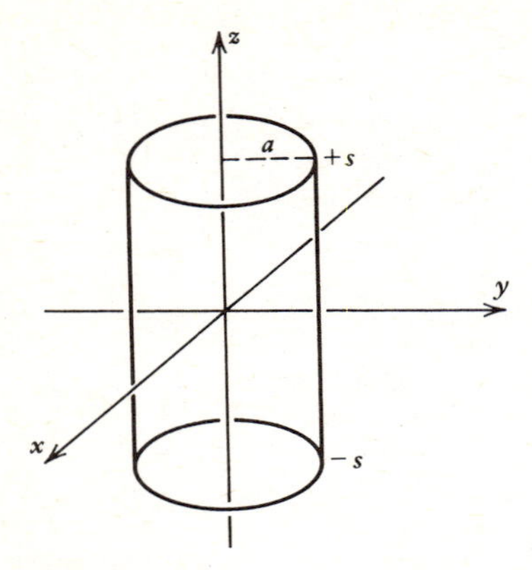

Fig. 12.6

12.8. Two discs of radius a each carry a uniform surface charge density of magnitude λ and are located at $z = \pm s$. The disc at $z = +s$ is positive and the disc at $z = -s$ is negative. The geometry is the same as in Fig. 12.6 with the ends of the cylinder replaced by the discs. If r is large compared to either a or s, show that the approximate potential is

$$\psi = \frac{sa^2\lambda}{2\varepsilon_0}\left[\frac{\cos\theta}{r^2} + \frac{4s^2 - 3a^2}{8r^4}(5\cos^3\theta - 3\cos\theta) + \cdots\right] \tag{12.61}$$

All the even multipoles are absent from the expansion. Note that the octupole term is zero if $s = \sqrt{3}\,a/2$. This result can be used as the magnetic scalar potential for a solenoid consisting of n turns if $2s\lambda = \varepsilon_0\mu_0 nI$.

12.9. Show that the addition theorem (12.24) can be written as

$$P_l(\cos\gamma) = P_l(\cos\theta)P_l(\cos\theta') + 2\sum_{m=1}^{l}\frac{(l-m)!}{(l+m)!}P_{lm}(\cos\theta)P_{lm}(\cos\theta')\cos m(\phi - \phi') \tag{12.62}$$

See equation (4.35).

12.10. By consulting the list of hydrogen atom wave functions at the end of section 5, we see that the probability distribution function for the electron in the state $N = 2$, $l = 1$, $m = 1$ is

$$|u_{211}|^2 = \frac{r^2}{64\pi a^5}e^{-r/a}\sin^2\theta \tag{12.63}$$

If $-q$ is the electronic charge, then the charge density associated with this state is

$$\rho = -q|u_{211}|^2 \tag{12.64}$$

If $r \gg a$, show that this charge density gives the potential

$$\psi = \frac{3qa^2}{2\pi\varepsilon_0}\frac{P_2(\cos\theta)}{r^3} \tag{12.65}$$

Hint: There is no monopole term because of the presence of the proton. Write $\sin^2\theta$ as a linear combination of $P_0(\cos\theta)$ and $P_2(\cos\theta)$. Only $l = 0$ and $l = 2$ will give nonzero multiple moments on account of the orthogonality of the Legendre polynomials. It is all right to calculate the multipole moments from (12.55) by extending the r integral to infinity because the charge density is essentially zero if $r \gg a$.

12.11. Show that the spherical harmonics obey the sum rule

$$\sum_{m=-l}^{+l} |Y_{lm}(\theta,\phi)|^2 = \frac{2l+1}{4\pi} \tag{12.66}$$

12.12. It is possible that the sun has a small quadrupole moment due to an ellipsoidal shape caused by its rotation. Show that the orbit equation for a planet moving in the equatorial plane of the sun is

$$\frac{d^2x}{d\theta^2} + x - \frac{m\varepsilon x^2}{\ell^2} - \frac{mC}{\ell^2} = 0 \tag{12.67}$$

where the parameter ε is defined by

$$\varepsilon = -\frac{3QGm}{4} \tag{12.68}$$

Refer to section 4 of Chapter 8 for a discussion of the two-body central force problem. Note especially that the angle θ which appears in (12.67) is the angular position of the planet in the plane of its orbit and is not the same θ which appears in equation (12.53). In equation (12.53), $\theta = \pi/2$ for this problem. In (12.67), $C = GmM$ is the force constant, m is the mass of the planet, M is the mass of the sun, and ℓ is the orbital angular momentum. Show that the change of variable $x = y + h$ gives the differential equation

$$\frac{d^2y}{d\theta^2} + \left(1 - \frac{2m\varepsilon h}{\ell^2}\right) y - \frac{m\varepsilon}{\ell^2} y^2 = 0 \tag{12.69}$$

provided that

$$h = \frac{\ell^2}{2m\varepsilon} - \frac{\ell^2}{2m\varepsilon}\sqrt{1 - \frac{4m^2\varepsilon C}{\ell^4}} \tag{12.70}$$

Note that $r_0 = 1/h$ is the radius of a stable circular orbit. If ε is very small, show that

$$h \simeq \frac{mC}{\ell^2}\left(1 + \frac{m^2\varepsilon C}{\ell^4}\right) \tag{12.71}$$

Compare this result to equation (4.37) of Chapter 8. Solve (12.69) approximately by the Fourier series method as given in section 3 of Chapter 9. Show that to first order in the parameter ε, the advance of the perihelion is given by

$$\Delta\theta \simeq \frac{2\pi m^2 \varepsilon C}{\ell^4} = \frac{3\pi a^2(A^2 - B^2)}{5b^2} \tag{12.72}$$

where a and b are the semimajor and semiminor axes, respectively, of the almost elliptical orbit as discussed in problem 4.5 of Chapter 8 and A and B are the semimajor and semiminor axes of the ellipsoid of Fig. 12.5 with an obvious change in notation.

13. INTRODUCTION TO BESSEL FUNCTIONS

Written in cylindrical coordinates [Chapter 4, equation (10.21)], Laplace's equation is

$$\nabla^2\psi = \frac{1}{r}\frac{\partial}{\partial r}\left(r\frac{\partial\psi}{\partial r}\right) + \frac{1}{r^2}\frac{\partial^2\psi}{\partial\theta^2} + \frac{\partial^2\psi}{\partial z^2} = 0 \tag{13.1}$$

The variables can be separated. If $\psi(r,\theta,z) = F(r)G(\theta)H(z)$, we find

$$H''(z) - \kappa^2 H(z) = 0 \tag{13.2}$$

$$G''(\theta) + m^2 G(\theta) = 0 \tag{13.3}$$

$$r^2F''(r) + rF'(r) + (\kappa^2r^2 - m^2)F(r) = 0 \tag{13.4}$$

where κ^2 and m^2 are separation constants. If we set $x = \kappa r$ and $F(r) = w(x)$, the radial equation (13.4) becomes

$$w''(x)+\frac{1}{x}w'(x)+\left(1-\frac{m^2}{x^2}\right)w(x)=0 \qquad (13.5)$$

Reference to equation (1.31) reveals that (13.5) is Bessel's differential equation in terms of the real variable x. Equation (1.37) gives one of its possible solutions in series form. If m is not an integer, $J_m(x)$ and $J_{-m}(x)$ as found from the series (1.37) are two linearly independent solutions. If m is an integer, the series expansion is not defined for negative values of m, and only one solution is found by the series method. The Bessel functions $J_m(x)$ of integral order are analytic at $x=0$. As shown in problem 1.8, the second linearly independent solution is singular at $x=0$. There is a certain degree of arbitrariness in the choice of the separation constants. For example, we could change the sign of κ^2 and obtain

$$H''(z)+\kappa^2 H(z)=0 \qquad (13.6)$$

$$w''(x)+\frac{1}{x}w'(x)-\left(1+\frac{m^2}{x^2}\right)w(x)=0 \qquad (13.7)$$

in place of (13.2) and (13.3). As will be shown later in this section, the choice between these two possibilities is dictated by the boundary conditions that occur in a given problem. Formally, (13.7) can be obtained from (13.5) by making the change of variable $x \to ix$. It is usual to define one of the possible solutions of (13.7) by means of

$$I_m(x)=i^{-m}J_m(ix) \qquad (13.8)$$

This is called a *modified* or *hyperbolic* Bessel function. From (1.37), we see that its series expansion is

$$I_m(x)=\left(\frac{x}{2}\right)^m \sum_{n=0}^{\infty}\frac{1}{n!\Gamma(m+n+1)}\left(\frac{x}{2}\right)^{2n} \qquad (13.9)$$

Note that for integral values of m, $I_m(x)$ is analytic at $x=0$.

In section 3 of Chapter 7, we found that the time-independent part of the wave function for a vibrating membrane obeys the two-dimensional Helmholtz equation [Chapter 7, equation (3.8)]. In plane polar coordinates, this equation is

$$\frac{1}{r}\frac{\partial}{\partial r}\left(r\frac{\partial u}{\partial r}\right)+\frac{1}{r^2}\frac{\partial^2 u}{\partial\theta^2}+\kappa^2 u=0 \qquad (13.10)$$

If we write $u(r,\theta)=F(r)G(\theta)$, we find that separation of the variables is again possible. Equations (13.3) and (13.4) are again obtained. For a circular drumhead which has been clamped at $r=a$, the boundary conditions to be imposed on the solutions of (13.4) are $F(a)=0$. If the solutions of (13.3) are to be single-valued functions, m is required to be an integer. The appropriate solutions of the radial equation (13.4) are then Bessel functions of integral order. Two-dimensional Helmholtz equations were also encountered in connection with electromagnetic waves propagating in a hollow conducting pipe as equations (5.15) and (5.16) in Chapter 7 show. If the cross section of the pipe is circular, it is appropriate to express these equations in plane polar coordinates as was done for the circular drumhead. For TM waves, E_3 vanishes at the boundary so that the boundary conditions are the same as for the drumhead. For TE waves, it is the normal derivative of B_3 which vanishes on the boundary so that the appropriate boundary condition for the radial part of the solution in this case is $F'(a)=0$.

Bessel's equation expressed in Sturm–Liouville form is

$$\frac{d}{dr}(rF')-\frac{m^2}{r}F+\kappa^2 rF=0 \qquad (13.11)$$

In the notation of equation (10.1) in Chapter 7, $p(r)=r$, $q(r)=m^2/r$, $\lambda=\kappa^2$, and $\rho(r)=r$. For the drumhead or waveguide problem, the range of the variable r is $0\le r\le a$. Since $p(r)=0$ at $r=0$, Sturm–Liouville boundary conditions are satisfied at $r=0$ if $F(0)$ is finite. These conditions are satisfied at $r=a$ if either $F(a)=0$ or $F'(a)=0$. These are exactly the boundary conditions that are required for the waveguide and the drumhead. These problems therefore have Sturm–Liouville eigenfunctions as their solutions. The use of rectangular coordinates led to the same basic conclusions, and it is evident that other shapes for the drumhead and the waveguide and the use of other coordinate systems will give qualitatively similar solutions. Since they are finite at $r=0$, it is the Bessel functions $J_m(\kappa r)$ which are the appropriate solutions of the radial equation. If the boundary condition is $F(a)=0$, the Sturm–Liouville eigenvalues are to be determined from the condition

$$J_m(\kappa_{mn}a)=0 \qquad (13.12)$$

Thus, if x_{mn} are the roots of the Bessel functions, then $\kappa_{mn} = x_{mn}/a$. We know from the general discussion of the Sturm–Liouville eigenvalue problem in section 10 of Chapter 7 that the Bessel functions will be oscillatory in nature. Figure 13.2 at the end of this section shows graphs of the Bessel functions from which you could make crude estimates of the values of x_{mn}. More accurate values can be found from tables. For each value of m, the functions

$$F_{mn}(r) = N_{mn} J_m(\kappa_{mn} r) \tag{13.13}$$

are a complete set of eigenfunctions. The constant N_{mn} is a normalizing factor. Since the weight function is $\rho(r) = r$, these eigenfunctions obey the orthogonality condition

$$\int_0^a F_{mn}(r) F_{mn'}(r) r\, dr = \delta_{nn'} \tag{13.14}$$

If the angular dependence is included, the circular drumhead eigenfunctions can be expressed in the form

$$u_{mn}(r,\theta) = N_{mn} J_m(\kappa_{mn} r) \frac{1}{\sqrt{2\pi}} e^{im\theta} \tag{13.15}$$

where now m is an integer on account of the single-valuedness requirement for a function which represents the displacements of a vibrating drumhead. The factor $1/\sqrt{2\pi}$ is required to normalize the Fourier eigenfunctions $e^{im\theta}$. The eigenfunctions obey

$$\int_0^a \int_0^{2\pi} u^*_{mn}(r,\theta) u_{m'n'}(r,\theta) r\, dr\, d\theta = \delta_{nn'} \delta_{mm'} \tag{13.16}$$

Note that $r\, dr\, d\theta$ is an element of area in plane polar coordinates. A complete solution of the wave equation (3.3) of Chapter 7 for the circular drumhead is

$$\eta(r,\theta,t) = \sum_{n=1}^{\infty} \sum_{m=-\infty}^{\infty} u_{mn}(r,\theta) \times (A_{mn} \sin \omega_{mn} t + B_{mn} \cos \omega_{mn} t) \tag{13.17}$$

which should be compared to equation (3.19) of Chapter 7. Negative values of m are required if we want to use Fourier eigenfunctions in complex form for the θ dependence. It is then necessary to define

$$J_{-m}(x) = (-1)^m J_m(x) \tag{13.18}$$

for integer values of m. This is similar to what happened with the associated Legendre functions for negative integer values of m; see equation (4.41). It will be shown a little later in this section that (13.18) is consistent with the recurrence relations obeyed by the Bessel functions.

If we are interested in TE waves propagating in a waveguide of circular cross section, the analysis is the same, except that the Sturm–Liouville eigenvalues are to be found as the roots of the derivatives of the Bessel functions at $r = a$: $J'_m(\kappa_{mn} a) = 0$.

A recurrence formula for the Bessel functions can be found from the series expansion (1.37) by means of the following computations:

$$J_{m+1}(x) = \sum_{n=0}^{\infty} \frac{(-1)^n}{n!\Gamma(m+n+2)} \left(\frac{x}{2}\right)^{2n+m+1} = \sum_{n=1}^{\infty} \frac{(-1)^{n-1}}{(n-1)!\Gamma(m+n+1)} \left(\frac{x}{2}\right)^{2n+m-1} \tag{13.19}$$

$$J_{m-1}(x) = \sum_{n=0}^{\infty} \frac{(-1)^n}{n!\Gamma(m+n)} \left(\frac{x}{2}\right)^{2n+m-1} \tag{13.20}$$

$$\begin{aligned}
J_{m-1} + J_{m+1} &= \frac{1}{\Gamma(m)} \left(\frac{x}{2}\right)^{m-1} + \sum_{n=1}^{\infty} \left[\frac{(-1)^n}{n!\Gamma(m+n)} - \frac{(-1)^n}{(n-1)!\Gamma(m+n+1)} \right] \left(\frac{x}{2}\right)^{2n+m-1} \\
&= \left(\frac{x}{2}\right)^{m-1} \left\{ \frac{m}{\Gamma(m+1)} + \sum_{n=1}^{\infty} \left[\frac{(-1)^n (m+n)}{n!\Gamma(m+n+1)} - \frac{(-1)^n n}{n!\Gamma(m+n+1)} \right] \left(\frac{x}{2}\right)^{2n} \right\} \\
&= \left(\frac{x}{2}\right)^{m-1} \left\{ \frac{m}{\Gamma(m+1)} + \sum_{n=1}^{\infty} \frac{(-1)^n m}{n!\Gamma(m+n+1)} \left(\frac{x}{2}\right)^{2n} \right\} \\
&= \frac{2m}{x} \left(\frac{x}{2}\right)^m \sum_{n=0}^{\infty} \frac{(-1)^n}{n!\Gamma(m+n+1)} \left(\frac{x}{2}\right)^{2n} = \frac{2m}{x} J_m
\end{aligned}$$

$$J_{m-1} + J_{m+1} = \frac{2m}{x} J_m \tag{13.21}$$

Similarly,

$$J_{m-1} - J_{m+1} = 2J'_m \tag{13.22}$$

Other recurrence formulas which can be derived from (13.21) and (13.22) are

$$J_{m-1} = \frac{m}{x} J_m + J'_m \tag{13.23}$$

$$J_{m+1} = \frac{m}{x} J_m - J'_m \tag{13.24}$$

$$\frac{d}{dx}(x^m J_m) = x^m J_{m-1} \tag{13.25}$$

$$\frac{d}{dx}(x^{-m} J_m) = -x^{-m} J_{m+1} \tag{13.26}$$

The recurrence relations (13.21) through (13.26) are actually valid for all the linearly independent solutions of Bessel's equation. As noted in equation (1.49), Bessel and Neumann functions are frequently used as a pair of linearly independent solutions. Another possible pair are the Hankel functions. These functions will be discussed at some length in a later section.

The recurrence relations can be used to define Bessel functions of negative integer order. Equation (13.21) can be expressed in two forms as

$$J_{m-1} = \frac{2m}{x} J_m - J_{m+1} \tag{13.27}$$

$$J_{m+1} = \frac{2m}{x} J_m - J_{m-1} \tag{13.28}$$

Setting $m = 0$ in (13.27) gives

$$J_{-1} = -J_1 \tag{13.29}$$

Now set $m = -1$ in (13.27) and use (13.29) to get

$$J_{-2} = -\frac{2}{x} J_{-1} - J_0 = \frac{2}{x} J_1 - J_0 \tag{13.30}$$

If $m = 1$ in (13.28), then

$$J_2 = \frac{2}{x} J_1 - J_0 = J_{-2} \tag{13.31}$$

By continuing in this manner, we arrive at (13.18).

The integral required to determine the normalizing factor N_{mn} in equation (13.13) can be evaluated by first noting that

$$\frac{d}{dr}\left[r^2 F'^2 + (\kappa^2 r^2 - m^2) F^2\right] = 2F'\left[r^2 F'' + rF' + (\kappa^2 r^2 - m^2) F\right] + 2\kappa^2 r F^2 = 2\kappa^2 r F^2 \tag{13.32}$$

The quantity in brackets is zero because F is a solution of Bessel's equation as given by (13.4). By integrating (13.32), we find

$$\int_0^a F^2 r\, dr = \frac{1}{2\kappa^2}\left[r^2 F'^2 + (\kappa^2 r^2 - m^2) F^2\right]_0^a \tag{13.33}$$

If $F(r)$ is a Bessel function and if the boundary condition is $F(a) = 0$, then

$$\int_0^a \left[J_m(\kappa_{mn} r)\right]^2 r\, dr = \tfrac{1}{2} a^2 \left[J'_{mn}(\kappa_{mn} a)\right]^2 = \tfrac{1}{2} a^2 \left[J_{m+1}(\kappa_{mn} a)\right]^2 \tag{13.34}$$

where use has been made of the recurrence relation (13.24) and the fact that $J_m(\kappa_{mn} a) = 0$. The normalized eigenfunctions associated with the radial equation are therefore

$$F_{mn}(r) = \frac{\sqrt{2}}{a J_{m+1}(\kappa_{mn} a)} J_m(\kappa_{mn} r) \tag{13.35}$$

As an example of a boundary value problem in potential theory, suppose that the electrostatic potential is known as a function of r and θ over the base of a circular cylinder of radius a and height s, as shown in Fig. 13.1. The other parts of the cylinder are grounded, and it is required to find the solution of Laplace's equation (13.1) at all points inside. Solutions of (13.2) are required which vanish at $z = s$. Therefore,

$$H(z) = \sinh\left[\kappa_{mn}(s - z)\right] \tag{13.36}$$

The complete solution is

$$\psi = \sum_{n=1}^{\infty} \sum_{m=-\infty}^{\infty} c_{mn} u_{mn}(r, \theta) \sinh\left[\kappa_{mn}(s - z)\right] \tag{13.37}$$

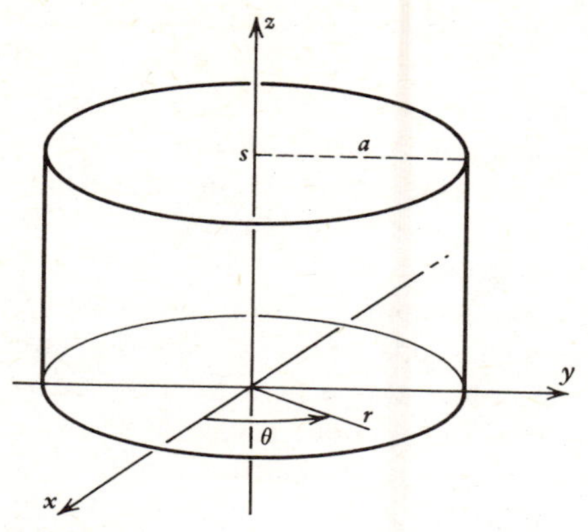

Fig. 13.1

where

$$u_{mn}(r,\theta)=\frac{1}{a\sqrt{\pi}\,J_{m+1}(\kappa_{mn}a)}J_m(\kappa_{mn}r)\,e^{im\theta} \tag{13.38}$$

are a complete set of orthonormal eigenfunctions with respect to the variables r and θ. Since $f(r,\theta)=\psi(r,\theta,0)$ is a known function, the coefficients in (13.37) are given by

$$c_{mn}\sinh(\kappa_{mn}s)=\int_0^a\int_0^{2\pi}u^*_{mn}(r,\theta)f(r,\theta)\,r\,dr\,d\theta \tag{13.39}$$

thus completing the problem.

Another boundary value problem involving the same geometry results if the potential is known as a function of θ and z over the cylindrical walls and the two ends are grounded. Eigenfunctions with respect to z are now needed, and we must replace the separation constant κ^2 by $-\kappa^2$ in (13.2) and (13.4). The required solutions of the z equation are the normalized eigenfunctions

$$u_n(z)=\sqrt{\frac{2}{s}}\sin\left(\frac{n\pi z}{s}\right)$$

$$\kappa_n=\frac{n\pi}{s}\qquad n=1,2,3,\ldots \tag{13.40}$$

The appropriate combined solutions of the θ and z equations are

$$u_{mn}(\theta,z)=\frac{1}{\sqrt{\pi s}}\sin\left(\frac{n\pi z}{s}\right)e^{im\theta} \tag{13.41}$$

The required solutions of the radial equation are now the hyperbolic Bessel functions given by (13.9). Therefore,

$$\psi(r,\theta,z)=\sum_{n=1}^{\infty}\sum_{m=-\infty}^{+\infty}c_{mn}u_{mn}(\theta,z)I_m(\kappa_n r) \tag{13.42}$$

If $f(\theta,z)=\psi(a,\theta,z)$ is the known potential over the cylindrical walls, then the constants in (13.42) are determined in the usual way by using the orthogonality property of (13.41):

$$c_{mn}I_m(\kappa_n a)=\int_0^s\int_0^{2\pi}u^*_{mn}(\theta,z)f(\theta,z)\,d\theta\,dz \tag{13.43}$$

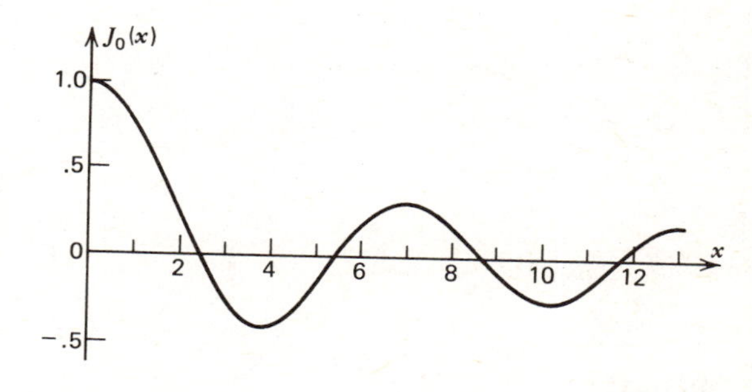

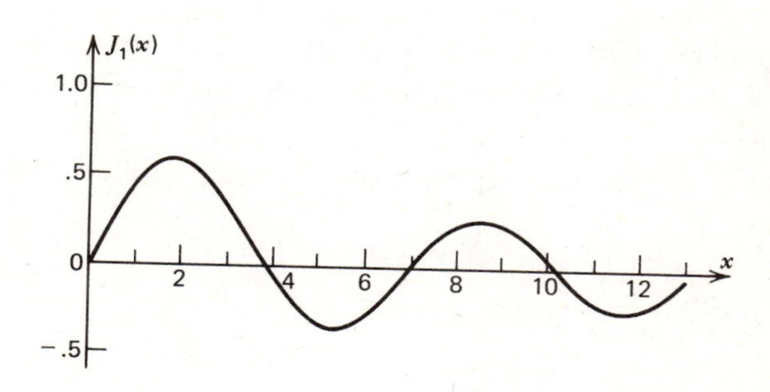

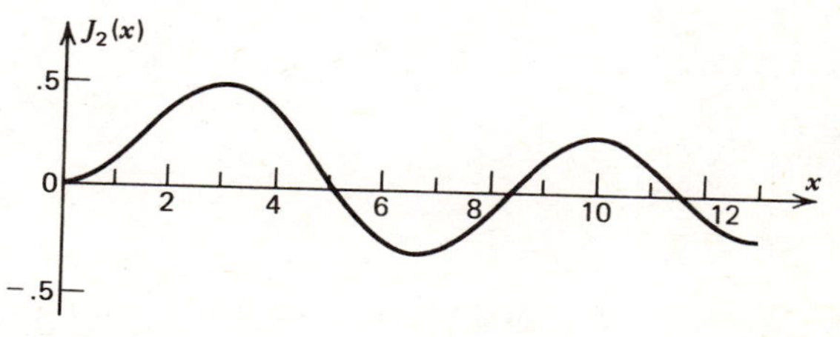

Fig. 13.2

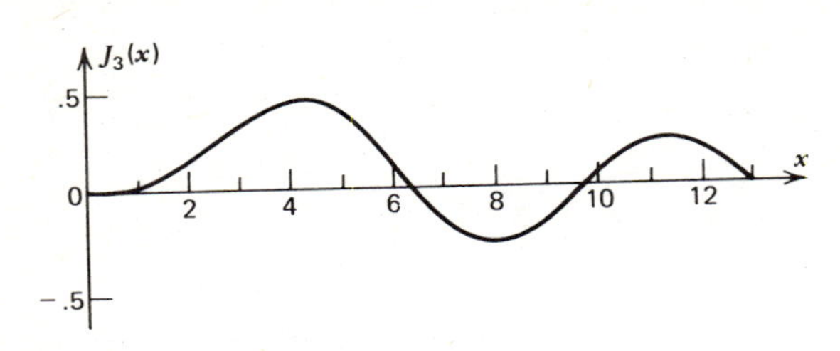

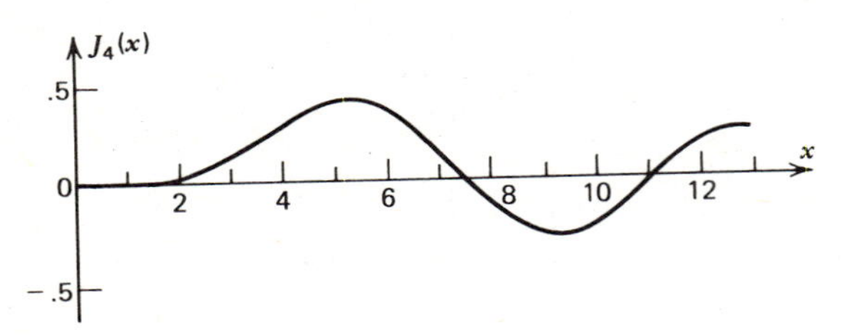

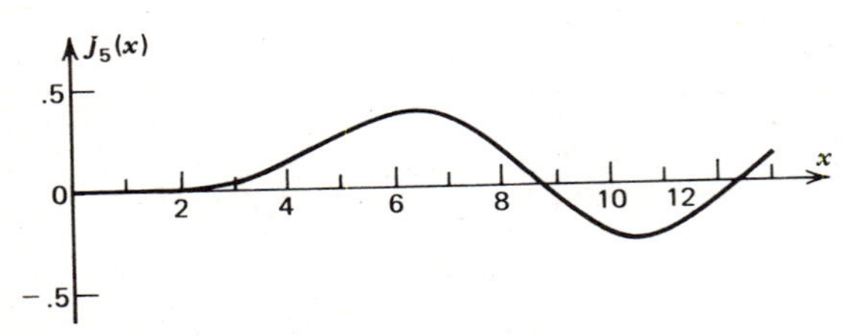

Fig. 13.2 (*continued*)

The reader should note that the choice of the sign of the separation constant κ^2 determines the type of solutions which are obtained. The hyperbolic Bessel functions are exponential in form, whereas the solutions of the z equation as given by (13.40) are oscillatory and are Sturm–Liouville eigenfunctions. The situation is turned around in the solution (13.37), where the eigenfunctions come out in terms of Bessel functions and the solutions of the z equation are exponential. If the potential is a prescribed function over both the base of the cylinder and its walls with only the end at $z=s$ grounded, then the potential at any point inside is found by superimposing (13.37) and (13.42).

Extensive tables giving numerical values of Bessel functions exist. The infinite series representation (1.37) is rapidly convergent and can be used as the basis for numerical evaluation of the Bessel functions. Figure 13.2 shows graphs of the first six Bessel functions of integer order. Note that all the Bessel functions are zero at $x=0$ except for $J_0(0)$, which is 1.

PROBLEMS

13.1. Use equation (12.12) of Chapter 5 to show that the Neumann function $N_0(x)$ has a logarithmic singularity at $x=0$ and that $N_m(x)$ is proportional to x^{-m} near $x=0$.

Hint: It is necessary to use only the leading term in the series expansion for the Bessel functions.

13.2. Prove the following recurrence formulas for the modified Bessel functions:

$$I_{m-1}-I_{m+1}=\frac{2m}{x}I_m \tag{13.44}$$

$$I_{m-1}-I_{m+1}=2I_m' \tag{13.45}$$

$$I_{m-1}=\frac{m}{x}I_m+I_m' \tag{13.46}$$

$$I_{m+1}=-\frac{m}{x}I_m+I_m' \tag{13.47}$$

$$\frac{d}{dx}(x^mI_m)=x^mI_{m-1} \tag{13.48}$$

$$\frac{d}{dx}(x^{-m}I_m)=x^{-m}I_{m+1} \tag{13.49}$$

13.3. Suppose it is established that $J_{-m}(x)=(-1)^mJ_m(x)$ for all positive integers less than or equal to m. Show that (13.27) and (13.28) imply that $J_{-(m+1)}=(-1)^{m+1}J_{m+1}$.

13.4. The cylinder in Fig. 13.1 has its base ($z=0$) at the constant potential V. The sides and the end at $z=s$ are grounded. Show that the potential at any point inside the cylinder is

$$\psi(r,z)=\sum_{n=1}^{\infty}\frac{2VJ_0(\kappa_n r)}{\kappa_n aJ_1(\kappa_n a)\sinh(\kappa_n s)}\sinh\kappa_n(s-z) \tag{13.50}$$

where the eigenvalues κ_n are determined by $J_0(\kappa_n a)=0$.

13.5. Refer to section 5 in Chapter 7. The eigenfunctions of equation (5.16) in Chapter 7 for TE waves propagating in a waveguide of circular cross section can be written as

$$B_3=B_0J_m(\gamma r)\cos m\theta \tag{13.51}$$

If a is the radius of the waveguide, then γ is determined for each mode by the condition $J_m'(\gamma a)=0$. We should really write γ_{mn} instead of γ, but let's use γ to economize the notation. There is degeneracy, and we could also use as eigenfunctions

$$B_3=B_0J_m(\gamma r)\sin m\theta \tag{13.52}$$

or a linear combination of (13.51) and (13.52). If (13.51) is used, show that the field components, including the time dependence, are given by

$$B_r=-\frac{\kappa_3 B_0}{\gamma}J_m'(\gamma r)\cos m\theta\sin(\kappa_3 z-\omega t) \tag{13.53}$$

$$B_\theta=\frac{m\kappa_3}{\gamma^2 r}B_0J_m(\gamma r)\sin m\theta\sin(\kappa_3 z-\omega t) \tag{13.54}$$

$$B_z=B_0J_m(\gamma r)\cos m\theta\cos(\kappa_3 z-\omega t) \tag{13.55}$$

$$E_r=\frac{\omega m B_0}{\gamma^2 r}J_m(\gamma r)\sin m\theta\sin(\kappa_3 z-\omega t) \tag{13.56}$$

$$E_\theta=\frac{\omega B_0}{\gamma}J_m'(\gamma r)\cos m\theta\sin(\kappa_3 z-\omega t) \tag{13.57}$$

Hint: First show from equations (5.10) through (5.14) in Chapter 7 that the transverse part of the time-independent electromagnetic field can be expressed as

$$\mathbf{E}_T=\frac{i\omega}{\gamma^2}(\nabla_T\times\hat{\mathbf{e}}_3B_3)\qquad \mathbf{B}_T=\frac{i\kappa_3}{\gamma^2}\nabla_T B_3 \tag{13.58}$$

where

$$\nabla_T=\hat{\mathbf{e}}_1\frac{\partial}{\partial x}+\hat{\mathbf{e}}_2\frac{\partial}{\partial y}\qquad \mathbf{E}_T=E_1\hat{\mathbf{e}}_1+E_2\hat{\mathbf{e}}_2 \tag{13.59}$$

Then express the curl and the gradient in plane polar coordinates.

13.6. In each of the following cases, show that the given function satisfies the indicated differential equation:

(a) $f(z)=J_m(\lambda z^q)$

$$z^2f''(z)+zf'(z)+(\lambda^2q^2z^{2q}-m^2q^2)f(z)=0 \tag{13.60}$$

(b) $w(z)=z^pJ_m(\lambda z^q)$

$$z^2w''(z)+(1-2p)zw'(z)+(\lambda^2q^2z^{2q}+p^2-m^2q^2)w(z)=0 \tag{13.61}$$

(c) $w(z)=\sqrt{z}\,J_{1/n}\left(\frac{2k}{n}z^{n/2}\right)$

$$w''(z)+k^2z^{n-2}w(z)=0 \tag{13.62}$$

13.7. By considering the order of the Bessel function as a continuous variable, show that

$$\frac{\partial}{\partial m} J_m(x) = J_m(x) \ln x - \left(\frac{x}{2}\right)^m \sum_{n=0}^{\infty} \frac{(-1)^n \psi(m+n+1)}{n!\Gamma(m+n+1)} \left(\frac{x}{2}\right)^{2n} \tag{13.63}$$

where ψ is the digamma function and is defined by equation (22.37) of Chapter 5.

13.8. Refer to the discussion of heat conduction in section 6 of Chapter 3. An infinitely long cylinder of radius a is initially at the temperature T_0 throughout. Starting at $t=0$, the cylinder walls are kept at a temperature $T=0$. Show that the temperature in the cylinder is given by

$$T(r,t) = \sum_{n=0}^{\infty} \frac{2T_0 J_0(\kappa_n r)}{\kappa_n a J_1(\kappa_n a)} e^{-\alpha\kappa_n^2 t} \tag{13.64}$$

The appropriate starting point is equation (6.8) in Chapter 3, with $\alpha = k/\rho c$ and $f=0$.

13.9. Evaluate the following integrals:

(a) $\int x^2 J_0(x)\,dx = x^2 J_1(x) + xJ_0(x) - \int J_0(x)\,dx$ (13.65)

(b) $\int x^3 J_0(x)\,dx = x(x^2-4)J_1(x) + 2x^2 J_0(x)$ (13.66)

13.10. If α is any root of $J_0(x)=0$, show that

(a) $\int_0^1 J_1(\alpha x)\,dx = \frac{1}{\alpha}$ (13.67)

(b) $\int_0^\alpha J_1(x)\,dx = 1$ (13.68)

(c) $\int_0^\infty J_1(x)\,dx = 1$ (13.69)

13.11. Show from the recurrence relation (13.24) that the zeros of $J_{m+1}(x)$ must interleave with those of $J_m(x)$. There is exactly one zero of $J_m(x)$ between two successive zeros of $J_{m+1}(x)$. The only exception is that all the Bessel functions have a common zero at $x=0$ except for $J_0(0)$, which is unity.

13.12. If κ denotes any positive root of $J_0(x)=0$ and $0 \le r \le 1$, obtain the following expansions:

(a) $1 = \sum_\kappa \frac{2J_0(\kappa r)}{\kappa J_1(\kappa)}$ (13.70)

(b) $r^2 = \sum_\kappa \frac{2(\kappa^2-4)J_0(\kappa r)}{\kappa^3 J_1(\kappa)}$ (13.71)

where the sums are over all the positive roots.

13.13. If α and β are two real numbers and $\alpha \ne \beta$, show that

$$\begin{aligned}\int J_\nu(\alpha r) J_\nu(\beta r)\, r\, dr &= \frac{r}{\alpha^2-\beta^2}\left[\beta J_\nu(\alpha r) J_\nu'(\beta r) - \alpha J_\nu'(\alpha r) J_\nu(\beta r)\right] \\ &= \frac{r}{\alpha^2-\beta^2}\left[\beta J_\nu(\alpha r) J_{\nu-1}(\beta r) - \alpha J_\nu(\beta r) J_{\nu-1}(\alpha r)\right] \\ &= \frac{\alpha\beta r^2}{2\nu(\alpha^2-\beta^2)}\left[J_{\nu+1}(\alpha r) J_{\nu-1}(\beta r) - J_{\nu+1}(\beta r) J_{\nu-1}(\alpha r)\right]\end{aligned} \tag{13.72}$$

Hint: Write Bessel's equation in Sturm–Liouville form and use equation (10.5) in Chapter 7.

13.14. Obtain the expansion

$$J_0(\alpha r) = \sum_\kappa \frac{2\kappa J_0(\alpha) J_0(\kappa r)}{(\kappa^2-\alpha^2) J_1(\kappa)} \qquad 0 \le r \le 1 \tag{13.73}$$

where the sum is over all the positive roots of $J_0(x)=0$ and α does not equal any of the roots.

13.15. Prove that

$$\frac{J_{n+1}(x)}{x^{n+1}} = -\left(\frac{1}{x}\frac{d}{dx}\right)\frac{J_n(x)}{x^n}$$

$$\frac{J_{n+k}(x)}{x^{n+k}} = (-1)^k\left(\frac{1}{x}\frac{d}{dx}\right)^k\frac{J_n(x)}{x_n}$$

$$x^{n-1}J_{n-1}(x) = \left(\frac{1}{x}\frac{d}{dx}\right)x^nJ_n(x) \qquad (13.74)$$

$$x^{n-k}J_{n-k}(x) = \left(\frac{1}{x}\frac{d}{dx}\right)^k x^nJ_n(x)$$

14. CONTOUR INTEGRAL SOLUTION OF BESSEL'S EQUATION

By separating Laplace's equation in rectangular coordinates, it is easily shown that a simple solution is

$$\psi = Ae^{i(ax+by)+\kappa z} \qquad (14.1)$$

where A, a, b, and κ are constants and

$$a^2 + b^2 = \kappa^2 \qquad (14.2)$$

If we set

$$a = -\kappa\sin\alpha \qquad b = \kappa\cos\alpha \qquad (14.3)$$

and change to cylindrical coordinates by means of the transformation

$$x = r\cos\theta \qquad y = r\sin\theta \qquad (14.4)$$

the result is

$$\psi = Ae^{-i\kappa r\sin(\alpha-\theta)}e^{\kappa z} \qquad (14.5)$$

A more general solution can be built up by integrating over the parameter α:

$$\psi = e^{\kappa z}\int_\alpha A(\alpha)e^{-i\kappa r\sin(\alpha-\theta)}\,d\alpha \qquad (14.6)$$

where, as indicated, A is to be regarded as a function of α. What we are doing is a little bit like taking a plane wave solution of the wave equation and then integrating over the wave number to obtain a Fourier integral representation of a more general solution. In general, the order of a Bessel function, or other solution of Bessel's equation, can be any real number, which we will now denote by the letter ν, reserving m or n for the case where the order is specifically an integer. Actually, it is possible for the order of a Bessel function to be a complex number, but we will not consider this generalization here. We know that in cylindrical coordinates the solutions of the θ equation are of the form $e^{i\nu\theta}$. This form of θ dependence can be separated out of (14.6) if we write

$$A(\alpha) = Ce^{i\nu\alpha} \qquad (14.7)$$

and then make the change of variable $\alpha = \theta + \zeta$ in the integral (14.6). The result is

$$\psi = Ce^{i\nu\theta}e^{\kappa z}\int_1^2 e^{-ix\sin\zeta + i\nu\zeta}\,d\zeta \qquad (14.8)$$

The integral is between two as yet unspecified points in the complex ζ plane. We have set $x = \kappa r$ in accordance with the notation of section 13. There is a notational problem here in that we just got through using x as one of the rectangular coordinates. For the rest of this section, x is the independent variable of Bessel's equation. Equation (14.8) has the correct form to be a solution of Laplace's equation in cylindrical coordinates, provided that

$$w(x) = C\int_1^2 e^{-ix\sin\zeta + i\nu\zeta}\,d\zeta \qquad (14.9)$$

is a solution of Bessel's equation (13.5). This will be the case if the limits of integration are appropriately chosen as we now show. By direct substitution of (14.9) into the differential equation (13.5), we find

$$\begin{aligned}0 &= \int_1^2 (x^2\cos^2\zeta - ix\sin\zeta - \nu^2)e^{-ix\sin\zeta + i\nu\zeta}\,d\zeta \\ &= \int_1^2 \frac{d}{d\zeta}\left[(ix\cos\zeta + i\nu)e^{-ix\sin\zeta + i\nu\zeta}\right]d\zeta \\ &= \left[(ix\cos\zeta - i\nu)e^{-ix\sin\zeta + i\nu\zeta}\right]_1^2 \end{aligned} \qquad (14.10)$$

In order for (14.10) to be valid, the quantity in square brackets must have the same value at the two end points. This can be accomplished if ζ is allowed to go to infinity in such a way that

$$e^{-ix\sin\zeta + i\nu\zeta} \to 0 \tag{14.11}$$

You may want to review the discussion of trigonometric and hyperbolic functions of a complex variable in section 14 of Chapter 5. If $\zeta = p + iq$, then

$$\begin{aligned}\sin\zeta &= \sin p\cos iq + \cos p\sin iq\\ &= \sin p\cosh q + i\cos p\sinh q\end{aligned}$$

$$e^{-ix\sin\zeta} = e^{-ix\sin p\cosh q}e^{x\cos p\sinh q} \tag{14.12}$$

If $q \to +\infty$, then (14.12) becomes zero if $\cos p < 0$. If $q \to -\infty$, then $\cos p > 0$ is required. Paths of integration in the complex ζ plane can therefore be chosen that go to infinity in the shaded portions of Fig. 14.1a. Three possible contours, C_0, C_1, and C_2, are shown. If the constant C in (14.9) is chosen as $1/(2\pi)$ and the contour C_0 is used, Bessel's functions are obtained:

$$J_\nu(x) = \frac{1}{2\pi}\int_{C_0} e^{-ix\sin\zeta + i\nu\zeta}d\zeta \tag{14.13}$$

The truth of this assertion will be verified in the subsequent discussion.

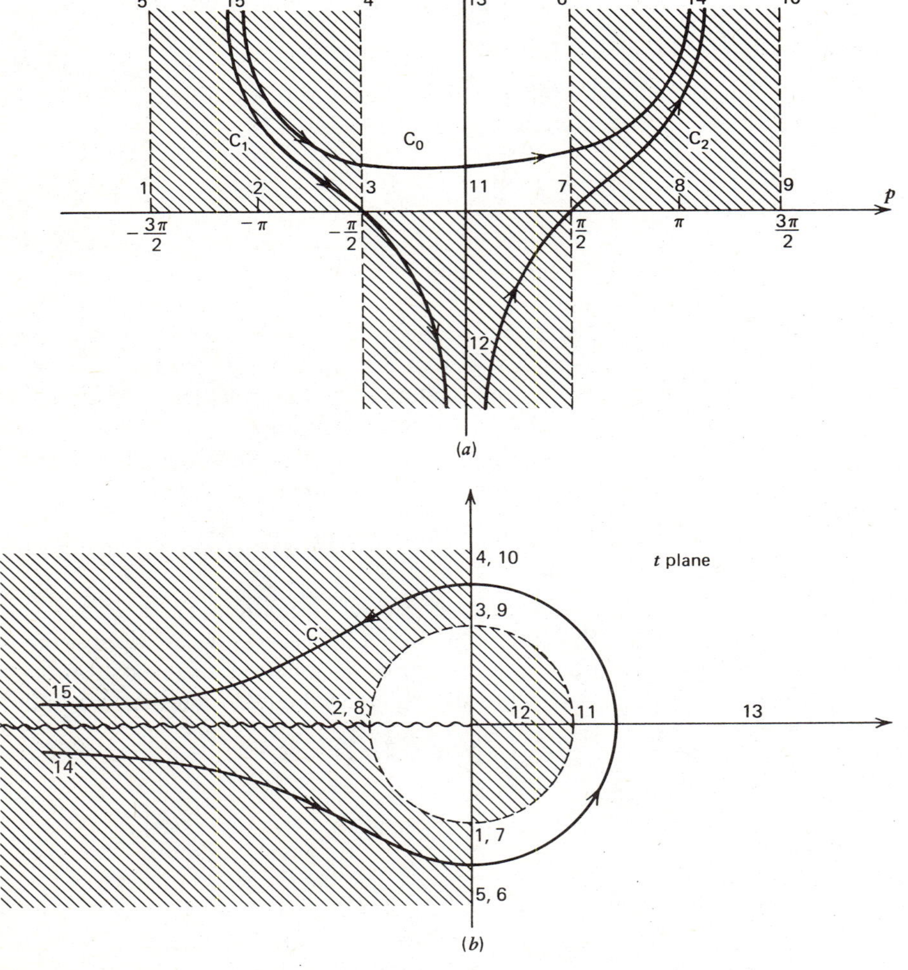

Fig. 14.1

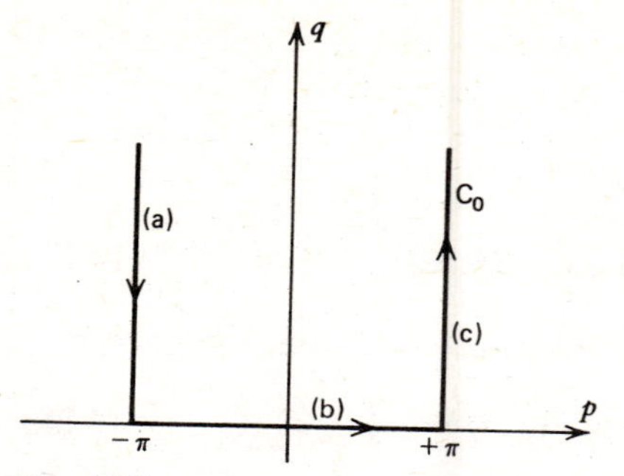

Fig. 14.2

The path of integration can be deformed as shown in Fig. 14.2. The variable of integration ζ along the three straight-line portions is to be written as (a) $\zeta = -\pi + iq$, (b) $\zeta = p$, and (c) $\zeta = \pi + iq$. The integrals along (a) and (c) can be combined into a single integral with the result that

$$J_\nu(x) = \frac{1}{2\pi}\int_{-\pi}^{+\pi} e^{-ix\sin p + i\nu p}\,dp - \frac{\sin \nu\pi}{\pi}\int_0^\infty e^{-x\sinh q - \nu q}\,dq \qquad (14.14)$$

If $\nu \geq 0$, we can now verify that (14.14) represents the solution of Bessel's equation which is finite at $x = 0$. Setting $x = 0$ and $\nu = 0$ in (14.14) gives

$$J_0(0) = \frac{1}{2\pi}\int_{-\pi}^{+\pi} dp = 1 \qquad (14.15)$$

If $x = 0$ and $\nu > 0$, then

$$J_\nu(0) = \frac{1}{2\pi}\int_{-\pi}^{+\pi} e^{i\nu p}\,dp - \frac{\sin \nu\pi}{\pi}\int_0^\infty e^{-\nu q}\,dq$$

$$= \left.\frac{e^{i\nu p}}{2\pi i\nu}\right|_{-\pi}^{+\pi} - \frac{\sin \nu\pi}{\pi\nu}$$

$$= \frac{e^{i\nu\pi} - e^{-i\nu\pi}}{2\pi i\nu} - \frac{\sin \nu\pi}{\pi\nu} = 0 \qquad (14.16)$$

For $\nu > 0$, equation (14.14) can therefore differ from a Bessel function by at most a multiplicative factor. For Bessel functions of integer order, (14.14) specializes to

$$J_m(x) = \frac{1}{2\pi}\int_{-\pi}^{+\pi} e^{-ix\sin p + imp}\,dp$$

$$= \frac{1}{2\pi}\int_{-\pi}^{+\pi} \cos(mp - x\sin p)\,dp$$

$$= \frac{1}{\pi}\int_0^{\pi} \cos(mp - x\sin p)\,dp \qquad (14.17)$$

The last two forms are true because the imaginary part of the integrand is an odd function of p whereas the real part is an even function.

The change of variable

$$t = e^{-i\zeta} \qquad d\zeta = \frac{i\,dt}{t} \qquad (14.18)$$

converts (14.13) into

$$J_\nu(x) = \frac{1}{2\pi i}\int_C e^{(x/2)[t-(1/t)]}\frac{dt}{t^{\nu+1}} \qquad (14.19)$$

The contour C in the complex t plane is shown in Fig. 14.1*b*. Reference to section 17 of Chapter 5 shows that a change of variable such as that given by (14.18) can be thought of as a mapping of the complex ζ plane onto the complex t plane. To explore the geometry of the mapping in detail, we have chosen several points in the ζ plane and have computed their images in the t plane. These are the numbered points in Fig. 14.1. For example, at point 5, $\zeta = -3\pi/2 + iq$ and

$$t = e^{-i\zeta} = e^{3\pi i/2}e^{q} = -ie^{q} \qquad (14.20)$$

which is a point of the negative imaginary axis in the t plane. The mapping is not one-to-one because more than one point in the ζ plane is mapped onto the same point in the t plane. Both the shaded regions, $-3\pi/2 < p < -\pi/2$, $q > 0$, and $\pi/2 < p < 3\pi/2$, $q > 0$, map onto the half-space $\mathrm{Re}\,t < 0$, excluding the region inside the unit circle centered at the origin. The shaded region, $-\pi/2 < p < \pi/2$, $q < 0$, maps onto points inside the unit circle given by $|t| < 1$, $\mathrm{Re}\,t > 0$. The contour C_0 maps onto the contour C, except that the direction of integration on C has been reversed. If you checked out the transformation leading to (14.19) and can't find a minus sign, this is what happened to it. If ν is not an integer, then $t^{-\nu-1}$ is a multiple-valued function with $t = 0$ as a branch point.

In Fig. 14.1*b*, a branch cut has been placed along the negative real t axis.

One of the uses to which (14.19) can be put is to derive a generating function for the Bessel functions of integer order. If $\nu = m$ is an integer, then the integrand of (14.19) becomes single-valued, and the contour C can be closed into a circle about the origin. Reference to equation (19.40) of Chapter 5 shows that the Laurent expansion of a function about $t = 0$ can be written as

$$f(t) = \sum_{m=-\infty}^{+\infty} A_m t^m \qquad A_m = \frac{1}{2\pi i}\oint \frac{f(t)}{t^{m+1}}\,dt \tag{14.21}$$

If we set

$$f(t) = e^{(x/2)[(t-1/t)]} \tag{14.22}$$

then $A_m = J_m(x)$ and

$$e^{(x/2)(t-1/t)} = \sum_{m=-\infty}^{+\infty} J_m(x) t^m \tag{14.23}$$

The infinite series for $J_m(x)$ can be recovered from (14.23). If we write the generating function as

$$\begin{aligned} e^{xt/2}e^{-x/(2t)} &= \left[1 + \frac{xt}{2} + \cdots + \frac{1}{m!}\left(\frac{x}{2}\right)^m t^m + \cdots\right] \\ &\quad \times \left[1 - \frac{x}{2t} + \cdots + \frac{(-1)^n}{n!}\left(\frac{x}{2t}\right)^n + \cdots\right] \end{aligned} \tag{14.24}$$

we see that the coefficient of t^m with the lowest power of x is $(1/m!)(x/2)^m$. This is in fact the leading term for the series expansion of the Bessel functions as reference to equation (1.38) shows. Thus, (14.13) gives the Bessel functions with the correct multiplicative factor.

The solutions of Bessel's equation which result if the contours C_1 and C_2 of Fig. 14.1*a* are used are called *Hankel functions* and are given by

$$H_\nu^1(x) = \frac{1}{\pi}\int_{C_1} e^{-ix\sin\zeta + i\nu\zeta}\,d\zeta \tag{14.25}$$

$$H_\nu^2(x) = \frac{1}{\pi}\int_{C_2} e^{-ix\sin\zeta + i\nu\zeta}\,d\zeta \tag{14.26}$$

It will become clear in section 15 that the Hankel functions are two linearly independent solutions of Bessel's equation. By appropriately deforming the contours C_0, C_1, and C_2 of Fig. 14.1*a*, it becomes evident that

$$J_\nu(x) = \frac{1}{2}\left[H_\nu^1(x) + H_\nu^2(x)\right] \tag{14.27}$$

A fourth solution of Bessel's equation is defined by

$$N_\nu(x) = \frac{1}{2i}\left[H_\nu^1(x) - H_\nu^2(x)\right] \tag{14.28}$$

and is called a *Neumann function*. There are of course only two linearly independent solutions of Bessel's equation. As possible pairs of linearly independent solutions we can use either J_ν and N_ν or H_ν^1 and H_ν^2. If ν is not an integer, then J_ν and $J_{-\nu}$ can also be used. Recall that pairs of linearly independent solutions of $w''(x) + w(x) = 0$ are $\sin x$ and $\cos x$ or e^{ix} and e^{-ix}. These pairs are related by

$$\cos x = \frac{1}{2}(e^{ix} + e^{-ix}) \qquad \sin x = \frac{1}{2i}(e^{ix} - e^{-ix}) \tag{14.29}$$

which are analogous to (14.27) and (14.28).

The possible solutions of Bessel's equation are sometimes referred to collectively as *cylinder functions*. They all obey the same set of recurrence relations, namely, those given by (13.21) through (13.26). For example,

$$\begin{aligned} H_{\nu+1}^1 + H_{\nu-1}^1 &= \frac{1}{\pi}\int_{C_1} e^{-ix\sin\zeta}\left[e^{i(\nu+1)\zeta} + e^{i(\nu-1)\zeta}\right] d\zeta \\ &= \frac{2}{\pi}\int_{C_1} e^{-ix\sin\zeta + i\nu\zeta}\cos\zeta\, d\zeta \\ &= \frac{2i}{\pi x}\int_{C_1} \frac{d}{d\zeta}\left(e^{-ix\sin\zeta + i\nu\zeta}\right) d\zeta \\ &\quad + \frac{2\nu}{\pi x}\int_{C_1} e^{-ix\sin\zeta + i\nu\zeta}\,d\zeta \\ &= \frac{2\nu}{x} H_\nu^1(x) \end{aligned} \tag{14.30}$$

Problem 13.1 reveals that the Neumann and Hankel functions are singular at $x = 0$.

The solution of the modified or hyperbolic Bessel equation (13.7) which is finite at $x = 0$ is given by

(13.8). A frequently used second linearly independent solution is defined by

$$K_\nu(x) = \frac{\pi}{2} i^{\nu+1} H_\nu^1(ix) \qquad (14.31)$$

As an example of the use of the generating function (14.23), we will derive the addition theorem for Bessel functions of integer order. Consider the identity

$$e^{[(x+y)/2][t-(1/t)]} = e^{(x/2)[t-(1/t)]} e^{(y/2)[t-(1/t)]} \qquad (14.32)$$

Expanding the exponential factors according to (14.23) leads to

$$\sum_{m=-\infty}^{+\infty} J_m(x+y) t^m = \sum_{n=-\infty}^{+\infty} \sum_{k=-\infty}^{+\infty} J_n(x) J_k(y) t^{n+k} \qquad (14.33)$$

By equating the like powers of t on the two sides of the equation we find

$$J_m(x+y) = \sum_{n=-\infty}^{+\infty} J_n(x) J_{m-n}(y) \qquad (14.34)$$

In particular, if $m=0$,

$$\begin{aligned} J_0(x+y) &= J_0(x)J_0(y) + J_1(x)J_{-1}(y) \\ &\quad + J_{-1}(x)J_1(y) + J_2(x)J_{-2}(y) \\ &\quad + J_{-2}(x)J_2(y) + \cdots \\ &= J_0(x)J_0(y) - 2J_1(x)J_1(y) \\ &\quad + 2J_2(x)J_2(y) + \cdots \\ &= J_0(x)J_0(y) + 2\sum_{n=1}^{\infty} (-1)^n J_n(x)J_n(y) \end{aligned} \qquad (14.35)$$

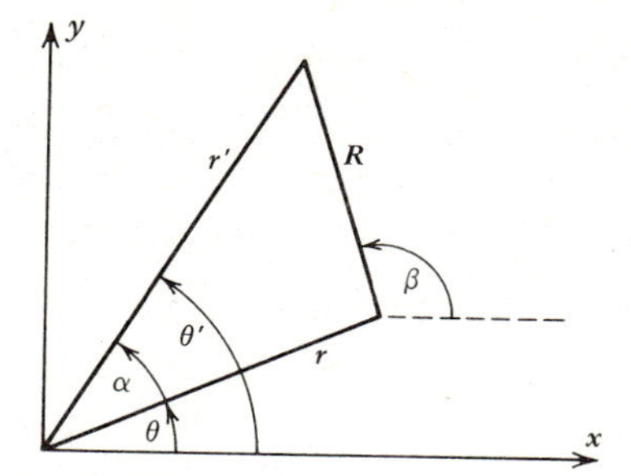

Fig. 14.3

This is a special case of what is known as the *addition theorem* for Bessel functions of integer order. A generalization can be found by first noting from Fig. 14.3 that

$$R = \sqrt{r^2 + r'^2 - 2rr'\cos\alpha} \qquad (14.36)$$

It is then possible to expand $J_0(R)$ in a Fourier cosine series:

$$J_0(R) = A_0(r,r') + \sum_{n=1}^{\infty} A_n(r,r')\cos n\alpha \qquad (14.37)$$

If we set $\alpha = \pi$, we find $J_0(R) = J_0(r+r')$, in which case (14.37) becomes identical to (14.35). The coefficients in (14.37) are then determined with the result that

$$J_0(R) = J_0(r)J_0(r') + \sum_{n=1}^{\infty} 2J_n(r)J_n(r')\cos n\alpha \qquad (14.38)$$

The analogous result for spherical harmonics is given by (12.24).

PROBLEMS

14.1. Show from the generating function (14.23) that

$$J_n(-x) = (-1)^n J_n(x) = J_{-n}(x) \qquad (14.39)$$

where n is an integer.

14.2. Draw a sketch showing the images of the Hankel function contours C_1 and C_2 of Fig. 14.1*a* in the complex t plane of Fig. 14.1*b*.

14.3. Derive the following:

$$e^{ix\sin\theta} = \sum_{n=-\infty}^{+\infty} J_n(x)e^{in\theta} \tag{14.40}$$

$$\cos(x\sin\theta) = J_0(x) + 2[J_2(x)\cos 2\theta + J_4(x)\cos 4\theta + \cdots] \tag{14.41}$$

$$\sin(x\sin\theta) = 2[J_1(x)\sin\theta + J_3(x)\sin 3\theta + J_5(x)\sin 5\theta + \cdots] \tag{14.42}$$

$$J_n(x) = \frac{1}{2\pi}\int_{-\pi}^{+\pi} \cos n\theta \cos(x\sin\theta)\,d\theta \qquad n \text{ even} \tag{14.43}$$

$$J_n(x) = \frac{1}{2\pi}\int_{-\pi}^{+\pi} \sin n\theta \sin(x\sin\theta)\,d\theta \qquad n \text{ odd} \tag{14.44}$$

Hint: Put $t = e^{i\theta}$ in (14.22). Separate into real and imaginary parts. Write as Fourier series in real form. See equation (1.45) in Chapter 8.

14.4. Show that

$$J_0(x) = \frac{1}{2\pi}\int_0^{2\pi} \cos(x\cos\theta)\,d\theta = \frac{1}{\pi}\int_0^{\pi} \cos(x\cos\theta)\,d\theta$$

$$= \frac{2}{\pi}\int_0^{\pi/2} \cos(x\cos\theta)\,d\theta \tag{14.45}$$

14.5. Evaluate the integral

$$\int_0^{\infty} e^{-ax}J_0(bx)\,dx = \frac{1}{\sqrt{a^2+b^2}} \qquad a > 0 \tag{14.46}$$

Hint: Start with the known integral

$$\int_0^{\infty} e^{-ax}\cos kx\,dx = \frac{a}{a^2+k^2} \tag{14.47}$$

Let $k = b\cos\theta$. Integrate θ from 0 to 2π. Refer to equation (21.34) of Chapter 5 to see how to evaluate the integral over θ.

14.6. Prove that

$$\int_0^{\infty} \frac{1}{x}J_1(x)\,dx = 1 \tag{14.48}$$

Hint: Write Bessel's equation of order zero as

$$J_0''(x) + \frac{1}{x}J_0'(x) + J_0(x) = 0 \tag{14.49}$$

Use $J_1(x) = -J_0'(x)$, which follows from (13.24). Equation (13.69) shows that $J_1(x) \to 0$ as $x \to \infty$.

14.7. Evaluate the following integrals:

(a) $$\int_0^a\int_0^{2\pi} \cos\left(\frac{r}{c}\cos\theta\right) r\,dr\,d\theta = 2\pi ac J_1\left(\frac{a}{c}\right) \tag{14.50}$$

(b) $$\int \ln x J_0(x) x\,dx = x\ln x J_1(x) + J_0(x) \tag{14.51}$$

(c) $$\int_0^a\int_0^{2\pi} \ln\left(\frac{a}{r}\right)\cos\left(\frac{r}{c}\cos\theta\right) r\,dr\,d\theta = 2\pi c^2\left[1 - J_0\left(\frac{a}{c}\right)\right] \tag{14.52}$$

14.8. Show that

(a) $$1 = [J_0(x)]^2 + 2\sum_{n=1}^{\infty}[J_n(x)]^2 \tag{14.53}$$

(b) $$J_1(x+y) = \sum_{n=0}^{\infty}(-1)^n[J_n(x)J_{n+1}(y) + J_{n+1}(x)J_n(y)] \tag{14.54}$$

14.9. Show that

(a) $$e^{ix\cos\theta} = J_0(x) + 2\sum_{n=1}^{\infty} i^n J_n(x)\cos n\theta \tag{14.55}$$

(b) $$\cos x = J_0(x) + 2\sum_{k=1}^{\infty} (-1)^k J_{2k}(x) \tag{14.56}$$

(c) $$\sin x = 2\sum_{k=0}^{\infty} (-1)^k J_{2k+1}(x) \tag{14.57}$$

14.10. From Fig. 14.3, show that

$$e^{iR\sin\beta} = e^{ir'\sin\theta'} e^{-ir\sin\theta} \tag{14.58}$$

Use (14.40) to expand the exponential factors involving $\sin\theta$ and $\sin\theta'$. Integrate β from $-\pi$ to $+\pi$ to obtain the addition theorem in the form

$$J_0(R) = \sum_{n=-\infty}^{+\infty} J_n(r')J_n(r)e^{in\alpha} \tag{14.59}$$

When the integration is carried out, the triangle of Fig. 14.3 rotates rigidly so that $d\theta = d\theta' = d\beta$ and $\alpha = \theta' - \theta = \text{constant}$.

14.11. Obtain the following expansions:

$$1 = J_0(x) + 2J_2(x) + 2J_4(x) + 2J_6(x) + \cdots \tag{14.60}$$

$$x = 2[J_1(x) + 3J_3(x) + 5J_5(x) + 7J_7(x)\cdots] \tag{14.61}$$

$$x^2 = 8[J_2(x) + 2^2J_4(x) + 3^2J_6(x) + 4^2J_8(x) + \cdots] \tag{14.62}$$

Hint: Differentiate (14.40) repeatedly with respect to θ and then set $\theta = 0$.

14.12. Show from (13.22) that

$$4J_\nu''(x) = J_{\nu-2}(x) - 2J_\nu(x) + J_{\nu+2}(x) \tag{14.63}$$

Higher derivatives can be similarly evaluated.

14.13. Prove that

$$J_0(x) = 2\frac{d}{dx}[J_1(x) + J_3(x) + J_5(x) + \cdots] \tag{14.64}$$

14.14. Show that if $w_\nu(x)$ is any solution of Bessel's equation (13.5), then

$$2xw_\nu^2 = \frac{d}{dx}\left[x^2\left(w_\nu^2 - w_{\nu+1}w_{\nu-1}\right)\right] \tag{14.65}$$

Hint: Use (13.32) and the recurrence relations (13.23) and (13.24), which have now been shown to be valid for all solutions of Bessel's equation and all real values of the order.

14.15. Establish the following definite integrals:

(a) $$\int_0^{\pi/2} J_0(x\cos\theta)\cos\theta\, d\theta = \frac{\sin x}{x} \tag{14.66}$$

(b) $$\int_0^{\pi/2} J_1(x\cos\theta)\, d\theta = \frac{1-\cos x}{x} \tag{14.67}$$

Hint: Use the series expansions for the Bessel functions and integrate term by term. You will need equations (22.16) and (22.29) in Chapter 5.

14.16. By using the power series for $J_\mu(xt)$ and integrating term by term, show that

$$J_\nu(x) = \frac{2x^{\nu-\mu}}{2^{\nu-\mu}\Gamma(\nu-\mu)}\int_0^1 J_\mu(xt)\,t^{\mu+1}(1-t^2)^{\nu-\mu-1}\,dt \tag{14.68}$$

provided that $\nu > \mu > -1$. This is called *Sonine's integral.*

14.17. By expanding $\cos(x\sin\theta)$ in a power series and integrating term by term, show that

$$J_\nu(x) = \frac{x^\nu}{2^{\nu-1}\sqrt{\pi}\,\Gamma(\nu+\frac{1}{2})}\int_0^{\pi/2}\cos(x\sin\theta)\cos^{2\nu}\theta\,d\theta \qquad \nu > -\tfrac{1}{2} \tag{14.69}$$

By setting $t = \sin\theta$, show that

$$J_\nu(x) = \frac{x^\nu}{2^{\nu-1}\sqrt{\pi}\,\Gamma(\nu+\frac{1}{2})}\int_0^{1}\cos xt(1-t^2)^{\nu-1/2}\,dt \qquad \nu > -\tfrac{1}{2} \tag{14.70}$$

15. SADDLE POINT METHOD

We consider in this section a method of obtaining the leading term of the asymptotic series representation of a function by means of a contour integral in the complex plane. A general discussion of asymptotic series is given in section 23 of Chapter 5. The technique to be used here is called the *saddle point method* or *method of steepest descents*. It is generally applicable to functions which have representations of the form

$$\int_C e^{xf(\zeta)}d\zeta \tag{15.1}$$

and is valid when x is large and positive. The representation of the Hankel functions by contour integrals as given by equations (14.25) and (14.26) is appropriate for finding their asymptotic formulas. The Hankel function $H_\nu^1(x)$ can be expressed as

$$H_\nu^1(x) = \frac{1}{\pi}\int_{C_1} e^{xf(\zeta)}e^{i\nu\zeta}d\zeta \tag{15.2}$$

where

$$\begin{aligned} f(\zeta) &= -i\sin\zeta = -i\sin(p+iq) \\ &= -i\sin p\cosh q + \cos p\sinh q \end{aligned} \tag{15.3}$$

The contour C_1 is illustrated in Fig. 14.1*a*. Figure 15.1 shows C_1 in the vicinity of the point $p = -\pi/2$. The magnitude of the integrand of (15.2) is given approximately by

$$|e^{xf(\zeta)+i\nu\zeta}| \simeq e^{x\cos p\sinh q} \tag{15.4}$$

If x is large, the integrand is small in the shaded portions of Fig. 15.1 and is large in the unshaded portions. The point $p = -\pi/2$ is therefore a saddle point. The factor in the exponent of the integrand of (15.2) is $F(\zeta) = xf(\zeta) + i\nu\zeta$. Formally, the location of

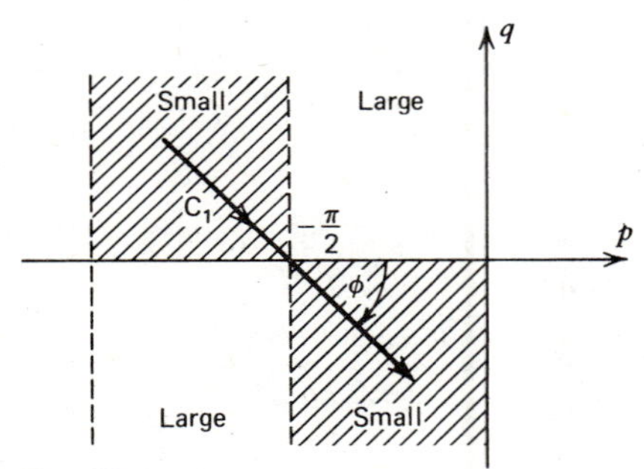

Fig. 15.1

the saddle point of the integrand is found from

$$F'(\zeta) = -ix\cos\zeta + i\nu = 0 \qquad \cos\zeta = \frac{\nu}{x} \tag{15.5}$$

If $x \gg \nu$, then $\zeta = -\pi/2$. If $f(\zeta)$ is expanded in a Taylor's series about the point $\zeta = -\pi/2$, the result is

$$\begin{aligned} f(\zeta) &= f\left(-\frac{\pi}{2}\right) + \frac{1}{2}f''\left(-\frac{\pi}{2}\right)\left(\zeta+\frac{\pi}{2}\right)^2 + \cdots \\ &= i - \frac{1}{2}i\left(\zeta+\frac{\pi}{2}\right)^2 + \cdots \end{aligned} \tag{15.6}$$

It is convenient to introduce the variables $u = p + \pi/2$ and $v = q$ so that the saddle point is given by $u = v = 0$. Polar coordinates (s, ϕ) with the saddle point as origin are also useful and can be defined by $u = s\cos\phi$ and $v = s\sin\phi$, or equivalently by $\zeta + \pi/2 = se^{i\phi}$. The angle ϕ then gives the direction with which the path of integration passes through the saddle point as illustrated in Fig. 15.1. The Taylor's series expansion of $f(\zeta)$ can be written in various ways as

$$\begin{aligned} f(\zeta) &= i + uv - i\frac{1}{2}(u^2 - v^2) = i - \frac{1}{2}is^2e^{2i\phi} \\ &= i + \frac{1}{2}s^2\sin 2\phi - \frac{1}{2}is^2\cos 2\phi \end{aligned} \tag{15.7}$$

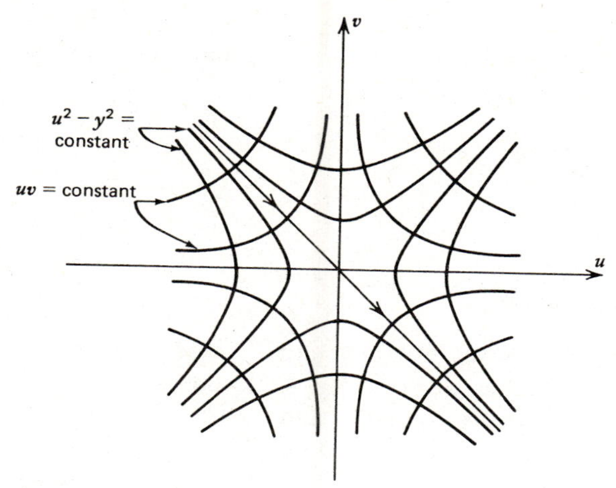

Fig. 15.2

Figure 15.2 shows the result of plotting the two sets of curves $\operatorname{Re} f(\zeta) = \text{constant}$ and $\operatorname{Im} f(\zeta) = \text{constant}$. The choice $\phi = -\pi/4$ results in the integration following a path given by $\operatorname{Im} f(\zeta) = i$ and going at right angles to the curves $\operatorname{Re} f(\zeta) = \text{constant}$. This is the path of steepest descent or most rapid change in the magnitude of the integrand of (15.2). In the limit as $x \to \infty$, all the contribution to the integral comes from the vicinity of the saddle point and (15.2) is approximately

$$H_\nu^1(x) \simeq \frac{1}{\pi} e^{ix - i\nu\pi/2 - i\pi/4} \int_{s=-\infty}^{+\infty} e^{-xs^2/2}\, ds \tag{15.8}$$

where we have set

$$d\zeta = ds\, e^{-i\pi/4} \qquad e^{+i\nu\zeta} \simeq e^{-i\nu\pi/2} \tag{15.9}$$

The path of steepest descent has the property that the greatest contribution to the integral is compressed into the shortest possible distance. Since the integrand forms a sharp spike at $s = 0$ and is nearly zero a very short distance away from $s = 0$, there is not much error in replacing the limits of integration by $\pm\infty$. Evaluation of the integral yields

$$H_\nu^1(x) \simeq \sqrt{\frac{2}{\pi x}}\, e^{ix - i\nu\pi/2 - i\pi/4} \tag{15.10}$$

Since $J_\nu(x)$ is real if x is real, (14.27) shows that $H_\nu^2(x) = H_\nu^1(x)^*$. Therefore,

$$H_\nu^2(x) \simeq \sqrt{\frac{2}{\pi x}}\, e^{-ix + i\nu\pi/2 + i\pi/4} \tag{15.11}$$

From (14.27) and (14.28), we find

$$J_\nu(x) \simeq \sqrt{\frac{2}{\pi x}} \cos\left(x - \frac{\nu\pi}{2} - \frac{\pi}{4}\right) \tag{15.12}$$

$$N_\nu(x) \simeq \sqrt{\frac{2}{\pi x}} \sin\left(x - \frac{\nu\pi}{2} - \frac{\pi}{4}\right) \tag{15.13}$$

We emphasize that the asymptotic formulas are the leading terms of asymptotic series and are valid if $x \gg 1$ and $x \gg \nu$. The qualitative nature of the Bessel functions for large values of the argument is clearly revealed by their asymptotic formulas. They are oscillatory, have approximately equally spaced roots, and decrease in magnitude in proportion to $1/\sqrt{x}$. The following table compares values of $J_0(x)$ as computed from (15.12) with accurate values and gives some idea of the accuracy of the asymptotic formulas.

x	$J_0(x)$ from Equation (15.12)	$J_0(x)$ from Tables
2	0.19674	0.22389
4	−0.39788	−0.39715
6	0.15680	0.15065
8	0.16833	0.17165
10	−0.24676	−0.24594

We now show how the complete asymptotic expansions for the cylinder functions can be found. By making the change of dependent variable

$$w(x) = x^{-1/2} e^{ix} f(x) \tag{15.14}$$

we find that Bessel's equation (13.5) becomes

$$x^2 f''(x) + 2x^2 i f'(x) + \left(\tfrac{1}{4} - \nu^2\right) f(x) = 0 \tag{15.15}$$

This effectively factors out the behavior of $w(x)$ at infinity. Since we are anticipating that $f(x)$ will be an asymptotic expansion of the form

$$f(x) = 1 + \frac{c_1}{x} + \frac{c_2}{x^2} + \cdots \tag{15.16}$$

we make the change of independent variable $x = 1/y$. The result is

$$y^2 f''(y) + (2y - 2i) f'(y) + \left(\tfrac{1}{4} - \nu^2\right) f(y) = 0 \tag{15.17}$$

In spite of the fact that $y = 0$ is an irregular singular

point of (15.17), a power series solution of the form

$$f(y)=\sum_{k=0}^{\infty} c_k y^k \tag{15.18}$$

is possible. Substituting (15.18) into (15.17) gives

$$\sum_{k=0}^{\infty} c_k\left[k(k+1)+\tfrac{1}{4}-\nu^2\right]y^k$$

$$-2i\sum_{k=-1}^{\infty} c_{k+1}(k+1)y^k=0 \tag{15.19}$$

Note that the case $k=-1$ is satisfied for all values of c_0 so that there is no inconsistency. The recurrence relation for the coefficients is

$$c_{k+1}=ic_k\frac{4\nu^2-1-4k(k+1)}{8(k+1)} \tag{15.20}$$

Setting $c_0=1$ and computing the coefficients one by one gives the result

$$H_\nu^1(x)=\sqrt{\frac{2}{\pi x}}\,e^{ix-i\nu\pi/2-i\pi/4}\left[P(x)+iQ(x)\right]$$

$$=H^2(x)^* \tag{15.21}$$

$$P(x)=1-\frac{(4\nu^2-1)(4\nu^2-9)}{8^2 2!\,x^2}$$

$$+\frac{(4\nu^2-1)(4\nu^2-9)(4\nu^2-25)(4\nu^2-49)}{4!\,(8x)^4}$$

$$-\cdots \tag{15.22}$$

$$Q(x)=\frac{4\nu^2-1}{8x}-\frac{(4\nu^2-1)(4\nu^2-9)(4\nu^2-25)}{3!\,(8x)^3}$$

$$+\cdots \tag{15.23}$$

Asymptotic series for $J_\nu(x)$ and $N_\nu(x)$ can be found from (14.27) and (14.28) if desired.

The asymptotic formulas are of course important for numerical computation, but they can also be used in some cases to find general relations among the various cylinder functions. For example, we know from Sturm–Liouville theory that the Wronskian for any two linearly independent cylinder functions is

$$w_1w_2'-w_1'w_2=\frac{C}{x} \tag{15.24}$$

The constant C can be found by a consideration of the limiting case $x\to\infty$. In this manner, the following Wronskians can be verified by using the appropriate asymptotic formulas in (15.24):

$$J_\nu N_\nu'-J_\nu'N_\nu=\frac{2}{\pi x} \tag{15.25}$$

$$H_\nu^1H_\nu^{2\prime}-H_\nu^{1\prime}H_\nu^2=-\frac{4i}{\pi x} \tag{15.26}$$

$$J_\nu J_{-\nu}'-J_\nu'J_{-\nu}=-\frac{2\sin\nu\pi}{\pi x} \tag{15.27}$$

Note that in the case of J_ν and $J_{-\nu}$, the Wronskian is zero if ν is an integer, thus confirming the assertion made in section 1 that J_ν and $J_{-\nu}$ are linearly independent solutions of Bessel's equation only if ν is not an integer.

So far, we have not developed a power series representation for the Neumann functions. This can be remedied by noting that for noninteger values of ν there must exist a relation of the form

$$N_\nu(x)=aJ_\nu(x)+bJ_{-\nu}(x) \tag{15.28}$$

By using the asymptotic forms (15.12) and (15.13), the constants a and b can be evaluated with the result that

$$N_\nu(x)=\frac{J_\nu(x)\cos\nu\pi-J_{-\nu}(x)}{\sin\nu\pi} \tag{15.29}$$

Since a and b are constants, their values, once found by using any value of x in (15.28), are then valid for all values of x. By means of the power series (1.38) for the Bessel functions, the dominant term of (15.29) near $x=0$ can be found:

$$N_\nu(x)\to-\left(\frac{x}{2}\right)^{-\nu}\frac{1}{\sin\nu\pi\,\Gamma(1-\nu)}$$

$$x\to 0\,(\nu>0) \tag{15.30}$$

Reference to equation (Ch. 5-22.23) shows that

$$N_\nu(x)\to-\left(\frac{x}{2}\right)^{-\nu}\frac{\Gamma(\nu)}{\pi}\qquad x\to 0\,(\nu>0) \tag{15.31}$$

Finding series representations for the Neumann functions of integer order is not so easy. As $\nu\to m$ in (15.29), the expression becomes indeterminate of the form $0/0$. It can however be evaluated by using l'Hospital's rule. It is possible to find the general series for $N_m(x)$ by this method, but the calculation is tedious and the resulting series is quite messy. In the following derivations, we will obtain the essential results without undue complication. We begin by finding the behavior of $N_0(x)$ near $x=0$. By taking the limit $\nu \to 0$ in (15.29), we find

$$N_0(x) = J_0(x) \lim_{\nu\to 0}\left[\frac{1-[J_{-\nu}(x)/J_\nu(x)]}{\pi\nu}\right]$$
$$= -J_0(x) \lim_{\nu\to 0}\left[\frac{\frac{\partial}{\partial \nu} J_{-\nu}(x)/J_\nu(x)}{\pi}\right] \tag{15.32}$$

In the limit $x \to 0$, the series (1.38) for the Bessel functions gives

$$\frac{J_{-\nu}(x)}{J_\nu(x)} \to \left(\frac{x}{2}\right)^{-2\nu} \frac{\Gamma(\nu+1)}{\Gamma(-\nu+1)} \tag{15.33}$$

By carrying out the indicated differentiation with respect to ν, we find

$$\frac{\partial}{\partial\nu}\left[\frac{J_{-\nu}(x)}{J_\nu(x)}\right]_{\nu=0} = -2\ln\left(\frac{x}{2}\right) + \frac{2\Gamma'(1)}{\Gamma(1)}$$
$$= -2[\ln x - \ln 2 - \psi(1)]$$
$$= -2[\ln x - \ln 2 + \gamma] \tag{15.34}$$

where ψ is the digamma function defined by equation (22.37) in Chapter 5 and $\gamma = 0.57722$ is Euler's constant defined by equation (22.38) in that chapter. Thus,

$$N_0(x) \to \frac{2}{\pi} J_0(x)[\ln x - \ln 2 + \gamma] \qquad x \to 0 \tag{15.35}$$

As expected (see problem 1.9), $N_0(x)$ has a logarithmic singularity at $x = 0$.

The behavior of the Neumann functions near $x = 0$ for any integer order can be found from the recurrence relations. Equation (13.24) shows that

$$N_1(x) = -N_0'(x) \to -\frac{2}{\pi x} \qquad x \to 0 \tag{15.36}$$

Now use (13.26) to get

$$N_2(x) = -x\frac{d}{dx}[x^{-1}N_1(x)]$$
$$\to -\frac{1}{\pi}\left(\frac{2}{x}\right)^2 \qquad x \to 0 \tag{15.37}$$

A continuation of this process shows that

$$N_m(x) \to -\left(\frac{2}{x}\right)^m \frac{(m-1)!}{\pi} \qquad x \to 0 \tag{15.38}$$

which is identical to (15.31) for noninteger values of the order.

We will now find the complete series for $N_0(x)$ and also show another method for obtaining a second linearly independent solution of a second-order differential equation when the power series method yields only one solution. Refer to the power series solution of (1.31). In the recurrence relation for the coefficients (1.34), set $m = 0$; but for the moment, leave s undetermined. The recurrence relation (1.34) then generates the function

$$w(x,s) = x^s\left[1 - \frac{x^2}{(s+2)^2} + \frac{x^4}{(s+2)^2(s+4)^2} - \frac{x^6}{(s+2)^2(s+4)^2(s+6)^2} + \cdots\right] \tag{15.39}$$

Note that $J_0(x)$ is obtained if $s = 0$. As it stands, (15.39) is not a solution of Bessel's equation because s has not been required to satisfy the indicial equation (1.33). This means that when the series is substituted into the differential equation, we do not get zero as in (1.32) but rather the left-over term corresponding to $k = 0$. Thus,

$$x^2 w''(x,s) + xw'(x,s) + x^2 w(x,s) = c_0 s^2 x^s \tag{15.40}$$

All the other terms drop out because the coefficients are required to obey the recurrence relation (1.34). If (15.40) is differentiated with respect to s and then s is put equal to zero, the result is

$$x^2\left[\frac{\partial w''}{\partial s}\right]_{s=0} + x\left[\frac{\partial w'}{\partial s}\right]_{s=0} + x^2\left[\frac{\partial w}{\partial s}\right]_{s=0} = 0 \tag{15.41}$$

Therefore, a second linearly independent solution can be found from (15.39) by means of

$$w_2(x) = \left[\frac{\partial}{\partial s} w(x,s)\right]_{s=0} \tag{15.42}$$

The required derivatives are

$$\left[\frac{d}{ds}x^s\right]_{s=0} = \ln x \qquad \left[\frac{d}{ds}\frac{1}{(s+2)^2}\right]_{s=0} = -\frac{1}{2^2}$$

$$\left[\frac{d}{ds}\frac{1}{(s+2)^2(s+4)^2}\right]_{s=0} = \frac{1}{2^2 4^2}\left(-1-\frac{1}{2}\right) \tag{15.43}$$

$$\left[\frac{d}{ds}\frac{1}{(s+2)^2(s+4)^2(s+6)^2}\right]_{s=0} = \frac{1}{2^2 4^2 6^2}\left(-1-\frac{1}{2}-\frac{1}{3}\right)$$

A second solution of Bessel's equation of order zero is therefore

$$w_2(x) = J_0(x)\ln x + \left(\frac{x}{2}\right)^2 - \frac{1}{2!\,2!}\left(1+\frac{1}{2}\right)\left(\frac{x}{2}\right)^4 + \frac{1}{3!\,3!}\left(1+\frac{1}{2}+\frac{1}{3}\right)\left(\frac{x}{2}\right)^6 - \cdots \tag{15.44}$$

The behavior of (15.44) at $x = 0$ is correct to qualify it as a second linearly independent solution of Bessel's equation of order zero, but it is not necessarily exactly $N_0(x)$. It must however be a linear combination of $N_0(x)$ and $J_0(x)$:

$$w_2(x) = aN_0(x) + bJ_0(x) \tag{15.45}$$

By taking the limit $x \to 0$ and using (15.35), we get

$$\ln x = \frac{2a}{\pi}(\ln x - \ln 2 + \gamma) + b \tag{15.46}$$

Thus, $a = \pi/2$, $b = \ln 2 - \gamma$, and

$$N_0(x) = \frac{2}{\pi}J_0(x)(\ln x - \ln 2 + \gamma) - \frac{2}{\pi}\sum_{k=1}^{\infty}\frac{(-1)^k}{k!\,k!}\left(\frac{x}{2}\right)^{2k}\sum_{n=1}^{k}\frac{1}{n} \tag{15.47}$$

Equation (13.24) shows that $N_1(x) = -N_0'(x)$. Therefore,

$$N_1(x) = \frac{2}{\pi}J_1(x)(\ln x - \ln 2 + \gamma) - \frac{2}{\pi x}J_0(x) + \frac{2}{\pi}\sum_{k=1}^{\infty}\frac{(-1)^k}{k!\,(k-1)!}\left(\frac{x}{2}\right)^{2k-1}\sum_{n=1}^{k}\frac{1}{n} \tag{15.48}$$

For the purposes of numerical computation, the series representations of the higher-order Neumann functions are not needed because their numerical values for a given x can be found by means of the recurrence relation (13.21) once $N_0(x)$ and $N_1(x)$ are known. Figure 15.3 shows graphs of the first three Neumann functions of integer order.

As an example of a boundary value problem in which the Neumann functions appear, suppose that the potential is wanted in the region $a < r < b$ between two coaxial cylinders of length s as shown in Fig. 15.4. Let the potential be specified as a function

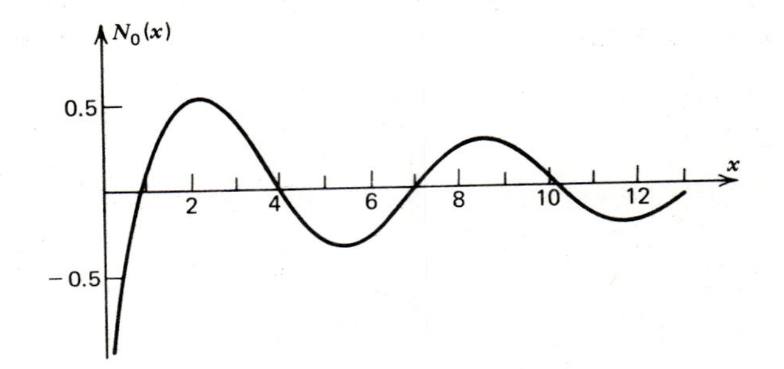

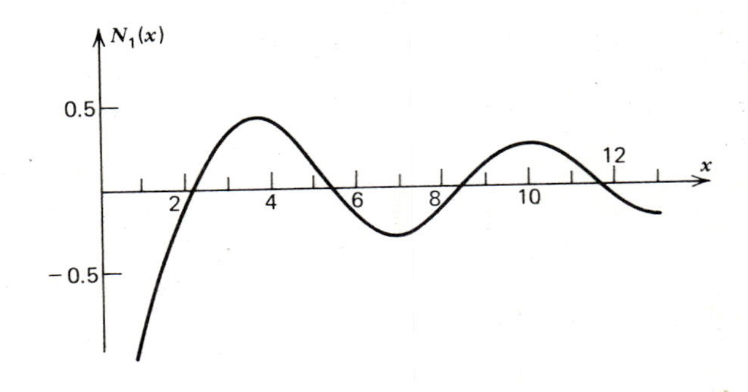

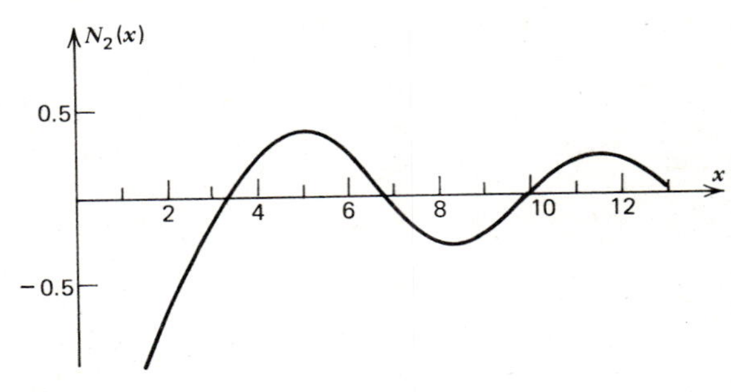

Fig. 15.3

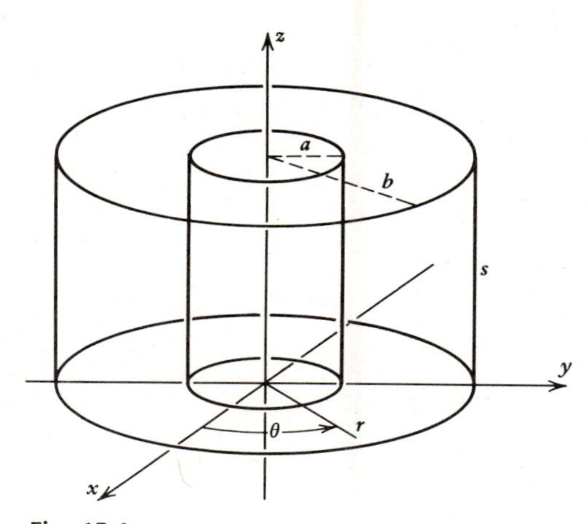

Fig. 15.4

of r over the base of the cylinder ($z=0$) and let all other boundaries of the enclosure be grounded. For simplicity, assume no θ dependence. The requirement that the potential vanish at $r=a$ and $r=b$ defines a Sturm–Liouville eigenvalue problem with respect to the differential equation (13.11). Since $r=0$ is not included in the range of allowed values of r, there is no reason to exclude the Neumann functions from the solution. The normalized eigenfunctions are of the form

$$u_\kappa(r) = C_\kappa F(r) \tag{15.49}$$

where C_κ is a normalization factor and $F(r)$ is a linear combination of $J_0(\kappa r)$ and $N_0(\kappa r)$:

$$F(r) = J_0(\kappa r) + B_\kappa N_0(\kappa r) \tag{15.50}$$

The boundary conditions require that

$$\begin{aligned} J_0(\kappa a) + B_\kappa N_0(\kappa a) &= 0 \\ J_0(\kappa b) + B_\kappa N_0(\kappa b) &= 0 \end{aligned} \tag{15.51}$$

This set of homogeneous equations will be consistent if

$$J_0(\kappa a) N_0(\kappa b) - J_0(\kappa b) N_0(\kappa a) = 0 \tag{15.52}$$

The eigenvalues are the values of κ which are roots of this equation. Each value of κ yields a value of the constant B_κ in (15.51). The normalization can be carried out by using (13.32). Since $F(a)=F(b)=0$,

$$\begin{aligned} \int_a^b rF^2\,dr &= \frac{1}{2\kappa^2}\left[r^2F'^2\right]_a^b \\ &= \frac{1}{2}\left[r^2\{J_0'(\kappa r) + B_\kappa N_0'(\kappa r)\}^2\right]_a^b \end{aligned} \tag{15.53}$$

By substituting in the limits and then using (15.51) to eliminate B_κ, we find

$$\begin{aligned} \int_a^b rF^2\,dr &= \frac{1}{2}b^2\left[\frac{N_0(\kappa b)\,J_0'(\kappa b) - J_0(\kappa b)\,N_0'(\kappa b)}{N_0(\kappa b)}\right]^2 \\ &\quad - \frac{1}{2}a^2\left[\frac{N_0(\kappa a)\,J_0'(\kappa a) - J_0(\kappa a)\,N_0'(\kappa a)}{N_0(\kappa a)}\right]^2 \end{aligned} \tag{15.54}$$

The expressions in square brackets are Wronskians; and by means of (15.25), the normalization integral is reduced to

$$\int_a^b rF^2\,dr = \frac{2}{\pi^2\kappa^2}\left[\frac{1}{[N_0(\kappa b)]^2} - \frac{1}{[N_0(\kappa a)]^2}\right] \tag{15.55}$$

The general solution of the problem is

$$\begin{aligned} \psi(r,z) &= \sum_\kappa A_\kappa u_\kappa(r)\sinh[\kappa(s-z)] \\ A_\kappa \sinh \kappa s &= \int_a^b u_\kappa(r)\,\psi(r,0)\,r\,dr \end{aligned} \tag{15.56}$$

where $\psi(r,0)$ is the known potential over the base of the cylinder and the sum is over all the roots of (15.52).

PROBLEMS

15.1. Refer to problem 1.2. Put Bessel's equation in normal form as defined by equation (1.41). In the limit as $|z| \to \infty$, show that the solutions of Bessel's equation are of the form

$$w(z) \simeq \frac{C}{\sqrt{z}}\sin(z-\phi) \tag{15.57}$$

This method correctly determines the asymptotic form of the solutions of Bessel's equation but does not determine the constants C and ϕ to correctly match the Bessel or Neumann functions as they are conventionally defined. Note that (15.57) holds if z is a complex variable. Since $z=0$ is a branch point, the asymptotic form can be made single-valued for $-\pi<\arg z<\pi$ if a branch cut is placed along the negative real axis.

15.2. The modified Bessel functions are defined by equations (13.8) and (14.31). Show that if $\nu>0$, then

$$K_\nu(x)\simeq\sqrt{\frac{\pi}{2x}}\,e^{-x}\qquad I_\nu(x)\simeq\frac{1}{\sqrt{2\pi x}}e^{x}\qquad x\to\infty$$

$$K_\nu(x)\simeq\frac{\Gamma(\nu)}{2}\left(\frac{2}{x}\right)^\nu\qquad I_\nu(x)\simeq\left(\frac{x}{2}\right)^\nu\frac{1}{\Gamma(\nu+1)}\qquad x\to0 \tag{15.58}$$

$$K_0(x)\simeq-\ln\left(\frac{x}{2}\right)-\gamma\qquad x\to0$$

15.3. If $w_1(x)$ and $w_2(x)$ are any two linearly independent solutions of the modified Bessel equation (13.7), show that the Wronskian is given by

$$W(x)=w_1(x)\,w_2'(x)-w_2(x)\,w_1'(x)=\frac{C}{x} \tag{15.59}$$

where C is a constant. For I_ν and K_ν, show that

$$I_\nu K_\nu'-I_\nu' K_\nu=-\frac{1}{x} \tag{15.60}$$

15.4. In Fig. 15.4, the inner cylinder ($r=a$) is held at a constant potential V. The outer cylinder ($r=b$) as well as the two ends are grounded. Show that the potential at any point between the cylinders ($a<r<b, 0<z<s$) is given by

$$\psi(r,z)=\sum_{n=1}^{\infty}\frac{4V}{n\pi}\left[\frac{I_0(\kappa r)\,K_0(\kappa b)-I_0(\kappa b)\,K_0(\kappa r)}{I_0(\kappa a)\,K_0(\kappa b)-I_0(\kappa b)\,K_0(\kappa a)}\right]\sin\kappa z \tag{15.61}$$

where $\kappa=n\pi/s$ and the sum is over odd values of n.

15.5. If $\nu>0$, show that

$$\int_0^\infty J_{\nu+1}(x)\,dx=\int_0^\infty J_{\nu-1}(x)\,dx \tag{15.62}$$

If n is a positive integer or zero, show that

$$\int_0^\infty J_n(x)\,dx=1\qquad\int_0^\infty J_n(bx)\,dx=\frac{1}{b}\qquad b>0 \tag{15.63}$$

Hint: For the case $n=0$, set $a=0$ in (14.46). The asymptotic forms show that the integrals are convergent.

15.6. Show that the modified Bessel functions defined by (14.31) obey the recurrence relations

$$\begin{aligned}K_{\nu-1}(x)-K_{\nu+1}(x)&=-\frac{2\nu}{x}K_\nu(x)\\K_{\nu-1}(x)+K_{\nu+1}(x)&=-2K_\nu'(x)\end{aligned} \tag{15.64}$$

Show that if m is an integer, then

$$I_{-m}(x)=I_m(x)\qquad K_{-m}(x)=K_m(x) \tag{15.65}$$

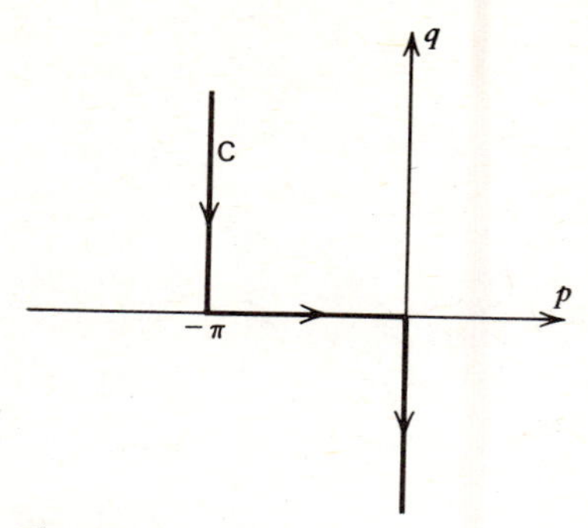

Fig. 15.5

15.7. The contour C_1 of Fig. 14.1*a* can be deformed into the contour C of Fig. 15.5. Express $H_0^1(x)$ as an integral along this path and show from this result that

$$N_0(x) = \frac{1}{\pi}\int_0^{\pi} \sin(x\cos\theta)\, d\theta - \frac{2}{\pi}\int_0^{\infty} e^{-x\sinh q}\, dq \tag{15.66}$$

Hint: Note from (14.27) and (14.28) that $H_0^1(x) = J_0(x) + iN_0(x)$. Thus, $N_0(x)$ is the imaginary part of the integral for $H_0^1(x)$.

15.8. Show that the transformation $t = -\sin\zeta$ maps the shaded region of the ζ plane in Fig. 15.6*a* into the entire half-plane $\text{Im}\, t > 0$ of Fig. 15.6*b*. If you will examine your solution of problem 15.7, you will see that if the contour C_1 of Fig. 14.1*a* is divided at $\zeta = -\pi/2$, then equal contributions to $H_0^1(x)$ come from the two portions. Thus,

$$H_0^1(x) = \frac{2}{\pi}\int_C e^{-ix\sin\zeta}\, d\zeta \tag{15.67}$$

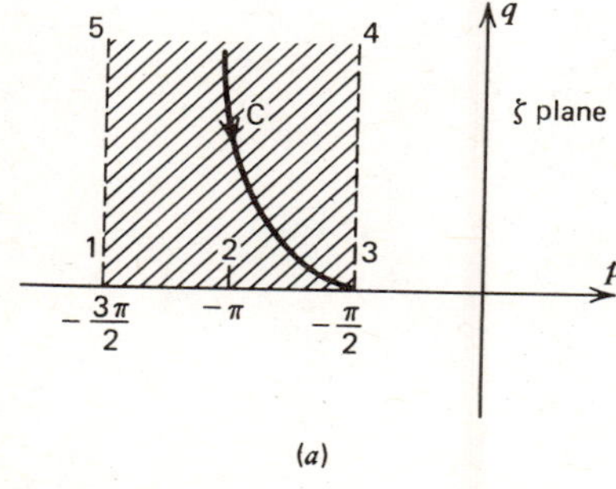

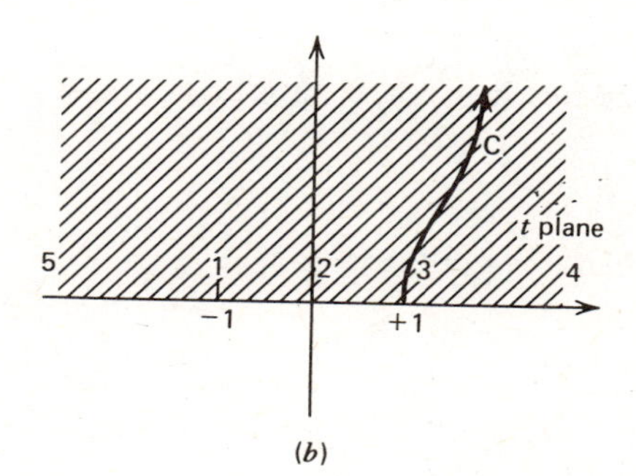

Fig. 15.6

where C is the contour shown in Fig. 15.6*a*. By making the change of variable $t = -\sin\zeta$, show that

$$H_0^1(x) = -\frac{2}{\pi}\int_1^{a+i\infty} \frac{e^{ixt}}{\sqrt{1-t^2}}\, dt \tag{15.68}$$

where a is any real number. The path of integration in the t plane starts at $t = 1$ and ends at infinity anywhere in the region $\text{Im}\, t > 0$ as shown in Fig. 15.6*b*. When you make the change of variable, you must use $\cos\zeta = \pm\sqrt{1-t^2}$. Can you see why the negative sign is appropriate? Similarly, you can show that the shaded region of Fig. 14.1*a* outlined by the points

6, 7, 8, 9, and 10 also maps onto $\operatorname{Im} t > 0$ and that

$$H_0^2(x) = \frac{2}{\pi} \int_{-1}^{a+i\infty} \frac{e^{ixt}}{\sqrt{1-t^2}}\, dt \tag{15.69}$$

15.9. For the closed contour of Fig. 15.7, it is true that

$$\frac{2}{\pi} \oint \frac{e^{ixt}}{\sqrt{1-t^2}}\, dt = 0 \tag{15.70}$$

This is because the integrand has no poles inside the contour. From (15.68), we see that the contribution to (15.70) from (a) is $H_0^1(x)$. Let (c) be a quarter circle arc defined by $t = 1 + Re^{i\theta}$. Show that the contribution to the integral from (c) is zero in the limit $R \to \infty$ and that

$$H_0^1(x) = -\frac{2}{\pi} \int_1^{\infty} \frac{e^{ixt}}{\sqrt{1-t^2}}\, dt \tag{15.71}$$

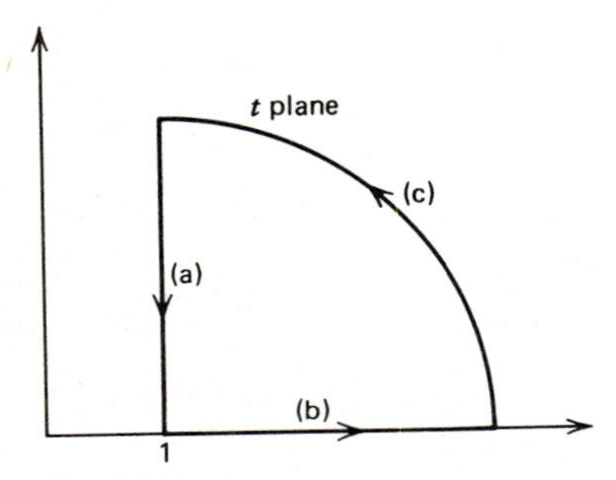

Fig. 15.7

15.10. Functions with branch points are treacherous and must be treated very carefully. In equation (15.71), we can write

$$\sqrt{1-t^2} = \pm i\sqrt{t^2-1} \tag{15.72}$$

How do we decide on the correct sign? In terms of the variable ζ,

$$\sin\zeta = -t \qquad \cos\zeta = -\sqrt{1-t^2} \tag{15.73}$$

Note that at $\zeta = -\pi$, $t = 0$ and $\cos\zeta = -1$, as it should. The line $p = -\pi/2$, $q > 0$ of Fig. 15.6a transforms into the real axis for $t > 1$, and this is the path of integration for (15.71). Along this path,

$$\begin{aligned} \cos\zeta &= \cos\left(-\frac{\pi}{2} + iq\right) = i \sinh q = -\sqrt{1-t^2} \\ \sinh q &= i\sqrt{1-t^2} \end{aligned} \tag{15.74}$$

Since $\sinh q > 0$, we must choose the negative sign in (15.72). Use (15.71) to show that

$$J_0(x) = \frac{2}{\pi} \int_1^{\infty} \frac{\sin xt}{\sqrt{t^2-1}}\, dt = \frac{2}{\pi} \int_0^{\infty} \sin(x\cosh u)\, du \tag{15.75}$$

$$N_0(x) = -\frac{2}{\pi} \int_1^{\infty} \frac{\cos xt}{\sqrt{t^2-1}}\, dt = -\frac{2}{\pi} \int_0^{\infty} \cos(x\cosh u)\, du \tag{15.76}$$

15.11. Show that the asymptotic expansions for the Bessel and Neumann functions can be written as

$$J_\nu(x) \simeq \sqrt{\frac{2}{\pi x}}\,[P(x)\cos\phi - Q(x)\sin\phi]$$
$$N_\nu(x) \simeq \sqrt{\frac{2}{\pi x}}\,[P(x)\sin\phi + Q(x)\cos\phi] \tag{15.77}$$

where $P(x)$ and $Q(x)$ are given by (15.22) and (15.23) and $\phi = x - \nu\pi/2 - \pi/4$.

15.12. Prove that if n is a positive integer,

$$\int_0^\infty \frac{1}{x} J_n(x)\,dx = \frac{1}{n} \tag{15.78}$$

Refer to problem 15.5.

15.13. Show that a generating function for the modified Bessel functions of integer order is

$$e^{(x/2)(s+1/s)} = \sum_{n=-\infty}^{+\infty} I_n(x) s^n \tag{15.79}$$

Refer to equations (13.8) and (14.23).

15.14. Obtain the following expansions:

$$1 = I_0(x) + 2\sum_{k=1}^{\infty} I_{2k}(x)(-1)^k \tag{15.80}$$

$$e^x = I_0(x) + 2\sum_{n=1}^{\infty} I_n(x) \tag{15.81}$$

15.15. Show that if $\nu > -\frac{1}{2}$,

$$I_\nu(x) = \frac{x^\nu}{2^{\nu-1}\sqrt{\pi}\,\Gamma(\nu+\frac{1}{2})}\int_0^{\pi/2} \cosh(x\sin\theta)\cos^{2\nu}\theta\,d\theta \tag{15.82}$$

Refer to equation (14.69).

15.16. Show from (14.70) that

$$J_\nu(x) = \frac{x^\nu}{2^\nu\sqrt{\pi}\,\Gamma(\nu+\frac{1}{2})}\int_{-1}^{+1} e^{ixt}(1-t^2)^{\nu-1/2}\,dt \tag{15.83}$$

More generally, if the solution of Bessel's equation is expressed as

$$w(x) = x^\nu \int_b^c e^{ixt}(1-t^2)^{\nu-1/2}\,dt \tag{15.84}$$

show that

$$x^2 w'' + xw' + (x^2-\nu^2)w = -ix^{\nu+1}\left[e^{ixt}(1-t^2)^{\nu+1/2}\right]_b^c \tag{15.85}$$

If $b = -1$ and $c = +1$, then (15.85) is zero and $w(x)$ is a solution of Bessel's equation. It is in fact Bessel's function. Other possibilities are $b = \pm 1$ and c any point at infinity in the upper half of the t plane. These two possibilities result in the Hankel functions:

$$-H_\nu^1(x)=\frac{x^\nu}{2^{\nu-1}\sqrt{\pi}\,\Gamma(\nu+\frac{1}{2})}\int_1^{a+i\infty} e^{ixt}(1-t^2)^{\nu-1/2}\,dt \tag{15.86}$$

$$H_\nu^2(x)=\frac{x^\nu}{2^{\nu-1}\sqrt{\pi}\,\Gamma(\nu+\frac{1}{2})}\int_{-1}^{a+i\infty} e^{ixt}(1-t^2)^{\nu-1/2}\,dt \tag{15.87}$$

These expressions are valid if $\nu > -\frac{1}{2}$ and are generalizations of the results of problem (15.8).

15.17. Show from (15.86) that

$$K_\nu(x)=\frac{\sqrt{\pi}\,x^\nu}{2^\nu\Gamma(\nu+\frac{1}{2})}\int_1^\infty e^{-xt}(t^2-1)^{\nu-1/2}\,dt \tag{15.88}$$

Hint: First replace x by ix, then use the contour of Fig. 15.7.

15.18. Show that the Hankel functions obey the relations

$$H_\nu^1(x)=J_\nu(x)+iN_\nu(x)=\frac{iJ_\nu(x)e^{-i\nu\pi}-iJ_{-\nu}(x)}{\sin\nu\pi} \tag{15.89}$$

$$H_\nu^2(x)=J_\nu(x)-iN_\nu(x)=\frac{iJ_{-\nu}(x)-iJ_\nu(x)e^{i\nu\pi}}{\sin\nu\pi} \tag{15.90}$$

$$H_{-\nu}^1(x)=e^{i\nu\pi}H_\nu^1(x) \qquad H_{-\nu}^2(x)=e^{-i\nu\pi}H_\nu^2(x) \tag{15.91}$$

This shows, for example, that $H_{-\nu}^1(x)$ and $H_\nu^1(x)$ are not linearly independent solutions of Bessel's equation.

15.19. Show that

$$N_\nu(x)N_{-\nu}'(x)-N_\nu'(x)N_{-\nu}(x)=\frac{-2}{\pi x}\sin\nu\pi \tag{15.92}$$

and therefore that $N_\nu(x)$ and $N_{-\nu}(x)$ are a set of linearly independent solutions of Bessel's equation if ν is not an integer. Show that

$$N_{-\nu}(x)\simeq-\left(\frac{x}{2}\right)^{-\nu}\frac{\cos\nu\pi\,\Gamma(\nu)}{\pi} \qquad x\to 0 \tag{15.93}$$

Compare this result to (15.31). Since both $N_\nu(x)$ and $J_\nu(x)$ obey the same recurrence relations,

$$N_{-n}(x)=(-1)^n N_n(x) \tag{15.94}$$

if n is an integer.

15.20. Show that

$$I_\nu(x)I_{-\nu}'(x)-I_\nu'(x)I_{-\nu}(x)=-\frac{2\sin\nu\pi}{\pi x} \tag{15.95}$$

This shows that $I_\nu(x)$ and $I_{-\nu}(x)$ are linearly independent solutions of the modified Bessel equation (13.7) if ν is not an integer. Show that

$$\begin{aligned} I_{-\nu}(x)&\simeq\left(\frac{x}{2}\right)^{-\nu}\frac{1}{\Gamma(1-\nu)} \qquad && x\to 0\\ I_{-\nu}(x)&\simeq\frac{1}{\sqrt{2\pi x}}e^x \qquad && x\to\infty \end{aligned} \tag{15.96}$$

15.21. Prove that

$$K_\nu(x) = \frac{\pi}{2}\frac{I_{-\nu}(x) - I_\nu(x)}{\sin \nu\pi} \qquad K_\nu(x) = K_{-\nu}(x) \tag{15.97}$$

This result extends (15.65) to noninteger values of ν in the case of $K_\nu(x)$ and shows that $K_\nu(x)$ and $K_{-\nu}(x)$ are not linearly independent.

15.22. Show that

$$\begin{aligned}
K_{\nu-1}(x) &= -\frac{\nu}{x}K_\nu(x) - K_\nu'(x)\\
K_{\nu+1}(x) &= \frac{\nu}{x}K_\nu(x) - K_\nu'(x)\\
\left(\frac{d}{x\,dx}\right)^n \frac{K_\nu(x)}{x^\nu} &= (-1)^n \frac{K_{\nu+n}(x)}{x^{\nu+n}}\\
\left(\frac{d}{x\,dx}\right)^n x^\nu K_\nu(x) &= (-1)^n x^{\nu-n} K_{\nu-n}(x)
\end{aligned} \tag{15.98}$$

It may have occurred to the reader that $I_\nu(x)$ and $K_\nu(x)$ obey different recurrence relations even though they are solutions of the same differential equation. The difference is superficial because $I_\nu(x)$ and $e^{i\nu\pi}K_\nu(x)$ do obey the same set of recurrence relations.

16. APPLICATIONS OF BESSEL FUNCTIONS

We have already indicated how Bessel functions arise in potential theory, heat conduction, electromagnetism, and mechanical waves. In this section, the list of applications is extended.

Fraunhofer Diffraction by a Circular Aperture

The basic theory of Fraunhofer diffraction is discussed in section 10 of Chapter 5 and should be reviewed at this time. In problem 10.4 of Chapter 5, you derived the intensity pattern which results after initially parallel light passes through an aperture in the shape of a long slit of width a. We will now derive the intensity pattern which is produced by a circular aperture of radius a under similar circumstances as pictured in Fig. 16.1a. The contribution to the electric field at point P, which comes from an element of area $d\sigma$ at point Q, is

$$dE = A\sin(\kappa R - \omega t)\,d\sigma \tag{16.1}$$

where A is a constant. Let r be the distance from the center 0 of the aperture to the point P. As shown in Fig. 16.1b, P is sufficiently distant from the aperture so that the light rays traveling to it from various parts of the aperture can be regarded as parallel. The path difference between the light rays departing from O and Q is $s\cos\gamma$. The vector $\mathbf{s}$ lies in the x, y plane and $\mathbf{r}$ lies in the x, z plane. The angle γ between $\mathbf{r}$ and $\mathbf{s}$ can be found by means of

$$\begin{aligned}
\mathbf{s} &= \hat{\mathbf{i}}s\cos\phi + \hat{\mathbf{j}}s\sin\phi\\
\mathbf{r} &= \hat{\mathbf{i}}r\sin\theta + \hat{\mathbf{k}}r\cos\theta\\
\mathbf{r}\cdot\mathbf{s} &= rs\sin\theta\cos\phi = rs\cos\gamma
\end{aligned} \tag{16.2}$$

Note that s and ϕ are the plane polar coordinates of points in the plane of the circular opening. By noting that $R = r - s\cos\gamma$ and defining $\chi = \kappa r - \omega t$, equation (16.1) can be written in complex form as

$$dE = Ae^{i(\chi - \kappa s\cos\gamma)}\,d\sigma \tag{16.3}$$

it being understood that the actual electric field is the imaginary part. The resultant electric field at P is found by integrating (16.3) over the aperture:

$$E = Ae^{i\chi}\int_{s=0}^{a}\int_{\phi=0}^{2\pi} e^{-i\kappa s\cos\phi\sin\theta} s\,ds\,d\phi \tag{16.4}$$

To many readers, this procedure may seem like a swindle because it ignores many of the complications predicted by Maxwell's theory of electromagnetic waves. In the simple theory of diffraction being given here, the light waves are treated simply as scalar waves which can interfere with one another. Empiri-

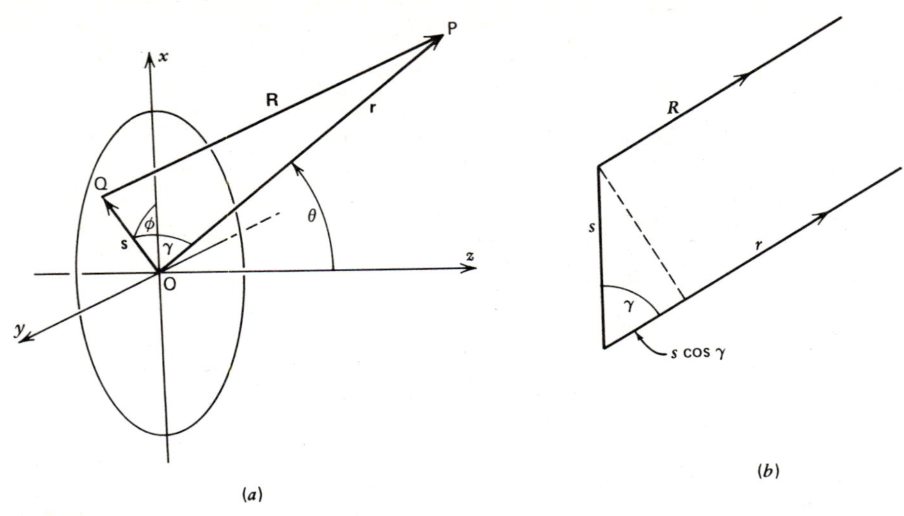

Fig. 16.1

cally, the theory works well and is useful because of its simplicity.

The change of variable $\phi \to \phi - \pi/2$ in (16.4) changes $\cos\phi$ into $\sin\phi$. Because of the periodic nature of the integrand, the integration can be carried out over any interval of length 2π. Equation (14.17) shows that the integral over ϕ gives a Bessel function of order zero:

$$E = Ae^{ix}2\pi\int_0^a J_0(\kappa s\sin\theta)\,s\,ds \tag{16.5}$$

The integral over s can be evaluated by using (13.25). The result is

$$\begin{aligned} E &= \frac{2\pi A e^{ix}}{\kappa^2\sin^2\theta}\int_0^{\kappa a\sin\theta} J_0(x)\,x\,dx \\ &= 2\pi a^2 A e^{ix}\frac{J_1(\kappa a\sin\theta)}{\kappa a\sin\theta} \end{aligned} \tag{16.6}$$

The intensity is therefore

$$I = C\left[\frac{J_1(\kappa a\sin\theta)}{\kappa a\sin\theta}\right]^2 \tag{16.7}$$

and is shown graphed in Fig. 16.2. Note that the secondary maxima are quite small compared to the central maximum. The points where the intensity is zero are found from the zeros of the Bessel function $J_1(x)$. From tables, the first zero is given by

$$\begin{aligned} \kappa a\sin\theta &= \frac{2\pi a\sin\theta}{\lambda} = 3.83171 \\ \sin\theta &= 0.6098\frac{\lambda}{a} \end{aligned} \tag{16.8}$$

where λ is the wavelength of the light.

Hanging Flexible Chain

Figure 16.3 shows a flexible string or chain of length s hanging vertically under its own weight. The x axis is the position of stable equilibrium of the chain and its lower end is at $x = 0$. We will study the possible small oscillations of the chain when its motion is

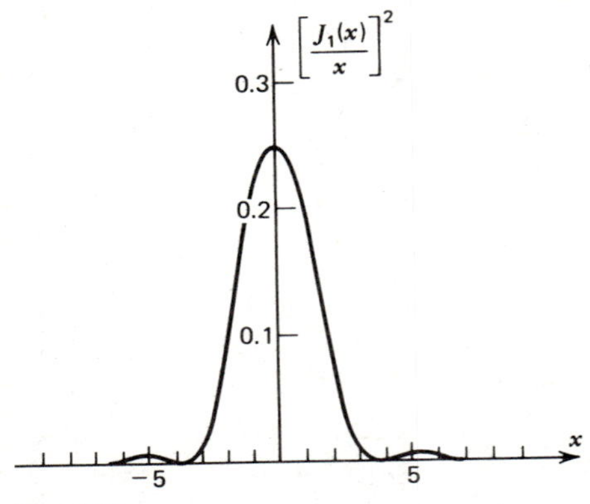

Fig. 16.2

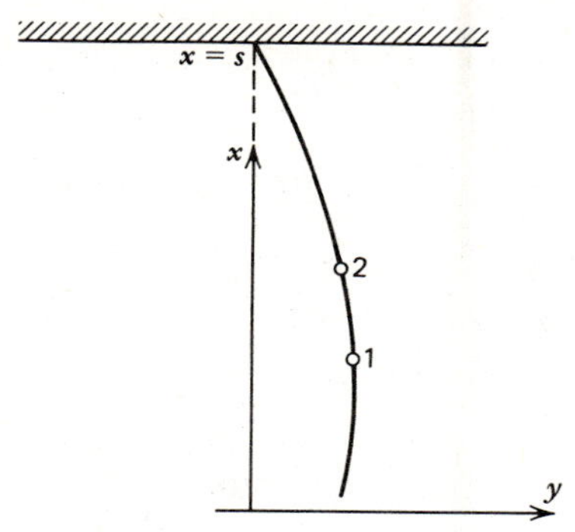

Fig. 16.3

confined to the x, y plane. The tension in the string at a given point equals the weight of the string which hangs below it and is given by $F = \mu g x$, where μ is the mass per unit length of the chain and $g = 9.8$ m/sec^2 is the acceleration of a freely falling object. The analysis of section 2 in Chapter 7 holds except that we must now replace equation (2.8) in Chapter 7 by

$$\frac{\left(F\frac{\partial y}{\partial x}\right)_2 - \left(F\frac{\partial y}{\partial x}\right)_1}{\Delta x} = \mu \frac{\partial^2 y}{\partial t^2} \tag{16.9}$$

to account for the fact that the tension is now variable. In the limit $\Delta x \to 0$,

$$\frac{\partial}{\partial x}\left(gx\frac{\partial y}{\partial x}\right) = \frac{\partial^2 y}{\partial t^2} \tag{16.10}$$

The first step in the solution is to separate the variables. If $y(x,t) = u(x)f(t)$, then

$$f''(t) + \omega^2 f(t) = 0$$

$$xu''(x) + u'(x) + \frac{\omega^2}{g}u(x) = 0 \tag{16.11}$$

where ω^2 is the separation constant. The differential equation for $u(x)$ is not a recognizable form as it stands. The change of variable

$$x = \tfrac{1}{4}g\zeta^2 \qquad w(\zeta) = u(x) \tag{16.12}$$

yields the differential equation

$$\zeta w''(\zeta) + w'(\zeta) + \omega^2 \zeta w(\zeta) = 0 \tag{16.13}$$

which is Bessel's equation of order zero. The general solution is

$$w(\zeta) = AJ_0(\omega\zeta) + BN_0(\omega\zeta) \tag{16.14}$$

Since the solution is required to be finite at $\zeta = 0$, we set $B = 0$. The end of the string at $x = s$ is fixed, requiring that

$$J_0\left(2\omega\sqrt{\frac{s}{g}}\right) = 0 \tag{16.15}$$

Thus, if the roots of $J_0(x)$ are denoted by α_n, the eigenfrequencies of the normal modes of the chain are given by

$$2\omega_n\sqrt{\frac{s}{g}} = \alpha_n \tag{16.16}$$

The general solution of (16.10) can now be expressed

$$y(x,t) = \sum_{n=0}^{\infty} u_n(\zeta)(A_n \cos\omega_n t + B_n \sin\omega_n t) \tag{16.17}$$

where for convenience the eigenfunctions are written in normalized form as

$$u_n(\zeta) = \frac{\sqrt{2}\,J_0(\omega_n\zeta)}{aJ_1(\alpha_n)} \qquad a = 2\sqrt{\frac{s}{g}} \qquad \omega_n a = \alpha_n \tag{16.18}$$

The normalization factor can be found by reference to equation (13.35).

Suppose that at $t = 0$, the shape of the chain is known to be $y(x,0) = F(x) = G(\zeta)$ and that its initial velocity is zero. The coefficients A_n are then determined by

$$G(\zeta) = \sum_{n=0}^{\infty} A_n u_n(\zeta) \qquad A_n = \int_0^a u_n(\zeta)\,G(\zeta)\,\zeta\,d\zeta \tag{16.19}$$

The coefficients B_n are all zero. For instance, if the initial shape is

$$y(x,0) = b(s - x) = b(s - \tfrac{1}{4}g\zeta^2) \tag{16.20}$$

then

$$A_n = \frac{b\sqrt{2}}{aJ_1(\alpha_n)}\left(sI_1 - \frac{g}{4}I_2\right) \tag{16.21}$$

where I_1 and I_2 denote the two integrals

$$I_1 = \int_0^a J_0(\omega_n \zeta)\zeta\, d\zeta = \frac{1}{\omega_n^2}\int_0^{\alpha_n} J_0(x)x\, dx = \frac{\alpha_n}{\omega_n^2} J_1(\alpha_n) \tag{16.22}$$

$$I_2 = \int_0^a J_0(\omega_n \zeta)\zeta^3\, d\zeta = \frac{1}{\omega_n^4}\int_0^{\alpha_n} J_0(x)x^3\, dx$$

$$= \frac{\alpha_n^3 - 4\alpha_n}{\omega_n^4} J_1(\alpha_n) \tag{16.23}$$

The integral I_2 is evaluated in problem 13.9. The coefficients reduce to

$$A_n = \frac{16bs^2}{\alpha_n^3\sqrt{2gs}} \tag{16.24}$$

The final form of the solution is

$$y(x,t) = \sum_{n=0}^{\infty} \frac{8bs}{\alpha_n^3 J_1(\alpha_n)} J_0\left(\alpha_n\sqrt{\frac{x}{s}}\right)\cos\omega_n t \tag{16.25}$$

Alternating Current in a Wire of Circular Cross Section

Suppose that an alternating current exists in a wire of circular cross section and that we want to know how the current is distributed over a cross section of the wire. We will assume that the frequency is low enough so that radiation effects are not important. To put it another way, the finite time for signals to propagate along the wire can be neglected so that the instantaneous current through all cross sections of the wire is essentially the same at a given time. This is basically the quasi-static approximation of AC circuit theory. We therefore neglect the displacement current term in the Maxwell equation (12.11), Chapter 3, and write the laws of Faraday and Ampere as

$$\nabla\times\mathbf{E} + \frac{\partial\mathbf{B}}{\partial t} = 0 \qquad \nabla\times\mathbf{B} = \mu_0\mathbf{j} \tag{16.26}$$

Cylindrical coordinates will be used with the z axis directed along the axis of the wire. The electric field has only a z component and the magnetic field only a θ component:

$$\mathbf{E} = E(r,t)\hat{\mathbf{k}} \qquad \mathbf{B} = B(r,t)\hat{\boldsymbol{\theta}} \tag{16.27}$$

By expressing (16.26) in cylindrical coordinates, equation (10.22) in Chapter 4, we find

$$-\frac{\partial E}{\partial r} + \frac{\partial B}{\partial t} = 0 \qquad \frac{1}{r}\frac{\partial}{\partial r}(rB) = \mu_0 j \tag{16.28}$$

The elimination of B between these two equations and the use of $j = \sigma E$ yields

$$\frac{1}{r}\frac{\partial}{\partial r}\left(r\frac{\partial j}{\partial r}\right) = \mu_0\sigma\frac{\partial j}{\partial t} \tag{16.29}$$

The constant σ is the conductivity of the wire. Since the current is known to vary sinusoidally with time, we write

$$j(r,t) = F(r)e^{i\omega t} \tag{16.30}$$

and find that $F(r)$ obeys the differential equation

$$F''(r) + \frac{1}{r}F'(r) - \kappa^2 F(r) = 0 \tag{16.31}$$

where κ^2 is an imaginary quantity and is given by

$$\kappa^2 = i\mu_0\sigma\omega \qquad \kappa = \frac{1+i}{\sqrt{2}}\sqrt{\mu_0\sigma\omega} \tag{16.32}$$

Reference to (13.7) shows that (16.31) is a modified Bessel equation of order zero. The solution which is finite at $r = 0$ is

$$F(r) = AI_0(\kappa r) \qquad j = AI_0(\kappa r)e^{i\omega t} \tag{16.33}$$

it being understood that the actual current density is the real part. The constant A can be regarded as real since any imaginary part would only introduce an inconsequential constant phase factor into the solution.

An important qualitative consequence of the solution can be found by calculating the total current through a coaxial cross section of radius $r < a$:

$$C(r) = \int_0^r j(r,t)2\pi r\, dr = \frac{2\pi A}{\kappa} rI_1(\kappa r)e^{i\omega t} \tag{16.34}$$

The ratio of $C(r)$ to the total current is

$$\frac{C(r)}{C} = \frac{rI_1(\kappa r)}{aI_1(\kappa a)} \tag{16.35}$$

If the frequency is high, κ will have a large magnitude, and the asymptotic form of the modified Bessel functions as given by (15.58) can be used to give

$$\frac{C(r)}{C} \simeq \sqrt{\frac{r}{a}}\, e^{\kappa(r-a)} \tag{16.36}$$

For large values of κ (high frequencies), this ratio is small if $r < a$. Most of the current is therefore near the surface of the wire and constitutes what is known as the *skin effect*. At the other extreme of very low frequencies, $\kappa \to 0$ and the current density is constant over the cross section of the wire.

For more precise calculations, it is necessary to separate the expression for the current density into its real and imaginary parts. For this purpose, the *Kelvin functions* are defined by

$$I_0\left(\frac{1+i}{\sqrt{2}}x\right) = \operatorname{ber} x + i \operatorname{bei} x$$

$$K_0\left(\frac{1+i}{\sqrt{2}}x\right) = \operatorname{ker} x + i \operatorname{kei} x \tag{16.37}$$

Handbooks giving numerical values and properties of these functions exist. Kelvin functions of higher order are also defined.

Quantum Constant Force Problem

The radial equation for the two-body central force problem in quantum mechanics is given by equation (5.19). If the change of dependent variable $F(r) = u(r)/r$ is made, we find

$$u'' - l(l+1)\frac{u}{r^2} + \frac{2M}{\hbar^2}[W - V(r)]u = 0 \tag{16.38}$$

Since $F(r)$ is required to be finite at $r = 0$, one of the conditions on $u(r)$ is $u(0) = 0$. The potential

$$V(r) = gr \tag{16.39}$$

describes a force of constant strength g between two particles and has been proposed as a possible force that exists between the elementary particles called quarks. We will examine here in detail the case of zero angular momentum which is given by

$$u'' + \frac{2M}{\hbar^2}(W - gr)u = 0 \tag{16.40}$$

This differential equation also occurs in the one-dimensional treatment of the constant force problem and is important in establishing the connecting formulas in the WKB approximation. The change of variable $r = \alpha x$ leads to

$$w'' + \frac{2Mg\alpha^3}{\hbar^2}\left(\frac{W}{\alpha g} - x\right)w = 0 \tag{16.41}$$

where we have put $u(r) = u(\alpha x) = w(x)$. We therefore define the parameter α and a new parameter a by

$$\frac{2Mg\alpha^3}{\hbar^2} = 1 \qquad a = \frac{W}{\alpha g} \tag{16.42}$$

The differential equation is reduced to

$$w'' + (a - x)w = 0 \tag{16.43}$$

The significance of the parameter a is that it is the value of x where the total energy W and the potential energy are equal. In other words, it is the location of the classical turning point in the motion. If you will look at equation (13.62), you will see that the possible solutions of (16.43) are

$$\begin{aligned} &\sqrt{a-x}\,J_{\pm 1/3}\left(\tfrac{2}{3}[a-x]^{3/2}\right) \qquad x < a \\ &\sqrt{x-a}\,I_{\pm 1/3}\left(\tfrac{2}{3}[x-a]^{3/2}\right) \qquad x > a \end{aligned} \tag{16.44}$$

In the classically forbidden region $x > a$, it is necessary to choose a linear combination of $I_{1/3}$ and $I_{-1/3}$ that decreases exponentially as $x \to \infty$. As reference to (15.58) shows, such a combination is given by $K_{1/3}$. The relation of $K_{1/3}$ to $I_{\pm 1/3}$ is found from (15.97). The proposed solution is therefore

$$\begin{aligned} w(x) &= c_1 Ai(x-a) \\ Ai(x-a) &= \frac{1}{\sqrt{3}\,\pi}\sqrt{x-a}\,K_{1/3}(\zeta) \\ &= \frac{1}{3}\sqrt{x-a}\left[I_{-1/3}(\zeta) - I_{1/3}(\xi)\right] \\ \zeta &= \tfrac{2}{3}(x-a)^{3/2} \end{aligned} \tag{16.45}$$

The new function $Ai(x-a)$ defined in (16.45) is called an *Airy function*. It has the asymptotic form

$$Ai(x-a) \simeq \frac{1}{2\sqrt{\pi}\sqrt{x-a}}e^{-\zeta} \qquad x \to \infty \tag{16.46}$$

which follows from the asymptotic form of $K_{1/3}(\zeta)$

as given by (15.58). The solution for $x < a$ is

$$w(x) = c_2\sqrt{a-x}\,J_{-1/3}(\eta) + c_3\sqrt{a-x}\,J_{1/3}(\eta)$$
$$\eta = \tfrac{2}{3}(a-x)^{3/2} \qquad (16.47)$$

The solutions given by (16.45) and (16.47) must join smoothly at $x = a$. The required values of the constants c_2 and c_3 can be found from the series expansions of the functions around the point $x = a$. By using the known expansions for the Bessel functions (1.37) and (13.9), we find that

$$\sqrt{a-x}\,J_{1/3}(\eta) = \frac{(a-x)3^{2/3}}{\Gamma(1/3)}\sum_{n=0}^{\infty} s_n(x)$$
$$= \frac{(a-x)3^{2/3}}{\Gamma(1/3)}\left[1 - \frac{(a-x)^3}{12} + \cdots\right]$$
$$\frac{s_{n+1}(x)}{s_n(x)} = \frac{-(a-x)^3}{3(n+1)(3n+4)} \qquad (16.48)$$

$$\sqrt{a-x}\,J_{-1/3}(\eta) = \frac{3^{1/3}}{\Gamma(2/3)}\sum_{n=0}^{\infty} s_n(x)$$
$$= \frac{3^{1/3}}{\Gamma(2/3)}\left[1 - \frac{(a-x)^3}{6} + \cdots\right]$$
$$\frac{s_{n+1}(x)}{s_n(x)} = \frac{-(a-x)^3}{3(n+1)(3n+2)} \qquad (16.49)$$

We have stated the general terms of the series in terms of recurrence formulas since they are useful in numerical computation of the series by computer. The series for $I_{\pm 1/3}(\zeta)$ are the same except that $a - x$ is replaced by $x - a$ and the terms of the series are all positive and do not alternate in sign. The first two terms in the series expansions of the solutions (16.45) and (16.47) are

$$w(x) = c_1\left[\frac{1}{3^{2/3}\Gamma(2/3)} - \frac{x-a}{3^{1/3}\Gamma(1/3)} + \cdots\right]$$
$$x > a \qquad (16.50)$$

$$w(x) = c_2\frac{3^{1/3}}{\Gamma(2/3)} + c_3\frac{3^{2/3}(a-x)}{\Gamma(1/3)} + \cdots$$
$$x < a \qquad (16.51)$$

By equating the solutions and their first derivatives at $x = a$, we find

$$c_2 = c_3 = \tfrac{1}{3}c_1 \qquad (16.52)$$

The solution for $x < a$ is therefore

$$w(x) = c_1\tfrac{1}{3}\sqrt{a-x}\left[J_{-1/3}(\eta) + J_{1/3}(\eta)\right]$$
$$= c_1 Ai(x-a) \qquad (16.53)$$

As indicated, equation (16.53) provides an extension of the Airy function to negative values of its argument. Since $w(0) = 0$, the possible values of the turning point a are the roots of $Ai(-a) = 0$. From tables, we find

$$a_1 = 2.33811 \qquad a_3 = 5.52056$$
$$a_2 = 4.08795 \qquad a_4 = 6.78671 \qquad (16.54)$$

The possible energy eigenvalues are then found from (16.42). Numerical values of the Airy function can be found from tables or computed from the series for the Bessel functions. The wave function as given by $w_n(x)$ is graphed in Fig. 16.4 for the first four energy levels. Each graph is a portion of $Ai(x-a)$. Successive wave functions are found by shifting $Ai(x-a)$ along the positive x axis until the next zero coincides with the origin. No attempt has been made to normalize the wave functions.

PROBLEMS

16.1. A uniform flexible chain of length s is suspended from one end as shown in Fig. 16.3 and rotates around the x axis at a constant angular velocity. Show that if the chain is in equilibrium so that each point rotates around the x axis in a circle of fixed radius, the possible angular velocities are given by (16.16). Assume that the displacements of the chain from the x axis are small.

16.2. The complex impedance of a length s of wire can be defined by $\mathscr{E} = ZC$, where $\mathscr{E}$ is the applied emf and C is the total complex current. If $j(a, t)$ is the current density at the surface

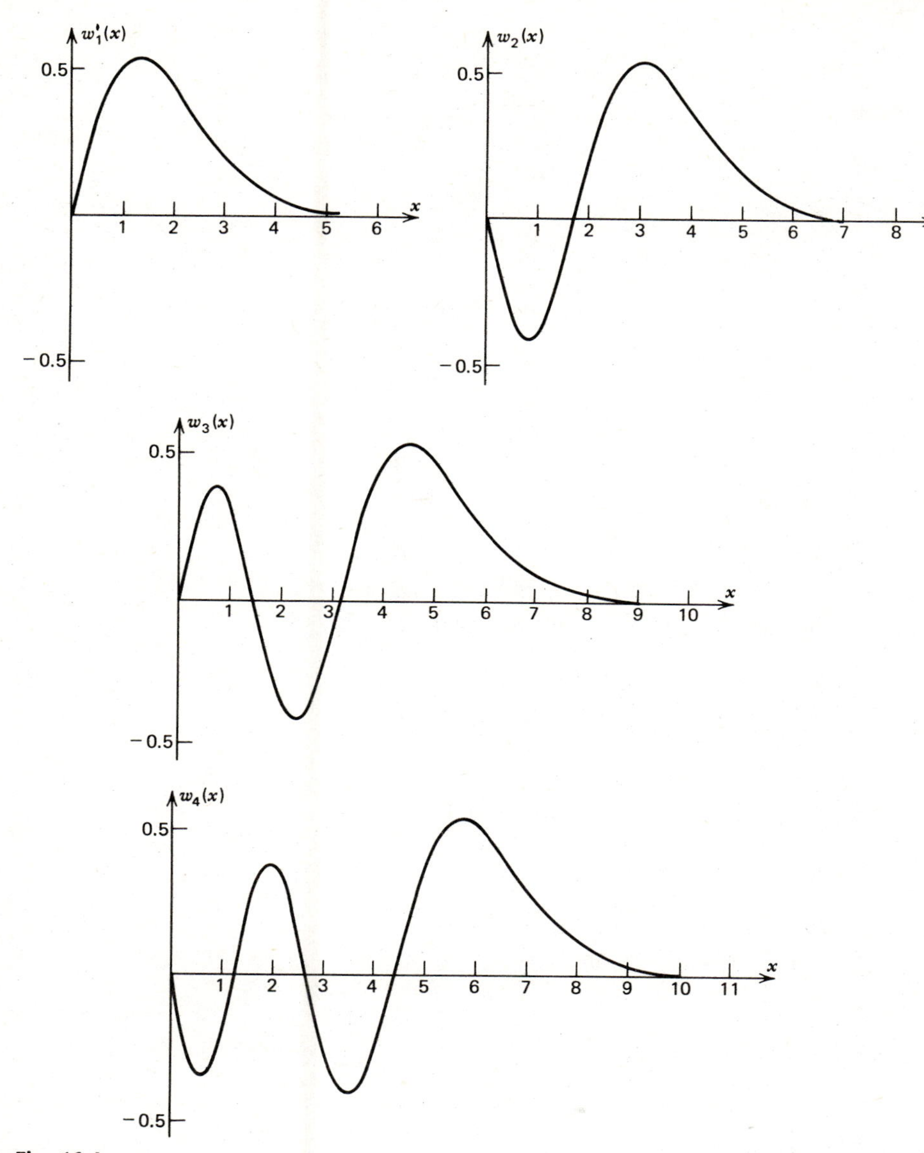

Fig. 16.4

of the wire, then $\mathscr{E} = sj(a, t)/\sigma$. Show that

$$Z = \frac{s\kappa}{2\pi\sigma a}\frac{I_0(\kappa a)}{I_1(\kappa a)} \tag{16.55}$$

By setting $Z = R + i\omega L$, show that at the low-frequency limit the resistance and self-inductance of a length s of wire are

$$R = \frac{s}{\pi a^2 \sigma} \qquad L = \frac{\mu_0 s}{8\pi} \tag{16.56}$$

16.3. Show that

$$\operatorname{ber} x = \sum_{k=0}^{\infty} \frac{(-1)^k}{(2k)!(2k)!}\left(\frac{x}{2}\right)^{4k}$$

$$\operatorname{bei} x = \sum_{k=0}^{\infty} \frac{(-1)^k}{(2k+1)!(2k+1)!}\left(\frac{x}{2}\right)^{4k+2} \tag{16.57}$$

16.4. Show that as $x \to \infty$,

$$\operatorname{ber} x \simeq \frac{1}{\sqrt{2\pi x}} e^{x/\sqrt{2}} \cos\left(\frac{x}{\sqrt{2}} - \frac{\pi}{8}\right)$$

$$\operatorname{bei} x \simeq \frac{1}{\sqrt{2\pi x}} e^{x/\sqrt{2}} \sin\left(\frac{x}{\sqrt{2}} - \frac{\pi}{8}\right) \tag{16.58}$$

Hint: First show that

$$\operatorname{ber} x + i \operatorname{bei} x = J_0(x e^{3\pi i/4}) \tag{16.59}$$

16.5. Show that as $x \to -\infty$,

$$Ai(x-a) \simeq \frac{1}{\sqrt{\pi}(a-x)^{1/4}} \cos\left(\frac{2}{3}[a-x]^{3/2} - \frac{\pi}{4}\right) \tag{16.60}$$

16.6. Show that the first four terms of the power series expansion of the Airy function are given by

$$Ai(x-a) = \frac{1}{3^{2/3}\Gamma(\frac{2}{3})} - \frac{x-a}{3^{1/3}\Gamma(\frac{1}{3})} + \frac{(x-a)^3}{6 \cdot 3^{2/3}\Gamma(\frac{2}{3})} - \frac{(x-a)^4}{12 \cdot 3^{1/3}\Gamma(\frac{1}{3})} + \cdots \tag{16.61}$$

and that this series is valid for both $x > a$ and $x < a$.

16.7. Show that the generalization of (16.43) for arbitrary angular momentum states is

$$w'' + \left[a - x - \frac{l(l+1)}{x^2}\right] w = 0 \tag{16.62}$$

Note that the wave functions are asymptotically Airy functions in the limit $x \to \infty$.

16.8. Show from Bessel's differential equation that

$$(\nu^2 - \mu^2)\frac{J_\mu J_\nu}{x} = \frac{d}{dx}\left[x\left(J_\mu J_\nu' - J_\mu' J_\nu\right)\right] \tag{16.63}$$

If $\mu + \nu > 0$, then

$$(\nu^2 - \mu^2)\int_0^\infty \frac{J_\mu J_\nu}{x} dx = \lim_{x \to \infty} \left[x\left(J_\mu J_\nu' - J_\mu' J_\nu\right)\right] \tag{16.64}$$

By using the asymptotic form (15.12), show that

$$\int_0^\infty \frac{J_\mu J_\nu}{x} dx = \frac{2 \sin\left[\frac{\pi}{2}(\nu - \mu)\right]}{\pi(\nu^2 - \mu^2)} \tag{16.65}$$

By an appropriate limiting process, show that

$$\int_0^\infty \frac{(J_\mu)^2}{x} dx = \frac{1}{2\mu} \tag{16.66}$$

From (16.7), the total amount of energy which passes through the circular aperture is proportional to

$$\int_0^\infty \frac{(J_1)^2}{x^2} 2\pi x\, dx = \pi \tag{16.67}$$

17. SPHERICAL BESSEL FUNCTIONS

A crude model of the force which exists between nucleons is described by the three-dimensional square well potential given by

$$\begin{aligned} V(r) &= -V_0 \qquad r<a \\ V(r) &= 0 \qquad r>a \end{aligned} \tag{17.1}$$

Bound states are characterized by negative total energy. The deuteron is a two-body system consisting of a proton and a neutron, and an elementary discussion of it can be based on the square well potential as given by (17.1). The radial part of the wave function obeys the differential equation (5.19). Square well potential bound states are therefore described by

$$F''(r) + \frac{2F'(r)}{r} - \frac{l(l+1)F(r)}{r^2} + \frac{2M}{\hbar^2}(V_0 - |W|)F(r) = 0 \qquad r<a \tag{17.2}$$

$$F''(r) + \frac{2F'(r)}{r} - \frac{l(l+1)F(r)}{r^2} - \frac{2M|W|}{\hbar^2}F(r) = 0 \qquad r>a \tag{17.3}$$

Consider first the solution inside the well. The change of variable

$$r = \frac{x}{\alpha} \qquad \alpha = \sqrt{\frac{2M(V_0 - |W|)}{\hbar^2}} \qquad F(r) = w(x) \tag{17.4}$$

converts (17.2) to

$$x^2 w''(x) + 2xw'(x) + \left[x^2 - l(l+1)\right] w(x) = 0 \tag{17.5}$$

Equation (17.5) matches equation (13.61) if

$$p = -\tfrac{1}{2} \qquad q = 1 \qquad \lambda = 1 \qquad \nu = l + \tfrac{1}{2} \tag{17.6}$$

Consequently, we are led to define *spherical Bessel*, *Neumann*, and *Hankel functions* by means of

$$\begin{aligned} j_l(x) &= \sqrt{\frac{\pi}{2x}}\, J_{l+1/2}(x) \\ n_l(x) &= \sqrt{\frac{\pi}{2x}}\, N_{l+1/2}(x) \\ h_l^1(x) &= \sqrt{\frac{\pi}{2x}}\, H_{l+1/2}^1(x) = j_l(x) + i n_l(x) \\ h_l^2(x) &= \sqrt{\frac{\pi}{2x}}\, H_{l+1/2}^2(x) = j_l(x) - i n_l(x) \end{aligned} \tag{17.7}$$

Any two of the functions given in (17.7) can be used as linearly independent solutions of (17.5). For the potential well problem, the appropriate solution inside the well is $j_l(x)$, but we will examine the mathematical properties of all the functions defined in (17.7). You will be pleased to know that spherical Bessel functions can be represented in terms of elementary functions. For example, from the series (1.37) for the Bessel functions, we find

$$j_0(x) = \sqrt{\frac{\pi}{2x}}\sqrt{\frac{x}{2}} \sum_{n=0}^{\infty} \frac{(-1)^n}{n!\Gamma(n+3/2)}\left(\frac{x}{2}\right)^{2n} \tag{17.8}$$

If we set $\Gamma(n+3/2) = (n+1/2)\Gamma(n+1/2)$ and use the Legendre duplication formula (22.29) in Chapter 5, the result is

$$j_0(x) = \frac{1}{x}\sum_{n=0}^{\infty} \frac{(-1)^n}{(2n+1)!} x^{2n+1} = \frac{\sin x}{x} \tag{17.9}$$

For $\nu = l + \frac{1}{2}$, equation (15.29) gives for the Neumann functions

$$N_{l+1/2}(x) = (-1)^{l+1} J_{-l-1/2}(x) \tag{17.10}$$

The spherical Neumann functions can therefore be expressed as

$$\begin{aligned} n_l(x) &= (-1)^{l+1}\sqrt{\frac{\pi}{2x}}\, J_{-l-1/2}(x) \\ &= (-1)^{l+1} j_{-l-1}(x) \end{aligned} \tag{17.11}$$

By again employing the series (1.37) for the Bessel functions, we get

$$n_0(x) = -\frac{\cos x}{x} \tag{17.12}$$

The spherical Bessel functions obey a number of recurrence relations which can be derived from the recurrence relations (13.21) through (13.26) obeyed by the Bessel functions. Recall that $J_\nu(x)$, $N_\nu(x)$, $H_\nu^1(x)$, and $H_\nu^2(x)$ all obey the same recurrence relations for any real value of the order ν. If $f_l(x)$ denotes any one of the four functions defined in (17.7), then

$$f_{l-1}(x) + f_{l+1}(x) = \frac{2l+1}{x} f_l(x) \tag{17.13}$$

$$l f_{l-1}(x) - (l+1) f_{l+1}(x) = (2l+1) f_l'(x) \tag{17.14}$$

$$\frac{l}{x} f_l(x) - f_{l+1}(x) = f_l'(x) \tag{17.15}$$

$$f_{l-1}(x) - \frac{l+1}{x} f_l(x) = f_l'(x) \tag{17.16}$$

$$\left(\frac{d}{x\,dx}\right)^m x^{l+1} f_l(x) = x^{l-m+1} f_{l-m}(x) \tag{17.17}$$

$$\left(\frac{d}{x\,dx}\right)^m x^{-l} f_l(x) = (-1)^m x^{-l-m} f_{l+m}(x) \tag{17.18}$$

Once equations (17.13) and (17.14) are established, the other equations can be derived from them. By setting $m = 1$ and $l = 0$ in (17.18), we get

$$j_1(x) = -j_0'(x) = \frac{\sin x}{x^2} - \frac{\cos x}{x} \tag{17.19}$$

$$n_1(x) = -n_0'(x) = -\frac{\cos x}{x^2} - \frac{\sin x}{x} \tag{17.20}$$

If $l = 1$ in (17.13), the result is

$$j_2(x) = \left(\frac{3}{x^3} - \frac{1}{x}\right) \sin x - \frac{3}{x^2} \cos x \tag{17.21}$$

$$n_2(x) = -\left(\frac{3}{x^3} - \frac{1}{x}\right) \cos x - \frac{3}{x^2} \sin x \tag{17.22}$$

The limiting forms of the spherical Bessel functions for large and small values of x can be found from the corresponding results for the Bessel functions:

$$j_l(x) \simeq \frac{1}{x} \cos\left(x - \frac{\pi l}{2} - \frac{\pi}{2}\right) = \frac{1}{x} \sin\left(x - \frac{\pi l}{2}\right) \qquad x \to \infty \tag{17.23}$$

$$j_l(x) \simeq \frac{x^l 2^l l!}{(2l+1)!} \qquad x \to 0 \tag{17.24}$$

$$n_l(x) \simeq \frac{1}{x} \sin\left(x - \frac{\pi l}{2} - \frac{\pi}{2}\right) = -\frac{1}{x} \cos\left(x - \frac{\pi l}{2}\right) \qquad x \to \infty \tag{17.25}$$

$$n_l(x) \simeq -\frac{(2l)!}{x^{l+1} 2^l l!} \qquad x \to 0 \tag{17.26}$$

In the differential equation (17.3) which applies to the classically forbidden region $r > a$, the change of variable

$$r = \frac{x}{\beta} \qquad \beta = \sqrt{\frac{2M|W|}{\hbar^2}} \qquad w(x) = F(r) \tag{17.27}$$

results in the differential equation

$$x^2 w''(x) + 2x w'(x) - \left[x^2 + l(l+1)\right] w(x) = 0 \tag{17.28}$$

Solutions are the modified spherical Bessel functions

$$\iota_l(x) = \sqrt{\frac{\pi}{2x}}\, I_{l+1/2}(x)$$

$$\kappa_l(x) = \sqrt{\frac{2}{\pi x}}\, K_{l+1/2}(x) \tag{17.29}$$

$$\upsilon_l(x) = \sqrt{\frac{\pi}{2x}}\, I_{-l-1/2}(x)$$

Note that the numerical factor in the definition of $\kappa_l(x)$ is not the same as it is for $\iota_l(x)$ and $\upsilon_l(x)$. This definition is not universal, and you may find $\kappa_l(x)$ defined with a factor of $\sqrt{\pi/2}$ in some references. Any two of the functions defined in (17.29) can be used as linearly independent solutions. The definitions of the modified Bessel functions as given by

(13.8) and (14.31) are

$$I_\nu(x) = i^{-\nu} J_\nu(ix) \qquad K_\nu(x) = \frac{\pi}{2} i^{\nu+1} H_\nu^1(ix) \tag{17.30}$$

By combining (17.7), (17.10), and (17.30), we can show that

$$\begin{aligned} \iota_l(x) &= i^{-l} j_l(ix) \\ \kappa_l(x) &= -i^l h_l^1(ix) \\ \upsilon_l(x) &= i^{-l-1} n_l(ix) \end{aligned} \tag{17.31}$$

From (15.97) it follows that

$$\begin{aligned} \kappa_l(x) &= \sqrt{\frac{\pi}{2x}}(-1)^l \left[I_{-l-1/2}(x) - I_{l+1/2}(x) \right] \\ &= (-1)^l \left[\upsilon_l(x) - \iota_l(x) \right] \end{aligned} \tag{17.32}$$

By using the series expansion (13.9) for $I_\nu(x)$, we find that

$$\iota_0(x) = \frac{\sinh x}{x}$$

$$\upsilon_0(x) = \frac{\cosh x}{x} \qquad \kappa_0(x) = \frac{e^{-x}}{x} \tag{17.33}$$

If $f_l(x)$ is either $\iota_l(x)$ or $\upsilon_l(x)$, then

$$f_{l-1}(x) - f_{l+1}(x) = \frac{2l+1}{x} f_l(x) \tag{17.34}$$

$$l f_{l-1}(x) + (l+1) f_{l+1}(x) = (2l+1) f_l'(x) \tag{17.35}$$

$$f_{l+1}(x) + \frac{l}{x} f_l(x) = f_l'(x) \tag{17.36}$$

$$f_{l-1}(x) - \frac{l+1}{x} f_l(x) = f_l'(x) \tag{17.37}$$

$$\frac{f_{l+n}(x)}{x^{l+n}} = \left(\frac{d}{x\,dx} \right)^n \frac{f_l(x)}{x^l} \tag{17.38}$$

$$x^{l-n+1} f_{l-n}(x) = \left(\frac{d}{x\,dx} \right)^n x^{l+1} f_l(x) \tag{17.39}$$

The recurrence relations for $\kappa_l(x)$ are slightly different:

$$\kappa_{l-1} - \kappa_{l+1}(x) = -\frac{2l+1}{x} \kappa_l(x) \tag{17.40}$$

$$l \kappa_{l-1}(x) + (l+1) \kappa_{l+1}(x) = -(2l+1) \kappa_l'(x) \tag{17.41}$$

$$\kappa_{l+1}(x) - \frac{l}{x} \kappa_l(x) = -\kappa_l'(x) \tag{17.42}$$

$$\kappa_{l-1}(x) + \frac{l+1}{x} \kappa_l(x) = -\kappa_l'(x) \tag{17.43}$$

$$\frac{\kappa_{l+n}(x)}{x^{l+n}} = (-1)^n \left(\frac{d}{x\,dx} \right)^n \frac{\kappa_l(x)}{x^l} \tag{17.44}$$

$$x^{l-n+1} \kappa_{l-n}(x) = (-1)^n \left(\frac{d}{x\,dx} \right)^n x^{l+1} \kappa_l(x) \tag{17.45}$$

Setting $l = 0$ and $n = 1$ in (17.38) and (17.44) yields

$$\iota_1(x) = \frac{\cosh x}{x} - \frac{\sinh x}{x^2} \tag{17.46}$$

$$\upsilon_1(x) = \frac{\sinh x}{x} - \frac{\cosh x}{x^2} \tag{17.47}$$

$$\kappa_1(x) = e^{-x} \left(\frac{1}{x} + \frac{1}{x^2} \right) \tag{17.48}$$

Application of the recurrence relations (17.34) and (17.40) yields

$$\iota_2(x) = \left(\frac{1}{x} + \frac{3}{x^3} \right) \sinh x - \frac{3 \cosh x}{x^2} \tag{17.49}$$

$$\upsilon_2(x) = \left(\frac{1}{x} + \frac{3}{x^3} \right) \cosh x - \frac{3 \sinh x}{x^2} \tag{17.50}$$

$$\kappa_2(x) = \left(\frac{1}{x} + \frac{3}{x^2} + \frac{3}{x^3} \right) e^{-x} \tag{17.51}$$

The limiting forms of the modified spherical Bessel functions for large and small values of x are

$$\iota_l(x) \simeq \upsilon_l(x) \simeq \frac{e^x}{x} \qquad x \to \infty \tag{17.52}$$

$$\kappa_l(x) \simeq \frac{e^{-x}}{x} \qquad x \to \infty \tag{17.53}$$

$$\iota_l(x) \simeq \frac{l! 2^l x^l}{(2l+1)!} \qquad x \to 0 \tag{17.54}$$

$$\upsilon_l(x) \simeq \frac{(-1)^l (2l)!}{2^l l! x^{l+1}} \qquad x \to 0 \tag{17.55}$$

$$\kappa_l(x) \simeq \frac{(2l)!}{2^l l! x^{l+1}} \qquad x \to 0 \tag{17.56}$$

In completing the square well problem, we require a solution inside the well which is finite at $r = 0$ and an exponentially decaying solution outside the well.

The appropriate choices are

$$F(r) = Aj_l(\alpha r) \qquad r < a$$
$$= B\kappa_l(\beta r) \qquad r > a \tag{17.57}$$

where A and B are constants. The requirement that the interior and exterior solutions join smoothly at $x = a$ gives the conditions

$$\begin{aligned} Aj_l(\alpha a) - B\kappa_l(\beta a) &= 0 \\ A\alpha j_l'(\alpha a) - B\beta\kappa_l'(\beta a) &= 0 \end{aligned} \tag{17.58}$$

A nontrivial solution for A and B exists if, and only if, the determinant of the coefficients vanishes:

$$\beta j_l(\alpha a)\kappa_l'(\beta a) - \alpha j_l'(\alpha a)\kappa_l(\beta a) = 0 \tag{17.59}$$

The recurrence relations (17.15) and (17.42) can be used to eliminate the derivatives with the result that

$$\beta j_l(\alpha a)\kappa_{l+1}(\beta a) - \alpha j_{l+1}(\alpha a)\kappa_l(\beta a) = 0 \tag{17.60}$$

It is convenient to introduce two new dimensionless parameters v and ε defined by

$$v = \sqrt{\frac{2MV_0a^2}{\hbar^2}} \qquad \varepsilon = \frac{|W|}{V_0} \tag{17.61}$$

Note that for bound states, $0 < \varepsilon < 1$. Reference to (17.4) and (17.27) shows that

$$\alpha a = v\sqrt{1-\varepsilon} \qquad \beta a = v\sqrt{\varepsilon} \tag{17.62}$$

The parameter v contains all the required information about the radius and depth of the potential well. Equation (17.60) can be expressed as

$$\sqrt{\varepsilon}\, j_l(v\sqrt{1-\varepsilon})\kappa_{l+1}(v\sqrt{\varepsilon}) - \sqrt{1-\varepsilon}\, j_{l+1}(v\sqrt{1-\varepsilon})\kappa_l(v\sqrt{\varepsilon}) = 0 \tag{17.63}$$

The possible energy eigenvalues are found from the values of ε which are roots of (17.63). The equation is transcendental and must be solved by numerical methods. For $l = 0$, you can show that (17.63) reduces to the simpler form

$$\sqrt{\varepsilon} = -\sqrt{1-\varepsilon}\cot(v\sqrt{1-\varepsilon}) \tag{17.64}$$

The number of roots which (17.63) has depends on the value of the well parameter v. By setting $\varepsilon = 0$ in (17.63), we can find the critical values of v at which new bound states appear. For this purpose, we must use the limiting form of $\kappa_l(v\sqrt{\varepsilon})$ which is valid as $\varepsilon \to 0$. By using (17.56), we find that (17.63) reduces to

$$(2l+1)j_l(v) - vj_{l+1}(v) = 0 \tag{17.65}$$

A still greater simplification can be achieved by using the recurrence relation (17.13). The final result is

$$j_{l-1}(v) = 0 \tag{17.66}$$

For the special case $l = 0$, (17.11) shows that (17.66) becomes $n_0(v) = 0$.

As an example, consider the case $l = 1$. Reference to (17.9) shows that (17.66) is satisfied if $v = \pi, 2\pi, 3\pi, \ldots$. This means that if $0 < v < \pi$, equation (17.63) has no roots over the range $0 < \varepsilon < 1$ and there are no $l = 1$ bound states. If $\pi < v < 2\pi$, there is one $l = 1$ bound state. If $2\pi < v < 3\pi$, there are two $l = 1$ bound states, and so on. Equation (17.63) is really not as formidable as it looks. A computer can find its roots quite easily. Let us define $x = v\sqrt{\varepsilon}$ and $y = v\sqrt{1-\varepsilon}$ and write, for $l = 1$,

$$F(\varepsilon) = j_1(y) - \frac{y\kappa_1(x)}{x\kappa_2(x)}j_2(y) \tag{17.67}$$

This is just (17.63) divided by the factor $x\kappa_2(x)$. It is easier numerically to deal with the ratio $\kappa_1(x)/\kappa_2(x)$ because the exponential factor e^{-x} is divided out. Also, this ratio never becomes infinite over the range of x values $0 < x < v$. The energy levels are the values of ε such that $F(\varepsilon) = 0$. Figure 17.1 shows graphs of $F(\varepsilon)$ for various values of v. In Fig. 17.1*a*, $v = \pi$ and the first bound state is just making its appearance at the top of the well. In Fig. 17.1*b*, $\pi < v < 2\pi$ and the function has a single zero between $\varepsilon = 0$ and $\varepsilon = 1$. Note that $F(\varepsilon)$ is always zero at $\varepsilon = 1$. This is the exact bottom of the well and is not an energy level. If $\varepsilon = 1$, $\alpha = 0$ and there is no wave function inside the well. In Fig. 17.1*c*, $v = 2\pi$ and a second energy level is making its appearance. Finally, in Fig. 17.1*d*, $5\pi < v < 6\pi$ and there are five bound states.

A quantum mechanical particle inside a sphere of radius a with perfectly reflecting walls is a limiting case of the potential well. The walls of the sphere become perfectly reflecting if the potential is zero inside and infinite outside. Equation (17.2) is replaced

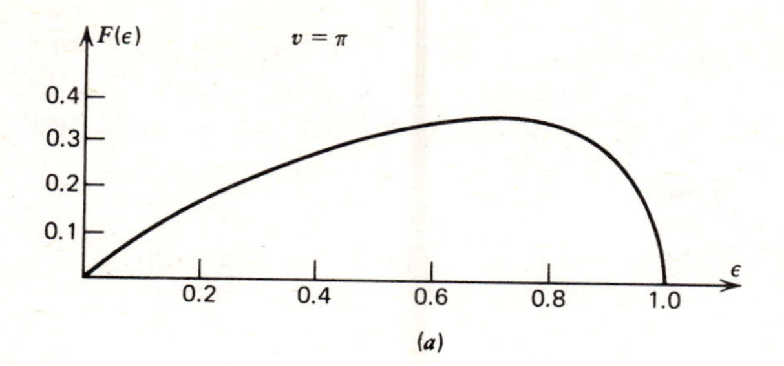

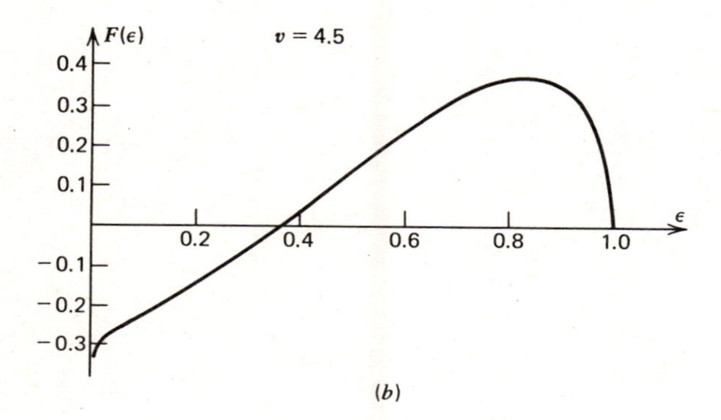

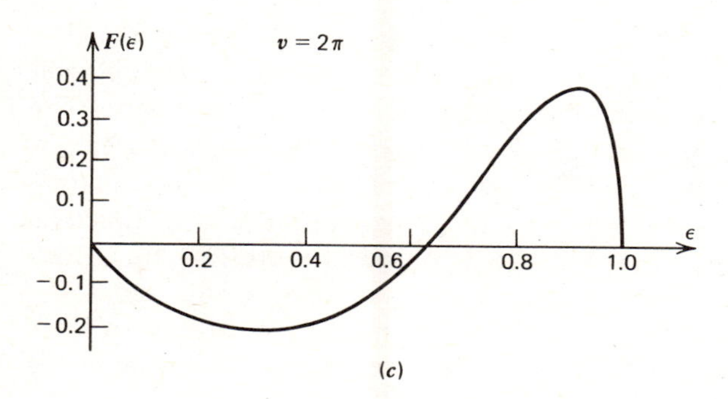

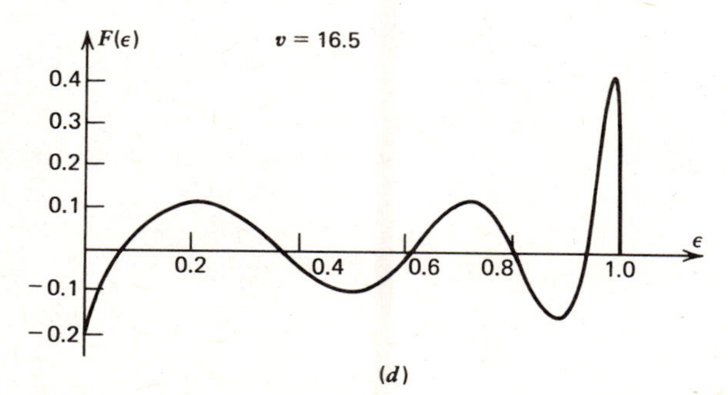

Fig. 17.1

by

$$\frac{d}{dr}(r^2F') - l(l+1)F + \alpha^2 Fr^2 = 0$$

$$\alpha^2 = \frac{2MW}{\hbar^2} > 0 \tag{17.68}$$

The wave function is now required to vanish at $r = a$. The solution and the eigenvalue condition is therefore

$$F(r) = Aj_l(\alpha_{nl}r) \qquad j_l(\alpha_{nl}a) = 0 \tag{17.69}$$

There are now an infinite number of eigenfunctions and eigenvalues. Equation (17.68) is written in Sturm–Liouville form; and by looking at (10.1) and (10.17) of Chapter 7, you can see that the general form of the Wronskian for two linearly independent solutions of (17.68) is

$$F_1F_2' - F_1'F_2 = \frac{C}{r^2} \tag{17.70}$$

where C is a constant. For example, the Wronskian relations

$$j_l(x)h_l^{1\prime}(x) - j_l'(x)h_l^1(x) = \frac{i}{x^2} \tag{17.71}$$

$$j_l(x)h_l^{2\prime}(x) - j_l'(x)h_l^2(x) = -\frac{i}{x^2} \tag{17.72}$$

can be established by using the asymptotic form (17.23) for $j_l(x)$ and

$$h_l^1(x) \simeq -\frac{i}{x}e^{ix - il\pi/2} \qquad h_l^2(x) \simeq \frac{i}{x}e^{-ix + il\pi/2} \tag{17.73}$$

for the spherical Hankel functions. The normalization factor for the radial part of the wave function can be found from the identity

$$\frac{d}{dr}\left[r^3F'^2 + r^2FF' + \alpha^2r^3F^2 - l(l+1)rF^2\right]$$

$$= 2\alpha^2r^2F^2 \tag{17.74}$$

which is true for any solution of (17.68). In particular, if $F(r)$ is given by (17.69),

$$\int_0^a [j_l(\alpha r)]^2 r^2\,dr = \frac{1}{2}a^3[j_l'(\alpha a)]^2 = \frac{1}{2}a^3[j_{l+1}(\alpha a)]^2 = \frac{1}{2}a^3[j_{l-1}(\alpha a)]^2 \qquad (17.75)$$

where use has been made of the recurrence relations (17.15) and (17.16). Note also that $F'(r) = \alpha j_l'(\alpha r)$.

In the remainder of this section, we are going to derive an addition theorem obeyed by the spherical Bessel functions which is a generalization of the expansions given by equations (12.20) and (12.21). Also relevant to the discussion is the development of the Green's function for the classical wave equation given in section 7 of Chapter 8; see especially equations (7.30) and (7.32). When the wave equation

$$\nabla^2\psi - \frac{1}{c^2}\frac{\partial^2\psi}{\partial t^2} = 0 \qquad (17.76)$$

is separated into spatial and temporal parts by means of $\psi(\mathbf{r}, t) = u(\mathbf{r})f(t)$, the result is

$$\nabla^2 u + \kappa^2 u = 0 \qquad f''(t) + \kappa^2 c^2 f(t) = 0 \qquad (17.77)$$

where κ^2 is the separation constant. We are of course now talking about the wave equation of classical physics such as would describe the propagation of electromagnetic, mechanical, or acoustical waves. If you will separate $u(\mathbf{r})$ into its radial and angular parts, you will find that

$$u_{lm}(\mathbf{r}) = F_l(r)Y_{lm}(\theta, \phi) \qquad (17.78)$$

where $Y_{lm}(\theta, \phi)$ are spherical harmonics and $F_l(r)$ are spherical Bessel functions of argument κr. The Green's function for the wave equation obeys the nonhomogeneous Helmholtz equation

$$\nabla^2 G(\mathbf{r}, \mathbf{r}') + \kappa^2 G(\mathbf{r}, \mathbf{r}') = -4\pi\delta(\mathbf{R}) \qquad (17.79)$$

When the region in which the wave equation is to be solved consists of all space and the sources are known functions of time and coordinates, the appropriate Green's function is

$$G(\mathbf{r}, \mathbf{r}') = \frac{e^{\pm i\kappa R}}{R} \qquad (17.80)$$

The only difference between equation (12.1) and (17.79) is the term $\kappa^2 G(r, r')$. By following exactly the same steps that led to equation (12.13), we find that

$$\sum_{l=0}^{\infty}\sum_{m=-\infty}^{+l}\left[\frac{1}{r^2}\frac{d}{dr}(r^2F_l') - \frac{F_l}{r^2}l(l+1) + \kappa^2 F_l\right]A_{lm}Y_{lm}(\theta, \phi) = -4\pi\frac{\delta(r-r')}{r^2}\sum_{l=0}^{\infty}\sum_{m=-l}^{+l}Y_{lm}^*(\theta', \phi')Y_{lm}(\theta, \phi) \qquad (17.81)$$

Therefore,

$$A_{lm} = Y_{lm}^*(\theta', \phi') \qquad (17.82)$$

$$\frac{d}{dr}(r^2F_l') - F_l l(l+1) + \kappa^2 F_l r^2 = -4\pi\delta(r-r') \qquad (17.83)$$

which should be compared to equations (12.14) and (12.15). For $r \neq r'$, the possible solutions of (17.83) are $j_l(\kappa r)$, $n_l(\kappa r)$, $h_l^1(\kappa r)$, $h_l^2(\kappa r)$, or a linear combination of these functions. For $r < r'$, the appropriate choice is $j_l(\kappa r)$ because this is the solution which is finite at $r = 0$. The two choices of sign in equation (17.80) represent outgoing and incoming spherical waves. If we agree to use the positive sign, then according to (17.73) it is the Hankel function $h_l^1(\kappa r)$ which has the appropriate asymptotic form for large values of r. We therefore express the solution of (17.73) as

$$\begin{aligned} F_l(r) &= Aj_l(\kappa r) \qquad r < r' \\ F_l(r) &= Bh_l^1(\kappa r) \qquad r > r' \end{aligned} \qquad (17.84)$$

Equation (12.17) remains valid; and by the same line of argument that led to (12.18), we find

$$\begin{aligned} A\kappa j_l'(\kappa r') - B\kappa h_l^{1\prime}(\kappa r') &= \frac{4\pi}{r'^2} \\ Aj_l(\kappa r') - Bh_l^1(\kappa r') &= 0 \end{aligned} \qquad (17.85)$$

The solution for A and B is

$$A = \frac{4\pi h_l^1(\kappa r')}{\kappa r'^2\left[j_l'(\kappa r')h_l^1(\kappa r') - j_l(\kappa r')h_l^{1\prime}(\kappa r')\right]} = 4\pi i\kappa h_l^1(\kappa r') \qquad (17.86)$$

$$B = 4\pi i\kappa\, j_l(\kappa r')$$

where use has been made of the Wronskian relation (17.71). Thus,

$$F_l(r) = 4\pi i\kappa h_l^1(\kappa r')\, j_l(\kappa r) \qquad r < r'$$
$$= 4\pi i\kappa h_l^1(\kappa r)\, j_l(\kappa r') \qquad r > r' \qquad (17.87)$$

Note the complete symmetry of these relations with respect to the exchange of r and r'. We therefore have the result

$$\frac{e^{i\kappa R}}{R} = 4\pi i\kappa \sum_{l=0}^{\infty}\sum_{m=-l}^{+l} h_l^1(\kappa r')\, j_l(\kappa r) \times Y_{lm}^*(\theta', \phi')Y_{lm}(\theta, \phi) \qquad r < r'$$

$$\frac{e^{i\kappa R}}{R} = 4\pi i\kappa \sum_{l=0}^{\infty}\sum_{m=-l}^{+l} h_l^1(\kappa r)\, j_l(\kappa r') \times Y_{lm}^*(\theta', \phi')Y_{lm}(\theta, \phi) \qquad r > r' \qquad (17.88)$$

This is a mathematically interesting result, and it is also important in the development of the theory of time-dependent multipole fields in electrodynamics.

PROBLEMS

17.1. Show that

$$h_{-l-1}^1(x) = i(-1)^l h_l^1(x) \qquad h_{-l-1}^2(x) = -i(-1)^l h_l^2(x) \qquad \kappa_{-l-1}(x) = \kappa_l(x) \qquad (17.89)$$

See equations (15.91) and (15.97).

17.2. Let $F_1(r)$ be a solution of (17.68) and let $F_2(r)$ be a solution of the same differential equation with α replaced by β. Show that if $\alpha \neq \beta$, then

$$(\beta^2 - \alpha^2)\int_a^b F_1(r)F_2(r)r^2\,dr = \left[r^2(F_1'F_2 - F_1F_2')\right]_a^b \qquad (17.90)$$

17.3. The most general wave function for a quantum mechanical particle inside a sphere of radius a with perfectly reflecting walls is

$$\psi(r,t) = \sum_{n=1}^{\infty}\sum_{l=0}^{\infty}\sum_{m=-l}^{+l} A_{nlm}u_{nl}(r)Y_{lm}(\theta,\phi)e^{-iW_{nl}t/\hbar}$$
$$u_{nl}(r) = \frac{\sqrt{2}\, j_l(\alpha_{nl}r)}{a^{3/2}j_{l+1}(\alpha_{nl}a)} \qquad (17.91)$$

A particle is inside a sphere of radius $a/2$ and is in its ground state. The sphere is suddenly enlarged so that its new radius is a. Show that the wave function is then

$$\psi = \sum_{\substack{n=1\\ n\neq 2}}^{\infty} \frac{-4}{\sqrt{a}\,\pi^{3/2}(n^2-4)r}\sin\left(\frac{n\pi}{2}\right)\sin\left(\frac{n\pi r}{a}\right)e^{-iW_nt/\hbar} + \frac{1}{2\sqrt{\pi a}}\frac{1}{r}\sin\left(\frac{2\pi r}{a}\right)e^{-iW_2t/\hbar}$$
$$W_n = \frac{\hbar^2 n^2\pi^2}{2Ma^2} \qquad (17.92)$$

17.4. Show that the function

$$w(x)=x^pI_\nu(\lambda x^q) \tag{17.93}$$

satisfies the differential equation

$$x^2w''(x)+(1-2p)xw'(x)-(q^2\lambda^2x^{2q}+q^2\nu^2-p^2)w(x)=0 \tag{17.94}$$

Verify that (17.28) is obtained if $p=-\frac{1}{2}$, $\nu=l+\frac{1}{2}$, $\lambda=1$, and $q=1$.

17.5. Prove the following relations:

$$j_l(x)=(-1)^lx^l\left(\frac{d}{x\,dx}\right)^l\frac{\sin x}{x} \tag{17.95}$$

$$n_l(x)=(-1)^{l+1}x^l\left(\frac{d}{x\,dx}\right)^l\frac{\cos x}{x} \tag{17.96}$$

$$\iota_l(x)=x^l\left(\frac{d}{x\,dx}\right)^l\frac{\sinh x}{x} \tag{17.97}$$

$$\upsilon_l(x)=x^l\left(\frac{d}{x\,dx}\right)^l\frac{\cosh x}{x} \tag{17.98}$$

$$\kappa_l(x)=(-1)^lx^l\left(\frac{d}{x\,dx}\right)^l\frac{e^{-x}}{x} \tag{17.99}$$

17.6. Diffusion and heat conduction are discussed in section 6 of Chapter 3. The diffusion equation can be written as

$$\nabla^2\psi-\frac{1}{D}\frac{\partial\psi}{\partial t}=-\frac{f}{D} \tag{17.100}$$

where ψ is the number of particles per unit volume, D is the diffusion constant, and f is the number of particles per second per unit volume which are being created in the medium. In uranium 235, neutrons are created at a rate which is proportional to the number of neutrons already present: $f=\lambda\psi$. Suppose that the uranium is in the shape of a sphere of radius a and that the initial density of neutrons depends only on r. As a boundary condition, assume that the density of neutrons at the surface is zero. Show that the solution of (17.100) is

$$\psi(r,t)=\sum_{n=1}^{\infty}A_nj_0(\beta_nr)e^{-\alpha_nt} \tag{17.101}$$

where the constants β_n are determined by $j_0(\beta_na)=0$ and

$$\alpha_n=D\beta_n^2-\lambda \tag{17.102}$$

Show that the density of neutrons will increase without limit if the radius of the sphere is larger than $\pi\sqrt{D/\lambda}$.

17.7. Show that

$$h_0^1(x)=-\frac{i}{x}e^{ix}$$

$$h_1^1(x)=-\left(\frac{1}{x}+\frac{i}{x^2}\right)e^{ix} \tag{17.103}$$

$$h_2^1(x)=\left(\frac{i}{x}-\frac{3}{x^2}-\frac{3i}{x^3}\right)e^{ix}$$

The corresponding relations for $h_l^2(x)$ can be found from $h_l^2(x)=h_l^1(x)^*$.

17.8. What happens to (17.88) in the limit $\kappa\to 0$?

17.9. By setting $F(\varepsilon)=0$ in equation (17.67), show that the eigenvalue condition for the $l=1$ states of the square well potential can be expressed as

$$\frac{\cot y}{y}-\frac{1}{y^2}=\frac{1}{x}+\frac{1}{x^2} \tag{17.104}$$

17.10. Show that

$$\frac{\sin \kappa R}{\kappa R}=4\pi\sum_{l=0}^{\infty}\sum_{m=-l}^{+l} j_l(\kappa r')\, j_l(\kappa r)\, Y_{lm}^*(\theta',\phi')\, Y_{lm}(\theta,\phi) \tag{17.105}$$

$$\frac{\cos \kappa R}{\kappa R}=-4\pi\sum_{l=0}^{\infty}\sum_{m=-l}^{+l} n_l(\kappa r')\, j_l(\kappa r)\, Y_{lm}^*(\theta',\phi')\, Y_{lm}(\theta,\phi) \tag{17.106}$$

where $r<r'$ for (17.106). Note that (17.105) is correct for both $r<r'$ and $r>r'$.

17.11. By setting $\theta'=0$ and $r'=a$ in (17.105) and (17.106), show that

$$\frac{\sin \kappa R}{\kappa R}=\sum_{l=0}^{\infty}(2l+1)\, j_l(\kappa a)\, j_l(\kappa r)\, P_l(\cos\theta) \tag{17.107}$$

$$\frac{\cos \kappa R}{\kappa R}=-\sum_{l=0}^{\infty}(2l+1)\, n_l(\kappa a)\, j_l(\kappa r)\, P_l(\cos\theta) \qquad a>r \tag{17.108}$$

The geometry is shown in Fig. 17.2.

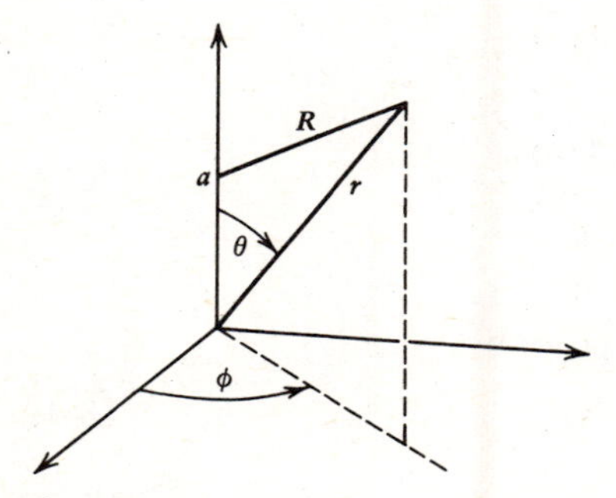

Fig. 17.2

17.12. In terms of the geometry of Fig. 17.2, equation (17.88) can be expressed as

$$\frac{e^{i\kappa R}}{\kappa R}=\sum_{l=0}^{\infty}(2l+1)\, ih_l^1(\kappa a)\, j_l(\kappa r)\, P_l(\cos\theta) \qquad r<a \tag{17.109}$$

Show that in the limit $a\to\infty$,

$$e^{ix\cos\theta}=\sum_{l=0}^{\infty}(2l+1)\, e^{il\pi/2} j_l(x)\, P_l(\cos\theta) \tag{17.110}$$

17.13. Each of the following expansions can be found as a special case of (17.107):

$$\frac{\sin \kappa(a-r)}{\kappa(a-r)}=\sum_{l=0}^{\infty}(2l+1)\, j_l(\kappa r)\, j_l(\kappa a) \tag{17.111}$$

$$1=\sum_{l=0}^{\infty}(2l+1)[\, j_l(x)]^2 \tag{17.112}$$

$$\frac{\sin 2x}{2x} = \sum_{l=0}^{\infty} (2l+1)(-1)^l [j_l(x)]^2 \tag{17.113}$$

To prove (17.113), it will be necessary to first show that $j_l(-x) = (-1)^l j_l(x)$. See equation (17.95).

17.14. Show from equation (14.69) that

$$j_l(x) = \frac{x^l}{2^l l!} \int_0^{\pi/2} \cos(x \sin\theta) \cos^{2l+1}\theta \, d\theta \tag{17.114}$$

17.15. Use (17.109) to show that

$$j_l(x) = \frac{1}{2} e^{-il\pi/2} \int_0^{\pi} e^{ix\cos\theta} P_l(\cos\theta) \sin\theta \, d\theta \tag{17.115}$$

17.16. Show from equation (16.65) that

$$\int_0^{\infty} j_n(x) j_l(x) \, dx = \frac{\sin\left[\frac{\pi}{2}(n-l)\right]}{n(n+1) - l(l+1)} \qquad n \neq l$$

$$= \frac{\pi}{2(2n+1)} \qquad n = l \tag{17.116}$$

17.17. Establish the expansions

$$\frac{e^{-\alpha R}}{\alpha R} = \sum_{l=0}^{\infty} (2l+1) \kappa_l(\alpha a) \iota_l(\alpha r) P_l(\cos\theta) \qquad r < a \tag{17.117}$$

$$e^{-x\cos\theta} = \sum_{l=0}^{\infty} (2l+1)(-1)^l \iota_l(x) P_l(\cos\theta) \tag{17.118}$$

The geometry for (17.117) is shown in Fig. 17.2. See equation (17.31).

18. HANKEL TRANSFORMS

The steps leading from the differential equation (13.11) to the orthogonality condition (13.14) do not depend on the order of the Bessel functions being an integer. We therefore write the complete orthonormal set given by (13.35) as

$$F_{\nu n}(r) = \frac{\sqrt{2}}{a J_{\nu+1}(\kappa_{\nu n} a)} J_\nu(\kappa_{\nu n} r) \tag{18.1}$$

requiring only that $\nu \geq 0$ to prevent the functions from becoming singular at $r = 0$. The values of $\kappa_{\nu n}$ are the positive roots of

$$J_\nu(\kappa_{\nu n} a) = 0 \tag{18.2}$$

Any function of r defined over $0 \leq r \leq a$ can be expanded as

$$F(r) = \sum_{n=0}^{\infty} A_n F_{\nu n}(r) \tag{18.3}$$

The coefficients are given by

$$A_n = \int_0^a F_{\nu n}(r) F(r) r \, dr \tag{18.4}$$

In section 5 of Chapter 8, we found that it is possible to take a Fourier series defined over the finite interval $-s \leq x \leq s$ and allow s to go to infinity. In this manner, a Fourier integral representation of a function defined over $-\infty < x < \infty$ is obtained. Similarly, it is possible to obtain an integral transform from (18.3) by allowing a to become infinite.

Reference to the asymptotic form (15.12) shows that when a is large, the roots of (18.2) are determined by

$$\cos\left(\kappa_{\nu n} a - \frac{\nu\pi}{2} - \frac{\pi}{4}\right) = 0$$

$$\kappa_{\nu n} a = \left(n + \frac{1}{2}\right)\pi + \frac{\pi}{4} + \frac{\nu\pi}{2} \tag{18.5}$$

The values of the first few roots given by (18.5) are

inaccurate, but this will not matter in the limit $a \to \infty$. By again using (15.12), we find that

$$J_{\nu+1}(\kappa_{\nu n} a) = (-1)^n \sqrt{\frac{2}{\pi \kappa_{\nu n} a}} \tag{18.6}$$

The approximate form of the eigenfunctions (18.1) which is valid if a is large then becomes

$$F_{\nu n}(r) = (-1)^n \sqrt{\frac{\pi \kappa_{\nu n}}{a}}\, J_\nu(\kappa_{\nu n} r) \tag{18.7}$$

Equation (10.43) in Chapter 7 shows that the eigenfunctions obey the closure relation

$$\sum_{n=0}^{\infty} F_{\nu n}(r) F_{\nu n}(r') = \frac{\delta(r-r')}{r} \tag{18.8}$$

For large values of a, this is

$$\sum_{n=0}^{\infty} J_\nu(\kappa_{\nu n} r) J_\nu(\kappa_{\nu n} r') \frac{\pi \kappa_{\nu n}}{a} = \frac{\delta(r-r')}{r} \tag{18.9}$$

The values of $\kappa_{\nu n}$ as given by (18.5) are closely spaced if a is large. In the limit as $a \to \infty$, $\kappa_{\nu n}$ becomes a continuous variable κ. Note that

$$\kappa_{\nu, n+1} - \kappa_{\nu n} = \frac{\pi}{a} \tag{18.10}$$

We therefore write $d\kappa = \pi/a$ and replace the sum in (18.9) by an integral to get

$$\int_0^\infty J_\nu(\kappa r) J_\nu(\kappa r') \kappa \, d\kappa = \frac{\delta(r-r')}{r} \tag{18.11}$$

By interchanging the roles of r and κ, we get

$$\int_0^\infty J_\nu(\kappa r) J_\nu(\kappa' r) r \, dr = \frac{\delta(\kappa-\kappa')}{\kappa} \tag{18.12}$$

The analogous results for Fourier integrals are given by equations (5.11) and (5.12) in Chapter 8.

It is apparent without going through the details that the limiting form of (18.4) is

$$A(\kappa) = \int_0^\infty J_\nu(\kappa r) F(r) r \, dr \tag{18.13}$$

The inverse of (18.13) can be found by means of (18.11). If (18.13) is multiplied by $J_\nu(\kappa r')\kappa\, d\kappa$ and then integrated with respect to κ, the result is

$$\begin{aligned} \int_0^\infty J_\nu(\kappa r') A(\kappa) \kappa \, d\kappa &= \int_0^\infty F(r) \int_0^\infty J_\nu(\kappa r) J_\nu(\kappa r') \kappa \, d\kappa \, r \, dr \\ &= \int_0^\infty F(r) \delta(r-r') \, dr = F(r') \end{aligned} \tag{18.14}$$

Equation (18.13) and its inverse

$$F(r) = \int_0^\infty J_\nu(\kappa r) A(\kappa) \kappa \, d\kappa \tag{18.15}$$

are called *Hankel transforms*. They should be compared to the Fourier integral transform pair given by equations (5.6) and (5.7) of Chapter 8. We could now proceed to establish the validity of the Hankel transform pair more rigorously as we did with Fourier transforms in section 5 of Chapter 8, but we will not carry out this program here.

Hankel transforms occur in the solution of Laplace's equation in cylindrical coordinates when the range of values of r is $0 \le r < \infty$. The appropriate solution of Laplace's equation (13.1) in the region $0 \le z < \infty$, $0 \le r \le a$ is

$$\psi(r, \theta, z) = \sum_{m=-\infty}^{+\infty} \sum_{n=0}^{\infty} u_{mn}(r, \theta) e^{-\kappa_{mn} z} \tag{18.16}$$

If the boundary conditions are $\psi = 0$ at $r = a$, then the appropriate eigenfunctions are given by (13.38). The summation over n is over all the positive roots of $J_m(\kappa_{mn} a) = 0$. In the limit $a \to \infty$, the solution (18.16) is replaced by

$$\psi(r, \theta, z) = \sum_{m=-\infty}^{+\infty} e^{im\theta} \int_0^\infty A_m(\kappa) J_m(\kappa r) e^{-\kappa z} \kappa \, d\kappa \tag{18.17}$$

If the potential is given as a known function $F(r, \theta)$ over the x, y plane, then by setting $z = 0$ in (18.17), we find

$$F(r, \theta) = \sum_{m=-\infty}^{+\infty} e^{im\theta} \int_0^\infty A_m(\kappa) J_m(\kappa r) \kappa \, d\kappa \tag{18.18}$$

In (18.18), the angular dependence of $F(r, \theta)$ is expanded in terms of a complex Fourier series and the radial dependence is represented by a Hankel transform. The orders of the Bessel functions are now integers because the potential is required to be single-valued with respect to increases in θ by 2π. Recall that $J_{-m}(x) = (-1)^m J_m(x)$. If (18.18) is multiplied by the complex conjugate of the Fourier eigenfunctions, $e^{-in\theta}$, and then integrated with respect to θ, the result is

$$\int_0^{2\pi} F(r,\theta) e^{-in\theta}\, d\theta = 2\pi \int_0^\infty A_n(\kappa) J_n(\kappa r)\kappa\, d\kappa \tag{18.19}$$

We can now use (18.13) to obtain

$$A_n(\kappa) = \frac{1}{2\pi}\int_0^{2\pi} e^{-in\theta}\int_0^\infty J_n(\kappa r) F(r,\theta) r\, dr\, d\theta \tag{18.20}$$

which formally completes the solution of the potential problem. The solution is valid in the half-space $0 \le z < \infty$.

If the potential is given as a function of r only over the x, y plane, then only $m = 0$ appears in the solution with the result that

$$\psi(r,z) = \int_0^\infty A(\kappa) J_0(\kappa r) e^{-\kappa z}\kappa\, d\kappa \tag{18.21}$$

$$A(\kappa) = \int_0^\infty J_0(\kappa r) F(r) r\, dr \tag{18.22}$$

For example, suppose that the circular region of the x, y plane given by $0 \le r \le b$ is maintained at the potential V and that the rest of the x, y plane is grounded:

$$F(r) = V \qquad 0 \le r \le b$$
$$F(r) = 0 \qquad r > b \tag{18.23}$$

Equation (18.22) gives

$$A(\kappa) = V\int_0^b J_0(\kappa r) r\, dr = \frac{V}{\kappa^2}\int_0^{\kappa b} J_0(x) x\, dx$$
$$= \frac{V}{\kappa^2}\int_0^{\kappa b}\frac{d}{dx}[xJ_1(x)]\, dx = \frac{VbJ_1(\kappa b)}{\kappa} \tag{18.24}$$

where use has been made of (13.25) with $m = 0$. The combination of (18.21) and (18.24) gives

$$\psi(r,z) = Vb\int_0^\infty J_1(\kappa b) J_0(\kappa r) e^{-\kappa z}\, d\kappa \tag{18.25}$$

PROBLEMS

18.1. Prove that

$$\int_0^\infty \frac{r}{\sqrt{1+r^2}} J_0(\kappa r)\, dr = \frac{e^{-\kappa}}{\kappa} \tag{18.26}$$

Hint: Equation (14.46) is helpful.

18.2. Suppose that the electrostatic potential is the known function $F(x, y)$ over the x, y plane and that the potential is wanted everywhere in the half-space $z > 0$. The x and y dependence of the potential can be represented as a two-dimensional Fourier transform:

$$\psi(x,y,z) = \frac{1}{2\pi}\int_{-\infty}^{+\infty}\int_{-\infty}^{+\infty} A(\alpha,\beta,z) e^{i(\alpha x+\beta y)}\, d\alpha\, d\beta \tag{18.27}$$

Show that

$$A(\alpha,\beta,z) = f(\alpha,\beta) e^{-\kappa z} \qquad \kappa = \sqrt{\alpha^2+\beta^2} \tag{18.28}$$

$$F(x,y) = \frac{1}{2\pi}\int_{-\infty}^{+\infty}\int_{-\infty}^{+\infty} f(\alpha,\beta) e^{i(\alpha x+\beta y)}\, d\alpha\, d\beta \tag{18.29}$$

$$f(\alpha,\beta) = \frac{1}{2\pi}\int_{-\infty}^{+\infty}\int_{-\infty}^{+\infty} F(x,y) e^{-i(\alpha x+\beta y)}\, dx\, dy \tag{18.30}$$

See section 5 of Chapter 8, especially equations (5.33) and (5.34). Suppose that the square region $-s \leq x \leq s$, $-s \leq y \leq s$ of the x, y plane is maintained at a potential V and the rest of the x, y plane is grounded. Show that

$$A(\alpha, \beta, z) = \frac{2V}{\pi\alpha\beta} \sin\alpha s \sin\beta s\, e^{-\kappa z} \tag{18.31}$$

18.3. If the variables are changed by means of

$$x = r\cos\theta \qquad y = r\sin\theta \qquad \alpha = \kappa\sin\theta' \qquad \beta = \kappa\cos\theta' \tag{18.32}$$

show that equation (18.29) becomes

$$F(r,\theta) = \frac{1}{2\pi}\int_0^\infty \int_0^{2\pi} f(\kappa,\theta')\, e^{i\kappa r\sin(\theta+\theta')}\kappa\, d\kappa\, d\theta' \tag{18.33}$$

Now make the change of variable $\phi = \theta + \theta'$ and use the expansion

$$f(\kappa,\theta') = \sum_{m=-\infty}^{+\infty} A_m(\kappa)\, e^{-im\theta'} \tag{18.34}$$

to obtain another derivation of (18.18). In this way, Hankel transforms can be derived from Fourier transforms.

18.4. The problem of the isolated electrified conducting disc of radius c is discussed from the point of view of spheroidal coordinates in section 12 of Chapter 4. This problem can also be treated as a mixed boundary value problem. The boundary conditions in the x, y plane are

$$\begin{aligned} \psi &= V \qquad 0 < r < c \\ \frac{\partial\psi}{\partial z} &= 0 \qquad r > c \end{aligned} \tag{18.35}$$

The second part of the condition results because of the fact that the electric field has no z component in the x, y plane if $r > c$. Show that the function $A(\kappa)$ of equation (18.21) must satisfy

$$\begin{aligned} V &= \int_0^\infty A(\kappa) J_0(\kappa r)\kappa\, d\kappa \qquad r < c \\ 0 &= \int_0^\infty A(\kappa) J_0(\kappa r)\kappa^2\, d\kappa \qquad r > c \end{aligned} \tag{18.36}$$

18.5. Equation (14.46) remains valid if it is analytically continued into the complex a plane as long as the integral remains convergent. Show that

$$\int_0^\infty J_0(\kappa r)\, e^{-i\kappa c}\, d\kappa = \frac{1}{\sqrt{r^2 - c^2}} \tag{18.37}$$

and from this deduce that

$$\begin{aligned} \int_0^\infty J_0(\kappa r)\cos\kappa c\, d\kappa &= \frac{1}{\sqrt{r^2-c^2}} \qquad r > c \\ &= 0 \qquad r < c \end{aligned} \tag{18.38}$$

$$\begin{aligned} \int_0^\infty J_0(\kappa r)\sin\kappa c\, d\kappa &= 0 \qquad r > c \\ &= \frac{1}{\sqrt{c^2-r^2}} \qquad r < c \end{aligned} \tag{18.39}$$

18.6. Combine the results of problems 18.4 and 18.5 to show that the potential due to an electrified disc is

$$\psi(r,z) = \frac{2V}{\pi}\int_0^\infty \frac{\sin \kappa c}{\kappa} J_0(\kappa r) e^{-\kappa z}\, d\kappa \qquad z > 0 \tag{18.40}$$

By using the solution given in section 12 of Chapter 4, show that

$$\int_0^\infty \frac{\sin \kappa c}{\kappa} J_0(\kappa r)\, d\kappa = \sin^{-1}\left(\frac{c}{r}\right) \qquad r > c$$

$$= \frac{\pi}{2} \qquad r < c \tag{18.41}$$

19. GREEN FUNCTIONS

The propagator used to find the general solution of the free-particle Schrödinger equation as given by (6.25) and (6.26) in Chapter 8 is really a type of Green's function. A similar treatment of the equation of heat conduction is given by (6.40), Chapter 8. In equation (7.35), Chapter 8, we find a Green's function suitable for solving the classical wave equation with time-dependent sources when the boundaries of the region of interest are at infinity. We consider in this section the formal solution of Poisson's equation in a region Σ of space in which the charge density is known and where the boundary conditions are either of the Dirichlet or Neumann type. See theorem 9 and problem 10.12 of Chapter 3 for a discussion of these boundary conditions.

If the boundaries of the region are at infinity, then the Green's function is $G(\mathbf{r},\mathbf{r}') = 1/R$ and the solution of Poisson's equation is given by (10.29), Chapter 3. Review the distinction between field point variables and source point variables given at the beginning of section 10 of Chapter 3. The formal solution of Poisson's equation begins with Green's second identity as given by (9.17), Chapter 3:

$$\int_\Sigma (\psi \nabla'^2 \phi - \phi \nabla'^2 \psi)\, d\Sigma' = \int_\sigma \left(\psi \frac{\partial \phi}{\partial n'} - \phi \frac{\partial \psi}{\partial n'}\right) d\sigma' \tag{19.1}$$

The variables of integration have been taken to be the source point variables (x', y', z'). If $\phi = G(\mathbf{r},\mathbf{r}')$ is the Green's function and $\psi(\mathbf{r}')$ is the potential, then these functions obey the Poisson equations

$$\nabla'^2 G(\mathbf{r},\mathbf{r}') = -4\pi\delta(\mathbf{R}) \tag{19.2}$$

$$\nabla'^2 \psi(\mathbf{r}') = -\frac{1}{\varepsilon_0}\rho(\mathbf{r}') \tag{19.3}$$

Equation (19.3) is of course the Poisson equation that we have set out to solve and (19.2) is to be regarded as part of the definition of the Green's function. In order to complete the definition, we must give the boundary conditions which $G(\mathbf{r},\mathbf{r}')$ satisfies on the bounding surface of the region Σ. By combining (19.1), (19.2), and (19.3) and utilizing the properties of the Dirac delta function, we find

$$\psi(\mathbf{r}) = \frac{1}{4\pi\varepsilon_0}\int_\Sigma G(\mathbf{r},\mathbf{r}')\rho(\mathbf{r}')\, d\Sigma'$$

$$= \frac{1}{4\pi}\int_\sigma \left[G(\mathbf{r},\mathbf{r}')\frac{\partial}{\partial n'}\psi(\mathbf{r}') - \psi(\mathbf{r}')\frac{\partial}{\partial n'}G(\mathbf{r},\mathbf{r}')\right] d\sigma' \tag{19.4}$$

We know that the Dirichlet boundary value problem requires the potential to be specified over the boundary of the region. This means that we should try to specify the boundary conditions on the Green's function in such a way that only ψ itself and not its normal derivative appears in the surface integral. If we complete the definition of the Green's function by requiring it to obey the boundary condition $G(\mathbf{r},\mathbf{r}') = 0$ on σ, then (19.4) specializes to

$$\psi(\mathbf{r}) = \frac{1}{4\pi\varepsilon_0}\int_\Sigma G(\mathbf{r},\mathbf{r}')\rho(\mathbf{r}')\, d\Sigma' - \frac{1}{4\pi}\int_\sigma \psi(\mathbf{r}')\frac{\partial}{\partial n'}G(\mathbf{r},\mathbf{r}')\, d\sigma' \tag{19.5}$$

which provides a formal solution of the Dirichlet problem. Note that all the conditions of the Dirichlet boundary value problem are met by both the functions $\psi(\mathbf{r})$ and $G(\mathbf{r},\mathbf{r}')$ and that these two functions are therefore uniquely specified. From the practical point of view, it is sometimes easier to solve the boundary value problem involved with finding the Green's function than to solve directly for the desired potential function. In other words, (19.5) gives the solution of a complicated potential problem in terms of a simpler one. Physically, the Green's function is the potential which is created by a point charge of magnitude $4\pi\varepsilon_0$ located at $\mathbf{r}$ subject to the boundary condition $G(\mathbf{r},\mathbf{r}')=0$ on the surface of the region.

In the Neumann problem, the normal derivative of the potential is specified over the bounding surface. It would seem that an appropriate boundary condition on the Green's function is $\partial G/\partial n'=0$, but this is incorrect. The reason is that the Green's function has a point charge source inside the region of interest and this produces a net outward flux across the bounding surface given by

$$\int_\sigma \frac{\partial G}{\partial n'}\,d\sigma' = -4\pi \tag{19.6}$$

A possible choice of boundary condition, but by no means the only one, is

$$\frac{\partial G}{\partial n'} = -\frac{4\pi}{\sigma} \tag{19.7}$$

where σ is the total area of the boundary surface. We then find from (19.4) that

$$\psi(\mathbf{r}) = \frac{1}{4\pi\varepsilon_0}\int_\Sigma G(\mathbf{r},\mathbf{r}')\rho(\mathbf{r}')\,d\Sigma' + \frac{1}{4\pi}\int_\sigma G(\mathbf{r},\mathbf{r}')\frac{\partial}{\partial n'}\psi(\mathbf{r}')\,d\sigma + \bar{\psi} \tag{19.8}$$

where

$$\bar{\psi} = \frac{1}{\sigma}\int_\sigma \psi(\mathbf{r}')\,d\sigma' \tag{19.9}$$

is the average value of the potential over the surface. It is of no consequence since it is a constant and does not affect the electric field which is ultimately calculated from the potential.

Green functions have a symmetry property. If $\psi = G(\mathbf{r},\mathbf{r}')$ and $\phi = G(\mathbf{r}'',\mathbf{r}')$, then Green's second identity (19.1) gives

$$\int_\Sigma \{G(\mathbf{r},\mathbf{r}')[-4\pi\delta(\mathbf{r}''-\mathbf{r}')] - G(\mathbf{r}'',\mathbf{r}')[-4\pi\delta(\mathbf{r}-\mathbf{r}')]\}\,d\Sigma' = -4\pi G(\mathbf{r},\mathbf{r}'') + 4\pi G(\mathbf{r}'',\mathbf{r})$$

$$= \int_\sigma \left\{G(\mathbf{r},\mathbf{r}')\frac{\partial}{\partial n'}G(\mathbf{r}'',\mathbf{r}') - G(\mathbf{r}'',\mathbf{r}')\frac{\partial}{\partial n'}G(\mathbf{r},\mathbf{r}')\right\}d\sigma' \tag{19.10}$$

For the Dirichlet problem, the surface integral is zero and

$$G(\mathbf{r},\mathbf{r}'') = G(\mathbf{r}'',\mathbf{r}) \tag{19.11}$$

The Green's function is therefore invariant with respect to exchange of source and field point variables. This symmetry property is sometimes helpful in finding the Green's function. If we have a Neumann problem and we use the boundary condition (19.7), then

$$-4\pi G(\mathbf{r},\mathbf{r}'') + 4\pi G(\mathbf{r}'',\mathbf{r}) = -\frac{4\pi}{\sigma}\int_\sigma [G(\mathbf{r},\mathbf{r}') - G(\mathbf{r}'',\mathbf{r}')]\,d\sigma' \tag{19.12}$$

The symmetry property (19.11) is therefore not automatic in the case of the Neumann problem but can usually be imposed as a separate condition.

As an example of a Dirichlet problem, we will find the potential in the half-space $z>0$ when the charge density is known and the potential is specified over the x, y plane. In order to find the Green's function, imagine that a positive point charge $q=4\pi\varepsilon_0$ is placed at (x, y, z) and that the x, y plane is a grounded conductor. The potential due to this arrangement can be found by placing an *image charge* of strength $-4\pi\varepsilon_0$ at $(x, y, -z)$ as shown in Fig. 19.1. The potential due to the actual charge and its image is

$$G(\mathbf{r},\mathbf{r}') = \frac{1}{r_1} - \frac{1}{r_2} = \frac{1}{\sqrt{(x'-x)^2+(y'-y)^2+(z'-z)^2}} - \frac{1}{\sqrt{(x'-x)^2+(y'-y)^2+(z'+z)^2}} \tag{19.13}$$

Since obviously $G(\mathbf{r},\mathbf{r}')=0$ when $r_1=r_2$, the x, y plane is at zero potential, meaning that (19.13) also gives the potential in the half-space $z>0$ due to a single point charge at (x, y, z) and a grounded conductor which coincides with the x, y plane. Part of

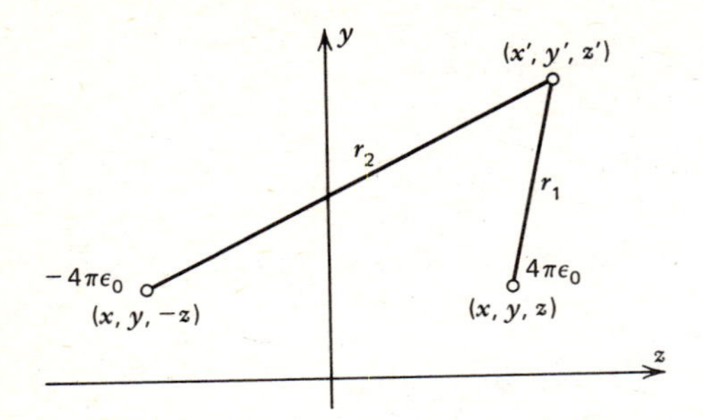

Fig. 19.1

the bounding surface of the region is at infinity, and we must suppose that the potential becomes zero there. It is necessary to compute the normal derivative of the Green's function over the remaining portion of the bounding surface, which is the x, y plane in this example:

$$\frac{\partial G}{\partial n'} = -\left(\frac{\partial G}{\partial z'}\right)_{z'=0} = -\frac{2z}{\left[(x'-x)^2+(y'-y)^2+z^2\right]^{3/2}} \tag{19.14}$$

The minus sign appears because the unit normal to the surface points outward from the region of interest, which is the negative z direction in this case. If $F(x, y)$ is the known potential over the x, y plane, then the final solution for the potential in the region $z > 0$ is

$$\psi(\mathbf{r}) = \frac{1}{4\pi\varepsilon_0}\int_\Sigma G(\mathbf{r},\mathbf{r}')\rho(\mathbf{r}')\,d\Sigma' + \frac{1}{2\pi}\int_{-\infty}^{+\infty}\int_{-\infty}^{+\infty}\frac{zF(x',y')\,dx'dy'}{\left[(x'-x)^2+(y'-y)^2+z^2\right]^{3/2}} \tag{19.15}$$

If there is no charge in the region $z > 0$ and if the potential over the x, y plane is given by equation (18.23), then the potential on the axis of the disc $(x = y = 0)$ can be found from (19.15) as

$$\psi(z) = \frac{1}{2\pi}\int_0^b \frac{zV2\pi r'\,dr'}{(r'^2+z^2)^{3/2}} = V\left(1-\frac{z}{\sqrt{b^2+z^2}}\right) \tag{19.16}$$

It sometimes happens that solving the same problem by different methods leads to interesting mathematical results. For example, by setting $r = 0$ in (18.25), we find that

$$\int_0^\infty J_1(\kappa b)e^{-\kappa z}\,d\kappa = \frac{1}{b}\left(1-\frac{z}{\sqrt{b^2+z^2}}\right) \tag{19.17}$$

Green's function for a sphere can also be found by the method of images. Green's function required for solving the Dirichlet problem in the region exterior to a sphere of radius a is found by placing a point charge q at a distance r from the center of the sphere as shown in Fig. 19.2. We then attempt to find an appropriate image charge q' inside the sphere that is a distance b from its center such that the potential at every point of the sphere is zero. The potential at point P due to the two point charges is

$$\psi = \frac{1}{4\pi\varepsilon_0}\left(\frac{q}{r_1}+\frac{q'}{r_2}\right) = \frac{1}{4\pi\varepsilon_0}\left[\frac{q}{\sqrt{r^2+r'^2-2rr'\cos\gamma}} + \frac{q'}{\sqrt{b^2+r'^2-2br'\cos\gamma}}\right] \tag{19.18}$$

By setting $\psi = 0$ and $r' = a$, we find

$$0 = \frac{q}{\sqrt{1+\dfrac{a^2}{r^2}-\dfrac{2a}{r}\cos\gamma}} + \frac{rq'}{a}\frac{1}{\sqrt{1+\dfrac{b^2}{a^2}-\dfrac{2b}{a}\cos\gamma}} \tag{19.19}$$

This expression will hold for all values of the angle γ provided that b and q' are determined by

$$\frac{rq'}{a} = -q \qquad \frac{b}{a} = \frac{a}{r} \tag{19.20}$$

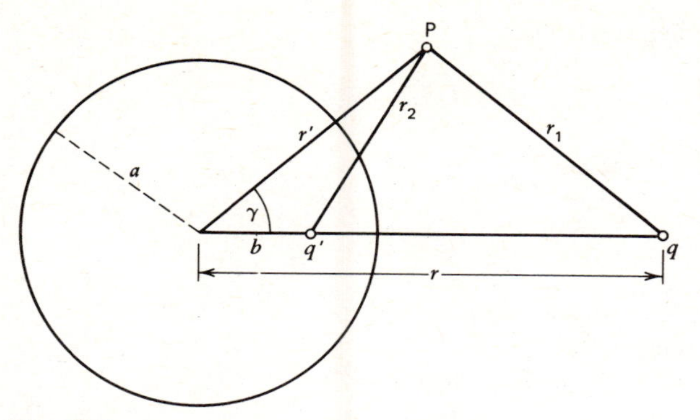

Fig. 19.2

The Green's function is found by setting $q = 4\pi\varepsilon_0$ and eliminating q' and b from the potential (19.18). The result is

$$G(\mathbf{r},\mathbf{r}') = \frac{1}{\sqrt{r^2 + r'^2 - 2rr'\cos\gamma}} - \frac{a}{\sqrt{r^2r'^2 + a^4 - 2rr'a^2\cos\gamma}} \tag{19.21}$$

In order to complete the problem, we require

$$\frac{\partial G}{\partial n'} = -\left(\frac{\partial G}{\partial r'}\right)_{r'=a} = -\frac{r^2 - a^2}{a\left(a^2 + r^2 - 2ra\cos\gamma\right)^{3/2}} \tag{19.22}$$

The final result is found from equation (19.5) to be

$$\psi(r,\theta,\phi) = \frac{1}{4\pi\varepsilon_0}\int_\Sigma G(\mathbf{r},\mathbf{r}')\rho(\mathbf{r}')\,d\Sigma' + \frac{1}{4\pi a}\int_0^\pi\int_0^{2\pi}\frac{F(\theta',\phi')(r^2 - a^2)a^2\sin\theta'}{\left(a^2 + r^2 - 2ra\cos\gamma\right)^{3/2}}\,d\theta'\,d\phi' \tag{19.23}$$

where $F(\theta',\phi')$ is the known potential over the surface of the sphere. In setting up the integral, we have represented an element of area on the sphere by $d\sigma' = a^2\sin\theta'\,d\theta'\,d\phi'$. Before the integral can be carried out, it is necessary to express $\cos\gamma$ in terms of the polar angles of the source and field point variables. This is done by equation (12.22).

PROBLEMS

19.1. A point charge q is on the z axis at $z = s$. If the (x, y) plane is a grounded conducting sheet, show that it has an induced surface charge density given by

$$\lambda = \frac{-qs}{2\pi\left(x^2 + y^2 + s^2\right)^{3/2}} \tag{19.24}$$

Show that the total induced charge is $-q$.

19.2. If the region $z > 0$ is free of charge, show that the Fourier integral solution given in problem 18.3 is the same as (19.15).

19.3. Evaluate the following integrals:

$$\int_0^\infty J_0(bx)\,xe^{-ax}\,dx = \frac{a}{\left(a^2 + b^2\right)^{3/2}} \tag{19.25}$$

$$\int_0^\infty J_2(bx)\,e^{-ax}\,dx = \frac{b^2 + 2a^2}{b^2\sqrt{b^2 + a^2}} - \frac{2a}{b^2} \tag{19.26}$$

19.4. Show that the Green's function for the Dirichlet problem inside a sphere of radius a is also given correctly by (19.21) and that the potential is given by

$$\psi(r,\theta,\phi) = \frac{1}{4\pi\varepsilon_0}\int_\Sigma G(\mathbf{r},\mathbf{r}')\rho(\mathbf{r}')\,d\Sigma' + \frac{1}{4\pi a}\int_0^\pi\int_0^{2\pi}\frac{F(\theta',\phi')(a^2 - r^2)a^2\sin\theta'}{\left(a^2 + r^2 - 2ra\cos\gamma\right)^{3/2}}\,d\theta'\,d\phi' \tag{19.27}$$

19.5. Show from the generating function (9.8) for the Legendre polynomials that

$$\frac{1-h^2}{(1+h^2-2xh)^{3/2}}=\sum_{l=0}^{\infty}(2l+1)h^l P_l(x) \tag{19.28}$$

Hint: First show that the generating function obeys

$$G(x,h)+2h\frac{\partial}{\partial h}G(x,h)=\frac{1-h^2}{(1+h^2-2xh)^{3/2}} \tag{19.29}$$

19.6. For the case where there is no charge inside the sphere, set $h=r/a$ in (19.29) and use the addition theorem (12.24) to show that (19.27) and (10.6) are equivalent.

19.7. Two functions $\psi(x)$ and $\phi(x)$ are defined over the interval $a \le x \le b$. Show that the one-dimensional version of Green's second identity is

$$\int_a^b\left(\psi\frac{d^2\phi}{dx^2}-\phi\frac{d^2\psi}{dx^2}\right)dx=\left[\psi\frac{d\phi}{dx}-\phi\frac{d\psi}{dx}\right]_a^b \tag{19.30}$$

19.8. A function $\psi(x)$ obeys the one-dimensional Poisson equation

$$\frac{d^2\psi}{dx^2}=-\frac{1}{\varepsilon_0}\rho(x) \qquad a \le x \le b \tag{19.31}$$

and satisfies the boundary conditions $\psi=\psi_a$ at $x=a$ and $\psi=\psi_b$ at $x=b$. Show that the general solution of (19.31) is

$$\psi(x)=\frac{1}{\varepsilon_0}\int_a^b G(x,x')\rho(x')\,dx'+\frac{\psi_a(b-x)+\psi_b(x-a)}{b-a} \tag{19.32}$$

where the Green's function is given by

$$\begin{aligned} G(x,x')&=\frac{(x'-a)(b-x)}{b-a} \qquad x'<x \\ G(x,x')&=\frac{(b-x')(x-a)}{b-a} \qquad x'>x \end{aligned} \tag{19.33}$$

Draw a graph of the Green's function.

20. GREEN'S FUNCTION EXPANSIONS

Expansion of Green's Function in Spherical Coordinates

The solution of the Dirichlet problem in the spherical region $0 \le r \le a$ is given by (19.27). It is not necessarily the best idea to try to carry out the integration with the Green's function in closed form, and we therefore seek an expansion of it in terms of spherical harmonics. This can be done by the same method that was used to obtain the expansion of $1/R$ given by (12.20) and (12.21). In fact, the only real difference is that the solutions of (12.15) must now be chosen to be

$$\begin{aligned} F_l(r)&=Ar^l \qquad r<r' \\ F_l(r)&=Br^{-l-1}+Cr^l \qquad r>r' \end{aligned} \tag{20.1}$$

rather than (12.16). The reason is that there is no basis for excluding r^l from the solution valid for $r>r'$ if the boundaries are not at infinity. We remark that we can solve either (12.1) or (12.4) for the Green's function with exactly the same final result because of the symmetry property $G(\mathbf{r},\mathbf{r}')=G(\mathbf{r}',\mathbf{r})$. In other words, we can consider either $\mathbf{r}$ or $\mathbf{r}'$ as the variable point and the other point as fixed. There are now three conditions on the solutions (20.1). One is the condition (12.17) on the derivative, a second is the requirement that the solution be continuous at $r=r'$, and the third is that the solution vanish at $r=a$. These three conditions give

$$(l+1)B(r')^{-l}+lA(r')^{l+1}-Cl(r')^{l+1}=4\pi$$

$$B(r')^{-l-1}-A(r')^{l}+C(r')^{l}=0 \tag{20.2}$$

$$Ba^{-l-1}+Ca^{l}=0$$

The solution for the three constants is

$$A=\frac{4\pi}{2l+1}\left[\frac{1}{(r')^{l+1}}-\frac{(r')^{l}}{a^{2l+1}}\right]$$

$$B=\frac{4\pi(r')^{l}}{2l+1} \tag{20.3}$$

$$C=-\frac{4\pi(r')^{l}}{(2l+1)a^{2l+1}}$$

The Green's function is therefore

$$G(\mathbf{r},\mathbf{r}')=\sum_{l=0}^{\infty}\sum_{m=-l}^{+l}\frac{4\pi}{2l+1}\left[\frac{(r')^{l}}{r^{l+1}}-\frac{(rr')^{l}}{a^{2l+1}}\right]$$
$$\times Y_{lm}^{*}(\theta',\phi')Y_{lm}(\theta,\phi)\qquad r>r' \tag{20.4}$$

The form of the expansion which is valid for $r<r'$ is found by exchanging r and r'. The same result can also be obtained by direct expansion of (19.21).

As an example, suppose that a total charge q is uniformly distributed over a thin wire in the shape of a ring of radius b. The ring is placed at the center of a grounded conducting sphere of radius a where $a>b$. The ring and the sphere both have their centers at the origin of coordinates and the ring lies in the x, y plane. The axis of the ring coincides with the z axis. The problem is to find the potential everywhere inside the sphere. The charge density can be conveniently represented by a delta function as

$$\rho(\mathbf{r}')=C\delta(r'-b)\delta(\cos\theta') \tag{20.5}$$

where the constant C is to be determined by the condition that

$$q=\int_{\Sigma}\rho(\mathbf{r}')\,d\Sigma'=C\int_{r'=0}^{a}\int_{\theta'=0}^{\pi}\delta(r'-b)\delta(\cos\theta')$$
$$\times 2\pi r'^{2}\sin\theta'\,d\theta'\,dr'$$
$$=C2\pi b^{2}\int_{-1}^{+1}\delta(x)\,dx=C2\pi b^{2} \tag{20.6}$$

Therefore,

$$\rho(\mathbf{r}')=\frac{q}{2\pi b^{2}}\delta(r'-b)\delta(\cos\theta') \tag{20.7}$$

Because of the axial symmetry, only $m=0$ terms will appear in the final solution. The potential in the region $b<r<a$ is

$$\psi(\mathbf{r})=\frac{1}{4\pi\varepsilon_0}\int_{\Sigma}G(\mathbf{r},\mathbf{r}')\rho(\mathbf{r}')\,d\Sigma'$$
$$=\frac{1}{4\pi\varepsilon_0}\sum_{l=0}^{\infty}\frac{4\pi}{2l+1}\int_0^a\int_0^\pi\left[\frac{(r')^{l}}{r^{l+1}}-\frac{(rr')^{l}}{a^{2l+1}}\right]$$
$$\times\frac{2l+1}{4\pi}P_l(\cos\theta')P_l(\cos\theta)\frac{q}{2\pi b^{2}}$$
$$\times\delta(r'-b)\delta(\cos\theta')2\pi r'^{2}\sin\theta'd\theta'dr'$$
$$=\frac{q}{4\pi\varepsilon_0}\sum_{l=0}^{\infty}\left[\frac{b^{l}}{r^{l+1}}-\frac{(rb)^{l}}{a^{2l+1}}\right]P_l(0)P_l(\cos\theta)$$
$$=\frac{q}{4\pi\varepsilon_0}\sum_{n=0}^{\infty}\left[\frac{b^{2n}}{r^{2n+1}}-\frac{(rb)^{2n}}{a^{4n+1}}\right]\frac{(-1)^n(2n)!}{2^{2n}(n!)^2}$$
$$\times P_{2n}(\cos\theta)\qquad b<r<a \tag{20.8}$$

We have used equation (4.45) to evaluate $P_l(0)$. The potential for $0<r<b$ is found by simply exchanging b and r in (20.8). The potential for a uniformly charged ring with boundaries at infinity can be found by setting $a=\infty$. See equation (10.37). [In (10.37), a is the radius of the ring.]

Expansion of Green's Function in Cylindrical Coordinates

We will find an expansion for the Green's function of Poisson's equation in cylindrical coordinates for the case where the boundaries are at infinity. This of course yields another expansion for $1/R$. First note that the expression for the delta function in cylindrical coordinates is

$$\delta(\mathbf{R})=\frac{\delta(r-r')}{r}\delta(\theta-\theta')\delta(z-z') \tag{20.9}$$

The factor of r in the denominator is required to cancel the factor of r in the volume element so that

$$\int_{\Sigma}\delta(\mathbf{R})\,d\Sigma$$

$$=\int_{\Sigma}\delta(r-r')\delta(\theta-\theta')\delta(z-z')\,dr\,d\theta\,dz=1 \tag{20.10}$$

The Poisson equation to be solved is

$$\nabla^2 G=\frac{1}{r}\frac{\partial}{\partial r}\left(r\frac{\partial G}{\partial r}\right)+\frac{1}{r^2}\frac{\partial^2 G}{\partial\theta^2}+\frac{\partial^2 G}{\partial z^2}$$

$$=-4\pi\frac{\delta(r-r')}{r}\delta(\theta-\theta')\delta(z-z') \tag{20.11}$$

The θ and z delta functions can be represented as

$$\delta(z-z')=\frac{1}{2\pi}\int_{-\infty}^{+\infty}e^{i\kappa(z-z')}\,d\kappa$$

$$=\frac{1}{\pi}\int_0^{\infty}\cos\kappa(z-z')\,d\kappa \tag{20.12}$$

$$\delta(\theta-\theta')=\frac{1}{2\pi}\sum_{m=-\infty}^{+\infty}e^{im(\theta-\theta')} \tag{20.13}$$

To justify (20.12), we can argue that $\sin\kappa(z-z')$ is an odd function and therefore contributes nothing to the integral. The same result can also be obtained by combining equations (5.40) and (5.43) in Chapter 8. The reason for expressing the delta function as an integral over positive values of κ is that it will result in Bessel functions of positive argument in the final answer. The θ and z dependence of the Green's function can be expanded in terms of the same eigenfunctions used to expand the delta functions:

$$G(\mathbf{r},\mathbf{r}')=\frac{1}{2\pi^2}\int_0^{+\infty}\sum_{m=-\infty}^{+\infty}F_m(r)\cos\kappa(z-z')$$

$$\times e^{im(\theta-\theta')}\,d\kappa \tag{20.14}$$

The differential equation (20.11) gives

$$\frac{1}{2\pi^2}\int_0^{+\infty}\sum_{m=-\infty}^{+\infty}\left[\frac{1}{r}\frac{d}{dr}\left(r\frac{dF_m}{dr}\right)-\left(\frac{m^2}{r^2}+\kappa^2\right)F_m\right]$$

$$\times\cos\kappa(z-z')e^{im(\theta-\theta')}$$

$$=-\frac{4\pi}{2\pi^2}\frac{\delta(r-r')}{r}$$

$$\times\int_0^{+\infty}\sum_{m=-\infty}^{+\infty}\cos\kappa(z-z')e^{im(\theta-\theta')}\,d\kappa \tag{20.15}$$

Since the eigenfunctions are complete, (20.15) implies that

$$\frac{d}{dr}\left(r\frac{dF_m}{dr}\right)-\left(\frac{m^2}{r}+\kappa^2 r\right)F_m=-4\pi\delta(r-r') \tag{20.16}$$

For $r\neq r'$, the solutions are modified Bessel functions of argument κr. Equation (15.58) shows that $I_m(\kappa r)$ is finite at $r=0$ and that $K_m(\kappa r)$ decays exponentially as $r\to\infty$. The appropriate form of the solution is therefore

$$\begin{aligned} F_m(r)&=AI_m(\kappa r) \qquad r<r' \\ F_m(r)&=BK_m(\kappa r) \qquad r>r' \end{aligned} \tag{20.17}$$

Because of the delta function in (20.16), the first derivative of $F_m(r)$ has a discontinuity at $r=r'$ given by

$$\left(\frac{dF_m}{dr}\right)_{r'+\varepsilon}-\left(\frac{dF_m}{dr}\right)_{r'-\varepsilon}=-\frac{4\pi}{r'} \tag{20.18}$$

This condition, coupled with the requirement that $F_m(r)$ itself be continuous at $r=r'$, gives

$$\begin{aligned} BK_m'(\kappa r')-AI_m'(\kappa r')&=\frac{-4\pi}{\kappa r'} \\ BK_m(\kappa r')-AI_m(\kappa r')&=0 \end{aligned} \tag{20.19}$$

Reference to the Wronskian relation (15.60) shows that

$$A=4\pi K_m(\kappa r') \qquad B=4\pi I_m(\kappa r') \tag{20.20}$$

The expansion for $1/R$ is therefore

$$\frac{1}{R}=\frac{2}{\pi}\sum_{m=-\infty}^{+\infty} e^{im(\theta-\theta')}\int_0^\infty K_m(\kappa r)I_m(\kappa r')$$
$$\times \cos\kappa(z-z')\,d\kappa \qquad r>r' \tag{20.21}$$

By using (15.65), we can write the expansion in the form

$$\frac{1}{R}=\frac{2}{\pi}\int_0^\infty K_0(\kappa r)I_0(\kappa r')\cos\kappa(z-z')\,d\kappa$$
$$+\frac{4}{\pi}\sum_{m=1}^{\infty}\cos m(\theta-\theta')\int_0^\infty K_m(\kappa r)I_m(\kappa r')$$
$$\times\cos\kappa(z-z')\,d\kappa \qquad r>r' \tag{20.22}$$

The imaginary part of the expansion in (20.21) must sum to zero because $1/R$ is real. As usual, the form of the expansion valid for $r<r'$ is found by simply exchanging r and r'. In cylindrical coordinates, R is given by

$$R=\sqrt{r^2+r'^2-2rr'\cos(\theta-\theta')+(z-z')^2} \tag{20.23}$$

Other significant results can be derived from (20.22). In the following development, keep in mind the limiting forms (15.58) of $K_m(x)$ and $I_m(x)$. For example, if we set $r'=z'=0$ in (20.22), we get the integral

$$\frac{1}{\sqrt{r^2+z^2}}=\frac{2}{\pi}\int_0^\infty K_0(\kappa r)\cos\kappa z\,d\kappa \tag{20.24}$$

The analogous result for $J_0(\kappa r)$ is given by (14.46). If r^2 in (20.24) is replaced by $r^2+r'^2-2rr'\cos(\theta-\theta')$ and the resulting expression equated to (10.22), then

$$\frac{2}{\pi}\int_0^\infty K_0\left(\kappa\sqrt{r^2+r'^2-2rr'\cos(\theta-\theta')}\right)\cos\kappa z\,d\kappa$$
$$=\frac{2}{\pi}\int_0^\infty K_0(\kappa r)I_0(\kappa r')\cos\kappa z\,d\kappa$$
$$+\frac{4}{\pi}\int_0^\infty\sum_{m=1}^{\infty}\cos m(\theta-\theta')K_m(\kappa r)$$
$$\times I_m(\kappa r')\cos\kappa z\,d\kappa \tag{20.25}$$

where we have set $z'=0$. If the Fourier cosine transforms of two functions are equal, then the functions themselves are equal. Therefore,

$$K_0\left(\kappa\sqrt{r^2+r'^2-2rr'\cos(\theta-\theta')}\right)$$
$$=K_0(\kappa r)I_0(\kappa r')+2\sum_{m=1}^{\infty}\cos m(\theta-\theta')$$
$$\times K_m(\kappa r)I_m(\kappa r') \qquad r>r' \tag{20.26}$$

This is an addition theorem for the modified Bessel functions and should be compared to (14.38). Finally, by taking the limit $\kappa\to 0$ in (20.26), we arrive at

$$\ln\left[\frac{r}{\sqrt{r^2+r'^2-2rr'\cos(\theta-\theta')}}\right]$$
$$=\sum_{m=1}^{\infty}\frac{1}{m}\left(\frac{r'}{r}\right)^m\cos m(\theta-\theta') \qquad r>r' \tag{20.27}$$

which is actually a Green's function for Poisson's equation in two-dimensional plane polar coordinates.

Expansion of Green's Function in Terms of Eigenfunctions

Many of the partial differential equations which we have studied are of the form

$$\nabla^2\psi+[\lambda-q(\mathbf{r})]\psi=0 \tag{20.28}$$

For example, reference to (7.3) of Chapter 7 shows that (20.28) is of the form of the time-independent Schrödinger equation if $q(\mathbf{r})$ is identified with the potential energy and the constant λ is the total energy. If $q(\mathbf{r})=0$, then (20.28) becomes a Helmholtz equation. Let us suppose that (20.28) is defined in some region Σ. We know that an eigenvalue problem is defined if the solutions are required to obey certain kinds of boundary conditions on the surface σ which bounds Σ. For example, the vanishing of the wave function at infinity gave us the energy eigenfunctions and eigenvalues for the hydrogen atom. To be specific, let us suppose that we have a set of eigenfunctions and eigenvalues which obey

$$\nabla^2u_n(\mathbf{r})+[\lambda_n-q(\mathbf{r})]u_n(\mathbf{r})=0 \tag{20.29}$$

subject to the boundary conditions $u_n(\mathbf{r})=0$ on σ. We will define a Green's function by requiring it to satisfy

$$\nabla^2 G(\mathbf{r},\mathbf{r}') + [\lambda - q(\mathbf{r})]\,G(\mathbf{r},\mathbf{r}') = -4\pi\delta(\mathbf{R}) \tag{20.30}$$

and the same boundary conditions as the eigenfunctions, namely, $G(\mathbf{r},\mathbf{r}') = 0$ on σ. It is specifically required that the constant λ in (20.30) *not* be one of the eigenvalues of (20.29). Both the Green's function and the delta function can be expanded in terms of the eigenfunctions:

$$G(\mathbf{r},\mathbf{r}') = \sum_{n=1}^{\infty} a_n(\mathbf{r}')u_n(\mathbf{r})$$
$$\delta(\mathbf{R}) = \sum_{n=1}^{\infty} u_n^*(\mathbf{r}')u_n(\mathbf{r}) \tag{20.31}$$

By combining (20.29), (20.30), and (20.31), we find that

$$\sum_{n=1}^{\infty} a_n(r')(\lambda - \lambda_n)u_n(\mathbf{r}) = -4\pi \sum_{n=1}^{\infty} u_n^*(\mathbf{r}')u_n(\mathbf{r}) \tag{20.32}$$

Since the eigenfunctions are complete, we have

$$a_n(\mathbf{r}')(\lambda - \lambda_n) = -4\pi u_n^*(\mathbf{r}') \tag{20.33}$$

The Green's function is therefore

$$G(\mathbf{r},\mathbf{r}') = 4\pi \sum_{n=1}^{\infty} \frac{u_n^*(\mathbf{r}')u_n(\mathbf{r})}{\lambda_n - \lambda} \tag{20.34}$$

You can see the reason at this point for requiring $\lambda \neq \lambda_n$. This gives us a way of calculating a Green's function directly as an expansion in terms of eigenfunctions and is an alternative to the methods we have used up to this point. As an example, suppose we want the Green's function for Poisson's equation with Dirichlet boundary conditions inside the rectangular box defined by the three coordinate planes and the planes $x = a$, $y = b$, and $z = c$. If we set $q(\mathbf{r}) = 0$ and $\lambda_n = \kappa_{lmn}^2$, then the eigenfunctions are the solutions of Helmholtz's equation which vanish on the boundaries:

$$u_{lmn}(\mathbf{r}) = \sqrt{\frac{8}{abc}}\sin\left(\frac{l\pi x}{a}\right)\sin\left(\frac{m\pi y}{b}\right)\sin\left(\frac{n\pi z}{c}\right) \tag{20.35}$$

The eigenvalues are

$$\kappa_{lmn}^2 = \pi^2\left(\frac{l^2}{a^2} + \frac{m^2}{b^2} + \frac{n^2}{c^2}\right) \tag{20.36}$$

To get the Green's function for Poisson's equation, we set $\lambda = 0$ in (20.34) with the result

$$G(\mathbf{r},\mathbf{r}') = 4\pi \sum_{l=1}^{\infty}\sum_{m=1}^{\infty}\sum_{n=1}^{\infty} \frac{u_{lmn}^*(\mathbf{r}')u_{lmn}(\mathbf{r})}{\kappa_{lmn}^2} \tag{20.37}$$

The main difference between (20.37) and the Green's functions which we have previously found for Poisson's equation is that all three coordinates are put on an equal footing and are represented by eigenfunction expansions. For example, the method used to obtain both (20.4) and (20.21) required that two of the three delta functions be represented in terms of eigenfunctions. Because of the nature of the equation relating the separation constants, the differential equation for the remaining variable then gives solutions which are exponential in nature and are not eigenfunctions. To put it another way, (20.37) gives an expansion for the Green's function of Poisson's equation in terms of the eigenfunctions of the Helmholtz equation.

If the boundaries are at infinity, the eigenvalue spectrum becomes continuous and the sum in (20.34) is replaced by an integral. To illustrate the point, the Helmholtz equation when expressed in rectangular coordinates yields the plane wave eigenfunctions

$$u(\boldsymbol{\kappa},\mathbf{r}) = \left(\frac{1}{2\pi}\right)^{3/2} e^{i\boldsymbol{\kappa}\cdot\mathbf{r}} \tag{20.38}$$

They are *delta function normalized*, meaning that

$$\int_{\Sigma_r} u^*(\boldsymbol{\kappa}',\mathbf{r})u(\boldsymbol{\kappa},\mathbf{r})\,d\Sigma_r = \delta(\boldsymbol{\kappa} - \boldsymbol{\kappa}') \tag{20.39}$$

See equations (5.36) and (5.37) in Chapter 8. The Green's function for the Helmholtz equation is

$$G(\mathbf{r},\mathbf{r}') = 4\pi \int_{\Sigma_\kappa} \frac{u^*(\boldsymbol{\kappa},\mathbf{r}')u(\boldsymbol{\kappa},\mathbf{r})}{\kappa^2 - \lambda}\,d\Sigma_\kappa \tag{20.40}$$

where the integral extends over all of κ space. If $\lambda = 0$, we get the Green's function for the Poisson equation:

$$\frac{1}{R} = \frac{1}{2\pi^2}\int_{\Sigma_\kappa} \frac{e^{i\boldsymbol{\kappa}\cdot(\mathbf{r}-\mathbf{r}')}}{\kappa^2}\,d\Sigma_\kappa \tag{20.41}$$

This is the Fourier integral representation of the function $1/R$.

PROBLEMS

20.1. Obtain (20.4) by direct expansion of (19.21).

20.2. Show that the Green's function suitable for solving the Dirichlet problem in the rectangular region bounded by the six planes $x=0$, $y=0$, $z=0$, $x=a$, $y=b$, and $z=c$ is

$$G(\mathbf{r},\mathbf{r}')=4\pi\sum_{l=1}^{\infty}\sum_{m=1}^{\infty}\frac{\sinh\kappa_{lm}(c-z)\sinh\kappa_{lm}z'}{\kappa_{lm}\sinh(\kappa_{lm}c)}u_{lm}(x,y)u_{lm}(x',y') \qquad z>z' \tag{20.42}$$

where

$$\begin{aligned}u_{lm}(x,y)&=\frac{2}{\sqrt{ab}}\sin\left(\frac{l\pi x}{a}\right)\sin\left(\frac{m\pi y}{b}\right)\\ \kappa_{lm}^2&=\left(\frac{l\pi}{a}\right)^2+\left(\frac{m\pi}{b}\right)^2\end{aligned} \tag{20.43}$$

The choice to represent the x and y dependence of the Green's function in terms of eigenfunctions and to single out the z dependence for special treatment is arbitrary. Other versions of the Green's function can be found by choosing the separation constants in different ways. The expansion valid for $z<z'$ is found by interchanging z and z'. By comparing (20.37) and (20.42), obtain the expansion

$$\frac{\sinh\kappa(c-z)\sinh\kappa z'}{\kappa\sinh\kappa c}=\frac{2}{c}\sum_{n=1}^{\infty}\frac{\sin\left(\frac{n\pi z}{c}\right)\sin\left(\frac{n\pi z'}{c}\right)}{\kappa^2+\left(\frac{\pi n}{c}\right)^2} \qquad z>z' \tag{20.44}$$

20.3. Two equal positive point charges are inside a grounded conducting sphere of radius a. The center of the sphere is at the origin and the charges are on the z axis at $z=\pm b$. Show that the potential inside the sphere is given by

$$\psi(r,\theta)=\frac{q}{2\pi\varepsilon_0}\sum_{n=0}^{\infty}\left[\frac{b^{2n}}{r^{2n+1}}-\frac{(rb)^{2n}}{a^{4n+1}}\right]P_{2n}(\cos\theta) \qquad b<r<a \tag{20.45}$$

The expression for the potential which is valid for $0<r<b$ is found by exchanging b and r. First solve the problem by using the Green's function (20.4), then find an exact solution in closed form by the method of images and expand it.

20.4. In potential problems where the range of values of θ is $0\le\theta\le 2\pi$, the requirement that the potential be a single-valued function of θ automatically leads to Fourier eigenfunctions for the θ dependence. In developing the expansion for $1/R$ in cylindrical coordinates by direct solution of (20.11), either $\delta(r-r')/r$ or $\delta(z-z')$ can be represented in terms of eigenfunctions. Represent $\delta(r-r')/r$ as a Hankel transform (18.11) and show that an expansion for $1/R$ which is an alternative to (20.21) is

$$\frac{1}{R}=\sum_{m=-\infty}^{+\infty}\int_0^{\infty}e^{-\kappa|z-z'|}J_m(\kappa r)J_m(\kappa r')\,d\kappa\, e^{im(\theta-\theta')} \tag{20.46}$$

Derive (14.46) and (14.38) from this result. Remember that if two functions have the same Laplace transform, they are essentially the same function.

20.5. Verify (20.41) by direct evaluation of the integral.

Hint: Write $\boldsymbol{\kappa}\cdot(\mathbf{r}-\mathbf{r}')=\kappa R\cos\gamma$ and express the volume element in κ space as $d\Sigma_\kappa=2\pi\kappa^2\sin\gamma\,d\kappa\,d\gamma$.

20.6. Show that the Green's function for solving the Dirichlet problem in the region between two infinite parallel planes at $z=0$ and $z=s$ has the two possible forms

$$G(\mathbf{r},\mathbf{r}') = \frac{4}{s}\sum_{n=1}^{\infty}\sum_{m=-\infty}^{+\infty} I_m\left(\frac{n\pi r'}{s}\right)K_m\left(\frac{n\pi r}{s}\right)\sin\left(\frac{n\pi z}{s}\right)\sin\left(\frac{n\pi z'}{s}\right)e^{im(\theta-\theta')} \qquad r>r' \tag{20.47}$$

$$G(\mathbf{r},\mathbf{r}') = 2\sum_{m=-\infty}^{+\infty}\int_0^{\infty}\frac{\sinh\kappa(s-z)\sinh\kappa z'}{\sinh\kappa s}J_m(\kappa r)J_m(\kappa r')\,d\kappa\, e^{im(\theta-\theta')} \qquad z>z' \tag{20.48}$$

If a point charge q is placed on the z axis at $z=a$ between two infinite parallel grounded conducting planes at $z=0$ and $z=s$, show that the resulting potential is

$$\psi(r,z) = \frac{q}{\pi\varepsilon_0 s}\sum_{n=1}^{\infty}K_0\left(\frac{n\pi r}{s}\right)\sin\left(\frac{n\pi z}{s}\right)\sin\left(\frac{n\pi a}{s}\right) \tag{20.49}$$

20.7. A uniform flexible string is stretched between two fixed supports at $x=0$ and $x=s$. If the string is driven at the point $x=a$ by a harmonic driving force of frequency ω, equation (2.9) of Chapter 7 shows that the equation of motion is

$$\frac{\partial^2 y}{\partial x^2} - \beta\frac{\partial y}{\partial t} - \frac{1}{v^2}\frac{\partial^2 y}{\partial t^2} = -N\delta(x-a)e^{i\omega t} \tag{20.50}$$

where β, v, and N are constants. If there is no friction ($\beta=0$), show that the steady-state motion is given by

$$y(x,t) = \frac{2Nv^2}{s}\sum_{n=1}^{\infty}\frac{\sin\left(\frac{n\pi a}{s}\right)\sin\left(\frac{n\pi x}{s}\right)e^{i\omega t}}{\omega_n^2-\omega^2} \qquad \omega_n = \frac{n\pi v}{s} \tag{20.51}$$

Note that the solution becomes undefined if the driving frequency matches one of the natural frequencies of the system. This is a typical resonance phenomenon. The singularity is removed if friction is included in the solution. Equation (20.51) represents only the steady-state motion; any so[illegible] of the homogeneous equation can also be present.

20.8. In order to solve the problem of the driven string including friction as given by (20.50), we first separate out the time dependence by writing

$$y(x,t) = F(x)e^{i\omega t} \tag{20.52}$$

The function $F(x)$ then obeys a nonhomogeneous equation of the Helmholtz type given by

$$F''(x) + \left(\frac{\omega^2}{v^2} - i\omega\beta\right)F(x) = -N\delta(x-a) \tag{20.53}$$

The eigenfunctions of the freely vibrating string without a driving force or friction are given by

$$u_n(x) = \sqrt{\frac{2}{s}}\sin\left(\frac{n\pi x}{s}\right) \tag{20.54}$$

Even though these functions are not solutions of (20.53), they are nevertheless complete and conform to the boundary conditions of the problem. It is therefore possible to expand $F(x)$ in terms of them as

$$F(x) = \sum_{n=1}^{\infty}A_n u_n(x) \tag{20.55}$$

where the coefficients A_n are now expected to be complex. Show that the final form of the steady-state solution is

$$y(x,t) = \frac{2Nv^2}{s} \sum_{n=1}^{\infty} \frac{\sin\left(\frac{n\pi a}{s}\right)\sin\left(\frac{n\pi x}{s}\right)e^{i(\omega t-\phi_n)}}{\sqrt{\left(\omega_n^2-\omega^2\right)^2+v^4\omega^2\beta^2}}$$

$$\tan\phi_n = \frac{v^2\omega\beta}{\omega_n^2-\omega^2} \tag{20.56}$$

20.9. A vibrating string fixed at its two ends has a continuously distributed harmonic driving force so that its equation of motion is

$$\frac{\partial^2 y}{\partial x^2} - \beta\frac{\partial y}{\partial t} - \frac{1}{v^2}\frac{\partial^2 y}{\partial t^2} = -f(x)e^{i\omega t} \tag{20.57}$$

Show that it is possible to express the steady state solution as

$$y(x,t) = F(x)e^{i\omega t} \qquad F(x) = \int_0^s G(x,x')f(x')\,dx' \tag{20.58}$$

where the Green's function obeys

$$\frac{\partial^2 G}{\partial x^2} + \left[\frac{\omega^2}{v^2} - i\beta\omega\right]G = -\delta(x-x')$$

$$G(0,x') = G(s,x') = 0 \tag{20.59}$$

20.10. The completion of the problem of the solution of the nonhomogeneous wave equation as exemplified by (20.57) requires that we find the transient solution. This is the general solution of the homogeneous equation

$$\frac{\partial^2 y}{\partial x^2} - \beta\frac{\partial y}{\partial t} - \frac{1}{v^2}\frac{\partial^2 y}{\partial t^2} = 0 \tag{20.60}$$

Show that the transient solution is

$$y(x,t) = \sum_{n=1}^{\infty}\sqrt{\frac{2}{s}}\sin\left(\frac{n\pi x}{s}\right)e^{-\alpha t}\left(A_n e^{r_n t} + B_n e^{-r_n t}\right) \tag{20.61}$$

where

$$\alpha = \frac{\beta v^2}{2} \qquad r_n = \sqrt{\alpha^2 - \omega_n^2} \tag{20.62}$$

Note that a given term can be *overdamped* (r_n real), *critically damped* ($r_n = 0$), or *underdamped* (r_n imaginary). The complete solution of (20.57) is the superposition of the steady-state solution and the transient solution.

20.11. A flexible string is stretched between $x = 0$ and $x = s$. The end at $x = 0$ is fixed and the end at $x = s$ moves according to

$$y(s,t) = h\cos\omega t \tag{20.63}$$

The initial shape of the string is given by

$$y(x,0) = \frac{hx}{s} \tag{20.64}$$

There is no friction. By making the change of dependent variable

$$y(x,t)=u(x,t)+\frac{hx}{s}\cos\omega t \tag{20.65}$$

show that

$$\frac{\partial^2 u}{\partial x^2}-\frac{1}{v^2}\frac{\partial^2 u}{\partial t^2}=-\frac{xh\omega^2}{sv^2}\cos\omega t \tag{20.66}$$

Show that the final solution is

$$y(x,t)=\frac{xh}{s}\cos\omega t+\sum_{n=1}^{\infty}\frac{2h\omega^2(-1)^{n+1}}{n\pi(\omega_n^2-\omega^2)}\sin\left(\frac{n\pi x}{s}\right)(\cos\omega t-\cos\omega_n t) \tag{20.67}$$

One approach to the solution is to start by substituting the expansions

$$u(x,t)=\sum_{n=1}^{\infty}A_n(t)u_n(x) \qquad x=\sum_{n=1}^{\infty}c_n u_n(x) \tag{20.68}$$

directly into the differential equation (20.66) and then determine $A_n(t)$. You can also find the steady-state solution by the Green's function method according to the scheme

$$\begin{aligned}
&\frac{\partial^2 u}{\partial x^2}-\frac{1}{v^2}\frac{\partial^2 u}{\partial t^2}=-f(x)\cos\omega t \qquad u(x,t)=F(x)\cos\omega t\\
&F''(x)+\kappa^2F(x)=-f(x) \qquad \kappa=\frac{\omega}{v}\\
&F(x)=\int_0^s G(x,x')f(x')\,dx' \qquad \frac{d^2G}{dx^2}+\kappa^2G=-\delta(x-x')\\
&G(x,x')=\sum_{n=1}^{\infty}\frac{u_n(x')u_n(x)}{\kappa_n^2-\kappa^2} \qquad \kappa_n=\frac{n\pi}{s}
\end{aligned} \tag{20.69}$$

You must then add in the general solution of the homogeneous wave equation and determine the constants to fit all initial conditions. Show also that the Green's function has the closed form

$$G(x,x')=\frac{1}{\kappa\sin\kappa s}\begin{cases}\sin\kappa x\sin\kappa(s-x') & x<x'\\ \sin\kappa x'\sin\kappa(s-x) & x>x'\end{cases} \tag{20.70}$$

20.12. If $\psi(\mathbf{r})$, $p(\mathbf{r})$, and $\phi(\mathbf{r})$ are any three functions, show that

$$\psi(\nabla\cdot p\nabla\phi)-\phi(\nabla\cdot p\nabla\psi)=\nabla\cdot[p(\psi\nabla\phi-\phi\nabla\psi)] \tag{20.71}$$

Prove that

$$\int_\Sigma[\psi(\nabla\cdot p\nabla\phi)-\phi(\nabla\cdot p\nabla\psi)]\,d\Sigma=\int_\sigma p\left(\psi\frac{\partial\phi}{\partial n}-\phi\frac{\partial\psi}{\partial n}\right)d\sigma \tag{20.72}$$

This is a generalized form of Green's second identity.

20.13. The partial differential equation

$$\nabla\cdot p\nabla u-qu+\lambda\rho u=0 \tag{20.73}$$

is a generalization of the one-dimensional Sturm–Liouville differential equation given in section 10 of Chapter 7. Assume that (20.73) is defined in some region Σ and that $p(\mathbf{r})$, $q(\mathbf{r})$, and $\rho(\mathbf{r})$ are real functions. The function $\rho(\mathbf{r})$ is a weight function and is required to be positive throughout the region Σ. Let $u_1(\mathbf{r})$ and $u_2(\mathbf{r})$ be any two solutions which conform to the boundary conditions

$$\int_\sigma p\left(u_1^* \frac{\partial u_2}{\partial n} - u_2 \frac{\partial u_1^*}{\partial n} \right) d\sigma = 0 \tag{20.74}$$

Show that an eigenvalue problem is defined and that the eigenvalues are real. Show that the eigenfunctions corresponding to distinct eigenvalues are orthogonal, that is,

$$\int_\Sigma \rho u_1^* u_2 \, d\Sigma = 0 \tag{20.75}$$

If a Green's function is required to satisfy

$$\nabla \cdot p \nabla G - qG + \lambda \rho G = -4\pi\delta(\mathbf{R}) \tag{20.76}$$

show that it can be expressed as

$$G(\mathbf{r},\mathbf{r}') = 4\pi \sum_{n=1}^{\infty} \frac{u_n^*(\mathbf{r}') u_n(\mathbf{r})}{\lambda_n - \lambda} \tag{20.77}$$

provided that it satisfies the same boundary conditions as the eigenfunctions.

$$\frac{d}{dx}\left(\frac{\partial F}{\partial y'}\right) - \frac{\partial F}{\partial y} = 0$$

10 The Calculus of Variations

1. THE BRACHISTOCHRONE PROBLEM

A child's slide is to be constructed between the points (1) and (2) as illustrated in Fig. 1.1. There is a uniform gravitational field in the y direction. We seek the solution of the problem of finding the shape of the slide which will minimize the time of travel between the two end points. This is the content of the *brachistochrone problem*. The technique used to solve it and other similar problems is called *the calculus of variations*. The name comes from the Greek *brachisto* (shortest) and *chronos* (time). To put it more succinctly, a particle is acted on by a uniform gravitational field and is constrained to move without friction along a path between fixed end points, and we are required to find the shape of the path which will minimize the time of travel. If ds represents a line element measured along the path and if u is the speed of the particle at any point, then the time is given by

$$t_{12} = \int_1^2 \frac{ds}{u} \tag{1.1}$$

The line element can be expressed as

$$ds = \sqrt{dx^2 + dy^2} = \sqrt{1 + y'^2}\, dx \tag{1.2}$$

where $y' = dy/dx$. The speed of the particle at any point is found from conservation of energy to be

$$u = \sqrt{u_1^2 + 2g(y - y_1)} \tag{1.3}$$

where u_1 is the initial speed at point (1). The problem is therefore to choose the function $y(x)$ such that the integral

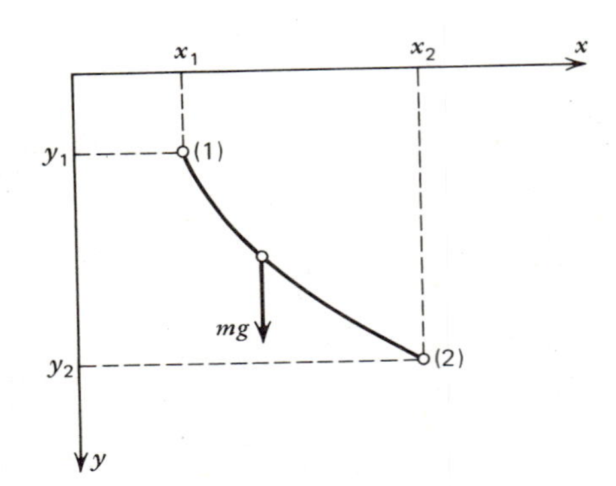

Fig. 1.1

$$t_{12} = \int_1^2 \frac{\sqrt{1 + y'^2}}{\sqrt{u_1^2 + 2g(y - y_1)}}\, dx \tag{1.4}$$

is a minimum. You are familiar with the problem of finding the points where a given function has maxima or minima. By contrast, minimizing the integral given by (1.4) involves finding an entire unknown function. In the following discussion, we show how this problem can be reduced to that of finding the minimum of a function which depends on a single parameter.

To formulate the problem in somewhat more general terms, we are given a function $F(x, y, y')$ and are required to determine $y(x)$ such that the integral

$$I = \int_1^2 F(x, y, y')\, dx \tag{1.5}$$

is an extremum. We use the term *extremum* because there is the possibility that the procedure we are about to develop will yield a maximum rather than a minimum. It is also possible that relative maxima or minima or even an inflexion point will be found. Before proceeding further, we will prove a basic theorem.

Theorem 1. If $G(x)$ is assumed to be piecewise continuous and if

$$\int_1^2 G(x)\eta(x)\,dx = 0 \tag{1.6}$$

for all functions $\eta(x)$ which are restricted only by the condition $\eta(x_1) = \eta(x_2) = 0$, then $G(x) = 0$.

Proof: Figure 1.2 shows two possible piecewise continuous functions $G(x)$ defined over $x_1 \le x \le x_2$. If $G(x) \ne 0$, then in each case a function $\eta(x)$ can be chosen so as to make the integrand of (1.6) positive over its entire range. The integrand, and hence the integral, can be made zero only if $G(x) = 0$. By contrast, if $G(x)$ is not assumed to be piecewise continuous, then it could be nonzero at a finite number of isolated points and still have (1.6) be true.

To proceed with the main problem, let $y(x)$ be the function which extremizes (1.5). Other functions which represent curves passing through the same end points can be expressed as

$$Y(x, \varepsilon) = y(x) + \varepsilon\eta(x) \tag{1.7}$$

provided that $\eta(x)$ satisfies the condition $\eta(x_1) = \eta(x_2) = 0$. The function $Y(x, \varepsilon)$ is called a *comparison function*. If the parameter ε is zero, then the comparison function becomes identical to the extremizing function. The integral (1.5) is replaced by

$$I(\varepsilon) + \int_1^2 F(x, Y, Y')\,dx \tag{1.8}$$

where $Y' = \partial Y/\partial x$. The function $y(x)$ is to be determined by the condition

$$\left(\frac{dI}{d\varepsilon}\right)_{\varepsilon=0} = 0 \tag{1.9}$$

which reduces the problem to that of finding the possible extrema of an ordinary function of a single parameter. By differentiating (1.8) with respect to ε and then setting $\varepsilon = 0$, we get

$$0 = \int_1^2 \left[\frac{\partial F}{\partial y}\eta(x) + \frac{\partial F}{\partial y'}\eta'(x)\right] dx \tag{1.10}$$

If the second term is integrated by parts and the condition that $\eta(x)$ vanish at the end points is used, the result is

$$0 = \int_1^2 \left[\frac{\partial F}{\partial y} - \frac{d}{dx}\left(\frac{\partial F}{\partial y'}\right)\right]\eta(x)\,dx \tag{1.11}$$

Since the only condition on $\eta(x)$ is that it be zero at the end points, theorem 1 gives

$$\frac{d}{dx}\left(\frac{\partial F}{\partial y'}\right) - \frac{\partial F}{\partial y} = 0 \tag{1.12}$$

The first thing that should strike you is that (1.12) is in exactly the same form as Lagrange's equations as we developed them in section 13 of Chapter 4. In fact, (1.12) is called the *Euler–Lagrange equation*. The possibility suggests itself that the basic equations of classical mechanics can be derived from variational principles, and this is in fact true.

Given the "Lagrangian" $F(x, y, y')$, the Euler–Lagrange equation provides a second-order differential equation the solution of which is the required function $y(x)$. We will next show that the quantity

$$h = \frac{\partial F}{\partial y'}y' - F \tag{1.13}$$

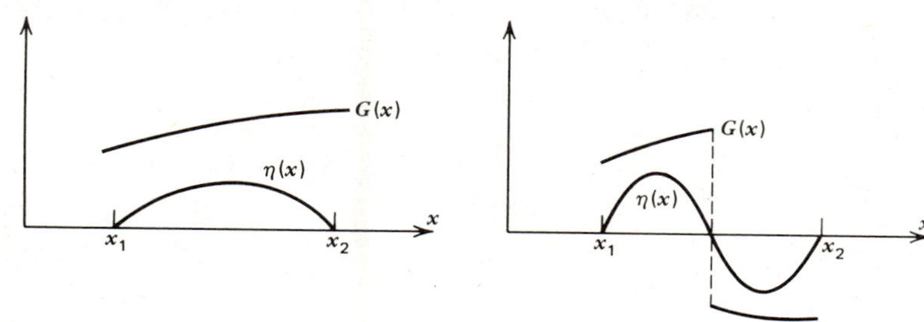

Fig. 1.2

is a constant is those cases where F has no explicit dependence on x, meaning that $F = F(y, y')$ and $\partial F/\partial x = 0$. To prove this, we differentiate (1.13) totally with respect to x:

$$\frac{dh}{dx} = \frac{d}{dx}\left(\frac{\partial F}{\partial y'}\right) y' + \frac{\partial F}{\partial y'} y'' - \frac{\partial F}{\partial x} - \frac{\partial F}{\partial y} y' - \frac{\partial F}{\partial y'} y'' \tag{1.14}$$

The Euler–Lagrange equation (1.12) reduces (1.14) to

$$\frac{dh}{dx} = -\frac{\partial F}{\partial x} \tag{1.15}$$

Thus, h is a constant if $\partial F/\partial x = 0$. This fact is useful in the solution of some problems.

For the brachistochrone problem, the integrand of (1.5) is

$$F(y, y') = \sqrt{\frac{1 + y'^2}{u_1^2 + 2gy}} \tag{1.16}$$

This is an example where h as defined by (1.13) is a constant. We find that

$$h = -\frac{1}{\sqrt{1 + y'^2}\sqrt{u_1^2 + 2gy}} \tag{1.17}$$

This is a not too pleasant-looking first-order differential equation for determining $y(x)$. Since obviously curves connecting the end points of Fig. 1.1 can be constructed which will make the time of transit arbitrarily long, the solution of (1.17) will represent a minimizing curve. The method of solution is to use a clever substitution. We define a parameter ϕ by means of

$$y' = -\tan\frac{\phi}{2} \qquad \sqrt{1 + y'^2} = \frac{1}{\cos\frac{\phi}{2}} \tag{1.18}$$

If (1.17) is solved for y, the result is

$$y = \frac{1}{2gh^2}\cos^2\frac{\phi}{2} - \frac{u_1^2}{2g} \tag{1.19}$$

By differentiating with respect to ϕ and combining the result with (1.18), we get

$$dy = -\frac{1}{2gh^2}\cos\frac{\phi}{2}\sin\frac{\phi}{2}\,d\phi = -\tan\frac{\phi}{2}\,dx \tag{1.20}$$

Now solve for dx and integrate to get

$$dx = \frac{1}{2gh^2}\cos^2\frac{\phi}{2}\,d\phi = \frac{1}{4gh^2}(1 + \cos\phi)\,d\phi$$

$$x = \frac{1}{4gh^2}(\phi + \sin\phi) + a \tag{1.21}$$

where a is a constant of integration. If we set $b = 1/(4gh^2)$, then (1.19) and (1.21) can be written as

$$x = b(\phi + \sin\phi) + a \qquad y = b(1 + \cos\phi) - \frac{u_1^2}{2g} \tag{1.22}$$

which represents the solution of the problem in parametric form. The curve generated by (1.22) is a *cycloid*. It is shown plotted in Fig. 1.3 for the case where $a = 0$ and $u_1 = 0$. In an actual application, we want a portion of a cycloid which connects the given end points (x_1, y_1) and (x_2, y_2). This gives a total of four conditions which permit the determination of the constants a and b and the initial and final values of the parameter ϕ. The cycloid can be generated by a point on the rim of a circle of radius b which rolls along the x axis. The parameter ϕ is the angle through which the circle turns.

We close this section by recapitulating the basic procedure and at the same time introducing a more compact notation. The derivative of (1.8) with respect

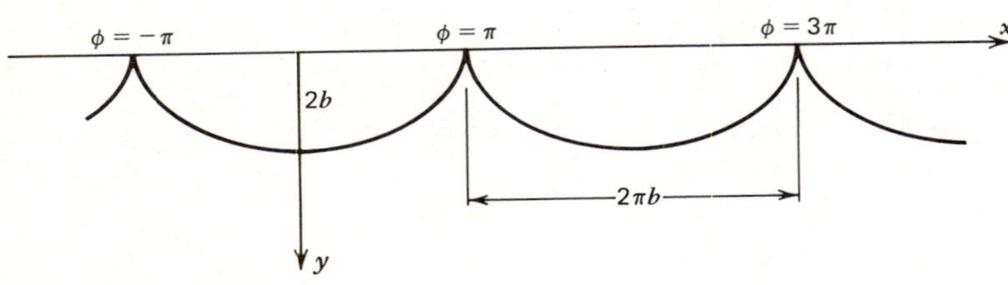

Fig. 1.3

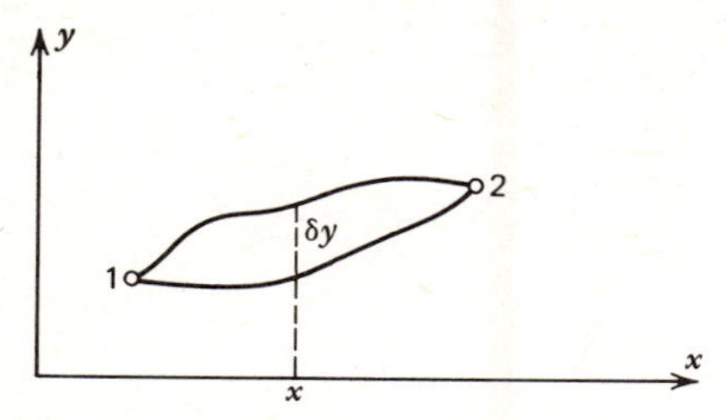

Fig. 1.4

to ε calculated at $\varepsilon = 0$ can be written

$$\left(\frac{dI}{d\varepsilon}\right)_{\varepsilon=0} d\varepsilon = \int_1^2 \left[\frac{\partial F}{\partial y}\eta(x)\, d\varepsilon + \frac{\partial F}{\partial y'}\eta'(x)\, d\varepsilon\right] dx \tag{1.23}$$

We define the *symbol of variation* by

$$\delta y = \eta(x)\, d\varepsilon \qquad \delta y' = \eta'(x)\, d\varepsilon \tag{1.24}$$

In using this notation, we visualize that a value of δy is calculated at each x, thus generating the difference between two possible comparison functions as shown in Fig. 1.4. To put it another way, δy is a function of x and represents at each point the difference between two functions. The graphs of these two functions both pass through the same end points. Consequently, $\delta y_1 = \delta y_2 = 0$. From (1.24), it is evident that

$$\delta y' = \frac{d}{dx}\delta y \tag{1.25}$$

We therefore write (1.23) as

$$\delta I = \int_1^2 \left[\frac{\partial F}{\partial y}\delta y + \frac{\partial F}{\partial y'}\delta y'\right] dx = 0 \tag{1.26}$$

Because of (1.25), we can integrate the second term in (1.26) by parts, giving

$$\int_1^2 \left[\frac{\partial F}{\partial y} - \frac{d}{dx}\left(\frac{\partial F}{\partial y'}\right)\right] \delta y\, dx = 0 \tag{1.27}$$

We now argue that since (1.27) holds for all possible variations δy, the integrand must vanish, giving again the Euler–Lagrange equation (1.12). In (1.26), we have written

$$\delta I = \left(\frac{dI}{d\varepsilon}\right)_{\varepsilon=0} d\varepsilon \tag{1.28}$$

This is to be thought of as a variation in the value of the integral brought about by varying the function $y(x)$. The condition $\delta I = 0$ means that the value of the integral is *stationary* for slight variations of $y(x)$ from the extremizing function. The variational notation is convenient because it is more compact; but if a question as to the validity of a procedure arises, it is best to fall back on the comparison function notation as given by (1.7).

PROBLEMS

1.1. A particle starts from rest at the origin and is constrained to move on the cycloid given by (1.22). Show that the equation of the path can be put into the alternative form

$$x = b(\theta - \sin\theta) \qquad y = b(1 - \cos\theta) \tag{1.29}$$

If the path ends at $x = y$, show that $0 \le \theta \le 2.412$ approximately. If $x = 10$ m, then $b = 5.729$ m.

1.2. A particle starts from rest at the origin and follows the cycloidal path given by (1.29). If the path ends at (x, y) and we define $R = y/x$, show that the final value of θ is determined by the root of

$$f(\theta) = 1 - \cos\theta - R(\theta - \sin\theta) \tag{1.30}$$

which lies in the interval $0 < \theta < 2\pi$. Assume that both x and y are positive. Show that $f(\theta)$ has precisely one root over this range. The significance of this result is that (1.29) gives one, and only one, cycloid passing through the points $(0, 0)$ and (x, y).

Hint: Show that $f'(\theta) = 0$ at 0, 2π, and at one point between these values. This means that $f(\theta)$ can cross the θ axis only once between 0 and 2π. Figure 1.5 shows $f(\theta)$ for several values

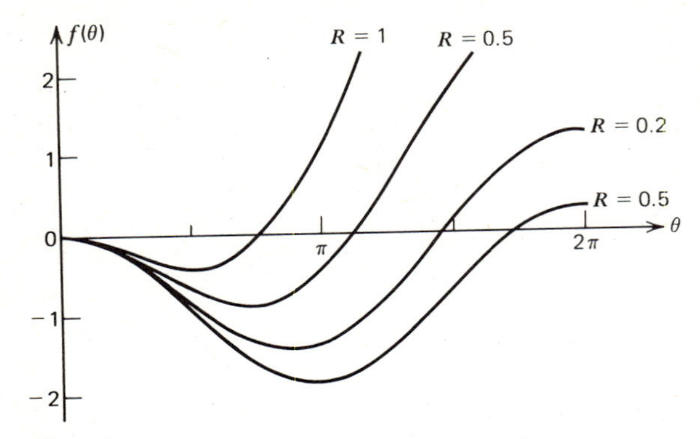

Fig. 1.5

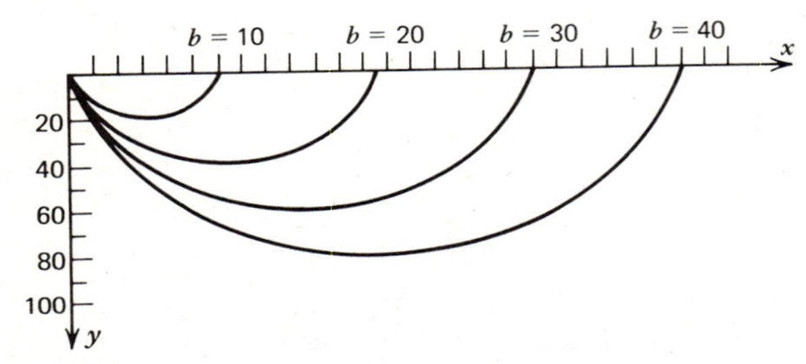

Fig. 1.6

of the ratio R. The point is further illustrated by Fig. 1.6, which shows plots of (1.29) for various values of the parameter b.

1.3. As shown in Fig. 1.7, the x, y plane separates two optical media with indices of refraction n_1 and n_2. A light ray starts from the z axis at $z = a$ and follows a straight-line path to the point (x, y) on the x, y plane. It then continues on in a straight line to the point $y = b$, $z = -c$ on the y, z plane. Derive the laws of refraction by assuming that the actual path followed by the light ray between the fixed end points $(0, 0, a)$ and $(0, b, -c)$ extremizes the time of transit. This is called *Fermat's principle*. Recall that the index of refraction of a medium is defined by $n = c/u$, where c is the speed of light in vacuum and u is its speed in the medium.

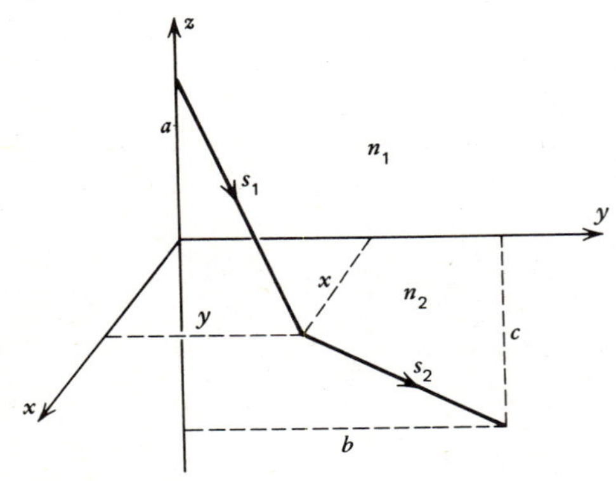

Fig. 1.7

1.4. A particle is acted on by a uniform gravitational field in the y direction and is constrained to move on the portion of the cycloid of Fig. 1.3 between the cusps at $\phi = \pm\pi$. Find the Lagrangian for the motion of the particle in terms of the parameter ϕ as the generalized coordinate. Find the second-order differential equation satisfied by $\phi(t)$. Make the change of variable $\theta = \theta(\phi)$ and determine the function $\theta(\phi)$ by the condition that the $\dot{\theta}$ term be absent from the differential equation for $\theta(t)$. Show that the solution is

$$\theta(t) = 2\sin\frac{\phi}{2} = A\cos\left(\sqrt{\frac{g}{4b}}\,t - \alpha\right) \tag{1.31}$$

where A and α are constants. As long as $A < 2$ so that the motion does not go beyond the cusps, the oscillations are strictly isochronous, meaning that the frequency does not depend on the amplitude. This is the solution of the *tautochrone problem*. (Greek: *tauto*, the same; *chronos*, time.)

1.5. A plane pendulum consists of a point particle on the end of a massless string of length $4b$. The string is tied to the origin which is also the location of the cusp of two cycloidal cylinders. The equation of the cylinders is given by (1.29). As the particle swings back and forth, the string must wrap around the cylinders as shown in Fig. 1.8. Show that the path of the particle is also a cycloid and that the pendulum is therefore isochronous.

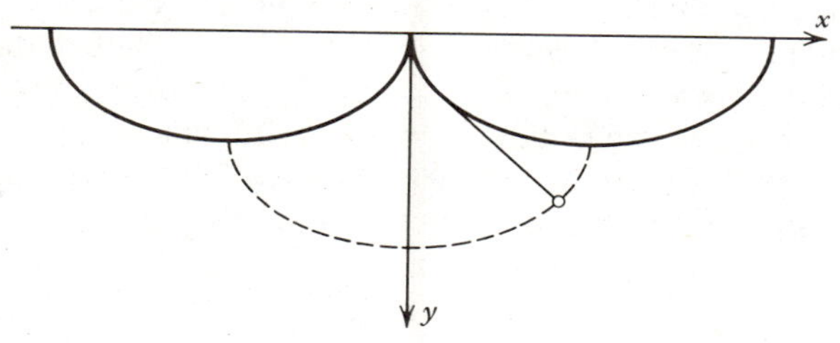

Fig. 1.8

1.6. Consider the class of curves $y(x)$ which pass through the points (x_1, y_1) and (x_2, y_2) in the x, y plane each having the same slope at these points. Out of this class of admissible curves, it is desired to pick the one which extremizes the integral

$$I = \int_1^2 F(x, y, y', y'')\,dx \tag{1.32}$$

Show that

$$\frac{d^2}{dx^2}\left(\frac{\partial F}{\partial y''}\right) - \frac{d}{dx}\left(\frac{\partial F}{\partial y'}\right) + \frac{\partial F}{\partial y} = 0 \tag{1.33}$$

Use the variational notation in the derivation.

1.7. A tunnel is to be bored through the Earth that connects two points on its surface. Assume that a vehicle moves through the tunnel without friction and has no source of power except that provided by the gravitational field of the Earth. Show that the path of least time is a hypocycloid. Show also that the time required for a particle to move through a straight-line tunnel is the same for all such tunnels connecting any two points on the Earth's surface. Use this fact to construct a proof that the hypocycloid in fact gives a path of minimum time and not some other kind of extremum. A hypocycloid is a plane curve that is generated by a circle rolling on the inside of the circumference of a larger circle.

2. SURFACES OF REVOLUTION OF MINIMUM AREA

Figure 2.1 shows two circular coaxial wire loops of radii a and b. The loops are separated by a distance s. The x axis is the common axis of the two loops and the loop of radius a lies in the y, z plane. When a soap film is suspended between the loops, it forms a surface of revolution, provided that the effects of gravity are neglected. The film adjusts itself in such a way that the surface area is a minimum. The area of the surface of revolution is

$$\sigma = 2\pi \int_0^s y\sqrt{dx^2 + dy^2} = 2\pi \int_0^s y\sqrt{1 + y'^2}\, dx \qquad (2.1)$$

In order to find the actual shape of the surface, we must determine the function $y(x)$ which minimizes the integral (2.1).

Since the integrand of (2.1) does not contain x explicitly, a first-order differential equation can be found from (1.13). It is convenient here to replace h by $-h$. The result is

$$y = h\sqrt{1 + y'^2} \qquad (2.2)$$

If we set $y' = \sinh u$, we find that (2.2) is satisfied, provided that

$$\frac{du}{dx} = \frac{1}{h} \qquad u = \frac{x - c}{h} \qquad (2.3)$$

where c is a constant of integration. The solution is therefore

$$y = h \cosh\left(\frac{x - c}{h}\right) \qquad (2.4)$$

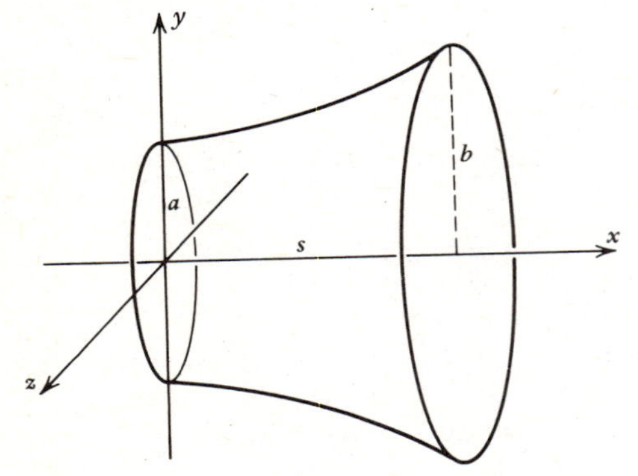

Fig. 2.1

and is called a *catenary*. The surface of revolution obtained from it is a *catenoid*. By setting $h = 0$ in equation (2.2), we get the special solution $y = 0$, which means that there is no surface at all between the two loops. In the example of the soap film, $y = 0$ corresponds to two separate plane surface films in the form of discs, one bounded by each wire loop.

The condition that the catenary (2.4) pass through the point $(0, a)$ gives

$$a = h \cosh \frac{c}{h} \qquad \frac{c}{h} = \cosh^{-1} \frac{a}{h}$$

$$= \ln\left(\frac{a \pm \sqrt{a^2 - h^2}}{h}\right) \qquad (2.5)$$

In the discussion of inverse functions in section 16 of Chapter 5, it was the general practice to use the branches of the functions which correspond to the principal values. However, there is no reason to exclude negative values of c from the solution (2.4), and we therefore retain both roots of (2.5). When the constant c is determined by (2.5), the catenaries given by (2.4) form a one-parameter family of curves passing through the point $(0, a)$, as shown in Fig. 2.2. In

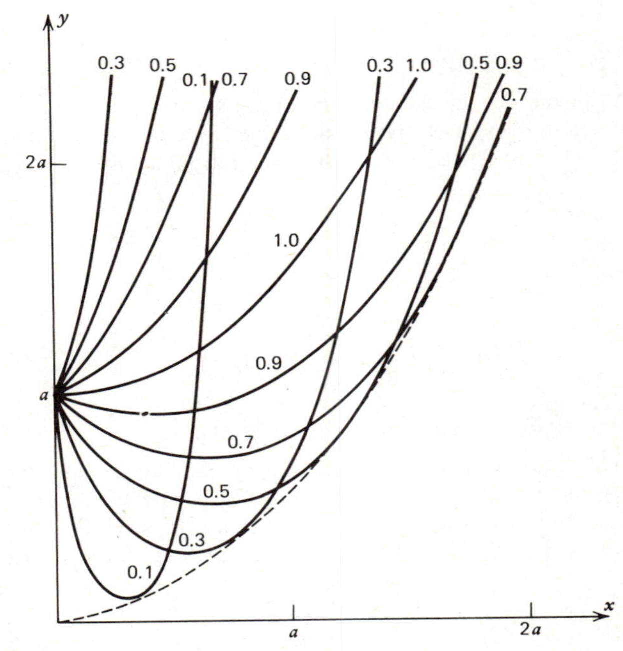

Fig. 2.2

the following graphical analysis, it is convenient to set $a = 1$. All quantities with dimensions of length are then measured in units of a. Equation (2.5) shows that the range of h values is $0 < h \leq 1$. Since $y > 0$, equation (2.2) shows that h is positive. Each curve in Fig. 2.2 results from a different choice of h. The curves below the $h = 1$ curve are for $c > 0$ [plus sign in (2.5)] and those above the $h = 1$ curve are for $c < 0$ [minus sign in (2.5)]. The catenaries cover only a portion of the x, y plane and have an *envelope* indicated by the dashed curve. If the loop separation s and radius b of the second loop are such that the point (s, b) lies outside the envelope, there is no catenary joining $(0, a)$ and (s, b) and hence no minimum surface of revolution. The only possible solution is then given by $h = 0$ and $y = 0$. Inside the envelope, there are two catenaries joining $(0, a)$ and (s, b). Later on in this section, it will be shown that the catenary which touches the envelope at a point between the loops is ruled out as a possible solution.

In general, a one-parameter family of curves can be expressed as

$$\phi(x, y, h) = 0 \tag{2.6}$$

Consider two curves given by h and $h + dh$ which touch the envelope of the family at the neighboring points 1 and 2, as shown in Fig. 2.3. By calculating the total differential of (2.6), we get

$$\frac{\partial \phi}{\partial x} dx + \frac{\partial \phi}{\partial y} dy + \frac{\partial \phi}{\partial h} dh = 0 \tag{2.7}$$

If $d\mathbf{s}$ is the infinitesimal displacement vector directed along the envelope between the points of contact 1 and 2, then (2.7) can be expressed as

$$\nabla \phi \cdot d\mathbf{s} + \frac{\partial \phi}{\partial h} dh = 0 \tag{2.8}$$

Since $\nabla \phi$ is perpendicular to the surfaces $\phi = 0$, it is at right angles to $d\mathbf{s}$. This gives the *envelope condition*

$$\frac{\partial \phi}{\partial h} = 0 \tag{2.9}$$

By expressing (2.6) as $\phi = y - f(x, y) = 0$, the envelope condition can be expressed as

$$\frac{\partial f}{\partial h} = 0 \tag{2.10}$$

Given that (s, b) lies inside the envelope, the two values of h which give the two catenaries connecting $(0, a)$ and (s, b) are found by setting $y = b$ in (2.4) and then eliminating c by means of (2.5):

$$b = h \cosh\left[\frac{s}{h} - \ln\left(\frac{a \pm \sqrt{a^2 - h^2}}{h}\right)\right] \tag{2.11}$$

The only way that this equation can be solved for h is by numerical methods. Figure 2.4 shows a graph of b vs. h for $a = 1.0$ and $s = 1.0$. The lower part of the graph for $0 < h \leq 1.0$ results if the plus sign is used in (2.11), and the upper part results if the negative sign is used. The minimum in the graph occurs at $h = 0.4044$ and $b = 0.587$. For $b < 0.587$, there is no value of h and no solution (other than $y = 0$), meaning that the point (s, b) is outside the envelope of Fig. 2.2. If $b = 0.587$, then (s, b) is exactly on the envelope and

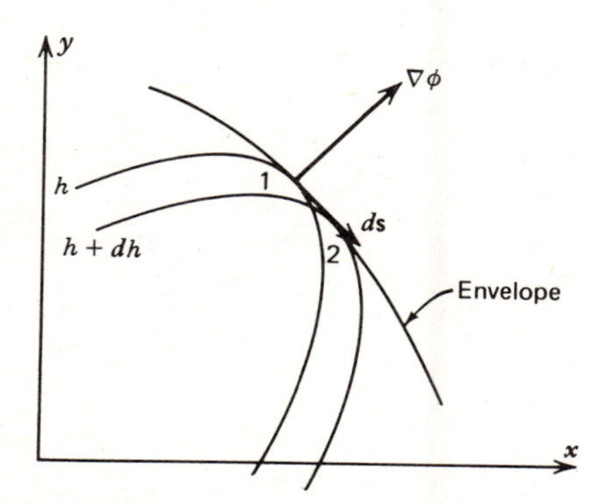

Fig. 2.3

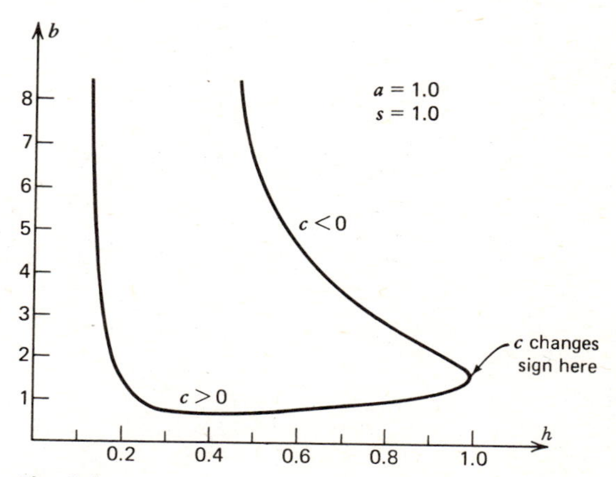

Fig. 2.4

there is a single catenary joining the points $(0, a)$ and (s, b). As will be shown later this is not a minimizing curve. If $b > 0.587$, two values of h are possible. The value of h on the lower curve in the range $0 < h \leq 0.4044$ gives a catenary which touches the envelope between $x = 0$ and $x = s$ and is not a minimizing curve. The other value of h gives a solution which may be either a relative or an absolute minimum.

The actual area of the minimizing catenoid as calculated from (2.1) is

$$\sigma_1 = 2\pi \int_0^s y\sqrt{1 + y'^2}\, dx = 2\pi h \int_0^s \cosh^2\left(\frac{x - c}{h}\right) dx$$
$$= \pi h^2 \left[\frac{1}{2}\sinh\frac{2(s - c)}{h} + \frac{1}{2}\sinh\frac{2c}{h} + \frac{s}{h}\right] \tag{2.12}$$

For the solution $h = 0$ and $y = 0$, the area of the soap film is the sum of the areas of the discs:

$$\sigma_2 = \pi(a^2 + b^2) \tag{2.13}$$

If a flat surface is allowed to bulge slightly, it can only increase its area. It follows that σ_2 provides a possible minimum area. Figure 2.5 shows graphs of σ_1 and σ_2 as functions of b for $a = 1.0$ and $s = 1.0$, which are the same values of these parameters used to construct Fig. 2.4. The graph of σ_1 begins at the values of h and b corresponding to the minimum of the graph in Fig. 2.4. The significant feature of Fig. 2.5 is that it shows that $\sigma_1 > \sigma_2$ for $0.589 < b < 0.900$, meaning that σ_1 is relative minimum and σ_2 is an absolute minimum. For $b > 0.900$, the situation is reversed.

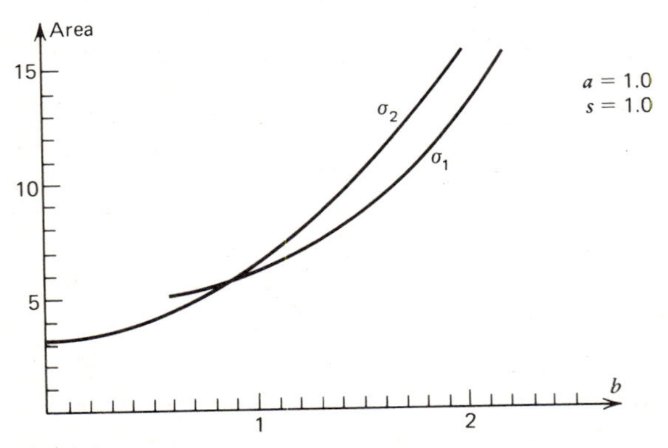

Fig. 2.5

In order that $y(x)$ be a function which extremizes the integral (1.5), it is necessary but not sufficient that it be a solution of the Euler–Lagrange equation (1.12). In the remainder of this section, conditions which are both necessary and sufficient for a minimum will be developed. The results can be easily modified if the function in question corresponds to a maximum rather than a minimum. We begin by computing the *second variation* of the integral (1.8):

$$\frac{dI}{d\varepsilon} = \int_1^2 \left(\frac{\partial F}{\partial Y}\eta + \frac{\partial F}{\partial Y'}\eta'\right) dx$$

$$\frac{d^2 I}{d\varepsilon^2} = I''(\varepsilon)$$
$$= \int_1^2 \left(\frac{\partial^2 F}{\partial Y^2}\eta^2 + 2\frac{\partial^2 F}{\partial Y\,\partial Y'}\eta\eta' + \frac{\partial^2 F}{\partial Y'^2}\eta'^2\right) dx \tag{2.14}$$

$$I''(0) = \int_1^2 \left[F_{yy}\eta^2 + F_{yy'}\frac{d}{dx}(\eta^2) + F_{y'y'}\eta'^2\right] dx$$

In the last line of (2.14), we have set $\varepsilon = 0$ and used the abbreviations

$$F_{yy} = \frac{\partial^2 F}{\partial y^2} \qquad F_{yy'} = \frac{\partial^2 F}{\partial y\,\partial y'} \qquad F_{y'y'} = \frac{\partial^2 F}{\partial y'^2} \tag{2.15}$$

The middle term of (2.14) can be integrated by parts to give

$$I''(0) = \int_1^2 (Q\eta^2 + P\eta'^2)\, dx$$

$$Q(x) = F_{yy} - \frac{d}{dx}F_{yy'} \qquad P(x) = F_{y'y'} \tag{2.16}$$

Remember that $\eta(x)$ can be any continuous and differentiable function which vanishes at the two end points. A necessary condition for $y(x)$ to be a minimizing function is $I''(0) \geq 0$ for all comparison functions in the neighborhood of $y(x)$, and a sufficient condition is $I''(0) > 0$ for all such functions. The comparison functions as given by (1.7) can all be made to lie in as small a neighborhood of $y(x)$ as we like by choosing ε sufficiently small.

Theorem 1. Legendre's Necessary Condition. A necessary condition for $I''(0) \geq 0$ is $P(x) \geq 0$ at each point of

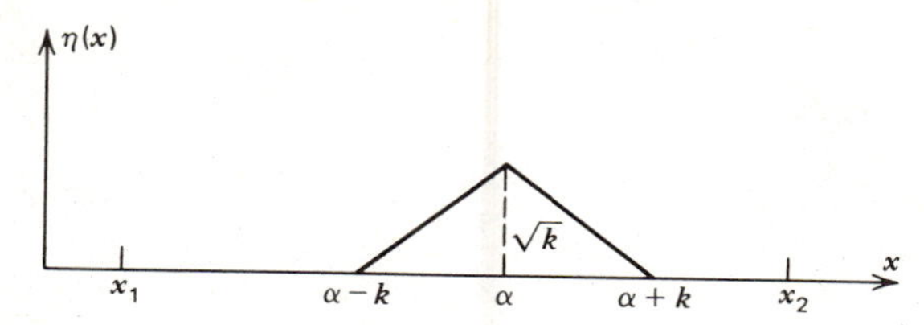

Fig. 2.6

$[x_1, x_2]$. (Recall that $[x_1, x_2]$ is an abbreviation of the closed interval $x_1 \le x \le x_2$.)

Proof: A possible choice for $\eta(x)$ is given by

$$\eta(x) = \frac{1}{\sqrt{k}}(k + x - \alpha) \qquad \alpha - k \le x \le \alpha$$

$$\eta(x) = \frac{1}{\sqrt{k}}(k - x + \alpha) \qquad \alpha \le x \le \alpha + k \tag{2.17}$$

$$\eta(x) = 0 \qquad x_1 \le x \le \alpha - k \quad \text{or} \quad \alpha + k \le x \le x_2$$

This function is shown in Fig. 2.6. Let M be the maximum value of $|Q(x)|$ in $[\alpha - k, \alpha + k]$. Then

$$\left|\int_1^2 Q\eta^2\, dx\right| \le M\int_{\alpha-k}^{\alpha+k} \eta^2\, dx = \frac{2}{3}Mk^2 \tag{2.18}$$

The contribution to $I''(0)$ from this term can be made arbitrarily small by choosing k small. For the remaining term in $I''(0)$, we find

$$\int_1^2 P\eta'^2\, dx = \frac{1}{k}\int_{\alpha-k}^{\alpha+k} P\, dx = 2\bar{P} \tag{2.19}$$

where $\bar{P}$ is the average value of $P(x)$ over $\alpha - k, \alpha + k$. If $P(x)$ is continuous at $x = \alpha$ and if $P(\alpha) < 0$, then (2.19) can be made negative if k is sufficiently small. It is therefore necessary that $P(\alpha) \ge 0$. Since α can be any point in $[x_1, x_2]$, the theorem is established. Note that if $P(\alpha) = 0$, it is also necessary that $P'(\alpha) = 0$. Otherwise, $P(x)$ would cross the axis at $x = \alpha$ and become negative.

It is of interest to consider the class of functions which obey the Euler–Lagrange equation associated with the integral (2.16). If $u(x)$ is such a function, then

$$\frac{d}{dx}\frac{\partial}{\partial u'}(Qu^2 + Pu'^2) - \frac{\partial}{\partial u}(Qu^2 + Pu'^2) = 0 \tag{2.20}$$

This can be simplified to

$$\frac{d}{dx}(Pu') - Qu = 0 \tag{2.21}$$

which is called the *Jacobi equation*. We emphasize here that $u(x)$ is not necessarily one of the functions $\eta(x)$ used to construct the comparison functions. Let α be a point which lies in the interval $x_1 < x \le x_2$. If a nonzero solution of the Jacobi equation exists which satisfies $u(x_1) = 0$ and $u(\alpha) = 0$, the point α is said to be *conjugate* to x_1. The concept of a conjugate point is of fundamental importance in the following theorems.

Theorem 2. A necessary condition for $I''(0) > 0$ is that $x_1 < x \le x_2$ contain no points conjugate to x_1.

Proof: Assume that $u(x)$ is a solution of (2.21) and that the point $x = \alpha$ is conjugate to x_1. The condition $I''(0) > 0$ must hold for all $\eta(x)$ that are continuous and differentiable and that satisfy $\eta(x_1) = \eta(x_2) = 0$. In particular, it must hold if $\eta(x)$ is given by

$$\eta(x) = u(x) \qquad x_1 \le x \le \alpha$$

$$\eta(x) = 0 \qquad \alpha \le x \le x_2 \tag{2.22}$$

We find that

$$I''(0) = \int_{x_1}^{\alpha} (Qu^2 + Pu'^2)\, dx \tag{2.23}$$

If the second term is integrated by parts, the result is

$$I''(0) = uu'P|_{x_1}^{\alpha} - \int_{x_1}^{\alpha}\left[u\frac{d}{dx}(Pu') - Qu^2\right] dx = 0 \tag{2.24}$$

The result $I''(0) = 0$ is obtained because u is a solution of (2.21) and vanishes at $x = x_1$ and $x = \alpha$. Since $I''(0) > 0$ is violated, there can be no points conjugate to x_1 in $x_1 < x \le x_2$.

Theorem 3. Sufficient conditions for $y(x)$ to be a minimizing function of the integral (1.5) are

1. $y(x)$ is a solution of the Euler–Lagrange equation (1.12).
2. There are no points in $x_1 < x \le x_2$ conjugate to x_1.
3. $P(x) > 0$ at all points of $[x_1, x_2]$.

Proof: Let $w(x)$ be a continuous and differentiable function defined at each point of $[x_1, x_2]$ which is required to obey the differential equation

$$P(Q + w') = w^2 \tag{2.25}$$

This is a special case of what is known as a *Riccati*

differential equation. Note that

$$Q\eta^2 + P\eta'^2 + \frac{d}{dx}(w\eta^2) = P\eta'^2 + 2w\eta\eta' + (Q + w')\eta^2$$

$$= P\left(\eta' + \frac{w\eta}{P}\right)^2 \tag{2.26}$$

Then, since

$$\int_1^2 \frac{d}{dx}(w\eta^2)\, dx = 0 \tag{2.27}$$

we find that

$$I''(0) = \int_1^2 P\left(\eta' + \frac{w\eta}{P}\right)^2 dx \tag{2.28}$$

Since we are trying to show that $I''(0) > 0$ for all possible $\eta(x)$, we must rule out the possibility that a nonzero $\eta(x)$ exists such that

$$\eta' + \frac{w\eta}{P} = 0 \tag{2.29}$$

The solution of this differential equation is

$$\eta(x) = \eta(x_1)\exp\left[-\int_{x_1}^{x} \frac{w}{P}\,dx\right] \tag{2.30}$$

Since $\eta(x_1) = 0$, there is no solution of (2.29) other than $\eta(x) = 0$. The function $w(x)$ is to be found by solving (2.25). By making the change of variable

$$w = -\frac{u'P}{u} \tag{2.31}$$

we find that

$$\frac{d}{dx}(Pu') - Qu = 0 \tag{2.32}$$

which is just the Jacobi equation. Since there are no conjugate points in $x_1 < x \leq x_2$, (2.31) defines $w(x)$ at each point. Equation (2.28) shows that $P(x) > 0$ is certainly a sufficient condition for $I''(0) > 0$, and the theorem is established.

In the literature on the subject, it is generally stated that $P(x) > 0$ at all points of $[x_1, x_2]$ is a necessary condition for $y(x)$ to be a minimizing function. It then follows that the conditions of theorem 3 are both necessary and sufficient for the existence of a minimum. In theorem 1, we established the necessity of the weaker condition, $P(x) \geq 0$. We will not pursue this question further here.

We will now explore the relevance of theorem 3 to the soap film problem discussed at the beginning of this section. The one-parameter family of catenaries $y(x, h)$ as given by (2.4) and all passing through the point $(0, a)$ are solutions of the Euler–Lagrange equation

$$\frac{d}{dx}F_{y'} - F_y = 0 \tag{2.33}$$

By differentiating partially with respect to h, we find

$$\frac{d}{dx}\left(F_{yy'}\frac{\partial y}{\partial h} + F_{y'y'}\frac{\partial y'}{\partial h}\right) - F_{yy}\frac{\partial y}{\partial h} - F_{yy'}\frac{\partial y'}{\partial h} = 0 \tag{2.34}$$

Now differentiate the first term with respect to x but leave the second term alone:

$$\frac{d}{dx}\left(F_{y'y'}\frac{\partial y'}{\partial h}\right) + \left(\frac{d}{dx}F_{yy'}\right)\frac{\partial y}{\partial h} + F_{yy'}\frac{\partial y'}{\partial h} - F_{yy}\frac{\partial y}{\partial h} - F_{yy'}\frac{\partial y'}{\partial h} = 0 \tag{2.35}$$

Cancel out the common term and rewrite to get

$$\frac{d}{dx}\left(F_{y'y'}\frac{\partial y'}{\partial h}\right) - \left(F_{yy} - \frac{d}{dx}F_{yy'}\right)\frac{\partial y}{\partial h} = 0 \tag{2.36}$$

Reference to (2.16) and (2.21) shows that (2.36) is the Jacobi equation with $\partial y/\partial h$ as its solution. Since all members of the family of catenaries pass through the point $(0, a)$, the function $y(0, h)$ is independent of h, meaning that $\partial y(0, h)/\partial h = 0$. We have shown in equation (2.10) that $\partial y/\partial h = 0$ at the point where the catenary touches the envelope. This point is therefore conjugate to $(0, a)$; and if it lies in the interval $0 < x \leq s$, the catenary is not a miminizing function. Note that the end point $x = s$ is included, meaning that if (s, b) lies exactly on the envelope, theorem 3 excludes the catenary passing through it as a minimizing function. By calculating $P(x)$ from the integrand of (2.1), we find

$$P(x) = F_{y'y'} = \frac{y}{(1 + y'^2)^{3/2}} = \frac{h}{\cosh^2\left(\dfrac{x - c}{h}\right)} > 0 \tag{2.37}$$

This proves the assertion made earlier that it is the catenary which passes through $(0, a)$ and (s, b) with no point of contact with the envelope between these points that is a minimizing function.

We add here a postscript on terminology. An integral such as (1.5) is frequently written $I[y]$ and referred to as a *functional*, meaning that it is an integral the value of which is a function of the function $y(x)$.

PROBLEMS

2.1. Solve $x = \cosh y$ for y and show that

$$y = \cosh^{-1} x = \ln\left(x \pm \sqrt{x^2 - 1}\right) \tag{2.38}$$

2.2. Consider the solution of the brachistochrone problem for the case $u_1 = 0$ as given by (1.29). Show that

$$F_{y'y'} = \frac{\sin^2 \dfrac{\theta}{2}}{2\sqrt{gb}} \tag{2.39}$$

and therefore that this function is positive everywhere on the cycloid except at the cusps, where it is zero. Note that y' becomes infinite at the cusps.

2.3. For the family of cycloids given by (1.29), show that

$$\frac{\partial y}{\partial b} = 2\left(\frac{\sin \dfrac{\theta}{2} - \dfrac{\theta}{2}\cos\dfrac{\theta}{2}}{\sin\dfrac{\theta}{2}}\right) \tag{2.40}$$

This function vanishes at $\theta = 0$ and at values of θ which are roots of

$$\tan\frac{\theta}{2} = \frac{\theta}{2} \tag{2.41}$$

If you will graph (1.29) for several closely spaced values of b, you will see that an envelope is generated corresponding to the root of (2.41) which lies in the range $2\pi < \theta < 4\pi$. This has been done in Fig. 2.7. There are no zeros of (2.40) in $0 < \theta \leq 2\pi$ and hence no conjugate points in this interval. We have already shown in problem 1.2 that there is a unique cycloid connecting the origin with any point below it. Problems 2.2 and 2.3 show conclusively that this cycloid provides a path of minimum time for a particle that starts from rest at the origin. *Hint:* You can't find an equation for y as a function of x and b because it is too hard to solve for θ in terms of x and b and then eliminate θ from the y equation. You have to think of θ as a function of x and b and then differentiate (1.29) partially with respect to b as it stands. Since x and b are treated as the independent variables, $\partial x/\partial b = 0$.

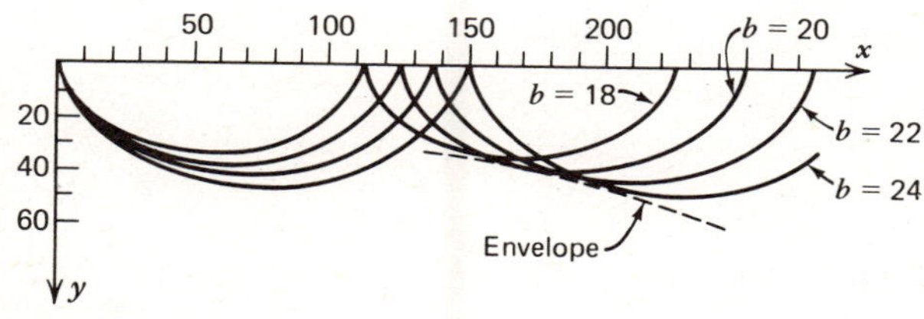

Fig. 2.7

2.4. By using the cycloids given by (1.29), show that in the case where a particle starts from rest at the origin and follows a path of least time,

$$Q(x) = \frac{1}{16\sqrt{g}\, b^{5/2} \sin^4 \dfrac{\theta}{2}} \tag{2.42}$$

where $Q(x)$ is defined by (2.16). Show that the Jacobi equation (2.32) is

$$\frac{d^2u}{d\theta^2} - \frac{u}{2\sin^2\frac{\theta}{2}} = 0 \tag{2.43}$$

Reference to section 1 of Chapter 9 reveals that (2.43) has regular singular points at $\theta = 0$ and $\theta = 2\pi$. Show that the limiting forms of the two linearly independent solutions in the vicinity of $\theta = 0$ are θ^2 and $1/\theta$. Since (2.43) has no singular points for $0 < \theta < 2\pi$, the function $u(x)$ is defined over this entire interval.

2.5. How does theorem 3 have to be modified if $y(x)$ is a maximizing function?

2.6. Suppose that the integral

$$I = \int_1^2 F(x, y')\, dx \tag{2.44}$$

is to be minimized. Show that a necessary and sufficient condition for the existence of a minimum is $P(x) = F_{y'y'} \geq 0$. Show moreover that $P(x)$ cannot vanish at all points of $[x_1, x_2]$.

Hint: If $P(x)$ vanishes for a continuous range of values of x contained as a subinterval of $[x_1, x_2]$, show that an $\eta(x)$ exists which will make the second variation of (2.44) zero.

2.7. In the soap film problem, show that $c = \frac{1}{2}s$ if $a = b$. Show that no catenary exists if the separation s between the loops is such that

$$s > \frac{2ra}{\cosh r} \tag{2.45}$$

where r is the positive root of

$$\cosh r - r\sinh r = 0 \tag{2.46}$$

2.8. If $f(x)$ is a given function, show that the curves $y(x)$ which pass through fixed end points and which minimize

$$I = \int_1^2 f^2 y'^2\, dx \tag{2.47}$$

are solutions of the differential equation

$$y' = \frac{A}{f^2} \tag{2.48}$$

where A is a constant. Reference to equation (11.24) in Chapter 7 shows that the Schwartz inequality for two real functions $f(x)$ and $g(x)$ is

$$\int_1^2 f^2\, dx \int_1^2 g^2\, dx \geq \left[\int_1^2 fg\, dx\right]^2 \tag{2.49}$$

Use it to show that the functions which satisfy (2.48) make I an absolute minimum and that this minimum is given by

$$I_{\min} = \frac{(y_2 - y_1)^2}{\int_1^2 \frac{1}{f^2}\, dx} \tag{2.50}$$

Assume that $f(x) > 0$ so that the integral exists.

3. VARIATION OF END POINTS

Suppose that the brachistochrone problem is modified by requiring that the path of least time be found between a fixed point (x_1, y_1) and a curve given by $H(x, y)=0$ as illustrated in Fig. 3.1. In other words, of all the functions $y(x)$ which can be constructed between the fixed point (x_1, y_1) and the given curve, we want the one along which the time of travel of a particle is minimized. As usual, the particle moves without friction and there is a uniform gravitational field in the y direction. If $y(x)$ is the actual minimizing curve, all other possible comparison curves can be represented by

$$Y(x)=y(x)+\varepsilon\eta(x) \tag{3.1}$$

where $\eta(x)$ is an arbitrary continuous and differentiable function which vanishes at (x_1, y_1) but *not* at the second end point. Let (X_2, Y_2) be the coordinates of the intersection of the comparison curve with the given curve $H(x, y)=0$. For a given $\eta(x)$, X_2 and Y_2 depend only on the parameter ε. In other words, by allowing ε to vary, the point at which $Y(x)$ intersects $H(x, y)=0$ changes. Differentiating $H(X_2, Y_2)=0$ with respect to ε yields

$$0=\frac{\partial H}{\partial X_2}\frac{dX_2}{d\varepsilon}+\frac{\partial H}{\partial Y_2}\frac{dY_2}{d\varepsilon} \tag{3.2}$$

At the point of intersection,

$$Y_2(\varepsilon)=y(X_2)+\varepsilon\eta(X_2) \tag{3.3}$$

Therefore,

$$\frac{dY_2}{d\varepsilon}=y_2'\frac{dX_2}{d\varepsilon}+\varepsilon\eta_2'\frac{dX_2}{d\varepsilon}+\eta_2 \tag{3.4}$$

where $y_2'=dy/dX_2$, $\eta_2'=d\eta/dX_2$, and $\eta_2=\eta_2(X_2)$. By eliminating $dY_2/d\varepsilon$ between (3.2) and (3.4) and

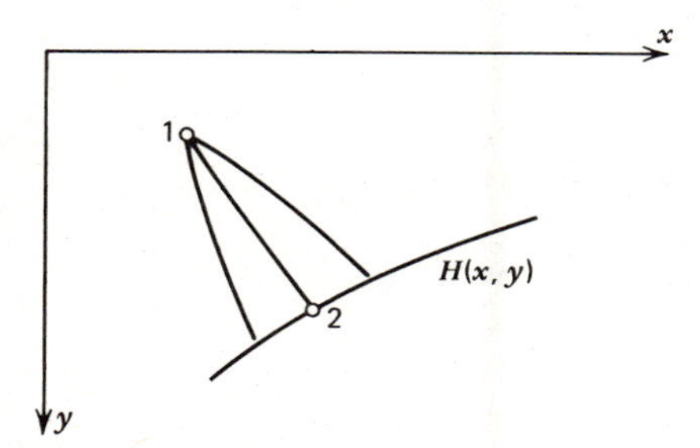

Fig. 3.1

then setting $\varepsilon=0$, we find that

$$\left(\frac{dX_2}{d\varepsilon}\right)_{\varepsilon=0}=-\frac{\dfrac{\partial H}{\partial y_2}\eta_2}{\dfrac{\partial H}{\partial x_2}+\dfrac{\partial H}{\partial y_2}y_2'} \tag{3.5}$$

The integral to be extremized is

$$I(\varepsilon)=\int_1^{X_2(\varepsilon)}F(x, Y, Y')\,dx \tag{3.6}$$

where, as indicated, the upper limit is a function of ε. Differentiation with respect to ε gives

$$\frac{dI}{d\varepsilon}=\frac{dX_2}{d\varepsilon}F_2+\int_1^2\left(\frac{\partial F}{\partial Y}\eta+\frac{\partial F}{\partial Y'}\eta'\right)dx \tag{3.7}$$

where $F_2=F(X_2, Y_2, Y_2')$. We next integrate the second term in the integral by parts and then set $\varepsilon=0$ to get

$$0=\left(\frac{dX_2}{d\varepsilon}\right)_{\varepsilon=0}F_2+\frac{\partial F}{\partial y_2'}\eta_2+\int_1^2\left(\frac{\partial F}{\partial y}-\frac{d}{dx}\frac{\partial F}{\partial y'}\right)\eta\,dx \tag{3.8}$$

Note that $\eta_1=0$ because all comparison functions pass through the initial point (x_1, y_1). If (3.5) and (3.8) are combined, the result is

$$0=\left[\frac{\partial F}{\partial y_2'}-\frac{\dfrac{\partial H}{\partial y_2}F_2}{\dfrac{\partial H}{\partial x_2}+\dfrac{\partial H}{\partial y_2}y_2'}\right]\eta_2+\int_1^2\left(\frac{\partial F}{\partial y}-\frac{d}{dx}\frac{\partial F}{\partial y'}\right)\eta\,dx \tag{3.9}$$

Included in the comparison functions which connect the initial and final points are those which satisfy the condition $\eta_2=0$ but which are otherwise arbitrary. Thus, (3.9) implies that the Euler–Lagrange equation

$$\frac{d}{dx}\frac{\partial F}{\partial y'}-\frac{\partial F}{\partial y}=0 \tag{3.10}$$

is still valid but it must be supplemented by the subsidiary condition

$$\frac{\partial F}{\partial y_2'}\frac{\partial H}{\partial x_2}-\left(F_2-\frac{\partial F}{\partial y_2'}y_2'\right)\frac{\partial H}{\partial y_2}=0 \tag{3.11}$$

which is known as the *end point condition* or, in some references, as the *transversality condition*. For the brachistochrone problem, the minimizing curve is still a cycloid because the Euler–Lagrange equation (3.10) holds. For the function $F(x, y, y')$, which is the integrand of (1.4), the end point condition reduces to

$$\frac{\partial H}{\partial x_2} dy_2 - \frac{\partial H}{\partial y_2} dx_2 = 0 \tag{3.12}$$

where dx_2 and dy_2 are components of displacement tangent to the minimizing curve $y(x)$. The interpretation of (3.12) is that the minimizing cycloid must intersect the given curve $H(x, y) = 0$ at right angles.

It is not hard to generalize the procedure to include variations of both end points. Suppose, as shown in Fig. 3.2, that an extremizing curve connecting two given curves $G(x, y) = 0$ and $H(x, y) = 0$ is desired. Since $\eta(x)$ does not vanish at either end point, we have, in addition to (3.5),

$$\left(\frac{dX_1}{d\varepsilon}\right)_{\varepsilon=0} = -\frac{\dfrac{\partial G}{\partial y_1}\eta_1}{\dfrac{\partial G}{\partial x_1} + \dfrac{\partial G}{\partial y_1} y_1'} \tag{3.13}$$

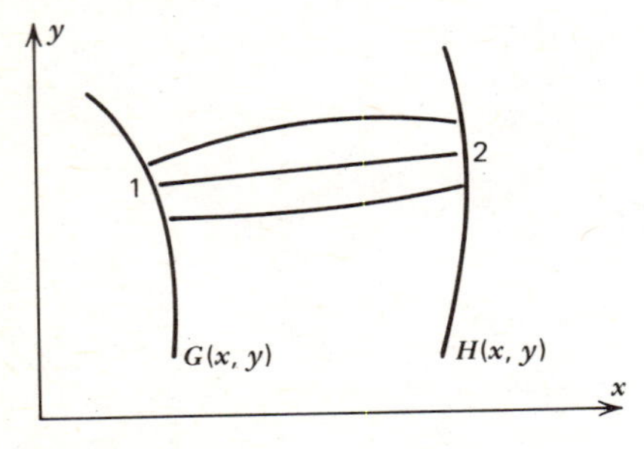

Fig. 3.2

The integral which is to be varied is

$$I(\varepsilon) = \int_{X_1(\varepsilon)}^{X_2(\varepsilon)} F(x, Y, Y')\, dx \tag{3.14}$$

This leads to

$$\frac{dI}{d\varepsilon} = \frac{dX_2}{d\varepsilon} F_2 - \frac{dX_1}{d\varepsilon} F_1 + \int_1^2 \left(\frac{\partial F}{\partial Y}\eta + \frac{\partial F}{\partial Y'}\eta' \right) dx$$

$$0 = \left(\frac{dX_2}{d\varepsilon}\right)_0 F_2 + \frac{\partial F}{\partial y_2'}\eta_2 - \left[\left(\frac{dX_1}{d\varepsilon}\right)_0 F_1 + \frac{\partial F}{\partial y_1'}\eta_1 \right]$$

$$+ \int_1^2 \left(\frac{\partial F}{\partial y} - \frac{d}{dx}\frac{\partial F}{\partial y'} \right) \eta\, dx \tag{3.15}$$

By combining (3.5), (3.13), and (3.15) and again using the arbitrariness of $\eta(x)$, we are led to the end point condition

$$\frac{\partial F}{\partial y_1'}\frac{\partial G}{\partial x_1} - \left(F_1 - \frac{\partial F}{\partial y_1'} y_1' \right) \frac{\partial G}{\partial y_1} = 0 \tag{3.16}$$

which applies to the initial point (x_1, y_1). Both equations (3.10) and (3.11) remain valid.

You may have noticed that the derivations of this section assume that the integrand $F(x, y, y')$ contains no explicit dependence on the coordinates of the end points. For the brachistochrone problem, this is true of the second end point but not the first. Thus, (3.16) is not valid for the brachistochrone problem. In place of (3.14), we should write

$$I(\varepsilon) = \int_{X_1(\varepsilon)}^{X_2(\varepsilon)} F(x, Y_1, Y, Y')\, dx \tag{3.17}$$

and then take into account the fact that Y_1 depends on ε.

PROBLEMS

3.1. A particle starts from rest at the origin and follows the cycloidal path given by (1.29). Show that the path of least time between the origin and the straight line $y = -mx + c$ occurs when

$$m = \tan\frac{A}{2} \qquad b = \frac{c}{(1 - \cos A) + (A - \sin A)\tan\dfrac{A}{2}} \tag{3.18}$$

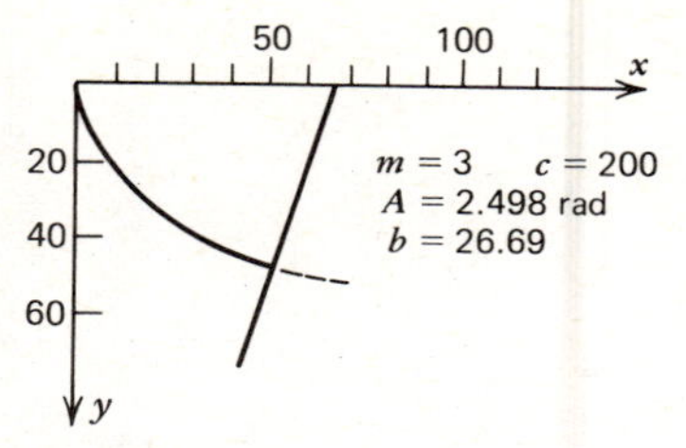

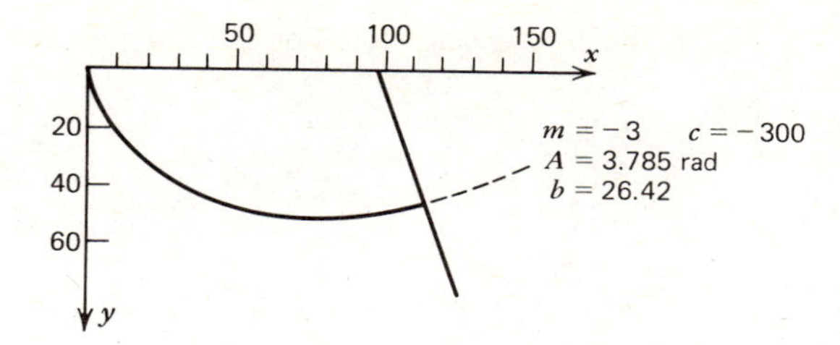

Fig. 3.3

where A is the value of θ where the cycloid and the straight line intersect. Two examples are shown in Fig. 3.3.

3.2. Suppose that the integrand of the functional to be minimized is of the form

$$F(x, y, y') = f(x, y)\sqrt{1+y'^2} \tag{3.19}$$

Show that end point condition (3.11) specializes to the form (3.12). As an example, two glass funnels are coated with soap solution and a soap film is caused to form between the rim of the smaller funnel and the walls of the larger one as shown in Fig. 3.4. The soap film forms a catenoid surface which intersects the larger funnel at right angles.

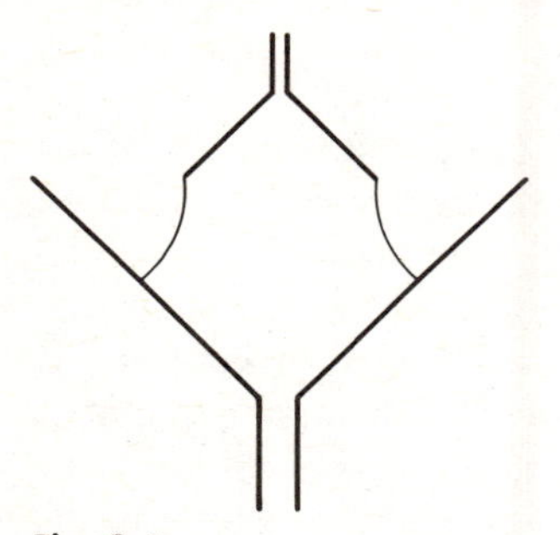

Fig. 3.4

3.3. If the initial point of an extremizing curve is allowed to vary along the curve $G(x, y) = 0$, show that

$$\left(\frac{dY_1}{d\varepsilon}\right)_0 = \frac{\dfrac{\partial G}{\partial x_1}\eta_1}{\dfrac{\partial G}{\partial x_1} + \dfrac{\partial G}{\partial y_1}y_1'} \tag{3.20}$$

3.4. For the brachistochrone problem, the integrand of the integral for the time is

$$F(y_1, y, y') = \sqrt{\frac{1+y'^2}{2g(y-y_1)+u_1^2}} \tag{3.21}$$

This shows explicitly its dependence on the initial value of y. In this problem, you are to find the conditions for minimum time from a given curve $G(x, y) = 0$ to a *fixed* point. As a first step, differentiate (3.17) with respect to ε for the case where $X_2(\varepsilon) = x_2 =$ constant; then, use equations (3.13) and (3.20). Show that the Euler–Lagrange equation is valid and that

$$\int_1^2 \frac{\partial F}{\partial y_1} dx = -\int_1^2 \frac{\partial F}{\partial y} dx = -\left(\frac{\partial F}{\partial y_2'} - \frac{\partial F}{\partial y_1'}\right) \tag{3.22}$$

$$\frac{\partial F}{\partial y_1'} y_1' - F_1 = \frac{\partial F}{\partial y_2'} y_2' - F_2 \tag{3.23}$$

$$\frac{\partial G}{\partial x_1}\frac{\partial F}{\partial y_2'} + \frac{\partial G}{\partial y_1}\left(\frac{\partial F}{\partial y_2'} y_2' - F_2\right) = 0 \tag{3.24}$$

$$\frac{\partial G}{\partial x_1} dy_2 - \frac{\partial G}{\partial y_1} dx_2 = 0 \tag{3.25}$$

Equation (3.25) shows that the tangent to the minimizing cycloid at its termination point is perpendicular to the curve $G(x, y) = 0$ at the point where the cycloid originates.

4. ISOPERIMETRIC PROBLEMS

In this section, problems in the calculus of variations are considered in which there is a subsidiary condition or constraint. The Lagrange multiplier method as introduced in section 6 of Chapter 6 will be used.

If a flexible chain is hung between fixed supports as shown in Fig. 4.1, its equilibrium shape can be determined from the condition that its potential energy is a minimum. If μ is the mass density of the chain per unit length, then the potential energy of an element of length ds is

$$dV = \mu g y\, ds = \mu g y\sqrt{1 + y'^2}\, dx \tag{4.1}$$

The total potential energy of the chain is

$$V = \mu g \int_1^2 y\sqrt{1 + y'^2}\, dx \tag{4.2}$$

This is the same functional which occurs in the surface of minimum area problem, as reference to

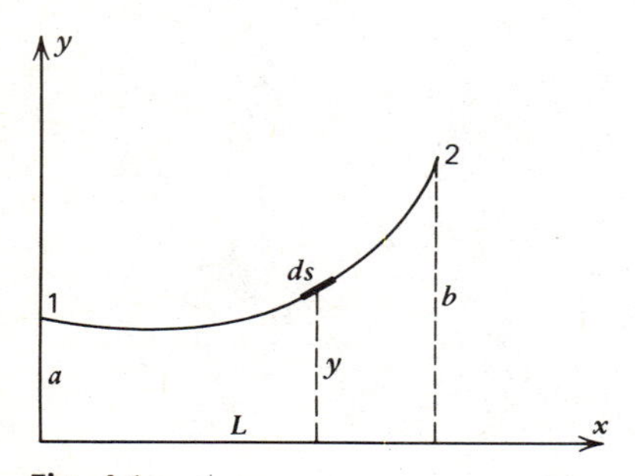

Fig. 4.1

equation (2.1) shows. The problem is however not the same because there is a constraint. The chain has a fixed length given by

$$s = \int_1^2 ds = \int_1^2 \sqrt{1 + y'^2}\, dx \tag{4.3}$$

This means that out of all the possible curves *of the same length* which connect the end points 1 and 2, we must pick the one which minimizes the potential energy. There is no such constraint in the soap film problem because the film can expand or contract. It is now not possible to express the comparison curves as $Y(x) = y(x) + \varepsilon\eta(x)$ because, for a given $\eta(x)$, the variation of the parameter ε produces a change in the length of the chain. The comparison curves are therefore made to depend on two parameters, ε_1 and ε_2:

$$Y(x) = y(x) + \varepsilon_1\eta(x) + \varepsilon_2\zeta(x) \tag{4.4}$$

where $\eta(x)$ and $\zeta(x)$ are two functions which are continuous and differentiable and vanish at the end points but are otherwise arbitrary. For a given choice of $\eta(x)$ and $\zeta(x)$, the two parameters ε_1 and ε_2 can be varied in such a way that the length of the chain is maintained. These two parameters are therefore not independent of one another because a variation of one of them requires a compensating variation in the other.

To state the problem in general terms, we are required to find the curve $y(x)$ passing through given end points which extremizes

$$I = \int_1^2 F(x, Y, Y')\, dx \tag{4.5}$$

subject to the constraint

$$J=\int_1^2 K(x, Y, Y')\,dx = \text{constant} \tag{4.6}$$

This is called an *isoperimetric problem*. The use of comparison curves of the form (4.4) means that I and J can be regarded as ordinary functions of the two variables ε_1 and ε_2. The total differentials of these functions are

$$dI=\frac{\partial I}{\partial \varepsilon_1}d\varepsilon_1+\frac{\partial I}{\partial \varepsilon_2}d\varepsilon_2 \tag{4.7}$$

$$0=\frac{\partial J}{\partial \varepsilon_1}d\varepsilon_1+\frac{\partial J}{\partial \varepsilon_2}d\varepsilon_2 \tag{4.8}$$

The condition for an extremum is $dI=0$ when $\varepsilon_1=\varepsilon_2=0$. If (4.8) is multiplied by the factor λ and then added to (4.7), the result is

$$0=\left(\frac{\partial I}{\partial \varepsilon_1}+\lambda\frac{\partial J}{\partial \varepsilon_1}\right)_0 d\varepsilon_1+\left(\frac{\partial I}{\partial \varepsilon_2}+\lambda\frac{\partial J}{\partial \varepsilon_2}\right)_0 d\varepsilon_2 \tag{4.9}$$

As explained in section 6 of Chapter 6, the Lagrange multiplier λ can be chosen in such a way that

$$\left(\frac{\partial I}{\partial \varepsilon_1}+\lambda\frac{\partial J}{\partial \varepsilon_1}\right)_0=0 \qquad \left(\frac{\partial I}{\partial \varepsilon_2}+\lambda\frac{\partial J}{\partial \varepsilon_2}\right)_0=0 \tag{4.10}$$

It is convenient to define

$$\bar{I}=I+\lambda J \qquad \bar{F}=F+\lambda K \tag{4.11}$$

By differentiating $\bar{I}$ with respect to ε_1 and then integrating by parts, we find

$$\frac{\partial \bar{I}}{\partial \varepsilon_1}=\int_1^2\left(\frac{\partial \bar{F}}{\partial Y}\eta+\frac{\partial \bar{F}}{\partial Y'}\eta'\right)dx$$

$$=\int_1^2\left(\frac{\partial \bar{F}}{\partial Y}-\frac{d}{dx}\frac{\partial \bar{F}}{\partial Y'}\right)\eta\,dx \tag{4.12}$$

Setting $\varepsilon_1=\varepsilon_2=0$ and using (4.10) gives

$$0=\int_1^2\left(\frac{\partial \bar{F}}{\partial y}-\frac{d}{dx}\frac{\partial \bar{F}}{\partial y'}\right)\eta\,dx \tag{4.13}$$

Because of the arbitrariness of $\eta(x)$, the function $\bar{F}$ obeys the Euler–Lagrange equation

$$\frac{d}{dx}\frac{\partial \bar{F}}{\partial y'}-\frac{\partial \bar{F}}{\partial y}=0 \tag{4.14}$$

After a reshuffling of constants, the function $\bar{F}$ for the hanging chain problem is essentially

$$\bar{F}=(y+\lambda)\sqrt{1+y'^2} \tag{4.15}$$

Since there is no explicit dependence on x, we can use

$$\frac{\partial \bar{F}}{\partial y'}y'-\bar{F}=-h=\text{constant} \tag{4.16}$$

in place of the Euler–Lagrange equation. This gives the differential equation

$$-h\sqrt{1+y'^2}+y+\lambda=0 \tag{4.17}$$

Its solution is

$$y=-\lambda+h\cosh\left(\frac{x-c}{h}\right) \tag{4.18}$$

where c is another constant. If, as shown in Fig. 4.1, the chain hangs between the points $(0, a)$ and (L, b), then its length is found from (4.3) to be

$$s=\int_0^L\cosh\left(\frac{x-c}{h}\right)dx=h\sinh\left(\frac{L-c}{h}\right)+h\sinh\frac{c}{h} \tag{4.19}$$

At the end points, (4.18) gives

$$a=-\lambda+h\cosh\frac{c}{h} \tag{4.20}$$

$$b=-\lambda+h\cosh\left(\frac{L-c}{h}\right) \tag{4.21}$$

Equations (4.19), (4.20), and (4.21) are three equations for the determination of the three constants λ, h, and c.

The name *isoperimetric* was originally given to the problem of finding the shape of a closed curve of given length which encloses the greatest possible area. To solve this problem, we start with Stokes' theorem, equation (9.19) in Chapter 3, applied to an area σ lying in the x, y plane bounded by a closed curve C:

$$\int_\sigma \nabla\times\mathbf{a}\cdot d\boldsymbol{\sigma}=\oint_C \mathbf{a}\cdot d\mathbf{s} \tag{4.22}$$

If we choose the components of the vector **a** to be

$$a_x = -\tfrac{1}{2}y \qquad a_y = \tfrac{1}{2}x \tag{4.23}$$

then $\nabla \times \mathbf{a} \cdot d\boldsymbol{\sigma} = d\sigma$ and (4.22) gives

$$\sigma = \oint_C \frac{1}{2}(x\,dy - y\,dx) \tag{4.24}$$

It is convenient in this problem to represent x and y parametrically as functions of t and express (4.24) as

$$\sigma = \oint_C \frac{1}{2}(x\dot{y} - y\dot{x})\,dt \tag{4.25}$$

The problem is to find functions $x(t)$ and $y(t)$ which will maximize (4.25) subject to the constraint that the perimeter of the area remain constant in length:

$$s = \oint_C ds = \oint_C \sqrt{\dot{x}^2 + \dot{y}^2}\,dt = \text{constant} \tag{4.26}$$

Stated in general terms, the problem is to choose functions $x(t)$ and $y(t)$ that will extremize the functional

$$I = \int_1^2 F(x, y, \dot{x}, \dot{y}, t)\,dt \tag{4.27}$$

subject to the constraint

$$J = \int_1^2 K(x, y, \dot{x}, \dot{y}, t)\,dt \tag{4.28}$$

According to the Lagrange multiplier method, the extremizing functions are to be found by setting the variation of the functional

$$\bar{I} = I + \lambda J = \int_1^2 (F + \lambda K)\,dt = \int_1^2 \bar{F}\,dt \tag{4.29}$$

equal to zero. We will use the somewhat less precise and less rigorous but more compact variational notation introduced in equation (1.24). The variation of (4.29) is

$$\delta\bar{I} = \int_1^2 \left(\frac{\partial\bar{F}}{\partial x}\delta x + \frac{\partial\bar{F}}{\partial\dot{x}}\delta\dot{x} + \frac{\partial\bar{F}}{\partial y}\delta y + \frac{\partial\bar{F}}{\partial\dot{y}}\delta\dot{y} \right) dt \tag{4.30}$$

Since $x(t)$ and $y(t)$ are required to pass through fixed end points, integrating the $\dot{x}$ and $\dot{y}$ terms by parts gives

$$\delta\bar{I} = \int_1^2 \left(\frac{\partial\bar{F}}{\partial x} - \frac{d}{dt}\frac{\partial\bar{F}}{\partial\dot{x}} \right) \delta x\,dt + \int_1^2 \left(\frac{\partial\bar{F}}{\partial y} - \frac{d}{dt}\frac{\partial\bar{F}}{\partial\dot{y}} \right) \delta y\,dt \tag{4.31}$$

The variations δx and δy are not independent of one another on account of the constraint. We argue that the Lagrange multiplier λ can be chosen so that the coefficient of δx is zero. Then, since y can be varied arbitrarily, the coefficient of δy is also zero. This gives

$$\frac{d}{dt}\frac{\partial\bar{F}}{\partial\dot{x}} - \frac{\partial\bar{F}}{\partial x} = 0 \qquad \frac{d}{dt}\frac{\partial\bar{F}}{\partial\dot{y}} - \frac{\partial\bar{F}}{\partial y} = 0 \tag{4.32}$$

We find that an Euler–Lagrange equation holds for each of the functions $x(t)$ and $y(t)$.

For the example of the isoperimetric problem, the integrand of the functional $\bar{I}$ is

$$\bar{F} = \tfrac{1}{2}(x\dot{y} - y\dot{x}) + \lambda\sqrt{\dot{x}^2 + \dot{y}^2} \tag{4.33}$$

The Euler–Lagrange equations (4.32) give

$$\frac{d}{dt}\left(\frac{\lambda\dot{x}}{\sqrt{\dot{x}^2 + \dot{y}^2}} - y \right) = 0 \qquad \frac{d}{dt}\left(\frac{\lambda\dot{y}}{\sqrt{\dot{x}^2 + \dot{y}^2}} + x \right) = 0 \tag{4.34}$$

Therefore,

$$\frac{\lambda\dot{x}}{\sqrt{\dot{x}^2 + \dot{y}^2}} - y = -a \qquad \frac{\lambda\dot{y}}{\sqrt{\dot{x}^2 + \dot{y}^2}} + x = b \tag{4.35}$$

where a and b are constants. The two equations in (4.35) can be combined to give

$$dy(y - a) + dx(x - b) = 0 \tag{4.36}$$

Integration gives

$$(y - a)^2 + (x - b)^2 = c^2 \tag{4.37}$$

where c is another constant. As you probably expected, the answer is a circle.

PROBLEMS

4.1. If the function $\bar{F}$ which appears in equation (4.32) has no explicit dependence on the parameter t, show that the quantity

$$H = \frac{\partial \bar{F}}{\partial \dot{x}}\dot{x} + \frac{\partial \bar{F}}{\partial \dot{y}}\dot{y} - \bar{F} \tag{4.38}$$

is a constant, that is, $dH/dt = 0$.

4.2. A polygon can be made to approximate a circle arbitrarily closely. Show that the area of a polygon with n equal sides and perimeter s is

$$\sigma_n = \frac{s^2}{4n\tan\dfrac{\pi}{n}} \tag{4.39}$$

If σ represents the area of a circle of circumference s, show that $\sigma/\sigma_n > 1$ if n is finite and that $\sigma/\sigma_n \to 1$ as $n \to \infty$.

4.3. Equations (4.19), (4.20), and (4.21) are transcendental and must be solved numerically for the constants λ, c, and h for given values of a, b, s, and L. Show that h is positive and can be found from

$$\sqrt{s^2 - (b-a)^2} = 2h\sinh\frac{L}{2h} = f(h) \tag{4.40}$$

Show that there is one value of h that will satisfy (4.40) provided that

$$s^2 > (b-a)^2 + L^2 \tag{4.41}$$

and that there is no solution otherwise. What does (4.41) mean geometrically? Show that the value of c is given by

$$c = \frac{L}{2} - \frac{h}{2}\ln\left(\frac{s+b-a}{s-b+a}\right) \tag{4.42}$$

Figure 4.2 shows a plot of $f(h)$ as defined by (4.40) for the case $L = 1$.

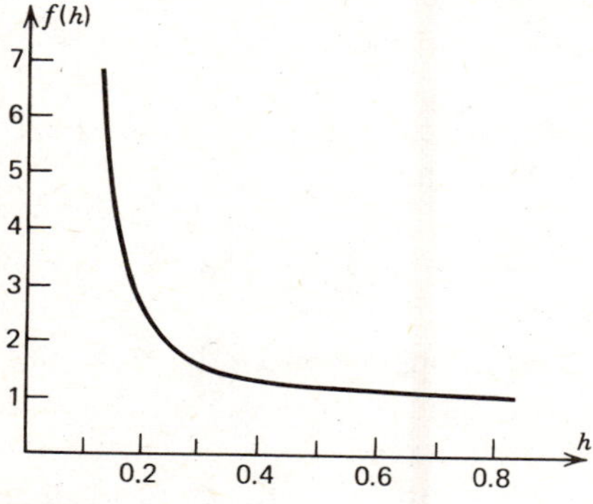

Fig. 4.2

4.4. Find the shape of a solid of given area that encloses the largest possible volume. To solve the problem, let $y(x)$ be a function which passes through the points $(0,0)$ and $(x_2, 0)$. Maximize the volume of the solid of revolution generated by rotating this function about the x axis subject to the constraint that the area remain a constant. *Caution:* The end point x_2 is a variable.

5. HAMILTON'S PRINCIPLE

The connection between the Lagrange formulation of classical mechanics and the calculus of variations is probably obvious to the reader by now. Suppose that a Lagrangian for a system with f degrees of freedom is known. Imagine that the motion of the system is represented by a single point in the configuration space made up out of the f independent coordinates q_k. Then, Hamilton's principle states that out of all the possible paths connecting two fixed points in the configuration space, the one actually followed by the system point is the one that extremizes the integral

$$I = \int_1^2 L(q_k, \dot{q}_k, t)\, dt \tag{5.1}$$

The Lagrangian generally depends on the f generalized coordinates $q_1, \ldots, q_f$, the f generalized velocities $\dot{q}_1, \ldots, \dot{q}_f$, and the time. Lagrange's equations are discussed in section 13 of Chapter 4. By varying the functional (5.1) and integrating by parts in the usual way, we find

$$\delta I = \int_1^2 \sum_{k=1}^{f} \left(\frac{\partial L}{\partial q_k} \delta q_k + \frac{\partial L}{\partial \dot{q}_k} \delta \dot{q}_k \right) dt$$

$$= \int_1^2 \sum_{k=1}^{f} \left(\frac{\partial L}{\partial q_k} - \frac{d}{dt} \frac{\partial L}{\partial \dot{q}_k} \right) \delta q_k\, dt = 0 \tag{5.2}$$

This is really the same process that we used to obtain equation (4.31), except that there are now f independent coordinates instead of just two. When we integrated by parts to obtain (5.2), we used the fact that the variations of the functions $q_k(t)$ vanish at the two end points. Since there are no constraints, the f variations δq_k are independent of one another, and (5.2) implies that

$$\frac{d}{dt} \frac{\partial L}{\partial \dot{q}_k} - \frac{\partial L}{\partial q_k} = 0 \tag{5.3}$$

which are the familiar Lagrange equations.

Knowing that Lagrange's equations can be obtained from a variational principle may not be of any particular value in the practical solution of a problem. There are however problems involving certain types of constraints, the solution of which is greatly facilitated by the calculus of variations. Suppose that a system is described by N generalized coordinates which are not independent but are connected by differential constraints of the form

$$\sum_{k=1}^{N} a_{\alpha k}\, \delta q_k = 0 \tag{5.4}$$

where $a_{\alpha k}$ are functions of the coordinates and the time and $\alpha = 1, 2, \ldots, n$. Necessarily, $n < N$. For example, the constraint connecting the coordinates x and ϕ of the disc rolling without slipping down the inclined plane of Fig. 13.1 in Chapter 4 can be expressed as

$$\delta x - b\, \delta \phi = 0 \tag{5.5}$$

One technique is to try to integrate (5.4) and put the equations of constraint in the form

$$f_\alpha(q_k, t) = 0 \tag{5.6}$$

We then solve for n of the coordinates in terms of the remaining $N - n$ and eliminate them from the Lagrangian. Lagrange's equations in the form (5.3) are then valid for the remaining coordinates. There are however cases where this can't be done because the differential equations of constraint are not exact and are therefore nonintegrable. It is still possible to treat the problem by the method of Lagrange multipliers. Each of the equations represented by (5.4) is multiplied by a function $\lambda_\alpha(t)$. We then sum over α and integrate with respect to the time to get

$$\int_1^2 \sum_{\alpha=1}^{n} \sum_{k=1}^{N} \lambda_\alpha a_{\alpha k}\, \delta q_k\, dt = 0 \tag{5.7}$$

If we run the sum from 1 to N in equation (5.2) and then add it to (5.7) we obtain

$$\int_1^2 \sum_{k=1}^{N} \left(\frac{\partial L}{\partial q_k} - \frac{d}{dt} \frac{\partial L}{\partial \dot{q}_k} + \sum_{\alpha=1}^{n} \lambda_\alpha a_{\alpha k} \right) \delta q_k\, dt \tag{5.8}$$

Only $N - n$ of the variations δq_k are independent of one another; but by appropriate choices for the Lagrange multipliers, the coefficients of each variation δq_k can be made to vanish with the result that

$$\frac{d}{dt} \frac{\partial L}{\partial \dot{q}_k} - \frac{\partial L}{\partial q_k} = \sum_{\alpha=1}^{n} \lambda_\alpha a_{\alpha k} \tag{5.9}$$

A somewhat trivial example is provided by the disc rolling down the incline. The Lagrangian is

$$L = \tfrac{1}{2} M \dot{x}^2 + \tfrac{1}{2} I \dot{\phi}^2 + M g x \sin\theta \tag{5.10}$$

There is a single equation of constraint given by (5.5) and therefore a single Lagrange multiplier. The two Lagrange equations are

$$\frac{d}{dt}\frac{\partial L}{\partial \dot{x}} - \frac{\partial L}{\partial x} = \lambda \qquad \frac{d}{dt}\frac{\partial L}{\partial \dot{\phi}} - \frac{\partial L}{\partial \phi} = -b\lambda \tag{5.11}$$

The equations of motion are found to be

$$M\ddot{x} - Mg\sin\theta = \lambda \qquad I\ddot{\phi} = -b\lambda \tag{5.12}$$

Elimination of the Lagrange multiplier gives

$$M\ddot{x} - Mg\sin\theta = -\frac{I\ddot{\phi}}{b} \tag{5.13}$$

If the original equation of constraint is differentiated we get $\ddot{x} - b\ddot{\phi} = 0$. Thus, (5.13) gives

$$\ddot{x} = \frac{Mg\sin\theta}{M + \dfrac{I}{b^2}} \tag{5.14}$$

as expected.

In the remainder of this section, we will assume that Lagrange's equations in the form (5.3) are valid. As a generalization of equations (1.13) and (4.38), the *Hamiltonian* of a mechanical system with f degrees of freedom is defined by

$$H = \sum_{k=1}^{f} p_k \dot{q}_k - L \tag{5.15}$$

where p_k are the *canonical momenta* defined by

$$p_k = \frac{\partial L}{\partial \dot{q}_k} \tag{5.16}$$

Canonical momenta are introduced in problem 13.2 of Chapter 4. By calculating the total derivative of H with respect to the time and utilizing Lagrange's equations, we find that

$$\begin{aligned}\frac{dH}{dt} &= \sum_k \dot{p}_k \dot{q}_k + \sum_k p_k \ddot{q}_k - \sum_k \frac{\partial L}{\partial q_k}\dot{q}_k - \sum_k p_k \ddot{q}_k - \frac{\partial L}{\partial t} \\ &= -\frac{\partial L}{\partial t}\end{aligned} \tag{5.17}$$

Thus, if L has no explicit dependence on the time so that $\partial L/\partial t = 0$, the motion of the system is such that $H =$ constant. When you did problem 6.5 in Chapter 7, you verified that for a conservative mechanical system, the Hamiltonian is identical to the total energy.

The importance of the Hamiltonian goes far beyond the fact that it is sometimes a constant of the motion. Let us consider the effect on the Hamiltonian of a variation of the coordinates and the canonical momenta:

$$\begin{aligned}\delta H &= \sum_k \dot{q}_k \delta p_k + \sum_k p_k \delta \dot{q}_k - \sum_k \frac{\partial L}{\partial q_k}\delta q_k - \sum_k p_k \delta \dot{q}_k \\ &= \sum_k \dot{q}_k \delta p_k - \sum_k \frac{\partial L}{\partial q_k}\delta q_k\end{aligned} \tag{5.18}$$

The variations δq_k and δp_k are to be understood in the sense defined by equation (1.24) and are not necessarily the actual changes in the coordinates and momenta that result from the motion of the system. The variations are to be performed at a fixed time in much the same way that the variation δy of Fig. 1.4 is carried out at a fixed x. Sometimes, such variations are called *virtual displacements* of the system. By contrast, equation (5.17) involves time rates of change in the various quantities which result from the actual motion of the system. If the Hamiltonian is expressed in terms of generalized coordinates and canonical momenta rather than coordinates and generalized velocities, then

$$\delta H = \sum_k \frac{\partial H}{\partial p_k}\delta p_k + \sum_k \frac{\partial H}{\partial q_k}\delta q_k \tag{5.19}$$

The direct comparison of (5.17) and (5.18) reveals that

$$\frac{\partial H}{\partial p_k} = \dot{q}_k \qquad \frac{\partial H}{\partial q_k} = -\dot{p}_k \tag{5.20}$$

These are Hamilton's canonical equations and can be used in place of Lagrange's equations to find the motion of a mechanical system. Thus, the Hamiltonian leads to a new formulation of classical mechanics. One of the basic features of the Hamiltonian formulation is that it puts the coordinates and momenta on an equal footing and enlarges the configuration space to a $2f$-dimensional space consisting of the coordinates and their canonically conjugate momenta. This $2f$-dimensional space is called *phase space*. In problem 7.6 of Chapter 6, we found that the phase space concept is useful in statistical mechanics. Hamilton's equations (5.20) are $2f$ first-order differential equations in the $2f$ variables

$q_1, \ldots, q_f$, $p_1, \ldots, p_f$. We can visualize the system as being represented by the motion of a single point in the phase space.

To illustrate the idea, the Lagrangian for a one-dimensional simple harmonic oscillator is

$$L = \tfrac{1}{2}m\dot{x}^2 - \tfrac{1}{2}kx^2 \tag{5.21}$$

The canonical momentum is

$$p_x = \frac{\partial L}{\partial \dot{x}} = m\dot{x} \tag{5.22}$$

In order to use Hamilton's equations, we must first express the Hamiltonian as a function of canonical momenta and coordinates:

$$H = p_x\dot{x} - L = \frac{p_x^2}{2m} + \frac{1}{2}kx^2 \tag{5.23}$$

In this example, the Lagrangian has no explicit time dependence and H is a constant of the motion. It is in fact the total mechanical energy. We see from equation (5.23) that the path of the system in the two-dimensional phase space is an ellipse. The equations of motion follow from

$$\frac{\partial H}{\partial p_x} = \dot{x} \qquad \frac{\partial H}{\partial x} = -\dot{p}_x \tag{5.24}$$

We find

$$\frac{p_x}{m} = \dot{x} \qquad kx = -\dot{p}_x \tag{5.25}$$

By eliminating p_x, we find the familiar result $kx = -m\ddot{x}$.

PROBLEMS

5.1. From the form of Lagrange's equations (5.9), it can be inferred that the Lagrange multipliers are related to the forces of constraint. For the example of the disc rolling down the inclined plane, show that the frictional force responsible for the rolling is given by $f = -\lambda$.

5.2. The problem of a bead moving without friction on the spoke of a wheel is discussed in section 13 of Chapter 4 and is illustrated in Fig. 13.2 of that chapter. Solve this problem by using Hamilton's equations. There are several interesting features. The Hamiltonian is a constant of the motion. The Hamiltonian is *not* equal to the total energy. The total energy is *not* a constant of the motion because the constraint is time-dependent.

5.3. Obtain the equation of motion of a plane pendulum by the method of Lagrange multipliers. Show that the Lagrange multiplier is essentially the tension in the string.

5.4. The Lagrangian for a charged particle moving in an electromagnetic field is given in problem (13.7) of Chapter 4. Show that the canonical momentum is

$$\mathbf{p} = m\mathbf{u} + q\mathbf{A} \tag{5.26}$$

and that the Hamiltonian can be expressed as

$$H = \frac{1}{2}mu^2 + q\psi = \frac{1}{2m}\left[p^2 - 2q(\mathbf{p}\cdot\mathbf{A}) + q^2A^2\right] + q\psi \tag{5.27}$$

5.5. If the equations of constraint (5.4) are integrable, then functions $f_\alpha(q_k, t)$ exist such that

$$a_{\alpha k} = \frac{\partial f_\alpha}{\partial q_k} \tag{5.28}$$

Show that under this circumstance it is possible to express the equations of motion (5.9) as

$$\frac{d}{dt}\frac{\partial \bar{L}}{\partial \dot{q}_k} - \frac{\partial \bar{L}}{\partial q_k} = 0 \qquad \bar{L} = L + \sum_{\alpha=1}^{n} \lambda_\alpha(t) f_\alpha \tag{5.29}$$

6. GEODESICS

In Chapter 4, we learned that the magnitude of the line element can be expressed in arbitrary curvilinear coordinates as

$$ds = \sqrt{g_{ij}\,dq_i\,dq_j} \tag{6.1}$$

where g_{ij} are elements of the metric tensor. The summation convention is in force meaning that sums over i and j are implied in (6.1). The sums run from 1 to n if the space is n-dimensional. We have suppressed the superscript notation for contravariant quantities because the development of this section does not require any transformations between different curvilinear coordinate systems. The space that we are talking about can be quite general and need not be Euclidean. The distance between two points in the space is

$$s = \int_1^2 ds = \int_1^2 \sqrt{g_{ij}q_i'q_j'}\,d\lambda \tag{6.2}$$

where λ is a parameter and $q_i' = dq_i/d\lambda$. The functions $q_i(\lambda)$ give a parametric representation of the path connecting the initial and final points. By definition, the path of shortest length between the two end points is a *geodesic*, and it is found by requiring that the functions $q_i(\lambda)$ satisfy the Euler–Lagrange equations

$$\frac{d}{d\lambda}\left(\frac{\partial}{\partial q_k'}\sqrt{g_{ij}q_i'q_j'}\right) - \frac{\partial}{\partial q_k}\sqrt{g_{ij}q_i'q_j'} = 0 \tag{6.3}$$

It is possible to put (6.3) in a more convenient form. If the differentiations with respect to q_k and q_k' are done, the result is

$$\frac{d}{d\lambda}\left[\frac{1}{\left(\frac{ds}{d\lambda}\right)}g_{ki}\left(\frac{dq_i}{d\lambda}\right)\right] - \frac{1}{2}\frac{1}{\left(\frac{ds}{d\lambda}\right)}\frac{\partial g_{ij}}{\partial q_k}\frac{dq_i}{d\lambda}\frac{dq_j}{d\lambda} = 0 \tag{6.4}$$

where we have set

$$\frac{ds}{d\lambda} = \sqrt{g_{ij}q_i'q_j'} \tag{6.5}$$

It is very often convenient to use s itself as the parameter. Thus, setting $\lambda = s$ in (6.4) gives

$$\frac{d}{ds}\left(g_{ki}\frac{dq_i}{ds}\right) - \frac{1}{2}\frac{\partial g_{ij}}{\partial q_k}\frac{dq_i}{ds}\frac{dq_j}{ds} = 0 \tag{6.6}$$

It is interesting to compare (6.6) with the covariant form of the acceleration of a particle as given by (9.27) in Chapter 4. If we replace s by t in (6.6), the two expressions become exactly the same if the covariant components of the acceleration are zero.

According to relativity theory, the proper time τ of a particle can be identified with "arc length" in the four-dimensional space-time continuum. The properties of the gravitational field are contained in the metric tensor; and when a particle moves under the influence of gravitational forces alone, its covariant acceleration is zero. The equation of motion of the particle is given by (6.6), with s replaced by τ. The indices $i, j, k, \ldots$ all run from 1 to 4, and dq_i/ds is replaced by the four-dimensional generalized velocity components $\dot{q}_i = dq_i/d\tau$. The path of the particle is a geodesic.

It is possible to put (6.6) in a still more convenient form. If a scalar function ψ is defined by

$$\psi = \frac{1}{2}g_{ij}q_i'q_j' \qquad q_i' = \frac{dq_i}{ds} \tag{6.7}$$

then (6.6) is equivalent to

$$\frac{d}{ds}\frac{\partial\psi}{\partial q_k'} - \frac{\partial\psi}{\partial q_k} = 0 \tag{6.8}$$

PROBLEMS

6.1. Show that the shortest distance between two points in a three-dimensional Cartesian space is a straight line. Find the shortest distance in the x, y plane from the point (4,0) to the parabola

$$y = \frac{1}{2\sqrt{2}}x^2$$

6.2. If

$$F(q_k, q_k') = \sqrt{g_{ij} q_i' q_j'} \qquad q_i' = \frac{dq_i}{d\lambda} \tag{6.9}$$

and the Euler–Lagrange equation (6.3) is satisfied, then

$$h = \frac{\partial F}{\partial q_k'} q_k' - F \tag{6.10}$$

is a constant. Show that $h = 0$. If ψ is given by (6.7) and the Euler–Lagrange equation (6.8) is satisfied, then

$$H = \frac{\partial \psi}{\partial q_k'} q_k' - \psi \tag{6.11}$$

is a constant. Show that

$$H = \psi = \tfrac{1}{2} \tag{6.12}$$

6.3. Suppose that we want to find the functions $q_k(\lambda)$ which will extremize the integral

$$I = \int_1^2 F(q_k, q_k', \lambda)\, d\lambda \tag{6.13}$$

where it is stipulated that the extremizing curve starts at a fixed point and terminates on the surface

$$H(q_1, q_2, q_3) = 0 \tag{6.14}$$

To solve the problem, we write

$$I(\varepsilon) = \int_1^{\lambda_2(\varepsilon)} F(Q_k, Q_k', \lambda)\, d\lambda \tag{6.15}$$

where the comparison curves are represented in parametric form as

$$Q_k(\lambda) = q_k(\lambda) + \varepsilon\eta_k(\lambda) \tag{6.16}$$

Over the surface given by (6.15), the end point of a comparison curve obeys

$$Q_{k,2}(\varepsilon) = q_k(\lambda_2) + \varepsilon\eta_k(\lambda_2) \tag{6.17}$$

Show that

$$\left(\frac{d\lambda_2}{d\varepsilon}\right)_{\varepsilon=0} = \left(-\frac{\dfrac{\partial H}{\partial q_k}\eta_k}{\dfrac{dH}{d\lambda}}\right)_2 \qquad \frac{dH}{d\lambda} = \frac{\partial H}{\partial q_j}\frac{dq_j}{d\lambda} \tag{6.18}$$

and that the appropriate end point condition is

$$\left(\frac{\partial F}{\partial q_k'}\frac{dH}{d\lambda}\right)_2 = \left(\frac{\partial H}{\partial q_k}F\right)_2 \tag{6.19}$$

In particular, if $F = \psi$ where ψ is given by (6.7),

$$\left(g_{kn} q_n' \frac{dH}{d\lambda}\right)_2 = \left(\frac{\partial H}{\partial q_k}\psi\right)_2 \tag{6.20}$$

If the space is Cartesian and we set $\lambda = s$, then

$$\left(\frac{dx_k}{ds}\frac{dH}{ds}\right)_2 = \left(\frac{\partial H}{\partial x_k}\psi\right)_2 \tag{6.21}$$

This shows that the components of ∇H are proportional to dx_k/ds, meaning that the extremizing line intersects $H=0$ at right angles. The same result is true even if the space is not Cartesian.

6.4. Find the function ψ of equation (6.7) in terms of spherical coordinates appropriate for finding the geodesics on the surface of a sphere. Show that the differential equation for finding the geodesics is

$$\frac{d\theta}{\sin\theta\sqrt{a\sin^2\theta-1}}=d\phi \tag{6.22}$$

where a is a constant. Complete the solution and show that the geodesics are great circles. *Hint:* The substitution $u=\cot\theta$ is useful.

7. CONTINUOUS SYSTEMS

Equation (2.54) of Chapter 7 gives as the expression for the energy density of a vibrating string

$$w=\frac{1}{2}\mu\left(\frac{\partial\psi}{\partial t}\right)^2+\frac{1}{2}F\left(\frac{\partial\psi}{\partial x}\right)^2 \tag{7.1}$$

To standardize the notation, we have called the displacement of the string ψ instead of y. By analogy with particle mechanics, it might be supposed that the quantity

$$L=\frac{1}{2}\mu\left(\frac{\partial\psi}{\partial t}\right)^2-\frac{1}{2}F\left(\frac{\partial\psi}{\partial x}\right)^2 \tag{7.2}$$

is a *Lagrange density* and that the equations of motion of the string can be obtained from it by means of an appropriate Euler–Lagrange equation. This is in fact the case.

More generally, we want to derive the basic equations of motion for a continuous system by requiring that the function $u(\mathbf{r})$ extremize the integral

$$I=\int_\Sigma L(x_k,u,\partial_k u)\,d\Sigma \tag{7.3}$$

where x_k stands for the three rectangular coordinates and we have used the shorthand notation

$$\partial_k u=\frac{\partial u}{\partial x_k} \tag{7.4}$$

The integral is taken over some region Σ of space in which the problem is defined. We will consider the appropriate generalization of (7.3) to include time dependence a little later. Comparison functions can be defined by means of

$$U(\mathbf{r})=u(\mathbf{r})+\varepsilon\eta(\mathbf{r}) \tag{7.5}$$

We then write

$$I(\varepsilon)=\int_\Sigma L(x_k,U,\partial_k U)\,d\Sigma \tag{7.6}$$

and determine u by the condition that $dI/d\varepsilon=0$ when $\varepsilon=0$. The derivative of (7.6) with respect to ε is

$$\frac{dI}{d\varepsilon}=\int_\Sigma\left[\frac{\partial L}{\partial U}\eta+\sum_{k=1}^{3}\frac{\partial L}{\partial(\partial_k U)}(\partial_k\eta)\right]d\Sigma \tag{7.7}$$

If we set $\varepsilon=0$ and define the components of a vector $\mathbf{a}$ by means of

$$a_k=\frac{\partial L}{\partial(\partial_k u)} \tag{7.8}$$

then (7.7) gives

$$0=\int_\Sigma\left[\frac{\partial L}{\partial u}\eta+(\mathbf{a}\cdot\nabla\eta)\right]d\Sigma \tag{7.9}$$

The second term is transformed by means of the vector identity (9.1) and Gauss' theorem (9.13), both of Chapter 3, as follows:

$$\begin{aligned}\int_\Sigma(\mathbf{a}\cdot\nabla\eta)\,d\Sigma&=\int_\Sigma[\nabla\cdot\eta\mathbf{a}-\eta(\nabla\cdot\mathbf{a})]\,d\Sigma\\&=\int_\sigma\eta\mathbf{a}\cdot d\boldsymbol{\sigma}-\int_\Sigma\eta(\nabla\cdot\mathbf{a})\,d\Sigma\end{aligned} \tag{7.10}$$

The boundary conditions are assumed to be such that the surface integral is zero. For example, if all of the comparison functions are required to have the same prescribed values on the boundary, then $\eta=0$. Another possibility is $\mathbf{a}=0$ at all points of σ. The result is then

$$0=\int_\Sigma\left[\frac{\partial L}{\partial u}-\nabla\cdot\mathbf{a}\right]\eta\,d\Sigma \tag{7.11}$$

By an obvious extension of theorem 1 in section 1, (7.11) implies that

$$\frac{\partial L}{\partial u} - \nabla \cdot \mathbf{a} = 0 \tag{7.12}$$

which is the required Euler–Lagrange equation obeyed by the function u. Written out in detail, the divergence term is

$$\nabla \cdot \mathbf{a} = \sum_{k=1}^{3} \frac{\partial}{\partial x_k} \left[\frac{\partial L}{\partial(\partial_k u)} \right] \tag{7.13}$$

As a simple example, suppose that the Lagrange density is

$$L = \frac{1}{2} \nabla u \cdot \nabla u = \frac{1}{2} \sum_{j=1}^{3} (\partial_j u)^2 = \frac{1}{2} |\nabla u|^2 \tag{7.14}$$

We find that

$$a_k = \frac{\partial L}{\partial(\partial_k u)} = \partial_k u \qquad \frac{\partial L}{\partial u} = 0 \tag{7.15}$$

The Euler–Lagrange equation (7.12) then gives

$$\nabla^2 u = 0 \tag{7.16}$$

which is Laplace's equation. Thus, if the boundary conditions on u are such that the surface integral in (7.10) vanishes, then the functional

$$I = \frac{1}{2} \int_\Sigma |\nabla u|^2 \, d\Sigma \tag{7.17}$$

is extremized if u is a solution of Laplace's equation.

We can show that the extremum is actually a minimum. Let ϕ be any function which vanishes on the surface σ of the region in question and let $\psi = u + \phi$. Consider

$$\int_\Sigma |\nabla \psi|^2 \, d\Sigma = \int_\Sigma |\nabla u|^2 \, d\Sigma + 2 \int_\Sigma \nabla u \cdot \nabla \phi \, d\Sigma + \int_\Sigma |\nabla \phi|^2 \, d\Sigma \tag{7.18}$$

The cross term can be evaluated by noting that

$$\int_\Sigma \nabla \cdot \phi \nabla u \, d\Sigma = \int_\Sigma \nabla \phi \cdot \nabla u \, d\Sigma + \int_\Sigma \phi \nabla^2 u \, d\Sigma \tag{7.19}$$

The first term can be converted into a surface integral by means of Gauss' theorem. It is therefore zero because $\phi = 0$ on σ. Then, since u is a solution of Laplace's equation,

$$\int_\Sigma \nabla \phi \cdot \nabla u \, d\Sigma = 0 \tag{7.20}$$

Therefore

$$\int_\Sigma |\nabla \psi|^2 \, d\Sigma = \int_\Sigma |\nabla u|^2 \, d\Sigma + \int_\Sigma |\nabla \phi|^2 \, d\Sigma > \int_\Sigma |\nabla u|^2 \, d\Sigma \tag{7.21}$$

which shows that u is a minimizing function. Since the electric field is given by $\mathbf{E} = -\nabla u$, the functional (7.17) is proportional to the electrostatic energy stored in the field. Thus, for given boundary conditions, the actual field which exists is such as to minimize the electrostatic potential energy.

It frequently happens that a problem is formulated by requiring that the functional (7.3) be extremized subject to a constraint of the form

$$J = \int_\Sigma K(x_k, u, \partial_k u) \, d\Sigma = \text{constant} \tag{7.22}$$

It is then necessary to write the comparison functions in the form

$$U(\mathbf{r}) = u(\mathbf{r}) + \varepsilon_1 \eta(\mathbf{r}) + \varepsilon_2 \zeta(\mathbf{r}) \tag{7.23}$$

The procedure is the same as that used in section 4 and leads to the result that u obeys the Euler–Lagrange equation

$$\frac{\partial \bar{L}}{\partial u} - \sum_{k=1}^{3} \frac{\partial}{\partial x_k} \left[\frac{\partial \bar{L}}{\partial(\partial_k u)} \right] = 0 \qquad \bar{L} = L + \lambda K \tag{7.24}$$

If the problem is time-dependent, then we seek a function $\psi(\mathbf{r}, t)$ that will extremize the functional

$$I = \int_\Sigma \int_{t_1}^{t_2} L(x_k, t, \psi, \partial_t \psi, \partial_k \psi) \, dt \, d\Sigma \tag{7.25}$$

where

$$\partial_t \psi = \frac{\partial \psi}{\partial t} \tag{7.26}$$

Essentially, the inclusion of the time just adds one more variable which can be treated on an equal footing with the three spatial coordinates. The integral is over a four-dimensional region made up of the three coordinates and the time. The variation of ψ must vanish at t_1 and t_2, and the surface integral which appears in (7.10) must vanish. The resulting Euler–Lagrange equation is

$$\frac{\partial L}{\partial \psi} - \frac{\partial}{\partial t}\frac{\partial L}{\partial(\partial_t \psi)} - \sum_{k=1}^{3} \frac{\partial}{\partial x_k}\frac{\partial L}{\partial(\partial_k \psi)} = 0 \qquad (7.27)$$

A simple example is provided by the Lagrange density (7.2) for the vibrating string. Note that

$$\frac{\partial L}{\partial \psi} = 0 \qquad \frac{\partial L}{\partial(\partial_t \psi)} = \mu \frac{\partial \psi}{\partial t} \qquad \frac{\partial L}{\partial(\partial_x \psi)} = -F\frac{\partial \psi}{\partial x} \qquad (7.28)$$

The Euler–Lagrange equation (7.27) then gives

$$\frac{\partial}{\partial x}\left(F\frac{\partial \psi}{\partial x}\right) - \mu \frac{\partial^2 \psi}{\partial t^2} = 0 \qquad (7.29)$$

This is the familiar one-dimensional wave equation and is valid if the linear mass density μ and the tension F are functions of x.

Another example is provided by the stretched flexible membrane. Equation (3.36) of Chapter 7 governs the flow of mechanical energy from point to point on the surface of the membrane. The term

$$w = \frac{1}{2}F(\nabla\psi \cdot \nabla\psi) + \frac{1}{2}\nu\left(\frac{\partial \psi}{\partial t}\right)^2 \qquad (7.30)$$

is the mechanical surface energy density. In (7.30), F is the tension in the membrane in newtons per meter and ν is the surface mass density in kilograms per square meter. The implication is that the Lagrange density is

$$L = \frac{1}{2}\nu\left(\frac{\partial \psi}{\partial t}\right)^2 - \frac{1}{2}F(\nabla\psi \cdot \nabla\psi) \qquad (7.31)$$

Application of the Euler–Lagrange equation (7.27) leads to the two-dimensional wave equation

$$\frac{\partial}{\partial x}\left(F\frac{\partial \psi}{\partial x}\right) + \frac{\partial}{\partial y}\left(F\frac{\partial \psi}{\partial y}\right) - \nu\frac{\partial^2 \psi}{\partial t^2} = 0 \qquad (7.32)$$

We will close this section by showing that the energy eigenvalues of a conservative system in quantum mechanics are the extrema of an appropriately defined functional. If $\psi(\mathbf{r}, t)$ is the wave function for the system, then the expectation value of the Hamiltonian is

$$\langle H \rangle = \int_\Sigma \psi^* H \psi \, d\Sigma \qquad (7.33)$$

A discussion of this concept is given in section 12 of Chapter 7; see equation (12.46). If the Hamiltonian is independent of the time, stationary-state energy eigenfunctions u_k and corresponding energy eigenvalues W_k exist which are found from $Hu_k = W_k u_k$. A general wave function for the system can be expanded in terms of these eigenfunctions as

$$\psi = \sum_{k=1}^{\infty} c_k u_k e^{-iW_k t/\hbar} \qquad (7.34)$$

Through the use of the orthonormality property of the eigenfunctions the expectation value of H is found to be

$$\langle H \rangle = \sum_{k=1}^{\infty} |c_k|^2 W_k \qquad (7.35)$$

If the system is in a pure state described by the eigenfunction u_k, then the expectation value of H is $\langle H \rangle = W_k$. Let u be a time-independent function. Consider the functional

$$I = \int_\Sigma u^* H u \, d\Sigma = \int_\Sigma u^* \left[-\frac{\hbar^2}{2m}\nabla^2 u + Vu \right] d\Sigma \qquad (7.36)$$

In (7.36), H has been assumed to be the Hamiltonian for a conservative system described by a potential V. Note that

$$\int_\Sigma \nabla \cdot (u^* \nabla u) \, d\Sigma = \int_\Sigma \nabla u^* \cdot \nabla u \, d\Sigma + \int_\Sigma u^* \nabla^2 u \, d\Sigma \qquad (7.37)$$

The first term can be converted into a surface integral. The system under consideration is localized and the surface of integration is sufficiently far removed that the wave function vanishes at each of its points. This gives the result

$$\int_\Sigma u^* \nabla^2 u \, d\Sigma = -\int_\Sigma \nabla u^* \cdot \nabla u \, d\Sigma \qquad (7.38)$$

The functional (7.36) is then

$$I=\int_{\Sigma}\left[\frac{\hbar^2}{2m}\nabla u^*\cdot\nabla u+u^*Vu\right]d\Sigma \tag{7.39}$$

It depends on both u and u^*, but there is no difficulty in generalizing the derivation of (7.12) to include the case where the Lagrangian depends on two (or more) functions. We will now extremize (7.39) subject to the constraint

$$\int_{\Sigma}u^*u\,d\Sigma=1 \tag{7.40}$$

The functions u and u^* are therefore required to obey the Euler–Lagrange equations

$$\frac{\partial\bar{L}}{\partial u}-\sum_{k=1}^{3}\frac{\partial}{\partial x_k}\frac{\partial\bar{L}}{\partial(\partial_k u)}=0 \tag{7.41}$$

$$\frac{\partial\bar{L}}{\partial u^*}-\sum_{k=1}^{3}\frac{\partial}{\partial x_k}\frac{\partial\bar{L}}{\partial(\partial_k u^*)}=0 \tag{7.42}$$

where

$$\bar{L}=\frac{\hbar^2}{2m}\nabla u^*\cdot\nabla u+u^*Vu+\lambda uu^* \tag{7.43}$$

Since $\bar{L}$ is real if the potential function V is real, (7.41) and (7.42) are merely complex conjugates of each other. Equation (7.42) gives

$$-\frac{\hbar^2}{2m}\nabla^2u+Vu+\lambda u=0 \tag{7.44}$$

This shows that the extremizing functions are eigenfunctions of the familiar time-independent Schrödinger equation and that the possible Lagrange multipliers are related to the energy eigenvalues by means of $\lambda=-W_k$. Equation (7.36) shows that the actual extrema of the original functional are the energy eigenvalues. We will encounter this same result again in section 8 in connection with the Sturm–Liouville eigenvalue problem. We will show that the Sturm–Liouville eigenvalues are actually minima of an appropriately defined functional and that the lowest eigenvalue is an absolute minimum.

The reader may think that it is an idle exercise to derive results that are already known from a variational principle. Approaching the eigenvalue problem through the calculus of variations leads to both theoretical and practical results of some importance. The variational aspects of the Sturm–Liouville problem are useful in the construction of a completeness proof for the eigenfunctions. The knowledge that the eigenfunctions minimize a certain functional and that the minima so obtained are the eigenvalues forms the basis for an important technique, usually called the Rayleigh–Ritz method, for finding approximate eigenfunctions and eigenvalues.

PROBLEMS

7.1. The functional to be extremized in the case of a one-dimensional time-dependent system such as the vibrating string is

$$I=\int_{t_1}^{t_2}\int_0^s L(x,t,\psi,\partial_x\psi,\partial_t\psi)\,dx\,dt \tag{7.45}$$

By using the variational notation introduced in equation (1.24), show that

$$\delta I=\int_{t_1}^{t_2}\int_0^s\left[\frac{\partial L}{\partial\psi}-\frac{\partial}{\partial x}\frac{\partial L}{\partial(\partial_x\psi)}-\frac{\partial}{\partial t}\frac{\partial L}{\partial(\partial_t\psi)}\right]\delta\psi\,dx\,dt \tag{7.46}$$

provided that

$$\left.\frac{\partial L}{\partial(\partial_t\psi)}\delta\psi\right|_{t_1}^{t_2}=0\qquad\left.\frac{\partial L}{\partial(\partial_x\psi)}\delta\psi\right|_0^s=0 \tag{7.47}$$

If the Lagrangian is given by (7.2), show that (7.47) becomes

$$\frac{\partial\psi}{\partial t}\delta\psi\bigg|_{t_1}^{t_2}=0 \qquad \frac{\partial\psi}{\partial x}\delta\psi\bigg|_{0}^{s}=0 \tag{7.48}$$

The second condition in (7.48) can be met if the string is fixed at both ends, in which case $\delta\psi=0$ at $x=0$ and $x=s$. It can also be met if $\partial\psi/\partial x=0$ at $x=0$ and $x=s$. The latter condition is called the *free end condition* and is valid if the ends of the string are free. It is of course possible to have one end of the string fixed and the other free. The free end condition can be realized physically by having the string attached to a small massless ring which can move without friction on a wire which is perpendicular to the x axis as shown in Fig. 7.1. An equivalent situation occurs in the case of longitudinal sound waves in a tube which is open at one or both ends. See problem 9.3 in Chapter 7.

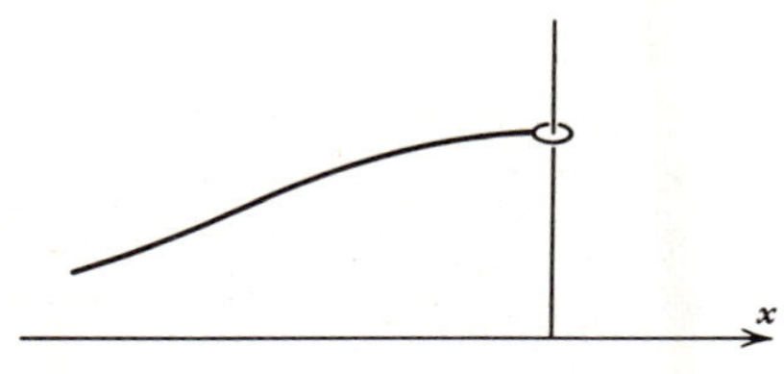

Fig. 7.1

7.2. Suppose that the functional I depends on the three rectangular coordinates x_k, the two functions u and v, and the six partial derivatives $\partial_k u$ and $\partial_k v$:

$$I=\int_\Sigma L(x_k, u, v, \partial_k u, \partial_k v)\, d\Sigma \tag{7.49}$$

If the components of **a** and **b** are defined to be

$$a_k=\frac{\partial L}{\partial(\partial_k u)} \qquad b_k=\frac{\partial L}{\partial(\partial_k v)} \tag{7.50}$$

show that u and v obey the Euler–Lagrange equations

$$\frac{\partial L}{\partial u}-\nabla\cdot\mathbf{a}=0 \qquad \frac{\partial L}{\partial v}-\nabla\cdot\mathbf{b}=0 \tag{7.51}$$

provided that the boundary conditions on u and v are such that

$$\int_\sigma \delta u\,\mathbf{a}\cdot d\boldsymbol{\sigma}=0 \qquad \int_\sigma \delta v\,\mathbf{b}\cdot d\boldsymbol{\sigma}=0 \tag{7.52}$$

In (7.52), we have used the variational notation instead of the comparison function notation.

7.3. Suppose that the potential has the constant value V over a sphere of radius a and that it is zero at infinity. Assume that the potential is of the form

$$u=V\left(\frac{r}{a}\right)^p \qquad a\le r\le\infty \tag{7.53}$$

where p is an unknown parameter. Calculate directly the functional (7.17). Note that convergence of the integral requires $p<-\frac{1}{2}$. Now show that I is minimized if $p=-1$. This is a trivial problem, but it illustrates the possibility of using a trial function which depends on one or more parameters. If the parameters are then chosen so as to minimize the functional, we may hope to obtain an approximation to the actual solution. In this example, the solution so obtained is exact, but in more complicated cases this is usually not true.

7.4. Show that the Lagrange density

$$L=\frac{1}{2}\varepsilon_0 E^2-\frac{1}{2\mu_0}B^2 \tag{7.54}$$

coupled with the Lorentz gauge condition leads to the homogeneous wave equations obeyed by the vector and scalar potentials. Refer to equations (12.35) through (12.46) in Chapter 3. If (7.54) is written in terms of the potentials, it contains four unknown functions each of which obeys an Euler–Lagrange equation. For example, the component A_1 of the vector potential obeys

$$\frac{\partial L}{\partial A_1}-\frac{\partial}{\partial t}\frac{\partial L}{\partial(\partial_t A_1)}-\sum_{k=1}^{3}\frac{\partial}{\partial x_k}\frac{\partial L}{\partial(\partial_k A_1)}=0 \tag{7.55}$$

7.5. Find the ground-state energy of a one-dimensional harmonic oscillator by using as a trial function the normalized Gaussian

$$u=\sqrt{\frac{\beta}{\sqrt{\pi}}}\,e^{-\beta^2x^2/2} \tag{7.56}$$

Calculate the functional (7.36) and then minimize it by using β as the variational parameter. The harmonic oscillator is discussed in section 2 of Chapter 9.

8. VARIATIONAL ASPECTS OF THE EIGENVALUE PROBLEM

We began the discussion of the Sturm–Liouville eigenvalue problem in section 10 of Chapter 7 and have referred to it repeatedly in the intervening chapters. It is appropriate that we return once more to this important topic and settle a bit of unfinished business, namely, the proof of the completeness of the eigenfunctions. In this section, we will establish the completeness of the Sturm–Liouville eigenfunctions when the problem is defined over the finite interval $a\le x\le b$. In problem 2.13 of Chapter 9, we were able to establish the completeness of at least one example of a set of Sturm–Liouville eigenfunctions defined over an infinite interval by a special technique. In the following discussion, all functions are assumed to be real.

We first show that the familiar Sturm–Liouville differential equation is obtained by extremizing the functional

$$I=\int_a^b\left(pu'^2+qu^2\right)dx \tag{8.1}$$

subject to the constraint

$$\int_a^b \rho u^2\,dx=1 \tag{8.2}$$

All of the quantities which appear in (8.1) and (8.2) have the same meaning as they did in section 10 of Chapter 7. According to the Lagrange multiplier method, the variation of the functional

$$\begin{aligned}\bar{I}=I-\lambda&=\int_a^b\left(pu'^2+qu^2-\lambda\rho u^2\right)dx\\&=\int_a^b\bar{L}(x,u,u')\,dx\end{aligned} \tag{8.3}$$

is to be set equal to zero. By using the variational notation we find that

$$\begin{aligned}\delta\bar{I}&=\int_a^b\left(\frac{\partial\bar{L}}{\partial u}\delta u+\frac{\partial\bar{L}}{\partial u'}\delta u'\right)dx\\&=\left.\frac{\partial\bar{L}}{\partial u'}\delta u\right|_a^b-\int_a^b\left(\frac{d}{dx}\frac{\partial\bar{L}}{\partial u'}-\frac{\partial\bar{L}}{\partial u}\right)\delta u\,dx=0\end{aligned} \tag{8.4}$$

The boundary conditions are required to be such that

$$\left.\frac{\partial\bar{L}}{\partial u'}\delta u\right|_a^b=\left.2pu'\delta u\right|_a^b=0 \tag{8.5}$$

For example, (8.5) holds if $p(a)=p(b)=0$, $u'(a)=u'(b)=0$, or $u(a)=u(b)=0$. Each of these possibilities satisfies equation (10.6) of Chapter 7 and means that the Sturm–Liouville operator is Hermitian and that an eigenvalue problem is defined. Equation (8.4) implies that

$$\frac{d}{dx}\frac{\partial \bar{L}}{\partial u'} - \frac{\partial \bar{L}}{\partial u} = 0 \tag{8.6}$$

which gives the Sturm–Liouville differential equation

$$\frac{d}{dx}(pu') - qu + \lambda \rho u = 0 \tag{8.7}$$

as expected. Notice that the Lagrange multiplier is the eigenvalue.

We next inquire as to the meaning of the functional (8.1). If the first term in the integrand is integrated by parts, the result is

$$I = puu'\Big|_a^b - \int_a^b u\left(\frac{d}{dx}pu' - qu\right) dx \tag{8.8}$$

By means of (8.5) and (8.7), (8.8) is reduced to

$$I = \lambda \tag{8.9}$$

We have the result that the functions which are normalized according to (8.2) and which extremize the functional (8.1) are the Sturm–Liouville eigenfunctions. The actual values of the extrema are the eigenvalues.

We will now show that the lowest eigenvalue is an absolute minimum with respect to the class of normalized functions which can be represented by the finite sum

$$u = \sum_{k=1}^{N} c_k u_k \tag{8.10}$$

It is assumed at this point that the completeness of the eigenfunctions is not yet established and that the convergence of (8.10) as $N \to \infty$ is not yet known. On account of the orthogonality of the eigenfunctions,

$$1 = \int_a^b u^2 \rho \, dx = \sum_{k=1}^{N}\sum_{j=1}^{N} c_k c_j \int_a^b u_k u_j \rho \, dx$$
$$= \sum_{k=1}^{N}\sum_{j=1}^{N} c_k c_j \delta_{kj} = \sum_{k=1}^{N} c_k^2 \tag{8.11}$$

The functional (8.1) with respect to any function of the form (8.10) is

$$I[u] = \int_a^b \left(p \sum_{k=1}^{N}\sum_{j=1}^{N} c_k c_j u_k' u_j' + q \sum_{k=1}^{N}\sum_{j=1}^{N} c_k c_j u_k u_j \right) dx \tag{8.12}$$

Integration by parts gives

$$I[u] = \sum_{k=1}^{N}\sum_{j=1}^{N} c_k c_j p u_k' u_j \Big|_a^b - \sum_{k=1}^{N}\sum_{j=1}^{N} c_k c_j \int_a^b u_j \frac{d}{dx}(pu_k')\, dx + \sum_{k=1}^{N}\sum_{j=1}^{N} c_k c_j \int_a^b q u_k u_j \, dx \tag{8.13}$$

The eigenfunctions conform to the boundary conditions (8.5) and obey the Sturm–Liouville differential equation. Therefore,

$$I[u] = \sum_{k=1}^{N}\sum_{j=1}^{N} c_k c_j \int_a^b u_j(\lambda_k \rho u_k - q u_k)\, dx + \sum_{k=1}^{N}\sum_{j=1}^{N} c_k c_j \int_a^b q u_k u_j \, dx$$
$$= \sum_{k=1}^{N}\sum_{j=1}^{N} c_k c_j \lambda_k \delta_{kj} = \sum_{k=1}^{N} c_k^2 \lambda_k \tag{8.14}$$

By using (8.11), this can be rewritten as

$$I[u] = \lambda_1 + \sum_{k=1}^{N} c_k^2(\lambda_k - \lambda_1) \tag{8.15}$$

Theorem 6 in section 10 of Chapter 7 shows that the eigenvalues form a monotonically increasing sequence. Equation (8.15) therefore shows that

$$I[u] \geq \lambda_1 \tag{8.16}$$

with respect to the class of normalized functions of the form (8.10). The equality holds if $u = u_1$.

Now consider the smaller subclass of functions which are of the form (8.10) but which are also orthogonal to the first eigenfunction u_1. Such functions can be represented as

$$u = \sum_{k=2}^{N} c_k u_k \tag{8.17}$$

Note the omission of u_1 from the sum. A calculation similar to the one in (8.11) shows that

$$\sum_{k=2}^{N} c_k^2 = 1 \tag{8.18}$$

By going through the same sequence of steps that led to (8.15), we get

$$I[u] = \lambda_2 + \sum_{k=2}^{N} c_k^2(\lambda_k - \lambda_2) \tag{8.19}$$

which implies that

$$I[u] \geq \lambda_2 \tag{8.20}$$

The equality holds if $u = u_2$. The meaning of (8.16) and (8.20) is that the first eigenvalue λ_1 is the absolute minimum of $I[u]$ with respect to functions of the form (8.10) and λ_2 is the first relative minimum beyond λ_1.

We now repeat the procedure, this time working in the even smaller subspace of all possible functions of the form (8.10) which are orthogonal to both u_1 and u_2. With respect to these functions, we find that $I[u] \geq \lambda_3$. In this manner, the entire sequence of eigenfunctions $u_1, u_2, \ldots, u_{N-1}$ are shown to be minima of $I[u]$ with respect to functions of the form (8.10). Once the completeness of the eigenfunctions has been established, this will include all piecewise continuous functions.

If u is any real function whatsoever which conforms to the boundary conditions (8.5), then (8.8) shows that

$$I[u] = -\int_a^b uS(u)\, dx \tag{8.21}$$

where S is the Sturm–Liouville operator

$$S = \frac{d}{dx} p(x) \frac{d}{dx} - q(x) \tag{8.22}$$

In Chapter 7, the Sturm–Liouville operator was denoted by L. The notation change is necessary to avoid confusion with the Lagrangian. If u_k is an eigenfunction, then

$$S(u_k) + \rho\lambda_k u_k = 0 \tag{8.23}$$

Let $f(x)$ be an arbitrary real function. We define

$$\Delta_n(x) = f - \sum_{k=1}^{n} c_k u_k \tag{8.24}$$

$$\delta_n^2 = \int_a^b \Delta_n^2 \rho\, dx \tag{8.25}$$

where $n < N$. Equation (8.25) is the same as equation (10.33) in Chapter 7 if all the functions are real. According to equation (10.38) of Chapter 7, completeness is established if

$$\lim_{n\to\infty} \delta_n^2 = 0 \tag{8.26}$$

We require the evaluation of $I[\Delta_n]$. In the following computation, the function $f(x)$ is assumed to conform to the same boundary conditions as the eigenfunctions:

$$\begin{aligned} I[\Delta_n] &= -\int_a^b \left[f - \sum_{k=1}^{n} c_k u_k\right]\left[S(f) - \sum_{j=1}^{n} c_j S(u_j)\right] dx \\ &= I[f] + \int_a^b \sum_{j=1}^{n} c_j f S(u_j)\, dx \\ &\quad + \int_a^b \sum_{k=1}^{n} c_k u_k S(f)\, dx \\ &\quad - \int_a^b \sum_{k=1}^{n} \sum_{j=1}^{n} c_k c_j u_k S(u_j)\, dx \end{aligned} \tag{8.27}$$

In the next step, we use the Hermitian property of the Sturm–Liouville operator as expressed by

$$\int_a^b u_k S(f)\, dx = \int_a^b S(u_k) f\, dx \tag{8.28}$$

Also needed is (8.23) and the expression for the expansion coefficients as given by

$$c_k = \int_a^b u_k f \rho\, dx \tag{8.29}$$

The result is

$$I[\Delta_n] = I[f] - \sum_{k=1}^{n} c_k^2 \lambda_k \tag{8.30}$$

Next consider

$$\int_a^b \Delta_n(x) u_j \rho\, dx = \int_a^b \left(f - \sum_{k=1}^{n} c_k u_k \right) u_j \rho\, dx$$
$$= c_j - \sum_{k=1}^{n} c_k \delta_{kj} \tag{8.31}$$

Equation (8.31) is zero if $j = 1, 2, \ldots, n$. The function Δ_n is therefore orthogonal to all eigenfunctions $u_1, u_2, \ldots, u_n$. If we define

$$\psi_n(x) = \frac{\Delta_n(x)}{\delta_n} \tag{8.32}$$

then according to (8.25), ψ_n is normalized. We therefore have the result that $\psi_n(x)$ is normalized and orthogonal to $u_1, u_2, \ldots, u_n$. This shows that $I[\psi_n] \geq \lambda_{n+1}$ for all $n < N$. Equation (8.30) can be written as

$$\delta_n^2 I[\psi_n] = I[f] - \sum_{k=1}^{n} c_k^2 \lambda_k \tag{8.33}$$

According to equation (10.32) in Chapter 7, the eigenvalues become arbitrarily large as n increases. Therefore, (8.33) implies

$$I[f] \geq \sum_{k=1}^{n} c_k^2 \lambda_k \tag{8.34}$$

If the function f is such that the functional $I[f]$ exists, then (8.34) puts an upper limit on the sum

$$\sum_{k=1}^{n} c_k^2 \lambda_k \tag{8.35}$$

which is independent of n. The eigenvalues become large and positive as n increases, which means that (8.35) cannot diverge to $-\infty$. We can therefore let N, and hence n, become arbitrarily large and arrive at the conclusion that

$$\sum_{k=1}^{\infty} c_k^2 \lambda_k \tag{8.36}$$

converges. For k sufficiently large, $c_k^2 \lambda_k > c_k^2$. Therefore, by comparison, Σc_k^2 converges. Since (8.11) is independent of N, we conclude that

$$\sum_{k=1}^{\infty} c_k^2 = 1 \tag{8.37}$$

Reference to equations (10.37) through (10.39) of Chapter 7 shows that the completeness of the eigenfunctions is established. We can allow N to become infinite in all equations of this section. In particular, (8.10) now includes all possible normalized functions that conform to the boundary conditions of the problem, and the sequence of eigenvalues is established as minima of the functional (8.1) with respect to all such functions. Our proof assumes that the class of functions which can be expanded in terms of the eigenfunctions conforms to the same boundary conditions as the eigenfunctions, but this requirement can in fact be relaxed. Also, we have not addressed ourselves to the general problem of establishing completeness if the problem is defined over an infinite interval.

As an example of how the results of this section can be used, we will pretend that we don't know the solution of

$$\frac{d^2u}{dx^2} + \kappa^2 u = 0 \qquad u(0) = u(1) = 0 \tag{8.38}$$

and attempt to estimate the lowest eigenvalue. The lowest eigenfunction will have no nodes between $x = 0$ and $x = 1$ and will be symmetric about $x = \frac{1}{2}$. Such a function is

$$u = Nx(1 - x) \tag{8.39}$$

Normalization gives

$$\int_0^1 u^2\, dx = \frac{N^2}{30} = 1 \qquad N = \sqrt{30} \tag{8.40}$$

The value of the functional (8.1) is

$$I = \int_0^1 u'^2\, dx = 10 \tag{8.41}$$

If λ_1 is the lowest eigenvalue, then $\lambda_1 \leq 10$. The actual value is $\lambda_1 = \pi^2 = 9.87$. This shows that a trial function which is correctly normalized, conforms to the boundary conditions of the problem, and resembles the actual eigenfunction is capable of producing a good estimate of the actual eigenvalue.

Problems (20.12) and (20.13) of Chapter 9 are concerned with the generalization of Sturm–Liouville theory to three dimensions. We will now continue with this development and show that extremizing the functional

$$I = \int_\Sigma (p\nabla u \cdot \nabla u^* + quu^*)\, d\Sigma \tag{8.42}$$

subject to the normalizing constraint

$$\int_\Sigma uu^* \rho\, d\Sigma = 1 \tag{8.43}$$

leads to the Sturm–Liouville differential equation

$$\nabla \cdot p\nabla u - qu + \lambda\rho u = 0 \tag{8.44}$$

The functions $p(\mathbf{r})$ and $q(\mathbf{r})$ are real. The weight function $\rho(\mathbf{r})$ is real and positive throughout the region Σ in which the problem is defined. For the sake of generality, we have allowed for the possibility that $u(\mathbf{r})$ is complex. We look for stationary values of the functional

$$\bar{I} = \int_\Sigma (p\nabla u \cdot \nabla u^* + quu^* - \lambda\rho uu^*)\, d\Sigma$$

$$= \int_\Sigma L(x_k, u, u^*, \partial_k u, \partial_k u^*)\, d\Sigma \tag{8.45}$$

This is really a generalization of the treatment of the Schrödinger eigenvalue problem given in section 7. The function u and its complex conjugate are to be treated as independent functions. The variation of (8.45) gives

$$\delta\bar{I} = \int_\Sigma \left[\frac{\partial L}{\partial u}\delta u + \frac{\partial L}{\partial(\partial_k u)}\delta(\partial_k u) + \frac{\partial L}{\partial u^*}\delta u^* + \frac{\partial L}{\partial(\partial_k u^*)}\delta(\partial_k u^*) \right] d\Sigma \tag{8.46}$$

If the components of a vector $\mathbf{a}$ are defined by

$$a_k = \frac{\partial L}{\partial(\partial_k u)} \tag{8.47}$$

then an integration by parts can be carried out in a manner similar to that given in equation (7.10). The result is

$$0 = \int_\Sigma \left(\frac{\partial L}{\partial u} - \nabla \cdot \mathbf{a} \right) \delta u\, d\Sigma + \int_\Sigma \left(\frac{\partial L}{\partial u^*} - \nabla \cdot \mathbf{a}^* \right) \delta u^*\, d\Sigma \tag{8.48}$$

provided that the boundary conditions are such that

$$\int_\sigma \delta u \mathbf{a} \cdot d\boldsymbol{\sigma} = \int_\sigma \delta u p \nabla u \cdot d\boldsymbol{\sigma} = 0 \tag{8.49}$$

The Euler–Lagrange equation

$$\frac{\partial L}{\partial u} - \nabla \cdot \mathbf{a} = 0 \tag{8.50}$$

then gives the Sturm–Liouville differential equation (8.44) as required. The boundary conditions on $u(\mathbf{r})$ in order to satisfy (8.49) are consistent with equation (20.74) as is required to define an eigenvalue problem.

Just as in the one-dimensional case, the stationary values of the functional (8.42) are the eigenvalues as we now show. By means of the identity

$$\nabla \cdot u^*(p\nabla u) = \nabla u^* \cdot p\nabla u + u^* \nabla \cdot (p\nabla u) \tag{8.51}$$

the divergence term can be converted into a surface integral which is zero on account of (8.49). Since u obeys (8.44) and the normalization condition (8.43), the final result is

$$I = \lambda \tag{8.52}$$

as expected. The eigenfunctions u_k form a sequence of functions which minimize the functional (8.42). The actual values of the minima are the eigenvalues λ_k.

The eigenvalue problem has been a central theme of this book. Let us look once more at it in matrix form, this time from a variational point of view. Let A and G be Hermitian matrices and let G be positive definite. Consider the problem of choosing vectors X such that the central quadric

$$\phi = X^\dagger A X \tag{8.53}$$

is extremized subject to the constraint

$$1 = X^\dagger G X \tag{8.54}$$

The idea of a central quadric is discussed at the end of section 10 in Chapter 2. The matrix G is required to be positive definite in order that $X^\dagger G X > 0$ for any X. This permits X to be normalized so that (8.54) is obeyed. In the theory of linear vibrations given in section 14 of Chapter 4, the matrix G played the role of a metric tensor. By varying the vector X in (8.53)

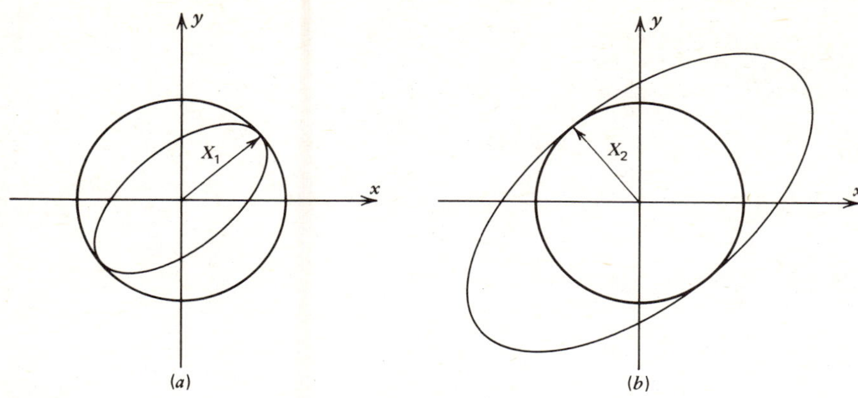

Fig. 8.1

and (8.54), we get

$$\delta\phi = \delta X^{\dagger}AX + X^{\dagger}A\,\delta X \tag{8.55}$$

$$0 = \delta X^{\dagger}GX + X^{\dagger}G\,\delta X \tag{8.56}$$

Now multiply (8.56) by a Lagrange multiplier λ and subtract the result from (8.55). Then set $\delta\phi = 0$ to get

$$0 = \delta X^{\dagger}(A - \lambda G)X + X^{\dagger}(A - \lambda G)\,\delta X \tag{8.57}$$

Since the variation δX is arbitrary, (8.57) implies that

$$(A - \lambda G)X = 0 \qquad X^{\dagger}(A - \lambda G) = 0 \tag{8.58}$$

Both equations in (8.58) are identical because A and G are Hermitian. We have therefore obtained the standard eigenvalue problem in matrix form from a variational principle. For a given normalized eigenvector X, the central quadric (8.53) has the value

$$\phi = X^{\dagger}\lambda GX = \lambda \tag{8.59}$$

The close analogy between the matrix eigenvalue problem and the Sturm–Liouville problem is obvious.

From the matrix version of the eigenvalue problem, we can construct a simple geometrical interpretation of the variational procedure. Suppose that the problem is two-dimensional. Assume that G is the identity matrix and that A is a real symmetric matrix with two positive eigenvalues. If x and y are the two components of X, then the normalization condition (8.54) is

$$x^2 + y^2 = 1 \tag{8.60}$$

and represents a unit circle in the x, y plane as shown in Fig. 8.1. If the curves obtained from setting $\phi =$ constant are plotted, a series of ellipses is obtained. Imagine starting out with a very small value of ϕ and plotting the corresponding ellipses as ϕ is allowed to increase. Finally, an ellipse is found the major axis of which just touches the unit circle (8.60) as shown in Fig. 8.1*a*. This represents the minimum possible value of ϕ that is consistent with the constraint. The coordinates of the point of contact of the major axis of the ellipse and the unit circle give the components of the eigenvector X_1. If ellipses are now plotted for larger values of ϕ, eventually the minor axis will just touch the unit circle as shown in Fig. 8.1*b*. This is the largest possible value of ϕ that is consistent with the constraint and represents the other eigenvalue. Again, the coordinates of the point of contact give the components of the eigenvector X_2. This analysis reveals that the constraint is absolutely necessary because extremizing ϕ subject to no constraint would give $\phi = \infty$.

PROBLEMS

8.1. Write down the expression for functions of the class (8.10) which are orthogonal to both u_1 and u_2. Demonstrate the orthogonality explicitly.

8.2. Show that the equality holds in (8.34) in the limit $n \to \infty$.

8.3. It is not always convenient to normalize a trial eigenfunction before beginning a calculation. Show that if u is not normalized but is orthogonal to the eigenfunctions $u_1, u_2, \ldots, u_{k-1}$ of a one-dimensional Sturm–Liouville system, then the kth eigenvalue obeys the inequality

$$\lambda_k \le \frac{\int_a^b (pu'^2 + qu^2)\, dx}{\int_a^b u^2 \rho\, dx} \tag{8.61}$$

8.4. Show that for the Helmholtz equation $\nabla^2 u + \kappa^2 u = 0$ the eigenvalues are to be estimated by means of

$$\kappa^2 \le \frac{\int_\Sigma \nabla u \cdot \nabla u\, d\Sigma}{\int_\Sigma u^2\, d\Sigma} \tag{8.62}$$

where u is a suitable trial eigenfunction and is not necessarily normalized. By assuming a trial eigenfunction of the form

$$u = 1 - \left(\frac{r}{a}\right)^n \tag{8.63}$$

show that the lowest eigenvalue for the vibrations of a circular membrane satisfies

$$\kappa^2 \le \frac{1}{a^2}\left(n + 3 + \frac{2}{n}\right) \tag{8.64}$$

The membrane is clamped at its perimeter $r = a$. *Caution:* The problem is two-dimensional so that the volume element $d\Sigma$ in (8.62) becomes an element of area. Find the optimum value of the parameter n and show that $\kappa a \le 2.412$. Compare this to the exact value. See section 13 of Chapter 9.

11 Numerical Methods

Most of the subject matter of this chapter deals with examples of computer programs which can be used along with other parts of the text. By limiting the treatment in this manner, we can introduce the reader to at least some of the aspects of the very extensive subject of numerical methods. Many readers will have had experience with computers and will want to try out a few programs using the mathematical material they have just learned. The programs are written in Basic language and are suitable for most small home computing systems. They were developed on one of the most popular of these, the Apple II computer. Many of the programs will be analyzed in detail but no attempt will be made to explain the fundamentals of programming. It is therefore assumed that the reader has some familiarity with Basic language. References to other parts of the text will be given where appropriate. For example, in the heading of section 1 below, (1.3.3) refers the reader to Chapter 1, equation (3.3), where a matrix as an array of numbers is first introduced.

1. THE MATRIX AS AN ARRAY OF NUMBERS (1.3.3)

Many computers have a special "MAT" command for matrices, but some small home computers do not. For this reason, the programs on matrices will be written using rectangular arrays. The following program loads a rectangular array into the computer which can then be used in various matrix operations.

```
1 DATA 1,2,3
2 DATA 4,5,6
10 PRINT ``ENTER THE NUMBER OF ROWS'':
INPUT N
20 PRINT ``ENTER THE NUMBER OF COLUMNS
'': INPUT M
22 PRINT
23 PRINT ``A='': PRINT
25 DIM A(N,M)
30 FOR R=1 TO N
40 FOR C=1 TO M
50 READ E
60 A(R,C)=E
70 PRINT A(R,C); SPC(3)
80 NEXT C: PRINT: PRINT
90 NEXT R
100 END
RUN
ENTER THE NUMBER OF ROWS
?2
ENTER THE NUMBER OF COLUMNS
?3
A=
1 2 3
4 5 6
```

Lines 1 through 9 have been reserved for the matrix elements to be entered as data. It is suggested that each line be used as one row of the matrix. In principle, the size of the matrix that can be entered is limited only by the memory of the computer. If you want to enter a matrix with 100 rows, you will of course have to leave 100 lines for that purpose and start the rest of the program at, say, line 110 instead of 10. Lines 10 and 20 tell the computer how many rows and columns the matrix has. Line 22 causes a space to be left between printed material. Line 25 reserves enough memory for the array. Lines 20 through 90 contain two loops, one nested inside the other. One loop is for the rows and the other is for the columns. In other words, for a given value of R, the second loop records the column entries. Line 50 reads the data and line 60 assigns the data just read to a location in the array. Lines 70 and 80 cause the array to be printed. In line 80, the first print statement causes the next row to be printed on a new line and the second print statement results in a space between the rows. The use of the colon permits more than one statement to be put on the same line. Line 80 is therefore equivalent to the three lines:

```
80 NEXT C
81 PRINT
82 PRINT
```

In line 70, the statement SPC(3) tells the computer to leave three spaces between entries in the array when it is printed. Running the program causes the array to be printed and also to be stored in the computer for use in various matrix operations that would be performed in a larger program. If you want the array to be stored and not printed, omit line 70 and the two print statements in line 80. Most of the commands used in the programs in this chapter are common to all computers using Basic language, but you will find a few that are peculiar to the Apple Computer.

2. EXCHANGE OF ROWS IN AN ARRAY

The computation of the inverse of a matrix by Gauss elimination (1.10.18) and the evaluation of determinants (1.5.1) involves various row and column operations on rectangular arrays. The following program exchanges the first two rows of a 3×3 array. It can easily be generalized to exchange any two rows (or columns) of an array of any size.

```
1 DATA 1,2,3
2 DATA 4,5,6
2 DATA 7,8,9
5 FOR R=1 TO 3
10 FOR C=1 TO 3
15 READ E
20 A(R,C)=E
25 PRINT A(R,C) SPC(3)
30 NEXT C: PRINT: PRINT
35 NEXT R
36 PRINT
40 FOR K=1 TO 3
45 X=A(1,K)
50 A(1,K)=A(2,K)
55 A(2,K)=X
60 NEXT K
65 FOR R=1 TO 3
70 FOR C=1 TO 3
75 PRINT A(R,C) SPC(3)
80 NEXT C: PRINT: PRINT
85 NEXT R
90 END
RUN
1 2 3
4 5 6
7 8 9

4 5 6
1 2 3
7 8 9
```

The exchange is performed in lines 40 through 60. For a given value of K, line 45 temporarily assigns the value of $A(1, K)$ to a variable X. Line 50 replaces the value of $A(1, K)$ with that of the corresponding element $A(2, K)$ of the second row. Finally, line 55 reassigns the original value of $A(1, K)$ to $A(2, K)$, thus completing the exchange. The statement RUN after the last line of the program causes the program to be executed with the result shown just after the program.

3. MULTIPLICATION OF A ROW BY A FACTOR

The following program multiplies the first row of the original array by a factor of two. Lines 1 through 36 are identical to those of program 2 and have been omitted.

```
40 FOR K=1 TO 3
45 A(1,K)=2*A(1,K)
50 NEXT K
100 FOR R=1 TO 3
110 FOR C=1 TO 3
120 PRINT A(R,C); TAB(1+8*C)
130 NEXT C: PRINT: PRINT
```

```
140 NEXT R
200 END
RUN
1 2 3
4 5 6
7 8 9

2    4    6
6    9    12
7    8    9
```

For a given value of K, line 45 replaces $A(1, K)$ by $2A(1, K)$. Line 120 illustrates the use of the TAB function to line up the elements of the different columns when they are printed. By contrast, SPC(N) leaves N spaces between the elements of a given row regardless of how many digits a given element has.

4. ADDITION OF A MULTIPLE OF ONE ROW TO ANOTHER ROW

If line 45 in program 3 is replaced by

```
45 A(2,K)=A(2,K)+2*A(1,K)
```

the result of running the program will be to replace the elements $A(2, K)$ of row 2 by $A(2, K) + 2A(1, K)$.

5. TRANSPOSE OF AN ARRAY (1.5.25)

Lines 1 through 36 of program 2 can be used. The portion of the program which calculates the transpose of $A(R, C)$ is

```
40 FOR R=1 TO 3
45 FOR C=1 TO 3
50 T(R,C)=A(C,R)
55 PRINT T(R,C); TAB(1+8*C)
60 NEXT C: PRINT: PRINT
65 NEXT R
100 END
RUN
1 2 3
4 5 6
7 8 9

1    4    7
2    5    8
3    6    9
```

Note that the TAB function TAB($1 + T * C$) produces columns in the printed array which are $T - 1$ spaces apart.

6. MATRIX MULTIPLICATION (1.3.19)

The following program multiplies together any two square matrices.

```
5 PRINT ``ENTER N'': INPUT N
10 DATA 1,2,3,5
11 DATA 2,1,0,6
12 DATA 1,0,9,2
13 DATA 5,1,2,1
20 PRINT ``A='': PRINT
30 FOR R=1 TO N
40 FOR C=1 TO N
50 READ M
60 A(R,C)=M
70 PRINT A(R,C) SPC(3)
80 NEXT C: PRINT: PRINT
90 NEXT R
100 DATA -1,2,0,-2
101 DATA 3,0,-6,2
102 DATA 1,5,-8,3
103 DATA 6,0,1,2
120 PRINT ``B='': PRINT
130 FOR R=1 TO N
140 FOR C=1 TO N
150 READ M
160 B(R,C)=M
165 PRINT B(R,C) SPC(3)
170 NEXT C: PRINT: PRINT
180 NEXT R
200 FOR J=1 TO N
210 FOR L=1 TO N
220 S=0
230 FOR K=1 TO N
240 S=S+A(J,K)*B(K,L)
250 NEXT K
260 P(J,L)=S
270 NEXT L
280 NEXT J
```

```
300 PRINT ``AB=´´: PRINT
310 FOR R=1 TO N
320 FOR C=1 TO N
330 PRINT P(R,C); TAB(1+8*C)
340 NEXT C: PRINT: PRINT
350 NEXT R
360 END
RUN
ENTER N
?4
A=
1 2 3 5
2 1 0 6
1 0 9 2
5 1 2 1
B=
-1 2 0 -2
3 0 -6 2
1 5 -8 3
6 0 1 2
AB=
38      17      -31     21
37      4       0       10
20      47      -70     29
6       20      -21     0
```

This program has no DIM $A(M, N)$ or DIM $B(M, N)$ statement. These are generally not needed unless $N > 10$. The matrix multiplication takes place in lines 200 through 280. There are three nested loops. Each value of J corresponds to a row and each value of L corresponds to a column. For each value of J and L, the K-loop multiplies row J of A by column L of B and then sums the products. Line 260 records the result in the appropriate location of the product array $P(J, L)$.

Many of the programs to follow involve calculating sums, and it may be well to remind the reader how the computer does this. For a given value of J and L, line 220 assigns the value 0 to the sum. For $K=1$, line 240 replaces the value $S=0$ by $S=A(J,1)B(1, L)$. Line 250 sends the computer back to line 230 with the value $K=2$. Line 240 then replaces S by the new value $S=A(J,1)B(1, L)+A(J,2)B(2, L)$. This process continues until $K=N$ at which point the value of S is assigned to $P(J, L)$ in line 260.

The following program calculates the product of three 3×3 matrices. Only that portion of the program which does the multiplication is shown. You can add the missing parts of the program which load in the three arrays $A(R, J)$, $B(J, K)$, and $D(K, C)$.

```
300 FOR R=1 TO 3
310 FOR C=1 TO 3
320 S=0
330 FOR J=1 TO 3
340 FOR K=1 TO 3
350 S=S+A(R,J)*B(J,K)*D(K,C)
360 NEXT K
370 NEXT J
380 P(R,C)=S
390 PRINT P(R,C); TAB(1+8*C)
400 NEXT C: PRINT: PRINT
410 NEXT R
500 END
```

Note that there are now four nested loops. For given values of R and C, the J and K loops perform the required double sum.

7. ORTHOGONAL TRANSFORMATION OF A VECTOR (2.2.13)

It is possible to include a matrix in a program that has elements which are functions of one or more parameters. To illustrate the point, the following program calculates the transformed components of a vector which are produced by a rotation about the z axis. Since the z component of the vector is unchanged, the program is written using a two-component vector and a 2×2 transformation matrix.

```
10 PRINT ``ENTER THE ANGLE OF ROTATION
IN DEGREES´´: INPUT A
20 PRINT ``ENTER THE COMPONENTS OF THE
VECTOR´´: INPUT C1,C2
25 PRINT
30 P=3.141592654
40 A=2*P*A/360
50 S(1,1)=COS(A): S(1,2)=SIN(A)
51 S(2,1)=-SIN(A): S(2,2)=COS(A)
60 X(1)=C1: X(2)=C2
70 FOR J=1 TO 2
80 T=0
```

```
90 FOR K=1 TO 2
100 T=T+S(J,K)*X(K)
110 NEXT K
120 Y(J)=T
130 NEXT J
140 PRINT ``THE TRANSFORMED COMPONENTS
OF THE VECTOR ARE'': PRINT
150 PRINT ``X'='';Y(1);SPC(3);``Y'='';
Y(2)
160 END
RUN
ENTER THE ANGLE OF ROTATION IN DEGREES
?30
ENTER THE COMPONENTS OF THE VECTOR
?1,2
THE TRANSFORMED COMPONENTS OF THE VECTOR
ARE X'=1.8660254 Y'=1.23205081
```

In line 10, the computer asks for the angle of rotation in degrees. In the example given above, 30° was entered. The components of the vector in the original coordinates are requested by line 20. Line 40 converts the angle of rotation to radian measure. In lines 50 and 51, the appropriate numerical values are assigned to the elements of the orthogonal transformation. Line 60 constructs a column matrix that represents the vector in the original coordinates. The required matrix multiplication is carried out in lines 70 through 130. The transformed vector is represented by a column matrix with elements $Y(1)$ and $Y(2)$. Line 140 prints the components of the transformed vector relabeled as X' and Y'.

8. SECOND-RANK TENSOR TRANSFORMATION (2.6.7)

The following program performs a second-rank tensor transformation on an arbitrary two-dimensional tensor. The transformation used is orthogonal.

```
1 PRINT ``ENTER ROW 1 OF THE TENSOR''
2 INPUT X1,X2
3 PRINT ``ENTER ROW 2 OF THE TENSOR''
4 INPUT X3,X4
5 PRINT ``ENTER THE ROTATION ANGLE IN
DEGREES''
6 INPUT D
7 PRINT
10 P=3.14159264
20 D=2*P*D/360
30 S(1,1)=COS(D): S(1,2)=SIN(D)
40 S(2,1)=-SIN(D): S(2,2)=COS(D)
50 A(1,1)=X1: A(1,2)=X2
60 A(2,1)=X3: A(2,2)=X4
70 PRINT: PRINT ``A='': PRINT
80 PRINT X1; TAB(10); X2
90 PRINT: PRINT X3; TAB(10); X4
95 PRINT: PRINT ``SA(TR S)='': PRINT
100 FOR R=1 TO 2
110 FOR C=1 TO 2
120 T=0
130 FOR J=1 TO 2
140 FOR K=1 TO 2
150 T=T+S(R,J)*A(J,K)*S(C,K)
160 NEXT K
170 NEXT J
180 B(R,C)=T
190 PRINT B(R,C); TAB(1+14*C)
200 NEXT C: PRINT: PRINT
210 NEXT R
220 END
RUN
ENTER ROW 1 OF THE TENSOR
?1,2
ENTER ROW 2 OF THE TENSOR
?2,1
ENTER THE ROTATION ANGLE IN DEGREES
?30
A=
1      2
2      1
SA(TR S)=
2.7320508  1.00000001
1.00000001 -7.32050803
```

There are some minor differences between this and previous programs. In lines 2 and 4, the elements of the tensor are fed into the computer as input rather than data. Note that variables can be labeled as $X1$, $X2$,... rather than X, Y,... if desired. In lines 50 and 60, values are assigned to the 2×2 array which represents the tensor. This is appropriate for tensors or matrices of small dimension. Lines 100 through 210 perform the necessary matrix multiplications.

Note carefully that in line 150, the final factor is written as $S(C, K)$ rather than $S(K, C)$ in order to represent the elements of $\tilde{S}$ rather than S.

9. MATRIX INVERSE (1.10.21)

Before giving the program for matrix inversion, we discuss a short program that examines a list of numbers and determines the location of the largest member. This program is needed as part of the matrix inversion program.

```
10 DATA 300, 40, 25, 700, 2, 500
20 N=6
30 FOR K=1 TO N
40 READ E
50 C(K)=E
60 NEXT K
70 J=1
80 FOR K=J TO N
90 IF C(J)<C(K) THEN 110
100 NEXT K
105 GOTO 130
110 J=K
120 GOTO 80
130 PRINT J
RUN
4
```

The entries of the list are entered as data in line 10. Lines 30 through 60 assign the entries to a one-subscript variable $C(K)$. In line 70, J is given the initial value 1. Lines 80 and 90 compare $C(1)$ with successive members of the list until one is found which is larger than $C(1)$. The program then goes to line 110 where J is assigned a new value corresponding to this location. Line 120 directs the computer back to line 80 where the whole process is repeated until another entry in the list is found which is still larger. When no larger value is found, we are sent to line 130 which then prints the location of the largest entry in the list. If all members of the list are equal, the program gives $J = 1$. If the maximum value appears more than once, then we are given the location where the maximum first occurs.

The following program is suitable for finding the inverse of a matrix of any size:

```
1 DATA 1,1,1,1
2 DATA 1,2,4,8
3 DATA 1,3,9,27
4 DATA 1,4,16,64
15 N=4
16 DIM A(N,N): DIM B(N,N); DIM C(N)
20 FOR R=1 TO N
25 FOR C=1 TO N
30 READ E
35 A(R,C)=E
40 PRINT A(R,C) SPC(3)
45 NEXT C: PRINT: PRINT
50 NEXT R
60 FOR R=1 TO N
70 B(R,R)=1
80 NEXT R
95 FOR S=1 TO N
100 GOSUB 500
110 NEXT S
115 PRINT: PRINT ``THE INVERSE IS'':
PRINT
120 FOR R=1 TO N
130 FOR C=1 TO N
140 PRINT B(R,C) SPC(3)
150 NEXT C: PRINT: PRINT
160 NEXT R
220 END
500 FOR J=S TO N
505 C(J)=ABS(A(J,S))
510 NEXT J
515 J=S
520 FOR K=J TO N
525 IF C(J)<C(K) THEN 540
530 NEXT K
535 GOTO 550
540 J=K
545 GOTO 520
550 IF C(J)=0 THEN 900
580 FOR K=1 TO N
585 X=A(S,K): Y=B(S,K)
590 A(S,K)=A(J,K): B(S,K)=B(J,K)
595 A(J,K)=X: B(J,K)=Y
600 NEXT K
```

```
700 D=A(S,S)
705 FOR J=1 TO N
710 A(S,J)=A(S,J)/D
715 B(S,J)=B(S,J)/D
720 NEXT J
725 FOR J=1 TO N: IF J=S THEN 755
730 M=A(J,S)
735 FOR K=1 TO N
740 A(J,K)=A(J,K) - M*A(S,K)
745 B(J,K)=B(J,K) - M*B(S,K)
750 NEXT K
755 NEXT J
760 RETURN
900 PRINT ``NO INVERSE'': END
RUN
1 1 1  1
1 2 4  8
1 3 9  27
1 4 16  64
THE INVERSE IS
4  -6  4  -1
-4.33333334  9.5  -7  1.83333333
1.5  -4  3.5  -1
-.16666667  .5  -.5  .16666667
```

The program is based on Gauss elimination. The matrix $A(R,C)$ to be inverted is entered in lines 1 through 50. Lines 60 through 80 load a matrix $B(R,C)$ into the computer as the identity matrix.* The program is designed to perform row operations on $A(R,C)$ and $B(R,C)$ until $A(R,C)$ is reduced to the identity and $B(R,C)$ becomes the inverse of $A(R,C)$. For a given value of S as defined by line 95, line 100 sends us to a subroutine beginning with line 500. We will follow the subroutine through with $S=1$. If $A(1,1)\neq 0$, we can divide row 1 by $A(1,1)$ and then subtract appropriate multiples of it from other rows in such a way that the first column of $A(R,C)$ is converted to $(1,0,0,\ldots,0)$. If $A(1,1)=0$, the procedure fails. Another possibility is that $A(1,1)$ is very small and possibly inaccurate. The situation is improved by exchanging row 1 with the row with the largest first element. Lines 500 through 550 perform a search of all the elements of column 1 to find the entry with the largest magnitude. This part of the program is identical to the short program given at the beginning of this section. Note that line 505 assigns to $C(J)$ the absolute values of $A(J,1)$, since we want to find the element of largest magnitude irrespective of sign. If it is determined that the element of largest magnitude is zero, then all elements of column 1 are zero. Line 550 sends the computer to line 900 where "NO INVERSE" is printed and the program ends. If this is not the case, lines 580 through 600 exchange rows 1 and J of both $A(R,C)$ and $B(R,C)$. In the new version of $A(R,C)$, element $A(1,1)$ is the element of largest magnitude in column 1. Lines 700 through 720 divide the first rows of both $A(R,C)$ and $B(R,C)$ by $A(1,1)$. Lines 725 through 755 subtract the appropriate multiples of row 1 from each of the other rows so that the final form of the first column of $A(R,C)$ is $(1,0,0,\ldots,0)$ as required. The same operations are done on $B(R,C)$. We now return to line 110 where S is given the value 2. The whole process repeats, this time working with column 2. Note that row 1 is left out of the search-and-exchange procedures of lines 500 through 600. We must not move row 1 because to do so would disrupt the first column. The search procedure may discover that all elements of the second column are zero with the possible exception of $A(1,2)$ which it does not examine. Since this means that column 2 is a multiple of column 1, we are sent to line 900 where the "NO INVERSE" message is printed. Assuming that this does not happen, lines 725 through 760 reduce column 2 of $A(R,C)$ to $(0,1,0,\ldots,0)$. Note that line 725 is necessary to prevent row 2 from being subtracted from itself.

*Since the off-diagonal elements of $B(R,C)$ are not specified, the computer automatically assigns them the value zero.

It may have occurred to the reader that the use of a subroutine in this program is unnecessary. It is a remnant of the program development. It is sometimes a good idea to write segments of the program and try them out separately before they are integrated into the main body of the program.

10. MATRIX EIGENVALUES

For a general discussion of the matrix eigenvalue problem, refer to Chapter 2, section 10, and Chapter 7, section 14. The general problem of finding the eigenvalues of an $n \times n$ matrix is difficult. We first offer a program that will find the dominant eigenvalue and corresponding eigenvector of a symmetric matrix. To start the program off, we first express a suitable starting vector X as a linear combination of the eigenvectors of the matrix A in which we are interested:

$$X = a_1 X_1 + a_2 X_2 + \cdots + a_n X_n \tag{10.1}$$

This is possible because the eigenvectors are a complete set of n linearly independent vectors. For convenience, we assume that they are normalized. If we operate on (10.1) by the matrix A a total of p times, the result is

$$A^p X = a_1 \lambda_1^p X_1 + a_2 \lambda_2^p X_2 + \cdots + a_n \lambda_n^p X_n \tag{10.2}$$

where $\lambda_1, \lambda_2, \ldots$ are the eigenvalues. Suppose that λ_1 is the eigenvalue of largest magnitude. Equation (10.2) can be expressed as

$$A^p X = \lambda_1^p \left[a_1 X_1 + a_2 \left(\frac{\lambda_2}{\lambda_1} \right)^p X_2 + \cdots + a_n \left(\frac{\lambda_n}{\lambda_1} \right)^p X_n \right] \tag{10.3}$$

If p is sufficiently large, then

$$A^p X = \lambda_1^p a_1 X_1 = N X_1 \tag{10.4}$$

Normalization of (10.4) then gives the eigenvector X_1. The eigenvalue then follows from $AX_1 = \lambda_1 X_1$. The starting vector X should have all nonzero components. In the following program, it is chosen to be $(1, 1, \ldots, 1)$.

```
1 DATA 1,2,3,4
2 DATA 2,1,5,6
3 DATA 3,5,1,7
4 DATA 4,6,7,1
15 N=4
16 DIM A(N,N): DIM X(N): DIM Y(N)
20 FOR R=1 TO N
30 FOR C=1 TO N
40 READ E
50 A(R,C)=E
60 PRINT A(R,C); SPC(3)
70 NEXT C: PRINT: PRINT
80 NEXT R
81 PRINT: PRINT ``THE NORMALIZED
EIGENVECTOR IS'': PRINT
85 FOR K=1 TO N
90 X(K)=1
100 NEXT K
105 P=20
110 FOR J=1 TO P
120 FOR K=1 TO N
130 S=0
140 FOR L=1 TO N
150 S=S+A(K,L)*X(L)
160 NEXT L
170 Y(K)=S
180 NEXT K
190 S=0
200 FOR K=1 TO N
210 S=S+Y(K)↑2
220 NEXT K
230 FOR K=1 TO N
240 X(K)=Y(K)/SQR(S)
250 NEXT K
260 NEXT J
270 FOR K=1 TO N
280 S=0
290 FOR L=1 TO N
300 S=S+A(K,L)*X(L)
310 NEXT L
320 Y(K)=S
330 PRINT X(K)
335 NEXT K
340 J=1
350 FOR K=J TO N
360 IF ABS (X(J))<ABS(X(K)) THEN 390
370 NEXT K
380 GOTO 410
390 J=K
400 GOTO 350
410 E=Y(J)/X(J)
415 PRINT
420 PRINT ``THE EIGENVALUE IS''; E
430 S=0
440 FOR K=1 TO N
450 S=S+Y(K)↑2
460 NEXT K
465 PRINT: PRINT ``THE MAGNITUDE OF AX
IS''
470 PRINT: PRINT SQR(S)
```

```
480 PRINT: PRINT ``WHERE A IS THE MATRIX
AND X IS THE NORMALIZED
EIGENVECTOR''
RUN
1 2 3 4
2 1 5 6
3 5 1 7
4 6 7 1
THE NORMALIZED EIGENVECTOR IS
.35249192
.492878785
.542032059
.582255267
THE EIGENVALUE IS 15.0169875
THE MAGNITUDE OF AX IS
15.0169875
WHERE A IS THE MATRIX AND X IS THE
NORMALIZED EIGENVECTOR
```

The p multiplications indicated by equation (10.4) are carried out in lines 105 through 260. Lines 190 through 250 normalize the vector after each multiplication to keep the components from becoming unduly large. We assume that when line 270 is reached, $X(K)$ represents the actual eigenvector corresponding to the dominant eigenvalue. Lines 270 through 320 carry out the multiplication AX one more time to produce the eigenvalue by means of $AX = \lambda X$. Lines 340 through 410 find the eigenvalue by comparing the largest component of X to the corresponding component of λX. Lines 430 through 460 find the magnitude of λX. Since X is normalized, this should be the magnitude of the eigenvalue.

If the dominant eigenvalue is either degenerate, or nearly degenerate, the program produces an accurate eigenvalue, but the eigenvector is questionable. In the case of degeneracy, there is more than one eigenvector corresponding to the eigenvalue,and the program gives one eigenvector which is some mix of these possibilities. Suppose that the dominant eigenvalue is almost degenerate and that λ_2 is slightly less than λ_1. This means that $(\lambda_2/\lambda_1)^p$ goes to zero very slowly with increasing p. The vector A^pX then still has a large X_2 component. Entering a larger value of p in line 105 may improve the situation.

An algorithm for finding the remaining eigenvalues can be based on the proof of theorem 3 in section 14 of Chapter 7. Once the eigenvector corresponding to the eigenvalue of largest magnitude is in hand, the matrix T_1 of equation (7.14.8) can be constructed. A possibility is

$$T_1 = \begin{pmatrix} x_1 & 0 & 0 & 0 \\ x_2 & 1 & 0 & 0 \\ x_3 & 0 & 1 & 0 \\ x_4 & 0 & 0 & 1 \end{pmatrix} \tag{10.5}$$

To simplify the equations a bit, the ideas will be illustrated with 4×4 matrices. It may turn out that $x_1 = 0$, in which case (10.5) is singular and therefore unacceptable. Suppose that the component of the eigenvector with the largest magnitude is x_3. A better choice for T_1 is then

$$T_1 = \begin{pmatrix} x_1 & 1 & 0 & 0 \\ x_2 & 0 & 1 & 0 \\ x_3 & 0 & 0 & 0 \\ x_4 & 0 & 0 & 1 \end{pmatrix} \tag{10.6}$$

If we make the assumption that the component of the eigenvector with the largest magnitude is also the most accurate component, then (10.6) is an improvement over (10.5) even if x_1 is not zero.

Since our concern at the moment is with real symmetric matrices, it would be even better to use an orthogonal transformation in place of (10.6). To accomplish this, columns two, three, and four of (10.6) are replaced by unit vectors which are orthogonal to one another and also to the first column, which must remain as the original eigenvector. The required mutually orthogonal vectors can be found by the Schmidt orthogonalization procedure given in equations (7.11.31) through (7.11.35). By setting $x_1 = S_{11}$, $x_2 = S_{12}, \ldots,$ we have an orthogonal transformation S_1 such that

$$A' = S_1 A \tilde{S}_1 = \begin{pmatrix} \lambda_1 & 0 & 0 & 0 \\ 0 & 22 & 23 & 24 \\ 0 & 32 & 33 & 34 \\ 0 & 42 & 43 & 44 \end{pmatrix} \tag{10.7}$$

It has been the custom in this book to write an orthogonal similarity transformation as $SA\tilde{S}$ and a general similarity transformation as $T^{-1}AT$. Thus, $\tilde{S}$ corresponds to T and the first row of S_1 is the eigenvector (x_1, x_2, x_3, x_4). In many references, you will find an orthogonal similarity transformation written as $\tilde{S}AS$.

The process is now repeated by finding the dominant eigenvalue and corresponding eigenvalue of the 3×3 submatrix appearing in (10.7). Once this is done, we can construct an orthogonal transformation of the form

$$S_2 = \begin{pmatrix} 1 & 0 & 0 & 0 \\ 0 & 22 & 23 & 24 \\ 0 & 32 & 33 & 34 \\ 0 & 42 & 43 & 44 \end{pmatrix} \tag{10.8}$$

such that

$$A'' = S_2 A' \tilde{S}_2 = \begin{pmatrix} \lambda_1 & 0 & 0 & 0 \\ 0 & \lambda_2 & 0 & 0 \\ 0 & 0 & 33 & 34 \\ 0 & 0 & 43 & 44 \end{pmatrix} \tag{10.9}$$

Repeating the process once more completes the diagonalization and yields an overall transformation

$$S = S_3 S_2 S_1 \tag{10.10}$$

The eigenvectors are displayed as the rows of S. A program for carrying out this procedure is given in section 12. Schmidt orthogonalization, which is one of the required ingredients, is discussed in section 11.

The program that we have discussed in this section is called the *power method* for obtaining the dominant eigenvalue of a matrix. It works for symmetric matrices and also nonsymmetric matrices if the eigenvectors form a complete set.

11. SCHMIDT ORTHOGONALIZATION (7.11.31)

Let the original complete set of vectors form the rows of the matrix T and the new orthogonal vectors form the rows of the matrix U. In terms of the notation used in the program given in this section, equations (7.11.31) through (7.11.34) translate into the following:

$$U(1, K) = T(1, K) \tag{11.1}$$

$$U(2, K) = T(2, K) + Q(2,1)U(1, K) \tag{11.2}$$

$$U(3, K) = T(3, K) + Q(3,1)U(1, K) + Q(3,2)U(2, K) \tag{11.3}$$

$$\vdots$$

$$U(N, K) = T(N, K) + Q(N, K)U(1, K) + \cdots + Q(N, N-1)U(N-1, K) \tag{11.4}$$

$$Q(2,1) = -\frac{1}{M(1)} \sum_K U(1, K)T(2, K) \tag{11.5}$$

$$Q(3,1) = -\frac{1}{M(1)} \sum_K U(1, K)T(3, K) \tag{11.6}$$

$$Q(3,2) = -\frac{1}{M(2)} \sum_K U(2, K)T(3, K) \tag{11.7}$$

$$M(1) = \sum_K U(1, K)^2 \quad M(2) = \sum_K U(2, K)^2 \tag{11.8}$$

where $Q(J, K)$ replaces α_{jk}. The program follows:

```
1 DATA 1,2,3,1
2 DATA 1,0,0,0
3 DATA 0,1,0,0
4 DATA 0,0,0,1
15 N=4
16 DIM T(N,N): DIM U(N,N)
17 DIM M(N): DIM Q(N,N)
18 DIM S(N,N): DIM I(N,N)
19 PRINT: PRINT ``ORIGINAL VECTORS
ARE ROWS OF'': PRINT
20 FOR R=1 TO N
25 FOR C=1 TO N
30 READ E
35 T(R,C)=E
40 PRINT T(R,C) SPC(3)
45 NEXT C: PRINT: PRINT
50 NEXT R
55 FOR K=1 TO N
60 U(1,K)=T(1,K)
65 NEXT K
70 FOR R=2 TO N
80 M=0
90 FOR K=1 TO N
100 M=M+U(R-1,K)↑2
110 NEXT K
120 M(R-1)=M
130 FOR C=1 TO R-1
```

```
140 S=0
150 FOR K=1 TO N
160 S=S+U(C,K)*T(R,K)
170 NEXT K
180 Q(R,C)=-S/M(C)
185 NEXT C
190 FOR K=1 TO N
195 S=T(R,K)
200 FOR J=1 TO R-1
210 S=S+Q(R,J)*U(J,K)
220 NEXT J
230 U(R,K)=S
240 NEXT K
250 NEXT R
260 M=0
270 FOR K=1 TO N
280 M=M+U(N,K)↑2
290 NEXT K
300 M(N)=M
305 PRINT ``THE ORTHONORMAL SET IS'':
PRINT
310 FOR R=1 TO N
320 FOR C=1 TO N
330 S(R,C)=U(R,C)/SQR(M(R))
340 PRINT S(R,C) SPC(3)
350 NEXT C: PRINT: PRINT
360 NEXT R
490 PRINT ``ROWS OF S ARE THE ORTHO
NORMAL SET'': PRINT
495 PRINT ``S*TRANS(S)='': PRINT
500 FOR R=1 TO N
510 FOR C=1 TO N
520 S=0
530 FOR K=1 TO N
540 S=S+S(R,K)*S(C,K)
550 NEXT K
560 I(R,C)=S
570 PRINT I(R,C) SPC(3)
580 NEXT C: PRINT: PRINT
590 NEXT R
RUN
```

```
ORIGINAL VECTORS ARE ROWS OF
1 2 3 1
1 0 0 0
0 1 0 0
0 0 0 1
THE ORTHONORMAL SET IS
.25819889 .516397779 .774586669 .25810889
.966091783 -.138013112 -.207019668 -.0690065559
-1.07612864E-10 .845154254 -.507092552 -.169030851
-6.71084115E-11 -9.82658882E-11 -.316227766 .948683298
THE ROWS OF S ARE THE ORTHONORMAL SET
S*TRANS(S)=
.999999999 7.50901563E-11 6.3096195E-11 1.32786226E-10
7.50901563E-11 .999999999 -1.40602197E-10 -1.00499165E-10
6.3096195E-11 -1.40602197E-10 .999999999 -2.48974175E-10
1.32786226E-10 -.100499165E-10 -2.48974175E-10 .999999999
```

Lines 55 through 65 evaluate the first row of U as dictated by equation (11.1). In line 70, we enter a loop in which R varies from 2 to N. Each value of R corresponds to a row of U as given by equations (11.2) through (11.4). For example, $R = 2$ constructs row two of U according to equation (11.2). We will follow the R loop through for the case $R = 2$. Lines 80 through 120 compute $M(1)$ as given by (11.8). Lines 130 through 180 evaluate $Q(2, 1)$ according to (11.5). Lines 190 through 240 then utilize (11.2) to evaluate $U(2, K)$. At line 250, the R loop repeats with $R = 3$ resulting in the evaluation of $U(3, K)$ as given by (11.3). At the completion of the R loop, $M(1)$ through $M(N-1)$ have been evaluated. Lines

260 through 300 evaluate the remaining $M(N)$. Finally, lines 310 through 360 normalize the new vectors to produce an orthogonal matrix with elements $S(R, C)$. The remaining part of the program calculates $S\tilde{S}$ as a check on the final result.

12. THE EIGENVALUES OF A SYMMETRIC MATRIX

All the ingredients are now in hand for carrying out the complete diagonalization of a real symmetric matrix. The algorithm to be used is explained in section 10. The program follows:

```
1 DATA 1,2,3,4
2 DATA 2,1,5,6
3 DATA 3,5,1,7
4 DATA 4,6,7,1
15 N=4
16 DIM A(N,N): DIM B(N,N)
17 DIM Y(N): DIM X(N)
18 DIM S(N,N): DIM Q(N,N)
19 DIM O(N,N): DIM F(N,N)
20 DIM D(N,N)
40 FOR R=1 TO N
41 O(R,R)=1
42 NEXT R
49 PRINT: PRINT ``THE ORIGINAL MATRIX
IS'': PRINT
50 FOR R=1 TO N
55 FOR C=1 TO N
60 READ E
65 B(R,C)=E
70 PRINT B(R,C) SPC(3)
75 NEXT C: PRINT: PRINT
80 NEXT R
81 RESTORE
100 FOR M=1 TO N-1
105 FOR R=M TO N
110 FOR C=M TO N
120 A(R,C)=B(R,C)
130 NEXT C
140 NEXT R
150 FOR R=1 TO N
160 FOR C=1 TO N
170 IF R<M OR C<M THEN A(R,C)=0
180 NEXT C
190 NEXT R
200 FOR K=1 TO M
205 X(K)=0: Y(K)=0
210 NEXT K
215 FOR K=M TO N
220 X(K)=1
225 NEXT K
230 P=40
240 FOR J=1 TO P
250 FOR K=M TO N
260 S=0
270 FOR L=M TO N
280 S=S+A(K,L)*X(L)
290 NEXT L
300 Y(K)=S
310 NEXT K
320 S=0
330 FOR K=M TO N
340 S=S+Y(K)↑2
350 NEXT K
360 FOR K=M TO N
370 X(K)=Y(K)/SQR(S)
380 NEXT K
390 NEXT J
400 FOR K=M TO N
410 S=0
420 FOR L=M TO N
430 S=S+A(K,L)*X(L)
440 NEXT L
450 Y(K)=S
460 NEXT K
470 J=M
480 FOR K=J TO N
490 IF ABS (X(J))<ABS(X(K)) THEN 520
500 NEXT K
510 GOTO 540
520 J=K
530 GOTO 480
540 E=Y(J)/X(J)
550 PRINT: PRINT ``EIG VAL'';M;
`` = '';E: PRINT
560 FOR R=1 TO N
565 FOR C=1 TO N
570 T(R,C)=0
575 NEXT C
580 NEXT R
```

```
585 FOR R=1 TO M
590 T(R,R)=1
595 NEXT R
600 FOR C=M TO N
610 T(M,C)=X(C)
620 NEXT C
630 E=1
640 FOR C=M TO N
650 IF C=J THEN 680
660 T(C+E,C)=1
670 GOTO 690
680 E=0
690 NEXT C
700 FOR R=1 TO N
705 FOR C=1 TO N
710 U(R,C)=0
715 NEXT C
720 NEXT R
725 FOR R=1 TO M
730 U(R,R)=1
735 NEXT R
740 FOR K=M TO N
750 U(M,K)=T(M,K)
760 NEXT K
770 FOR R=M+1 TO N
780 S=0
790 FOR K=M TO N
800 S=S+U(R-1,K)↑2
810 NEXT K
820 M(R-1)=S
830 FOR C=M TO R-1
840 S=0
850 FOR K=M TO N
860 S=S+U(C,K)*T(R,K)
870 NEXT K
880 Q(R,C)=-S/M(C)
885 NEXT C
890 FOR K=M TO N
895 S=T(R,K)
900 FOR J=M TO R-1
910 S=S+Q(R,J)*U(J,K)
920 NEXT J
930 U(R,K)=S
940 NEXT K
950 NEXT R
960 S=0
970 FOR K=M TO N
980 S=S+U(N,K)↑2
990 NEXT K
1000 M(N)=S
1010 FOR R=1 TO N
1020 FOR C=1 TO N
1030 IF R=C THEN S(R,C)=1
1040 IF R<>C THEN S(R,C)=0
1050 NEXT C
1060 NEXT R
1070 FOR R=M TO N
1080 FOR C=M TO N
1090 S(R,C)=U(R,C)/SQR(M(R))
1100 NEXT C
1110 NEXT R
1120 GOSUB 2000
1130 FOR R=1 TO N
1140 FOR C=1 TO N
1150 S=0
1160 FOR K=1 TO N
1170 FOR L=1 TO N
1180 S=S+S(R,K)*A(K,L)*S(C,L)
1190 NEXT L
1200 NEXT K
1210 B(R,C)=S
1220 NEXT C
1230 NEXT R
1240 NEXT M
1250 FOR R=1 TO N
1260 FOR C=1 TO N
1270 READ E
1280 A(R,C)=E
1290 NEXT C
1300 NEXT R
1310 PRINT: PRINT ``THE ORTHOGONAL
TRANSFORMATION IS'': PRINT
1320 FOR R=1 TO N
1330 FOR C=1 TO N
1340 PRINT O(R,C) SPC(3)
1350 NEXT C: PRINT: PRINT
1360 NEXT R
1370 PRINT: PRINT ``O(R,K)*A(K,L)
*O(C,L)='': PRINT
1380 FOR R=1 TO N
1390 FOR C=1 TO N
1400 S=0
```

```
1410 FOR K=1 TO N
1420 FOR L=1 TO N
1430 S=S+O(R,K)*A(K,L)*O(C,L)
1440 NEXT L
1450 NEXT K
1460 D(R,C)=S
1470 PRINT D(R,C) SPC(3)
1480 NEXT C: PRINT: PRINT
1490 NEXT R
1495 PRINT: PRINT ``EIG VAL'';N; `` = '';
D(N,N)
1500 END
2000 FOR R=1 TO N
2010 FOR C=1 TO N
2020 S=0
2030 FOR L=1 TO N
2040 S=S+S(R,L)*O(L,C)
2050 NEXT L
2060 F(R,C)=S
2070 NEXT C
2080 NEXT R
2090 FOR R=1 TO N
2100 FOR C=1 TO N
2100 O(R,C)=F(R,C)
2120 NEXT C
2130 NEXT R
2140 RETURN
RUN
```

```
THE ORIGINAL MATRIX IS
1 2 3 4
2 1 5 6
3 5 1 7
4 6 7 1
EIG VAL 1=15.0169875
EIG VAL 2=-6.30105553
EIG VAL 3=-4.04123361
THE ORTHOGONAL TRANSFORMATION IS
.35249192 .492878784 .542032058 .582255268
.149643512 .24975683 .530656766 -.796009887
.210385861 .728428624 -.633516592 -.154227936
.899496555 -.405072548 -.152516792 -.059672473
O(R,K)*A(K,L)*O(C,L)=
15.0169875 9.05583875E-09 8.6693035E-09 8.55550298E-10
9.17225408E-09 -6.30105552 2.22935341E-08 1.02289732E-09
6.25368557E-09 2.34867912E-08 -4.04123361 -2.43431941E-10
2.16522267E-09 1.21207222E-09 -3.60671493E-11 -.674698311
EIG VAL 4=-.674698311
```

Lines 40 through 42 define a matrix $O(R, C)$ as the identity matrix for use in the subroutine starting at line 2000. The RESTORE statement of line 81 permits the original data to be read again later. Starting at line 100, each value of M corresponds to a submatrix of size $N - M + 1$. For example, if $M = 1$, the M loop finds the dominant eigenvalue and the corresponding eigenvector of the original $N \times N$ matrix and constructs an orthogonal matrix which performs the transformation given by (10.7). (The value of N can be anything—the examples of section 10 use $N = 4$.) Lines 105 through 190 define a matrix $A(R, C)$, which has the elements of its first $M - 1$ rows and columns set equal to zero and its remaining submatrix set equal to $B(R, C)$. The matrix $B(R, C)$ is the transformed matrix from the previous pass through the M loop. For example, if $M = 1$, then $A(R, C)$ is the original matrix entered as data in lines 1 through 4. If $M = 2$, then $A(R, C)$ represents the transformed matrix of equation (10.7) with the exception that $A(1,1)$ has been set equal to zero rather than λ_1. It is best to do this in order to prevent the possibility that the second pass through the M loop will again pick up λ_1 as the dominant eigenvalue. Lines 200 through 210 represent a purging operation designed to get rid of any values assigned to the vectors $X(K)$ and $Y(K)$ during the previous pass through the M loop. In lines 215 through 550, the dominant eigenvalue of $A(R,C)$ is found by the method of section 10. Lines 560 through 595 represent another purging procedure to ensure that $T(R,C)$ starts each new pass through the M loop as the identity matrix. If $M=2$, then lines 600 through 690 construct a matrix of the form

$$T(R,C)=\begin{pmatrix} 1 & 0 & 0 & 0 \\ 0 & x_1 & x_2 & x_3 \\ 0 & 1 & 0 & 0 \\ 0 & 0 & 0 & 1 \end{pmatrix} \qquad (12.1)$$

Equation (12.1) is written for the case where x_2 is the component of the eigenvector (x_1, x_2, x_3) with the largest magnitude. Lines 700 through 1110 put $T(R,C)$ through the Schmidt orthogonalization procedure. For example, if $M=2$, then (12.1) is transformed into a matrix of the form (10.8). Note that lines 700 through 735 and 1010 through 1080 perform purges on $U(R,C)$ and $S(R,C)$. At line 1120, we are sent to a subroutine beginning at line 2000 which combines the orthogonal transformation just found with those of the previous passes through the M loop. In other words, the subroutine carries out equation (10.10). Note that if $M=1$, then $O(R,C)$ of line 2040 is the identity matrix defined by lines 40 through 42. At the end of the subroutine, $O(R,C)$ has been redefined as S_1 of equation (10.10). After we return from the subroutine, lines 1130 through 1230 transform $A(R,C)$. Specifically, these lines carry out (10.7) if $M=1$ and (10.9) if $M=2$. Each time this is done, a new matrix $B(R,C)$ is defined, and at line 1240 we are sent back to the beginning of the M loop at line 100. After the M loop has been completed, lines 1250 through 1500 reread the original matrix and then transform it by the overall transformation that has been accumulated by the subroutine.

The program works reasonably well for degenerate matrices. For a degenerate eigenvalue, the power method constructs an eigenvector which is a linear combination of all the possible eigenvectors. The Schmidt orthogonalization procedure then constructs a set of unit vectors which are orthogonal to this choice. Apparently, near-degeneracy causes more problems than exact degeneracy. Trouble may also result if too many of the eigenvalues are zero. For example, the program is defeated by the matrix

$$\begin{pmatrix} 1 & 1 & 1 & 1 \\ 1 & 1 & 1 & 1 \\ 1 & 1 & 1 & 1 \\ 1 & 1 & 1 & 1 \end{pmatrix} \qquad (12.2)$$

This matrix has eigenvalues 4, 0, 0, and 0. The program successfully finds $\lambda_1 = 4$ and $\lambda_2 \simeq 0$ but is then stopped by a "DIVISION BY ZERO ERROR." The reason for this is that at some stage of the reduction, the transformed matrix has become the zero matrix. The trial eigenvector is then annihilated and the computer stops when it tries to normalize the resulting zero vector.

It is possible that the program given in this section would not work well for large matrices due to error accumulation in the many stages of the reduction to diagonal form. It seems to be quite satisfactory for matrices of modest size. See problems 5 through 7 for some important modifications and refinements of the procedure.

13. THE CHARACTERISTIC FUNCTION OF A MATRIX (2.10.10), (7.12.34)

The algorithm for finding the coefficients in the characteristic function of a matrix is provided by theorem 5 in section 14 of Chapter 7. The following program computes the coefficients as given by equations (7.14.23):

```
1 DATA 1,2,3,4
2 DATA 2,1,5,6
3 DATA 3,5,1,7
4 DATA 4,6,7,1
15 N=4
16 DIM A(N,N): DIM P(N,N)
17 DIM T(N): DIM B(N,N)
18 DIM C(N)
20 FOR R=1 TO N
25 FOR C=1 TO N
26 READ E
```

```
30 A(R,C)=E
35 P(R,C)=E
40 PRINT A(R,C) SPC(3)
45 NEXT C: PRINT: PRINT
50 NEXT R
55 T=0
60 FOR R=1 TO N
65 T=T+A(R,R)
70 NEXT R
75 T(1)=T
80 FOR M=2 TO N
85 FOR J=1 TO N
90 FOR L=1 TO N
95 S=0
100 FOR K=1 TO N
105 S=S+A(J,K)*P(K,L)
110 NEXT K
115 B(J,L)=S
120 NEXT L
125 NEXT J
130 T=0
135 FOR J=1 TO N
140 T=T+B(J,J)
145 NEXT J
150 T(M)=T
155 FOR J=1 TO N
160 FOR K=1 TO N
165 P(J,K)=B(J,K)
170 NEXT K
175 NEXT J
180 NEXT M
185 FOR J=1 TO N
190 PRINT ``T('';J;``)='';T(J)
195 NEXT J: PRINT
200 C(0)=1
210 FOR K=1 TO N
215 S=0
220 FOR L=0 TO K-1
221 U=K-L
225 S=S-(1/K)*T(U)*C(L)
230 NEXT L
235 C(K)=S
240 NEXT K
245 PRINT ``F(X)=X↑N+C(1)*X↑N-1+C(2)
*X↑N-2+···+C(N)''
246 PRINT: PRINT ``N='';N; PRINT
250 FOR K=1 TO N
255 PRINT ``C('';K;``)='';C(K)
260 NEXT K
RUN
1 2 3 4
2 1 5 6
3 5 1 7
4 6 7 1
T(1)=4
T(2)=282
T(3)=3070
T(4)=52698
F(X)=X↑N+C(1)*X↑N-1+C(2)*X↑N-2+···
-C(N)
N=4
C(1)=-4
C(2)=-133
C(3)=-470
C(4)=-258
```

Lines 55 through 75 calculate the trace $T(1)$ of the original matrix A. The remaining traces are calculated in lines 80 through 175. For example, if $M=2$, then $B(J,L)$ defined by line 115 are the elements of a matrix B which is the square of A. Lines 130 through 150 calculate the trace $T(2)$ of B. Lines 155 through 175 redefine the product matrix P as A^2 to be used for $M=3$. Note that no purging operation is needed before P is redefined because line 165 reassigns a new value to each element of P. Lines 200 through 240 carry out the computations indicated by equations (7.14.23). We have run the same example here that was used in section 12. Figure 13.1 shows a plot of the characteristic function $f(\lambda)$. This plot was made on the computer, but we will not give the graphing program here because there are too many differences in the methods of doing graphics from computer to computer. As you can see, approximate eigenvalues can be read from the graph as roots of $f(\lambda)$. Note that $f(\lambda)$ becomes negative and of large magnitude between $\lambda_4=-0.675$ and $\lambda_1=15.017$. This portion of $f(\lambda)$ is beyond the range of Fig. 13.1. Experts on numerical methods do not recommend the use of $f(\lambda)$ as a way to find the eigenvalues of a matrix. Apparently, error buildup in the computation of the coefficients becomes too great for large matrices.

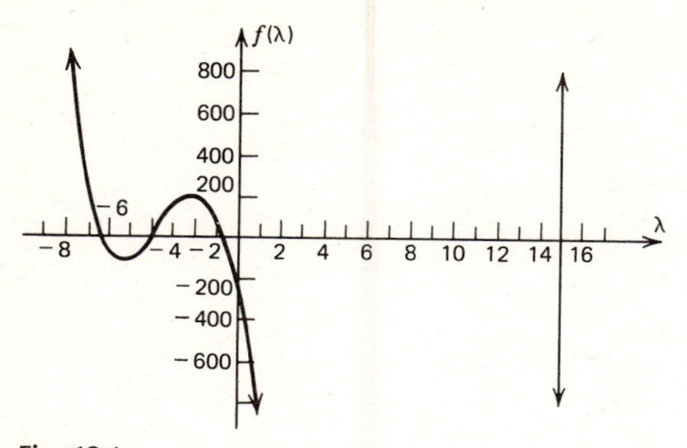

Fig. 13.1

There is however nothing wrong with finding the eigenvalues of matrices of modest size as roots of $f(\lambda)$. If you wish, you can program your computer to find these roots. A program for doing this is offered in the next section. Note that the characteristic function provides a method of finding the eigenvalues of any matrix, symmetric or not. As a side benefit, the determinant of the original matrix can be found by means of equation (7.14.31).

14. FINDING THE ROOTS OF A FUNCTION

The roots of a function $F(x)$ are the values of x for which $F(x)=0$. There are many applications in which the numerical values of the roots of a function are required. Of immediate interest is the problem of finding the roots of the characteristic function of a matrix as discussed in section 13. The program given in this section works by determining the points at which $F(x)$ changes sign.

```
1 A=-8
2 B=.16
3 S=.1
4 N=5
5 R=0
10 DEF FN F(X)=X↑4-4*X↑3-133*X↑2
-470*X-258
20 FOR X=A TO B STEP S
25 IF FN F(X)=0 THEN 70
40 IF FN F(X+S)*FN F(X)<0 THEN GOSUB
500
50 NEXT X
60 END
70 R=R+1
71 PRINT ``ROOT'';R;`` = '';X: PRINT
75 GOTO 50
500 Y=X
510 FOR K=0 TO N
520 U=S/(10↑(K+1))
530 FOR Z=Y TO Y+S/(10↑K) STEP U
540 IF FN F(Z)=0 THEN 600
560 IF FN F(Z+U)*FN F(Z)<0 THEN 580
570 NEXT Z
580 Y=Z
590 NEXT K
600 R=R+1
610 PRINT ``ROOT'';R;`` = '';Z: PRINT
620 RETURN
RUN
ROOT 1=-6.3010556
ROOT 2=-4.04123361
ROOT 3=-.674698322
ROOT 4=15.0169875
```

Lines 1 and 2 define the interval over which the function is to be examined. The computer plot of Fig. 13.1 assists us in this choice. In general, you will have to make some such preliminary analysis to find the range over which the roots can be expected to be found. Line 10 defines the function which is to be examined. Once a function is defined in this manner, it can be used throughout the remainder of the program. Lines 20 through 50 evaluate the function at the points $x=A, A+S, A+2S, \ldots, B$, where S is given by line 3. If it should happen that $F(X)=0$ exactly for one of these values, then line 25 causes this value of X to be printed as a root before going to next X. If this does not happen, then line 40 alerts the computer when a sign change occurs in the function indicating the presence of a root between X and $X+S$. If you make S too large, you might get into a situation like that pictured in Fig. 14.1a, where the sign change in $F(X)$ between $X=S$ and $X=2S$ would not be detected. This results in two of the roots of $F(X)$ being missed. Another hazard occurs when the graph of the function is like that pictured in Fig. 14.1b. This circumstance indicates that at least two of the roots of the function are equal.

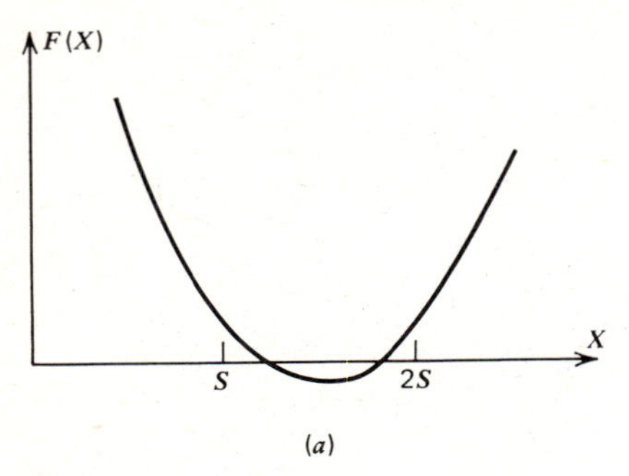

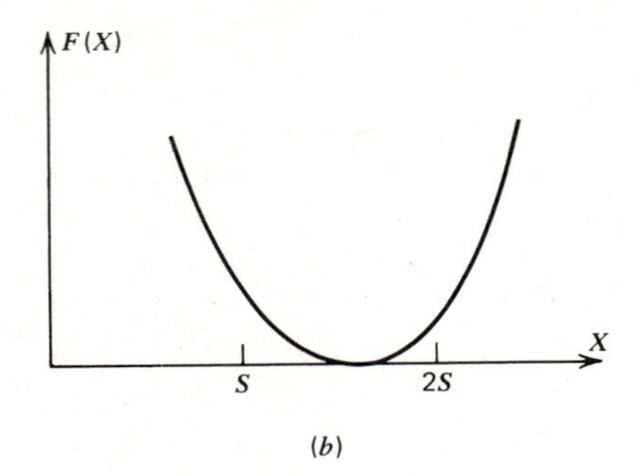

Fig. 14.1

Once it is determined that there is a root between X and $X+S$, we go to a subroutine which examines this interval in finer detail. For example, if $K=0$, this interval is divided into subintervals of length $S/10$ and examined by the same technique that was used in lines 20 through 50. The root is then confined to an interval of length $S/10$. For $K=1$, the process is repeated, this time dividing the interval of length $S/10$ into still smaller subintervals of length $S/100$. The value of N in line 4 determines how often this process is to be repeated and hence determines the accuracy of the final result. If you make N too large, the computer will get stuck in the Z loop because the difference between $F(Z)$ and $F(Z+U)$ becomes less than the least count of the machine and a sign change may not be detected.

If you think you might have a multiple root, you can first find the locations of the maxima and minima of $F(X)$ and then determine if any of these points are also zeros of the function. You can do this separately or you can write a more elaborate program which includes this contingency.

It is comforting to observe that we have duplicated the eigenvalues found in section 12 to a high accuracy. Having more than one method for achieving the same answer is valuable as a check on our procedure.

15. MAXIMA AND MINIMA

The program given in this section, locates the maxima and minima of a function. It is quite similar to the program in section 14. As an exercise, try to write the program yourself before you read on.

```
1 A=-8
2 B=16
3 S=1
4 N=4
5 M1=0
6 M2=0
10 DEF FN F(X)=X↑4-4*X↑3-133*X↑2-470
*X-258
20 FOR X=A TO B STEP S
30 D1=FN  F(X+S)-FN F(X)
40 D2=FN F(X+2*S)-FN F(X+S)
50 IF D1*D2<=0 THEN GOSUB 500
60 NEXT X
70 END
500 Y=X
510 FOR K=0 TO N
520 U=S/(10↑(K+1))
530 FOR Z=Y TO Y+2*S/(10↑K) STEP U
540 D3=FN F(Z+U)-FN F(Z)
550 D4=FN F(Z+2*U)-FN F(Z+U)
560 IF D3*D4<=0 THEN 580
570 NEXT Z
580 Y=Z
590 NEXT K
600 IF D2>0 OR D1<0 THEN 640
610 M1=M1+1
620 PRINT ``MAX'';M1;`` = '';Z;``F(X)='';
FN F(Z): PRINT
630 RETURN
640 M2=M2+1
650 PRINT ``MIN'';M2;`` = '';Z;``F(X)='';
FN F(Z); PRINT
660 RETURN
```

```
RUN
MIN 1=-5.33600001 F(X)=-118.543871
MAX 1=-2.10848001 F(X)=198.967825
MIN 2=10.4442 F(X)=-12332.9239
```

Lines 20 through 50 essentially determine whether or not a change in the sign of the derivative of $F(X)$ has occurred over the interval X to $X+2S$. If a sign change does occur, the subroutine examines this interval in finer detail. The accuracy of the final answer depends on the choice of N in line 4. For $S=1$ and $N=4$, the answers are not accurate beyond the fourth decimal place. Each loop of the subroutine essentially gives us another order of magnitude in the accuracy.

16. DETERMINANTS

The method of evaluating a determinant to be presented in this section is based on reducing it to triangular form using the theorems of section 5 in Chapter 1. By triangular form is meant that there are all zeros below the main diagonal. The value of the determinant of the triangular form is the product of the diagonal elements. The method of carrying this out is similar to that used in finding the inverse of a matrix, and we will therefore not make a detailed analysis. You should try to write the program yourself before reading on.

```
1 DATA 1, 2, 3, 4
2 DATA 2, 1, 5, 6
3 DATA 3, 5, 1, 7
4 DATA 4, 6, 7, 1
15 N=4
16 DIM A(N,N): DIM C(N)
25 PRINT ``THE ORIGINAL MATRIX IS'':
PRINT
30 FOR R=1 TO N
40 FOR C=1 TO N
50 READ E
60 A(R,C)=E
70 PRINT A(R,C) SPC(3)
80 NEXT C: PRINT: PRINT
90 NEXT R
100 P=1
110 FOR S=1 TO N
120 FOR J=S TO N
130 C(J)=ABS(A(J,S))
140 NEXT J
150 J=S
160 FOR K=J TO N
170 IF C(J)<C(K) THEN 200
180 NEXT K
190 GOTO 220
200 J=K
210 GOTO 160
220 IF C(J)=0 THEN 900
230 IF J=S THEN 300
240 P=-P
250 FOR K=1 TO N
260 X=A(S,K)
270 A(S,K)=A(J,K)
280 A(J,K)=X
290 NEXT K
300 FOR J=1 TO N
310 IF J=S THEN 360
320 M=A(J,S)/A(S,S)
330 FOR K=S TO N
340 A(J,K)=A(J,K)-M*A(S,K)
350 NEXT K
360 NEXT J
370 NEXT S
380 FOR J=1 TO N
390 P=P*A(J,J)
400 NEXT J
410 PRINT ``DET=''; P: PRINT
420 END
900 PRINT ``DET=0'': END
RUN
THE ORIGINAL MATRIX IS
1 2 3 4
2 1 5 6
3 5 1 7
4 6 7 1
DET=-258
```

As an example, we have used the same matrix that was used in section 13. The determinant equals the coefficient $C(4)$ of the characteristic function as expected.

17. ARRAYS WITH MORE THAN TWO SUBSCRIPTS

It is possible to define arrays which depend on more than two indices or subscripts. The following program assigns the numbers 1 through 27 to the elements of the three-index symbol $B(J, K, L)$.

```
10 A=1
20 FOR J=1 TO 3
30 PRINT ``J=''; J: PRINT
40 FOR K=1 TO 3
50 FOR L=1 TO 3
60 B(J,K,L)=A
70 A=A+1
80 PRINT B(J,K,L) SPC(3)
90 NEXT L: PRINT: PRINT
100 NEXT K: PRINT
110 NEXT J
RUN
J=1
1 2 3
4 5 6
7 8 9
J=2
10 11 12
13 14 15
16 17 18
J=3
19 20 21
22 23 24
25 26 27
```

The dimension and size of the arrays that can be accommodated are limited only by the memory of the computer. You can therefore program your computer to do any numerical computations involving tensors of higher rank than two.

18. THE PERMUTATION SYMBOL

The permutation symbol is discussed in section 4 of Chapter 1. The following program determines if a given permutation of $1, 2, 3, \ldots, n$ is even or odd.

```
1 DATA 1, 3, 2, 6, 5, 4
10 N=6
20 FOR K=1 TO N
30 READ E
40 E(K)=E
45 PRINT E(K) SPC(3)
50 NEXT K
60 C=0
70 FOR J=1 TO N-1
80 FOR K=J+1 TO N
90 IF E(J)<E(K) THEN 140
100 X=E(J)
110 E(J)=E(K)
120 E(K)=X
130 C=C+1
140 NEXT K
150 NEXT J
154 PRINT: PRINT
155 FOR K=1 TO N
156 PRINT E(K) SPC(3)
157 NEXT K
160 IF C/2=INT(C/2) THEN 190
170 PRINT ``ODD''
180 END
190 PRINT ``EVEN''
RUN
1 3 2 6 5 4
1 2 3 4 5 6 EVEN
```

Lines 60 through 150 represent a sorting procedure which examines the sequence of numbers given as data in line 1 and rearranges them in the order $1, 2, 3, \ldots, n$ by a series of transpositions. The index C keeps track of how many transpositions are made. For example, suppose $J = 1$ in line 70. Line 90 then compares $E(1)$ with $E(K)$ until a value of K is found such that $E(1) > E(K)$. Lines 100 through 120 then exchange $E(1)$ and $E(K)$ so that a new list is defined in which $E(1) < E(K)$. We then go on to next K and the process repeats. At the end of the K loop, $E(1) = 1$ and the number of required transpositions to achieve this result equals the value of C. We now go on to $J = 2$. This time when the K loop is complete, $E(2) = 2$. Ultimately, the list is restored to $1, 2, 3, \ldots, n$ as required. Lines 154 through 157 print the list in restored order and may be omitted if desired. They were put in as assurance that the program was working properly. In line 160, INT($C/2$) is the largest integer less than or equal to $C/2$. If $C/2 = \text{INT}(C/2)$, then C is even and the original permutation is even. If not, the original permutation is odd.

The sorting procedure used in this program is suitable for arranging any list of numbers in either ascending or descending order.

19. NUMERICAL INTEGRATION

In this section, we turn our attention to the problem of finding the definite integral of a function by a numerical method known as *Simpson's rule*. Consider a function $y(x)$ defined over $0 \le x \le 4h$ as pictured in Fig. 19.1. The definite integral of $y(x)$ is the area between its graph and the x axis, and we will find the contribution to the integral from $0 \le x \le 2h$ by approximating the function over this interval by the parabola

$$y = ax^2 + bx + c \tag{19.1}$$

The area is

$$S_1 = \int_0^{2h} y\,dx = \tfrac{8}{3}ah^3 + 2bh^2 + 2ch \tag{19.2}$$

The constants a, b, and c are to be determined by the requirement that the parabola match the actual function exactly at $x = 0$, h, and $2h$. This gives

$$y_0 = c$$

$$y_1 = ah^2 + bh + c \qquad y_2 = 4ah^2 + 2bh + c \tag{19.3}$$

The idea is that, if h is sufficiently small, the parabola approximates the actual function accurately over the entire interval $0 \le x \le 2h$. By solving (19.3) for the constants a, b, and c and then substituting into (19.2), we find

$$S_1 = \frac{h}{3}(y_0 + 4y_1 + y_2) \tag{19.4}$$

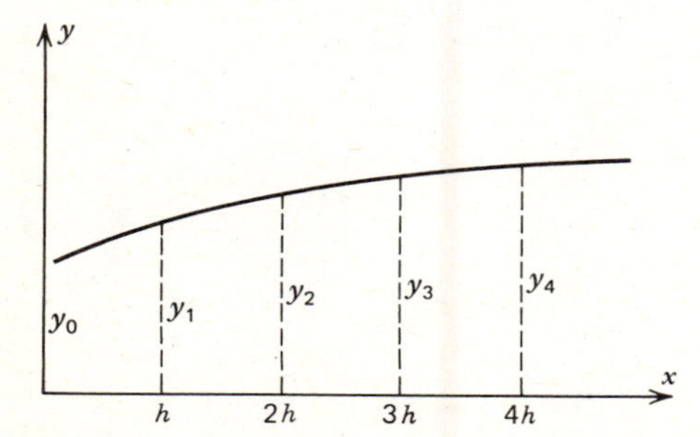

Fig. 19.1

The integral between $x = 0$ and $x = 4h$ is

$$S_2 = \frac{h}{3}(y_0 + 4y_1 + y_2) + \frac{h}{3}(y_2 + 4y_3 + y_4)$$

$$= \frac{h}{3}(y_0 + 4y_1 + 2y_2 + 4y_3 + y_4) \tag{19.5}$$

Finally, if we divide the arbitrary interval $A \le x \le B$ up into an even number of subintervals of length h, then

$$\int_A^B y\,dx \simeq \frac{h}{3}(y_0 + 4y_1 + 2y_2 + 4y_3 + 2y_4 + \cdots + 2y_{2k-2} + 4y_{2k-1} + y_{2k})$$

$$= \frac{h}{3}(y_0 - y_{2k}) + \frac{4h}{3}(y_1 + y_3 + \cdots + y_{2k-1}) + \frac{2h}{3}(y_2 + y_4 + \cdots + y_{2k}) \tag{19.6}$$

where k is an integer and $h = (B - A)/(2k)$. The following program uses (19.6) to compute the definite integral of any function:

```
10 PRINT ``THE NUMBER OF INTERVALS IS
2K''
20 PRINT ``ENTER AN INTEGER FOR K'':
INPUT K
30 PRINT ``NUMBER OF INTERVALS='';2*K
40 PRINT ``ENTER THE LIMITS OF
INTEGRATION''
50 INPUT A,B
60 DEF FN F(X)=1/X
70 H=(B-A)/(2*K)
80 S=0: T=0
90 FOR J=1 TO K
100 S=S+FN F(A+2*J*H-H)
110 T=T+FN F(A+2*J*H)
120 NEXT J
125 PRINT
130 I=(H/3)*(FN F(A)-FN F(B)+4*S+2*T)
140 PRINT ``THE INTEGRAL IS''; I
RUN
THE NUMBER OF INTERVALS IS 2K
ENTER AN INTEGER FOR K
?50
NUMBER OF INTERVALS=100
ENTER THE LIMITS OF INTEGRATION
?1,2
THE INTEGRAL IS .693147181
```

The function used as an example is $1/x$. Many of the functions of mathematical physics are defined by integrals or have an integral representation. For example, we can define $\ln x$ by

$$\int_1^x \frac{1}{x} dx = \ln x \tag{19.7}$$

If we divide $1 \leq x \leq 2$ into 100 subintervals, the program of this section gives $\ln 2$ accurate to nine decimal places.

The parabolic approximation does not work in the vicinity of a point where the function or its derivative is discontinuous. We can get around this difficulty by performing separate integrations between such points. In this manner, the limits of integration fall exactly at the points of discontinuity.

20. RUNGE–KUTTA METHOD

We offer in this section a widely used numerical procedure for integrating a first-order differential equation of the form

$$\frac{dy}{dx} = f(x, y) \tag{20.1}$$

Suppose that starting values x_0 and y_0 are known. The crudest numerical procedure for calculating y at $x_0 + h$ that we can imagine is

$$y_1 = y(x_0 + h) = y_0 + f(x_0, y_0)h \tag{20.2}$$

We next use $x_1 = x_0 + h$ and y_1 as starting values for the next interval to get

$$y_2 = y(x_0 + 2h) = y_1 + f(x_1, y_1)h \tag{20.3}$$

and so on. This method is inaccurate unless h is taken very small. A much more accurate procedure is the Runge–Kutta method given by

$$\begin{aligned}
k_1 &= hf(x_0, y_0) \\
k_2 &= hf(x_0 + \tfrac{1}{2}h, y_0 + \tfrac{1}{2}k_1) \\
k_3 &= hf(x_0 + \tfrac{1}{2}h, y_0 + \tfrac{1}{2}k_2) \\
k_4 &= hf(x_0 + h, y_0 + k_3) \\
y(x_0 + h) &= y_0 + \tfrac{1}{6}(k_1 + 2k_2 + 2k_3 + k_4)
\end{aligned} \tag{20.4}$$

In the above formulas, x_0 and y_0 are the values of x and y at the beginning of the interval, h is the length of the interval, and $y(x_0 + h)$ is the value of y at the end of the interval. The derivation of these formulas is too complicated to be presented here, but we can make them seem plausible by expressing the integral of (20.1) as

$$y(x_0 + h) - y_0 = \int_0^h f(x, y)\, dx \tag{20.5}$$

If we approximate the integral by Simpson's rule and use (19.4), the result is

$$y(x_0 + h) = y_0 + \frac{h}{6}\left[f(x_0, y_0) + 4f\left(x_0 + \frac{1}{2}h, y_1\right) + f(x_0 + h, y_2)\right] \tag{20.6}$$

where y_1 is the value of y at $x_0 + \frac{1}{2}h$ and y_2 is the value of y at $x_0 + h$. Note that the interval of integration is h rather than $2h$ as in (19.4). There is at least a resemblance between (20.6) and (20.4). They become identical in those cases where $f(x, y)$ depends on x only. In problem 13, it is shown that the error in Simpson's rule is of the order h^5. The same is true of the Runge–Kutta method. Writing a program that will perform the calculations indicated by (20.4) is fairly easy:

```
10 PRINT ``ENTER THE LIMITS OF INTEGRATION''
15 INPUT A,B
20 PRINT ``ENTER THE NUMBER OF
INTERVALS''
25 INPUT N
30 PRINT ``ENTER THE INITIAL VALUE OF
Y''
35 INPUT YO
40 H=(B-A)/N
50 DEF FN F(X)=X*Y
60 PRINT ``X''; TAB(14); ``Y'': PRINT
70 PRINT A; TAB(14); YO
80 XO=A
90 FOR J=1 TO N
100 Y=YO
110 K1=H*FN F(XO)
120 Y=YO+K1/2
130 K2=H*FN F(XO+H/2)
140 Y=YO+K2/2
150 K3=H*FN F(XO+H/2)
160 Y=YO+K3
```

```
170 K4=H*FN F(XO+H)
180 Y=YO+(K1+2*K2+2*K3+K4)/6
200 XO=XO+H
210 YO=Y
220 NEXT J
230 PRINT A+(J - 1)*H; TAB(14); Y
240 END
RUN
ENTER THE LIMITS OF INTEGRATION
?0,2
ENTER THE NUMBER OF INTERVALS
?50
ENTER THE INITIAL VALUE OF Y
?1
X   Y
0   1
2   7.3890545
```

As line 50 shows, it is possible to define a function in Basic which depends on more than one variable. Lines 90 through 120 perform the computations indicated by (20.4). For each value of J, the loop starts with values of $X0$ and $Y0$ that were obtained from the previous value of J. If a graph of the solution is wanted, an appropriate plot statement could be inserted as line 190. In this way, a point would be plotted at the end of each interval. We have used as an example

$$\frac{dy}{dx} = xy \tag{20.7}$$

The solution is easily found to be

$$y = y_0 e^{x^2/2} \tag{20.8}$$

If $y_0 = 1$ and $x = 2$, then (20.8) gives $y = 7.38905610$, in close agreement with the value produced by the numerical integration.

The Runge–Kutta method can be extended to a system of simultaneous first-order differential equations of the form

$$\frac{dy}{dx} = f(x, y, z) \qquad \frac{dz}{dx} = g(x, y, z) \tag{20.9}$$

The integration formulas are

$$\begin{aligned}
\Delta_1 &= hf(x_0, y_0, z_0) \\
\Delta_2 &= hf\left(x_0 + \tfrac{1}{2}h,\ y_0 + \tfrac{1}{2}\Delta_1,\ z_0 + \tfrac{1}{2}\delta_1\right) \\
\Delta_3 &= hf\left(x_0 + \tfrac{1}{2}h,\ y_0 + \tfrac{1}{2}\Delta_2,\ z_0 + \tfrac{1}{2}\delta_2\right) \\
\Delta_4 &= hf\left(x_0 + h,\ y_0 + \Delta_3,\ z_0 + \delta_3\right) \\
\delta_1 &= hg(x_0, y_0, z_0) \\
\delta_2 &= hg\left(x_0 + \tfrac{1}{2}h,\ y_0 + \tfrac{1}{2}\Delta_1,\ z_0 + \tfrac{1}{2}\delta_1\right) \\
\delta_3 &= hg\left(x_0 + \tfrac{1}{2}h,\ y_0 + \tfrac{1}{2}\Delta_2,\ z_0 + \tfrac{1}{2}\delta_2\right) \\
\delta_4 &= hg\left(x_0 + h,\ y_0 + \Delta_3,\ z_0 + \delta_3\right) \\
y &= y_0 + \tfrac{1}{6}(\Delta_1 + 2\Delta_2 + 2\Delta_3 + \Delta_4) \\
z &= z_0 + \tfrac{1}{6}(\delta_1 + 2\delta_2 + 2\delta_3 + \delta_4)
\end{aligned} \tag{20.10}$$

where x_0, y_0, and z_0 are the initial values of the variables. One important consequence of this result is that it allows the Runge–Kutta method to be used for second-order differential equations of the form

$$\frac{d^2y}{dx^2} = g\left(x, y, \frac{dy}{dx}\right) \tag{20.11}$$

If we write $dy/dx = z$, then (20.11) is equivalent to

$$\frac{dy}{dx} = z \qquad \frac{dz}{dx} = g(x, y, z) \tag{20.12}$$

The integration of (20.12) can be done by means of (20.10).

21. SUMMING SERIES

The study of the convergence properties of infinite series is greatly facilitated by the use of the computer. For example, consider the series given by (5.21.56). The following program calculates partial sums of $\Sigma(1/k^2)$:

```
10 PRINT ``ENTER THE NUMBER OF TERMS''
20 INPUT N
30 PRINT ``ACTUAL VALUE=1.644934067''
35 PRINT
40 PRINT ``NUMBER''; TAB(10); ``SUM''
41 PRINT ``OF TERMS'': PRINT
50 S=0
60 FOR K=1 TO N
70 S=S+1/K↑2
80 IF K/200=INT(K/200) THEN 100
90 NEXT K
95 END
100 PRINT K; TAB(10); S
110 GOTO 90
```

```
RUN
ENTER THE NUMBER OF TERMS
?2000
ACTUAL VALUE=1.644934067
NUMBER        SUM
OF TERMS
200           1.63994655
400           1.64243719
600           1.64326879
800           1.64368485
1000          1.64393456
1200          1.64410108
1400          1.64422004
1600          1.64430926
1800          1.64437866
2000          1.64443419
```

We have calculated a total of 2000 terms of the series. Note that line 80 of the program provides a means of interrupting the K loop every 200 terms and causing the value of the partial sum at that point to be printed. It is evident that this series converges fairly slowly. A better result could be obtained by using one of the other series given by (5.21.56)

As another example, the series for sin x is given by

$$\sin x = \sum_{k=0}^{\infty} \frac{(-1)^k x^{2k+1}}{(2k+1)!} = \sum_{k=0}^{\infty} u_k \tag{21.1}$$

We already studied the convergence properties of this series in section 5 of Chapter 5. See Table (5.5.1) and Fig. (5.5.1). We will now show how the computer does the calculations. Of great utility is the recurrence formula for the terms of the series:

$$u_{k+1} = -\frac{x^2}{(2k+3)(2k+2)} u_k \tag{21.2}$$

The following program utilizes (21.2) to print a short table of the values of sin x for radian arguments:

```
10 N=5
15 PRINT ``NUMBER OF TERMS=''; N+1:
PRINT
20 PRINT ``X''; TAB(8); ``SUM'';
TAB(22); ``LAST TERM''; TAB(42); ``SIN(X)''
25 PRINT
30 FOR X=0 TO 1.8 STEP .2
40 U=X: S=U
50 FOR K=0 TO N
60 U=-U*(     )/((2*K+2)*(2*K+3))
70 S=S+U
80 NEXT K
90 PRINT X; TAB(8); S; TAB(22); U;
TAB(42); SIN(X)
100 NEXT X
RUN
  X        SUM            LAST TERM          SIN(X)
  0          0                0                 0
  .2       .198669331     1.31555688E-19     1.98669331
  .4       .389418342      1.0777042E-15     .389418342
  .6       .564642474     2.09742258E-13     .564642473
  .8       .717356091     8.82855278E-12     .717356091
  1        .841470985     1.60590438E-10     .841470985
  1.2      .932039086     1.71820858E-09     .932039086
  1.4       .98544973     1.27463004E-08      .98544973
  1.6      .999573604     7.23235044E-08     .999573603

NUMBER OF TERMS=6
```

In addition to the partial sum, the program prints the value of the last term included in the partial sum and the values of $\sin x$ for comparison. We know from theorem 1 in section 4 of Chapter 5 that the value of the last term is a good measure of the error in the partial sum approximation. The series is rapidly convergent and great accuracy is obtained by using only six terms. Because of the periodic property of $\sin x$, the entire function is determined by its values for $0 \leq x \leq \pi/2$.

22. COMPUTING WITH RECURRENCE FORMULAS

Many of the special functions that we have studied obey recurrence formulas, and these are very convenient in computing. Reference to equation (9.9.16) shows that the Legendre polynomials obey the recurrence relation

$$P_{l+1}(x) = \frac{x(1+2l)}{1+l} P_l(x) - \frac{l}{1+l} P_{l-1}(x) \tag{22.1}$$

Once it is known that $P_0(x) = 1$ and $P_1(x) = x$, it is easy to write a program that utilizes (22.1a) to compute the numerical values of $P_l(x)$ for any order:

```
10 PRINT ``ENTER THE ORDER''
20 INPUT N
25 PRINT ``ORDER=''; N: PRINT
26 PRINT ``X''; TAB(12); ``P(X)'':
PRINT
30 DIM P(N)
40 FOR X=0 TO 1 STEP .2
50 P(0)=1: P(1)=X
60 FOR L=1 TO N-1
70 A=(1+2*L)/(L+1): B=L/(L+1)
80 P(L+1)=A*X*P(L)-B*P(L-1)
90 NEXT L
100 PRINT X; TAB(12); P(N)
110 NEXT X
RUN
ENTER THE ORDER
?9
ORDER=9
X         P(X)
0         0
.2        .24595712
.4        -.18876356
.6        -.0461030397
.8        .187855279
1         .999999997
```

PROBLEMS

1. Write a program that will add two matrices.

2. Generalize program 2 so that it will exchange any two rows of an array of any size.

3. Transform a three-component vector by means of an arbitrary proper orthogonal transformation which depends on the three Euler angles as defined by equations (15.3) through (15.5) of Chapter 4. Write the program in such a way that it performs the matrix multiplication indicated by equation (4.15.6).

4. Suppose that the n homogeneous algebraic equations represented by $AX = Y$ are complex. By writing $A = A_1 + iA_1$, $X = X_1 + iX_2$, and $Y = Y_1 + iY_2$, show that $AX = Y$ can be replaced by the $2n$ real equations

$$\begin{pmatrix} A & -A_2 \\ A_2 & A_1 \end{pmatrix} \begin{pmatrix} X_1 \\ X_2 \end{pmatrix} = \begin{pmatrix} Y_1 \\ Y_2 \end{pmatrix} \tag{1}$$

If $A^{-1} = B_1 + iB_2$, show that

$$\begin{pmatrix} A_1 & -A_2 \\ A_2 & A_1 \end{pmatrix} \begin{pmatrix} B_1 & -B_2 \\ B_2 & B_1 \end{pmatrix} = \begin{pmatrix} I & 0 \\ 0 & I \end{pmatrix} \tag{2}$$

5. If A is symmetric and nonsingular, show that the smallest eigenvalue of A is the dominant eigenvalue of A^{-1}. If A^{-1} does not exist, then at least one eigenvalue of A is zero.

6. If H is a complex matrix, then it can be written as $H = A + iB$, where A and B are real. If H is Hermitian, show that A is symmetric and that B is antisymmetric. The eigenvalues and eigenvectors of H are determined by $HZ = \lambda Z$. Recall that the eigenvalues of a Hermitian matrix are real. By writing the eigenvector as $Z = X + iY$, show that

$$\begin{pmatrix} A & -B \\ B & A \end{pmatrix}\begin{pmatrix} X \\ Y \end{pmatrix} = \lambda \begin{pmatrix} X \\ Y \end{pmatrix} \tag{3}$$

Note that the square matrix appearing in (3) is symmetric. Thus, a Hermitian eigenvalue problem of dimension n can be replaced by an equivalent $2n$-dimensional real symmetric eigenvalue problem. If the eigenvector appearing in (3) is normalized, then $\tilde{X}X + \tilde{Y}Y = 1$. Show that Z is also normalized in the sense $Z^{\dagger}Z = 1$. If λ is not a degenerate eigenvalue of $HZ = \lambda Z$, then it must be a twofold degenerate eigenvalue of (3). If

$$\begin{pmatrix} X_1 \\ Y_1 \end{pmatrix} \quad \text{and} \quad \begin{pmatrix} X_2 \\ Y_2 \end{pmatrix} \tag{4}$$

are the two possible eigenvectors corresponding to the eigenvalue λ, then it must in fact be true that $Z_2 = aZ_1$, where a is a constant. If Z_1 and Z_2 are both normalized, show that $aa^* = 1$. Let us assume that the eigenvectors (4) are orthogonal:

$$\tilde{X}_1 X_2 + \tilde{Y}_1 Y_2 = 0 \tag{5}$$

Show that as a consequence

$$a = i(\tilde{X}_1 Y_2 - \tilde{Y}_1 X_2) \tag{6}$$

Since we have already shown that a has unit magnitude, it must be true that $\tilde{X}_1 Y_2 - \tilde{Y}_1 X_2 = 1$, $X_2 = -Y_1$, and $Y_2 = X_1$. Confirm these results by finding the eigenvalues and normalized eigenvectors of

$$\begin{pmatrix} 1 & 2 & 0 & -1 \\ 2 & 3 & 1 & 0 \\ 0 & 1 & 1 & 2 \\ -1 & 0 & 2 & 3 \end{pmatrix} \tag{7}$$

7. If X is an eigenvector of A corresponding to the eigenvalue λ, show that X is also an eigenvector of $A' = A - \alpha I$ corresponding to the eigenvalue $\lambda - \alpha$. This simple fact is quite useful in the numerical computation of eigenvalues. The transformation $A' = A - \alpha I$ is sometimes called a *shift of origin*. It was pointed out that the program of section 12 runs into difficulty if an eigenvalue is almost degenerate. Suppose that there are two eigenvalues of A given by $\lambda_1 = \lambda_0$ and $\lambda_2 = \lambda_0 + \varepsilon$, where ε is small compared to λ_0. The eigenvalues of the shifted matrix A' are $\lambda'_1 = \lambda_0 - \alpha$ and $\lambda'_2 = \lambda_0 + \varepsilon - \alpha$. The ratio of these eigenvalues is

$$\frac{\lambda'_1}{\lambda'_2} = \frac{\lambda_0 - \alpha}{\lambda_0 - \alpha + \varepsilon} \tag{8}$$

If we choose $\alpha \simeq \lambda_0$, then (8) can be made small and the power method of section 10 converges rapidly.

Shift of origin also cures another even more serious difficulty. If there are two eigenvalues $\lambda_1 = +\lambda_0$ and $\lambda_2 = -\lambda_0$, then there is no way for the power method to separate out the eigenvectors corresponding to these eigenvalues. In the shifted problem, $\lambda'_1 = \lambda_0 - \alpha$, $\lambda'_2 = -\lambda_0 - \alpha$, and λ'_1/λ'_2 can be made small by an appropriate choice of α, thus removing the difficulty. If any of the difficulties just mentioned are present, they will make themselves evident by an imperfect diagonalization when you try to run the program of section 12. In fact, the degree to which the final transformation diagonalizes the original matrix is generally a good indication of the accuracy of the method. You can also discover potential problems by examining a plot of the characteristic function of the matrix in which you are interested. Knowing approximate eigenvalues allows you to make an appropriate shift of origin before attempting to calculate accurate eigenvalues. An example is provided by the matrix

$$\begin{pmatrix} 1 & 2 & 3 & 0 & -1 & -2 \\ 2 & -1 & 1 & 1 & 0 & -3 \\ 3 & 1 & -2 & 2 & 3 & 0 \\ 0 & 1 & 2 & 1 & 2 & 3 \\ -1 & 0 & 3 & 2 & -1 & 1 \\ -2 & -3 & 0 & 3 & 1 & -2 \end{pmatrix} \tag{9}$$

Note that this matrix could have arisen from a three-dimensional Hermitian eigenvalue problem as discussed in problem (6). Each eigenvalue is therefore at least twofold degenerate. The eigenvalues are in fact

$$\begin{aligned} \lambda_1 = \lambda_2 &= -5.887675657 \\ \lambda_3 = \lambda_4 &= 5.1032956 \\ \lambda_5 = \lambda_6 &= -1.23561992 \end{aligned} \tag{10}$$

Note that λ_1 and λ_3 are approximately equal in magnitude. To get a good result from the program in section 12, we must set $P = 100$ in line 230. The situation is much improved if we add 5 to each diagonal element of (9). You will find that $P = 20$ is quite sufficient for the shifted matrix. When you compare the eigenvectors of (9) with those of the shifted matrix, they are apparently different. The reason for this is the exact twofold degeneracy of each eigenvalue. The computer chooses different linear combinations in the two cases. Try out a shift of origin on the matrix given by equation (12.2). As another example, find the eigenvalues and the normalized eigenvectors of

$$\begin{pmatrix} 1 & 2 & 3 & 4 & 5 & 6 \\ 2 & 7 & 8 & 9 & 10 & 11 \\ 3 & 8 & 12 & 13 & 14 & 15 \\ 4 & 9 & 13 & 16 & 17 & 18 \\ 5 & 10 & 14 & 17 & 19 & 20 \\ 6 & 11 & 15 & 18 & 20 & 21 \end{pmatrix} \tag{11}$$

8. Show that the total energy of a spherical pendulum can be written as

$$W = \tfrac{1}{2} m r^2 \dot{\theta}^2 + V_e(\theta) \tag{12}$$

where

$$V_e(\theta) = mgr\left(\frac{\mu}{\sin^2\theta} - \cos\theta\right) \qquad \mu = \frac{\ell_z^2}{2m^2 g r^3} \tag{13}$$

The spherical pendulum is discussed in problem 9.5 of Chapter 4. The function $V_e(\theta)$ is an effective potential. An effective potential was used in section 15 of Chapter 4 in discussing the motion of a spinning top and also in section 4 of Chapter 8 in connection with the classical two-body central force problem. The total energy W and the z component of the angular momentum ℓ_z are determined by the initial conditions. Suppose that the initial conditions are such that $\mu = 0.5$ and $W/mgr = 0.6$. Use whatever graphics capability your computer has to plot the function $f(\theta) = V_e(\theta)/mgr$. The condition $f(\theta) - 0.6 = 0$ gives the turning points of the motion, that is, the maximum and minimum values of θ. Show that these points are at 36.67° and 95.48°. Show that the minimum of $f(\theta)$ occurs at 58.34° and that this corresponds to the pendulum moving on an exactly circular path with energy $W/mgr = 0.16525$.

9. Show that the locations of the maxima of the single-slit diffraction pattern found in problem 10.4 of Chapter 5 are the roots of $\tan\alpha = \alpha$. Find the numerical values of the first 10 positive roots and show that they approach $(n+\frac{1}{2})\pi$ for large values of α. *Caution:* You might be tempted to use the program of section 14 to find the roots of $F(\alpha) = \tan\alpha - \alpha$. This is not a good idea because $\tan\alpha$ is undefined at $\pi/2, 3\pi/2, \ldots$. How can you get around this problem?

10. Find the determinant of the matrix appearing at the end of problem (7) and compare its value to the product of the eigenvalues.

11. Write a program that defines and prints out the three-dimensional permutation symbol in the manner illustrated in section 17. Write the program in such a way that the computer assigns the correct numerical values to the elements of the permutation symbol. In other words, don't plug in the values one by one yourself.

12. Modify the program of section 19 so that it will print out a table of the values of $\ln x$ for integer values of x between $x = 1$ and $x = 12$.

13. According to equations (5.28) and (5.29) of Chapter 5, a continuous and differentiable function can be represented by four terms of its Taylor's series plus a remainder term:

$$y(x) = y(0) + y'(0)x + \frac{1}{2!}y''(0)x^2 + \frac{1}{3!}y^{(3)}(0)x^3 + \frac{1}{4!}y^{(4)}(\xi)x^4 \tag{14}$$

where $0 \le \xi \le x$. By applying Simpson's rule to (14) over the interval $0 \le x \le 2h$, show that

$$\int_0^{2h} y\,dx - S_1 = -\frac{1}{90}y^{(4)}(\xi)h^5 \tag{15}$$

where S_1 is given by (19.4). Thus, the error in Simpson's rule is of the order of h^5, where h is the interval size. Since Simpson's rule depends on a parabolic approximation, we might have expected a contribution to the error from the cube term in (14). This does not happen, and the approximation is therefore better than anticipated.

14. Consider the differential equation

$$\frac{dy}{dx} = xy^2 \tag{16}$$

If we attempt a numerical integration for $y_0 = 1$ and $0 \le x \le 2$, the Runge–Kutta method of section 20 fails. Can you see why? The program works fine for $y_0 = -1$.

15. The value of π is defined by the equation $\sigma = \pi r^2$, where σ is the area of a circle and r is its radius. A circle of radius $\sqrt{2}$ has a square inscribed in it. By considering the difference between the area of the circle and the area of the square, show that

$$\pi = 4\int_0^1 \sqrt{2 - x^2}\,dx - 2 \tag{17}$$

Now use Simpson's rule to numerically evaluate π. It is also true that

$$\pi = 2\int_0^{\sqrt{2}} \sqrt{2 - x^2}\,dx \tag{18}$$

Why won't this work as well?

16. Write a computer program for integrating differential equations based on (20.10). Try it out on the second-order differential equation

$$\frac{d^2y}{dx^2} = -6\frac{dy}{dx} - 5y \tag{19}$$

Compare the numerical integration with values found from the known solution of (19).

17. The recurrence relation obeyed by the binomial coefficients is given by equation (5.42) of Chapter 5. Write a program based on it that calculates the numerical values of the binomial coefficients for integer values of k.

18. The recurrence relation obeyed by the Bernoulli numbers is given by equation (14.39) of Chapter 5. Write a program that computes these numbers.

19. The orbit equation for a charged particle moving in a radial electric field is given by equation (4.25) of Chapter 8. Integrate this differential equation by the Runge–Kutta method. Let the limits of integration run from the inner turning point at A to slightly past the outer turning point at B as shown in Fig. 4.3 of Chapter 8. For convenience, choose $r_0 = 1$, but otherwise use the same numerical values that were used in equation (4.31) of Chapter 8. Write the program in such a way that at least ten values of the variables are printed over the range of integration. For comparison purposes, include in the program a computation based on the Fourier series solution given by equation (4.31) of Chapter 8.

20. Asymptotic series are discussed in section 23 of Chapter 5. From equation (23.3) of Chapter 5, we see that the asymptotic series for the exponential integral can be written as

$$xe^xE_1(x) = \sum_{k=0}^{\infty} u_k = \sum_{k=0}^{\infty} \frac{(-1)^k k!}{x^k} \tag{20}$$

Write a program that computes the terms in the series one by one and then plots their magnitudes as illustrated in Fig. 1. The terms in an asymptotic series typically decrease in magnitude up to a certain point, after which an increase occurs. The location of the minimum term in the series depends on the value of the variable x. Confirm this by setting up your program so you can make plots like Fig. 1 for various values of x.

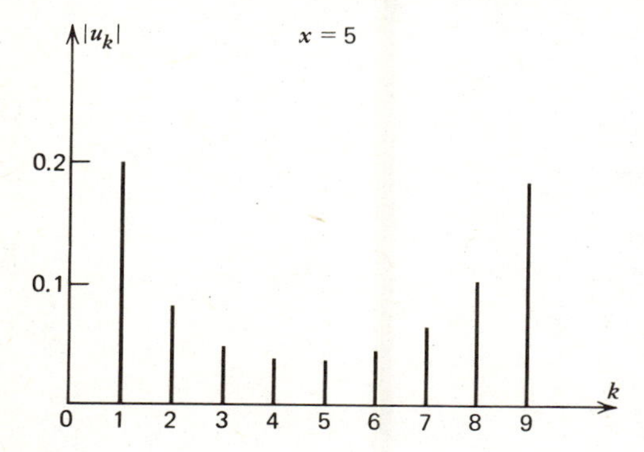

Fig. 1

The error in approximating a function by its asymptotic series is less than the first neglected term. Write a program that sums equation (20) and automatically truncates the series when the term of smallest magnitude is reached. Design the program so that it prints x, $xe^xE_1(x)$, the number of terms summed, and the value of the last term used. Compare the calculated values with tables. Remember that the asymptotic series can be expected to be accurate only for large values of x.

If you want values of $E_1(x)$ for other than large values of x, a couple of possibilities suggest themselves. You can find the contribution to the integral (5.23.1) from $x \le t \le a$ by expanding e^{-t} in the integrand and then integrating term by term. This would be good for small values of x because there is a logarithmic singularity at $x = 0$. A portion of the integral for, say, $a \le t \le b$ could be computed by Simpson's rule. Finally, the asymptotic series is good for $b \le t < \infty$. You have to investigate the convergence properties of the series in order to choose values for a and b that are consistent with the accuracy desired.

21. Hamilton's canonical equations are discussed in section 5 of Chapter 10. See equation (5.20) of Chapter 10. Set up the plane pendulum problem using the Hamilton formulation. The result is the two first-order differential equations

$$\dot{q} = \frac{p}{mr^2} \qquad \dot{p} = -mgr \sin q \tag{21}$$

where $q = \theta$. The solution of equation (21) gives the path of the pendulum in phase space in parametric form. The Hamiltonian is a constant of the motion. If it is written as

$$H = \frac{p^2}{2mr^2} - mgr \cos q \tag{22}$$

then periodic motion occurs if $-1 < H/(mgr) < 1$. If $H/(mgr) > 1$, the pendulum goes over the top and q increases indefinitely. By making the changes of variable

$$W = \frac{H}{mgr} \qquad y = q \qquad z = \frac{p}{m\sqrt{g}\, r^{3/2}} \qquad x = \sqrt{\frac{g}{r}}\, t \tag{23}$$

show that

$$W = \frac{1}{2} z^2 - \cos y \qquad \frac{dy}{dx} = z \qquad \frac{dz}{dx} = -\sin y \tag{24}$$

The differential equations are of the form (20.9) and can be integrated by the Runge–Kutta method. Carry this out and graph the possible trajectories of the pendulum in phase space. You can obtain p and q as a function of t from the integration if you want. Use as initial conditions

$$x_0 = 0 \qquad y_0 = 0 \qquad z_0 = \sqrt{2(W+1)} \tag{25}$$

Figure 2 was obtained by this method. Each choice of W gives a possible path; $W = 1$ gives the sticking solution. The trajectories can also be found directly from $W =$ constant by using W as given by (24). You can use this as a check on your method.

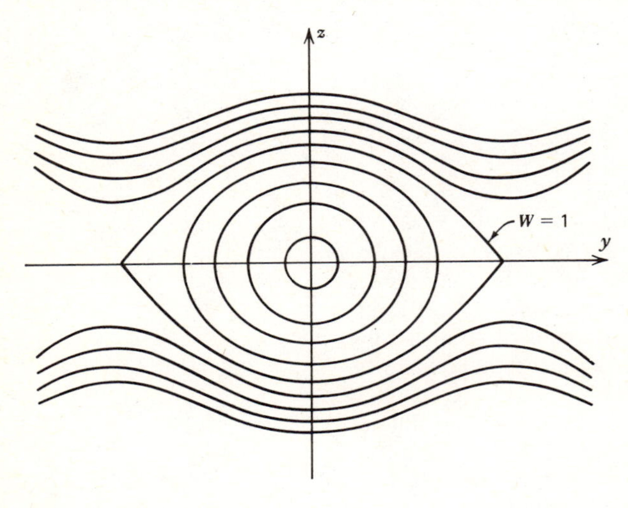

Fig. 2

22. This problem deals with finding the eigenvalues and eigenfunctions of Sturm–Liouville differential equations by numerical methods. As a specific example, we will use Bessel's equation of order zero:

$$F''(r)+\frac{1}{r}F'(r)+\kappa^2F(r)=0 \tag{26}$$

See section 13 of Chapter 9. As boundary conditions, we will use $F(a)=F(b)=0$. This eigenvalue problem occurs in connection with finding the potential between two concentric grounded conducting cylinders as pictured in Fig. (15.4) of Chapter 9. Analytically, the solutions are given by equation (15.50) of Chapter 9. The problem with carrying out a numerical integration of equation (26) is that it contains an unknown parameter, namely, the eigenvalue κ. A little consideration of equation (15.50) of Chapter 9 will convince you that the requirement $F(a)=0$ determines the mix of the two functions $J_0(\kappa r)$ and $N_0(\kappa r)$ in the solution. Since eigenfunctions are determined only up to a multiplicative factor, it makes no difference what we use for the first derivative of the solution at $r=a$. In the numerical work of this problem, we will use $F'(a)=1$. Let us suppose that we want the lowest eigenvalue and corresponding eigenfunction. The problem, then, is to choose κ so that the solution passes through $r=b$ without having crossed the axis anywhere between a and b. We will start by making a guess as to the value of κ. If a rather bad guess is made, the solution passes through the point $r=r_1$ as shown in Fig. (3). How shall we then choose a better value of κ? Consider the transformation $x=\alpha r+\beta$ and determine the constants α and β by the requirement that $x=a$ when $r=a$ and $x=b$ when $r=r_1$. The result is

$$\alpha=\frac{b-a}{r_1-a} \qquad \beta=a(1-\alpha) \tag{27}$$

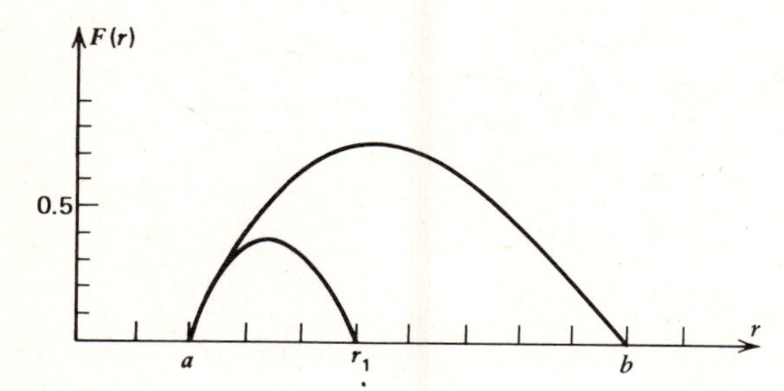

Fig. 3

If the differential equation (26) is transformed by means of $F(r)=u(x)$, we find

$$u''+\frac{u'}{x-\beta}+\left(\frac{r_1-a}{b-a}\kappa\right)^2u=0 \tag{28}$$

The implication is that a reasonable way to try to correct the eigenvalue is by means of

$$\kappa'=\frac{r_1-a}{b-a}\kappa \tag{29}$$

Write a program that uses the Runge–Kutta method to integrate equation (26) between $r=a$ and the next zero of the solution. Include a provision in the program whereby at the end of the integration the eigenvalue is corrected by means of (29) and the integration performed again. This can obviously be repeated as often as desired. As numerical values, use $a=1$ and $b=5$.

Solution:

```
10 HGR: HCOLOR=3
20 HPLOT 0,0 TO 0,180
30 HPLOT 0,90 TO 279,90
40 FOR X=0 TO 279 STEP 10
50 HPLOT X, 88 TO X, 92
60 NEXT X
70 FOR Y=0 TO 180 STEP 10
80 HPLOT 0, Y TO 2, Y
90 NEXT Y
100 S=40
101 A=1
102 B=5
103 N=200
104 T=100
105 K=2
106 M=3
107 FOR L=1 TO M
110 Y0=0
115 Z0=1
120 H=(B-A)/N
130 DEF FNF(X)=Z
140 DEF FNG(X)=-Z/X-Y*K↑2
150 X0=A
160 FOR J=1 TO 5*N
170 Y=Y0: Z=Z0
180 D1=H*FNF(X0)
190 E1=H*FNG(X0)
200 Y=Y0+D1/2: Z=Z0+E1/2
210 D2=H*FNF(X0+H/2)
220 E2=H*FNG(X0+H/2)
230 Y=Y0+D2/2: Z=Z0+E2/2
240 D3=H*FNF(X0+H/2)
250 E3=H*FNG(X0+H/2)
260 Y=Y0+D3: Z=Z0+E3
270 D4=H*FNF(X0+H)
280 E4=H*FNG(X0+H)
290 Y=Y0+(D1+2*D2+2*D3+D4)/6
295 X=X0+H
300 IF Y*Y0<0 THEN 500
310 Z=Z0+(E1+2*E2+2*E3+E4)/6
320 IF S*X>279 THEN 340
321 IF T*Y>90 THEN 340
330 HPLOT S*X, 90-T*Y
340 X0=X0+H
350 Y0=Y: Z0=Z
```

```
360 NEXT J
370 NEXT L
380 PRINT ``K=''; K: PRINT
390 PRINT ``X=''; X: PRINT
410 END
500 K=K*(X - A)/(B - A)
510 GOTO 370
RUN
K=.764400013
X=5.00000002
```

A more accurate value of K as found from tables is $K = 0.76319$. The final value of X is printed so that we may see how well the final eigenfunction has hit the mark at $X = 5$.

We have included in the program a plotting routine which may make sense only to owners of Apple Computers. The eigenvalue can of course be found without the graphics, but it is useful (and fun) to be able to see the solution graphed as the program proceeds. Lines 100 and 104 define parameters S and T which are useful as scaling factors when the plotting is done in line 330. If the graph is too small or if part of it is off the screen, you can easily change the values of S and T. Line 103 determines the number of intervals used in the Runge–Kutta integration. This can be made larger if you want more accuracy. The value of M in line 106 determines the number of times that the integration is repeated. Again, the accuracy can be improved by making M larger. Line 105 gives the initial trial eigenvalue. The program works well even if it is inaccurate. The Runge–Kutta integration as given by (20.10) is carried out in lines 110 through 360. Line 300 interrupts the integration when the solution crosses the x (or r) axis. The eigenvalue is corrected in line 500. We then go on to next L, and the integration is repeated.

23. Modify the program of problem 22 appropriately so that higher eigenvalues and eigenfunctions can be found. Remember that the second eigenfunction crosses the axis one time between $r = a$ and $r = b$. The third eigenfunction crosses twice, and so on.

Bibliography

Abramowitz, M., and I. A. Stegun, eds., *Handbook of Mathematical Functions*. U.S. National Bureau of Standards, Washington, D.C. (1965).

Arfken, G., *Mathematical Methods for Physicists*. Academic Press, New York (1970).

Becker, R., and F. Sauter, *Electromagnetic Fields and Interactions*, Vols. I and II. Blaisdell Publishing Co., New York (1964).

Bennett, A. A., W. E. Milne, and H. Bateman, *Numerical Integration of Differential Equations*. Dover Publications, Inc., New York (1956).

Bennett, W. R., *Scientific and Engineering Problem Solving With the Computer*. Prentice-Hall, Englewood Cliffs, New Jersey (1976).

Bent, R. J., and G. C. Sethares, *Basic, An Introduction to Computer Programming*. Brooks/Cole Publishing Co., Monterey, California (1978).

Bliss, G. A., *Calculus of Variations*. Open Court Publishing Co., La Salle, Illinois (1925).

Boas, M. L., *Mathematical Methods in the Physical Sciences*, 2nd ed. John Wiley and Sons, Inc., New York (1982).

Bowman, F., *Introduction to Bessel Functions*. Dover Publications, Inc., New York (1958).

Bradbury, T. C., *Theoretical Mechanics*. John Wiley and Sons, Inc., New York (1971).

Byrd, P. F., and M. D. Friedman, *Handbook of Elliptic Integrals for Engineers and Physicists*. Springer-Verlag, Berlin (1954).

Churchill, R. V., *Fourier Series and Boundary Value Problems*. McGraw-Hill Book Co., Inc., New York (1978).

Churchill, R. V., *Introduction to Complex Variables and Applications*. McGraw-Hill, Inc., New York (1948).

Churchill, R. V., *Operational Mathematics*. McGraw-Hill Book Co., Inc., New York (1971).

Coan, J. S., *Advanced Basic*. Hayden Book Co., Inc., Rochelle Park, New Jersey (1977).

Copson, E. T., *An Introduction to the Theory of Functions of a Complex Variable*. Oxford University Press, London (1935).

Courant, R., and D. Hilbert, *Methods of Mathematical Physics*, Vols. I and II. Interscience Publishers, New York (1962).

Courant, R., *Differential and Integral Calculus*. Interscience Publishers, New York (1952).

Dahlquist, G., and A. Björck, *Numerical Methods*. Prentice-Hall, Inc., Englewood Cliffs, New Jersey (1974).

Feller, W., *An Introduction to Probability Theory and its Applications*, Vols. I and II. John Wiley and Sons, Inc., New York (1966).

Ferrar, W. L., *Algebra, A Text Book of Determinants, Matrices, and Algebraic Forms*. Oxford University Press, New York (1953).

Gelefand, I. M., and S. V. Fomin, *Calculus of Variations*. Prentice-Hall, Inc., Englewood Cliffs, New Jersey (1963).

Goldstein, H., *Classical Mechanics*, 2nd ed. Addison-Wesley Publishing Co., Cambridge, Massachusetts (1980).

Gourlay, A. R., and G. A. Watson, *Computational Methods for Matrix Eigenproblems*. John Wiley and Sons, New York (1973).

Hamermesh, M., *Group Theory and its Application to Physical Problems*. Addison-Wesley, Reading, Massachusetts (1966).

Hamming, R. W., *Numerical Methods for Scientists and Engineers*. McGraw-Hill Book Co., Inc., New York (1973).

Hancock, H., *Elliptic Integrals*. Dover Publications, New York (1958).

Hardy, G. H., *A Course of Pure Mathematics*. Cambridge University Press, New York (1952).

Jackson, J. D., *Classical Electrodynamics*, 2nd ed. John Wiley and Sons, Inc., New York (1975).

Knopp, K., *Elements of the Theory of Functions*. Dover Publications, Inc., New York (1952).

Kreyszig, E., *Advanced Engineering Mathematics*, 2nd ed. John Wiley and Sons, Inc., New York (1979).

Kunz, K. S., *Numerical Analysis*. McGraw-Hill Book Co., New York (1957).

Ledermann, W., *Introduction to the Theory of Finite Groups*. Interscience Publishers, New York (1953).

Levy, H., and E. A. Baggott, *Numerical Solutions of Differential Equations*. Dover Publications, Inc., New York (1950).

Lighthill, M. J., *Introduction to Fourier Analysis and Generalized Functions*. Cambridge University Press, New York (1959).

Margenau, H., and G. M. Murphy, *The Mathematics of Physics and Chemistry*, Vols. I and II. D. Van Nostrand Co., Inc., New York (1964).

Marion, J. B. and M. Heald, *Classical Electromagnetic Radiation*. Academic Press, New York (1980).

Mathews, J., and R. L. Walker, *Mathematical Methods of Physics*, 2nd ed. W. A. Benjamin, Inc., New York (1970).

Mayer, J. E., and M. G. Mayer, *Statistical Mechanics*, 2nd ed. John Wiley and Sons, Inc., New York (1977).

McLachlan, N. W., *Theory of Vibrations*, Dover Publications, Inc., New York (1951).

Morse, P. M. and K. U. Ingard, *Vibration and Sound*. McGraw-Hill Book Co., Inc., New York (1968).

Park, D., *Introduction to the Quantum Theory*. McGraw-Hill Book Co., New York (1964).

Pauling, L., and E. B. Wilson, *Introduction to Quantum Mechanics*. McGraw-Hill Book Co., New York (1935).

Perlis, S., *Theory of Matrices*. Addison-Wesley Publishing Co., Cambridge, Massachusetts (1952).

Phillips, H. B., *Vector Analysis*. John Wiley and Sons, Inc., New York (1953).

Rainville, E. D. and P. E. Bedient, *Elementary Differential Equations*. Macmillan Co., New York (1981).

Rainville, E. D., *Intermediate Differential Equations*. Macmillan Co., New York (1964).

Ross, S. L., *Differential Equations*. Xerox College Publishing, Lexington, Massachusetts (1974).

Schied, F., *Numerical Analysis*. Schaum's Outline Series, McGraw-Hill Book Co., Inc., New York (1968).

Schiff, L. I., *Quantum Mechanics*, 3rd ed. McGraw-Hill Book Co., New York (1968).

Sears, F. W. and G. L. Salinger, *An Introduction to Thermodynamics, the Kinetic Theory of Gases, and Statistical Mechanics*. Addison-Wesley Publishing Co., Reading, Massachusetts (1975).

Spiegel, M. R., *Mathematical Handbook*. Shaum's Outline Series, McGraw-Hill Book Co., Inc., New York (1968).

Sokolnikoff, I. S., *Tensor Analysis*. John Wiley and Sons, Inc., New York (1951).

Sommerfeld, A., *Thermodynamics and Statistical Mechanics*. Academic Press, New York (1964).

Sommerfeld, A., *Electrodynamics*. Academic Press, New York (1952).

Sommerfeld, A., *Partial Differential Equations in Physics*. Academic Press, New York (1949).

Symon, K. R., *Mechanics*. Addison-Wesley Publishing Co., Reading, Massachusetts (1971).

Temple, G., *Cartesian Tensors: An Introduction*. John Wiley and Sons, Inc., New York (1960).

Tolman, R. C., *The Principles of Statistical Mechanics*. Oxford University Press, London (1950). Reprint. Dover Publications, Inc., New York (1980).

Weinstock, R., *Calculus of Variations*. McGraw-Hill Book Co., Inc., New York (1952).

Whittaker, E. T., and G. N. Watson, *A Course of Modern Analysis*. Cambridge University Press, New York (1944).

Wigner, E. P., *Group Theory*. Academic Press, New York (1959).

Wilson, E. B., *Advanced Calculus*, Dover Publications, Inc., New York (1958).

Winter, R. G., *Quantum Physics*. Wadsworth Publishing Co., Belmont, California (1979).

Young, H. D., *Statistical Treatment of Experimental Data*. McGraw-Hill Book Co., Inc., New York (1962).

Index